Herbert Busemann
Selected Works II

Herbert Busemann (probably in 1934)
With kind permission of the archives of the Mathematisches Forschungsinstitut Oberwolfach, and Konrad Jacobs, Erlangen

Herbert Busemann
Selected Works II

Edited by
Athanase Papadopoulos

Springer

Herbert Busemann
(1905 Berlin – 1994 Santa Ynez)

Editor
Athanase Papadopoulos
Institut de Recherche Mathématique
University of Strasbourg, CNRS
Strasbourg Cedex
France

ISBN 978-3-319-65623-6

Library of Congress Control Number: 2017950015

Mathematics Subject Classification (2010): 01A75, 53C70, 53C23, 58B20, 53C60

Printed on acid-free paper

This Springer imprint is published by Springer Nature
The registered company is Springer International Publishing AG
The registered company address is: Gewerbestrasse 11, 6330 Cham, Switzerland

PREFACE

Herbert Busemann is an eminent representative of the remarkable Göttingen school of geometry that was active until the end of the first third of the twentieth century. During 50 years, as a faithful heir of Hilbert, Busemann was deeply involved in fundamental questions of geometry and its foundations. He developed, in relative isolation, an extremely rich theory of geodesic metric spaces, extending to such a general setting a theory of parallels, perpendiculars, asymptotes, horofunctions, boundaries, and many other notions from our familiar Euclidean and non-Euclidean geometries. Busemann was the first to proclaim loud and clear that the major theorems of Riemannian and Finsler geometry are not exclusive to these fields, but they belong to a larger field concerned with spaces he called G-spaces (G stands for geodesic). These are metric spaces that satisfy certain axioms on the existence of geodesics between two points and the uniqueness of local prolongation. Let us quote Busemann from the Preface of his *Geometry of geodesics* (1955):

> In spite of the great success of Finsler's thesis, the later development of this aspect lacks simple geometric facts to the extent that their existence in non-Riemannian geometry has been doubted.
> Here we take the attitude that Finsler spaces are point spaces of an essentially new character which we must approach directly without being prejudiced by euclidean or Riemannian methods. The euclidean tradition is much too strong to be shaken off readily, and it may take several generations of mathematicians to get real freedom.
> However, the direct approach has produced many new results and has materially generalized many known phenomena, thereby revealing two startling facts: much of Riemannian geometry is not truly Riemannian and much of differential geometry requires no derivatives.

Talking about the later methods he introduced, Busemann declares in the same preface:

> The present methods also offer the aesthetical and methodological advantage of a geometric approach to geometric problems and they extend to differential geometry the trend of replacing calculations by syllogisms which has so deeply permeated modern mathematics.
> Some differential geometers find this tendency unnatural. We need not take them seriously: the term "differential geometry" carries no obligation to use derivatives, just as names like "calculus of variations" or "algebra" do not confine us to operating within their original meanings.

In the Preface of his *Spaces with distinguished geodesics*, written more than 30 years after the *Geometry of geodesics*, Busemann writes:

> In Riemannian geometry the assertions of many theorems in the large with a beautiful intuitive geometric content do not depend, or can be formulated so that they do not depend, either on the Riemannian character of the metric or on the analytic methods inherent in the standard approach. G-spaces theory shows that these properties are in fact very often superfluous and that the theorems are accessible to a direct, purely geometric approach.
> Generalizations lack interest if they solve only the problems that gave rise to them, for example, if complex numbers led only to the fundamental theorem of algebra. G-spaces lead to many novel results, of which we mention here only one: the theory of parallels or asymptotes, which is almost completely neglected in the theory of Riemann and Finsler spaces.

Busemann set a research program, in the form of problems and conjectures concerning G-spaces, and his program is more active today than ever. He also contributed in a substantial way to the development of convexity theory and integral geometry. The major part of the problems he formulated in his paper with Petty, *Problems on convex bodies* (1956), are still unsolved. Examining Busemann's *Selected Works* will give an idea of the vastness of the task that Busemann set for himself and the extreme richness of his work.

This *Selected Works* edition is divided into four parts:

1. The foundations of geodesic spaces and the metric geometry of Finsler spaces
2. Convexity and integral geometry
3. Hilbert's problem IV
4. Miscellaneous articles

The first part constitutes Volume I of these *Selected Works*, and the other three parts constitute Volume 2. This division has been made for practical purposes, and, in fact, these four parts are related to each other. For instance, Finsler geometry is a generalization of convexity theory in the sense that it is a geometry obtained by assigning a convex body to each point on a manifold, namely, the unit ball of the norm in the tangent space at that point. In this way, a Finsler manifold becomes a parametrized family of convex bodies. One aspect of Hilbert's Problem IV asks for the characterization of Minkowski geometry, which is precisely the local geometry of Finsler spaces. The work on this problem also involves the study of the Hilbert metrics, which are among the main examples that Busemann used in his metric theory. Integral geometry has strong applications in convexity theory, and it also plays a major role in the solution of Hilbert's Problem IV by Pogorelov who was inspired by Busemann. Thus, the division into 4 parts is not completely sharp, but it will hopefully permit an easier localisation of the main ideas. In any case, the collection of articles by Busemann that are published in these two volumes constitutes a single profound œuvre whose influence on modern research will continue to grow, with metric geometry as a guiding line. These two volumes will not only promote Busemann's highly original work – which in the editor's opinion is still very much under-estimated – but they will also be a useful tool to the researchers in the areas that Busemann founded. Presenting all these articles available in two volumes will make them easily accessible and will highlight the unity of his work.

During the last few years, I talked with several people about my conviction that an edition of Busemann's Selected Works was a necessity, and Juan-Carlos Alvarez-Paiva was the most enthusiastic supporter of this idea. I am thankful to him for his help at the beginning of this project. I would also like to thank Marcel Berger, Yuri Burago and Vitali Milman who encouraged the project before it started, Annette A'Campo-Neuen and Dmitry Slutskiy who translated several of Busemann's articles from the German and the Russian, Manfred Karbe for his translation of Busemann's correspondence and Hubert Goenner for the search he made in the University of Göttingen archives, and Elena Griniari, Manfred Karbe and Irene Zimmermann for their editorial help and support.

Athanase Papadopoulos
Strasbourg and Providence, June 2017

Contents

Preface v

Contents of Volume I xv

Introduction xix

Publications of Herbert Busemann xxv

Acknowledgements xxxi

Contributors xxxv

Part I Essays, Correspondence and Biographical Material

Documents Concerning Busemann from the library of the University of Göttingen 3
Hubert Goenner

Correspondence related to the paper by Busemann and Feller *Zur Differentiation der Lebesgueschen Integrale* (On the differentiation of Lebesgue's integrals), Fundamenta Mathematicae 22 (1934) 226–256. Translated by Manfred Karbe 5
Hubert Goenner and Willy Feller

Letter to Richard Courant, May 12, 1935 13
Herbert Busemann

***Outline of research*, document written around 1935** 23
Herbert Busemann

Letter to Solomon Lefschetz, November 16, 1935. Translated by M. Karbe 25
Herbert Busemann

A report on the Busemann–Courant Correspondence (New York University; Elmer Holmes Bobst Library) 29
Manfred Karbe

Herbert Busemann, private assistant for mathematics. From: Anikó Szábo, Vertreibung, Rückkehr, Wiedergutmachung, Wallstein Verlag, Göttingen, 2000, pp. 482–484. English translation by M. Karbe 57

A letter from Busemann to V. Pambuccian 61
Herbert Busemann

Meeting Herbert Busemann 65
Michael Waterman

Busemann and Feller on curvature properties of convex surfaces 67
Annette A'Campo Neuen and Athanase Papadopoulos

Herbert Busemann and Convexity 89
Valerii N. Berestovskii

On Several Problems Posed by H. Busemann and Related to Geometry in Normed Spaces 119
Dmitri Burago

Busemann's Work on Hilbert Geometry 123
Athanase Papadopoulos

Busemann's Work on the Characterization of Minkowski Geometries . . 145
Athanase Papadopoulos

Part II Busemann's Papers on Convexity and Integral Geometry

Bemerkungen zur Differentialgeometrie der konvexen Flächen. I 163
Herbert Busemann and William Feller

Bemerkungen zur Differentialgeometrie der konvexen Flächen. II 175
Herbert Busemann and William Feller

Bemerkungen zur Differentialgeometrie der konvexen Flächen. III 205
Herbert Busemann and William Feller

Curvature Properties of Convex Surfaces 235
Herbert Busemann and William Feller

Regularity Properties of a Certain Class of Surfaces 271
Herbert Busemann and William Feller

Intrinsic Area 287
Herbert Busemann

Note on a Theorem on Convex Sets 291
Herbert Busemann

Intrinsic Area 295
Herbert Busemann

The Isoperimetric Problem in the Minkowski Plane 329
Herbert Busemann

A Theorem on Convex Bodies of the Brunn-Minkowski Type 339
Herbert Busemann

The Isoperimetric Problem for Minkowski Area 345
Herbert Busemann

Volume in Terms of Concurrent Cross-Sections 365
Herbert Busemann

Problems on Convex Bodies 377
Herbert Busemann and C.M. Petty

Minkowski's and Related Problems for Convex Surfaces with Boundaries 385
Herbert Busemann

Volumes and Areas of Cross-Sections 393
Herbert Busemann

Corrections to Volumes and Areas of Cross-Sections 397
Herbert Busemann

Area and Normality 399
Herbert Busemann and Ernst Gabor Straus

Areas in Affine Space I. II 437
Herbert Busemann

Areas in Affine Spaces III 447
Herbert Busemann

Geometries in Which the Planes Minimize Area 463
Herbert Busemann

Convexity on Grassmann Manifolds 483
Herbert Busemann

Convex Bodies and Convexity on Grassmann Cones I-IV 497
Herbert Busemann and G.C. Shephard

Convex Bodies and Convexity on Grassmann Cones 513
Herbert Busemann, G. Ewald and G.C. Shephard

Convexity problems arising in the theory of area 555
Herbert Busemann

Convex Bodies and Convexity on Grassmann Cones (*X*): Projection Functions of Parallel Convex Bodies 563
Herbert Busemann and G.C. Shephard

Un Nuovo Tipo di Problemi nella Teoria dei Corpi Convessi 587
Herbert Busemann

Convex bodies and Convexity on Grassmann Cones XI: 615
Sublinear Functions
Herbert Busemann

Part III Busemann's Papers on Hilbert's Problem IV

On Hilbert's fourth problem 627
Herbert Busemann

Spaces with homothetic spheres 637
Herbert Busemann

Planes with Analogues to Euclidean Angular Bisectors 649
Herbert Busemann

Remark on "Planes with Analogues to Euclidean Angular Bisectors" . . . 657
Herbert Busemann

Problem IV: Desarguesian Spaces 659
Herbert Busemann

Minkowskian Geometry, Convexity Conditions and the Parallel Axiom 671
Herbert Busemann

Locally Desarguesian Spaces 689
Herbert Busemann and B.B. Phadke

Book Review of "Hilbert's Fourth problem" by A.V. Pagorelov 705
Herbert Busemann

A General Version of Beltrami's Theorem in the Large 709
Herbert Busemann and B.B. Phadke

Part IV Miscellaneous Papers

Zur Differentiation der Lebesgueschen Integrale 729
Herbert Busemann and W. Feller

Lokale Eigenschaften der zu Variationsproblemen gehörigen metrischen Räume 761
Herbert Busemann

Non-Euclidean Geometry 785
Herbert Busemann

The Role of Geometry for the Mathematics Student 801
Herbert Busemann

Two Applications of Geometry 807
Herbert Busemann

Irreducible Sums of Simple Multivectors 841
Herbert Busemann and D. Edward Glassco

Contents of Volume I

Preface v

Contents of Volume II xiii

Introduction xix

Publications of Herbert Busemann xxi

Acknowledgements xxvii

Contributors xxxi

Part I Essays and Biographical Material

Herbert Busemann (1905–1994) 3
Athanase Papadopoulos

On Three Early Papers by Herbert Busemann on the Foundations of Geometry 15
Athanase Papadopoulos and Marc Troyanov

On Pasch's Axiom and Desargues' Theorem in Busemann's Work 29
Marc Troyanov

Busemann's Results, Ideas, Questions and Locally Compact Homogeneous Geodesic Spaces 41
Valerii N. Berestovskii

Busemann's Problems on *G*-spaces 85
Pavel D. Andreev

Busemann's Metric Theory of Timelike Spaces 115
Athanase Papadopoulos and Sumio Yamada

Chronogeometry 133
Athanase Papadopoulos

Book review of "Metric Methods in Finsler Spaces and in the Foundations of Geometry" 143
Frederick A. Ficken

Book review of "The Geometry of Geodesics" 149
William M. Boothby

Part II Busemann's Papers on the Foundations of Geodesic Spaces and on the Metric Geometry of Finsler Spaces

On the Geometries Where Circles of Infinite Radius are the Shortest Lines 155
Herbert Busemann

On Spaces With Convex Spheres and the Parallel Postulate 173
Herbert Busemann

Pasch's Axiom and Two-Dimensionality 193
Herbert Busemann

On Translations in General Plane Geometries 199
Herbert Busemann

Two-Dimensional Metric Spaces with Prescribed Geodesics 229
Herbert Busemann

On the Foundations of Calculus of Variations 241
Herbert Busemann and Walther Mayer

Metric Conditions for Symmetric Finsler Spaces 267
Herbert Busemann

On Leibniz's Definition of Planes 271
Herbert Busemann

On Spaces in which Two Points Determine a Geodesic 283
Herbert Busemann

Local Metric Geometry 297
Herbert Busemann

Metrically Homogeneous Spaces 373
Herbert Busemann

Two-Dimensional Geometries with Elementary Areas 379
Herbert Busemann

Spaces with Non-Positive Curvature 385
Herbert Busemann

Angular Measure and Integral Curvature 437
Herbert Busemann

The Geometry of Finsler Spaces 455
Herbert Busemann

On Geodesic curvature in two-dimensional Finsler spaces 467
Herbert Busemann

The Foundations of Minkowskian Geometry 483
Herbert Busemann

Metrics on the Torus Without Conjugate Points 515
Herbert Busemann

Motions with Maximal Displacements 533
Herbert Busemann

Tori With One-Parameter Groups of Motions 541
Herbert Busemann and Flemming P. Pedersen

On Normal Coordinates in Finsler Spaces 553
Herbert Busemann

Quasihyperbolic Geometry 561
Herbert Busemann

Groups of Motions Transitive on Sets of Geodesics 575
Herbert Busemann

Metrizations of Projective Spaces 581
Herbert Busemann

Similarities and Differentiability 585
Herbert Busemann

The Synthetic Approach to Finsler Spaces in the Large 597
Herbert Busemann

Length-Preserving Maps 667
Herbert Busemann

Metric Collineations and Inverse Problems 689
Herbert Busemann and Helmut Salzmann

Axioms for Indefinite Metrics 717
Herbert Busemann and John K. Beem

Timelike Spaces 741
Herbert Busemann

Transitive Geodesics in Minkowski Planes 787
Herbert Busemann

Spaces with Distinguished Shortest Joins 791
Herbert Busemann

Nonconvex Spheres in G-Spaces 805
Herbert Busemann and B.B. Phadke

Two Theorems on General Symmetric Spaces 817
Herbert Busemann and B.B. Phadke

Symmetric Spaces and Ellipses 827
Herbert Busemann and B.B. Phadke

Peakless and Monotone Functions on *G*-spaces 837
Herbert Busemann and B.B. Phadke

Novel Results in the Geometry of Geodesics 869
Herbert Busemann and B.B. Phadke

INTRODUCTION

The present volume of Busemann's *Selected Works*, the second of the two volumes set, contains his papers on convexity and integral geometry, those on Hilbert's Fourth Problem, and some other papers on various topics. The papers on convexity cover the period 1935–1969, whereas those on Hilbert's Problem IV were published later, between 1966 and 1984.

Busemann's papers are preceded by introductory material including biographical elements, correspondence, and five essays, by various authors, which may be considered as reading guides for his work.

The correspondence is edited, presented and translated by Manfred Karbe. It is based on a collection of letters that are kept at the Elmer Holmes Bobst Library of New York University, written by Herbert Busemann, Alfred Busemann, William Feller and Richard Courant. These letters give a lively picture of the period 1933–1936 which Busemann spent in Denmark, including his doubts and hesitations concerning his project of emigrating to the US. The letters also provide interesting information on the scientists (mostly mathematicians, but also physicists) with whom Busemann interacted during that period. Besides the letters, the present volume contains a biography of Herbert Busemann extracted from Anikó Szábo's *Vertreibung, Rückkehr, Wiedergutmachung* (Expulsion, return, reparation), Wallstein Verlag, Göttingen, 2000, related to his request to the German State of a compensation for the damages caused to his career by Nazi Germany. The volume also includes two reports written by Busemann on his own work, a letter from Busemann to Victor Pambuccian, written in 1989, and a short text on memories of Busemann written by Michael Waterman.

The first of the five essays included in this volume is written by Annette A'Campo-Neuen and the author of this introduction. It consists in a commentary to the paper *Krümmungseigenschaften konvexer Flächen* (Curvature properties of convex surfaces), written in 1935, by Busemann and William Feller.[1] This paper is important for several reasons: First, it is one of the first papers of Busemann on convexity theory, a topic that became one of Busemann's main subjects of interest during more than 35 years. The paper is also important because we see in it Busemann's first attempts to do geometry with a minimum amount of differentiability, which became also one of Busemann's lifelong mathematical themes. The questions addressed in this paper belong to the main subject of Busemann's monograph *Convex surfaces*, published in 1958.

---

[1]William Feller (1906–1970) was, like Busemann, a former doctoral student of Courant (PhD obtained in 1926). He left Germany in 1933 for Denmark, the same year as Busemann, then he moved to Sweden in 1934, where he stayed 5 years, before emigrating to the United States. Busemann and Feller wrote seven joint papers, published between 1934 and 1945.

The second essay is written by Valerii Berestovskii, and it is a survey of Busemann's work on convexity theory. This work is included by Berestovskii in a larger context. In particular, it is compared to other works that appeared in the same period on the same subject, starting with the works of Bonnesen-Fenchel, Steinitz-Rademacher, and especially, the later works by A. D. Alexandrov and his School. Particular attention is given in this article to the problems that were posed by Busemann and his collaborators, and which became the foundations of a large portion of the research conducted on convexity theory. They concern area and volume, the Brunn-Minkowski theory and isoperimetric problems in Minkowski spaces. Most of all, Berestovskii discusses the famous Busemann-Petty problems, several of which remain until today among the most difficult and interesting open problems in convexity theory and integral geometry. Berestovskii provides detailed comments on these problems, their (partial) solutions and their ramifications, with a review of the modern literature on this subject.

Let us stress on the fact that like for the work on metric geometry that is contained in Volume I of the present *Selected Works*, there are strong connections between Busemann's work on convexity and that of A. D. Alexandrov and the school he founded on the same subject. As a matter of fact, Busemann's monograph *Convex surfaces* (1958) was initially conceived as a book on the work of the Russian School on convexity. The correspondence kept at the Elmer Holmes Bobst Library includes a copy of a letter to Busemann, dated July 23, 1956 from Interscience Publishers, signed by E. S. Proskauer, Editor-in-Chief, with a copy to Courant, Stocker, and Bers, saying:

> Dear Dr. Busemann,
> It is with very special pleasure that we learn from Dr. Stocker of your willingness to present in our Tracts series an account of the important recent development of the work of certain Russian geometers.

This book was translated into Russian in 1962, and it is reviewed in detail in Berestovskii's essay in the present volume.

The third essay is written by Dmitri Burago, and it concerns more specifically Busemann's problems on convexity and integral geometry. As the author explains, several of these problems have many facets, and they constituted the basis of his joint research with Sergei Ivanov. In particular, Burago and Ivanov gave an affirmative answer to one aspect of Busemann's problem asking whether flats are area-minimizing in normed spaces. They considered the case of two-flats, both for the Hausdorff measure and for the Holmes-Thompson volume. A detailed and interesting review of different points of view on this problem is presented in Burago's essay.

The fourth and the fifth essay are written by the author of this introduction. They concern two topics that accompanied Busemann during many years, and which are related to Hilbert's Problem IV, asking for a characterization of the geometries on subsets of Euclidean n-space or n-dimensional real projective space for which the (Euclidean, or projective) lines are geodesics. The problem also asks for the study of these individual metrics. The two essays concern respectively Busemann's work on

Minkowski geometries (the geometries of finite-dimensional normed spaces) and Hilbert geometries. These two classes of metrics satisfy Hilbert's requirements and are the two classes that are mentioned by Hilbert in his statement of Problem IV. Busemann's results on these two topics are spread in several of his books and papers, and we thought that it would be useful to collect in these two papers his work on the subject. The characterization of Minkowski geometries attracted Busemann since his early works, and, as a matter of fact, was the subject of his doctoral dissertation. Regarding Hilbert geometry, the works of Busemann and his collaborators constitute, with two exceptions (a paper by Birkhoff and another one by Samelson) the only work done on this topic during the hundred years that followed Hilbert's discovery of that geometry. Hilbert geometry, since a couple of decades, became an important subject of research, in particular as part of the study of projective structures on manifolds. Busemann's work on this subject remains poorly known, with several of his results rediscovered with an ignorance of his original results. This is one of the main reasons that led us to writing this survey.

Let us mention that Busemann was the main promoter of Hilbert's Problem IV, and his work acted as the main catalyst to the solution of that problem that was given by A. V. Pogorelov in 1973. Let us quote the latter, from the introduction to his monograph *Hilbert's fourth problem*:[2]

> The occasion for the present investigation is a remarkable idea due to Herbert Busemann, which I learned from his report to the International Congress of Mathematicians at Moscow in 1966. Busemann gave an extremely simple and very general method of constructing Desarguesian metrics by using a nonnegative completely additive set function on the set of planes and defining the length of a segment as the value of the function of the set of planes intersecting the segment.

Busemann's construction mentioned by Pogorelov is based on an integral formula for distances defined as a measure on the set of planes, contained in earlier work of Blaschke, which in turn is based on an integral formula due to Crofton. The formula was promoted by Busemann in order to construct metrics satisfying Hilbert's problem IV.

Let us also mention that Hilbert's Fourth Problem has several interpretations, and in some respect it is still an open problem (in particular for what concerns non-symmetric metrics).

The collection of papers by Busemann on convexity and on Hilbert's Problem IV that are contained in the present volume include several survey articles which the reader can take advantage of. We mention in particular his paper with Feller, *Regularity properties of a certain class of surfaces* (1945), his paper with Petty, *Problems on convex bodies* (1956), his surveys *On Hilbert's fourth problem* (1964), *Desarguesian spaces* (1976), *Locally Desarguesian spaces*, with B. B. Phadke (1980) and his book review on Pogorelov's *Hilbert's fourth problem* (1981).

[2]Translated by R. A. Silverman, V. H. Winston & Sons, Washington D. C., 1979. Original by Nauka

The last section of Busemann's papers which are reproduced in the present volume contains 6 papers on various topics, some of them on specialized subjects, like his paper with Feller, *Zur Differentiation der Lebesgueschen Integrale*, (1934) and others on elementary topics, like his survey of *non-Euclidean geometry* (1950) or his paper *Two applications of geometry* (1967). The reason of the inclusion of a paper in this last section is that it was not obvious to which of the preceding sections it belongs (some of them could belong to more than one section).

I would like to make a few comments on the paper *The Role of Geometry for the Mathematics Student* in which Busemann reports on geometry in college curricula, in 1960. His first observation is that "the standard course 'Analytic geometry and calculus' originated from the waning interest in geometry and almost always proceeds at the expense of geometry." Differential geometry is taught at a reasonable number of colleges, but, Busemann says, "there are influential mathematicians doubting the value of any of the remaining courses." The list of remaining courses is: analytic geometry (which is *not* taught in the "Analytic geometry and calculus" course), projective geometry, foundations of geometry, non-Euclidean geometry, higher geometry and modern Euclidean geometry (triangle and circle). The usual objection for including these subjects in the curricula is that "the topic is no longer an active subject of research." Busemann's defense for a subject like projective geometry is that "the subject is beautiful, it required major efforts of some of the best mathematicians during and after the last century to understand its structure thoroughly, it still gives a deep insight into geometry ... But most of all, do we have the right to completely disrupt historical continuity whenever a subject moves out of the focus of contemporary interests? Do we really expect or agree that our present mathematical efforts will be altogether junked (at least from courses) as soon as the interests change?"

Busemann would have certainly be pleased to know that today projective geometry, the subject of a book he wrote with Kelly in 1953, is today, like his metric geometry, at the forefront of research, in topology and geometry. The same holds for non-Euclidean geometry, after the vast horizons opened up by Thurston. In his last book, written with Phadke, *Spaces with distinguished geodesics* (1987), Busemann writes (p. 111) about the importance of using metric methods in the study of Teichmüller spaces. This is also a topic that is extremely active today.

The return to basic geometry is, contrary to what one may think, a difficult matter, and discovering something new, using basic geometry, requires a deep understanding of mathematics. One is reminded here of A. D. Alexandrov's Slogan: "Retreat to Euclid,"[3] and of René Thom's apology of returning to elementary geometry: "For Euclidean geometry, I keep a soft spot which my colleagues and my fellow mathematicians do not forgive me."[4]

---

[3] S. S. Kutateladze, Alexandrov of Ancient Hellas, Siberian Electronic Mathematical Reports, 9 (2012), p. A6–A11.

[4] "J'ai conservé pour la géométrie euclidienne une faiblesse que mes collègues et confrères mathématiciens ne me pardonnent pas." In: *Logos et Théorie des Catastrophes*, actes du colloque international de Cerisy de 1982, dir. J. Petitot, ed. Patiño, Genève, 1989.

One is also reminded, of course, of Bill Thurston's approach to mathematics: everything is based on elementary geometry.

Busemann worked on fundamental questions, independently of the fashion of the day; this is probably the seal of a great mathematician.

Athanase Papadopoulos
Strasbourg, July 2017

PUBLICATIONS OF HERBERT BUSEMANN

Books

(1) H. Busemann, Methods in Finsler Spaces and in the Foundations of Geometry. Annals of Mathematics Studies, no. 8. University Press, Princeton, N. J., 1942.
(2) H. Busemann and P. J. Kelly, Projective geometry and projective metrics. Academic Press Inc., New York, N. Y., 1953.
(3) H. Busemann, The geometry of geodesics. Academic Press Inc., New York, N. Y., 1955. Reprinted by Dover, Mineola, NY, 2005. Russian edition: Geometriya geodezicheskikh, translated by M. Z. Kainer, edited by I. M. Jaglom Gosudarstv. Izdat. Fiz.-Mat. Lit., Moscow 1962.
(4) H. Busemann, Convex surfaces. Interscience Tracts in Pure and Applied Mathematics, no. 6 Interscience Publishers, Inc., New York; Interscience Publishers Ltd., London 1958. Reprinted by Dover, Mineola, NY, 2008. Russian edition: Vypuklye poverkhnosti. translated by V. A. Zalgaller and S. I. Zalgaller, Izdat. Nauka, Moscow 1964.
(5) H. Busemann, Recent synthetic differential geometry. Ergebnisse der Mathematik und ihrer Grenzgebiete, Band 54. Springer Verlag, New York-Berlin, 1970.
(6) H. Busemann and B. B. Phadke, Spaces with distinguished geodesics. Monographs and Textbooks in Pure and Applied Mathematics, 108. Marcel Dekker, Inc., New York, 1987.

Articles

(1) H. Busemann, Die Vollständigkeit der Minimalfolgen von Eigenwertproblemen. Nachrichten Göttingen 1930, 295–307 (1930).
(2) H. Busemann, Über die Geometrien, in denen die "Kreise mit unendlichem Radius" die kürzesten Linien sind. Math. Ann. 106, no. 1, 140–160 (1932).
(3) H. Busemann, Über Räume mit konvexen Kugeln und Parallelenaxiom. Nachr. Ges. Wiss. Göttingen, Math.-Phys. Kl. (I, No.38) 1933, 116–140 (1933).
(4) H. Busemann, Paschsches Axiom und Zweidimensionalität. Math. Ann. 107, no. 1, 324–328 (1933).
(5) H. Busemann and W. Feller, Zur Differentiation der Lebesgueschen Integrale. Fundam. Math. 22, p. 226–256 (1934).

(6) H. Busemann, and W. Feller, Krümmungseigenschaften konvexer Flächen. Acta Math. 66, p. 1–47 (1935).
(7) H. Busemann and W. Feller, Bemerkungen zur Differentialgeometrie der konvexen Flächen. I. Kürzeste Linien auf differenzierbaren Flächen. Matematisk Tidsskrift. B. p. 25–36 (1935).
(8) H. Busemann and W. Feller, Bemerkungen zur Differentialgeometrie der konvexen Flächen. II. Über die Krümmungsindikatrizen. Matematisk Tidsskrift. B. p. 87–115 (1935).
(9) H. Busemann and W. Feller, Bemerkungen zur Differentialgeometrie der konvexen Flächen. III. Über die Gauss'sche Krümmung. Matematisk Tidsskrift. B. p. 41–70 (1936).
(10) H. Busemann, On translations in general plane geometries. Am. J. Math. 60, 227–256 (1938).
(11) H. Busemann, Lokale Eigenschaften der zu Variationsproblemen gehörigen metrischen Räume. Fundam. Math., Warszawa, 32, 265–287 (1939).
(12) H. Busemann, Two-dimensional metric spaces with prescribed geodesics. Ann. Math., (2) 40, 129–140 (1939).
(13) H. Busemann and W. Mayer, On the foundations of calculus of variations. Trans. Am. Math. Soc. 49, 173–198 (1941).
(14) H. Busemann, Metric conditions for symmetric Finsler spaces. Proc. Natl. Acad. Sci. USA 27, 533–535 (1941).
(15) H. Busemann, On Leibniz's definition of planes. Amer. J. Math. 63, 101–111 (1941).
(16) H. Busemann, On spaces in which points determine a geodesic. Trans. Am. Math. Soc. 54, 171–184 (1943).
(17) H. Busemann, Local metric geometry. Trans. Am. Math. Soc. 56, 200–274 (1944).
(18) H. Busemann, Book review of: H. Weyl and J. Weyl, *Meromorphic functions and analytic curves*, Annals of Mathematics Studies, no. 12, Princeton University Press; London, Humphrey Milford and Oxford University Press, 1943. Bull. Amer. Math. Soc. Volume 50, Number 9, 655–657 (1944).
(19) H. Busemann and W. Feller, Regularity properties of a certain class of surfaces. Bull. Am. Math. Soc. 51, 583–598 (1945).
(20) H. Busemann, Intrinsic area. Proc. Natl. Acad. Sci. USA 32, 5–8 (1946).
(21) H. Busemann, Metrically homogeneous spaces. Am. J. Math. 68, 340–344 (1946).
(22) H. Busemann, Intrinsic area. Ann. of Math. (2) 48, 234–267 (1947).
(23) H. Busemann, Two-dimensional geometries with elementary areas. Bull. Am. Math. Soc. 53, 402–407 (1947).

(24) H. Busemann, Note on a theorem on convex sets. Matematisk Tidsskrift. B. 1947, 32–34 (1947).
(25) H. Busemann, The isoperimetric problem in the Minkowski plane. Am. J. Math. 69, 863–871 (1947).
(26) H. Busemann, Spaces with non-positive curvature. Acta Math., 80, 259–311 (1948).
(27) H. Busemann, On the problem of Dido, Studies and essays, presented to R. Courant on his 60th birthday January 8, 1948, Interscience Publishers, Inc., New York, 63–73 (1948).
(28) H. Busemann, Angular measure and integral curvature, Canad. J. Math. 1, 279–296 (1949).
(29) H. Busemann, A theorem on convex bodies of the Brunn–Minkowski type. Proc. Natl. Acad. Sci. USA 35, 27–31 (1949).
(30) H. Busemann, The isoperimetric problem for Minkowski area. Am. J. Math. 71, 743–762 (1949).
(31) H. Busemann, The geometry of Finsler spaces. Bull. Am. Math. Soc. 56, 5–16 (1950).
(32) H. Busemann, Noneuclidean geometry. Math. Mag. 24, 19–34 (1950).
(33) H. Busemann, On geodesic curvature in two–dimensional Finsler spaces. Ann. Mat. Pura Appl., IV. Ser. 31, 281–295 (1950).
(34) H. Busemann, The foundations of Minkowskian geometry. Comment. Math. Helv. 24, 156–187 (1950).
(35) H. Busemann, Metrics on the torus without conjugate points. Bol. Soc. Mat. Mexicana 10, . nos. 3–4, 1–18 (1953).
(36) H. Busemann, Volume in terms of concurrent cross-sections. Pac. J. Math. 3, 1–12 (1953).
(37) H. Busemann, Motions with maximal displacements. Comment. Math. Helv. 28, 1–8 (1954).
(38) H. Busemann and P. F. Pedersen, Tori with one-parameter groups of motions. Math. Scand. 3, 209-220 (1955).
(39) H. Busemann, On normal coordinates in Finsler spaces. Math. Ann. 129, 417–423 (1955).
(40) H. Busemann, Quasihyperbolic geometry. Rend. Circ. Mat. Palermo, II. Ser. 4, 256–269 (1955).
(41) H. Busemann, Groups of motions transitive on sets of geodesics. Duke Math. J. 23, 539–544 (1956).
(42) H. Busemann, and C. M. Petty, Problems on convex bodies. Math. Scand. 4, 88–94 (1956).
(43) H. Busemann, Metrizations of projective spaces. Proc. Am. Math. Soc. 8, 387–390 (1957).
(44) H. Busemann, Similarities and differentiability. Tôhoku Math. J., II. Ser. 9, 56–67 (1957).

(45) H. Busemann, Spaces with finite groups of motions. J. Math. pur. appl., IX. Sér. 37, 365-373 (1958).
(46) H. Busemann, Axioms for geodesics and their implications. The axiomatic method. With special reference to geometry and Physics. Proceedings of an International Symposium held at the Univ. of Calif., Berkeley, Dec. 26, 1957-Jan. 4, 1958 (edited by L. Henkin, P. Suppes and A. Tarski) pp. 146–159 Studies in Logic and the Foundations of Mathematics North-Holland Publishing Co., Amsterdam, 1959.
(47) H. Busemann, Minkowski's and related problems for convex surfaces with boundaries. Mich. Math. J. 6, 259–266 (1959).
(48) H. Busemann, Volumes and areas of cross-sections. Am. Math. Mon. 67, 248–250 (1960).
(49) H. Busemann, Corrections to *Volumes and areas of cross-sections* Am. Math. Mon. 67, 671 (1960).
(50) H. Busemann, and E. G. Straus, Area and normality. Pac. J. Math. 10, 35–72 (1960).
(51) H. Busemann, Areas in affine spaces. III: The integral geometry of affine area. Rend. Circ. Mat. Palermo, II. Ser. 9, 226–242 (1960).
(52) H. Busemann, The role of geometry for the mathematics student. Am. Math. Mon. 67, 281–285 (1960).
(53) H. Busemann, On spaces with one-parameter groups of motions. Tensor, New Ser. 10, 21–25 (1960).
(54) H. Busemann, The synthetic approach to Finsler spaces in the large. C.I.M.E., Geometria del Calcolo delle Variazioni 1 (No.2), 70 p. (1961).
(55) H. Busemann, Geometries in which the planes minimize area. Ann. Mat. Pura Appl., IV. Ser. 55, 171–189 (1961).
(56) H. Busemann, Convexity on Grassmann manifolds. Enseign. Math., II. Sér. 7, 139–152 (1962).
(57) H. Busemann, G. Ewald and G. C. Shephard, Convex bodies and convexity on Grassmann cones. Arch. Math. 13, 512–526 (1962).
(58) H. Busemann, The theory of parallels with applications to closed geodesics. Algebr. Topol. Foundations Geom., Proc. Colloq. Utrecht, August 1959, 7–14 (1962).
(59) H. Busemann and G. C. Shephard, Convex bodies and convexity on Grassmann cones. I-IV. Math. Ann. 151, 1–14 (1963).
(60) H. Busemann, Length-preserving maps. Pac. J. Math. 14, 457–477 (1964).
(61) H. Busemann, Convexity problems arising in the theory of area (Russian), Proc. Fourth All-Union Math. Congr. (Leningrad, July 3–12 1961), Vol. II, Leningrad: Nauka, p. 177–182 (1964).

(62) H. Busemann, Convexity problems arising in the theory of area, Proc. Fourth All-Union Math. Congr. Leningrad, July 3–12 1961, Vol. II, Leningrad: Nauka, p. 177–182 (1964).

(63) H. Busemann and G. C. Shephard, Convex bodies and convexity on Grassmann cones. X: Projection functions of parallel convex bodies. Ann. Mat. Pura Appl., IV. Ser. 70, 271–293 (1965).

(64) H. Busemann and H. Salzmann, Metric collineations and inverse problems. Math. Z. 87, 214–240 (1965).

(65) H. Busemann, Un nuovo tipo di problemi nella teoria dei corpi convessi. Seminari 1962/63 Anal. Alg. Geom. e Topol., Vol. 1, Ist. Naz. Alta Mat. pp. 178–205, Ediz. Cremonese, Rome, 1965.

(66) H. Busemann, Extremals on closed hyperbolic space forms. Tensor, New Ser. 16, 313-318 (1965).

(67) H. Busemann, On Hilbert's fourth problem (Russian) Uspehi Mat. Nauk 21, no. 1, pp. 155-164 (1966).

(68) H. Busemann and J. K. Beem, Axioms for indefinite metrics. Rend. Circ. Mat. Palermo, II. Ser, 15, 223–246 (1966).

(69) H. Busemann, Timelike spaces. Diss. Math. 53, 52 p. (1967).

(70) H. Busemann and G. C. Shephard, Convexity on nonconvex sets. Proc. Colloq. Convexity, Copenhagen 1965, 20-33 (1967).

(71) H. Busemann, Two applications of geometry, Report of the CUMP Geometry conference, Part III, proceedings, US Department of health, education and welfare, office of Education, Mathematical Association of America, p. 73–105 (1967).

(72) H. Busemann, Transitive geodesics in Minkowski planes. Math. Chronicle 1, 27–29 (1969).

(73) H. Busemann, Convex bodies and convexity on Grassmann cones. XI: Sublinear functions. Math. Scand. 24, 93–101 (1969).

(74) H. Busemann, Synthetische Differentialgeometrie. Jahresber. Dtsch. Math.-Ver. 71, 1–24 (1969).

(75) H. Busemann, Spaces with distinguished shortest joins. A spectrum of mathematics (Essays presented to H. G. Forder), pp. 108–120. Auckland Univ. Press, Auckland, 1971.

(76) H. Busemann and E. D. Glassco II, Irreducible sums of simple multivectors. Pac. J. Math. 49, 13-32 (1973).

(77) H. Busemann, Spaces with homothetic spheres. J. Geometry 4, p. 175–186, (1974).

(78) H. Busemann, Planes with analogues to Euclidean angular bisectors. Collection of articles dedicated to Werner Fenchel on his 70th birthday. Math. Scand. 36, p. 5–11 (1975).

(79) H. Busemann, Remark on: "Planes with analogues to Euclidean angular bisectors" (Math. Scand. 36 (1975), 5–11). Math. Scand. 38, no. 1, p. 81–82, (1976).

(80) H. Busemann, Problem IV: Desarguesian spaces. Math. Dev. Hilbert Probl., Proc. Symp. Pure Math. 28, De Kalb 1974, 131-141 (1976).
(81) H. Busemann, Minkowskian geometry, convexity conditions and the parallel axiom. J. Geom. 12, no. 1, p. 17–33, (1979).
(82) H. Busemann and B. B. Phadke, Locally Desarguesian spaces. J. Geom. 14, no. 2, p. 123–138, (1980).
(83) H. Busemann and B. B. Phadke, Nonconvex spheres in G-spaces. J. Indian Math. Soc., New Ser. 44, 39-50 (1980).
(84) H. Busemann, Book review of: A. V. Pogorelov *Hilbert's fourth problem*, Holt, Rinehart and Winston, New York, 1979. Bull. Amer. Math. Soc. 4, p. 87–90, (1981).
(85) H. Busemann and B. B. Phadke, Two theorems on general symmetric spaces. Pac. J. Math. 92, 39–48 (1981).
(86) H. Busemann and B. B. Phadke, Symmetric spaces and ellipses. In: E. B. Christoffel: the influence of his work on mathematics and the physical sciences, int. Symp., Aachen 1979, 626–635 (1981).
(87) H. Busemann, New developments in the theory of G-spaces. Differential geometry, Budapest 1979, Colloq. Math. Soc. Janos Bolyai 31, 125-140 (1982).
(88) H. Busemann and B. B. Phadke, Peakless and monotone functions on G–spaces. Tsukuba J. Math. 7, 105–135 (1983).
(89) H. Busemann and B. B. Phadke, A general version of Beltrami's theorem in the large. Pacific J. Math. 115, no. 2, p. 299–315 (1984).
(90) H. Busemann and B. B. Phadke, Novel results in the geometry of geodesics. Adv. Math. 101, No. 2, 180–219 (1993).

ACKNOWLEDGEMENTS

The editor is grateful to all the individuals, editors, publishing houses, societies and institutions who granted permission for reproducing Busemann's papers and other material related to his life. In particular, he is grateful to Wallstein Verlag for the permission to publish an article on Herbert Busemann as well as an English translation of the German original, to the library of the University of Göttingen for material concerning Busemann published in the present volume, and to the Elmer Holmes Bobst Library of New York University for giving is access and permission to publish part of correspondence between Herbert Busemann, Willy Feller, Alfred Busemann and Richard courant.

The following papers are reprinted by permission of the Mathematical Association of America and the American Math. Monthly:

- H. Busemann, Noneuclidean geometry. Math. Mag. 24, 19–34 (1950).
- H. Busemann, Volumes and areas of cross-sections. Am. Math. Mon. 67, 248–250 (1960).
- H. Busemann, Corrections to *Volumes and areas of cross-sections* Am. Math. Mon. 67, 671 (1960).
- H. Busemann, The role of geometry for the mathematics student. Am. Math. Mon. 67, 281–285 (1960).
- H. Busemann, Two applications of geometry, Report of the CUMP Geometry conference, Part III, proceedings, US Department of health, education and welfare, office of Education, Mathematical Association of America, p. 73–105 (1967).

The following papers are reprinted by permission of the American Mathematical Society:

- H. Busemann and W. Feller, Regularity properties of a certain class of surfaces. Bull. Am. Math. Soc. 51, 583-598 (1945).
- H. Busemann, Problem IV: Desarguesian spaces. Math. Dev. Hilbert Probl., Proc. Symp. Pure Math. 28, De Kalb 1974, 131-141 (1976).
- H. Busemann, Book review of: A. V. Pogorelov "Hilbert's fourth problem", Holt, Rinehart and Winston, New York, 1979. Bull. Amer. Math. Soc. 4, p. 87–90 (1981).

The following papers are reprinted by permission of the Institute of Mathematics of the Polish Academy of Sciences:

- H. Busemann and W. Feller, Zur Differentiation der Lebesgueschen Integrale. Fundam. Math. 22, p. 226–256 (1934).
- H. Busemann, Lokale Eigenschaften der zu Variationsproblemen gehörigen metrischen Räume. Fundam. Math., Warszawa, 32, 265–287 (1939).

The following papers are reprinted by permission of the Johns Hopkins University Press:

- H. Busemann The isoperimetric problem in the Minkowski plane. Am. J. Math. 69, 863-871 (1947).
- H. Busemann, The isoperimetric problem for Minkowski area. Am. J. Math. 71, 743-762 (1949).

The following paper is reprinted by permission of the European Mathematical Society:

- H. Busemann, Convexity on Grassmann manifolds. Enseign. Math., II. Sér. 7, 139-152 (1962).

The following papers are reprinted by permission of the Danish Mathematical Society and Mathematica Scandinavica:

- H. Busemann and W. Feller, Bemerkungen zur Differentialgeometrie der konvexen Flächen. I. Kürzeste Linien auf differenzierbaren Flächen. Matematisk Tidsskrift. B. p. 25–36 (1935).
- H. Busemann and W. Feller, Bemerkungen zur Differentialgeometrie der konvexen Flächen. II. Über die Krümmungsindikatrizen. Matematisk Tidsskrift. B. p. 87–115 (1935).
- H. Busemann and W. Feller, Bemerkungen zur Differentialgeometrie der konvexen Flächen. III. Über die Gauss'sche Krümmung. Matematisk Tidsskrift. B. p. 41–70 (1936).
- H. Busemann, Note on a theorem on convex sets, Matematisk Tidsskrift. B, p. 32–34 (1947).
- H. Busemann, and C. M. Petty, Problems on convex bodies. Math. Scand. 4, 88-94 (1956).
- Busemann, Convex bodies and convexity on Grassmann cones. XI: Sublinear functions. Math. Scand. 24, 93-101 (1969).
- H. Busemann, Planes with analogues to Euclidean angular bisectors. Collection of articles dedicated to Werner Fenchel on his 70th birthday. Math. Scand. 36, p. 5–11 (1975).
- H. Busemann, Remark on: "Planes with analogues to Euclidean angular bisectors" (Math. Scand. 36 (1975), 5–11). Math. Scand. 38, no. 1, p. 81–82 (1976).

The following papers are reprinted by permission of the Pacific Journal of Mathematics:

- H. Busemann, Volume in terms of concurrent cross-sections. Pac. J. Math. 3, 1-12 (1953).
- H. Busemann and E. G. Straus, Area and normality. Pac. J. Math. 10, 35-72 (1960).
- H. Busemann and E. D. Glassco II, Irreducible sums of simple multivectors. Pac. J. Math. 49, 13–32 (1973).
- H. Busemann and B. B. Phadke, A general version of Beltrami's theorem in the large. Pacific J. Math. 115, no. 2, p. 299–315 (1984).

The following paper is reprinted by permission of the Princeton University Press:

- H. Busemann, Intrinsic area. Ann. of Math. (2) 48, 234–267 (1947).

The following papers are reprinted by permission of Springer Verlag:

- H. Busemann, Intrinsic area. Ann. of Math. (2) 48, 234–267 (1947).
- H. Busemann, Areas in affine space. I, II. Rend. Circ. Mat. Palermo, II. Ser. 9, 81-90 (1960).
- H. Busemann, Areas in affine spaces. III: The integral geometry of affine area. Rend. Circ. Mat. Palermo, II. Ser. 9, 226-242 (1960).
- H. Busemann, Geometries in which the planes minimize area. Ann. Mat. Pura Appl., IV. Ser. 55, 171-189 (1961).
- H. Busemann, E. Ewald, G. and G. C. Shephard, Convex bodies and convexity on Grassmann cones. Arch. Math. 13, 512-526 (1962).
- H. Busemann, and G. C. Shephard, Convex bodies and convexity on Grassmann cones. I-IV. Math. Ann. 151, 1-14 (1963).
- Busemann, and G. C. Shephard, Convex bodies and convexity on Grassmann cones. X: Projection functions of parallel convex bodies. Ann. Mat. Pura Appl., IV. Ser. 70, 271-293 (1965).
- H. Busemann, Spaces with homothetic spheres. J. Geometry 4, p. 175–186 (1974).
- H. Busemann, Minkowskian geometry, convexity conditions and the parallel axiom. J. Geom. 12), no. 1, p. 17–33 (1979).
- H. Busemann and B. B. Phadke, Locally Desarguesian spaces. J. Geom. 14, no. 2, p. 123–138 (1980).

The following paper is translated by permission of Springer Verlag:

- H. Busemann and W. Feller, Krümmungseigenschaften konvexer Flächen. Acta Math. 66, p. 1–47 (1936).

The following paper is reprinted by permission of the Michigan Mathematical Journal.

- H. Busemann, Minkowski's and related problems for convex surfaces with boundaries. Mich. Math. J. 6, 259-266 (1959).

The following papers are translated by permission of the Uspekhi Matematicheskikh Nauk and the Nauka Editions of the Russian Academy of Sciences:

- H. Busemann, Convexity problems arising in the theory of area, Proc. Fourth All-Union Math. Congr. Leningrad, July 3–12 1961, Vol. II, Leningrad: Nauka, p. 177–182 (1964). (Translated from the Russian.)
- H. Busemann, On Hilbert's fourth problem. Uspehi Mat. Nauk 21 (1966) no. 1 (127), 155–164 (1966). (Translated from the Russian.)

CONTRIBUTORS

- Annette A'Campo-Neuen, Universität Basel, Department Mathematik und Informatik, Basel, Switzerland.
  Email: annette.acampo@unibas.ch

- Valerii N. Berestovskii, Sobolev Institute of Mathematics SB RAS, Novosibirsk, Russia.
  Email: vberestov@inbox.ru

- Dmitri Burago, Department of Mathematics,The Pennsylvania State University University Park, USA.
  Email: burago@math.psu.edu

- Hubert Goenner, Institute for Theoretical Physics, University of Göttingen, Germany.
  Email: goenner@theorie.physik.uni-goettingen.de

- Manfred Karbe, Freiburg, Germany.
  Email: Manfred_Karbe@t-online.de

- Victor Pambuccian, School of Mathematical and Natural Sciences, Arizona State University, Phoenix AZ, 85069-7100, USA.
  Email: victor.pambuccian@asu.edu

- Michael Waterman, Unisverity of Southern California, Los Angeles, USA.
  Email: msw@usc.edu

- Athanase Papadopoulos, Institut de Recherche Mathématique Avancée, Strasbourg, France.
  Email : papadop@math.unistra.fr

- Dmitry Slutskiy, Institut de Recherche Mathématique Avancée, Université de Strasbourg, Strasbourg, France.
  Email: dmitry.slutsky@gmail.com

PART I
ESSAYS, CORRESPONDENCE AND BIOGRAPHICAL MATERIAL

Documents concerning Busemann from the library of the University of Göttingen*

Collected and translated by Hubert Goenner

Busemann bequested his papers to the Deutsche Nationalbibliothek, but the material can be seen only after prior arrangement of a date in Deutsches Exilarchiv Frankfurt. There exists a catalog of this Deutsches Exilarchiv, Franfurt/Main.

In the Handschriftenabteilung (Department for handwritten manuscripts) of the University of Göttingen, the file on Busemann's PhD is available (code: UAG.Kur.Alt.V.c.340). It contains:

– The minutes of the PhD examination on 25. Feb. 1931. Examiners were Courant (geometry), Landau (mathematical analysis), and Max Born (theoretical physics). The grades achieved by Busemann were (in the same order); "with honors" (best grade!); "very good to good" (very good is the second best grade); "good". The dissertation received the grade "very good". The minutes contain catchwords of the questions asked.

The title of the thesis is:
"Ueber die Geometrien, in denen die 'Kreise mit unendlichem Radius' die kürzesten Linien sind." ("On geometries in which the 'circles with infinite radius' are the shortest lines.")
In full, the report by Courant on Busemann's thesis is:
"Die Arbeit von Herrn Busemann ist ein Beitrag zur mengentheoretischen Geometrie, welche Grundlagenfragen im Rahmen der modernen Mengenlehre in Angriff nimmt. Herr Busemann betrachtet einen metrischen dreidimensionalen Raum, d.h. einen Raum mit einer symmetrischen, definiten, dem Dreiecksaxiom genügenden Metrik, für welche ausserdem noch die Häufungsstelleneigenschaft (Kompaktheit) vorausgesetzt wird. Man fordert ferner, dass es zu jedem Punktepaar des Raumes genau einen Halbierungspunkt und zwei Verdoppelungspunkte gibt; die so definierten, schon von Menger untersuchten "Geradenräume" erfüllen sämtliche Hilbertschen Axiome. Herr Busemann fügt nun das weitere Axiom hinzu, dass die gerade Verbindungsstrecke zweier Punkte als Limes eines Kreises mit

*We would like to thank Martin Liebetruth from the University Library for providing a cassette with an interview with H. Busemann.

unendlich werdendem Radius durch die beiden Punkte aufgefasst werden kann. Er gelangt zu dem Resultat, dass durch dieses "Grenzkreisaxiom" die klassische Minkowskische Geometrie mit einem konvexen Eichkörper entsteht."

"Bei der Bewertung der Arbeit - welche übrigens von zwei hervorragenden Sachkennern des Gebiets Prof. Alexandroff und Prof. Kolmogoroff bis in alle Einzelheiten durchgesehen worden ist - ist zu berücksichtigen, dass Herr Busemann sich das Thema selbst gestellt und es in absolut selbständiger Weise ohne jede Hilfe und Anregung durchgeführt hat. Es handelt sich also um eine durchaus selbständige wissenschaftliche Leistung. Jedoch kann ich für die Arbeit nicht das beste Prädikat vorschlagen mit Rücksicht darauf, dass es sich um kein wirklich zentrales Problem handelt.

Ich schlage daher vor, die Arbeit mit der Note II (sehr gut) zu bewerten."

Englisch translation of Courant's report on Busemann's thesis:

The work by Mr. Busemann contributes to set-theoretical geometry, which tackles fundamental questions in the framework of modern set theory. Mr. Busemann considers a metric three-dimensional space with a symmetric, definite metric which satisfies the triangle-axiom and for which furthermore the accumulation-point property (compactness) is assumed. Moreover, it is demanded that for each pair of points of the space exactly one midpoint and two doubling points do exist; the thus defined spaces, "straightline spaces" already studied by Menger, satisfy all of Hilbert's axioms. Mr. Busemann now adds the further axiom that the straight connecting line of two points can be understood as the limit of a circle with infinite radius. He arrives at the result that by this "limitcircle-axiom" the classical Minkowski geometry with a convex gauge field originates.

For the evaluation of the work - which by the way has been looked through up to all particulars by two outstanding experts, Prof Alexandroff and Prof. Kolmogoroff -one must bear in mind that Mr. Busemann has picked the topic he himself and has carried it out in an absolutely independent manner without any help and stimulation. Therefore, it is a definitely independent scientific performance. However, I cannot suggest the best predicate for the work by taking into account that it deals not with a really fundamental problem.

Therefore, I suggest to valuate the work with grade II (very good).

- A handwritten cv of Busemann. Before beginning his studies (München 1 Sem. 1925/26, Göttingen 1926/27, Paris 1927/28, Göttingen 1928/1930), for 2 1/2 years he was working in commercial business in Hamburg and Berlin. Vater Dr. Alfreed Busemann, stellvertr. Direktor der Krupp A.-G. in Essen-Ruhr. Mutter: Olga, geb. Wetzlar. In his application for the examination, he states that he was stimulated to pick the topic of his thesis by the courses of Prof. Alexandroff. He also lists 26 courses attended during his studies. "Ich hörte Vorlesungen bei den Herren Professoren und Priatdozenten: Alexandroff, Bessel-Hagen, Bohr, Borel, Born, Caratheodory, Cartan, Courant, Franck, Graetz, Grandjot, Herglotz, Hilbert, Julia, Landau, Montel, Noether, Nordheim, Ostrowski, Perron, Walther, Van der Waerden, Wegner, Weyl, Wien."

- The doctoral diploma.

Correspondence related to the paper by Busemann and Feller, Zur Differentiation der Lebesgueschen Integrale [On the differentiation of Lebesgue's integrals]. Fundamenta Mathematicae 22 (1934) 226–256.

The English translation is by Manfred Karbe.

Kopenhagen, den 19.XI.33.

Lieber Herr Professor.

Über die Göttinger Ereignisse pflegen Sie ja besser orientiert zu sein als ich. Vor einigen Tagen schrieb mir Mesmer etwas, was Sie vielleicht noch nicht gehört haben und Sie vielleicht interessiert. Der Fall Nikuradse ist Ihnen wohl an sich bekannt, vielleicht auch, daß in einer Art Wutanfall verschiedene Leute aus dem Prandtlschen Institut ein Pamphlet gegen Nikuradse losgelassen hatten, das leider neben wahren auch einige offenbar blödsinnige Beschuldigungen enthielt. Im Juli war eine Verhandlung mit Zeugenaussagen, im September kam die Mitteilung, daß die Zeugen, welche gegen N. etwas ausgesagt hatten, zu entlassen wären. Die Begründung hierfür war ihrerseits blödsinnig, sodaß ein Protest Erfolg hatte, und am 26.X. kam, wie Mesmer schreibt, der Generaldirektor (wer das ist, weiß ich nicht) aus Berlin und fällte ein Urteil: 3 fristlos entlassen, Mesmer ab 1.XII. unter Gehaltskürzung wieder auf Bewährung eingestellt, zwei weitere ab 1.XI. wieder auf Bewährung eingestellt. Das Skandalöse ist aber, daß Prandtl sich hat zwingen lassen, das Urteil zu unterschreiben, worin man lesen kann, daß die Verurteilten das Ansehen des Deutschen Reiches geschädigt hätten.

Nachträglich steigt mein Respekt vor Levi-Civita unbeschränkt, welcher nach achtjährigem Faschismus nicht einmal den Beamteneid leisten wollte. Damals konnten wir alle noch nicht ganz beurteilen, was ein Druck bedeuten kann. Levi-Civita ist übrigens, wie er schreibt, sehr betrübt, daß er garnichts tun kann, aber: dato il regime e la scarsità di mezzi ....

Mir geht es hier ausgezeichnet. H. Bohr ist unglaublich freundlich, alle Augenblicke muß man dort essen, wo er irgendwie behilflich sein kann, ist er es, ohne daß man ihn darum gebeten hätte. Das ist für einen Emigranten, selbst wenn man, wie ich, finanziell gesichert ist, äußerst wichtig, zumal man in jeder Hitlerrede noch einige Tritte hinterher bekommt. Glücklicherweise ist mein Selbstvertrauen noch durch eine andere Sache gestärkt worden; Feller und mir ist es nämlich gelungen eine in einer berühmten Arbeit von Lebesgue (1910) gestellte und seitdem von verschiedenen bekannteren Mathematikern ohne Erfolg behandelte Frage zu lösen. Es handelt sich um folgendes: Wenn man irgend eine meßbare Menge K in der Ebene hat und betrachtet zu jedem Punkt P eine Folge sich auf ihn zusammenziehender achsenparalleler Würfel W_n, so ist (bis auf eine Nullmenge) für $P \subset K$ $\lim \frac{m(W_n K)}{m W_n} = 1$ und für $P \not\subset K$ $\lim \frac{m(W_n K)}{m W_n} = 0$. Die Frage ist, welche Mengen an Stelle der Würfel treten können, ohne daß dieser „Dichtesatz" verloren geht, insbesondere, ob man beliebige achsenparallele Rechtecke nehmen kann. Wir haben gezeigt, daß letzteres geht, daß man aber merkwürdigerweise nicht mehr beliebig gelegene Rechtecke nehmen darf. Man kann das Resultat noch verschiedentlich anders wenden und bekommt so Sätze über die Vertauschbarkeit der Differentiationsreihenfolge bei gemischten partiellen Ableitungen.

Ich wollte Sie fragen, an welchen Tagen um Weihnachten herum Sie in Göttingen sind. Ich werde ein oder zwei Tage dort sein, um sozusagen aus sentimentalen Gründen die letzte Phase des alten Göttingen noch kurz sehen und dann für immer Lebwohl zu sagen.

Herzliche Grüße

Ihr Herbert Busemann

Copenhagen, 19.11.33

Dear Professor.

You seem to be more informed than I am about the events in Göttingen. A few days ago, Mesmer wrote me something you might not yet have heard and you might be interested in. The case of Nikuradze as such is probably known to you, perhaps also that in a kind of tantrum various people from the Prandtl Institute had released a pamphlet against Nikuradze, which unfortunately contained, along with true accusations, some apparently stupid accusations. In July there was a trial with testimony, and in September came the announcement that the witnesses who had said something against N. were to be dismissed. The reasoning was, on their part, stupid, so that a protest had been successful, and on 10/26 came, as Mesmer writes, the director general (who that is I do not know) from Berlin and made a judgment: 3 dismissed without notice, Mesmer from 12/1 back on probation with salary cut, two more back on probation from 11/1. But the scandalous thing is that Prandtl let himself be forced to sign the verdict, in which it can be read that the condemned had damaged the reputation of the German Reich.

My respect for Levi-Civita continues to rise unreserved, who, after eight years of fascism, did not even want to swear the civil servant oath. At that time, we all could not yet fully assess what a pressure could mean. Levi-Civita, by the way, writes, he is very sorrowful that he can do nothing whatsoever, but: dato il regime e la scarsità di mezzi ....

I am doing very well here. H. Bohr is incredibly friendly, every moment one must eat there, where he can somehow be of help, he is, without being asked. This is extremely important for an emigrant, even if, like me, one is financially secure, especially since in each of the speeches of Hitler one still gets a few kicks from behind. Fortunately, my self-confidence has been strengthened by another thing; Feller and myself have succeeded in solving a question which has been put up in a famous work by Lebesgue (1910) and has since been unsuccessfully treated by various well-known mathematicians. This is the following: If we have any measurable set K in the plane, and, for each point P, consider a sequence of axis-parallel cubes W_n contracting to P, it is true that (except for a null set) $\lim \frac{m(W_n K)}{mW_n} = 1$ for $P \subset K$, and $\lim \frac{m(W_n K)}{mW_n} = 0$ for $P \not\subset K$. The question is which sets can be substituted for the cubes without loss of this "density theorem", in particular whether arbitrary axis-parallel rectangles can be taken. We have shown that the latter is the case, but, strangely enough, one cannot take arbitrary rectangles. The result can still be formulated differently, thus giving theorems on the interchangeability of the differentiation order in mixed partial derivatives.

I wanted to ask you on which days around Christmas you are in Göttingen. I will be there for a day or two, so to say for the sake of the sentimental reasons, to catch a brief glimpse of the last phase of the old Göttingen, and then to say good-bye forever.

Best regards

Yours, Herbert Busemann

* * *

Kopenhagen-Hellerup, den 18.November 33.
Onsgaardsvej 25, Pension Olsen.

Lieber Herr Courant,

haben Sie recht herzlichen Dank für Ihren Luftpostbrief und für Ihr Interesse! Ich habe für Leeds alles getan, was ich nur konnte, und habe mich auch s o f o r t beworben, ohne auf irgendwelche Informationen zu warten. Was ich Ihnen darüber schrieb, war eine fast wörtliche Wiedergabe dessen, was mir die Notgemeinschaft in Zürich geschrieben hat: ich weiss selbst kein Wort mehr darüber. Wie weit die Sache also ernst zu nehmen ist, kann ich nicht beurteilen. Ich bin mir auch absolut im klaren darüber, dass ich wissenschaftlich weniger geleistet habe, als andere, und dass ich den Lehrerfolg, den ich einigermassen hatte, nicht mitnehmen kann, so dass ich meine Chance (hier wie überhaupt) in der geringen Anziehungskraft der Stelle sehe. Wenn wieder nichts daraus wird, so bin ich vorläufig doch noch nicht am Verhungern, und freue mich sogar, endlich wieder in Ruhe Mathematik lernen zu können, und hoffe, dass sich noch rechtzeitig etwas hinreichend Schlechtes finden wird.

Zu Ihrer Frage! Ueber den Deutschen Unterstützungsfond in Leeds, der die Mittel für die drei Stellen herbeischaffen soll, weiss ich nichts mehr, als eben diesen Satz. Ich nehme an, dass es sich um eine rein örtliche Angelegnheit [sic] handelt. Vielleicht bezieht sich aber Ihre Frage auf die "Notgemeinschaft" – wenn nicht, so wird das folgende doch nicht beissen.

Das Unternehmen heisst "Notgemeinschaft Deutscher Wissenschaftler im Ausland", Zürich, Uraniastr. 40 II. Ich habe inzwischen gehört, dass mein Name dort durch Frau Stenzel, bzw. Anna Stenzel, bekannt wurde, die die Leiterin des Unternehmens zufällig kennt. Von Frau Stenzel habe ich folgendes erfahren:

Leiter war früher ein Mediziner Schwarz, der jetzt in Konstantinopel ist. Jetzt leitet ein andrer Professor die Sache, doch soll die Arbeit hauptsächlich Frau Thea Wolf leisten. Es soll ein gross aufgezogener Betrieb sein, dessen Vermittlung auch viele noch amtierende Professoren benutzen. Für Mathematiker sein ein Stipendium ausgeschrieben worden, für das freilich bereits 5 Bewerber vorhanden seien (ich nicht!). Wer bei allen den Sachen berücksichtigt werden will, muss einen Lebenslauf mit einer Liste der Publikationen und Referenzen in dreifacher Ausführung einsenden. – Uebrigens soll es zur Zeit für Mathematiker Professuren in Russland und Indien geben. Die Stelle in Leeds – also der mysteriöse Unterstützungsfond – scheint finanziell von diesem Unternehmen unabhängig zu sein, denn die Bewerbung sollte direkt bei der Universität eingereicht werden, und die Notgemeinschaft hat mir nur die Tatsache mitgeteilt.

Das ist alles, was ich über die Angelegenheit weiss.

Mir geht es hier ganz ausgezeichnet. Busemann und ich haben soeben eine Arbeit beendet, die ich noch vor Toresschluss des Zweiten Reiches begonnen hatte. Wir sind beim Niederschreiben, wobei Busemann darauf achtet, dass hinreichend viele Worte ohne Bedeutung vorkommen. Ich dachte immer, man sollte mit möglichst wenig Worten sachlich sein. Das war Bescheidenheit. Busemann behauptet aber, es erinnere an Minkowski und werde anders ausgelegt. Zum Glück ist der Inhalt – über die Differenzierbarkeit der Lebesgueschen Integrale – so nüchtern, dass nicht viele Konjunktive Platz haben. Ich hatte mich dem nüchternen Thema zugewandt, als Reaktion auf die Feller–Tornier Arbeit, (un)seligen Andenkens.

Nochmals besten Dank und viele herzliche Grüsse

Ihr Willy Feller

[...] I feel very well here. Busemann and I have just finished a work which I had begun in the nick of time of the Second Reich. We are in the process of writing down, and Busemann pays attention to the fact that there are sufficiently many words without meaning. I always thought one should be factual with as few words as possible. That was modesty. Busemann, however, claims it reminds of Minkowski and is interpreted differently. Fortunately, the content – about differentiability of Lebesgue integrals – is such that not many subjunctives have room. I had turned to the dry theme, in response to the Feller–Tornier work, with (un)fortunate memory.

Thanks again and many kind regards.

Yours, Willy Feller

Kopenhagen-Hellerup, den 18.November 33.
Onsgaardsvej 25, Pension Olsen.

Lieber Herr Courant,

haben Sie recht herzlichen Dank für Ihren Luftpostbrief und für Ihr Interesse! Ich habe für Leeds alles getan, was ich nur konnte, und habe mich auch s o f o r t beworben, ohne auf irgendwelche Informationen zu warten. Was ich Ihnen darüber schrieb, war eine fast wörtliche Wiedergabe dessen, was mir die Notgemeinschaft in Zürich geschrieben hat: ich weiss selbst kein Wort mehr darüber. Wie weit die Sache also ernst zu nehmen ist, kann ich nicht beurteilen. Ich bin mir auch absolut im klaren darüber, dass ich wissenschaftlich weniger geleistet habe, als andere, und dass ich den Lehrerfolg, den ich einigermassen hatte, nicht mitnehmen kann, so dass ich meine Chance (hier wie überhaupt) in der geringen Anziehungskraft der Stelle sehe. Wenn wieder nichts daraus wird, so bin ich vorläufig doch noch nicht ~~ganz~~ am Verhungern, und freue mich sogar, endlich wieder in Ruhe Mathematik lernen zu können,und hoffe, dass sich noch rechtzeitig etwas hinreichend Schlechtes finden wird.

Zu Ihrer Frage! Ueber den Deutschen Unterstützungsfonds in Leeds, der die Mittel für die drei Stellen herbeischaffen soll, weiss ich nichts mehr, als eben diesen Satz. Ich nehme an, dass es sich um eine rein örtliche Angelegnheit handelt. Vielleicht bezieht sich aber Ihre Frage auf die "Notgemeinschaft" - wenn nicht, so wird das folgende doch nicht beissen.

Das Unternehmen heisst "Notgemeinschaft Deutscher Wissenschaftler im Ausland", Zürich, Uraniastr. 40 II. Ich habe inzwischen gehört, dass mein Name dort durch Frau Stenzel, bzw. Anna Stenzel, bekannt wurde, die die Leiterin des Unternehmens zufällig kennt. Von Frau Stenzel habe ich folgendes erfahren:

Leiter war früher ein Mediziner Schwarz, der jetzt in Kostantinopel ist. Jetzt leitet ein andrer Professor die Sache, doch soll die Arbeit hauptsächlich Frau Thea Wolf leisten. Es soll ein gross aufgezogener Betrieb sein, dessen Vermittlung auch viele noch amtierende Professoren benutzen. Für Mathematiker sei ein Stipendium ausgeschrieben worden, für das freilich bereits 5 Bewerber vorhanden seien (ich nicht!).
Wer bei allen den Sachen brücksichtigt werden will, muss einen Lebenslauf mit einer Liste der Publikationen und Referenzen in dreifacher

Ausführung einsenden. - Uebrigens soll es zur Zeit für Mathematiker Professuren in Russland und Indien geben. Die Stelle in Leeds - also der mysteriöse Unterstützungsfond$ - scheint finanziell von diesem Unternehmen unabhängig zu sein, denn die Bewerbung sollte direkt bei der Universität eingereicht werden, und die Notgemeinschaft hat mir nur die Tatsache mitgeteilt.

Das ist alles, was ich über die Angelegenheit weiss.

Mir geht es hier ganz ausgezeichnet. Busemann und ich haben soeben eine Arbeit beendet, die ich noch vor Toresschluss des Zweiten Reiches begonnen hatte. Wir sind beim Niederschreiben, wobei Busemann darauf achtet, dass hinreichend viele Worte ohne Bedeutung vorkommen. Ich dachte immer, man sollte mit möglichst wenigen Worten sachlich sein. Das war Bescheidenheit. Busemann behauptet aber, es erinnere an Minkowski und werde anders ausgelegt. Zum Glück ist der Inhalt - über die Differenzierbarkeit der Lebesgueschen Integrale - so nüchtern, dass nicht viele Konjunktive Platz haben. Ich hatte mich dem nüchternen Thema zugewandt, als Reaktion auf die Feller-Tornier Arbeit, (un)seligen Andenkens.

Nochmals besten Dank und viele herzliche Grüsse
Ihr
Willy Feller

Kopenhagen, den 19. XI. 33.

Lieber Herr Professor.

Über die Göttinger Ereignisse pflegen Sie ja besser orientiert zu sein als ich. Vor einigen Tagen schrieb mir Messner etwas, was Sie vielleicht noch nicht gehört haben und Sie vielleicht interessiert. Der Fall Nikuradse ist Ihnen wohl an sich bekannt, vielleicht auch, daß in einer Art Unterfalle verschiedene Leute aus dem Prandtlschen Institut ein Pamphlet gegen Nikuradse losgelassen hatten, daß leider neben wahren auch einige offenbar blödsinnige Beschuldigungen enthielt. Im Juli war eine Verhandlung mit Zeugenaussagen, im September kam die Mitteilung, daß die Zeugen, welche gegen N. etwas ausgesagt hatten, zu entlassen wären. Die Begründung hierfür war ihrerseits blödsinnig, sodaß ein Protest Erfolg hatte, und am 26. X. kam, wie Messner schreibt, der Generaldirektor (wer das ist, weiß ich nicht) aus Berlin und fällte ein Urteil: 3 fristlos entlassen, Messner ab 1. XII unter Gehaltskürzung wieder auf Bewährung eingestellt, zwei weitere ab 1. XI. wieder auf Bewährung eingestellt. Das Skandalöse ist also, daß Prandtl sich hat zwingen lassen, das Urteil zu unterschreiben, worin man lesen kann, daß die Verurteilten das Ansehen des Deutschen Reiches geschädigt hätten.

Nachträglich steigt mein Respekt vor Levi-Civita unbeschränkt, welcher nach achtjährigem Faschismus nicht einmal den Beamteneid leisten wollte. Damals konnten wir alle noch nicht ganz beurteilen, was ein Druck bedeuten kann. Levi-Civita ist übrigens, wie

er schreibt, sehr betrübt, daß er garnicht turnen kann, aber
: dato il regime e la scarsità di mezzi...

M.- geht es hier ausgezeichnet. H. Bohr ist unglaublich freundlich, alle Augenblicke muß man dort essen, wo er irgendwie behilflich sein kann, ist er es, ohne daß man ihn darum gebeten hätte. Das ist für einen Emigranten, selbst wenn man, wie ich, finanziell gesichert ist, äußerst wichtig, zumal man in jeder Hitlerrede noch einige Tritte hinterher bekommt. Glücklicherweise ist mein Selbstvertrauen noch durch eine andere Sache gestärkt worden; Feller und mir ist es nämlich gelungen eine in einer berühmten Arbeit von Lebesgue (1910) gestellte und seitdem von verschiedenen bekannten Mathematikern ohne Erfolg behandelte Frage zu lösen. Es handelt sich um folgendes: Wenn man irgend eine meßbare Menge K in der Ebene hat und betrachtet zu jedem Punkt P eine Folge sich auf ihn zusammenziehender achsenparalleler Würfel W_n, so ist (bis auf eine Nullmenge) für $P \subset K$ $\lim \frac{m(W_n K)}{m W_n} = 1$ und für $P \not\subset K$ $\lim \frac{m(W_n K)}{m W_n} = 0$. Die Frage ist, welche Mengen an Stelle der Würfel treten können, ohne daß dieser „Dichtesatz" verloren geht, insbesondere, ob man beliebige achsenparallele Rechtecke nehmen kann. Wir haben gezeigt, daß letzteres geht, daß man aber merkwürdigerweise nicht mehr beliebig gelegene Rechtecke nehmen darf. Man kann das Resultat noch verschiedentlich anders wenden und bekommt so Sätze über die Vertauschbarkeit der Differentiationsreihenfolge bei gemischten partiellen Ableitungen.

Ich wollte Sie fragen, an welchen Tagen um Weihnachten herum Sie in Göttingen sind. Ich werde ein oder zwei Tage dort sein, um sozusagen aus sentimentalen Gründen die letzte Phase des alten Göttingen noch kurz sehen und dann für immer Lebewohl zu sagen.

Herzliche Grüße

Ihr Herbert Busemann.

Kopenhagen, 12.V.1935.

Lieber Herr Professor.

Ihren Brief, für den ich herzlich danke, finde ich jetzt erst, nach einer längeren Reiseperiode, vor; die letzte Zeit war ich in Deutschland, wohin ich mir Briefe prinzipiell nicht nachschicken lasse. Ich kann daher heute erst antworten. Die ungebührliche Länge der Antwort empfinden Sie hoffentlich nicht als lästig, da Sie mich so sehr auffordern, sowohl über die allgemeinen wie auch über meine persönlichen Verhältnisse ausführlich zu berichten.

Wie ich wohl damals ankündigte, bin ich Mitte März für 10 Tage nach Stockholm und von dort nach Lappland gefahren, $68^\circ\,30'$ nördlicher Breite. Die Landschaft ist dort ungeheuer weitzügig; wenige, weit auseinanderliegende Berge, große Seen und Wasserfälle, letztere habe ich natürlich nur in erstarrter Form gesehen. Dem Prinzip nach habe ich an Lappland dasselbe auszusetzen wie an Arosa, nur daß man in Lappland keinen berechtigten Anspruch auf alpines Gelände machen kann. Die Schweden sind steilere Hänge nicht gewöhnt, laufen dagegen auf flachem Gelände mit unglaublichem Tempo, sodaß ich kaum mitkam. Bei etwas steileren, für alpine Begriffe aber sehr harmlosen Abfahrten glänzte ich dagegen, obwohl ich in den Alpen keineswegs als Meister auftreten kann. Den größten Eindruck hat mir in Lappland das Nordlicht gemacht.

Von der Skistation aus fuhr ich für ein paar Tage in die Lofoten. Selten habe ich so etwas Schönes gesehen. Denken Sie sich die Berge des Vierwaldstättersees ins Meer abfallend, dabei manchmal ganz enge Durchfahrten, manchmal weite Wasserflächen umrahmend, schließlich ganz ins offene, schwarzblaue Meer hinausgehend, dann haben Sie ungefähr eine Vorstellung. Großenteils sind die Bergformen ganz bizarr, das Gestein ist ein rötlicher Granit, auf dem Ganzen etwas Schnee. Einen Tag habe ich mich bei der Kabeljaufischerei herumgetrieben. Die Fischerflotte, ca. 6000 Boote, folgt dem Schwarm der Fische, der jedes Jahr die Lofoten passiert, durch zwei Monate, und zwar liegen die vielen Boote ganz dicht beieinander auf einem ganz kleinen Raum. Die Fische schwimmen ca. 25m tief und werden (ohne Stock direkt über dem Bordrand) geangelt, an Land ihrer Köpfe und Schwänze beraubt, aufgehängt und getrocknet, worauf sie als Stockfische die Welt durchwandern.

Von den Lofoten aus bin ich eine Reihe von Nächten durchfahrend, von einer Hotelnacht in Stockholm abgesehen (ich kann hier vielleicht für den Fall, daß Sie es noch nicht wissen, einfügen, daß Feller sein Stipendium bis Ende des Jahres verlängert bekommen hat), schließlich in Marburg bei Rellich gelandet. Rellich und Reidemeister scheinen sich den Umständen nach wohl zu befinden. Die Studentenzahl ist, wie überall und in allen Fächern, gleich $\frac{\varepsilon}{2}$. Rellich erzählte folgendes über W. W. Herr Ziegenbein, wie Sie wohl wissen, der neue Zeus des Göttinger Mathematischen Instituts, der auf Grund eines Parteibuchs geringerer Ordnungszahl Herrn Tornier den Rang abgelaufen hat, hielt einen Vortrag über Hilbert und Brouwer (in der analytischen Fortsetzung der Mafa). Er kannte nicht mehr davon als die beiden Namen Formalismus und Intuitionismus und erklärte demgemäß die Stellungnahme der neuen Jugend als gegeben. Arnold Schmidt ging mangels Diskussion zu ihm hin und meinte, so einfach sei das doch nicht, und wurde zu seinem Erstaunen aufgefordert, das nächste Mal seinerseits darüber zu reden. Der Arme, er hatte nicht bedacht, daß die Volkswut das nicht zuzulassen hatte. Herr Weber brachte sie zum Kochen, indem er sich entsann, daß Schmidt im Jahre 1930 (!) gesagt hatte, die Sozialdemokraten hätten besser den Kriegsdienst verweigert. Der Vortrag fand natürlich nicht statt, und Schmidt war längere Zeit damit beschäftigt, hinter Weber herzulaufen und ihn um Rücknahme der Behauptung zu bitten.

Von Marburg fuhr ich nach Essen; ich kam nach Stresa und vor Genf nach Essen. Ich hörte, daß man nach Erklärung der Wehrpflicht zunächst sehr pessimistisch war und mit Repressalien rechnete, nach Stresa war man ziemlich optimistisch, da man die Tragweite nicht erkannte, und wurde nach Genf wieder sehr gedrückt und zwar zum Schluß in allen Schichten, da merkwürdigerweise alles Außenpolitische, z. B. auch der Mac-Donald-Artikel in den deutschen Zeitungen zu finden war. Man hat das Bewußtsein, in ähnlicher Lage zu sein wie 1914.

Was die Wirtschaft angeht, ist folgendes zu sagen: Es herrscht eine starke Binnenkonjunktur, aber doch kein eigentlicher Optimismus, da man nicht glaubt, daß sich eine reine Binnenkonjunktur à la longue aufrecht erhalten läßt. Private Berechnungen sollen auch zeigen, das von früheren Regierungen her geheime Fonds dagewesen sein müssen, da sonst der Devisenbestand schon lange negativ sein müßte. Ferner werden von privater Seite die Staatsschulden (durch Finanzierungswechsel) auf $8\frac{1}{2}$ statt offiziell $4\frac{1}{2}$ Milliarden Mark berechnet. Stimmungsmäßig schien mir im Rheinland, da man die Mißerfolge der Außenpolitik mehr der Böswilligkeit der andren als den eigenen Fehlern zuschreibt, eigentlich eher eine Verschiebung zugunsten Hitlers eingetreten zu sein, abgesehen von den ernsthaft katholischen Kreisen. Der Kampf gegen letztere ist sehr heftig, zur Illustration kann vielleicht folgendes dienen: Auf einer Kruppschen Zeche sollten vier Lehrlinge (spätere Bergfachstudenten) eingestellt werden, drei waren aus der H. J. hervorgegangen, der vierte gehörte einem katholischen Verband an und war offenbar der tüchtigste. Es wurde gegen Einstellung dieses letzteren Einspruch erhoben, und obwohl die Zechenleitung sich wegen der Qualitäten des jungen Mannes sehr für ihn einsetzte, konnte sie ihn nicht behalten. Personelle Eingriffe kommen überhaupt sehr weitgehend vor, besonders auf folgende Art (z. B. bei meinem Frankfurter Onkel). Man erklärt plötzlich, irgend ein Werk brauche einen Direktor mehr und zwar den und den Herrn. In dem Frankfurter Fall ließ sich der betreffende P. G. wegen Unzulänglichkeit völlig kaltstellen, sodaß das Ganze nur eine Belastung des Spesenkontos bedeutet.

Von Essen fuhr ich nach Leipzig zu v. d. Waerden, mit dem ich einiges über algebraische Geometrie besprechen wollte. Dort scheinen, abgesehen von der minimalen Studentenzahl die Verhältnisse ganz angenehm zu sein; v. d. W. will sich sogar in Leipzig ein Haus bauen. Er sagt aber trotzdem, für einen deutschen Beamten, selbst als Holländer, sei es doch sehr bedrückend, zu sehen, wie durch den Kampf jedes gegen jeden alles systematisch zugrunde gerichtet werde. „Wenn A. und B. ein geordnetes Paar von Kollegen sind, such[t] Herr A. Herrn B. herauszuschmeißen.“ In Leipzig sei (im Gegensatz zu Dresden) die Stimmung ziemlich nazifeindlich infolge des stark reduzierten, für Leipzig sehr wichtigen internationalen Geschäfts.

Der letzte Haltepunkt in Deutschland war Berlin, wo ich 8 Tage war. Dort war ich aufs höchste erstaunt, wie offen sich die nazifeindliche Stimmung bemerkbar macht. Am 1. Mai wurde die Kundgebung überallhin übertragen, man konnte ihr nicht entgehen. Niemand muckste sich, nicht einen ε-Beifall habe ich vernommen. Das Ende erlebte ich in einem Restaurant, wo ich zu Mittag aß. Genau zum Schluß, als Dr. Ley sagte: hiermit schließen wir die diesjährige Kundgebung, ertönte ostentativer Beifall – ich hätte es nicht für möglich gehalten, vor allem den Ton nicht. Ich war mit einem zufällig getroffenen Göttinger (arischen) Philologen in einer Bar, es war allerdings schon $\frac{1}{3}3$, als man Elisabeth Bergener [Bergner] hochleben ließ, lauter ähnliche Erlebnisse hatte ich (soeben lese ich übrigens, daß eine witzelnde Truppe, die ich auch gehört habe, ins Kola gewandert ist).

Landau habe ich auch besucht, er wird immer schrulliger. Er wohnt bei seiner Mutter in einem Minimalzimmer, wo er gleichzeitig schläft und arbeitet, hat im Ganzen nur 20 Bücher in Berlin und die nicht mal in seinem Zimmer. Frau Landau wohnt bei Susi, dreimal die Woche (genau) sehen sich die beiden. Interessant ist, daß Susi nun ihr chemisches Examen machen darf, wegen starker Unterschreitung des numerus clausus. Sonst macht der Antisemitismus in Deutschland Fortschritte, in verschiedenen Städten wurden an jüdische Geschäfte beleidigende Plakate angeschlagen und einige Scheiben gingen in Stücke, sogar in Düsseldorf.

Sie hatten die Freundlichkeit, sich nach einem Studienzuschuß für mich erkundigen zu wollen. Unabhängig davon hatte ich schon lange die Absicht, Sie bezüglich meiner um Rat zu fragen. Meine Idee war bisher folgende: Ich wollte zuallererst die politische Entwicklung in Deutschland abwarten; aber was das Regierungssystem wie die Arierfrage angeht, ist wohl sobald keine Änderung zu erwarten. Ferner wollte ich sozusagen unter Aufrechterhaltung des vermögensmäßigen status quo so lange hier warten, bis wieder ein halbwegs regelmäßiger Zahlungsverkehr Deutschlands mit dem Auslande hergestellt ist, sodaß ich auf eine gewisse monatliche Summe bestimmt rechnen kann. Aber auch in diesem Punkt ist noch keine Änderung abzusehen. Wegen des niedrigen Standes der Krone kann ich hier trotz der Unregelmäßigkeit der Geldüberweisungen allein von diesen leben. Sollten sie ganz aufhören, so könnte ich hier ohne Einschränkungen 6 Jahre leben. Bei einem monatlichen Verbrauch von $ 130.– könnte ich in Amerika von demselben Geld etwa 3 Jahre leben. Augenblicklich bekomme ich aus Deutschland monatlich durchschnittlich etwa $ 70.–, es droht aber, wie gesagt, völlig aufzuhören (Neugebauer erzählt auch von großen Schwierigkeiten bei Hirschwald und Springer). Ich habe es also trotz allem an sich sehr gut und schäme mich beinahe, daran zu denken, andren einen Studienzuschuß fortzunehmen, andererseits möchte ich mir natürlich für den Fall einer völligen deutschen Abriegelung mein Geld nach Möglichkeit als Reserve verwahren. Zunächst wäre ich Ihnen dankbar, wenn Sie mir mitteilen könnten, ob die von mir angegebene Summe von $ 130.– pro Monat eine vernünftige Zahl ist. Ich möchte nicht bei jedem Cent überlegen müssen, ob ich ihn ausgeben darf oder nicht, nicht nur aus Bequemlichkeit sondern auch weil ich bemerkt habe, daß man dann sofort mit dem höchsten Mißtrauen wie jemand behandelt wird, der bald um Hilfe zu suchen droht.

Nun die mathematische Situation. Hier wesentliche Anregungen zu bekommen, halte ich für unwahrscheinlich. Nielsen hat sich seit längerer Zeit mit anderen Dingen beschäftigt und scheint wohl zu wünschen, wieder in topologischen Schwung zu kommen, es aber nicht zu tun. Ich habe mich daher sozusagen selbständig gemacht und einen lang gehegten Wunsch durchgeführt, nämlich algebraische Geometrie gelernt, d. h. v. d. Waerdens Arbeiten studiert und modulo dieser den italienischen Normalkursus in Mathematik verwandelt (und wie berichtet, hier daüber Vorlesungen gehalten). Das hat mich den ganzen Winter intensiv in Anspruch genommen. Von v. d. Waerden hoffte ich einige Tips zu bekommen, wie man solche Schwierigkeiten leichter überwindet, er meinte jedoch, ich sei der einzige Mensch in Europa, der sich diese Mühe gemacht habe, und auch er könne die italienischen Arbeiten absolut nicht ohne weiteres in Ordnung bringen. Vielleicht ist es vom praktischen Standpunkt aus leichtsinnig, seine Zeit damit zu verbringen, statt zu publizieren, da ja in andren Ländern die Zahl der Arbeiten das Entscheidende zu sein scheint. Vom überirdischen Standpunkt aus glaube ich jedoch, mit ungewöhnlichem Ertrag für mich gearbeitet zu haben, da ich durch die dauernden Lücken und Unexaktheiten, mit denen ich mich auseinandersetzen mußte, in die Sache ganz anders eingedrungen bin, als ich das sonst bei rezeptiver Aufnahme zu tun pflege. Von demselben Standpunkt aus würde ich mir selbst etwa noch ein halbes Jahr Zurückgezogenheit verschreiben, in welchem ich mich mit den Lefschetzschen Arbeiten befassen würde. Diese sind bekanntlich besonders schwer verständlich, aber ich glaube, sie führen an die heute aussichtsreichste Seite der algebraischen Geometrie.

Daß das „wirkliche" Interesse, das die Princetoner damals für mich gehabt haben, heute noch besteht, halte ich leider für unwahrscheinlich. Es entsprang zunächst einer persönlichen Sympathie von Lefschetz für mich, die wegen der sehr kurzen Bekanntschaft, gewiß längst verschwunden ist, und dann war damals der Zug nach dem Westen ja noch nicht so groß und selbstverständlich. Inzwischen ist eine große Zahl erstklassiger Leute nach Amerika gekommen und die nicht nur von Ihnen bemerkte Fremdenfeindlichkeit hat zum Teil sicher in der Saturierung ihren Grund.

Meine Lage kennen Sie jetzt genau; was ich tun soll, läßt sich schwer sagen. Aber eine genaue Antwort auf die Frage, was Sie an meiner Stelle täten, würde mir sehr am Herzen liegen. Dafür, hier zu bleiben, spricht vom mathematischen Standpunkt aus eigentlich nichts und sonst zunächst die billigen Lebensverhältnisse und die angenehmen persönlichen Beziehungen. Für Amerika spricht vom mathematischen Gesichtspunkt aus viel. Princeton sieht von hier aus bereits etwas zu groß aus, und ich habe mal an Baltimore gedacht, weil mir das Buch von Zariski einen so großen Eindruck gemacht hat. (Es hat nur den einen Fehler, in den Ergebnissen statt in der gelben Sammlung erschienen und so zu komprimiert zu sein). Aber Sie kennen die Verhältnisse ja besser und könnten mir vielleicht auch darüber Auskunft geben. Was bei Ihrer Beurteilung meiner Lage vielleicht auch wichtig ist: heute bin ich 30 Jahre alt geworden.

Mit einem Schluß haben Sie vielleicht schon garnicht mehr gerechnet, trotzdem bin ich jetzt soweit. Mit vielen Grüßen an Sie und Ihre Familie

Ihr Herbert Busemann

Kopenhagen, May 12, 1935

Dear Professor.

Your letter, for which I sincerely thank you, I just find now, after a longer travel period; the last time I was in Germany, whereto in principle I do not let letters redirect. So I can reply only today. Hopefully you will not feel the unusual length of the letter as annoying since you are asking me so much to report on the general as well as my personal circumstances in detail.

As I announced then, I went to Stockholm in mid-March for ten days, and from there to Lapland, at latitude 68° 30′ N. The landscape is immensely extensive; few, widely separated mountains, great lakes and waterfalls, the latter of course I have only seen in a frozen form. In principle, I have the same objections against Lapland as against Arosa, only that one is not qualified to make any claim to alpine terrain in Lapland. The Swedish are not accustomed to steeper slopes, but run on flat ground with incredible speed, so I could hardly keep up. In the case of somewhat steeper but for alpine terms very harmless descents, however I shone although I can by no means appear as a master in the Alps. For me the most important impression in Lapland were the Northern Lights.

From the ski station I drove for a few days to the Lofoten. Seldom have I seen something so beautiful. Imagine the mountains of the Lake Lucerne were falling into the sea, sometimes narrow passages, thereby sometimes encircling wide water surfaces, finally going out into the open, black and blue sea, then you have a rough idea. For the most part the mountain shapes are very bizarre, the rock is reddish granite, on the whole some snow. One day, I've been hanging out at cod fishery. The fishing fleet, about 6,000 boats, follows the swarm of fish passing through the Lofoten Islands every two months, and the many boats are quite serried in a localized manner. The fish swim about 25m deep and are fished (without rod directly over the edge of the board), ashore robbed of their heads and tails, hung and dried, wandering through the world as stockfish.

From the Lofoten Islands, I have been traveling through a series of nights, apart from one hotel night in Stockholm (in the event that you do not know it yet I can perhaps add that Feller got his scholarship extended up to the end of the year), to finally arrive at Rellich in Marburg. Rellich and Reidemeister seem to be well according to circumstances. The number of students, as everywhere and in all disciplines, is equal to $\frac{\varepsilon}{2}$. Rellich told the following about W. W.[1] As you may well know, Mr. Ziegenbein, the new Zeus of the Göttingen

[1] W. W. stands for Werner Weber, Edmund Landau's assistant since Dec. 1, 1928. Weber joined both the NSDAP and SA in July resp. August 1931, and replaced Rellich as acting director of the mathematical institute.

Mathematical Institute, who has outstripped Mr. Tornier on the grounds of a party membership book of lower ordinal number, gave a lecture on Hilbert and Brouwer (in the analytic continuation of Mafa[2]). He knew not more about it than the two names formalism and intuitionism, and thus declared the comments of the new youth as given. Arnold Schmidt went to him for lack of discussion and said that it was not so easy, and was asked to his own surprise to speak himself about it next time. The poor man, he had not considered that the rage of the people had not to admit this. Mr. Weber brought them to the boil by recalling that Schmidt had said in 1930 that the Social Democrats had better refused the military service. The lecture, of course, did not take place, and Schmidt was for a long time busy running after Weber and asking him to take back the assertion.

From Marburg I went to Essen; I came after Stresa[3] and before Geneva[4] to Essen. I heard that, after the declaration of compulsory military service, one was very pessimistic at first and reckoned with reprisals, after Stresa one was fairly optimistic as one didn't see the consequences, and after Geneva depressed again, in fact eventually in all social classes, as strangely all the foreign policy, for example the MacDonald article[5], was to be found in the German newspapers. One has the consciousness to be in a similar situation as in 1914.

As far as the economy is concerned, the following is to be said: There is a strong domestic economy, but there is no real optimism, since one does not believe that a purely domestic economy can be sustained in the long run. Private calculations are also supposed to show that secret funds originating from previous governments must have existed since otherwise foreign currency holdings would have been negative for a long time. Furthermore, from the private side the state debt (by means of finance bills) is calculated at $8\frac{1}{2}$ instead of officially $4\frac{1}{2}$ billion Marks. Atmospherically, in the Rhineland, as the failure of foreign policy is attributed to the malice of others rather than to one's own mistakes, it seemed to me to be a shift in favor of Hitler, apart from the seriously Catholic circles. The struggle against the latter is very vehement, the following may serve the as an illustration: On a Krupp coal mine, four apprentices (later mining students) were to be recruited, three were from the H. J.[6], the fourth belonged to a Catholic Union and was evidently the most capable. There was an objection to

---

[2]Abbreviation of **Ma**thematische **Fa**kultät, Department of mathematics.

[3]A city at Lake Maggiore, where, in response to Hitler's renouncement on March 16, 1935, of the disarmament clauses of the Treaty of Versailles and the reintroduction of conscription, a conference was held on April 11–14, 1935, resulting in a resolution between French foreign minister Pierre Laval, British prime minister Ramsay MacDonald, and Italian prime minister Benito Mussolini, opposing "any unilateral repudiation of treaties which may endanger the peace of Europe".

[4]Council Session in Geneva on April 15–17, 1935, of the League of Nations, the resulting resolution contained a formal condemnation of the German government's decision to unilaterally renounce the disarmament clauses of the Versailles Treaty.

[5]In an article entitled "Peace, Germany and Stresa" and published in the *News Letter* on April 25, 1935, Ramsay MacDonald, in one of the strongest indictments of Hitler's Germany ever made by a member of the British government, accused Germany of destroying European confidence in the outlook for continued peace.

[6]"Hitlerjugend", Hitler youth.

the appointment of this latter, and although the coal mine's management was very much in favor of him because of the young man's qualification, it could not keep him. Personnel interventions are very far-reaching, especially in the following way (for example with my uncle from Frankfurt). One suddenly declares that some plant needs an additional director, namely such and such gentleman. In the case of Frankfurt, the P. G.[7] in question let completely sideline himself because of his inadequacy, so that the whole thing was only a burden on the account expenses.

From Essen I went to Leipzig to see v. d. Waerden, with whom I wanted to discuss some aspects of algebraic geometry. There, except for the minimal number of students, the conditions seem quite pleasant; v. d. W. even wants to build a house in Leipzig. But he nevertheless says that for a German official, even as a Dutchman, it is very depressing to see how everything is systematically ruined by the struggle of everyone against everyone. "If A. and B. are an ordered pair of colleagues, Mr. A. seeks to kick Mr. B. out." In Leipzig (as opposed to Dresden), the atmosphere is quite Nazi-hostile as a result of the greatly reduced international business, which is very important to Leipzig.

The last stop in Germany was Berlin, where I was 8 days. There I was wonderstruck how open the Nazi-hostility makes its presence felt. The May 1st rally was broadcast everywhere, you could not miss it. Nobody muttered, I did not hear an ε-applause. I ended up at a restaurant where I ate lunch. At the very end, when Dr. Ley[8] said: "We are closing this year's demonstration", there was an ostentatious applause – I would not have considered it possible, especially not the sound. I was in a bar with a randomly met (Aryan) philologist from Göttingen, however it was already $\frac{1}{3}3$, when Elisabeth Bergner[9] was given three cheers. I had very similar experiences (I have just read, by the way, that a joking company, which I have also heard, has migrated into the Kola[10]).

I have also visited Landau, he is getting more and more quirky. He lives with his mother in a minimal room where he both sleeps and works, he has a total of only 20 books in Berlin, and these not even in his room. Mrs. Landau lives with Susi[11], the two meet each other (exactly) three times a week. It is interesting that Susi is now allowed to make her chemical examination because of the severe lower deviation of the numerus clausus. Otherwise, anti-Semitism is progressing in Germany, in various cities, insulting bills were posted on Jewish shops, and some panes came asunder, even in Düsseldorf.

You had the kindness to inquire about a study grant for me. Independently, I long had the intention to ask you for advice regarding me. My idea was the following so far: I first of all wanted to await the political development in Germany. But as far as the government system as well as the Aryan question is concerned, it is likely that at the moment no change can be expected. Moreover, while, so to speak, maintaining the status quo of the financial situation, I

[7] "Parteigenosse", party member.
[8] A Nazi politician and head of the German Labour Front.
[9] A famous (Jewish) actress (1897–1986).
[10] In addition to "KZ" another abbreviation for "concentration camp".
[11] Landau's second daughter (1908–1982).

wanted to wait here until regular payment transactions between Germany and foreign countries can be executed again to a certain degree, so that I can count on a certain monthly sum. But no change can yet be foreseen on this point either. Because of the low level of the crown, I can live here from money transfers alone, in spite of their irregularity. If they stop completely, I could live here without restrictions for 6 years. With a monthly consumption of $ 130, I could live in America for about 3 years from the same money. Currently I get from Germany a monthly average of $ 70, but, as I said, it threatens to stop completely (Neugebauer also tells of great difficulties for Hirschwald and Springer). Nevertheless, in principle I am doing very well, and feel almost ashamed to think of taking away a study grant from others, on the other hand, I would, of course, like to keep my money as a reserve, if possible, in the event of a complete German lockdown. First of all, I would be grateful if you could tell me whether the indicated sum of $ 130 per month is a reasonable figure. I do not want to have to think twice about every cent whether or not I can spend it, not only for convenience, but also because I have noticed that one is immediately treated with greatest distrust like someone who is soon asking for help.

Now the mathematical situation. Here I think it is unlikely to obtain a major source of inspiration. Nielsen has been concerned with other things for a long time and seems to wish to return to topological momentum, but he does not do so. I have, therefore, made myself, so to speak, independent, and carried out a long-felt wish, namely, I learned algebraic geometry, that is, I studied v. d. Waerden's papers, and modulo these I turned the Italian standard course into mathematics (and, as reported, held lectures on this here). That took me hard all the winter. From v. d. Waerden I hoped to get some tips on how to overcome such difficulties more easily, but he said that I was the only person in Europe who had made this effort, and that he could absolutely not readily put the Italian works in order without further ado. Perhaps, from a practical point of view, it is imprudent to spend time instead of publishing, since in other countries the number of papers seems to be decisive. From a supramundane point of view, however, I believe I have worked with unusual gain for myself, as in view of the permanent gaps and inaccuracies I had to deal with I have penetrated the matter quite differently than I usually proceed within a receptive context. From the same point of view, I would prescribe myself another half year of sequestration, in which I would deal with Lefschetz's works. These are, as is well known, particularly difficult to understand, but I believe they lead to the most promising side of algebraic geometry today.

Unfortunately, I think it is unlikely that the "real" interest which the Princetonians have had for me at that time still exists today. It was primarily personal sympathy from Lefschetz for me, which has surely disappeared by now because of the very short acquaintance, and then the migration to the west was not yet so vast and natural. In the meantime, a large number of first-class people have come to America, and the xenophobia, which is not just noticed by you, has for sure partly its reason in saturation.

You now know my situation exactly; what I am to do is difficult to say. But an exact answer to the question of what you would do in my place would be very important to me. From the mathematical standpoint nothing speaks for staying here, except at first the cheap living conditions and the pleasant personal relationships. For America speaks a lot from a mathematical point of view. Princeton looks a little too big from here, and I once thought of Baltimore, because the book of Zariski made so much impression on me. (It has only one error, being published in the Ergebnisse series instead of in the yellow collection and hence being so compressed.) But you know the circumstances better and could perhaps also give me information about this. What is perhaps also important in your assessment of my situation: today I am 30 years old.

Perhaps you have no longer expected a closing, but I am now finished.

With best regards to you and your family.

Yours, Herbert Busemann

Outline of research, document written around 1935

My first paper deals with a question raised by Prof. Courant concerning the direct methods of the calculus of variations. The dissertation and the papers connected with it belong to the domain of the foundations of geometry and treat more especially the question to relate intersection properties of the geodetical lines of metrical spaces (as the theorem of Desargues) with simpler metrical properties as the convexity of the spheres. In collaboration with Dr. Feller, Stockholm, I treated the problem proposed by Lebesgue relatively to which systems of point sets the Lebesgue integrals may be differentiated. Later on I worked, also jointly with Dr. Feller, at the curvature properties of convex surfaces without assumptions regarding differentiability. Among others we study completely the distribution of the curvatures of the normal intersections at an arbitrary point with tangential plane and prove moreover that at almost all points of an arbitrary convex surface the classical Euler theorem holds. In the last time I have applied myself to algebraic geometry and topology, especially to Prof. Nielsen's research. The results of Nielsen's show strong analogies with certain theorems of algebraic geometry. The origins of these analogies are not yet cleared up thoroughly and I intend to take up this matter.

(List of papers before)

Hellerup, 16. XI. 35.

Lieber Herr Professor.

Herzlichen Dank für Ihren ermunternden Brief, der sich mit meinem kreuzte. Sie haben diesen, da er durch Ihre Ausführungen bereits überholt war, hoffentlich nur noch als Stimmungsbild betrachtet.

Mit jedem Ihrer Briefe wird mein Gesicht dümmer. Wie Sie es fertig gebracht haben, in Princeton jemand für mich zu interessieren, ist mir ein Rätsel. Meine Situation stellt sich für mich selbst so dar: 1932 hat Lefschetz auf Grund einer flüchtigen (einen Abend in Ihrem Haus dauernden) Bekanntschaft eine rein persönliche Zuneigung zu mir gefaßt. Damals war es nicht so trivial, daß jemand nach Princeton kam und deshalb bestand ein relatives Interesse. Jetzt da man absolut jeden nach Princeton haben kann und einen wesentlichen Teil der gesamten Weltkraft dort bereits konzentriert hat, schien mir, daß dort nur die Extraklasse Aussicht hat, nicht lästig zu fallen. Ich kann mir ja nicht einreden, auch nur approximativ dazu zu gehören. Von meiner bisherigen mathematischen Produktion war eigentlich nur die Fundamenta [A]rbeit ein Beweis dafür, daß ich imstande bin, mit objektiven Schwierigkeiten, d. h. Problemen, mit denen sich gute Mathematiker erfolglos beschäftigt haben, fertig zu werden. Auch konnte ich die anderen hier allzu leicht von meiner Chancenlosigkeit überzeugen. Wenn ich glaube, daß es in irgendeiner Hinsicht nicht so viele mir gleichwertige Exemplare gibt, so geht das in Richtung auf die außermathematische Bildung, wo ich das Gefühl habe, mich infolge eines guten Gedächtnisses und der Lust, mich mit meinen Büchern allein zu befinden, über das normale Maß hinaus ausgebreitet zu haben. Ganz klar, wie sich diese Teilnahme an andren Dingen einmal für andre nutzbar gestalten sollte, war mir nicht, doch habe ich immer fest an ihren Wert, nicht nur für mich, geglaubt. Aber das zählt ja sozusagen nicht.

Also so sehe ich mich selbst im Rahmen der übrigen Welt und bekomme nun Ihren vorigen Brief, worin stand, daß man mir sicher gerne schreiben wolle, daß man sich freue, wenn ich zu wissenschaftlicher Zusammenarbeit kommen wolle. Schon das setzte mich, wie Sie an meinem letzten Brief bemerkt haben werden, in Erstaunen, und nun schreiben Sie „... (Lefschetz) war *sehr* freundlich, äußerte sich mit größter Wärme und Anerkennung und würde Sie sehr gerne dort haben". Nicht nur ich habe Mund und Nase aufgesperrt. Nachträglich ist mir klar, wieso der Beginn unsrer Korrespondenz über Amerika mißglücken mußte; und ich habe Ihnen die Selbstbetrachtungen geschrieben, damit auch Sie meine anfänglich ablehnende Haltung verstehen, und auch, wenn ich das schreiben darf, weil ich Sie bitten möchte, sie mit Ihrer Ansicht über mich zu vergleichen, da ich etwas Angst habe, man wird mich an Hand des Bildes, das Sie von mir entworfen haben, nicht wiedererkennen.

Aber da Sie nun Möglichkeiten geschaffen haben, an die ich nicht glaubte, habe ich schon im letzten Brief mein Schicksal Ihnen anvertraut und tue es auch weiterhin.* Ihrer Aufforderung gemäß habe ich daher einen Brief an Lefschetz

*Insbesondere auch über den Zeitpunkt der Amerikareise. Zwar scheint mir jetzt klar, daß es nicht zweckmäßig ist, unabhängig von dem etw. Grant of Aid noch dieses Schuljahr zu

geschrieben, von dem ich eine Kopie einlege. Weiter wäre nur, was damit aber bereits gesagt ist, mitzuteilen, daß ich den von Ihnen vorgeschlagenen Weg sehr gern beschreite, d.h. mich um einen Grant of Aid bewerben möchte.

Ich danke Ihnen herzlich, um so mehr, da Sie zunächst alles gegen meinen passiven Widerstand unternommen haben, den Sie hoffentlich nun verstehen.

Herzliche Grüße an Sie und Ihre Familie

Ihr Herbert Busemann

Hellerup, November 16, 1935

Dear Professor.

Thank you very much for your encouraging letter, which crossed mine [from November 3, 1935]. You have hopefully considered it, already superseded by your remarks, only as an atmospheric picture.

With each of your letters my face becomes dumber. How you managed to get someone interested in me at Princeton is a mystery to me. My situation presents itself to me as follows: In 1932 Lefschetz, on the basis of a casual acquaintance (an evening in your house), conceived a purely personal affection for me. At that time, it was not so trivial that someone came to Princeton, and so there was a relative interest. Now that you can have absolutely everyone in Princeton, and have already concentrated a substantial part of the entire world elite there, it seemed to me that there only top-class people have good prospects of not being a nuisance to somebody. I cannot persuade myself to belong to them even approximately. Of my previous mathematical production, only the Fundamenta paper was a proof that I am capable of dealing with objective difficulties, that is, problems with which good mathematicians have been unsuccessful. Also I was able to convince the others here all too easily of my lack of opportunities. If I believe that there are not so many copies tantamount to me, this is is heading for an extra-mathematical education, where I have the feeling that I have spread beyond the normal range as a result of a good memory and the desire to be alone with my books. It was not clear to me how this participation in other things could be used by others, but I have always believed in their value, not only for myself. But that does not count so to speak.

Anyway, this way I perceive myself within the framework of the rest of the world and now get your previous letter, saying that one would certainly like to write to me, that one would be glad if I wanted to come to scientific cooperation. This, as you have noticed in my last letter, already astonished me, and now you write "...(Lefschetz) was *very* friendly, expressed himself with great warmth and appreciation, and would very much like to have you there". Not only I

kommen, doch wäre es mir angenehm, falls Sie mir über Ihre Ansicht in diesem Punkt, sobald sie endgültig ist, Mitteilung machen könnten.

have opened my mouth and nose. It is, after all, clear to me why the beginning of our correspondence on America must have failed; and I have written to you these meditations so that you may understand my initially negative attitude, and also, if I may write this, because I would like to ask you to compare them with your view of me, since I am a bit afraid one will not recognize me by the image you have drawn of me.

But since you have now created opportunities that I did not have faith in, I have already entrusted my fate to you in the last letter and continue to do so.* I therefore wrote a letter to Lefschetz, of which I enclose a copy. Further on, but this is already said, would be only to communicate that I am very fond of the path you have proposed, that is, I would like to apply for a grant of aid.

I thank you very much, especially since you have done everything against my passive resistance, which I hope you will now understand.

Kind regards to you and your family.

Yours, Herbert Busemann

*In particular also about the date of the trip to America. It seems to me that it is not appropriate to come independently of the possible Grant of Aid still this school year, but I would be pleased if you could let me know your opinion on this point as soon as it is final.

On the Busemann–Courant Correspondence, New York University (Elmer Holmes Bobst Library)

The archive contains some 50 letters between Herbert Busemann and Richard Courant, and 24 letters between Alfred Busemann (Herbert's father) and Richard Courant. Busemann's father (March 24, 1878 – September 7, 1946) was Director of Finance at the legendary Krupp steel producer in Essen, Germany. He became acquainted with Richard Courant via his son Herbert, who was then a PhD student of Courant in Göttingen. A first meeting took place on August 13, 1929, at Hannover Main Station, where Alfred Busemann interrupted a train journey from Essen to Berlin. On this occasion Busemann promised Courant a 35,000 Reichsmark personal interest-free loan for several years, possibly with a later donation, for the new mathematical institute in Göttingen.[1] Most of the correspondence between Busemann father and Courant falls into the late period of the Weimar Republic, from 1929 to 1932, and is concerned with things related to the loan. Beginning in December 1932, Alfred Busemann also agreed to arrange for a monthly payment of 50 Reichsmark to Courant as support for distressed students.[2]

The correspondence between Herbert Busemann and Richard Courant begins on August 9, 1930, when Courant sent a letter to Essen informing Busemann that, on the basis of the judgements, in all probability his geometric work is sufficient for a decent dissertation. On December 23, 1930, Courant let Busemann know that his work found grace before the eyes of the criticism and could then be accepted with a perfectly reasonable predicate as a dissertation. He received the doctorate on February 25, 1931. For Courant's dissertation report see [1]. The main part of the Busemann–Courant correspondence relates to the years 1935–36, with emphasis on Busemann's move from Denmark to America as well as the situation in Germany. The first letter (February 20) in 1935 from Courant to Busemann starts as follows: "May I remind you that you have promised to tell me occasionally how you look at the world in general and in particular, and

[1] In a letter dated Sep. 27, 1929, Alfred Busemann reports that Gustav Krupp von Bohlen und Halbach, Chairman of Krupp's Supervisory Board, declined to provide financial support for the institute since he has to reject all those requests for "nowadays, industry is forced to continually issue bonds and borrow loans in order to keep the plants competitive". In fact, since 1925 Krupp was in a financially precarious position, see Karl Heinrich Pohl and O. Wiedfeldt, Die Finanzkrise bei Krupp und die Sicherheitspolitik Stresemanns, Vierteljahrschrift für Sozial- und Wirtschaftsgeschichte, Bd. 61 (1974), pp. 505–525.

[2] From a letter of Alfred Busemann to Courant dated November 29, 1932: "Perhaps you use the December amount for the stark case you mentioned and give the person in question a further amount of RM 50,–, so that he can possibly drive home at Christmas." [Vielleicht verwenden Sie den Dezember-Betrag für den von Ihnen erwähnten krassen Fall und geben dem Betreffendem zu Weihnachten noch einen weiteren Betrag von RM 50,–, damit er evtl. nach Hause fahren kann.]

how you are doing. ... What are your plans? Don't you want to visit us?"[3] At this time, Courant has been in New York for exactly six months, Busemann emigrated to Copenhagen in May 1933. In his reply to Courant (March 13, 1935) he writes:[4]

> or people still remain in the Mathematical Society, after it is explicitly stated that not every member may write what he wants on politics in his own country in the newspapers, which is, it seems, based on a completely different attitude. This is so strange to me that I have quitted membership right after the Bieberbach article and have not even come up with the idea that someone could stay in it. When Nielsen, whom I met afterwards as one of the first in Sweden, told me that he did not want to break off relations with Germany, I absolutely did not and do not understand the conceivable content of this statement, I would not have responded differently if he had told me that the current position of Sirius in the universe was against it.
>
> Unfortunately I cannot write you anything real about Germany; how imaginary everything is, is perhaps best seen in the fact that after the victory of Schmeling, spontaneously the German national hymn was sung.[5] It does not seem to me to be more insane when I read about a flagellant procession in the Middle Ages.

Courant replies (March 30, 1935):[6]

---

[3]"Darf ich Sie daran erinnern, dass Sie mir versprochen haben, mir gelegentlich einmal zu berichten, wie Sie die Welt im allgemeinen und im besonderen ansehen und wie es Ihnen geht. ... Wie sind Ihre Pläne? Haben Sie nicht Lust, uns einmal hier zu besuchen?"

[4]"oder Leute noch in der Mathematikervereinigung bleiben, nachdem explizit gesagt wird, nicht jedes Mitglied dürfe in seinem Lande in den Zeitungen über Politik schreiben was es wolle, liegt dem offenbar eine ganz andere Einstellung zugrunde. Diese ist mir so fremd, daß ich postwendend nach dem Bieberbach-Artikel ausgetreten bin und nicht einmal auf die Idee gekommen bin, es könnte jemand drin bleiben. Als Nielsen, den ich als ersten nachher in Schweden traf mir sagte, er wolle die Beziehungen zu Deutschland doch nicht abbrechen, verstand und verstehe ich absolut nicht den denkbaren Inhalt dieser Aussage, ich hätte nicht anders reagiert, wenn er mir gesagt hätte, die augenblickliche Stellung des Sirius im Weltall spräche dagegen.

Ich kann Ihnen leider garnichts Reelles über Deutschland schreiben; wie imaginär alles ist, sieht man vielleicht am besten daran, daß nach dem Sieg von Schmeling spontan das Deutschlandlied angestimmt wurde. Es kommt mir nicht irrsinniger vor, wenn ich von einer Flagellantenprozession im Mittelalter lese."

[5]Max Schmeling was a German boxer who was heavyweight champion of the world between 1930 and 1932. Busemann probably refers to the fight Steve Hamas vs. Max Schmeling at the Hanseatenhalle, Hamburg, on March 10, 1935, which resulted in the ninth-round knockout and was the end of Hamas' boxing career.

[6]"Wie ist es denn mit Ihren Absichten, nach Amerika zu kommen? Wenn Sie wirklich Lust dazu haben und wenigstens teilweise den Aufenthalt hier aus eigenen Mitteln bestreiten können, würde ich einmal sondieren, ob nicht beim Princeton Institute ein gewisser Zuschuss als Stipendium für Sie gewährt werden könnte. So weit ich weiss, haben sich doch damals die Leute in Princeton für Sie wirklich interessiert. ... Hier in Amerika scheint mir im Moment die Situation der emigrierten Wissenschaftler unbehaglich zu werden, weil entschieden eine Art fremdenfeindlicher Stimmung gärt, die Reappointments nach Ablauf der meist zweijährigen Fristen in vielen Fällen schwierig macht. Z. B. fürchte ich sehr, dass Lewy wenigstens temporäre Schwierigkeiten haben wird. In Brown scheint er wenigstens wegen Umorganisation der dortigen Einrichtungen nicht bleiben zu können."

> What about your intentions to come to America? If you really want to do this and, at least in part, you are able to stay here at your own expenses, I would like to probe if the Princeton Institute can award a grant as a scholarship for you. As far as I know, at that time the people at Princeton were really interested in you.... Here in America, the situation of the emigrated scientists seems to me to become uncomfortable at the moment since a kind of xenophobic mood seethes, which makes reappointments difficult in many cases after expiration of the mostly two-year deadlines. E.g., I am very much afraid that Lewy will have at least temporary difficulties. At least at Brown, he does not seem to be able to stay because of the reorganization of the institutions there.

In the above letter Courant encourages Busemann to report more details next time, which results in the longest reply by Busemann (9 handwritten pages, May 12, 1935) available in the collection. Busemann wrote this letter on his 30th birthday. It is subsequently reproduced in full [2]. In his reply (May 28, 1935) Courant states that it is too late to arrange a stay for the coming winter, so that one should rather consider the next academic year starting in autumn 1936. Courant mentions that both in Baltimore and in Princeton there is a chance to get a scholarship for the year 1936–37. He continues:[7]

> $ 130 a month is quite plentiful in view of your humble claims. In Baltimore, even $ 100 is considered abundant. ...On the whole, if you can get a small aid in Baltimore, I would consider a stay there as convenient since you can certainly get in touch there with Zariski. Princeton is actually overcrowded with strangers but would of course be very good as well.
>
> As to the principal question of whether you should be encouraged to come to America, I will still look as accurately as possible before I give you definite advice. However, I am inclined to a positive attitude in your case because the risk does not seem so high to me because of your reserves. Once the depression will be over, there is undoubtedly very well place for people of your qualities, and the fact that you can wait a little and can also appear on the scene as an independent person here is very essential.

[7] "130 $ im Monat sind durchaus reichlich bei Ihren bescheidenen Ansprüchen. In Baltimore gelten sogar 100 $ als reichlich. ...Im ganzen würde ich, wenn Sie in Baltimore eine kleine Beihilfe bekommen können, einen Aufenthalt dort für günstig halten, da Sie dort mit Zariski sicherlich gut Kontakt bekommen können. Princeton ist mit Fremden schon eigentlich überfüllt, würde aber natürlich auch sehr gut sein.

Was die prinzipielle Frage anbetrifft, ob man Ihnen zu Amerika zureden soll, so will ich, bevor ich Ihnen einen definitiven Rat gebe, mich noch möglichst genau umsehen. Ich neige aber gerade in Ihrem Falle zu einer positiven Einstellung, weil mir aufgrund Ihrer Reserven das Risiko nicht so hoch zu sein scheint. Wenn einmal die Depression hier vorbei sein wird, ist zweifellos sehr gut Platz für Leute von Ihren Qualitäten, und die Tatsache, dass Sie ein wenig abwarten und auch gewissermasser als unabhängiger Mensch hier auftreten können, ist sehr wesentlich.

Hätten Sie nicht Lust, sich die Geschichte einfach einmal selber anzusehen? Sie können mit deutschem Geld ein relativ billiges Rückreisebillett kaufen und hier in N.Y., in Baltimore, in Princeton, eventuell auch in Harvard usw. sich mit ziemlich wenig Geld ein paar Wochen umsehen, Kontakt anknüpfen und dann von Europa aus viel einfacher und mit viel klarerer Vorstellung die Dinge betreiben."

> Wouldn't you like to have a look at the situation yourself? You can buy a relatively cheap return ticket with German money and here in NY, in Baltimore, in Princeton, possibly also in Harvard, etc. you can look around with very little money for a few weeks, make new contacts and then do things much more easily from Europe with a clearer idea.

In the next letters Busemann's transition to America is further discussed as well as the case of Rudolf Lüneburg, which is covered in Siegmund-Schultze, Mathematicians Fleeing from Nazi Germany, Princeton University Press, 2009, pp. 180–185. Busemann obviously cannot get along with the idea of initially just visiting America, without having any kind of official association. He explains (September 11, 1935):[8]

> Finally, as far as my own journey to America is concerned, I must confess that I had not taken very seriously your proposal of an information journey. With my temperament, the whole would only result in a reversal of the room here with one in America. Fenchel also thinks that if I do not belong "somewhere", the stay in America were a failure; especially in view my joy of solitude, which has here (unfortunately, or fortunately, as the case may be) strengthened, it is out of the question becoming some kind of burden or maneuvering myself without a concrete cause, a kind of well-defined affiliation.

Courant is obviously somewhat irritated about the harsh rejection of an information trip to America (and, possibly, Busemann's attitude in the Lüneburg case) and expounds his thoughts in a $2\frac{1}{2}$ page typewritten letter (September 26, 1935):[9]

[8] "Was nun schließlich meine eigene ev. Fahrt nach Amerika betrifft, so muß ich gestehen, daß ich Ihren Vorschlag einer Informationsreise garnicht sehr ernst genommen hatte. Bei meinem Temperament würde das Ganze nur auf eine Vertauschung des Zimmers hier mit einem in Amerika hinauslaufen. Auch Fenchel meint, wenn ich nicht „irgendwozu gehöre", sei der Aufenthalt in Amerika verfehlt; zumal bei meiner Freude am Alleinsein, die sich hier (leider oder glücklicherweise, je nachdem) noch verstärkt hat, kommt irgend eine Art von Lästigfallen oder sich Einbugsieren ohne einen konkreten Anlaß, eben eine Art wohldefiniertes Zugehören nicht in Frage."

[9] "Ich habe Ihnen auf Ihre Anfrage wegen Amerika damals folgendes als meine Meinung geschrieben: Auf längere Sicht halte ich die Chancen in Amerika für grösser als in den anderen mir bekannten Ländern. Im Moment allerdings und für die nächste Zeit wird die Situation noch sehr schwierig bleiben. Um hier hereinzukommen, ist es äusserst vorteilhaft, wenn man nicht von vornherein als jüdischer Emigrant, sozusagen durch die Not gezwungen, um jeden Preis eine Stelle sucht, sondern gewissermassen als unabhängiger Mensch auftreten, sich akklimatisieren und seine Chance abwarten kann. Eine Sicherheit, nach einiger Zeit etwas zu bekommen, besteht natürlich auch dann nicht, aber ich hatte das Gefühl, dass in Ihrem Falle, falls Sie sich akklimatisieren und mit den richtigen Leuten in Kontakt kommen und, wie wir alle, Ihre früheren europäischen Ansprüche zurückschrauben, nach einiger Zeit sich schon ein Weg öffnen wird. Ich halte es nicht für ausgeschlossen, dass Sie für das im Herbst beginnende akademische Jahr 1936–37, wenn rechtzeitig vorbereitende Schritte getan werden, in Baltimore oder sonstwo irgend eine Art von Stipendium erhalten können, welches allerdings nur einen Teil Ihrer Ausgaben decken würde. Ich habe mich bereit erklärt, in dieser Hinsicht weitere Sondierungen vorzunehmen. Nachdem die Ferienunterbrechung Anfang Oktober aufhört,

On your request I then wrote to you [May 28, 1935; cf. p. 3] concerning America as my opinion: In the longer term, I consider the chances in America to be greater than in the other countries I know. At the moment, however, and for the time being, the situation will still remain very difficult. In order to get a foothold here, it is extremely beneficial not to be forced from the outset as a Jewish emigrant, so to say forced by necessity, to look for a job at any cost, but to be able to stand as an independent man, to acclimatize and wait for one's chance. Of course, even then a guarantee to get something after some time does not exist, but I felt that in your case, if you get acclimatized and come into contact with the right people and, like all of us, lower back your previous European expectations, after a while a path will open. I do not think it is impossible for you to receive some kind of scholarship, which would however only cover part of your expenses, in Baltimore or elsewhere for the academic year 1936–37 beginning in autumn if timely preparatory steps are taken. I have agreed to carry out further explorations in this direction. Since the holiday break ends at the beginning of October, the right moment would be to start with such inquiries now. However, according to your last letter, I do not quite know how far you are inclined to seriously consider the risk that is actually involved in such a transfer to America. If you were already partly present in Baltimore or Princeton for the academic year starting now and made contact with Zariski or Lefschetz, in my opinion the psychological and factual situation would be made easier for the following year. I had not expected you to come here as a tourist and resident of a hotel room; on the contrary, I was very willing to prepare everything in Princeton and Baltimore for your stay and the start of cooperation with Lefschetz and Zariski, and I was therefore anxiously waiting for your answer until yesterday. It would not make sense to come three months before the start of the academic year, because then here are holidays. Only a longer visit between October and mid-May would make sense.

I admit that you probably feel more comfortable with your temperament in Scandinavia than in America; but I am afraid that things will become more

---

wäre an sich jetzt der richtige Moment, mit solchen Nachfragen zu beginnen. Ich weiss aber nach Ihrem letzten Briefe nicht recht, wie weit Sie Neigung haben, das Risiko, das in einer Transferierung nach Amerika tatsächlich liegt, ernstlich ins Auge zu fassen. Wenn Sie schon einen Teil des jetzt beginnenden akademischen Jahres in Baltimore oder in Princeton wären und dort den Kontakt mit Zariski bzw. Lefschetz herstellten, so würde nach meinem Gefühl psychologisch und sachlich die Situation für das folgende Jahr sehr erleichtert werden. Bloss als Tourist und Bewohner von Hotelzimmern hierherzukommen, hatte ich Ihnen keineswegs zugemutet; ich war vielmehr sehr bereit, in Princeton und Baltimore alles für Ihren Aufenthalt und die Aufnahme der Zusammenarbeit mit Lefschetz bzw. Zariski vorzubereiten und habe deswegen bis gestern gespannt auf Ihre Antwort gewartet. Drei Monate vor Beginn des akademischen Jahres zu kommen, hätte keinen Sinn, weil dann hier Ferien sind. Einen Sinn hätte nur ein längerer Besuch zwischen Oktober und Mitte Mai.

Ich gebe zu, dass Sie sich bei Ihrem Temperament wahrscheinlich in Skandinavien wohler fühlen würden als in Amerika; aber ich fürchte, dass in Skandinavien oder sonst in Europa die Dinge noch schwieriger werden als sie schon jetzt sind, und dass auch der Andrang von Leuten jeder Art von Abstammung hierher erheblich steigen wird. Ich möchte Sie nicht zu irgend einem Entschluss drängen, der Ihrem Temperament nicht entspricht; aber bitte schreiben Sie mir doch möglichst bald, ob ich versuchen soll, in der Frage Ihrer Transferierung irgend welche weiteren Sondierungen vorzunehmen, und ob Sie Ihre ursprüngliche Idee, gegebenenfalls auf einen längeren Zeitraum hier auch ohne offizielle Stellung zu sein, endgültig aufgegeben haben."

> difficult in Scandinavia or elsewhere in Europe than they are now, and that the crowds of people of all sorts of lineage will also rise considerably. I do not want to push you to any decision that does not correspond to your temperament; but please write to me as soon as possible, whether I should try to carry out any further explorations in the question of your transfer, and whether you have finally abandoned your original idea to stay here, if necessary, for a longer period of time without an official position.

Busemann registers the slight resentment and closes his reply (October 6, 1935): "I can say that for a long time nothing impresses me so much as your willingness to do something for me, even after I have obviously annoyed you with my last letter."[10] He further explains in this letter:[11]

> I got your letter yesterday, and I've been talking last night, very extensively, about $2\frac{1}{2}$ hours, with Bohr about everything, and today pondered on it with Fenchel. Both of them have strengthened me in my attitude and have essentially come to their judgment independently of me.
>
> The risk of a relocation to America without a fixed "reference point" (it is exactly this what we are talking about just to make sure you understand what I mean), I consider not so much on the financial but mainly on the psychological side: Here my life as a private man is authentical. If I now go to America with an invitation, an epsilontic scholarship, or something solid in my pocket, and it turns out that the possibility, which is also to be considered in your opinion,

[10] "Ich kann sagen, daß mich lange nichts so beeindruckt, als Ihre Bereitwilligkeit, für mich etwas zu tun, selbst nachdem ich Sie offenbar mit meinem letzten Brief verstimmt habe."

[11] "Ich bekam Ihren Brief gestern und habe mich noch gestern abend, sehr ausführlich, so $2\frac{1}{2}$ Stunden, mit Bohr über alles unterhalten und heute mit Fenchel überlegt. Beide haben mich in meiner Haltung bestärkt und waren im wesentlichen sogar unabhängig von mir zu ihrem Urteil gelangt.

Das Risiko einer Übersiedlung nach Amerika ohne festen „Anhaltspunkt" (gerade darum, damit Sie richtig verstehen, was ich damit meine, dreht es sich), sehe ich nicht so sehr auf finanziellem, sondern hauptsächlich auf psychologischem Gebiet: Hier hat mein Leben als Privatmann eine sozusagen glaubhafte Form angenommen. Wenn ich nun nach Amerika gehe mit einer Aufforderung, einem epsilontischen Stipendium oder irgend etwas Festem in der Tasche und es zeigt sich, daß die auch nach Ihrer Ansicht durchaus in Betracht zu ziehende Möglichkeit eintritt, daß eine Dauerperspektive sich nicht ergibt, so ist der formale Rückzug gesichert: ich habe dann eben, psychologisch gesehen, etwa dasselbe gemacht wie Magnus, nämlich ein bestimmtes Programm durchgeführt und mich dann z.B. wieder hierher begeben. Wenn ich mich dagegen ohne ein solches Programm (Anhaltspunkt oder Zugehörigkeit, wie ich es das letzte Mal ausdrückte) nach Amerika begebe und der ins Auge gefaßte Fall tritt ein, so scheint mir meine Situation völlig unmöglich, da eben dort mein Dasein wohl doch nicht so glaubhaft privat ist, wie es hier einfach auf Grund der Entwicklung geworden ist. Und weiter folgendes: Wenn ich erst einmal aufgefordertermaßen drüben bin, kann man mir aus meiner bloßen Anwesenheit keinen Vorwurf machen. Es wird sich dann, wie mir scheint, leichter als jetzt die Form finden lassen, in einen Privatmann überzugehen; denn mir ist ganz klar, daß das für eine Zeitlang trotz allem nötig sein wird und sinnvoll ist, sobald sich nur irgend eine Chance für die Zukunft sehen läßt. Hoffentlich ist Ihnen nun klar geworden, was ich meine. Um zu resumieren: ich würde mit Enthusiasmus mit irgend einer Zugehörigkeit (in dem hoffentlich jetzt geklärten Sinn) kommen, halte es aber ohne eine solche für bedenklich. Wie gesagt, ist Fenchel derselben Ansicht und Bohr schien sie auch vernünftig."

that a permanent perspective does not arise, the formal retraction is secured: I have then, psychologically speaking, made the same thing as Magnus, namely I have done a certain program, and then, for example, gone back here. If, on the other hand, I go to America without such a program (a reference point or affiliation, as I put it last time), and the envisaged case occurs, my situation seems quite impossible to me, since at that very place my presence is probably not so credibly private, as it has become here simply due to the development. And the following: Once I have been asked to come to America, one cannot reproach me for my mere presence. It will then, as it seems to me, be easier than now to find the form of passing into private man's existence; for it is quite clear to me that this will be necessary for a while, in spite of everything, and is useful as soon as there is any chance for the future to be seen. I hope you now understand what I mean. To summarize: I would come with enthusiasm under the condition of any affiliation (in the now hopefully clarified sense) but consider it to be questionable without such a guarantee. As I said, Fenchel is of the same opinion, and it also seemed reasonable to Bohr.

Courant replies to Busemann (October 17, 1935):[12]

As far as your matter is concerned, I understand your own psychological point of view quite well, even if I had still considered an orientational visit a simplification of your situation. For the current year here or there [Baltimore or Princeton] to get a scholarship is impossible, however, a written inquiry whether, at your own expense, you want to come to Baltimore or Princeton for scientific collaboration with Zariski or Lefschetz would probably always be kindly addressed to you by the relevant authorities.

Busemann responds (November 3, 1935):[13]

Needless to say, if an invitation comes from Princeton to apply, I will do it. Bohr and Nielsen of course want to support me as much as possible.

An invitation to scientific cooperation would suffice for me; honestly, I have not dared to ask you to do such a thing as it seemed to me more demanding

[12] "Was nun Ihre Angelegenheit betrifft, so verstehe ich Ihren eigenen psychologischen Standpunkt ganz gut, wenn ich auch nach wie vor einen orientierenden Besuch für eine Vereinfachung Ihrer Situation angesehen hätte. ... Für das laufende Jahr hier oder dort [Baltimore oder Princeton] ein Stipendium zu bekommen, ist unmöglich, dagegen eine schriftliche Anfrage, ob Sie zur wissenschaftlichen Zusammenarbeit mit Zariski oder Lefschetz nach Baltimore bzw. Princeton auf eigene Kosten kommen wollten, würde wahrscheinlich immer mit grosser Freude von den betreffenden Stellen an Sie gerichtet werden."

[13] "Falls eine Aufforderung, mich zu bewerben, von Princeton an mich gerichtet wird, bewerbe ich mich selbstverständlich. Bohr und Nielsen wollen mich natürlich dabei soviel wie möglich unterstützen.

Eine Aufforderung zu wissenschaftlicher Zusammenarbeit würde mir an sich genügen; ehrlich gesagt, habe ich nur nicht gewagt, Sie zu bitten, so etwas zu veranlassen, da es mir anspruchsvoller vorkam, als ein ε-Stipendium, wie Sie es mir zu Beginn der ganzen Korrespondenz schilderten. ...

Mein weiteres Schicksal lege ich vertrauensvoll in Ihre Hände."

> than an ε-scholarship, as you described it to me at the beginning of the whole correspondence. ...
>
> I put my future fate confidently in your hands.

Meanwhile Courant came up with a concrete plan to apply for Princeton (October 29, 1935):[14]

> I have now talked with both the people of Baltimore (but not directly with Zariski, who lives a very secluded life) and Lefschetz. My impression is the following: Baltimore does not seem so favorable because of Zariski's isolation and the difficulty of having personal contact with him. Otherwise, maybe in Baltimore, the possibility of getting a scholarship for the year 1936–37 would indeed exist. However, under the present circumstances, I would greatly prefer Princeton. Lefschetz, with whom I spoke for a long time about the case, was *very* friendly, expressed himself with great warmth and appreciation, and would very much like to have you there. He suggested that you should contact him directly by letter. You can correspond with Lefschetz completely openly about all things. However, I should point out an issue that may be uncomfortable to

[14]"Ich habe jetzt sowohl mit den Leuten in Baltimore (allerdings nicht direkt mit Zariski, welcher sehr zurückgezogen lebt) als auch mit Lefschetz gesprochen. Danach ist mein Eindruck folgender: Baltimore scheint nicht so günstig zu sein wegen der Isoliertheit von Zariski und der Schwierigkeit, mit ihm persönlichen Kontakt zu haben. An und für sich wäre sonst in Baltimore eventuell die Möglichkeit für ein Stipendium für das Jahr 1936–37 durchaus vorhanden. Jedoch würde ich unter den jetzigen Verhältnissen Princeton erheblich vorziehen. Lefschetz, mit dem ich lange über den Fall sprach, war *sehr* freundlich, äusserte sich mit grösster Wärme und Anerkennung und würde Sie sehr gern dort haben. Er schlug vor, dass Sie direkt brieflich mit ihm in Verbindung treten sollten. Sie können mit Lefschetz vollständig offen über alle Dinge korrespondieren. Nur auf einen Punkt, der Ihnen vielleicht unbehaglich ist, muss ich Sie hinweisen: Lefschetz ist begeistert von der Entwickelung in Russland und gibt gern jüngeren Leuten den Rat, nach Russland zu gehen. Er nimmt es aber durchaus nicht übel, wenn man persönlich anders denkt. – Unter der Voraussetzung, dass Sie überhaupt den amerikanischen Plan weiter verfolgen wollen, scheint mir jetzt der beste modus procedendi folgender: Sie bewerben sich um einen "grant of aid" beim Instute for Advanced Study. Bohr, Nielsen und andere Leute unterstützen Sie dabei durch direkte Briefe an Veblen usw. Ebenso wird Lefschetz Sie unterstützen und erklären, dass man Sie gern dort haben möchte. Es ist dann ziemlich wahrscheinlich, dass Sie einen grant bekommen, der zum mindesten einen Teil Ihrer hiesigen Ausgaben für ein Jahr deckt (der eventuell auch auf ein zweites Jahr verlängert werden kann). Wenn Sie diesen Weg beschreiten wollen, so schreiben Sie mir bitte, ich werde dafür sorgen, dass Sie entweder direkt eine Aufforderung zur Ausfüllung der Antragsformulare erhalten, oder ich werde einen solchen Schritt gleich mit Harald Bohr gemeinsam tun.

Ich habe über die Frage nachgedacht, ob und wie Sie in Amerika von Anfang an in einen Lehrbetrieb "eingeordnet" werden können. In Princeton ist dies wahrscheinlich schwierig wegen der ungeheuer vielen Professoren und Dozenten und der geringen Anzahl von Studenten. Es ist aber nicht ganz ausgeschlossen, dass ich im Rahmen eines noch sehr vagen Planes der Zusammenarbeit zwischen New York University und Princeton irgend ein Seminar oder dergleichen hier in New York einrichten kann, an welchem Sie sich regelmässig in irgend einer aktiven Eigenschaft beteiligen können. ...

Bitte fassen Sie diesen Brief nicht etwa so auf, als wollte ich Sie weiter zu einem Entschluss in Richtung Amerika drängen. Ich könnte es durchaus verstehen, wenn Sie lieber die europäische Entwickelung an Ort und Stelle abwarten wollen. Ich wollte aber, dass Sie die gegenwärtige Situation möglichst klar sehen, bevor Sie sich zu dem Risiko eines längeren Aufenthalts hier entschliessen."

> you: Lefschetz is enthusiastic about the development in Russia and likes to give younger people the advice to go to Russia. But he does not take it badly at all if one thinks differently personally. – On the assumption that you want to pursue the American plan at all, the best modus procendi now seems to me the following: You apply for a "grant of aid" at the Institute for Advanced Study. Bohr, Nielsen and other people support you by direct letters to Veblen, etc. Likewise, Lefschetz will support you and declare that one would like to have you there. It is then quite likely that you will get a grant that covers at least part of your expenses for one year (which may be extended to a second year). If you wish to pursue this path, please write to me, I will ensure that you either receive a direct request to complete the application forms, or I will even do so together with Harald Bohr.
>
> I've been thinking about the question of whether and how you could be "brought into line" with teaching in America from the very beginning. In Princeton this is probably difficult because of the immense number of professors and lecturers and the small number of students. It is, however, not altogether impossible that, in the framework of a still vague plan of collaboration between New York University and Princeton, I can set up a seminar or the like here in New York, at which you can regularly participate in some active form.
>
> Please do not take this letter as if I wanted to push you further towards America. I would certainly understand if you would rather stay where you are and await the European development. However, I wanted you to see the current situation as clearly as possible before you decide to take the risk of a longer stay here.

Busemann's response (November 16, 1935) [3] is subsequently reproduced in full because it provides an excellent account of his self-perception at this time. Towards the end of the year, after half a year of extensive discussions on the relocation of Busemann to America, things seem to have turned to the good. At the beginning of December, Courant returned from a visit to Princeton, where he had talked to Weyl and Lefschetz (Veblen was traveling). Courant reports to Busemann about full support from both and announces receipt of the forms so that the application process can start. He concludes (December 9, 1935):[15]

> On the whole, I would advise not to think too much about oneself and the world in this context, but, at most, to prepare oneself factually and linguistically.

In this letter, however, light clouds appear in the sky. Recently, Courant explains, a "stupid" decision has been taken not to award any smaller subsidiary grants, which are below the subsistence minimum. Moreover, it is primarily Americans who have to be taken into account when awarding grants. Nevertheless, Busemann is now in a position to declare "mission accomplished" (January, 13, 1936):[16]

---

[15] Im Ganzen würde ich raten, nicht mehr zu viel über sich selbst und die Welt in diesem Zusammenhange nachzudenken, sondern höchstens, sich sachlich vorzubereiten, und sprachlich.

[16] "Die Bewerbung ist am 4.I. abgegangen, als Hilfskräfte wurden v. d. Waerden, Levi-Civita,

The application[17] was sent on January 4, 1936. As assistants were mobilized v. d. Waerden, Levi-Civita, Alexandroff, Bohr and Nielsen, and besides these, of course, you and Lefschetz as references. The others have written directly to Princeton. After this had been done with the exception of the letters of Bohr and Nielsen, a letter from Veblen to Bohr arrived in which my chances for 1936/7 were described as "less than ε". Bohr and Nielsen have nevertheless still written, in fact Bohr wrote to Veblen that, after the application had been released, it would be unnatural if one were not supported from where I was best known.

Previously I had received an unusually friendly letter from Lefschetz because of the warm tone (of which Veblen had seen a copy and sent it to Bohr). This has induced my wish to come to the United States regardless of the scholarship in autumn. In order not to let this enterprise fail because of the negative scholarship, I wrote to Veblen, in fact absolutely in the same way as the whole thing has developed: Your advice to come to the United States, my doubts that, according to my local experience, I would not have thought it possible to be welcome as an emigrant, even as an independent one, that, on the other hand, here I felt not to disturb, and therefore I wanted to be sure that they wanted me. I also wrote how the idea of the ε-scholarship developed from this, and I had intended to apply for it, and finally something else would have come out.

In Veblen's letter to Bohr, it was made clear that I would not get the scholarship, but that he would like to have me there too, and that he probably "could give an official status" to me. I do not have the slightest idea what I can imagine, and therefore I have sent my letter to Veblen also with the aim to get an answer

---

Alexandroff, Bohr und Nielsen mobilisiert, und außer diesen natürlich Sie und Lefschetz als Referenzen angegeben. Die anderen haben direkt nach Princeton geschrieben. Nachdem dies gerade bis auf die Briefe von Bohr und Nielsen erledigt war, kam ein Brief von Veblen an Bohr, in dem meine Chancen für 1936/37 als „less than ε" bezeichnet werden. Bohr und Nielsen haben dann doch noch geschrieben, und zwar schrieb Bohr an Veblen, daß das deshalb geschehe, weil es unnatürlich wäre, wenn man von dort, wo man mich am besten kennt, nicht unterstützt würde, nachdem die Bewerbung mal losgelassen wäre.

Nun hatte ich vorher einen wegen des warmen Tones ungewöhnlich freundlichen Brief von Lefschetz bekommen (von dem Veblen eine Kopie gesehen und an Bohr geschickt hat). Dies hatte mich bewogen, unabhängig von dem Stipendium im Herbst nach U.S.A. kommen zu wollen. Um nun diese Unternehmung nicht an dem negativen Stipendium stranden zu lassen, habe ich an Veblen geschrieben, und zwar absolut genau so, wie sich die ganze Sache entwickelt hat: Ihr Rat nach U.S.A. zu kommen, meine Bedenken, daß ich nach meinen hiesigen Erfahrungen es für keineswegs gegeben angesehen hätte, daß man als Emigrant, selbst als zunächst unabhängiger, willkommen wäre, daß ich andrerseits hier das Gefühl hätte, nicht zu stören und daher erst sicher sein wollte, daß man mich auch haben wolle. Ferner schrieb ich, wie sich dann daraus die Idee des ε-Stipendiums entwickelt hat und ich die Absicht gehabt hätte, mich um so eines zu bewerben und schließlich etwas andres herausgekommen wäre.

In Veblens Brief an Bohr stand, daß ich zwar das Stipendium nicht kriegen werde, daß er mich aber auch gerne dort haben wolle und er mir wohl „could give an offical status". Ich weiß nicht im geringsten, was ich mir darunter vorzustellen habe, und habe daher meinen Brief an Veblen auch mit der Hinterabsicht geschickt, eine Antwort zu bekommen, aus der ich sehen kann a.) ob er das überhaupt nur Bohr zuliebe geschrieben hat oder ob er ernstlich an meiner Anwesenheit interessiert ist und b.) wenn ja, was darunter zu verstehen ist.

Jedenfalls kurz und gut, ich dachte, nachdem die Sache mal den jetzigen Zustand erreicht hat, ist es am besten, ich trete selbst in Aktion."

[17]For a brief outline of previous and intended research see [4], reproduced on p. ?? of this volume.

from which I can see a.) whether he wrote it for the sake of Bohr, or whether he was seriously interested in my presence, and b.) if so, what is meant by this.

In any case, in short, I thought, after the matter has reached the present state, it is best, I take action myself.

The same letter contains impressions about the situation of Germany. For those who are interested, some samples are provided in the following.[18]

Now about Germany. The best way is to express it negatively: If you have not felt the pressure, you cannot imagine it. It seems to be the main pleasure to give each one a kick who could possibly fancy himself as something special, of course, from revenge for one's feelings of inferiority. For example, on December 7 all bonzes, at the universities the deans and the rector, in industry the directors (my uncle and father as well) had to sway the collection box, they were directed to some place and had to stand under control for five hours. If you have money

[18] "Nun über Deutschland. Am besten wendet man es negativ: wenn man den Druck nicht gefühlt hat, kann man ihn sich nicht ausmalen. Jedem einen Tritt zu versetzen, der sich selbst für etwas halten könnte, scheint das Hauptpläsir zu sein, natürlich aus Rache für die eigenen Minderwertigkeitsgefühle. Z. B. mußten am 7.12. alle Bonzen, an den Universitäten die Dekane und der Rektor, in der Industrie die Direktoren (mein Onkel und Vater auch) die Sammelbüchse schwenken, sie wurden an irgend einen Platz dirigiert und mußten kontrolliert 5 Stunden stehen. Wenn man selbst Geld hat, geniert man sich natürlich, und so haben mein Vater wie sein Bruder heimlich die Büchsen selbst mit 10 Pfennigstücken gefüllt.

Butter und Eier sind kaum zu haben, selbst wir hatten zeitweise keine frischen Eier. Die Zeitungen und alles, was geistig sein sollte, haben allmählich ein Niveau, oder besser Nivöchen gekriegt, das man mal ein paar Wochen hintereinander gesehen haben muß, um den geistigen Zusammenbruch erfassen zu können; ...

Das neue Flugministerium ist fertig, nur 2600 Zimmer, mit Schwimmbad für die Offiziere, als Versammlungslokal wurde das alte Herrenhaus mit einbezogen, und das alles in der teuersten Gegend von Berlin: zwischen der Anhalter und der Leipziger Straße ist die halbe Wilhelmstraße abgerissen. Bild in der Berliner Illustrierten: Göring führt Hitler im neuen Ministerium rund, u. a. in Görings Arbeitszimmer, an der Wand ein kolossales Portrait, von wem meinen Sie wohl? Von Göring selbst. Göring hat bereits die 5-te Uniform für die Flieger entworfen, das ganze Geld der armen Leute geht in die Uniformen. Der Wahnsinn hat Formen angenommen, von denen Sie sich keine Vorstellungen machen können.

Manche schleichen umher, indem sie versuchen, die Zeit nicht zu beachten, z. B. Haupt, der mich zu einem Besuch aufgefordert hatte, was ich auch getan habe.

Er liest niemals eine Zeitung, dreht niemals eine deutsche Station am Radio an, hat ein altes Lesekränzchen, das die Zeit überdauert und aus lauter Leuten zu bestehen scheint, die nach einer ziemlich flachen, bequemen Erbauung schnappen und sich damit sozusagen die Ohren zukleistern. Ich muß gestehen, mich empört es, wenn jemand nicht rebelliert, und war dem Haupt daher ziemlich unbequem, ich wollte ihm seine Ruhe nicht lassen. (Mathematisch war es übrigens recht fruchtbar, da wir uns gegenseitig ziemlich viel erzählen konnten) Ferner gab es in Erlangen den schleichendsten aller Schleicher, Herrn Nöbeling. Er ließ es sich nicht nehmen, mich zur Bahn zu bringen. ...

Wenn man die ganze Misere in Deutschland sieht, so ekelt man sich und ist traurig. Die Judenfrage ist, von der Not des einzelnen Betroffenen abgesehen, nichts als ein Wahnsinnsanfall unter vielen. Die einzigen Leute, die wirklich eine gewisse Kontinuität in ihrer geistigen Haltung bewahrt haben und dabei auch ihre Würde, sind die Juden. ... Vom ersten Tag an sehnte ich mich sofort aus Deutschland wieder weg. Ich mußte fortwährend an die Apokalypse denken, besonders das eine Bild von Dürer mit den 4 apokalyptischen Reitern hatte ich dauernd vor Augen. Man hat den Eindruck, daß alles unaufhaltsam stürzt, und zwar nur stürzt, ohne daß man sich irgendwie ausmalen könnte, wohin."

> yourself, you feel embarrassed, and so my father like his brother secretly filled the boxes themselves with 10 pence coins.
>
> Butter and eggs are hard to get, even we had intermittently no fresh eggs. The newspapers, and everything that should be intellectual, have gradually attained a niveau, or better, miniature niveau, which you must have seen in a row for a few weeks in order to grasp the intellectual collapse; ...
>
> The new Ministry of Flight is ready, only 2,600 rooms, with a pool for the officers, the old mansion was included as a meeting place, and all this in the most expensive area of Berlin: Between Anhalter and Leipziger Strasse half the Wilhelmstrasse is demolished. Photograph in the Berliner Illustrierte: Göring is leading Hitler in the new Ministry, among others in Göring's office, on the wall a colossal portrait, from whom do you suppose? Of Göring himself. Göring has already designed the 5th uniform for the pilots, the whole money of the poor people goes into uniforms. Madness has assumed forms which can not be imagined.
>
> Some sneak around, trying not to pay attention to the time, for instance Haupt [Otto Haupt (1887–1988), mathematician at the University of Erlangen, Germany], who had asked me to visit, which I did.
>
> He never reads a newspaper, never sets the radio to a German station, has an old reading circle that stands the test of time and seems to consist of a few people who are snapping at a rather flat, comfortable entertainment, and, so to speak, are pasting their ears. I must confess, I am indignant when someone is not rebellious, and was therefore rather uncomfortable to Haupt, I did not want to leave him his peace. (Mathematically, it was quite fruitful since we could tell each other a lot.) In Erlangen, too, there was the most insidious of all sneakers, Herr Nöbeling. He insisted to take me to the train. ...
>
> When one sees the whole misery in Germany, one is disgusted and sad. The Jewish question is, apart from the distress of the individual concerned, nothing but a madness attack among many. The only people who have really maintained a certain continuity in their spiritual attitude, and thereby also their dignity, are the Jews. ... From the very first day, I immediately longed to leave Germany. I had to think of the Apocalypse continually, especially the one painting of Dürer with the four apocalyptic riders I had constantly in mind. One has the impression that everything is falling unstoppably, and only falls, without being able to imagine whereto.

Things now develop rapidly. Courant comes with good news (February 4, 1936):[19]

> You will be treated as a fully qualified member of the Institute just as the people with scholarships and you are really welcome everywhere. The business with N.Y.U. will give you "teaching experience" and will definitely improve your chances for later.

[19] "Sie werden im Institute als vollberechtigtes Mitglied genau so wie die Leute mit Stipendien behandelt werden und sind wirklich überall willkommen. Die Sache mit N.Y.U. wird Ihnen "teaching experience" geben und dadurch Ihre Chancen für später ganz entschieden verbessern."

The last sentence in the above excerpt refers to what had been officially confirmed to Busemann by Courant almost three months later, cf. p. 9 and [5]: For the academic year 1936–37, Busemann is invited to give a course on function theory at New York University.

The rest of the story is nasty paperwork, application for immigrant visa, travel arrangements, and other things. Busemann is carefully planning all the necessary steps, from trying to hire one of Courant's students "where I can learn English and get an impression of his mathematical education" up to "Do you agree that I will deposit the paddle boat at home with you until I move to Princeton. I will send it to you immediately by railway express company." After having crossed the Atlantic with the passenger steamer "Berlin", Busemann finally reached New York on September 9, 1936.

On December 22, 1936, Courant sends New Year's greetings to Busemann's parents:[20]

> Above all, I would like to confirm to you today that the stay of your son has been a really great success so far. I am very glad that I have pushed him so much to the decision to emigrate. For it really could not have come any better. As a scientist as well as a personality, he has prevailed here from the very first moment; one feels everywhere that he is a particularly valuable member, and this will undoubtedly make for him the paths here very flat.

For a brief account of Busemann's vita in connection with his claim for reparation, see the excerpt from [6] on pp. 57–60.

References

[1] R. Courant, Report on Busemann's dissertation, Selected Works, Herber Busemann, Vol. II. p. 29–55.

[2] Letter from H. Busemann to R. Courant dated May 12, 1935.

[3] Letter from H. Busemann to R. Courant dated November 16, 1935.

[4] H. Busemann, Outline of previous and intended research, [Jan. 1936?].

[5] Letter from R. Courant to H. Busemann dated May 26, 1936.

[6] A. Szábo, Vertreibung, Rückkehr, Wiedergutmachung, Wallstein Verlag, Göttingen, 2000, Selected Works, Herbert Busemann, Vo. II, p. 57–60.

[20] "Vor allem möchte ich Ihnen heute gern bestätigen, dass der Aufenthalt Ihres Sohnes bisher ein wirklich grosser Erfolg ist. Ich bin sehr froh, ihn seinerzeit so sehr zu dem Entschluss gedrängt zu haben. Denn es hätte wirklich garnicht besser auskommen können. Sowohl als Wissenschaftler als auch Persönlichkeit hat er sich vom ersten Moment ab hier durchgesetzt; man fühlt überall, dass es sich hier um ein besonders wertvolles Mitglied handelt und dies wird ihm zweifellos die Wege hier sehr ebenen."

Kopenhagen, 12. V. 1935.

Lieber Herr Professor.

Ihren Brief, für den ich herzlich danke, finde ich jetzt erst, nach einer längeren Reiseperiode, vor; die letzte Zeit war ich in Deutschland, wohin ich mir Briefe prinzipiell nicht nachschicken lasse. Ich kann daher heute erst antworten. Die ungebührliche Länge der Antwort empfinden Sie hoffentlich nicht als lästig, da Sie mich so sehr auffordern, sowohl über die allgemeinen wie über meine persönlichen Verhältnisse ausführlich zu berichten.

Wie ich wohl damals ankündigte, bin ich Mitte März für 10 Tage nach Stockholm und von dort nach Lappland gefahren, 68°30' nördlicher Breite. Die Landschaft ist dort ungeheuer weitzügig; wenige, weit auseinander liegende Berge, große Seen und Wasserfälle, letztere habe ich natürlich nur in erstarrter Form gesehen. Dem Prinzip nach habe ich an Lappland dasselbe auszusetzen wie an Arosa, nur daß man in Lappland keinen berechtigten Anspruch auf alpines Gelände machen kann. Die Schweden sind steilere Hänge nicht gewöhnt, laufen dagegen auf flachem Gelände mit unglaublichem Tempo, sodaß ich kaum mitkam. Bei etwas steileren, für alpine Begriffe aber sehr harmlosen Abfahrten glänzte ich dagegen, obwohl ich in den Alpen keineswegs als Meister auftreten kann. Den größten Eindruck hat mir in Lappland das Nordlicht gemacht.

Von der Skistation aus fuhr ich für ein paar

Tage in die Lofoten. Selten habe ich so etwas Schönes gesehen. Denken Sie sich die Berge des Vierwaldstättersees ins Meer abfallend, dabei manchmal ganz enge Durchfahrten, manchmal weite Wasserflächen umrahmend, schließlich ganz in offene, schwarzblaue Meer hinausgehend, dann haben Sie ungefähr eine Vorstellung. Großartig sind die Bergformen ganz bizarr, das Gestein ist ein rötlicher Granit, auf dem Ganzen etwas Schnee. Einen Tag hab' ich mich bei der Kabeljaufischerei herumgetrieben. Die Fischerflotte, ca. 6000 Boote, folgt dem Schwarm der Fische, der jedes Jahr die Lofoten passiert, durch zwei Monate, und zwar liegen die vielen Boote ganz dicht beieinander auf einem ganz kleinen Raum. Die Fische schwimmen ca 25 m tief und werden (ohne Stock direkt über den Bordrand) geangelt, an Land ihrer Köpfe und Schwänze beraubt, aufgehängt und getrocknet, worauf sie als Stockfische die Welt durchwandern.

Von den Lofoten aus bin ich eine Reihe von Nächte durchfahrend, von einer Hotelnacht in Stockholm abgesehen (ich kann hier vielleicht, für den Falle, daß Sie es noch nicht wissen, einfügen, daß Feller sein Stipendium bis Ende des Jahres verlängert bekommen hat), schließlich in Marburg bei Rellich gelandet. Rellich und Reidemeister scheinen sich den Umständen nach wohl zu befinden. Die Studentenzahl ist, wie überall und in allen Fächern, gleich $\frac{\varepsilon}{2}$. Rellich erzählte folgendes über W.W. Herr Ziegenbein, wie Sie wohl wissen, der neue Zeus des Göttinger Mathematischen Instituts, der auf

Grund eines Parteibuchs geringerer Ordnungszahl Herrn Tornier den Rang abgelaufen hat, nicht einen Vortrag über Hilbert und Brouwer (in der analytischen Fortsetzung der Maße). Er kannte nicht mehr davon als die beiden Namen Formalismus und Intuitionismus und erklärte demgemäß die Stellungnahme der neuen Jugend als gegeben. Arnold Schmidt ging mangels Diskussion zu ihm hin und meinte, so einfach sei das doch nicht, und wurde zu seinem Erstaunen aufgefordert, das nächste Mal seinerseits darüber zu reden. Der Arme, er hatte nicht bedacht, daß die Volksart das nicht zu erlassen hatte. Herr Ucher brachte sie zum Kochen, indem er sich entsann, daß Schmidt im Jahre 1930 (!) gesagt hatte, die Sozialdemokraten hätten besser den Kriegsdienst verweigert. Der Vortrag fand natürlich nicht statt, und Schmidt war längere Zeit damit beschäftigt, hinter Ucher herzulaufen und ihn um Rücknahme der Behauptung zu bitten.

Von Marburg fuhr ich nach Essen; ich kam nach Stresa und von Genf nach Essen. Ich hörte, daß man nach Erklärung der Wehrpflicht zunächst sehr pessimistisch war und mit Repressalien rechnete, nach Stresa war man ziemlich optimistisch, da man die Tragweite nicht erkannte, und wurde nach Genf wieder sehr gedrückt und war zum Schluß in allen Schichten, da merkwürdigerweise alles Außenpolitische, z.B. auch der Mac-Donald-Artikel in den deutschen Zeitungen zu finden war. Man hat das Bewußtsein, in ähnlicher Lage zu sein wie 1914.

Was die Wirtschaft angeht, ist folgendes zu sagen:

Es herrscht eine starke Binnenkonjunktur, aber doch kein eigentlicher Optimismus, da man nicht glaubt, daß sich eine reine Binnenkonjunktur à la longue aufrecht erhalten läßt. Private Berechnungen wollen auch zeigen, daß von früheren Regierungen her geheime Fonds dagewesen sein müssen, da sonst der Devisenbestand schon lange negativ sein müßte. Ferner werden von privater Seite die Staatsschulden (durch Finanzierungswechsel) auf $8\frac{1}{2}$ statt offiziell $4\frac{1}{2}$ Milliarden Mark berechnet. Stimmungsmäßig scheint mir im Rheinland, da man die Mißerfolge der Außenpolitik mehr der Böswilligkeit der anderen als den eigenen Fehlern zuschreibt, eigentlich eher eine Verschiebung zugunsten Hitlers eingetreten zu sein, abgesehen von den ernsthaft katholischen Kreisen. Der Kampf gegen letztere ist sehr heftig, zur Illustration kann vielleicht folgendes dienen: Auf einer Kruppschen Zeche sollten vier Lehrlinge (spätere Bergfachstudenten) eingestellt werden, drei waren aus der H. J. hervorgegangen, der vierte gehörte einem katholischen Verband an und war offenbar der tüchtigste. Es wurde gegen Einstellung dieses letzteren Einspruch erhoben, und obwohl die Zechenleitung sich wegen der Qualitäten des jungen Mannes sehr für ihn einsetzte, konnte sie ihn nicht behalten. Personelle Eingriffe kommen überhaupt sehr weitgehend vor, besonders auf folgende Art (z. B. bei meinem Frankfurter Onkel). Man erklärt plötzlich, irgend ein Werk brauche einen Direktor mehr und zwar den und den Herrn. In dem Frankfurter Falle ließ sich der betreffende P.G.

wegen Unzulänglichkeit völlig kaltstellen, sodass das Ganze nur eine Belastung des Spesenkontos bedeutet.

Von Essen fuhr ich nach Leipzig zu v. d. Waerden, mit dem ich einiges über algebraische Geometrie besprechen wollte. Dort scheinen, abgesehen von der minimalen Studentenzahl die Verhältnisse ganz angenehm zu sein; v. d. W. will sich sogar in Leipzig ein Haus bauen. Er sagt aber trotzdem, für einen deutschen Beamten, selbst als Holländer, sei es doch sehr bedrückend, zu sehen, wie durch den Kampf jedes gegen jeden alles systematisch zugrunde gerichtet werde. „Wenn A. und B. ein geordnetes Paar von Kollegen sind, sucht Herr A. Herrn B. herauszuschmeißen." In Leipzig sei (im Gegensatz zu Dresden) die Stimmung ziemlich nazifeindlich infolge des stark reduzierten, für Leipzig sehr wichtigen internationalen Geschäfts.

Der letzte Haltepunkt in Deutschland war Berlin, wo ich 8 Tage war. Dort war ich aufs höchste erstaunt, wie offen sich die nazifeindliche Stimmung bemerkbar macht. Am 1. Mai wurde die Kundgebung überallhin übertragen, man konnte ihr nicht entgehen. Niemand muckste sich, nicht einen Beifall habe ich vernommen. Das Ende erlebte ich in einem Restaurant, wo ich zu Mittag aß. Gerade zum Schluß, als Dr. Ley sagte: hiermit schließen wir die diesjährige Kundgebung, ertönte ostentativer Beifall – ich hätte es nicht für möglich gehalten, vor allem den Ton nicht. Ich war mit einem zufällig getroffenen Göttinger (arischen) Philologen in einer Bar, es war allerdings schon ½3, als man Elisabeth

Bergner hochleben ließ, lauter ähnliche Erlebnisse hatte ich (soeben lese ich übrigens, daß eine witzelnde Truppe, die ich auch gehört hatte, ins Kola gewandert ist).

Landau habe ich auch besucht, er wird immer schrulliger. Er wohnt bei seiner Mutter in einem Minimalzimmer, wo er gleichzeitig schläft und arbeitet, hat im Ganzen nur 20 Bücher in Berlin und die nicht mal in seinem Zimmer. Frau Landau wohnt bei Susi, dreimal die Woche (genau) sehen sich die beiden. Interessant ist, daß Susi nun ihr chemisches Examen machen darf, wegen starker Unterschreitung des numerus clausus. Sonst macht der Antisemitismus in Deutschland Fortschritte, in verschiedenen Städten wurden an jüdische Geschäfte beleidigende Plakate angeschlagen und einige Scheiben gingen in Stücke, sogar in Düsseldorf.

Sie hatten die Freundlichkeit, sich nach einem Studienzuschuß für mich erkundigen zu wollen. Unabhängig davon hatte ich schon lange die Absicht, Sie bezüglich meiner um Rat zu fragen. Meine Idee war bisher folgende: Ich wollte zuallererst die politische Entwicklung in Deutschland abwarten; aber was das Regierungssystem wie die Arierfrage angeht, ist wohl sobald keine Änderung zu erwarten. Ferner wollte ich sozusagen unter Aufrechterhaltung des vermögensmäßigen status quo so lange hier warten, bis wieder ein halbwegs regelmäßiger Zahlungsverkehr Deutschlands mit dem Auslande hergestellt ist, sodaß ich auf eine gewisse monatliche Summe bestimmt rechnen kann. Aber auch in diesem Punkte ist noch keine Änderung abzusehen. Wegen des niedrigen

Standes der Krone kann ich hier trotz der Unregelmäßigkeit der Geldüberweisungen allein von diesen leben. Sollten sie ganz aufhören, so könnte ich hier ohne Einschränkungen 6 Jahre leben. Bei einem monatlichen Verbrauch von $ 130.– könnte ich in Amerika von demselben Geld etwa 3 Jahre leben. Augenblicklich bekomme ich aus Deutschland monatlich durchschnittlich etwa $ 70.–, es droht aber, wie gesagt, völlig aufzuhören (Neugebauer erzählt auch von großen Schwierigkeiten bei Hirschwald und Springer) Ich habe es also trotz allem an sich sehr gut und schäme mich beinahe, daran zu denken, andren einen Studien-zuschuß fortzunehmen, andrerseits möchte ich mir natürlich für den Fall einer völligen deutschen Abriegelung mein Geld nach Möglichkeit als Reserve verwahren. Zunächst wäre ich Ihnen dankbar, wenn Sie mir mitteilen könnten, ob die von mir angegebene Summe von $ 130.– pro Monat eine vernünftige Zahl ist. Ich möchte nicht bei jedem Cent überlegen müssen, ob ich ihn ausgeben darf oder nicht, nicht nur aus Bequemlichkeit sondern auch weil ich bemerkt habe, daß man dann sofort mit dem höchsten Mißtrauen wie jemand behandelt wird, der bald um Hilfe zu suchen droht.

Nun die mathematische Situation. Hier wesentliche Anregungen zu bekommen, halte ich für unwahrscheinlich. Nielsen hat sich seit längerer Zeit mit andren Dingen beschäftigt und scheint wohl zu wünschen, wieder in topologischen Schwung zu kommen, es aber nicht zu tun. Ich habe mich daher sozusagen selbständig gemacht und einen lang gehegten Wunsch durchgeführt, nämlich algebraische

Geometrie gelernt, d.h. v.d. Waerdens Arbeiten studiert und insbesondere diese den italienischen Normalkursus in Mathematik verwandelt (und wie berichtet, hier darüber Vorlesungen gehalten). Das hat mich den ganzen Winter intensiv in Anspruch genommen. Von v.d. Waerden hoffte ich einige Tips zu bekommen, wie man solche Schwierigkeiten leichter überwindet, er meinte jedoch, ich sei der einzige Mensch in Europa, der sich diese Mühe gemacht habe, und auch er könne die italienischen Arbeiten absolut nicht ohne weiteres in Ordnung bringen. Vielleicht ist es vom praktischen Standpunkt aus leichtsinnig, seine Zeit damit zu verbringen, statt zu publizieren, da ja in anderen Ländern die Zahl der Arbeiten das Entscheidende zu sein scheint. Vom überirdischen Standpunkt aus glaube ich jedoch, mit ungewöhnlichem Ertrag für mich gearbeitet zu haben, da ich durch die dauernden Lücken und Unexaktheiten, mit denen ich mich auseinandersetzen musste, in die Sache ganz anders eingedrungen bin, als ich das sonst bei rezeptiver Aufnahme zu tun pflege. Von demselben Standpunkt aus würde ich mir selbst etwa noch ein halbes Jahr Zurückgezogenheit verschreiben, in welchem ich mich mit den Lefschetzschen Arbeiten befassen würde. Diese sind bekanntlich besonders schwer verständlich, aber ich glaube, sie führen auch die heute aussichtsreichste Seite der algebraischen Geometrie.

Daß das „wirkliche" Interesse, das die Princetoner damals für mich gehabt haben, heute noch besteht, halte ich leider für unwahrscheinlich. Es entsprang zunächst einer persönlichen Sympathie von Lefschetz für mich, die wegen der sehr kurzen Bekanntschaft, gewiß längst verschwunden ist, und dann war damals der Zug nach dem Westen ja noch nicht so groß und selbverständlich. Inzwischen ist eine große Zahl erstklassiger Leute nach Amerika gekommen und die nicht nur von Ihnen bemerkte Fremdenfeindlichkeit hat zum Teil sicher in der Saturierung ihren Grund.

Meine Lage kennen Sie nun genau; was ich tun soll, läßt sich schwer sagen. Aber eine genaue Antwort auf die Frage, was Sie an meiner Stelle täten, würde mir sehr am Herzen liegen. Dafür, hier zu bleiben, spricht vom mathematischen Standpunkt aus eigentlich nichts und sonst zunächst die billigen Lebensverhältnisse und die angenehmen persönlichen Beziehungen. Für Amerika spricht vom mathematischen Gesichtspunkt aus viel. Princeton sieht von hier aus bereits etwas zu groß aus, und ich habe mal an Baltimore gedacht, weil mir das Buch von Zariski einen so großen Eindruck gemacht hat. (Es hat nur den einen Fehler, in den Ergebnissen statt in der gelben Sammlung erschienen und so zu komprimiert zu sein). Aber Sie kennen die Verhältnisse ja besser und könnten mir vielleicht auch darüber Auskunft geben. Was bei Ihrer Beurteilung meiner Lage vielleicht auch wichtig ist: heute bin ich 30 Jahre alt geworden.

Mit einem Schluß haben Sie vielleicht schon gar nicht mehr gerechnet, trotzdem bin ich jetzt soweit.
Mit vielen Grüßen an Sie und Ihre Familie

Ihr Herbert Busemann.

Hellerup, 16. XI. 55.

Lieber Herr Professor.

Herzlichen Dank für Ihren ermunternden Brief, der sich mit meinem kreuzte. Sie haben diesen, da er durch Ihre Ausführungen bereits überholt war, hoffentlich nur noch als Stimmungsbild betrachtet.

Mit jedem Ihrer Briefe wird mein Gesicht dümmer. Wie Sie es fertig gebracht haben, in Princeton jemand für mich zu interessieren, ist mir ein Rätsel. Meine Situation stellt sich für mich selbst so dar: 1932 hat Lefschetz auf Grund einer flüchtigen (einen Abend in Ihrem Hause dauernden) Bekanntschaft eine rein persönliche Zuneigung zu mir gefasst. Damals war es nicht so trivial, daß jemand nach Princeton kam und deshalb bestand ein relatives Interesse. Jetzt da man absolut jeden nach Princeton haben kann und einen wesentlichen Teil der gesamten Weltkraft dort bereits konzentriert hat, schien mir, daß dort nur die Extraklasse Aussicht hat, nicht lästig zu fallen. Ich kann mir ja nicht einreden, auch nur approximativ dazu zu gehören. Von meiner bisherigen mathematischen Produktion war eigentlich nur die Fundamentalarbeit ein Beweis dafür, daß ich imstande bin, mit objektiven Schwierigkeiten, d.h. Problemen, mit denen sich gute Mathematiker erfolglos beschäftigt haben, fertig zu werden. Auch konnte ich die andern hier allzuleicht von meiner Chancenlosigkeit überzeugen. Wenn ich glaube, daß es in irgendeiner Hinsicht nicht so viele mir gleichwertige Exemplare gibt, so geht das in Richtung auf die ~~allgemeine~~ außermathematische Bildung, wo ich das Gefühl habe, mich infolge eines guten Gedächtnisses und der Lust, mich mit meinen Büchern allein zu befinden, über das normale Maß hinaus ausgebreitet zu haben. Ganz klar, wie sich diese Teilnahme an andern Dingen, einmal für andere nutzbar gestalten sollte, war mir nicht, doch habe ich immer fest an ihren Wert, nicht nur für mich, geglaubt. Aber das zählt ja sozusagen nicht.

Also so sehe ich mich selbst im Rahmen der übrigen Welt und bekomme nun Ihren vorigen Brief, worin stand, daß man mir sicher gerne schreiben wolle, daß man sich freue, wenn ich zu wissenschaftlicher Zusammenarbeit kommen wolle. Schon das setzte mich, wie Sie aus meinem letzten Brief bemerkt haben werden, in Erstaunen, und nun schreiben Sie „... (Lefschetz) war <u>sehr</u> freundlich, äußerte sich mit größter Wärme und Anerkennung und würde Sie sehr gerne dort haben." Nicht nur ich habe Mund und Nase aufgesperrt. Nachträglich ist mir klar, wie der Beginn unserer Korrespondenz über Amerika auffallen mußte; und ich habe Ihnen die Selbstbetrachtungen geschrieben, damit auch Sie meine anfänglich ablehnende Haltung verstehen, und auch, wenn ich das schreiben darf, weil ich Sie bitten möchte, sie mit Ihrer Ansicht über mich zu vergleichen, da ich etwas Angst habe, man wird mich an Hand des Bildes, das Sie von mir entworfen haben, nicht wiedererkennen.

Aber da Sie nun Möglichkeiten geschaffen haben, an die ich nicht glaubte, habe ich schon im letzten Brief mein Schicksal Ihnen anvertraut und tue es auch weiterhin.*) Ihrer Aufforderung gemäß habe ich daher einen Brief an Lefschetz geschrieben, von dem ich eine Kopie einlege. Weiter wäre nur, was damit aber bereits gesagt ist, mitzuteilen, daß ich den von Ihnen vorgeschlagenen Weg sehr gern beschritte, d.h. mich um einen Grant of Aid bewerben möchte.

Ich danke Ihnen herzlich, um so mehr, da Sie zunächst alles gegen meinen passiven Widerstand unternommen haben, den Sie hoffentlich nun verstehen.

Herzliche Grüße an Sie und Ihre Familie

Ihr Herbert Busemann.

*) Insbesondere auch über den Zeitpunkt der Ausreise. Wie ~~ist~~ scheint mir jetzt klar, daß es nicht zweckmäßig ist, unabhängig von dem etw. Grant of Aid noch dieses Schuljahr zu kommen, doch wäre es mir angenehm, falls Sie mir über Ihre Ansicht in diesem Punkte, sobald sie endgültig ist, Mitteilung machen könnten.

Kopenhagen, den 19. XI. 33.

Lieber Herr Professor.

Über die Göttinger Ereignisse pflegen Sie ja besser orientiert zu sein als ich. Vor einigen Tagen schrieb mir Messner etwas, was Sie vielleicht noch nicht gehört haben und Sie vielleicht interessiert. Der Fall Nikuradse ist Ihnen wohl an sich bekannt, vielleicht auch, daß in einer Art Unterfall verschiedene Leute aus dem Prandtlschen Institut ein Pamphlet gegen Nikuradse losgelassen hatten, daß leider neben wahren auch einige offenbar blödsinnige Beschuldigungen enthielt. Im Juli war eine Verhandlung mit Zeugenaussagen, im September kam die Mitteilung, daß die Zeugen, welche gegen N. etwas ausgesagt hatten, zu entlassen wären. Die Begründung hierfür war ihrerseits blödsinnig, sodaß ein Protest Erfolg hatte, und am 26. X. kam, wie Messner schreibt, der Generaldirektor (wer das ist, weiß ich nicht) aus Berlin und fällte ein Urteil: 3 fristlos entlassen, Messner ab 1. XII unter Gehaltskürzung wieder auf Bewährung eingestellt, zwei weitere ab 1. XI. wieder auf Bewährung eingestellt. Das Skandalöse ist also, daß Prandtl sich hat zwingen lassen, das Urteil zu unterschreiben, worin man lesen kann, daß die Verurteilten das Ansehen des Deutschen Reiches geschädigt hatten.

Nachträglich steigt mein Respekt vor Levi-Civita unbeschränkt, welcher nach achtjährigem Faschismus nicht einmal den Beamteneid leisten wollte. Damals konnten wir alle noch nicht ganz beurteilen, was ein Druck bedeuten kann. Levi-Civita ist übrigens, wie

er schreibt, sehr betrübt, daß er garnichts tun kann, aber
: dato il regime e la scarsità di mezzi . . .

Mir geht es hier ausgezeichnet. H. Bohr ist unglaublich freundlich, alle Augenblicke muß man dort essen, wo er irgendwie behilflich sein kann, ist er es, ohne daß man ihn darum gebeten hätte. Das ist für einen Emigranten, selbst wenn man, wie ich, finanziell gesichert ist, äußerst wichtig, zumal man in jeder Hitlerrede noch einige Tritte hinterher bekommt. Glücklicherweise ist mein Selbstvertrauen noch durch eine andere Sache gestärkt worden; Feller und mir ist es nämlich gelungen eine in einer berühmten Arbeit von Lebesgue (1910) gestellte und seitdem von verschiedenen bekannten Mathematikern ohne Erfolg behandelte Frage zu lösen. Es handelt sich um folgendes: Wenn man irgend eine meßbare Menge K in der Ebene hat und betrachtet zu jedem Punkt P eine Folge sich auf ihn zusammenziehender achsenparalleler Würfel W_n, so ist (bis auf eine Nullmenge) für $P \subset K$ $\lim \frac{m(W_n K)}{m W_n} = 1$ und für $P \not\subset K$ $\lim \frac{m(W_n K)}{m W_n} = 0$.
Die Frage ist, welche Mengen an Stelle der Würfel treten können, ohne daß dieser „Dichtesatz" verloren geht, insbesondere, ob man beliebige achsenparallele Rechtecke nehmen kann. Wir haben gezeigt, daß letzteres geht, daß man aber merkwürdigerweise nicht mehr beliebig gelegene Rechtecke nehmen darf. Man kann das Resultat noch verschiedentlich anders wenden und bekommt so Sätze über die Vertauschbarkeit der Differentiationsreihenfolge bei gemischten partiellen Ableitungen.

Ich wollte Sie fragen, an welchen Tagen um Weihnachten herum Sie in Göttingen sind. Ich werde ein oder zwei Tage dort sein, um sozusagen aus sentimentalen Gründen die letzte Phase des alten Göttingen noch kurz sehen und dann für immer Lebewohl zu sagen.

Herzliche Grüße

Ihr Herbert Busemann.

Herbert Busemann, Privatassistent für Mathematik. Extracted from Anikó Szábo, *Vertreibung, Rückkehr, Wiedergutmachung* (Expulsion, return, reparation), Wallstein Verlag, Göttingen, 2000, pp. 482–484. Dem Wallstein Verlag sei für die Genehmigung zur Wiedergabe herzlich gedankt.[1]

Herbert Busemann wurde am 12. Mai 1905 in Berlin geboren.[2] Nach seinem Schulbesuch in Frankfurt und Essen studierte er Mathematik in München, Göttingen, Paris und Rom. Die Promotion legte er am 25. Februar 1931 in Göttingen ab, wo er sich auch habilitieren wollte.[3] Am Mathematischen Institut war er für Richard Courant tätig. Als dieser vorschlug, auf eine bezahlte Stelle zugunsten eines anderen Kommilitonen zu verzichten, stimmte Busemann zu. Weil er aus vermögendem Verhältnissen kam, war er auf diese Einkünfte nicht angewiesen. So arbeitete er bei Courant und Herglotz als Privatassistent und hatte die Aufgabe, Übungen und Seminare zu leiten und einen Teil der Korrespondenz des Göttinger Mathematischen Instituts zu führen.[4]

Durch seinen Aufenthalt in Italien hatte Busemann den Faschismus bereits kennengelernt. In seinem Wiedergutmachungsverfahren berichtete er, daß er „ähnliche Verhältnisse unter Hitler erwartete".[5] Nach den Erlebnissen der ersten Monate im nationalsozialistischen Deutschland verließ er bereits im Mai 1933 als 28jähriger das Land, um nach Dänemark zu emigrieren. Dort arbeitete er als Lecturer an der Universität in Kopenhagen. Sein Aufenthalt in dem Land war als vorübergehende Lösung gedacht, denn anfangs hoffte er noch, Deutschland würde bald „zur Demokratie zurückkehren".[6] Während der Zeit in Dänemark wurde er von seinem Vater unterstützt, weil er kein Gehalt bezog. Erst in Dänemark wurde ihm klar, daß er nach den nun geltenden Gesetzen in Deutschland seine akademische Laufbahn dort nicht fortsetzen konnte, weil er einen jüdischen Großvater hatte.[7] Im Jahr 1936 emigrierte Busemann in die USA und wurde Assistent für Mathematik beim Institute for Advanced Study in Princeton. Vertretungsweise lehrte er 1939 als Instructor am Swarthmore College und an der

[1]Special thanks are due to Wallstein Verlag for the permission to publish this article on Herbert Busemann as well as an English translation of the German original.

[2]BHdE II/I, S. 175.

[3]HStAH, Nds. 401, Acc. 92/85, Nr. 390, Bl. 5, Wiedergutmachungsantrag vom 15.12.1957.

[4]HStAH, Nds. 401, Acc. 92/85, Nr. 390, Bl. 9, Wiedergutmachungsantrag vom 15.12.1957, Anlage Lebenslauf; C. Reid, Courant 1888–1972, Springer-Verlag, Berlin 1979, S. 157.

[5]HStAH, Nds. 401, Acc. 92/85, Nr. 390, Bl. 9, Wiedergutmachungsantrag vom 15.12.1957, Anlage Lebenslauf; HStAH, Nds. 110 W, Acc. 119/94. Entschädigungsakte Herbert Busemann, 629/1, Bl. 15–18, Schilderung seiner Verfolgung, 21.11.1957.

[6]HStAH, Nds. 401, Acc. 92/85, Nr. 390, Bl. 9, Wiedergutmachungsantrag vom 15.12.1957, Anlage Lebenslauf; HStAH, Nds. 110 W, Acc. 119/94. Entschädigungsakte Herbert Busemann, 629/1, Bl. 15–18, Schilderung seiner Verfolgung, 21.11.1957.

[7]HStAH, Nds. 401, Acc. 92/85, Nr. 390, Bl. 9, Wiedergutmachungsantrag vom 15.12.1957, Anlage Lebenslauf; HStAH, Nds. 110 W, Acc. 119/94. Entschädigungsakte Herbert Busemann, 629/1, Bl. 15–18, Schilderung seiner Verfolgung, 21.11.1957.

Johns Hopkins University in Baltimore. Im folgenden Jahr ging er zunächst als Instructor, dann als Assistant Professor an das Illinois Institute of Technology in Chicago. Im Jahr 1945 wurde er als Assistant Professor an das Smith College in Northampton berufen. Weitere zwei Jahre später erhielt er die gesicherte Position als Full Professor an der University of Southern California in Los Angeles, wo er 1964 zum Distinguished Professor ernannt wurde.[8]

Im Dezember 1957 übersandte das Auswärtige Amt in Bonn die Unterlagen von Herbert Busemann, die zur Bearbeitung der Wiedergutmachung notwendig waren, an das Niedersächsische Kultusministerium.[9] Obgleich Busemann Gutachten von Courant aus New York und Siegel aus Göttingen eingereicht hatte sowie Werner Fenchel seine Assistententätigkeit bezeugte, erhielt Busemann keine Wiedergutmachung nach dem BWGöD. Als Privatassistent ohne Habilitation hatte er keine Anspruchsberechtigung. Seine Klage gegen diesen Bescheid wurde am 27. August 1958 von der Ersten Kammer des Landesverwaltungsgerichts in Hannover abgewiesen: „Daß der Kläger aus achtenswerten Gründen seinerzeit auf ein Anstellungsverhältnis verzichtet hat, konnte bei meiner Entscheidung leider nicht berücksichtigt werden [...].“[10]

Nach dem BEG erhielt Busemann eine Entschädigung: Durch einen Vergleich vom Oktober 1960 erhielt er 776,05 DM zum Ausgleich für seinen Auswanderungsschaden und durch Teilvergleich vom 15. September 1960 einen Ausgleich wegen seines Schadens im beruflichen Fortkommen 5.000 DM. Dieser erhöhte sich nach dem BEG-Schlußgesetz von 1965 um weitere 5.000 DM.[11] Aufgrund seines damaligen großzügigen Verhaltens, nicht auf einer regulären Assistentenstelle zu bestehen, bekam Busemann im Gegensatz zu Rupert Wild, Werner Fenchel und Heinrich Hauptmann keine Entschädigung nach dem BWGöD zugesprochen.[12] Ob Busemann mit einem weitergehenden Klageverfahren erfolgreich gewesen wäre, ist nur spekulativ zu beantworten. Busemann verstarb im Jahr 1994.[13]

[8]HStAH, Nds. 401, Acc. 92/85, Nr. 390, Bl. 9, Wiedergutmachungsantrag vom 15.12.1957, Anlage Lebenslauf; BHdE II/1, S. 176.

[9]HStAH, Nds. 401, Acc. 92/85, Nr. 390, AA, Kunisch, an das Nds. Kultusministerium, 15.12.1957.

[10]HStAH, Nds. 401, Acc. 92/85, Nr. 390, II A (1) 3289/58 an die Erste Kammer des Landesverwaltungsamtes Hannover. Busemann zog die Klage daraufhin zurück.

[11]HStAH, Nds. 110 W, Acc. 119/94. Entschädigungsakte Herbert Busemann, 629/1, Bl. E39.

[12]Sein Vater war Direktor bei den Krupp-Werken. Vgl. C. Reid, Courant 1888–1972, Springer-Verlag, Berlin 1979, S. 150.

[13]R. Siegmund-Schultze, Mathematiker auf der Flucht vor Hitler, Vieweg, Braunschweig 1998, S. 156.

Herbert Busemann, Private assistant for Mathematics. English translation by Manfred Karbe.*

Herbert Busemann was born in Berlin on May 12, 1905.[1] After attending school in Frankfurt and Essen, he studied mathematics in München, Göttingen, Paris and Rome. He took his doctorate in Göttingen on February 25, 1931, where he also wanted to complete habilitation.[2] At the Mathematical Institute he worked for Richard Courant. When the latter proposed to waive a paid position in favor of another fellow student, Busemann agreed. Because he came from wealthy circumstances, he was not dependent on these revenues. Thus he worked with Courant and Herglotz as a private assistant, and had the task of conducting exercises and seminars and managing part of the correspondence of the Göttingen Mathematical Institute.[3]

Because of his stay in Italy, Busemann already learned about fascism. In his restitution procedures, he reported that he "expected similar conditions under Hitler".[4] After the experiences of the first months in National Socialist Germany he, by then 28 years old, already left the country in May 1933 to emigrate to Denmark. There he worked as a lecturer at the University of Copenhagen. His stay in the country was thought of as a temporary solution, for at first he hoped that Germany would soon "return to democracy".[5] During his time in Denmark he was supported by his father because he did not receive any salary. Only in Denmark did he realized that he could not continue his academic career in Germany because he had a Jewish grandfather.[6] In 1936 Busemann emigrated to the USA and became an assistant for mathematics at the Institute for Advanced Study in Princeton. In 1939, as a replacement, he taught as an instructor at Swarthmore College and Johns Hopkins University in Baltimore. The following year, he held a position as instructor, then as assistant professor at the Illinois Institute of Technology in Chicago. In 1945, he was appointed Assistant Professor at Smith College in Northampton. Two years later, he obtained the secure position of a full professor at the University of Southern

*Translated from Anikó Szábo, Vertreibung, Rückkehr, *Wiedergutmachung, Wallstein Verlag* (Expulsion, return, reparation), Göttingen, 2000, pp. 482–484.

[1]BHdE II/I, S. 175.

[2]HStAH, Nds. 401, Acc. 92/85, Nr. 390, Bl. 5, Request for restitution of Dec. 15, 1957.

[3]HStAH, Nds. 401, Acc. 92/85, Nr. 390, Bl. 9, Request for restitution of Dec. 15, 1957, enclosure résumé; Reid, Courant, pp. 132–33.

[4]HStAH, Nds. 401, Acc. 92/85, Nr. 390, Bl. 9, Request for restitution of Dec. 15, 1957, enclosure résumé; HStAH, Nds. 110 W, Acc. 119/94. restitution-claim file Herbert Busemann, 629/1, Bl. 15–18, description of his persecution, Nov. 21, 1957.

[5]HStAH, Nds. 401, Acc. 92/85, Nr. 390, Bl. 9, Request for restitution of Dec. 15, 1957, enclosure résumé; HStAH, Nds. 110 W, Acc. 119/94. restitution-claim file Herbert Busemann, 629/1, Bl. 15–18, description of his persecution, Nov. 21, 1957.

[6]HStAH, Nds. 401, Acc. 92/85, Nr. 390, Bl. 9, Request for restitution of Dec. 15, 1957, enclosure résumé; HStAH, Nds. 110 W, Acc. 119/94. restitution-claim file Herbert Busemann, 629/1, Bl. 15–18, description of his persecution, Nov. 21, 1957.

California, Los Angeles, where in 1964 he became Distinguished Professor.[7]

In December 1957, the Foreign Office at Bonn sent to the Lower Saxon Ministry of Culture the documents of Herbert Busemann, which were necessary to deal with the restitution.[8] Although Busemann had submitted expert opinions from Courant in New York and Siegel in Göttingen, and Werner Fenchel testified his assistant activity, Busemann did not receive any compensation according to BWGöD [law regulating the restitution for former employees of public service institutions]. As a private assistant without habilitation, he did not have any entitlement to benefits. On August 27, 1958, his action against this decision was dismissed by the First Chamber of the State Administrative Court at Hannover: "The fact that the applicant had renounced an employment relationship for the sake of good reason could unfortunately not be taken into account in my decision [...]."[9]

According to the BEG [Federal Compensation Law], Busemann received a compensation: By a compromise agreement of October 1960, he received DM 776.05 in compensation for his emigration damage and by a partial agreement of September 15, 1960, compensation for DM 5,000 because of damage to his professional career. This increased by additional DM 5,000 according to the BEG final act of 1965.[10] Because of his then generous attitude, not to insist on a regular assistant position, Busemann, in contrast to Rupert Wild, Werner Fenchel and Heinrich Hauptmann, was not awarded any compensation according to the BWGöD.[11] Whether Busemann would have succeeded in a more extensive legal proceeding is to be answered only speculatively. Busemann passed away in 1994.[12]

Abbreviations

AA	Auswärtiges Amt
BEG	Bundesentschädigungsgesetz
BHdE	Biographisches Handbuch der deutschsprachigen Emigration
BWGöD	Gesetz zur Wiedergutmachung für die Angehörigen des öffentlichen Dienstes
HStAH	Hauptstaatsarchiv Hannover
	Nds. 110 W Wiedergutmachungsbehörden
	Nds. 401 Ministerium für Wissenschaft und Kunst

[7]HStAH, Nds. 401, Acc. 92/85, Nr. 390, Bl. 9, Request for restitution of Dec. 15, 1957, enclosure résumé; BHdE II/1, p. 176.

[8]HStAH, Nds. 401, Acc. 92/85, Nr. 390, AA, Kunisch, to the Lower Saxon Ministry of Culture, Dec. 15, 1957.

[9]HStAH, Nds. 401, Acc. 92/85, Nr. 390, II A (1) 3289/58 to the First Chamber of the State Administrative Office Hannover. A s a result Busemann withdrew his case.

[10]HStAH, Nds. 110 W, Acc. 119/94. restitution-claim file Herbert Busemann, 629/1, Bl. E39.

[11]His father was a director of the Krupp concern. Cf. C. Reid, Courant in Göttingen and New York, Springer-Verlag, New York 1976, p. 106.

[12]R. Siegmund-Schultze, Mathematicians fleeing from Nazi Germany, Princeton University Press, 2009, p. 183.

26. November 1989

Sehr geehrter Herr Pambuccian

Besten Dank für Ihren Brief, Ihre Klage ber die Vernachlässigung der synthetischen Geometrie in den USA ist leider gerechtfertigt. Über die mutmaßlichen Gründe äußere ich mich später.

Zunächst sind meine *The Geometry of Geodesics* (GG) und *Recent synthetic Differential Geometry* noch im Handel, allerdings zu lächerlich hohen Preisen. Aber beide sind in sehr bekanten Serien erschienen, auf welche Ann Arbeit [sic] sicher abonniert ist,

Die Interessenlosigkeit in den USA ist allein daran erkenntlich, daß jedes Jahr ungefähr doppelt so viele Exemplare im Ausland verkauft werden wie in den USA. Dabei ist der Faktor zwei vermutlich kleiner als in Wirklichkeit, weil 3 meiner Bücher ins Russische übersetzt wurden und in großen, gut gedruckten, sehr billigen Auflagen erschienen sind, z. B. kostet GG nur 1,88 Rubel.

Ich hätte Ihnen gerne als Ph D Vater gedient, aber das ist unmöglich, weil ich 84 Jahre alt bin und seit 1970 pensioniert bin. Bis zu etwa 80 Jahren hatte ich das Glück, noch gute neue Ideen zu haben, wie Sie an meinem (und Phadkes) Buch sehen können, aber meine Produktivität ist jetzt vorbei. Ich hätte nicht mehr die Kraft, mich ganz einer neuen Idee zu widmen.

Die Exemplare meiner Arbeiten, die ich noch habe, schicke ich Ihnen. Aber sie bilden natürlich eine zufällige Auswahl, weil ich die meisten längst weggeben habe.

Einige der neueren Arbeiten lassen viele interessante Fragen offen, z.B. wäre die Theorie der Peakless Functions einer viel weitergehenden Untersuchung zugänglich.

Der Grund für die Vernachlässigung der synthetischen Geometrie scheint mir zunächst daran zu liegen, daß jeder Professor an dem Weiterleben seiner Ideen via seiner Schüler interessiert ist. Zuzugeben, daß seine Methoden weitgehend unnötige Voraussetzungen verlangen, ist unbequem und wird einfach verschwiegen. Tatsächlich kenne [ich] außer einigen meiner Schüler niemand in den USA, der GG wirklich gelesen hat.

Seit ich 1985 den Internationalen Lobachevski Preis für GG bekommen habe, ist Ignorieren unmöglich geworden, weil vor mir nur sehr große Mathematiker den Preis bekommen haben, wie Lie, Hilbert, Klein, E. Cartan, Weyl, H. Hopf, und ich außerdem der erste so geehrte USA Bürger bin.

Mit freundlichem Gruß,

Ihr Herbert Busemann

26. November 1989

Sehr geehrter Herr Pambuccian

Besten Dank für Ihren Brief. Ihre Klage über die Vernachlässigung der synthetischen Geometrie in den USA ist leider gerechtfertigt. Über die vermutlichen Gründe äussere ich mich später.

Zunächst sind meine The Geometry of Geodesics (GG) und Recent synthetic Differential Geometry noch im Handel, allerdings zu lächerlich hohen Preisen. Aber beide sind in sehr bekannten Serien erschienen, auf welche Ann Arbor sicher abonniert ist.

Die Interessenlosigkeit in den USA ist allein daran erkenntlich, daß jedes Jahr ungefähr doppelt so viele Exemplare im Ausland verkauft werden wie in den USA. Dabei ist der Faktor zwei wesentlich ~~größer~~ kleiner als in Wirklichkeit, weil 3 meiner Bücher ins Russische übersetzt worden und in großen, gut gedruckten, sehr billigen Auflagen erschienen sind, z. B. kostet GG nur 1.88 Rubel.

Ich hätte Ihnen gerne als Ph.D. Vater gedient, aber das ist unmöglich, weil ich 84 Jahre alt und seit 1970 pensioniert bin. Bis zu etwa 80 Jahren hatte ich das Glück, noch gute neue Ideen zu haben, wie Sie an meinem (und Phadkes) Buch sehen können, aber meine Produktivität ist jetzt vorbei. Ich hätte nicht mehr die Kraft, mich ganz einer neuen Idee zu widmen.

Die Exemplare meiner Arbeiten, die ich noch habe, schicke ich Ihnen. Aber sie bilden natürlich eine zufällige Auswahl, weil ich die meisten längst weggegeben habe.

Einige der neueren Arbeiten lassen viele interessante Fragen offen, z.B. wäre die Theorie der Peakless Funktions einer viel weitergehenden Untersuchung zugänglich.

Der Grund für die Vernachlässigung der Synthetischen Geometrie scheint mir zunächst darin zu liegen, daß jeder Professor an dem Weiterleben seiner Ideen via seiner Schüler interessiert ist. Zuzugeben, daß seine Methoden weitgehend unnötige Voraussetzungen verlangen, ist unbequem und wird einfach verschwiegen. Tatsächlich kenne außer einigen meiner Schüler niemand in den USA, der GG wirklich gelesen hat.

Seit ich 1985 den Internationalen Lobatchevski Preis für GG bekommen habe, ist Ignorieren unmöglich geworden, weil vor mir nur sehr große Mathematiker den Preis bekommen haben, wie Lie, Hilbert, Klein, E. Cartan, Weyl, H. Hopf und ich außerdem der erste so geehrte USA Bürger bin.

Mit freundlichen Grüßen,

Ihr Herbert Busemann

HERBERT BUSEMANN
3839 OAK TRAIL ROAD
SANTA YNEZ
CALIFORNIA 93460

SAN LUIS OBISPO CA
PM
28 NOV
1989

ALWAYS
USE ZIP CODE

Mr. Victor Pambuccian
Department of Mathematics
University of Michigan
Ann Arbor, MI 48109-1003

November 26, 1989

Dear Mr. Pambuccian

Many thanks for your letter. Your complaint regarding the neglect of synthetic geometry in the USA is unfortunately justified. I'll return later to the probable causes.

First, my *The Geometry of Geodesics* (GG) and *Recent synthetic Differential Geometry* are still available, however for ridiculously high prices. But both appeared in very well known series, to which Ann Arbor certainly subscribes.

The absence of interest in the USA is recognizable from the very fact that every year there are approximately twice as many copies sold abroad than in the USA. And the factor two is probably lower than the real one, for three of my books were translated in Russsian, and appeared in well printed, large, very cheap editions, e.g. GG costs 1.88 rubles.

I would have gladly served as your Ph. D. adviser, but that is impossible, for I am 84 years old and retired since 1970. Until I was about 80 I was fortunate to still have good new ideas, as you can see from my (and Phadke's) book, but my productivity is now gone. I wouldn't have the energy to devote myself to a completely new idea.

I am sending the copies of those of my papers that I still have. However, they naturally form a random selection, for I have long ago given away most of them.

Some of the newer papers leave many interesting questions open, for example the theory of peakless functions would be open to a much more advanced investigation.

The reason for the neglect of synthetic geometry appears to me to lie first in the fact that every professor cares about seeing his ideas live on via his students. To admit that his methods use to a large extent unnecessary hypotheses is inconvenient and is simply swept under the rug. In fact, I do not know, except for some of my students, anybody in the USA who has actually read GG.

Since I got in 1985 the International Lobachevsky Prize for GG, ignoring me has become impossible, as only great mathematicians, such as Lie, Hilbert, Klein, E. Cartan, Weyl, H. Hopf, have gotten the prize before me, and besides, I am the first thus honored US citizen.

With best regards,

Your Herbert Busemann

MEETING HERBERT BUSEMANN

MICHAEL WATERMAN

In 1984 Hugo Martinez, the editor of the Bulletin of Mathematical Biology, asked me to assemble articles for a special issue focused on DNA and protein sequence analysis. I wrote three of the fourteen articles and one of them was titled "Line Geometries for Sequence Comparisons." On seeing the manuscript, my geometer colleague Paul Yang told me that what I had discovered was an example of Straight Spaces that Herbert Busemann had developed. Busemann had from 1947 to 1970 been on the faculty in the Mathematics Department of the University of Southern California and he had retired in 1970. I tracked down his address and mailed him a copy of my paper with a letter explaining the circumstances. It was one of those salutes one makes to the past and I expected at most a note thanking me.

However Herbert wrote back, graciously inviting me to visit him at his home. After a couple of phone calls we arranged a date and I drove to Santa Barbara which is 95 miles from Los Angeles to see him and have dinner. At Santa Barbara I drove up a steep road and then along the Santa Inez Valley. That is where many wineries produce excellent Chardonnay; it is higher and drier than nearby costal Santa Barbara. After asking directions at least once, I finally located the turnoff to Herbert's home. The road went up a steep ridge and I had to stop to allow a covey of quail to cross. Herbert's California-modern home was located on the narrow ridge top.

I was greeted at the door by Herbert and his wife and was even more impressed by the house as I was taken through it. The house, as I mentioned, straddled the ridge and the landscape determined its footprint. There were big windows with views of the valley and the canyon behind, bright abstract paintings, books everywhere, Middle Eastern rugs, and confortable furniture. Out of sight was the regulation length pool in which Herbert swam in each day for what seemed to me a heck of a long time for a man 80 years old. We toured the house, had coffee, and then his wife went to see to dinner preparations. Herbert and I went through his painting room, where large canvases of bright geometrical abstract paintings, finished and unfinished, were stacked everywhere. His retirement occupation of painting was serious and it is possible to find some of his striking work for sale online with a simple Internet search.

Then Herbert said, "Now let's talk mathematics." He began to ask me about my paper. There is a subtlety that makes sequence alignment difficult which anyone not expert with the subject usually misses, and I assumed Herbert was making the same mistake. So I began to gently steer him away from that. "No! No! I understand that," he said impatiently and then dove into deeper waters. I already had high regard for this intense older man but this bumped that to another level. I told him the department was building in geometry with Paul Yang and Robert Brooks. I had even read a couple of Brook's lighter papers which I thought were beautiful and I could say something about them to Herbert.

If only I had a tape of our dinner conversation! The Busemanns came to USC in 1947 so they just missed the brief unfortunate stay of my friend Stan Ulam. They bought a fine home in Baldwin Hills to the southwest of USC. It had a great view of the Los Angeles Basin, facing toward Griffith Park and the Hollywood Hills. But over the years the area became predominantly African American and, I believe, less comfortable for them. So when he retired in 1970, they bought the property in the Santa Inez Valley and Herbert designed and built their home.

It was not long after my visit that it was announced that Herbert Busemann had received the Soviet Union's Lobachevsky Prize, the first American mathematician ever to do so. In my head I sketched an article about Herbert and his award. Since I knew Lee Dembart, a Los Angeles Times science writer, I gave him my ideas and he wrote an article very much on those lines. In 1992 the Busemanns made a substantial donation to the University of Southern California ($770,000 in 1992 dollars) that created the Herbert and Ruth Busemann Assistant Professorship in Mathematics. This, with two of his donated paintings, keep his name and legacy in the minds of all of us in the Mathematics Department.

BUSEMANN AND FELLER ON CURVATURE PROPERTIES OF CONVEX SURFACES

ANNETTE A'CAMPO-NEUEN AND ATHANASE PAPADOPOULOS

Contents

1. Introduction	67
2. Euler and Meusnier	69
3. Curvature of space curves	73
4. Meusnier's Theorem	74
5. Normal curvatures and Dupin's indicatrix	76
6. Conjugate directions and Olinde Rodriguez Formula	78
7. Blaschke's interpretation of Gaussian curvature	80
8. Euler's Theorem holding almost everywhere	80
9. The Umbilical Theorem	80
10. A special example	82
11. Geodesics	82
12. Later developments	83
References	86

1. Introduction

In their article [5],[1] Busemann and Feller study curvature properties of convex surfaces contained in three-dimensional Euclidean space. They generalize classical results of the differential geometry of surfaces to this setting. In such a setting, geometric methods are needed instead of differential calculus and the familiar notions of tangent line, tangent plane, osculating circle, radius of curvature, etc. take their full original meaning. At the same time, Busemann and Feller prove important results on convex surfaces, among which the fact that such a surface has almost everywhere second-order derivatives. (It was known before that such a surface is almost everywhere differentiable; cf. [39]). Another result they obtain is the fact that there is a well defined nearest point projection of curves lying in the complement of a convex body bounded by a convex surface, onto this

[1]The paper *Krümmungseigenschaften konvexer Flächen* appeared in Acta Math. **66**, 1936, p. 1-47, but it carries the note "Imprimé le 26 juillet 1935."

convex surface, which is length non-increasing. They also study geodesics (shortest lines between two points) on convex surfaces.

This early work of Busemann and Feller is the starting of a lifelong program conducted by Busemann on two different fronts: on the one hand, to generalize results and phenomena that are known to hold in the setting of Riemannian or differential geometry to a setting where a minimal amount of differentiability (or no differentiability at all) is needed, and on the other hand, to investigate the general field of convexity (convex surfaces, convex bodies, volumes, Brunn–Minkowski theory, convexity in general metric spaces, etc.). Busemann and Feller's goal is also to build a complete geometrical system of local differential geometry starting with curvature, lines, osculating circles, etc. with a minimum assumption on differentiability. We shall see that the criterion, to see whether this new setting is reasonable, is to show that some classical results such as Euler and Meusnier's theorems are valid in the present general setting.

The main result is that a convex surface in $\mathbb{R}^3$, without any differentiability hypothesis, has sufficient properties so that one can define a notion of curvature and obtain theorems on the normal curvature which are analogous to the classical theorems of Euler, Meusnier and others on the differential geometry of surfaces. These results hold at almost all points.

Busemann and Feller present conceptual proofs of classical results, reducing the arguments to a synthetic essence. As a consequence, they obtain generalizations of these results to a setting with a minimum amount of regularity conditions. They prove that any convex surface is twice differentiable almost everywhere, with existence at almost all points of curvatures of normal cross sections satisfying Euler's theorem, and admitting a Dupin indicatrix of ordinary type: ellipse or pair of parallel lines, or is lying at infinity. In the latter case, the curvature of a normal cross section is zero. The indicatrix, in this setting, is defined as follows. Consider the tangent plane T at a point x of the surface where such a plane exists, and choose the coordinates in such a way that this tangent plane is not vertical. Take the plane parallel to T and lying above T, at distance ϵ. It intersects the surface in a curve, call it c_ϵ. Busemann and Feller prove that for almost every x, the curve c_ϵ converges, as $\epsilon \to 0$, to a curve, which they call the indicatrix at that point. They also prove a version of the theorem saying that a differentiable convex surface where almost all points are umbilics and the upper normal curvatures are bounded is either a piece of a sphere or a plane. On the other hand, they construct an example of a closed differentiable convex surface where almost all points are umbilics even with the same curvature and that is yet no sphere. This shows that the boundedness condition on the upper normal curvatures may not be omitted.

The theorems of Meusnier and Euler are recurrent in Busemann's work. In a paper he wrote 15 years later ([7], 1950) Busemann, introduced a new notion of curvature in Minkowski spaces (finite-dimensional normed vector spaces), and to justify the relevance of this notion, he showed that analogues of the theorems of Meusnier and Euler hold in this new setting as well. We shall say more on these developments in §12. Before Busemann's paper [7], Paul Finsler, in his 1940 paper [25], introduced a notion of curvature valid for (what we call today) n-dimensional Finsler spaces. He also introduced a function which is the analogue of the cosine function in Euclidean geometry. He showed that with these definitions, the theorem of Meusnier is satisfied. In some sense, the theorems of Meusnier and Euler were again used as tests for the suitability of the generalization of the notion of curvature. We also elaborate on this in §12 of the present article.

2. Euler and Meusnier

The study of the differential geometry of curves and surfaces, based on the notion of tangent, normal, order of contact, etc. of plane curves finds its origin in the work of Archimedes and Apollonius (third and second centuries B.C.) and other Greek geometers. It would be interesting to talk about these results, but this would take us too far.[2] Rather, we shall start our exposition with the work of Euler, to which Busemann and Feller refer in their article.

We owe to Euler the first systematic study of the differential geometry of surfaces in the modern sense of the word (curvature defined in terms of second derivatives) and the first important theorems concerning this notion. One should mention though that the ground for Euler's work had been prepared by Kepler, Fermat, Leibniz, Descartes, Johann and Jacob Bernoulli, and that at the same time when Euler was working on the subject, Clairaut (at the age of 16), published another important systematic study of curves and surfaces, see *Recherches sur les courbes à double courbure* [14].

Volume II of Euler's treatise *Introductio in analysin infinitorum* [20] (1748) is concerned with the differential geometry of space curves and surfaces. A curve is given there by a parametric equation of the form $x = x(t), y = y(t), z = z(t)$, and a surface by a parametric equation of the form $x = x(u,v), y = y(u,v), z = z(u,v)$. Chap XIV of this volume concerns the osculating circle and the curvature of a curve. Euler also discusses convexity and concaveness of curves in relation with the sign of the curvature radius. Algebraic curves are classified, and transcendental curves are studied in Chapter XXI. A long appendix is concerned with surfaces in 3-space. In Chapter II of this appendix, Euler studies the intersection of a surface with an arbitrary plane. In Chapter III, he considers sections of cylinders, cones and spheres. In Chapter V, he studies second-degree surfaces and in Chapter VI intersections of two surfaces.

About twenty years after his *Introductio*, Euler wrote a memoir entitled *Recherches sur la courbure des surfaces* (Researches on the curvature of surfaces) [21] (1767) in which he introduced the notion of curvature at a point of a differentiable surface. The definition is based on the collection of curvatures of the curves that pass through that point. Euler noticed that instead of studying the totality of curves passing through a point, it suffices to consider the curves that are intersections of planes containing the normal vector to the surface at that point. Such a plane is called a *normal plane*. He considered the collection of osculating circles of a curve associated with these normal planes and he showed that the collection of radii of these circles contains all the information about the curvature of the surface at

[2] On the notion of angle, tangents, and contact in Greek mathematics, see the remarkable recent editions by R. Rashed, [34], [35], [36], [37] [38].

that point. In particular, Euler proved that at any given point of the surface, the maximal and minimal curvatures associated with these normal planes determine all the other normal curvatures.

More precisely, given a point on the surface and a tangent vector v at that point, we call *normal curvature* through v the curvature of the curve obtained by intersecting the surface with the plane containing the vector v and the normal vector at that point. The *maximal* and *minimal normal curvature* at the given point are the maximum and minimum of the normal curvatures taken over all the normal planes passing through that point. Likewise, the *normal curvature radius* at the given point in the direction of the vector v is the curvature radius of the associated curve. There is a similar notion of *maximal* and *minimal normal curvature radii* at the given point. Euler showed that the directions of the planes that realize the extremal curvatures (except in very special cases) are orthogonal to each other, and he proved that at a given point, if ρ_1 and ρ_2 are the maximal and minimal normal curvature radii respectively, then the normal curvature radius ρ of the normal section through an arbitrary tangent vector v is given by the formula

$$\rho = \frac{2\rho_1\rho_2}{(\rho_1+\rho_2)-(\rho_1-\rho_2)\cos(2\varphi)},$$

where φ is the angle between v and the tangent vector to the normal plane with maximal curvature radius.

The following is a convenient way of re-writing Euler's equation:

$$\frac{1}{\rho} = \frac{\cos^2\varphi}{\rho_1} + \frac{\sin^2\varphi}{\rho_2}.$$

Notice that the values ρ_1 and ρ_2 may also be negative, and that Euler's equation has also a meaning when ρ_1 or ρ_2 is infinite: in this case, the curvature $\frac{1}{\rho}$ is zero for all φ. Euler concludes his memoir [21] as follows (Réflexion VI, p. 143):

> Thus, the judgement of the curvature of surfaces, however complicated it seems at the beginning, is reduced for each point to the knowledge of two osculating radii, one of which is the largest and the other the smallest at that element. These two objects determine entirely the nature of the curvature, displaying for us the curvature of all the possible sections that are perpendicular to the proposed element.[3]

To summarize, Euler's idea is the following: To define the curvature at a point on a surface, one looks for the curves that pass through that point. Not all of them need to be considered. In general, these curves have infinitely many different curvature radii, but there are relations among these magnitudes, and in fact, only finitely many of them are needed to describe the curvature of a surface.

---

[3]Ainsi le jugement sur la courbure des surface, quelque compliqué qu'il ait paru au commencement, se réduit pour chaque élément à la conaissance de deux rayons osculateurs, dont l'un est le plus grand et l'autre le plus petit dans cet élément ; ces deux choses déterminent entièrement la nature de la courbure en nous découvrant la courbure de toutes les sections possibles qui sont perpendiculaires sur l'élément proposé.

There are other memoirs by Euler on differential geometry. In his memoirs *De linea brevissima in superficie quacunque duo quaelibet puncta jungente* (On the shortest line joining two points on a surface) [18] and *Accuratior evolutio problematis de linea brevissima in superficie quacunque ducenda* (A more accurate treatment of the problem of drawing the shortest line on a surface) [24], Euler studies geodesic lines on surfaces. The memoir *De curvis triangularibus* (On triangular curves) [22] is concerned with curves of constant breadth. The memoir *Methodus facilis omnia symptomata linearum curvarum non in eodem plano sitarum investigandi* (Easy method to investigate all points of intersection of curves which do not lie in the same plane) [23] is concerned with the angle of elevation of the tangents, on the equation of the bending planes and on other notions for space curves. Euler's important essay on the calculus of variations, *Methodus inveniendi lineas curvas maximi minimive proprietate gaudentes, sive solutio problematis isoperimetrici lattissimo sensu accepti* (A method for finding curved lines enjoying properties of maximum or minimum, or solution of isoperimetric problems in the broadest accepted sense) [19] contains a study of a minimal surface, etc.

Another important result on which the paper by Busemann and Feller relies is due to Meusnier.[4] The result is contained in the article [30] which Meusnier published at the age of 22, the only mathematical article he wrote. This paper contains an important complement to Euler's paper [21]. Indeed, whereas Euler had obtained a formula for the curvature radius of a curve obtained by intersecting the surface with an arbitrary plane (and not only the ones containing a normal vector), Meusnier gives a simple relation between that magnitude and the curvature radius of a curve obtained as a normal section that has the same tangent vector. This relation is the so-called Meusnier formula. Meusnier's article carries almost the same title as Euler's [21], *Mémoire sur la courbure des surfaces* [30] (1785). It is interesting to read how Meusnier surveys Euler's work ([30] p. 478):

> The curvature of a line is evidently the reason for which the tangent changes its position from a point to another, and the more this change is substantial, the more we can say that the curvature is large: therefore if we conceive two tangents at two infinitely close points, the angle formed by these tangents measures the curvature of the element contained between these two points, and the expression of this angle contains all the theory of curvature. But it has been observed that this formula depends only on the first and

[4]Jean-Baptiste Marie Charles Meusnier de la Place (1745–1793) was a student of Gaspard Monge (1746–1818) at the Military Academy of Mézières, from whom he learnt about Euler's work on curvature of surfaces. He became a general in the French army. Besides that, he was a mathematician, physicist, chemist, and engineer. A detailed and interesting biography of Meusnier was written by Darboux [15]. We learn from the latter that Meusnier presented at the Académie des Sciences the discoveries he made at the Military School on the 14th and 21st of February 1776 and that d'Alembert, who was present at the Academy, declared: "Meusnier starts while I am finishing." On the 12th of June of the same year, and despite his very young age, Meusnier was made corresponding member of the Academy. Monge taught at Mézières and at another military Academy, the famous École Polytechnique. He had several students and descendants, and he is considered as the main founder of the French schools of differential and projective geometry.

second difference equations of the given element. Therefore it follows that a curve which is tangent to that element and that has at its contact the same second difference equation will also have the same curvature. Yet it is always possible to assign a circle which has this property; therefore one can regard any element of a curve as a portion of a certain circle.

Likewise, we say that the curvature of a surface constitutes the fact that a tangent plane varies from one point to the other: thus the curvature of a surface element depends on the formula that expresses the change in the tangent plane in the extent of that element. But we shall see shortly that this formula depends itself on the first and second difference equations: therefore, a surface which is tangent to the element will have at its contact the same curvature as this element, if it has the same second difference equation, and the given element may be regarded as part of that surface. [...]

Mr. Euler treated the same matter in a very beautiful memoir, printed in 1760 among those of the Berlin Academy. This famous geometer addresses the question in a way different from the one which we just presented. He makes the curvature of an element of the surface dependent on that of the various sections that one can perform on it by cutting it with planes. This is why he starts by determining the curvature radius of a section to an element of surface made by an arbitrary plane. He also restricts this determination to the case where the cutting plane is perpendicular to the tangent to the element which is considered, and he discovers the following beautiful property: that among all such cutting planes, the one which gives the section of largest curvature makes with the one which gives the smallest curvature a right angle, regardless of the nature of the surface considered. Finally, he shows that radii of curvature of these two sections are sufficient to determine all the others. whence he concludes that the curvature of a surface element will be known provided we are given this curvature in the two directions where it is the smallest and the greatest.[5]

---

[5]La courbure d'une ligne est évidemment ce qui fait que d'un point à un autre la tangente change de position, et plus ce changement est considérable, plus on peut dire que la courbure est grande : donc si l'on conçoit deux tangentes en deux points infiniment voisins, l'angle formé par ces tangentes, mesure la courbure de l'élément compris entre ces deux points, et l'expression de cet angle contient toute la théorie de la courbure ; mais on a observé que cette formule dépend uniquement des équations aux différences premières et secondes de l'élément dont il s'agit ; d'où il suit qu'une courbe tangente à cet élément et ayant au contact la même équation aux différences secondes aura aussi la même courbure. Or il est toujours possible d'assigner un cercle qui ait cette propriété ; on peut donc regarder tout élément de courbe comme une portion d'un certain cercle.

De même, disons-nous, la courbure d'une surface constitute en ce que d'un point à un autre le plan tangent varie : ainsi la courbure d'un élément de surface dépend de la formule qui exprime le changement du plan tangent dans l'étendue de cet élément ; mais nous verrons bientôt que cette formule dépend elle-même des équations aux différences premières et secondes : donc une surface tangente à l'élément aura au contact la même courbure que lui, si elle a la même équation aux différences secondes, et l'élément en question pourra être regardé comme faisant portion de cette surface. [...]

M. Euler a traité la même matière dans un fort beau mémoire, imprimé en 1760 parmi ceux de l'Académie de Berlin. Cet illustre géomètre envisage la question d'une manière différente de celle que nous venons d'exposer ; il fait dépendre la courbure d'un élément de surface de celle des différentes sections qu'on peut y faire en le coupant par des plans.

We shall quote Meusnier's results that were generalized by Busemann and Feller in §4. In the same article, Meusnier obtained several other results; for instance, he determined the surfaces for which the two extremal radii of curvature are always equal, he solved an isoperimetric problem, a problem on developable surfaces and a problem on minimal surfaces.

3. Curvature of space curves

The starting point of Busemann and Feller's investigation is a purely geometric definition of tangent line, curvature and osculating plane, generalized from the setting of curves in space to that of sequences of points converging to a given point (see Section 1 of [5]):

Definition 3.1 (Tangent line, curvature radius, curvature and osculating plane of sequences of points). Let $P \neq P_n$ be a sequence of points in three-dimensional Euclidean space converging to a point P. If the lines joining P_n to P converge to a line t, this limit is called the *tangent line* of $\{P_n\}$. The pair (P, t) is said to be a *line element*. If $\{P_n\}$ has tangent line t, then the circle in the plane (t, P_n) through P_n that is tangent to t is called the *approximating circle* of P_n with respect to the line element (P, t). If ρ_n denote the radius of this circle and if the limit ρ of ρ_n exists, then this limit is called the *curvature radius*, and its inverse $\frac{1}{\rho}$ the *curvature* of the sequence of points $\{P_n\}$. If the planes through P_n and t converge to a plane E, then this plane E is called the *osculating plane* of $\{P_n\}$.

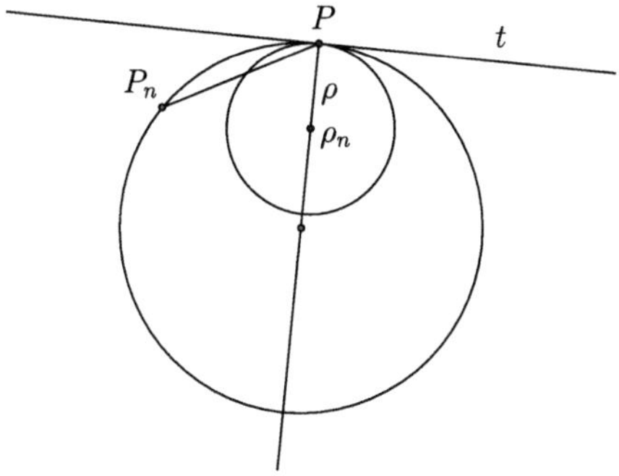

In the same paper, other metric notions are introduced, such as the *foot of the perpendicular*, *normal segment* and *tangent segment* from P_n to P. Formulae for curvature radii, etc. are given in terms of limits of normal

C'est pourquoi il commence par déterminer le rayon de courbure d'une section faite dans un élément de surface par un plan quelconque. Il restreint aussi cette détermination au cas où le plan coupant est perpendiculaire sur le plan tangent à l'élément qu'on considère, et découvre cette belle propriété: qu'entre tous les plans coupant qui sont dans ce cas, celui qui donne la section de plus grande courbure fait, avec celui qui donne la section de moindre courbure, un angle droit, quelle que soit la nature de la surface dont il s'agit. Il fait voir enfin que les rayons de courbure de ces deux sections suffisent pour déterminer tous les autres ; d'où il conclut qu'on connaîtra la courbure d'un élément de surface, pourvu qu'on ait cette courbure dans les deux sens où elle est la moindre et la plus grande.

segments, tangent segments, etc. generalizing the classical formulae given in terms of derivatives.

From this, Busemann and Feller give the following definition of curvature of a space curve:

Definition 3.2 (tangent, curvature and osculating plane of space curves). Let κ be a continuous space curve through a point P (that is, κ is the topological image of a closed segment of the real line). If for any sequence P_n of points on κ converging to P the tangent and curvature exist and are independent of the choice of P_n, then these tangent and curvature are defined to be the *tangent* and *curvature of* κ at P. The osculating plane of κ at P is defined similarly.

Several other definitions of the same metric nature are given. In particular, if the tangent and the osculating plane exist but the curvature depends on the chosen sequence, then the upper limit of all the possible curvatures is called the *upper curvature of* κ. Choosing the sequence P_n to be on the same side of P on the curve, one obtains notions of right-sided and left-sided tangent and curvature.

Busemann and Feller compare these definitions with the classical ones by taking derivatives in the special case of plane curves and they show that their concept is broader. Namely, if κ is given as the graph of a continuous function f of one variable x and at some point x_0 of the interval of definition the right second derivative of f exists, then the usual notion of right curvature of κ at x_0 coincides with the one they introduce. They show that the converse is not true: there exists a C^1-function f with right curvature at x_0 (in the sense of the authors) admitting no right second derivative at x_0.

The case of convex plane curves is of particular interest. Here, at each point of such a curve, a left and a right tangents exist. Because the slope of these one-sided tangents is monotone with respect to the point, the curve has a two-sided tangent almost everywhere (up to countably many exceptions). This monotonicity property also implies that up to a set of measure zero, f has also almost everywhere a second derivative.

In the same trend, the authors give a geometric definition of a strict tangent plane at a point of a surface (see § 2 of [5]):

Definition 3.3 (Strict tangent plane). Let P be a point in the interior of a continuous surface Φ in the Euclidean space. A plane Π is called *strict tangent plane* of Φ at P if for every sequence of secants Q_nP_n of Φ converging to P all the limits of the lines joining Q_n and P_n are contained in Π.

Busemann and Feller indicate in the introduction of their paper that their notion of strict tangent plane coincides with that of *paratangent* introduced by Bouligand [4].

Again, the case of a convex surface is a typical example: any tangent plane is automatically strict.

4. Meusnier's Theorem

Meusnier, in the paper [30], completed Euler's theorem on the curvature of surfaces by proving that the curvature radius at a point of a curve on a surface obtained by intersecting it with an arbitrary (oblique) plane is equal to the product of the curvature radius of the normal section containing the tangent to the same curve multiplied by the cosine of the angle between the plane of the normal section and the osculating plane of the curve. More precisely, if the oblique plane makes an angle ϕ with the normal plane containing the same tangent vector at the given point, then Meusnier's formula is

$$\kappa_\phi \cos\phi = \kappa$$

where κ is the curvature of the curve cut out by the normal plane and κ_ϕ the curvature of the curve cut out by the oblique plane.

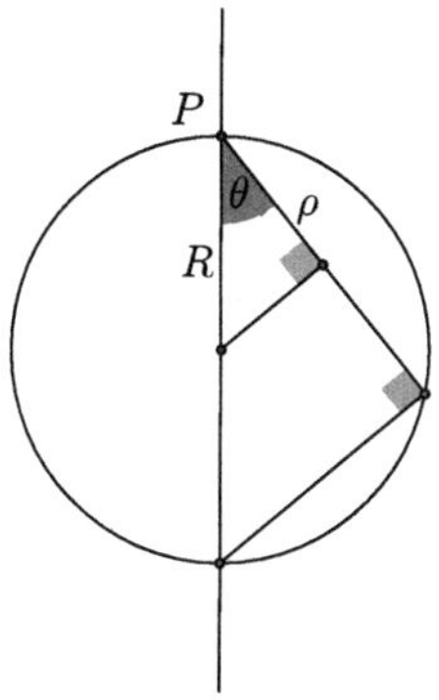

Meusnier's statement can be rephrased as follows (see [3], p. 89):

Theorem 4.1. *Suppose that R is the curvature radius of the normal section of an analytic*[6] *surface through the tangent t at a point P. Let E be a plane through t making an angle θ with the normal of the surface at P. Then the curvature radius ρ of the curve κ on Φ cut out by E is related to R by:*

$$\frac{\rho}{\cos\theta} = R.$$

Thus, the theorem reduces the curvature of a curve cut out on the surface Φ by an plane intersecting it and tangent to the direction t, to the curvature of the curve cut out by the normal plane section.

Meusnier proved at the same time that at any point on a surface the osculating circles to a curve on this surface of all the oblique sections having the same tangent are on the same sphere; in fact, they are the small circles that are the intersections of the sphere with the given planes. (This is the sphere whose center is the center of the osculating circle of a normal section and whose radius is the curvature radius of the normal section.)

Busemann in [6] p. 15 says that there are several equivalent forms of Meusnier's theorem for smooth surfaces, but that for convex surfaces, there is just one natural form, which is given in Bonnesen and Fenchel [5], but which essentially goes back to Hjelmslev [26].

[6] Editor's note: Blaschke works in the setting of analytic surfaces, but for what concerns us here and for the rest of the theorems we quote from [3], surfaces of class C^2 suffice.

In Section 2 of [5], Busemann and Feller give a generalization of Meusnier's theorem to the setting of converging sequences of points on a (non-necessarily convex) surface.

More precisely, Busemann and Feller prove the following:

Let Φ be a surface having a strict tangent plane Π at the point P; let P_n and P'_n be two sequences of points on the surface, both converging to P and having the same tangent t and whose osculating planes make the angles $\theta \neq \frac{\pi}{2}$ and $\theta' \neq \frac{\pi}{2}$ respectively with the normal of Φ at P. Assume moreover that the angles formed by the lines $\overline{P_nP'_n}$ with t are bounded from below and that $\{P_n\}$ has curvature $\frac{1}{\rho}$. Then $\{P'_n\}$ has curvature $\frac{1}{\rho'}$ where

$$\frac{\rho}{\cos\theta} = \frac{\rho'}{\cos\theta'}. \tag{1}$$

In the case where the sequences of points are replaced by curves, one gets the results of Hjelmslev [26] and Bouligand [4].

5. Normal curvatures and Dupin's indicatrix

In Section 3 of [5], Busemann and Feller, following Euler, consider the curvature of normal sections of a convex surface at a given point. In his article on the curvature of surfaces [21] (1767), Euler already emphasizes the significance of normal curvatures (see [21], p. 120):

> The set of all the curvature radii [of the normal sections] will give us the right measure of the curvature of the surface at a given point.[7]

For a surface given as the graph of an analytic[8] function in two variables, as described in §2, Euler calculates the normal curvatures at a given point and proves the following:

Theorem 5.1. *At any given point of an analytic surface, the maximal normal curvature and the minimal normal curvature determine all the other normal curvatures.*

More precisely, Euler shows (see 2):

Theorem 5.2. *If ρ_1 and ρ_2 denote the maximal and minimal normal curvature radius respectively, then the normal curvature radius ρ of the normal section through a tangent t satisfies*

$$\frac{1}{\rho} = \frac{\cos^2\varphi}{\rho_1} + \frac{\sin^2\varphi}{\rho_2},$$

where φ is the angle between t and the tangent of the normal section with maximal curvature radius. Here ρ_1 or ρ_2 may also be negative.

Euler's equation has still a meaning when ρ_1 or even ρ_2 is infinite. In the latter case it means that the curvature $\frac{1}{\rho} = 0$ for every φ.

---

[7] L'assemblage de tous ces rayons [osculateurs des sections perpendiculaires à la surface] nous donnera la juste mesure de la courbure de la surface au point donné.

[8] The word "analytic" in Euler's sense refers to a function given by some formula, that is, a function to which the methods of analysis apply.

In 1813, Ch. Dupin[9] in his collection of articles [17] observes that Euler's equation in fact defines a conic section in the tangent plane at the given point P written in polar coordinates. For this conic section, Dupin introduces the name *indicatrix*, since it indicates by its type the complete nature of the curvature of the surface at P.[10]

If the surface is convex and if not all normal curvatures at P vanish, the indicatrix can be either an ellipse or a pair of two parallel lines and P lies at its center. Written in suitable polar coordinates in the tangent plane at P, the indicatrix is the curve given by $r = \sqrt{\rho(\varphi)}$, where $\rho(\varphi)$ is the normal curvature in the direction of φ.

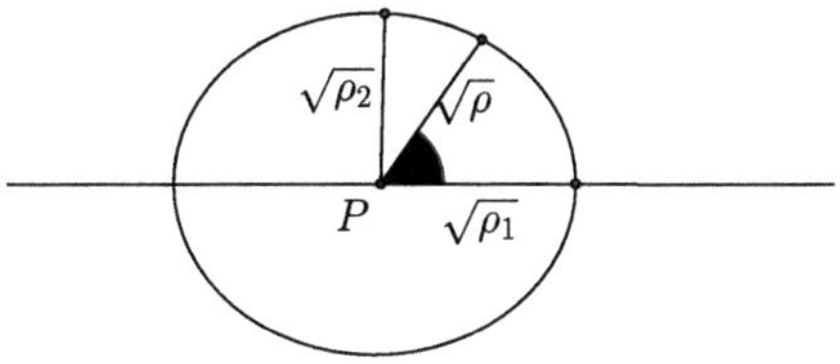

Now Busemann and Feller use this observation to define an indicatrix of a point P on an arbitrary convex surface provided it has a tangent plane at P and certain curvatures exist (see Section 3 of [5]):

Definition 5.3 (indicatrix). In the tangent plane at P, for every angle φ, the ray t emanating from P in the direction of φ defines a normal half-plane section of Φ through t. Let us assume that it has a (one-sided) curvature $\frac{1}{\rho_\varphi}$. Then by associating $\sqrt{\rho_\varphi}$ to φ, we obtain a curve in the tangent plane called the *indicatrix* of P. Here ρ_φ may also be infinite. In this case, the corresponding point of the indicatrix in the direction of φ lies at infinity.

By definition, if all normal sections have two-sided curvatures at P, then the indicatrix is centrally symmetric with center P.

There is also another approach to the indicatrix generalizing the following remark. As Blaschke notes in [3], p. 91, the indicatrix of a point on an analytic[11] surface is similar to the intersection of the surface with any plane parallel to the tangent plane at sufficiently small distance (and on the appropriate side of the tangent plane). Using this idea, Busemann and Feller show (see Section 3 of [5]) the following:

---

[9]Charles Dupin (1784–1873) was a student of G. Monge at the École Polytechnique in Paris and went to Corfu as Naval engineer in 1807 where he took part in the Greek science revival by teaching mathematics and mechanics. During that time, he wrote his treatises on geometry that he sent to his teacher, Monge, for inspection and then published it in [17].

[10]See p. xvii, introductory report of Monge et al. of [17].

[11]See Footnote 6.

Proposition 5.4. *For any small number $h > 0$, consider the plane parallel to the tangent plane Π at distance $h/2$ (in the direction of the inner normal) and intersect it with Φ. The resulting curve is projected to Π and rescaled at scale $\sqrt{h} : 1$ to obtain a curve σ_h in Π. Taking the limit of those curves for $h \to 0$, we obtain exactly the indicatrix of P (if the limit exists).*

To see this, the authors choose a coordinate system for the space with P as origin, the tangent plane Π as (x, y)-plane and the inner normal as z-axis. Then one can describe the convex surface near P as the graph of a nonnegative convex function f in polar coordinates $z = f(r, \varphi)$. If the normal section in the direction of φ has curvature, it is given by

$$\frac{1}{\rho_\varphi} = \lim_{r \to 0} \frac{2\, f(r, \varphi)}{r^2}.$$

The curve σ_h is given by the equation $f(\sqrt{h}\, r, \varphi) = \frac{h}{2}$.

Now let $P_n = (r_n, \varphi_0, h_n/2)$ be a sequence of points on the normal section in direction of φ_0 converging to P. Projecting to Π and rescaling as above, we obtain the sequence of points $P_n' = (\frac{r_n}{\sqrt{h_n}}, \varphi_0, 0)$ and $\lim_{n \to \infty} \frac{r_n}{\sqrt{h_n}} = \sqrt{\rho_{\varphi_0}}$.

Busemann and Feller prove the following properties of the indicatrix:

Theorem 5.5. *Suppose that at a point P on a convex surface for almost all directions two-sided normal curvatures exist and are finite. Then an indicatrix exists and is a convex curve. Conversely, any convex plane curve together with a chosen point P in its interior or on the curve itself may occur as the indicatrix of P for a suitable convex surface through P.*

It is easy to see the second claim: For a convex plane curve κ with P in its interior, choose P as the origin, the plane as the (x, y)-plane, and describe κ in polar coordinates by an equation of the form $\lambda = \lambda(\varphi)$. Then the surface defined by $z = \frac{r^2}{2\lambda^2(\varphi)}$ has a tangent plane at the origin, and its indicatrix at the origin is precisely κ.

Finally, Busemann and Feller consider the case where there is no tangent plane at P but the tangent cone contains a flat piece. Then the indicatrix is just defined in the corresponding section of angles.

6. Conjugate directions and the Olinde Rodriguez Formula

Dupin in [17], further investigated the properties of the indicatrix and introduced a notion of conjugate directions in the tangent plane of a point on an analytic surface. His definition is the following:

Definition 6.1. Let P be a point on an analytic surface with tangent plane Π. For a tangent t of the surface at P let Q be one of the points on the indicatrix of P in the direction of t. The line $\tilde{t}$ through P in Π parallel to the tangent of the indicatrix at Q is called *conjugate to* t. Since the indicatrix is a conic section, the conjugate is well-defined and conjugacy is a symmetric relation on the lines in the tangent plane Π. In the directions t, where the normal curvature takes an extremal value, i.e. the principal directions, the conjugate line is perpendicular to t.

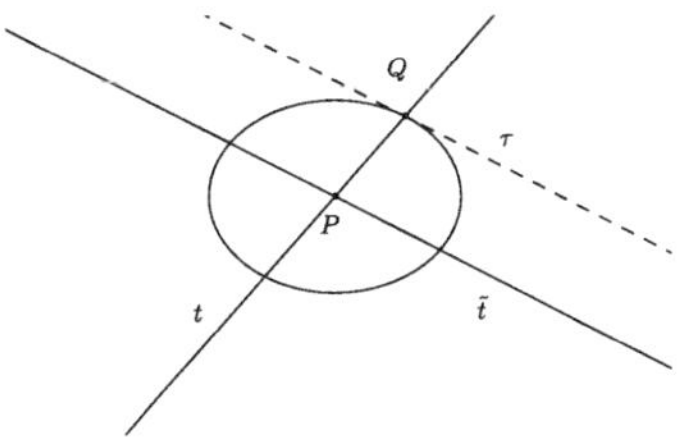

Busemann and Feller define an analogue of this notion for the following setting. They assume P to be a point on a convex surface such that locally near P tangent planes exist, and that the tangent plane at P touches the surface only in P. Moreover, they assume that the indicatrix of P exists, contains P in its interior and that not all the normal curvatures vanish. In the case of a sufficiently differentiable surface, it would just be an ellipse.

Definition 6.2 (conjugate line). Let P' be a point on a surface curve κ through P with tangent t at P and bounded upper curvature at P. Let $g(P')$ denote the line obtained by intersecting the tangent planes at P and P'. If for any sequence of such points P' converging to P the lines $g(P')$ converge to a common line g, the line g is called *conjugate to* t.

Under these assumptions Busemann and Feller prove the following result:

Theorem 6.3. *Suppose that the indicatrix of P has a tangent τ at the point Q, then the conjugate of the line joining P and Q exists and it is parallel to τ. In this more general setting however, the conjugacy relation need not be symmetric. If the line joining P and Q is orthogonal to τ, it is called a principal direction. The normal curvature in any principal direction takes an extremal value. However, the converse need not be true since the indicatrix may have singularities.*

In this context there is an analogue of the following statement usually attributed to Olinde Rodriguez[12] (see Blaschke, [3], p. 95):

Theorem 6.4. *If κ is a curve tangent to a principal direction at P with principal curvature $\frac{1}{r}$, then*

$$\frac{d\xi}{ds} = \frac{1}{r}\eta ,$$

where ξ denotes the inner unit normal vector to the surface at P, η is the unit tangent vector of κ at P and s the arclength of κ.

As Busemann and Feller show, this theorem remains true also for any curve κ on Φ emanating from P with finite upper curvature and tangent to one of the principal directions.

[12]Benjamin Olinde Rodriguez (1795–1851) was a French banker also known as utopian socialist supporting Saint-Simon. He wrote a thesis on differential geometry at Paris Faculty of Science, which was published in 1816, where he refined results of Monge and Dupin.

7. Blaschke's interpretation of Gaussian curvature

Theorem 7.1. *Let Φ be a (not necessarily differentiable) convex surface and P a point on Φ with bounded indicatrix. Let $\Omega(h)$ denote the area of the piece of the surface cut out by the plane parallel to the tangent plane at P at distance h. Furthermore, let $\rho(\varphi)$ denote the normal curvature at P in the direction of a line making an angle φ with some fixed line chosen as a reference in the tangent plane. Then*

$$\lim_{h\to 0} \frac{\Omega(h)}{h} = \int_0^{2\pi} \rho(\varphi)\, d\varphi\,.$$

In the special case, where the indicatrix is an ellipse given by

$$\frac{1}{\rho} = \frac{\cos^2\varphi}{\rho_1} + \frac{\sin^2\varphi}{\rho_2}\,,$$

this amounts to

$$\lim_{h\to 0} \frac{\Omega(h)}{h} = \int_0^{2\pi} \rho(\varphi)\, d\varphi = 2\pi\sqrt{\rho_1\rho_2}\,,$$

which is equivalent to

$$\rho_1\rho_2 = \lim_{h\to 0} \left(\frac{\Omega(h)}{2\pi h}\right)^2 .$$

This was noted by Blaschke in [3], Exercise 10, p. 120.

8. Euler's Theorem holding almost everywhere

Using their version of Meusnier's theorem, Busemann and Feller can directly conclude that on any convex surface Φ, at almost all points (up to a subset of Lebesgue measure zero) a tangent plane exists and all plane sections through P have finite curvatures. At these points, the indicatrix is a convex curve having P as its center. They proceed to show even more, namely the following analogue of Euler's theorem 5.2 (see Section 4 of [5]):

Theorem 8.1. *For almost all points on a convex surface (up to a set of Lebesgue measure zero) the indicatrix exists and is a conic section with P as center (more precisely, it is either an ellipse or a pair of parallel lines).*

For the proof, Busemann and Feller assume that the surface is locally given as the graph of a continuous convex function f in two variables. Since the normal sections are convex curves, as noted in §3, the existence of a normal curvature of the surface at P implies the existence of first and second derivatives of f in the corresponding direction. Then, they investigate the second directional derivative $f_{\alpha\alpha}$ as a function of α and show that for almost all points it is in fact a quadratic form in $\cos\alpha$ and $\sin\alpha$. The claim can be concluded with the help of Meusnier's theorem.

9. The Umbilical Theorem

Section 5 of [5] is devoted to a generalization of the classical theorem on umbilical points. An *umbilical point* on a surface is a point P such that the normal sections at P all have the same curvature. Thus, at such a point, the principal directions are indeterminate.

The classical statement on umbilical points is the following (see Blaschke [3], p. 97):

Theorem 9.1. *A surface in Euclidean space consisting only of umbilical points is a part of a sphere or of a plane.*

For the argument given by Blaschke, it is sufficient to assume that the surface is of class C^3. This is because he uses the Olinde Rodrigues formula (6.4) mentioned above for two orthogonal principal directions η_1 and η_2

$$\frac{d\xi}{ds_j} = \frac{1}{r_j}\eta_j .$$

Taking derivatives again with respect to the other direction respectively and subtracting, he finds that the normal curvature of the surface in any direction and at any point is indeed constant and then the claim follows.

Busemann and Feller prove a generalization of this theorem, in which the differentiability condition of the function is relaxed, and under another hypothesis saying that almost all (instead of all) points on the surface are umbilics.

Theorem 9.2. *Let Φ be a convex surface given as a graph of a differentiable function f in the two variables x, y, such that the partial derivatives f_x and f_y are absolutely continuous*[13] *as functions of both variables. If almost all points of Φ are umbilics, then Φ is part of a sphere or of a plane.*

The regularity assumptions on the partial derivatives of f are indeed necessary and cannot be omitted. This is shown by a counterexample constructed in Section 6 of [5].

The proof of this result is based on this observation following from the Olinde Rodriguez formula 6.4:

Lemma 9.3. *Let Φ be a convex surface, given as a graph of a differentiable function f in the two variables x, y. Let κ be a normal section of Φ at some point P. Moreover, suppose that f_x and f_y are absolutely continuous along κ (parametrized by arclength). If almost all points of κ are umbilics of Φ then κ is a normal section of Φ at all its points.*

Furthermore, the authors of [5] show the following criterion for the regularity condition in the theorem:

Lemma 9.4. *If the upper normal curvatures of the surface are bounded, then the functions $f_x(x, y)$ and $f_y(x, y)$ are absolutely continuous with respect to both variables.*

This lemma follows from the observation that f_x and f_y satisfy uniform Lipschitz conditions. The observation is needed to see that in the example constructed in the Section 6 of [5] the regularity condition is in fact not fulfilled.

---

[13]Here, referring to Lebesgue [29], Busemann and Feller call a function h on an interval absolutely continuous if h is differentiable almost everywhere and if the Lebesgue integral of h' over any subinterval $[x_0, x_1]$ equals $h(x_1) - h(x_0)$.

10. A special example

In Section 6 of [5], Busemann and Feller construct an example of a surface illustrating the fact that in the umbilical theorem the boundedness condition for the upper normal curvature is indeed needed.

Example 10.1. For any given curvature $0 \le c < 1$, Busemann and Feller construct a closed convex differentiable surface Φ contained in the unit ball such that almost all points are umbilics with curvature c and yet Φ is neither a piece of a sphere nor a plane. More precisely, up to a set of measure zero, Φ consists of a countable union of regions congruent to pieces of the sphere of radius $\frac{1}{c}$ (or flat pieces if $c = 0$).

A sequence of convex closed surfaces Φ_n containing a union Γ_n of finitely many pieces of the sphere of radius $\frac{1}{c}$ (or flat pieces if $c = 0$) is constructed recursively. One starts from the unit sphere Φ_1 (and the empty set Γ_1) and then iterates the following two steps:

Step 1: Choose a finite number of sufficiently small disjoint caps in $\Phi_n \setminus \Gamma_n$, each bounded by a simply closed curve obtained by intersecting the surface with the sphere of radius $1/c$. Cut off each cap and glue back a piece of the sphere of radius $1/c$ instead. Here one has to make sure that the curves are so small that the area of each new cap is still more than half the area of the original cap. Moreover, the curves are chosen in such a way that the union of all the original caps cover more than two thirds of the area of $\Phi_n \setminus \Gamma_n$.

Step 2: Smooth out the edges along the small curves by inserting sufficiently small patches that are parts of canal surfaces with normal curvatures $> c$. Here the patches are chosen so that they are disjoint, do not intersect Γ_n, and replace at most one quarter of the total area of the new pieces inserted in Step 1.

The choice of the curvature of the canal patches guarantees that each Φ_n is again convex. Moreover, the area of $\Phi_n \setminus \Gamma_n$ for $n \to \infty$ converges to zero, because the ratio of the area of Γ_n to the area of Φ_n converges to 1. Therefore the limit surface Φ has all the required properties.

11. Geodesics

In the final Section 7 of [5], Busemann and Feller consider shortest paths, or geodesics, on convex surfaces. They refer to Lebesgue for the existence of a shortest rectifiable curve on a given convex surface between any two points,[14] and show that such a curve at a point with tangent plane is in fact differentiable. More precisely they prove:

---

[14]In his dissertation, entitled *Intégrale, longueur, aire*, [28], where Lebesgue introduces his concept of integration, he also defines the notion of a rectifiable curve and its length. In chapter VI on Plateau's problem, among other things, he addressed the problem of whether the integral of a given function f taken over curves in space with some extra properties such as given its endpoints or lying on a given surface, attains a minimum. He proves this under some very general conditions on the extra properties and on the function f (see [28], p. 115–118). This general result implies the existence of at least one shortest rectifiable path between any two points on a given closed, convex surface, although this fact is never explicitly stated.

Theorem 11.1. *If a convex surface has a tangent plane Π at P, then any shortest line passing through P has a tangent at P and an osculating plane which is perpendicular to Π (that is, passing through the normal at that point). So in particular, its geodesic curvature at P is zero.*

This also holds for points where there are no normal curvatures or if certain normal curvatures are infinite.

The proof is again purely synthetic. It involves several lemmas of a purely metric character, like their Lemma 7.1, saying that a closest-point projection of a rectifiable curve in the complement of a closed convex surface onto that surface is distance non-increasing.

The authors point out that even if the surface has a tangent plane and an indicatrix at every point, there need not be a shortest line emanating from a given point in every possible direction. However, if the upper normal curvatures are bounded then this holds.

12. Later developments

In 1935 and 1936, Busemann and Feller published additions to their paper [5] on curvature properties of convex surfaces in three separate short articles entitled *Remarks on the Differential Geometry of Convex Surfaces, I, II and III* ([9] [10] [11]).

The first article [9], carrying the subtitle *Shortest Lines on Differentiable Surfaces*, deals with the question mentioned in the previous section, namely whether for a given point P on a convex surface there exists a geodesic emanating from P in any possible direction. Busemann and Feller first construct an example of a differentiable convex surface where this property does not hold even though the surface at each point has finite one-sided normal curvatures. More precisely, starting from a sphere and then iteratively glueing smaller and smaller rounded cones along some chosen arc near the North pole P, they construct a surface such that there is no geodesic emanating from P in the direction of the arc.

Busemann and Feller then proceed to prove the following existence result: *If Φ is a convex differentiable surface such that the upper one-sided normal curvatures at any point are bounded by $\frac{1}{R}$, then through every point on Φ there is a unique geodesic emanating in any given direction. Moreover, if two points A and B on Φ have at least two different shortest paths joining them, then their distance is $\geq \pi R$.* The proof is based on an approximation of Φ (using a theorem by Tonelli) by analytic surfaces whose normal curvatures satisfy the same conditions as the normal curvatures of Φ.

The second article [10] carries the subtitle *On the Curvature Indicatrices*. It is devoted to a more general notion of indicatrix. As mentioned above (see Theorem 5.5), in [5] Busemann and Feller show that at almost every point of a convex surface an indicatrix exists and is a convex curve with P in its interior or on its boundary, and any such convex curve may occur as an indicatrix. However, there may be exceptional points on the surface where the normal curvatures do not exist. At such points, for a given half-tangent t, Busemann and Feller consider the upper and lower one-sided normal curvatures, i.e. the limes superior or the limes inferior respectively of all one-sided curvatures in the direction of the ray t. Replacing these upper and lower normal curvatures in the definition of an indicatrix, they arrive at the concept of an upper and lower indicatrix respectively at P. In their article [10], Busemann and Feller describe the possible sets that can occur as upper and lower indicatrix and illustrate all those types with suitable examples.

In the third article [11], carrying the subtitle *On Gaussian Curvature*, the authors aim at a definition of Gaussian curvature at some point that only depends on the indicatrix and its position relative to the point. For a point P of the surface Φ such that an indicatrix exists, they consider the surface cap Ω_h around P, obtained by intersecting Φ with a plane parallel to the tangent plane at P at distance h. They consider then its spherical image Ω_h^*, i.e. the set of points on the unit sphere whose tangent planes are parallel to a tangent plane at some inner point of Ω_h. Now the Gaussian curvature at P is defined as the limit

$$\kappa(P) = \lim_{h \to 0} \frac{|\Omega_h^*|}{|\Omega_h|},$$

where $|\Omega_h^*|$ and $|\Omega_h|$ denotes the Lebesgue measure of the set respectively. Busemann and Feller prove that if the indicatrix of P is bounded, then the limit $\kappa(P)$ exists. Moreover, it is finite if P lies in the interior and infinite if P lies on the boundary of the indicatrix. For the finite case they also give an explicit formula in terms of integrals over the indicatrix. In the special case where Euler's theorem holds and the indicatrix is an ellipse centered around P, then

$$\kappa(P) = \frac{1}{a^2 b^2},$$

where a, b are the semi-axes of the ellipse. If the indicatrix is unbounded, the limit $\kappa(P)$ in general does not exist, even if Euler's theorem holds and the indicatrix is a pair of parallel lines.

There is another article of Busemann and Feller [8] where they further investigate a question that has come up during their work on curvature properties of convex surfaces. It is entitled *On differentiation of Lebesgue integrals*. In Section 4 of [5], as a tool for their proof, they use the variation of a function f in two variables on an axis-parallel rectangle (an "interval") to associate to f an additive, absolutely continuous function on intervals and differentiate this function (almost everywhere) with respect to intervals. Now in [8], generalizing the system of intervals, they consider a system R of measurable sets of positive measure in n–dimensional space such that for each point Q in space there is a sequence of sets ρ_n in R contracting to Q. They show that such a system R has the property that for every bounded, Lebesgue-integrable function f, we have almost everywhere

$$\lim_{n \to \infty} \frac{1}{m(\rho_n)} \int_{\rho_n} f(P) dP = f(Q)$$

if and only if the system R fulfils the density theorem, namely, if the following holds: If κ is a measurable set and ρ_n a sequence of sets in R contracting to Q, then at almost all points Q the limit $\lim_{n \to \infty} \frac{m(\rho_n \cap \kappa)}{m(\rho_n)} = 1$ if $Q \in \kappa$ and $\lim_{n \to \infty} \frac{m(\rho_n \cap \kappa)}{m(\rho_n)} = 0$ otherwise. If one allows possibly unbounded integrands f, this is no longer true. For example, the system of intervals R satisfies the density theorem, but the contraction property of the integrals for unbounded integrands f does not hold in general.

Let us now mention some other developments of the work of Busemann and Feller in [5]. A. D. Alexandrov, in his paper [1], generalized to any dimension $n \geq 3$ some of the major results of Busemann and Feller. He proved in particular that for any convex body in $\mathbb{R}^n$, second-order differentiability holds almost everywhere on the boundary with respect to Hausdorff $(n-1)$-dimensional measure, that the notion of normal curvature may be defined almost everywhere and that theorems analogous to those of Euler and Meusnier hold. (We note that the existence of a derivative almost everywhere for a convex function of one variable was already known; it follows from the fact that a monotone function of one variable is almost everywhere differentiable). For dimension $n \geq 3$, the existence almost everywhere of a second derivative for a convex function is stated in Alexandrov's paper in the form saying that the gradient of a convex function is almost everywhere differentiable.

Busemann and Feller's paper was also the starting point for general investigations on other subjects. In their foundational paper on the metric theory of Finsler geometry [13] (1941), Busemann and Mayer write:

> If the underlying space is two–dimensional we shall prove that already under this condition all minimizing arcs are continuously differentiable. Since we do not even require the existence of first partial derivatives of $F(x, \dot{x})$, our method has to be quite different from the usual ones. Our procedure was suggested by the treatment of geodesics on arbitrary convex surfaces in [5]. So far we have not succeeded in getting a similar result for more–dimensional spaces. In the two-dimensional case three questions arise next, namely, whether under the same conditions the minimizing curves have second derivatives, whether every line element is contained in at least one minimizing arc, and whether a given line element occurs in at most one such arc. Simple examples will show that the answer to all three questions is negative.

We mentioned in the introduction of the present paper the fact that Busemann generalized some of the ideas in the paper [5] to the setting of Minkowski spaces. In this context, he defines a function $\mathrm{sm}(L, R)$ which he calls a "Minkowski sine" and he proves the following analogue of Euler's theorem in [7] (p. 179):

Theorem 12.1. *Let M be a hypersurface of class C^2 in a Minkowski space, p a point of M, and L a line not in the tangent hyperplane H of M at p. If we take on each ray R in H with origin p a segment of Minkowski length $\big(\rho(R)\mathrm{sm}(L, R)\big)^{1/2}$ a segment starting at p, where $\rho^{-1}(R)$ is the curvature κ of the intersection of the plane through L and R with M, then the endpoints of these segments traverse a pair of conjugate quadrics in H.*

The sine function that is used in this paper is different from the sine curvature which Finsler introduced in [25] in the setting of Finsler spaces, which we mentioned in the introduction.

In the same article (p. 178), Busemann proves an analogue of Meusnier's theorem, in a form that says that a certain factor $\kappa_s \mathrm{sm}(t_{r+1}, t^*_{r+2})$ is constant in the pencil of planes.

In their joint paper [12], Busemann and Feller prove that a curve of finite total curvature has almost everywhere a unique tangent and a well defined curvature. They highlight a condition for a curve or a hypersurface in Euclidean space for having finite total curvature and they show that this condition is essentially more general than convexity. They also obtain new general forms of the theorems of Euler and Meusnier. In fact, in the paper [12], Busemann and Feller extend some results they obtain for convex surfaces in [5] to the more general setting of surfaces with total extrinsic curvature.

Finally, we mention that in his monograph on convex surfaces [6], Busemann makes a comprehensive review of several results he obtained with Feller in [5], stating them in a more general setting.

References

[1] A. D. Alexandrov, Almost everywhere existence of the second differential of a convex function and some properties of convex surfaces connected with it, Leningrad State Univ. Annals [Uchenye Zapiski] Math. Ser. 6 (1939), 3–35.

[2] S. Altmann and E. L. Ortiz, Mathematics and social utopias in France. Olinde Rodrigues and his times. History of Mathematics Series, No. 28. American Mathematical Society, Providence, RI and London Mathematical Society, London, 2005

[3] Wilhelm Blaschke, Vorlesungen über Differentialgeometrie, Band I. Elementare Differentialgeometrie. Springer–Verlag, Berlin, Fourth Edition 1945, Unveränderter Nachdruck der dritten erweiterten Auflage, bearbeitet von Gerhard Thomsen, von 1930.

[4] G. Bouligand, Introduction à la géométrie infinitésimale directe, Paris, Gauthier-Villars, 1931.

[5] Herbert Busemann, Willy Feller, Krümmungseigenschaften konvexer Flächen, Acta Math. **66**, 1936, p. 1-47. English translation by A. A'Campo Neuen: Curvature Properties of Convex Surfaces. Selected Works, Herbert Busemann, Vol. II, p. 235–269.

[6] H. Busemann, Convex surfaces, Interscience tracts in pure and applied mathematics, NY, 1958.

[7] H. Busemann, The foundations of Minkowskian geometry. Comment. Math. Helv. 24, (1950). 156-187

[8] Herbert Busemann, Willy Feller, Zur Differentiation der Lebesgueschen Integrale. Fundamenta Mathematicae 22 (1934), p. 226–256.

[9] Herbert Busemann, Willy Feller, Bemerkungen zur Differentialgeometrie der konvexen Flächen. I: Kürzeste Linien auf differenzierbaren Flächen, in: Mat. Tidsskr. B, København, 1935, p. 25-36.

[10] Herbert Busemann, Willy Feller, Bemerkungen zur Differentialgeometrie der konvexen Flächen. II: Über die Krümmungsindikatrizen, in: Mat. Tidsskr. B, København, 1935, p. 87–115.

[11] Herbert Busemann, Willy Feller, Bemerkungen zur Differentialgeometrie der konvexen Flächen. III: Über die Gaußsche Krümmung, in: Mat. Tidsskr. B, København, 1936, p. 41–70.

[12] H. Busemann, and W. Feller, Regularity properties of a certain class of surfaces. Bull. Amer. Math. Soc. 51, (1945). 583–598.

[13] H. Busemann and W. Mayer, On the foundations of calculus of variations. Trans. Amer. Math. Soc. 49, (1941). 173–198.

[14] A.-C. Clairaut, Recherches sur les courbes à double courbure, Nyon, Didot et Quillau, Paris, 1731.

[15] G. Darboux, Notice historique sur le général Meusnier, membre de l'Ancienne Académie des Sciences, lue dans la séance publique annuelle du 20 décembre 1909,

Paris, Gauthier-Villars, 19019 and Mém. Acad. Sciences, t. 51, 2e série, 1910, p. 1—40.

[16] Ch. Dupin. Essai sur la description des lignes et les surfaces du second degré. Journal de l'École Polytechnique, Série 1, 7 (1807), Cahier 14, 45–83.

[17] Ch. Dupin, Développements de géométrie. Paris, Coursier, 1813.

[18] L. Euler, De linea brevissima in superficie quacunque duo quaelibet puncta jungente, Commentarii academiae scientiarum Petropolitanae 3, 1732, pp. 110–124, Opera Omnia,Series 1, Volume 25, pp. 1–12.

[19] L. Euler, Methodus inveniendi lineas curvas maximi minimive proprietate gaudentes, sive solutio problematis isoperimetrici lattissimo sensu accepti, 1744. In: Opera Omnia, Series 1, Volume 24.

[20] L. Euler, *Introductio in analysin infinitorum*, First edition: Lausannae: Apud Marcum-Michaelem Bousquet & socios., 1748. *Opera omnia*, Series 1, vol. VIII. English translation by J. T. Blanton, 2 vol., Springer-Verlag, New York, 1988, 1990.

[21] L. Euler, Recherches sur la courbure des surfaces. Memoires de l'académie de science de Berlin, **16** (1767), p. 119–143.

[22] L. Euler, De curvis triangularibus, Acta Academiae Scientarum Imperialis Petropolitinae 1778, pp. 3–30, Opera Omnia, Series 1, Volume 28, pp. 298–321.

[23] L. Euler, Methodus facilis omnia symptomata linearum curvarum non in eodem plano sitarum investigandi, Acta Academiae Scientarum Imperialis Petropolitinae 1782, pp. 19–57 Opera Omnia, Series 1, Volume 28, pp. 348–381.

[24] L. Euler, Accuratior evolutio problematis de linea brevissima in superficie quacunque ducenda, Nova Acta Academiae Scientarum Imperialis Petropolitinae 15, 1806, pp. 44–54, Opera Omnia, Series 1, Volume 25, pp. 269–279

[25] P. Finsler, Über eine Verallgemeinerung des Satzes von Meusnier, Vierteljahrsschr. Naturf. Ges. Zürich, vol. 85 (1940) 155–164.

[26] J. Hjelmslev, Grundlag for Fladerness Geometri, Mem. Acad. Roy. Sc. let. Danemark. Copenhague, (7) XII (1914).

[27] J.-L. de Lagrange, Théorie des fonctions analytiques contenant les principes du calcul différentiel dégagés de toute considération d'infiniment petits, d'évanouissans, de limites et de fluxions, et réduits à l'analyse algébrique des quantités finies. Journal de l'École Polytechnique, 9e cahier, Tome III, Bernard, Paris, 1796. Œuvres complètes, Tome 9, p.13–413.

[28] H. Lebesgue, Intégrale, longueur, aire, Annali di Matematica (3), VII (1902).

[29] H. Lebesgue, Leçons sur l'Intégration, Paris 1928.

[30] J.-B. Meusnier, Mémoire sur la courbure des surfaces, Mém. div. sav. Paris, t. X, 1785, p. 477–510.

[31] G. Monge, Mémoire sur les développées, les rayons de courbure, et les différens genres d'inflexions des courbes à double courbure, Mem. Div. Sav.,10 (1784), 551–550.

[32] G. Monge, Applications de l'analyse appliquée à la géométrie, à l'usage de l'École Polytechnique, 1th ed.1795 with the title *Feuilles d'analyse appliquée à la géométrie*, 4th ed. 1809, Bernard, Paris.

[33] S.D. Poisson, Mémoire sur la courbure des surfaces, Journal für die reine und angewandte Mathematik (1832) Volume: 8, page 280-297.

[34] R. Rashed, *Angles et grandeur : d'Euclide à Kamāl al-Dīn al-Fārisī*, Scientia Graeco-Arabica , Band 17. De Gruyter, Boston/Berlin, 2015

[35] R. Rashed, Apollonius : Les Coniques, tome 1.1 : Livre I, commentaire historique et mathématique, édition et traduction du texte arabe, de Gruyter, 2008, 666 p.

[36] R. Rashed, Apollonius : Les Coniques, tome 2.2 : Livre IV, commentaire historique et mathématique, édition et traduction du texte arabe, de Gruyter, 2009, 319 p.; Apollonius : Les Coniques, tome 3 : Livre V, commentaire historique et mathématique, édition et traduction du texte arabe, de Gruyter, 2008, 550 p.

[37] R. Rashed, Apollonius : Les Coniques, tome 4 : Livres VI et VII, commentaire historique et mathématique, édition et traduction du texte arabe, Scientia Graeco-Arabica, vol. 1.4, de Gruyter, 2009, 572 p.;

[38] R. Rashed, Apollonius : Les Coniques, tome 2.1 : Livres II et III, commentaire historique et mathématique, édition et traduction du texte arabe, de Gruyter, 2010, 682 p.

[39] K. Reidemester, Über die singulären Randpunkte eines konvexen Körpers. Math. Ann. 83(1921), 116-118.

[40] O. Rodrigues, Recherches sur la théorie des lignes et des rayons de courbure des surfaces, et sur la transformation d'une classe d'intégrales doubles qui ont un rapport direct avec les formules de cette théorie, Correspondance de l'Ecole Polytechnique, 3 (1815), 162–182.

Annette A'Campo-Neuen, Universität Basel, Departement Mathematik und Informatik, Fachbereich Mathematik, Spiegelgasse 1, CH-4051 Basel, Schweiz

Athanase Papadopoulos, Institut de Recherche Mathématique Avancée, Université de Strasbourg and CNRS, 7 rue René Descartes, 67084 Strasbourg Cedex, France, and Brown University, Mathematics Department, 151 Thayer Street Providence, RI 02912, USA.

HERBERT BUSEMANN AND CONVEXITY

VALERII. N. BERESTOVSKII

Introduction

It is not possible to give, even briefly, a complete account of the diverse, interesting, and very active investigations on convexity theory that were made by Herbert Busemann and around his work.

In the bibliography, at the end of this essay, we have collected the numerous books and some of the interesting and important articles on convex geometry in order to present, essentially with no comments, the basic trends in the development of this subject at the time of Busemann's investigations, but also before and after his work on the subject.

We start the bibliography with references to Busemann's publications. Then the literature is disposed in a way that distinguishes (in a rather conventional manner) the main directions of research in convex geometry. More precisely, the plan is the following:

(1) Busemann's book and his papers on convexity [1] – [33].
(2) Extrinsic geometry of convex sets and convex (hyper)surfaces [34]–[76].
(3) Intrinsic geometry of convex surfaces and generalizations [77]–[79].
(4) Extremal problems of convex geometry, isoperimetry, mixed volumes [37], [39], [45], [51], [64], [68], [80] – [104], [118], [122], [125].
(5) Combinatorial convexity, convex polytopes [34], [40], [104] – [128].
(6) Asymptotic convex geometry [45], [119], [129]–[130].

In the bibliography, only the unconventional and interesting book [71] by L. Hörmander represents the important field of convex analysis.

The References in each direction are generally given in ascending chronological order.

I thank Professor Athanase Papadopoulos for his help in preparing this text.

1. Busemann's articles before 1960

Two important books appeared in the same year, 1934: the one by Bonnesen and Fenchel [37], and the one by Steinitz and Rademacher [105]. These books laid a firm basis for all the subsequent systematic investigations in the respective theories of convex sets/hypersurfaces and combinatorial convexity. Branko Grünbaum begins his excellent book [110] with the following dedication: "*In humble homage dedicated to the memory of the outstanding geometer Ernst Steinitz (1871–1928).*"

The book [37], even though it is not large, presents the subject in an encyclopedic manner. It contains a very complete information about all the important results in the theory of convex bodies obtained before 1934. The proofs are often omitted, but

Supported by the Minist of Education and Science of the Russian Federation, Project Number 1.8126.2017/8.9.

the book is supplied with a highly extensive bibliography. The following titles of section in [37] will give a precise idea of the content on this book: 1. Main definitions, 1–6. 2. Barycentres and convex hulls, 7–13. 3. Classification of boundary points and support planes of a convex body, 13–18. 4. Representation of a convex body by convex functions, 18–28. 5. Linear combinations of convex bodies. Linear and concave families, 28–34. 6. Approximation of a convex body, 34–37. 7. Numbers and figures corresponding to convex bodies, 37–55. 8. Integral formulas for volumes and mixed volumes, 55–69. 9. Symmetrizations and related modifications of convex bodies, 69–74. 10. Inequalities, extremals and cover problems, 74–87. 11. The Brunn-Minkowski Theorem and Minkowski inequalities, 87–104. 12. Special cases and applications of the Brunn-Minkowski Theorem and Minkowski inequalities, 105–114. 13. Definability of convex bodies by curvature functions, 114–124. 14. Central convex bodies, 124–127. 15. Bodies of constant width, 127–141. 16. Characteristic properties of second order forms, 141–143. 17. Differential geometry of convex curves and surfaces, 143–149. References, 150–164.

It is remarkable that the papers [1] and [2] by Busemann are quoted in this book. The paper [1] is a published version of Busemann's PhD dissertation defended in 1931 at the University of Göttingen under the supervision of Richard Courant (1888–1972). In the preface, the authors of [37] thank Busemann and H. Kneser for their help in the proof-reading and error-correction. We recall that Helmut Kneser (1898–1973) was a doctoral student of David Hilbert and specialist in topology and group theory. His most famous result is probably his theorem on the existence of a *prime decomposition for 3-manifolds.* In his proof of this result lies the origin of the concept of *normal surface*, a fundamental cornerstone of the theory of 3-manifolds. But for us, more important is his joint paper with Wilhelm Süss (1895–1958) on the proof of the Brunn-Minkowski inequality [85]. Süss was a doctoral student of Ludwig Bieberbach (1886–1982), and the founder and first director of the Oberwolfach Mathematical Institute.

The next five papers [3]–[7] by Busemann are joint with William Feller (1906–1970), a doctoral student of Courant and future outstanding probabilist. Like the book [37], the first four of these papers were written in Copenhagen. Busemann notes in the preface to the book [19] that Chapter I (Extrinsic Geometry) and Chapter II (The Brunn-Minkowski Theory and Its Applications) form, in many respects, a natural complement to [37]. Results of all papers [3]–[7] are used in the first chapter of [19]. The co-author of the paper [18], where was posed the famous *Busemann-Petty problem*, is Clinton Petty, a doctoral student of Busemann. The next paper by Busemann [23] is co-authored by Ernst Gabor Straus (1922-1983), who was an assistant of Albert Einstein in 1950–1953. Straus's extensive list of co-authors, besides Busemann, includes Einstein, Paul Erdös, Richard Bellman, Béla Bollobás, Sarwadaman Chowla, Ronald Graham, Lászlo Lovász, and George Szekeres. Straus's interests ranged widely over his career, beginning with his early work on relativity with Einstein and counting deep work on analytic number theory, extremal graph theory, and combinatorics. He is known for the *Erdös-Straus conjecture.* Günther Ewald (1929–2014) and Geoffrey Colin Shephard, Busemann's co-authors of the papers [29]–[30]

and [31], [32], are accomplished and known specialists in combinatorial convexity, as it is seen from their books [122] and [114]. Shephard also worked on reflection groups. He addressed the *Shephard problem* on the volumes of projected convex bodies and another problem on polyhedral nets. He began the study of complex polytopes, and classified complex reflection groups. Tudor Zamfirescu, the author of the book [74], is a doctoral student of Ewald (1968). The book [128] contains the proceedings of a geometry conference in Mulhouse (France), in 2014, in honor of Zamfirescu, on the occasion of his 70th anniversary.

K. Reidemeister showed in [36] that a convex hypersurface in Euclidean space E^n is almost everywhere differentiable. Busemann and Feller proved in the paper [3] that a convex surface in E^3 has almost everywhere approximative second differential and that the Dupin indicatrix has the usual form. In 1939, A. D. Alexandrov extended this result to an arbitrary convex hypersurface in E^n [39]. Roughly speaking, it is proved in [7] that similar statements hold not only in the convex case but for curves in E^2 and surfaces in E^3 with finite total extrinsic curvatures. Some analogues of classical theorems of Euler and Meusnier for such surfaces are proved as well.

The papers [8] and [9] are dedicated to the natural definition of intrinsic area for parametrized surfaces in Finsler spaces using Hausdorff measure. Busemann begins his paper [9] by saying that examples of Nöbeling and Besicovitch convinced many mathematicians that Lebesgue area does not represent the geometric idea of area in this setting. This paper frees the concept of area from both its Euclidean and its infinitesimal character.

There is a general agreement that area must

(1) be a generalization of arclength;
(2) coincide with the elementary notion of area for piecewise Euclidean or Minkowskian spaces;
(3) have the usual value for smooth surfaces in Riemannnian spaces;
(4) be *intrinsic*, that is, depends only on the geodesic (intrinsic) distances on the surface and not on the way the surface is embedded.
Busemann also thinks that the following principle, called Kolmogoroff Principle, is geometrically indispensable:
(5) The surface with the greater intrinsic distances must have the greater area.

It is shown in [9] that conditions (2) to (5) select among the many area elements suggested for Finsler spaces the one indicated by Choquet (see [9]), so that the question of area may be considered settled for smooth surfaces in Finsler spaces.

Although Hausdorff measure is defined only for non-parametrized surfaces, Busemann proved that for birectifiable surfaces in Finsler spaces (corresponding to rectifiable surfaces in non-parametric form in E^3), area coincides with Hausdorff measure.

Busemann proved in [10] the following main theorems.

I. In a two-dimensional Busemann G-space, a local notion of area exists for which triangles with equal sides have equal areas if and only if the space is locally isometric to either the Euclidean plane, or a hyperbolic plane, or a sphere.

II. In a two-dimensional Busemann G-space, a local notion of area exists such that the area of a triangle pab depends only on p, the local branch of the geodesic $\mathfrak{g}$

that contains the segment $[ab]$ *and the distance* ab*, if and only if the space is locally isometric to a Minkowski plane.*

Roughly speaking, in [11], Busemann solves the isoperimetric problem in a Minkowski plane with a non-necessarily convex or symmetric indicatrix.

The results are the following:

Let $F(u)$ be a continuous function of vectors $u \in E^2$ which is positive for $u \neq 0$, and positive homogeneous of order 1. The problem of finding among all simple closed curves $z = z(t) \in E^2$ with a given orientation and a given Minkowski length $L = \int F(z'(t))dt$ one which bounds the greatest (Euclidean) area *has a unique solution* (up to translation) *no matter whether the indicatrix* $C : F(z) = 1$ is *convex or not.* The solution *is homothetic to the polar reciprocal* (figuratrix) *of* $\overline{C}$*, the boundary of the convex closure of* C*, with respect to the unit Euclidean circle rotated through* $\pm\pi/2$.

At the end of [11] Busemann proves that the intrinsic (not necessarily symmetric) metrics defined by F and defined by $\overline{F}$ with the indicatrix $\overline{C}$ coincide.

To prove these results, Busemann uses the *Brunn-Minkowski Theory* (see below). Notice that this may be done by other methods.

It is clear that if $F(\cdot)$ is the usual norm, then $\overline{C} = C$ and the orientation of curves plays no role. In this case, the Minkowski area differs from the Euclidean area by a constant factor. Therefore *the solution of the Minkowskian isoperimetric problem is the same as for the above problem.* For the theory of Finsler spaces, it is interesting to find in what case the Minkowski circles solve the isoperimetric problem. A necessary and sufficient condition for this is the symmetry of perpendicularity relation for non-oriented Euclidean straight lines with respect to Minkowski distance. J. Radon in 1916 (respectively, M. M. Day in 1947) determined some (respectively, all) Minkowski circles with this property [11].

In the note [12], Busemann characterized Minkowski (or finite dimensional Banach) spaces X with strictly convex closed unit balls B as those in which every closed convex set C is a *Chebyshev set*, i.e. any point $x \in X$ has a unique nearest point in C. Notice that B is strictly convex if and and only if the unit ball B^* in its dual Minkowski space X^* has smooth boundary (unit sphere).

A problem considered in [13] is a direct generalization of the problem in [11]: A positive continuous function $\sigma(u)$ is defined on the set S^{n-1} of all unit vectors u of E^n:

To find among all simple closed surfaces D *in a certain class* Γ *and with a given "area"* $\int_D \sigma(u)dS$ *those which bound the maximal (Euclidean) volume* $V(D)$.

Here dS means the (Euclidean) surface element of D and u the exterior normal of D "at dS."

For the rather restricted class Γ which consists of all convex surfaces and non-convex surfaces whose first partial derivatives satisfy piecewise Lipschitz conditions, the problem is solved by the *isoperimetric inequality*

$$(1) \qquad \int_D \sigma(u)dS \geq nV^{(n-1)/n}(D)V^{1/n}(K),$$

where K is the polar reciprocal (with respect to the Euclidean unit sphere) of the boundary of the convex closure of the surface $C := \{\sigma^{-1}(u)u : u \in S^{n-1}\}$. Furthermore, the equality sign holds only for D homothetic to K.

Inequality (1) solves the isoperimetric problem for the intrinsic area in Minkowski spaces (Section 6 of [13]). The solution is in general not a Minkowski sphere, but a new convex surface which proves to be of great importance for the theory of Finsler spaces [15].

The paper [13] shows that Minkowski's definition of area in Euclidean spaces by means of parallel sets carriers over to Minkowski spaces if the parallel sets consist of solutions of the isoperimetric problem instead of spheres. Section 7 of the paper completes the paper [9] by proving the much more general theorem that $\int \sigma(u)dS$ *represents the intrinsic area of any rectifiable manifold in a Finsler space.*

The form of (1) indicates a close connection with the Brunn-Minkowski Theory. In fact, the area $\int \sigma(u)dS$ was studied by Minkowski in [35], and (1) follows for convex C and D directly from Minkowski's inequality (Section 1).

The *Brunn-Minkowski inequality* for convex sets C and D in E^n and n-dimensional volume V is

$$(2) \qquad V^{1/n}((1-\lambda)C + \lambda D) \geq (1-\lambda)V^{1/n}(C) + \lambda V^{1/n}(D) \quad \text{for} \quad 0 \leq \lambda \leq 1,$$

where equality holds if and only if either one of the sets C, D is a point, or C and D have no interior points and lie in parallel hyperplanes, or both C and D have interior points and are positive homothetic.

The Brunn-Minkowski inequality was first proved by Brunn [81, 82] with a clever but rather vague proof. The equality case, in particular, was not settled satisfactorily, as was pointed out by Minkowski. Later on, both Brunn [83] and Minkowski [104] provided correct proofs. A different proof is due to Blaschke [84]. It uses Steiner symmetrizations [80]. At present several essentially different proofs are known, as well as a series of extensions and applications [125].

The note [14] establishes a theorem on convex bodies in E^n which is the counterpart of the Brunn-Minkowski Theorem, or Brunn Theorem, by [37]. The latter follows from the Brunn-Minkowski inequality. Busemann gives the following reformulation of this theorem to exhibit the analogy with his theorem:

In E^n, let K be a convex body with interior points, P a two-dimensional half-plane bounded by the line L, and H a hyperplane normal to L intersecting K in the non-empty set $H \cap K$ with $(n-1)$-dimensional volume $V(H \cap K)$. For any such H, we take the segment of length $V^{1/(n-1)}(H \cap K)$ with origin $H \cap L$ on the ray $H \cap P$. Then the ends of these segments form a convex curve in P turning its concavity towards L.

Busemann's theorem is concerned with the intersection of K by pencils of non-parallel half-hyperplanes:

I. In E^n, let K be a convex body with interiors points, L an $(n-2)$-dimensional linear space which intersects K, and P a 2-dimensional space normal to L at a point O. Any half-hyperplane H bounded by L intersects K in a non-empty set $H \cap K$. For any such H, we take the segment of length $V(H \cap K)$ with origin O on the ray $H \cap P$. Then the ends of these segments form a convex curve in P.

The following result is a corollary of I:

II. Let K be a convex body in E^n with O as interior point and center. If for any hyperplane H through O we take a segment with origin O and length $V(H \cap K)$ on the normal to H at O (in both directions), then the ends of these segments form a convex surface S.

The result II is decisive for the theory of area in Finsler spaces and more specifically in Minkowski spaces.

Let r and f be non-parallel rays in P with origin O, q a ray with origin O in the angle with legs r, f, let α, β be angles between r and q, and q and f respectively. Let H_r, H_q, H_f denote the half-hyperplanes bounded by L and containing r, q, f respectively. Then Theorem I above is equivalent to the inequality

$$V(H_q \cap K) \geq \frac{\sin(\alpha+\beta)V(H_r \cap K)V(H_f \cap K)}{\sin\alpha V(H_r \cap K) + \sin\beta V(H_f \cap K)}. \tag{3}$$

III. Equality holds in (3) if and only if $H_r \cap K$ can be transformed into $H_f \cap K$ by a rotation about L and a subsequent dilation in the direction of f, and $H_q \cap K$ is the intersection of H_q with the convex closed hull of $H_r \cap K$ and $H_f \cap K$.

Statement *III* implies the following addendum to *II*:

IIa. The surface S in II *is strictly convex if K is strictly convex.*

But III shows that the strict convexity of K is not a necessary condition.

A proof of these results of Busemann, in a convenient reformulation of W. Barthel [64], is also given in [64].

Busemann essentially used results of his paper [13] in his programmed papers [15] and [16] on the study of Finsler geometry. We shall give here only a short extract from the introduction of [16]. Busemann considered the paper [15] as a useful introduction to [16] for the readers not familiar with Finsler spaces.

The exploration of Minkowski spaces is a necessary preliminary step for any progress in the theory of Finsler spaces. By Busemann's unconventional opinion, these are the local spaces which belong to any finitely compact space with geodesics and a minimum amount of differentiability properties.

Contrary to the common point of view, Busemann thinks that one needs only a few number of invariants to determine simple geometric objects in Finsler geometry: a curve in a plane is determined by two curvatures instead of one (Section 5) and more generally, a hypersurface is determined by three fundamental forms instead of two (Section 7); however, the problem of characterizing a k-dimensional submanifold in an n-dimensional space is open for $k < n - 1$ and seems to be difficult.

The new tools in [16] are

(1) the basic theorem (19) on convex bodies;
(2) an *adequate concept of area,*
(3) a function which corresponds to the *sine function* in Euclidean geometry (Section 2), and which has many applications. Its definition does not require differentiability properties of the unit sphere.

The reason for the success of the suggested concepts is that they are *integro-differential expressions instead of pure differential expressions.* As a consequence, they depend continuously on the metric. The sine function is used in Section 3 to discuss *normality of linear sub-spaces* in Minkowski spaces. The solutions of the isoperimetric problem in Minkowski spaces (see [13]) is closely connected with the normality of hyperplanes to straight lines (Section 4). They are significant because *many Euclidean results remain valid if the unit sphere is replaced by a properly normalized isoperimetrix.* One of the two curvatures of curves is a Minkowski invariant and an integro-differential expression. The other permits to associate intrinsically a Euclidean space with a given Minkowski space. In Section 6, the *theorems of Meusnier and Euler* are proved for curves on a surface. Section 7 illustrates that the *methods of relative differential geometry* (see [37], pp. 64,65) *become significant* for surfaces in 3-dimensional Minkowski space if the isoperimetrix, instead of the unit sphere, is taken as carrier of the spherical image.

Busemann proved in [17] some important results in integral geometry which have applications to Finsler spaces. We shall reproduce here the main results stated in Section 1 of [17], together with two applications of them at the end of the paper.

There are known two expressions

$$|M| = \frac{1}{2}\int_0^{2\pi} r^2(\omega)d\omega = \frac{1}{2}\int_0^{2\pi}\left(\int_{-r(\omega-\pi/2)}^{r(\omega+\pi/2)} |\rho|d\rho\right)d\omega \tag{4}$$

for the area $|M|$ of a plane domain M, given in polar coordinates ρ, ω by the inequalities $0 \leq \rho \leq r(\omega)$, $0 \leq \omega \leq 2\pi$.

The well-known extension to dimension n of the first expression in (4) is

$$|M| = \frac{1}{n}\int_{S^{n-1}} r^n(u)d\omega_{(n-1)}(y). \tag{5}$$

Here S^{n-1} is the unit sphere in E^n centered at z, $d\omega_{(n-1)}(u)$ is its volume (area) element at the point $u \in S^{n-1}$, and M is given by $0 \leq \rho \leq r(u)$, $u \in S^{n-1}$.

In the second expression, $|\rho|$ may be interpreted as the (1-dimensional) volume of the simplex with one vertex at the origin z and the other at a variable point $p = (\rho, \omega \pm \pi/2)$ in the cross-section (intersection) of M with the line normal to ω.

The purpose of the note [17] is *the proof and the application of the following extension of this second expression to* $n-1$ *sets* $M_1, \ldots, M_{n-1}$ *in* E^n:

$$|M_1|...|M_{n-1}| = \tag{6}$$

$$\frac{(n-1)!}{2}\int_{S^{n-1}}\left(\int_{M_1(u)}\cdots\int_{M_{n-1}(u)} T(p_1,...,p_{n-1},z)dV_{n-1}(p_1)\ldots dV_{n-1}(p_{n-1})\right)d\omega_{(n-1)}(u).$$

Here, $M_j(u)$ is the cross-section of M_j with the hyperplane $H(u)$ through z normal to the unit vector u, the point p_j varies in $M_j(u)$, the differential $dV_{n-1}(p_j)$ is the $(n-1)$-dimensional volume element of $M_j(u)$ at p_j, and $T(p_1,...,p_{n-1},z)$ is the volume of the simplex with vertices $p_1,...,p_{n-1},z$.

Replacing the sets $M_{n-r+1},...,M_{n-1}$ by the unit ball B^n with center z yields expressions for $|M_1|...|M_{n-r}|$ in terms of the volume $T(p_1,...,p_{n-1},z)$, in particular (5) for $r = n-1$.

With the notation
$$\kappa_n = |B^n| = \pi^{n/2}/\Gamma\left(\frac{n}{2}+1\right),$$
Steiner's symmetrization [80] leads from (6) to the following result:

If $M_1, ..., M_{n-1}$ are convex bodies in E^n ($n \geq 3$) with interior points, z is a given point in E^n, and $M_j(u)$ is the cross-section of M_j with the plane normal to u and through z, then

$$|M_1|...|M_n| \geq \frac{1}{n}\frac{\kappa_n^{n-2}}{\kappa_{n-1}^n}\int_{S^{n-1}} (|M_1(u)|...|M_{n-1}(u)|)^{n/(n-1)}\, d\omega_{(n-1)}(u), \tag{7}$$

with equality holding only when the M_j are homothetic ellipsoids with center z.

It follows in particular for a convex body M that for $n \geq 3$,

$$|M|^{n-1} \geq \frac{1}{n}\frac{\kappa_n^{n-2}}{\kappa_{n-1}^n}\int_{S^{n-1}} |M(u)|^n d\omega_{(n-1)}(u), \tag{8}$$

with equality (if $|M| > 0$) only for ellipsoids with center z.

The last result may also be formulated as follows:

Among all convex bodies M with a given volume, the ellipsoids with center z (and only these) maximize $\int_{S^{n-1}} |M(u)|^n d\omega_{(n-1)}(u)$.

In the special case $M_{n-1} = M$, $M_j = B^n$ for $j < n-1$, the inequality (7) gives a weaker estimate due to L. A. Santalo

$$|M| \geq \frac{1}{n}\kappa_n^{-n/(n-1)}\int_{S^{n-1}} |M(u)|^{n/n-1} d\omega_{(n-1)}(u), \tag{9}$$

with equality (if $|M| > 0$) only for the balls with center z. Finally (9) shows the following:

Among all convex bodies with center z, the balls (and only these) yield the maximum of $\min_u |M(u)|^n |M|^{1-n}$. The corresponding minimum maximum problem seems quite difficult.

2. The Busemann-Petty problems

In the paper [18] are suggested ten interesting and difficult problems on convex bodies in the Euclidean space E^n with unit sphere S^{n-1} with center z at the origin of E^n. Usually, the results require $n \geq 3$ but in some instances one may have $n = 2$. We shall quote here all of them but then discuss only part of the problems.

The theory of Minkowski spaces leads to various problems concerning plane sections through the center z and interior point of a symmetric convex body K. Any one of these problem has an interpretation in terms of Minkowskian geometry but also can be considered directly, without this interpretation.

Throughout the paper [18] (see also [22]), the authors use the following notation. An arbitrary element in S^{n-1} is denoted as u. If f is a positive function on S^{n-1}, then the "surface $uf(u)$" is given by the equation $x = uf(u)$, $u \in S^{n-1}$. The hyperplane in E^n through z which is orthogonal to $u \in S^{n-1}$ is denoted by $H(u)$ or $H(-u)$; V is the volume of K, $K(u) := K \cap H(u)$ and $A(u)$ is the $(n-1)$-dimensional measure or area of $K(u)$.

Let K, K' be two such convex bodies, V' and $A'(u)$ be defined for K' as V and $A(u)$ for K.

The *Busemann-Petty problem* is usually understood as the following:

Problem 1. *Does $A'(u) \leq A(u)$ for all $u \in S^{n-1}$ imply $V' \leq V$?*

The answer is affirmative if K' is an ellipsoid. The only other special case known at the time of the paper [18] was when $A'(u) = cA(u)$ for all $u \in S^{n-1}$.

The ellipsoid maximizes $V^{1-n} \int_{S^{n-1}} A(u)^n d\omega(u)$, where $d\omega$ is the area element of S^{n-1} at u. This suggests

Problem 2. *To find a convex body K (with center z) which minimizes the quantity $V^{1-n} \int_{S^{n-1}} A(u)^n d\omega(u)$.*

The greatest lower bound of this expression is positive (see [17], p. 11) and the existence of K can be deduced from an affine form of Blaschke's Selection theorem [84]. The following modification is important:

Problem 3. *To find an estimate from above for V in terms of $A(u)$ such that the equality sign characterizes the ellipsoids.*

Problem 4. *Which K yields $\min_K \max_u A(u)V^{1-n}$?*

An affirmative answer to Problem 1 in the case where K is a ball would imply that K is a solution of Problem 4.

The answer to Problem 4 would in particular decide the question of whether the so-called isoperimetrix, i.e. the solution of the isoperimetric problem with center z and Minkowski area equal to n times its Minkowski volume, can be properly contained in the unit ball K. Problem 6, which is related to Problem 4, would also settle this question.

For a given u is constructed a cone of maximal volume $C(u)$ with base $K(u)$ and apex in K.

Problem 5. *Are the ellipsoids characterized by the property that $C(u)$ is constant (i.e. independent of u for fixed K)?*

Problem 6. *To find a convex body for which $V^{-1} \max_u C(u)$ is minimal.*

The corresponding min min and max max problems are simple and are solved by cylinders and double cones. The max min problem in the plane is solved by regular affine hexagons [88]. It is shown in [88] that the ellipse solves Problem 6 in two dimensions.

Problem 7. *Find a convex body K for which the integral $\int_{\partial K} A(u)^{-1} ds(u)$ is minimal, where $ds(u)$ is the area element of the boundary ∂K of K at a point of contact of a tangent plane parallel to $H(u)$.*

The Minkowskian interpretation of Problem 7 is to find a Minkowski metric for which the area of the unit sphere is minimal.

Problem 8. *Are the ellipsoids characterized by the fact that the Gauss curvature at a point of contact with a tangent plane parallel to $H(u)$ is proportional to $A(u)^{-n+1}$?*

The answer is affirmative for dimension 2.

For a given point p on the boundary ∂K of K, determine u such that the cone with apex p and base $K(u)$ has maximal volume (in general p is not an apex for $C(u)$). Call $P(u)$ the $(n-1)$-dimensional area of the projection of K on $H(u)$ parallel to the line pz. Clearly $P(u) \geq A(u)$.

Problem 9. *Can the ratio $P(u)/A(u)$ be constant for a K which is not an ellipsoid?*

If the value of this constant is 1, a positive answer to this problem would contradict Blaschke's theorem in [84] that the ellipsoids are the only convex bodies with plane shadow boundaries.

It is proved in [14] that the surface $uA(u)$ is convex. Calling an r-dimensional linear subspace of E^n briefly an r-flat, the last problem is the following:

Problem 10. *If $n \geq 4$ and $2 \leq r < n-1$, does a result analogous to the convexity of the surface $uA(u)$ hold for the sections of K by r-flats through z?*

Busemann's time period before 1960 ended by two reviews [20], [21] of the interesting books [42] and [43] by H. G. Eggleston.

3. The book "Convex surfaces" [19]

Busemann writes in the preface to the book [19]: "The purpose of the tract is to acquaint the mathematical public with a subject, convex surfaces, which during the past 25 years has experienced a striking and beautiful development, principally in Russia, but has remained largely unknown, at least in the USA [...] The major part of the book, also of Chapters I and II, is in fact due to A. D. Alexandrov, and the great deal of the rest was stimulated by him."

Earlier, we mentioned the titles of Chapters I and II of [19] and said something about the content of Chapter I. The titles of the other chapters are the following: III. Intrinsic Geometry, IV. Realization of Intrinsic Metrics, V. Conclusions.

We add to what we mentioned earlier a few results by Alexandrov discussed in Chapter I.

Let us consider only *closed convex hypersurfaces* C in E^n. The *spherical image* $n'(p)$ of a point $p \in C$ consists of all the outward unit normals to the supporting planes of C at p. The spherical image of a set $M \subset C$ is the union $n'(M) = \bigcup_{p \in M} n'(p)$.

It is easy to see that $n'(p)$ is convex (in S^{n-1}) and that $n'(M)$ is closed if M is closed. Busemann sets, for any set $M \subset C$ for which $n'(M)$ is measurable, $n(M) =$ *measure of* $n'(M) =$ *integral curvature of* M.

Theorem 1. *$n'(M)$ is measurable for any Borel set $M \subset C$ and the integral curvature $n(M)$ is completely additive on the Borel sets of C.*

Theorem 2. *Let C_1 and C_2 be two closed convex hypersurfaces containing a point z in their interior. If $n(M_1) = n(M_2)$ for any two Borel sets $M_j \subset C_j$ which are projections of each other from z, then C_2 is obtained from C_1 by a dilation with center z.*

Denote by $A(M)$ the measure (or area) of the Borel set M on C. The ratio $n(M)/A(M)$ is called the *specific curvature* of M.

Theorem 3. *If the specific curvature of C is bounded (on all Borel sets of positive measure on C), then C is continuously differentiable.*

In Section 6 of Chapter II, Busemann considered *mixed volumes.*

If K_i are convex bodies in E^n and $\lambda_i \geq 0$, $i = 1, \ldots, r$, then $K = \lambda_1 K_1 + \cdots + \lambda_r K_r$ is a convex body (by [37], Section 20). We come now to the *basic ideas of Minkowski,* Section 29 in [37] (or [35] for $n = 3$).

The n-dimensional measure or volume of K is denoted by $V(K)$. By *Minkowski's theorem* [122], [125], for variable $\lambda_i \geq 0$, the volume of $K = \sum_{i=1}^{r} \lambda_i K_i$ is a homogeneous polynomial

$$V(K) = \sum_{i_1=1}^{r} \cdots \sum_{i_n=1}^{r} \lambda_{i_1} \ldots \lambda_{i_n} V_{i_1 \ldots i_n}$$

of degree n in the λ_i, where the coefficients $V_{i_1 \ldots i_n}$ are uniquely determined by requiring that they are symmetric in their subscripts.

Then, $V_{i_1 \ldots i_n}$ *depends only on the bodies* $K_{i_1}, \ldots, K_{i_n}$. *Hence we may write*

$$V(K_{i_1}, \ldots, K_{i_n})$$

instead of $V_{i_1 \ldots i_n}$ *and call it the mixed volume of* $K_{i_1}, \ldots, K_{i_n}$.

Although i runs originally from 1 to r we see that studying mixed volumes in E^n amounts to studying $V(K_{i_1}, \ldots, K_{i_n})$ where the K_{i_j} are not necessarily distinct.

It is clear that $V(K_{i_1}, \ldots, K_{i_n})$ is invariant under translations of any K_{i_j} and $V(K, \ldots, K) = V(K)$ for any convex body $K \subset E^n$. A very important case is that with only two distinct K_i, so that it makes sense to use the notation $V_m(K, L) = V_{n-m}(L, K)$ when K and L enter $V(\ldots)$ m and $n - m$ times respectively. Then:

$$V(\lambda_1 K + \lambda_2 L) = \sum_{m=0}^{n} C_m^n \lambda_1^{n-m} \lambda_2^m V_m(K, L). \tag{10}$$

The support function of a body K in E^n is defined by formula

$$h_K(u) = \sup_{x \in K} (u \cdot x), \quad u \in E^n.$$

Let suppose that h_K is of class C^2 on $E^n - \{0\}$.

A *normal coordinate system for* $u \in S^{n-1}$ is any orthogonal coordinate system $u_1, \ldots, u_n$ in E^n with origin 0 such that $u = (0, \ldots, 0, 1)$. Then, for $h = h_K$, we have

$$d^2 h(u) = \sum_{i,j=1}^{n-1} \frac{\partial^2 h}{\partial u_i \partial u_j}(u) du_i du_j := \sum_{i,j=1}^{n-1} h_{ij}(u) du_i du_j. \tag{11}$$

Let $D_m(h, u)$ denote the sum of all principal minors of the size m of the matrix $h_{ij}(u)$. In particular, $D_{n-1}(h, u) := D(h, u)$ is the determinant of the matrix $h_{ij}(u)$; it is called the *discriminant of the quadratic form* (11). It is clear that $D_m(u) = \{R_1 \ldots R_m\}$, where $\{R_1 \ldots R_j\}$ is the *elementary symmetric function of the* R_i of which $R_1 \dots R_j$ *is one summand* and $R_1, \dots, R_{n-1}$ *are principal radii of curvature* of ∂K at a point $v \in \partial K$ such that $h(u) = u \cdot v$. In particular, $D(h, u) = R_1 \cdot \dots \cdot R_{n-1}$.

Assume that $K_1, \dots, K_n$ are convex bodies in E^n with supporting functions $h_1, \dots, h_n$ of class C^2 on $E^n - \{0\}$. Denote by $D(h_1, \dots, h_{n-1}, u)$ the factor of $\lambda_1 \dots \lambda_{n-1}$ in $D_{n-1}(\lambda_1 h_1 + \dots + \lambda_{n-1} h_{n-1}, u)$. Then

$$V(K_1, \dots, K_n) = n^{-1} \int_{S^{n-1}} h_n(u) D(h_1, \dots, h_{n-1}, u) d\omega_{(n-1)}(u). \tag{12}$$

Let B is the unit Euclidean ball in E^n. Then (12) implies the following statements:

(1) $nV_1(K, B)$ is the $(n-1)$-volume of ∂K.

(2) $V_{n-1}(K, B) = (2n)^{-1} \int_{S^{n-1}} (h_K(u) + h_K(-u)) d\omega_{(n-1)}(u)$, $h_K(u) + h_K(-u)$ is the so-called width of K in the direction $\pm u$.

(3) If h_K is of class C^2 on S^{n-1} then for $1 < m < n-1$,

$$V_m(K, B) = (nC_{n-m}^{n-1})^{-1} \int_{S^{n-1}} \{R_1 \dots R_{n-m}\}(u) d\omega_{(n-1)}(u).$$

In Section 7 of [19], Busemann considers the *general Brunn-Minkowski Theorem.*

The Brunn-Minkowski inequality (2) and (10) imply that the function

$$g(\rho) := V^{1/n}((1-\rho)K + \rho L) = \sum_{m=0}^{n} C_m^n (1-\rho)^{n-m} \rho^m V_m(K, L)$$

is for $0 \le \rho \le 1$ a concave function of ρ. Therefore the tangent of the graph of $g(\rho)$ at $\rho = 0$ lies over the the graph and passes through $(1, g(1))$ only when $g(\rho)$ is linear. This yields *Minkowski's inequalities* [35]:

$$V_1^n(K, L) \ge V^{n-1}(K) V(L), \quad V_{n-1}^n(K, L) \ge V(K) V^{n-1}(L). \tag{13}$$

If K and L do not lie in parallel hyperplanes the equality sign holds only when K and L are homothetic.

The fact that the second derivative at 0 of $g(\rho)$ is non-positive yields the further, so-called *quadratic, inequalities of Minkowski* [35]:

$$V_1^2(K, L) \ge V(K) V_2(K, L), \quad V_{n-1}^2(K, L) \ge V_{n-2}(K, L) V(L). \tag{14}$$

Assume that K and L do not lie in parallel hyperplanes. Then equality holds in the first of these inequalities not only when K and L are homothetic but also when K is homothetic to a "cap" of L, i.e. the convex hull of L and a point not lying in L, see [37], pp. 17, 92. Minkowski conjectured in [35], and Bol proved in 1943 (see [19]), that this is the only case where equality holds.

These inequalities solve many extremal and uniqueness problems, see [37]. *Busemann will use them only to prove in Section 8 of* [19] *uniqueness of the solution of Minkowski's problem.*

There were conjectures on extensions of these results to other mixed volumes, see [37], pp. 92,93. These conjectures were proved independently in 1936 by Fenchel in two short notes (each of them three pages long) (see [19]) and in 1937, 1938 by Alexandrov in four papers [39], pp. 31-59, 61-97, 99-117, 119-144. Both established first the following generalization of (14), later called the *Alexandrov-Fenchel inequality*:

(15) $$V^2(C_1, \dots, C_{n-2}, K, L) \geq V(C_1, \dots, C_{n-2}, K, K)V(C_1, \dots, C_{n-2}, L, L),$$

where $C_1, \dots, C_{n-2}$, K, L *are convex bodies in* E^n.

Busemann noticed the following: "Three proofs of the basic inequality (15) are available, one by Fenchel and two by Alexandrov. Fenchel's proof is very sketchy, a detailed version has never appeared and it is not quite clear what it would involve. The simplest one, at present, seems to be the second proof of Alexandrov which generalizes Hilbert's approach to (14)", see [37], pp. 102-104. It uses the corresponding inequality for *mixed discriminant* of quadratic forms, see [39], pp. 119-144. Busemann discuss the essential steps of the second proof by Alexandrov; the proof is based on (12).

Inequality (15) contains the *General Brunn-Minkowski Theorem*:

Theorem 4. *If* $K_\rho = (1-\rho)K + \rho L$, *then*

(16) $$f(\rho) = V^{1/m}(C_1, \dots, C_{n-m}, K_\rho, \dots, K_\rho), \quad m \geq 2,$$

is for $0 \leq \rho \leq 1$ *a concave function of* ρ.

After that, Busemann writes that Alexandrov noticed that (15) and the General Brunn-Minkowski Theorem leads to an easily remembered more *general inequality*

(17) $$V^m(C_1, \dots, C_n) \geq \prod_{k=0}^{m-1} V(C_1, \dots, C_{n-m}, C_{n-k}, \dots, C_{n-k}).$$

Alexandrov proved the following theorem calling a convex body K *regular* if K contains interior points and ∂K is a surface of class C^2 with positive curvature:

If K *and* L *are convex bodies of dimension at least* m *and if* $C_1, \dots, C_{n-m}$ *are regular, then the function (16) is linear if and only if* K *and* L *are homothetic.*

In [19], Busemann proves this theorem only under the assumption that K and L are also regular.

Consider r real quadratic forms

$$q_k = \sum_{i,j=1}^{n} a_{ij}^{(k)} x_i x_j, \quad k = 1, \dots, r, \quad a_{ij}^{(k)} = a_{ji}^{(k)}$$

in n variables. For any real $\lambda_1, \dots, \lambda_r$,

$$q = \lambda_1 q_1 + \dots + \lambda_r q_q = \sum_{i,j=1}^{n} a_{ij} x_i x_j, \quad a_{ij} = \sum_{k=1}^{r} a_{ij}^{(k)}$$

is also a quadratic form. Its *discriminant*, i.e. the determinant of the matrix (a_{ij}), can be written in the form:

(18) $$D(q) = \sum_{i_1=1}^{r} \cdots \sum_{i_n=1}^{r} \lambda_{i_1} \dots \lambda_{i_n} D(q_{i_1}, \dots, q_{i_n}),$$

where $D(q_{i_1}, \dots, q_{i_n})$ is independent of the order of the q_{i_k}, and depends only on $q_{i_1}, \dots, q_{i_n}$. $D(q_i, \dots, q_i) = D(q_i)$, since $\sum \lambda_{i_1} \dots \lambda_{i_n} = (\lambda_1 + \dots + \lambda_r)^n$. In analogy with mixed volumes, $D(q_{i_1}, \dots, q_{i_n})$ is called the *mixed discriminant* of $q_{i_1}, \dots, q_{i_n}$. Studying its properties amounts to investigating $D(q_1, \dots, q_n)$ where the q_i are not necessarily distinct. It is important that *if* $q_1, \dots, q_n$ *are positive definite, then* $D(q_1, \dots, q_n) > 0$.

The following is a crucial fact on mixed discriminants, analogous to (15):

Theorem 5. [39] *If the forms* $q_1, \dots, q_{n-1}$ *are positive definite and* $Q = \sum b_{ik}x_ix_k$, $b_{ik} = b_{ki}$, *is any form then*

$$D^2(q_1, \dots, q_{n-1}, Q) \geq D(q_1, \dots, q_{n-1}, q_{n-1})D(q_1, \dots, q_{n-2}, Q, Q); \tag{19}$$

the equality sign holds only when $Q = \lambda q_{n-1}$.

In Section 8 of Chapter II, Busemann considers the *Minkowski problem*. This problem, in its original formulation, asks whether, given a continuous positive-valued function $f(u)$, $u \in S^{n-1}$, there is a closed convex surface F in E^n with $f^{-1}(u) = \{R_1 \dots R_{n-1}\}(u)$ as the Gaussian curvature of F at the point of F where the outward unit normal is the vector u. The function f must necessarily satisfy the exactness condition expressed by the vector equation

$$\int_{S^{n-1}} uf(u)d\omega_{n-1}(u) = 0. \tag{20}$$

This problem was solved originally by Minkowski only in the following "weak" sense. Given a function f satisfying (20), there exists a closed convex hypersurface $F \subset E^n$, unique up to translation, such that for any given closed region $G \subset S^{n-1}$ the integral $\int_G f(u)d\omega_{n-1}(u)$ equals the area of the region of F whose spherical image under the Gauss map (generalized, if F is not differentiable) is G.

According to Fenchel-Jessen (see [19]) a weak solution exists in the same sense as for the Minkowski problem extended by replacing the regular measure density $f(u)d\omega_{n-1}(u)$ on S^{n-1} by any non-negative absolutely additive Borel measure whose support is not contained in any hyperplane section of S^{n-1}. A similar result was proved by Alexandrov in [39].

The following theorem holds:

Theorem 6. *If* $n = 3$ *and* $f = f(u)$ *is positive and of class* C^m, $m \geq 2$, *then* F *is at least of class* C^{m+1}. *If* $f = f(u)$ *is analytic, then* F *is also analytic.*

Theorem 6 for $m \geq 3$ is due to Pogorelov (see [53]). The improvement from $m \geq 3$ to $m \geq 2$ is obtained by applying the methods of Nirenberg (see [19]). Pogorelov in [54] (the Russian original is from 1975) and Cheng and Yau in [55] proved that for arbitrary $n \geq 3$, F is of class $C^{m+1,\alpha}$ for every $0 < \alpha < 1$, if $f = f(u)$ is of class C^m, $m \geq 3$; if $f = f(u)$ is analytic, then F is also analytic. The case $m = 1$, which in some respect is the most interesting case, is not covered.

In Section 9 of Chapter II, Busemann discusses *uniqueness for given* $\{R_1 \dots, R_m\}$. In [39], Alexandrov deduces from Theorem 5 that a regular convex hypersurface is uniquely determined up to translation when one of the functions $\{R_1, \dots, R_m\}(u)$, $u \in S^n$, $1 \leq m \leq n-1$, is known as a function of the normal $u \in S^{n-1}$. In Section

9, Busemann extends this statement to general convex hypersurfaces. In this case, one must have

$$\int_{S^{n-1}} u\{R_1 \dots R_m\}(u) d\omega(u) = 0. \tag{21}$$

Alexandrov (see [39]) constructed *positive analytic functions* $f(u)$, $u \in S^n$, *satisfying Equality (21) which are not the curvature functions* $\{R_1 \dots R_m\}$ *of a convex hypersurface for any* $m < n-1$. This implies that *if* $m < n-1$, $f(u)$, $u \in S^{n-1}$, *must satisfy other conditions in order to arise from a convex hypersurface*, but so far there is not even a plausible conjecture as to what these might be. A sufficient but not necessary condition is found in Pogorelov [54] for all n. So that *there is no known natural analogue to the existence statement for the Minkowskii problem, when the curvature functions* $\{R_1 \dots R_m\}(u)$, $u \in S^{n-1}$, *with* $m < n-1$ *are given.*

Possibly, a reason of this failure is that for $m < n-1$ in a natural setting of the problem instead of $\{R_1 \dots R_m\}(u)$, $u \in S^{n-1}$, must be the elementary symmetric function $\{k_1 \dots k_m\}(u)$ of main curvatures $k_i(u) = 1/R_i(u)$. See in this connection Section 5.

Most of the last two chapters in [19] can be found in books, mainly in Alexandrov's [77], the English translation of the Russian 1948 original, and [39], [40]. Many theorems, in particular in Chapter III, dealing with general forms of well-known classical theorems, require usually quite long proofs. Therefore these proofs are omitted. Here Busemann's own contributions is restricted to a discussion of the intrinsic measure of Borel sets on convex hypersurfaces in E^n, pp. 86, 87, his joint theorem with Feller (12.5), p. 92, published in [4], and remarks on his joint paper with W. Mayer on calculating length of curves in Finsler manifolds, p. 93.

We shall discuss only complete convex surfaces C without boundary in E^3.

The first important notion is the *intrinsic metric* ρ on C. The distance $\rho(x, y)$ for $x, y \in C$ is defined as the infimum of lengths of curves in C (calculated in E^3), joining the points x, y. This infimum is always attained, (C, ρ) is a complete metric space whose metric topology coincides with the induced one from E^3, and the length of any curve in C, calculated in E^n and in ρ, coincide. This implies that (C, ρ) is a so-called *inner metric space* (or *length space* in M. Gromov's terms). The latter means for a metric space (M, ρ) that for any two points $x, y \in M$, $\rho(x, y)$ is equal to the infimum of lengths of paths in (M, ρ), calculated in the usual manner, joining the points x, y. The very important *Cohn-Vossen theorem* states that for any two points x, y in an arbitrary complete locally compact inner metric space (M, ρ) this infimum is always attained by some (not necessarily unique) path (curve). This path is usually called *shortest arc* (joining the points x and y). Busemann calls it a *segment.* We shall always suppose that a shortest arc is parametrized by arc length. A path in (M, ρ) parametrized by arc length is called *geodesic* if it is locally a shortest arc.

We shall not discuss in detail Section 12, on *Differentiability properties of geodesics.* We only mention that an arbitrary geodesic $\gamma = \gamma(t)$ in C has always one-sided tangents $\gamma_l'(t)$ and $\gamma_r'(t)$. Moreover, $\gamma_l'(t) = \gamma_r'(t)$ except for an at most countable set of

t. At any point $c \in C$, there exists a *tangent cone* C_c to C. It may be of one of the following three types:

(1) a hyperplane;
(2) a union of two half-hyperplanes with a common line;
(3) a cone with total angle $\theta < 2\pi$ around the vertex c.

In all cases it is convenient to understand C_c as a cone with total angle $0 < \theta \leq 2\pi$.

If $c = \gamma(t_0)$ is an interior point of a geodesic γ, then Case (3) is impossible, while the inequality $\gamma_l'(t_0) \neq \gamma_r'(t_0)$ holds only in the case (2).

Section 13 of the book carries the title *Angles. The convexity condition.* Let $\gamma_1 = \gamma_1(t)$, $0 \leq t \leq a_1$; $\gamma_2 = \gamma_2(t)$, $0 \leq t \leq a_2$, be two segments in C with common origin $c = \gamma_1(0) = \gamma_2(0)$. The *convexity condition* on C means that the *angle function*

$$5) \quad \alpha(t_1, t_2) := \arccos \frac{\rho^2(c, \gamma_1(t_1)) + \rho^2(c, \gamma_2(t_2)) - \rho^2(\gamma_1(t_1), \gamma_2(t_2))}{2\rho(c, \gamma_1(t_1)) \cdot \rho(c, \gamma_2(t_2))}$$

is a non-increasing function both on $0 < t_1$ *and on* $0 < t_2$. This implies that

(1) there exists a so-called *angle* $\alpha(\gamma_1, \gamma_2) := \lim \alpha_{t_1,t_2 \to 0}(t_1, t_2)$ *between the segments* γ_1 *and* γ_2 *at* c;
(2) $\alpha(\gamma_1, \gamma_2) = 0$ if and only if $\gamma_1'(0) = \gamma_2'(0)$;
(3) $\gamma_1'(0) = \gamma_2'(0)$ if and only if $\gamma_1(t) = \gamma_2(t)$ on $[0, \min(a_1, a_2)]$.
The following is clear:
(4) $\alpha(\gamma_1, \gamma_2) = \alpha(\gamma_2, \gamma_3)$.
Moreover, if γ_1, γ_2, γ_3 are segments in C with common origin $c \in C$ then
(5) $\alpha(\gamma_1, \gamma_3) \leq \alpha(\gamma_1, \gamma_2) + \alpha(\gamma_2, \gamma_3)$, *the triangle inequality for angles*;
(6) For some θ, where $0 < \theta \leq 2\pi$, $\alpha(\gamma_1, \gamma_2) + \alpha(\gamma_2, \gamma_3) + \alpha(\gamma_3, \gamma_1) \leq \theta$ and inequality turns to equality for some γ_1, γ_2, γ_3.
The inequalities (2), (3), and (6) imply the *inclusion property* of geodesics in C:
(7) If $\gamma_1(t) = \gamma_2(t)$ for $0 \leq t \leq t_1$, where $0 < t_1 < \min(a_1, a_2)$, then $\gamma_1(t) = \gamma_2(t)$ for $0 \leq t \min(a_1, a_2)$.

The above statements imply that we have a metric space with metric α after identifying locally coinciding segments with origin c. In the general case, this metric space is incomplete or disconnected. In this case there are points, arbitrarily close to c, joined to c by more than one segment [39]. After applying the Hausdorff completion to this space we get the so-called *space of directions* (D_c, α) *at* c, an inner metric space of diameter $\theta/2$ and homeomorphic to the circle S^1.

The convexity condition means in other words that (C, ρ) has Alexandrov curvature ≥ 0. In 1986, Berestovskii gave the following short and general definition-theorem for an arbitrary real number K: A locally compact inner metric space (M, ρ) has Alexandrov curvature $\geq K$ if and only if for any point $x \in M$ there exists an open subset $U \subset M$ containing x such that every quadruple of points $\{a, b, c, d\}$ in U embeds isometrically (as a metric space with the induced metric) in $S^3_{K'}$, a simply connected complete three-dimensional Riemannian manifold of constant sectional curvature K', for some $K' \geq K$ depending on the given quadruple.

Section 14 of Busemann's book carries the title *Triangulations. Intrinsic curvature.* A *triangle* $T = \Delta ABC$ *with vertices* A, B, C in an inner metric space (M, ρ) is a union of segments (sides) $[AB]$, $[BC]$, $[CA]$, having no other common points besides the vertices. A triangle is called *proper* if the sum (of lengths) of any two sides is greater than the third. A set $V \subset M$ is called *convex* if any two points in V can be joined by at least one segment entirely lying in V. A *polygon* in an inner metric space is a curve consisting of a finite number of segments. It is closed if its end coincides with its origin and *simple* when it is has no self-intersection (and in this case it is called a Jordan curve). A compact set in C with interior points is a *polygonal domain* if its boundary consists of a finite number of disjoint simple closed polygons. On the base of the inclusion property of geodesics, Alexandrov proved in [77] that for any point $c \in C$ there exists an arbitrarily small proper triangle bounding an open convex neighborhood U of c. In this manner, he obtained a triangulation of an arbitrary polygonal domain P on C (the domain can coincide with C) with arbitrarily small proper triangles bounding convex open sets.

Let T be a triangle in C with three angles α_1, α_2, α_3 at its three vertices. We set $\delta(T) = \alpha_1 + \alpha_2 + \alpha_3 - \pi$ and call it the *excess of (the interior of) T*. For a point p of C, we denote by $\delta(p)$ the complete angle of C_p minus 2π. Consider a polygonal domain P on C with interior D and boundary B and denote by $\alpha_1, \ldots \alpha_n$ those (interior sector) angles in P at points of the boundary B of P which are different from π. If P is triangulated into the triangles $T_1, \ldots, T_r$ with vertices $p_1, \ldots p_s$ *inside* D, then

$$\sum_{i=1}^{r} \delta(T_i) + \sum_{j=1}^{s} \delta(p_j) = 2\pi\chi(P) - \sum_{k=1}^{k} (\pi - \alpha_k), \tag{22}$$

where $\chi(P)$ is the *Euler characteristic* of P. This is the *Gauss-Bonnet theorem for the polygonal domain P.*

If $P = C$ and C is a closed convex surface in E^3 (necessarily homeomorphic to the sphere S^2) then the formula (22) becomes the following formula:

$$\sum_{i=1}^{r} \delta(T_i) + \sum_{j=1}^{s} \delta(p_j) = 4\pi. \tag{23}$$

The right hand side of (22) shows that the left hand side is independent of a triangulation of P. It is called the *excess $\delta(P)$ of (the interior of) P.* By definition, a primitive set on (C, ρ) is a point, a segment without its ends, or a triangle without its boundary. An *elementary set* is the union of a finite number of disjoint primitive sets. For any elementary set W its *excess* $\delta(W)$ is defined as the left hand side of (22). Then *on C the excess $\delta(W)$ is* ≥ 0. Then the excess can be extended to all Borel sets on C. We have the following *Gauss theorem*:

Theorem 7. *For any Borel set $M \subset C$, the excess $\delta(M)$ is equal to the area $\nu(M)$ of the spherical image of M.*

Section 15 of the book carries the title *The Gauss-Bonnet theorem. Quasi-geodesics.* In this section, Busemann discusses mainly an extension of the Gauss-Bonnet theorem from polygonal domains to "general" domains, replacing segments on boundaries of domains by curves of an appropriate class. For these curves he needs the notion of *angle between curves* and the *swerve ("povorot" in Russian) of curve*, which were defined by Alexandrov in [77].

Much of the theory of convex surfaces, that is, of two-dimensional manifolds of non-negative curvature, extends to a wide class of *two-dimensional manifolds with inner metric of bounded curvature* [78], [79]. Unlike the condition of curvature $\geq K$, there is no analogue here of boundedness of curvature for locally compact inner metric spaces or manifolds of higher dimension with inner metric. Yu. G. Reshetnyak obtained a purely analytical characterization of the linear element of a two-dimensional manifold with bounded curvature in terms of subharmonic functions, see [79].

We need to skip many details of Chapter IV of the book [19]. We only state briefly the following facts. Any simply connected two-dimensional manifold M (without boundary) with complete length metric ρ of curvature ≥ 0 in the sense of Alexandrov is isometric to a complete convex surface C in E^3. The surface C is unique up to congruence if and only if (M, ρ) is closed or has total excess equal to 2π [53]. Alexandrov first proved this for polyhedral metrics ρ of curvature ≥ 0, where all the curvature is concentrated in points. In fact, the methods of the paper [34] by Cauchy work here. Then, using arbitrary small triangulations of (M, ρ) and replacing domains bounded by triangles by flat domains bounded by Euclidean triangles with the same side lengths, Alexandrov obtains polyhedral metrics on M of curvature ≥ 0 which are arbitrarily close to ρ. Using isometric embeddings of these polyhedral metrics in E^3 in the form of complete convex surfaces and going to a limit, Alexandrov gets a complete convex surface $C \subset E^3$ isometric to (M, ρ). If ρ is defined by a sufficiently smooth metric tensor, then C can be chosen smooth enough [53].

4. Busemann papers after 1959 [22]–[33]

We already mentioned the paper [22] in connection with the Busemann-Petty problems.

Fortunately, unlike all his other papers, Busemann's paper [26] is supplied with a summary. So we may simply reproduce it: It is shown that for $1 \leq a \leq n$, an open convex set in n-dimensional affine space can be provided with a complete a-dimensional area for which the a-flats minimize area, and that any such area can be extended to an area with the same properties in an open convex set in $(n+1)$-space.

All ten papers by Busemann in this section, together with the papers [47], [48] by G. Ewald and [49], [50] by G.C. Shephard form a series with a great unity, dedicated to one subject. The title of Busemann's paper [27] reveals this subject: *convexity on Grassmann manifolds*, unless the reader prefers the ironical title of another of his paper [32], joint with Shephard: *convexity on non-convex sets.* This series begins with the joint paper [23] by Busemann and E. G. Straus and indicates, together with [24], a connection with the paper [38] by V. V. Wagner (1908-1981), the 1937 Lobachevsky prize laureate. Busemann wrote in his paper [24], concerning [23], the following: "Recently E. G. Straus and the author showed in [23] that there are indeed many concrete problems concerning the simplest area, many of which are still open, while others have quite unexpected answers." All ten papers deserve a serious study, but a review of them is beyond the scope of this essay.

5. Notes on related results

The first notes below concern the Busemann-Petty problems, discussed in Section 2, and preserves the numbering of these problems.

Problem 1. This problem has attracted a good deal of interest, achieved a certain notoriety, and appears in several articles, as well as in the books of Burago and Zalgaller [94], Croft, Falconer, and Gay [68] (Problem a9, p. 22), Schneider [96], Gardner [97], and Gruber [125]. Its history is quite involved and described in detail in the book [97].

Evidently, the problem has a positive answer if $n = 2$. This is also true when $n = 3$ [97], and $n = 4$, see the paper [99] by G. Zhang.[1] The problem has a negative answer when $n > 4$, see Papadimitrakis 1992, Gardner [102] and Zhang 1994. Previously, negative answers were given by Larman and Rodgers for $n > 11$ in 1975, by Ball for $n > 9$ in 1986, and by Giannopoulos and Bourgain for $n > 6$ in 1990 and in 1991 respectively. D. Larmann and C. Rogers obtained their result by a probabilistic argument. Ball found the maximal volume, $\sqrt{2}$, of $(n-1)$-slices of the unit cube. Then he observed that from this it follows easily that for $n > 9$, a ball in E^n, whose central slices have areas $\varepsilon + \sqrt{2}$ (where ε is sufficiently small), has smaller volume than the unit cube. A counterexample for $n > 6$ was given by A. Giannapoulous. All references for the above results are given in [125].

An important role for the above solutions of the Busemann-Petty problem was played by the notion of *intersection body* introduced by Erwin Lutwak in the paper [95]. Let us give the necessary definitions. A set $K \subset E^n$ is *star-shaped* at the origin if it contains the origin and if every line through the origin meets K in a line segment. If K is star-shaped at the origin, its *radial function* is defined by

$$\rho_K(u) = \sup\{c \geq 0 : cu \in K\}, \quad u \in S^{n-1}. \tag{24}$$

A *star-body* is a compact set which is star-shaped at the origin and whose radial function is continuous. We say that a set is *centered* if it is centrally symmetric with centre at the origin. The star-body K is called *intersection body* if there is another star-body M such that $\rho_K(u) = V_{n-1}(M \cap H(u))$ for all $u \in S^{n-1}$. In this case, K is said to be the intersection body of M, and we write $K = IM$. It is shown in [95] that in this case there is a unique such M which is also centered. Clearly, any intersection body is centered. A star-body is an intersection body if and only if

[1] It is interesting that the same author, G. Zhang, published earlier papers in Duke Math. J. (1993) and in Math. Ann. (1994) where he "proved" the opposite statement.

$\rho_K = Rg$ for some nonnegative continuous function g on S^{n-1} where

$$Rg(u) = \int_{S^{n-1}\cap H(u)} g(v)d\omega_{n-2}(v) \tag{25}$$

is the *spherical Radon transform* of g. In this case we have $g = \rho_M^{n-1}/(n-1)$.

If K is a ball then $K = IM$, where M is a ball. If K is an ellipsoid then $K = IM$, where M is an ellipsoid. Hence, *every ellipsoid is an intersection body.* Let us recall that in [18] it was said that the answer to Problem 1 is affirmative if K' is an ellipsoid. These two statements imply that the following theorem by Lutwak (see [95]) generalizes the second of these statements:

Suppose that K is an intersection body and L an arbitrary star body in E^n such that $V_{n-1}(K\cap H(u)) \leq V_{n-1}(L\cap H(u))$ for all $u \in S^{n-1}$. Then $V(K) \leq V(L)$.

This theorem together with the counterexample of Ball we already mentioned imply that *any (even centered) cube in E^n, $n > 9$, is not an intersection body.* The following two theorems by Gardner (see [97]) give a characterization of E^n, $n \geq 3$, having a positive answer for the Busemann-Petty problem:

1) The Busemann-Petty problem has a positive answer in E^n if and only if the following holds: each centrally symmetric convex body K (with inner points) in E^n, with $\rho_K \in C^\infty(S^{(n-1)})$, whose boundary hypersurface has everywhere positive Gaussian curvature, is an intersection body.

2) A right spherical cylinder in E^n is an intersection body if and only if $n \leq 4$.

Note that Statement 1) is true for the family of all convex centrally symmetric bodies K (with inner points) in E^n.

As a corollary, the Busemann-Petty problem has a negative answer for $n > 4$. Thus, $n = 2$, $n = 3$ ([97]), $n = 4$ ([99]) are all dimensions of Euclidean spaces with positive answers to this problem.

A unified analytic solution for the Busemann-Petty problem was given in [100] by Gardner, Koldobsky, and Schlumprecht. For all the connected notions and statements, given below, we refer the reader to [100].

It turned out to be quite difficult to calculate the inverse spherical Radon transform of a radial function (24) in order to check whether a given body is an intersection body. The first step for this is the formula

$$2^n\pi^{n-1}R^{-1}\rho_K = \widehat{\rho_K} \tag{26}$$

when the Fourier transform $\widehat{\rho_K}$ of ρ_K is a continuous function on S^{n-1}. For the calculation of the right hand side of (26) in [100] is used *the parallel section function of K in the direction of $\xi \in S^{n-1}$*:

$$A_\xi(z) = V_{n-1}(K \cap \{x \in E^n : (x,\xi) = z\}).$$

Theorem 1 in [100] implies the following

Corollary 1. *Let K be an origin-symmetric star body in E^n with C^∞-boundary.*

(a) If n is even, then

$$\widehat{\rho_K}(\xi) = (-1)^{(n-2)/2}\pi A_\xi^{(n-2)}(0).$$

(b) If $n \geq 3$ is odd, then

$$\widehat{\rho_K}(\xi) = c_n \int_0^\infty \frac{A_\xi(z) - A_\xi(0) - A_\xi''(0)(z^2/2) \cdots - A_\xi^{(n-3)}(z^{(n-3)}/(n-3)!)}{z^{n-1}} dz$$

where $c_n = (-1)^{(n-1)/2} 2(n-2)!$ (and $\widehat{\rho_K}$ is the Fourier transform in the sense of distributions).

Then, by Corollary 1 and (26),

$$R^{-1}\rho_K(\xi) = \frac{1}{4} A_\xi(0) > 0, \quad \xi \in S^1, \quad \text{if} \quad n = 2,$$

$$R^{-1}\rho_K(\xi) = -\frac{1}{16\pi^2} A_\xi^{(2)}(0) \geq 0, \quad \xi \in S^3, \quad \text{if} \quad n = 4$$

by the Brunn-Minkowski inequality, if K is a convex body. Similarly,

$$R^{-1}\rho_K(\xi) = -\frac{1}{(2\pi)^2} \int_0^\infty \frac{A_\xi(z) - A_\xi(0)}{z^2} dz \geq 0, \quad \xi \in S^2, \quad \text{if} \quad n = 3,$$

if K is a convex body. Note that all the derivatives of A_ξ at 0 of odd order vanish since A_ξ is an even function. This, together with the Brunn-Minkowski inequality, implies that the integral in the last formula is finite and the function under the integral is non-positive. Then we get the last inequality. These inequalities and Theorem 1 from [97], stated above, imply that the Busemann-Petty problem has a positive answer for $2 \leq n \leq 4$. On the other hand, while convexity controls the second derivative of A_ξ, it does not control the derivatives of higher order. Then the authors of [100] use Corollary 1 to present a simple example that confirms the negative answer to the Busemann-Petty problem when $n \geq 5$.

F. Barth, M. Fradelizi, and B. Maurey gave in [101] a simple proof of Theorem 1 from [100].

The *generalized Busemann-Petty problem* asks the following: If K and L are convex bodies in E^n which are symmetric with respect to the origin, and the volume satisfies $V_i(K \cap H) \leq V_i(L \cap H)$ for every i-dimensional subspace H, $1 < i < n$, with fixed i, does it follows that $V_n(K) \leq V_n(L)$?

The problem has a negative answer for $n > 4$ and a positive answer for $n = 3, 4$. The paper [98] by Bourgain and Zhang gives a negative answer to the generalized Busemann-Petty problem for $3 < i < n$ in the stronger sense where i is not fixed. For $i = 2, 3$, it is shown in the paper [103] by E. Milman that when ρ_K is an $(n-i)$-th root of the radial function of a convex body, the answer to the generalized Busemann-Petty problem is positive for any star-body L.

Problem 7. *The corresponding maximum problem is solved by the parallelepiped.* The reason is that one can easily prove that any convex body has a circumscribed parallelepiped. In dimension 2, the answer to Problem 7 is a *regular affine hexagon* [88]. Earlier, Yu. G. Reshetnyak proved the last two statements for dimension 2 in [87]. For this, he proved that each convex body in E^2 admits at least one inscribed affine-regular hexagon. It is possible that the first person who proved this was A. S. Besikovitch [86].

G. P. Egorychev proved in 1980 and published in the paper [91] his proof of the famous *van der Waerden conjecture on permanents of doubly stochastic matrices.* The permanent per(A) is defined for any $(m \times n)$-matrix (a_{ij}) where $m \leq n$, but we need only the case $m = n$. In this case,

$$\mathrm{per}(A) = \sum_{s \in S_n} a_{1s(1)} a_{2s(2)} \dots a_{ns(n)},$$

where the summation is over all permutations s on $\{1, \dots, n\}$. The matrix A with non-negative elements is called doubly stochastic if the sum of the elements in any row and column is equal to 1. The simplest and most important examples are the $(n \times n)$-matrices J_n whose elements are equal to $1/n$ and the unit matrix E_n.

We denote by Ω_n the set of all doubly stochastic $(n \times n)$-matrices. First, we note that it is not difficult to see that for any matrix $A \in \Omega_n$, $\mathrm{per}(A) \leq 1$ and that equality holds if and only if $A = E_n$ or A is a result of transpositions of columns in E_n. Contrary to this, the following conjecture about the minimum, suggested by van der Waerden in 1926, is highly non-trivial.

Conjecture 1. *([90]). If A is a doubly stochastic $(n \times n)$-matrix then* $\mathrm{per}(A) \geq n!/n^n$ *with equality only in the case $A = J_n$.*

A matrix $A \in \Omega_n$ is called *minimizing* if $\mathrm{per}(A) = \min_{X \in \Omega_n} \mathrm{per}(X)$. Egorychev used in his proof only two known results: part of a theorem by D. London on minimizing a matrix A and Alexandrov's inequality (19) for mixed discriminants of quadratic forms in the special case where the matrices of the quadratic forms are diagonal. London proved that *if $A \in \Omega_n$ is a minimizing matrix, then*

$$\text{(27)} \qquad \mathrm{per}(A) \leq \mathrm{per}(A(i/j)) \quad \textit{for all} \quad i, j \in \overline{1, n},$$

where $A(i/j)$ is the $(n-1) \times (n-1)$-matrix obtained from A by deleting its i-th row and j-th column.

Egorychev observed that for any real $(n \times n)$-matrix A, $\mathrm{per}(A) = n! D(q_1, \dots, q_n)$, where q_i, $i \in \overline{1, n}$, is the quadratic n-form with the diagonal matrix whose diagonal is a disposition of the i-th column of A preserving the row numbers. Then, Inequality (19) implies: *if A is a real $(n \times n)$-matrix whose first $n-1$ columns are nonnegative, we have*

$$\text{(28)} \qquad (\mathrm{per}(A))^2 \geq \mathrm{per}(A^{(1)} \dots A^{(n-2)} A^{(n-1)} A^{(n-1)})\, \mathrm{per}(A^{(1)} \dots A^{(n-2)} A^{(n)} A^{(n)})$$

where $A^{(j)}$ is the j-th column of A. If the first $n-1$ columns of A are positive, then equality holds if and only if $A^{(n-1)}$ and $A^{(n)}$ are linearly dependent. Incidentally, in the course of the proof, Egorychev decides positively Conjecture 11 in [90] by Marcus and Newman: *if for $A \in \Omega_n$ all inequalities (27) hold, then all these inequalities must be equalities.*

Note also that after the proof of van der Waerden's conjecture, the investigation of some problems connected with the conjecture become redundant since one gets their solutions automatically. For example, this concerns Conjectures 2, 16, and Problem 9 in [90].

Another independent proof of the van der Waerden conjecture (without proof of uniqueness of the minimizing matrix J_n) was given by D. I. Falikman in [92]. This proof, like the one in [91], is based on ideas of geometric character. For their works, Egorychev and Falikman won the Fulkerson prize in 1982.

In the paper [93], B. Gyires proves that an inequality for permanents proved by him in 1977 in a paper, quoted in [93], is equivalent to the van der Waerden conjecture. On the basis of this equivalence, Gyires considers that an equivalent of the van der Waerden conjecture was solved already in 1977.

Let Θ_n be the set of all real $(n \times n)$-matrices where the row and column sums are equal to 1, so $\Omega_n \subset \Theta_n$. Let A^* denote the transpose of the matrix A, and let $(AA^*)^{1/2}$ and $(A^*A)^{1/2}$ denote the positive semidefinite roots of AA^* and A^*A respectively. Assume that $p \geq 0$, $q \geq 0$, $p+q=1$. Then the Gyires inequality which we mentioned is, for any $A \in \Theta_n$,

$$p^2 \operatorname{per}((AA^*)^{1/2}) + q^2 \operatorname{per}((A^*A)^{1/2}) + 2pq \operatorname{per}(A) \geq \frac{n!}{n^n} \tag{29}$$

with equality if and only if $A = J_n$.

A natural complement to Theorem 3 is one of the main results by I. G. Nikolaev and S. Z. Shefel in [56], [57] which can be formulated as follows:

Theorem 8. *If F is any (not necessarily complete) convex surface in E^3 with the specific curvature bounded from below by $c > 0$ and from above by $c_1 \geq c$ then F is of classes $C^{1,\alpha}$, $0 < \alpha < 1$ and W_p^2, $p \geq 1$. Moreover, α and p depend only on the ratio c_1/c and $\alpha \to 1$ and $p \to \infty$ when $c_1/c \to 1$.*

Roughly speaking, Theorem 2 from the paper [58] means that if at some point on a convex surface $F \subset E^3$ there exists an approximate differential of order (n, α), where $n = 0, 1, 2, \ldots$, $0 < \alpha < 1$, then at the same point F has approximate differentials up to order $(n+2, \alpha)$. This theorem implies corresponding earlier results by A. V. Pogorelov [53] and by I. Kh. Sabitov (see References in [58]).

V. Bangert, using some results by Yu. D. Burago and B. V. Kulikov from the paper [59], proved in [60] the following theorem (Proposition (1.3) from [60]):

Theorem 9. *Let $C \subset E^{n+1}$ be a closed convex set and assume that the interior of C is not empty and does not coincide with E^{n+1}. If ∂C has (first) mean (measurable) curvature density bounded above by $\alpha > 0$ (and defining a absolutely additive measure) then ∂C is a $C^{1,1}$-hypersurface of E^{n+1}; more precisely, ∂C is a C^1-hypersurface in E^{n+1} and the inner unit normal $N : \partial C \to E^{n+1}$ is Lipschitz with constant α (relative to the inner metric on ∂C).*

For more clarity we added the words in parentheses in this theorem. Bangert notes that his proof of Theorem 9 is essentially local and so a corresponding local statement is true. I. G. Nikolaev presented in [62] a simple proof of Theorem 9.

Now we shall mainly reproduce the abstract from the paper [61]. In this paper Nikolaev considered a convex hypersurfaces in E^n. The integral scalar curvature of any such hypersurface is defined as a *weak* limit of the integral scalar curvatures of regular convex hypersurfaces approximating a given hypersurface. Nikolaev proved in [61] that a convex hypersurface with bilaterally bounded strictly positive specific scalar curvature and absolutely continuous elementary symmetric curvature measures can be approximated by $C^{1,1}$-smooth hypersurfaces with uniformly bounded from both sides strictly positive specific scalar curvatures. If, in addition, the specific Gaussian curvature is bilaterally bounded and strictly positive then Nikolaev proved that the hypersurface itself is of class $C^{1,1}$.

No paper or book by Busemann concerns combinatorial convexity. Nevertheless, we discuss briefly the famous *Borsuk problem* formulated in 1933 (see [45]):

Is it possible to partition every bounded set $A \subset E^n$, with $\operatorname{diam}(A) > 0$*, into $n+1$ sets, each of diameter less than that of A?*

One of the most interesting aspects of this subject is the great discrepancy between the elementary formulation of the problem and the difficulty of its solution. Its setting and solution were connected not only with combinatorial geometry but also with discrete and asymptotic geometries, number theory and graph theory.

Due to the well-known fact that the diameter of any set is the same as that of its convex hull, we may assume that A is a convex body (with inner points). Using a homothety, it is possible also to assume that $\operatorname{diam}(A) = 1$. One more possible reduction of the problem follows from the following result by Pál: Every set $A \subset E^n$ is a subset of a convex body K of constant width equal to $\operatorname{diam}(A)$, [45].

The Borsuk problem is trivial for $n = 1$. It was solved positively for $n = 2$ by Borsuk, in 1932 and for $n = 3$ by Julian Perkal in 1947, and independently by H. G. Eggleston in 1955 [42]. Its answer for $n = 4$ is unknown. Affirmative answers to the Borsuk problem (for all n) were given for smooth convex bodies by Hugo Hadwiger in 1946, for centrally symmetric convex bodies by A. S. Riesling in 1971, and for convex bodies of revolution by Boris Dekster in 1995.

It seems that generally speaking Borsuk's conjecture is believed to be true, and various stronger conjectures have been proposed. The possibility of a counterexample based on combinatorial configurations was suggested by Erdös in 1981 and by D. Larman in 1984, see the references in [119]. In fact, using such configurations, Jeff Kahn and Gil Kalai showed in 1993 that the general answer to Borsuk's question is no [119]. Thus, the problem was finally solved 60 years after its setting.

Let $\alpha(n)$ be the smallest number so that every set in E^n of diameter 1 can be partitioned into $\alpha(n)$ sets of diameter smaller than 1. The regular simplex with $n+1$ vertices in E^n shows that $\alpha(n) \geq n+1$. Another example showing this is, by the Borsuk-Ulam theorem, the n-dimensional Euclidean ball. Kahn and Kalai proved that $\alpha(n) \geq (1.2)^{\sqrt{n}}$ for sufficiently large n. Their construction shows that Borsuk's conjecture is false for $n = 1325$ and for every $n \geq 2015$. In their proof, Kahn and Kalai used the prime number theorem and the following result by Frankl and Wilson published in 1981:

Theorem 10. *Let s be a prime power and $n = 4s$. Let S be a family of $n/2$-subsets of $\{1, \ldots, n\}$, so that no two sets in the family have intersection of size $n/4$. Then*

$$|S| \leq 2C^{n-1}_{n/4-1}.$$

Oded Schramm showed in 1988 that $\alpha(n) \leq (\sqrt{3/2} + \varepsilon)^n$ for every $\varepsilon > 0$ and sufficiently large n. A different proof of Schramm's bound was given by Bourgain and Lindenstrauss in 1991, see [119]. It is conjectured that $\alpha(n) \geq c^n$ for some $c > 1$ and for all $n \geq 1$.

It follows from the papers [126] by A. V. Bondarenko and [127] by T. Jenrich and A. E. Brouwer that Borsuk's conjecture is false for $n \geq 64$. Bondarenko answered Larman's question on Borsuk's conjecture for two-distance sets. He found a two-distance set consisting of 416 points on $S^{64} \subset E^{65}$, showing that $\alpha(65) \geq 84$. He also proved that for $n \geq 1$ and $k \geq 0$, $\alpha(66n + k) \geq 84n + k + 1$ which implies that for all $n \geq 65$, $\alpha(n) \geq n + 19$. It is shown in [127] that Bondarenko's 65-dimensional counterexample contains a 64-dimensional counterexample to Borsuk's conjecture which is a two-distance set of 325 points. This counterexample implies that $\alpha(64) \geq 71$. Both counterexamples are closely connected with special strongly regular graphs.

The books [118] and [122] by T. Oda and G. Ewald and the Appendix by A. G. Khovanskii in the book [94] by Yu. D. Burago and V. A. Zalgaller demonstrate interesting connections of convex geometry with algebraic geometry. In particular, "the theory of toric varieties, which are also called torus embeddings, relates algebraic geometry to the geometry of convex figures in real affine spaces" [118]. In considering toric varieties, many of the general notions of algebraic geometry occur. For this reason it is natural "to construct toric varieties as complex analytic spaces, so that they can be understood more easily without much prior knowledge of algebraic geometry. Not only can some of the important complex analytic properties of these spaces be translated into easily visualized elementary geometry of conves figures, but many interesting examples of complex analytic spaces can be easily constructed by means of this theory" [118]. Notice that mixed volumes and the Alexandrov-Fenchel inequalities naturally appear in [94], [118], [122].

References

[1] H. Busemann, Über die Geometrien, in denen die "Kreise mit unendlichem Radius" die kürzesten Linien sind. Math. Ann. 106(1932), 140-160.

[2] H. Busemann, Über Räume mit konvexen Kugeln und Parallelenaxiom. Nachr. Ges. Wiss. Göttingen (1933), 116-140.

[3] H. Busemann and W. Feller, Krümmungseigenschaften konvexer Flächen. Acta Math. 66(1935), 1-47.

[4] H. Busemann and W. Feller, Bemerkungen zur Differentialgeometrie der konvexen Flächen. I. Kürzeste Linien auf differenzierbaren Flächen. Mat. Tidsskr., **B 1935**, 25-36.

[5] H. Busemann and W. Feller, ibid. II. Über die Krümmungsindikatrizen. *ibid*, **B 1935**, 87-115.

[6] H. Busemann and W. Feller, ibid. III. Über die Gaussche Krümmung. *ibid*, **B 1936**, 41-70.

[7] H. Busemann and W. Feller, Regularity properties of a certain class of surfaces. Bull. Amer. Math. Soc., 51(1945), No. 8, 583-598.

[8] H. Busemann, Intrinsic area. Proc. Nac. Acad. Sci. 32(1946), 5-8.

[9] H. Busemann, Intrinsic area. Ann. Math. 48(1947), No. 2, 234-267.

[10] H. Busemann, Two-dimensional geometries with elementary areas. Bull. Amer. Math. Soc. 53(1947), No. 4, 402-407.

[11] H. Busemann, The isoperimetric problem in the Minkowski plane. Amer. J. Math. 69(1947), 863-871.

[12] H. Busemann, Note on a theorem on convex sets. Matematisk Tidsskrift. (1947), 32-34.

[13] H. Busemann, The isoperimetric problem for Minkowski area. Amer. J. Math. 71(1949), 743-762.

[14] H. Busemann, A theorem on convex bodies of the Brunn-Minkowski type. Proc. Nac. Acad. Sci. 35(1949), 27-31.
[15] H. Busemann, The geometry of Finsler spaces. Bull. Amer. Math. Soc. 56(1950), 5-16.
[16] H. Busemann, The foundations of Minkowskian geometry. Comment. Math. Helv. 24(1950), 156-186.
[17] H. Busemann, Volume in terms of concurrent cross-sections. Pacific J. Math. 3(1953), 1-12.
[18] H. Busemann and C.M. Petty, Problems on convex bodies. Math. Scand. 4(1956), 88-94.
[19] H. Busemann, Convex surfaces. New York: Interscience Publishers, Inc., 1958.
[20] H. Busemann, Book review of "Convexity" by H.G.Eggleston. Bull. Amer. Math. Soc. 64(1958), No. 5, 295-296.
[21] H. Busemann, Book review of "Problems in Euclidean Space: Application of convexity" by H.G.Eggleston. Bull. Amer. Math. Soc. 64(1958), No. 5, 298-299.
[22] H. Busemann, Volumes and areas of cross-sections. Amer. Math. Monthly. 67(1960), 248-250 and 671.
[23] H. Busemann and E.G. Straus, Area and Normality. Pacific J. Math. 10(1960), No. 1, 35-72.
[24] H. Busemann, Areas in affine space. I, II. Rend. Circ. Mat. Palermo (2). 9 (1960), 81-90.
[25] H. Busemann, ibid. III. The integral geometry of affine area. Rend. Circ. Mat. Palermo (2). 9 (1960), 226-242.
[26] H. Busemann, Geometries in which the planes minimize area. Annali di Matematica 55 (1961), No. 1, 171-189.
[27] H. Busemann, Convexity on Grassmann manifolds. Enseign. Math. (2) 7 (1961), 139-152.
[28] H. Busemann, Un nuovo tipo di problemi nella teoria dei corpi convessi. Seminari dell'Istituto Nazionale di Alta Matematica 1962-1963, 178-205.
[29] H. Busemann, G. Ewald and G.C. Shephard, Convex bodies and convexity on Grassmann cones. I– IV. Math. Ann. 151(1963), 1-41.
[30] H. Busemann, G. Ewald and G.C. Shephard, Convex bodies and convexity on Grassmann cones. V. Arch. Math. 13(1962), 512-526.
[31] H. Busemann and G.C. Shephard, Convex bodies and convexity on Grassmann cones (X): Projection functions of parallel convex bodies. Annali di Matematica 70(1965), No. 1, 271-293.
[32] H. Busemann and G.C. Shephard, Convexity on nonconvex sets. Proc. Colloq. Convexity (Copenhagen, 1965). Math. Inst. Univ. Copenhagen, 1967, 20-33.
[33] H. Busemann, Convex bodies and convexity on Grassmann cones. XI. Sublinear functions. Math. Scand. 24(1969), 93-101.
[34] A. Cauchy, Sur les polygones et les polyèdres. Second mémoire. J. École polytechn. Bd. 9 (1813).
[35] H. Minkowski, Gesammelte Abhadlungen von Hermann Minkowski. Band II. B.G. Teubner, Leipzig – Berlin, 1911. Chelsea 1967.
[36] K. Reidemester, Über die singulären Randpunkte eines konvexen Körpers. Math. Ann. 83(1921), 116-118.
[37] T. Bonnesen und W. Fenchel, Theorie der konvexen Körper. Ergeb. Math. Grenzg. V. 3. No. 1. Berlin: Springer, 1934; New York, 1948. Engl. transl.: Theory of convex bodies. BCS Associates, Moscow, Idaho, 1987.
[38] V.V. Wagner, Geometry of a space with areal metric and its applications to the calculus of variations (Russian). Mat. Sb. N.S. 19(41)(1946), 341-44.
[39] A.D. Alexandrov, Selected Works. Part 1: Selected Scientific Papers. Ed. Yu.G. Reshetnyak and S.S. Kutateladze. Amsterdam: Gordon and Breach Publishers, 1996.
[40] A.D. Alexandrov, Convex Polyhedra. Berlin etc.: Springer-Verlag, 2005.
[41] H. Hadwiger, Altes und Neues über konvexen Körper. Birkhäuser, Basel etc., 1955.
[42] H.G. Eggleston, Problems in Euclidean Space. Application of Convexity. Pergamon Press, New York, 1957. Dover republication, 2007.

[43] H.G. Eggleston, Convexity. Cambridge Tracts in Mathematics and Mathematical Physics. No. 47. Cambridge University Press. Cambridge - London - New York - Melbourne, 1958. Reprinted 1963, 1966. Reprinted with corrections 1969, 1977. Digitally printed version 2008.
[44] I.M. Yaglom and V.G. Boltyanskii, Convex figures. Holt, Rinehart and Winston, New York, 1961.
[45] V.L. Klee, (Ed.), Convexity. Proceedings of Symposia in Pure Mathematics. VII. Amer. Math. Soc., Providence, RI, 1963.
[46] L.A. Lyusternik Convex Figures and Polyhedra. Dover Publ., New York, 1963.
[47] G. Ewald, Über die Schattengrenzen konvexer Körper (Konvexe Körper und konvexität auf Grassmann-Kegeln VII). Abh. Math. Sem. Univ. Hamburg 27(1964), 167-170.
[48] G. Ewald, Konvexe Körper und Konvexität auf Grassmann Kegeln IX. Stark konvexe Funktionen. Math. Ann. 157(1964), No. 3, 219-230.
[49] G.C. Shephard, Convex bodies and convexity on Grassmann cones (VI). The projection functions of a simplex. London Math. Soc. 39(1964), 307-319.
[50] G.C. Shephard, Convex bodies and convexity on Grassmann cones (VIII): Projection functions of vector sums of convex sets. J. London Math. Soc. 39(1964), 417-423.
[51] F.A. Valentine, Convex sets. New York, Inc.: McGraw-Hill Book Company, 1964.
[52] R.V. Benson, Euclidean Geometry and Convexity. McGraw-Hill Book Company, New York, 1966.
[53] A.V. Pogorelov, Extrinsic Geometry of Convex Surfaces. Transl. of Math. Monogr, V. 35. Amer. Math. Soc. Providence, RI, 1973.
[54] A.V. Pogorelov, The Minkowski multidimensional problem. Wiley, New York, Toronto, London, Sydney, 1978.
[55] S.-Y. Cheng, S.-T. Yau, On the Regularity of the Solution of the n-Dimensional Minkowski Problem. Comm. Pure Appl. Math. 29(1976), 495-516.
[56] I.G. Nikolaev, S.Z. Shefel, Smoothness of convex surfaces on the basis of differential properties of quasiconformal mappings. Soviet Math. Dokl., 26(1982), 599-602.
[57] I.G. Nikolaev, S.Z. Shefel, Convex surfaces with positive bounded specific curvature and a priori estimates for Monge-Ampère equation. Siberian Math. J., 26(1985), 572-586.
[58] I.G. Nikolaev, S.Z. Shefel, Smoothness of convex surfaces and generalized solutions of the Monge-Ampère equation based on differential properties of quasiconformal mappings. Siberian Math. J., 26(1985), 841-851.
[59] Yu.D. Burago, Kalinin B.V., Smoothness and singularities of convex hypersurfaces. St. Petersburg Mathematical Journal, 8:3(1997), 399-411.
[60] V. Bangert, Convex hypersurfaces with bounded first mean curvature measure. Calc. Var. 8(1999), 259-278.
[61] I.G. Nikolaev, Approximation of a convex hypersurface with bounded positive specific scalar curvature. Siberian Adv. Math., 13:4(2003), 54-69.
[62] I.G. Nikolaev, On convex hypersurfaces with bounded specific mean curvature. Proceedings on geometry and analysis. Novosibirsk, Sobolev Institute of Mathematics SB RAS (2003) (Russian), 332-342.
[63] L.M. Kelly, M.L. Weiss, Geometry and Convexity. Wiley/Interscience Publ., New York etc., 1979.
[64] K. Leichtweiss, Konvexen Mengen. Berlin: VEB Deutscher Verlag der Wissenschaften, 1980.
[65] K. Leichtweiss, Affine geometry of convex bodies. Heidelberg: Johann Ambrosius Barth Verlag, 1998.
[66] S.R. Lay, Convex Sets and Their Applications. John Wiley & Sons, Inc., New York, 1982. Dover Publications, Inc., Mineola, N.Y., 2007.
[67] P. Gruber and J. Willis (Eds.), Convexity and Its Applications. Birkhäuser, Basel, 1983.
[68] H.T. Croft, K.J. Falconer, and R.K. Gay, Unsolved Problems in Geometry. Springer, New York, 1991.

[69] P. Gruber and J. Willis (Eds.), Handbook of Convex Geometry. North-Holland, Amsterdam, 1993. two volumes.
[70] M.L.J. Van de Vel, Theory of Convex Structures. Elsevier Science Publishers B.V., North-Holland, Amsterdam, etc., 1993.
[71] L. Hörmander, Notions of Convexity. Birkhäuser, Basel etc., 1994.
[72] R.J. Webster, Convexity. Oxford University Press, Oxford, 1994.
[73] A. Barvinok, A Course in Convexity. A.M.S., Providence, RI: 2002.
[74] T. Zamfirescu, The Majority in Convexity. Part I: Smoothness, Curvature and Consequences. Editura Univ. Bucuresti, 2009.
[75] V. Soltan, Lectures on Convex Sets. World Scient. Publ. Co. Pte. Ltd., Singapore, 2015.
[76] I.E. Leonard and J.E.Lewis, Geometry of Convex Sets. John Wiley & Sons, Inc., Hoboken, New Jersey, 2016.
[77] A.D. Alexandrov, Intrinsic Geometry of Convex Surfaces. Selected Works. Part II. Ed. S.S. Kutateladze. Boca Raton: Chapman & Hall/CRC, 2006.
[78] A.D. Alexandrov and V.A. Zalgaller, Intrinsic Geometry of Surfaces. A.M.S., Providence, RI: 1967. (Transl. Math. Monogr. Vol. 15).
[79] Yu.G. Reshetnyak (Ed.), Geometry IV. Non-Regular Riemannian geometry. Encycl. Math. Sci., 70. Berlin, Inc.: Springer-Verlag, 1993.
[80] J. Steiner, Ges. Werke. Bd. 2. Berlin, 1882.
[81] H. Brunn, Über Ovale und Eiflächen. PhD Thesis. U. München, 1887.
[82] H. Brunn, Über Kürven ohne Wendepunkte. Habilitationsschrift. U. München, 1889.
[83] H. Brunn, Referat über eine Arbeit: Exacte Grundlagen für eine Theorie der Ovale. S.-ber. Bayer. Akad. Wiss. (1894), 93-111.
[84] W. Blaschke, Kreis und Kugel. Leipzig: Verlag von Veit & Comp., 1916.
[85] H. Kneser, W. Süss, Die Volumina in linearen Scharen konvexer Körper. Mat. Tidsskr. B 1 (1932), 19-25.
[86] A.S. Besikovitch, Measure of asymmetry of convex curves. J. London Math. Soc., 23(1945), 237-240.
[87] Yu.G. Reshetnyak, One extremal problem from the theory of convex curves (Russian). Usp. Mat. Nauk, 8:6(58)(1953), 125-126.
[88] C.M. Petty, On the geometry of the Minkowski plane. Riv. Mat. Univ. Parma, 6(1955), 269-292.
[89] H. Hadwiger, Vorlesungen über Inhalt, Oberfláche un Isoperimetrie. Springer-Verlag, Berlin-Göttingen- Heidelberg, 1957.
[90] H. Minc, Permanents. Addison-Wesley Publishing Company. London-Amsterdam-Don Mills-Ontarion-Sydney-Tokyo, 1978.
[91] G.P. Egorychev, The solution of van der Waerden's problem for permanents. Adv. Math. 42(1981), No. 3, 299-305.
[92] D.I. Falikman, Proof of the van der Waerden conjecture regarding the permanent of doubly stochastic matrix. Mathemarical Notes 29:6(1981), 475-479.
[93] B. Gyires, Elementary Proof for a Van der Waerden's conjecture and Related Theorems. Computers Math. Applic. 31:10(1996), 7-21.
[94] Y.D. Burago and V.A. Zalgaller, Geometric Inequalities. Grundlehren der Math. Wiss. 285. Berlin etc., Springer, 1988.
[95] E. Lutwak, Intersection Bodies and Dual Mixed Volumes. Adv. Math. 71(1988), 232-261.
[96] R. Schneider, Convex Bodies: The Brunn-Minkowski Theory. Encycl. Math., 44. Cambridge University Press, Cambridge, 1993. 2nd expanded edition, 2014.
[97] R.G. Gardner, Geometric tomography. Encyclopaedia Mathematics and its Applications 58. Cambridge University Press, Cambridge. First published 1995. Second edition, 2006, 2012. Reprinted 2013.
[98] J. Bourgain and G. Zhang, On a generalization of the Busemann-Petty problem. Convex Geometric Analysis. 34(1998), 65-76.

[99] G. Zhang, A positive solution to the Busemann-Petty problem in $\mathbb{R}^4$. Ann of Math. 149(1999), 535-543.

[100] R.G. Gardner, A. Koldobsky, and T. Schlumprecht, An analytic solution to the Busemann-Petty problem on sections of convex bodies. Ann. of Math. 149(1999), 691-703.

[101] F. Barthe, M. Fradelizi, and B.Maurey, A Short Solution to the Busemann-Petty problem. Positivity 3(1999), 95-100.

[102] R.G. Gardner, The Brunn-Minkowski inequality. Bull. Amer. Math. Soc., New Ser. 39(2002), 355-405.

[103] E. Milman, A comment on the low-dimensional Busemann-Petty problem. Lect. Notes Math. 1910, 2007, 245-253.

[104] H. Minkowski, Geometrie der Zahlen. Teubner, Leipzig 1896. Chelsea, New York, 1953. Johnson Reprint Corp., New York, 1968.

[105] E. Steinitz / H.Rademacher, Vorlesungen über die Theorie der Polyeder unter Einschluss der Elemente der Topologie. Grundlehren der mathematischen Wissenschaften 41. Berlin, Springer-Verlag, 1934. Reprint 1976.

[106] H.S.M. Coxeter, Regular Polytopes. Third edition. Dover Publications, Inc., New York, 1973.

[107] H. Hadwiger and H.Debrunner, Combinatorial geometry in the plane. Holt, Rinehart & Winston, Inc., New York, 1964. Reprinted Dover Publications, Inc., Mineola, New York, 2015.

[108] V.J.D. Baston, Some properties of polyhedra in Euclidean space. Pergamon Press, Oxford etc., 1965.

[109] W.T. Wu, A theory of Embedding, Immersion, and isotopy of Polytopes in a Euclidean Space. Science Press, Beijing, 1965.

[110] B. Grünbaum, Convex Polytopes. 2nd ed. GTM, 221. New York, Inc.: Springer-Verlag, 2003.

[111] B. Grünbaum and G.C. Shephard, Convex Polytopes. Bull. London Math. Soc., 1(1969), 257-300.

[112] B. Grünbaum, Polytopes, graphs, and complexes. Bull. Amer. Math. Soc., New Ser. 76(1970), 1131-1201.

[113] J. Stoer and C. Witzgall, Convexity and Optimization in Finite Dimensions I. Grundllehren der mathematischen Wissenschaften, 163. Berlin, Inc.: Springer-Verlag, 1970.

[114] P. McMullen and G.C.Shephard, Convex Polytopes and the Upper Bound Conjecture. vol. 3 of London Math. Soc. Lecture Notes Series, Cambridge University Press, Cambridge, 1971.

[115] A. Bronsted, An introduction to convex polytopes. New York Inc.: Springer-Verlag, 1983.

[116] V.A. Yemelichev, M.M. Kovalev and M.K. Kravtsov, Polytopes, Graphs and Optimization. Cambridge University Press, Cambridge, 1984.

[117] J.E. Goodman, E. Lutwak, J. Malkevitch, and R. Pollak (Eds.), Discrete Geometry and Convexity (New York 1982). Annals of the New York Academy of Sciences, 440, 1985.

[118] T. Oda, Convex bodies and algebraic geometry: An introduction to toric varieties. Ergebnisse der Mathematic und ihrer Grenzgebiete 3, Folge, Band 15. Springer-Verlag, Berlin, Heidelberg, New York, 1988.

[119] J. Kahn and G. Kalai, A counterexample to Borsuk's conjecture. Bull. Amer. Math. Soc., New Ser. 29(1993), 60-62.

[120] T. Bisztriczky, P. McMullen, and A. Weiss (Eds.), Polytopes: Abstract, Convex and Computational (Proc. NATO Advanced Study Institute, Toronto). Kluwer Academic Publishers, Dordrecht, 1994.

[121] G.M. Ziegler, Lectures on polytopes. GTM, 152. New York, Inc.: Springer-Verlag, 1995; revised 6th printing, 2006.

[122] G. Ewald, Combinatorial Convexity and Algebraic Geometry. GTM, 168. New York, Inc.: Springer-Verlag, 1996.

[123] C. Zong, Strange Phenomena in Convex and Discrete Geometry. New York, Inc.: Springer-Verlag, 1996.

[124] V. Boltyanski, H. Martini, and P.S. Soltan, Excursion into Combinatorial Geometry. Springer-Verlag, Berlin etc., 1997.

[125] P.M. Gruber, Convex and Discrete Geometry. Grundlehren der mathematischen Wissenschaften. V. 36. Springer-Verlag, Berlin - Heidelberg- New York, 2007.

[126] A.V. Bondarenko, On Borsuk's conjecture for two-distance sets. Discrete and Computational Geometry. 51(2014), 509-515.

[127] T. Jenrich, A.E. Brouwer, A 64-dimensional counterexample to Borsuk's conjecture. The Electronic Journal of Combinatorics. 21(4)(2014), # P. 4.29.

[128] K.A. Adiprasio, I. Barany, and C. Vilcu (Eds.), Convexity and Discrete Geometry Including Graph Theory: Mulhouse, France, September 2014. Springer International Publishing Switzerland, 2016.

[129] V.D. Milman and G.Schechtman (Eds.), Asymptotic Theory of Finite Dimensional Normed Spaces. Lect. Notes Math. 1200, 1986.

[130] A.A. Giannopoulos and V.D. Milman, Asymptotic Convex Geometry. Short Overview. In.: Different Faces of Geometry. Eds.: S.K. Donaldson, Ya. Eliashberg, M. Gromov. New York, Boston, Dordrecht, London, Moscow: Kluwer Academic/ Plenum Publishers, 2004. 87-162.

V.N. Berestovskii, Sobolev Institute of Mathematics SB RAS, Acad. Koptyug avenue, 4, Novosibirsk, 630090, Russia

E-mail address: vberestov@inbox.ru

ON SEVERAL PROBLEMS POSED BY H. BUSEMANN AND RELATED TO GEOMETRY IN NORMED SPACES

DMITRI BURAGO

The Author believes that an extremely important part of H. Busemann's contribution to Science was his extraordinary vision of the future of Mathematics expressed through the problems he formulated. They still keep many mathematicians at unrest. A lot of problems remain unsolved or only partially solved. They also inspired a lot of new research. In this short article we briefly discuss a few problems by H. Busemann with a hope that this would help to read original Busemann's papers and appreciate their importance. Let me mention that a large part of the joint research I conducted together with S. Ivanov came from our attempts to understand these problems. The reader may also want to have a look at [1] where some issues discussed here are mentioned.

At some point Busemann got interested in the following question: In which geometries planes minimize area ([8, 9, 10, 11])? He was especially concerned with normed spaces. A general formulation of the problem is as follows. By "definition" planes minimize k-area in $(V^n, || \cdot ||)$ if and only if for every k-dimensional affine subspace of $L \subset V^n$ and any topological ball $B \subset V$ with a smooth boundary ∂B, and for every Lipschitz (Lip in the sequel) "surface" $\Sigma \subset V^n$ with $\partial\Sigma = \partial B$ one has "$Area_k$"$(\Sigma) \geq$"$Area_k$"(∂B). We call this Σ a *competing surface.* Now the questions are: What do the words "surface" and "$Area_k$" (k-dimensional area) mean?

There are a number of possible clear answers to the first question. The word "surface" could mean a Lip immersed ball or a manifold with prescribed topology, a Lip immersed manifold, a Lip chain with coefficients in $\mathbb{Z}$ or in $\mathbb{R}$. As it turned out not so long ago, the problems we discuss heavily depend on the choice of the interpretation. Nowadays we do know that in some cases even the answers are different too [4].

The situation with "$Area_k$" is yet more involved. Up to a multiplicative constant, there is one translation invariant volume on $(V, || \cdot ||)$. Not the same with surface k-areas. The choice of a surface area depends on the choice of a normalization constant in each k-space tangent to Σ. There is an affine-invariant approach to the notion of k-area, but it is easier to comprehend the notion using an axillary Euclidean structure. Then the surface k-area is given by a choice of its *density function* $A : G(k, n) \to \mathbb{R}+$ from the Grassmannian to non-negative reals. The area is then found in the standard way: by integration of A against the Euclidean surface area.

Every k-surface in a normed space carries the structure of a Finsler manifold. By now we know that every smooth compact Finsler manifold can be isometrically embedded into some normed space [2]. Thus, without loss of generality, one can think of Finsler manifolds as sub-manifolds of normed spaces. Our first requirement

The Author was partially supported by NSF grant DMS-1205597.

is that k-areas depend only on the intrinsic metrics of sub-manifolds. That is, we need a notion of volume for Finsler manifolds. We also want it to strictly increase when the metric strictly increases, and to change smoothly if we smoothly vary the metric. Of course, we also want it to coincide with the Riemannian volume in the Riemannian case.

There are two main examples of k-areas satisfying these properties: The Busemann-Haudorff and Holmes-Thompson (symplectic) surface areas. Of course, any positive linear combination of the two also satisfies the above properties. There are other examples but they look artificial. These two are of the outmost importance.

The Busemann-Hausdorff volume is just the Hausdorff measure. Busemann however showed that it can be obtained from a certain density A which is given by a very nice expression. He brought this notion of what we know now as the Busemnn-Hausdorff surface area to the context of surfaces in normed spaces.

The Holmes-Thompson volume is also given by a nice expression. Both expressions (for the Busemann-Hausdorff and Holmes-Thomson surface areas) in Euclidean coordinates are written via the volumes of the unit balls of duals to the unit balls of the norms. The Holmes-Thompson volume, just as the Busemann-Hausdorff volume, also has a nice invariant description. It originates from Symplectic Geometry. The co-tangent bundle to a manifold carries a canonical symplectic structure and therefore a canonical volume form. If one considers the set of unit co-balls of the norm over a set in the manifold, its "symplectic volume" (that is, the nth power of the symplectic form) is the Holmes-Thompson area of the set.

We have already defined what it means that "planes minimize area". This property is nowadays often referred to as *semi-ellipticity* of the surface area. This notion is of outmost importance in a number of areas of Geometry and PDEs.

We need yet another notion. One says that a volume density A is semi-convex if it can be extended to a convex function on the vector space of all k-vectors, that is, to the space of linear combinations of elements of $G(n, k)$.

It is easy to see that *convexity* $\Longrightarrow$ *semi-ellipticity*. Busemann made a remarkable observation: The Holmes-Thompson density is NOT convex! He gave an explicit example of four indecomposable 2-vectors e_1, e_2, e_3, e_4 in a certain four-dimensional normed space with $e_1 + e_2 + e_3 = e_4$ but $||e_1|| + ||e_2|| + ||e_3|| < ||e_4||$. Here $|| \cdot ||$ stands for the norm corresponding to the Holmes-Thompson area (that is, $||v \wedge w||$ is the Holmes-Thompson area of the parallelogram spanned by v and w).

This first made Busemann think that the Holmes-Thomson surface area is not elliptic in any sense. He had tried to construct an example for a number of years and then realized that the reverse implication *semi-ellipticity* $\Longrightarrow$ *convexity* is not at all obvious. He left this question as an open problem specifically for the Holmes-Thompson area. It attracted a lot of attention since then but still remains widely open and very intriguing. In my humble opinion, the largest progress in this direction is the following result: The 2-dimensional surface area is elliptic over $\mathbb{Z}$ if the competing filling surface is assumed to have the topology of a disc; and it is not semi-elliptic for Lip chains over $\mathbb{R}$ ([6, 4]). This progress is very modest, but it appeared in the Annals, which indicates the importance of this problem posed by Busemann. Even if the filling film is allowed to have handles, the answer is not known. The higher dimensional case is widely open.

There is another related problem suggested by Busemann: Is the Busemann-Hausdorff surface k-area at least convex? This is a classical result in co-dimension 1, that is if $k = n - 1$ ([12]). Therefore, in co-dimension 1 it is semi-elliptic. One is however really interested in higher co-dimensions. It has been proven recently that the 2-dimensional Hausdorff-Busemann surface area is semi-elliptic (and, therefore, convex) ([5]). In higher dimensions this remains a complete mystery.

The Author of this brief article believes that this circle of problems posed by Busemann opened a can of worms, a new direction or perhaps even a new area of research which is likely to be very active for many years to come.

References

[1] J.C. Alvarez Paiva., *Some problems in Finsler Geometry*, Handbook on Differential Geometry, Vol. 2, pp. 1-33, Elsevier 2006.

[2] D. Burago, S. Ivanov, *Isometric embeddings of Finsler manifolds'*, Algebra & Analysis, vol. 5, no. 1, 159-169, 1993.

[3] D. Burago, S. Ivanov, *On asymptotic volume of Finsler tori, minimal surfaces in normed spaces, and symplectic filling volume*, Ann. of Math. (2) **156** (2002), no. 3, 891–914.

[4] D. Burago, S. Ivanov, *Gaussian images of surfaces and ellipticity of surface area functionals*, Geom. Funct. Anal. **14** (2004), no. 3, 469–490.

[5] D. Burago, S. Ivanov. *Minimality of planes in normed spaces*. Geom. Funct. Anal. 22 (2012), no. 3, 627638.

[6] D. Burago, S. Ivanov, *Minimal surfaces in normed spaces, symplectic fillings, and asymptotic volume growth of Finsler tori*. Annals of Mathematics, 2002, vol.156, n. 3.

[7] H. Busemann, *Intrinsic area*, Ann. of Math. (2) **48** (1947), 234–267.

[8] H. Busemann, *A theorem on convex bodies of the Brunn-Minkowski type*, Proc. Nat. Acad. Sci. U.S.A. **35** (1949), 27–31.

[9] H. Busemann, *Geometries in which the planes minimize area*, Ann. Mat. Pura Appl. (4) **55** (1961), 171–189.

[10] H. Busemann, G. Ewald, G. C. Shephard, *Convex bodies and convexity on Grassmann cones. I–IV*, Math. Ann. **151** (1963), 1–41.

[11] H. Busemann, E. G. Straus, *Area and normality*, Pacific J. Math. **10** (1960), 35–72.

[12] K.Leichtweiss, *Convex Sets*, Springer, 1980.

Dmitri Burago: Pennsylvania State University, Department of Mathematics, University Park, PA 16802, USA

E-mail address: burago@math.psu.edu

BUSEMANN'S WORK ON HILBERT GEOMETRY

ATHANASE PAPADOPOULOS

Contents

1. Introduction 123
2. Definitions 125
3. The Hilbert metric of a simplex 127
4. Geodesics and convexity properties 128
5. Curvature 131
6. Perpendicularity 134
7. Asymptotes in Hilbert geometry 138
8. The Finsler structure, Hilbert area and Finsler volume 140
9. Isometries 142
References 143

1. Introduction

The Hilbert metric was introduced by David Hilbert in a letter he addressed to Klein on August 14, 1894. The mathematical part of the letter is reproduced as an article that was published in Mathematische Annalen in 1895 [16], and also as an appendix to Hilbert's *Grundlagen der Geometrie* (see p. 108 ff. of the English edition [17]). This metric is present in the background in several papers and books of Busemann that are spread all along his work (cf. [3], [6], [7] and the articles [8], [9], [11]). Busemann used the Hilbert metric as a class of examples that illustrate several notions that are of fundamental importance in his work. He writes in his *Geometry of geodesics* [7] (p. 106):

> The Hilbert metric will not be studied for its own sake. It will merely be used
> (1) to illustrate convexity questions;
> (2) to construct other Desarguesian spaces;
> (3) to provide examples for the theory of parallels.

The notion of Desarguesian space, referred to in the second item above, is one of the main geometric concepts introduced in [7], and it is used extensively in Busemann's work on Hilbert's Problem IV. For the definition, the reader may refer to Busemann's book or to the survey [24] in the present volume.

Beyond the circle of Busemann and his students and collaborators, Hilbert geometry did not attract much attention, until it was revived in the 1990s. This topic constitutes today a very active field of research (see the surveys in the recent volume [23]). Thus, during the first hundred years that followed its birth (roughly speaking, from 1895 to beginning of the 1990s), most of the work done on the Hilbert metric is due to Busemann, his collaborators and his students. The exception is a version of the Hilbert metric which was used to prove a theorem in algebra, due to Birkhoff [1] and Samelson [25].

Because of the dispersion of the work of Busemann on the Hilbert metric, and since Busemann's books are not included in the present *Selected Works* edition, we shall provide in this essay a rather detailed review of his work on the subject. Surveying this topic in Busemann's writings is at the same time the occasion of reviewing several metric notions that played an essential role in his work: geodesics, horospheres, perpendicularity, reflections, motions, horofunctions, bounded curvature, parallelism, etc. This is also the occasion to see some aspects of Busemann's comprehensive theory of metric spaces that includes all these notions.

Busemann, in studying Hilbert geometry, was largely motivated by his interest in Hilbert's Problem IV, a problem which asks for the study of the metrics on subsets of Euclidean (or projective) space for which the Euclidean (or projective) lines are geodesics. Hilbert, in his formulation of the problem, gave essentially two examples of such metrics, one of them being the class of Hilbert metrics, the other one being the class of Minkowski metrics, on which Busemann also worked extensively. We review Busemann's work on Minkowski metrics in the essay [24] in the present volume.

The plan of the present article is the following. In §2, we recall the basic definitions concerning Hilbert geometry. In §3, we consider the special case of the Hilbert metric of a simplex. Busemann showed that this is the only Hilbert geometry which is isometric to a Minkowski space. In §4, we survey Busemann's work on the geodesics and the convexity of balls in Hilbert geometry. In §5, we study the various metrical notions of bounded and zero curvature that were introduced by Busemann, in the setting of Hilbert geometry. In §6, we study perpendicularity in metric spaces and more specifically the construction of perpendiculars in Hilbert geometry. This construction generalizes the construction of perpendiculars in the Cayley-Klein model of hyperbolic geometry (which is the Hilbert geometry of an ellipsoid). The symmetry of perpendicularity characterizes hyperbolic geometry among all Hilbert geometries. §7 is concerned with asymptotes. A criterion is given for the differentiability at a point in $\partial\Omega$ in terms of a metric property involving two asymptotic lines having this point as a common point at infinity. §8 concerns the fact that a Hilbert metric is locally Finslerian. Formulae are given for the associated Finsler norm. Its length, area and volume elements are discussed. §9 is concerned with isometries. It contains a characterization of the isometry group of a Hilbert geometry of a strictly convex domain. We also discuss reflections in Hilbert geometries. A theorem of Busemann and Kelly gives a characterization of hyperbolic geometry as the unique Hilbert geometry for which reflections about all hyperplanes exist.

Some of the results obtained by Busemann and his collaborators were rediscovered later, and they are often attributed to other mathematicians. This includes the characterization of the Hilbert geometries whose isometry groups are Abelian and act transitively (cf. Theorem 3.1 below), the characterization of the isometries of Hilbert geometries associated to strictly convex bodies (Theorem 9.1 below), and there are others. Highlighting this fact is a way of repairing an injustice.

To end this introduction let us note that Hilbert geometries of non-strictly convex sets are important examples of the so-called "chord spaces" introduced by Busemann, a topic we shall not consider in this essay; cf.[14].

2. Definitions

The title of §18 of Busemann's *Geometry of geodesics* [7] (p. 105ff.) is "Hilbert's geometry." We shall make a self-contained review of this section. The definition of the distance function that defines this metric is based on the cross ratio, and we start by recalling this notion.

Let A^n be n-dimensional affine space and let a, b, c, d be four distinct points lying in that order on a line in A^n. The cross ratio of the ordered quadruple (a, b, c, d) is defined as

$$[a,b,c,d] = \frac{|b-d|}{|c-d|} \cdot \frac{|c-a|}{|b-a|}. \tag{1}$$

In this formula, a choice of a Euclidean metric on A^n is made, $|x-y|$ denotes the distance between the two points x and y with respect to this metric, and the result does not depend on the choice of such a metric.

Let Ω be a relatively compact open convex subset of A^n and $\partial\Omega$ its boundary.

An *affine line in* Ω is the intersection of Ω with an affine line of the ambient affine space A^n. We shall sometimes use the terminology "line" instead of "affine line."

For $x \neq y$ in Ω, let $L(x, y)$ denote the affine line in Ω containing these two points. Since we assumed that Ω is relatively compact, $L(x, y)$ intersects $\partial\Omega$ in two points; let a and b be the these points, ordered in such a way that the four points a, x, y, b lie in that order on the line $L(x, y)$ (Figure 1).

We set

$$h_\Omega(x,y) = h(x,y) = \log[a,x,y,b]. \tag{2}$$

For $x = y$, we set $h(x, y) = 0$.

This function $h(x, y)$ defined on $\Omega \times \Omega$ satisfies all the properties of a metric and is called the *Hilbert metric* on Ω. Hilbert, in the letter to Klein that we mentioned, gave a proof of the triangle inequality satisfied by this metric, based on properties of the cross ratio. A similar proof is recalled by Busemann in [7], p. 106. This proof relies more specifically on three properties of the cross ratio: the invariance under perspectivities, an addition formula and a monotonicity property. In the next proposition, we state precisely

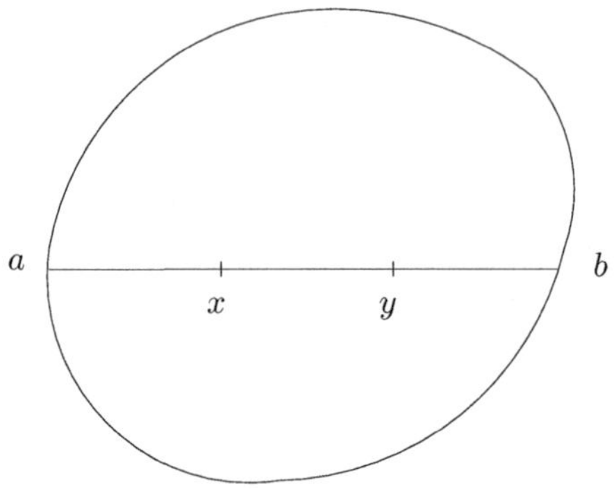

FIGURE 1.

these three properties since they are used in the later developments of the Hilbert metric. The three properties refer to Figure 2. In the figure to the left, the four ordered points a', b', c', d' and a, b, c, d are said to be obtained from each other by a *perspectivity*, or *projectivity of center* p.

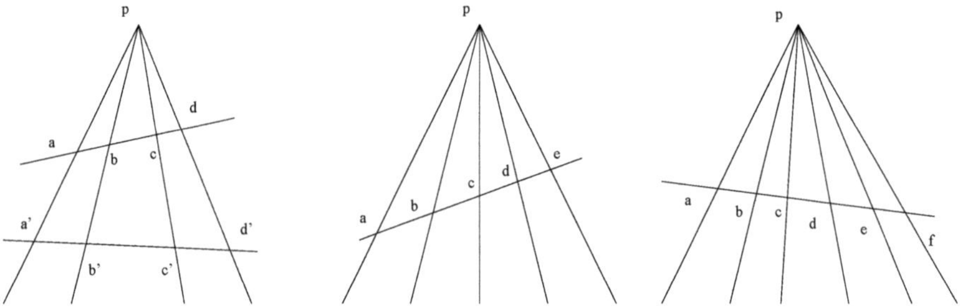

FIGURE 2. The three properties of Proposition 2.1

Proposition 2.1. *The cross ratio satisfies the following properties, see* Figure 2.

(1) (invariance) $[a, b, c, d] = [a', b', c', d']$,
(2) (product) $[a, b, c, e].[a, c, d, e] = [a, b, d, e]$,
(3) (monotonicity) $[b, c, d, e] \geq [a, c, d, f]$.

Let us now prove that the Hilbert metric satisfies the triangle inequality. We use the notation of Figure 3. We show the following:

$$h(x, y) + h(y, z) \geq h(x, z)$$

with equality if and only if x, y, z are aligned.

If x, y, z are aligned, the equality is trivial.

Suppose they are not. We produce ac and db until they meet at a point p. (This point may be at infinity, that is, the lines may be parallel and in this case a special argument is needed.)

By the invariance and the monotonicity properties of the cross ratio (Proposition 2.1), we have

$$[a, x, y, b] = [u, x, v, w] \geq [e, x, v, f]$$

and

$$[c,y,z,d] = [u,v,z,w] \geq [e,v,z,f].$$

Multiplying the two inequalities and using the product property of the cross ratio, we obtain

$$[a,x,y,b].[c,y,z,d] \geq [e,x,z,f]$$

which implies, by taking logarithms, the triangle inequality.

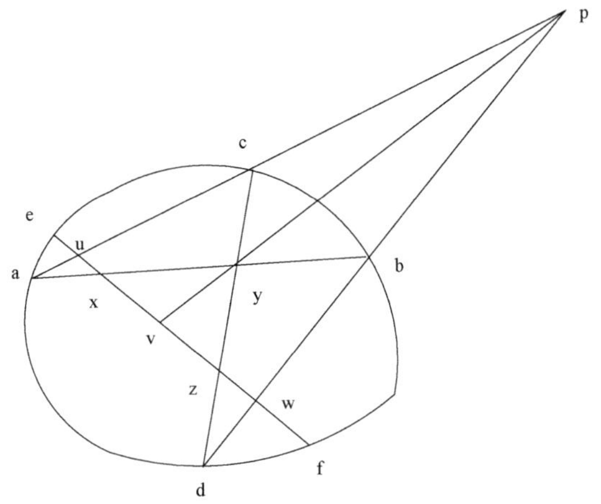

FIGURE 3. The triangle inequality

At some places in his work, Busemann includes a multiplicative positive factor k in the definition of the Hilbert metric (2). (Hilbert did the same.) In the case where Ω is a ball, the Hilbert metric is a model of hyperbolic geometry, usually called the Cayley-Klein projective ball model. In fact, this metric of the ball gives different models, if we allow k to vary. In each model, the curvature of the space (which is constant) depends on the factor k. In the present survey, we shall stick to the case $k = 1$.

To end this section, let us remark that the definition of the Hilbert metric is naturally extended to the case of non-compact convex domains Ω. For this, one uses an extension of the definition of the cross ratio (Equation (1)): if the point d (respectively a) is at infinity, then the ratio $\frac{|b-d|}{|c-d|}$ (respectively $\frac{|c-a|}{|b-a|}$) is taken to be equal to 1. In this case, the metric h_Ω is non-degenerate (that is, it satisfies $f(x,y) \neq 0$ if $x \neq y$) if and only if the convex set Ω does not contain an affine line of A^n.

3. THE HILBERT METRIC OF A SIMPLEX

Busemann and Phadke, in their paper [12], discuss the Hilbert geometry of a triangle. They note that all triangles are affinely equivalent and, therefore their Hilbert metrics are isometric. Furthermore, such a Hilbert metric is isometric to the metric on the positive quadrant in $\mathbb{R}^2$,

$$\{x = (x_1, x_2) | x_1 > 0, x_2 > 0\}$$

defined by the formula

$$d(x,y) = \max\{|\log \frac{x_1}{y_1}|, |\log \frac{x_2}{y_2}|, |\log \frac{x_3}{y_3}\}|$$

where $x = (x_1, x_2)$ and $y = (y_1, y_2)$. Busemann and Phadke recall that this metric was already discussed in the paper [21] but that at the time that paper was written, it was not realized that this metric can be identified with the Hilbert metric of the triangle. A construction for the circles of the Hilbert geometry of a quadrant and of a triangle is provided in Phadke's paper [21]; these are the polygons $bcdefgb$ in Figures 4 and 5. In both figures, the center of the circle is the point a.

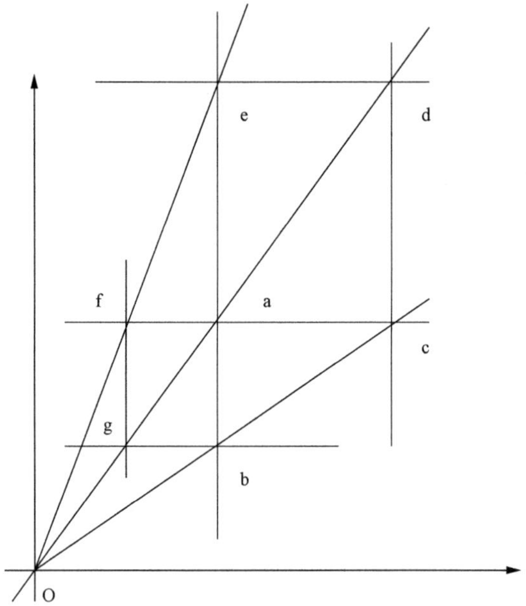

FIGURE 4.

A more important result of Busemann concerning the Hilbert geometry of the triangle, and more generally, of the simplex, is the following:

Theorem 3.1 (Busemann, [9], p. 35; cf. also Busemann and Phadke, [13], p. 313.). *The class of simplices in Euclidean spaces is exactly the class of convex subsets whose associated Hilbert geometry has the property that its isometry group has an Abelian subgroup of order two which acts transitively. Furthermore, this geometry is isometric to a Minkowski geometry (a finite-dimensional normed space).*

This theorem of Busemann is usually attributed to others who rediscovered it.

4. Geodesics and convexity properties

We start this section by fixing a few definitions and some notation that we shall use throughout the rest of this survey.

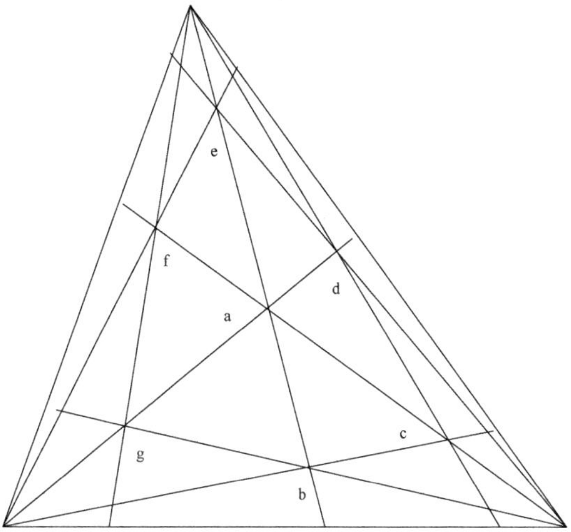

FIGURE 5.

Let (M, d) be a metric space and $I \subset \mathbb{R}$ an arbitrary nonempty interval. A *geodesic* $\gamma : I \to (M, d)$ is a map such that for any triple of points x_1, x_2, x_3 in I satisfying $x_1 \leq x_2 \leq x_3$, we have

$$d(x_1, x_2) + d(x_2, x_3) = d(x_1, x_3).$$

In the case where I is a closed interval $[a, b]$, we say that the geodesic γ *joins the points* $\gamma(a)$ *and* $\gamma(b)$. A *geodesic line* (or, more simply, a *line*) is a geodesic whose domain of definition is $\mathbb{R}$.

Let O be a point in Ω and ρ a nonnegative real number. The *sphere* in Ω of center O and radius ρ is the subset

$$S(O, \rho) = \{x \in \Omega \ ; \ h(O, x) = \rho\}.$$

The *open (respectively closed) ball* in Ω of center O and radius ρ is the subset

$$B(O, \rho) = \{x \in \Omega \ ; \ h(O, x) < \rho\},$$

respectively

$$B(O, \rho) = \{x \in \Omega \ ; \ h(O, x) \leq \rho\}.$$

More generally, for any subset Q of M and for any $\rho \geq 0$, we define the open (respectively closed) ρ-neighborhood of Q by

$$B(Q, \rho) = \{p \in \Omega \mid h(p, Q) < \rho\},$$

respectively

$$B(Q, \rho) = \{p \in \Omega \mid h(p, Q) \leq \rho\}.$$

In these formulae, the distance $h(p, Q)$ from a point h to a subset Q of M is defined as

$$h(p, Q) = \inf_{x \in Q} h(p, x).$$

For any convex subset Ω of A^n, a *support hyperplane* of Ω at a point $a \in \partial\Omega$ is an affine hyperplane in A^n that has nonempty intersection with the boundary $\partial\Omega$ of Ω and that has empty intersection with Ω. Support hyperplanes exist at every point of the boundary of a convex set, but they may be not unique. If $\partial\Omega$ is smooth then the tangent hyperplane of $\partial\Omega$ at a is a support hyperplane. If Ω is strictly convex, then the support hyperplane at any boundary point of Ω is unique. If each point of the boundary of a subset Ω of A^n has a support hyperplane, then Ω is convex. For these classical results we refer the reader to the books by Bonnesen and Fenchel [2], Eggleston [15] and Valentine [26].

We now study the convexity of balls in a Hilbert metric. We refer to Figure 6 in which H_a and H_b are support hyperplanes at the points a and b respectively of $\partial\Omega$, and H_p and H_f are hyperplanes in the same pencil.

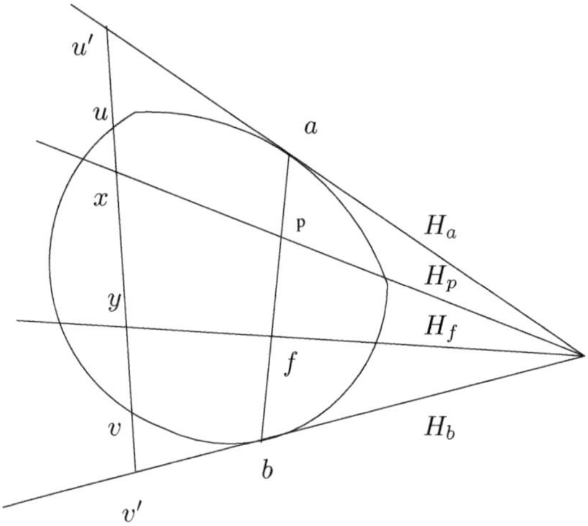

FIGURE 6. Convexity of balls

We have, by the invariance and the monotonicity of the cross ratio (Proposition 2.1),

$$[a, p, f, b] = [u', x, y, v'] \leq [u, x, y, v].$$

Therefore, $h(p, f) \leq h(x, y)$. Taking $x = p$, we obtain $h(p, f) \leq h(p, y)$, for every $y \in H_p$. This implies that H_f is a support hyperplane for the ball of center p passing through f. Thus, every point on the boundary of that ball has a support hyperplane, which implies that the ball is convex.

The next proposition includes this result together with some of other results of Busemann on the Hilbert metric:

Proposition 4.1 ([7], § 18). *The Hilbert metric $h(a, b)$ defined on Ω satisfies the following properties:*

(1) *Closed balls are compact.*
(2) *The affine lines in Ω are lines for the Hilbert metric.*
(3) *The balls are convex.*
(4) *More generally, for any convex subset Q of Ω and for any $\rho \geq 0$, the (open or closed) ρ-neighborhood $B(Q, \rho)$ is convex. Furthermore, if Ω is strictly convex, then $B(Q, \rho)$ is strictly convex for every $\rho > 0$.*
(5) *Any h-geodesic is contained in a Euclidean line of A^n if and only if $\partial\Omega$ does not contain two coplanar but non-collinear segments.*

(6) *If* (Ω, h) *satisfies one of the equivalent conditions in Property (5), then the balls in this space are strictly convex; equivalently; their boundary does not contain any nontrivial affine segment.*

(7) *Any geodesic in* Ω *can be extended to a geodesic line.*

Remarks 4.2. 1.– Property (2) implies that any Euclidean segment in Ω is the image of a geodesic joining its endpoints. Thus, for any two points in Ω, there is a geodesic connecting them.

2.– Convexity of the subsets in properties (3) and (4) refers to lines in Ω: for any two points in the subset Q of Ω, the affine segment joining them is contained in Ω. There exists another notion of convexity for subsets of a metric space, but we are not referring to it here.

3.– When one of the equivalent conditions of Property (5) is satisfied, Ω becomes an example of what Busemann calls a *straight Desarguesian space* (see the definition in Busemann [7] §11, and also in the essay [24] in the present volume).

4.–The convexity property of $S(Q, \rho)$ for any convex set Q, expressed in (4), follows from the convexity of $S(T, \rho)$ for any segment T, cf. [7], (36.21). This property in general metric spaces is much stronger than the convexity of balls expressed in (3), cf. Chapter V of [7].

5.— The convexity of balls in Hilbert geometries (and in particular the convexity of discs, in the two-dimensional case) does not imply the convexity of ellipses. Busemann and Phadke in [12] give an example of a Hilbert geometry (this is the geometry of the quadrant which we discussed in §3) in which the ellipses are not convex; cf. Figure 7 which represents the ellipse of foci f_1 and f_2. In this example, the two arcs cb and gf are concave.

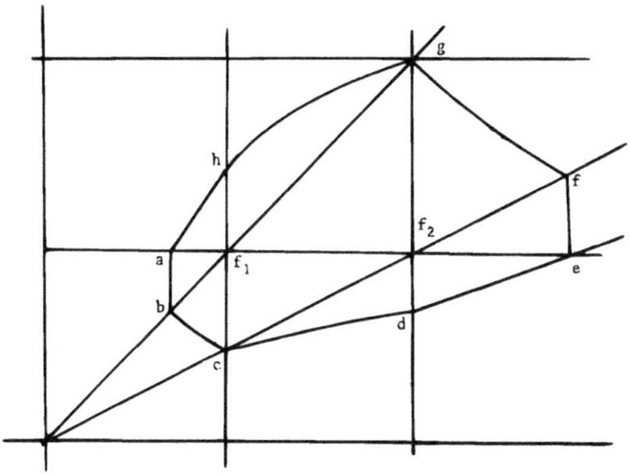

FIGURE 7. An ellipse in the Hilbert metric of the quadrant (from the paper [12]).

5. Curvature

In this section, we discuss three notions introduced by Busemann that express the boundedness of the curvature in general metric spaces. We shall recall these notions in the more specific setting of Hilbert geometry. One of these properties is a property of triangles which is known under the names

Busemann nonpositive curvature, respectively *nonnegative curvature*, *positive curvature*, *negative curvature* or *zero curvature*, depending on the inequalities that are involved in the definition. We note that Busemann introduced other notions of curvature in metric spaces than those which we discuss in section. We mention in particular a notion that invokes the *angle excess* of triangles in a G-surface, developed in §42 of his book *The geometry of geodesics* [7].

We start with a general property introduced by Busemann in [7], p. 109, which he called Peaklessness. It generalizes the notion of convexity of a function.

Definition 5.1 (Peakless function)**.** Let f be a continuous function defined on an arbitrary interval I of the real line. Then f is said to be *peakless* if for all $t_1 < t_2 < t_3$ in I we have

$$f(t_2) \leq \max(f(t_1), f(t_3))$$

with equality if and only if $f(t_1) = f(t_2)$.

Equivalently, the function f is peakless if and only if it satisfies one of the following three properties:

(1) f is constant;
(2) f is strictly increasing or decreasing;
(3) f takes a minimum value at all points of some interval (which may be reduced to a point), is strictly decreasing to the left of that interval and strictly increasing to the right.

Convex functions are peakless, but the converse is not true in general, since any increasing function is peakless.

It is a classical fact that if $f(t)$ is a convex positive function, then for every $\alpha > 1$, the function f^α is convex. The converse is not true: the fact that the function f^α for some $\alpha > 0$ is convex does not imply in general that f is convex. In fact, the convexity property of f^α becomes weaker and weaker as α increases: for any $\epsilon > 0$, one can find functions which are $\alpha + \epsilon$-convex and not α-convex. Taking the limit as $\alpha \to \infty$, the convexity property of f^α tends to the peaklesness property; cf. [7], p. 109. In this case, peaklnessness is a limiting convexity property.

A peakless function is said to be *strictly peakless* if it is not constant on any interval.

We now turn back to Hilbert geometry.

Proposition 5.2 ([7] (18.12))**.** *Let (Ω, h) be a Hilbert geometry and let $\gamma : I \to \Omega$ be a geodesic whose image lies on an affine line. Then, for any convex subset Q of Ω, the distance function*

$$t \mapsto h(x(t), Q)$$

is peakless. If Ω is strictly convex then this distance function is strictly peakless.

Next, we discuss the notions of Busemann bounded or zero curvature developed in [7], p. 110 and in other places. The notion known as Busemann non-positive curvature which we recall below is one among several properties that express non-positive curvature in metric spaces. It is more restrictive than peaklessness in the sense that there are spaces in which the distance function is peakless but which are not Busemann non-positively curved. One class of examples is precisely the class of Hilbert metrics. We shall review this below. Let us first recall the definition.

The setting is that of spaces which are *midpoint convex*, that is, metric spaces whose distance function d satisfies the following property: for any two points x and z there exists a point y such that $d(x,y) = d(y,z) = 1/2d(x,z)$. Such a point z is said to be a *midpoint* of x and y.

A midpoint convex geodesic metric space which is complete is geodesic (that is, any two points may be joined by a geodesic). A geodesic metric space is always midpoint convex.

Definition 5.3 (Busemann non-positively curved space, [7] §36.6). A midpoint convex metric space (M,d) is said to be *Busemann non-positively curved* if for any three points a,b,c in M if a' is a midpoint of a and c and b' a midpoint of b and p, then

$$d(a',b') \leq 1/2(d(a,b)). \tag{3}$$

There are variations of this definition, namely, the space is called *Busemann negatively curved, non-negatively curved, or positively curved* depending on whether the inequality in (3) is replaced by a strict inequality, or the reverse large or strict inequality respectively. If the inequality is replaced by an equality, the space is said to be *zero curved.* The last property appears in the setting of Minkowski spaces, cf. [7], p. 261 and the report [24] in the present volume.

Busemann gives examples in [7] §18 to show the following:

Proposition 5.4 ([7], p. 110). *A Hilbert geometry is not necessarily Busemann non-positively curved.*

Busemann shows ([7] (36.15)) that the property of being non-positively curved implies the convexity of the distance functions to geodesics. More precisely, it implies that for any two geodesics $x(t)$ and $y(t)$, if H denotes the image of $y(t)$, then the two functions

$$t \mapsto d(x(t), y(t))$$

and

$$t \mapsto d(x(t), H)$$

are convex. He also shows that in a Hilbert geometry (Ω, h_Ω) (even in the case of a strictly convex Ω), these functions are not necessarily convex ([7], p. 110).

On p. 406 of [7], Busemann asks (Problem 34) whether a Hilbert geometry with nonpositive curvature is necessary the hyperbolic metric (that is, the metric of constant negative curvature on the disc). The problem was solved affirmatively by Kelly and Straus, who proved in [22] a more general result which we quote, after making another definition.

In a midpoint convex metric space (M,d), the curvature is said to be *determinate* at a point if it is either non-positive or non-negative in the sense of Definition 5.3 in a neighborhood of that point. The curvature is said to be *indeterminate* at a point if it is neither non-positive nor non-negative in any neighborhood of that point.

Kelly and Straus obtained the following characterization of hyperbolic geometry using the notion of curvature of Hilbert geometries:

Theorem 5.5 (Kelly–Straus [22]). *Let Ω be a bounded open convex subset of A^n ($n \geq 2$) equipped with its Hilbert metric. Then Ω is an ellipsoid if and only if the curvature is determinate everywhere in Ω.*

Kelly and Strauss stated their theorem for $n = 2$, but their result is valid in higher dimensions. In fact, if the curvature of (Ω, h_Ω) is determinate everywhere, the curvature of any plane section of Ω is determinate everywhere. Thus, by Theorem 5.5, the boundary of any plane section of Ω is an ellipse, which implies that the boundary of Ω is an ellipsoid.

Finally, we talk about capsules and their convexity.

Definition 5.6 (Capsules, [7] Chapter V). If T is a segment in a metric space (M, d) and ρ a nonnegative number, a subset of the form

$$\{x \in M \mid d(x, T) \leq \rho\}$$

is said to be a *capsule* of axis T and radius ρ. This is also the closed ρ-closed neighborhood of T.

Busemann proves a local-implies-global result for convexity of capsules (see the precise statement, Theorem 36.20 in [7]). At the same time, he proves the following:

Theorem 5.7 ([7] Theorem 36.20). *A space which has Busemann non-positive (resp. negative) curvature has convex (resp. strictly convex) capsules.*

Busemann shows that in a Hilbert geometry, capsules are convex:

Theorem 5.8 ([7] §18 and 41). *For any segment T in a Hilbert geometry and for any $\rho > 0$, the capsules of axis T and radius ρ are convex.*

To see the relevance of this notion, we note that for Riemannian manifolds, convexity of capsules is equivalent to nonpositive curvature. Indeed, Busemann proves the following:

Theorem 5.9 ([7], Theorem 41.6). *If the metric space (M, d) is a Riemannian manifold, then the following three properties are equivalent:*

(1) The space is locally Busemann non-positively curved;
(2) capsules are convex
(3) the sectional curvatures at any point are non-positive.

6. Perpendicularity

There are large sections on perpendicularity in general metric spaces, including its relation with convexity of spheres and other important notions, in Busemann's monograph *Metric methods in Finsler spaces and in the foundations of geometry* (1942) [3], in his book with Kelly, *Projective geometry and projective metrics* (1953) [6], in his monograph *The geometry of geodesics* (1955) [7], as well as in other papers of Busemann and his collaborators. In this section, following the rule we set at the beginning of the present survey, we shall not look for generality, but stick to the notion of perpendicularity in Hilbert geometry.

The notion of perpendicularity that will be used in this paper is essentially two-dimensional. Therefore, we restrict ourselves, as in [6] and [7] to the case of convex subsets of the affine plane $\mathbb{A}^2$.

Let Ω be an open convex subset of A^2 equipped with its Hilbert metric h_Ω.

The next definitions were made by Busemann in a general metric space, see [6] and [7], but we restrict their use here to the case of the Hilbert metric h_Ω.

Definition 6.1 (Foot). Let X be a subset of Ω and let p be a point in Ω. A *foot* of p on X is a point $f \in X$ satisfying $d(p,f) \leq d(p,x)$ for all x in X.

Equivalently, f is a foot of p on X if we have $d(p,X) = d(p,f)$.

The existence and uniqueness of feet is closely related to convexity. Jessen proved in [19] that a subset M of $\mathbb{R}^n$ is closed and convex if and only if every point in $\mathbb{R}^n$ has a unique foot on M. Busemann, in [4], generalized this result to more general spaces he calls "straight line spaces." He proved the following:

(1) Consider the following property: *Every point has exactly one foot on a given closed convex set.*

A straight line space satisfies this property if and only if the spheres in that space are convex:

(2) Consider the following property: *If every point has exactly one foot on a given set, then the set is closed and convex.*

A straight line space satisfies this property if and only if its limit spheres are flat (that it, if it contains two distinct points then it contains the whole straight line joining them):

Recall that an *affine line* (or, more concisely, a *line*) in Ω is a nonempty subset which is the intersection with Ω of an affine line in A^2. By Proposition 4.1, a line in Ω is a geodesic for the metric h_Ω.

Theorem 6.2 ([3], p. 111). *In a Hilbert geometry of a strictly convex surface, every point P has a unique foot on a straight line.*

The construction is given in Figure 8, in which G is the point on the prolongation of g outside the strictly convex set Ω such that P belongs to the line that connects the two support lines from G to Ω.

Definition 6.3 (Perpendicularity). Let L and L' be two lines in Ω that intersect at a point $f \in \Omega$. The line L is said to be perpendicular to L' (and we write $L \perp L'$) if f is a foot on L' of any point on L.

Proposition 6.4 (Perpendiculars in two-dimensional Hilbert geometry, [6] §28). *Let Ω be an open convex subset of A^2 equipped with its Hilbert metric h_Ω, let L be a line in Ω and let x and y be the two intersection points of L with $\partial\Omega$. Assume that at these two points there are two support lines ξ and η to Ω that intersect at a point $w \in A^2 \setminus \Omega$. (If the lines ξ and η are parallel, we consider that their intersection point w is at infinity.) If L' be a line in Ω whose extension in A^2 passes through w, then, $L \perp L'$.*

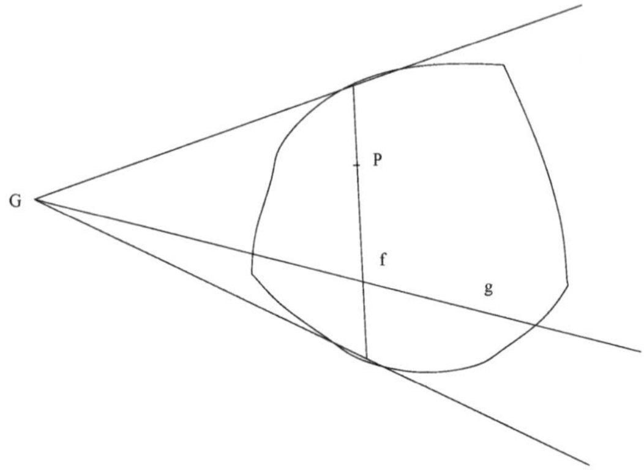

FIGURE 8. The point f is the foot of P on the line g

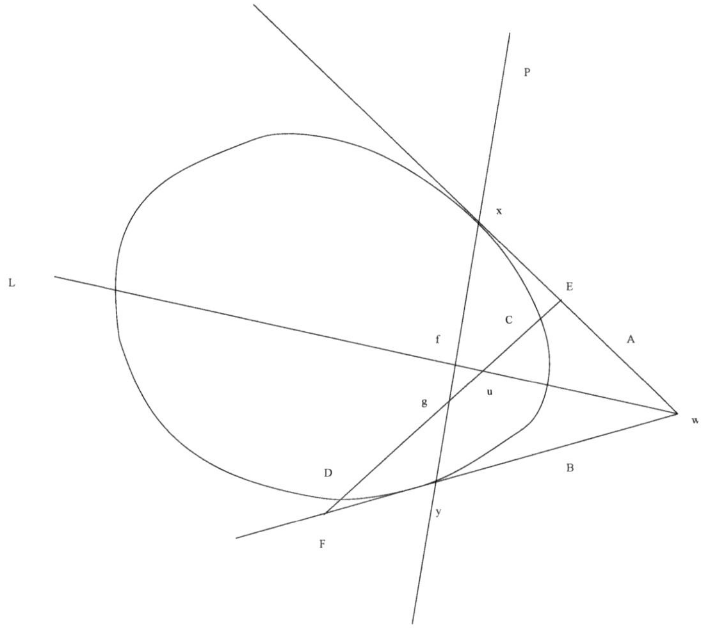

FIGURE 9. The line L is perpendicular to the line L'. (The converse is not true)

Proof: To spare notation, we denote by the same letters affine lines in A and their extensions as affine lines in A^2.

The situation is represented in Figure 9.

By elementary Euclidean geometry, the line L in A intersects the line P. Let $f = P \cap L$. We show that for all g on P with $g \neq f$ and for all $u \in L$ with $u \neq f$, we have $h_A(g, u) > h_A(g, f)$.

We assume, as in Figure 9, that the points y, f, g, x lie in that order on P.

We use the following notation:

- x' is the intersection with ∂A of the ray starting at g and containing u;

• y' is the intersection with ∂A of the ray starting at u and containing g;
• x_0 is the intersection with ξ of the ray starting at g and containing u;
• y_0 is the intersection with ξ of the ray starting at u and containing g.

The quadruple (g, f, x, y) is obtained from the quadruple (g, u, x_0, y_0) by a perspectivity of center w. Therefore,

$$[g, f, x, u] = [g, u, x_0, y_0].$$

By the monotonicity property of the cross ratio (Proposition 2.1), we have

$$[g, u, x', y'] \leq [g, u, x_0, y_0].$$

Therefore, $h_A(g, f) \leq h_A(g, u)$, which proves that $P \perp L$. □

Busemann and Kelly prove a more general result saying that such a construction gives all the perpendiculars to P (see [6] §28).

From Proposition 6.4, one can easily construct examples which show that, in general, perpendicularity between lines in a Hilbert geometry is not a symmetric relation (see [6], p. 173, Exercise 28.8). We have:

Proposition 6.5 ([6], Proposition 28.12). *If Ω is an ellipse, perpendicularity is symmetric, and two intersecting lines in Ω are perpendicular if and only if they are conjugate with respect to Ω.*

Busemann and Kelly note, as a corollary of Proposition 6.5, that the Euclidean parallel axiom (the fact that through a point there are infinitely many lines which do not intersect a given line) fails in Hilbert geometry.

Kelly and Paige show in [20] that in dimension 2, the symmetry of perpendicularity almost characterizes hyperbolic geometry among Hilbert geometries. More precisely, they prove the following:

Proposition 6.6 (Kelly and Paige, [20]). *If Ω is a convex open subset of A^2, then (Ω, h) has the symmetry of perpendicularity if and only if Ω is the interior of an ellipse or the interior of a triangle.*

Thus, the symmetry of perpendicularity in a Hilbert geometry implies that the space is standard in some sense. It is tempting to make the analogy with the situation of perpendicularity in Minkowski geometries. In both cases, Busemann shows that perpendicularity is symmetric if and only if the space is standard:

Theorem 6.7 ([7], Theorem 17.25). *If, in a Minkowski G-space of dimension greater than two, perpendicularity between lines is symmetric, then the metric is Euclidean.*

Figure 10 shows perpendicular lines in the Klein model of hyperbolic space.

The following proposition is a reformulation of Proposition 6.4 (cf. Figure 11).

Proposition 6.8 ([6] (28.11)). *In the Hilbert geometry of a strictly convex domain Ω of the plane, a line P is perpendicular to the line K if and only if Ω has, at its intersections with P, support lines concurrent with K.*

Busemann and Kelly prove in [6], §21 that for the Hilbert geometry of a strictly convex domain Ω of the plane, the existence of perpendiculars is equivalent to the fact that the circles are strictly convex.

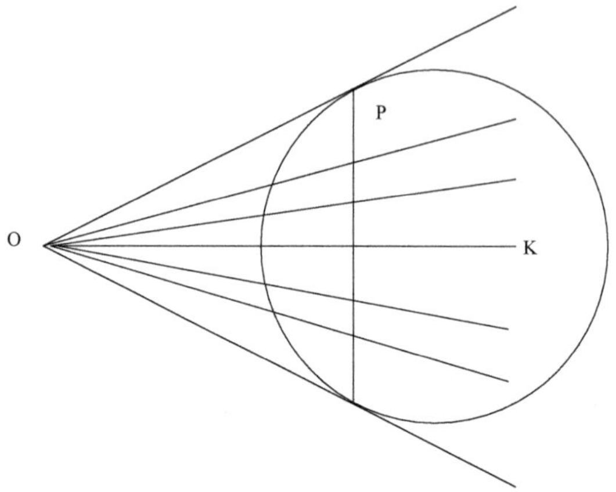

FIGURE 10.

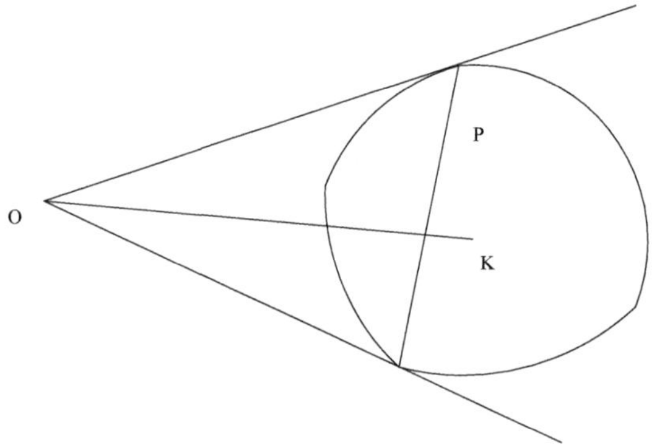

FIGURE 11.

We end this section by the following proposition generalizing a property of perpendicularity in hyperbolic geometry.

Proposition 6.9 ([6] Problem 28.7). *In a Hilbert geometry of a convex curve Ω in the plane, two lines which do not intersect in Ω or on its boundary have exactly one common perpendicular. If the intersection of the two lines is in Ω or on its boundary then they do not have a common perpendicular.*

After perpendiculars, we talk, in the next section, about asymptotes.

7. Asymptotes in Hilbert geometry

Asymptotes are defined by Buseman and Kelly in [6] as follows. Let Ω be a convex set in A^n, L an affine line in Ω and P a point in Ω which is not on L. It is clear that from P, one can always draw a line that intersects L and a line that is disjoint from it. An *asymptote* from P to L is a line that contains P and that separates the set of lines through P from those that separate P. If Ω is bounded, then from the point P there are exactly two asymptotes to L containing P (Figure 12). Asymptoticity is a symmetric relation, but it is not transitive.

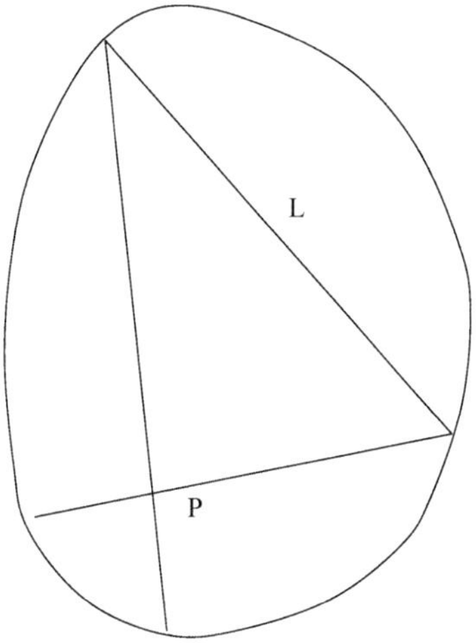

FIGURE 12. The two asymptotes to L containing P.

This notion of asymptote is defined in the same way as in hyperbolic geometry. It is easy to see that a two-dimensional Hilbert geometry shares the following property with hyperbolic geometry: the set of points at constant distance from a given straight line does not consist of two straight lines (see Buseman and Kelly, p. 172, problem 28.3).

After the above set-theoretic definition of asymptote, Busemann and Kelly give the following metric characterization for two lines to be asymptotic in Hilbert geometry:

Proposition 7.1 ([6] (28.13)). *Let L_1 and L_2 be two distinct lines in Ω. Then L_1 is asymptotic to L_2 if and only if for some variable point x on L_1, if f is the foot of x on L_2, there exists a subray of L_1 such that for x on this subray, the set of distances $\{h_\Omega(x, f)\}$ is bounded.*

Busemann and Kelly then give the following result which relates asymptoticity with differentiability.

Proposition 7.2 ([6] (28.14)). *Let L_1 be asymptotic to L_2 and let $g \in \partial\Omega$ be their common endpoint. Let x be a point on L_1 having f as foot on L_2. Then, $\partial\Omega$ is differentiable at g if and only if $h_\Omega(x, f) \to 0$ as $f \to g$.*

We mention another fact related to the structure at infinity of the Hilbert metric: In [7], Busemann gives a characterization of hyperbolic geometry among Hilbert geometries in terms of flatness of loci associated to pairs of Busemann functions associated to opposite geodesic rays. In the following statement, $\alpha(r, .)$ and $\alpha(s, .)$ are the Busemann functions associated to the geodesic rays r and s, and in this context, a subspace R of a Hilbert geometry is said to be *flat* if for any two points in R, any geodesic containing them is also in R.

Theorem 7.3 ([7], p. 404, Theorem 18). *In a Hilbert geometry, assume that for any two opposite geodesic rays (that is, rays that have only their origin in common and that lie on the same line) r and s, the locus*

$$\{x \mid \alpha(r, x) = \alpha(s, x)\}$$

is flat. Then, the metric is hyperbolic.

8. The Finsler structure, Hilbert area and Finsler volume

The formula for the Finsler norm of a Hilbert metric is given by Busemann and Kelly in [6], p. 165, although the name "Finsler norm" is not used there. We recall the main steps for this definition.

Let Ω be a bounded convex subset of the Euclidean plane with Euclidean coordinates x_1, x_2 and equipped with the distance function

$$e(x,y) = \sqrt{(x_1-y_1)^2+(x_2-y_2)^2}.$$

Let L be a line in Ω and p a point on L. The pair (L,d) is usually called a line element. To this line element, Busemann and Kelly associate the magnitude

$$\Phi(p,L) = \frac{1}{2}\left(\frac{1}{e(p,x_0)} + \frac{1}{e(p,y_0)}\right) \tag{4}$$

where x_0 and y_0 are the two intersection points of L with $\partial\Omega$. They show that for any two variable distinct points a and b tending to p in such a way that the line joining them tends to L, we have

$$\Phi(p,L) = \lim_{a,b\to p} \frac{h_\Omega(a,b)}{e(a,b)}. \tag{5}$$

Busemann and Kelly then associate to each point p in Ω a closed curve K_p defined as follows: On each line of the pencil of lines at p, two distinct points z_1 and z_2 are chosen so that $e(p,z_i) = 1/\Phi(p,L)$. The curve K_p is the locus of such points z_1 and z_2. They show the following:

Proposition 8.1 (Busemann–Kelly [6] (29.4))**.** *The closed curve K_p is convex and symmetric with respect to p. It is strictly convex if Ω is strictly convex.*

For each p, the closed curve K_p defines a Minkowski metric (a metric arising from a norm) on the affine plane (x_1,x_2), which has K_p as unit ball. More precisely, this metric, denoted by $m(x,y)$, is defined by the formula ([6] (29.6)):

$$m_p(x,y) = e(x,y)\Phi(p,L_{xy})$$

where L_{xy} is the line through p which is parallel to the affine line through x and y.

Busemann and Kelly then prove the following:

Theorem 8.2 (Busemann–Kelly [6] (29.7))**.** *The Minkowski metric m is the infinitesimal form of the Hilbert metric h_Ω. More precisely, for $a \neq b$, we have*

$$\lim_{a,b\to p} \frac{h_\Omega(a,b)}{m_p(a,b)} = 1.$$

In other words, the norm associated with the Minkowski metric m is the infinitesimal Finsler structure associated to the Hilbert metric h_Ω. Busemann and Kelly call the function $m_p(a,b)$ the *local Minkowski metric belonging to the point p*.

We also have the following ([6] Exercice 29.5):

Proposition 8.3. *Let L_1 and L_2 be two lines in a Hilbert geometry (Ω, d_Ω) intersecting at a point p. Then L_1 is perpendicular to L_2 if and only if it is perpendicular to it in the sense of the Minkowski geometry defined by the closed curve K_p.*

We now recall the definition of the *Hilbert area.*

Let $B \subset \Omega$ be a bounded region in the sense of the Hilbert metric. Subdivide B into smaller regions B_i. In each B_i choose a point p_i and to such a point associate its local Minkowski metric. The area of B_i with respect to this local metric is given by the following formula ([6] (29.8)):

$$A_{p_i}(B_i) = \sigma(p_i) \iint_{B_i} dx_1 dx_2$$

where

$$\sigma(p_i) = \pi \left(\int_{-\pi/2}^{\pi/2} \Phi^{-2}(p_i, L) d\omega \right)^{-1}$$

where ω is the angle measure in the coordinates (x_1, x_2). Then we have:

Proposition 8.4 (Busemann–Kelly [6] (29.9) and (29.10))**.** *$\sum_i A_{p_i}(B_i)$ is an approximate expression of the Hilbert area. As the number of regions B_i increases while their diameter goes to zero, the sums $\sum_i A_{p_i}(B_i)$ tend to the following limit:*

$$A(B) = \iint_B \sigma(x_1, x_2) dx_1 dx_2.$$

The magnitude $A(B)$ is, by definition, the *Hilbert area* of B, and it is invariant under isometries ([6], p. 168).

In §50 of [6], Busemann and Kelly generalize the construction of the Finsler norm associated to a Hilbert metric to the case of three dimensions. The definition of $\Phi(p, L)$ in (4) and the formula in (5) remain the same for every line L in the bundle of lines through a point p where $e = e(x, y)$ is now the three-dimensional Euclidean metric

$$e(x, y) = \sqrt{(x_1 - y_1)^2 + (x_2 - y_2)^2 + (x_3 - y_3)^2}.$$

Two points z_1 and z_2 are selected on L so that $e(p, z_1) = e(p, z_2) = \Phi(p, L)$ and the locus of z_1, z_2 is a (strictly) convex surface K_p symmetric with respect to p (cf. [6] (50.4); the analogue of Proposition 8.1 above). The analogue of Proposition 8.2 is also valid ([6] (50.4)).

Busemann and Kelly then give formulae for the Hilbert line element at a point x in the direction L ([6], (50.6):

$$ds = \Phi(x, L)(dx_1^2 + dx_2^2 + dx_3^2)^{1/2}. \tag{6}$$

In other words, the length of a C^1 curve in Ω given in coordinates as $x(t) = (x_1(t), x_2(t), x_3(t))$ is equal to

$$\int \Phi(x(t), L(t))(x_1'^2 + x_2'^2 + x_3'^2)^{1/2} dt. \tag{7}$$

Finally, Busemann and Kelly give a formula for the Hilbert volume. We refer the interested reader to [6], p. 299.

9. Isometries

Busemann and Kelly in §29 of [6] discuss isometries of the Hilbert metric. We mention the following result, which was rediscovered several years Busemann and Kelly proved it, and which is often quoted without any reference to them:

Theorem 9.1 (Busemann–Kelly [6], p. 164). *Let Ω_1 and Ω_2 be two strictly convex domains in projective space, with associated Hilbert metrics $h_1(.,)$ and $h_2(.,.)$. Suppose there is a mapping $f : \Omega_1 \to \Omega_2$ and a constant k such that $h_1(x, y) = kh_2(f(x), f(y))$ for every x and y in Ω_1. Then there exists a projective map that carries Ω_1 to Ω_2.*

Some properties that concern isometries of a Hilbert geometry are given by Busemann and Kelly as exercises in [6].

Starting from §23 of their book [6], Busemann and Kelly study motions in general metric spaces, and in particular reflections. These authors are essentially interested in metrics on subsets of two-dimensional affine space, but most of their developments apply to general metric spaces. We highlight some of their results on reflections in Hilbert geometry. In order not to be too technical, shall not be very precise for what concerns the hypotheses and we refer the interested reader to the treatise by Busemann and Kelly.

Proposition 9.2 ([6] (23.4)). *In the Hilbert geometry of a convex subset of the plane, suppose a reflection ϕ in a line L exists. Then every point a not on L has exactly one foot on this line, and this foot is the center of a and $\phi(a)$.*

Given two points $a \neq b$ in a metric space (M, d), their *bisector* is the set of points x satisfying $d(x, a) = d(x, b)$.

Proposition 9.3 ([6] (23.5)). *Suppose a reflection ϕ in a line L exists. Then L is the bisector of a and $\phi(a)$. Furthermore, perpendiculars to L exist and ϕ maps each perpendicular to L to itself.*

Proposition 9.4 ([6] (23.8)). *If a motion ϕ which is not the identity acts as the identity on a line L, then ϕ is the reflection in L.*

Proposition 9.5 ([6] (23.9)). *If L and L' are two lines, if the circles in (Ω, d_Ω) are convex and the reflection in L exists, then L' is perpendicular to L if and only if L is perpendicular to L'.*

Busemann and Kelly then prove the following, in dimension two:

Theorem 9.6 ([6] (29.3)). *The domain Ω is the interior of an ellipse if and only if its Hilbert geometry admits reflections about all lines.*

They also prove the following, for any $n \geq 2$:

Theorem 9.7 ([6] (50.1)). *For any bounded open convex set Ω in $\mathbb{R}^n$ ($n \geq 2$) equipped with its Hilbert metric, reflections in all the hyperplanes in Ω through one fixed point exist and are isometries if and only if Ω is enclosed by an ellipsoid (and, thus, its Hilbert metric is a model of hyperbolic geometry).*

In his 1970 monograph [10] (p. 38), Busemann proves the following theorem which gives a joint characterization of Hilbert geometries and (non-necessarily symmetric) Minkowski geometries among Desarguesian spaces in terms of isometries:

Theorem 9.8 ([10], p. 38). *Among all noncompact and not necessarily symmetric Desarguesian spaces with compact spheres $B^{\pm}(p,\rho)$, the Hilbert and Minkowski geometries are characterized by the property that an isometry between any two geodesics is a projectivity.*

In this statement, $B^{\pm}(p,\rho)$ are the left and right closed balls with center p with radius ρ. We recall that in the setting of a non-symmetric metric, one has to distinguish between left and right spheres:

$$B^{+}(p,\rho) = \{x \mid d(p,x) \leq \rho\}$$

and

$$B^{-}(p,\rho) = \{x \mid d(x,p) \leq \rho\}.$$

Talking about non-symmetric metrics, we note that Busemann studied in some detail the so-called Funk metric, a non-symmetric version of the Hilbert metric; see Busemann's monograph [10], his article [11] and the survey [8].

References

[1] G. Birkhoff, Extensions of Jentzsch's theorem. Trans. Amer. Math. Soc. 85 (1) (1957) 219–227.

[2] T. Bonnesen and W. Fenchel, Theorie der konvexen Körper, Julius Springer, 1934, several later editions. English translation, BCS Assciates, Moscow, Idaho USA, 1987.

[3] H. Busemann, Metric methods in Finsler spaces and in the foundations of geometry, Annals of Mathematical tudies, No. 8, Princeton University Press, 1942.

[4] H. Busemann, Note on a theorem on convex sets, Matematisk Tidsskrift. B, p. 32-34 (1947).

[5] H. Busemann and P. J. Kelly, Projective geometry and projective metrics, Academic Press, New York, 1953.

[6] H. Busemann, *The geometry of geodesics*, Academic Press (1955), reprinted by Dover in 2005.

[7] H. Busemann, On Hilbert's fourth problem. (Russian) Uspehi Mat. Nauk 21 (1966) no. 1 (127), 155–164 (1966). Translated from the Russian by D. Slutskiy, Selected Works, Herbert Busemann, Vol. II, p. 629–638.

[8] H. Busemann, Timelike spaces. Dissertationes Math. Rozprawy Mat. 53 (1967) 52 pp.

[9] H. Busemann, Recent synthetic differential geometry. Berlin-Heidelberg-New York: Springer-Verlag 1970.

[10] H. Busemann, Spaces with homothetic spheres. J. Geometry 4, 175–186 (1974).

[11] H. Busemann and B. B. Phadke, Minkowskian geometry, convexity conditions and the parallel axiom. J. Geom. 12 (1979), no. 1, 17-33.

[12] H. Busemann and B. B. Phadke, A general version of Beltrami's theorem in the large. Pac. J. Math. 115, 299-315 (1984).

[13] H. Busemann and B. B. Phadke, Spaces with distinguished geodesics. Monographs and Textbooks in Pure and Applied Mathematics, 108. Marcel Dekker, Inc., New York, 1987.

[14] H. G. Eggleston, *Convexity*, Cambridge Tracts in Mathematics and Mathematical Physics No. 47, Cambridge University Press, 1958.

[15] D. Hilbert, Über die gerade Linie als kürzeste Verbindung zweier Punkte, *Math. Ann.*, 46 (1895), 91-96.

[16] D. Hilbert, Foundations of geometry. Second edition. Translated from the tenth German edition by Leo Unger Open Court, LaSalle, Ill. 1971.

[17] D. Hilbert, Mathematische Probleme, *Göttinger Nachrichten*, 1900, pp. 253–297, reprinted in *Archiv der Mathematik und Physik*, 3d. ser., vol. 1 (1901) pp. 44–63 and 213–237. English version, "Mathematical problems", reprinted also in the *Bull. Amer. Math. Soc.* (N.S.) 37 (2000), no. 4, 407-436.

[18] B. Jessen, To Sætninger om konvekse Punktmængder, Matematisk Tidsskrift. B, pp. 66-70 (1940).

[19] P. J. Kelly and L. J. Paige, Symmetric perpendicularity in Hilbert geometries, Pacific Jour. of Math. 2 (1952), p. 319-322.

[20] B. B. Phadke, A triangular world with hexagonal circles. Geometriae Dedicata 3 (1974/75), 511–520.

[21] P. J. Kelly and E. Straus, Curvature in Hilbert geometries. *Pacific J. Math.* 8 (1958), 119–125.

[22] A. Papadopoulos, On Hilbert's Fourth Problem, Handbook of Hilbert geometry, (ed. A. Papadopoulos and M. Troyanov), European Mathematical Society, Zürich, 2014, 375-431.

[23] A. Papadopoulos, Busemann's work on the characterization of Minkowski geometries, Selected Works, Herbert Busemann, Vol. II, p. 145–159.

[24] H. Samelson, On the Perron–Frobenius theorem, Michigan Math. J., 4 (1957) 57– 59.

[25] F. A. Valentine, Convex Sets. McGraw-Hill, New York 1964.

Athanase Papadopoulos, Université de Strasbourg and CNRS, 7 rue René Descartes, 67084 Strasbourg Cedex, France

E-mail address: athanase.papadopoulos@math.unistra.fr

BUSEMANN'S WORK ON THE CHARACTERIZATION OF MINKOWSKI GEOMETRIES

ATHANASE PAPADOPOULOS

Contents

1. Introduction 145
2. G-space, straight space, Desarguesian space 147
3. Limit sphere, parallelism and asymptotes 149
4. Minkowski geometry 151
5. On the characterization of Minkowski geometries 153
6. Non-symmetric Minkowski geometries 157
References 158

1. Introduction

Hilbert's fourth problem, from the collection of problems he presented at the occasion of the Second International Congress of Mathematicians, held in Paris (1900), is entitled: *Problem of the straight line as the shortest distance between two points* [27]. The title gives an idea of the topic, and Hilbert's comments show that this problem has many facets. In particular, it is not limited to the question of characterizing the metrics on subsets of Euclidean space for which Euclidean straight lines are geodesics. One of the questions raised by this problem asks for the characterization of Minkowski geometries (metrics induced by a norm on a finite-dimensional vector space). Another aspect of this problem is the inclusion of the triangle inequality – which, in the classical literature, is a theorem – as an axiom. Let us quote part Hilbert (the translation is from the article published in the 1902 in the Bulletin of the AMS [27]):

> [...] We are asking then, for a geometry in which all the axioms of ordinary Euclidean geometry hold, and in particular all the congruence axioms except the one of the congruence of triangles (or all except the theorem of the equality of the base angles in the isosceles triangle), and in which, besides, the proposition that in every triangle the sum of two sides is greater than the third, is assumed as a particular axiom.
>
> One finds that such a geometry really exists and is no other than that which Minkowski constructed in his book, Geometrie

> der Zahlen,[1] and made the basis of his arithmetical investigations. Minkowski's is therefore also a geometry standing next to the ordinary Euclidean geometry; it is essentially characterized by the following stipulations:
>
> 1. The points which are at equal distances from a fixed point O lie on a convex closed surface of the ordinary Euclidean space with O as a center.
>
> 2. Two segments are said to be equal when one can be carried to the other by a translation of the ordinary Euclidean space.
>
> In Minkowski's geometry the axiom of parallels also holds. By studying the theorem of the straight line as the shortest distance between two points, I arrived[2] at a geometry in which the parallel axiom does not hold, while all other axioms of Minkowski's geometry are satisfied. The theorem of the straight line as the shortest distance between two points and the essentially equivalent theorem of Euclid about the sides of a triangle, play an important part not only in number theory but also in the theory of surfaces and in the calculus of variations. For this reason, and because I believe that the thorough investigation of the conditions for the validity of this theorem will throw a new light upon the idea of distance, as well as upon other elementary ideas, *e.g.*, upon the idea of the plane, and the possibility of its definition by means of the idea of the straight line, *the construction and systematic treatment of the geometries here possible seems to me desirable.*[3]

Busemann, in his review of Hilbert's problem in [13], has a section on Minkowski geometries. A.V. Pogorelov, in his monograph [33] in which he accounts for his solution of Hilbert's fourth problem, notes that the Minkowski geometries are those that realize the system of axioms of Euclidean geometry after we drop from them all the axioms of congruence involving the concept of angle and we add instead the triangle inequality as an axiom. This is in the spirit of Hilbert's Problem. Pogorelov also notes that if instead of modifying in this way the axioms of Euclidean geometry, we modify in the same manner those of Lobachevsky geometry, then we get the Hilbert geometries. In his report [17] on Hilbert's problem, Busemann writes:

> The fourth problem concerns the geometries in which the ordinary lines, i.e. lines of an n-dimensional (real) projective space $\mathbb{P}^n$ or pieces of them, are the shortest curves or geodesics. Specifically, Hilbert asks for the construction of these metrics and the study of the individual geometries. It is clear from Hilbert's comments that he was not aware of the immense number of these metrics, so that the second part of the problem is not at all well posed and has inevitably been replaced by the investigation of special, or special classes of, interesting geometries.

---

[1] Leipzig, 1896.

[2] *Math. Annalen*, vol. 46, p. 91.

[3] The emphasis is Hilbert's.

Minkowski geometry is one such class.

Thus, the question of studying the class of Minkowski geometries is considered in more than one way as part of Hilbert's problem.

Hilbert metrics constitute another class of geometries that are solutions of Hilbert's Fourth Problem to which Busemann also dedicated a large place in his work; see the report [29] in the present volume.

Finally, we note that besides his interest in Hilbert's Problem IV, Busemann was interested in Minkowski geometries as part of his work on the metric foundations of Finsler geometry, another topic which haunted him during his lifetime. Minkowski metrics play, in Finsler spaces, the role that Euclidean metrics play in Riemannian manifolds: they are the infinitesimal models of these spaces.

Our aim here is to highlight some of Busemann's work on Minkowski geometry.

The content of the rest of this article is the following:

In §2, we recall the notions of G-space, straight space and Desarguesian space, which are essential in Busemann's work and which will be used throughout the survey. In §3, we recall the notions of limit sphere, parallelism and asymptotes. The material in this section is developed in §22 and 23 of Busemann's *Geometry of geodesics* [11]. In §4, we recall the notions that form the bases of Minkowski geometry. §5 is the longest section of this article. We survey there several results of Busemann on the characterization of Minkowski geometry. §6 concerns non-symmetric Minkowski geometries.

2. G-space, straight space, Desarguesian space

The notion of G-space, like most of the other notions that we introduce in this section, is already contained in Busemann's early book [4] (1942) and in his paper [5] (1943). We shall nevertheless rely on Busemann's 1955 exposition in his monograph *The geometry of geodesics* [11] which is the standard reference for the subject.

Let (X, d) be a metric space. Following Busemann, for any ordered triple of points x, y, z in X, we shall write (xyz) and we shall say that *y lies between x and z* if the three points are distinct and satisfy $d(x, y) + d(y, z) = d(x, z)$.

In the metric space (X, d), a theory of length of curves (i.e. continuous images of a compact intervals) is developed, as in Euclidean space: the length of a curve is defined as the supremum over all lengths of polygonal curves joining the endpoints of the curve. Length satisfies the usual additivity and lower-semicontinuity properties, and the length of a curve is bounded below by the distance between its endpoints. This is contained in [11], Chapter 1, §6. The length of a curve c is denoted by $\lambda(c)$.

If x and y are two points in X, a *segment* joining them is a continuous map c from a closed interval $[a, b]$ of $\mathbb{R}$ into X satisfying $c(a) = x$, $c(b) = y$ and $\lambda(c) = d(x, y)$.

A *straight line* in X is an isometric map from $\mathbb{R}$ into X, that is, a map $g : \mathbb{R} \to X$ satisfying $d(a, b) = |a - b|$ for every a and b in X.

A *geodesic* in X is a locally isometric map from $\mathbb{R}$ into X.

Definition 2.1 (G-space; [11] p. 37). A G-space is a metric space (X,d) satisfying the following properties:

(1) (Finite compactness)[4] Every bounded infinite set has an accumulation point.
(2) (M-convexity)[5] For any x and z in X, there exists a point y such that (xyz).
(3) (Local extendability) For every point p of X there exists an open ball $B(p,\rho_p)$ of center p and radius $\rho_p > 0$ ($\rho = \infty$ is allowed) such that for every x and y in $B(p,\rho_p)$ there exists a point z in $B(p,\rho_p)$ such that (xyz).
(4) (Uniqueness of extension) If (xyz_1) and (xyz_2) and if $d(y,z_1) = d(y,z_2)$ then $z_1 = z_2$.

The following basic properties of G-spaces are proved by Busemann in the first chapter of [11].

(1) In a finitely compact M-convex metric space, any two points can be joined by a segment ([11], 6.8). Thus, the same property holds in a G-space.
(2) In a G-space, every segment can be extended to a geodesic ([11], 7.9).

Definition 2.2 (Straight space; [11] p. 37). A *straight space* is a G-space such that in Property (3) of Definition 2.1, we have $\rho_p = \infty$ for each p in X.

In a straight space, any segment can be extended to a straight line ([11], 7.10). By property (4) of Definition 2.1, this extension is unique. Busemann proves in [11] (7.5) that in a straight space, all geodesics are straight lines.

Definition 2.3 (Desarguesian space [11] p. 65 and [20]). A Desarguesian space is a metric space (X,d) satisfying the following properties:

(1) X is an open nonempty subset of the projective space $\mathbb{P}^n$ equipped with a metric d whose associated topology is the one induced from its inclusion in $\mathbb{P}^n$.
(2) The closed balls in X are compact.
(3) Any two points in X can be joined by a geodesic segment, that is, a curve of length (x,y).
(4) For any three points x,y,z in $X \subset \mathbb{P}^n$ that are not collinear, the strict triangular inequality $d(x,y)+d(y,z) < d(x,z)$ holds; that is, the three points do not belong to a geodesic.

Busemann, in his survey [13] on Hilbert's problem and in his book with Phadke [24], declares that Minkowski spaces are by far the most important Desarguesian spaces. If its spheres are strictly convex, a Minkowski space is a G-space. If its spheres are not strictly convex, a Minkowski space is an important example of chord space. The theory of chord spaces is developed in [24].

---

[4] Busemann attributes this terminology to Morse.

[5] The letter M stands for Menger.

3. Limit sphere, parallelism and asymptotes

The definition of limit spheres and the theory of parallels, as developed by Busemann in [11], are based on the introduction of a class of functions which were called later on *Busemann functions*. We recall this definition.

Let (X, d) be a metric space.

A *ray* r in (X, d) is a "half-straight line," that is, an isometric map $r : [0, \infty[\to X$. (We shall sometimes call *ray* the image of such a map).

To a ray r and a point p in X, Busemann associated the function

$$\alpha(r, p) = \lim_{t \to \infty} (d(p, r(t)) - t).$$

For any r and p, this limit exists and is finite.

The function $\alpha(r, p)$ is continuous in p. In fact we have

$$|\alpha(r, p) - \alpha(r, q)| \leq d(p, q)$$

for any p and q in X.

Given two rays r_1 and r_2 in X and a point p in X, we define the *distance between* r_1 *and* r_2 *at infinity* as the quantity

$$\delta_\infty(r_1, r_2) = \liminf_{\substack{t_1 \to \infty \\ t_2 \to \infty}} d(r_1(t_1), r_2(t_2)).$$

For any two rays r_1 and r_2 in X and for any two points p and q in X, we have ([11] p. 131):

$$|\alpha(r_1, p) + \alpha(r_2, q) - \alpha(r_1, q) - \alpha(r_2, p)| \leq \delta_\infty(r_1, r_2).$$

Busemann also proves the following ([11] p. 132):

If $\delta_\infty(r_1, r_2) = 0$ then $\alpha(r_1, q) - \alpha(r_2, p)$ is a constant.

Definition 3.1 (Limit sphere [11] p. 132)**.** If p is a point in X and r a ray in X, then a subset of X defined by

$$K_\infty(r, p) = \{p \in X | \alpha(r, p) \ = \ \text{constant}\}$$

is called the *limit sphere through* p *with central ray* r.

For any pair of points p and q in X, we denote by $K(p, q)$ the sphere with center p passing by q, that is,

$$K(p, q) = \{x \in X | d(p, x) = d(p, q)\}.$$

In a straight space, a limit sphere is always a limit of spheres (cf. Busemann [11] p. 134). More precisely, if (X, d) is a straight space, if $r : [0, \infty[\to X$ is a ray in X, if $t_n \to \infty$ and $p_n \to p$, then we have

$$\lim K(r(t_n), p_n) = K_\infty(r, p).$$

In order to introduce the theory of parallelism, we first need to talk about oriented geodesics and asymptotes. Depending on whether a geodesic g (identified with its image in X) is traversed in the sense of its parametrization or in the opposite sense, we shall talk about the geodesic g^+ or g^-, and we shall consider it as *an oriented geodesic*. We shall also talk about *oriented straight lines*, or, more simply, *oriented lines* or *lines* (if the orientation is understood).

We say that a sequence g_n^+ of oriented straight lines converges to an oriented straight line g^+ if, in terms of suitable parametrizations, we have $g_n(t) \to g(t)$ for any $t \geq 0$.

Busemann proves the following:

Theorem 3.2 (Busemann [11] Theorem 23.2). *Let (X, d) be a straight space, g^+ an oriented line in X, q_n a sequence of points converging to a point q and t_n a sequence of positive numbers converging to infinity. Then the line $g^+(q_n, x(t_n))$ so oriented that $x(t_n)$ follows q_n converges to an oriented line a^+ which is independent of the choice of the sequences q_n and t_n.*

Definition 3.3 (Asymptote). The line a^+ in Theorem 3.2 is called the *asymptote* to g^+ through q.

Busemann proves the following complements to Theorem 3.2:

Theorem 3.4 (Busemann [11] Theorem 23.2). *The asymptote to g^+ through any point q' on a^+ coincides with a^+. Furthermore, the feet of q on the limit spheres $K_\infty(r, x)$ are unique and a^+ is their locus.*

In other words, the line a is perpendicular to all the limit spheres $K_\infty(r, x)$ where r is a positive subray of g^+. We recall the notion of perpendicularity in metric spaces, used by Busemann's in his book [11] and in other papers and books, since this notion will be used below.

Let X be a subset of a metric space (M, d) and let p be a point in M. We say that a point f in X is a *foot* of p on X if we have $d(p, f) \leq d(p, x)$ for all x in X. Let L be a line in M that intersects X at a point f. Then L is said to be *perpendicular* to X if f is a foot on X of any point on L.

Theorem 3.5 (Busemann [11] Theorem 23.3). *In a straight Desarguesian space, asymptotes have the following two properties:*

(1) (Symmetry) If a^+ is asymptotic to b^+, then b^+ is asymptotic to a^+.
(2) (Transitivity) If a^+ is asymptotic to b^+ and b^+ is asymptotic to c^+, then a^+ is asymptotic to c^+.

Theorem 3.6 (Busemann [11] Theorem 23.4). *(1) In a straight space, the transitivity property of asymptotes implies the symmetry property.*

(2) In a two-dimensional straight space, transitivity and symmetry are equivalent.

Busemann gave examples that show that neither symmetry nor transitivity hold in general in straight spaces.

This notion of asymptoticity is a generalization of the notion of parallelism in hyperbolic geometry. In Euclidean geometry, this notion coincides with the usual notion of parallelism.

Finally, the *Parallel Axiom* is stated as follows:

The asymptotes have the symmetry property and furthermore, if a^+ is asymptotic to g^+ then a^- is asymptotic to g^-.

Busemann proves:

Theorem 3.7 (Busemann [11] Theorem 23.7). *In a straight two-dimensional space, the parallel Axiom holds if and only if for any line g and for any point p not on g, there exists exactly one line through p which does not intersect g.*

4. Minkowski geometry

In §17 of his *Geometry of geodesics* [11], Busemann makes the following definition:

Definition 4.1 (Minkowski metric). A metric d on the affine space A^n is called *Minkowskian* if its restriction to each affine line of A^n is equivalent to the natural topology induced by A^n and if the following property holds: for any two points x, y in A^n, if $z = (x+y)/2$ then $d(x,z) = d(z,y) = d(x,y)/2$

Another way of formulating the property in the preceding definition is to say that any affine midpoint of two points is their metric midpoint.

Note that only Minkowski spaces whose associated metrics are symmetric (that is, $d(x,y) = d(y,x)$ for every x and y) are considered in [11]. In some other publications, Busemann considers non-symmetric Minkowski spaces.

Busemann in [11] gives several properties and equivalent definitions which show that the notion of Minkowski metric coincides with the familiar notion of finite-dimensional normed vector space. We mention the following

Theorem 4.2 ([11], Theorem 17.9). *A Minkowski metric on A^n is finitely compact and M-convex. Prolongation in property (3) of a G-space (Definition 2.1) is possible in the large, that is, $\rho_p = \infty$ for all p. Furthermore, all spheres $K(p,\rho)$ of center p and radius ρ for $\rho > 0$ are convex and homothetic as subsets of A^n, and they have p as their affine center. The metric space (A^n, d) is a G-space (and therefore straight) if and only if the spheres are strictly convex.*

The following theorem is a form of the so-called Borsuk-Ulam theorem:

Theorem 4.3 ([11], Theorem 17.10). *Two Minkowski metrics on A^n are isometric if and only if for some point $p \in A^n$ there exists an affine map of A^n which maps a sphere of center p and radius 1 for one metric to a sphere of center p and radius 1 for the other metric.*

The following result makes the relation with a metric associated with a norm:

Theorem 4.4 ([11], Theorem 17.17). *A Minkowski metric $d(x,y)$ on A^n with affine coordinates $x_1, \ldots, x_n$ is of the form $d(x,y) = F(y-x)$ where*

(1) $F(x) > 0$ for any $x > 0$.
(2) $F(\mu x) = |\mu|$ for any real number μ;
(3) $F(x)$ is convex.

Conversely, any function F satisfying these three properties has an associated metric $d(x,y) = F(y-x)$ which is Minkowski. The affine space A^n associated with this metric is a G-space if and only if the function $F(x)$ is strictly convex.

Using the notion of G-space and Desarguesian space, Busemann gave two intrinsic characterizations of finite-dimensional Minkowski spaces among abstract metric spaces. The following result is a characterization of Minkowski geometries among Desarguesian spaces:

Theorem 4.5 ([11], Theorem 24.1). *A metric space (X,d) is isometric to a Minkowski space if and only if it is a Desarguesian space in which the parallel axiom holds and the spheres are strictly convex.*

The second result involves the notion of *Busemann zero curvature.* We recall that a geodesic metric space is said to have *zero curvature in the sense of Busemann* if the distance between the midpoints of two sides of an arbitrary triangle is equal to half the length of the remaining side. Busemann formulates the following:

Theorem 4.6 ([11], Theorem 39.12). *A simply connected finite-dimensional G-space of zero curvature is isometric to a Minkowski space.*

The fact that a Minkowski metric on A^n is Euclidean if and only if its spheres are ellipsoids (Theorem 17.11 of [11]) can be taken as the definition of a Euclidean metric.

Busemann shows (cf. [11] p. 102) that in a Minkowski space, if $x(t)$ is a straight line, then the sphere $K(x(t), x(0))$ of center $x(t)$ passing through $x(0)$ converges as $t \to \infty$ to a limit sphere. He also proves the following:

Theorem 4.7 ([11], Theorem 17.21). *In a Minkowski G-space, the tangent cones of the spheres are the limit spheres. Furthermore, the limit spheres are hyperplanes if and only if the spheres are differentiable.*

The following theorem is attributed by Busemann to Blaschke [1]:

Theorem 4.8 ([11], Theorem 17.25). *If, in a Minkowski G-space of dimension greater than two, perpendicularity between lines is symmetric, then the metric is Euclidean.*

This result may be compared with a result by Kelly and Paige saying that in dimension 2, the symmetry of perpendicularity in Hilbert geometry implies that the space is standard in some sense (see [29] for a report on Hilbert geometry in Busemann's work):

Proposition 4.9 (Kelly and Paige, [28]). *If Ω is a convex open subset of A^2, then (Ω, h) has the symmetry of perpendicularity if and only if Ω is the interior of an ellipse or the interior of a triangle.*

Finally, we mention that Section 25 of Busemann and Kelly's book [21] is concerned with reflections in Minkowski plane geometry. The authors obtain the following

Proposition 4.10 ([21] (25.2)). *In a Minkowski geometry, if the reflection in a line L exists, then reflections in all lines parallel to L exist.*

Proposition 4.11 ([21] (25.2)). *In a Minkowski geometry, reflection in a line L exists if and only if in every circle of center L there exists a family of parallel chords bisected by L.*

Proposition 4.12 ([21] (25.3)). *A Minkowski geometry admits a reflection in every line if and only if the circles are ellipses.*

The Minkowskian geometries satisfying the preceding proposition are the Euclidean ones. Busemann and Kelly prove the following special case of the Borsuk-Ulam theorem (cf. Theorem 4.3):

Proposition 4.13 ([21] (25.6)). *All Euclidean geometries are congruent.*

In the book [11], Theorem 52.7, Minkowski planes, together with some other surfaces, are characterized by the fact that they are the only G-spaces of dimension 2 having transitive groups of motions. The other surfaces are: the quasi-hyperbolic metric (see the paper [12]), the cylinder and torus with Minkowskian metrics, and the sphere and projective plane with a spherical metric.

Busemann's paper [6] concerns the isoperimetric problem in Minkowski geometry: to find, among all given simple closed curves of a given length those that bound a disc of greatest area. In this paper, the notion of mixed area as well as the Brunn-Minkowski theory are used. The paper [9] is concerned with the isoperimetric problem in Minkowski geometry of higher dimensions. Busemann studies there the question of finding among all closed surfaces in a certain class and with a given area those which bound a maximal volume. In general, the solution is not a Minkowski sphere but a new convex surface which plays an important role in the theory of Finsler spaces. In the first part of the paper [10], Busemann develops the general theory of Minkowski spaces as local models for Finsler spaces. He studies in particular the so-called Loewner ellipsoid, that is, the ellipsoid of smallest Minkowskian volume containing the Minkowski unit sphere. Using this ellipsoid, he introduces a natural Euclidean metric on the given Minkowskian space. He then proposes a generalization of the sine function which is defined in this setting on pairs of flats and he uses this function to study the notion of normality of linear subspaces. In the second part of the same paper, Busemann studies the differential geometry of curves and surfaces, introducing for them a notion of curvature adapted to this Minkowskian setting.

5. On the characterization of Minkowski geometries

The problem of providing axioms for Minkowski geometries is the main subject of Busemann's early papers [2] and [3]. We report on these two papers in the article [30]. In the book [11], this subject is thoroughly considered in §§24 and 25. Busemann gave several variations and improvements of the result of [3], and we shall mention them in what follows. He proves:

Theorem 5.1 (Theorem 24.1 of [11]). *A Desarguesian space in which the parallel axiom holds and the spheres are convex is a Minkowski space.*

In this statement, unlike in the early papers [2] and [3], there is no mention of the differentiability of the spheres and there is no restriction on the dimension. Busemann states in [11], after Theorem 24.1 (p. 146):

> Theorem (24.1) shows that the main problem is to find conditions under which a straight space with the parallel axiom and convex spheres is Desarguesian. It may be that no additional conditions are necessary but this seems hard to prove. In higher dimensions we will see that it suffices to add a condition, which in smooth spaces amounts to differentiability of the spheres.

In the paper [11], the hypothesis on the differentiability of spheres is used to insure the differentiability of the spheres in the resulting Minkowski space. Busemann proves:

Theorem 5.2 (Theorem 24.6 of [11]). *A straight space of dimension ≥ 3 in which the spheres are convex and differentiable and the parallelism axiom holds is a Minkowski space with differentiable spheres.*

Busemann says in [11] (beginning of §25) that it is an open question whether Theorem 5.2 holds or not in dimension two.

A subspace K of a metric space X is called *flat* if for any two distinct points in K, the entire line joining them is contained in K. Busemann proves the following:

Theorem 5.3 (Theorem 24.7 of [11]). *A straight space of dimension ≥ 3 is a Minkowski space with differentiable spheres if and only if the limit spheres are flat.*

Theorem 5.2 is obtained in [11] as a consequence of Theorem 5.3.

Busemann noticed that the existence and uniqueness of feet of points on closed convex subsets is closely related to the convexity of spheres, he proves:

Theorem 5.4 (Theorems 24.10 and 24.12 of [11]). *A straight space of dimension ≥ 3 is Euclidean if and only if it satisfies the parallel axiom and the following condition: if L is a straight line, if p a point in X which is not on L and if f is a foot of p on L, then f is the foot on the line connecting p and f of any point x on L.*

Busemann proved the following result:

Theorem 5.5 ([7] and Theorem 24.14 of [11]). *Among the straight spaces of dimension ≥ 3, Minkowski spaces with differentiable spheres are characterized by the property that a non-empty set on which every point has exactly one foot is convex and closed.*

The following result gives a characterization of Minkowski geometry in terms of the convexity of the α-th power of the distance function. It is a revised version of a theorem in [3] whose statement contains an additional differentiability condition.

Theorem 5.6 (Theorem 25.6 of [11]). *A two-dimensional straight space (X, d) is Minkowskian if and only if it satisfies the parallel axiom and if there exists $\alpha \geq 1$ such that for any p, x, y in X, we have*

$$d(p,y)^\alpha \leq \frac{1}{2}(d(p,x)^\alpha + d(p,z)^\alpha)$$

where y is the midpoint of x and z.

Busemann also provides characterizations of Minkowski spaces that use the notion of maximal displacement of motions. We recall that a point z in a metric space (E, d) is said to have maximal displacement with respect to a motion Φ if

$$d(z, \Phi(z)) = \sup_{x \in E} d(x, \Phi(x)).$$

Theorem 5.7 (Theorem 50.5 of [11]). *If a straight space possesses a transitive group of motions and for each such motion points with maximal displacement exist, then the space is Minkowskian.*

The next characterization of Minkowski spaces involves the area of triangles.

In the Euclidean plane, the area of a triangle is the product of the base with half of the altitude. This property does not characterize Euclidean geometry, but it characterizes Minkowskian geometries with symmetric perpendicularity. The reference is to a notion of area (and more generally of volume) in Minkowski spaces that is thoroughly studied by Busemann in [11]. The statement is the following:

Theorem 5.8 (Theorem 50.9 of [11]). *A straight space possesses the property that a triangle ABC has area which depends only on the distances BC and the distance from A to the segment BC if and only if the space is a Minkowski plane with symmetric perpendicularity.*

A result of the same kind is also proved in [8].

The next theorem gives a common characterization of Minkowskian and hyperbolic planes:

Theorem 5.9 (Theorem 51.5 of [11]). *If in a straight plane R there exists two lines g and h with h not asymptotic to g^+ or g^- and such that all translations along g and h exist as motions of the space, then R is either Minkowskian or hyperbolic.*

In §30 of [11], Busemann introduces *locally Minkowskian* spaces as G-spaces in which every point has an open disk neighborhood which is isometric to an open disk in a Minkowski space. He proves (Theorem 30.1) that the universal covering of a locally Minkowskian space is Minkowskian. He provides several characterizations of *locally* Minkowski geometries, after having introduced a notion of curvature in terms of angle excess of triangles in a G-space. He introduces spaces of positive, nonpositive, negative, nonnegative, and zero curvature using the so-called Busemann property for midpoints of triangles. In particular, a straight space (X, d) is said to be of zero curvature in the sense of Busemann if and only if for any triple of points a, b, c , if m is the midpoint of ab and n the midpoint of ac, we have $d(m, n) = d(b, c)/2$. He shows (Theorems 36.2 and 39.12 of [11]) that spaces of curvature zero under any of these notions are locally Minkowskian.

In the paper [22], Busemann and Phadke consider again the problem of characterizing Minkowski geometry, and more specifically, they ask whether the assumption that the space is Desarguesian is necessary or not. They declare that this question has remained unsolved for over forty years. The goal in their paper is to replace the fact that a space is Desarguesian by a property which is stronger than the convexity of spheres. They derive such a condition from functions which measure the distance between lines and points on lines. They prove the following:

Theorem 5.10 (Theorem 3 of [22]). *A straight space (X, d) is a Minkowskian space with differentiable spheres if and only if it satisfies the following condition:*

For any pair of points $a \neq b$, if m is the midpoint of a, b then for p traversing a ray not on $L(a, b)$ with origin m, we have

$$\lim_{d(p,m)\to\infty} (2d(p, m) - d(p, a) - d(p, b)) = 0.$$

In the book [15], written 15 years after [11], Busemann updates his results and moreover, in some sections of the book, he considers the setting of non-symmetric metrics. At several places, he addresses the question of characterizing Minkowski geometry. Theorem 13 (p. 32) in that book gives a condition under which the universal covering of a G-space that possesses local similarities is Minkowskian. Busemann writes in [15] p. 25 that "one would expect that a G-space is Minkowskian if similarities exist for all factors k. Actually, the space may have infinitely many points each of which is the fixed point of similarities with given k without being Minkowskian." Rather, he proves the following:

Theorem 5.11 (Theorem 13 of [15]). *Let R be a G-space with a local similarity ϕ with factor $k \neq 1$, and assume that R is continuously differentiable at a fixed point of ϕ. When $k < 1$ the R is Minkowskian. When $k > 1$ then the universal covering space of R is Minkowskian.*

Theorem 5.12 (Theorem 14 of [15]). *If a straight space possesses a similarity with factor $\neq 1$ which is continuously differentiable at a fixed point, then the space is Minkowskian.*

In the paper [18] Busemann gives a characterization of Minkowski planes by the Desarguesian spaces which satisfy a property on angle bisectors which is defined in a purely metrical way, avoiding angular measures. The idea is based on the fact that an angle bisector in a Euclidean triangle is a segment which joins the given vertex to a point on basis which divides it by a ratio equal to the ratio of the two legs of the triangle. More precisely, the property is the following:

> *In a Desarguesian plane with distance function d, inside a (nonstraight) convex angle with legs N_1, N_2 and vertex v there is a ray M with origin v such that any segment $[a_1, a_2]$ with $a_i \in N_i, a_i \neq v$, intersects M in a point $b = b(a_1, a_2)$ such that we have*
>
> $$\frac{d(v, a_1)}{d(v, a_2)} = \frac{d(b, a_1)}{d(b, a_2)}.$$

Busemann proves the following:

Theorem 5.13. *The above property characterizes the Minkowski planes among the Desarguesian planes. Furthermore, if we assume that in this geometry the circles are differentiable then the property characterizes the Minkowski planes among all the two-dimensional straight G-spaces.*

The proofs use the parallel axiom and the convexity of circles. The result is improved in the paper [19].

In their paper [23], Busemann and Phadke study symmetric and locally symmetric spaces in a general metric setting. Here, a space is called *symmetric* if for every point p and every positive number ρ, the open ball $B(p, \rho)$ admits an isometric involution, that is, a map $\phi \neq \mathrm{Id}$ such that $\phi^2 = \mathrm{Id}$. The space is said to be *locally symmetric* if for some positive continuous function $\sigma(p)$ defined on this space, the open ball $B(p, \sigma(p))$ is symmetric. Busemann and Phadke prove the following:

Theorem 5.14 ([23]). *Minkowskian spaces are characterized by the following: they are locally symmetric simply connected without conjugate points and they satisfy the parallel axiom.*

6. Non-symmetric Minkowski geometries

In the paper [14], Busemann considers Minkowskian planes whose distance function is not necessarily symmetric and whose circles are convex polygons. He shows that any two oriented affine segments can be joined by a polygonal "partial geodesic" in the sense of a local geodesic. In particular, the space has closed geodesics. He also shows that there exist in such a space polygonal geodesics whose oriented segments are dense in the space of oriented affine segments.

In his PhD thesis [31], written under Busemann, and in his paper [32], B. Phadke studied the characterization of Minkowski geometries with a non-necessarily symmetric metric. This characterization involves equidistant loci. Since one deals here with non-symmetric metrics, in defining the distance from a point p to a set M, one has to distinguish all the possibilities for the variable point: it could be on the left or on the right of the distance formula. Phadke proves the following three theorems in which the term "equidistant locus" refers to all such possibilities in the definition of equidistant loci: The space is Minkowskian if and only if one of these three conditions is satisfied:

(1) the equidistant loci to hyperplanes are hyperplanes;
(2) the equidistant loci to lines are convex sets which are unions of lines;
(3) the balls are compact, the equidistant loci to lines are convex, and the equidistant sets to hyperplanes are hyperplanes.

The fact that Minkowski planes are characterized by the equidistant loci to lines being lines was already known to P. Funk who obtained this result under strong differentiability and regularity conditions (see [25] and [26]). In fact, Funk considered the case of a Finsler metric. Using the techniques of the calculus of variations, he proved in [26] that such a space is either Minkowskian or Funk if, locally, the equidistant loci to lines are lines.

The Funk geometry, an asymmetric version of the Hilbert geometry, is one of the main examples of non-symmetric geometries that Busemann studied. Funk and Minkowski geometries share a few basic formal properties; for instance, two arbitrary points are joined by a unique geodesic if and only if the convex set associated to this metric is strictly convex.

We mention the following common characterizations of the class of Minkowski/Funk geometries:

Theorem 6.1 (Busemann [16]). *A Desarguesian space in which all the right spheres of positive radius around any point are homothetic is either a Funk space or a Minkowski space.*

Theorem 6.2 (Busemann [15]). *Among noncompact and nonnecessarily symmetric Desarguesian spaces in which all the right and left spheres of positive radius around any point are compact, the Hilbert and Minkowski geometries are characterized by the property that any isometry between two (distinct or not) geodesics is a projectivity.*

References

[1] W. Blaschke, Räumliche Variationsprobleme mit symmetrischer Transversalitätsbedingung, Ber. d. math.-phys. Kl. d. Sächs. Ges. Wiss. Leipzig 68, 50–55 (1916).

[2] H. Busemann, *Über die Geometrien, in denen die "Kreise mit unendlichem Radius" die kürzesten Linien sind.* Math. Ann. 106 (1932), no. 1, 140–160. English translation by A. A'Campo-Neuen: On the geometries where circles of infinite radius are the shortest lines, Busemann's *Selected Works*, Volume 1, p. 155–172.

[3] H. Busemann, *Über Räume mit konvexen Kugeln und Parallelenaxiom.* Nachrichten Göttingen 1933, 116-140 (1933). English translation by A. A'Campo-Neuen: On spaces with convex spheres and the parallel postulate, Busemann's *Selected Works*, Volume 1, p. 173–191.

[4] H. Busemann, *Metric Methods in Finsler Spaces and in the Foundations of Geometry.* Annals of Mathematics Studies, no. 8. Princeton University Press, Princeton, N. J., 1942.

[5] H. Busemann, On spaces in which points determine a geodesic. *Trans. Am. Math. Soc.* 54 (1943), 171-184.

[6] H. Busemann, The isoperimetric problem in the Minkowski plane. Amer. J. Math. 69, (1947), 863–871.

[7] H. Busemann, Note on a theorem on convex sets, Matematisk Tidsskrift. B, p. 32-34 (1947).

[8] H. Busemann, Two-dimensional geometries with elementary areas. Bull. Amer. Math. Soc. 53, (1947). 402–407.

[9] H. Busemann, The isoperimetric problem for Minkowski area. Amer. J. Math. 71, (1949), 743–762.

[10] H. Busemann, The foundations of Minkowskian geometry. Comment. Math. Helv. 24, (1950). 156-187

[11] H. Busemann, *The geometry of geodesics*, Academic Press (1955), reprinted by Dover in 2005.

[12] H. Busemann, Quasihyperbolic geometry. Rend. Circ. Mat. Palermo (2) 4 (1955), 256–269.

[13] H. Busemann, On Hilbert's fourth problem. *Uspehi Mat. Nauk* 21 (1966) no. 1 (127), 155-164. Translated from the Russian by D. Slutskiy, Busemann's *Selected Works*, Volume II, p. 629–638.

[14] H. Busemann, Transitive geodesics in Minkowski planes. Math. Chronicle 1 1969/1970 part 1, 27-29.

[15] H. Busemann, Recent synthetic differential geometry. Berlin-Heidelberg-New York: Springer-Verlag 1970.

[16] H. Busemann, Spaces with homothetic spheres. J. Geometry 4, 175–186 (1974).

[17] H. Busemann, Problem IV: Desarguesian spaces, in: Mathematical Developments arising from Hilbert's Problems, Proc. Symp. Pure Math. 28, De Kalb 1974, 131–141 (1976).

[18] H. Busemann, Planes with analogues to Euclidean angular bisectors. Collection of articles dedicated to Werner Fenchel on his 70th birthday. Math. Scand. 36 (1975), 5–11.

[19] H. Busemann, Remark on: "Planes with analogues to Euclidean angular bisectors" (Math. Scand. 36 (1975), 5–11). Math. Scand. 38 (1976), no. 1, 81–82.

[20] H. Busemann, Review of: A. V. Pogorelov, Hilbert's fourth problem. *Bull. Amer. Math. Soc.* (N.S.) Volume 4, Number 1 (1981), 87–90.

[21] H. Busemann and P. J. Kelly, Projective geometry and projective metrics, Academic Press, New York, 1953.

[22] H. Busemann and B. B. Phadke, Minkowskian geometry, convexity conditions and the parallel axiom. J. Geom. 12 (1979), no. 1, 17–33.

[23] H. Busemann and B. B. Phadke, Symmetric spaces and ellipses. In: E. B. Christoffel: the influence of his work on mathematics and the physical sciences, int. Symp., Aachen 1979, 626–635 (1981).

[24] H. Busemann and B. B. Phadke, Spaces with distinguished geodesics. Monographs and Textbooks in Pure and Applied Mathematics, 108. Marcel Dekker, Inc., New York, 1987.
[25] P. Funk, Über die Geometrien bei denen die Geraden die Kürzesten sind, Math. Ann. 101 (1929), 226–237.
[26] P. Funk, Über Geometrien, in denen die Kürzesten sind, und die Äquidistanten zu einer Geraden wieder Geraden sind, Monatshefte Math. Phys. 37 (1930), 153–158.
[27] D. Hilbert, Mathematische Probleme, *Göttinger Nachrichten*, 1900, pp. 253–297, reprinted in *Archiv der Mathematik und Physik*, 3d. ser., vol. 1 (1901) pp. 44–63 and 213–237. English version, "Mathematical problems", translated by M. Winston Newson, *Bulletin of the AMS*, vol. 8, 1902, pp. 437– 445 and 478–479. Reprinted also in the Bull. Amer. Math. Soc. (N.S.) 37 (2000), no. 4, 407-436.
[28] P. J. Kelly and L. J. Paige, Symmetric perpendicularity in Hilbert geometries, Pacific Jour. of Math. 2 (1952), p. 319-322.
[29] A. Papadopoulos, Busemann's work on Hilbert geometry, Busemann's *Selected Works*, Volume 2, p. 123–144.
[30] A. Papadopoulos and M. Troyanov, On three early papers by Herbert Busemann on the foundations of geometry, Busemann's *Selected Works*, Volume 1, p. 15–28.
[31] B. B. Phadke, Equidistant loci, PhD dissetration, University of Southern California. 1970. 113 p.
[32] B. B. Phadke, Equidistant loci and the Minkowskian geometries. Canad. J. Math. 24 (1972), 312–327.
[33] A. V. Pogorelov, Hilbert's fourth problem. Translated by R. A. Silverman, edited by I. Kra, in cooperation with E. Zaustinskyi, V. H. Winston and Sons, Washington DC, 1979. Original published by Nauka.

Athanase Papadopoulos, Université de Strasbourg and CNRS, 7 rue René Descartes, 67084 Strasbourg Cedex, France
E-mail address: athanase.papadopoulos@math.unistra.fr

PART II
BUSEMANN'S PAPERS ON CONVEXITY AND INTEGRAL GEOMETRY

Bemerkungen zur Differentialgeometrie der konvexen Flächen. I.

Kürzeste Linien auf differenzierbaren Flächen.

Von Herbert Busemann und Willy Feller.

1. Ist Φ eine beliebige konvexe Fläche im dreidimensionalen Euklidischen Raum, so gibt es zu je zwei Punkten auf ihr mindestens eine kürzeste Verbindung. An anderer Stelle*) haben wir gezeigt, dass jede solche kürzeste Linie, die von einem Punkt P ausgeht, in welchem Φ eine Tangentialebene Π hat, dort eine Halbtangente und eine Schmiegebene besitzt, und dass letztere senkrecht auf Π steht. In jedem inneren Punkt der Kurve, in dem die Fläche differenzierbar ist, fallen die beiden Halbtangenten zusammen. Jedoch ist noch ungeklärt, ob in jeder durch P gehenden Richtung t eine kürzeste Linie von P ausgeht. In Beantwortung dieser Frage zeigen wir zunächst (Nr. 2) an einem Beispiel, dass dies sogar für durchweg differenzierbare konvexe Flächen nicht der Fall zu sein braucht, selbst dann nicht, wenn die Fläche in jedem Punkt endliche (einseitige) Normalkrümmungen**) besitzt.

Geht durch ein Linienelement (P, t) keine kürzeste Linie, so sieht man leicht, dass die kürzeste Verbindung in der Umgebung von P nicht eindeutig sein kann, genauer, dass es in beliebiger Nähe von P Flächenpunkte gibt, mit denen P in verschiedener Weise durch kürzeste Linien verbunden werden kann. Es zeigt sich, dass diese Vorkommnisse nicht dem Fehlen der höheren Ableitungen, sondern der Unbeschränktheit der oberen Normalkrümmungen zuzuschreiben sind. Es gilt nämlich sogar:

Ist Φ eine konvexe differenzierbare Fläche und sind alle oberen (einseitigen) Normalkrümmungen auf Φ kleiner als $1/R$, so gibt es durch jedes tangentielle

*) Krümmungseigenschaften konvexer Flächen, Acta Mathematica (im Druck), § 7. im folgenden zur Abkürzung mit K. k. F. zitiert.

**) Darunter verstehen wir die einseitigen Krümmungen der Normalschnitte. Allgemein definieren wir die Krümmungen der ebenen Kurve k im Punkte P mit der Tangente t vermöge der Radien $r(Q)$ der Kreise mit der Tangente t in P durch einen weiteren Kurvenpunkt Q. Den limes superior von $\frac{1}{r(Q)}$ bei Annäherung von Q an P „von rechts" nennen wir die obere rechte Krümmung von k in P. Endliche einseitige Normalkrümmungen bedeuten in der Sprechweise von K. k. F. die Existenz einer den Flächenpunkt im Inneren enthaltenden Indikatrix. Vgl. § 1 und 3.

Linienelement von Φ genau eine kürzeste Linie. Sind A und B zwei Punkte auf Φ, zu denen es zwei verschiedene kürzeste Verbindungen gibt, so ist deren Länge mindestens gleich πR.

Unser Beweis beruht auf einer Approximation von Φ gemäss einem Satz von *Tonelli* durch analytische Flächenstücke, auf denen die Normalkrümmungen im Wesentlichen in denselben Schranken liegen wie auf Φ. (Nr. 3).

2. Wir beginnen mit dem Beispiel. (x, y, z) seien Cartesische Koordinaten, Φ_1 die Einheitskugel $x^2+y^2+z^2=1$, k der Kreisquadrant $x=0$, $y \geq 0$, $z \geq 0$ auf Φ_1. Wir konstruieren eine konvergente Folge $\{\Phi_\nu\}$ konvexer Flächen, die aus Φ_1 durch Aufsetzen von Kegeln und Abrundung hervorgehen. Auf der Grenzfläche Φ wird es keine kürzeste Linie durch $P(0, 0, 1)$ mit der Richtung von k geben.

Z_2 sei ein innerer Punkt von k. Um Z_2 schlagen wir auf Φ_1 einen Kreis ζ_2, der k in zwei inneren Punkten $D_1{}^2$ und $D_2{}^2$ trifft, und zwar sei $D_1{}^2$ weiter von P entfernt. Wir ersetzen nun das Innere von ζ_2 durch den Tangentialkegel an Φ_1, der Φ_1 längs ζ_2 berührt, und lassen Φ_1 im übrigen ungeändert; die so entstehende Fläche heisse $\Phi_2{}^*$. Die kürzeste Verbindung von P mit $D_1{}^2$ ist, wie leicht zu sehen, genau zweideutig; $\gamma_1{}^2$ und $\gamma_2{}^2$ seien die beiden Verbindungen; sie liegen symmetrisch zu k und treffen ζ_2 ausser in $D_1{}^2$ in je einem weiteren Punkt $E_1{}^2$ bzw. $E_2{}^2$. Die Teilbögen $\widehat{PE_1}{}^2$, $\widehat{PE_2}{}^2$ von $\gamma_1{}^2$ bzw. $\gamma_2{}^2$ sind Grosskreisbögen auf Φ_1, sie mögen in P den Winkel α_2 bilden. Der von $\widehat{PE_1}{}^2$, $\widehat{PE_2}{}^2$ und dem Bogen $\widehat{E_1{}^2 D_2{}^2 E_2{}^2}$ von ζ_2 begrenzte Bereich heisse w_2. Der von $\gamma_1{}^2$ und $\gamma_2{}^2$ begrenzte Bereich enthält die Spitze des aufgesetzten Kegels im Innern. Dieser lässt sich daher so durch ein kleines einbeschriebenes Kugelstück abrunden, dass auf der entstehenden Fläche Φ_2 die Kurven $\gamma_1{}^2$ und $\gamma_2{}^2$ auch noch die einzigen kürzesten Verbindungen zwischen P und $D_1{}^2$ sind.

Wir setzen nun dieses Verfahren fort: Um den Mittelpunkt Z_3 des Bogens $\widehat{PD_2}{}^2$ von k schlagen wir einen so kleinen Kreis ζ_3, dass er ganz im Inneren von w_2 verläuft und die von P aus an ihn gezogenen berührenden Grosskreise auf Φ_1 höchstens den Winkel $\frac{\alpha_2}{2}$ bilden. Wir ersetzen das Innere von ζ_3 durch den Tangentialkegel an Φ_2 (oder, was dasselbe ist, an Φ_1) längs ζ_3 und erhalten so Φ_3^*. Der Kreisbogen k treffe ζ_3 in D_1^3 und D_2^3, und zwar sei D_1^3 der weiter von P entfernte Punkt. Die kürzeste Verbindung von P mit D_1^3 ist wieder genau zweideutig, γ_1^3 und γ_2^3 seien diese Verbindungen. Sie definieren wie eben die Grössen α_3, E_1^3, E_2^3 und w_3. Die Spitze von Φ_3^* wird wieder durch eine kleine Kugel so abgerundet, dass γ_1^3 und γ_2^3 die einzigen kürzesten Verbindungen von P mit D_1^3 bleiben. Die abgerundete Fläche nennen wir Φ_3.

Durch unbegrenzte Fortsetzung dieser Konstruktion bekommt man eine Folge konvexer Flächen Φ_ν mit folgenden Eigenschaften: $\{\Phi_\nu\}$ konvergiert wegen der Monotonie gegen eine Grenzfläche Φ. Diese ist überall differenzierbar, auf ihr liegt eine Folge von Kurvenpaaren γ_1^ν, γ_2^ν die P mit den Punkten D_1^ν verbinden. Da γ_1^ν und γ_2^ν die einzigen kürzesten Verbindungen von P mit D_1^ν auf Φ_ν sind, folgt leicht*), dass sie dies auch auf Φ sind. Der von γ_1^ν und γ_2^ν auf Φ begrenzte Bereich liegt ganz im Inneren des von $\gamma_1^{\nu-1}$ und $\gamma_2^{\nu-1}$ begrenzten, und zieht sich mit wachsendem ν gegen P zusammen. Ferner liegt der Bereich symmetrisch zu k. Die Tangenten an γ_1^ν und γ_2^ν in P streben gegen die Tangente t an k in P. Die Fläche Φ besitzt offenbar überall endliche einseitige Normalkrümmungen (in P ergibt sich das daraus, dass in jeder von t verschiedenen Richtung eine Normalkrümmung existiert und gleich 1 ist, woraus folgt**), dass auch für die Richtung t dasselbe gilt). Es kann aber auf Φ keine von P ausgehende kürzeste Linie mit der Halbtangente t geben; denn jede von P ausgehende Kurve h mit dieser Tangente verläuft für jedes ν zunächst zwischen γ_1^ν und γ_2^ν, trifft also für grosses ν alle $\gamma_1^\nu + \gamma_2^\nu$. Ist etwa der Punkt Q^ν von γ_1^ν ein solcher Schnittpunkt, so ist der Bogen $\widehat{Q^\nu P}$ von h keine kürzeste Verbindung von P mit Q^ν, da der Bogen $\widehat{PQ^\nu}$ von γ_1^ν jedenfalls kürzer ist. Φ liefert also ein Beispiel der gewünschten Art.

3. Der Approximationssatz, auf dem das Folgende beruht,

*) Z. B. unmittelbar aus folgendem Hilfsatz (K. k. F. § 7): Durchläuft P eine rektifizierbare Kurve λ im Aeusseren einer konvexen Fläche ψ, so durchläuft der Fusspunkt F_P von P auf ψ eine rektifizierbare Kurve, deren Länge höchstens so gross ist wie die von λ.

**) K. k. F. § 3.

lautet: Φ sei eine differenzierbare konvexe Fläche, Φ^* eine abgeschlossene geodätische Kreisscheibe*) auf Φ, die nicht ganz Φ überdeckt. Die oberen Normalkrümmungen von Φ seien beschränkt, etwa kleiner als $\frac{1}{R}$. Dann gibt es bei vorgegebenem $\varepsilon > 0$ $(\varepsilon < R)$ eine Folge analytischer Flächenstücke, die mit ihren Tangentialebenen gegen Φ^* konvergieren, und auf denen die Normalkrümmungen kleiner als $\frac{1}{R-\varepsilon}$ sind.

Wir leiten das aus folgendem Spezialfall eines Satzes von *Tonelli***) ab.

Es sei $\zeta = f(u, v)$ eine in einem messbaren Teilbereich des Quadrates $q = \left(0 \leq x \leq \frac{1}{\sqrt{2}},\ 0 \leq y \leq \frac{1}{\sqrt{2}}\right)$ definierte, stetig differenzierbare Funktion. Für alle in Frage kommenden reellen a und b mögen in q die zweiten oberen Derivierten von $f(u, au+b)$ nach u der Ungleichung***)

$$|D''f(u, au+b)| \leq h(u, a, b) \tag{1}$$

genügen, wo h in allen Variablen stetig ist. Die Polynome

$$\begin{aligned} P_n(u, v) &= \frac{1}{K_n} \iint_q f(\alpha, \beta) \cdot [1-(u-\alpha)^2-(v-\beta)^2]^n \, d\alpha \, d\beta \\ K_n &= 4 \iint_q [1-\alpha^2-\beta^2]^n \, d\alpha \, d\beta \end{aligned} \tag{2}$$

streben dann mit ihren ersten Ableitungen gegen $f(u, v)$ und dessen Ableitungen, und bei vorgegebenem $\varepsilon > 0$ gilt für hinreichend grosses n

$$|P_n''(u, au+b)| \leq h(u, a, b) + \varepsilon. \tag{3}$$

Um Betrachtungen nicht lokaler Natur anstellen zu können (bei denen man sich auf ein in der Form $z = f(x, y)$ darstell-

---

*) d. h. die Menge aller Punkte von Φ, deren auf Φ gemessener Abstand von einem festen Punkt eine gewisse Zahl nicht übersteigt.

**) Rend. Circ. mat. Palermo 29 (1910), S. 1—36.

***) Bekanntlich (vgl. etwa *H. Lebesgue:* Leçons sur l'intégration, Paris 1928, S. 75) sind die Grenzen für die oberen und unteren und damit für alle Derivierten einer stetigen Funktion $\varphi(x)$ in jedem Intervall dieselben. Daher genügen wegen der Stetigkeit von $h(u, a, b)$ mit einer zugleich alle zweiten Derivierten von f der Ungleichung (1).

bares Flächenstück beschränken könnte), benutzen wir zum Beweise besondere Koordinaten u, v, ζ, in denen Φ durch eine Gleichung

$$\zeta = f(u, v)$$

dargestellt wird, mit stetig differenzierbarem f, wobei ferner der Definitionsbereich von f im Quadrat q liegt und die Gleichungen $v = au + b$ ebene Schnitte der Fläche liefern. Ein solches Koordinatensystem erhält man z. B. so:

Wir legen eine Kugel Σ ins Innere der Fläche Φ und nehmen als Südpol S von Σ einen solchen Punkt, von dem aus sich Φ^* auf einen Bereich $\Sigma^* \subset \Sigma$ projiziert, der S frei lässt. Wir beziehen Σ auf stereographische Parameter*) (u', v'), sodass die einzige Singularität des Parameternetzes in S liegt. Dem Bereich Σ^* entspricht ein beschränkter Bereich Γ' der (u', v')-Ebene; statt u', v' führen wir auf Σ Parameter $u = cu'$, $v = cv'$ ein, wo c so gewählt ist, dass Γ' in einen Teilbereich Γ des Quadrats q der (u, v)-Ebene übergeht. Dem variablen Punkt P von Σ^* ordnen wir als Koordinaten ein Wertetripel u, v, ζ, zu, wo u, v die Parameterwerte des Schnittpunkts der Strecke $\overline{PS}$ mit Σ sind und ζ die Länge dieser Strecke bezeichnet. Dann stellt sich Φ^* dar durch eine Gleichung $\zeta = f(u, v)$, wobei u, v in Γ variiert, und diese Darstellung hat die gewünschten Eigenschaften.

Wir definieren jetzt die Polynome $P_n(u, v)$ durch (2). Bei derselben geometrischen Deutung von u, v, ζ liefern dann die Gleichungen

$$\zeta = P_n(u, v)$$

eine Folge analytischer Flächenstücke Φ_n, die mit ihren Tangentialebenen gegen Φ^* konvergieren. Nun sei $Q[u_0, v_0, \zeta_0 = P_n(u_0, v_0)]$ irgend ein Punkt von Φ_n. Einer durch (u_0, v_0) gehenden Geraden $v = au + b$ der u, v-Ebene entspricht ein ebener Schnitt von Φ_n durch Q. Infolge des Meusnierschen Satzes berechnet sich die Krümmung des Normalschnittes von Φ_n mit derselben Tangente in der Form

$$P_n''(u, au + b)\, G_n(u, a, b)\,, \tag{4}$$

*) Darunter verstehen wir, wie üblich, folgendes: Es sei (u', v') ein Cartesisches Koordinatensystem in der Tangentialebene an Σ im Nordpol. Dem variablen Kugelpunkt Q werden als Parameter die Koordinaten u', v' des Schnittpunktes der Verbindungsgeraden SQ mit dieser Ebene zugeordnet.

wo G_n ein von n unabhängiger positiver analytischer Ausdruck in P_n und dessen ersten Ableitungen ist. Nun gilt aber der Meusniersche Satz bereits für die oberen Krümmungen der ebenen Schnitte jeder Fläche mit stetiger Tangentialebene*). Die oberen Normalkrümmungen von Φ^* berechnen sich demgemäss durch die (4) entsprechende Formel

$$D'' f(u, au+b)\; G(u, a, b),$$

wobei D'' eine bestimmte zweite Derivierte**) von $f(u, au+b)$ bedeutet, und g von f und dessen Ableitungen genau so abhängt, wie G_n von P_n. Daher strebt G_n gleichmässig gegen G und wegen (3) ergibt sich somit, dass bei vorgegebenem $\eta > 0$ $(\eta < R)$ die Normalkrümmungen von Φ_n für hinreichend grosses n kleiner sind als $\frac{1}{R-\eta}$.

4. Durch diese Konstruktion ist die Fragestellung weitgehend auf analytische Flächen reduziert, und wir wollen daher vorweg die später für diese benutzten Abschätzungen ausführen.

Ψ sei ein analytisches Flächenstück, dessen sämtliche Normalkrümmungen kleiner als $\frac{1}{R}$ sind. In jeder geodätischen Kreisscheibe um einen Flächenpunkt P mit einem kleineren Radius als πR kann man geodätische Polarkoordinaten mit P als Ursprung einführen. In diesen nimmt das Linienelement die Form***)

$$ds^2 = d\varrho^2 + g(\varrho, \vartheta)\cdot d\vartheta^2 \tag{5}$$

an, mit

$$\sqrt{g(0, \vartheta)} = 0 \qquad \frac{\partial}{\partial\varrho}\sqrt{g(0, \vartheta)} = 1. \tag{5a}$$

Die Gaussche Krümmung berechnet sich aus

$$K = -\frac{1}{\sqrt{g}}\;\frac{\partial^2 \sqrt{g}}{\partial\varrho^2}. \tag{6}$$

---

*) *Hjelmslev:* Grundlag for Fladernes Geometri, Mém. Acad. Sc. Let. Danemark, Copenhague, (7) XII, (1914). Vgl. auch K. k. F. § 2.

**) Das sieht man z. B. so: Stellt $y = \varphi(x)$ eine konvexe differenzierbare Kurve dar, welche im Nullpunkt die x-Achse berührt, so ist der Mittelpunkt jedes die x-Achse im Nullpunkt berührenden Kreises durch einen weiteren Kurvenpunkt zugleich Schnittpunkt der y-Achse mit einer Kurvennormale. Daher ist in diesem Falle jede Krümmung im Nullpunkt gleich einer zweiten Derivierten von $\varphi(x)$.

***) vgl. z. B. *W. Blaschke:* Vorl. über Differentialgeometrie I, Zweite Aufl., Berlin 1924, § 57.

Daraus folgt für $\varrho < R\pi$ wegen $K < \frac{1}{R^2}$

$$\int_0^\varrho \frac{\partial^2 \sqrt{g}}{\partial \varrho^2} \cdot R \sin \frac{\varrho}{R}\, d\varrho \geq -\int_0^\varrho \sqrt{g} \cdot \frac{1}{R} \sin \frac{\varrho}{R}\, d\varrho$$

und hieraus durch zweimalige Produktintegration unter Beachtung von (5a)

$$\frac{\partial \sqrt{g(\varrho, \vartheta)}}{\partial \varrho} \cdot R \sin \frac{\varrho}{R} - \sqrt{g(\varrho, \vartheta)} \cos \frac{\varrho}{R} \geq 0$$

oder

$$\frac{\partial}{\partial \varrho} \frac{\sqrt{g(\varrho, \vartheta)}}{R \sin \frac{\varrho}{R}} \geq 0. \tag{7}$$

Aus (5a) folgt aber

$$\lim_{\varrho \to 0} \frac{\sqrt{g(\varrho, \vartheta)}}{R \sin \frac{\varrho}{R}} = 1$$

und somit aus (7)

$$\sqrt{g(\varrho, \vartheta)} \geq R \sin \frac{\varrho}{R}. \tag{8}$$

Genau so ergibt sich wegen $K > -\frac{1}{R^2}$ auch

$$\sqrt{g(\varrho, \vartheta)} \leq R \operatorname{sh} \frac{\varrho}{R}. \tag{9}$$

Wir wollen nun noch eine Folgerung aus (8) ziehen. Es sei Σ die Kugel vom Radius R und P' irgend ein Punkt auf ihr. Wir betrachten wie eben eine geodätische Kreisscheibe um P mit einem Radius $< \pi R$ und bilden sie auf Σ ab, indem wir dem Punkt (ϱ, ϑ) von Ψ den Punkt mit den gleichen sphärischen Polarkoordinaten auf Σ mit P' als Zentrum zuordnen. Den Bildpunkt eines Punktes X von Ψ bezeichnen wir mit X'. Es bedeute $g(\ ,\ )$ den geodätischen Abstand auf Ψ, $g'(\ ,\ )$ denjenigen auf Σ. Q sei ein Punkt auf Ψ mit $g(P, Q) = \varrho_0 < \pi R$ und γ die (eindeutige) geodätische Verbindung von P mit Q. Ferner sei C ein nicht auf γ gelegener Punkt von Ψ mit $g(P, C) = \varrho_1 < \varrho_0$ und ψ der Winkel zwischen γ und der kürzesten Verbindung von P mit C. Nun ist

$$g'(P', Q') = g(P, Q) = \varrho_0, \quad g'(P', C') = g(P, C) = \varrho_1,$$

und auch der Winkel zwischen den kürzesten Verbindungen von P' mit Q' und C' ist gleich ψ. Nach (8) ist das Bild des kürzesten Bogens $\widehat{CQ}$ höchstens so lang wie dieser, und daher hat man

$$g(P, C) + g(C, Q) \geq g'(P', C') + g'(C', Q') = f(\varrho_0, \varrho_1, \psi) \cdot \varrho_0$$
$$= f(\varrho_0, \varrho_1, \psi)\, g(P, Q)$$

mit

$$f(\varrho_0, \varrho_1, \psi) = \frac{\varrho_1 + R \cdot \arccos\left\{\cos\frac{\varrho_0}{R}\cos\frac{\varrho_1}{R} + \sin\frac{\varrho_0}{R}\sin\frac{\varrho_1}{R}\cos\psi\right\}}{\varrho_0}.$$

Bei festem ϱ_0 und ϱ_1 und für $0 \leq \psi \leq \pi$ wächst f monoton und ist für $\psi = 0$ gleich 1. Man hat daher auf Ψ für $\psi \geq \varphi$

$$g(P, C) + g(C, Q) \geq f(\varrho_0, \varrho_1, \varphi)\; g(P, Q),$$

wo $f(\varrho_0, \varrho_1, \varphi) > 1$ für $0 < \varrho_1 < \varrho_0 < \pi R$, $0 < \varphi \leq \pi$ und f nur von den angegebenen Argumenten, nicht aber von der Gestalt von Ψ abhängt.

5. Es sei nun Φ eine beliebige differenzierbare konvexe Fläche, deren Normalkrümmungen durchweg kleiner sind als $\frac{1}{R}$ und P irgend ein Punkt auf ihr. Unter Φ^* verstehen wir eine abgeschlossene geodätische Kreisscheibe um P, die nicht ganz Φ überdeckt. Φ_n bedeute bei vorgegebenem $\varepsilon > 0$ eine Folge gegen Φ^* konvergierender analytischer Flächenstücke im Sinne des Approximationssatzes aus Nr. 3. Insbesondre seien also die Normalkrümmungsradien aller Φ_n grösser als $R - \varepsilon$. Die auf Φ, Φ^*, Φ_n gemessenen Entfernungen bezeichnen wir mit $g(\ ,\ )$, $g^*(\ ,\ )$ $g_n(\ ,\ )$. Es ist stets

$$g(A, B) \leq g^*(A, B).$$

Es sei nun A, B ein Punktepaar auf Φ^*, für das eine kürzeste Verbindung η auf Φ^* existiert, die ganz im Inneren von Φ^* verläuft. Ist A_n, B_n ein gegen A, B konvergierendes Punktepaar auf Φ_n, so ist zunächst wegen der Halbstetigkeit der Bogenlänge

$$\underline{\lim}\; g_n(A_n, B) \geq g(A, B).$$

Andererseits gibt es wegen der gleichmässigen Approximation der ersten Ableitungen von Φ^* durch die von Φ_n eine A_n mit B_n verbindende Kurve η_n auf Φ_n, die mit ihren Tangenten gegen η konvergiert; also ist wegen

$$g_n(A_n, B_n) \leq \text{Länge } \eta_n \to g(A, B)$$

weiter

$$\lim g_n(A_n, B_n) = g(A, B). \tag{11}$$

Hieraus folgt: Jede gegen eine im Inneren von Φ^* verlaufende Kurve konvergente Teilfolge von kürzesten Verbindungen von A_n mit B_n auf Φ_n strebt gegen eine kürzeste Verbindung von A mit B auf Φ^*.

Es sei nun q eine ganz im Inneren von Φ^* gelegene abgeschlossene Kreisscheibe um P vom Radius $\varrho < \pi(R-\varepsilon)$ und P_n ein gegen P konvergierender Punkt auf Φ_n; schliesslich q_n die geodätische Kreisscheibe um P_n auf Φ_n mit dem Radius ϱ (für hinreichend grosse n verläuft q_n im Inneren von Φ_n). Unter einem *Radius* von q verstehen wir irgend eine kürzeste Verbindung von P mit einem Randpunkt von q. Diejenigen Radien von q, die Grenzlagen einer Folge von Radien von q_n sind, wollen wir für einen Augenblick *Grenzradien* nennen. (In Wirklichkeit sind, wie wir sehen werden, alle Radien Grenzradien). Ist Q irgend ein Punkt von q und Q_n eine gegen Q konvergierende Punktfolge aus q_n, so liegt Q_n auf genau einem Radius γ_n von q_n. Wählen wir aus der Folge $\{\gamma_n\}$ eine konvergente Teilfolge aus, so enthält die Grenzkurve eine kürzeste Verbindung von Q mit P und ist ein Grenzradius. Daher: Jeder Punkt von q liegt auf einem Grenzradius. γ sei nun der Bogen PQ irgend eines Grenzradius durch Q. Wir wollen beweisen, dass jeder in q verlaufende, P mit Q verbindende, rektifizierbare Bogen s, der einen nicht auf γ liegenden Punkt C enthält, länger ist als γ.

Nach Definition des Grenzradius können wir eine Teilfolge Φ_{n_i} und auf Φ_{n_i} einen Punkt Q_{n_i} so wählen, dass die Teilfolge der kürzesten Verbindungen γ_{n_i} von P_{n_i} mit Q_{n_i} gegen γ und insbesondre Q_{n_i} gegen Q konvergiert. Es werde $g(P, Q) = g^*(P, Q) = \varrho_0$ gesetzt; es darf dann $g(P, C) = \varrho_1 < \varrho_0$ vorausgesetzt werden. C_{n_i} sei so gewählt, dass $C_{n_i} \to C$ und

$g_{n_i}(C_{n_i}, P_{n_i}) = \varrho_1$ gilt. Schliesslich sei φ_{n_i} der Winkel zwischen γ_{n_i} und der geodätischen Verbindung von P_{n_i} mit C_{n_i} und l_{n_i} der auf dem Kreis um P_{n_i} durch C_{n_i} gemessene Abstand zwischen C_{n_i} und γ_{n_i}. Da die Normalkrümmungen von Φ_{n_i} kleiner sind als $\frac{1}{R-\varepsilon}$, folgt aus (9)

$$l_{n_i} \leq (R-\varepsilon) \operatorname{sh} \frac{\varrho_1}{R-\varepsilon} \varphi_{n_i}.$$

Die φ_{n_i} sind also nach unten beschränkt, etwa grösser als $\psi > 0$. Daher hat man nach (10)

$$g_{n_i}(P_{n_i}, C_{n_i}) + g_{n_i}(C_{n_i}, Q_{n_i}) \geq f(\varrho_0, \varrho_1, \psi) g_{n_i}(P_{n_i}, Q_{n_i}).$$

Zu s lassen sich Kurven s_{n_i} auf Φ_{n_i} durch C_{n_i} angeben, die P_{n_i} mit Q_{n_i} verbinden und mit ihren Längen gegen s konvergieren. Da

$$\text{Länge } s_{n_i} \geq g_{n_i}(P_{n_i}, C_{n_i}) + g_{n_i} C_{n_i}, Q_{n_i})$$

ist, folgt

$$\text{Länge } s \geq f(\varrho_0, \varrho_1, \psi) g(P, Q), \qquad \text{w. z. b. w.}$$

Die Radien von q überdecken also q (abgesehen von P) einfach und lückenlos. Da über den Radius von q nur vorausgesetzt war, dass er kleiner als $\pi(R-\varepsilon)$ sein sollte, und ε beliebig war, liegt hierin, dass in jedem Kreis in Φ^* um P mit einem kleineren Radius als πR die Radien den Kreis einfach und lückenlos überdecken, oder auch dass für Punktepaare P, Q auf Φ^* mit einem kleineren geodätischen Abstand als πR die geodätische Verbindung eindeutig ist. Wenn Φ nicht geschlossen ist, ist die letzte Behauptung des Satzes aus Nr. 1 damit bewiesen, da Φ^* beliebig gross gewählt werden konnte. Ebenso im Fall eines geschlossenen Φ, wenn Q nur nicht einer der von P am weitesten entfernten Punkte ist, da dann Φ^* so gewählt werden kann, dass Q ein innerer Punkt von Φ^* wird.

6. Den noch ausstehenden Fall kann man durch eine Sonderbetrachtung erledigen, doch gibt es eine einfache Beziehung, mit der man das ganze Resultat samt der Existenz und Eindeutigkeit einer kürzesten Linie durch ein Linienelement sofort ableitet.

Der Beweisgang aus Nr. 5 zeigt: Ist A_n enthalten in q_n und $A_n \to A \neq P$, so konvergieren die Radien von q_n durch A_n gegen den Radius von q durch A. Nun seien γ und γ' zwei Radien von q und $A(s)$ bzw. $A'(s)$ die Punkte auf γ bzw. γ' mit

$$g(P, A(s)) = g(P, A'(s)) = s,$$

und $A_n(\varrho)$ bzw. $A'_n(\varrho)$ seien auf dem Rand von q_n so gewählt, dass $A_n(\varrho) \to A(\varrho)$ und $A'_n(\varrho) \to A'(\varrho)$ strebt. Nach der eben gemachten Feststellung gilt dann für die Radien γ_n bzw. γ'_n von q_n durch $A_n(\varrho)$ bzw. $A'_n(\varrho)$ die Beziehung

$$\gamma_\nu \to \gamma \quad \text{bzw.} \quad \gamma'_n \to \gamma';$$

insbesondre strebt also bei analoger Bezeichnungsweise wie für γ und γ' auch

$$A_n(s) \to A(s) \quad \text{und} \quad A'_n(s) \to A'(s).$$

Wird noch die Länge des auf dem geodätischen Kreise um P gemessenen kleineren Bogens $\widehat{A_n(s) A'_n(s)}$ mit $l_n(s)$ bezeichnet, so gilt nach (9)

$$l_n(s') \leq (R-\varepsilon) \operatorname{sh} \frac{s'}{R-\varepsilon} \sphericalangle(\gamma_n, \gamma'_n)$$

und nach (8)

$$l_n(s) \geq (R-\varepsilon) \sin \frac{s}{R-\varepsilon} \sphericalangle(\gamma_n, \gamma'_n),$$

also

$$l_n(s) \geq \frac{\sin \dfrac{s}{R-\varepsilon}}{\operatorname{sh} \dfrac{s'}{R-\varepsilon}} l_n(s')$$

und damit bei entsprechenden Bezeichnungen auf Φ

$$l(s) \geq \frac{\sin \dfrac{s}{R-\varepsilon}}{\operatorname{sh} \dfrac{s'}{R-\varepsilon}} l(s'). \tag{12}$$

Das ist die gesuchte Beziehung.

Verlaufen zwischen den Punkten P und Q zwei kürzeste Verbindungen γ und γ' so gilt für die Längen der entsprechenden Kreisbögen $\widehat{A_n(s)A_n'(s)}$ offenbar

$$l(s) \to 0 \quad \text{für} \quad s \to g(P, Q).$$

Ist Φ^* die abgeschlossene Kreisscheibe um P mit dem Radius $g(P, Q)-\eta$, wo $0<\eta<g(P, Q)$, so gilt (12) für jedes $s< g(P, Q)-\eta$. Für $\eta \to 0$ und $s = g(P, Q)-2\eta$ folgt demnach wegen $l(s)\to 0$

$$g(P, Q) \geq \pi(R-\varepsilon).$$

Damit ist der noch ausstehende Fall aus Nr. 5 erledigt.

Um zu sehen, dass verschiedene Radien von q in P verschiedene Tangenten haben, wenden wir (12) an auf irgend ein festes $s' \leq \varrho$ und $s \to 0$. Wir bekommen dann

$$\lim_{s\to 0} \frac{l(s)}{s} \geq \frac{(R-\varepsilon)\, l(s')}{\operatorname{sh} \dfrac{s'}{R-\varepsilon}} > 0$$

und daraus unmittelbar (z. B. durch Projektion auf die Tangentialebene von Φ) $\sphericalangle(\gamma, \gamma') > 0$.

Nun ordnen wir jedem Randpunkt X von q die Tangente t_X in P des Radius von P nach X zu. Wenn X den Rand von q durchläuft, dreht sich t_X monoton, weil die Radien q einfach überdecken. Die Abbildung $X \to t_X$ ist, wie wir eben sahen, eineindeutig, und aus der Monotonie und der lückenlosen Ueberdeckung von q durch seine Radien folgt, dass die Zuordnung $t_X \to X$ stetig ist. Wegen der Eineindeutigkeit ist also auch $X \to t_X$ stetig und damit ist gezeigt, dass es zu jeder Tangentialrichtung an Φ in P genau eine von P ausgehende kürzeste Linie mit dieser Richtung als Halbtangente gibt.

Bemerkungen zur Differentialgeometrie der konvexen Flächen. II.

Über die Krümmungsindikatrizen.

Von Herbert Busemann und Willy Feller.

1. Trägt man von einem Punkt O eines zweimal stetig differenzierbaren Flächenstücks Φ in jeder von O ausgehenden Tangentialrichtung an Φ nach beiden Seiten die Wurzel aus dem Krümmungsradius des zugehörigen Normalschnitts ab, so entsteht die sog. *Dupinsche Indikatrix* (der Normalkrümmungen von Φ in O). Sie ist stets ein Kegelschnitt mit O als Mittelpunkt, und wenn Φ konvex ist, ein Paar paralleler Geraden oder eine Ellipse (die auch ganz im Unendlichen liegen kann).

Ähnliches gilt, wie wir an anderer Stelle gezeigt haben[1]), auch noch für fast alle Punkte eines beliebigen konvexen Flächenstücks Φ: Es gibt auf Φ stets eine Punktmenge Λ vom Lebesgueschen Mass Null derart, dass in jedem Punkt O von $\Phi - \Lambda$ eine Tangentialebene an Φ existiert und sämtliche Normalschnitte in O endliche Krümmungen besitzen; die Dupinsche Indikatrix in O besteht wieder aus einer (evtl. ausgearteten) Ellipse mit O als Mittelpunkt. Im einzelnen Flächenpunkt braucht es dagegen keine Krümmungen der Normalschnitte zu geben: An deren Stelle treten vier Zahlen, nämlich die einseitigen unteren und oberen Krümmungen[2]). Zweck dieser Note ist es, eine Uebersicht über alle möglichen Wertverteilungen dieser vier Grössen im einzelnen Punkt einer konvexen Fläche zu geben, und zwar legen wir weniger den sich ergebenden Sätzen selbst als dem Umstand Wert bei, dass wir zu jedem als möglich erkannten Typus wirklich eine Fläche angeben und dadurch ein vollständiges Beispielmaterial liefern.

---

[1]) Krümmungseigenschaften konvexer Flächen, Acta Mathematica Bd. 66, im folgenden zur Abkürzung mit K. k. F. zitiert.

[2]) Stellt $y = f(x)$ eine die x-Achse im Nullpunkt O berührende Kurve dar, so verstehen wir etwa unter ihrem rechten oberen Krümmungsradius $\bar{\varrho}$ in O den limes superior für $x \to 0+$ der Radien $R(x)$ der Kreise, die in O die x-Achse berühren und durch den Kurvenpunkt $(x, f(x))$ gehen. Es ist

$$R(x) = \frac{x^2 + f^2(x)}{2f(x)}, \quad \bar{\varrho} = \overline{\lim_{x \to 0+}} \frac{x^2}{2f(x)}.$$

Die so definierten Krümmungsradien unterscheiden sich im allgemeinem von denjenigen, die man vermöge der Grenzlage des Schnittpunktes benachbarter Kurvennormalen definiert, vgl. K. k. F. § 1.

Es sei Φ ein beliebiges konvexes Flächenstück, welches im Punkte O die Tangentialebene Π besitzt. In jeder von O ausgehenden Richtung[1]) in Π tragen wir von O aus die Wurzel aus dem oberen bzw. unteren Krümmungsradius des zugehörigen Normalhalbschnitts ab. Es entstehen so zwei Punktmengen, die wir a. a. O. als die *„obere"* bzw. *„untere" Indikatrix* (von Φ in O) bezeichnet haben; wenn beide zusammenfallen, so sprechen wir von einer Indikatrix schlechthin. Es wurde dort darüber folgendes festgestellt:

I. Die untere Indikatrix i liegt ganz auf einer konvexen Kurve j, die O im Inneren oder auf dem Rande enthält. Im letzten Fall kann j eine oder zwei von O ausgehende Strecken enthalten; abgesehen von diesen Strecken, von denen natürlich höchstens je ein von O verschiedener Punkt zu i gehören kann, gehört jeder Punkt von j zu i. Man kann das auch so ausdrücken: Es seien r, φ ebene Polarkoordinaten in Π mit O als Ursprung, und $g(\varphi)$ sei die Wurzel aus dem zur Richtung φ gehörenden unteren Normalkrümmungsradius. Dann gibt es einen (evtl. leeren) Winkelraum (φ_0, φ_1) derart, dass in seinem offenen Inneren $g(\varphi) > 0$ ist und $r = g(\varphi)$ einen konvexen Kurvenbogen darstellt, während im Inneren des Komplementärwinkelraumes $g(\varphi) = 0$ ist. Unstetigkeiten von $g(\varphi)$, d. h. i, können höchstens in den Richtungen φ_0 und φ_1 vorkommen.

II. Es sei $r = g(\varphi)$ die Gleichung der unteren Indikatrix i, $r = G(\varphi)$ die der oberen I, und es bezeichne γ bzw. Γ die Menge der Punkte (r, φ) mit $r \leq g(\varphi)$ bzw. $r \leq G(\varphi)$. Ist dann P irgendein Punkt von γ und Q einer von Γ, so enthält Γ die ganze Strecke $\overline{PQ}$.

Aus II ergeben sich gewisse Stetigkeitseigenschaften von $G(\varphi)$, von denen wir a. a. O. nur die folgende festgestellt haben: Falls j (in diesem Fall $= i$) den Punkt O im Inneren enthält, so ist $G(\varphi)$ stetig. Es zeigt sich aber (vgl. No. 5), dass die allgemeinste in II enthaltene Aussage die folgende ist: Ist für einen Wert φ_0

[1]) Hier und im folgenden ist stets zwischen entgegengesetzten Richtungen (φ und $\varphi+\pi$) zu unterscheiden. Es sei gleich bemerkt, dass wir das Wort Normalschnitt, wo es zu keinen Verwechslungen Anlass gibt, auch für den Schnitt der Fläche mit einer durch die Flächennormale gehenden Halbebene statt Normalhalbschnitt verwenden.

die Funktion $g(\varphi_0)$ positiv, so existieren für alle Richtungen, ausser vielleicht für $\varphi = \varphi_0$ und $\varphi = \varphi_0 + \pi$, die Limites $G(\varphi+)$ und $G(\varphi-)$, und es ist

$$G(\varphi+) \leq G(\varphi) \leq G(\varphi-) \quad \text{für } \varphi_0 < \varphi < \varphi_0 + \pi,$$
$$G(\varphi-) \leq G(\varphi) \leq G(\varphi+) \quad \text{für } \varphi_0 + \pi < \varphi < \varphi_0 + 2\pi.$$

Wenn also für zwei Richtungen φ_0 und φ_1 mit $0 < \varphi_0 - \varphi_1 < \pi$ $g(\varphi_0) \neq 0$ und $g(\varphi_1) \neq 0$ ist, so ist $G(\varphi)$ in den Winkelräumen $\varphi_0 < \varphi < \varphi_1$ und $\varphi_0 + \pi < \varphi < \varphi_1 + \pi$ stetig, woraus man den erwähnten Stetigkeitssatz wiedergewinnt.

Die in I und II enthaltenen Aussagen lassen sich nicht verschärfen. Es wird nämlich im folgenden gezeigt:

(a) Zu jeder beliebigen konvexen Kurve j der Ebene Π, die O im Inneren oder auf dem Rande enthält, kann eine konvexe Fläche angegeben werden, die Π in O berührt und dort eine Indikatrix i hat, die ganz auf j liegt. (No. 3). Enthält j eine von O ausgehende Strecke in der Richtung φ_0, so kann insbesondre erreicht werden, dass ein beliebiger Punkt dieser Strecke (die Endpunkte mitgerechnet) zu i gehört. (No. 4).

Es wird dabei *nicht* vorausgesetzt, dass j überhaupt ein Inneres hat; j kann z. B. aus einer einzigen von O ausgehenden Strecke bestehen ($g(\varphi_0) \neq 0$, $g(\varphi) = 0$ sonst).

(b) Durch die Gleichungen $r = g(\varphi)$ und $r = G(\varphi)$ mögen zwei Punktmengen i und I der Ebene Π definiert sein, von denen die erste nach I als untere Indikatrix einer konvexen Fläche in O auftreten kann, und es bezeichne γ bzw. Γ die Gesamtheit der Punkte (r, φ) mit $r \leq g(\varphi)$ bzw. $r \leq G(\varphi)$. Ferner sei nicht überall $g(\varphi) = 0$. Wenn dann Γ jede Strecke $\overline{PQ}$ mit $P \subset \gamma$ und $Q \subset \Gamma$ enthält, so gibt es eine konvexe Fläche, welche in O die Tangentialebene Π, und dort i zur unteren, I zur oberen Indikatrix hat. (No. 7).

Wenn i mit O zusammenfällt, so wird die Aussage II leer; trotzdem kann nicht jede in der Form $r = G(\varphi)$ darstellbare Punktmenge als obere Indikatrix auftreten, sondern $G(\varphi)$ muss eine Bedingung erfüllen, die im Falle $i \neq O$ in II enthalten ist:

(c) Stellt $r = G(\varphi)$ die obere Indikatrix einer konvexen Fläche im Nullpunkt dar, so ist $G(\varphi)$ der Limes einer monoton fallenden Folge von nach unten halbstetigen Funktionen. (No. 8).

(c′) Ist $G(\varphi)$ eine solche Funktion, so gibt es eine konvexe Fläche durch O, mit O als unterer und $r = G(\varphi)$ als oberer Indikatrix. (No. 9).

Darin ist insbesondre enthalten, dass es obere Indikatrizen gibt, welche in keiner Richtung φ stetig sind, d. h. $G(\varphi)$ braucht nicht von erster Bairescher Klasse zu sein.

Für das Verständnis des Inhalts der letzten Sätze ist es nützlich, sich die folgenden Zusammenhänge ins Gedächtnis zurückzurufen (von denen übrigens weiter kein Gebrauch gemacht wird):

$f(x)$ heisst nach unten bzw. nach oben halbstetig, wenn für jedes x

$$\underline{\lim}_{\xi \to x} f(\xi) \geq f(x) \quad \text{bzw.} \quad \overline{\lim}_{\xi \to x} f(\xi) \leq f(x)$$

ist. Eine Funktion is dann und nur dann halbstetig nach unten (oben), wenn sie der Limes einer monoton wachsenden (fallenden) Folge stetiger Funktionen ist.[1])

Eine monoton wachsende (fallende) Folge nach unten (oben) halbstetiger Funktionen konvergiert gegen eine nach unten (oben) halbstetige Funktion.

Wir wollen die Gesamtheit der Funktionen, die sich als Limites monoton fallender (wachsender) Folgen nach unten (oben) halbstetiger Funktionen darstellen lassen, mit γ (bzw. Γ) bezeichnen.

Nach einem Satz von *Young*[2]) gilt dann: Die Funktionen, welche Grenzfunktionen konvergenter (aber nicht notwendig monotoner) Folgen von stetigen Funktionen sind — d. h. die Funktionen der 0-ten oder ersten Baireschen Klasse — sind genau diejenigen Funktionen, die gleichzeitig zu γ und Γ gehören.

Ein bekannter Satz von *Baire*[3]) erlaubt dies auch so zu formulieren: Eine Funktion gehört dann und nur dann gleichzeitig γ und Γ an, wenn sie auf jeder perfekten Menge E höchstens punktweise unstetig ist, d. h.: wenn man die Funktion bloss auf E als Definitionsbereich betrachtet, so liegen ihre Stetigkeitspunkte (auf E) überall dicht.

Hieraus folgt insbesondre, dass eine zu Γ gehörende Funktion $G(\varphi)$, die in allen Punkten eines Teilintervalls von $(0, 2\pi)$ unstetig ist, nicht auch zu γ gehören kann.

Zwei Beispiele mögen die Anwendungsmöglichkeit dieser Sätze erläutern.

Es sei $G(\varphi) = 1$ für rationale φ, und $G(\varphi) = 0$ sonst. Wir setzen

$$G_n(\varphi) = 1 \text{ für } \varphi = \varphi_1, \ \varphi = \varphi_2, \cdots\cdots, \ \varphi \doteq \varphi_n,$$
$$G_n(\varphi) = 0 \text{ sonst,}$$

---

[1]) Vgl. z. B. *F. Hausdorff:* Mengenlehre, 2. Aufl. 1927. S. 248; oder *H. Hahn:* Reelle Funktionen, 1. Aufl. 1921, S. 162.

[2]) *H. Hahn:* a. a. O. S. 346.

[3]) *Hausdorff*, a. a. O. S. 255: *Hahn*, a. a. O. S. 363 oder auch *H. Lebesgue:* Leçons sur l'intégration, 2. Aufl. 1928, S. 202.

wobei $\{\varphi_n\}$ die rationalen φ-Werte durchlaufen möge. $G_n(\varphi)$ ist dann halbstetig nach oben, und die Folge $\{G_n(\varphi)\}$ strebt monoton wachsend gegen $G(\varphi)$. Daher gehört $G(\varphi)$ zu Γ und ist andrerseits in jedem Punkte unstetig. Nach den obigen Feststellungen kann also $G(\varphi)$ nicht zu γ gehören, d. h. $r = G(\varphi)$ kann nach (c) nicht eine obere Indikatrix einer konvexen Fläche darstellen.

Hingegen ist die Funktion: $G(\varphi) = 0$ für rationale φ, $G(\varphi) = 1$ sonst, der Limes der fallenden Folge $G_n(\varphi)$ mit

$$G_n(\varphi) = 0 \text{ für } \varphi = \varphi_1,\ \varphi = \varphi_2,\ \cdots\cdots,\ \varphi = \varphi_n,$$
$$G_n(\varphi) = 1 \text{ sonst};$$

$G_n(\varphi)$ ist offenbar nach unten halbstetig, und daher gehört $G(\varphi)$ zu γ. Es gibt also nach (c') konvexe Flächen, deren obere Indikatrix in einem Punkte O durch $r = G(\varphi)$ dargestellt wird.

Ferner folgt aus den zitierten Sätzen, dass jede Funktion $G(\varphi)$ mit nur abzählbar vielen Unstetigkeiten zu γ (nämlich sogar zu $\gamma \cdot \Gamma$) gehört. Wenn $r = g(\varphi)$ und $r = G(\varphi)$ die beiden Indikatrizen einer konvexen Fläche in O darstellen, und wenn $g(\varphi)$ nicht überall verschwindet, hat $G(\varphi)$ diese Eigenschaft (vgl. No. 5), womit die erwähnte Tatsache, dass (c) für $i \neq O$ in II enthalten ist, bewiesen ist.

2. Wir beziehen unsere Betrachtungen auf rechtwinklige kartesische Koordinaten x, y, z, und gleichzeitig auf ein System von Zylinderkoordinaten r, φ, z, das mit den ersten durch die Gleichungen $x = r\cos\varphi$, $y = r\sin\varphi$ verbunden ist. Bei den zu konstruierenden Flächen wird stets der Ursprung O der kritische Punkt sein, ferner wird die Fläche stets im Halbraum $z \geq 0$ liegen und in O die x, y-Ebene berühren. Letzteres, d. h. die Differenzierbarkeit in O, wird sich stets unmittelbar aus der Bemerkung ergeben, dass es wegen der Konvexität des Tangentialkegels in jedem Punkt einer konvexen Fläche genügt zu wissen, dass zwei Normalschnitte (nicht Halbschnitte!) in O differenzierbar sind. Wir werden daher die Tatsache der Differenzierbarkeit nicht jedesmal ausdrücklich erwähnen.

Wir beginnen nun mit einigen Bemerkungen über konvexe Hüllen. Unter der konvexen Hülle einer Punktmenge μ verstehen wir die kleinste konvexe Menge, in welcher μ enthalten ist.[1]

---

[1] Der Gebrauch des Wortes „konvexe Hülle" scheint nicht ganz festgelegt zu sein, insofern manche Definitionen in der Literatur (die allerdings dann gewöhnlich nur auf abgeschlossene Mengen angewandt werden) auf die abgeschlossene Hülle dessen, was wir als konvexe Hülle bezeichnen, führen. In der Formulierung von Satz II haben wir hier daher, im Gegensatz zu K. k. F. § 3, nicht gesagt, dass Γ mit einem Punkt Q auch die konvexe Hülle von $Q + \gamma$ enthält, da II falsch wird, wenn darunter die kleinste abgeschlossene konvexe Menge verstanden wird, die $Q + \gamma$ enthält; z. B. dann, wenn eine Indikatrix i existiert, welche zu einer Kurve j gehört, die eine von O ausgehende Strecke enthält, wobei nicht der von O verschiedene Endpunkt derselben zur Indikatrix gehört.

Uns kommt es gewöhnlich nicht auf die ganze konvexe Hülle an, sondern bloss auf ihren Rand; um einen kurzen Ausdruck zu haben, werden wir den Rand der konvexen Hülle von μ als Haube von μ bezeichnen.

Die Bestimmung der konvexen Hülle wird oft sehr vereinfacht, wenn man folgenden Satz benutzt.[1])

Die konvexe Hülle einer beschränkten abgeschlossenen zusammenhängenden Menge μ ist gleich der Vereinigungsmenge aller (evtl. auch ausgearteten) Dreiecksflächen, deren Ecken auf μ liegen.

Wir werden insbesondre eine Folgerung aus diesem Satze benutzen, die sich durch Diskussion der Lage der Eckpunkte der Dreiecke sofort ergibt: Sind k_1 und k_2 zwei beschränkte abgeschlossene konvexe Mengen mit nicht leerem Durchschnitt, so ist die konvexe Hülle von $k_1 + k_2$ gleich der Vereinigungsmenge aller Strecken, deren Endpunkte auf k_1 oder k_2 liegen.

Besonders wichtig für das Weitere ist folgende Feststellung: k_1 und k_2 seien zwei in den Viertelebenen $\varphi = \varphi_1$ bzw. $\varphi = \varphi_2$, $z \geq 0$, $0 < \varphi_2 - \varphi_1 < \pi$, gelegene konvexe Kurvenstücke, die von O ausgehen und dort die x, y-Ebene berühren. In einer Umgebung von O setzt sich dann die Haube von $k_1 + k_2$ zusammen aus ebenen Stücken und aus den Strecken, welche diejenigen Punkte von k_1 und k_2 verbinden, in denen es komplanare Stützgeraden gibt.

Der *Beweis* ist ganz leicht: Die angegebenen Strecken gehören zunächst jedenfalls der konvexen Hülle von $k_1 + k_2$ an; andrerseits ist eine Ebene, die je eine Stützgerade an k_1 und k_2 enthält, Stützebene von $k_1 + k_2$, so dass unsre Strecken in der Tat auf dem Rande der konvexen Hülle liegen.

Das aus diesen Strecken bestehende Stück der Haube von $k_1 + k_2$ wollen wir weiterhin mit $H(k_1, k_2)$ bezeichnen. $H(k_1, k_2)$ berührt die x,y-Ebene im Nullpunkt, und ist im übrigen nur im Winkelraum $\varphi_1 \leq \varphi \leq \varphi_2$ definiert. Trotzdem werden wir auch hier die allgemeine Terminologie anwenden, und in unmittelbar verständlicher Weise von Normalschnitten bzw. von

---

[1]) Vgl. *W. Fenchel:* Ueber Krümmung und Windung geschlossener Raumkurven, Math. Ann. *101*, (1929), § 2.

einer Indikatrix von $H(k_1, k_2)$ in O und diesem Winkelraum sprechen.

Sind k_1 und k_2 insbesondre Stücke der Parabeln $z = \frac{r^2}{2a_1}, \varphi = \varphi_1$, bzw. $z = \frac{r^2}{2a_2}, \varphi = \varphi_2$, $0 < a_1, a_2 < \infty$, so entsteht $H(k_1, k_2)$, indem man die Punkte von k_1 und k_2 mit gleichen z-Koordinaten verbindet, d. h. $H(k_1, k_2)$ besteht aus einem Stück des parabolischen Zylinders

$$z = \frac{r^2}{2a_1 a_2 \sin^2(\varphi_2 - \varphi_1)} \left\{ \sqrt{a_1} \sin(\varphi - \varphi_1) + \sqrt{a_2} \sin(\varphi_2 - \varphi) \right\}^2,$$
$$\varphi_1 \leq \varphi \leq \varphi_2.$$

Man sieht unmittelbar, (vgl. die Formel von Fussnote [2]), S. 87), dass hier eine Indikatrix existiert und aus der Strecke mit den Endpunkten $(\sqrt{a_1}, \varphi_1, 0)$ und $(\sqrt{a_2}, \varphi_2, 0)$ besteht. Dies bleibt auch noch für den Fall richtig, dass eine oder beide Parabeln in Geraden ausarten, wobei die Punkte $(\sqrt{a_i}, \varphi_i, 0)$ als uneigentliche Punkte zu interpretieren sind. Aus dieser Bemerkung wollen wir das Analogon für den allgemeinen Fall ableiten:

Besitzen die Kurven k_i des letzten Satzes (endliche oder unendliche) Krümmungsradien a_i, $i = 1, 2$, so hat $H(k_1, k_2)$ im fraglichen Winkelraum eine Indikatrix, und diese besteht aus der Strecke (bzw. Halbgeraden usf.) von $(\sqrt{a_1}, \varphi_1, 0)$ nach $(\sqrt{a_2}, \varphi_2, 0)$.

Beweis. Die Aussage von Satz I (S. 88) über die Konvexität der unteren Indikatrix bleibt auch noch für den Fall richtig, dass letztere nur in einem Winkelraum definiert ist[1]). Daher enthält der von der unteren Indikatrix von $H(k_1, k_2)$ zusammen mit den Strecken von O nach $(\sqrt{a_i}, \varphi_i, 0)$, $i = 1, 2$, begrenzte abgeschlossene Bereich das abgeschlossene Dreieck mit den (evtl. uneigentlichen) Ecken O, $(\sqrt{a_i}, \varphi_i, 0)$, $i = 1, 2$, d. h., für den Radiusvektor in der Richtung φ der unteren Indikatrix gilt[2])

$$(*) \qquad \underline{\varrho}_\varphi \geq \frac{\sqrt{a_1 a_2} \sin(\varphi_2 - \varphi_1)}{\sqrt{a_1} \sin(\varphi - \varphi_1) + \sqrt{a_2} \sin(\varphi_2 - \varphi)}$$

---

[1]) K. k. F., S. 16.

[2]) Im Falle $a_i = \infty$ ist die Gleichung natürlich sinngemäss durch Grenzübergang zu interpretieren.

Nun betrachten wir bei beliebigem $\varepsilon > 0$ die (evtl. ausgearteten) Parabeln k_1^ε und k_2^ε mit den Gleichungen $z = \frac{r^2}{2(a_1+\varepsilon)}$, $\varphi = \varphi_1$ bzw. $z = \frac{r^2}{2(a_2+\varepsilon)}$, $\varphi = \varphi_2$. Bei endlichem a_i hat k_i^ε einen grösseren Krümmungsradius als k_i, und verläuft daher zufolge der Definition der Krümmung in einer Umgebung von O unterhalb von k_i. Für $a_i = \infty$ ist k_i^ε eine Gerade und hat erst recht keinen Punkt oberhalb k_i. In allen Fällen liegt also (nach unsren Feststellungen über die Konstruktion der Haube zweier Kurven) in einer gewissen Umgebung von O die Fläche $H(k_1^\varepsilon, k_2^\varepsilon)$ unterhalb oder auf $H(k_1, k_2)$, und daher ist die obere Indikatrix von $H(k_1, k_2)$ ganz enthalten im abgeschlossenen Bereich, der von der Indikatrix von $H(k_1^\varepsilon, k_2^\varepsilon)$ und Stücken der Geraden $\varphi = \varphi_i$, $i = 1, 2$, begrenzt wird. Man hat also für den Radiusvektor $\bar{\varrho}_\varphi$ dieser oberen Indikatrix

$$\bar{\varrho}_\varphi \leq \frac{\sqrt{(a_1+\varepsilon)(a_2+\varepsilon)}\sin(\varphi_2 - \varphi_1)}{\sqrt{a_1+\varepsilon}\sin(\varphi - \varphi_1) + \sqrt{a_2+\varepsilon}\sin(\varphi_2 - \varphi)}.$$

Für $\varepsilon \to 0$ ergibt das zusammen mit (*) gerade die Behauptung.

3. Wir kommen nun zum Beweis der Behauptung (a), S. 89. Es sei j eine konvexe Kurve der x, y-Ebene, die entweder durch O geht, oder O im Inneren enthält. In dieser Nummer beweisen wir zunächst den ersten Satz von (a), d. h. wir konstruieren eine konvexe Fläche, die in O die x, y-Ebene berührt und dort eine Indikatrix i hat, die, abgesehen von etwaigen von O ausgehenden Strecken, ganz mit j zusammenfällt.

Wir betrachten zunächst den Fall, dass j den Punkt O im Inneren enthält. Der Radiusvektor von j sei durch die Gleichung $r = g(\varphi)$ gegeben; dann ist $g(\varphi) > 0$, und die Fläche Φ:

$$z = \frac{r^2}{2g^2(\varphi)}$$

hat j in O als Indikatrix. Dass diese Fläche konvex ist, erkennt man aus dem, was über die Konstruktion der konvexen Hülle von Parabeln gesagt wurde.

Will man erreichen, dass die Fläche mit der x, y-Ebene nur den Punkt O gemeinsam hat, so kann man etwa so verfahren: Die Fläche $z = r^3$ ist konvex. Statt Φ betrachten wir die Haube Φ^* des Durchschnitts der konvexen Mengen, die von dieser

Fläche und von Φ begrenzt werden. Φ^* ist eine Fläche der gesuchten Art. Der ebene Schnitt (φ) von Φ^* hat die Gestalt

$$(\lambda_{g(\varphi)}):\qquad z = f(r, g(\varphi)) \begin{cases} = \dfrac{r^2}{2g^2(\varphi)} & \text{für}\quad r \leq \dfrac{1}{2g^2(\varphi)} \\ = r^3 & \text{für}\quad r \geq \dfrac{1}{2g^2(\varphi)}. \end{cases}$$

Da diese Kurven noch öfter vorkommen werden, wollen wir sie generell mit $\lambda_{g(\varphi)}$ bezeichnen. Der Grenzfall $g(\varphi) = \infty$ ist dabei zugelassen.

Wir betrachten nun den Fall, dass die Kurve j ein Inneres hat, aber durch den Nullpunkt geht[1]). Abgesehen von höchstens zwei Richtungen ist der Radiusvektor von j wieder eine eindeutige Funktion der $r = g(\varphi)$; aus der Konvexität von j folgt unmittelbar, dass $g(\varphi)$ in einem Winkelraum mindestens der Grösse π verschwindet, und wir denken uns diesen Winkelraum symmetrisch zur negativen x-Achse gelegen. Es ist dann mit einem passenden $0 < \eta \leq \frac{\pi}{2}$.

$$\begin{aligned} g(\varphi) &= 0 \quad \text{für}\quad \eta < \varphi < 2\pi - \eta \\ g(\varphi) &> 0 \quad \text{für}\quad |\varphi| < \eta. \end{aligned}$$

Es sei dann Φ^* die Haube der durch

$$|\varphi| < \eta,\quad z > f(r, g(\varphi))$$

definierten offenen konvexen Menge (Φ^* besteht also aus dem Flächenstück $z = f(r, g(\varphi))$, der z-Achse und evtl. aus Stücken der beiden Halbebenen $\varphi = \pm\eta$). Um aus Φ^* eine in O differenzierbare Fläche zu machen, fügen wir die beiden kongruenten Kurven

$$\left.\begin{aligned} k:&\qquad \varphi = \pi - \frac{\eta}{2} \\ k':&\qquad \varphi = \pi + \frac{\eta}{2} \end{aligned}\right\}\ z = r^{\frac{3}{2}}$$

hinzu, und bilden die Haube Φ von $\Phi^* + k + k'$. Wir wollen zeigen, dass ihre Indikatrix auf j liegt.

---

[1]) Ausserdem gibt es nur noch den Fall, dass j kein Inneres hat, also aus einer durch O gehenden Strecke besteht. Dann ist die Behauptung dieser Nummer trivial, da z. B. die Fläche $z = r^{\frac{8}{3}}$ eine Indikatrix hat, die mit O zusammenfällt, und daher auf j liegt. Dass aber die Endpunkte der Strecke zur Indikatrix gehören können, wird in Nr. 4 bewiesen werden.

Zunächst ist klar, dass Φ im Winkelraum $|\pi-\varphi|\leq\frac{\eta}{2}$ mit der Haube von $k+k'$ zusammenfällt, also ein Stück des k und k' umhüllenden Zylinders mit horizontalen Erzeugenden ist. Alle Normalhalbschnitte von Φ in diesem Winkelraum haben also in O einen verschwindenden Krümmungsradius.

Im Winkelraum $\eta<\varphi<\pi-\frac{\eta}{2}$ besteht Φ, wie man leicht erkennt (vgl. den Beweis, S. 92), aus einer einparametrigen Schar von Strecken $\overline{P_1 P_2}$, mit P_1 auf Φ^*, P_2 auf k. Diese Endpunkte entsprechen sich eineindeutig, und es ist klar, dass P_1 mit P_2 gegen O strebt. Es seien $(r_1, \varphi_1, z_1=f(r_1, g(\varphi_1))$ die Koordinaten von P_1, und $(r_2, \pi-\frac{\eta}{2}, z_2=r_2^{\frac{3}{2}})$ die von P_2. Der Punkt P des Normalschnitts (φ) von Φ, $(\eta<\varphi<\pi-\frac{\eta}{2})$, der auf der Strecke $\overline{P_1 P_2}$ liegt, hat dann die Koordinaten

$$r=\frac{r_1 r_2 \sin(\varphi_1+\frac{\eta}{2})}{r_1 \sin(\varphi-\varphi_1)+r_2\sin(\varphi+\frac{\eta}{2})},\ \varphi,\ z=\frac{z_1 r_2 \sin(\varphi+\frac{\eta}{2})+z_2 r_1 \sin(\varphi-\varphi_1)}{r_1\sin(\varphi-\varphi_1)+r_2\sin(\varphi+\frac{\eta}{2})}.$$

Man erhält also für den oberen Krümmungsradius $\overline{\varrho\varphi}$ dieses Normalschnitts

$$\overline{\varrho\varphi}=\overline{\lim}\frac{r^2}{2z}=$$

$$=\overline{\lim}\frac{r_1^2 r_2^2 \sin^2(\varphi_1+\frac{\eta}{2})}{2\left\{z_1 r_2 \sin(\varphi+\frac{\eta}{2})+z_2 r_1\sin(\varphi-\varphi_1)\right\}\left\{r_1\sin(\varphi-\varphi_1)+r_2\sin(\varphi+\frac{\eta}{2})\right\}}.$$

Hierin wird die rechte Seite durch Fortlassung der Glieder mit r_2 im Nenner höchstens vergrössert; ferner hat $\sin(\varphi-\varphi_1)$ wegen $\varphi-\varphi_1>\varphi-\eta>0$ und $\pi-(\varphi-\varphi_1)>\frac{3}{2}\eta$ eine positive untere Grenze. Man erhält also

$$\overline{\varrho\varphi}\leq\overline{\lim}\frac{r_2^2}{2z_2}\cdot\frac{\sin^2\left(\varphi_1+\frac{\eta}{2}\right)}{\sin^2(\varphi-\varphi_1)}=0.$$

Die gleiche Rechnung gilt natürlich auch für den symmetrischen

Winkelraum $\pi+\frac{\eta}{2}<\varphi<2\pi-\eta$, und wir haben somit, dass die Indikatrix von Φ im ganzen Winkelraum $\eta<\varphi<2\pi-\eta$ mit O zusammenfällt.

Um zu beweisen, dass Φ auch im Winkelraum $|\varphi|<\eta$ eine Indikatrix hat und diese mit dem Bogen von j zusammenfällt, bemerken wir, dass die Koordinate φ_1 von P_1 bei Annäherung an O gegen η streben muss. Nach Definition ist nämlich zunächst $0\leq\varphi_1\leq\eta$. Wäre nun etwa für eine Teilfolge $\{P_1^\nu\}$ der Punkte P_1 mit $P_1^\nu\to O$ etwa $\lim\varphi_1^\nu=\psi<\eta$, so hätte jeder Normalschnitt (φ) von Φ mit $\psi<\varphi<\eta$ eine gegen O konvergierende Folge von Schnittpunkten mit den Strecken $\overline{P_1^\nu P_2^\nu}$; die soeben durchgeführte Rechnung zeigt[1]), dass diese Folge in O einen verschwindenden Krümmungsradius hätte, das ist aber deshalb unmöglich, weil Φ nach Definition Φ^* im Inneren oder auf dem Rande enthält, so dass die Normalschnitte von Φ für $|\varphi|<\eta$ unterhalb oder auf den entsprechenden Schnitten von Φ^* verlaufen und daher nur positive Krümmungsradien haben.

Somit ist $\lim\varphi_1=\eta$, d. h. die Schnittpunkte der Normalschnitte von Φ mit $0<\varphi<\eta$ mit den Strecken $\overline{P_1P_2}$ können O nicht beliebig nahe kommen. Diese Normalhalbschnitte fallen also ein Stück weit (von O gerechnet) mit denjenigen von Φ^* d. h. mit den Kurven $\lambda_{g(\varphi)}$ zusammen und haben daher in O dieselben Krümmungen wie die $\lambda_{g(\varphi)}$, w. z. b. w..

4. Wir wollen nun für den Fall, dass j eine oder zwei von O ausgehende Strecken enthält, beweisen, dass es konvexe Flächen mit einer ganz auf j liegenden Indikatrix gibt, bei denen ein beliebig vorgeschriebener Punkt dieser (abgeschlossenen) Strecken zu i gehört.

Wir betrachten zunächst den Sonderfall, das j kein Inneres hat, d. h. aus einer durch O gehenden Strecke besteht. Es ist also zu zeigen, dass der Normalkrümmungsradius ϱ_φ den Bedingungen

$$\varrho_\varphi=0 \text{ für } \varphi\neq 0 \text{ und } \varphi\neq\pi;\ \varrho_0=a\geq 0;\ \varrho_\pi=b\geq 0$$

genügen kann. Dies ist nach S. 93 offenbar der Fall für die Haube der vier Kurven:

[1]) Die Voraussetzung $\varphi>\eta$ wurde nur zur Feststellung verwandt, dass $\sin(\varphi-\varphi_1)$ eine positive untere Schranke hat.

$$z = r^{\frac{3}{2}},\ \varphi = \frac{\pi}{2} \text{ und } \varphi = \frac{3\pi}{2},$$

$$z = \frac{r^2}{2a} \text{ bzw. } z = \frac{r^2}{2b} \text{ für } \varphi = 0 \text{ bzw. } \varphi = \pi \text{ falls}$$

a bzw. b positiv sind, $z = r^{\frac{3}{2}}$ im Falle $a = 0$ bzw. $b = 0$.

Wenn die vorgeschriebene Indikatrix i ein Inneres hat, so denken wir uns die x-Achse wieder so orientiert, dass i die Gleichung $r = g(\varphi)$ erhält, mit

$$(i):\quad \begin{array}{ll} g(\varphi) > 0 & \text{für } |\varphi| < \eta,\quad 0 < \eta \leq \frac{\pi}{2} \\ g(\varphi) = 0 & \text{für } \eta < \varphi < 2\pi - \eta\ . \end{array}$$

Dieser allgemeine Fall lässt sich nun unmittelbar auf den Spezialfall der Indikatrix $\bar{i}$: $r = g(\varphi)$ zurückführen, mit

$$(\bar{i}):\quad \begin{array}{ll} g(\varphi) = \infty & \text{für } |\varphi| < \eta \\ g(\varphi) = 0 & \text{für } \eta \leq \varphi \leq 2\pi - \eta\ . \end{array}$$

Es sei nämlich $\bar{\Phi}$ eine Fläche mit der Indikatrix $\bar{i}$ in O, und Φ sei die in No. 3 definierte Fläche, deren Indikatrix für $|\varphi| \neq \eta$ mit $r = g(\varphi)$ übereinstimmt. Dann hat die Haube Φ^* des Durchschnitts der Inneren[1]) von Φ und $\bar{\Phi}$ in O offenbar eine Indikatrix mit der Gleichung

$$\begin{array}{lll} r = 0 & \text{für} & \eta \leq \varphi \leq 2\pi - \eta\ . \\ r = g(\varphi) & \text{für} & |\varphi| < \eta, \end{array}$$

und damit ist der Fall $g(\pm\eta) = 0$ erledigt. Ist aber etwa $g(\eta) > 0$, so genügt es die Haube H von Φ^* und der Parabel

$$z = \frac{r^2}{2g^2(\eta)}\ ,\quad \varphi = \eta$$

zu betrachten. Durch eine fast wörtliche Uebertragung der Betrachtung von S. 97 erkennt man wieder,[2]) dass für $|\varphi| < \eta$ die Normalschnitte von Φ^* und H von O gerechnet ein Stückweit übereinstimmen, und dass die obige Parabel Normalschnitt der Fläche ist.

[1]) Sind $z = U(r, \varphi)$ und $z = V(r, \varphi)$ die Gleichungen von Φ bzw. $\bar{\Phi}$, so ist Φ^* durch $z = \min \{ U(r, \varphi), V(r, \varphi) \}$ gegeben.

[2]) Wesentlich benutzt wird an beiden Stellen nur, dass $g(\eta) \leq \lim\limits_{\varphi \to \eta -} g(\varphi)$ ist.

Zur Konstruktion eine Fläche mit der Indikatrix $\bar{i}$ betrachten wir wieder die beiden Kurven

$$\left.\begin{aligned} k: \quad & \varphi = \pi - \frac{\eta}{2} \\ k': \quad & \varphi = \pi + \frac{\eta}{2} \end{aligned}\right\} \quad z = r^{\frac{3}{2}}$$

und ausserdem eine geschlossene konvexe zur x-Achse symmetrische Kurve γ der x,y-Ebene, die von O ausgeht und dort die beiden Halbgeraden $\varphi = \pm \eta$ berührt; sie habe die Gleichung $r = \mu(\varphi)$, so dass

$$(\gamma): \quad \begin{aligned} & \mu(\varphi) > 0 \quad && \text{für} \quad 0 \leq \varphi < \eta, \\ & \mu(-\varphi) = \mu(\varphi) \quad , && \lim_{\varphi \to \eta -} \mu(\varphi) = 0 \end{aligned}$$

ist. Es bezeichne ferner Z den im Halbraum $z \geq 0$ gelegenen Halbzylinder mit der Leitkurve γ, dessen Erzeugende parallel zur z-Achse sind. Wir wollen zeigen, dass die Haube von $z + k + k'$ in O die Indikatrix i hat, vorausgesetzt bloss, dass $\mu(\varphi)$ hinreichend klein gewählt vurde; es genügt z. B., dass

$$\mu(\varphi) \leq \sin^6 \frac{(\eta - \varphi)}{2} \cdot \sin^3 \frac{\eta}{4}$$

ist. (Das Gleichheitszeichen hierin liefert übrigens bereits eine konvexe Kurve, und man kann diese als γ nehmen).

Es ist leicht, sich über das Aussehen von $\bar{\Phi}$ Rechenschaft abzulegen. Im Winkelraum $|\pi - \varphi| < \frac{\eta}{2}$ besteht $\bar{\Phi}$ natürlich wieder aus einem Stück des horizontalen Zylinders durch k und k', hat dort also verschwindende Krümmungsradien. Ferner enthält $\bar{\Phi}$ natürlich den von γ begrenzten ebenen Bereich; für $|\varphi| < \eta$ enthält also der Normalschnitt (φ) von $\bar{\Phi}$ die Strecke $0 < r < \mu(\varphi)$, $z = 0$, und hat daher in O einen unendlichen Krümmungsradius. Wir haben also nur den übrigen Teil von $\bar{\Phi}$ zu untersuchen[1]), wobei wir uns aus Symmetriegründen auf den Halbraum $\varphi \geq 0$ beschränken können.

[1]) Dass die Normalschnitte für $\varphi > \eta$ in O verschwindende Krümmungsradien haben, folgt aus dem Beweis von No. 3, und gilt natürlich ohne Einschränkung für $\mu(\varphi)$. Diese ist nur für den Schnitt $\varphi = \eta$ notwendig, sowie für eine Verschärfung unsrer Behauptung, die wir für spätere Zwecke brauchen.

Nun sieht man leicht (vgl. die Schlussweise von S. 92), dass der restliche Teil von $\overline{\Phi}$ aus einem Stück der Einhüllenden Λ der Ebenen besteht, welche gleichzeitig Stützebenen an Z und k sind. Man erhält Λ folgendermassen: Es sei $P_1 = (r_1 = \mu(\varphi_1), \varphi_1, 0)$ ein Punkt der Kurve γ (natürlich mit $\varphi_1 \geqq 0$), in dem es eine Stützgerade t an γ gibt, welche die Halbgerade $\varphi = \pi - \frac{\eta}{2}$ in Q_1 schneidet. Von Q_1 aus zieht man die Tangente an k, ihr Berührungspunkt sei $P_2 = \left(r_2, \pi - \frac{\eta}{2}, z_2 = r_2^{\frac{3}{2}}\right)$. Dann ist die Ebene $\underline{P_1 Q_1 P_2}$ offenbar Stützebene sowohl an Z als auch an k und daher auch an $\overline{\Phi}$. Die Strecke $\overline{P_1 P_2}$ gehört daher Φ an. Bezeichnet also P^* den Punkt von γ (mit $\varphi \geqq 0$), in welchem es eine zu $\varphi = \pi - \eta$ parallele Stützhalbgerade gibt, so entspricht jedem Punkt P_1 des Bogens $\widehat{OP^*}$ von γ ein Punkt P_2 von k (und falls γ differenzierbar ist, auch nur ein solcher Punkt), und $\overline{\Phi}$ besteht aus den Strecken $\overline{P_1 P_2}$. Für $P_1 \to P^*$ strebt diese Strecke gegen die Zylindererzeugende durch P, und der Zylinderteil zwischen P^* und dem dazu in bezug auf $\eta = 0$ symmetrischen Punkt gehört $\overline{\Phi}$ an.

Wir erhalten also den Normalschnitt (φ) von $\overline{\Phi}$ für $\eta \leqq \varphi \leqq \pi - \frac{\eta}{2}$, indem wir die Punkte der Strecken $\overline{P_1 P_2}$ berechnen, deren Winkelkoordinate φ ist. Für deren Koordinaten erhalten wir wieder (vgl. S. 96, wo jetzt $z_1 = 0$, $r_1 = \mu(\varphi_1)$ zu setzen ist):

$$r = \frac{r_2 \mu(\varphi_1) \sin(\varphi_1 + \frac{\eta}{2})}{\mu(\varphi_1) \sin(\varphi - \varphi_1) + r_2 \sin(\varphi + \frac{\eta}{2})}, \ \varphi, \ z = r_2^{\frac{3}{2}} \frac{\mu(\varphi_1) \sin(\varphi - \varphi_1)}{\mu(\varphi_1) \sin(\varphi - \varphi_1) + r_2 \sin(\varphi + \frac{\eta}{2})}.$$

Hieraus ergibt sich

$$r = \frac{z^2}{r^2} \frac{\mu(\varphi_1) \sin^3(\varphi_1 + \frac{\eta}{2})}{\mu(\varphi_1) \sin^3(\varphi - \varphi_1) + \frac{z^2}{r^2} \sin^2(\varphi_1 + \frac{\eta}{2}) \sin(\varphi + \frac{\eta}{2})}.$$

Zur Abschätzung der rechten Seite zerlegen wir das φ-Intervall in zwei Teile:

1) Es sei $\varphi \geq \eta + \frac{\pi}{8}$. Dann ist wegen $\varphi_1 < \eta \leq \frac{\pi}{2}$

$$\frac{\pi}{8} \leq \varphi - \varphi_1 < \pi - \left(\frac{\eta}{2} - \varphi_1\right),$$

und daher

$$\sin(\varphi - \varphi_1) \geq \min\left\{\sin\frac{\pi}{8},\ \sin\left(\varphi_1 + \frac{\eta}{2}\right)\right\} \geq \sin\frac{\pi}{8}\sin\left(\varphi_1 + \frac{\eta}{2}\right).$$

Man erhält also

$$r < \frac{z^2}{r^2}\frac{1}{\sin^3\frac{\pi}{8}},$$

oder

$$z > \left(r \sin\frac{\pi}{8}\right)^{\frac{3}{2}}.$$

2) Es sei $\eta \leq \varphi \leq \eta + \frac{\pi}{8}$. Dann ist

$$\eta - \varphi_1 \leq \varphi - \varphi_1 \leq \frac{5\pi}{8} < \pi - \frac{\eta - \varphi_1}{2}$$

und

$$\varphi_1 + \frac{\eta}{2} < \varphi + \frac{\eta}{2} \leq \frac{3\eta}{2} + \frac{\pi}{8} \leq \pi - \frac{\eta}{4},$$

und daher

$$\sin(\varphi - \varphi_1) > \sin\frac{\eta - \varphi_1}{2},\ \sin\left(\varphi + \frac{\eta}{2}\right) > \sin\frac{\eta}{4},\ \sin\left(\varphi_1 + \frac{\eta}{2}\right) > \sin\frac{\eta}{4}.$$

Man erhält damit

$$r < \frac{z^2}{r^2}\frac{\mu(\varphi_1)}{\mu(\varphi_1)\sin^3\frac{\eta - \varphi_1}{2} + \frac{z^2}{r^2}\sin^3\frac{\eta}{4}}.$$

Da die rechte Seite mit $\mu(\varphi_1)$ monoton wächst, wird hieraus für $\mu(\varphi_1) \leq \sin^6\frac{\eta - \varphi_1}{2}\sin^3\frac{\eta}{4}$

$$r < \frac{z^2}{r^2}\frac{\sin^6\frac{\eta - \varphi_1}{2}}{\sin^9\frac{\eta - \varphi_1}{2} + \frac{z^2}{r^2}} \leq \frac{z^2}{r^2}\max_{t \geq 0}\frac{t^{\frac{2}{3}}}{t + \frac{z^2}{r^2}} = \frac{1}{3}2^{\frac{2}{3}}\left(\frac{z}{r}\right)^{\frac{4}{3}},$$

oder

$$z > r^{\frac{7}{4}}.$$

Sämtliche Normalschnitte von $\overline{\Phi}$ im betrachteten Winkelraum liegen also ganz oberhalb entweder von $z =$ konst. $r^{\frac{3}{2}}$ oder von

$z = r^{\frac{7}{4}}$, und da diese beiden Kurven in O verschwindende Krümmungsradien besitzen, gilt dasselbe auch von den Normalschnitten, w. z. b. w..

Damit ist nicht nur die Behauptung (a) der Einleitung vollständig beweisen, sondern es hat sich sogar eine Gleichmässigkeitseigenschaft der Normalschnitte von $\overline{\Phi}$ ergeben, die wir zusammen mit einer bereits erwähnten weiteren Eigenschaft dieser Schnitte zwecks späterer Anwendung noch besonders vermerken. Da die zu Anfang dieser Nummer mit Φ^* bezeichnete Fläche ganz oberhalb oder auf $\overline{\Phi}$ verläuft, haben wir:

Stellt

$$r = g(\varphi) > 0 \quad \text{für } |\varphi| < \eta$$
$$r = 0 \quad \text{für } \eta \leq \varphi \leq 2\pi - \eta$$

eine nach I mögliche Indikatrix i dar, so gibt es eine in O differenzierbare Fläche mit der Indikatrix i und folgenden Eigenschaften:

(1) Für $|\varphi| < \eta$ fällt der Normalschnitt (φ) von Φ^* in einem Bogen der Länge $\delta(\varphi) > 0$ von O aus gerechnet mit der Kurve $\lambda_{g(\varphi)}$ zusammen, und zwar hat $\delta(\varphi)$ in jedem Intervall $|\varphi| \leq \eta' < \eta$ eine positive untere Grenze.

(2) Alle Normalschnitte (φ) im abgeschlossenen Winkelraum $\eta \leq \varphi \leq 2\pi - \eta$ verlaufen ganz oberhalb der durch

$$z = \min\left\{\left(r \sin\frac{\pi}{8}\right)^{\frac{3}{2}}, r^{\frac{7}{4}}\right\}$$

definierten Fläche.

Wir machen noch eine Anmerkung, die später Verwendung finden wird. Es sei Φ die soeben konstruierte Fläche mit der Indikatrix i der Gleichung $r = g(\varphi)$, und es stelle $r = g_1(\varphi)$ eine andere mögliche Indikatrix i_1 dar, wobei durchweg

$$g(\varphi) \leq g_1(\varphi)$$

ist. Dann gibt es eine in O differenzierbare konvexe Fläche Φ_1 mit der Indikatrix i_1 in O, die ganz unterhalb oder auf Φ verläuft (und natürlich ebenfalls im Halbraum $z \geq 0$). Konstruiert man nämlich nach dem beschriebenen Verfahren eine Fläche mit der Indikatrix i_1, so gelangt man offenbar höchstens beim zweiten Schritt von No. 3 — beim Haubenbilden von $\Phi^* + k + k'$ — ins Innere von Φ. Nun geht aber aus der Konstruktion hervor, dass die Lage der Ebenen von k und k' ganz gleichgültig ist, so dass man zur Konstruktion auch von Φ_1 etwa dieselben Kurven k, k' verwenden kann, wie für Φ. Es liegt auf der Hand, dass man so zu einer Fläche Φ_1 mit der gewünschten Eigenschaft gelangt.

5. Wir wollen nun die auf S. 89 angegebenen Stetigkeitseigenschaften der oberen Indikatrix bestätigen, die z. T. auch zum Beweise von (b) gebraucht werden.

Wir legen der Betrachtung die Bezeichnungen von II, S. 88 zugrunde und denken uns dabei das Koordinatensystem so orientiert, dass $g(0) > 0$ wird. Es sei dann P ein Punkt der positiven x-Achse im Abstand $p \leq g(0)$ von O (das Ungleichheitszeichen wird zugelassen, um den Fall $g(0) = \infty$ miteinzubegreifen). Wir betrachten etwa die Halbebene $0 < \varphi < \pi$ und bezeichnen den laufenden Punkt $(G(\varphi), \varphi)$ von I mit R_φ. Schliesslich werde der Winkel OPR_φ mit $\vartheta = \vartheta(\varphi)$ bezeichnet; zufolge des Sinussatzes hängt er mit $G(\varphi)$ durch

$$p = G(\varphi)(\cos\varphi + \sin\varphi \cdot \cot\vartheta)$$

zusammen.

Ist nun etwa $0 < \varphi < \psi < \pi$, so folgt aus II, S. 88 unmittelbar, dass R_φ nicht im Inneren des Dreiecks OPR_φ liegen kann, dass also $\vartheta(\varphi) \geq \vartheta(\psi)$: Für $0 < \varphi < \pi$ fällt also $\vartheta(\varphi)$ mit wachsendem φ monoton. Daher existieren für jedes φ dieses Intervalls die Grenzwerte $\vartheta(\varphi-)$ und $\vartheta(\varphi+)$, und es ist

$$\vartheta(\varphi-) \geq \vartheta(\varphi) \geq \vartheta(\varphi+).$$

Andrerseits ist im betrachteten Winkelraum $0 < \vartheta < \pi$, und in diesem Winkelraum fällt $\cot\vartheta$ monoton. Aus dem obigen Zusammenhang zwischen ϑ und $G(\varphi)$ folgt also unmittelbar, dass für $0 < \varphi < \pi$ auch die Grenzwerte $G(\varphi-)$ und $G(\varphi+)$ existieren mit

$$G(\varphi-) \geq G(\varphi) \geq G(\varphi+).$$

Die entsprechenden Beziehungen für das Intervall $\pi < \varphi < 2\pi$ lauten natürlich

$$G(\varphi+) \geq G(\varphi) \geq G(\varphi-), \qquad \text{w. z. b. w.}$$

Wie schon in der Einleitung erwähnt, liegt hierin, dass sobald $g(\varphi_1) > 0$ und $g(\varphi_2) > 0$ für zwei Richtungen mit $0 < \varphi_2 - \varphi_1 < \pi$,

$G(\varphi)$ stetig ist für $\varphi_1 < \varphi < \varphi_2$ und $\varphi_1 + \pi < \varphi < \varphi_2 + \pi$. Insbesondre ist also $G(\varphi)$ durchwegs stetig, wenn i den Punkt O im Inneren enthält. – Bemerkenswert ist vielleicht noch eine unmittelbare Folgerung aus unsren Ungleichungen: Wenn sowohl $g(\varphi_0) > 0$ als auch $g(\varphi_0 + \pi) > 0$ ist, so ist $G(\varphi)$ stetig für alle φ mit Ausnahme höchstens der Richtungen φ_0 und $\varphi_0 + \pi$. Zum Beweise genügt es die positive x-Achse einmal in die Richtung φ_0, dann in die Richtung $\varphi_0 + \pi$ zu legen.

Wir schliessen nun noch einige Bemerkungen für den Fall $g(0) > 0$ an. Die Existenz der Grenzwerte $G(\varphi -)$ und $G(\varphi +)$ besagt, dass $G(\varphi)$ (ausser für $\varphi = 0$ und $\varphi = \pi$) höchstens sog. Unstetigkeiten erster Art haben kann. Hieraus folgert man aber leicht, dass $G(\varphi)$ höchstens abzählbar viele Unstetigkeitsstellen besitzt. Denn die Stellen φ, in denen $G(\varphi -) - G(\varphi +) > \frac{1}{n}$ ist, können infolge der Existenz einseitiger Grenzwerte und der Monotonität von $\vartheta(\varphi)$ offenbar keine von 0 und π verschiedene Häufungsstellen haben, d. h. es gibt für jedes n höchstens abzählbar viele solche Stellen. Für $n \to \infty$ strebt aber diese Menge gegen die Gesamtheit aller Unstetigkeitsstellen ausser evtl. 0 und π. Weiter folgt, dass $G(\varphi)$ vollständig bestimmt ist, sobald man dessen Werte in einer abzählbaren überall dicht liegenden Menge $\{\varphi_n\}$ kennt, in der die Unstetigkeitsstellen vorkommen. Denn die Werte $G(\varphi_n)$ bestimmen alle Grenzwerte $G(\varphi +)$ und $G(\varphi -)$ eindeutig. Wir notieren dieses Ergebnis in der Form, in welcher es später verwandt wird: i und I mögen den Bedingungen von II, S. 88 genügen und i falle nicht mit O zusammen. Hat dann die konvexe in O differenzierbare Fläche Φ dort i zur unteren Indikatrix, und fällt ihre obere Indikatrix in sämtlichen – höchstens abzählbar vielen – Unstetigkeitsstellen von $G(\varphi)$ und ausserdem in einer überall dicht liegenden Menge von φ-Werten mit $G(\varphi)$ zusammen, so stellt $r = G(\varphi)$ für alle φ-Werte die obere Indikatrix von φ dar.

Dass $G(\varphi)$ im allgemeinem keine weitergehenden Stetigkeitseigenschaften hat, folgt aus dem nun zu beweisenden Satz (b).

6. Den Beweis, dass die Bedingungen des Satzes (b), S. 89 hinreichend sind, werden wir wieder durch Konstruktion der entsprechenden Fläche erbringen. Das allgemeine Verfahren ist jedoch so verwickelt, dass der einfache Grundgedanke nicht mehr klar hervortritt; der grösseren Anschaulichkeit wegen führen wir daher in dieser Nummer zunächst die Konstruktion für den Spezialfall durch, dass die Kurve i den Punkt O im Inneren enthält. Dabei benutzen wir wesentlich eine Idee, die uns Prof. *B. Jessen*, Kopenhagen, gesprächsweise mitgeteilt hat.

i und I seien durch $r = g(\varphi) > 0$ bzw. $r = G(\varphi)$ gegeben, und der Bereich $r \leq G(\varphi)$ umfasse mit jedem Punkt Q zugleich die konvexe Hülle von $Q + i$ (vgl. II und Fussnote [1]) auf S. 91). Dann sind nach dem Vorangehendem i und I geschlossene stetige Kurven.

Wir bilden die beiden Flächen

$$\omega: \qquad z = f(r, g(\varphi)),$$
$$\Omega: \qquad z = f(r, G(\varphi)),$$

deren Schnitte $\varphi =$ Konst. Kurven $\lambda_{g(\varphi)}$ bzw. $\lambda_{G(\varphi)}$ sind (Definition S. 95). (Ω braucht nicht konvex zu sein). Die konvexe Hülle von ω und irgend einem Normalschnitt $\varphi = \varphi_0$ von Ω liegt dann ganz im Inneren oder auf dem Rande von Ω; es sei nämlich $r = h(\varphi)$ die Gleichung des Randes der konvexen Hülle von i und dem Punkt $r = G(\varphi_0)$, $\varphi = \varphi_0$, $z = 0$. Dann ist nach Voraussetzung $h(\varphi) \leq G(\varphi)$ für alle φ, und daher liegt die Fläche
$$z = f(r, h(\varphi))$$
oberhalb oder auf Ω. Andrerseits ist diese Fläche — da $r = h(\varphi)$ eine konvexe Kurve darstellt — konvex, und enthält daher die konvexe Hülle von ω und dem betrachteten Normalschnitt. Insbesondre verläuft also auch die Haube von ω und einem beliebigen Punkt von Ω in oder auf Ω.

Es sei nun $\{\varphi_n\}$ eine abzählbare, in $(0, 2\pi)$ dichte Folge von Richtungen und $\{\varphi^n\}$ eine solche Anordnung derselben, dass jedes φ_n unter den φ^n unendlich oft vorkommt (etwa $\{\varphi^n\} = \varphi_1, \varphi_2, \varphi_1, \varphi_2, \varphi_3, \varphi_1, \cdots$). Q^1 sei ein von O verschiedener Punkt des Normalschnitts $\varphi = \varphi^1$ von Ω. Der Tangentialkegel von Q^1 an ω berührt ω längs einer Kurve, die einen positiven Abstand von O hat; sie verlaufe etwa oberhalb der Ebene $z = z_1 > 0$. Dann wählen wir auf den Normalschnitt $\varphi = \varphi^2$ von Ω einen Punkt $Q^2 \neq O$ derart, dass die Berührungskurve des Tangentialkegels von Q^2 an ω ganz unterhalb der Ebene $z = \frac{z_1}{2}$ verläuft. So fortfahrend erhalten wir eine Punktfolge $\{Q^n\}$, bei welcher Q^n auf dem Normalschnitt $\varphi = \varphi^n$ von Ω liegt, und derart, dass die Berührungskurve des Tangentialkegels von Q^n an ω ganz zwischen zwei Ebenen $z = \frac{z_{n-1}}{2}$ und $z = z_n < \frac{z_{n-1}}{2}$ verläuft.

Die einzelnen Tangentialkegel haben daher keine Punkte gemeinsam, und die konvexe Hülle von ω mit allen Punkten Q^n besteht daher aus der Vereinigungsmenge der von ω und allen diesen Tangentialkegeln (die wir natürlich stets nur bis zur Berührungskurve mit ω rechnen) begrenzten abgeschlossenen Bereiche. Ihr Rand ist eine konvexe in O differenzierbare Fläche Φ, die nach der Vorbemerkung zwischen ω und Ω verläuft; stellt also $r = g^*(\varphi)$ die untere, $r = G^*(\varphi)$ die obere Indikatrix von Φ in O dar, so ist

$$g^*(\varphi) \geq g(\varphi) \qquad \text{und } G^*(\varphi) \leq G(\varphi).$$

Andrerseits fallen etwa die Schnitte $z = \frac{3}{4} z_n$ von Φ und ω zusammen, und daher ist $g^*(\varphi) = g(\varphi)$. Ferner hat für jedes n der Normalschnitt $\varphi = \varphi_n$ von Φ mit demjenigen von Ω unendlich viele Punkte gemeinsam, und daher ist für alle diese Richtungen $G^*(\varphi_n) = G(\varphi_n)$. Wegen der Stetigkeit der beiden Funktionen ist also durchweg $G^*(\varphi) = G(\varphi)$.

7. Wir gehen nun zum allgemeinem Fall des Satzes (b) über. Wir haben zu zeigen: Wenn Γ mit jedem Punkt Q zugleich die konvexe Hülle von $Q + i$ enthält, so gibt es eine in O differenzierbare Fläche Φ mit i als unterer und I als oberer Indikatrix.

Es bezeichne für den Augenblick wieder R_φ den laufenden Punkt $(G(\varphi), \varphi)$ von I. Für $0 \leq \psi \leq 2\pi$ bezeichne σ_ψ die konvexe Hülle von $R_\psi + i$, und i_ψ ihren Rand, abgesehen von den inneren Punkten der auf ihm etwa enthaltenen von O ausgehenden Strecken (in deren Richtung also der von O verschiedene Endpunkt mitzählt).

$$(i_\psi): \qquad r = g_\psi(\varphi)$$

sei die Gleichung von i_ψ, so dass speziell

$$g_\psi(\psi) = G(\psi)$$

wird. Nach Voraussetzung ist für jedes ψ

$$\gamma \subset \sigma_\psi \subset \Gamma,$$

und daher gilt für alle φ und ψ

$$g(\varphi) \leq g_\psi(\varphi) \leq G(\varphi).$$

Schliesslich bezeichne H die nach den Vorschriften der Nr. 3 – 4 konstruierte Fläche mit der Indikatrix i in O.

Wir beginnen mit der Konstruktion einer Fläche H_ψ zu jedem ψ, und ausserdem einer festen Fläche H mit folgenden Eigenschaften:

(1) H_ψ ist konvex und berührt in O (von oben) die (x, y)-Ebene.

(2) In O hat H_ψ die Indikatrix i_ψ.

(3) Das Stück $0 \leq z \leq \zeta$ (mit $\zeta = \zeta(\psi) > 0$) von H liegt ganz oberhalb oder auf H_ψ.

(4) In endlich vielen beliebig vorgegebenen Richtungen der (x, y)-Ebene $\chi_1, \chi_2, \cdots, \chi_N$ verläuft der Schnitt $\varphi = \chi_n$ ($n = 1, \cdots, N$) von H_ψ für $z \leq \zeta$ im Falle $g_\psi(\chi_n) > 0$ ganz oberhalb oder auf der Kurve $\lambda_{g_\psi(\chi_n)}$ (Definition vgl. S. 95), und falls $g_\psi(\chi_n) = 0$ ist, oberhalb einer nur von der Richtung χ_n (nicht aber von ψ) abhängigen Kurve λ'_{χ_n}, welche in O die x, y-Ebene berührt und dort einen verschwindenden Krümmungsradius bezitzt.

Nach der Schlussbemerkung von No. 4, S. 102, gibt es zunächst eine Fläche H'_ψ, welche den Forderungen (1)—(3) genügt, und zwar für alle z. Diese Fläche wollen wir nun in N Schritten so abändern, dass auch (4) erfüllt wird.

Wir betrachten zuerst eine solche Richtung χ_n, für welche $g_\psi(\chi_n) > 0$ ist. Wenn χ_n eine Unstetigkeitsstelle von $g_\psi(\varphi)$ ist (so dass die konvexe Kurve j_ψ, auf der i_ψ liegt, in der Richtung χ_n eine von O ausgehende Strecke enthält), so liegt nach No. 4, S. 98 die Kurve $\lambda_{g_\psi(\chi_n)}$ der Halbebene $\varphi = \chi_n$ von selbst auf H'_ψ, und es ist nichts zu beweisen. Sonst gibt es im Punkte $(g_\psi(\chi_n), \chi_n)$ von i_ψ eine Stützgerade t an i_ψ (d. h. an j_ψ), die nicht durch O geht. Es sei dann Z der Zylinder, dessen Erzeugende parallel zu t sind, der die Halbebene $\varphi = \chi_n$ längs der Kurve $\lambda_{g_\psi(\chi_n)}$ schneidet und ferner eine Halbebene von $z = 0$ enthält, die von der Geraden t begrenzt wird. Z ist demnach in O differenzierbar und hat dort die Gerade t zur Indikatrix. Wegen $g(\varphi) \leq g_\psi(\varphi)$, liegt i ebenso wie i_ψ ganz auf einer Seite von t, und aus der Konstruktion von H und Z geht ohne weiteres hervor, dass H in einer Umgebung von O ganz oberhalb oder auf Z liegt. Bildet man also den Durchschnitt der Inneren von H_ψ und Z (vgl. Fussnote [1]), S. 98) so ist ihr Rand eine in O differenzierbare konvexe Fläche, die natürlich i_ψ zur Indikatrix hat und in der erwähnten Umgebung von O nicht ins Innere von H dringt; ihr Schnitt mit der Halbebene $\varphi = \chi_n$ verläuft aber ganz oberhalb oder auf $\lambda_{g_\psi(\chi_n)}$. Entsprechend verfahren wir der Reihe nach mit allen Richtungen χ_n, $n \leq N$, für welche $g_\psi(\chi_n) > 0$ ist.

Nun betrachten wir eine Richtung χ_n, für welche $g_\psi(\chi_n) = 0$ und daher erst recht $g(\chi_n) = 0$ ist. Wenn nun $g(\varphi)$ durchweg stetig ist, oder aber in den Unstetigkeitspunkten verschwindet — d. h. wenn $g(\pm\eta) = 0$ in der Bezeichnung von S. 98 — so verläuft der Schnitt von H mit der Halbebene $\varphi = \chi_n$ mindestens für $z \leq \sin^{\frac{21}{2}} \frac{\pi}{8}$ oberhalb von $z = r^{\frac{7}{4}}$ (vgl. S. 102). Wir bezeichnen dann die Kurve $z = r^{\frac{7}{4}}$, $\varphi = \chi_n$, mit λ'_{χ_n}. Hat aber $g(\varphi)$ Unstetigkeitsstellen φ_ν $(\nu = 1, 2)$ mit $g(\varphi_\nu) > 0$, so betrachten wir die Haube der Rotationsfläche $z = r^{\frac{7}{4}}$ mit den Kurven (oder der Kurve) $z = f(r, g(\varphi_\nu))$, $\varphi = \varphi_\nu$ und ihre Schnittkurve mit der Halbebene $\varphi = \chi_n$ bezeichnen wir mit λ'_{χ_n}. Sie hängt, wie ausdrücklich bemerkt sei, nur von i (sogar nur von den Unstetigkeitspunkten von i) ab, nicht von ψ. Aus der Konstruktion von H folgt unmittelbar (vgl. S. 102), dass die Schnittkurve von H mit $\varphi = \chi_n$ mindestens in der Umgebung $z < \sin^{\frac{21}{2}} \frac{\pi}{8}$ oberhalb oder auf λ'_{χ_n} verläuft. In beiden Fällen können wir also wieder das für den Fall $g_\psi(\chi_n) > 0$ beschriebene Verfahren anwenden, indem wir diesmal die Erzeugenden des Zylinders Z parallel zu einer Stützgeraden t von i_ψ in O nehmen, und als Leitkurve die Kurve λ'_{χ_n} zusammen mit der Halbgeraden $\varphi = \pi + \chi_n$, $z = 0$. (Der Radiusvektor der Indikatrix von Z in O wird diesmal unendlich in dem einen von t begrenzten Winkelraum und verschwindet in dem andern). — Wendet man dieses Verfahren für alle Richtungen an, so kommt man schliesslich zu einer Fläche H_ψ, für welche auch (4) erfüllt ist mindestens in der kleinsten der Umgebungen, die sich bei den einzelnen Schritten ergeben haben.

Nach diesen Vorbemerkungen bestimmen wir zur Konstruktion der gesuchten Fläche eine Folge von paarweise verschiedenen Richtungen $\{\psi_n\}$ der x, y-Ebene, welche alle Unstetigkeitspunkte von $G(\varphi)$ enthält, und darüber hinaus eine in $(0, 2\pi)$ dicht liegende Menge von Richtungen. Zu jedem ψ_n konstruieren wir nach dem eben beschriebenen Verfahren eine Fläche $H_{\psi_n} = H_n$, die (1)—(4) erfüllt, und zwar seien $\chi_i = \psi_i$, $i = 1, \cdots, n-1 = N$ die endlich vielen vorgegebenen Richtungen und etwa $0 \leq z \leq \zeta_n$ der Bereich, in dem (3) und (4) bestehen.

Wie in No. 6 werden nun die ψ_n in eine Folge $\{\varphi^n\}$ geordnet, so dass jedes ψ_n unter den φ^n unendlich oft vorkommt, und es sei etwa

$$\varphi^n = \psi s_n$$

für jedes n. Schliesslich werde der Schnitt der Fläche Hs_n mit der Ebene $z = h > 0$ mit $q_n(h)$ bezeichnet. Für $h < \zeta s_n$ ist $q_n(h)$ eine geschlossene Kurve die ganz ausserhalb von H verläuft: die Haube von $q_n(h) + H$ unterscheidet sich von H nur in einem Streifen $h' < z < h''$, und zwar gehen h' und h'' mit h gegen 0 (was z. B. daraus folgt, dass alle unsre Flächen in der Rotationsfläche $z = r^3$ enthalten sind). Ferner liegt diese Haube für hinreichend kleine h (nämlich jedenfalls sobald auch $h'' < \zeta s_n$ wird) ganz in oder auf Hs_n. Wir können also ähnlich wie in No. 6 eine Folge $h_n \to 0$ derart bestimmen, das sich die Haube von $q_n(h_n) + H$ von H nur in einem Streifen $h'_n < z < h''_n$ unterscheidet, etwa mit

$$h''_{n+1} < \frac{1}{2} h'_n .$$

Ferner sei h_n so gewählt, dass auch noch $h'_n < \zeta s_n$ wird. Die einzelnen Hauben stören sich gegenseitig nicht, und daher ist die Haube von H und der Vereinigungsmenge aller Schnitte $q_n(h_n)$ gleich der Haube der Vereinigungsmenge aller konvexen Hüllen von H mit den einzelnen Schnitten: wir bezeichnen diese Haube mit Φ, und wollen zeigen, dass sie in O die Kurve i zur unteren und I zur oberen Indikatrix hat.

Zur diesem Zwecke bezeichnen wir die beiden Indikatrizen von Φ mit i^* bzw. I^*. Da nun Φ einerseits H im Inneren oder auf dem Rande enthält, andrerseits in einer Folge von Schnitten (etwa $z = \frac{3}{4} h_n''$) mit H übereinstimmt, schliesst man genau wie in Nr. 6 unmittelbar dass $i^* = i$ ist.

Wenn nun $r = G^*(\varphi)$ die Gleichung von I^* ist, so genügt es nach S. 104 zu zeigen, dass $G^*(\varphi)$ und $G(\varphi)$ an den Stellen $\varphi = \psi_n$ gleich sind. Da aber unendlich viele Punkte des Schnittes $\varphi = \psi_n$ von H_n auf Φ liegen, ist trivialerweise

$$G(\psi_n) \leq G^*(\psi_n).$$

Zum Nachweis der umgekehrten Richtung werde N beliebig aber fest gewählt, und es bezeichne P einen Punkt von Φ mit der Winkelkoordinate $\varphi = \psi_N$. Gehört dann für irgend ein n mit $s_n > N$ der Punkt P der Haube von $H + q_n(h_n)$ an, so liegt P im Falle, dass $g_{\psi_{s_n}}(\psi_N) > 0$, oberhalb der Kurve $\lambda g_{s_n}(\psi_N)$, und im Falle, dass $g_{\psi_{s_n}}(\psi_N) = 0$, oberhalb der Kurve λ'_{ψ_N} (welche in O einen verschwindenden Krümmungsradius hat). Wegen $g_\psi(\varphi) \leq G(\varphi)$ liegt im ersten Falle P erst recht oberhalb von $\lambda_{G(\psi_n)}$. – Wenn schliesslich P der Haube von $H + q_n(h_n)$ mit $s_n \leq N$ angehört, so liegt P oberhalb des Schnittes $\varphi = \psi_N$ der Fläche $H s_n$, und auch dessen Krümmungsradius in O kann $G^2(\psi_N)$ nicht übersteigen. Demnach liegt P in allen Fällen oberhalb einer der erwähnten $N+2$ Kurven, mit Krümmungsradien $\leq G^2(\psi_N)$, und der Schnitt $\varphi = \psi_N$ von Φ enthält daher keinen Punkt, der unterhalb des von diesen Kurven zusammen mit der z-Achse begrenzten Bereiches liegt. Damit ist auch

$$G^*(\psi_N) \leq G(\psi_N)$$

gezeigt.

8. Der Beweis der Behauptung (c) der Einleitung ist äusserst einfach. Es sei

$$z = F(r, \varphi) \geq 0$$

die Gleichung der zu untersuchenden in O differenzierbaren konvexen Fläche Φ; ohne Beschränkung der Allgemeinheit kann vorausgesetzt werden, dass $F(r, \varphi) > 0$ ist für $r > 0$, da man sonst durch Durchschnittbildung (vgl. Fussnote [1]), S. 98) von Φ mit der Rotationsfläche $z = r^3$ zu einer Fläche gelangt, die mit der (x, y)-Ebene nur den Punkt O gemeinsam hat und in O dieselbe obere (und untere) Indikatrix besitzt, wie Φ. — Die Gleichung der oberen Indikatrix von Φ in O sei wieder

$$r = G(\varphi),$$

und Γ bezeichne den Bereich $r \leq G(\varphi)$.

Nach K. k. F. S. 16 erhält man Γ folgendermassen: Es bezeichne σ_h die Kurve

$$(\sigma_h): \qquad F(\sqrt{h}\, r, \varphi) = \frac{h}{2}, \quad z = 0.$$

σ_h ist die von O aus ähnlich und im Masstab $\sqrt{h}:1$ vergrösserte Projektion der Schnittkurve von Φ mit der Ebene $z = \frac{h}{2}$ auf die x, y-Ebene, und daher ist σ_h konvex und enthält O im Inneren. Γ_n sei die Vereinigungsmenge der durch die Kurven σ_h für $0 < h < \frac{1}{n}$ begrenzten abgeschlossenen Bereiche. Mit wachsendem n fällt Γ_n monoton: der Limes dieser fallenden Mengenfolge ist gerade Γ.

Wen nun ein Punkt (r_0, φ_0) zu Γ_n gehört, so liegt offenbar die ganze Strecke $\varphi = \varphi_0$, $r < r_0$ in Γ_n. Es gibt daher eine eindeutige Funktion $G_n(\varphi)$ der Eigenschaft, dass alle Punkte $\varphi = \varphi_0$, $r < G_n(\varphi_0)$ zu Γ_n gehören, während die Punkte $\varphi = \varphi_0$, $r = G_n(\varphi_0)$ ausserhalb von Γ_n liegen (für die Punkte mit $r = G_n(\varphi)$ bleibt die Frage unentschieden). Aus dem Gesagten folgt, dass die Folge $\{G_n(\varphi)\}$ monoton fällt, und dass

$$G(\varphi) = \lim_{n \to \infty} G_n(\varphi)$$

ist. Für (c) ist also nur zu beweisen, dass $G_n(\varphi)$ für jedes n halbstetig nach unten ist (Definition s. S. 90).

Es sei nun $P = (a, \varphi)$ ein Punkt von Γ_n mit $a < G_n(\varphi)$. Dann liegt jeder Punkt (a', φ) mit $a < a' < G_n(\varphi)$ im Inneren oder auf dem Rande einer Kurve $\sigma_{h'}$ mit $h' < \frac{1}{n}$, und P ist dann innerer Punkt von $\sigma_{h'}$. Es gehört also eine ganze Umgebung von P auch zu Γ_n, und P ist somit nicht Häufungspunkt von Punkten der Komplementärmenge von Γ_n. Das ergibt aber unmittelbar die Beziehung

$$\varliminf_{\psi \to \varphi} G_n(\psi) \geq G_n(\varphi), \qquad \text{w. z. b. w..}$$

9. Den Beweis von (c′) (S. 90) beginnen wir mit einer Vorbemerkung über $G(\varphi)$.

Es sei $\{G_n(\varphi)\}$ eine monoton fallende Folge von nicht negativen, nach unten halbstetigen Funktionen, $0 \leq \varphi \leq 2\pi$, $G(0) = G(2\pi)$, und

$$G(\varphi) = \lim_{n\to\infty} G_n(\varphi).$$

Für irgend eine Zahl $a > 0$ bezeichne $(G > a)$ bzw. $(G \geq a)$ die Menge aller φ-Werte für welche $G(\varphi) > a$ bzw. $G(\varphi) \geq a$ ist; Die Mengenfolge $\{G_n > a\}$ fällt nach Voraussetzung mit wachsendem n für jedes feste a monoton, und es existiert daher $\lim_{n\to\infty}(G_n > a)$. Offenbar gilt für jedes a

$$(*) \qquad (G \geq a) \supset \lim_{n\to\infty}(G_n > a) \supset (G > a).$$

Durch diese Relation ist, wie man unmittelbar einsieht, $G(\varphi)$ eindeutig bestimmt: Es gibt nur eine Funktion $G(\varphi)$, die (*) für jedes a erfüllt. In Wirklichkeit ist nun $G(\varphi)$ bereits durch das Bestehen von (*) für eine überall dichte Menge von a-Werten eindeutig bestimmt: Ist $\{a_k\}$ eine in $(0, \infty)$ dichte Folge, $a_k > 0$, und gilt für eine Funktion $G^*(\varphi)$ und jedes a_k

$$(**) \qquad (G^* \geq a_k) \supset \lim_{n\to\infty}(G_n > a_k) \supset (G^* > a_k),$$

so ist[1] $G^*(\varphi) = G(\varphi)$. Zum Beweise von (c′) genügt es somit eine in O differenzierbare konvexe Fläche zu konstruieren, deren untere Indikatrix mit O zusammenfällt, und deren obere Indikatrix in O durch die Gleichung $r = G^*(\varphi)$ gegeben ist, wobei $G^*(\varphi)$ die Beziehung (**) erfüllt.

Wegen der Halbstetigkeit nach unten von $G_n(\varphi)$ ist die Menge $(G_n > a_k)$ für jedes n und k offen, d. h. Vereinigungsmenge von höchstens abzählbar vielen paarweise fremden (offenen) φ-Intervallen. Wir stellen nun $(G_n > a_k)$ in doppelter Weise so als Vereinigungsmenge von abzählbar vielen (nicht notwendig fremden) offenen Intervallen $I_\nu^{n,k}$ und $i_\nu^{n,k}$ dar:

$$(G_n > a_k) = \sum_\nu I_\nu^{n,k} = \sum_\nu i_\nu^{n,k},$$

dass:

1) Die Länge aller $I_\nu^{n,k}$ höchstens gleich π wird.

[1]) Ein Beweis dieses an sich naheliegenden Satzes etwa bei *Carathéodory*, Vorlesungen über reelle Funktionen, 1. Aufl. S. 371.

2) Ein Punkt gleichzeitig höchtens zwei Intervallen $I_\nu^{n,k}$ angehört (n, k fest gedacht!)

3) $i_\nu^{n,k}$ mit seinen Endpunkten ganz im Inneren von $I_\nu^{n,k}$ liegt. Hierzu genügt es offenbar in der ursprünglichen Darstellung von $(G_n > a_k)$ als Vereinigungsmenge von paarweise fremden Intervallen diese einzeln so in Teilintervalle zu zerlegen, dass die Bedingungen 1) bis 3) erfüllt sind. Ein einzelnes Intervall ist aber leicht in dieser Form darzustellen.[1])

H bezeichne nun die Rotationsfläche $z = r^{\frac{5}{4}}$. Zu jedem Intervall $I_\nu^{n,k}$ mit $n > k$ konstruieren wir eine in O differenzierbare konvexe Fläche $H_\nu^{n,k}$ mit folgenden Eigenschaften:

a) Wenn φ nicht dem Intervall $I_\nu^{n,k}$ angehört, so verläuft der Normalschnitt (φ) von $H_\nu^{n,k}$ etwa für $z > \sin^{2\frac{1}{2}} \frac{\pi}{8}$ ganz oberhalb der Kurve $z = r^{\frac{7}{4}}$.

b) Wenn φ dem Intervall $I_\nu^{n,k}$ angehört, so verläuft der Schnitt (φ) von $H_\nu^{n,k}$ ganz oberhalb oder auf der Kurve λ_{a_k} (Definition S. 95).

c) Wenn φ dem Intervall $i_\nu^{n,k}$ angehört, so fällt der Schnitt (φ) von $H_\nu^{n,k}$ für $z < \zeta_\nu^{n,k}$ mit der Kurve λ_{a_k} susammen,[2]) $\zeta_\nu^{n,k} > 0$.

d) Für $z < \zeta_\nu^{n,k}$ verläuft $H_\nu^{n,k}$ ganz unterhalb oder auf H.

Die Existenz solcher Flächen steht nach dem Ergebnis von No. 4 (vgl. den Satz auf S. 102) fest.[3]).

Zur Konstruktion der Fläche mit der vorgeschriebenen oberen Indikatrix verfahren wir nun ähnlich wie in No. 7. Der Schnitt $q_\nu^{n,k}(h)$ von $H_\nu^{n,k}$ mit der Ebene $z = h > 0$ verläuft für $h < \zeta_\nu^{n,k}$

---

[1]) Etwa folgendermassen: Es sei $\alpha < \varphi < \beta$ das vorgegebene Intervall. Wir zerlegen es zunächst durch Einschaltung von unendlich vielen Teilpunkten φ_ν, $\nu = 0, \pm 1, \pm 2, \pm 3, \cdots$, die nur α und β als Häufugspunkte haben: $0 < \varphi_{\nu+1} - \varphi_\nu < \frac{\pi}{3}$, $\lim_{\nu \to -\infty} \varphi_\nu = \alpha$, $\lim_{\nu \to +\infty} \varphi_\nu = \beta$. Dann setzen wir I_ν gleich $\varphi_\nu < \varphi < \varphi_{\nu+2}$ nnd i_ν gleich $\frac{2}{3}\varphi_\nu + \frac{1}{3}\varphi_{\nu+1} < \varphi < \frac{1}{3}\varphi_{\nu+1} + \frac{2}{3}\varphi_{\nu+2}$.

[2]) Im Folgenden wird $\zeta_\nu^{n,k} < \sin^{2\frac{1}{2}} \frac{\pi}{8}$ angenommen.

[3]) Man kann als Indikatrix von $H_\nu^{n,k}$ z. B. $g(\varphi) = a_k$ für φ aus $I_\nu^{n,k}$ und $g(\varphi) = 0$ sonst vorschreiben, und $H_\nu^{n,k}$ nach dem in No. 3—4 beschriebenen Verfahren konstruieren. Die Bedingung d) ist dann natürlich von selbst erfüllt. b) wurde zwar nicht besonders erwähnt, ist aber stets vermöge einer Durchschnittbildung (vgl. Fussnote [1]), S. 98) mit dem Paraboloid $z = \frac{r^2}{2a_k^2}$ zu erreichen.

ganz ausserhalb oder auf H, und die Haube von $H + q_\nu^{n,k}(h)$ unterscheidet sich von H nur in einem Streifen $h' \leq z \leq h''$, wobei h' und h'' mit h gegen 0 gehen. Wir können daher zu jedem $H_\nu^{n,k}$ ein $h_\nu^{n,k} < \zeta_\nu^{n,k}$ so bestimmen, dass die zugehörigen Streifen paarweise einen positiven Abstand von einander haben, und die $h_\nu^{n,k}$ Null als einzigen Häufungswert besitzen. Wir ordnen jedoch in dieser Weise nur Flächen $H_\nu^{n,k}$ mit $n > k$ Schnitte zu[1]): die Haube Φ von H und der Vereinigungsmenge aller dieser $q_\nu^{n,k}(h_\nu^{n,k})$ ist wieder (vgl. S. 106) gleich der Haube der Vereinigungsmenge der Hauben von H und der einzelnen $q_\nu^{n,k}(h_\nu^{n,k})$. Es soll nun gezeigt werden, dass die obere Indikatrix von Φ in O durch $r = G(\varphi)$ dargestellt wird. (Dass die untere Indikatrix mit O zusammenfällt, ist trivial, da Φ mit H eine Folge von Schnitten $z = h_s$ gemeinsam hat, mit $h_s \to 0$.)

Es sei $r = G^*(\varphi)$ die Gleichung der oberen Indikatrix von Φ in O.

α) Es sei die Stelle $\varphi = \psi$ enthalten in $\lim\limits_{n\to\infty}(G_n > a_k)$. Dann gehört die Stelle $\varphi = \psi$ für jedes n, k einem Intervall $i_{\nu_{n,k}}^{n,k}$ an, und der Punkt $\varphi = \psi$ von $q_{\nu_{n,k}}^{n,k}(h_{\nu_{n,k}}^{n,k})$ liegt auf der Kurve λ_{a_k}. Der Schnitt $\varphi = \psi$ von Φ enthält also in jeder Umgebung von O unendlich viele Punkte auf der Kurve λ_{a_k}, und daher ist $G^*(\psi) \geq a_k$. Wir haben damit die Beziehung

$$\lim_{n\to\infty}(G_n > a_k) \subset (G^* \geq a_k).$$

β) $\varphi = \psi$ sei enthalten in $(G^*(\varphi) > a_k)$, d.h. es sei $G^*(\psi) > a_k$. Dann enthält der Schnitt $\varphi = \psi$ von Φ eine Folge von Punkten $P_s \to O$, die sämtlich unterhalb der Kurve λ_{a_k} gelegen sind. P_s möge etwa der Haube von $H + q_{\nu_s}^{n_s,k_s}(h_{\nu_s}^{n_s,k_s})$ angehören. Nach Konstruktion von $H_\nu^{n,k}$ muss dann für hinreichend grosse s stets $a_{k_s} > a_k$ sein, und ausserdem ψ enthalten im Intervall $I_{\nu_s}^{n_s,k_s}$. Wäre nämlich ψ nicht enthalten in $I_{\nu_s}^{n_s,k_s}$, so läge P_s nach a) oberhalb der Kurve $z = r^{\frac{7}{4}}$, und daher erst recht oberhalb λ_{a_k}. Wenn aber ψ in $I_{\nu_s}^{n_s,k_s}$ enthalten ist, so liegt P_s nach b) oberhalb oder auf der Kurve $\lambda_{a_{k_s}}$; wäre nun $a_{k_s} \leq a_k$, so verliefe $\lambda_{a_{k_s}}$ ganz oberhalb oder auf λ_{a_k}, und P_s könnte nicht unterhalb von λ_{a_k} liegen.

[1]) Von jeder Fläche $H_\nu^{n,k}$ kommt diesmal nur ein einziger Schnitt vor!

Wir sehen also, dass die Stelle $\varphi = \psi$ enthalten ist in der Menge $(G_{n_s} > a_k)$. Nun kann wegen $P_s \to O$ die Haube von $H + q_\nu^{n,j}(h_\nu^{n,j})$ bei festen n, j und ν höchstens endlich viele Punkte P_s enthalten. Mindestens einer der drei Indizes n_s, k_s, ν_s muss demnach unendlich viele Werte durchlaufen. Bei festem n_s, k_s gehört aber ψ höchstens zwei Intervallen $I_\nu^{n_s, k_s}$ an (Bedingung 2), so dass ν_s durch n_s und k_s zweideutig bestimmt ist. Ferner kamen bei der Konstruktion nur Kombinationen mit $n_s > k_s$ vor, so dass bei festem n_s auch bloss endlich viele Werte k_s in Frage kommen. Somit strebt $n_s \to \infty$, und ψ gehört unendlich vielen Mengen $(G_n > a_k)$ (k fest!) an, also wegen der Monotonie auch sämtlichen. Damit ist auch

$$(G^* > a_k) \subset \lim_{n \to \infty} (G_n > a_k)$$

gezeigt, so dass (**) besteht.

Bemerkungen zur Differentialgeometrie der konvexen Flächen. III.

Über die Gaussssche Krümmung.

Von Herbert Busemann und Willy Feller.

1. Diese Note soll für konvexe Flächen einige mit der vermittels des sphärischen Bildes definierten Gaussschen Krümmung zusammenhängende Fragen klären und bei dieser Gelegenheit überhaupt den Mechanismus des dabei vollzogenen Grenzübergangs deutlicher, als es gewöhnlich geschieht, zu Tage treten lassen.

Wir gehen von folgendem Sachverhalt aus: Auf jedem konvexen Flächenstück Φ gibt es[1]) drei Nullmengen

$$\Lambda \subset \Lambda' \subset \Lambda''$$

derart, dass in den Punkten von $\Phi - \Lambda$ die Fläche überall eine Tangentialebene besitzt; in den Punkten von $\Phi - \Lambda'$ existiert eine Indikatrix, d. h. alle Normalhalbschnitte durch einen Punkt P von $\Phi - \Lambda'$ haben in P eine (endliche oder unendliche) Krümmung (oder: die Normalschnitte haben einseitige Krümmungen); die Indikatrix i selbst ist die Kurve die entsteht, wenn man von P aus in der Tangentialebene an Φ in jeder Richtung die Quadratwurzel aus dem Krümmungsradius des zugehörigen Normalhalbschnitts abträgt. i liegt auf einer konvexen Kurve j, welche P im Inneren oder auf dem Rande enthält, und zwar ist i mit j identisch, ausser wenn j eine oder zwei von P ausgehende Strecken enthält, von welchen natürlich nur höchstens je ein von P verschiedener Punkt zu i gehören kann. Alle konvexen Kurven, die P im Inneren oder auf dem Rande enthalten, können wirklich als Träger der Indikatrix in P auftreten[2]). Schliesslich ist in jedem Punkt P von $\Phi - \Lambda''$ die Indikatrix sogar eine Ellipse mit P als Mittelpunkt oder ein zu P symmetrisch liegendes Paar paralleler Geraden[3]), d. h. in P gilt der Eulersche Satz.

Unter dem sphärischen Bild eines inneren Punktes P von Φ verstehen wir die Gesamtheit aller Punkte der Einheitskugel E, deren Tangentialebenen einer Stützebene an Φ in P parallel sind.

---

[1]) Vgl. „Krümmungseigenschaften konvexer Flächen", Acta math. *66*, (1936), § 4; im folgenden zitieren wir diese Arbeit zur Abkürzung mit K. k. F.

[2]) „Bemerkungen zur Differentialgeometrie der konvexen Flächen II, Ueber die Krümmungsindikatrizen". Matem. Tidsskr. *B 1935*.

[3]) K. k. F. § 4.

Für eine beliebige Menge Ω nennen wir die Vereinigungsmenge der sphärischen Bilder der Punkte P von Ω das sphärische Bild von Ω und bezeichnen es allgemein mit Ω^*.

Es sei nun P ein Punkt von $\Phi - \Lambda'$ und i die Indikatrix von Φ in P. Die Frage, die zunächst behandelt werden soll, ist: existiert in P eine nur von der Indikatrix i und von deren Lage relativ zu P abhängende Gausssche Krümmung von Φ? Hierbei ist noch zu präzisieren, durch welchen Grenzübergang die Gausssche Krümmung definiert sein soll. Ist $P \subset \Phi - \Lambda$ und Π die Tangentialebene an Φ in P, so betrachten wir die Φ schneidende zu Π parallele Ebene Π_h im Abstand h von Π; die von Π_h auf Φ ausgeschnittene (P enthaltende) Flächenkappe werde mit Ω_h bezeichnet. Unsere Frage ist dann, ob in allen Punkten von $\Phi - \Lambda'$

$$K(P) = \lim_{h \to 0} \frac{|\Omega_h^*|}{|\Omega_h|} \tag{1}$$

existiert (Betragstriche bezeichnen hier und im folgenden stets das Lebesguesche Mass der betreffenden Menge). Zur Beantwortung dieser Frage untersuchen wir gesondert die beiden Grenzübergänge

$$\frac{|\Omega_h|}{h} \quad \text{und} \quad \frac{|\Omega_h^*|}{h}$$

für $h \to 0$. Es gilt dann:

Es sei P ein Flächenpunkt mit einer Indikatrix i, die in Polarkoordinaten r, φ um P durch

$$r = g(\varphi) \tag{i}$$

dargestellt ist.

a) In P existiert stets $\lim \frac{|\Omega_h|}{h}$ und ist gleich dem doppelten Flächeninhalt von i (evtl. ∞):

$$\lim_{h \to 0} \frac{|\Omega_h|}{h} = \int_0^{2\pi} g^2(\varphi)\, d\varphi. \tag{2}$$

(vgl. Nr. 2).

b1) Wenn i beschränkt ist und nicht durch P geht, so existiert ein endlicher $\lim \frac{|\Omega_h^*|}{h}$, und zwar ist (vgl. Nr. 4)

$$\lim_{h \to 0} \frac{|\Omega_h^*|}{h} = \int_0^{2\pi} \frac{g^2(\varphi) - g'^2(\varphi)}{g^4(\varphi)}\, d\varphi \tag{3}$$

(die Ableitung $g'(\varphi)$ existiert wegen der Konvexität bis auf abzählbar viele φ-Werte überall, und daher hat das Integral einen Sinn).

b2) Wenn i beschränkt ist und durch P geht, so ist stets

$$\lim_{h\to 0}\frac{|\Omega_h^*|}{h}=\infty. \qquad \text{(vgl. Nr. 5).}$$

b3) Wenn i unbeschränkt ist, braucht $\lim\frac{|\Omega_h^*|}{h}$ nicht zu existieren.

Durch Kombination dieser Resultate folgt:

Unter den Voraussetzungen von b1) und b2) existiert in P die Gausssche Krümmung $K(P)$, und zwar ist im Falle b1)

$$K(P)=\lim_{h\to 0}\frac{|\Omega_h^*|}{|\Omega_h|}=\frac{\int\limits_0^{2\pi}\frac{g^2(\varphi)-g'^2(\varphi)}{g^4(\varphi)}\,d\varphi}{\int\limits_0^{2\pi}g^2(\varphi)\,d\varphi}, \tag{4}$$

und im Falle b2) $K(P)=\infty$.

Ist i speziell eine Ellipse mit P als Mittelpunkt:

$$g(\varphi)=\frac{1}{\sqrt{\frac{\cos^2\varphi}{a^2}+\frac{\sin^2\varphi}{b^2}}},$$

so ergibt sich aus (4)

$$K(P)=\frac{\frac{2\pi}{ab}}{2ab\pi}=\frac{1}{a^2b^2}.$$

Im Falle b3) kann man trotz (2) keine unmittelbaren Schlussfolgerungen ziehen. Entgegen dem zu vermutenden Resultat (dass nämlich, wenn i nicht durch P geht, $\frac{|\Omega_h^*|}{|\Omega_h|}\to 0$ strebt) gilt hier:

c) Wenn i unbeschränkt ist, existiert $K(P)$ im allgemeinen nicht, selbst wenn der Eulersche Satz gilt, d. h. i aus zwei parallelen zu P symmetrisch gelegenen Geraden besteht oder ganz im Unendlichen liegt; (vgl. Nr. 6).

Diese Erscheinung liegt, roh gesprochen, an der u. U. ungleichmässigen Approximation der unbeschränkten Indikatrizen durch die ähnlich vergrösserten Kurven s_h (vgl. S. 46); man kann sie durch die folgende Forderung ausschliessen:

Es sei γ_h die kleinste geodätische Kreisscheibe[1]) um P, die Ω_h enthält; dann soll

$$(5) \qquad \overline{\lim_{h\to 0}} \frac{|\gamma_h|}{|\Omega_h|}$$

endlich sein. Wir sagen in diesem Fall, dass die Gebiete Ω_h im Punkte P bezüglich der geodätischen Kreisscheiben zugelassen sind. Man hat dann:

d) Die Gebiete Ω_h seien im Punkte P bezüglich der geodätischen Kreisscheiben zugelassen; wenn dann die Indikatrix i unbeschränkt ist, aber nicht durch P geht, so ist (vgl. Nr. 7)

$$K(P) = \lim_{h\to 0} \frac{|\Omega_h^*|}{|\Omega_h|} = 0.$$

Geht jedoch i durch P (und ist unbeschränkt) so kann man sogar Flächen mit derselben Indikatrix i in P angeben, für welche die Gebiete Ω_h bezüglich der geodätischen Kreisscheiben zugelassen sind, und für welche die Limites $\lim \frac{|\Omega_h^*|}{|\Omega_h|}$ zwar existieren, aber verschieden ausfallen; (vgl. Nr. 8).

Es sei übrigens bemerkt, dass wenn i beschränkt ist und innere Punkte enthält, die Gebiete Ω_h stets bezüglich der geodätischen Kreisscheiben zugelassen sind[2]), und dass diese Bedingung nie erfüllt ist, wenn die Indikatrix ein Paar paralleler im Endlichen gelegener Geraden ist, sodass die obige Aussage nur für solche Punkte mit Eulerschem Satz etwas Neues liefert, die Nabelpunkte der Krümmung 0 sind.

Diese Ergebnisse können zur Untersuchung der Gaussschen Totalkrümmung (curvatura integra) von Φ verwandt werden. Unter der Totalkrümmung einer offenen Menge Γ von Φ versteht man bekanntlich das Mass ihres sphärischen Bildes, also

[1]) d. h. die Menge aller Punkte von Φ, deren auf Φ gemessener Abstand von P eine gewisse Zahl nicht übersteigt.

[2]) Es sei nämlich Π die Tangentialebene an Φ in P, ferner $\bar{s}_h$ der Rand der Projektion von Ω_h auf Π und g_h der Rand der Projektion auf Π der kleinsten geodätischen Kreisscheibe um P, welche Ω_h im Inneren enthält. Die von P aus im Masstab $\sqrt{2h} : 1$ ähnlich vergrösserte Kurve $\bar{s}_h$ strebt nach K. k. F. § 3 für $h \to 0$ gegen i; wegen der Differenzierbarkeit von Φ in P strebt daher die in gleicher Weise vergrösserte Kurve g_h gegen den kleinsten Kreis um P, der i im Inneren enthält, woraus (5) folgt.

$|\Gamma^*|$. Aus einem bekannten Satz der Lebesgueschen Theorie folgt dann, dass in unsrer obigen Bezeichnung in fast allen Punkten P von Φ

$$\lim_{h \to 0} \frac{|\gamma_h^*|}{|\gamma_h|} = \lambda(P) \tag{6}$$

existiert (hierin kann man natürlich unter γ_h auch einfach die geodätische Kreisscheibe um P vom Radius h verstehen). Es fragt sich nun, was der Grenzwert $\lambda(P)$ mit der oben definierten Gaussschen Krümmung zu tun hat. Wir zeigen:

In fast allen Punkten von Φ, in denen der Eulersche Satz gilt (also fast überall auf Φ) ist

$$\lambda(P) = \frac{1}{\varrho_1(P)\varrho_2(P)}$$

wobei $\varrho_1(P)$ und $\varrho_2(P)$ die beiden Hauptkrümmungen bezeichnen; (vgl. Nr. 9).

Es sei schon hier bemerkt, dass sich dieser Satz für fast alle Punkte, in denen die Indikatrix eine nicht ausgeartete Ellipse ist, unmittelbar aus der Lebesgueschen Theorie ergibt, weil dort die Gebiete Ω_h bezüglich der geodätischen Kreisscheiben zugelassen sind. Eine Sonderüberlegung ist nur für den Fall notwendig, wo i unbeschränkt ist.

Es sei nun insbesondre die sphärische Abbildung von Φ totalstetig, d. h. für jede Menge μ von Φ mit $|\mu| = 0$ sei auch $|\mu^*| = 0$. Dann gilt nach bekannten Sätzen für jede messbare Menge μ von Φ

$$|\mu^*| = \iint_{\mu} \lambda(P)\, dP,$$

wobei $\lambda(P)$ nach dem Vorangehenden fast überall auf μ definiert ist und fast überall durch das Produkt der beiden Hauptkrümmungen ersetzt werden kann. Damit ist eine in K. k. F. S. 34 gestellte Frage gelöst:

Ist die sphärische Abbildung von Φ totalstetig, so gilt

$$|\Phi^*| = \iint_{\Phi - \Lambda''} \frac{1}{\varrho_1(P)\varrho_2(P)}\, dP, \tag{7}$$

wo die $\varrho_i(P)$ die Hauptachsen der Indikatrix in P bezeichnen.

2. Φ sei nun eine konvexe Fläche, welche im Punkte P eine Tangentialebene Π und eine Indikatrix i besitzt. Wir beziehen uns im folgenden auf ein System von Zylinderkoordinaten r, φ, z mit dem Ursprung in P, wobei Π zur Ebene $z = 0$ wird. In einer Umgebung von P werde Φ durch

$$z = f(r, \varphi) \geq 0$$

dargestellt. Der Bereich Ω_h ist dann durch $z < h$ gegeben; wir bezeichnen die Ebene $z = h$ mit Π_h, ihre Schnittkurve mit Φ durch s_h. Die Projektion der letzteren auf Π hat die Gleichung

$$f(r, \varphi) = h.$$

Vergrössern wir diese ähnlich von P aus im Massstab $\sqrt{2h} : 1$, so entsteht eine Kurve σ_h mit der Gleichung

$$\frac{1}{h} f(r\sqrt{2h}, \varphi) = 1. \tag{8}$$

Für $h \to 0$ strebt σ_h gegen eine (evtl. ausgeartete) konvexe Kurve j, auf welcher die Indikatrix i in P liegt[1]). Wir bezeichnen der Einfachheit halber mit $|j|$ den Flächeninhalt des von j begrenzten Bereiches. Wenn s_h beschränkt ist — was bei beschränkter Indikatrix immer der Fall ist — so bezeichnen wir entsprechend mit $|s_h|$ den Flächeninhalt des von der Projektion von s_h auf Π begrenzten ebenen Bereiches. Wegen der Differenzierbarkeit von Φ in P ist offenbar

$$\lim_{h \to 0} \frac{|\Omega_h|}{|s_h|} = 1. \tag{9}$$

j habe nun die Gleichung

$$r = g(\varphi). \tag{10}$$

Dann folgt aus (8) und (9) wegen $\sigma_h \to j$ unmittelbar

$$\lim_{h \to 0} \frac{|\Omega_h|}{h} = \lim_{h \to 0} \frac{|s_h|}{h} = 2 \lim_{h \to 0} |\sigma_h| = 2\,|j| = \int_0^{2\pi} g^2(\varphi)\, d\varphi \tag{11}$$

(wenn für ein h_0 die Kurve s_{h_0} unbeschränkt ist, so gilt dasselbe von allen Kurven s_h mit $h < h_0$, und daher ist (11) in diesem Falle trivial).

Der Grenzwert (11), den im regulären Falle Blaschke[2])

---

[1]) K. k. F. § 3; dort wurde s_h mit s_{2h} bezeichnet.

[2]) Differentialgeometrie I, Berlin 1924, 2. Aufl. S. 85.

angegeben hat, existiert somit immer, wenn Φ im Punkte P eine Indikatrix besitzt. Er steht im regulären Fall zur Gaussschen Krümmung $K(P)$ in der Beziehung

$$K(P) = \frac{4\pi^2}{\left(\lim \frac{|\Omega_h|}{h}\right)^2}.$$

Wenn aber j keine Ellipse ist, wird im allgemeinen

$$\frac{4\pi^2}{\left(\lim \frac{|\Omega_h|}{h}\right)^2} \neq \lim \frac{|\Omega_h^*|}{|\Omega_h|}$$

sein, wie sich im folgenden ergeben wird[1]).

Zur Untersuchung der Gausschen Krümmung genügt es nach (11), das Verhalten von $\frac{|\Omega_h^*|}{h}$ für $h \to 0$ zu untersuchen; hierzu brauchen wir zwei Gleichmässigkeitseigenschaften, die wir zuerst ableiten wollen.

3. Aus der Definition der Krümmung und der Indikatrix[2]) folgt für jedes feste φ

$$(12) \qquad \frac{1}{g^2(\varphi)} = \lim_{r \to 0} \frac{2f(r, \varphi)}{r^2} = \lim_{r \to 0} \frac{f_r(r, \varphi)}{r}.$$

Hierbei ist unter $f_r(r, \varphi)$ folgendes zu verstehen: wenn n_φ den Normalhalbschnitt (φ) von Φ in P bezeichnet, so ist $f_r(r, \varphi)$ der Richtungstangens irgend einer Stützgeraden an n_φ im Punkte $(r, \varphi, f(r, \varphi))$; mit Ausnahme von höchstens abzählbar vielen Punkten auf n_φ ist demnach $f_r(r, \varphi)$ eine eindeutige Funktion von r. (Analytisch ausgedrückt ist $f_r(r, \varphi)$ irgend eine Derivierte von f nach r).

$g(\varphi)$ kann natürlich höchstens für zwei Werte φ unstetig werden (d. h. j kann höchstens zwei von P ausgehende Strecken enthalten). Wir zeigen, dass abgesehen evtl. von einer Umgebung dieser Stellen, die Grenzübergänge in (12) gleichmässig erfolgen. Wir beweisen nämlich:

Es sei B ein abgeschlossener Teilbereich von $(0, 2\pi)$, der keine Unstetigkeitsstelle von $g(\varphi)$ enthält; bei vorgegebenem $\delta > 0$ sei ferner $B'(\delta)$ die Teilmenge von B,

[1]) Vgl. auch K. k. F. S. 23, wo ein explizites Beispiel dafür gegeben wird.
[2]) K. k. F. § 1.

auf welcher $g^2(\varphi) \geq \delta$ ist, $B''(\delta)$ diejenige mit $g^2(\varphi) < \delta$. Dann gibt es zu jedem $\delta > 0$ und $\varepsilon > 0$ ein $R = R(\varepsilon, \delta) > 0$ derart, dass für $0 < r < R$ und $\varphi \subset B'(\delta)$

$$(13) \qquad \left|\frac{1}{g^2(\varphi)} - \frac{2f(r,\varphi)}{r^2}\right| < \varepsilon, \quad \left|\frac{1}{g^2(\varphi)} - \frac{f_r(r,\varphi)}{r}\right| < \varepsilon$$

wird, und ebenso für $0 < r < R$ und $\varphi \subset B''(\delta)$

$$(14) \qquad \frac{r^2}{2f(r,\varphi)} < 2\delta \quad \text{und} \quad \frac{r}{f_r(r,\varphi)} < 2\delta .$$

Soweit die Behauptung die Grösse $\frac{f}{r^2}$ betrifft, wurde sie bereits in K. k. F. § 3 beweisen, und folgt übrigens unmittelbar aus $\sigma_h \to j$. — Der andere Teil der Behauptung lässt sich hierauf zurückführen.

Die Neigung einer Stützgeraden an n_φ im Punkte $(r, \varphi, f(r, \varphi))$ ist wegen der Konvexität von n_φ nicht kleiner als die der Sehne von P nach diesem Punkt, d. h. es ist

$$f_r(r,\varphi) \geq \frac{f(r,\varphi)}{r}.$$

Die zweite Gleichmässigkeitseigenschaft in (14) folgt also aus der ersten, und wir haben nur noch die zweite der Ungleichungen (13) für den Bereich $B'(\delta)$ zu beweisen. Hierbei kann $\delta, \varepsilon < 1$ vorausgesetzt werden.

Wir wählen ein $0 < \eta < \left(\frac{\varepsilon\delta}{6}\right)^2$ und setzen für jedes φ aus $B'(\delta)$

$$\varrho(\varphi) = g^2(\varphi)(1-\eta) \quad \text{wenn} \quad \delta \leq g^2(\varphi) \leq \frac{5}{\varepsilon},$$
$$\varrho(\varphi) = \frac{5}{\varepsilon} - \eta \qquad \text{wenn} \quad g^2(\varphi) > \frac{5}{\varepsilon}$$

ist. Ferner bestimmen wir ein $R_1 > 0$ derart, dass für alle φ aus $B'(\delta)$ und $0 < r < R_1$

$$\left|\frac{1}{g^2(\varphi)} - \frac{2f(r,\varphi)}{r^2}\right| < \frac{\varepsilon^2\eta}{25}$$

wird. Dann ist für $0 < r < R_1$ und φ aus $B'(\delta)$

$$\frac{2f(r,\varphi)}{r^2} < \frac{\varepsilon^2\eta}{25} + \frac{1}{g^2(\varphi)} < \frac{1}{\varrho(\varphi)},$$

und daher verläuft Φ im angegebenen Bereiche ganz unterhalb der (nicht notwendig konvexen) Fläche

$$z = \frac{r^2}{2\varrho(\varphi)}.$$

Nun ist $\varrho(\varphi) > \frac{\delta}{2}$; für $0 < r < R_2 = \min\left[R_1, \frac{\delta}{2}\right]$ und jedes φ aus $B'(\delta)$ verläuft daher der Normalhalbschnitt n_φ von Φ erst recht unterhalb der Fläche

$$(15) \qquad z = \frac{r^2}{\varrho(\varphi) + \sqrt{\varrho^2(\varphi) - r^2}} = \varrho(\varphi) - \sqrt{\varrho^2(\varphi) - r^2}.$$

Für konstantes φ liefert (15) einen Kreisbogen k_φ. Es seien t_φ und t'_φ die beiden von dem Flächenpunkt $(r, \varphi, f(r, \varphi))$ aus an k_φ gezogenen Tangenten; für $r < \frac{1}{2} R_2$ fallen die Abszissen ihrer Berührungspunkte in den Bereich $0 \leq r < R_2$, liegen also oberhalb Φ. Wegen der Konvexität liegt also in dem Bereich $0 < r < R_2$, $\varphi \subset B'(\delta)$ jede Stützgerade an einen Normalschnitt n_φ im Flächenpunkte $(r, \varphi, f(r, \varphi))$ in dem äusseren von t_φ und t'_φ gebildeten Winkelraum: Der Richtungstangens der Stützgeraden liegt also zwischen den Richtungstangenten von t_φ und t'_φ. Dies ergibt ausgerechnet die Ungleichung

$$(16) \qquad \frac{r(\varrho - f) - \varrho\sqrt{r^2 + (\varrho - f)^2 - \varrho^2}}{\varrho^2 - r^2} \leq f_r \leq \frac{r(\varrho - f) + \varrho\sqrt{r^2 + (\varrho - f)^2 - \varrho^2}}{\varrho^2 - r^2}.$$

Nun geht im Bereiche $B'(\delta)$ gleichmässig

$$\frac{f_r}{r} \to 0, \qquad \frac{2f}{r^2} \to \frac{1}{g^2(\varphi)}.$$

Es existiert also ein R derart, dass für $0 < r < R$ und diejenigen Werte φ aus $B'(\delta)$, für welche $g^2(\varphi) \leq \frac{5}{\varepsilon}$ ausfällt,

$$\frac{1}{\varrho}\left\{1 - \sqrt{1 - \frac{\varrho^2}{g^2}}\right\} - \frac{\varepsilon}{2} < \frac{f_r}{r} < \frac{1}{\varrho}\left\{1 + \sqrt{1 - \frac{\varrho^2}{g^2}}\right\} + \frac{\varepsilon}{2}$$

wird. Setzt man hierin den Wert für $\varrho(\varphi)$ ein, so erhält man

$$\frac{1 - \sqrt{\eta}}{1 - \eta} \cdot \frac{1}{g^2(\varphi)} - \frac{\varepsilon}{2} < \frac{f_r}{r} < \frac{1 + \sqrt{\eta}}{1 - \eta} \cdot \frac{1}{g^2(\varphi)} + \frac{\varepsilon}{2}$$

oder

$$\left|\frac{1}{g^2(\varphi)} - \frac{f_r(r,\varphi)}{r}\right| < \frac{1}{g^2(\varphi)} \cdot \frac{\eta + \sqrt{\eta}}{1-\eta} + \frac{\varepsilon}{2} < \varepsilon.$$

Für diejenigen φ Werte aus $B'(\delta)$, für welche $g^2(\delta) > \frac{5}{\varepsilon}$ ausfällt, erhält man aber aus (16) für hinreichend kleine r gleichmässig

$$0 \leq \frac{f_r}{r} \leq \frac{2\varrho - f}{\varrho^2 - r^2} < \frac{3}{\varrho} = \frac{3}{\frac{5}{\varepsilon} - \eta} < \frac{3\varepsilon}{4},$$

oder

$$\left|\frac{1}{g^2(\varphi)} - \frac{f_r(r,\varphi)}{r}\right| < \frac{\varepsilon}{5} + \frac{3\varepsilon}{4} < \varepsilon, \qquad \text{w. z. b. w.}$$

4. Wir betrachten nun einen Flächenpunkt P mit einer Indikatrix i der Gleichung $r = g(\varphi)$, welche beschränkt ist und nicht durch P geht. Wir wollen den Grenzwert von $\frac{1}{h}|\Omega_h^*|$ für $h \to 0$ berechnen.

Hierzu denken wir uns auf der Einheitskugel E sphärische Polarkoordinaten u, v eingeführt, mit dem Bild P^* von P als Nordpol, u als Poldistanz. s_h^0 sei der Rand des Bildes Ω_h^*.

Dann ist jeder Punkt von s_h^0 Bildpunkt eines Punktes von s_h (doch kann, weil Φ nicht differenzierbar zu sein braucht, das Bild von s_h auch innere Punkte enthalten). Wenn s_h eine Strecke σ enthält, so muss jede Stützebene an Φ in einem inneren Punkte von σ die Strecke σ enthalten, so dass sämtliche Bildpunkte von inneren Punkten von σ auf demselben Meridian von E liegen. Die andren Meridiane schneiden s_h^0 offenbar je in einem Punkt (der Bildpunkt desjenigen Punktes von s_h ist, in welchem es eine zur Meridianhalbebene gleichsinnig parallele Hauptnormale an s_h gibt).

Ω_h^* ist daher ein Sternbereich, d. h. wenn ein Punkt (u_0, v_0) in Ω_h^* liegt, so enthält Ω_h^* auch den ganzen Meridianbogen $0 < u < u_0$, $v = v_0$. Man hat also

$$|\Omega_h^*| = \int_{s_h} (1 - \cos u)\, dv; \tag{15}$$

hierbei ist u längs s_h^0 eine eindeutige und stetige Funktion von v, abgesehen bloss von den (höchstens abzählbar vielen) Stellen v, die Strecken auf s_h entsprechen, und die für die Integration nichts ausmachen. Jedem anderen ν-Wert entspricht auch eindeutig ein Punkt von s_h, den wir mit $Q(\nu)$ bezeichnen, und der die Koordinaten $r = r(\nu)$, $\varphi = \varphi(\nu)$, $z = h$ haben möge. Es bezeichne dann $\alpha = \alpha(\nu)$ den Neigungswinkel der Stützgeraden im Punkte $Q(\nu)$ an den Normalschnitt $\varphi(\nu)$ von Φ, welche parallel ist zur Tangentialebene an E im betrachteten Punkt von s_h^0. Schliesslich sei $\boldsymbol{\nu}(\nu)$ diejenige äussere Normale an s_h in $Q(\nu)$, die zur Meridianhalbebene (ν) parallel ist, $\psi = \psi(\nu)$ ihr Winkel mit dem Radiusvektor $\varphi = \varphi(\nu)$. Dann ist

$$\operatorname{tg} u = \frac{\operatorname{tg}\alpha}{\cos\psi}. \tag{18}$$

(Der Zusammenhang zwischen den erwähnten Grössen wird, wie man leicht nachrechnet, durch

$$\operatorname{tg} u = \sqrt{f_r^2 + \frac{1}{r^2} f_\varphi^2}, \qquad \operatorname{tg} \nu = \frac{f_r \sin\varphi + f_\varphi \dfrac{\cos\varphi}{r}}{f_r \cos\varphi - f_\varphi \dfrac{\sin\varphi}{r}},$$

$$\operatorname{tg}\alpha = f_r \qquad , \qquad \operatorname{tg}\psi = -\frac{f_\varphi}{r f_r}$$

geliefert).

(17) und (18) ergeben

$$|\Omega_h^*| = \int_0^{2\pi} \frac{\cos^2 u}{1+\cos u} \operatorname{tg}^2 u \, d\nu = \int_0^{2\pi} \frac{\cos^2 u}{1+\cos u} \frac{\operatorname{tg}^2\alpha}{\cos^2\psi} d\nu,$$

wobei sämtliche unter dem Integralzeichen vorkommenden Grössen bis auf höchstens abzählbar viele ν-Werte eindeutige Funktionen von ν sind). Man hat also

$$\frac{|\Omega_h^*|}{h} = 2\int_0^{2\pi} \frac{\cos^2 u}{1+\cos u} \cdot \frac{r^2}{2h} \cdot \frac{\operatorname{tg}^2\alpha}{r^2} \cdot \frac{d\nu}{\cos^2\psi}.$$

Die hierin auftretenden Grössen können wir statt auf s_h auf den ähnlich vergrösserten Kurven σ_h (vgl. No. 2, S. 46) interpretieren. Diese Kurven streben für $h \to 0$ gegen die Indikatrix i, und zwar konvergeiren wegen der Konvexität die Stützgeraden mit. Für $h \to 0$ strebt daher $\varphi(\nu)$ gegen eine Richtung φ, in welcher i eine zu $\boldsymbol{\nu}$ gleichsinnig parallele äussere Normale besitzt: abgesehen von höchstens abzählbar vielen ν-Werten, die zu auf i liegenden Strecken gehören, existiert somit ein Grenzwert $\lim \varphi(\nu)$.

Für diese ν-Werte hat man nach dem unter No. 3 Gesagten wegen der gleichmässigen Konvergenz

$$\lim_{h\to 0}\frac{r^2}{2h}=g^2(\varphi),\quad \lim_{h\to 0}\frac{\operatorname{tg}\alpha}{r}=\lim_{h\to 0}\frac{f_r(r,\varphi)}{r}=\frac{1}{g^2(\varphi)},$$

wobei natürlich $\varphi=\lim\varphi(\nu)$ gemeint ist. Ferner strebt $\psi(\nu)$ gegen den Winkel zwischen der äusseren Kurvennormale und dem Radiusvektor der Indikatrix i. Da schliesslich u gegen 0 strebt, konvergiert der Integrand in (19) für alle ν mit Ausnahme von höchstens abzählbar vielen. Dieser Integrand ist auch gleichmässig beschränkt. Denn zunächst hat $\cos\psi$ längs i eine positive untere Schranke (weil der Radiusvektor von i nirgends mit einer Stützgeraden zusammenfallen kann), und wegen der gleichmässigen Konvergenz von ψ ist daher $\frac{1}{\cos^2\psi}$ beschränkt; ferner gibt es nach Voraussetzung zwei positive Konstanten A, B, mit $A<g^2(\varphi)<B$, und wegen der unter 3 bewiesenen Gleichmässigkeit hat man daher für hinreichend kleine h

$$\frac{A}{2}<\frac{r^2}{2h}<2B,\quad \frac{1}{2B}<\frac{\operatorname{tg}\alpha}{r}<\frac{2}{A}.$$

Somit erhält man aus (19)

$$\lim_{h\to 0}\frac{|\Omega_h^*|}{h}=\int_0^{2\pi}\frac{d\nu}{g^2(\varphi)\cos^2\psi}, \tag{20}$$

wobei der Integrand längs i ausserhalb der evtl. auf i liegenden Strecken eine eindeutige Funktion von ν ist. Nun ist aber $g(\varphi)\cos\psi$, wie man unmittelbar sieht, die Projektion des Radiusvektors von P zum Kurvenpunkt auf die btr. Kurvennormale, und diese Grösse bleibt längs einer auf i liegenden Strecke offenbar konstant: so sieht man nachträglich, dass der Integrand in (20) eine eindeutige Funktion von ν ist, selbst wenn zu einem ν-Wert eine ganze Strecke auf i gehören sollte.

Damit ist die Existenz der Gaussschen Krümmung in unserem Fall beweisen. Wir wollen jedoch in (20) φ als Integrationsvariable einführen, um einen zweckmässigeren Ausdruck für die Krümmung zu erhalten.

Dazu sei k eine beliebige konvexe Kurve der Ebene Π, die nicht durch P geht und beschränkt ist; ihre Gleichung sei

$$r=G(\varphi).$$

Wir setzen dann

$$F(k) = \int_k \frac{dv}{r^2 \cos^2 \psi},$$

wobei ψ wieder den Winkel zwischen dem Radiusvektor und der Kurvennormale, v den Winkel zwischen der Kurvennormale und einer festen Richtung bezeichnet. Das Funktional $F(k)$ hat offenbar für jede solche Kurve einen endlichen Wert.

Wenn nun insbesondre $G(\varphi)$ zweimal stetig differenzierbar ist, so hat man

$$\operatorname{tg} \varphi = \frac{G'(\varphi)}{G(\varphi)}$$

und

$$dv = d\varphi - d\psi = \left(1 - \frac{GG'' - G'^2}{G^2 + G'^2}\right) d\varphi,$$

also

$$F(k) = \int_0^{2\pi} \frac{G^2(\varphi) + 2G'^2(\varphi) - G(\varphi)\, G''(\varphi)}{G^4(\varphi)} d\varphi.$$

Wegen

$$0 = \int_0^{2\pi} \left(\frac{G'(\varphi)}{G^3(\varphi)}\right)' d\varphi = \int_0^{2\pi} \frac{G(\varphi)\, G''(\varphi) - 3G'^2(\varphi)}{G^4(\varphi)} d\varphi$$

folgt hieraus schliesslich

$$F(k) = \int_0^{2\pi} \frac{G^2(\varphi) - G'^2(\varphi)}{G^4(\varphi)} d\varphi. \tag{21}$$

Das Integral (21) hat nun auch dann einen Sinn, wenn k nicht zweimal differenzierbar ist, nur dass dann der Integrand an abzählbar vielen Stellen nicht definiert zu sein braucht. Approximiert man eine beliebige beschränkte, nicht durch P gehende konvexe Kurve k durch zweimal stetig differenzierbare solche Kurven k_n, so sieht man unmittelbar, dass $F(k_n) \to F(k)$ strebt, und ebenso auch die entsprechenden Integrale (21) konvergieren. Die Gleichung (21) besteht daher für jede beschränkte, nicht durch P gehende konvexe Kurve der Ebene Π.

(21) ergibt somit vermöge (11):

Wenn die konvexe Fläche Φ im Punkte P eine Tangentialebene und eine beschränkte, nicht durch P gehende Indikatrix i mit der Gleichung $r = g(\varphi)$ besitzt, so existieren in P

$$\lim_{h \to 0} \frac{|\Omega_h^*|}{h} \quad \text{und} \quad \lim_{h \to 0} \frac{|\Omega_h^*|}{|\Omega_h|}$$

und hängen nur von i und dessen Lage zu P ab, nicht aber von der speziellen Gestalt von Φ, und zwar ist

$$\lim_{h\to 0}\frac{|\Omega_h^*|}{h}=\int_0^{2\pi}\frac{g^2(\varphi)-g'^2(\varphi)}{g^4(\varphi)}\,d\varphi,$$

$$\lim_{h\to 0}\frac{|\Omega_h^*|}{|\Omega_h|}=\frac{\displaystyle\int_0^{2\pi}\frac{g^2(\varphi)-g'^2(\varphi)}{g^4(\varphi)}\,d\varphi}{\displaystyle\int_0^{2\pi}g^2(\varphi)\,d\varphi}.$$

Wenn i insbesondre eine Ellipse mit P als Mittelpunkt ist, ergibt sich, wie schon in der Einleitung erwähnt, der bekannte Ausdruck für die Gausssche Krümmung.

5. Wir wollen nun beweisen:

Wenn Φ im Punkte P eine Indikatrix i hat, die beschränkt ist und durch P geht (oder mit P zusammenfällt), so ist

$$\lim_{h\to 0}\frac{|\Omega_h^*|}{h}=\lim_{h\to 0}\frac{|\Omega_h^*|}{|\Omega_h|}=\infty$$

Wir betrachten zunächst den Fall, dass i nicht mit P zusammenfällt. i sei wieder durch die Gleichung $r=g(\varphi)$ gegeben. Die kleinste konvexe Kurve, auf der i liegt, nennen wir wieder j. Wir denken uns das Koordinatensystem so gelegt, dass die Halbgerade $\varphi=0$ eine Halbtangente von j wird, und dass j im Winkelraum $0\leq\varphi\leq\pi$ verläuft. Wir verstehen nun unter v den

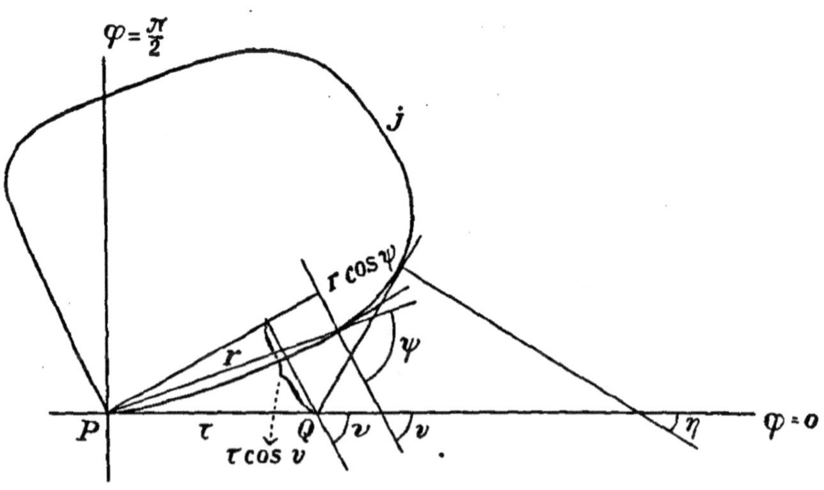

Winkel der äusseren Normalen von j mit der Halbgeraden $\varphi = 0$, und unter ψ den Winkel zwischen dem Radiusvektor und der äusseren Normalen von j.

Wir wählen dann eine Folge von Zahlen

$$0 < \eta < \eta_1 < \eta_2 < \cdots < \frac{\pi}{2},$$

so dass die $\eta_n \to \frac{\pi}{2}$ strebt, und dass die den Werten $v = -\eta$, $v = -\eta_1, \cdots$ entsprechenden Kurvennormalen von j zu Punkten mit eindeutiger Tangente gehören. Die zur Normalenrichtung $-\eta$ gehörende Stützgerade an j treffe die Halbgerade $\varphi = 0$ im Punkte $Q = (\tau, 0)$ (so dass $\tau > 0$ ist). Für jedes v zwischen $-\eta$ und π, zu welchem ein einziger Punkt von j gehört, ist dann, wie ein Blick auf die Figur zeigt,

$$\tau \cos v \geq r \cos \psi .$$

Man hat daher für alle n

$$\int_{-\eta}^{-\eta_n} \frac{dv}{r^2 \cos^2 \psi} \geq \frac{1}{\tau^2} \int_{-\eta}^{-\eta_n} \frac{dv}{\cos^2 v} = \frac{1}{\tau^2} (\operatorname{tg}^2 \eta_n - \operatorname{tg}^2 \eta). \tag{22}$$

Nun zeigen aber die Schlüsse der vorigen Nummer unmittelbar, dass

$$\lim_{h \to 0} \frac{|\Omega_h^*|}{h} \geq \int_{-\eta}^{-\eta_n} \frac{dv}{r^2 \cos^2 \psi}$$

ist; aus (22) folgt also

$$\lim_{h \to 0} \frac{|\Omega_h^*|}{h} = \infty,$$

und wegen (11) hat man daher auch

$$\lim_{h \to 0} \frac{|\Omega_h^*|}{|\Omega_h|} = \infty .$$

Den noch ausstehenden Fall, dass i mit P zusammenfällt, kann man folgendermassen erledigen. ϱ_h sei das Maximum der Abstände der Punkte von s_h von der z-Achse. Die unter Nr. 3 bewiesene Gleichmässigkeitseigenschaft liefert dann

$$\frac{\varrho_h^2}{h} \to 0 .$$

Wir schlagen um den Punkt $r = 0$, $z = h$ in der Ebene $z = h$

einen Kreis vom Radius ϱ_h und betrachten die beiden Kegel K_h und S_h mit der Spitze in P durch diesen Kreis, bzw. durch s_h. Dann ist offenbar

$$\Omega_h^* \supset S_h^* \supset K_h^*.$$

Nun ist K_h ein gewöhnlicher Kreiskegel, K_h^* also eine sphärische Kreisscheibe, deren Radius β_h zu ϱ_h in der Beziehung

$$\cot \beta_h = \frac{\varrho_h}{h}$$

steht. Dabei ist

$$|K_h^*| = 2\pi(1 - \cos \beta_h),$$

also

$$\lim_{h \to 0} \frac{|K_h|}{h} = \pi \lim_{h \to 0} \frac{\operatorname{tg}^2 \beta_h}{h} = \pi \lim_{h \to 0} \frac{h}{\varrho_h^2} = \infty,$$

und a fortiori

$$\lim_{h \to 0} \frac{|\Omega_h^*|}{h} = \lim_{h \to 0} \frac{|\Omega_h^*|}{|\Omega_h|} = \infty \qquad \text{w. z. b. w.}$$

6. Wir konstruieren nun eine konvexe Fläche Φ, welche in P die Ebene $z = 0$ berührt und dort eine Indikatrix besitzt, welche aus zwei parallelen, zu P symmetrisch gelegenen (evtl. „unendlich fernen") Geraden besteht (so dass P ein „normaler" Punkt von Φ ist); für dieses Φ wird aber in P

$$(23) \qquad \underline{\lim}_{h \to 0} \frac{|\Omega_h^*|}{|\Omega_h|} = 0, \qquad \overline{\lim}_{h \to 0} \frac{|\Omega_h^*|}{|\Omega_h|} = \infty$$

sein. – Dabei wird Φ einfach die konvexe Hülle von zwei konvexen Kurven c und k sein, welche in P die Ebene $z = 0$ berühren und in den Ebenen $\varphi = \pm\frac{\pi}{2}$ bzw. $\varphi = 0, \pi$ liegen.

Wenn k in P eine verschwindende Krümmung und c die endliche Krümmung $a > 0$ hat, so ist wegen der Konvexität der Indikatrix klar, dass die konvexe Hülle von c und k in P eine Indikatrix besitzt, welche aus den beiden zu $\varphi = 0$ parallelen Geraden im Abstand a besteht. Hat auch c eine verschwindende Krümmung in P, so liegt die Indikatrix ganz im Unendlichen.

Wir werden speziell für c die Kurve mit der Gleichung

$$z = r^q, \quad \varphi = \pm\frac{\pi}{2},$$

nehmen, wobei q eine feste Zahl > 1 ist; für k werden wir eine passende Kurve mit der Gleichung

$$z = f(r), \quad \varphi = 0, \pi$$

bestimmen, welche in $r = 0$ eine verschwindende Krümmung hat. Setzt man dann insbesondre $q = 2$, so hat die konvexe Hülle von c und k zwei parallele Geraden zur Indikatrix, während die Indikatrix für $q > 2$ ganz ins Unendliche fällt (und für $1 < q < 2$ mit der Geraden $\varphi = 0, \pi$ zusammenfällt). Wir müssen also bloss $f(r)$ so bestimmen, dass die Kurve $z = f(r)$ konvex wird, für $r = 0$ eine verschwindende Krümmung hat und derart, dass (23) gilt.

Die konvexe Hülle von c und k entsteht dadurch, dass man diejenigen Punkte von c und k verbindet, deren Stützgeraden sich in demselben Punkt der z-Achse treffen. Bezeichnet nun $f'(r_0)$ irgend eine Derivierte von $f(r)$ in $r = r_0$, so schneidet die Stützgerade in $r = r_0$ mit der Neigung $f'(r_0)$ die z Achse im Punkte $f(r_0) - r_0 f'(r_0)$. Demnach ist zur Bildung von Φ stets der Punkt $(r, 0, f(r))$ von k mit den Punkten $(\bar{r}, \pm\frac{\pi}{2}, \bar{r}^q)$ zu verbinden, mit

$$\bar{r}^q = \frac{rf'(r) - f(r)}{q-1}. \tag{24}$$

Wenn insbesondre $\dfrac{df}{dr}$ in $r = r_0$ nicht existiert, so muss dieser Punkt von k mit zwei ganzen Bögen von c verbunden werden.

Die Fläche Φ ist Einhüllende der erwähnten Stützgeraden und setzt sich daher aus vier Torsenstücken zusammen, die oberhalb der vier Quadranten der Ebene $z = 0$ liegen. Zur Berechnung der Flächennormalen ist es bequem, mit Cartesischen Koordinaten zu operieren, und wir setzen daher $x = r\cos\varphi$, $y = r\sin\varphi$.

Wir betrachten etwa das Torsenstück im Quadranten $0 < \varphi < \frac{\pi}{2}$. Es sind

$$-r, \quad \sqrt[q]{\frac{rf'(r) - f(r)}{q-1}}, \quad \frac{rf'(r) - qf(r)}{q-1}$$

die Komponenten des Vektors in der Torsenerzeugenden, der den Punkt $(r, 0, f(r))$ mit $(0, \bar{r}, \bar{r}^q)$ verbindet. Ferner sind $1, 0, f(r)$ die Komponenten eines Tangentialvektors an k im selben Punkt. Man erhält demnach den Normalenvektor an Φ längs der genannten Erzeugenden durch Bilden des vektoriellen Produkts; der Einheitsvektor in der Normalenrichtung ξ_1 hat also die Komponenten

$$(\xi_1):\quad \frac{f'(r)}{\sqrt{1+f'^2(r)+q^2H^{\frac{2q-2}{q}}}},\quad \frac{qH^{\frac{q-1}{q}}}{\sqrt{1+f'^2(r)+q^2H^{\frac{2q-2}{q}}}},$$

$$\frac{-1}{\sqrt{1+f'^2(r)+q^2H^{\frac{2q-2}{q}}}}$$

wobei zur Abkürzung $H=\dfrac{rf'(r)-f(r)}{q-1}$ gesetzt wurde.

Spiegelt man ξ_1 an der (x, z)-Ebene, so entsteht der Normalenvektor ξ_2 in demselben Punkt $(r, 0, f(r))$ an das andere an k angrenzende Torsenstück von Φ. Alle Vektoren des Winkelraums zwischen ξ_1 und ξ_2 gehören offenbar dem sphärischen Bilde des Punktes $(r; 0, f(r))$ an. Als Sternbereich (vgl. S. 50) muss demnach $\Omega^*_{f(r)}$ das gesamte sphärische Dreieck enthalten, dessen Eckpunkte der Bildpunkt P^* von P und die Endpunkte der vom Kugelmittelpunkt aus abgetragenen Vektoren ξ_1 und ξ_2 sind. Der Flächeninhalt dieses sphärischen Dreiecks ist grösser als der seiner ebenen Projektion auf die Tangentialebene in P, und dieses enthält noch das ebene Dreieck, dessen Eckpunkte die Projektionen der Eckpunkte des sphärischen Dreiecks sind. Dies ergibt die Abschätzung

$$|\Omega^*_{f(r)}| \geq 2\,\frac{qf'(r)H^{\frac{q-1}{q}}}{1+f'(r)+q^2H^{\frac{2q-2}{q}}}$$

Zur Berechnung von $|\Omega_{f(r)}|$ bedienen wir uns der Beziehung (9), und schätzen daher jetzt $|s_{f(r)}|$ ab. Als konvexe Hülle von c und k liegt Φ ganz im Inneren jedes der beiden konvexen Zylinder durch c senkrecht zur (y, z)-Ebene, bzw. durch k senkrecht zur (x, z)-Ebene. Daher liegt Φ auch im Inneren der konvexen Fläche, die das gemeinsame Innere dieser beiden Zylinder berandet. Der Schnitt dieser Fläche mit der Ebene $\Pi_{f(r)}$ enthält also $s_{f(r)}$ ganz im Inneren oder auf dem Rande, woraus man

$$|s_{f(r)}| \leq 4\,r\sqrt[q]{f(r)}$$

schliesst. Da weiter $f'(r)$ und daher auch H für $r \to 0$ gegen 0 streben, hat man vermöge (9)

$$\overline{\lim_{r\to 0}} \frac{|\Omega^*_{f(r)}|}{|\Omega_{f(r)}|} = \overline{\lim_{r\to 0}} \frac{|\Omega^*_{f(r)}|}{|\Omega_{f(r)}|} \geq \frac{q}{2}\left(\frac{1}{q-1}\right)^{\frac{q-1}{q}} \overline{\lim_{r\to 0}} \frac{f'(r)}{r f^{\frac{1}{q}}(r)} \{rf'(r) - f(r)\}^{\frac{q-1}{q}}$$

$$= \frac{q}{2}\left(\frac{1}{q-1}\right)^{\frac{q-1}{q}} \overline{\lim_{r\to 0}} \left\{\frac{f'^{\,2q-1}(r)}{rf(r)}\right\}^{\frac{1}{q}} \left\{1 - \frac{f(r)}{rf'(r)}\right\}^{\frac{q-1}{q}}.$$

Wir haben also zunächst zu zeigen, dass sich $f(r)$ so wählen lässt, dass

$$\overline{\lim_{r\to 0}} \frac{f'^{\,2q-1}(r)}{rf(r)} = \infty \tag{25}$$

wird; da K für $r = 0$ eine verschwindende Krümmung haben soll, muss ferner

$$\lim_{r\to 0} \frac{f(r)}{r^2} = \lim_{r\to 0} \frac{f'(r)}{r} = 0$$

sein. Dann ist für $q \geq 2$ und jede Folge $r_n \to 0$ mit $\frac{f'^{\,2q-1}(r_n)}{r_n f(r_n)} \to \infty$

$$\lim \frac{f(r_n)}{r_n f'(r_n)} = \lim \frac{r_n f(r_n)}{f'^{\,3}(r_n)} \cdot \frac{f'^{\,2}(r_n)}{r_n^2} \leq \lim \frac{r_n f(r_n)}{f'^{\,2q-1}(r_n)} \cdot \frac{f'^{\,2}(r_n)}{r_n^2} = 0,$$

also

$$\overline{\lim_{r\to 0}} \frac{|\Omega^*_{f(r)}|}{|\Omega_{f(r)}|} = \frac{q}{2}\left(\frac{1}{q-1}\right)^{\frac{q-1}{q}} \overline{\lim_{r\to 0}} \left\{\frac{f'^{\,2q-1}(r)}{rf(r)}\right\}^{\frac{1}{q}} = \infty,$$

so dass die zweite der Beziehungen (23) sicher erfüllt ist.

Eine Funktion $f(r)$ mit (25) erhält man aber z. B., wenn man setzt

$$f'(r) = \frac{1}{2^{(2q)^n}} \quad \text{für} \quad \frac{1}{2^n} < r < \frac{1}{2^{n-1}},$$

$$f(r) = \int_0^r f'(t)\,dt;$$

bei dieser Definition hat die Kurve k, wie verlangt war, in $r = 0$ eine verschwindende Krümmung, denn man hat

$$\frac{f(r)}{r^2} < \frac{f'(r)}{r} \to 0.$$

Ferner hat man an den Stellen $r = \frac{1}{2^n}$

$$\frac{f'^{2q-1}(r+)}{rf(r)} = \frac{f'^{2q-1}\left(\frac{1}{2^n}+\right)}{\frac{1}{2^n}f\left(\frac{1}{2^n}\right)} > \frac{1}{2^{(2q)^n(2q-1)}} 2^{2n} 2^{(2q)^n+1}$$

$$= 2^{(2q)^n+2n} \to \infty,$$

so dass (25) erfüllt ist.

Wir haben also nur noch zu zeigen, dass bei unsrer Definition von $f(r)$ auch

$$\lim_{r\to 0} \frac{|\Omega^*_{f(r)}|}{|\Omega_{f(r)}|} = 0$$

wird. Hierzu kann man so verfahren: Es ist offenbar für jedes n

$$f\left(\frac{1}{2^n}\right) > \frac{1}{2^{n+1}} f'\left(\frac{1}{2^n}-\right),$$

und daher erhält man für $r = \frac{1}{2^n}$ und $q \geq 2$

$$\frac{rf'(r-)-f(r)}{q-1} < f(r),$$

und

$$\frac{f'^{2q-1}(r-)}{rf(r)} < \frac{2^n 2^{n+1} 2^{(2q)^n+1}}{2^{(2q)^{n+1}(2q-1)}} = \frac{1}{2^{(2q)^{n+1}(2q-2)-(2n+1)}} \to 0.$$

Es lässt sich daher r_n zwischen $\frac{1}{2^{n+1}}$ und $\frac{1}{2^n}$ nahe an $\frac{1}{2^n}$ so wählen dass

(26) $$\frac{r_n f'(r_n) - f(r_n)}{q-1} < f(r_n)$$

wird, und

(27) $$\frac{f'^{2q-1}(r_n)}{r_n f(r_n)} \to 0$$

geht. Wegen (24) und (27) liegt dann die Kappe $\Omega_{f(r)}$ ganz in dem Bereich $\Sigma_{f(r)}$ von Φ, der durch die Verbindungsstrecken der Punkte $(\pm r, 0, f(r))$ (Cartesische Koordinaten!) von k mit den entsprechenden Punkten $(0, \pm \bar{r}, \bar{r}^q)$ von c begrenzt wird. $\Sigma_{f(r)}$ besteht aber aus vier Torsenstücken, und die Flächennormalen längs der begrenzenden Erzeugenden haben wir für den einen Quadranten bereits ausgerechnet; in den übrigen Quadranten erhält man dann die Flächennormale aus ξ_1 einfach durch Spiegelungen an den Koordinatenebenen. Bezeichnet man diese Vektoren mit ξ_2, ξ_3, ξ_4, so besteht das sphärische Bild von $\Sigma_{f(r)}$ einfach aus dem spärischen Viereck, dessen vier Eckpunkte die Endpunkte X_i der vom Kugelmittelpunkt aus abgetragenen Vektoren ξ_i sind. Nun sieht man leicht, dass für hinreichend kleine r_n der Flächeninhalt jedes dieser Dreiecke, z. B. $P^* X_1 X_2$ höchstens doppelt so gross ist wie das geradlinige Dreieck, dessen Ecken P^* und die Projektionen der Eckpunkte X_1, X_2 auf die Kugeltangentialebene in P^* sind. Man erhält daher aus der oben gefundenen Darstellung von ξ_1

$$|\Omega^*_{f(r)}| \leq |\Sigma^*_{f(r)}| \leq 8q \frac{f'(r) H^{\frac{q-1}{q}}}{1 + f'^2(r) + q^2 H^{\frac{2q-2}{q}}}.$$

Andrerseits schneidet $s_{f(r)}$ die y, z-Ebene im Punkte $(0, \sqrt[q]{f(r)}, f(r))$, und daher ist wegen der Konvexität

$$|s_{f(r)}| \geq r \sqrt[q]{f(r)}.$$

Man erhält demnach insgesamt

$$\lim_{r\to 0} \frac{|\Omega^*_{f(r)}|}{|\Omega_{f(r)}|} \leq \lim_{n\to\infty} \frac{|\Omega^*_{f(r_n)}|}{|\Omega_{f(r_n)}|} \leq \lim_{n\to\infty} \frac{|\Sigma^*_{f(r_n)}|}{|s_{f(r_n)}|} \leq$$

$$\leq 8q \frac{1}{(q-1)^{\frac{q-1}{q}}} \lim_{n\to\infty} \frac{f'(r_n)}{r_n f^{\frac{1}{q}}(r_n)} \{r_n f(r_n) - f(r_n)\}^{\frac{q-1}{q}} \leq$$

$$\leq \frac{8q}{(q-1)^{\frac{q-1}{q}}} \lim_{n\to\infty} \left\{\frac{f'^{\frac{2q-1}{q}}(r_n)}{r_n f(r_n)}\right\}^{\frac{1}{q}} = 0,$$

w. z. b. w.

7. Hier stellt sich die Frage, ob sich einfache Bedingungen angeben lassen, unter welchen

$$\lim_{h\to 0} \frac{|\Omega^*_h|}{|\Omega_h|} = 0 \tag{28}$$

ist. Eine solche ist im folgenden Satz gegeben:

Wenn die Gebiete Ω_h bezüglich der geodätischen Kreisscheiben γ_h in P zugelassen sind (vgl. S. 44), d. h. wenn

$$\overline{\lim} \frac{|\gamma_h|}{|\Omega_h|}$$

endlich ist und die Indikatrix von Φ im Punkte P unbeschränkt ist, aber nicht durch P geht, so gilt (28).

Beweis. Wir projizieren den Rand von γ_h auf die Ebene $z=0$ und dilatieren die Projektion von P aus im Massstab $\varrho_h : 1$, wo ϱ_h der Radius von γ_h ist. Es sei Γ_h das bei dieser Dilatation entstehende Gebiet, und t_h die Kurve, die entsteht, wenn man die Projektion von s_h auf die Ebene $z=0$ ebenfalls im Massstab $\varrho_h : 1$ von P aus dilatiert. t_h ist dann wieder konvex, enthält P im Inneren und ist in dem abgeschlossenen Bereich Γ_h enthalten und hat ferner mindestens einen Punkt mit dem Rand von Γ_h gemeinsam. Nun hat man nach (9)

$$\lim_{h\to 0}\frac{|s_h|}{|\Omega_h|}=\lim_{h\to 0}\frac{\varrho_h^2\,|t_h|}{|\Omega_h|}=0,$$

und ebenso (wegen der Differenzierbarkeit von Φ in P)

$$\lim_{h\to 0}\frac{\varrho_h^2\,|\Gamma_h|}{|\gamma_h|}=1.$$

Die Voraussetzung des Satzes besagt also, dass

$$\overline{\lim_{h\to 0}}\,\frac{|\Gamma_h|}{|t_h|}<\infty \tag{29}$$

ist. Nach einem Satz von Blaschke[1]) gibt es nun in jeder Folge $h_n\to 0$ mit $h_n>0$ eine Teilfolge $h_n'\to 0$ derart, dass $t_{h_n'}$ gegen eine konvexe Kurve t konvergiert; diese muss wegen (29) innere Punkte besitzen.

Wir werden nun zeigen, dass

$$\frac{|\Omega^*_{h_n'}|}{|\Omega^*_{h_n'}|}\to 0$$

strebt; damit sind wir dann fertig, da dies ja besagt, dass es in jeder Folge $h_n\to 0$ eine Teilfolge gibt, für welche (28) gilt.

t_h habe die Gleichung

$$r=\mu_h(\varphi).$$

Es ist dann

$$\frac{1}{\varrho_h^2}s_h=|t_h|=\frac{1}{2}\int_0^{2\pi}\mu_h^2(\varphi)\,d\varphi,$$

und daher

---

[1]) K. k. F. § 1, S. 7. T. Bonnesen–W. Fenchel: Theorie der konvexen Körper; Ergebnisse der Math., III, 1, Berlin 1934, S. 34.

$$\lim_{n\to\infty} \frac{|s_{h'_n}|}{\varrho^2_{h'_n}} = |t|.$$

Ferner hat man nach (19)

$$\frac{1}{\varrho_h^2}|\Omega_h^*| = \int_{s_h} \frac{\cos^2 u}{1+\cos u}\,\frac{\operatorname{tg}^2\alpha}{r^2}\,\frac{r^2}{\varrho_h^2}\,\frac{dv}{\cos\psi}. \tag{30}$$

Hierin ist $r = r(\varphi)$ der Radiusvektor der Kurve s_h, also $\frac{r}{\varrho_h}$ derjenige von t_h, d. h. es ist

$$\frac{r}{\varrho_h} = \mu_h(\varphi).$$

Wegen $t_{h'_n} \to t$ ist $\mu_{h'_n}$ beschränkt. Genau so wie auf S. 52 folgert man nun aus der zweiten Gleichmässigkeitseigenschaft wieder, dass $\frac{\operatorname{tg}\alpha}{r}$ gleichmässig gegen die Normalkrümmung $\frac{1}{g^2(\varphi)}$ in derjenigen Richtung φ strebt, in welcher die Indikatrix eine Stützgarade in der Richtung v hat; $\psi(v)$ strebt gegen den Winkel $\bar\psi(v)$ zwischen diesem Radiusvektor und der entsprechenden Normale an die Indikatrix. Man hat also wegen $u \to 0$ für die Kurven $s_{h'_n}$

$$\lim_{n\to\infty} \frac{\cos^2 u}{1+\cos u}\,\frac{\operatorname{tg}^2\alpha}{r^2}\,\frac{1}{\cos^2\psi} = \frac{1}{2}\,\frac{1}{g^4(\varphi)\cos^2\bar\psi}, \tag{31}$$

und aus der geometrischen Deutung folgt wieder, dass der Integrand beschränkt ist.

Nun erhält man die Kurven t und i als Limites von Kurven, die durch Dilatation aus denselben Kurven $s_{h'_n}$ entstehen; da t beschränkt und i unbeschränkt ist, hat man daher

$$\lim_{n\to\infty} \frac{h'_n}{\varrho^2_{h'_n}} = 0.$$

In jeder Richtung, in der $g(\varphi)$ endlich ist, wird also der Radiusvektor $\mu(\varphi) = \lim_{n\to\infty} \mu_{h'_n}(\varphi) = 0$. Dies zusammen mit (31) ergibt, dass der Integrand in (30) für die Folge h'_n überall gegen 0 strebt, ausgenommen höchstens zwei Richtungen φ, die zu in t enthaltenen, von P ausgehenden Strecken gehören. Somit ist

$$\lim_{n\to\infty} \frac{1}{\varrho^2_{h'_n}}|\Omega^*_{h'_n}| = 0,$$

also auch

$$\lim_{n\to\infty}\frac{|\Omega^*_{h'_n}|}{|\Omega_{h'_n}|}=\lim_{n\to\infty}\frac{|\Omega^*_{h'_n}|}{\varrho^2_{h'_n}}\cdot|t_{h'_n}|=0,$$

w. z. b. w.

8. Es bleibt nur noch der Fall zu untersuchen, dass die Indikatrix i unbeschränkt ist und durch P geht. Hier existiert $\lim\frac{|\Omega^*_h|}{|\Omega_h|}$ im allgemeinen natürlich erst recht nicht (vgl. Nr. 6), und es kann höchstens der Fall interessieren, wo die Bedingung des Satzes von Nr. 7 erfüllt ist. Es ist aber leicht zu sehen, dass selbst dann der Limes, falls er überhaupt existiert, von der Fläche und nicht nur von der Indikatrix abhängt. Als Beispiel betrachten wir eine Indikatrix, deren Inneres eine Halbebene ist: Unter den konvexen Flächen, deren Indikatrix in P durch $g(\varphi)=\infty$ für $\frac{\pi}{2}\leq\varphi\leq\frac{3}{2}\pi$, $g(\varphi)=0$ für $|\varphi|<\frac{\pi}{2}$ gegeben ist, und welche die Bedingung des Satzes von Nr. 7 erfüllen, gibt es solche, für die $\lim\frac{|\Omega^*_h|}{|\Omega_h|}=0$ ist, aber auch solche, bei denen dieser Grenzwert unendlich wird.

Genauer zeigen wir das folgende:

Es sei $1<a<2$ und die Fläche $\Phi=\Phi(a)$ werde für $x\leq 0$ definiert durch

$$z=(x^2+y^2)^{\frac{1+a}{2}},\quad x\leq 0,$$

und für $x\geq 0$ als Rand der konvexen Hülle der beiden Kurven

$$(c):\qquad z=y^{a+1},\quad y=0$$

und

$$(k):\qquad z=x^a,\quad y=0,\quad x\geq 0.$$

Dann ist

$$\lim_{h\to 0}\frac{|\Omega^*_h|}{|\Omega_h|}=0,$$

wenn $a>\frac{1+\sqrt{3}}{2}$, und $\lim\limits_{h\to 0}\frac{|\Omega^*_h|}{|\Omega_h|}=\infty$, wenn $1<a<\frac{1+\sqrt{3}}{2}$ ist.

Das „Innere" der Indikatrix von Φ besteht für alle a aus der Halbebene $x<0$, und die Gebiete Ω_h sind bezüglich der geodätischen Kreisscheiben in P zugelassen.

Um letzteres zu beweisen genügt es zu bemerken, dass

$$\lim_{h\to 0}\frac{|\gamma_h|}{|\Omega_h|}=\lim_{h\to 0}\frac{|\gamma_h|}{|s_h|}=2$$

ist; die von P aus im Masstab $h^{\overline{1+a}}:1$ ähnlich vergrösserten Projektionen der Kurven s_h auf die x, y-Ebene streben nämlich offenbar gegen den Halbkreis $x^2+y^2=1$, $x<0$ und dessen Durchmesser; die in gleicher Weise vergrösserten Projektionen der γ_h streben aber gegen den ganzen Einheitskreis. — Die übrigen Behauptungen beweisen wir ähnlich wie den Satz von Nr. 6, wobei sich jedoch einige Vereinfachungen ergeben.

Zunächst sieht man unmittelbar, dass es für die Rechnungen nur auf den Teil $x>0$ von Φ ankommt; genauer: bezeichnet man den über $x>0$ gelegenen Teil von Ω_h (also den Bereich $x>0$, $z<h$ von Φ) mit ω_h, so gilt

$$\overline{\lim_{h\to 0}}\frac{|\Omega_h^*|}{|\Omega_h|}=\overline{\lim_{h\to 0}}\frac{|\omega_h^*|}{|\Omega_h|}. \tag{32}$$

Der über der Halbebene $x>0$ gelegene Teil von Φ besteht (vgl. S. 57) aus zwei Torsenstücken die man erhält, indem man den Punkt

$$(x,\ 0,\ x^a),\quad x>0$$

von k mit den Punkten

$$\left(0,\ \pm\left(\frac{a-1}{a}\right)^{\frac{1}{a+1}},\ \frac{a-1}{a}x^a\right)$$

von c verbindet. Es bezeichne nun $\Sigma(x)$ den Bereich von Φ, der von diesen beiden Strecken und von c begrenzt wird. Setzt man dann

$$x_h=h^{\frac{1}{a}},\quad \bar{x}_h=\left(\frac{a}{a-1}h\right)^{\frac{1}{a}},$$

so wird offenbar

$$\Sigma(x_h)\subset\omega_h\subset\Sigma(\bar{x}_h). \tag{33}$$

Die Flächennormale längs unsrer Verbindungsstrecken hat die Komponenten

$$\xi_1:\ \xi_2: \qquad -\frac{ax^{a-1}}{H},\quad \mp\left(\frac{a-1}{a}\right)^{\frac{a}{a+1}}\frac{(a+1)\,x^{\frac{a^2}{a+1}}}{H},\quad \frac{1}{H},$$

wo zur Abkürzung

$$H^2 = 1 + x^{2a-2}\left\{a^2 + \left(\frac{a-1}{a}\right)^{\frac{2a}{a+1}} \left(a+1\right)^2 x^{\frac{2}{a+1}}\right\}$$

gesetzt wurde. Das sphärische Bild von $\Sigma(x)$ besteht also aus einem Kugelrechteck, dessen eine Seite (das Bild der Bogens von c) in der zur y, z-Ebene parallelen Ebene durch das Bild P^* liegt, und die anderen drei Seiten in den zur x, z-Ebene bzw. y, z-Ebene senkrechten Ebenen durch die vom Kugelmittelpunkt aus abgetragenen Vektoren ξ_1 und ξ_2. Der Flächeninhalt von Σ^* ist natürlich asymptotisch gleich dem Inhalt des Rechtecks, das in der Tangentialebene an die Kugel in P^* durch die Projektionen seiner Eckpunkte gebildet wird, d. h. man hat

$$(34)\quad 1 = \lim_{x\to 0} \frac{1}{|\Sigma^*(x)|} \cdot \frac{2}{H^2}(a+1)\,a\left(\frac{a-1}{a}\right)^{\frac{a}{a+1}} x^{a-1+\frac{a^2}{a+1}}$$

$$= 2(a+1)\,a^{\frac{1}{a+1}}(a-1)^{\frac{a}{a+1}} \lim_{x\to 0} \frac{x^{\frac{2a^2-1}{a+1}}}{|\Sigma^*(x)|}.$$

Um schliesslich auch noch $|\Omega_h|$ d. h. $|s_h|$ abzuschätzen, bemerke man, dass Φ ganz in der Rotationsfläche $z = (x^2+y^2)^{a+1}$ liegt, und für $x<0$ mit ihr zusammenfällt. Dies ergibt

$$\frac{\pi}{2}\, h^{\frac{2}{a+1}} < |s_h| < \pi\, h^{\frac{2}{a+1}}.$$

Man erhält daher aus (32) und (33) vermöge (34) durch Einsetzen der Werte für x_h und $\bar{x}_h$

$$\varliminf_{h\to 0} \frac{|\Omega_h^*|}{|\Omega_h|} = \varliminf_{h\to 0} \frac{|\omega_h^*|}{|s_h|} \geq \frac{2}{\pi}(a+1)\,a^{\frac{1}{a+1}}(a-1)^{\frac{a}{a+1}} \lim_{h\to 0} h^{\frac{2a^2-1}{a(a+1)}-\frac{2}{a+1}}$$

$$\varlimsup_{h\to 0} \frac{|\Omega_h^*|}{|\Omega_h|} = \varlimsup_{h\to 0} \frac{|\omega_h^*|}{|s_h|} \leq \frac{4}{\pi}(a+1)\,a\left(\frac{a-1}{1}\right)^{\frac{a-1}{a}} \lim_{h\to 0} h^{\frac{2a^2-1}{a(a+1)}-\frac{2}{a+1}}$$

Wenn also $2a^2-2a-1<0$ ist, erhält man den Grenzwert ∞, und wenn $2a^2-2a-1>0$ ist, den Grenzwert 0, w. z. b. w.

9. Wir wollen nun sehen, welche Aussagen die Theorie der additiven Mengenfunktionen über die Gausssche Krümmung liefert.

Die Abbildung der Fläche Φ auf die Einheitskugel E durch parallele Normalen heisst bekanntlich t o t a l s t e t i g, wenn jede

Nullmenge von Φ auf eine Nullmenge von E abgebildet wird, und das ist zugleich eine notwendige und hinreichende Bedingung dafür, dass jede messbare Menge von Φ in eine messbare Menge von E übergeht. Für die Totalstetigkeit der Abbildung ist die Differenzierbarkeit von Φ offenbar notwendig, aber sie ist, wie wir in K. k. F. § 6 gezeigt haben, nicht hinreichend. Um die Theorie der additiven Mengenfunktionen anwenden zu können, muss man sich daher zunächst einen Mengenkörper[1]) $K = \{\mu\}$ auf Φ verschaffen, der nur Mengen μ enthält, deren Bilder auf E messbar sind, und der noch solche Teilsysteme enthält, nach denen man differenzieren kann. Da das Bild γ^* einer offenen geodätischen Kreisscheibe γ von Φ messbar ist, besteht der aus den Kreisscheiben durch wiederholte Summen- und Durchschnittbildungen abgeleitete Körper K nur aus Mengen, deren Bilder auf E messbar sind. Andrerseits gilt für die geodätischen Kreisscheiben der Vitalische Ueberdeckungssatz, so dass nach ihnen differenziert werden kann, d. h. folgendes:

Es gibt auf Φ eine Nullmenge $\bar{A}$ derart, dass für jeden nicht in $\bar{A}$ gelegenen Punkt P von Φ und jede Folge von P (nicht notwendig als Mittelpunkt) enthaltenden geodätischen Kreisscheiben γ_n mit gegen Null konvergierendem Radius

$$\lim_{n \to \infty} \frac{|\gamma_n^*|}{|\gamma_n|} = \lambda(P)$$

existiert (und daher von der Wahl der Folge γ_n unabhängig ist).

Ordnet man nun jedem Punkt P einer beliebigen Teilmenge von Φ eine sich mit $n \to \infty$ auf P zusammenziehende Mengenfolge $\mu(P, n)$ $(n = 1, 2, \cdots)$ zu mit den folgenden Eigenschaften:

a) $\mu(P, n)$ und $\mu^*(P, n)$ sind messbar,

b) Die Mengen $\mu(P, n)$ sind im Punkte P bezüglich der geodätischen Kreisscheiben zugelassen, d. h. es gibt eine Folge geodätischer Kreisscheiben $\gamma(P, n)$ um P, die $\mu(P, n)$ enthalten und für die

---

[1]) Darunter verstehen wir ein System von Mengen, das mit zwei Mengen Γ_1 und Γ_2 zugleich die Summe $\Gamma_1 + \Gamma_2$ und den Durchschnitt $\Gamma_1 \Gamma_2$ enthält, und, falls Γ_2 in Γ_1 enthalten ist, auch die Differenz $\Gamma_1 - \Gamma_2$; schliesslich soll mit abzählbar vielen paarweise fremden Mengen auch ihre Vereinigungsmenge zum Körper gehören. Manchmal wird bei der Definition von Körpern die letzte Forderung fortgelassen; Körper, bei denen sie erfüllt ist, werden dann σ-Körper genannt (Terminologie von F. Hausdorff).

$$\lim_{n\to\infty}\frac{|\gamma(P,n)|}{|\mu(P,n)|}$$

endlich ist (vgl. S. 44, diese Zahl darf mit P beliebig varieren), so besagt die Lebesguesche Theorie, dass man ohne Aenderung der Ableitung $\lambda(P)$ auch nach diesen Mengen differenzieren kann, d. h. es gibt eine Nullmenge $\overline{\Lambda}_1$ auf Φ derart, dass für alle nicht in $\overline{\Lambda}_1$ liegenden Punkte von Φ

$$\lim\frac{|\mu^*(P,n)|}{|\mu(P,n)|}=\lambda(P)$$

wird[1]).

Nun sei P ein Punkt von Φ, in dem eine Indikatrix existiert, und $\Omega_h(P)$ die Flächenkappe von Φ, die durch die zur Tangentialebene an Φ in P parallelen Ebene im Abstand h abgeschnitten wird. Für $\Omega_{\frac{1}{n}}(P)$ $(n=1,2,\cdots)$ als Menge betrachtet gilt die Bedingung a) stets; die Bedingung b) gilt, wie wir gesehen haben (S. 44 [2])), jedenfalls dann, wenn die Indikatrix beschränkt ist. Wir haben damit:

In fast allen Punkten von Φ existiert die durch die geodätischen Kreisscheiben definierte Gausssche Krümmung und stimmt in fast allen Punkten in denen die Bedingung des Satzes von Nr. 7 S. 61 erfüllt ist, mit der durch die ebenen Schnitte $\Omega_h(P)$ definierten Gaussschen Krümmung überein.

Schwierigkeiten könnten somit höchstens die Punkte bereiten, in denen die Indikatrix aus zwei parallelen zu P symmetrisch gelegenen Geraden besteht (die anderen Punkte mit unbeschränkter Indikatrix bilden ja eine Nullmenge). Alle Behauptungen der Einleitung werden bewiesen sein, wenn wir das Folgende zeigen können:

Es sei Φ eine konvexe Fläche, die im Punkte P eine aus zwei parallelen zu P symmetrisch gelegenen Geraden bestehende Indikatrix hat, und es sei γ_ϱ die geodätische Kreisscheibe vom Radius ϱ um P. Dann ist

$$\lim_{\varrho\to 0}\frac{|\gamma_\varrho^*|}{|\gamma_\varrho|}=0.$$

[1]) Der Begriff einer zugelassenen Mengenfolge ist der Lebesgueschen Theorie bekanntlich sehr geläufig. Es ist interessant zu bemerken, dass sich uns dieser Begriff in Nr. 7 unabhängig davon aus einer geometrischen Betrachtung lokaler Natur ergeben hat. Dies zeigt wieder, wie naturgemäss die Begriffsbildungen der Lebesgueschen Theorie sind. Wir hoffen durch diese Ausführungen (sowie durch K. k. F. § 4) auch für Geometer, denen die Theorie der Reellen Funktionen ferner liegt, dargelegt zu haben, ein wie kräftiges Hilfsmittel in dieser Theorie auch für die Geometrie zur Verfügung steht.

Zum Beweise[1]) legen wir die Fläche Φ so, dass P in den Koordinatenursprung fällt und die Indikatrix aus den beiden Geraden $y = \pm b_0$ $(b_0 > 0)$ besteht. Dabei soll Φ im Halbraum $z \geq 0$ liegen. Ohne Einschränkung der Allgemeinheit kann ferner angenommen werden, dass Ψ mit der Ebene $z = 0$ nur den Punkt P gemeinsam hat. Es sei dann Ψ das Paraboloid mit der Gleichung

$$z = \frac{1}{2}\left(\frac{x^2}{a^2} + \frac{y^2}{b^2}\right)$$

mit $0 < b < b_0 < a$ für $b_0 \neq \infty$ und $0 < b = a$ für $b_0 = \infty$. In einer gewissen Umgebung von P verläuft dann Ψ ganz oberhalb von Φ.

Wir bilden nun die konvexe Hülle der Kreisscheibe $x^2 + y^2 \leq \varrho^2$ der x,y-Ebene mit der Fläche Ψ; ihr Rand sei $\overline{\Psi}$. Dann besteht $\overline{\Psi}$ aus der Kreisscheibe, einem Torsenstück das längs einer Kurve k das Paraboloid Ψ berührt, und schliesslich aus dem ausserhalb k gelegenen Teil von Ψ. k ist eine geschlossene Kurve, die sich mit $\varrho \to 0$ auf P zusammenzieht. Es seien ω bzw. $\overline{\omega}$ die von k auf Ψ bzw. $\overline{\Psi}$ begrenzten (P enthaltenden) Gebiete. Dann ist

$$\omega^* = \overline{\omega}^*.$$

Da Ψ in einer Umgebung von P ganz oberhalb von Φ verläuft, schneidet Ψ die Fläche Φ für hinreichend kleine ϱ längs des Torsenstückes, d. h. unterhalb der Kurve k. Es sei μ das von der Torse auf Φ ausgeschnittene (P enthaltende) Gebiet. Da sich γ_ϱ auf ein Teilgebiet der Kreisscheibe $x^2 + y^2 < \varrho^2$, $z = 0$ projiziert, hat man $\gamma_\varrho \subset \mu$; andrerseits sieht man mühelos, dass

$$\mu^* \subset \overline{\omega}^* = \omega^*.$$

Man hat demnach

$$|\gamma_\varrho^*| \leq |\omega^*|.$$

Nun hat die Kurve k, wie man leicht nachrechnet, in Zylinderkoordinaten ($x = r\cos\varphi$ usf.) die Gleichungen

$$r = 2\varrho\,\frac{\dfrac{\cos^2\varphi}{a^4} + \dfrac{\sin^2\varphi}{b^4}}{\dfrac{\cos^2\varphi}{a^2} + \dfrac{\sin^2\varphi}{b^2}}, \qquad z = 2\varrho^2\left(\frac{\cos^2\varphi}{a^4} + \frac{\sin^2\varphi}{b^4}\right),$$

[1]) Wesentliche Punkte der folgenden Überlegung verdanken wir einem Gespräch mit Herrn Dr. Fenchel, Kopenhagen.

sie verläuft deshalb ganz unterhalb der Ebene $z = \frac{2\varrho^2}{b^4}$. Es sei Ω der Bereich $z \leq \frac{2\varrho^2}{b^4}$ von Ψ. Dann ist

$$\lim_{\varrho \to 0} \frac{|\Omega^*|}{\varrho^2} = \frac{4\pi}{ab^5},$$

also

$$\overline{\lim_{\varrho \to 0}} \frac{|\gamma_\varrho^*|}{|\gamma_\varrho|} = \frac{1}{\pi} \overline{\lim_{\varrho \to 0}} \frac{|\gamma_\varrho^*|}{\varrho^2} \leq \frac{1}{\pi} \lim_{\varrho \to 0} \frac{|\Omega^*|}{\varrho^2} = \frac{4}{ab^5}.$$

Ein Grenzübergang $a \to \infty$ liefert hieraus die Behauptung.

Hier liegt es nahe, zu fragen, ob man die im Falle der Totalstetigkeit der sphärischen Abbildung aus diesen Betrachtungen folgende Formel (7) (S. 45) nicht ohne lokale Überlegungen direkt aus der Theorie der reellen Funktionen gewinnen kann. Das ist in der Tat möglich mit Hilfe eines Approximationssatzes von Tonelli[1]). Dagegen gelang es uns nicht, ohne Benutzung der lokalgeometrischen Ergebnisse zu zeigen, dass auch bei nicht totalstetiger sphärischer Abbildung in fast allen Punkten, wo der Eulersche Satz gilt, der Grenzwert $\lim \frac{|\gamma_n^*|}{|\gamma_n|}$ von S. 67 das Produkt der Hauptkrümmungen ist.

---

[1]) Sulla rappresentazione analitica delle funzioni di più variabli reali. Rend. Circ. mat. Palermo 29 (1910), S. 1–36. Die Fläche wird dabei durch analytische, aber im allgemeinen nicht konvexe Flächen approximiert und so die Formel (7) auf die gleiche Beziehung für analytischen Flächen zurückgeführt.

CURVATURE PROPERTIES OF CONVEX SURFACES

HERBERT BUSEMANN (COPENHAGEN) AND
WILLY FELLER (STOCKHOLM)

Translated from the German original by Annette A'Campo-Neuen[1]

CONTENTS

	Introduction	235
1.	Preliminaries on Curves	238
2.	Meusnier's Theorem	240
3.	The Curvatures of the Normal Sections at a Point of the Surface	244
4.	Curvature Properties of Normal Sections Holding Almost Everywhere	251
5.	On the Theorem on Umbilical Points	256
6.	Example of a Surface that up to a Null Set Consists only of Pieces of Spheres of the Same Radius	259
7.	Geodesics	263

INTRODUCTION

The aim of this paper is to develop the elements of differential geometry of convex surfaces without making the usual regularity assumptions. Hereby necessarily, purely geometric considerations often replace the usual analytic arguments so that our method approaches the one of HJELMSLEV in his foundations of planar geometry.[2] Our method is less related to the work of BOULIGAND and his school, since as far as our subject is concerned they give geometric formulations of analytic processes rather than presenting directly geometric approaches.

Our main topics are: Analoga of the theorems by Meusnier, Olinde Rodriguez and Euler; the theorem stating that the sphere is the only closed surface with only umbilical points; finally the differential properties of geodesics.

In Section 1, we will compile some definitions and properties of curves. Section 2, which also holds for *non–convex surfaces*, is devoted to Meusnier's theorem. Here it is useful to introduce the following notion. The plane Π is called *strict tangent plane* at the point P of the surface Φ (paratangent

---

Krümmungseigenschaften Konvexer Flächen, in: Acta Math. 66, 1936, p. 1–47.

[1]Department of Mathematics and Computer Science, University of Basel, Spiegelgasse 1, 4051 Basel, Switzerland, email: annette.acampo@unibas.ch

[2]Grundlag for Fladernes Geometri, Mém. Acad. Roy. Sc. Let. Danemark, Copenhague, (7) XII, (1914).

plane in the terminology of Bouligand), if for *each* sequence of secants $\overline{Q_nR_n}$ converging to P, all the limits of the lines $\underline{Q_nR_n}$ are contained in Π. For example, *every* tangent plane of a *convex* surface is automatically a strict tangent plane. If at some point P, where the surface has a strict tangent plane, *some* plane section (other than the tangent plane) has curvature,[3] then *all* plane sections with the same tangent at P have curvature, and the corresponding curvature circles lie on a common sphere. This statement constitutes a minor, but nevertheless in the case of convex surfaces important strengthening of Meusnier's theorem compared to the formulation given by HJELMSLEV.[4] – The well-known connection between the curvature of a curve on a surface and its projection to the tangent plane is valid even under these very general assumptions.

In Section 3, we study the curvatures of normal sections of a convex surface at a point P where the surface has a tangent plane. If we mark the square root of the smallest (one–sided) radius of curvature of each normal section on the corresponding tangent ray starting from P, then the endpoint describes a curve that we call the *lower indicatrix* of P. *This curve is always convex* (however it may lie entirely at infinity). Then follow statements on the analogously defined upper indicatrix. We simply speak of an *indicatrix* if both curves coincide. *Every convex curve containing P in its interior or its boundary may occur as indicatrix of a surface point P.*

Now let the convex surface have a tangent plane in the neighborhood of P and at P itself an indicatrix. We consider the mapping from the surface to the unit sphere given by parallel normals. If Q is a point on the indicatrix of P where that has the tangent τ, then the spherical image of every plane section of the surface with tangent $\underline{PQ}$ has a tangent at the image point of P, and this tangent is orthogonal to τ. Therefore the lines of intersection of neighboring tangent planes along these sections have a limit when going to P, and the limit is parallel to τ. This corresponds to the theorem on *conjugate directions* in the regular case: *If the indicatrix does not have singular points then for every direction through P there is a conjugate one; this relation may even be symmetric without the indicatrix having to be an ellipse.* The normal is conjugate to those directions PQ, where the radius vector of the indicatrix takes an extremal value. In these directions (of which in general there can be infinitely many), the formula of *Olinde Rodrigues* for principal curvature directions holds.

In Section 4, we put aside statements on curvature of normal sections that have to be valid at *every* point. Instead, we obtain stronger results that however are only valid *almost everywhere* (i.e., except at a set of measure zero). Of course, theorems of this type can only be obtained using the

[3]Here the center of curvature is not meant to be the limit of the points of intersection of neighboring normals to the curve, but defined as the limit of the centers of the circles touching the curve at P and passing through a neighboring point on the curve.

[4]as above. Recently, also BOULIGAND proved Meusnier's theorem under weaker assumptions than the classical ones, yet his assumptions are much stronger than the ones used by HJELMSLEV. See also p. 10.

theory of real functions. Convex surfaces have a tangent plane almost everywhere,[5] and convex curves have finite curvature almost everywhere.[6] Using Meusnier's theorem and the results of Section 3, one can conclude that *at almost all points of the surface all the plane sections have finite curvature.* Moreover, at almost every point Euler's theorem holds, i.e., *almost everywhere, the indicatrix is a conic section with the given point as center.* It may be of interest that the latter points can be characterized analytically as follows: We consider a piece of the surface Φ that can be projected uniquely onto a plane Π. In Π we introduce cartesian coordinates x, y in all possible ways. If in one of those systems we represent Φ in the form $z = f(x, y)$, then $f(x, y)$ is of bounded variation and hence almost everywhere differentiable along rectangles.[7] Then the set of points in question is the set of points where all plane sections have finite curvature and all the functions $f(x, y)$ are simultaneously differentiable.

In Section 5, without any further assumptions on regularity we will prove that *a differentiable piece of a surface consisting entirely of umbilical points, must be a piece of a sphere.* It is an obvious question whether here it would suffice to assume that *almost every point is an umbilic.* Under certain assumptions this is true, for example if the curvature of the normal sections is bounded, or if the first surface derivatives are absolutely continuous. However, Section 6 shows that the theorem is not true if we omit any such assumptions. Namely, we will construct a *closed differentiable convex surface such that almost every point is an umbilic of the same curvature, and yet it is not a sphere. Here the curvature can also be zero.* At the same time, this surface serves as an example showing that the spherical mapping of differentiable convex surfaces does *not need to be absolutely continuous*, i.e., that sets of measure zero on the surface may be mapped to sets of positive measure on the sphere. If the curvature of normal sections is bounded, this cannot happen, as we have already proved previously applying a relation occuring in Section 5.

The last section is independent of Sections 3–6. According to LEBESGUE[8], on every convex surface there is a shortest path between any two points. In Section 7, we show that such a geodesic is differentiable at each point P (including the singular points) where the surface has a tangent plane, and

---

[5]Compare T. BONNESEN and W. FENCHEL: Theorie der konvexen Körper. Ergebnisse der Math. III,1, Berlin 1934, p. 13. – We will refer to this report whenever possible, and we will cite it in the sequel as B.F.

[6]B.F., p. 144.

[7]This means that at almost all points the following limit exists

$$\lim_{n\to\infty} \frac{f(x+h_n, y+k_n) + f(x-h_n, y-k_n) - f(x+h_n, y-k_n) - f(x-h_n, y+k_n)}{4h_n k_n},$$

where h_n and k_n are sequences converging to zero such that

$$0 < \liminf \frac{h_n}{k_n} \leq \limsup \frac{h_n}{k_n} < \infty .$$

[8]Annali di Matematica (3), VII (1902). Compare also O. Bolza: Variationsrechnung, Leipzig and Berlin 1909, p. 422.

that it admits an osculating plane passing through the normal to the surface. *The projection of the line onto the tangent plane has curvature zero at such points, even if the normal sections (and in particular the geodesic) do not have curvature.*

1. Preliminaries on Curves

All the geometric objects that will be considered in the following are contained in three–dimensional Euclidean space. Capital Latin letters will denote points, $\overline{PQ}$ the closed segment from P to Q and $\underline{PQ}$ for $P \neq Q$ the line through P and Q. If R does not lie on the line $\underline{PQ} = t$ then $\underline{PQR} = \underline{Rt}$ shall denote the plane through R and t. The distance between two sets of points α and β will be denoted by $d(\alpha, \beta)$.

It is convenient to extend some notions that are usually defined only for curves to convergent sequences of points. If for a sequence of points $\{P_n\}$, converging to P, the lines $\underline{PP_n}$ converge, then the limit line t shall be called the *tangent* of $\{P_n\}$. Let F_{P_n} denote the foot of the perpendicular from P_n onto t. We call $\sigma_n = d(P_n, F_{P_n})$ the *normal segment* and $\tau_n = d(F_{P_n}, P)$ the *tangent segment* of P_n on t. Then

$$\lim \frac{\sigma_n}{\tau_n} = 0\,. \tag{1}$$

The circle through the points P and P_n with tangent t at P shall be called *approximating circle* of P_n with respect to the line element (P, t). Its radius is given by

$$\rho(P_n) = \frac{d^2(P, P_n)}{2d(P_n, F_{P_n})} = \frac{\sigma_n^2 + \tau_n^2}{2\sigma_n}\,, \tag{2}$$

where the right hand side is meant to be ∞ if $\sigma_n = 0$. If $\rho = \lim \rho(P_n)$ exists then we call ρ the *curvature radius* of the sequence of points $\{P_n\}$. Using (1.1) and (1.2) we find

$$\rho = \lim \frac{\tau_n^2}{2\sigma_n}\,. \tag{3}$$

The number $1/\rho$ or 0 if $\rho = \infty$ respectively is called the *curvature* of $\{P_n\}$.

If the convergent sequence $P_n \to P$ has a tangent t and the planes $\underline{P_n t}$ converge, then we call the limit the *osculating plane* of $\{P_n\}$. Furthermore, we speak of an osculating plane even if $\{P_n\}$ does not have a tangent but if all the subsequences with tangent admit the same osculating plane. For example this is true for all plane curves. If a sequence of points $\{P_n\}$ has an osculating plane Π as well as curvature $1/\rho$, then *the sequence of footpoints from P_n onto Π has the same curvature.* If Π' and Π'' are two orthogonal planes through t, then the sequences of footpoints from P_n onto these planes also have curvature $1/\rho'$ and $1/\rho''$, and

$$\frac{1}{\rho^2} = \frac{1}{(\rho')^2} + \frac{1}{(\rho'')^2}\,. \tag{4}$$

Now let κ be the topological image of a closed segment and let P be its endpoint. If all the sequences of points converging to P on κ admit a tangent

t (and hence always the same), then t is called the *tangent* to κ at P. If for *every* sequence of secants $\overline{Q_nR_n}$ of κ converging to P the lines $\underline{Q_nR_n} \to t$, then we call t a *strict tangent*.

If we choose P as the origin and the strict tangent t as the x-axis of a coordinate system then in a half-plane, say for $x \geq 0$, κ has a unique description of the form $y = f(x)$, $z = g(x)$. If the x-axis at the origin is not a strict but just an ordinary tangent, then such a description may not exist. The fact that the x-axis is a strict tangent at the origin, can be expressed analytically by the relations

$$\lim_{x\to 0^+, h\to 0} \frac{f(x+h)-f(x)}{h} = \lim_{x\to 0^+, h\to 0} \frac{g(x+h)-g(x)}{h} = 0 .$$

Therefore the difference quotients of $f(x)$ and $g(x)$ are bounded in a neighborhood of the origin, so that there the derivatives $f'(x)$ and $g'(x)$ exist almost everywhere. For every sequence $\{x_n\}$ converging to zero such that $f'(x_n)$ and $g'(x_n)$ exist, this implies $f'(x_n) \to 0$ and $g'(x_n) \to 0$. These conditions are also sufficient for the x-axis being a strict tangent at the origin.[10]

Now let κ again have a tangent t at P (not necessarily strict). We consider all the sequences of points on κ converging to P having curvature. The lower (upper) bound of all those curvatures will be called the *lower (upper) curvature* of κ at P. Analogously, we define the notions *curvature, curvature radius and osculating plane*.

Every inner point Q decomposes κ into two paths κ_r and κ_l that we will call the right or left side respectively of Q on κ. If κ_r and κ_l have tangents at Q, then we call these the *right* or *left tangent* respectively of κ at Q. If both coincide, then we simply speak of a *tangent*. The other notions are defined analogously.

Now let us consider a plane curve κ given by the equation $y = f(x)$. If κ admits a right tangent at $x = x_0$, then the right derivative $f^r(x_0)$ exists, and for the right upper and lower curvature we find the values

$$(5) \qquad \frac{2}{(\sqrt{1+[f^r(x_0)]^2})^3} = \lim_{h\to 0^+} \begin{matrix}\inf\\ \sup\end{matrix} \frac{1}{h}\left\{\frac{f(x_0+h)-f(x_0)}{h} - f^r(x_0)\right\} .$$

One can see from this description that κ even with continuous first derivative $f'(x)$ may have curvature at $x = x_0$ although the second derivative $f''(x_0)$ does not exist.[11] An example for this feature is given by $y = x^3 \sin(\frac{1}{x})$ at the origin. However, if the second right derivative $f^{rr}(x_0)$ exists then κ has a right curvature at $x = x_0$,[12] namely

$$(6) \qquad \frac{1}{\rho_0} = \frac{f^{rr}(x_0)}{(\sqrt{1+[f^r(x_0)]^2})^3} .$$

---

[10] Compare Section 2.

[11] Only if we define the curvature radius via the limit of the point of intersection of neighboring normals then the existence of curvature is equivalent to the existence of a second derivative.

[12] Compare B. JESSEN: Om konvekse Kurvers Krumning. Matematisk Tidsskrift B, 1929, Kopenhagen.

In particular, if $f''(x_0)$ exists then κ has curvature at $x = x_0$. If we apply (1.5) and (1.6) to the right and the left curvature then we find

$$f''(x_0) = \lim_{h\to 0} \frac{f(x_0+h)+f(x_0-h)-2f(x_0)}{h^2}. \tag{7}$$

For *convex* curves the situation is particularly easy. Here at each point there is a right and a left tangent, and these are monotonous. Hence the curve admits a (two-sided) tangent at every point up to at most countably many exceptions. It is always automatically a strict tangent. The monotonicity of the tangents implies that *almost everywhere a second derivative*[13] of the function describing the curve exists. Note that in contrast to non–convex curves here *the existence of a right (left) curvature implies the existence of the second right (left) derivative.*[14] Finally, let us mention *Blaschke's theorem of choice:*[15] *Every uniformly bounded sequence of convex curves* $\{\kappa_n\}$ *has a subsequence* $\{\kappa_{\nu_n}\}$ *converging to a (convex) curve* κ. Here the curves converge *together with their tangents* in the following sense: Let P_n be a point on κ_{ν_n} and let P_n converge to a point where κ has a tangent. Then the tangents of κ_{ν_n} at P_n converge to this tangent.

2. Meusnier's Theorem

Let Φ be a topological image of the disc and let P be an inner point of Φ. A plane Π is called a *strict tangent plane of* Φ at P if for *every* sequence of secants $\overline{Q_nR_n}$ of Φ converging to P all the limits of the lines $\underline{Q_nR_n}$ are contained in Π. If Φ has a strict tangent plane at P and this plane is not perpendicular to the xy-plane then one can describe Φ in a neighborhood of P uniquely in the form $z = f(x,y)$. If the plane is given in this form then one can rephrase the geometric definition as follows: Φ has a strict tangent plane at (x_0, y_0) if and only if the following limits exist:

$$\lim_{\substack{(x,y)\to(x_0,y_0)\\ h\to 0}} \frac{f(x+h,y)-f(x,y)}{h} = f_x(x_0,y_0) \quad \text{and}$$

$$\lim_{\substack{(x,y)\to(x_0,y_0)\\ k\to 0}} \frac{f(x,y+k)-f(x,y)}{k} = f_y(x_0,y_0). \tag{8}$$

In particular, if Φ has a tangent plane at *every* point then it is in fact a strict tangent plane whereever it is continuous.

For the sake of completeness, we mention what the strict tangent plane means in terms of the function $f(x,y)$. However, we will *not use* the following criterion later:

---

[13]This means more precisely that the derivatives of the right and the left derivative coincide outside of a set of measure zero.

[14]Compare Jessen, as above.

[15]Compare B. F., p. 34.

[17]According to Lebesgue, $g(x)$ is absolutely continuous on some interval, if $g'(x)$ exists almost everywhere in that interval and for every subinterval $\int_{x_0}^{x_1} g'(x)\,dx = g(x_1) - g(x_0)$ holds.

The surface $z = f(x, y)$ has a strict tangent plane at (x_0, y_0) that is not perpendicular to the xy-plane if and only if

(1) $f_x(x_0, y_0)$ and $f_y(x_0, y_0)$ exist;
(2) there is a neighborhood of (x_0, y_0) such that $f(x, y)$ for almost all values of one of the two variables is absolutely continuous in the other variable,[17] and
(3) the hence almost everywhere existing derivatives $f_x(x, y)$ and $f_y(x, y)$ are continuous at (x_0, y_0).

Proof. (a) If there is a strict tangent plane at (x_0, y_0) then the difference quotients occuring in (2.1) are bounded in a neighborhood of (x_0, y_0). Therefore $f(x, y)$ as a function of one variable is even absolutely continous at *all* values of the other variable.[18] The continuity of the partial derivatives at (x_0, y_0) follows from (2.1).

(b) Let us assume that all three conditions are satisfied. We have to show that for all sequences $(x_n, y_n) \to (x_0, y_0)$ and $h_n \to 0$ such that

$$\lim_{n\to\infty} \frac{f(x_n + h_n, y_n) - f(x_n, y_n)}{h_n} \tag{9}$$

exists, this limit is equal to $f_x(x_0, y_0)$. To see this we choose y'_n sufficiently close to y_n such that the sequence

$$\frac{f(x_n + h_n, y'_n) - f(x_n, y'_n)}{h_n}$$

has the same limit as (2.2) and that $f(x, y'_n)$ is absolutely continuous. Then

$$f(x_n + h_n, y'_n) - f(x_n, y'_n) = \int_{x_n}^{x_n+h_n} f_x(x, y'_n)\, dx = h_n \Delta_n\,,$$

where Δ_n is a number lying between the maximum and the minimum of $f_x(x, y'_n)$ on the integration interval. Because of the third assumption, we have in fact

$$\lim \frac{f(x_n + h_n, y'_n) - f(x_n, y'_n)}{h_n} = \lim \Delta_n = f_x(x_0, y_0)\,.$$

□

Now let us consider the special case of *convex surfaces*. Here the tangent planes at a point P always enfold a convex half–cone generated by the tangent rays of all plane sections at P. The surface lies entirely on one side of each of its tangent planes, so that every limit of a tangent plane is again a tangent plane. On the other hand, for every secant of the surface there is a parallel tangent plane. This immediately implies that *every tangent plane of a convex surface is automatically a strict tangent plane.*

The notion of a strict tangent plane provides the natural assumptions for MEUSNIER's theorem:

Let the surface Φ have a strict tangent plane Π at the point P. Let $P_n \to P$ and $P'_n \to P$ be two sequences of points on the surface with the same tangent t and with osculating planes that are inclined with respect to the normal of

[18]Compare e.g. H. LEBESGUE: Leçons sur l'Intégration, Paris 1928, p. 77 and p. 183.

Φ at P with the angle $\theta \neq \frac{\pi}{2}$ and $\theta' \neq \frac{\pi}{2}$ respectively. If moreover the angles formed by the lines $\underline{P_nP_n'}$ with t are bounded from below and $\{P_n\}$ has curvature $\frac{1}{\rho}$ then $\{P_n'\}$ has curvature $\frac{1}{\rho'}$ where

$$\frac{\rho}{\cos\theta} = \frac{\rho'}{\cos\theta'}\,. \tag{10}$$

Proof. For the proof we choose the coordinate system such that P is in the origin, t the x-axis and Π the xy-plane. Let $P_n = (x_n, y_n, z_n)$ and $P_n' = (x_n', y_n', z_n')$. Moreover, let θ_n and θ_n' be the angles of the z-axis with the planes through t and P_n or P_n' respectively, so that

$$z_n = y_n \tan\theta_n \quad z_n' = y_n' \tan\theta_n'\,. \tag{11}$$

Since the angles between $\underline{P_nP_n'}$ and the x-axis are bounded from below, the direction factor

$$\frac{x_n - x_n'}{y_n - y_n'}$$

can only have finite accumulation values. Because of

$$\frac{y_n}{x_n} \to 0 \quad \frac{y_n'}{x_n'} \to 0\,,$$

this implies

$$\lim \frac{x_n'}{x_n} = 1\,. \tag{12}$$

Moreover, since the xy-plane is a strict tangent plane, we obtain from (2.4)

$$\lim \frac{z_n - z_n'}{y_n - y_n'} = \lim \frac{z_n - z_n'}{\frac{z_n}{\tan\theta_n} - \frac{z_n'}{\tan\theta_n'}} = 0\,,$$

and hence also

$$\lim \frac{z_n'}{z_n} = 1\,. \tag{13}$$

Now the tangent segments and the normal segments (compare Section 1, p. 4) of the two sequences of points take the following values:

$$\tau_n = x_n \quad \sigma_n = \frac{z_n}{\cos\theta_n}\,; \quad \tau_n' = x_n' \quad \sigma_n' = \frac{z_n'}{\cos\theta_n'}\,.$$

Hence by assumption

$$\lim \frac{x_n^2 \cos\theta_n}{2z_n} = \rho\,.$$

Therefore (2.5) and (2.6) imply the claim

$$\lim \frac{(\tau_n')^2}{2\sigma_n'} = \lim \frac{(x_n')^2 \cos\theta_n'}{2z_n'} = \frac{\rho\cos\theta'}{\cos\theta}\,. \qquad \square$$

For curves on surfaces the theorem means the following:[20]

[20]The following theorem in principle is equivalent to the theorem of Hjelmslev (as above), who assumes a continuous tangent plane in the neighborhood of P. A remark of Bouligand (Géométrie infinitésimale directe, Paris 1932, p. 173–174) on the necessity of further, much stronger assumptions could lead to misunderstandings since in his example the partial derivatives are not continuous.

Let the surface Φ have a strict tangent plane at P. Let κ and κ' be two curves on Φ emerging from P with the same tangent ray t at P and with osculating planes forming the angle $\theta \neq \frac{\pi}{2}$ and $\theta' \neq \frac{\pi}{2}$ respectively with the normal of Φ at P. Then for the upper as well as the lower curvatures of κ and κ' at P we have

$$\frac{\rho}{\cos\theta} = \frac{\rho'}{\cos\theta'}.$$

To see this let $\{P_n\}$ be any sequence of points on κ converging to P with curvature radius ρ. Let P'_n be a point on κ' lying in the same plane orthogonal to t as P_n. Then by the theorem we just proved, $\{P'_n\}$ has curvature $\frac{1}{\rho'}$ with (2.3). Now the claim follows from the fact that the sequence $\{P_n\}$ was chosen arbitrarily.

For a later application, we also mention a *corollary* of Meusnier's theorem. If two surfaces Φ_1 and Φ_2 at a common point P have distinct strict tangent planes then in a neighborhood of P there are further points common to both surfaces. These intersection points form a curve κ having a strict tangent t at P. *If the normal sections of Φ_1 and Φ_2 in the direction of t at P have finite positive curvature radii ρ_1 and ρ_2 respectively, then κ has an osculating plane at P (and hence curvature).*

To see this let $\{P_n\}$ be any sequence of points on κ converging to P with curvature radius ρ and an osculating plane forming the angles θ_1 and θ_2 respectively with the normals of the surfaces at P. Then

$$\frac{\rho}{\cos\theta_i} = \rho_i \quad \text{for } i = 1, 2. \tag{14}$$

Now either $\theta_1 + \theta_2$ or $|\theta_1 - \theta_2|$ is one of the angles between the normals of the surfaces. If we denote it with ν, then $\nu \neq 0$ and in both cases we obtain from (2.7)

$$\cos\nu + \sin\nu \tan\theta_1 = \frac{\rho_1}{\rho_2}. \tag{15}$$

So there are only two possible values for θ_1, and thus we can conclude that κ must have an osculating plane at P. Because if on κ there were two sequences of points $P_n \to P$ and $P'_n \to P$ with distinct osculating planes Σ and Σ' then every plane in the angle range between Σ and Σ' would occur as osculating plane of a suitable sequence of points on κ.

The position of the osculating plane is determined by the respective values in question. In the case that we will be interested in later, we have

$$\frac{\rho_1}{\rho_2} < |\cos\nu|. \tag{16}$$

Since $\sin\nu$ and $\tan\theta_1$ are positive, in this case the osculating plane can *not* decompose the acute angle between the normals of the planes.

Note the following well–known relation which holds already under the assumptions made here. *Let the surface Φ have a strict tangent plane Π at P, and let the normal section with the tangent t at P have finite curvature*

$\frac{1}{\rho_\nu}$. *A curve κ on the surface with tangent t at P has finite curvature $\frac{1}{\rho}$ at P if and only if its projection to Π has curvature $\frac{1}{\rho_g}$. In this case the following holds*

$$\frac{1}{\rho^2} = \frac{1}{\rho_g^2} + \frac{1}{\rho_\nu^2}. \tag{17}$$

Proof. In order to prove this we choose P as the origin, t as the x-axis and Π as the xy-plane. Suppose that Φ is given by $z = f(x, y)$. Let the projection κ' be given by $y = g(x)$. Then

$$\frac{1}{\rho_\nu} = \lim_{x\to 0} \frac{2|f(x,0)|}{x^2},$$

and if it exists

$$\frac{1}{\rho_g} = \lim \frac{2|g(x,0)|}{x^2}, \tag{18}$$

$$\frac{1}{\rho} = \lim \frac{2\sqrt{(g(x,0))^2 + (f(x,g(x)))^2}}{x^2}. \tag{19}$$

Because the xy-plane is a strict tangent plane at the origin, we have

$$\lim_{x,y\to 0} \frac{f(x,y) - f(x,0)}{y} = 0,$$

and hence for every curve $y = g(x)$, for which $\frac{y}{x^2}$ is bounded,

$$\lim_{x\to 0} \frac{f(x,y) - f(x,0)}{x^2} = 0.$$

So if one of the equations (2.11) or (2.12) holds then the other one is true as well, and the equation (2.10) follows by insertion. □

3. The Curvatures of the Normal Sections at a Point of the Surface

Let the convex surface Φ have the tangent plane Π at P. Through each tangent ray t_φ at Φ emanating from P, we take the halfplane bounded by the normal of the surface. The resulting half normal section shall be denoted by x_φ. For short, we call the upper (lower) curvature of x_φ at P *the upper (lower) normal curvature belonging to* t_φ, an we denote it by $\frac{1}{\underline{\rho}_\varphi}$ or $\frac{1}{\overline{\rho}_\varphi}$ respectively. Hence there are *two* upper (lower) normal curvatures for every normal section. If all the normal sections have finite curvatures then all the normal curvatures exist and $\rho_\varphi = \rho_{\varphi+\pi}$.

If we mark $\sqrt{\underline{\rho}_\varphi}$ and $\sqrt{\overline{\rho}_\varphi}$ on t_φ from P, then the endpoints form two sets in Π that we will call the *upper* and *lower indicatrix*. If they coincide then we speak of the *indicatrix* at P. If all normal sections at P have (two-sided) curvatures then the indicatrix is a midpoint curve with P as center. Naturally, we think of every half tangent t_φ as being equipped with a point at infinity. If $\underline{\rho}_{\varphi_0} = \infty$ say, then we call the lower indicatrix for $\varphi = \varphi_0$ continuous if $\lim_{\varphi \to \varphi_0} \underline{\rho}_{\varphi} = \infty$. More generally, we also speak of an indicatrix if at P there is no tangent plane but the tangent cone contains a flat piece. In this case the two indicatrices are only defined in the corresponding flat sector of angles.

In order to be able to investigate the curvature properties even at this kind of points, we introduce the following notation:

Let r, φ be polar coordinates in the xy-plane. Let ω be a fixed angle, $0 < \omega \leq 2\pi$, and for $0 \leq \varphi \leq \omega$, let

$$z = f(r, \varphi), \quad f \geq 0,$$

describe a convex surface Φ such that the plane sections $\varphi =$ constant at the origin have tangents lying in the xy-plane.

So in particular, if $\omega = 2\pi$, then P is an inner point of Φ, and the xy-plane is the tangent plane of Φ at P. In the following we always think of the angle φ as being chosen from the interval $0 \leq \varphi \leq \omega$.

The lower and upper normal curvature belonging to the direction φ, equal

$$(20) \qquad \frac{1}{\overline{\rho}_{\varphi}} = \liminf_{r \to 0} \frac{2|f(r,\varphi)|}{r^2} \quad \text{and} \frac{1}{\underline{\rho}_{\varphi}} = \limsup_{r \to 0} \frac{2|f(r,\varphi)|}{r^2}.$$

Let s_h denote the intersection curve of Φ with the plane $z = \frac{h}{2}$. We project s_h onto the xy-plane and magnify the projection from P similarly on a scale of $\sqrt{h} : 1$. The resulting curve shall be called σ_h. The equations describing the curves are the following:

$$(21) \qquad \begin{array}{rll} s_h: & f(r,\varphi) = \frac{h}{2}, & z = \frac{h}{2}; \\ \sigma_h: & f(\sqrt{h}\,r,\varphi) = \frac{h}{2}, & z = 0; \end{array}$$

We can view the operation transfering the curve s_h into σ_h as a mapping of the surface to the considered sector of angles in the xy-plane: the point $Q(r, \varphi_0, \frac{h}{2})$ of the surface lying on s_h, is mapped to the point $Q'(r' = \frac{r}{\sqrt{h}}, \varphi_0, 0)$ lying on σ_h.

Now let $\{P_n\}$ be a sequence of points of the surface converging to P, that are all lying on the normal section $\varphi = \varphi_0$. If $(r_n, \varphi_0, \frac{h_n}{2})$ are the coordinates of P_n, then $P_n' = (r_n' = \frac{r_n}{\sqrt{h_n}}, \varphi_0, 0)$ is its image and by (3.1)

$$\sqrt{\underline{\rho}_{\varphi_0}} \leq \liminf r_n' \leq \limsup r_n' \leq \sqrt{\overline{\rho}_{\varphi_0}}\,.$$

One sees immediately that for *every* number r' between $\sqrt{\underline{\rho}_{\varphi_0}}$ and $\sqrt{\overline{\rho}_{\varphi_0}}$ there is such a sequence of points $\{P_n\}$ with

$$\lim r_n' = r'.$$

[22]This shows that all normal sections at P have one–sided curvatures if and only if the curves σ_h converge for $h \to 0$. The limit curve in that case is the indicatrix, and therefore it is always a convex curve.

[23]If we talk about the interior of the curve σ_h, then of course we think of the domain enclosed by σ_h together with the rays $\varphi = 0$ and $\varphi = \omega$.

Hence if for a sequence $h_n \to 0$ we consider the points of intersection of σ_{h_n} with the ray $\varphi = \varphi_0$, then the points of accumulation of these points for all the sequences $h_n \to 0$ cover the entire[22] interval $\sqrt{\underline{\rho}_{\varphi_0}} \le r' \le \sqrt{\overline{\rho}_{\varphi_0}}$ on $\varphi = \varphi_0$. In this way, we easily obtain the two indicatrices of P, and to simplify further analysis we introduce the following *notation*:

Let γ_n be the intersection and let Γ_n be the union of the closed domains bounded by the curves σ_h for $h < \frac{1}{n}$.[23] If $n \to \infty$ then γ_n increases monotonously whereas Γ_n decreases monotonously. We set

$$\gamma = \lim \gamma_n \quad \text{and} \quad \Gamma = \lim \Gamma_n\,.$$

(Of course, these domains may also be unbounded.)

3.1. **The Lower Indicatrix.** Obviously, the lower indicatrix lies entirely on the boundary of γ. On the other hand, γ_n being the intersection of convex domains, is convex as well and therefore the same is true for γ. The point P lies in the interior or on the boundary of γ. In the latter case the boundary of γ may contain one or two segments emanating from P: apart from the interior of these segments the boundary of γ coincides with the lower indicatrix. So we have:

In the interior of every plane piece of the tangent cone of a convex surface the lower indicatrix is contained in a convex curve. Hence the upper normal curvatures are continuous except at at most two points in the sense of the following equation

$$\lim_{\varphi\to\varphi_0} \underline{\rho}_{\varphi} = \underline{\rho}_{\varphi_0}\,.$$

Of course, the indicatrix may lie completely at infinity. The point P does not necessarily lie in the interior of the indicatrix, even if Φ has a tangent plane at P (i.e., if $\omega = 2\pi$). For example, the circle $x^2 + (y-1)^2 = 1$ might represent the indicatrix; then the upper normal curvatures belonging to the directions $\pi \le \varphi \le 2\pi$ are infinite. This is actually the case at the point $u = 0$ of the convex surface, given by

$$x = u^2(1+u)\cos v\,, \quad y = u^2(1+(1+u)\sin v)\,, \quad z = \frac{uv}{2}\,.$$

One can conclude that this surface is convex e.g. from the fact that its Gaussian curvature at all points except $u = 0$ is strictly positive.

3.2. **The Upper Indicatrix.** The set Γ_n being the union of convex domains containing P, is starlike with center P, i.e., if Q' is a point of Γ_n then all the points on the segment $\overline{PQ'}$ are also contained in Γ_n. Therefore also $\Gamma = \lim \Gamma_n$ is a starlike domain with center P. The endpoints of the segments emanating from P and filling Γ, obviously form the upper indicatrix. This statement can be sharpened. If $Q' = (r', \varphi_0)$ is a point of Γ contained in the finite part, then there exists a sequence $h_n \to 0$ such that the point of intersection of σ_{h_n} with $\varphi = \varphi_0$ converges to Q'. We can choose a converging subsequence from the sequence $\{\sigma_{h_n}\}$ (compare the end of Section 1). Of course, the limit curve is convex and by definition contains γ in its interior. *Hence if Q' is a point of Γ then the convex hull of Q' and the lower indicatrix is contained in Γ.*

Now let us assume that the upper normal curvatures at P are all finite. Then γ cuts off a segment from each ray φ = constant, and it follows immediately from the previous theorem that for $0 \leq \varphi \leq \omega$ we have

$$\liminf_{\varphi \to \varphi_0} \overline{\rho}_\varphi \geq \overline{\rho}_{\varphi_0} .$$

This means that the upper indicatrix is half continuous from below. However, for $0 < \varphi < \omega$, the inequality cannot occur. Because for every φ_1 with $0 < \varphi_1 < \omega$ there is a segment s contained in γ and perpendicular to $\varphi = \varphi_0$. The convex hull of s and each point of Γ is entirely contained in Γ, which directly implies

$$\limsup_{\varphi \to \varphi_1} \overline{\rho}_\varphi \leq \overline{\rho}_{\varphi_1} .$$

Thus we obtain:

Suppose that the upper normal curvatures belonging to the directions of a plane sector of angles in the tangent cone of Φ are finite. Then the lower normal curvatures with respect to each inner direction are continuous in the sense of the equation

$$\lim_{\varphi \to \varphi_0} \overline{\rho}_\varphi = \overline{\rho}_{\varphi_0} .$$

For the two directions $\varphi_0 = 0$ and $\varphi_0 = \omega$ bounding the plane sector of angles we have

$$\lim_{\varphi \to \varphi_0} \overline{\rho}_\varphi \geq \overline{\rho}_{\varphi_0} .$$

Whether the upper indicatrix in the case of finite upper normal curvatures has even more characteristic properties is left open here.[25]

3.3. **The Indicatrix.** Let $\{\varphi_n\}$ be a sequence of directions which is dense in $0 \leq \varphi \leq \omega$. Let the normal sections $\varphi = \varphi_n$ at P have finite (one–sided) curvatures. Then γ and Γ cut out the same segments on the ray $\varphi = \varphi_n$, and it follows from the properties of those domains that were just proved that both indicatrices of P coincide. Hence we obtain:

If at a point P of a surface the upper and the lower normal curvature coincide for a set of directions which is dense in $0 \leq \varphi \leq \omega$ and if they are finite then at P an indicatrix exists. Moreover, it is always a convex curve.

Every convex curve κ may occur as indicatrix of a point P on a convex surface, and moreover P can be any point in the interior[26] of κ. Namely, if

[25] Addition after proofreading: There are no such extra properties. Namely, as we shall prove elsewhere, the following theorem holds: Let γ be a convex curve containing P in its interior, and let Γ be a curve containing together with any point Q' also the convex hull of $Q' + \gamma$. *Then there is a convex surface having Γ as upper indicatrix at the point P, whereas γ occurs as its lower indicatrix at P.* Using the same tools, one can also study the points of discontinuity of the indicatrices, and it is easy to see that *every* convex curve containing P in its interior or *on the boundary* may occur as indicatrix. Therefore we have deleted the proof that we originally wanted to give here (compare the remark in the following subsection).

Translator's note: Here the authors refer to the following article: Herbert Busemann, Willy Feller. Bemerkungen zur Differentialgeometrie der konvexen Flächen. II. Über die Krümmungsindikatrizen, in: Mat. Tidsskr. B, København, 1935, p. 87-115.

[26] See previous footnote

κ in plane polar coordinates λ, φ is given by the equation $\lambda = \lambda(\varphi)$, then the surface

$$z = \frac{r^2}{2\lambda^2(\varphi)}$$

has a tangent plane at the point $r = 0$, and this point has κ as its indicatrix. The surface is convex since the plane sections $z =$ constant are similar convex curves in similar positions whereas the sections $\varphi =$ constant are parabolas.

3.4. **A Uniformity Property.** Let $\{P_n\}$ be a sequence of points on Φ converging to P with tangent t_{φ_0}. We set $P_n = (r_n, \varphi_n, \frac{h_n}{2})$ say, so that $\varphi_n \to \varphi_0$. The approximating circle of the normal halfsection with direction φ_n has radius

$$\rho(P_n) = \frac{r_n^2 + \frac{h_n^2}{4}}{h_n} = (r')_n^2 + \frac{h_n}{4},$$

where $(r'_n = \frac{r_n}{\sqrt{h_n}}, \varphi_n, 0)$ is the point on σ_{h_n} corresponding to P_n. Now obviously, if $\underline{\rho}_\varphi$ and $\overline{\rho}_\varphi$ are continuous at φ_0, then

$$\sqrt{\underline{\rho}_{\varphi_0}} \le \liminf r'_n \le \limsup r'_n \le \sqrt{\overline{\rho}_{\varphi_0}},$$

and hence also

$$\underline{\rho}_{\varphi_0} \le \liminf \rho(P_n) \le \limsup \rho(P_n) \le \overline{\rho}_{\varphi_0}.$$

If the point P has an indicatrix then $\underline{\rho}_{\varphi_0} = \overline{\rho}_{\varphi_0}$. Thus we have found the following:

Let the surface Φ have an indicatrix at P, and let $\{P_n\}$ be a sequence of points on Φ converging to P with tangent t_{φ_0}. Moreover, let ρ_φ be continuous at φ_0. Then the radii of the circles through P and P_n whose planes are perpendicular to Π and that touch Π in P, converge to the normal curvature radius belonging to the direction t_{φ_0}. *Hence the radii of the approximating circles of the normal half-sections approximate the normal curvature radii uniformly outside of arbitrarily small neighborhoods of possibly at most two points of discontinuity of the indicatrix.*

3.5. **Conjugate Directions and Principal Directions.** Now we want to investigate what facts connected with the notion of conjugate directions remain true. Let us assume that *the surface has a tangent plane in a neighborhood of P, and moreover that the tangent plane Π at P meets the surface only in P. Suppose that at P an indicatrix exists* that contains P in its interior and is not entirely at infinity (i.e., the normal curvatures are finite and not all of them vanish simultaneously).

Fiirst, we shall prove the following theorem on the spherical image of the surface Φ given by parallel normals.

Let Q be a point on the indicatrix of P where it has a tangent τ, and let κ be a curve on the surface emanating from P with tangent $\overline{PQ}$ and finite upper curvature. Then the spherical image of κ has a tangent at the image point of P, and moreover it is perpendicular to τ.

Proof. We take Π as the xy-plane and $\overline{PQ}$ as positive x-axis. Let the surface be given by $z = f(x, y)$. Let first κ be the intersection of the surface with the plane $y = z \tan\theta$ where $\theta \neq \frac{\pi}{2}$. We introduce orthogonal coordinates x and $u = \frac{z}{\cos\theta}$ in the plane of the curve. Then κ is given by the equation

$$u \cos\theta = f(x, u \sin\theta)\,.$$

Since κ is convex and P has a right curvature by assumption, for $x = 0$ the right second derivative of u with respect to x exists (compare Section 1, page 6). Moreover, by Meusnier's theorem

$$\frac{d^2u}{dx^2} = \lim_{x \to +0} \frac{-f_x(x,y)}{x(f_y(x,y)\sin\theta - \cos\theta)} = f_{xx}(0,0) \cdot \frac{1}{\cos\theta}\,,$$

where $f_{xx}(0,0)$ denotes the second right derivative. Since $f_y(0,0) = 0$, then along every plane section through the x-axis we have

$$\lim_{x \to +0} \frac{f_x(x,y)}{x} = f_{xx}(0,0)\,. \tag{22}$$

By (3.2), the tangents of the curves s_h and σ_h at corresponding points are parallel. But the slope of the tangent at s_h is

$$-\frac{f_x(x,y)}{f_y(x,y)}\,.$$

By the final remark of Section 1, the curves σ_h converge to the indicatrix along *with their tangents*, and hence along κ we have

$$\lim_{x \to +0} \frac{f_x(x,y)}{f_y(x,y)} = -\tan\alpha\,, \tag{23}$$

where α denotes the angle between the tangent τ of the indicatrix and the x-axis. Now (3.3) and (3.4) yield the existence of

$$\lim_{x \to +0} \frac{f_y(x,y)}{x} = -\frac{1}{\tan\alpha} \cdot f_{xx}(0,0)\,, \tag{24}$$

where (x, y) has to be taken along κ.

The coordinates of the spherical image of the surface point $(x, y, f(x, y))$, i.e., the cosines of the directions of the surface normals are

$$\frac{-f_x}{\sqrt{1 + f_x^2 + f_y^2}}\,, \quad \frac{-f_y}{\sqrt{1 + f_x^2 + f_y^2}}\,, \quad \frac{1}{\sqrt{1 + f_x^2 + f_y^2}}\,.$$

If we approach the origin along κ then by (3.3) and (3.5) these terms have derivatives with respect to x at $x = 0$ and hence also with respect to the arc length of κ. These derivatives equal

$$-f_{xx}(0,0)\,, \quad \frac{f_{xx}(0,0)}{\tan\alpha}\,, \quad 0\,. \tag{25}$$

The vector with these components is indeed perpendicular to τ, and hence the claim is proved for *plane* sections. Now by Meusnier's theorem, every surface curve of finite upper curvature with tangent $\overline{PQ}$ in a neighborhood of P runs between two plane sections with the same tangent. Therefore its spherical image must have the same tangent as the image of these plane sections. □

Of special interest is the case where $\underline{PQ}$ is perpendicular to τ. We will call such a direction a *principal direction. In a principal direction the normal curvature takes an extremal value in a weak sense.*[28] For a principal direction $\alpha = \pi/2$ and the terms (3.6) take the values $f_{xx}(0,0)$, 0, 0. Thus the following special case of our theorem holds:

Let κ be a curve on a surface emanating from P which has finite upper curvature at P and a tangent in one of the principal directions. Then for the associated normal curvature $1/r$ and the normal vector of the surface ξ at P the following holds:

$$\frac{d\xi}{ds} = \frac{1}{r} \cdot \eta \,, \tag{26}$$

where s denotes the arc length and η the unit vector of the tangent of κ. This is the precise analogon of the *formula of Olidne Rodriguez.*

Next we want to interpret our theorem geometrically. Let P' denote a variable point and let $g(P')$ be the line of intersection of the tangent plane at P' with the tangent plane at P. We let P' go to P along a curve κ with tangent t and finite upper curvature at P: If for every such curve $g(P')$ converges to the same limit then we call the limit line *conjugate to t*. Now we can rephrase our theorem as follows:

If the indicatrix of P at the point Q has a tangent τ then there exists a line that is conjugate to $\underline{PQ}$ and moreover it is parallel to τ. In particular, the lines that are conjugate to the principal directions are perpendicular to them.

The proof follows immediately from the fact that conjugate directions on the sphere are perpendicular to each other and that the tangent of the spherical image of κ is perpendicular to t.

If t is conjugate to t' then t' need not be conjugate to t. Those indicatrices where this relation is indeed symmetric can all be classified;[30] not all of them are ellipses.

3.6. **On a Limit of Blaschke.** Further we want to show how a geometric interpretation of Gaussian curvature introduced by BLASCHKE can be carried over to points with arbitrary indicatrix.

Let Φ be a (not necessarily differentiable) convex surface with a tangent plane at P. Moreover, at P there shall be a bounded indicatrix, i.e., the normal sections at P shall have one–sided non–vanishing curvatures. As before, we choose P as the origin and the normal of the surface as z-axis

[28] We have assumed that the indicatrix in Q has a tangent. But since singularities also may occur, not every direction where the normal curvature takes an extremal value, needs to be a principal direction. Therefore not in each point with indicatrix there is a principal direction, and if there is there may indeed be infinitely many.

[30] Compare RADON: Über eine besondere Art ebener konvexer Kurven, Ber. Verh. Sächs. Akad. Wiss., Leipzig, **68** (1916), p. 123.

[31] Compare B. F., p. 47.

and we describe the surface in cylindrical coordinates in the form $z = f(r, \varphi)$ with $f \geq 0$ say. Then the issue is an estimate of the area $\Omega(h)$ of the surface cap cut out by the plane $z = h$. Let $r = \overline{r}(\varphi)$ be the equation of the projection of the intersection s_{2h} of the surface with the plane $z = h$ and let $\Sigma(h)$ be the area enclosed by this convex curve. Since the area of a closed convex surface is always greater than the area of any convex surface contained in it,[31] $\Omega(h)$ is smaller than $\Sigma(h)$ increased by the area of the perpendicular cylinder of cross section s_{2h} between $z = 0$ and $z = h$. Hence we obtain

$$\Sigma(h) < \Omega(h) \leq \Sigma(h) + 2\pi h \cdot \max \overline{r}(\varphi)\,,$$

and therefore

$$\lim \genfrac{}{}{0pt}{}{\inf}{\sup} \frac{\Omega(h)}{h} = \lim \genfrac{}{}{0pt}{}{\inf}{\sup} \frac{\Sigma(h)}{h}\,.$$

Now

$$\Sigma(h) = \frac{1}{2} \int_0^{2\pi} (\overline{r}(\varphi))^2 \, d\varphi\,.$$

Here $\frac{(\overline{r}(\varphi))^2}{2h}$ converges to the normal curvature radius $\rho(\varphi)$ associated to the direction φ, and because of the uniformity property of this convergence (compare p. 16)

$$\lim \frac{\Omega(h)}{h} = \lim \frac{\Sigma(h)}{h} = \int_0^{2\pi} \rho(\varphi) \, d\varphi\,.$$

In other words:

Let the surface Φ have a tangent plane at P and a bounded indicatrix. Let $\Omega(h)$ denote the area of the cap cut out by the plane parallel to the tangent plane at P at distance h. Then $\lim 2\pi \cdot \frac{\Omega(h)}{h}$ exists and equals the arithmetic mean of the normal curvature radii at P.

In particular, if the indicatrix is an ellipse then one can describe ρ in the form

$$\frac{1}{\rho} = \frac{\cos^2 \varphi}{\rho_1} + \frac{\sin^2 \varphi}{\rho_2}\,,$$

and obtains

$$\lim \frac{\Omega(h)}{h} = \int_0^{2\pi} \rho(\varphi) \, d\varphi = 2\pi \sqrt{\rho_1 \rho_2}\,,$$

or

$$\rho_1 \rho_2 = \lim \left(\frac{\Omega(h)}{2\pi h} \right)^2\,.$$

This is BLASCHKE'S formula.[32]

Further note that the Gaussian curvature defined by the spherical mapping of the surface if it exists, *in the case of an arbitrary indicatrix may not*

---

[32] Compare W. BLASCHKE: Differentialgeometrie I, Berlin 1924, 2nd edition, p. 85.

[33] Translator's note: The authors treat these questions in detail in the following separate article: Herbert Busemann, Willy Feller. Bemerkungen zur Differentialgeometrie der konvexen Flächen. III: Über die Gaußsche Krümmung, in: Mat. Tidsskr. B, København, 1936, p. 41-70.

be equal to Blaschke's limit.[33] Namely, if we map the surface cap of area $\Omega(h)$ just defined for the point $r = 0$, onto the sphere using parallel normals and if $\Omega^*(h)$ denotes the area of its image, then for example for the convex surface given by

$$z = \frac{r^2}{1 + \sin^2 \varphi}$$

we find

$$\lim \left(\frac{\Omega(h)}{2\pi h} \right)^2 = \frac{9}{16} \quad \text{and} \quad \lim \frac{\Omega(h)}{\Omega^*(h)} = \frac{2\sqrt{2}}{5} .$$

4. Curvature Properties of Normal Sections Holding Almost Everywhere

Now we set aside the question of finding properties of normal curvatures holding at every single point, and using the theory of real functions, we look for stronger statements that however are true only almost everywhere.

Let $z = f(x, y)$ describe a convex piece of a surface Φ whose tangent planes are nowhere perpendicular to the xy-plane. Every plane section of the surface, being a convex curve, possesses a tangent almost everywhere. By the usual argument, for every fixed α, those points on the surface where the section of the surface with the plane $y = x \tan \alpha +$ constant has no tangent, form a set Λ_α of measure zero. Let α, β be two angles and $\Lambda = \Lambda_\alpha + \Lambda_\beta$. Then at each point of the surface that is not contained in Λ, there are two plane sections of the surface having a tangent: *Therefore the null set Λ consists of precisely those points where Φ does not have a tangent plane.*

Furthermore, all plane sections of the surface have a finite (two–sided) curvature at almost all points. If α is a fixed angle, then the points on the surface where the section of the surface with the plane $y = x \tan \alpha +$constant has no finite curvature, is a set Λ'_α of measure zero. Now let $\{\alpha_n\}$ be a dense sequence in the interval $(-\frac{\pi}{2}, \frac{\pi}{2})$, and let

$$\Lambda' = \sum_n \Lambda'_{\alpha_n} .$$

Of course, $\Lambda \subset \Lambda'$, and Λ' is again a set of measure zero. If the point $P(x_0, y_0)$ is not contained in Λ' then all the sections of the surface with the planes $y - y_0 = (x - x_0) \tan \alpha_n$ have finite curvature at P. Since on convex surfaces each tangent plane is automatically a strict tangent plane (compare Section 2, p. 8), we can apply Meusnier's theorem and conclude that the corresponding normal sections have finite curvature at P. Hence P has an indicatrix (compare Subsection 3.3), and because the right and left curvatures of the normal sections coincide, it is a curve with P as center. Conversely, Λ' consists *only* of points where not all normal sections have finite two–sided curvatures. Thus the following theorem holds:

[35] Let us note that the following proof uses the convexity of the surface only in a weak sense. The proof remains valid for every surface $z = f(x, y)$ where 1) f is of bounded variation and 2) at almost all points (x_0, y_0), the function $f(x, y_0 + (x - x_0) \tan \alpha)$ has a second derivative for *every* α.

[36] Compare S. Sacs, Théorie de l'Intégrale, Warszawa 1933, p. 121 (Tonelli's theorem).

On every convex surface Φ there is a set Λ' of measure zero such that at every point on the surface outside of Λ' all the plane sections have finite curvatures. At each such point, the indicatrix is a convex curve with the point of the surface as its center.

But we have even more. Let us call a point P a *normal point* of the surface if the indicatrix at P is a conic section (either an ellipse or a pair of lines) with P as center. We want to show that with the exception of a set Λ^* of measure zero, all surface points are *normal*, i.e., that Euler's theorem holds for them.[35]

First we will not work on the surface but in the xy-plane. For a set μ of points on the surface, let $\underline{\mu}$ denote its projection onto the xy-plane. In particular, $\underline{\Phi}$ is the domain of definition of $f(x, y)$. If μ is a set of measure zero in Φ then also $\underline{\mu}$ is a set of measure zero and vice-versa. This e.g. follows from the fact that for convex surfaces of the form $z = f(x, y)$ the well–known formula for its area holds.[36]

By Section 1, p. 6-7, for convex curves the existence of curvature implies the existence of the second derivative. Hence at all points of Φ outside of $\underline{\Lambda}'$, the sections of the surface with all planes $y = x\tan\alpha + \text{constant}$, considered as functions of the length on their trace lines in the xy-plane have first and second derivatives. We shall denote them by $f_\alpha(x, y)$ and $f_{\alpha\alpha}(x, y)$. For example,

$$f_\alpha(x_0, y_0) = \cos\alpha \cdot \frac{d}{dx} f(x, (x - x_0)\tan\alpha + y_0)\Big|_{x=x_0},$$

and in particular $f_{\pi/2} = f_y$. Our first aim is to deduce a relation between the derivatives $f_{\alpha\alpha}(x_0, y_0)$ as function of α at a fixed point P.

For this purpose, we introduce an additive interval function $F(I)$ associated to $f(x, y)$: If I is an interval in $\underline{\Phi}$, defined by the inequalities

$$x_0 - h \le x < x_0 + h, \quad y_0 - k \le y < y_0 + k,$$

then we associate to it the value[37]

$$\begin{aligned} F(I) = \Delta(x_0, y_0; h, k) = f(x_0 + h, y_0 + k) + \\ + f(x_0 - h, y_0 - k) - f(x_0 - h, y_0 + k) - f(x_0 + h, y_0 - k)\,. \end{aligned} \tag{27}$$

This function is absolutely additive: If $I = \sum I_n$ is a decomposition of I into finitely or countably many subintervals then because $f(x, y)$ as a convex function is of bounded variation

$$F(I) = \sum F(I_n)\,.$$

Therefore the function $F(I)$ can be differentiated with respect to intervals at almost every point which is meant as follows: Let $h_n \to 0$ and $k_n \to 0$ be two positive sequences converging to zero with

$$\frac{1}{N} \le \frac{h_n}{k_n} \le N\,,$$

[37]For sufficiently regular $f(x, y)$ obviously

$$F(I) = \int_{x_0-h}^{x_0+h} \int_{y_0-k}^{y_0+k} f_{xy}(x, y)\, dx\, dy\,.$$

Acta. Math. 66 (1936), 1–47

where $N \geq 1$ is given arbitrarily. Then for all points (x_0, y_0) that are not contained in a certain set $\underline{\Lambda_N^1}$ of measure zero, the limit

$$\lim_{n\to\infty} \frac{\Delta(x_0, y_0; h_n, k_n)}{4h_n k_n}$$

exists and is independent of the special choice of the sequences $\{h_n\}$ and $\{k_n\}$.[38] The set $\underline{\Lambda_N^1}$ grows monotonously with N, and its limit set is again a set of measure zero that we shall denote by $\underline{\Lambda^1}$. For each point (x_0, y_0) not in $\underline{\Lambda^1}$ and each pair of positive sequences $\{h_n\}$ and $\{k_n\}$ converging to zero with

$$0 < \liminf \frac{h_n}{k_n} \leq \limsup \frac{h_n}{k_n} < \infty$$

the limit

$$\lim \frac{\Delta(x_0, y_0; h_n, k_n)}{4h_n k_n} = a(x_0, y_0) \tag{28}$$

exists. For sufficiently regular $f(x, y)$ we have $a(x_0, y_0) = f_{xy}(x_0, y_0)$.

Now (4.2) yields the desired relation between the $f_{\alpha\alpha}(x_0, y_0)$. Let (x_0, y_0) be a point that lies neither in $\underline{\Lambda'}$ nor in $\underline{\Lambda^1}$. In (4.2) we set $k_n = h_n \tan(\alpha)$, and obtain for each $0 \leq \alpha < \frac{\pi}{2}$

$$\begin{aligned} &a(x_0, y_0) = \\ &\lim\nolimits_{h\to 0} \tfrac{1}{4h^2 \tan\alpha} \left[f(x_0 + h, y_0 + h\tan\alpha) + f(x_0 - h, y_0 - h\tan\alpha)\right. \\ &\qquad \left. - f(x_0 - h, y_0 + h\tan\alpha) - f(x_0 + h, y_0 - h\tan\alpha)\right]. \end{aligned} \tag{29}$$

Since (x_0, y_0) does not lie in $\underline{\Lambda'}$, the derivatives $f_{\alpha\alpha}(x_0, y_0)$ and $f_{-\alpha,-\alpha}(x_0, y_0)$ exist. By Section 1, (1.7) we have

$$f_{\alpha\alpha} = \cos^2(\alpha) \cdot \lim_{h\to 0} \frac{f(x_0 + h, y_0 + h\tan\alpha) + f(x_0 - h, y_0 - h\tan\alpha) - 2f(x_0, y_0)}{h^2}$$

and

$$f_{-\alpha,-\alpha} = \cos^2(\alpha) \cdot \lim_{h\to 0} \frac{f(x_0 + h, y_0 - h\tan\alpha) + f(x_0 - h, y_0 + h\tan\alpha) - 2f(x_0, y_0)}{h^2}.$$

If we insert that into (4.3) then we obtain

$$f_{\alpha\alpha}(x_0, y_0) - f_{-\alpha,-\alpha}(x_0, y_0) = 4\sin(\alpha)\cos(\alpha) a(x_0, y_0). \tag{30}$$

This is the desired relation.

Now we turn the coordinate system x, y by an angle ω with irrational ratio $\frac{\omega}{\pi}$. Let us call the new system ξ, η, and let $f(x, y)$ go over to $\varphi(\xi, \eta)$. As we have just shown, for $\varphi(\xi, \eta)$ there is set $\underline{\Lambda^2} + \underline{\Lambda'}$ of measure zero outside of which the derivative $\varphi_{\beta\beta}$ exists for every β and satisfies the equation

$$\varphi_{\beta\beta}(\xi_0, \eta_0) - \varphi_{\beta,-\beta}(\xi_0, \eta_0) = 4\sin(\beta)\cos(\beta) \cdot b. \tag{31}$$

Here of course the angle β is taken with respect to the ξ–axis. If instead of β we introduce the angle α relative to the x-axis and if we rewrite (4.5) in the old coordinates then we obatin the equation

[38] One can find a proof of this well–known theorem of Lebesgue's theory e.g. in S. Saks, as above, p. 49. Indeed, there it is assumed that $h_n = k_n$, i.e., one differentiates only with respect to squares. However, the proof can be carried over *word by word* to our assumptions, since it requires only that Vitali's theorem is valid.

(32) $f_{\alpha\alpha}(x_0, y_0) - f_{2\omega-\alpha,2\omega-\alpha}(x_0, y_0) = 4\sin(\alpha - \omega)\cos(\alpha - \omega) \cdot b(x_0, y_0)$.

The two relations (4.4) and (4.6) hold simultaneously at each point (x_0, y_0) that is not contained in the set $\underline{\Lambda}^* = \underline{\Lambda}^1 + \underline{\Lambda}^2 + \underline{\Lambda}'$ of measure zero. We would like to conclude that $f_{\alpha\alpha}(x_0, y_0)$ is a quadratic form in $\cos\alpha$ and $\sin\alpha$.

By Subsection 3.2, $f_{\alpha\alpha}(x_0, y_0)$ is a continuous function of α. We will show now that the two functional equations

$$(33)\qquad \begin{array}{ll} (a) & g(\alpha) - g(-\alpha) = 2a\sin(2\alpha) \\ (b) & g(\alpha) - g(2\omega - \alpha) = 2b\sin(2(\alpha - \omega)) \end{array}$$

for given a, b and $g(0)$ have one unique continuous solution, namely the function

$$(34)\qquad g(\alpha) = A\sin(2\alpha) + A'\cos(2\alpha) + A'' \quad , \text{ where}$$

$$A = a\,, \quad A' = \frac{a\cos(\omega) - b}{\sin(2\omega)}\,, \quad A'' = g(0) - A'\,.$$

Clearly, (4.8) is a quadratic function in $\cos\alpha$ and $\sin\alpha$. It is easy to verify that (4.8) solves the equations (4.7). On the other hand, by (4.7) the values $g(2n\omega)$ for $n = 1, 2, \ldots$ are already determined. Namely, it follows from (4.7(b)) that

$$g(2\omega) = g(0) + 2b\sin(2\omega)\,.$$

So if $g(2n\omega)$ is determined, then by (4.7(a)) also $g(-2n\omega)$ is determined and then taking (4.7(b)) for $\alpha = -2n\omega$ one obtains

$$g((2(n+1)\omega) = g(-2n\omega) + 2b\sin(2(2n+1)\omega)\,.$$

Since the ratio $\frac{\omega}{\pi}$ is irrational, the angles $2n\omega$ are everywhere dense and hence there is only *one unique* continuous solution of (4.7).

Thus at all points (x_0, y_0) except at the set of measure zero $\underline{\Lambda}^*$, a relation of the form

$$(35)\qquad f_{\alpha\alpha}(x_0, y_0) = A_1\cos^2(\alpha) + A_2\sin^2(\alpha) + A_3\sin(\alpha)\cos(\alpha)$$

holds, where the coefficients depend only on (x_0, y_0).

From this we can conclude that Euler's theorem holds at all points of the surface that do not lie above a point of the set $\underline{\Lambda}^*$. Let $P = (x_0, y_0, f(x_0, y_0))$ be such a point. For simplicity, let us assume that the coordinate system has been turned such that the x-axis is parallel to the line of intersection of the tangent plane at P with the xy-plane.[40] We may do this since by definition $f_{\alpha\alpha}$ is invariant under rotations of the coordinate system x, y in its plane and therefore the relation (4.9) remains valid. This implies

$$f_x(x_0, y_0) = 0 \quad \text{and} \quad f_\alpha(x_0, y_0) = f_y(x_0, y_0)\sin(\alpha)\,.$$

From now on, we omit the arguments x_0, y_0.

The curvature of the intersection of the surface with the plane $y - y_0 = (x - x_0)\tan\alpha$ at P is

[40] If the two planes are parallel, then the assertion is trivial.

$$\frac{1}{\rho_\alpha} = \frac{f_{\alpha\alpha}}{\sqrt{1+f_\alpha^2}} = \frac{f_{\alpha\alpha}}{\sqrt{1+f_y^2\,\sin^2(\alpha)}}. \tag{36}$$

Let t_α be the tangent of this intersection at P and $\varphi = \varphi(\alpha)$ the angle between t_α and t_0. Further, let ν be the inclination angle of the tangent plane at P relative to the xy-plane. Then

$$\cos(\nu) = \frac{1}{\sqrt{1+f_y^2}}, \qquad \tan(\alpha) = \cos(\nu)\tan(\varphi)$$

and hence

$$\sin\varphi = \frac{\sin\alpha}{\sqrt{1+f_y^2\,\sin^2(\alpha)}}\cdot\sqrt{1+f_y^2}, \qquad \cos\varphi = \frac{\cos\alpha}{\sqrt{1+f_y^2\,\sin^2(\alpha)}}. \tag{37}$$

For the angle $\theta = \theta(\alpha)$ between the plane of the curve and the normal of the surface we find

$$\cos\theta = \sqrt{\frac{1+f_y^2\,\sin^2(\alpha)}{1+f_y^2}}. \tag{38}$$

Using Meusnier's theorem, it follows from (4.10) and (4.12) that the curvature $\frac{1}{r}$ of the normal section with tangent t_α is

$$\frac{1}{\rho} = \frac{\cos\theta}{\rho} = \frac{1}{\sqrt{1+f_y^2}}\cdot\frac{f_{\alpha\alpha}}{1+f_y^2\,\sin^2(\alpha)}. \tag{39}$$

Now (4.9), (4.11) and (4.13) imply that

$$\frac{1}{\rho} = a_1\cos^2(\varphi) + a_2\sin^2(\varphi) + a_3\sin(\varphi)\cos(\varphi), \tag{40}$$

where the coefficients depend only on the point P. The equation (4.12) is Euler's theorem stating that the indicatrix is a conic section. Since in our context it must be convex, only an ellipse and a pair of lines are possible. Thus we have proved the following result:

Almost all points of a convex surface are normal, i.e., except on a set of measure zero, the indicatrix at every point P of the surface is either an ellipse or a pair of lines with P as center.

Summarizing, we can say: *On Φ there are three sets of measure zero $\Lambda \subset \Lambda' \subset \Lambda^*$ such that at all points on $\Phi \setminus \Lambda$ tangent planes exist and that at all points of $\Phi \setminus \Lambda'$ there are finite normal curvatures at all directions. Moreover, at the points of $\Phi \setminus \Lambda^*$ even Euler's theorem holds.* One can characterize these three sets analytically as follows: *$\Phi \setminus \Lambda$ consists of all points where f_α exists for every α. $\Phi \setminus \Lambda'$ consists of those points where*

[42]Indeed our proof shows that in a *fixed* coordinate system the relation (4.4) is necessary and sufficient for this. But if (4.4) holds in two coordinate systems then as we have seen $f_{\alpha\alpha}$ is a quadratic form in $\cos(\alpha)$ and $\sin(\alpha)$, so that relation (4.4) in this case is true for all coordinate systems.

[43]By an umbilical point we mean a point whose indicatrix is a concentric circle or lies entirely at infinity.

[44]Compare footnote 14.

even $f_{\alpha\alpha}$ exists. Now let us introduce orthogonal coordinate systems on the xy-plane in all possible ways. *Then $\underline{\Phi \setminus \Lambda^*}$ consists of those points in the xy-plane where the surface is differentiable in all these systems with respect to similar paraxial rectangles,* i.e., if (x_0, y_0) is a point of $\underline{\Phi \setminus \Lambda^*}$ then in any of these coordinate systems[42] for each α (compare (4.1)) the following limit exists

$$\lim_{h\to 0} \frac{\Delta(x_0, y_0; h, h\tan(\alpha))}{4h^2 \tan(\alpha)}$$

and is independent of α.

5. On the Theorem on Umbilical Points

Usually, for proving the theorem that a surface consisting only of umbilical points[43] is part of a sphere, one assumes at least threefold differentiability. Trivially, it is necessary to assume differentiability at least once. However, the other assumptions are not necessary. But there are closed differentiable convex surfaces where *almost* all points are umbilics even of the same curvature, and yet they are no spheres (compare Section 6). Therefore it seems natural to ask under which regularity assumptions it is enough to know that *almost* all points are umbilics to obtain a characterization of the sphere.

We first note the following *Lemma*: Let $z = f(x, y)$ describe a differentiable convex part of a surface Φ whose tangent planes are nowhere perpendicular to the xy-plane. Let k be a plane section of Φ which is the normal section at a point P. Moreover, let $f_x(x, y)$ and $f_y(x, y)$ along k be absolutely continuous[44] (as functions of the length say), and let us assume that almost all points of k are umbilics of the surface. Then k is the normal section of the surface at all its points.

Proof. By Section 3, (3.7), the derivative of the normal of the surface at all umbilical points of k is tangent to k. Hence the angle ω between the normal of the surface and the plane of k along k has a vanishing derivative almost everywhere on k. Since f_x and f_y are absolutely continuous the angle itself is also absolutely continuous and therefore constant. At P we have $\omega = 0$, so in fact $\omega = 0$ everywhere. □

In particular, if *all* the points of k are umbilics of the surface then the derivative of ω vanishes at *all* points of k. Again this implies $\omega = 0$, and in this case it is *not* necessary to explicitly assume that f_x and f_y are absolutely continuous. Thus if Φ consists only of umbilical points then *every* plane section that is a normal section at one of its points, is even a normal section of the surface at each of its points. From this one can easily conclude that Φ is a piece of a sphere. Hence:

A differentiable surface consisting only of umbilical points with positive curvature is a piece of a sphere.[45]

---

[45] It is not necessary to explicitly assume convexity because the tangent plane at such an umbilical point is also locally tangent. *A surface having a local tangent plane at each point is convex.* A proof of the corresponding fact for plane curves going back to E. Schmidt can be found in L. Bieberbach: Differentialgeometrie, Berlin 1932, p. 20f. The proof carries over to n dimensions immediately (compare B.-F., p. 6–7).

From our lemma, one can also derive different versions of the theorem on umbilics, but all of them are contained in the following one: *Let us assume that through two points P and P' of Φ there are each in Φ meaning ??? dense sets of normal sections on which almost all points are umbilics of the surface and where the functions f_x and f_y are absolutely continuous. Then Φ is a piece of a sphere.*

This is because the lemma implies that the normal sections through P and P' occuring in the theorem are in fact normal sections of the surface at all of their points. Since the normal of the surface is continuous, the same is true for *all* the normal sections through P and P'. It is easy to see that a differentiable surface where all the normal sections through a given point P are indeed normal sections at all points, is a surface of revolution with P as pole. Since we assumed that the tangent plane is nowhere perpendicular to the xy-plane, the points P and P' cannot be antipodal so that Φ can be considered as a surface of revolution in two different ways. This immediately implies the claim.

Thus a convex piece of a surface $z = f(x, y)$ is indeed a piece of a sphere if the following conditions are satisfied:

(1) almost all points are umbilics and
(2) $f_x(x, y)$ and $f_y(x, y)$ are absolutely continuous with respect to both variables.

These conditions are much more natural than the conditions given above, and we only have mentioned them to illustrate that in fact it is sufficient to dispose of a set of measure zero of umbilical points.

The indicatrix at a normal point of the surface is an ellipse or a pair of lines. Hence if there are three directions emanating from a normal point such that any two of them form an angle less than π and where the normal curvatures are equal then the indicatrix is a circle. On the other hand we know that almost all points of a convex surface are normal. *Hence condition 1) is certainly satisfied if from almost every point on the surface three pairwise non–opposite tangent directions are emanating whose corresponding lower and upper normal curvatures coincide.*

The following section will show that the second condition may not be omitted even if we assume that all umbilical points have the same curvature. In addition, we now want to prove that $f_x(x, y)$ and $f_y(x, y)$ *are absolutely continuous with respect to both variables if the upper normal curvatures of the surface are bounded.*

Namely, if we consider a piece of a surface on which f_x and f_y are bounded than by Meusnier's theorem the boundedness of the upper normal curvatures implies boundedness of the upper curvatures of the sections of the surface with all vertical planes $ax + by + c = 0$. Therefore, also the upper derivatives $\overline{f}_{\alpha\alpha}$ are uniformly bounded (compare Section 4, (4.10)), i.e.,

$$|\overline{f}_{\alpha\alpha}(x, y)| < M\,. \tag{41}$$

For $\alpha = 0$ and $\alpha = \frac{\pi}{2}$ this implies the two Lipschitz–conditions

$$(42) \qquad \begin{aligned} |f_x(x_1,y)-f_x(x_0,y)| &< M|x_1-x_0| \\ |f_y(x,y_1)-f_y(x,y_0)| &< M|y_1-y_0| \end{aligned}$$

For $\alpha = \frac{\pi}{4}$, we further obtain (compare Section 4, p. 23)

$$\frac{1}{4h^2}|f(x+h,y+h)+f(x-h,y-h)-f(x-h,y+h)-f(x+h,y-h)| < 2\sqrt{2}\,M\,.$$

Since each rectangle can be decomposed into disjoint sideaxial squares, we also have

$$\frac{1}{4hk}|f(x+h,y+k)+f(x-h,y-k)-f(x-h,y+k)-f(x+h,y-k)| < 2\sqrt{2}\,M\,.$$

Hence the upper derivatives $\overline{f}_{xy}$ and $\overline{f}_{yx}$ also fulfill

$$|\overline{f}_{xy}| \le 2\sqrt{2}\,M \quad \text{and} \quad |\overline{f}_{yx}| \le 2\sqrt{2}\,M\,.$$

Thus we have two more Lipschitz–conditions

$$(43) \qquad \begin{aligned} |f_x(x,y_1)-f_x(x,y_0)| &\le 2\sqrt{2}\,M|y_1-y_0|\,, \\ |f_y(x_1,y)-f_y(x_0,y)| &\le 2\sqrt{2}\,M|x_1-x_0|\,. \end{aligned}$$

Now (5.2) and (5.3) implies that f_x and f_y are absolutely continuous as claimed.[47]

Finally, we mention an application of these relations. It follows from (5.2) and (5.3) that also the components of the normal of the surface

$$\frac{f_x}{\sqrt{1+f_x^2+f_y^2}}\,, \quad \frac{f_y}{\sqrt{1+f_x^2+f_y^2}}\,, \quad \frac{-1}{\sqrt{1+f_x^2+f_y^2}}$$

fulfil certain Lipschitz–conditions in both variables. But these terms provide a representation of the spherical image of the surface with parameters x and y. The fact that the Lipschitz–conditions hold implies[48] that the well–known formula for the area is true here: If B is a domain in the xy-plane and $S(B)$ denotes the area of the spherical image of B then

$$(44) \qquad S(B) = \int\!\!\int_B \frac{f_{xx}f_{yy}-f_{xy}^2}{\sqrt{1+f_x^2+f_y^2}}\,dx\,dy$$

(where the integrand is indetermined only on a set of measure zero). For the area $O(B)$ of the piece of the surface lying above B we have

$$(45) \qquad O(B) = \int\!\!\int_B \sqrt{1+f_x^2+f_y^2}\,dx\,dy\,.$$

Since the integrands in (5.4) and (5.5) are bounded, one can differentiate the integrals with respect to similar rectangles. Hence for almost all points (x,y) of the domain of definition of $f(x,y)$, if $\{I_n\}$ is a sequence of similar concentric intervals contracting to (x,y), the following holds

---

[47]See Carathéodory, Vorlesungen über reelle Funktionen, 1st ed., Leipzig 1918, §568.

[48]T. Radó: Über das Fl"achenmass rektifizierbarer Flächen, Math. Ann. **100** (1928).

[49]With absolute value dashes, we shall denote the area or the measure of sets lying in the surface.

$$\lim_{n\to\infty} \frac{S(I_n)}{O(I_n)} = \frac{f_{xx}f_{yy} - f_{xy}^2}{(1 + f_x^2 + f_y^2)^2}\,. \tag{46}$$

Now let (x, y) in particular be a normal point. By introducing a coordinate system where f_{xy} vanishes, one can see immediately that the term on the right hand side of (5.6) is precisely the product of the principal curvatures. Hence:

If Φ is a differentiable convex surface with bounded upper normal curvatures then the Gaussian curvature exists at almost all normal points and equals the product of the principal curvatures. The integral of the Gaussian curvature over the normal points of Φ gives the area of the spherical image of Φ.

It is not clear under which general assumptions this theorem remains valid. In particular, the question arises whether the absolute continuity of the spherical mapping is sufficient here. The following section will show that this mapping may not be absolutely continuous even for differentiable convex surfaces, and that the integrand in (5.4) may vanish almost everywhere on a closed Φ contained in the unit ball.

6. Example of a Surface that up to a Null Set Consists only of Pieces of Spheres of the Same Radius

We now shall construct a surface Φ with the following properties: *Φ is a differentiable convex closed surface, entirely contained in the unit ball. On Φ there is a domain Γ with*[49]

$$|\Gamma| = |\Phi|\,,$$

(i.e., $\Phi \setminus \Gamma$ is set of measure zero). The set Γ is the union of domains that are each congruent to a piece of the sphere of radius r for a given $r > 1$. Such a surface exists for any choice of r. In particular, one might choose $r = \infty$.

Clearly, the area of the spherical image of Γ is at most $|\Gamma|/r^2$, hence since Φ is contained in the unit ball, at most $4\pi/r^2$. Therefore, the measure of the image of $\Phi \setminus \Gamma$ is positive. Thus the example shows two facts:

(1) *The spherical image of a continuously differentiable convex surface may not be absolutely continuous.* It can happen that almost the entire surface is mapped to a set of measure zero on the sphere, and a set of measure zero of the surface is mapped onto the rest of the sphere.

(2) *There are convex, differentiable surfaces contained in the unit ball whose points up to a set of measure zero are all umbilics of the same curvature $\frac{1}{r} < 1$.*

Our construction consists in translating the well–known construction of a monotonous, continuous, but not absolutely continuous function of one variable[51] to the setting of a sphere. We intersect the unit ball with a sphere Σ of radius r and thereby obtain a closed convex surface Φ' consisting of

[51] Compare e.g. Carathéodory as above, §§336/7.

one piece of each of the two spheres. Φ' has an edge that we smooth out so that we obtain a differentiable surface Φ_1 still containing a piece Γ_1 of the sphere Σ. Now we replace a piece of $\Phi_1 \setminus \Gamma_1$ by another piece of the sphere Σ, smooth the edge out, etc. Iterating this process, we obtain a monotonously decreasing sequence of surfaces $\{\Phi_n\}$ whose limit surface has the required properties. There are two points to take care of during the procedure:

(1) The area of the union of the pieces of the sphere of radius r contained in Φ_n shall get arbitrarily close to $|\Phi_n|$.
(2) The normal curvatures in $\Phi_n \setminus \Gamma_n$ must be greater than $\frac{1}{r}$ so that one can replace pieces of $\Phi_n \setminus \Gamma_n$ with pieces of the sphere of radius r.

The second point does not occur if $r = \infty$, i.e., if one replaces pieces of the surface with plane pieces. Therefore that case is somewhat simpler.

Before starting the construction in detail, we make a few auxiliary observations.

Let Φ be a closed, differentiable convex surface with bounded upper normal curvatures, let Δ be an open analytic piece on it, and let all the normal curvatures on Δ be greater than $\frac{1}{r}$. If P is a point of Δ then for every positive ϵ, let $\Sigma = \Sigma(P, \epsilon)$ be the sphere of radius r whose center lies on the *inner* normal of P at distance $r + \epsilon$ from P. The intersection of Δ and $\Sigma(P, \epsilon)$, for sufficiently small ϵ contains a closed curve $\kappa = \kappa(P, \epsilon)$ lying in an arbitrarily small neighborhood of P. The closed domain in Δ bounded by κ shall be denoted by $\lambda = \lambda(P, \epsilon)$, the smaller one in Σ bounded by κ shall be denoted by $\Lambda = \Lambda(P, \epsilon)$. Obviously, for sufficiently small ϵ we have

$$(*) \quad \frac{|\Lambda(P, \epsilon)|}{|\lambda(P, \epsilon)|} > \frac{1}{2}.$$

Now we replace the domain $\lambda(P, \epsilon)$ of Δ by $\Lambda(P, \epsilon)$. Hereby, we obtain a new closed convex surface Φ' contained in Φ and such that the curve κ is an edge. This edge we want to smooth out.

Lemma 6.1. *For sufficiently small ϵ there is a closed differentiable convex surface Φ_1 coinciding with Φ' outside of a given neighborhood of $\kappa(P, \epsilon)$. The part of Φ_1 not in Φ' is an analytic pipe surface whose normal curvatures are all bounded and greater than $\frac{1}{r}$.*

Proof. Let δ be sufficiently small such that all normal curvatures of Φ' are less than $\frac{1}{\delta}$. We shall denote the *inner* parallel surfaces of Δ and Σ at distance δ by Δ_δ and Σ_δ. Both surfaces are again convex and analytic. Their intersection curve shall be called κ'. We consider the envelope of the spheres of radius δ with centers on κ'. This pipe surface touches Φ' along two closed curves. One of them lies in the interior of $\Lambda(P, \epsilon)$ and the other one on Δ in the exterior of $\lambda(P, \epsilon)$, so that these two curves bound a strip around κ on Φ'. We now replace this strip by the corresponding strip on the pipe surface bounded by the same curves, and thereby we obtain a closed, differentiable, convex surface Φ_1. (The convexity can be seen as follows:

The interior of κ' on the parallel surface Σ_δ and the exterior on Δ_δ together form a convex piece of surface, and Φ_1 is obtained by constructing its *outer* parallel surface at distance δ.) Now Φ' and Φ_1 coincide except on the strip around κ, and this strip can be made arbitrarily small together with δ. So it just remains to consider the normal curvatures of the pipe surface.

Every point Q of the pipe surface lies on a circle γ of radius δ with center Q' on κ' (namely the directrix of the pencil of spheres). Let D and S be the points of contact of γ with Δ and Σ. The principal directions in Q are tangent to γ and perpendicular to it. Therefore the first principal curvature is $\frac{1}{\delta} > \frac{1}{r}$. The other one changes monotonously on γ along each quadrant that is cut out of γ by the main normal and the binormal of κ' at Q'. Now the angle between the normals of Δ and Σ with small ϵ is small as well. Hence the angle between the normals of Δ_ϵ and Σ_ϵ at Q' can be made arbitrarily small with ϵ. We choose ϵ sufficiently small such that this angle satisfies the inequality (2.9) from Section 2. Then the main normal of κ' cannot decompose the angle of the outer normals of the surfaces at Q' (Section 2, p. 11), and this implies that the points D and S both lie on *one* of the above–mentioned quadrants so that the second principal curvature varies *monotonously* between D and S. But at the points of contact D and S, all the normal curvatures of the pipe surface are at least as large as the corresponding ones of the touching surfaces (since it lies entirely in their interior), hence at S at least equal to $\frac{1}{r}$ and at D greater than $\frac{1}{r}$.

The boundedness of the normal curvatures of the inserted piece of the pipe surface follows e.g. from the fact that every point of the surface lies on a sphere of radius δ that is entirely contained in it. □

Lemma 6.2. *For the domains $\lambda(P,\epsilon)$, Vitali's covering theorem holds. This means, if $s(P)$ is any positive function then for every given $\alpha < 1$ and $\delta > 0$ there are finitely many pairwise disjoint domains $\lambda(P_\nu,\epsilon_\nu)$ on Δ with $\epsilon_\mu < \epsilon(P_\mu)$ whose diameters are smaller than δ and such that*

$$\sum_\nu |\lambda(P_\nu,\epsilon_\nu)| > \alpha|\Delta|\,.$$

Proof. For the geodesic discs $\beta(P,\epsilon)$ on Δ (i.e., for the set of all points of Δ having at most distance ϵ to P), Vitali's theorem holds.[52] Now if $\beta(P,\epsilon^*)$ is the smallest geodesic disc containing $\lambda(P,\epsilon)$, then as it is easy to see,

$$\lim_{\epsilon\to 0}\frac{|\beta(P,\epsilon^*)|}{|\lambda(P,\epsilon)|} < a(P)\,,$$

where a only depends on the ratio of the principal curvature radii at P. Therefore the domains $\lambda(P,\epsilon)$ are regular with respect to the discs and hence Vitali's theorem holds for them as well.[54]

After these preliminaries, we proceed to construct the surface. We will show the following: There is a sequence of convex surfaces $\{\Phi_n\}$ with the following properties:

[52]The proof is standard, compare for example CARATHÉODORY, as above, §288.
[54]Compare for example CARATHÉODORY, as above, §289.

(1) Φ_n is entirely contained in the closed region enclosed by Φ_{n-1}.
(2) On Φ_n there is a domain Γ_n consisting of pieces of spheres of fixed radius $r > 1$ so that Γ_{n-1} is contained in Γ_n.
(3) $\frac{|\Gamma_n|}{|\Phi_n|}$ grows monotonously and converges to 1.
(4) Φ_n is decomposed into analytic pieces of surfaces by finitely many rectifiable curves.
(5) The normal curvatures at inner points of $\Phi_n \setminus \Gamma_n$ are all greater than $\frac{1}{r}$ and bounded from above.

Proof. Let Φ_1 be the unit sphere, and let $\Phi_1, \ldots, \Phi_n$ already be defined. Let Δ be the union of the analytic domains contained in $\Phi_n \setminus \Gamma_n$. By Lemma (6.2), for every $\epsilon(P) > 0$, we can choose finitely many pairwise disjoint domains $\lambda(P_\nu, \epsilon_\nu)$ in Δ such that $\epsilon_\mu < \epsilon(P_\mu)$ and

$$\sum_\nu |\lambda(P_\nu, \epsilon_\nu)| > \frac{2}{3}|\Delta| = \frac{2}{3}|\Phi_n \setminus \Gamma_n|,$$

and such that the diameters are as small as the smoothing procedure described in Lemma (6.1) requires. Moreover, we choose $\epsilon(P)$ sufficiently small such that these domains satisfy equation $(*)$ above. The domains $\lambda(P_\nu, \epsilon_\nu)$ are closed and therefore any two of them have a positive distance. Therefore we can carry out the smoothing procedure for each of these domains in such a way that the others are not affected. Further, we choose the number δ sufficiently small that the part $\Lambda'(P_\nu, \epsilon_\nu)$ that remains from the sphere piece $\Lambda(P_\nu, \epsilon_\nu)$ after smoothing out, satisfies the following inequality

$$|\Lambda'(P_\nu, \epsilon_\nu)| \geq \frac{3}{4}|\Lambda(P_\nu, \epsilon_\nu)|.$$

Let us call the resulting surface Φ_{n+1}. Property (1) is trivially satisfied. (2) Φ_{n+1} contains Γ_n and moreover the union Γ' of all the sphere pieces $\Lambda'(P_\nu, \epsilon_\nu)$. We set $\Gamma_{n+1} = \Gamma_n + \Gamma'$. (3) We have

$$|\Gamma'| = \sum |\Lambda'(P_\nu, \epsilon_\nu)| \geq \frac{3}{4}\sum |\Lambda(P_\nu, \epsilon_\nu)|.$$

Now $\Lambda(P_\nu, \epsilon_\nu)$ lies on the inner side of $\lambda(P_\nu, \epsilon_\nu)$. Hence (compare footnote 23 on p. 19) $|\Lambda(P_\nu, \epsilon_\nu)| \leq |\lambda(P_\nu, \epsilon_\nu)|$. Thus if we set

$$a_n = \frac{\sum |\Lambda(P_\nu, \epsilon_\nu)|}{\sum |\lambda(P_\nu, \epsilon_\nu)|}$$

then because of $(*)$ we have $\frac{1}{2} < a_n \leq 1$ and

$$|\Gamma'| \geq \frac{3}{4} a_n \sum |\lambda(P_\nu, \epsilon_\nu)| \geq \frac{a_n}{2}|\Phi_n \setminus \Gamma_n|.$$

Therefore

$$\frac{|\Gamma_{n+1}|}{|\Phi_{n+1}|} \geq \frac{|\Gamma_{n+1}|}{|\Phi_n|} = \frac{|\Gamma_n|}{|\Phi_n|} + \frac{|\Gamma'|}{|\Phi_n|} \geq \frac{|\Gamma_n|}{|\Phi_n|} + \frac{a_n}{2}(1 - \frac{|\Gamma_n|}{|\Phi_n|}).$$

[55]See footnote 7

So $\frac{|\Gamma_n|}{|\Phi_n|}$ grows monotonously and clearly the limit is 1. The last two properties follow from the fact that Φ_{n+1} only consists of parts of Φ_n, parts of spheres and added strips of analytic pipe surfaces. The normal curvatures of the latter satisfy the requirement. □

The sequence of surfaces $\{\Phi_n\}$ converges to Φ say. If we set $\Gamma = \lim \Gamma_n$ then Γ is contained in Φ, and because of $|\Phi_n \setminus \Gamma_n| \to 0$ and $|\Gamma_n| < |\Gamma_{n+1}|$ we have $|\Gamma| = |\Phi|$. Hence, it only remains to show that Φ is differentiable. Through every point P of Φ there are infinitely many plane sections consisting of circular arcs up to a (linear) set of measure zero. Φ certainly has a tangent plane at P if two such plane sections at P are differentiable. We want to show that *every* such section π is differentiable at *all* its points. It is enough to see that the directions of the tangents of these arcs are dense in the set of all directions of the plane of π. We introduce an orientation on π. Let A_1 and A_2 be two points of π, each lying on an arc γ_1 and γ_2 respectively of π. We have to show that between A_1 and A_2 there is another point, also lying on an arc of π, and whose tangent direction lies between the tangent directions of A_1 and A_2. If $\gamma_1 = \gamma_2$ there is nothing to show. Otherwise γ_1 and γ_2 have a positive distance, and since the arcs are dense on π there is another arc γ between them. For a suitable n, the arcs γ_1, γ_2 and γ lie in Γ_n and hence on Φ_n. Since this surface is differentiable, all the tangents of γ lie between the ones of γ_1 and γ_2.

7. Geodesics

The considerations of this section are of a local nature again and they are independent from Sections 3–6.

We consider the collection of all rectifiable curves on a convex surface between two points P and Q. According to Lebesgue[55] there is at least one rectifiable arc g from P to Q whose length is equal to the lower bound of all lengths of such curves. Let us call g a *geodesic*. It is justified not to mention P and Q in this notation since any piece of g is again a geodesic. We study the differentiable properties of geodesics at those points of the surface where the surface has a tangent plane. Let us announce the result already:

Assume that the convex surface Φ has a tangent plane Π at the point P. Then every geodesic on Φ through P (emanating from P) has a tangent line at P (a tangent ray) and an osculating plane perpendicular to Π. By Section 1 (1.4), it follows from the position of the osculating plane that the projection of the geodesic onto Π at P has curvature zero. Hence we can also rephrase the theorem as follows:

On a convex surface the geodesics at every point with tangent plane have geodesic curvature zero. This also holds for points where there is no normal curvature or where the normal curvatures are infinite.

Acta. Math. 66 (1936), 1–47

For the proof[57] we need some auxiliary considerations.

1. Let Φ be a closed convex surface. We define the footpoint on Φ from a point P to be the uniquely defined point F_P of Φ which has the shortest distance to P. The footpoint depends continuously on P.

Lemma 7.1. *If P runs through a rectifiable curve κ in the exterior of Φ then the footpoint F_p of P on Φ runs through a rectifiable curve on Φ whose length is at most as big as the length of κ.*

Proof. It suffices to show: If A and B are any two points in the exterior of Φ then for the distances we have

$$d(F_A, F_B) \leq d(A, B)\,. \tag{47}$$

Φ at F_A and F_B has tangent planes Π_A and Π_B that are orthogonal to the perpendiculars $\underline{AF_A}$ and $\underline{BF_B}$ (compare Figure 1). If these planes are parallel then we are done. Otherwise, let h be their line of intersection. Let h_A and h_{F_A} denote the parallels of h through A and F_A respectively. Let the normal plane of h through B intersect h at D, and h_{F_A} at A_1 and h_A at A'. Then

$$d^2(A, B) = d^2(A, A') + d^2(A', B)\,, \tag{48}$$

$$d^2(F_A, F_B) = d^2(F_A, A_1) + d^2(A_1, F_B) = d^2(A, A') + d^2(A_1, F_B)\,, \tag{49}$$

Now let the parallel of $\underline{A_1F_B}$ through B intersect the segment $\overline{A_1A'}$ at A_2 say (otherwise we consider the parallel of $\underline{A_1F_B}$ through A'). Then

---

[57] At first sight, the following simple idea seems to imply that geodesics are differentiable. Suppose that the (not necessarily convex) surface Φ at P has a strict tangent plane, and let g be a geodesic emanating from P. Further, let $\{Q_n\}$ and $\{R_n\}$ be two sequences on g converging to P such that Q_n lies between P and R_n. Let us denote the distance measured on g by $g(A, B)$ and let (as in Section 1) $d(A, B)$ denote their Euclidean distance. Then

$$1 = \frac{g(P, Q_\nu) + g(Q_\nu, R_\nu)}{g(P, R_\nu)} \geq \frac{d(P, Q_\nu) + d(Q_\nu, R_\nu)}{g(P, R_\nu)}\,.$$

It is easy to see (for example with the criterion from Section 2) that

$$\lim \frac{g(P, R_\nu)}{d(P, R_\nu)} = 1\,,$$

and hence also

$$\lim \frac{d(P, Q_\nu) + d(Q_\nu, R_\nu)}{d(P, R_\nu)} = 1\,.$$

Now if the two sequences satisfy a relation of the form

$$\frac{1}{a} \leq \frac{d(P, Q_\nu)}{d(P, R_\nu)} \leq a\,, \tag{$*$}$$

then this implies that the angle between the secants $\overline{PQ_n}$ and $\overline{PR_n}$ converges to zero, so that if one of the two sequences has a tangent then the other one has the same line as tangent too.

However, this does not imply the existence of a tangent. If r, φ are polar coordinates then the curve $\varphi = \sin\sqrt{|\log r|} \pm \frac{\pi}{2}$ is rectifiable, but at the origin P it does not have a tangent. Yet, any two sequences of points on this curve satisfying a relation of the form $(*)$ have the property just described.

$$d(A_2, B) \geq d(A_1, F_B)\,. \tag{50}$$

Moreover, since the footpoint of the perpendicular from B onto $\underline{A'A_1}$ lies on the prolongation of $\overline{A'A_2}$ beyond A_2, we have

$$d(A_2, B) \leq d(A', B)\,.$$

Inserting this into (7.4), with (7.2) and (7.3) we obtain the claim (7.1). □

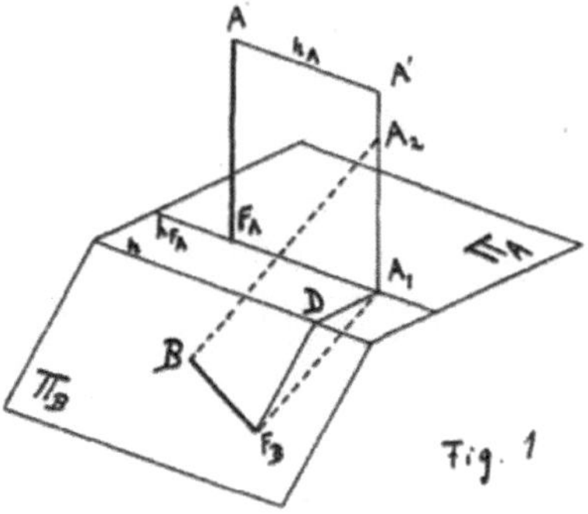

Fig. 1

2. Let the convex surface Φ at O have a tangent plane,[58] and let s be a skew plane section of Φ through O. Let the plane $\Pi(s)$ of s form the angle $\theta(s)$ with the normal of the surface at O with $0 < \theta(s) < \frac{\pi}{2}$. Let us call the *open* halfspace bounded by $\Pi(s)$ and containing the *inner* normal of Φ at O, the *positive* side of $\Pi(s)$. We choose a coordinate system such that the origin is placed at O and the x-axis lies in the intersection of $\pi(s)$ with the tangent plane at O and that the inner normal of the surface shows in the direction of the positive x-axis. The positive y-axis shall be on the positive side of $\Pi(s)$.

Then the plane $\Pi(s)$ is given by the equation

$$\Pi(s): \qquad z = -y \cot\theta(s)\,.$$

Now let $\gamma(s)$ be the line defined by the equations (compare Figure 2)

$$\gamma(s): \qquad x = 0\,, \quad z = y \tan\frac{\theta}{2}\,,$$

and let $\zeta(s)$ be the cylinder through the section s whose generators are parallel to $\gamma(s)$. Since the xy-plane is a strict tangent plane, the intersection of this cylinder with Φ defines a half–neighborhood on $\zeta(s)$ and Φ each. The first one lies entirely in the interior of Φ. The *open* half–neigbourhood of O on Φ shall be denoted by $h(s)$.

The cross section of the cylinder is a convex and hence rectifiable curve so that $\zeta(s)$ is developable to the plane.

[58]The following considerations partially assume that this tangent plane touches Φ only at O. If there were contacts at several points that would require certain changes but in fact they are all simplifications.

Acta. Math. 66 (1936), 1–47

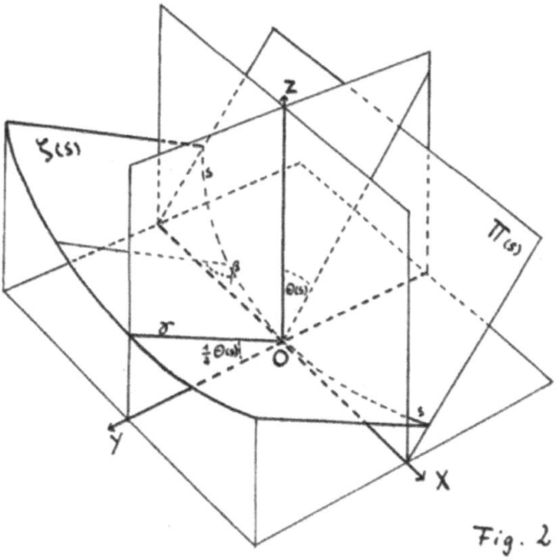

Fig. 2

Lemma 7.2. *On s there is a point $W_i(s)$ on each side of O $(i = 1, 2)$ such that the arc from $W_1(s)$ to $W_2(s)$ after developing the cylinder $\zeta(s)$ becomes a convex curve, and moreover its outer normals in the neighborhood of O are directed to the interior of $h(s)$.*

Proof. Let α be the angle of the tangent to s with the x-axis and β its angle with the generator of the cylinder. Then

$$\cos\beta = -\sin(\alpha)\sin(\frac{\theta(s)}{2}).$$

At the point O the angle α vanishes. If α runs from $-\frac{\pi}{2}$ to $\frac{\pi}{2}$ then $\cos\beta$ decreases monotonously, and after developing the cylinder β still remains to be the angle with the generator of the cylinder. Hence if for $W_i(s)$ we choose the two points of s where the tangent is perpendicular to the x-axis then the arc joining them fulfils the claim. □

Lemma 7.3. *For every positive δ, there is a neighborhood u_δ of O on Φ such that the points $W_i(s)$ for all plane sections s with $0 < \theta(s) < \frac{\pi}{2} - \delta$ lie outside of u_δ.*

Proof. Let $t(s)$ be the tangent of s at O, and let $l_i(s)$ be the perpendicular from $W_i(s)$ onto $t(s)$. If follows from the previous proof that $l_i(s)$ is also tangent to s and hence tangent to the surface Φ and that it forms the angle $\frac{\pi}{2} - \theta(s) > \delta$ with the xy-plane. Therefore the points of contact $W_i(s)$ of $l_i(s)$ with the surface cannot be arbitrarily close to the point O. □

Lemma 7.4. *For every positive δ, there is a neighborhood v_δ of O on Φ such that for all plane sections s with*

$$\delta < \theta(s) < \frac{\pi}{2} - \delta\,, \tag{51}$$

the part of v_δ lying on the positive side of $\Pi(s)$ is contained in $h(s)$. (This

Acta. Math. 66 (1936), 1–47

means that the half-neighborhoods $h(s)$ of O cut out of Φ by the cylinders $\zeta(s)$ can be extended uniformly.)

Proof. If v is a neighborhood of O for which its part v' lying on the positive side of $\Pi(s)$ is not contained in $h(s)$, then v' has points in common with the boundary of $h(s)$. By definition, v' does not contain any points of s. Hence there must be points of Φ on v' where the generators of the cylinder intersect the surface a second time. These points of intersection cannot be arbitrarily close to the point O. Thus there is a neighborhood v_δ of O not containing any such points of intersection. □

3. Let ω_δ be a geodesic disc with center in O that is contained in both neighborhoods u_δ and v_δ as in Lemma 7.3 and 7.4. Obviously every geodesic emanating from O and ending in a point of ω_δ lies entirely in ω_δ. Now we show first:

Every rectifiable path τ in ω_δ lying on the positive side of $\Pi(s)$ whose endpoints T_1, T_2 are in s, is at least as long as the arc from T_1 to T_2 in s. (Here of course we assume that s satisfies (7.5).)

For the proof we consider the set of all footpoints of τ on the cylinder $\zeta(s)$. It lies entirely on the positive side of $\Pi(s)$ and is at most as long as τ (by Lemma 7.1). We develop the cylinder and denote the objects that we obtain after developing with a dash. Then s' is convex between T_1' and T_2' (by Lemma 7.3), and τ' lies on the outside of s'. Every arc on this side is at least as long as the arc from T_1' to T_2' on s', and since lengths are preserved under development, this proves the proposition.

For geodesics the proposition means the following:

I. *Let s be any plane section through O, and choose δ such that (7.5) holds. Let R be a point of ω_δ that is not on the positive side of the plane $\Pi(s)$. Then none of the geodesics from O to R contains points on the positive side of $\Pi(s)$.*

Now every curve emanating from O having no tangent at O, intersects at least one skew section s through O infinitely often (here skew means: $0 < \theta(s) < \frac{\pi}{2}$). This means that the curve in any neighborhood of O visits the positive side of $\Pi(s)$ infinitely often. Therefore **I.** immediately implies:

II. *Every geodesic emanating from O has a half-tangent at O.*

Now we prove:

III. *Let the geodesic g emanating from O have half-tangent t at O, and let s be a skew section with tangent t at O. If moreover (7.5) holds then g within the neighborhood ω_δ lies entirely on the positive side of $\Pi(s)$.*

Proof. 1) The geodesic g cannot contain any point of ω_δ lying on the negative side of $\Pi(s)$ (outside of $\Pi(s)$). Otherwise we consider the plane section s^* through O and R with the same inclination $\theta(s^*) = \theta(s)$ at O as s (there are two such sections, but we take the one where the angle between $\Pi(s)$ and $\Pi(s^*)$ is small). The points of g that are sufficiently close to O then lie on the positive side of $\Pi(s^*)$, whereas R lies on s^*. This contradicts **I.**

2) The geodesic g does not meet s. Otherwise by **I**, g would not contain any point on the positive side of $\Pi(s)$, and by the previous argument no point on the negative side either. Hence g between O and R would coincide with s. But then there were a plane section s^* through O and R with

$$\theta(s) > \theta(s^*) > \delta\,.$$

And the arc between O and R would lie on the positive side of $\Pi(s^*)$, whereas R lies on s^*. This again contradicts **I**. □

IV. *Every geodesic emanating from O has an osculating plane and it contains the normal of the surface.*

Let $\delta > 0$ be arbitrarily chosen, and let s_1, s_2 be the two plane sections through the tangent t of g at O with

$$\theta(s_i) = 2\delta\,, \quad i = 1, 2.$$

Then by **III**, within the neighborhood ω_δ, g lies entirely on the positive side of both $\Pi(s_1)$ and $\Pi(s_2)$. Every plane through t and a point on the positive side of both $\Pi(s_1)$ and $\Pi(s_2)$, forms at most the angle 2δ with the z-axis. Hence all the limits of planes through t and a point of g converging to O, form at most the angle 2δ with the z-axis. Since δ can be chosen arbitrarily small, the claim follows.

Now from the results announced at the beginning of this section, it only remains to show the following:

V. *Every geodesic g through O on Φ has a tangent at O.*

To see this, let t and t' be the right or the left half–tangent of g at O respectively, and let $\{P_n\}$ and $\{P'_n\}$ be two sequences of points on g converging to O from the right or the left respectively. If A and B are points on a surface, let $l(A, B)$ denote the distance of their projections onto the tangent plane at O and let $g(A, B)$ denote their geodesic distance. Then clearly

$$\lim_{n\to\infty} \frac{g(P_n, P'_n)}{l(P_n, P'_n)} = 1\,.$$

On the other hand, we have

$$1 \le \frac{l(O, P_n) + l(O, P'_n)}{l(P_n, P'_n)} \le \frac{g(O, P_n) + g(O, P'_n)}{l(P_n, P'_n)} = \frac{g(P_n, P'_n)}{l(P_n, P'_n)}\,,$$

and therefore

$$\lim_{n\to\infty} \frac{l(O, P_n) + l(O, P'_n)}{l(P_n, P'_n)} = 1\,.$$

This immediately implies that the angle between t and t' vanishes and that means that the two half–tangents of g at O coincide.

Addition after Proofreading. Let us emphasize that it does not follow from the theorems proven here that on a surface from a point with tangent plane there is at least one geodesic emanating in every direction. More strongly, in general this is not true, not even if the surface possesses a tangent plane and an indicatrix everywhere. However, the claim is true for surfaces with bounded upper normal curvatures. For such surfaces, one can also make statements about the minimal distance that two points need to have in order to have more than one geodesic between them.[61]

Kopenhagen, 27th April 1934.

[61]Translator's note: These considerations were published by the authors in the following separate article: Herbert Busemann, Willy Feller. Bemerkungen zur Differentialgeometrie der konvexen Flächen. I: Kürzeste Linien auf differenzierbaren Flächen, in: Mat. Tidsskr. B, København, 1935, p. 25-36.

Acta. Math. 66 (1936), 1–47

REGULARITY PROPERTIES OF A CERTAIN CLASS OF SURFACES

H. BUSEMANN AND W. FELLER

Introduction. Convex curves in the plane and convex surfaces in space have many at first sight unexpected regularity properties.[1] The stringency of the convexity condition makes it desirable to find a geometric property of convex curves or surfaces which has similar analytical implications but applies to a wider class of geometrical objects. The first and most studied generalization of convexity is finiteness of the order: a curve or surface is said to be of order n if its intersection with any straight line consists of at most n connected sets[2] (segments, points, or rays). However, this condition proves to be much too weak.[3] On the other hand, the dual condition, finiteness of class, will turn out to be more restrictive than necessary.

The present paper will show that for curves *finiteness of the total curvature* is an entirely satisfactory condition. It includes in particular all curves of finite class. More generally, hypersurfaces in E^n with finite total curvature seem to share the most important differentiability properties with convex hypersurfaces.

The present results on curves with finite total curvature and the modern theory of functions of a real variable lead to the following surprising fact concerning surfaces in E^3: *If the paratingens*[4] *of* Φ *at a given point* p *leaves out at least one line* L, *and if, locally, the plane sections of* Φ *parallel to* L *have uniformly bounded total curvature, then the following facts hold simultaneously at almost all points* q *of* Φ: *the surface* Φ *has a tangent plane* Π *at* q. *All plane sections* ($\neq\Pi$) *of* Φ *through* p *have a curvature*[5] *at* p, *and these curvatures satisfy the theorems of Meusnier and Euler.*

Received by the editors April 13, 1945.

[1] For curves compare Jessen [1], for surfaces Busemann-Feller [1], and for hypersurfaces Alexandrov [1]. Numbers in brackets refer to the references cited at the end of the paper.

[2] This formulation is due to Hjelmslev [2] and has the obvious advantage not to exclude polygons, polyhedrons, ruled surfaces, $\cdots$.

[3] Compare Marchaud [1], Haupt [1], and §2 of the present paper.

[4] This concept is due to Bouligand, see Bouligand [1]. The paratingens of Φ represented parametrically by $p(u, v)$ at (u_0, v_0) consists of all non-oriented lines G which are limits of sequences of non-oriented lines $G(p(u'_\nu, v'_\nu), p(u_\nu, v_\nu))$ with $(u'_\nu, v'_\nu) \neq (u_\nu, v_\nu)$ and $u'_\nu, u_\nu \to u_0$, $v'_\nu, v_\nu \to v_0$. For the definition of total curvature see the following §1.

[5] More precisely the ordinary curvature as defined at the end of §1.

It will be seen that the hypotheses can be somewhat reduced. Theorems (4.5) and (4.8) are believed to be of interest for the theory of real variables.

1. **Curves of finite total curvature.** For $a \leqq t \leqq b$ let the point $p(t) = (x(t), y(t))$ of the (x, y)-plane describe a continuous curve C which is locally a Jordan arc: that is to say, for every t_0 let there be an $\eta(t_0) > 0$ such that the mapping $p(t) \to t$ is topological for $|t - t_0| \leqq \eta(t_0)$, or $a \leqq t \leqq a + \eta(a)$, or $b - \eta(b) \leqq t \leqq b$,[6] respectively.

The oriented straight line G is called a *paratangent* of C at t_0 if there exist sequences $t_k' \to t_0$, $t_k'' \to t_0$ with $t_k' < t_k''$ and such that the oriented line $G(p(t_k'), p(t_k'')) \equiv G(t_k', t_k'')$ tends to G. If, in particular, the two sequences can be chosen so that $t_k' \leqq t_0 \leqq t_k''$, then the paratangent G is called a *tangent* of C at t_0. If $t_k'' \leqq t_0$ ($t_k'' = t_0$), then G is called a *left-hand* paratangent (tangent) of C at t_0; right-hand tangents and paratangents are defined similarly.[7]

(1.1) *The paratangent (tangent) of C at t_0 is unique if and only if the one-sided paratangents (tangents) of C at t_0 are unique and coincide.*

The statement is obvious, but it should be noticed that it holds only for oriented tangents and paratangents.

The *mean value theorem* holds for these tangents in the following form:

(1.2) *If $t_1 < t_2$ and $p(t_1) \neq p(t_2)$, then a value t_0 between t_1 and t_2 exists such that a suitable tangent of C at t_0 is parallel either to $G^+ = G(t_1, t_2)$ or to its opposite orientation G^-.*[8]

A loop joining two points (for example, a strophoid) will show that a tangent parallel to G^+ need not exist. The statement (1.2) is easily proved by considering the points of C for which the distance from G^+ is maximal.

For every ϕ, $0 \leqq \phi < 2\pi$, define $N(\phi)$ (possibly $N(\phi) = \infty$) as the number of values t for which a tangent of C at t forms the angle ϕ with the positive x-axis. *If $N(\phi)$ is summable, $\kappa = \int_0^{2\pi} N(\phi)d\phi$ is called the total curvature of C, otherwise put $\kappa = \infty$.*

If the tangent $T(t)$ of C is unique for every t and varies continuously with t, then the end point of the unit vector with origin $(0, 0)$ and

[6] The end points will frequently require similar obvious changes, but these will in general not be mentioned.

[7] One-sided tangents or paratangents will be used as common word for right-hand and left-hand tangents or paratangents.

[8] For curves given in the form $y = f(x)$ this result (compare (3.1)) is due to Hjelmslev [1].

parallel to $T(t)$ traverses a continuous curve on the unit circle whose (finite or infinite) length equals κ.[9] This justifies the above nomenclature. It will now be shown that curves of finite total curvature have an important property of convex curves.

(1.3) THEOREM. *A curve of finite total curvature has everywhere unique one-sided paratangents.*

Suppose that there exist two left-hand paratangents T' and T'' at t_0. Then at least one of the two angular intervals between T' and T'' has the property that to every line L contained in it there exist infinitely many chords of $p(t)$ parallel to L. By the mean value theorem there exist therefore infinitely many points $t_\nu \to t$ such that a tangent at t_ν is parallel or opposite to L. Therefore, for every ϕ of one of the two angular intervals between T' and T'' either $N(\phi)=\infty$ or $N(-\phi)=\infty$ so that $N(\phi)$ can not be finite almost everywhere.

For any curve C let t_0 be a value for which the one-sided paratangents are unique. Then it is easily seen that the paratingens of C at t_0, that is, the set of all paratangents at t_0, is determined as follows: Let T_1 and T_2 be the two one-sided paratangents, T_1' and T_2' their positive subrays with origin $p(t_0)$. Then either the angle between T_1' and T_2' has measure π, in which case the paratingens is the whole plane and we say that C has at t_0 a *cusp*; or T_1' and T_2' bound exactly one angle of measure $\omega(t_0)<\pi$. This closed angle together with its vertical angle form the paratingens of C at t_0 whose measure is defined as $\omega(t_0)$.

The importance of the last theorem is shown by the following simple facts:

(1.4) *If the curve C: $p(t)$, $a \leqq t \leqq b$, has everywhere unique one-sided paratangents then:*

(a) *Every paratangent of C is a tangent.*

(b) *If $t_\nu \to t_0 - 0$ and T_ν is a tangent of C at t_ν, then T_ν tends to the left-hand tangent of C at t_0.*

(b′) *If $t_\nu \to t_0$, and the tangent T of C at t_0 is unique, then $T_\nu \to T$.*

(c) *For every $\eta > 0$ the set of those values t for which $\omega(t_0) > \eta$ is finite.*

(c′) *The number of t's for which the paratangent of C is not unique is at most countable.*

To see (a) let $t_\nu' < t_\nu''$ and $t_\nu' \to t_0$, $t_\nu'' \to t_0$, $G(t_\nu', t_\nu'') \to G$. If $t_\nu' \leqq t_0 \leqq t_\nu''$ for infinitely many ν nothing is to be proved. If $t_\nu' \leqq t_\nu'' \leqq t_0$, then

[9] This is a geometric formulation of a known theorem in the theory of real variables; compare, for example, Saks [1].

$G(t'_\nu, t''_\nu)$ tend by hypothesis to the same limit as $G(t'_\nu, t_0)$, so that C is a tangent.

To prove (b) choose t'_ν and t''_ν with $t'_\nu \leqq t_\nu \leqq t''_\nu < t_0$ such that $0 < t''_\nu - t'_\nu < 1/\nu$ and the smaller angle between $G(t'_\nu, t''_\nu)$ and T_ν has measure less than $1/\nu$. Then $t'_\nu \to t_0$ and $G(t'_\nu, t''_\nu)$ tends by (a) to the left-hand tangent of C at t_0.

If (c) were not true, then a sequence t_ν with $\omega(t_\nu) > \eta$ would exist. It may be assumed that $t_\nu \to t_0 - 0$. The two one-sided tangents of C at t_ν tend by (b) to the left-hand tangent of C at t_0, which contradicts the hypothesis $\omega(t_\nu) > \eta$. If C has the finite total curvature κ, the statement (c′) may be improved to $\sum \omega(t_\nu) \leqq \kappa$, where the t_ν are the values of t where the paratangent is not unique. In particular, there are fewer than κ/π cusps.

The uniqueness of the one-sided paratangents allows us also to associate with every tangent T of C at t an angle $\psi(t)$ between the tangent and the positive x-axis ($\psi(t)$ not necessarily between 0 and 2π) which, although multivalued, varies with t in very much the same manner as when C has a continuous tangent; let $\tau_1 < \tau_2 < \cdots < \tau_r$ denote the cusps of C; and put $a = \tau_0$, $b = \tau_{r+1}$. Then $\tau_0 < \tau_1$, $\tau_r < \tau_{r+1}$. Consider a subarc C_i: $\tau_i \leqq t \leqq \tau_{i+1}$ of C. Define as tangents of C_i at τ_i and τ_{i+1} the right-hand and left-hand tangents of C at τ_i and τ_{i+1}, respectively. By (1.4b) there is for every τ, $\tau_i \leqq \tau \leqq \tau_{i+1}$, a $\delta(\tau) > 0$ such that the smaller angle between any two paratangents of C for $|\tau - t| < \delta(\tau)$ $(0 \leqq \tau - \tau_i < \delta(\tau_i),\ 0 \leqq \tau_{i+1} - \tau < \delta(\tau_{i+1}))$ is less than π. A finite number of these intervals belonging to values $t_1 = \tau_i < t_2 < \cdots < t_m = \tau_{i+1}$ will cover the interval (τ_i, τ_{i+1}) so that only consecutive intervals $|t_i - t| < \delta(\tau_i)$ have common points. If we fix the value $\psi(\tau_i) = \psi(t_1)$ arbitrarily, then the angle $\psi(\tau)$ belonging to all t in $|t_1 - t| < \delta(\tau_i)$ is uniquely determined by the requirement that $|\psi(t) - \psi(\tau_i)| < \pi$. The same requirement and the already defined value of $\psi(t)$ for common points of $|t_i - t| < \delta(t_i)$, $i = 1, 2$, determine $\psi(t)$ uniquely for $|t_2 - t| < \delta(t_2)$. Continuing in the same way one defines $\psi(t)$ for all t on C_i. Notice that $\psi(\tau_i + 0) = \psi(\tau_i)$ and $\psi(\tau_{i+1} - 0) = \psi(\tau_{i+1})$.

Now choose $\psi(a)$ so that $0 \leqq \psi(a) < 2\pi$. Then the angles $\psi(t)$ are determined for $a \leqq t \leqq \tau_1$. As values $\psi(\tau_1)$ take those with $\psi(\tau_1 - 0) \leqq \psi(\tau_1) \leqq \psi(\tau_1 - 0) + \pi$. As initial value on C_1 choose $\psi(\tau_1 - 0) + \pi$, then $\psi(t)$ is determined on C_2, and so forth. A consequence of this determination of $\psi(t)$ is that for any t the limits $\psi(t-0)$ and $\psi(t+0)$ exist and that for any $\psi(t)$ belonging to t either

$$(1.5)\quad \psi(t-0) \leqq \psi(t) \leqq \psi(t+0) \text{ or } \psi(t-0) \geqq \psi(t) \geqq \psi(t+0).$$

(1.6) *If C has total curvature* κ, *then for any* $a=t_1<t_2<\cdots<t_n=b$ *and any choice of the values* $\psi(t_i)$

$$\sum_{\nu=1}^{n-1} |\psi(t_{i+1}) - \psi(t_i)| \leqq \kappa.$$

For by (1.5)

$$\sum |\psi(t_i) - \psi(t_{i+1})| \leqq \sum |\psi(t_i + 0) - \psi(t_{i+1} - 0)| + \sum |\psi(t_i + 0) - \psi(t_i - 0)|.$$

Now $\psi(t)$ takes on every value between $\psi(t_i+0)$ and $\psi(t_{i+1}-0)$ at least once in (t_i, t_{i+1}) and every value between $\psi(t_i-0)$ and $\psi(t_i+0)$ exactly once at t_i, which proves (1.6).

Let C_i, $\tau_i \leqq t \leqq \tau_{i+1}$, denote the same subarc of C as above. For every τ_0 the paratingens of C_i at t_0 leaves out at least one line L'. Choose the perpendicular L to L' as an x'-axis and L' as a y'-axis. Then for a suitable $\delta>0$, the subarc $|t_0-t| \leqq \delta$ has a representation $y'=f(x')$, $-\delta_1 \leqq x' \leqq \delta_2$, $\delta_i>0$, and the difference quotients $(f(x_1')-f(x_2'))/(x_1'-x_2')$ are bounded[10] in $(-\delta_1, \delta_2)$, since the paratingens of C at t_0 is a closed set. Therefore the arc $|t_0-t| \leqq \delta$ of C has the finite length $\int_{-\delta_1}^{\delta_2} \{1+f'^2(x')\}^{1/2}dx'$. A trivial application of the Heine-Borel theorem yields that C_i has finite length, so that C is also rectifiable. Introduce the arc length s as parameter on C, and let $\psi(t)$ become the function $\phi(s)$. Let $\phi^*(s)$ denote any of the values $\phi(s)$. Then $\phi^*(s)$ has by (1.6) bounded variation so that $\phi^{*\prime}(s)$ exists for almost all s. If $\bar{\Phi}(s)$ denotes any other value of $\phi(s)$, then because of (1.5), $\bar{\Phi}'(s)$ exists and equals $\phi^{*\prime}(s)$ whenever $\phi^{*\prime}(s)$ exists. Therefore we shall simply speak of $\phi'(s)$. Whenever $\phi'(s)$ exists, the tangent $T(s)$ of C is unique.

The geometric equivalent to the existence of $\phi'(s)$ is this: Let $\nu(s)$ be the normal to $T(s)$ at $p(s)$, and denote by $\nu(s+h)$ the normal to any tangent of C at $(s+h)$, and by $d_s(h)$ the distance (possibly ∞) of the intersection of $\nu(s)$ and $\nu(s+h)$ from $p(s)$. Then $1/d_s(h) \to |\phi'(s)|$, and $|\phi'(s)|$ is the so-called *strong curvature* of C at s.

(1.7) *A curve of finite total curvature has almost everywhere (with respect to the arclength s) a finite strong curvature.*

The (ordinary) curvature of C at a point s where C has a unique tangent $T(s)$ is defined as follows: Let $r_s(h)$ denote the radius (possibly ∞) of the circle which has tangent $T(s)$ at $p(s)$ and passes through $p(s+h)$. If $1/r_s(h)$ converges, then the limit is the curvature

[10] Compare Bouligand [1, p. 123].

of C at s. For any curve the existence of the strong curvature $\lim (1/d_s(h))$ implies the existence of the curvature $\lim (1/r_s(h))$; for convex curves the converse holds.[11] This is not necessarily true for curves of finite total curvature.

2. **Examples.** Before passing on to surfaces, we exhibit three examples of curves. (a) We show that curves of finite total curvature are really essentially more general than convex curves. This will be accomplished by constructing a curve with $N(\phi) \leqq 3$ which is not the union of a countable set of convex arcs and a set of measure 0. (b) We prove the statement at the end of §1. (c) We show that finiteness of the order is not an adequate generalization of convexity.

Let $\kappa_1, \kappa_2, \cdots$ be any finite or countable number of closed disjoint subarcs of a semicircle κ. Denote by κ_ν' the image of κ_ν under the reflection in the line $G(a_\nu, b_\nu)$ connecting the end points a_ν, b_ν of κ_ν, and replace κ_ν by κ_ν' for every ν. The curve C' thus obtained from κ has $N(\phi) \leqq 3$. If the κ_ν are chosen so that the set $\kappa - \sum \kappa_\nu$ is a nowhere dense subset of κ with positive measure, C' will not be the union of a countable number of convex arcs and a set of measure 0.

It requires a little more to find an example for (b). With the same notation as before but arbitrary κ_ν take a subarc κ_ν^* of κ_ν' whose end points a_ν' and b_ν' are close to but different from a_ν and b_ν, respectively, and replace κ_ν^* by its image κ_ν'' under reflection in $G(a_\nu', b_\nu')$. For the curve C'' thus obtained from C' the function $N(\phi)$ is at most 6.

Now let κ be the semicircle $x \geqq 0$ of the circle $x^2+y^2=1$, and E any curve which passes through (0, 1) and lies except for (0, 1) entirely in the unit circle and has there $y=1$ as tangent, and curvature 1. Then any curve between κ and E through (0, 1) will have curvature 1 at (0, 1). Now choose $a_\nu = (\cos(\pi/2^{2\nu-1}), \sin(\pi/2^{2\nu-1}))$, $b_\nu = (\cos(\pi/2^{2\nu}), \sin(\pi/2^{2\nu}))$ and a_ν' and b_ν' so close to a_ν and b_ν that the arc κ_ν'' lies between κ and E. Then C'' has curvature 1 at (0, 1). But it does not have a strong curvature at (0, 1); for a parallel P_ν through a_ν to the radius of κ from (0, 0) to b_ν is a normal to C'' at a_ν. The equation of P_ν is $y - \sin(\pi/2^{2\nu-1}) = \tan(\pi/2^{2\nu})(x - \cos(\pi/2^{2\nu-1}))$, hence P_ν intersects the x-axis, which is the normal to C'' at (0, 1), at

$$x_\nu = \cos(\pi/2^{2\nu-1}) - \frac{\sin(\pi/2^{2\nu-1})}{\tan(\pi/2^{2\nu})} \to -1.$$

Curves of order 3 are the union of at most 4 convex arcs[12] and have therefore all differentiability properties of convex curves.

[11] See Jessen [1].

[12] See Marchaud [1, p. 88, Théorème IV].

Haupt[13] constructed an example of a curve of order 4, where at a set μ of positive measure the paratingens is the whole plane. The strong curvature can, of course, not exist at any point of μ. An example of a curve of order at most 8 with the same properties can easily be constructed as follows: With the above notations choose the κ_ν again so that $\kappa-\sum\kappa_\nu$ is nowhere dense and has positive measure. Construct a semicircle $\bar{\kappa}_\nu$ with two suitably chosen points c_ν, d_ν of κ_ν as end points such that $\bar{\kappa}_\nu$ is contained in the circular segment S_ν bounded by κ_ν and the chord from a_ν to b_ν. Let $\overline{C}$ be the curve obtained from κ by replacing the arc from c_ν to d_ν by $\bar{\kappa}_\nu$. Since any straight line intersects at most two segments S_ν, and has at most 4 common points with the subarc of $\overline{C}$ in S_ν, it follows that $\overline{C}$ has at most order 8 (actually it has order 6). The paratingens is the whole plane at every point of $\kappa-\sum\kappa_\nu$. If the corners of $\overline{C}$ at c_ν and d_ν are smoothed off by changes taking place in a sufficiently small neighborhood of c_ν and d_ν, a curve $\overline{C}'$ is obtained, where the tangent exists and is continuous, hence the paratingens is everywhere a straight line, but the strong curvature exists at no point of $\kappa-\sum\kappa_\nu$. The question whether a curve of finite order has almost everywhere an ordinary curvature is open.

3. **Auxiliary facts on real variables.** The results of §1 can be applied to surfaces Φ: $p(u, v)=(x(u, v), y(u, v), z(u, v))$, $-\infty<u, v<\infty$, in a Cartesian (x, y, z)-space with the following properties:

I. *For every pair* (u_0, v_0) *there is a* $\delta(u_0, v_0)>0$ *so that the mapping* $p(u, v)\rightarrow(u, v)$ *is topological for* $|u-u_0|<\delta$ *and* $|v-v_0|<\delta$.

II. *For every* (u_0, v_0) *there is at least one line* L *not contained in the paratingens*[14] *of* Φ *at* (u_0, v_0).

III. *There is a positive* $\delta'\leqq\delta$ *such that planes parallel to* L *intersect the part* $|u-u_0|\leqq\delta'$, $|v-v_0|\leqq\delta'$ *of* Φ *in curves of uniformly bounded total curvature.*

Some observations to explain the content of conditions II and III are in order:

There is a positive $\delta^*\leqq\delta$ such that with L as z'-axis and the plane Π perpendicular to L as (x', y')-plane the part $|u-u_0|<\delta^*$, $|v-v_0|<\delta^*$ of Φ is represented in the form $z'=f(x', y')$, where (x', y') traverses some open set Γ in Π containing $x'=0$, $y'=0$. Clearly the number δ^* may in addition be chosen so that the difference quotients are bounded, that is,

$$|f(x_1', y_1')-f(x_2', y_2')|/\{(x_1'-x_2')^2+(y_1'-y_2')^2\}^{1/2}<M$$

[13] See Haupt [1].

[14] See footnote 4.

for $(x_i', y_i') \in \Gamma$. The planes parallel to L intersect therefore for $\delta' < \delta^*$ the part $|u-u_0| \leqq \delta'$, $|v-v_0| \leqq \delta'$ of Φ in Jordan arcs of the form $p(u(t), v(t))$ so that the concept of total curvature becomes applicable. The set $|u-u_0| < \delta'$, $|v-v_0| < \delta'$ corresponds to an open set Γ' in Π which contains $x'=0$, $y'=0$. Therefore an $\eta > 0$ exists such that the domain $|x'| < \eta$, $|y'| < \eta$ is contained in Γ'; it corresponds to some open set Δ containing (u_0, v_0) in the (u, v)-plane.

This leads to a function $f(x', y')$ with the following properties:

(a) $f(x', y')$ is defined and continuous for $|x'| \leqq \eta$, $|y'| \leqq \eta$.

(b) $|f(x_1', y_1') - f(x_2', y_2')| / \{(x_1' - x_2')^2 + (y_1' - y_2')^2\}^{1/2} < M < \infty$.

(c) All sections $ax' + by' + c = 0$ with $z' = f(x', y')$ have total curvature less than $\kappa < \infty$.

Before proceeding to prove the result mentioned in the introduction, some facts on functions of a real variable will be recalled. Let $f(x)$ be defined and continuous for $a \leqq x \leqq b$. The mean value theorem (1.2) becomes then:[9]

(3.1) *If $a \leqq x_1 < x_2 \leqq b$ then an x_0 between x_1 and x_2 and a tangent of $y = f(x)$ (tangent in the above sense) at x_0 exist whose slope $f^*(x_0)$ satisfies the relation $f^*(x_0) = \{f(x_2) - f(x_1)\} / \{x_2 - x_1\}$.*

The first and second *Schwarz derivatives* $D'f(x_0)$ and $D''f(x_0)$ of f at x_0 are defined by

$$D'f(x_0) = \lim_{h \to 0} \frac{f(x_0 + h) - f(x_0 - h)}{2h},$$

$$D''f(x_0) = \lim_{h \to 0} \frac{f(x_0 + h) + f(x_0 - h) - 2f(x_0)}{h^2}.$$

If $f'(x_0)$ exists, then $D'f(x_0)$ exists, but not conversely. However,[15] if $D'f(x_0)$ exists in a set of positive measure E then $f'(x)$ exists almost everywhere (a.e.) in E.

Let the difference quotients of $f(x)$ be bounded by M and let the total curvature of $y = f(x)$ be at most κ. If $f^*(x)$ denotes the slope of any tangent of $y = f(x)$ at x, it follows from §1 that $f^*(x)$ has variation $V \leqq \kappa(1 + M^2)^{1/2}$. The derivative $f'(x)$ exists a.e. in (a, b). At all points where $f'(x)$ exists, the existence of $f^{*\prime}(x)$ for one function $f^*(x)$ implies the existence of the derivative for every other slope function, and the derivatives are equal; they will be denoted by $f''(x)$.

[15] This is a special case of the general and deeper theorem of Marcinkiewicz and Zygmund [1] that the existence of the kth Riemann derivative of $f(x)$ in a set E of positive measure implies the existence a.e. in E of the kth de la Valée Poussin derivative, because the kth Riemann derivative coincides for $k = 1, 2$, with the Schwarz derivative.

If $f''(x)$ exists then $D''f(x_0)$ and the so-called *second de la Vallée Poussin derivative*

$$d''f(x_0) = \lim \frac{2}{h}\left\{\frac{f(x_0+h)-f(x_0)}{h} - f'(x_0)\right\}$$

exist and equal $f''(x_0)$. The existence of $f''(x_0)$ and $d''f(x_0)$ are equivalent to the existence of the strong and ordinary curvatures of $y=f(x)$ at x_0[16] respectively.

$d''f(x)$ may exist in a set E of positive measure, without $f''(x)$ existing anywhere in E.[17] If $d''f(x_0)$ exists, then $D''f(x_0)$ exists but not conversely, however:[15]

(3.2) *If $D''f(x)$ exists in a set E of positive measure, then $d''f(x)$ exists a.e. in E.*

If $f(x, y)$ is defined and continuous for $|x| \leqq \eta$, $|y| \leqq \eta$ all these concepts can be applied to derivatives of $f(x, y)$ in a given direction α and lead to concepts like $f_\alpha^*(x_0, y_0)$, $f_\alpha'(x_0, y_0)$, $D_\alpha' f(x_0, y_0)$, $f_\alpha''(x_0, y_0)$, $D_\alpha'' f(x_0, y_0)$ and $d_\alpha'' f(x_0, y_0)$, where, for example,

$$D_\alpha' f(x_0, y_0) = \lim \frac{1}{2h}\{f(x_0 + h\cdot\cos\alpha, y_0 + h\cdot\sin\alpha) - f(x_0 - h\cdot\cos\alpha, y_0 - h\cdot\sin\alpha)\},$$

$$d_\alpha' f(x_0, y_0) = \lim \frac{2}{h}\left\{\frac{f(x_0 + h\cdot\cos\alpha, y_0 + h\cdot\sin\alpha) - f(x_0, y_0)}{h} - f_\alpha'(x_0, y_0)\right\}.$$

For fixed α the set where any one of these derivatives exists is measurable.

For any interval I: $x'<x<x''$, $y'<y<y''$ put

$$\Delta(I, f) = f(x', y') + f(x'', y'') - f(x', y'') - f(x'', y').$$

The *variation of f* in the rectangle S: $|x| < \delta$, $|y| < \delta$ is defined as $V(f, S) = \sup_{[I_\nu]} \sum |\Delta(I_\nu, f)|$ where $[I_\nu]$ traverses all finite sets of non-overlapping intervals in S. To establish $V(f, S) \leqq M$ it suffices to show $V(f, S') < M$ for every rectangle S': $|x| < \delta' < \delta$, $|y| < \delta'' < \delta$. If $V(f, S)$ is finite, $f(x, y)$ is called of bounded variation in S in the sense of Lebesgue. According to a fundamental theorem of Lebesgue[18] $V(f, S) < \infty$ implies that almost everywhere in S

[16] Compare Busemann-Feller [1, p. 7].
[17] See Denjoy [1].
[18] See Lebesgue [1] or Saks [1].

$$\lim_{\nu\to\infty} \Delta(I_\nu, f)/|I_\nu| = F(x_0, y_0), \qquad |I_\nu| = \text{area of } I_\nu, \tag{3.3}$$

exists and is finite, provided $(x_0, y_0)\in\bar{I}_\nu$ and the sequence $\{I_\nu\}$ is regular, that is, if h_ν and k_ν are the sides of I_ν, then

$$0 < \liminf \frac{h_\nu}{k_\nu} \leqq \limsup \frac{h_\nu}{k_\nu} < \infty. \tag{3.4}$$

4. The existence of the second differential. After these preparations we show first:

(4.1) *If f is defined and continuous for $|x| \leqq \delta$, $|y| \leqq \delta$ and if for some α, $0<\alpha<\pi/2$, the slopes of the sections $x \sin\alpha \pm y\cos\alpha=$const. of $z=f(x, y)$ have variation at most B, then $f(x, y)$ has in S: $|x|<\delta$, $|y|<\delta$ at most variation $2\delta B/\cos\alpha$ and $z=f(x, y)$ has a tangent plane a.e., that is, the derivatives f'_β exist simultaneously for all β and*

$$f'_\beta = f'_0 \cos\beta + f'_{\pi/2}\sin\beta. \tag{4.2}$$

Proof. For given δ', $\delta''<\delta$ choose a rational number $r=p/q$ so that $\delta''\leqq r\delta'\cdot\tan\alpha<\delta$. For any N that is divisible by q let $I^N_{\mu\nu}$ denote the interval $[(\mu-1)/N-1]\delta'<x<(\mu/N-1)\delta$, $\{(\nu-1)/N-r\}\delta'\tan\alpha<y<(\nu/N-r)\delta'\tan\alpha$, $\mu=1,\cdots,2N$; $\nu=1,\cdots,2Nr=N'$. Then the diagonals of $I^N_{\mu\nu}$ are along lines $x\cdot\sin\alpha\pm y\cdot\cos\alpha=$const. Since $f(x, y)$ is continuous it suffices to show that

$$\sum_{\mu,\nu=1}^{2N,N'} |\Delta(I^N_{\mu\nu}, f)| \leqq 2\delta' B/\cos\alpha \text{ for every } N \text{ divisible by } q. \tag{4.3}$$

Let $p_{\mu\nu}(s, t)$ denote the point $x=\delta'[(2\mu-1-2N+s)/(2N)]$, $y=\delta'\tan\alpha[(2\nu-1-2rN+t)/(2N)]$ so that $p_{\mu\nu}(0, 0)$ is the center of $I^N_{\mu\nu}$. Then

$$\begin{aligned}\Delta(I^N_{\mu\nu}, f) = {} & \{[f(p_{\mu\nu}(1, 1)) - f(p_{\mu\nu}(0, 0))] - [f(p_{\mu\nu}(0, 0)) \\ & - f(p_{\mu\nu}(-1, -1))]\} - \{[f(p_{\mu\nu}(1, -1)) - f(p_{\mu\nu}(0, 0))] \\ & - [f(p_{\mu\nu}(0, 0)) - f(p_{\mu\nu}(-1, 1))]\}.\end{aligned}$$

Hence by (3.1)

$$\begin{aligned}\Delta(I^N_{\mu\nu}, f) = {} & (\delta'/(2N\cos\alpha))\{f^*_\alpha(p_{\mu\nu}(\theta^1, \theta^1)) \\ & - f^*_\alpha(p_{\mu\nu}(-\theta^2, -\theta^2))\} - (\delta'/(2N\cos\alpha)) \\ & \cdot\{f^*_{-\alpha}(p_{\mu\nu}(\theta^3, -\theta^3)) - f^*_{-\alpha}(p_{\mu\nu}(-\theta^4, \theta^4))\},\end{aligned}$$

where $0<\theta^i<1$ and the θ^i depend on μ and ν. Therefore

$$\sum_{\mu,\nu=1}^{2N,N'} |\Delta(I_{\mu\nu}^{N}, f)| \leqq (\delta'/(2N\cos\alpha)) \sum_{\lambda=1}^{2N} \sum_{\mu-\nu=\lambda} |f_\alpha^*(p_{\mu\nu}(\theta^1, \theta^1)) - f_\alpha^*(p_{\mu\nu}(-\theta^2, -\theta^2))| + (\delta'/(2N\cos\alpha)) \sum_{\lambda=1}^{N'} \sum_{\mu+\nu=\lambda} |f_\alpha^*(p_{\mu\nu}(\theta^3, -\theta^3)) - f_\alpha^*(p_{\mu\nu}(-\theta^4, \theta^4))|.$$

Since the variation of $f^*_{\pm\alpha}$ along the lines $x\cdot\sin\alpha \pm y\cdot\cos\alpha=\text{const.}$ is at most B and $\mu-\nu=\lambda$ implies that $p_{\mu\nu}(\theta^1, \theta^1)$ and $p_{\mu\nu}(-\theta^2, -\theta^2)$ lie on the line $x\cdot\sin\alpha - y\cdot\cos\alpha = \delta\lambda - \sin\alpha - r\cdot\cos\alpha$ the relation (4.3) follows.

The derivative $f_\alpha'(x, y)$ exists a.e. along a fixed line $x\cdot\sin\alpha - y\cdot\cos\alpha=\text{const.}$ Since the two-dimensional set where f_α' exists is measurable, it follows that $f_\alpha'(x, y)$ exists a.e. in $|x|\leqq\delta$, $|y|\leqq\delta$. Denote by N_1 the set of measure 0 in S where either f_α' or $f_{-\alpha}'$ or the set derivative $F(x, y)$ of $f(x, y)$ does not exist. If $(x_0, y_0)\in S-N_1$ and $k=h\cdot\cos\alpha$, $l=h\cdot\sin\alpha$, then

$$\begin{aligned} F(x_0, y_0) &= \lim \frac{1}{kl}\{f(x_0+k, y_0+l) + f(x_0, y_0) - f(x_0+k, y) + f(x_0, y_0+l)\} \\ &= \lim \frac{1}{kl}\{-f(x_0+k, y_0-l) - f(x_0, y_0) + f(x_0+k, y) + f(x_0, y_0-l)\}. \end{aligned} \tag{4.4}$$

Hence

$$2F(x_0, y_0) = \lim (k\sin\alpha)^{-1}\left\{\frac{f(x_0+k, y_0+l) - f(x_0, y_0)}{h} - \frac{f(x_0+k, y_0-l) - f(x_0, y_0)}{h} - 2\sin\alpha\frac{f(x_0, y_0+l) - f(x_0, y_0-l)}{2l}\right\}.$$

The first two terms in the braces tend to $f_\alpha'(x_0, y_0)$ and $f_{-\alpha}'(x_0, y_0)$ respectively, therefore the last term converges, that is, $D_{\pi/2}'f(x_0, y_0)$ exists. Similarly $D_0'f(x_0, y_0)$ exists. It follows from the last section that f_0' and $f_{\pi/2}'$ exist a.e. in $S-N_1$, so that f_0', $f_{\pi/2}'$, $f_{\pm\alpha}'$ and F exist in S except for a set $N_2\supset N_1$ of measure 0.

Let $(x_0, y_0) \in S - N_2$, $0 < \beta < \pi$, $\beta \neq \pi/2$, and put $h \cdot \cos \beta = k$, $h \cdot \sin \beta = l$. By transforming (4.4) it follows that

$$F(x_0, y_0) = \lim (h \cdot \sin \beta \cos \beta)^{-1} \left\{ \frac{1}{h} (f(x_0 + k, y_0 + l) - f(x_0, y_0)) \right.$$
$$- \frac{\cos \beta}{k} (f(x_0 + k, y_0) - f(x_0, y_0))$$
$$\left. - \frac{\sin \beta}{l} (f(x_0, y_0 + l) - f(x_0, y_0)) \right\}.$$

The last part in the braces tends to $-\cos \beta f'_0 - \sin \beta f'_{\pi/2}$, hence the first part converges and (4.2) follows.

This theorem leads to the following main result on functions $f(x, y)$ for the present paper.

(4.5) THEOREM. *If f is defined and continuous for $|x| \leqq \delta$, $|y| \leqq \delta$, and if the slopes of the sections $y = \text{const.}$ and $x \cdot \sin \alpha \pm y \cdot \cos \alpha = \text{const.}$, for some fixed α, $0 < \alpha < \pi/2$, have variation at most B, then the following facts hold simultaneously almost everywhere in S $(|x| < \delta, |y| < \delta)$.*

(1) *The set derivative $F(x, y)$ of $f(x, y)$ exists.*

(2) *$z = f(x, y)$ has a tangent plane.*

(3) *$d''_\beta f(x, y)$ exists for all β in the following strong sense:*

$$(4.6) \qquad d''_\beta f(x_0, y_0) = \lim_{\bar{\beta} \to \beta, h \to 0} \frac{2}{h} \left\{ \frac{f(x_0 + h \cdot \cos \bar{\beta}, y_0 + h \cdot \sin \bar{\beta}) - f(x_0, y_0)}{h} - f'_{\bar{\beta}}(x_0, y_0) \right\}$$

and

$$(4.7) \qquad d''_\beta f = d''_0 f \cos^2 \beta + 2F \cdot \cos \beta \sin \beta + d''_{\pi/2} \sin^2 \beta.$$

PROOF. By (4.1) and previous remarks the following statements are true except for a set N_3 of measure 0: f'_β exists for all β and $f'_\beta = f'_0 \cos \beta + f'_{\pi/2} \sin \beta$; the derivatives $d''_0 f$ and $d_{\pm\alpha} f$ exist. It is important that in the first part of this proof only the existence of $d''_0 f$ is used and not the fact that the slopes of $y = \text{const.}$ have variation at most B. Let $(x_0, y_0) \in S - N_3$. Put $f_0(s, t) = f(x_0 + s, y_0 + t)$, $f_0 = f(0, 0) = f(x_0, y_0)$ and $k = h \cdot \cos \alpha$, $l = h \cdot \sin \alpha$. Then

$$\sin \alpha \cos \alpha F(x_0, y_0) = \lim h^{-2} \{ f_0(k, l) + f_0 - f_0(k, 0) - f(0, l) \}$$
$$= \lim h^{-2} \{ f_0 + f_0(-k, -l)$$
$$- f_0(-k, 0) - f_0(0, -l) \},$$

hence

$$2\cdot\sin\alpha\cos\alpha\, F(x_0, y_0) = \lim \{ h^{-2}(f_0(k, l) + f_0(-k, -l) - 2f_0) \\ - k^{-2}\cos^2\alpha(f_0(k, 0) + f_0(-k, 0) - 2f_0) \\ - l^{-2}\sin^2\alpha(f_0(0, l) + f(0, l) - 2f_0) \}.$$

The first two terms in the braces tend to $D''_\alpha f = d''_\alpha f$ and $-\cos^2\alpha D''_0 f = -\cos^2\alpha d''_0 f$, hence the last term converges, so that $D''_{\pi/2}f(x_0, y_0)$ exists. By (3.2) the derivative $d''_{\pi/2}f$ will exist except at the set $N_4 \supset N_3$ of measure 0.

Let $(x_0, y_0) \in S - N_4$ and $\bar\beta \to \beta^*$, then by (4.2)

$$f'_{\bar\beta}(x_0, y_0) = f'_0 \cos\bar\beta + f'_{\pi/2}\sin\bar\beta \to f'_{\beta^*}(x_0, y_0).$$

If $\bar k = h\cdot\cos\bar\beta$, $\bar l = h\cdot\sin\bar\beta$, and $0 < \beta^* < \pi$, $\beta^* \neq \pi/2$, then the regularity condition (3.4) is satisfied for $h \to 0$ and $\bar\beta \to \beta^*$. Therefore

$$2\sin\bar\beta\cos\bar\beta F(x_0, y_0) = \lim_{\bar\beta\to\beta^*, h\to 0} \left\{ \frac{2}{h}\left(\frac{f_0(\bar k, \bar l) - f_0}{h} - f'_{\bar\beta}\right) \right. \\ - \frac{2\cos^2\bar\beta}{\bar k}\left(\frac{f_0(\bar k, 0) - f_0}{\bar k} - f'_0(x_0, y_0)\right) \\ \left. - \frac{2\sin^2\beta}{\bar l}\left(\frac{f_0(0, \bar l) - f_0}{\bar l} - f'_{\pi/2}(x_0, y_0)\right) \right\}.$$

The last part of the terms in braces tends to $d''_0 f(x_0, y_0)\cos^2\beta^* + d_{\pi/2}\sin^2\beta^*$, respectively. Therefore the first part has a limit, which proves (4.7) for all β^* and (4.6) for $\beta^* \neq 0, \pi/2$. The proof fails for these two values because the regularity condition (3.4) does not hold.

However, we now apply our result to the coordinate system $\xi = x\cdot\sin\alpha/2 + y\cdot\cos\alpha/2$, $\eta = -x\cdot\cos\alpha/2 + y\cdot\sin\alpha/2$. The lines $y = \text{const.}$ and $x\cdot\sin\alpha + y\cdot\cos\alpha = \text{const.}$ have then the equations $\xi\cdot\sin\alpha/2 \pm \eta\cdot\cos\alpha/2 = \text{const.}$ By hypothesis the slopes of these lines have variation at most B; moreover, it was proved that $d''f$ exists along the lines $\eta = \text{const.}$ As was pointed out in the beginning, the first part of the proof used only the existence a.e. of $d''f$ along $y = \text{const.}$, hence it follows that (4.6) holds for all directions except those parallel to the ξ- or η-axis; (4.6) holds therefore in particular in the direction of the old x- and y-axes. This completes the proof of (4.5). A simple application of (4.6) and (4.7) yields

(4.8) COROLLARY. *Under the assumptions of* (4.5) *the function* $f(x, y)$ *has almost everywhere a second differential in the sense of Stolz, that is, the relation*

$$(4.9)\quad \begin{aligned} f(x_0+h, y_0+k) &= f(x_0, y_0) + hf_0'(x, y) + kf_{\pi/2}' \\ &\quad + (h^2 d_0'' f + 2h\cdot k\cdot F + k^2 d_{\pi/2}'' f)/2 + o(h^2+k^2) \end{aligned}$$

holds for almost all x_0, y_0.

Relation (4.9) implies that

$$F(x, y) = \lim_{h\to 0, k\to 0} \{f(x_0+h, y_0+k) + f(x_0, y_0) - f(x_0+h, y_0) - f(x_0, y_0+k)\}$$

without any regularity condition. If f_0' and $f_{\pi/2}'$ exist in a neighborhood of (x_0, y_0) it follows that the mixed partial derivatives $\partial^2 f/\partial x\cdot\partial y$ and $\partial^2 f/\partial y\cdot\partial x$ exist at (x_0, y_0) and equal $F(x_0, y_0)$. A similar interpretation can be given to $F(x_0, y_0)$ in the general case, however this question is alien to the geometric purpose of the present paper.

The theorem of Euler on the curvatures of the normal sections of $z=f(x, y)$ at (x_0, y_0) follows from (4.2) and (4.7) alone.[19] In the present case both the theorems of Euler and of Meusnier follow immediately from (4.9). For it follows from II of §3 that the paratingens at (x_0, y_0) does not contain the normal to the tangent plane Π of $z=f(x, y)$ at (x_0, y_0). If $(\bar{x}, \bar{y}, \bar{z})$ are new coordinates with Π as $(\bar{x}, \bar{y})$-plane and (x_0, y_0) as origin, then $z=f(x, y)$ will in a neighborhood of $(0, 0)=(x_0, y_0)$ be represented in the form $\bar{z}=\bar{f}(\bar{x}, \bar{y})$. Then $o((x-x_0)^2+(y-y_0)^2)=o(\bar{x}^2+\bar{y}^2)$. The relation (4.9) is equivalent to the statement that $f(x, y)$ coincides, in a neighborhood of (x_0, y_0), except for $o((x-x_0)^2+(y-y_0)^2)$, with a paraboloid $a_1+a_2x+a_3y+a_4x^2+a_5xy+a_6y^2$. This same paraboloid written in the new coordinates will approximate $\bar{z}=f(\bar{x}, \bar{y})$ up to $o((x-x_0)^2+(y-y_0)^2)=o(\bar{x}^2+\bar{y}^2)$. In the new coordinates (4.9) reads as

$$(4.9')\quad \bar{f}(\bar{x}, \bar{y}) = (\bar{x}^2 d_0''\bar{f} + 2\cdot\bar{x}\cdot\bar{y}\cdot\bar{F} + \bar{y}^2 d_{\pi/2}''\bar{f})/2 + o(\bar{x}^2+\bar{y}^2)$$

from which it follows that

$$(4.10)\quad d_\beta''\bar{f}(0, 0) = \cos^2\beta d_0''\bar{f} + 2\cdot\cos\beta\sin\beta\bar{F} + \sin^2\beta d_{\pi/2}''\bar{f}.$$

But now $d_\beta'' f(0, 0)$ is the curvature of the section $\bar{x}\cdot\sin\beta+\bar{y}\cdot\cos\beta=0$ with $\bar{z}=f(\bar{x}, \bar{y})$, so that (4.10) is Euler's theorem.

Meusnier's theorem can easily be derived from (4.9′) in the following form (cf. Bouligand [1, p. 120]): Suppose that the surface $z=f(x, y)$ has a second differential in the sense of Stolz (cf. (4.9)) at $P=(x_0, y_0, f(x_0, y_0))$. *Then any two curves on the surface with the*

[19] Compare Busemann-Feller [1, pp. 28–29].

same tangent and same osculating plane at P have there the same curvature; and the circles of curvature of all plane sections with a given tangent at P lie on a sphere.

We apply these results to the neighborhood $|x'| \leqq \eta$, $|y'| \leqq \eta$ of (a), (b), (c) of §3. Because of (b) the Lebesgue area of the surface $z' = f(x', y')$ over $|x'| \leqq \eta$, $|y'| \leqq \eta$ is finite; it equals the area of the corresponding piece $\bar{\Delta}$ of the original surface Φ, where $\bar{\Delta}$ is the closure of the open set Δ in the (μ, ν)-plane corresponding to $|x'| < \eta$, $|y'| < \eta$. In order to avoid the known difficulties regarding the additivity of area, observe that Φ is, according to Lindelöf's theorem, covered by a countable number of sets $\bar{\Delta}$. This yields the result of the introduction, which may be stated more precisely as follows:

(4.11) THEOREM. *If the surface Φ satisfies the hypotheses* I, II, III *of* §3, *then Φ is the union of a countable number of pieces $\bar{\Delta}$ of finite area, such that for almost all points q of $\bar{\Delta}$ the following statements hold simultaneously: Φ has a tangent plane Π at q. All curves on $\bar{\Delta}$ through q which have an osculating plane, but different from π, have a curvature at q; and these curvatures satisfy the theorems of Euler and Meusnier.*

REFERENCES

A. ALEXANDROV

1. *Almost everywhere existence of second differentials of convex functions and some properties of convex manifolds connected with it* (Russian), Leningrad State University Annals (Uchenye Zapiski), Mathematics Series, vol. 6 (1939) pp. 3–35.

G. BOULIGAND

1. *Introduction à la géometrie infinitésimale directe*, Vuibert, Paris, 1932.

H. BUSEMANN AND W. FELLER

1. *Krümmungseigenschaften konvexer Flächen*, Acta Math. vol. 66 (1935) pp. 1–47.

A. DENJOY

1. *Sur l'intégration des coefficientes différentiels d'ordre supérieur*, Fund. Math. vol. 25 (1935) pp. 273–326.

O. HAUPT

1. *Ueber die Krümmung ebener Bogen endlicher Ordnung*, Sitzungsberichte der physikalisch-medicinischen Societät zu Erlangen vol. 71 (1939) pp. 219–227.

J. HJELMSLEV

1. *Contribution à la géometrie infinitésimale de la courbe réele*, Oversigt over det Kongelige Danske Videnskabernes Selskabs Forhandlinger, 1911, pp. 433–454.
2. *Die graphische Geometrie*, Comptes Rendus du Huitième Congrès des Mathématicians Scandinaves, Lund, 1935, pp. 2–12.

B. JESSEN

1. *Om konvekse Kurvers Krumming*, Mathematisk Tidsskrift B, 1929, pp. 50–62.

H. LEBESGUE

1. *Sur l'intégration des fonctions discontinues*, Ann. École Norm. (3) vol. 27 (1910) pp. 361–450.

A. MARCHAUD

1. *Sur les continus d'ordre borné*, Acta Math. vol. 55 (1930) pp. 67–115.

J. Marcinkiewicz and A. Zygmund
1. *On the differentiability of functions and summability of trigonometrical series*, Fund. Math. vol. 26 (1936) pp. 1–43.
S. Saks
1. *Theory of the integral*, Monografie Matematyczne, vol. 7, Warsaw and New York (Stechert), 1937.

Illinois Institute of Technology and
Brown University

Bull. Am. Math. Soc. 51 (1945), 583–598

INTRINSIC AREA

By Herbert Busemann

Department of Mathematics, Smith College

Communicated November 20, 1945

Many definitions of angle and area have been proposed for Finsler spaces. So far there is no criterion, according to which one definition of angle is more natural than the others. But there is such a criterion for area, as the present note will show. The corresponding definition of area is also new for Riemann or even Euclidean spaces.

In spite of the many available other definitions it seems to deserve attention in the latter case, in particular, for surfaces in parametric form; because it does not use special properties of the Euclidean space or the spaces of constant curvature (like the existence of polyhedra or projections on planes); and because it *depends only on the geometry on the surface.* In differential geometry this last property has always been considered an indispensable attribute of area.

The concept of a parametrized set S requires two spaces: a parameter space P and a space R in which S lies. To describe their properties denoted generally by M^* the (up to isometries) unique smallest complete metric space that contains the metric space M. It will then be required (α) *P is separable and metric, and P^* is arcwise connected.* Points of P^* are denoted by p, q, r, and the distance of p, q by pq.

(β) *R is metric.* x, y, z are used for points of R; xy is the distance of x and y.

(γ) *There is a uniformly continuous mapping $x = x(p)$ of P in R and $S = x(P)$.*

The non-parametric case is the special case where the sets P and S coincide and $x(p) = p$.

The mapping $x(p)$ can be extended to a uniformly continuous mapping of P^* in S^*. For two given points p, q of P^* let Λ be the set of all continuous curves $p(\tau)$ from q to r in P^* and put

$$\kappa(q, r, x(P)) = \inf_{p(\tau)\in\Lambda} \lambda[x(p(\tau))]$$

where $\lambda[x(p(\tau))]$ is the length of the image $x(p(\tau))$ of $p(\tau)$. The function κ satisfies the triangle inequality, but may be 0 or ∞. By identifying points q, r with $\kappa(q, r, x(P)) = 0$ new spaces $\bar{P}$ and $\bar{P}^*$ are obtained from P and P^* with the agreement that convergence in terms of κ means convergence in terms of $\kappa/(1 + \kappa)$ with $\infty/(1 + \infty) = 1$. Now the n-dimensional area of S is defined as

$$\alpha_n(x(P)) = \alpha_n(S) = h_n(\bar{P}, \bar{P})$$

where $h_n(M, E)$ denotes the n-dimensional Hausdorff measure of the set M in E. (That is, all coverings C of M by an at most countable number of spheres $S(p_i, \rho_i)$, with $p_i \in E$, $\rho_i < \epsilon$ are formed. Then

$$h_n^\epsilon(M, E) = \{\pi^{n/2}/\Gamma(n/2 + 1)\} \inf_C \sum_i \rho_i^n,$$

and $h_n^\epsilon(M, E) = \infty$, if no C exists. Finally

$$h_n(M, E) = \lim_{\epsilon\to 0} h_n^\epsilon(M, E).)$$

If $\kappa(q, r, x(P)) = \infty$ for some pair q, r, then it does not follow that $\alpha_n(S) = \infty$ for all n.

When $\dim P = m < \infty$, then $\alpha_m(S)$ is simply called the area $\alpha(S)$ of S. This agrees with the usual language in which the dimension of the parameter space decides whether we talk of a curve, a surface, ...

To deserve the name of an area $\alpha(S)$ must satisfy certain generally accepted requirements.

1. *$\alpha_n(x(P))$ is independent of the parametrization.* For obviously: If $q \to p(q)$ is a uniformly topological mapping of the metric space Q on P and $y(q) = x(p(q))$ then $\alpha_n(y(Q)) = \alpha_n(x(P))$.

2. *$\alpha(S) = \alpha_1(S)$ coincides with the arc length if S is a curve*, or $P = P^*$ is an interval $\alpha \leqslant \tau \leqslant \beta$ of the real axis.

3. If $\alpha_n(S) > 0$, then $\alpha_i(S) = \infty$ for $i < n$; if $\alpha_n(S) < \infty$ then $\alpha_i(S) = 0$ for $i > n$.[3]

4. *For an n-dimensional polyhedron $\alpha_n(S) = \alpha(S)$ coincides with the elementary area.* It is very easy to prove this in the following general form:

P is an arcwise connected locally finite n-dimensional euclidean polyhedron in some E^m consisting of at most countably many closed n-simplices $Q_1, Q_2, \ldots$ where different Q_i have at most boundary points in common. $x(p)$ is a continuous mapping of P in a metric space R, which is topological considered as mapping of a fixed Q_i, and $x(Q_i)$ is congruent to a euclidean n-simplex S_i. If $\mu(S_i)$ denotes the elementary volume of S_i, then $\alpha(x(P)) = \sum_i \mu(S_i)$.

5. *$\alpha_n(S)$ satisfies Kolmogoroff's Principle*[4] in this form:

If $y(p)$ is another uniformly continuous mapping of P in a metric space R' and $y(q)y(r) \leqslant \beta x(q)x(r)$ then

$$\alpha_n(y(P)) \leqslant \beta^n \alpha_n(x(P)).$$

The following property distinguishes $\alpha(P)$ from other areas, for which it is either hitherto unproved or false.

6. *$\alpha_n(S)$ depends only on the intrinsic geometry on the surface $x(P)$.* This becomes clearer in the non-parametric case. Then $\kappa(x, y)$ is simply the greatest lower bound of all curves from x to y on S or the geodesic distance of x and y on S.

In Riemann spaces of class C' the number $\alpha(S)$ coincides for surfaces of class C' with the usual integral, moreover $\alpha(S)$ is for surfaces of class C' the only area which satisfies the following three conditions:

(I) *$\alpha(S)$ is intrinsic,* i.e., depends only on the geodesic distances $\kappa(q, r, x(P))$.

(II) *$\alpha(S)$ satisfies Kolmogoroff's Principle* (5) *for $\beta = 1$.*

(III) *For piecewise Euclidean spaces (finite polyhedra) $\alpha(S)$ coincides with the elementary area.*

In Finsler spaces no generally accepted analytical expression for the area exists. But similarly as in Riemann spaces it is true that $\alpha(S)$ is for surfaces of class C' in Finsler spaces of class C' the only area which is intrinsic (I), satisfies Kolmogoroff's Principle for $\beta = 1$ (II) and is such that:

(III′) *For piecewise Minkowskian spaces $\alpha(S)$ coincides with the elementary area.* (There is only one natural elementary area.)

For surfaces of class C' in a three-dimensional Finsler space[1] (section 2) indicates a definition of area. Provided the author interprets that brief remark correctly, this area can be shown to coincide with α_2.

If $x(P)$ is rectifiable, or in the present terminology, if P is the unit cube of E^k, R is the E^n and the mapping $x(p)$ satisfies a Lipschitz condition $x(p)x(q) < \rho \cdot pq$, then $\alpha(S) = \alpha_k(S)$ coincides with the classical integral and, therefore,[2,5] with the various other areas defined by Peano Lebesgue, Radò and Federer. For non-rectifiable surfaces the relations are mostly unclear.

Proofs will be forthcoming in a comprehensive paper with the same title.

[1] Bouligand, Georges, and Choquet, Gustave, "Problèmes liés à des métriques variationnelles," *C. R. Acad. Sci. Paris*, **218**, 696–698 (1944).

[2] Federer, Herbert, "Surface Area I," *Trans. Amer. Math. Soc.*, **55**, 420–437 (1944).

[3] Hurewicz, Witold, and Wallman, Henry, *Dimension Theory*, Princeton, 1941.

[4] Kolmogoroff, A., "Beiträge zur Masstheorie," *Math. Ann.*, **107**, 351–366 (1932).

[5] Nöbeling, G., "Ueber den Flächeninhalt dehnungsbeschränkter Flächen," *Math. Zeitschrift*, **48**, 747–771 (1943).

Note on a theorem on convex sets.

By Herbert Busemann.

1. In an earlier volume of this journal B. Jessen[1]) gave a very short proof for the following fact:

A point set M in a euclidean space is closed and convex, if and only if every point of the space has exactly one foot[2]) *on M.*

This theorem may be decomposed into two statements:

I. *Every point of the space has exactly one foot on a given closed convex set.*

II. *If every point of the space has exactly one foot on a given set, then the set is closed and convex.*

Jessen gives only the proof for II, because I is trivial. In fact I holds in much more general spaces:

Theorem 1. Statement I holds in a straight line space[3]), if and only if the spheres are convex.

The proof given by Jessen for II rests on the fact that in euclidean spaces spheres become flat as the radius increases[4]). It is the main purpose of the present note to show, that the use of this property of spheres is not accidental, but that the property is actually equivalent to the validity of II:

Theorem 2. Statement II holds in a straight line space, if and only if the limit spheres[3]) are flat, that is contain with any two different points the whole straight line[3]) connecting them.

This permits to largely determine the geometries in which II holds. The dimension n of the space is always finite. *For* $n>2$ *the Minkowski spaces*[5]) *with differentiable spheres are the only spaces in which* II *holds.* For $n=2$, the question whether there are non-Minkowskian metrics for which the limit circles are

[1]) To Sætninger om konvekse Punktmængder, Matem. Tidsskrift B, 1940, pp. 66—70.

[2]) In any metric space the point f is called a foot of the point p on the set M, if $f \varepsilon M$ and pf equals the greatest lower bound of px for $x \varepsilon M$.

[3]) The concepts of straight line space, straight line, limit sphere are explained later. They are taken from the author's: Metric Methods in Finsler Spaces and in the Foundations of Geometry, Annals of Mathematics Studies No 8, Princeton 1942. This book will be frequently referred to in the sequel and is therefore simply quoted as B. – The term „straight line" in the present note is equivalent to „open straight line" in B.

[4]) This fact is used in the following form: Let S be a sphere that does not contain either of the points a or b, but whose center is an interior point of the segment ab; then the radii of the spheres that contain S but neither a nor b are uniformly bounded.

[5]) Minkowski spaces are discussed in detail in section II 1 of B.

straight lines, is undecided. For all dimensions II is equivalent to the requirement, that the spheres be convex and differentiable and that the euclidean parallel axiom holds.

2. In a metric space R with distance xy let (xyz) denote the statement that the three points x, y, z are different and the equation $xy + yz = xz$ holds. R is a *straight line space* (abbreviated as *SL-space*), if the following conditions are satisfied:

A. Each bounded infinite set has an accumulation point.

B. For any two different points x, z there is a point y with (xyz).

C. For any two different points x, y there is a point z with (xyz).

D. If (xyz_1), (xyz_2), and $yz_1 = yz_2$, then $z_1 = z_2$.

A *straight line* in a metric space is a continuous curve which possesses a parametrization $p(\varrho), -\infty < \varrho < \infty$ for which $p(\varrho_1)p(\varrho_2) = |\varrho_1 - \varrho_2|$. A *segment* from a to b is an arc $q(\varrho)$, $\alpha \leqq \varrho \leqq \beta$, with $q(\alpha) = a$, $q(\beta) = b$, and $|\varrho_1 - \varrho_2| = q(\varrho_1)q(\varrho_2)$; hence $\beta - \alpha = ab$. In an *SL*-space there is for any two different points a, b exactly one segment $S(a, b)$ from a to b, and exactly one straight line $G(a, b)$ through a and b (B. p. 16, Theorem I 2.3'). Of course $G(a, b)$ contains $S(a, b)$.

A set M in an *SL*-space is called *convex* if for any two points a, b the segment $S(a, b)$ lies in M. The sphere $K(p, q)$ is the locus of those points x, for which $px = pq$. Following the usual not quite consistent terminology we call $K(p, q)$ convex, when the set consisting of the points x with $px \leqq pq$ is convex.

Theorem 1 may now be proved as follows: The spheres of an *SL*-space are convex, if and only if a given point has exactly one foot on a given straight line (B. p. 132, Theorem IV 3.1 and p. 135, Theorem IV 3.3). Since straight lines are closed convex sets, it follows that the convexity of the spheres is necessary for the validity of I.

Let therefore the spheres of R be convex, p a given point, and M a given closed convex set. Assume that p has two different feet f_1 and f_2 on M. Then f_1, f_2 are also feet of p on $S(f_1, f_2)$ because this segment lies in M. Let f be the foot of p on $G(f_1, f_2)$. Since f is unique and $pf_1 = pf_2$ it follows that $pf < pf_i$ and that f is not on $S(f_1, f_2)$. Hence f_1 and f_2 lie on the same one, T, of the two rays into which f divides $G(f_1, f_2)$. But then $pf_1 = pf_2$ contradicts the fact (B. p. 132, Theorem IV 3.1 and Condition K_2), that px increases monotonically as x traverses T from p to infinity.

3. To prove Theorem 2 the concept of *limit spheres* must be explained. Let G^+ be an oriented straight line in the *SL*-space R, and p, x points on G^+. When x tends on G^+ in the positive direction to infinity, the spheres $K(x, p)$ converge to the limit sphere $L(G^+, p)$. It is important to notice that every point of R has exactly one foot on $L(G^+, p)$, no matter whether the spheres of R are convex or not (B. p. 102, Theorem III 4.5).

If II holds in R it follows therefore that a limit sphere L contains with any two different points a, b the segment $S(a, b)$. It is very easy to see that L must then contain the whole line $G(a, b)$. Conversely, if each limit sphere of R contains with any two different points a, b the whole line $G(a, b)$, then the space has finite dimension n and is for $n > 2$ a Minkowski space with differentiable spheres (B. p. 157, Theorem IV 6.2). To these spaces Jessen's proof (l. c. p. 67) applies with a trivial modification avoiding the use of small circles on spheres.

For $n = 2$ it is not known, whether the metric of an *SL*-plane whose limit circles are straight lines is necessarily Minkowskian. Nevertheless Jessen's proof carries over with even less change than in the higher dimensional case.

The final remark of the introduction regarding the equivalence of II with convexity and differentiability of spheres and the parallel axiom follows from B. p. 157, Theorems IV 6.2 and 3, and p. 163, Theorem 7.3.

INTRINSIC AREA

By Herbert Busemann

(Received February 20, 1946)

Introduction

Recent examples of Nöbeling and Besicovitch[1] convinced many mathematicians that Lebesgue area does not represent their geometric idea of area. A new start oriented by the requirements of geometry rather than analysis seems justified.

A survey of the literature on area reveals *two entirely different trends* which never influenced each other. One of these, using *real variables*, is interested in general surfaces but only in euclidean spaces. All definitions[2] of area for general parametrized surfaces use therefore such special properties of euclidean spaces as the existence of planes and perpendicular projection on them, and cannot even be applied to Riemann spaces. The other trend comes from *differential geometry* and is interested in area of smooth surfaces only but in much more general, namely, Finsler spaces. The definitions[3] of area appear in the form of analytic expressions for the element of area, which for Riemann spaces reduce to the standard area element. What is lacking is a criterion which selects one of these expressions as the "true" area element.

The present paper tries to combine the two trends by freeing the concept of area both from its euclidean and from its infinitesimal character, and by satisfying the basic requirements of both schools of thought. There is general agreement that area must 1) *be a generalization of arclength* 2) *coincide with the elementary area for piecewise euclidean spaces* 3) *have the usual value for smooth*

[1] See [19] and [2]. Whereas it has been known for a long time (see [22], p. 5) that a continuous surface $x_i(u, v)$, $i = 1, 2, 3$, $0 \leqq u \leqq 1$, $0 \leqq u \leqq 1$ in E^3 may have finite Lebesgue area while it is carried by a set of positive three dimensional Lebesgue measure, the new examples show that this may happen even when the mapping $(u, v) \rightarrow x(u, v)$ is *topological*. Besicovitch shows moreover, that a surface *homeomorphic* to a sphere may have arbitrarily small Lebesgue area and bound a (closed) set of arbitrarily large measure. The insolvability of the isoperimetric problem by Lebesgue area is very striking in view of the great rôle which Lebesgue area has played in the calculus of variations due to its lower semi-continuity. The results of the present paper have been partially announced in [6].

[2] See references [22], [18], [19], and [9]. Some of these definitions can be—and have been—modified to apply to Riemann Spaces. For instance, Lebesgue area may be defined by approximating a given surface by smooth surfaces whose area is assumed to be the classical integral. But then area of smooth surfaces remains without a geometric definition. Moreover this procedure breaks down entirely for Finsler spaces. It leads for a euclidean triangular surface imbedded in a Minkowski space to an area which is in general smaller than the euclidean area.

[3] The first definition is probably due to Bliss [3]. Other definitions are found in [1], [5], and [10]. In addition there are various definitions for special cases, mostly for the n-dimensional volume in n-dimensional spaces. In some of these definitions the area depends on a distinguished field of curves and is therefore not an area in the present sense.

surfaces in Riemann spaces 4) *be intrinsic*, that is, depend only on the geodesic distances on the surface and not on the way the surface is imbedded. It is true that the real variable trend has never taken this property into account, but it is doubtlessly fundamental for geometry.

There is less agreement on a last requirement, which seems to the author nevertheless geometrically indispensable: 5) *the surface with the greater intrinsic distances must have the greater area.* The non-intrinsic form of this principle goes back to the known paper [14] of Kolmogoroff.

For Finsler spaces 2) must be replaced by the analogous requirement for piecewise Minkowskian spaces. It will be shown that then conditions 2) to 5) select among the many area elements given for Finsler spaces the one indicated by Choquet[4] so that the question of area may be considered settled for smooth surfaces in symmetric[5] Finsler spaces.

The only k-dimensional measure in E^n which is not based on specific euclidean properties is Hausdorff measure. Although this measure is defined only for non-parametrized surfaces, it leads through the requirement of intrinsicness rather directly to the present definition of area for a general type of parametrized set in metric spaces, which satisfies the conditions 1) to 5) listed above. The use of Hausdorff measure implies that this area fits into a continuous scale of areas, a property whose importance was stressed by Besicovitch.

The area of a parametrized surface equals at least the area of the carrier of the surface considered as a non-parametrized set, and the latter value is at least the Hausdorff measure of the set.[6] Thus one of the main objections to Lebesgue area[1] does not apply to the present intrinsic area.[7]

Simple examples show that intrinsicness is incompatible with the requirement that area of a non-parametrized set shall always equal the Hausdorff measure of the set, so that *a disti ıction between area and measure is unavoidable.* However a measure cognate to intrinsic area can be defined for parametrized sets which for non-parametrized sets always coincides with Hausdorff measure. The one-dimensional measure coincides for continuous curves with the arclength.

The present paper emphasizes the basic general ideas and does not go far into the study of area for non-differentiable surfaces. It is only proved that for birectifiable surfces in Finsler spaces (corresponding to the rectifiable surfaces in non-parametric form in E^3) the area coincides with Hausdorff measure.

---

[4] Hausdorff measure yields immediately a volume element (compare section 6), and this element formed for the induced metric on a sub-manifold is the general area element. The equivalent form given by Choquet in [5] conceals the connection with Hausdorff measure and the validity of Komogoroff's Principle 5) which therefore remain unnoticed.

[5] Formally, the expression for area of smooth surfaces in Finsler spaces has a meaning also in non-symmetric spaces. But for reasons explained in sections 2 and 7 its geometric significance seems questionable.

[6] n-dimensional area and measure of a set are defined for every real $n \geqq 0$. Of course, area and measure are always compared for the same value of n.

[7] Besicovitch states that the area of Federer [9] may also be finite for a surface whose n-dimensional Hausdorff measure is infinite for $n < 3$.

This proof can easily be extended to show that for rectifiable surfaces in E^n the present area is given by the standard integral and coincides therefore with the area of Lebesgue, Radò, Peano and Federer. However, the extension of this result to Finsler spaces requires a previous study of these spaces by real variable methods, which the author intends to take up in a future paper.

1. Hausdorff measure

The word *metric space* will be applied to any point set R on whose pairs a distance $\delta(x, y)$ is defined which is either real and non-negative or ∞ and satisfies the conditions

$$(1.1)\qquad \begin{aligned} &a)\quad \delta(x, y) = 0 \text{ if } x = y\\ &b)\quad \delta(x, y) = \delta(y, x)\\ &c)\quad \delta(x, y) + \delta(y, z) \geqq \delta(x, z). \end{aligned}$$

R will be called an ordinary metric space when $\delta(x, y)$ is always finite and $x \neq y$ implies $\delta(x, y) > 0$.

The points of a metric space R divide into classes, two points x, y belonging to the same class if and only if $\delta(x, y) = 0$. The class axioms hold because of (1.1). The class to which the point x belongs will be denoted by x^*. Because of (1.1)

$$(1.2)\qquad \delta(x, y) = \delta(x', y') \text{ if } x' \in x^*, y' \in y^*.$$

The sphere $S(p, \rho)$ (or, when there is an ambiguity as to the distance considered $S(p, \rho; \delta)$) is the locus of those points x in R for which $\delta(p, x) < \rho$. With any point x, the whole class x^* belongs to $S(p, \rho)$. Open and closed sets are defined in the usual way: G is open when for every $x \in G$ a sphere $S(x, \rho(x)) \subset G$, $\rho(x) > 0$, exists. F is closed if any point x with $\delta(x_\nu, x) \to 0$ and $x_\nu \in F$ belongs to F. Hence both open and closed sets contain with any point x the whole class x^*.

To define n-dimensional Hausdorff measure of any set X in R denote by $\alpha(X)$ the diameter of X, that is

$$\alpha(X) = \sup_{y, z \in X} \delta(y, z).$$

For a given $\epsilon > 0$ let Λ be the class of all countable[8] coverings $\bigcup X_i = X$ of X with $\alpha(X_i) < \epsilon$ and put

$$(1.3)\qquad | x, \delta |_n^\epsilon = \omega^{(n)} \inf_\Lambda \sum \alpha^n(X_i), \quad \text{where} \quad \omega^{(n)} = \pi^{n/2} 2^{-n} \Gamma^{-1}(n + 2^{-1}),$$

and n is any non-negative real number. We put $| X, \delta |_n^\epsilon = \infty$ if Λ is empty. Obviously

$$(1.4)\qquad | X, \delta |_n^\epsilon \leqq | X', \delta |_n^{\epsilon'} \text{ if } X' \supset X \text{ and } \epsilon' \leqq \epsilon.$$

[8] "Countable" will always mean "denumerable or finite."

Therefore

$$|X, \delta|_n = \lim_{\epsilon \to 0+} |X, \delta|_n^\epsilon \tag{1.5}$$

exists and is called the *n-dimensional Hausdorff measure of X* (with respect to the distance δ). It satisfies the following two trivial relations:

$$|X, \delta|_n \leqq |X', \delta|_n \text{ if } X \subset X' \tag{1.6}$$

$$|\bigcup X_i, \delta|_n \leqq \sum |X_i, \delta|_n . \tag{1.7}$$

Besides dividing R into the classes x^* it is sometimes useful to divide R into finiteness classes, the two points x, y belonging to the same class if and only if $\delta(x, y) < \infty$. Again (1.1) guarantees the class axioms.

(1.8) If X has common points with non-countably many finiteness classes of R then $|X, \delta|_n = \infty$ for all n. If X has common points with the countable number of finiteness classes $R_1, R_2, \cdots$ only then $|X, \delta|_n = \sum_i |X \cap R_j, \delta|_n$.

Remember also the following well known property of Hausdorff measure

$$\begin{aligned} &\textit{If } |X, \delta|_n > 0, \textit{ then } |X, \delta|_m = \infty \textit{ for } m < n; \\ &\textit{if } |X, \delta|_n < \infty \textit{ then } |X, \delta|_m = 0 \textit{ for } m > n. \end{aligned} \tag{1.9}$$

To develop measure theory in R we asociate with R an ordinary metric space R^* whose elements are the classes x^* of R and whose distance function is

$$\delta^*(x^*, y^*) = \delta(x, y)[1 + \delta(x, y)]^{-1} \text{ for } x \in x^*, y \in y^* \tag{1.10}$$

where $\delta^*(x^*, y^*) = 1$ when $\delta(x, y) = \infty$. Because of (1.2) the distance (1.10) is independent of the choice of x in x^* and y in y^*. If X^* is the set of classes which contain points of X then

$$|X, \delta|_n = |X^*, \delta^*|_n \qquad \text{for all } n. \tag{1.11}$$

For if $\bigcup X_i = X$ and $\alpha(X_i) < \epsilon$, then $\bigcup X_i^* = X$ and $\alpha(X_i^*) < \epsilon$ hence $|X, \delta|_n^\epsilon \geqq |X^*, \delta^*|_n^\epsilon$. If $\bigcup X_i^* = X^*$ and $\alpha(X_i^*) < \epsilon$, let X_i denote the set of all points in X whose classes are in X_i^*. Then $X = \bigcup X_i$ and $\alpha(X_i) < \epsilon(1 - \epsilon)^{-1}$, hence $|X^*, \delta^*|_n^{\epsilon(1-\epsilon)^{-1}} \geqq |X, \delta|_n^\epsilon \geqq |X^*, \delta^*|_n^\epsilon$, which yields (1.11) when $\epsilon \to 0$. It follows in particular that $|X, \delta|_n = |Y, \delta|_n$ if $X^* = Y^*$.

A set X in R is called n-measurable when X^* is measurable, that is if for any set Y^* in R^*

$$|Y^*, \delta^*|_n = |X^* \cap Y^*, \delta^*|_n + |Y^* - X^* \cap Y^*, \delta^*|_n .$$

Because of the previous observations open and closed sets in R are measurable for every n, and so are all Borel sets since $(\bigcup X_i)^* = \bigcup X_i^*$ and $(\bigcap X_i)^* = \bigcap X_i^*$ when X_i contains with any point x the whole class x^*. However measure is in general not additive, only the following fact holds always:

(1.12) If the sets X_i, $i = 1, 2, \cdots$, are n-measurable and $X_i^* \cap X_j^* = 0$ for $i \neq j$ then $|\bigcup X_i, \delta|_n = \sum |X_i, \delta|_n$.

Many results can be carried over verbally. Because of its frequent occurrence the following is mentioned explicitely

(1.13) If $X_1 \subset X_2 \subset \cdots$ then $| \bigcup X_i, \delta |_n = \lim_i | X_i, \delta |_n$.

Hausdorff measure satisfies *Kolmogoroff's Principle* in the following general form:

(1.14) *For the mapping $x \to x'$ of the set X in a metric space R' with distance δ' let positive constants β and σ exist such that $\delta(x, y) \geqq \beta\delta'(x', y')$ for $x, y \in X$ and $\delta(x, y) < \sigma$. If X' denotes the image of X then*

$$| X, \delta |_n \geqq \beta^n | X, \delta' |_n.$$

For the proof it may be assumed that $| X, \delta |_n < \infty$. Then a sequence of coverings $\bigcup_i X_i^k = X$ with $\alpha(X_i^k) < \sigma k^{-1}$ and $\lim_k \omega^{(n)} \sum_i \alpha^n(X_i^k) = | X, \delta |_n$ exists. Then $\bigcup (X_i^k)' = X'$ and $\alpha((X_i^k)') \leqq \beta^{-1}\alpha(X_i^k) < \beta^{-1}\sigma k^{-1}$, hence $| X',\delta |^{k^{-1}\beta^{-1}} \leqq \omega^{(n)}\beta^{-n} \sum_i \alpha^n(X_i^k)$ which proves (1.14).

The measure $| X, \delta |_n$ does not depend on the ambient space. This is contained in the following more general statement.

(1.14a) If a mapping $x \to x'$ of X in a metric space R' with distance δ' exists such that $\delta(x, y) \leqq \delta'(x', y')$ for $x, y \in X$ and every point $p \in X$ has a neighborhood $S(p, \rho(p))$, $\rho(p) > 0$ with $\delta'(x', y') = \delta(x, y)$ for $x, y \in S(p, \rho(p) \cap X$ then the image X' of X has the same measure as X, that is $| X, \delta |_n = | X', \delta' |_n$ for all n.

PROOF. Denote by X_ν the subset of X for which $\rho(p) > \nu^{-1}$. Then $X_\nu \subset X_{\nu+1}$ and $\bigcup X_\nu = X$. If $\bigcup Y_i = X_\nu$ and $\alpha(Y_i) < \epsilon < \nu^{-1}$ then $\delta(x, y) = \delta(x', y')$ for $x, y \in Y_i$. Therefore $\bigcup Y_i' = X_\nu'$ and $\alpha(Y') < \epsilon < \nu^{-1}$, hence $| X_\nu, \delta |_n \geqq | X_\nu', \delta' |_n$ and by (1.13) $| X, \delta |_n \geqq | X', \delta' |_n$.

If x^* and x'^* denote the classes of x and x' in R and R' respectively, then the hypothesis yields $\delta'^*(x'^*, y'^*) \geqq \delta^*(x^*, y^*)$. Hence the mapping $x'^* \to x^*$ is single valued and by (1.11) and (1.14) $| X, \delta |_n = | X^*, \delta^* |_n \geqq | X', \delta'^* |_n = | X', \delta' |_n$.

The assumption $\delta(x, y) \leqq \delta'(x', y')$ is essential even for ordinary metric spaces as a locally congruent mapping of a sphere on an elliptic plane shows.

Besides $| X, \delta |_n$ there are two other similarly defined numbers which at first sight seem to have a better claim to be considered as measures:

Let Λ' and Λ'' be the classes of all countable coverings of the set X by spheres, $X \subset \bigcup S(p_i', \rho_i/2)$ and $X \subset \bigcup S(p_i'', \rho_i''/2)$ with $\rho_i' < \epsilon$ and no condition on the p_i', or with $\rho_i'' < \epsilon$ and $p_i'' \in X$ respectively. Then

$$\omega^{(n)} \inf_{\Lambda'} \sum \rho_i'^n = | X, \delta |_n'^\epsilon \leqq | X, \delta |_n''^\epsilon = \omega^{(n)} \inf_{\Lambda''} \sum \rho_i''^n$$

since $\alpha(X \cap S(p, \rho)) \leqq 2\rho$ we have

$$| X, \delta |_n^\epsilon \leqq | X, \delta |_n'^\epsilon \leqq | X, \delta |_n''^\epsilon.$$

As ϵ decreases none of these numbers can decrease, hence $| X, \delta |_n' = \lim_{\epsilon\to 0+} | X, \delta |_n'^\epsilon$ and $| X, \delta |_n'' = \lim_{\epsilon\to 0+} | X, \delta |_n''^\epsilon$ exist and satisfy the relation

(1.15) $| X, \delta |^n \leqq | X, \delta |_n' \leqq | X, \delta |_n'' \leqq 2^n | X, \delta |_n$.

The last inequality can be seen as follows: Let $\bigcup X_i = X$, $\alpha(X_i) < \epsilon$, and $\omega^{(n)} \sum \alpha^n(X_i) < |X, \delta|_n^\epsilon + \epsilon$. It may be assumed that each X_i contains at least one point p_i. Then $S(p_i, \alpha_i) \supset X_i$ for any $\alpha_i > \alpha(X_i)$, hence for suitable α_i

$$S(p_i, \alpha_i) \supset X \text{ and } |X, \delta|_n^{\prime\prime 2\epsilon} < 2^n \omega^{(n)} \sum \alpha_i^n < 2^n(|X, \delta|_n^\epsilon + 2\epsilon).$$

In general spaces (1.15) is the only relation between the three numbers. Their relative merits are these: $|X, \delta|_n$ and $|X, \delta|_n'$ *are Caratheodory measures*, but $|X, \delta|_n''$ is in general not additive. $|X, \delta|_n$ *and* $|X, \delta|_n''$ *do not depend on the ambient space*, that is only distances of points in X enter the definition, $|X, \delta|_n'$ does not have this property. *The factor* $\omega^{(n)}$ *has an obvious justification*, at least in Euclidean spaces, *in the case of* $|X, \delta|_n'$ *and* $|X, \delta|_n''$, but not for $|X, \delta|_n$. In fact, $|X, \delta|_n$ is in general not even a constant multiple of either $|X, \delta|_n'$ or $|X, \delta|_n''$. For instance according to a well known result of Besicovitch $|X, \eta_2|_1'$ is not a constant multiple of $|X, \eta_2|_1$, when R in the Euclidean plane with distance η_2. However in the most important special case the three numbers are equal:

(1.16) *If* E^m *is an m-dimensional euclidean space with distance* η_m *and exterior Lebesgue measure* $|X|_m^L$, *then for any set* X *in* E^m

$$(1.17) \qquad |X, \eta_m|_m = |X, \eta_m|_m' = |X, \eta|_m'' = |X|_m^L.$$

Although this fact is well known a brief proof will be given in a form which will be generalized to other spaces later on: By Vitali's Theorem there are for a given $\epsilon > 0$, spheres $S(p_i, \rho_i/2)$ with $p_i \in X$ and $\rho_i < \epsilon$ such that $\bigcup S(\rho_i, \rho_i/2) \supset X$ and

$$(1.18) \qquad \omega^{(m)} \sum \rho_i^m = \sum |S(p_i, \rho_i/2)|_m^L < |X|_m^L + \epsilon$$

hence $|X, \eta_m|_m^{\prime\prime\epsilon} \leqq |X|_m^L + \epsilon$.

If $\bigcup Y_i = X$ and $\alpha(Y_i) < \epsilon$, then $|X|_m^L \leqq \sum |Y_i|_m^L$. Among all sets of a given diameter the sphere has the greatest volume or measure,[9] hence $|Y_i|_m^L \leqq \omega^{(m)}\alpha^m(Y_i)$ and $|X|_m^L \leqq |X, \eta_m|_m^\epsilon$. Therefore $|X, \eta_m|_m'' \leqq |X|_m^L \leqq |X, \eta_m|_m$ and (1.17) follows from (1.15).

The question, under which conditions the three numbers $|X, \eta_m|_n$, $|X, \eta_m|_n'$, $|X, \eta_m|_n''$ are equal also for $n < m$ seems to be very difficult. In certain special cases a result of Kolmogoroff answers the question for integral n. For the theory of area the integral case will prove the most important.

(1.19) *Let a mapping* $p \to x(p)$ *of an analytical set* M *in* E^n *on an analytical set* X *in* E^m, $n < m$, *exist such that*

$$(1.20) \qquad \eta_n(p, q) \geqq \beta\eta_m(x(p), x(q)).$$

Then $|X, \eta_m|_n = |X, \eta_m|_n' = |X, \eta_m|_n''$

Proof. Let Y traverse all analytical sets in E^n which admit a mapping y —

[9] See [4] p. 76 and p. 109. The fact is also contained in theorem (2.2) whose proof is found at the end of section 2.

$x(y)$ on X with $\eta_n(y_1, y_2) \geqq \eta_m(x(y_1), x(y_2))$. Such Y exist because of (1.20), and since Kolmogoroff's Principle holds obviously for $|X, \delta|_n''$ it follows from (1.14), (1.15), and (1.17) that

$$|X, \eta_m|_n \leqq |X, \eta_m|_n' \leqq |X, \eta_m|_n'' \leqq |Y, \eta_n|_n'' = |Y|_n^L.$$

The measure $|X, \eta_m|_n$ is an n-dimensional Kolmogoroff measure in E^m[10]. The Kolmogoroff measure of X is unique because of (1.20) (compare [14], Eindeutigkeitssatz p. 361). Hence $|X, \eta_m|_n$ coincides with the maximal Kolmogoroff measure, or $|X, \eta_m|_n = \inf |Y|_n^L$ which proves (1.19).

2. Measure in Minkowski spaces

Since Finsler Spaces behave locally like Minkowski spaces the latter are very important for the present investigation, therefore their properties will be briefly reviewed: In E^n let K be a closed convex hyper-surface which has 0 as center. If 0 is the origin of a rectangular coordinate system, denote for any point $x \neq 0$ by x' the intersection of the ray from 0 through x with K and put $\varphi(x) = \eta_n(0, x)/\eta_n(0, x')$. Then $\varphi(-x) = \varphi(x)$ and φ is positively homogeneous and convex (compare [4], p. 23). Therefore the *Minkowski distance*

$$\mu_n(x, y) = \varphi(x - y) \tag{2.1}$$

satisfies the axioms for distance and makes E^n a new metric space M^n. The distance $\mu_n(x, y)$ is invariant under all translations $x' = x + a$ of the E^n. The relations $\eta_n(x, y) = (1 - t)\eta_n(x, z)$ and $\eta_n(y, z) = t\eta_n(x, z)$ imply the same relations for μ_n. Therefore a convex subset of E^n is also convex for M^n, but not conversely unless K is strictly convex. The spheres $S(p, \rho, \mu_n)$ are with respect to E^n all homothetic to the interior of K.

A Minkowski space R can be characterized intrinsically as an internally and externally convex, finitely compact metric space which admits a transitive abelian group T of notions. The equation $\mu_n(a\phi, a\Psi) = \mu_n(b\Phi, b\Psi)$ holds for any two elements Φ, Ψ of T and any two points a, b in R, so that $\mu(\Phi, \Psi) = \mu_n(a\Phi, a\Psi)$ is a metrization under which T becomes isometric to R. A fundamental theorem of Pontrjagin ([21], p. 170) shows that R can be mapped topologically on E^n with coordinates $x_1, \cdots, x_n$ in such a way that the group T becomes the group of translations $x' = x + a$.

Any *euclidean metrization* of R which is invariant under T is called *associated to* R. The associated E^n are determined up to a (non degenerate) affine transformation.

The triangle inequality and symmetry of the distance in R imply that in every associated euclidean metric the surface $K: \mu_n(0, x) = 1$ is convex and has 0 as center, and $\mu_n(x, y)$ is obtained from K by the relation (2.1).

Any reasonable n-dimensional measure in a Minkowski space M^n with dis-

[10] A proof that $|X, \eta_m|_n'$ is a Kolmogoroff measure, which also applies to $|X, \eta_m|_n$ is found in [18], p. 752.

tance μ must be invariant under the motions which M^n admits. If η is the euclidean distance of a euclidean space E^n associated with M^n, then measure in M^n must be invariant under the translations T of E^n. According to known results on Haar measure in groups or of Loomis [16] there is up to a constant factor only one non-trivial Caratheodory measure in M^n invariant under T. Since the exterior Lebesgue meaure $|X|_n^L$ in the E^n and the measures $|X, \mu|_n$, $|X, \mu|'_n$ are all Caratheodory measures in M^n and invariant under T it follows that they differ only by a constant factor and that *any reasonable measure in* M^n *is a constant multiple of* $|X|_n^L$.

Another property of Minkowski spaces leads according to Loomis [17] to the same conclusion. Loomis shows that a finitely compact space has, up to a constant factor, only one measure in which spheres with equal radii have equal measure ($\neq 0, \infty$), whenever the space satisfies this condition: If any closed sphere of radius ρ can be covered by m open spheres of radius ξ, then any closed sphere of radius $\alpha\rho$ can be covered by m open spheres of radius $\alpha\xi$. Minkowski spaces obviously satisfy this condition.

For a single Minkowski space the problem of volume can therefore be considered as settled. The value of the constant factor cannot be essential for the geometric properties of the space. The situation changes radically as soon as different Minkowski spaces are considered, as in Finsler spaces, where the Minkowski metric changes from point to point, and the constant factor becomes a function of the point. It is therefore of greatest importance to single out one value of the constant as "*natural.*"

Both measures $|X, \mu|_n$ and $|X, \mu|'_n$ are definite numbers and yields therefore definite values for the constant. But as pointed out in the preceding section the expression $\omega^{(n)} \sum \alpha^n(X_i)$ by means of which $|X, \mu|_n$ is evaluated is justified only if $|X, \mu|_n = |X, \mu|'_n$ and $|S(p, 2^{-1}), \mu|_n = \omega^{(n)}$.[11] The proof of (1.17) suggests that this will follow from the theorem:

(2.2) *For any set X in M^n with Minkowski diameter at most* α

$$|S(q, \alpha/2; \mu), \mu|_n \geqq |X, \mu|_n .$$

Since the measures $|X, \mu|_n$ and $|X|_n^L$ are proportional this relation is equivalent to

$$|S, (q, \alpha/2; \mu)|_n^L \geqq |X|_n^L . \tag{2.2a}$$

The rather long proof for (2.2a) will be given at the end of this section. (2.2a) implies that the measures $|X, \mu|_n$ and $|X|_n^L$ are proportional, so that the results of Loomis are not needed in this connection; for (2.2a) yields the following analogue to (1.17)

(2.3) *Let $|X|_n^L$ denote the exterior Lebesgue measure of a euclidean space*

[11] If this were not correct, using $|X, \mu|_n$ would mean acting as though a Minkowski space had a euclidean property which is actually does not have. This procedure has all too often been followed in Finsler spaces and has led to theories without geometric significance.

associated with a given Minkowski space M^n with metric μ. Then for any set X in M^n

$$|X, \mu|_n = |X, \mu|_n' = |X, \mu|_n'' = \omega^{(n)} |X|_n^L / S(q, 2^{-1}; \mu)|_n^L.$$

For the proof put

$$\Xi = \omega^{(n)}/|S(q, 2^{-1}; \mu)|_n^L.$$

Since the Minkowski spheres are homothetic it follows that for any point p and any radius ρ

$$(2.4) \qquad |S(p, \rho/2; \mu)|_n^L = \omega^{(n)}\rho^n\Xi^{-1}.$$

The Vitali Theorem holds for Lebesgue measure and the spheres $S(p, \rho; \mu)$. Therefore points $p_i \in X$ and $\rho_i < \epsilon$ exist such that $\bigcup S(p_i, \rho_i/2; \mu) \supset X$ and $\sum |S(p_i, \rho_i/2; \mu)|_n^L < |X|_n^L + \epsilon$. By (2.4) $\omega^{(n)} \sum \rho_i^n \leqq \Xi(|X|_n^L + \epsilon)$ hence $|X, \mu|_n'' \leqq \Xi |X|_n^L$. Next let $\bigcup X_i = X$, $\alpha(X_i) < \epsilon$, where $\alpha(X_i)$ is the Minkowski diameter of X_i. Then $|X|_n^L \leqq \sum |X_i|_n^L$. By (2.2a) the Minkowski sphere of diameter $\alpha(X_i)$ has at least Lebesgue measure $|X_i|_n^L$, so that by (2.4) $|X_i|_n^L \leqq \omega^{(n)}\alpha^n(X_i)\Xi^{-1}$. Hence $\Xi |X|_n^L \leqq \omega^{(n)} \sum \alpha^n(X_i)$ and $\Xi |X|_n^L \leqq |X, \mu|_n$.

A corollary of (2.3) and (2.4) is

$$(2.5) \qquad |S(p, \rho/2; \mu)|_n = \omega^{(n)}\rho^n.$$

These results justify the use of the measure $|X, \mu|_n$. But the question remains whether there are not other natural measures in Minkowski spaces. Since we know that any such measure has the form $\kappa |X|_n^L$, the question amounts to this: $|X, \mu|_n$ and $|X, \mu|_n'$ lead to the choice $\kappa = \omega^{(n)}$ (or to the normalization $|S(q, 2^{-1}), \mu|_n = \omega^{(n)}$). Are there geometric reasons which make this choice appear more natural than other choices? The following considerations will exhibit such reasons.

Let (μ) be the class of all n-dimensional Minkowski metrics referred to the *same* associated Euclidean space E^n. A natural measure for any of the metrics μ has the form $\kappa(\mu) \cdot |X|_n^L$. If the affine transformation $x \to x'$ of the E^n on itself with determinant $D \neq 0$ carries the unit sphere $\mu(p, x) = 1$ into the unit sphere $\mu(p', x') = 1$ ($\mu, \mu' \in (\mu)$) then $\mu(x, y) = \mu(x', y')$ so that $\kappa(\mu) |X|_n^L = \kappa(\mu') |X'|_n^L = \kappa(\mu') \cdot |X|_n^L \cdot |D|$ for any set X and its image X'. Hence

$$(2.6) \qquad \kappa(\mu):\kappa(\mu') = |D| \cdot |S(p', 1; \mu')|_n^L : |S(p, 1; \mu)|_n^L.$$

A very simple form of Kolmogoroff's Principle is

A.) If $\mu(x, y) \geqq \mu'(x, y)$ for all x, y then $\kappa(\mu) \geqq \kappa(\mu')$.

This condition is equivalent to the following

B.) If $S(p, 1; \mu) \subset S(p', 1; \mu')$ for suitable p, p', then $k(\mu) \geqq k(\mu')$.

For the hypotheses are equivalent: if $\mu(x, y) \geqq \mu'(x, y)$ then obviously $S(p, 1; \mu) \subset S(p, 1; \mu')$. If $S(p, 1; \mu) \subset S(p', 1; \mu')$, the fact that the $S(p, \rho; \mu)$ and

$S(p', \rho; \mu')$ considered as sets in E^n are convex and have p, p' as centers shows that $S(p, 1; \mu) \subset S(p, 1; \mu')$. Then the definition of Minkowski distance yields $\mu(x, y) \geqq \mu'(x, y)$.

A) or B) are not sufficient to determine $\kappa(\mu)$ up to a factor. *But the following stronger condition is sufficient*

B′.) *If* $\bigcup_{i=1}^{\lambda} S(q_i, 1; \mu') \supset S(p, 1; \mu) \supset \bigcup_{i=1}^{\nu} S(r_i, 1; \mu')$ *where* $\mu'(r_i, r_j) \geqq 2$ *for* $i \neq j$, *then*

$$\lambda\kappa(\mu) \geqq \kappa(\mu') \geqq \nu \cdot \kappa(\mu).$$

To prove this, consider the mesh in E^n defined by the planes $j \cdot x_i = l$; j a fixed positive integer, $i = 1, 2, \cdots, n$; $l = 1, 2, 3, \cdots$ and let μ^j be the metric in (μ) for which a cube of the mesh is a unitsphere. (2.6) shows that $\kappa(\mu^j) = j^n\kappa(\mu^1)$. Let μ be any element in (μ), p an arbitrary but fixed point, and denote by k'_j, k_j the number of open cubes of the mesh which are contained in, or contain points of, $S(p, 1; \mu)$ respectively. Then B' yields $\kappa(\mu') \geqq k'_j\kappa(\mu^j)$. The other part of B' cannot be applied directly to the k_j spheres because they do not cover $S(p, 1; \mu)$. Associate with each of these k_j cubes a homothetic cube with the same center and side $j^{-1}(1 + k_j^{-1})^{1/n}$. If $\bar{\mu}^j$ denotes the metric in (μ) with these cubes as unitspheres, then by (2.6) and B'

$$(1 + k_j^{-1})k_j j^{-n}\kappa(\mu) \geqq (1 + k_j^{-1})j^{-n}\kappa(\bar{\mu}^j) = j^{-n}\kappa(\mu^j) = \kappa(\mu') \geqq k'_j j^{-n}\kappa(\mu).$$

Because for convex bodies in E^n Jordan measure coincides with Lebesgue measure $\lim_{j\to\infty} k'_j j^{-n} = \lim k_j \cdot j^{-n} = | S(p, 1; \mu) |_n^L$, so that $\kappa(\mu) = | S(p, 1; \mu) |_n^L$ $\kappa(\mu')$ and fixing a value for $\kappa(\mu')$ determines all $\kappa(\mu)$.

The condition that the unitsphere K of a Minkowski metric has 0 as center, is not always required for a Minkowski metric. If 0 is not the center of K, then $\mu_n(x, y)$ will in general differ from $\mu_n(y, x)$ and there are two kinds of spheres: $S'(p, \rho)$ consisting of those points x for which $\mu_n(p, x) < \rho$ and $S''(p, \rho)$ consisting of the points x with $\mu_n(x, p) < \rho$. The set $S''(p, \rho)$ goes into $S'(p, \rho)$ by a reflection in p, therefore the two spheres have the same Lebesgue measure in any associated Euclidean metric, hence formula (2.3) may be used for measure without ambiguity. In fact the definition of the measure $| X, \mu |'_n$ can be applied and yields the same result for either type of sphere. The measure $| X, \mu |_n$ will in general be different and equal $| X, \bar{\mu} |_n^{\cdot}$, where $\bar{\mu}(x, y) = \max | \mu(x, y), \mu(y, x)]$ if $\alpha(X)$ is defined $\sup_{x,y\in X} \mu(x, y)$. But neither of these measures is acceptable, because both are independent of the orientation of the space, and are therefore no generalization of the length of intervals in the one dimensional case, which does depend on the orientation. In the case of intervals $| X, \mu |_1$ is the maximum of the two lengths of the interval and $| X, \mu |'_1$ is their arithmetic mean.

We now turn to the *proof of* (2.2) *or* (2.2a). A bounded set M with at least two points in an ordinary metric space R with distance δ will be called *entire*[12]

[12] This word is used to avoid the term "complete" which would be the translation of the German term "vollständig" used in this connection.

if it is impossible to add a point (of R) to M without increasing the diameter $\alpha(M)$ of M. Since $\alpha(\bar{M}) = \alpha(M)$ (where $\bar{M}$ is the closure of M) an entire set is closed. For any point $p \in R$ put $\beta(p, M) = \sup_{x \in M} \delta(p, x)$. Then obviously

(2.7) *The set M is entire if and only if every point p with $\beta(p, M) \leqq \alpha(M)$ belongs to M.*

(2.8) M is an entire set if and only if

$$M = \bigcap_{x \in M} \bar{S}(x, \alpha(M))$$

where $\bar{S}(x, \rho)$ denotes the set of points z with $\delta(x, z) \leqq \rho$.

For let M be etire and $x, y \in M$. Then $\delta(x, y) \leqq \alpha(M)$ hence $y \in \bar{S}(x, \alpha(M))$ and $M \subset \bigcap_{x \in M} \bar{S}(x, \alpha(M))$. If $y \in \bigcap_{x \in M} \bar{S}(x, \alpha(M))$ then $\delta(x, y) \leqq \alpha(M)$ for every $x \in M$, hence $\beta(y, M) \leqq \alpha(M)$ and $y \in M$ by (2.7). Conversely, if $M = \bigcap_{x \in M} S(x, \alpha(M))$ and $\beta(y, M) \leqq \alpha(M)$ then $\delta(x, y) \leqq \alpha(M)$ for all $x \in M$, hence $x \in \bigcap \bar{S}(x, \alpha(M)) = M$, so that M is by (2.6) entire.

(2.9) Every boundary point p of an entire set M satisfies the relation $\beta(p, M) = \alpha(M)$.

Assume there is a boundary point $p \in M$ with $\beta(p, M) < \alpha(M)$. Then $\beta(q, M) < \alpha(M)$ for all points q of a neighborhood U of p. But since p is on the boundary of M the set U contains points q not in M, so that M is by (2.6) not entire.

Let now M be a Minkowski space with distance μ. If M is any closed set in M containing more than one point and $\mu(x, y) = \alpha(M)$ for two points x, y in M, then any (hyper)plane π through y (or x) to which the line xy is perpendicular, in the sense of the Minkowski metric, is a supporting plane of M at y (or x). (For the elementary properties of Minkowskian geometry needed for the following considerations, see section II of the author's: *Metric methods in Finsler spaces and in the foundation of geometry*, Princeton 1942.) Otherwise there would be points z of M on the other side of π from x, but then $\mu(x, y) > \alpha(M)$. The number $\alpha(M)$ is therefore at the same time the greatest Minkowski distance that parallel supporting planes of M can have. (2.8) and (2.9) yield now:

(2.10) *An entire set in a Minkowski space M is convex* (with respect to any associated euclidean metric). *For every boundary point x of M there is another boundary point with $\mu(x, y) = \alpha(M)$. Planes to which the line xy is perpendicular at x or y are supporting planes of M, and $\alpha(M)$ is the maximal Minkowski distance of parallel supporting planes of M.*

Choose now a definite associated euclidean space to M with rectangular coordinates $x_1, \cdots, x_n$ and distance η. Let $E(u)$ be the supporting function (see [4], p. 23) of the sphere $S(0; 1; \mu)$, where 0 is the origin and let the convex body H with supporting function $H(u)$ contain 0 in its interior. Then $H(u)/E(u)$ is the Minkowski distance of the supporting plane π of H whose euclidean normal from 0 has direction u. For let π' be the supporting plane of $S(0, 1; \mu)$ with euclidean normal u, and p, p' the points where the euclidean normal from 0 to

π and π' intersects these planes. If a Minkowskian perpendicular to π and π' intersects these planes in q and q', then

$$\mu(0, q) = \eta(0, q)/\eta(0, q') = \eta(0, p)/\eta(0, p') = H(u)| u |^{-1}/E(u) | u |^{-1} = H(u)/E(u).$$

The supporting plane of H parallel to π has Minkowski distance $H(-u)/E(-u) = H(-u)/E(u)$ from 0, so that $(H(u) + H(-u))/E(u)$ is the Minkowski distance of the two parallel supporting planes to H whose euclidean normal has direction u or $-u$.

The convex body K with the supporting function $K(u) = (\frac{1}{2})(H(u) + H(-u))$ has 0 as center and $| K |_n^L \geqq | H |_n^L$ and actually $| K |_n^L > | H |_n^L$ unless H has already a center.[13] (2.10) shows that $\alpha(K) = \alpha(H)$. Hence

(2.11) *For every convex body H without center, there is a convex body K with center, $\alpha(K) = \alpha(H)$, and $| K |_n^L > | H |_n^L$.*

Next observe:

(2.12) If K is the convex closure of $\bar{N}$, then $\alpha(K) = \alpha(N)$.

For obviously $\alpha(K) \geqq \alpha(\bar{N}) = \alpha(N)$. Let $p, q \in K$ with $\mu(p, q) = \alpha(K)$. Then parallel planes π_p, π_q through p and q to which pq is a Minkowski perpendicular are supporting planes of K. If π_p were not a supporting plane of $\bar{N}$, then a plane π parallel to π_p would exist, which separates p from q and is a supporting plane of $\bar{N}$. But then neither p nor q would belong to the convex closure of N. Hence π_p and π_q contain points of $\bar{N}$ and $\alpha(\bar{N}) \geqq \mu(p, q) = \alpha(K)$.

Using an idea of Lebesgue (compare [4], p. 130] we prove next

(2.13) *A convex set K with center c is contained in an entire set K^* with the same center and diameter as K.*

K may be assumed closed. If K is not entire than a point p not in K with $\beta(p, K) \leqq \alpha(K)$ exists (see (2.7)). The convex closure K' of $p \cup K$ has diameter $\alpha(K)$. A point p' on the segment pc close to p is still outside of K but inside of K' therefore $\beta(p', K') < \alpha(K') = \alpha(K)$. If p'' is the point symmetric to p' with respect to c, then $\beta(p'', p' \cup K) = \beta(p', p'' \cup K)$ and this number depends continuously on p'. If q', q'' are the points where the line pc intersects the boundary of K, then $\beta(q', q'' \cup K) < \alpha(K)$. Therefore $\beta(p', p'' \cup K) < \alpha(K)$ holds also for suitable points p', p'' outside of K, and there is a $\rho > 0$ such that

$$\text{(2.14)} \quad \alpha[S(p', \rho) \cup S(p'', \rho) \cup K] = \alpha(K), \qquad S(p', \rho) \cap K = 0.$$

Among all pairs of spheres $S(p', \rho)$, $S(p'', \rho)$ symmetric to c which satisfy (2.14) choose one, $S(p_1', \rho_1)$, $S(p_1'', \rho_1)$ for which $\rho_1 > \sup \rho/2$ for all ρ for which (2.14) holds. Let K_1 be the convex closure of $S(p_1', \rho_1) \cup S(p_1'', \rho_1) \cup K$. If K_1 is entire (2.12) is proved, otherwise we proceed as before. There are points p', p'' with center c and a $\rho > 0$ such that

$$\text{(2.15)} \quad \alpha[S(p', \rho) \cup S(p'', \rho) \cup K_1] = \alpha(K_1) = \alpha(K), \qquad S(p', \rho) \cap K_1 = 0.$$

[13] Compare [4], pp. 73 and 106.

Among the pairs of spheres satisfying this condition we choose one, $S(p_2', \rho_2)$, $S(p_2'', \rho_2)$ such that $\rho_2 > (\frac{1}{2}) \sup \rho$ for all ρ that satisfy (2.15), and denote by K_2 the convex closure of $S(p_2', \rho_2) \cup S(p_2'', \rho_2) \cup K_1$, etc. Thus we either arrive at a symmetric entire set $K^* = K_n$ with $\alpha(K_n) = \alpha(K)$ or we obtain an infinite sequence $K_1, K_2, \cdots$. Then $\rho_\nu \to 0$ because of the Bolzano Weierstrass Theorem. Moreover $\bigcup K_\nu$ is convex and its closure K^* has diameter $\alpha(K^*) = \alpha(K)$ and is entire. For if K^* were not entire a pair of spheres with the property (2.14) but K replaced by K^* would exist which contradicts the definition of ρ_ν when $\rho_\nu < \rho/2$.

Finally we show

(2.16) *If K is an entire set with center c, then* $K = \bar{S}(c, \alpha(K)/2)$.

For let B be the boundary of K. By (2.9) there is for every $x' \in B$ at least one point $x_0 \in B$ with $\mu(x', x_0) = \alpha(K)$. The point x'' symmetric to x' with respect to c satisfies also the relation $\mu(x', x'') = \alpha(K)$. For if $x_0 \neq x''$ then planes π' and π_0 through x' and x_0 to which the line $x'x_0$ is a Minkowski perpendicular are by (2.10) supporting planes of K. There is a supporting plane π'' through x'' parallel to π' because K has c as center. Since there is only one supporting plane different from, and parallel to, π' the planes π'' and π_0 coincide. Because $x'x_0$ is perpendicular to π'' it follows that $\alpha(K) \geqq \mu(x', x'') \geqq \mu(x', x_0) = \alpha(K)$ and $\mu(x', c) = \mu(c, x'') = \alpha(K)/2$. This proves (2.16) because x' was arbitrary on B.

The statements (2.11), (2.13), and (2.16) prove (2.2a) and the Theorem (2.2a).

3. Intrinsic area

The intrinsic metric on a non-parametrized ordinary surface is determined by the length of the curves drawn on the surface. In the parametric case there are two spaces to consider, the parameter space P, which determines the topology of the surface with the intrinsic metric, and the space R in which the surface lies, which determines the numerical value of the distance. This leads to the following hypotheses and definitions.

I). *The parameter space P is metrizable and arcwise connected.*[14] Its points are denoted by p, q, r with or without sub, or super scripts.

II). *R is an ordinary metric space* with distance δ and points x, y, z.

III). *There is a continuous mapping $p \to x(p)$ of P in R.*

The image set $x(P)$ is the *set (Px) in R parametrized by P* whose area is now defined as follows:

For any two points q, r in P let Λ be the set of all continuous curves $p(\tau)$ from q to r in P. If λ denotes generally length of curves we put

$$\pi(q, r) = \inf_\Lambda \lambda[x(p(\tau))]. \tag{3.1}$$

[14] Instead of arcwise connectedness the apparently weaker condition is used, that any two points of P can be connected by a continuous curve in P. However the two conditions are equivalent, compare [24], p. 36. Therefore arcwise connectedness is preserved under continuous mapping.

$\pi(q, r)$ may be 0 or ∞ but is symmetric and satisfies the triangle inequality

$$\pi(r_1, r_2) + \pi(r_2, r_3) \geqq \pi(r_1, r_3), \qquad \pi(q, r) = \pi(r, q). \tag{3.2}$$

With the metric $\pi(q, r)$ the set P becomes a metric space in the sense of section 1. In the future $\bar{P}$ will denote the set P with the distance π, whenever it is not obvious that this distance is meant. Of course the distance π is in general not topologically equivalent to the original topology of P. The *n-dimensional* (n real and non-negative) *area of* (P, x) is now defined as

$$| P, x |_n = | P, \pi |_n . \tag{3.3}$$

The non-parametric case is the special case where $P \subset R$ and $x(p) = p$. In that case we simply use $| P |_n$ as symbol for area.

If P is separable and the (then unique, see [13]) dimension of P is finite, we put

$$| P, x |_{\dim P} = | P, x | \tag{3.4}$$

and call $| P, x |$ *the area of* (P, x). This agrees with the usual terminology in which the dimension of the parameter space decides whether we speak of a curve (Peano curve!), a surface, etc.

The mathematical justification for distinguishing $| P, x |$ among the $| P, x |_n$ is the fact that in normal cases $| P, x |$ is the only significant number because (1.9) yields

$$\begin{aligned} &\textit{If } | P, x |_n > 0, \textit{ then } | P, x |_m = \infty \textit{ for } m < n, \\ &\textit{if } | P, x |_n < \infty, \textit{ then } | P, x |_m = 0 \textit{ for } m > n. \end{aligned} \tag{3.5}$$

It is obvious that $| P, x |_n$ *does not depend on the parametrization*:

(3.6) If $p' \to p(p')$ is a topological mapping of the metrizable space P' on P, and $x'(p') = x(p(p'))$ then

$$| P', x' |_n = | P, x |_n \text{ for all } n.$$

The corresponding statement in the non-parametric case is this

(3.7) If $p' \to p(p')$ is a topologial mapping of the metric space P' on P then $| P', p |_n = | P |_n$.

This shows that the main difference between the parametric and the non-parametric case lies in the fact that in the latter case the mapping $x(p)$ of P on $x(P)$ is topological. In fact, (3.6) shows that

(3.8) $| P, x |_n = | x(P) |_n$ *if* $x(p)$ *maps* P *topologically on* $x(P)$.

It was mentioned in the introduction that with the present definition the area of (P, x) is at least as large as the area of $x(P)$. To prove this, notice that

$$\pi(q, r) \geqq \delta(x(q), x(r)) \tag{3.9}$$

and in the non-parametric case

$$\pi(q, r) \geqq \delta(q, r) \tag{3.10}$$

consequently by (1.14)

$$(3.11)\qquad | P |_n = | P, \pi |_n \geqq | P, \delta |_n$$

so that the area of a non-parametrized set is at least its Hausdorff measure. When (P, x) is a parametrized surface, then $x(P)$ with the metric δ may be considered a non-parametrized surface because it is also arcwise connected.[15] Then two points of $x(P)$ obtain, by (3.1), the distance

$$(3.12)\qquad \pi(x, y) = \inf \lambda(z(\tau)),$$

where $z(\tau)$ traverses all continuous curves from x to y in $x(P)$. If $x = x(q)$ and $y = x(r)$ then by (3.12), (3.1) and (3.9) $\pi(q, r) \geqq \pi(x, y) \geqq \delta(x, y)$, hence by (3.11), (3.3) and (1.14)

$$(3.13)\qquad | P, x |_n \geqq | x(P) |_n \geqq | x(P), \delta |_n .$$

This leads to a nice geometric property of area which will however not be used: If $x(P)$ is separable and metrizable and $\dim x(P) = n < \infty$, then $| P, x |_n > 0$ (compare [13], p. 104). $| P, x |_n$ is in general not additive. It is not even monotone; that is, there may be arcwise connected subsets Q of P with $| Q, x |_n > | P, x |_n$. But

(3.14) If $P = \bigcup P_i$ and the P_i are arcwise connected, then

$$| P, x |_n \leqq \sum | P_i, x |_n .$$

For if π_i denotes the distance (3.1) for (P_i, x) then obviously $\pi_i(q, r) \leqq \pi(q, r)$ when $q, r \in P_i$. Therefore by (1.14) $| P_i, x |_n = | P_i, \pi_i |_n \geqq | P_i, \pi |_n$. The assertion follows now from $\sum | P_i, \pi |_n \geqq | \bigcup P_i, \pi |_n = | P, \pi |_n$.

The following simple but fundametal form of *Kolmogoroff's Principle in its intrinsic form* is an immediate consequence of (1.14) and was implicitly used in the last proof.

K) *If (P, x) and (P, x') are both parametrized by P and the corresponding intrinsic distances π, π' satisfy the relation $\pi(q, r) \geqq \pi'(q, r)$ then $| P, x |_n \geqq | P, x' |_n$.*

There is a much more general form of this principle which is analogous to the following form of Kolmogoroff's Principle in non-intrinsic form.

(3.15) *Let $p \to p'(p)$ be a continuous mapping of P on P' and $p' \to x'(p')$ a continuous mapping of P' in a metric space R' with distance δ' such that every point p of P has a neighborhood $U(p)$ with $\delta[x(q), x(r)] \geqq \beta\delta'[x'(p'(q)), x'(p'(r))]$ for $q, r \in U(p)$, where β is positive and independent of p. Then*

$$| P, x |_n \geqq \beta^n | P', x' |_n .$$

According to (1.14) it suffices to show that $\pi(q, r) \geqq \beta\pi(q', r')$ for given points q', r' in P' and any points q, r in P with $p'(q) = q'$, $p'(r) = r'$ ($\pi(q', r')$ is the distance (3.1) for (P', x')). Let $p(\tau)$, $a \leqq \tau \leqq b$ be any continuous curve from

[15] Due to the invariance under change of parametrization this assumption is not more restrictive than requiring P to be a topological polyhedron.

q to r in P, with $\lambda[x(p(\tau))] < \pi(q, r) + \epsilon$. Then $p'(\tau) = p'(p(\tau))$ is a continuous curve from q' to r'. Denote by $\alpha(\tau)$ the least upper bound of all α such that

(3.16) $\delta[x(p(\tau_1)), x(p(\tau_2))] \geqq \beta\delta'[x'(p'(\tau_1)), x'(p'(\tau_2))]$ for $|\tau_i - \tau| \leqq \alpha$.

By hypothesis $\alpha(\tau) > 0$. Whenever $|\tau' - \tau_i| \leqq \alpha(\tau) - |\tau - \tau'|$ then (3.16) holds, hence $\alpha(\tau') \geqq \alpha(\tau) - |\tau - \tau'|$ and $|\alpha(\tau) - \alpha(\tau')| \leqq |\tau - \tau'|$. This shows that $\alpha(\tau)$ is continuous and has a positive minimum α^*. If $\tau_0 = a < \tau_1 < \cdots < \tau_n = b$ and $\max|\tau_i - \tau_{i+1}| < \alpha^*$, then

$$\sum_i \delta[x(p(\tau_i)), x(p(\tau_{i+1}))] \geqq \beta \sum_i \delta'[x'(p'(\tau_i)), x'(p'(\tau_{i+1}))]$$

so that

$$\pi(q, r) + \epsilon \geqq \lambda[x(p(\tau))] \geqq \beta\lambda[x'(p'(\tau))] \geqq \beta\pi(q', r').$$

Note the corollary

(3.17) Let $p \to x'(p)$ be a continuous mapping of P in a metric space R' such that every point of P has a neighborhood $U(p)$ with

$$\delta[x(p), x(r)] = \beta\delta'[x'(p')q)), x'(p'(r))] \text{ for } q, r \in U(p)$$

then $|P, x|_n = \beta^n(P, x'|_n$.

To deserve the name of area $|P, x|$ must coincide: for curves with the arc-length, for polyhedra with the elementary area, and for smooth surfaces in Riemann Spaces with the standard area.

The first two facts will be proved now, the last in section 7.

(3.18) *If P is the interval $a \leqq \tau \leqq b$ of the real axis with the metric $|\tau_1 - \tau_2|$, then $|P, x| = \lambda(x,(\tau))$.*

Proof. The segment $\tau_1 \leqq \tau \leqq \tau_2$ yields $\min \lambda[x(q(\sigma))]$ for all continuous curves $q(\sigma)$ from τ_1 to τ_2 in P, so that $\pi(\tau_1, \tau_2) = \lambda(x, \tau_1, \tau_2)$ where $\lambda(x, \tau_1, \tau_2)$ denotes the length of the subarc $\tau_1 \leqq \tau \leqq \tau_2$ of $p(\tau)$. Therefore, if $\lambda(x, a, b)$ is finite, P is congruent to the segment $0 \leqq \sigma \leqq \lambda(x, a, b)$ of the real axis and (3.18) follows from (1.17).

If $\lambda(p, a, b) = \infty$, then P divides into finiteness classes. If τ_1 and τ_2 belong to the same class, then all τ between τ_1 and τ_2 belong to this class. Hence every finiteness class is connected (possibly one point). If the number of classes is non-countable, then $|P, x| = |P, x|_1 = \infty$ (see (1.8)). If the number is countable, at least one class is not a closed set in P (see [24] p. 16). Let there be a finiteness class $\bar{P}_1$ of the form $\tau' \leqq \tau < \tau''$ say. Then $\lambda(p, \tau', \tau) \to \infty$ for $\tau \to \tau''-$ and $\bar{P}_1$ is congruent to the ray $\sigma \geqq 0$ of the real axis, so that $|\bar{P}_1, \pi|_1 = |P, x| = \infty$ (compare (1.8)).

The proof that $|P, x|$ is for polyhedra the elementary area rests on the following simple remark: Call the subset Q of a metric space R with distance δ *arc convex* if any points q, r in Q can for a given $\epsilon > 0$ be connected by a curve $p(\tau)$ in Q such that $\lambda(p(\tau)) < \delta(q, r) + \epsilon$. The intrinsic metric of Q coincides in that case with the original metric:

(3.19) *If Q is arc convex, then $\pi(q, r) = \delta(q, r)$ and $|Q|_n = |Q, \delta|_n$.*

Next the concept of a polyhedron must be fixed. Let $P = \bigcup Q_i^j$ be an arcwise connected locally finite euclidean polyhedron[15] of dimension n. More explicitly the Q_i^j are the open simplices of dimension j; $Q_i^j \cap Q_h^k = 0$ for $j \neq k$ or $i \neq h$; $j \leqq n$; and there is at least one n-dimensional simplex. $x(p)$ is a continuous mapping of P in a metric space R such that $x(p)$ maps each Q_i^j topologically on $x(Q_i^j)$ and $x(Q_i^j)$ is congruent to an open j-simplex in some Minkowski space M_i^j with metric μ_i^j (a simplex in a Minkowski space means a simplex in an associated euclidean space). Then (P, x) will be called an n-dimensional Minkowski Polyhedron.

(3.20) *If (P, x) is an n-dimensional Minkowski Polyhedron then $|P, x|_m = 0$ for $m > n$, $|P, x|_m = \infty$ for $m < n$ and $|P, x|_n$ is the elementary area $|P, x| = |P, x|_n = \sum_i |x(Q_i^n), \mu_i^n|_n$.*

By (3.8) and (3.19) $|Q_i^j, x|_m = |x(Q_i^j)|_m = |x(Q_i^j), \mu_i^j|_m$. Now $|x(Q_i^j), \mu_i^j|_j$ is different from 0 and ∞, so that by (3.5) $|Q_i^j, x|_m = 0$ for $m > j$, and $|Q_i^j, x|_m = \infty$ for $m < j$. If $q, r \in Q_i^j$ then $\pi(q, r) = \mu_i^j(x(q), x(r))$, hence $|Q_i^j, x|_n = |Q_i^j, \pi|_n$. Therefore $|P, x|_n = \sum |Q_i^n, \pi|_n = \sum |Q_i^n, x|_n$ which proves the assertion.

Any open set in a Minkowski space can be represented as a locally finite polyhedron, therefore (3.20) and (2.3) yield

(3.21) For a connected open set P in a Minkowski space M^n with metric μ

$$|P| = |P, \mu|_n = \omega^{(n)}(|S(q, 2^{-1}; \mu|_n^L)^{-1} \cdot |P|_n^L.$$

In particular, euclidean polyhedra in the usual sense have the elementary area as area; for connected open sets in E^n the "area" coincides with Lebesgue measure.

The present area is by its very definition intrinsic. But in the non-parametric case a question arises which needs some elucidation. If the finiteness class $\bar{P}_0$ of $\bar{P}$ is arcwise connected the procedure leading to the intrinsic metric can be repeated, that is the greatest lower bound of the lengths of all curves connecting two given points q, r in $\bar{P}_0$ can be introduced as a new distance (which is at least as large as $\pi(q, r)$) and this distance can also claim to be an intrinsic distance in P. To avoid uninteresting details we show that in the most important case, namely when P possesses a complete metrization, nothing new is gained (compare (3.19)).

(3.22) *If P is a complete metric space then each finiteness class Q of P is complete and arc convex. In particular, if any two points of P can be connected by a rectifiable curve, then $|P|_n = |\bar{P}|_n$.*

The proof requires a trivial lemma which holds also in the parametric case (for simplicity we put in the following proof $\pi(q, r) = \bar{q}\bar{r}$):

(3.23) Each finiteness class $\bar{P}_0$ of $\bar{P}$ has the following convexity property: For given points $q, r \in \bar{P}_0$ and given $\epsilon > 0$, $0 < \theta < 1$ there is a point p_0 such that

$$\bar{q}\bar{p}_0 \leqq \theta \bar{q}\bar{r} + \epsilon \quad \text{and} \quad \bar{p}_0\bar{r} \leqq (1 - \theta)\bar{q}\bar{r} + \epsilon.$$

For there is a sequence of curves $p_\nu(\tau)$, $a_\nu \leqq \tau \leqq b_\nu$ in P with $\lambda(x(p_\nu(\tau)) \to \bar{q}\bar{r}$. If τ_ν is such that $\lambda(x(p_\nu), a_\nu, \tau_\nu) = \theta\lambda(x(p_\nu)a_\nu, b_\nu)$ it is easily seen that $p_\nu(\tau_\nu)$ satisfies, for large ν, the conditions for p_0.

Under the assumptions of (3.22) we show first that Q is complete. Let $\{p_\nu\}$ be a fundamental sequence in Q. By (3.10) $\{p_\nu\}$ is also a fundamental sequence in P and converges therefore to a point p in P. It suffices to show that for a subsequence $p_{\mu(\lambda)}$ of p_μ also $\bar{p}\bar{p}_{\mu(\lambda)} \to 0$. Choose $\mu(\lambda)$ such that $\bar{p}_\nu\bar{p}_\mu < 2^{-\lambda}$ for $\nu, \mu \geqq \mu(\lambda)$ and let $q_\lambda(\tau)$, $0 \leqq \tau \leqq 1$ be a continuous curve from $p_{\mu(\lambda)}$ to $p_{\mu(\lambda+1)}$ with

$$0 \leqq \lambda(q(\tau)) - \bar{p}_{\mu(\lambda)}\bar{p}_{\mu(\lambda+1)} < 2^{-\lambda}.$$

Define $r(\sigma)$ for $0 \leqq \sigma < 1$ by $r(\sigma) = q_\lambda(\tau)$ if $\sigma = \sum_{i=0}^{\lambda-1} 2^{-i} + \tau\cdot 2^{-\lambda} - 1$. Since $q_\lambda(\tau) \in S(p, \epsilon)$ for large λ it follows that $r(\sigma) \to p$ for $\sigma \to 1$. With $r(1) = p$ the curve $r(\sigma)$ is continuous and connects $p_{\mu(1)}$ to p; since $r(1 - 2^{-\lambda}) = p_{\mu(\lambda+1)}$ and

$$\lambda(r, 1 - 2^{-\lambda}, 1) = \sum_{\kappa>\lambda} \lambda(q_\kappa(\tau)) < \sum_{\kappa>\lambda} \bar{p}_{\mu(\kappa)}\bar{p}_{\mu(\kappa+1)} + \sum_{\kappa>\lambda} 2^{-\lambda} < 2^{-\lambda+1}$$

hence $\bar{p}_{\mu(\lambda+1)}\bar{p} \to 0$.

That Q *is* arc convex means that for any two points q, r of Q a curve $q(\tau)$ from q to r with $\lambda(q(\tau)) < \bar{q}\bar{r} + \epsilon$ exists, where λ is now measured with the distance π. Put $q(0) = q, q(1) = r$ and choose by (3.23) $q(\frac{1}{2})$ such that $\bar{q}(0)\bar{q}(\frac{1}{2}) < \bar{q}(0)\bar{q}(1)/2 + \epsilon/4$ and by induction with respect to μ the points $q((2\nu - 1)2^{-\mu})$, $\nu = 1, 2, \cdots, 2^{\mu-1}$ such that

$$\left.\begin{matrix}\bar{q}((\nu - 1)2^{-\mu+1})\bar{q}((2\nu - 1)2^{-\mu}) \\ \bar{q}((2\nu - 1)2^{-\mu})\bar{q}(\nu 2^{-\mu+1})\end{matrix}\right\} < \bar{q}((\nu - 1)2^{-\mu+1})\bar{q}(\nu 2^{-\mu+1})2^{-1} + \epsilon\cdot 4^{-\mu} < \bar{q}(0)\bar{q}(1)\cdot 2^{-\mu} + \epsilon\cdot 2^{-\mu}.$$

If $|\tau - \sigma| = \nu\cdot 2^{-\mu}$ then

$$\bar{q}(\tau)\bar{q}(\sigma) < \nu q(0)q(1)\cdot 2^{-\mu} + \nu\cdot 2^{-\mu}\cdot\epsilon = |\tau - \sigma|\cdot[\bar{q}(0)\bar{q}(1) + \epsilon].$$

The mapping $\tau \to \bar{q}(\tau)$ is therefore uniformly continuous for dyadic τ with $0 \leqq \tau \leqq 1$. Because Q is complete the mapping can be extended to a continuous mapping of the whole interval $0 \leqq \tau \leqq 1$ in Q and $\lambda(q(\tau)) < \bar{q}\bar{r} + \epsilon$ follows from the last inequality.

4. Area and Measure

It seems plausible that area of a non-parametrized set and its measure are closely related only when there are not "too many" pairs of points that cannot be connected by a recifiable curve on the set. The following examples will bear out this conjecture.

With cylindrical coordinates ρ, θ, ζ in E^3 and the euclidean distance η, let P be the surface $\zeta = \rho\cdot\sin(\pi/\rho)$, $\zeta(0) = 0$, $0 \leqq \rho \leqq 1$. Since no point with $\rho > 0$ can be connected with the origin by a rectifiable curve on P, the space $\bar{P}$ (that is P with the metric π) has two finiteness classes, one, $\bar{P}_0$, consisting of the origin, and the other, $\bar{P}_1$ of all remaining points. If P_δ denotes the subset $\delta \leqq \rho \leqq 1$ of P then it will be seen later that $|P_\delta|_2$ is the ordinary area and at the same time the two-dimensional Hausdorff measure:

$$|P_\delta|_2 = |P_\delta, \eta_3|_2 = 2\pi \int_\delta^1 \rho(1 + \zeta'^2)^{\frac{1}{2}} \delta\rho$$

hence by (1.8) and (1.13)

$$0 < |P|_2 = |\bar{P}_0, \pi|_2 + |\bar{P}_1, \pi|_2 = |\bar{P}_2, \pi|_2 = |P, \eta_3|_2 < \infty$$

and by (1.9)

$$|P|_n = |P, \eta_3|_n = 0 \text{ for } n > 2, |P|_n = |P, \eta_3|_n = \infty \text{ for } n < 2.$$

With the same coordinates consider the surface $\zeta = (1 - \rho) \sin (\pi/(1 - \rho))$, $0 \leqq \rho \leqq 1$, $\zeta(1) = 0$. Then $\bar{P}$ has again two finiteness classes, $\bar{P}_0$ consisting of the circle $\rho = 1$, $\zeta = 0$ and $\bar{P}_1$ comprising the rest of $\bar{P}$. If P_δ denotes the subset $0 \leqq \rho \leqq \delta < 1$, then, as above

$$0 < |P_\delta|_2 = |P_\delta, \eta_3|_2 = 2\pi \int_0^\delta \rho(1 + \zeta'^2)^{\frac{1}{2}} d\rho < \infty$$

hence $|P_\delta|_n = |P_\delta, \eta_3|_n = 0$ when $n > 2$, $= \infty$ when $n < 2$ and by (1.8) and (1.13)

$$|P|_n = |\bar{P}_0, \pi|_n + |\bar{P}_1, \pi|_n = |\bar{P}_0, \pi|_n + \lim_{\delta \to 1} |P_\delta, \pi|_n = |\bar{P}_0, \pi|_n + |P, \eta_3|_n$$

so that $|P|_n = |P, \eta_3|_n = \infty$ for $n \leqq 2$, $|P|_n = |P, \eta_3|_n = 0$ for $n > 2$.

In these examples area and measure coincide for all n because there are only two finiteness classes. Consider however, in an E^2 with the distance η_2 and rectangular coordinates x, y a curve $y = f(x)$, $0 \leqq x \leqq 1$, $0 \leqq f(x) \leqq 1$ where $f(x)$ is continuous and nowhere differentiable. Let P be the graph of the curve with η_2 as distance. Then no two points of P can be connected by a rectifiable curve in P, so that $\pi(q, r) = \infty$ for any points $q \neq r$ in P. Each point of P forms an infinity class and $|P|_n = \infty$ for all n (see (1.8)) whereas $|P, \eta_2|_1 = \infty \cdot |P, \eta_2|_n = 0$ for $n \geqq 2$ because of (1.17), and probably $|P, \eta_2|_n = 0$ for $1 < n < 2$. In any case this example shows that Hausdorff measure and area do not coincide.

The example confirms also that the present area is not monotone: if Q denotes the square $0 \leqq x \leqq 1, 0 \leqq y \leqq 1$ with the distance η_2, then $|Q|_2 = |Q, \eta_2|_2 = 1$, $|Q|_n = 0$, for $n > 2$ hence $|P|_n > |Q|_n$ for $n \geqq 2$, although $P \subset Q$.

These occurrences are inherent in the use of intrinsic area. A surface, no two points of which can be connected by a rectifiable curve (like $z = f(x) + f(y)$ with the above f), leads to exactly the same $\bar{P}$, namely a set of isolated points with the power of the continuum, and can therefore not be distinguished intrinsically from the curve.

This is no objection to the use of intrinsic area, it only shows that in such cases an additional discussion may become necessary, and that *a k-dimensional measure in n-dimensional space rather than an area* will be appropriate. For non-parametrized sets we may simply use Hausdorff measure, but difficulties as in the preceding example are also encountered for parametrized sets. For instance, let Q be the set $0 \leqq u_1 \leqq 1$, $0 \leqq u_2 \leqq 1$ of an E^2 and consider the surface $x_1 = u_1$, $x_2 = f(u_1)$ (with same f as before) in another E^2. Then $| Q, x |_n \geqq | x(Q) |_n = \infty$ for all n.[16]

It is therefore desirable to dispose of a measure for parametrized sets which for non-parametrized sets coincides with Hausdorff measure. Guided by the definition of area we obtain such a measure under the following conditions.

I′. *P is ordinary metric with distance $\rho(q, r)$ and union of a countable number of compact sets.*

II′. *R is an ordinary metric space with distance $\delta(x, y)$.*

III′. *$x(p)$ is a continuous mapping of P in R.*

An ϵ-chain $\{p_i\}$ from q to r in P is a finite number of points $p_0 = q, p_1, \cdots, p_n = r$ such that $\rho(p_i, p_{i+1}) < \epsilon$. Let $\lambda\{p_i\}$ denote the number

$$\sum_{i=1}^{n} \delta[x(p_{i-1}), x(p_i)]$$

and define for $q, r \in P$

$$\pi^\epsilon(q, r) = \begin{cases} \inf \lambda\{p_i\} & \text{for all } \epsilon\text{-chains } \{p_i\} \text{ from } q \text{ to } r \\ \infty & \text{if no } \epsilon\text{-chain from } q \text{ to } r \text{ exists.} \end{cases}$$

Then it is very easy to see that

$$(4.1) \qquad \pi^\epsilon(p_1, p_2) + \pi^\epsilon(p_2, p_3) \geqq \pi^\epsilon(p_1, p_3)$$

$$(4.2) \qquad \pi^\epsilon(q, r) \geqq \delta[x(q), x(r)]$$

$$(4.3) \qquad \pi^\epsilon(q, r) \geqq \pi^{\epsilon'}(q, r) \text{ if } \epsilon < \epsilon',$$

hence $| P, \pi^\epsilon |_n \geqq | P, \pi^{\epsilon'} |_n$ for $\epsilon < \epsilon'$, so that

$$(4.4) \qquad | P, x |_n^H = \lim_{\epsilon \to 0+} | P, \pi^\epsilon |_n \geqq | P, \pi^\epsilon |_n$$

exists. *$| P, x |_n^H$ is called the n-dimensional Hausdorff measure of (P, x)* (n is again any real non-negative number). This is justified by the fact that

(4.5) *In the non-parametric case $P \subset R$ and $x(p) = p$*

$$| P, x |_n^H = | P, \pi^\epsilon |_n = | P, \delta |_n \textit{ for } \epsilon > 0.$$

For if $\delta(q, r) < \epsilon$, then $\pi^\epsilon(q, r) = \delta(q, r)$, hence $| P, \pi^\epsilon |_n^\eta = |P, \delta |_n^\eta$ for $\eta < \epsilon$. If $P_1 \subset P$ and π_1^ϵ is the distance π^ϵ formed for (P_1, x) then obviously $\pi_1^\epsilon(q, r) \geqq$

[16] That *Lebesgue measure is not intrinsic* is shown by the following example: Let C and C' be Cantor sets on $0 \leqq X \leqq 1$, $y = 0$ of measures 0 and ½ respectively, and $q \to q'$ a homeomorphic mapping of C on C'. Connect q to the point $r = (0, 3)$ and q' to $r' = (0, 1)$, and map the segment qr proportionally on $q'r'$. Then $\pi(x, y) > \pi(x', y')$ for any two corresponding pairs of points x, y in $P = \bigcup_q qr$ and x', y' in $P' = \bigcup_{q'} q'r'$, but $0 = | P |_2^L < | P' |_2^L$. The remainder of this section will not be used later on.

$\pi^\epsilon(q, r)$ for $q, r \in P_1$ and therefore $| P_1, \pi^\epsilon |_n \leqq | P_1, \pi_1^\epsilon |_n$. Actually the equality sign holds.

For if $q, r \in S(p, \epsilon/2; \rho)$, then $\rho(q, r) < \epsilon$, hence by (4.2) $\pi^\epsilon(q, r) = \delta[x(q), x(r)]$. The condition I' implies that P is separable. Therefore P can be covered by a countable number of spheres $S_1, S_2, \cdots$ of radius $\epsilon/2$. Denote by S_k^* the set of all points p in P for which a point $q \in S_k$ with $\pi^\epsilon(p, q) = 0$ exists and define the Borel sets V_k in P by $V_1 = S_1^*$, $V_k = S_k^* - \bigcup_{i=1}^{k-1} S_i^*$. Then $P = \bigcup V_k$ and $\pi^\epsilon(p_i, p_k) > 0$ for $p_i \in V_i$, $p_k \in V_k$ and $i \neq k$. It then follows from (1.12), (1.14a) and $\pi^\epsilon(q, r) = \delta[x(q), x(r)]$ for $q, r \in V_k$ that

$$| P, \pi^\epsilon |_n = \sum_k | V_k, \pi^\epsilon |_n = \sum_k | x(V_k), \delta |_n.$$

This equation and $\pi_1^\epsilon(q, r) \geqq \pi^\epsilon(q, r)$ yield

$$| P_1, \pi^\epsilon |_n \leqq | P_1, \pi_1^\epsilon |_n = \sum_k | x(P_1 \cap V_k), \delta |_n = | P_1, \pi^\epsilon |_n$$

$$\leqq \sum_k | x(V_k), \delta |_n = | P, \pi^\epsilon |_n.$$

Therefore

(4.6) *If* $P_1 \subset P$, *then* $| P_1, \pi_1^\epsilon |_n = | P_1, \pi^\epsilon |_n$ *and* $| P_1, x |_n^H \leqq | P, x |_n^H$ *so that Hausdorff measure is monotone.* A consequence of (4.6) is

(4.7) If $P_1 \subset P_2 \subset \cdots$ *and* $\bigcup P_\nu = P$, *then* $\lim_\nu | P_\nu, x |_n^H = | P, x |_n^H$.

For, if π^ϵ denotes the distance π^ϵ corresponding to (P_ν, x), then by (1.13), (4.4), and (4.6)

$$| P, \pi^\epsilon |_n = \lim | P_\nu, \pi^\epsilon |_n = \lim | P_\nu, \pi_\nu^\epsilon |_n \leqq \lim | P_\nu, x |_n^H \leqq | P, x |_n^H.$$

It can now be seen that $| P, x |_n^H$ is *invariant under change of the parametrization*:

(4.8) If $p' \rightarrow p(p')$ is a topological mapping of P' on P and $x(p(p')) = x'(p')$, then $| P, x |_n^H = | P', x' |_n^H$.

By I$'$ compact sets P_ν exist with $P_1 \subset P_2 \subset \cdots$ and $\bigcup P_\nu = P$. On $P'_\nu = p^{-1}(P_\nu)$ the mappings $p' \rightarrow p(p')$ is uniformly continuous. Hence a $\delta > 0$ exists such that $\rho'(q', r') < \delta$ (where ρ' is the distance in P') and $q', r' \in P'_\nu$ implies $\rho(q, r) < \epsilon$, where $q = p(q')$, $r = p(r')$. If now $\{p'_i\}$ is a δ-chain in P' from q' to r', then $\{p(p'_i)\}$ is an ϵ-chain in P from q to r, therefore $\pi^\delta(q', r') \geqq \pi^\epsilon(q, r)$ and $| P'_\nu, x' |_n \geqq | P_\nu, x |_n$; by symmetry $| P'_\nu, x' |_n^H = | P_\nu, x |_n^H$. The assertion follows now from (4.7). A consequence of (4.5) and (4.8) is

(4.9) If the mapping $p \rightarrow x(p)$ is topological, then

$$| P, x |_n^H = | x(P), \delta |_n.$$

A useful link between Hausdorff measure and intrinsic area, $| P, x |_n^{0}$, can be obtained as follows: By (4.3) the limit

$$\pi^0(q, r) = \lim_{\epsilon \rightarrow 0+} \pi^\epsilon(q, r)$$

exists, hence we may put

$$| P, x |_n^0 = | P, \pi^0 |_n \geqq | P, x |_n^H.$$

In a sense $|P, x|_n^0$ is an intrinsic measure. It has the advantage over $|P, x|_n$ that it is defined also for P which are not arcwise connected, but it will in general be invariant only under such changes $p' \to p(p')$, $p \to p'(p)$ of the parametrization which are one-to-one and uniformly continuous both ways. Similarly the relation $|P, x|_n^0 \geqq |x(P)|_n^0$ will in general only hold when $p \to x(p)$ is uniformly continuous. Consequently $|P, x|_n^0$ is really useful only when P is compact. Examples given later on will corroborate this from a slightly different angle.

If P is arcwise connected and satisfies I′ then all three numbers $|P, x|_n$, $|P, x|_n^H$, and $|P, x|_n^0$ are defined. Since for any continuous curve $p(t)$, $\alpha \leqq t \leqq \beta$ values $t_0 = \alpha < t_1 < \cdots < t_n = \beta$ can be found such that $(p(t_i))$ is an ϵ-chain, and because $\lambda[x(p(t))] \geqq \sum \delta[x(p(t_i)), x(p(t_{i+1}))]$ it follows that $\pi(q, r) \geqq \pi^0(q, r)$ and

$$(4.10) \qquad |P, x|_n \geqq |P, x|_n^0 \geqq |P, x|_n^H.$$

Cases where the three numbers coincide will be found later on. It is immediately seen that this is the case for the Minkowski polyhedra defined in section 1. Moreover in the one-dimensional case all three numbers coincide with the arclength for any continuous curve:

(4.11) *If P is the interval $\alpha \leqq t \leqq \beta$ of the real axis with distance $|t_1 - t_2|$, then*

$$|P, x|_1 = |P, x|_1^0 = |P, x|_1^H = \lambda(x(t)).$$

Proof. Because of (3.18) and (4.10) it suffices to prove that

$$(4.12) \qquad |P, x|_1^H \geqq \lambda(x(t)).$$

Every ϵ-chain $t_0 = t', t_1, \cdots, t_l = t''$, $|t_{i+1} - t_i| < \epsilon$ with $t' < t''$ contains a sub chain $(t^j) = (t_{i_j})$ of the form $t^0 = t' < t^1 < \cdots < t^k = t''$ with $t^{j+1} - t^j < \epsilon$, $i_{j+1} > i_j$. Then $\sum \delta[x(t_i), x(t_{i+1})] \geqq \sum \delta[x(t^j), x(t^{j+1})]$, hence only ascending ϵ-chains need be considered.

Since generally $\alpha(\bigcup_{y \in x_i} S(y, \delta)) < \alpha(X_i) + 2\delta$ we may restrict ourselves to decompositions $\bigcup X_i = P^\epsilon$ (that is P with the metric π^ϵ) where the X_i are open in P^ϵ. Choose therefore an open covering $\bigcup X_i = P^\epsilon$ with $|P, \pi^\epsilon|_1 + \epsilon > \sum \alpha_\epsilon(X_i)$, where $\alpha_\epsilon = \sup_{x,y \in X_i} \pi^\epsilon(x, y)$. Because P^ϵ is compact a finite number of X_i covers P^ϵ. Assume that $X_1, \cdots, X_k$ are such X_i. Because of the continuity of $x(t)$ the sets X_i are all open considered as subsets of P with the distance $|t_1 - t_2|$.

Let i_1 be the first subscript among $1, \cdots, k$ for which $X_{i_1} \supset \alpha = t^0$. If $X_{i_1} \supset \beta$ we stop, otherwise put $t_1' = \sup t$ for $t \in X_{i_1}$. Then $\pi^\epsilon(t_0, t_1') \leqq \alpha_\epsilon(X_{i_1})$. Proceeding in this manner we reach a first subscript i_m for which $X_i \supset t_{m-1}$ and $X_i \supset \beta = t_m$. Because of $\pi^\epsilon(t_j', t_{j+1}') \leqq \alpha_\epsilon(X_{i_{j+1}})$ there are ascending ϵ-chains from t_j' to t_{j+1}' of length at most $\alpha_\epsilon(X_{i_{j+1}})$. These chains form together an ascending ϵ-chain of length at most $\sum_j \alpha_\epsilon(X_{i_j}) \leqq \sum_i \alpha_\epsilon(X_i) \leqq |P, \pi^\epsilon|_1 + \epsilon$. For $\epsilon \to 0$ we obtain (4.12).

The higher dimensional measures may differ for continuous curves. For let P the interval $0 \leqq t \leqq 1$ and $f(t)$ the same function as in the beginning of this section, and (P, x) the curve $(t, f(t))$ in the euclidean (x_1, x_2)-plane. Then $|P, x|_n = |x(P)|_n = |P, x|_n^0 = |x(P)|_n^0 = \infty$ for all n, but $|P, x|_1^H = |x(P), \eta_2|_1 = \infty$, $|P, x|_n^H = |x(P), \eta_2|_n = 0$ for $n \geqq 2$.

Examples where $|P, x|_n$ and $|P, x|_n^0$ differ are easy to find. In a euclidean (x, y)-plane let P with the metric η_2 consist of the segments which connect $(0, 1)$ to the point $(x, 0)$ with irrational x between 0 and 1. Then $|P|_n = \infty$ for all n, because any two different points $q_i = (x_i, 0)$ in P have distance $\pi(q_1, q_2) > 2$. But $\pi^0(q_1, q_2) = \eta_2(q_1, q_2)$ for any two points of P, hence $|P|_n^0 = \infty$, for $n < 2$, $|P|_2^0 = 2^{-1}$, $|P|_n^0 = 0$ for $n > 2$, moreover $|P|_n^0 = |P, \eta_2|_n$ for all n.

If the points $(x, 0)$ traverse, instead of the irrational numbers, a perfect nowhere dense set in the interval $0 \leqq x \leqq 1$, then $|P|_n = |P|_n^0 = \infty$ for all n, whereas $|P, \eta_2|_1 = \infty$, $|P, \eta_2|_2 < \infty$, $|P, \eta_2|_n = 0$ for $n > 2$, and the values of $|P, \eta_2|_n$ depend for $1 < n < 2$ on the choice of the perfect nowhere dense set (compare [12]).

These examples confirm that *$|P, x|_n^0$ is for non-compact P too unstable to be interesting in itself.* How it can be used for comparison in a different way from before may be seen from the following

(4.13) If P is compact and $\pi^0(q, r) < \infty$ for any q, r in P, then $x(P)$ is arcwise connected and $|P, x|_n^0 \geqq |x(P)|_n$.

Proof. Let $y = x(q)$, $z = x(r)$. For every r there is a ν^{-1}-chain $(p_i^\nu) = (p_1^\nu, \cdots, p_{k_\nu}^\nu)$ from q to r in P such that $\sum \delta[x(p_i^\nu), x(p_{i+1}^\nu)] \to \pi^0(q, r) < \infty$. Put $\delta[x(p_i^\nu), x(p_{i+1}^\nu)] = \delta_i^\nu$. Let $\sigma_1, \sigma_2, \cdots$ be the different rational numbers between 0 and 1 and $q^\nu = p_j^\nu$ the last p_j^ν (ν fixed) for which $\sum_{i=0}^{j-1} \delta_i^\nu \leqq \sigma_\mu \sum_{i=1}^{k_\nu} \delta_i^\nu$. Choose a subsequent $\{1\nu\}$ of $\{\nu\}$ for which $\{q^{1\nu}\}$ converges, then a subsequence $\{2\nu\}$ of $\{1\nu\}$ for which $\{q_2^{2\nu}\}$ converges, and so forth. If $\{\kappa\}$ denotes the diagonal sequence $\{\nu\nu\}$ then $r_\mu = \lim_{\nu\to\infty} q_\mu^\nu$ exists for every ν moreover

$$\sum_{i=1}^{j(\mu,\kappa)-1} \delta_i^\kappa \leqq \sigma_\mu \sum_{i=1}^{\kappa\kappa} \delta_i^\kappa < \sum_{i=1}^{j(\mu,\kappa)} \delta_i^\kappa. \tag{4.14}$$

If $\sigma_\lambda < \sigma_\mu$ then $j(\lambda, \kappa) < j(\mu, \kappa)$ and because of (4.14)

$$\delta(x(q_\lambda^\kappa), x(q_\lambda^\kappa)) < \sum_{j(\lambda,\kappa)}^{j(\mu,\kappa)} \delta_i^\kappa \leqq (\sigma_\mu - \sigma_\lambda) \sum_1^{\kappa\kappa} \delta_i^\kappa + 2/\kappa.$$

Putting $x(r_\lambda) = x_\lambda$ it follows that $\delta(x_\lambda, x_\mu) \leqq (\sigma_\mu - \sigma_\lambda)\pi^0(q, r)$ so that $\sigma_\lambda \to x_\lambda$ is a uniformly continuous mapping of $\{\sigma_\lambda\}$ on $\{x_\lambda\}$ which because of the compactness of $x(P)$ can be extended to a continuous mapping of the interval $0 \leqq \tau \leqq 1$ on a set x_τ in $x(P)$. Then x_τ is a continuous curve from y to z in $x(P)$ of length at most $\pi^0(q, r)$. This shows that

$$\pi^0(q, r) \geqq \pi(x(q), x(r)) \tag{4.15}$$

and proves the assertion.

The hypothesis of (4.13) does not imply that P itself is arcwise connected, for instance because $x(p)$ is always continuous, when $x(P)$ consists of one point.

(The assumption $\pi^0(q, r) < \infty$ implies only that P is well chained, see [24], p. 13.)

In the non-parametric case (4.13) is significant. In that case compactness of P may be replaced by finite compactness, because it may be assumed that $\sum_i \delta_i^\nu < \pi^0(q, r) + 1$ for all ν, so that all p_i^ν lie in $S(q, \pi^0(q, r) + 1)$ whose closure is compact. Therefore (4.6) and (4.15) yield

(4.16) *If P is finitely compact and $\pi^0(q, r) < \infty$, then P is arcwise connected and $\pi^0(q, r) = \pi(q, r)$, so that $| P |_n^0 = | P |_n$ for all n.*

5. Definition of rectifiable surfaces

While the general theory of rectifiable surfaces is involved and will be left to another paper, their definition and simplest properties are useful even for the study of smooth surfaces in Finsler spaces.

The word rectifiable surface is ordinarily used as follows; P is the unit cube of the E^k, R is the E^n and the mapping $x(p)$ satisfies for a suitable $\beta > 0$ the condition $\eta_n(x(q), x(r)) \leqq \beta\eta_k(q, r)$, where η again denotes euclidean distance.

If the surface is given in the form $z = f(x_1, \cdots, x_k)$ then rectifiablility means $| z(x') - z(x'') | \leqq \beta\eta_n(x', x'')$. This can be reduced to the preceding case by considering the x-space as the E^k and the (x, z)-space as the E^n because then $\eta_{n+1}[(x', z(x')), (x'', z(x''))] \leqq (1 + \beta)\eta_n(x', x'')$. However, in this case also $\eta_n(x', x'') \leqq \eta_{n+1}[(x', z(x')), (x'', z(x''))]$ so that a Lipschitz condition holds both for the mapping $x \to (x, z(x))$ and for its inverse. This leads to the following definitions:

(P, x) is a *k-dimensional rectifiable surface* (in the metric space R) if a topological mapping $p' \to p(p')$ of a convex set P' with interior points in E^k on P exists, such that

(5.1) $$\delta[x(p(q')), \quad x(p(r'))] \leqq \beta\eta_k(q', r').$$

Putting $x'(p') = x(p(p'))$ (5.1) becomes

(5.1a) $$\delta[x'(p'), x'(r')] < \beta\eta_k(q', r').$$

By (3.6) $| P, x |_n = | P', x' |_n$. We call (P', x') a *rectifiable representation* of (P, x). Let (P, x) be rectifiable and (P, x) itself a rectifiable representation. If $p(\tau)$ traverses all curves from q to r in P then (5.1a) shows

(5.2) $$\pi(q, r) = \inf \lambda x(p(\tau)) \leqq \beta \inf \lambda(p(\tau)) = \beta\eta_k(q, r)$$

so that $\pi(q, r)$ is finite and

(5.3) $$| P, x |_k \leqq \beta^k | P |_k^L, | P, x |_n = 0 \text{ for } n > k.$$

The last follows from the first and (3.5) when $| P |_k^L < \infty$. If $| P |_k^L = \infty$ the set P can be represented as a union of convex sets P_i with $| P_i |_k^L < \infty$, then $| P, x |_n \leqq \sum | P_i, x |_n = 0$ because of (3.14).

The relation (5.3) generalizes with the help of (1.14) to a subset Q of P:

(5.4) $$| Q, \pi |_k \leqq \beta^k | Q |_k^L, | Q, \pi |_n = 0 \text{ for } n > k.$$

Hence $H(Q) = | Q, \pi |_k$ is an absolutely continuous set function. It is additive (on a properly chosen class of sets) obviously when $q \neq r$ implies $\pi(q, r) \neq 0$; in that case $| P, x |_k$ (or $H(Q)$) can be represented as a Lebesgue integral over the set derivative of $H(Q)$. It seems difficult to find conditions under which additivity holds in the general case.

To have a different name we call (P, x) *a birectifiable surface* and (P', x') (where $x'(p') = x(p(p'))$) a *birectifiable representation* of (P, x) if a topological mapping $p' \rightarrow p(p')$ on a convex set P' in E^k with interior points exists such that

$$(5.5) \qquad \alpha\eta_k(q', r') \leqq \delta[x(p(q')), x(p(r'))] < \beta\eta_k(q', r').$$

This is essentially a non-parametric case and by (3.8) $| P, x |_n = | P', x' |_n = | x'(P') |_n = | x(P) |_n$. Lebesgue measurable sets in P' correspond to sets measurable with respect to $|$ sic, $\pi |_k$ in P, and the above mentioned integral representation holds.

Finally (P, x) is called a k-dimensional rectifiable (birectifiable) manifold if P is a topological k-manifold (for the terminology compare [15] p. 331) and every point p of P has a coordinate neighborhood $U(p)$ for which $(U(p), x)$ is a rectifiable (birectifiable) surface. Since P can be covered by a countable number of $U(p_i)$ it follows from (3.14) and (5.3) that

(5.6) $| P, x |_n = 0$ *for* $n > k$ *when* (P, x) *is a rectifiable k-manifold.*

Every birectifiable manifold has an integral representation for its area.

6. Volume in Finsler Spaces

The general theory developed in the preceding sections will now be applied to the case where R is a Finsler space F of the following type:

F is an n-dimensional manifold of class C^1.[17] On the contravariant vectors of F a continuous function Φ is defined, which in a fixed coordinate neighborhood (x) is written as $\Phi(x_1, \cdots, x_n, \xi^1, \cdots, \xi^n) = \Phi(x, \xi)$ and satisfies the conditions

1) $\Phi(x, \xi) > 0$ *for* $\xi \neq (0, \cdots, 0)$,
2) $\Phi(x, k\xi) = k\Phi(x, \xi)$ *for* $k > 0$,
3) *For fixed* x *the surface* $\Phi(x, \xi) = 1$ *in* ξ*-space is convex.*
4) $\Phi(x, \xi) = \Phi(x, -\xi)$.

The distance of two points y, z (not necessarily in one coordinate neighborhood) is usually defined as

$$(6.1) \qquad \delta(y, z) = \inf \int \Phi(x(\tau), \dot{x}(\tau))\, d\tau$$

where $x(\tau)$ traverses all curves of class D' from y to z in F. But it is important to observe that $\delta(y, z)$ does not change if $x(\tau)$ is permitted to traverse all absolutely continuous curves (that is, the $x_i(\tau)$ are absolutely continuous functions of τ); moreover, for any absolutely continuous curve

$$x(\tau), \lambda(x(\tau)) = \int \Phi(x, \dot{x}) d\tau.^{18}$$

---

[17] Compare [15], pp. 330, 331. Instead of basic parameter neighborhood the term coordinate neighborhood will be used.

[18] This follows from the equivalence and convergence theorems of Tonelli. Compare [20], p. 51, where the pertaining original literature is quoted.

Hence F is already arc convex, and F is identical with F (compare (3.19)).

Let p be a point in the coordinate neighborhood (x). Then $\mu_p^{(x)}(y, z) = \Phi(p, y - z)$ defines because of 1) to 4) a Minkowski space $M_p^{(x)}$ depending on p and (x). If $\Psi_1 = \min_{|\alpha|=1} \Phi(p, \alpha)$, $\Psi_2 = \max_{|\alpha|=1} \Phi(p, \alpha)$, then

$$\Psi_1 \eta_n(y, z) \leqq \mu_p^{(x)}(y, z) \leqq \Psi_2 \eta_n(y, z) \tag{6.2}$$

where $\eta_n(y, z) = [\sum_i (y_i - z_i)^2]^{\frac{1}{2}}$. The metric $\mu_p^{(x)}$ approaches δ in the neighborhood of p in the following sense (compare [7] pp. 184, 185).

(6.3) For a given η between 0 and q there is a $\rho > 0$ such that for $y, z \in S(p, \rho; \eta_n) \subset (x)$ (or $y, z \in S(p, \rho; \delta) \subset (x)$)

$$(1 - \eta)\mu_p^{(x)}(y, z) \leqq \delta(y, z) \leqq (1 + \eta)\mu_p^{(x)}(y, z).$$

(6.2) and (6.3) show that F is a Lipschitz manifold, so that by (5.4) and (1.15) $| X, \delta |_m = | X, \delta |'_m = | X, \delta |''_m = 0$ for $m > n$ and $X \subset F$, in particular $|F|_m = 0$ for $m > n$. Since (1.14) holds for the three functions $| X, \delta |_n$, $| X, \delta |'_n$, and $| X, \delta |''_n$ it follows from (6.3) that

$$(1 - \eta)^n | X, \mu_p^{(x)} |_n < \begin{Bmatrix} | X, \delta |_n \\ | X, \delta |'_n \\ | X, \delta |''_n \end{Bmatrix} < (1 + \eta)^n | X, \mu_p^{(x)} |_n .$$

If $| X |_n^L$ denotes Lebesgue measure in (x) with metric $\eta_n = | x^1 - x^2 |$ and $(\Xi^{(x)}(p))^{-1} = | S(p, 2^{-1}, \mu_p^{(x)}) |_n^L$, then by (2.3) $| X, \mu_p^{(x)} |_n = \omega^{(n)}\Xi^{(x)}(p) \, | X |_n^L$. Therefore the last inequality shows that $| X, \delta |_n$, $| X, \delta |'_n$, and $| X, \delta |''_n$ are absolutely continuous set functions with respect to $| X |_n^L$ and have the same continuous derivative $\omega^{(n)}\Xi^{(x)}(p)$. Although $| X, \delta |''_n$ is in general spaces not additive it is easy to see that

$$| X, \delta |_n = | X, \delta |'_n = | X, \delta |''_n = \omega^{(n)} \int_X \Xi^{(x)}(x) \, dx_1 \cdots dx_n \tag{6.5}$$

for any measurable subset X of (x). But then (6.5) can in the known fashion be extended to any measurable set X in F. In particular

$$| F | = | F, \delta |_n = \omega^{(n)} \int_F \Xi^{(x)}(x) \, dx.$$

If F is a Riemann Space, that is, if $\Phi(x, \xi) = \sum [g_{ik}(x)\xi_i\xi_k]^{\frac{1}{2}}$ the ellipsoid $\sum g_{ik}\xi^i\xi^k = \frac{1}{4}$ has volume $\omega^{(n)} |g_{ik}(x)|^{-\frac{1}{2}}$, when $| g_{ik} |$ is the determinant of the g_{ik}, so that (6.5) takes the well known form

$$| X, \delta |_n = |X, \delta |'_n = | X, \delta |''_n = \int_X | g_{ik}(x) |^{\frac{1}{2}} \, dx_1 \cdots dx_n .$$

Let G be a connected open subset of F. Then G is a sub-manifold of class C^1 of F and becomes with the same integrand a Finsler Space. The formula (6.5) shows that

$$| G, \delta |_n = | G |_n \tag{6.6}$$

In section 4 it was observed that this is in general not true for other than open sets. If the subsets $G_1, G_2, \cdots$ of F are open and disjoint and $| F - \bigcup G_i, \delta |_n = 0$ then $| F | = \sum | G_i |$; in that case areas can be added. A simple case in which $| F - \bigcup G_i, \delta |_n = 0$, is $F - \bigcup G_i \subset \bigcup T_\nu$, where $T_\nu = y_\nu(P_\nu)$ and (P_ν, y_ν) is a rectifiable manifold of dimension $k_\nu < n$. For (3.13) and (5.6) show $| T_\nu |_n \leqq | P_\nu, y_\nu |_n = 0$ hence $| T_\nu, \delta |_n = 0$. The same argument extends (6.6) to arcwise connected sets Y which are not open but whose boundary $B(Y)$ is contained in a set $\bigcup T_\nu$ of the above type. For if $Y = G + B(Y)$, where G is either open or empty, then by (3.13) and (3.14)

$$| Y, \delta |_n \leqq | Y |_n \leqq | G |_n + \sum | T_\nu |_n = | G |_n = | G, \delta |_n \leqq | Y, \delta |_n .$$

(6.7) *If the set* $Y \subset F$ *is arcwise connected and its boundary* $B(Y) \subset \bigcup y_\nu(P_\nu)$, *where* (P_ν, y_ν) *is a rectifiable manifold of dimension less than* n, *then* $| Y |_n = | Y, \delta |_n = \int_Y \Xi^{(x)}(x)\, dx$. *If* $F = \bigcup Y_\nu$, $Y_\nu \cap Y_\mu = 0$ *and* Y_ν *has the properties of* Y, *then* $| F | = \sum | Y_\nu |$.

This additivity suffices amply for all purposes of differential geometry. As a manifold of class C^1 the set F can be triangulated or covered by C^1-complex $\{C_i^p\}$ (compare [15] p. 333), where the C_i^p may denote the simplices of dimension p without their faces. If D_i^n denotes any set containing C_i^n and contained in the closure of C_i^n, then $| F |_n = \sum | D_i^n |_n$.

7. Area in Finsler Spaces

The preceding results permit to determine the area of regular manifolds of class C^1 in F. To define them let P be a k-manifold of class C^1 and $x(p)$ a continuous mapping of P in F. Every point $q \in P$ has a neighborhood $U(p)$ whose image $x(U(p))$ lies entirely in one coordinate neighborhood (x) of F. A suitable subneighborhood $V(p) \subset U(p)$ of p will lie in one coordinate neighborhood (u) of P. Hence the basic covering of P by coordinate neighborhoods may be assumed to be such that the image of any coordinate neighborhood (u) in P lies entirely in one coordinate neighborhood (x) of F. In (u) the mapping $x(p)$ takes the form

$$x_i = x_i(u_1, \cdots, u_k), i = 1, 2, \cdots, n.$$

If the functions x_i have continuous first partial derivatives we call (P, x) a k-manifold of class C^1 in F.

If in addition the matrix $\left(\frac{\partial x_i}{\partial u_j}\right)$ has everywhere rank k (then $k \leqq n$) (P, x) is called regular.

Let (P, x) be k-manifold of class C^1 in F. If $u(\tau)$ is any absolutely continuous curve in P, then

$$\lambda[x(u(\tau))] = \int \Phi\left(x(u), \sum \frac{\partial x}{\partial u_j} \dot{u}_j\right) d\tau. \tag{7.1}$$

Since curves which have no absolutely continuous parametrization make the left side of (7.1) infinite, it follows that

$$\pi(q, r) = \inf \int \Phi\left(x(u), \sum \frac{\partial x}{\partial u_j} \dot{u}_j\right) d\tau \tag{7.2}$$

where $u(\tau)$ traverses all absolutely continuous curves from q to r in P.

If now

$$\Psi(u, \eta) = \Phi\left(x(u), \sum \frac{\partial x}{\partial u_j} \eta^j\right) \tag{7.3}$$

satisfies the conditions 1) to 4) of section 6, then P becomes with the integrand $\Psi(u, \eta)$ a Finsler space, and the right side of (7.2) (compare (6.1)) is the distance of q and r in that Finsler space.

Obviously $\Psi(u, k\eta) = k\Psi(u, \eta)$ for $k > 0$. But condition 1) is satisfied if and only if P is a regular C^1-manifold. For since Φ satisfies 1), $\Psi(u, \eta) = 0$ for $(\eta^1, \cdots, \eta^k) \neq (0, \cdots, 0)$ if and only if $\sum_j \frac{\partial x_i}{\partial u_j} \eta^j = 0$, which is equivalent to the rank of $\left(\frac{\partial x^i}{\partial u_j}\right)$ being less than k.

Let P be regular. Then it remains to show that $K: \Psi(u, \eta) = 1$ is for fixed u a convex surface in η-space. Under the affine transformation $\xi^i = \sum \frac{\partial x_i}{\partial u_j} \eta^j$ of the η-space on the k-plane E in ξ-space spanned by the k vectors $\frac{\partial x}{\partial u_j}$ the indicatrix K goes into the intersection K' of $\Phi(x, \xi) = 1$ with E. Since K' is convex K is convex. We have found

(7.4) *If P is a regular C^1-manifold in F, then $\bar{P}$ is the Finsler space defined by P and the integrand* (7.3).

This justifies considering $\pi(q, r)$ as the intrinsic distance on (P, x). If with the preceding notations $(\mathrm{H}^{(u)}(u))^{-1}$ and $(\Xi^{(u,x)}(u))^{-1}$ denote the k-dimensional areas bounded by K in η-space and by K' in E respectively, then [19]

$$\mathrm{H}^{(u)}(u) : \Xi^{(u,x)}(u) = \Delta^{(u,x)}(u), \quad \text{where} \quad \Delta^{(u,x)}(u) = \left[\sum_{j_1 < \cdots < j_k} \left(\frac{\partial(x_{j_1}, \cdots, x_{j_k})}{\partial(u_1, \cdots, u_k)}\right)^2\right]^{\frac{1}{2}}. \tag{7.5}$$

Consequently (6.5) yields for any measurable set Y in P

$$| Y, \pi |_k = \omega^{(k)} \int_Y \mathrm{H}^{(u)}_{(u)}\, du \cdots du_k = \omega^{(k)} \int_Y \Xi^{(u,k)}(u)\, \Delta^{(uk)}(u)\, du_1 \cdots du_k. \tag{7.6}$$

In the case of a regular k-manifold P of class C^1 in E^n the integrand $\Phi(x, \xi)$ is simply $|\xi| = (\sum (\xi^i)^2)^{\frac{1}{2}}$. The intersection of any k-plane through 0 is a

[19] A detailed proof for the general case in [11], p. 155.

unitsphere of area $\omega^{(k)}$ so that (7.6) *becomes the standard expression* $\int_P \Delta^{(u,x)}(u)\, du$.

If P is a regular k-manifold in a Riemann Space, then area of P is usually defined (compare for instance [7], p. 57) as volume of P with the induced metric (7.3). Since it was already seen that (6.5) gives the correct answer for volume, it follows that the middle term of (7.6) *coincides with the usual expression for area in Riemann Spaces.*

The right side of (7.6) was given by Choquet in [5] as area in the case of a two-dimensional manifold in a three-dimensional Finsler space. Since it equals $| Y, \pi |_k$ it has certain invariance properties which are rather laborious to establish directly: besides invariance under change of the x and the u, it must be kept in mind, that the evaluation of $\Xi^{(u,x)}(u)$ refers to the metric $\eta_n(\xi^1, \xi^2) = |\xi_1 - \xi_2|$ used in E as a subset of ξ-space; that is, the δ_{ik} ($\delta_{ii} = 1$, $\delta_{ik} = 0$ for $i \neq k$) are arbitrarily chosen as g_{ik} to establish a Euclidean metric. Any other choice of constant g_{ik} can claim the same right as the δ_{ik}, as long as $\sum g_{ik}\xi^i\xi^k$ is positive definite. Using the right side of (7.6) as definition of area would require an analytic proof of invariance under the choice of the g_{ik}.

As in section 2 it may be remarked that the analytic expressions in (7.6) remain meaningful even if Φ does not satisfy the condition $\Phi(x, \xi) = \Phi(x, -\xi)$, because the two different types of spheres in the Minkowski space $M^{(x)}_{x(u)}$ (compare section 6) through which $\Xi^{(u,x)}(u)$ is determined yield the same value for that number. But area would then not coincide with the generally adopted value for arclength for one-dimensional P.

(6.7) leads to a similar statement regarding additivity of area for k-manifolds in F. In many cases the assumption that $\left(\frac{\partial x_i}{\partial u_j}\right)$ has rank k is not satisfied throughout, it ceases to hold on certain singular loci. The following will take care of this possibility:

(7.7) *P is a k-dimensional manifold of class C^1, $x(p)$ maps P continuously in a Finsler space F in such a way that* 1) *P is the union of a set H and mutually disjoint connected open sets G_ν,* 2) *(G_ν, x) is a regular manifold of class C^1 (in F), and $H \subset \bigcup Q_i$ where (Q_i, x) is a rectifiable manifold of dimensions less than k.*[20] *Then* $| P, x | = \sum | G_\nu, x |$.

For (6.2) applied to G_ν and the integrand (7.3) shows that $\pi(q_\nu, q_\mu) > 0$ if $q_\nu \in G_\nu$, $q \in G_\mu$ and $\nu \neq \mu$. Let $p \in G_\nu$ and (u) a coordinate neighborhood of p which is contained in G_ν. Choose $\rho > 0$ such that (6.3) holds in $S(p, \rho; \eta_k)$ (always with respect to the integrand $\Psi(u, \eta)$). Since segments of an associated euclidean metric are shortest connections for a Minkowski metric it follows from (6.2) that for sufficiently small σ and $q, r \in S(p, \sigma; \eta_k)$ the distance $\pi(q, r)$ equals the distance of q and r as points of (G_ν, x). Therefore $| G_\nu, x | = | G_\nu, \pi |_k$, and by (3.14) and (5.6)

$$\sum | G_\nu, \pi |_k \leqq | P, \pi |_k = | P, x | \leqq \sum | Q_i, x |_k + \sum | G_\nu, x | = \sum | G_\nu, \pi |_k .$$

[20] This assumption is less restrictive than it appears at first sight, because it is not required that (Q_i, x) is a rectifiable representation.

A similar idea leads to the concept of a *piecewise Minkowskian space* which is important for the discussion of uniqueness. An n-dimensional manifold P of class C^1 is metrized as follows: $\{C_i^j\}$ is a C^1-covering of P. A function Φ is defined on the contravariant vectors with origins in $\bigcup C_i^n$ (the C_i^j are again the simplices without their faces), and for each C_i^n there is a coordinate neighborhood (x^i) in which $|x^i|$ ∴ bounded and which covers all of C_i^n and for which Φ does not depend on $(x_1^i, \cdots, x_n^i)$ but only on ξ. The distance of any two points y, z in P is defined as

$$\delta(y, z) = \inf \int \Phi(x, \dot{x})\, d\tau$$

where $x(\tau)$ traverses all curves of class D^1 from y to z in P such that $x(\tau) \in C_i^j$ with $j < n$ only for a finite number of τ's.

P is arc convex (hence $\bar{P} = P$). Each C_i^j with $j < n$ is a rectifiable j-manifold in P. As in the preceding proof every point of C_i^n has a neighborhood in which $\delta(y, z)$ coincides with the metric which C_i^n has as a Finsler space with the integrand Φ. But as such C_i^n is Minkowskian and has by (1.14) (applied twice) the elementary area $\omega^{(n)}\Xi^{(x^i)}\,|\,C_i^n, x^i\,|_n^L$, where $|\,C_i^n, x^i\,|_n^L$ denotes the Lebesgue measure of C_i^n evaluated in the x^i-space (in which $\Xi^{(x^i)}(x^i)$ is constant). We found

A) *An n-dimensional piecewise Minkowskian space P has the elementary area* $|\,P\,| = \omega^{(n)} \sum_i \Xi^{(x^i)}\,|\,C_i^n, x^i\,|_n^L$.

This statement postulated as an axiom together with the simple intrinsic form K) of Kolmogoroff's Principle (compare section 3) determine the area of C^1-manifolds in Finsler spaces completely. An axiom analogous to K) may be stated as follows:

B) *If (P, x) and (P, x') are two surfaces with intrinsic distances π and π' such that $\pi(q, r) \geqq \pi'(q, r)$ for $q, r \in P$, then the area of (P, x) is at least as large as the area of (P, x').*

The *Uniqueness Theorem* is this

If a surface area satisfies axioms A *and* B *then it coincides for regular manifolds of class C^1 in Finsler spaces with $|\,P, x\,|$.*

For let (P, x) b a k-dimensional regular manifold of class C^1 in a Finsler space F. It was proved before that $\pi(q, r)$ coincides with the distance which P has as Finsler space with the integrand $\Psi\,(u, \eta)$ of (7.3). For any $\epsilon > 0$ there is a C^1-covering (C_i^j) of P with these properties: each C_j^k lies entirely in one coordinate neighborhood (u^i), there is a point $\bar{u}^i$ in C_i^k such that

$$(1 - \epsilon)\Psi(\bar{u}^i, \eta) \leqq \Psi(u, \eta) \leqq (1 + \epsilon)\Psi(\bar{u}^i, \eta) \tag{7.8}$$

and

$$\left| \int_P \mathrm{H}^{(u)}(u)\, du - \sum_i \mathrm{H}^{(u^i)}(\bar{u}^i)\,|\,C_i^k, u^i\,|_k^L \right| < \epsilon. \tag{7.9}$$

With $(1 - \epsilon)\Psi(\bar{u}^i, \eta)$ (or $(1 + \epsilon)\Psi(\bar{u}^i, \eta)$) as integrands in C_i^k the manifold P becomes a piecewise Minkowski space P_ϵ^- (or P_ϵ^+) with distance δ^-(or δ^+). Although the definition of δ^- and δ^+ admits only curves which cross the C_i^j

with $j < k$ a finite number of times it is obvious that $\delta^-(q, r) \leqq \pi(q, r) \leqq \delta^+(q, r)$.

Therefore the area of (P, x) lies by B between the volume of P_ϵ^- and P_ϵ^+, but by A, (7.8) and (2.3)

$$|P_\epsilon^\pm| = \omega^{(k)} \sum_i (1 \pm \epsilon)^k \mathrm{H}^{(u^i)}(\bar{u}^i) \,|\, C_i^k, u^i \,|_k^L$$

so that by (7.9) the area of (P, x) lies between the numbers

$$(1 \pm \epsilon)^k \omega^{(k)} \left(\int_p \mathrm{H}^{(u)}(u)\, du \pm \epsilon \right).$$

The relation (7.6) shows now that the area of (P, x) must be $|P, x|$.

This result and repeated application of A) and B) allow to find more and more surfaces whose area is uniquely determined by A and B. The extent of the class of all these surfaces is not known.

As long as only such P are considered which can be covered by one coordinate system, it suffices to require A for ordinary euclidean triangulations and for piecewise Minkowski spaces in which the (x^i) always coincides with (x). For Riemann spaces the additional restriction may be added that the Minkowski metrics in the C_i^n are all euclidean. But even then A implies more than that euclidean polyhedra in E^n have the elementary area, since it is not required that the euclidean metrics in C_i^n which have a common face C_k^j induce the same metric on C_k^j.

8. Area of birectifiable surfaces

The definitions of birectifiable surface, birectifiable representation and birectifiable manifold are found in section 5. The purpose of this section is to prove the following theorem

(8.1) *If (P, x) is a k-dimensional birectifiable surface in a Finsler space with distance δ, then* $|P, x| = |x(P), \delta|_k$.

The following two theorems follow from (8.1).

(8.1a) *If (P, x) is a k-dimensional birectifiable manifold in a Finsler space and the mapping $p \to x(p)$ is topological then* $|P, x| = |x(P), \delta|_k$.

(8.2) *If (P, x) is birectifiable representation of a birectifiable surface in E^n, then*

$$|P, x| = \int_P \Delta(u)\, du, \quad \textit{where} \quad \Delta(u) = \left[\sum_{j_1 < \cdots < j_k} \left(\frac{\partial(x_{j_1}, \cdots, x_{j_k})}{\partial(u_1, \cdots, u_k)} \right)^2 \right]^{\frac{1}{2}}.$$

This formula extends to birectifiable manifolds in E^n (without the assumption that $p \to x(p)$ is topological), only $x(u)$ means then a local birectifiable representation.

First it will be shown that (8.1a) and (8.2) follow from (8.1). Under the assumption of (8.1a) every point p has a coordinate neighborhood U such that (U, x) is a birectifiable surface. In U choose a $(k-1)$-dimensional convex polyhedron T such that p lies in the interior V of T. Then for each of the closed $(k-1)$-faces T' of T the surface (T', x) is a $(k-1)$-dimensional birectifiable surface, so that $|T, x|_k = 0$. Now P can be covered by a countable number of V's; $V_1, V_2, \cdots$ with corresponding boundaries $T_1, T_2, \cdots$. Put $\bar{V}_i = V_i \cup T_i$ and $W_1 = V_1$, $W_2 = V_2 - V_2 \cap \bar{V}_1$ and generally $W_n = V_n - V_n \cap (\bigcup_{i=1}^{n-1} \bar{V}_i)$. The W_n are open and disjoint and $P = (\bigcup W_i) \cup (\bigcup T_i)$. Each W_i lies in one coordinate neighborhood and is an open set union of disjoint open cubes Q and their closed $(k-1)$-dimensional faces S. Each (Q, x) is a k-dimensional, and each (S, x) a $(k-1)$-dimensional birectifiable surface. Let $Q_1, Q_2, \cdots$ be all the cubes obtained from the different W_i and $S_1, S_2, \cdots$ all their $(k-1)$-dimensional faces. Then P is union of the Q_i, the S_j and the T_l. Because $p \to x(p)$ is topological the sets $x(Q_i)$ are disjoint, hence by (8.1)

$$|P, x| \leqq \sum |Q_i, x| + \sum |S_j, x|_k + \sum |T_l, x_k| = \sum |x(Q_i), \delta|_k \leqq |x(P), \delta|_k$$

and (8.1a) follows from (3.13).

This same consideration shows that the remark after (8.2) follows from (8.2). The assumption that $x \to x(p)$ is topological is not necessary because of the additivity of the integral.

(7.2) follows from (8.1) because it is well known that the integral representation holds for the Hausdorff measure $|x(p), \eta_n|_k$ compare [14], p. 362.

For the *proof of* (8.1) observe first that

$$(8.3) \qquad |P, x| = |x(P)|_k \geqq |x(P), \delta|_k,$$

compare section 5.

If (P, x) is a birectifiable representaion, the P is a convex domain with interior points in E^k, and if $u = (u_1, \cdots, u_k)$ are cartesian coordinates in E^k, then

$$(8.4) \qquad \beta_1 |u' - u''| \leqq \delta[x(u'), x(u'')] \leqq \beta_2 |u' - u''| \text{ for } u', u'' \in P$$

where $x(u)$ lies in some Finsler space F. The theorem is obvious when $|P|_k^L = \infty$. Assume therefore $|P|_k^L < \infty$, and let a positive $\epsilon < \min(1, |P|_k^L/4)$ be given.

If P_ν is the subset of those points in P for which $\eta_k(u, E^k - P) > \nu^{-1}$, then $|P_\nu|_k^L \to |P|_k^L$, hence m can be chosen so that $|P - P_m|_k^L < \epsilon$.

For any unitvector $\alpha = (\alpha_1, \cdots, \alpha_k)$ and $-m^{-1} < h < m^{-1}$ the point $u + \alpha h$ is in P when $u \in P_m$. Choose a finite number of unitvectors $\overset{1}{\alpha}, \overset{2}{\alpha}, \cdots, \overset{\theta}{\alpha}$ such that for any unitvector α an $\overset{i}{\alpha}$ with $|\alpha - \overset{i}{\alpha}| < \epsilon$ exists. For a constant vector γ put $\overset{i}{y}(\tau) = x(\overset{i}{\alpha}\tau + \gamma)$ for $\overset{i}{a}\tau + \gamma \in P$. Since $\beta_1 |h| \leqq \delta[\overset{i}{y}(\tau + h), \overset{i}{y}(\tau)] \leqq \beta_2 |h|$, the curve $\overset{i}{y}(\tau)$ is absolutely coninuous in τ so that for almost all τ

$$(8.5) \qquad \lim_{h \to 0} \frac{\lambda(\overset{i}{y}, \tau, \tau + h)}{\delta[\overset{i}{y}(\tau + h), \overset{i}{y}(\tau)]} = 1.$$[21]

Well known methods show then that a set P' in P_m with $|P'|_k^L = |P_m|_k^L$ exists such that (8.5) holds for $i = 1, \cdots, \theta$ and all γ for which $\overset{i}{\alpha}\tau + \gamma \in P'$.

[21] For the notation see the proof of (3.18) for the proof of (8.5) see footnote 18).

Since $\lambda(y^i, \tau, \tau + h) \geqq \pi(\alpha^i(\tau + h) + \gamma, \alpha^i\tau + \gamma) \geqq \delta[y^i(\tau + h), y^i(\tau)]$ it follows that for $u \epsilon P'$ also

$$(8.6) \qquad 1 \leqq \Delta(u, h) = \max_i \frac{\pi(u + \alpha^i h, u)}{\delta[x(u + \alpha^i h), x(u)]} \to 1.$$

Moreover $\Delta(u, h)$ is continuous for $u \epsilon P'$ and $0 < |h| < m^{-1}$. Let P'' be a compact subset of P' with $|P' - P''|_k^L < \epsilon$, then $\Delta(u, h)$ is uniformly continuous for $u \epsilon P''$ and $(\nu + 1)^{-1} \leqq |h| \leqq \nu^{-1} \leqq m^{-1}$. Following a known pattern we choose the values h_i^ν with $(\nu + 1)^{-1} = h_1^\nu < h_2^\nu < \cdots < h_{k_\nu}^\nu = \nu^{-1}$ such that

$$(8.7) \qquad |\Delta(u, h) - \Delta(u, h_i^\nu)| < \nu^{-1} \text{ for } h_i^\nu < |h| < h_{i+1}^\nu \text{ and } u \epsilon P''$$

and form the sequence $\{h'_\mu\} = h_1^m, -h_1^m, \cdots, h_{k_1}^m, -h_{k_1}^m, h_1^{m+1}, -h_1^{m+1}, \cdots$. Then $\Delta(u, h'_\mu) \to 1$ for $u \epsilon P''$. By Egoroff's Theorem there is a compact subset P^* of P'' with $|P' - P^*|_k^L < \epsilon$, hence $|P - P^*|_k^L < 3\epsilon$, on which $\Delta(u, h'_\mu)$ converges uniformly. It follows from (8.7) that $\Delta(u, h)$ tends on P^* uniformlv to 1 for $h \to 0$.

Choose $h_0 > 0$ such that $\Delta(u, h) - 1 < \epsilon$ for $0 < |h| < h_0$. If $u \epsilon P^*$, $0 < |h| < h_0$ and α is any unitvector, and α^i is such that $|\alpha - \alpha^i| < \epsilon$, then by (5.2) and (8.4)

$$\begin{aligned} \pi(u + \alpha h, u) &\leqq \pi(u + \alpha h, u + \alpha^i h) + \pi(u + \alpha^i h, u) \leqq \beta_2 |\alpha - \alpha^i| \cdot |h| \\ &+ (1 + \epsilon)\delta[x(u + \alpha^i h), x(u)] \leqq \beta_2\epsilon \cdot |h| + (1 + \epsilon)\{\delta[x(u + \alpha^i h), x(u + \alpha h) \\ &+ \delta[x(u + \alpha h), x(u)]\} \leqq (1 + \epsilon)\delta[x(u + \alpha h), x(u)] + \beta_2\epsilon |h|(2 + \epsilon) \end{aligned}$$

Therefore because of (8.4)

$$\pi(u + \alpha h, u) \leqq \delta[x(u + \alpha h), x(u)][1 + \epsilon(1 + 3\beta_2\beta_1^{-1})].$$

This relation, (8.4) and (1.14) show with $\sigma = 1 + 3\beta_2\beta_1^{-1}$ that

$$|P^*, \pi|_k < |x(P^*), \delta|_k(1 + \epsilon\sigma)^k.$$

Hence because of $|P - P^*|_k^L < 3\epsilon$.

$$|P, \pi|_k = |P, x| \leqq |x(P), \delta|_k(1 + \epsilon\sigma)^k + \beta_2^k \cdot 3\epsilon,$$

so that $|P, x| \leqq |x(P), \delta|_k$ and (8.1) follows from (8.3).

(8.2) contains the fact that *Lebesgue area* (or Peano area, $\cdots$) *of a birectifiable surface* (P, x) *in* E^n *is intrinsic* in spite of its non intrinsic definition. It is worthwhile to formulate the implications explicitly: If for the surface (P', x') in any metric space R a topological mapping $p' \to p(p')$ of P' on P exists such that $\pi(p'(q), p'(r)) = \pi(q, r)$ then $|P', x'|$ can be evaluted by approximating (P, x) in the known fashion by polyhedra in E^n.

Smith College

References

[1] Berwald, L. and P. Funk, *Flächeninhalt und Winkel in der Variationsrechnung.* Lotos 67/68, Prag (1919–1920), pp. 45–49.

[2] Besicovitvh, A. S., *On the definition and value of the area of a surface.* Quart. Jour. of Math., 16 (1945), pp. 86–102.

[3] Bliss, G. A., *Generalization of geodesic curvature and a theorem of Gauss concerning geodesic triangles.* Am. Jour. of Math., 37 (1915) pp. 1–18.
[4] Bonnesen, F. and W. Fenchel, *Theorie der konvexen Körper.* Ergeb. der Math., v. 3, part 1, 1934.
[5] Bouligand, G. and G. Choquet, *Problèmes liès a des mètriques variationnelles.* C. R., Acad. Sci. Paris, 218 (1944), pp. 696–698.
[6] Busemann, H., *Intrinsic area.* Proc., Nat. Acad. Sci., 32 (1946), pp. 5–8.
[7] Busemann, H. and W. Mayer, *On the foundations of calculus of variations*, Trans., Am. Math. Soc., 49 (1941), pp. 173–198.
[8] Duschek, A. and W. Mayer, *Lehrbuch der Differentialgeometrie*, vol. II. Leipzig, Teubner, 1930.
[9] Federer, H., *Surface area* I. Trans., Am. Math. Soc., 55 (1944), pp. 420–437.
[10] Grüss, G., *Beiträge zur Differentialgeometrie zweidimensionaler allgemein-metrisher Flächen*, Math. Ann., 103 (1930), pp. 162–184.
[11] Haupt, O., Differential- und Integralrechnung, vol. III. Berlin, De Gruyter, 1938.
[12] Hausdorff, F., *Dimension und äusseres Mass.* Math. Ann., 79 (1919), pp. 157–179.
[13] Hurewicz, W. and H. Wallman, Dimension theory, Princeton Univ. Press, 1941.
[14] Kolmogoroff, A., *Beiträge zur Masstheorie.* Math. Ann., 107 (1932), pp. 351–366.
[15] Lefschetz, S., *Algebraic topology.* Am. Math. Soc., Coloquium Publ., vol. 27, 1942.
[16] Loomis, L. H., *Abstract congruence and the uniqueness of Haar measure.* Ann. Math., 46 (1945), pp. 348–355.
[17] Loomis, L. H., *The intrinsic measure theory of Riemannian and euclidean metric spaces.* Ann. Math., 45 (1944), pp. 367–374.
[18] Nöbeling, G., *Über den Flächeninhalt dehnungsbeschränkter Flächen.* Math. Zeits., 48 (1943), pp. 747–771.
[19] Nöbeling, G., *Über Flächenmasse im Euklidischen Raum.* Math. Ann. 118 (1941/3), pp. 687–701.
[20] Pauc, C., *La mèthode directe en calcul des variations.* Actualities Sci., 885, Paris, 1941.
[21] Pontrjagin, L., Topological groups. Princeton Univ. Press, 1939.
[22] Radò, T., *On the problem of Plateau.* Ergeb. der Math., v. 2, part 2, Berlin, 1933.
[23] Saks, S., Theory of the integral. Warsaw, 1937.
[24] Whyburn, G. T., *Analytic topology.* Am. Math. Soc., Colloquium Publ. vol. 28, 1942.

Ann. Math. 48 (1947), 234–267

THE ISOPERIMETRIC PROBLEM IN THE MINKOWSKI PLANE.*

By Herbert Busemann.

The slow progress in the theory of Finsler spaces as compared to Riemann spaces is partly due to lack of information regarding the corresponding local, that is the Minkowskian, geometry. Those Minkowskian features will contribute most to an understanding of Finsler spaces which are not merely *verbal* generalizations of *known* euclidean statements.[1] The purpose of the present note was originally only to show that the isoperimetric problem (for any dimension) in Minkowski spaces leads to such a feature.

It turned out, however, that the plane problem can be solved in a general form—no longer significant for Finsler spaces—and then exhibits a phenomenon which is of interest for the theory of isoperimetric problems in the calculus of variations. The result seems to indicate that the standard methods may have followed too closely the pattern of the fixed endpoint problem. For that reason the plane case is here presented separately.

The following are the results: let $F(x,y)$ be continuous, positive for $x, y \neq 0$, and positive homogeneous of order 1. The problem, to find among all simple closed curves $x(t), y(t)$ with a given orientation and a given Minkowski length $L = \int F(\dot{x}, \dot{y})\,dt$ one which bounds the greatest (euclidean) area, *has a unique solution* (up to translations) *no matter whether the indicatrix* $C : F(x,y) = 1$ *is convex or not.* For non-convex C the solution is the same as for the boundary $\bar{C}$ of the convex closure of C as indicatrix and *is homothetic to the polar reciprocal* (figuratrix) *of* $\bar{C}$ *with respect to the unit circle rotated through* $\pm\,\pi/2$.

For Finsler spaces only the case is of interest where C is convex and has the origin as center. Since Finsler or Minkowski area differs from the euclidean area by a constant factor, *the solution of the Minkowskian isoperimetric problem is the same as for the above problem.* In intrinsic Minkowskian terms it may be described as the curve of length L for which as new

* Received January 22, 1947.

[1] Since the euclidean geometry is a special Minkowskian geometry, every theorem on general Minkowskian spaces is a generalization of some euclidean fact.—It has been stated that the so-called relative differential geometry is the differential geometry of Minkowski space. This is however not so. Relative length or area are not Minkowski length or area.

unit circle (indicatrix) perpendicularity (transversality) is the reverse of perpendicularity with respect to C. *It is in general not a Minkowski circle.*[2]

The result for non-convex C suggests the question: what is the relation between the lengths defined by C and $\bar{C}$ as indicatrices respectively. An answer is given in the final section for general n-dimensional Finsler spaces to the effect that *the corresponding Lebesgue lengths are equal.*

1. The form and the uniqueness of the solution for a *convex indicatrix* C can be derived in a few lines from the *Brunn-Minkowski Theory.* If $x = r\cos\theta$, $y = r\sin\theta$ and $F(\cos\theta, \sin\theta) = \rho(\theta)$, then $r = \rho^{-1}(\theta)$ is the equation of C in polar coordinates.

Let D be an analytic closed convex curve which contains the origin in its interior. If $h(x,y)$ is the supporting function of D then $h(\theta) = h(\cos\theta, \sin\theta)$ is the distance of the origin from the tangent of D at the point q where the exterior normal of D has direction θ. The radius of curvature of D at q is $h(\theta) + h''(\theta)$ (see [1, p. 65]) so that the euclidean line element of D at q equals $(h(\theta) + h''(\theta))d\theta$. Therefore the Minkowski length $L(D)$ of D is

$$L(D) = \int_0^{2\pi} (h(\theta) + h''(\theta))\rho(\theta + \delta\pi/2)\,d\theta \tag{1}$$

where $\delta = 1$ ($\delta = -1$) if the orientation of D is positive (negative). From now on we fix the orientation of D and give δ the corresponding value.

If C is convex then $F(x,y)$ is convex, hence $H(x,y) = r\rho(\theta + \delta\pi/2)$ is convex and therefore the supporting function of a certain convex curve K. Then $L(D)$ is by (1) twice the mixed area $A(D,K)$ of D and K (see [4, p. 245]),

$$L(D) = 2A(D,K). \tag{2}$$

Both $L(D)$ and $A(D,K)$ do not change when D or K undergo translations ([1, p. 40]) and depend continuously on D. Since every convex curve can be approximated by analytic convex curves ([1, section 27]), the relation (2) holds for any convex curve D. The Theorem of Brunn-Minkowski yields ([1, p. 97]) that

$$A^2(D,K) \geq A(D)A(K), \tag{3}$$

[2] This fact was first noticed by S. Ulam who found, and communicated to the author, that the solution for $F(x,y) = \max(|x|, |y|)$ is $|x| + |y| = \text{const.}$ In this connection it is of interest that *the Minkowski circle is the solution of the problem to find among all curves with a given Minkowski diameter one which bounds the greatest area,* compare [2, section 2].

where $A(D)$ and $A(K)$ are the areas bounded by D and K respectively, and the equality sign holds only when D and K are homothetic. The relations (2) and (3) yield the isoperimetric inequality

$$(4) \qquad L^2(D)/A(D) \geq 4A(K).$$

If D is a general simple closed curve with the given orientation, then $L(D)$ is to be defined by the Weierstrass sum. If D is not rectifiable in the euclidean sense, then $L(D) = \infty$. If D is rectifiable and suitably parametrized (for instance by the arc length) then the Weierstrass sum coincides with the Lebesgue integral $\int F(\dot{x}, \dot{y})\,dt$ along D (see [5, p. 51]). If C is convex the euclidean straight lines are shortest connections. Hence, if D is not convex the boundary of the convex closure of D has by the usual arguments at most the length of D and bounds a greater area, so that (4) holds for any simple closed curve with the given orientation, and the equality still holds only for D homothetic to K.

It is well known that the curve $\bar{K}$ with the supporting function $F(x, y)$ (the so called figuratrix of F) originates from C by a polar reciprocity with respect to the euclidean unit circle (see [4, pp. 146, 147]). Then K is obtained from $\bar{K}$ by a rotation through $-\delta\pi/2$ about O.

2. This way of generating K leads to the property of K which permits the discussion of *non-convex indicatrices.* Call tangent of K a supporting line of K which is at the same time a right hand or left hand tangent of K at one of its common points with K. Then (if C is convex)

(5) No tangent of K is parallel to a radius of C from O to an interior point of a (euclidean) segment ab which lies on C.

For the polars of the different points of ab with respect to $x^2 + y^2 = 1$ pass through one point p and are supporting lines of $\bar{K}$ at p. Hence a supporting line $\bar{S}_x$ of $\bar{K}$ which corresponds to an interior point x of ab cannot be a tangent of $\bar{K}$. After rotation of $\bar{K}$ through $-\delta\pi/2$ this fact becomes (5).

Let now C be not convex and call $\bar{F}(x, y)$ the positive, positive homogeneous function of degree 1 for which $\bar{F}(x, y) = 1$ is the boundary $\bar{C}$ of the convex closure of C. Then $\bar{F}(x, y) \leq F(x, y)$ and if $\bar{L}(D)$ denotes the length of D measured by $\bar{F}$,

$$(6) \qquad \bar{L}(D) \leq L(D).$$

If K denotes the curve originating from $\bar{C}$ by a polar reciprocity in $x^2 + y^2 = 1$ and rotation through $-\delta\pi/2$, then by (4) and (6)

$$L^2(D)/A(D) \geq \bar{L}^2(D)/A(D) \geq 4A(K)$$

and the equality signs can hold only if D is homothetic to K and $L(D) = \bar{L}(D)$. *The last relation holds always for D homothetic to K.*

For if $u(t), v(t)$ represents K with the euclidean arc length t as parameter, then u and v exist except for an at most countable number of t's and

$$L(D) = \int F(\dot{u}, \dot{v})\,dt, \qquad \bar{L}(D) = \int \bar{F}(\dot{u}, \dot{v})\,dt.$$

If $F(x,y) > \bar{F}(x,y)$, then the ray from O to (x,y) intersects C in a point where C has no supporting line, and therefore $\bar{C}$ in an interior point of a segment ab which is part of $\bar{C}$. The statement (5) shows now that $F(u,v) = \bar{F}(\dot{u}, \dot{v})$ for any t for which $\dot{u}$ and $\dot{v}$ exist, which proves $L(D) = \bar{L}(D)$. The following has been proved

Theorem 1. *Let $F(x,y)$ be continuous, positive for $x, y \neq 0$ and positive homogeneous of degree 1. If $r = \bar{\rho}^{-1}(\theta)$ is the polar equation of the boundary $\bar{C}$ of the convex closure of $F(x,y) = 1$, then for any simple closed curve D*

$$L^2(D)/A(D) \geq 2 \int_0^{2\pi} (\bar{\rho}^2(\theta) - \bar{\rho}'^2(\theta))\,d\theta \tag{7}$$

where $L(D)$ is length in terms of F and $A(D)$ is the euclidean area.

The equality sign holds for positively (negatively) traversed D only if D is homthetic to the polar reciprocal K of C with respect to $x^2 + y^2 = 1$ rotated about O through $-\pi/2\,(\pi/2)$.

The right side of (7) equals $4A(K)$ by a known integral representation of area in terms of the supporting function, see [1, p. 66]. It is true that this reference states the formula only for analytic curves, but its validity for general convex curves follows from the fact that if the analytic convex curves $r = \rho_\nu^{-1}(\theta)$ tend to $r = \rho(\theta)$, then their derivatives ρ'_ν tend automatically to ρ', see [1, p. 35].

3. Area is defined only for symmetric Finsler spaces [see 2, section 2]. Therefore, as far as Minkowski geometry is concerned, Theorem 1 is interesting only when C is convex and has O as center $(F(x,y) = F(-x,-y))$. In that case the (non-oriented euclidean straight) line g is said to be perpendicular to the line h (or h transversal to g) for the Minkowski geometry with unit circle C, in a formula: $g \perp_C h$, if the parallel to h through O intersects C in those points where C has supporting lines parallel to h. This terminology is due to the fact that the intersection of g and h minimizes the Minkowski

distance of a given point on g from a variable point on h. If K is **defined** as before, then the statements

(8) $g \perp_C h$ *and* $h \perp_K g$ *are equivalent*

This may be seen as follows: If p is a point of C and h is a supporting line of C at p let the euclidean normal to h from the origin O intersect h at f and $\bar{K}$ at $\bar{p}$. By the definition of $\bar{K}$ the line Op must intersect a supporting line $\bar{h}$ of $\bar{K}$ at $\bar{p}$ at a right angle. Denote the intersection by $\bar{f}$. If $\bar{p}$, $\bar{f}$, $\bar{h}$, are transformed into p_1, f_1, h_1 after rotation through $-\pi/2$ about O, then Op is parallel to h_1 and Op_1 is parallel to h. This means: if $Op = g$ is an arbitrary radius of C and h is a supporting line of C at p then the supporting line h_1 of K parallel to g contains a point p_1 for which $g_1 = Op_1$ is parallel to h, which is equivalent to (8).

Minkowski area (measure) is the euclidean area (Lebesgue measure) multiplied by π and divided by the euclidean area $A(C)$. Hence, if $M(D)$ denotes the Minkowski area bounded by D, then

$$M(D) = 2\pi A(D) / \int_0^{2\pi} \rho^{-2}(\theta)\, d\theta,$$

where $r = \rho^{-1}(\theta)$ is the polar equation of C. The relation (7) yields then the *Minkowskian Isoperimetric Inequality*

$$L^2(D)/M(D) \geq \pi^{-1} \int_0^{2\pi} (\rho^2 - \rho'^2)\, d\theta \int_0^{2\pi} \rho^{-2} d\theta, \tag{9}$$

and the equality sign holds only for the curves K which satisfy (8).

The right side of (9) is a Minkowski invariant, hence it does not change if the x, y undergo a non-degenerate affine transformation.

For the theory of Finsler spaces it is interesting to find out in which cases the Minkowski circles solve the isoperimetric problem. By (8) the symmetry of the relation $g \perp_C h$ is necessary and sufficient. The analytic equivalent to this statement is that the supporting function $h(\theta)$ of C must be proportional to $\rho(\theta + \pi/2)$. The factor of proportionality is easily evaluated by expressing $A(C)$ first in terms of the radius vector and then in terms of the supporting function of C. It is then found that

$$h(\theta) = \left[\int_0^{2\pi} \rho^{-2}(\theta)\, d\theta\right]^{1/2} \cdot \left[\int_0^{2\pi} (\rho^2(\theta) - \rho'^2(\theta))\, d\theta\right]^{-1/2} \rho(\theta + \pi/2). \tag{10}$$

All curves with a continuous non-vanishing curvature (the Legendre condition of the calculus of variations) which satisfy (10) have been determined by Radon in [6]. But there are other curves with this property, for instance the regular n-gons with $n \equiv 2 \bmod 4$.[3]

To have examples for the preceding theory it is useful to construct K for any convex polygon C, not necessarily with O as center. Since the convex closure of any polygon is again bounded by a polygon, this construction will yield K for arbitrary polygonal C. To obtain the counterclockwise solution corresponding to C let $p_1, p_2, \cdots, p_n$ be the vertices in counterclockwise order (see figure). Orient the radii

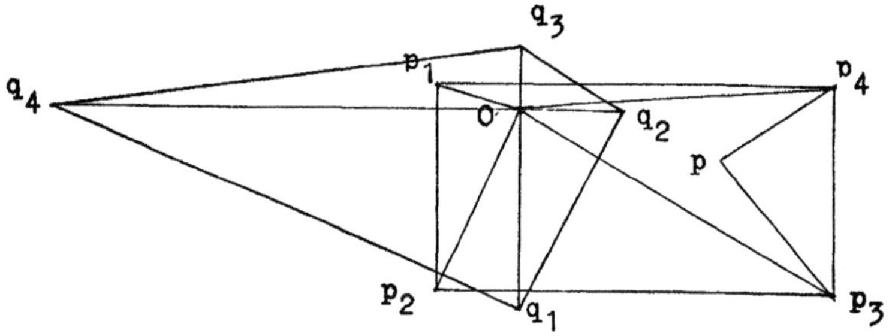

of C from p_i to O. Then K is the polygon $q_1, \cdots, q_n$ such that the oriented side $q_{i-1}q_i$ is parallel to p_iO and Oq_i is parallel to p_ip_{i+1} (subscripts are to be reduced mod n if necessary). In the figure the polygon K is also the solution for the polygon $C' = (p_1p_2p_3pp_4)$ which has C as boundary of its convex closure. If ab denotes also the euclidean distance of the points a, b and β_i the angle $p_{i-1}Op_i$, the following formula determines the lengths of the sides of K:

$$(11) \qquad \frac{L(K)}{2A(K)} = q_{i-1}q_i \frac{Op_{i-1}{}^{-1} \sin \beta_{i+1} - Op_i{}^{-1} \sin (\beta_i + \beta_{i+1}) + Op_{i+1}{}^{-1} \sin \beta_i}{\sin \beta_i \sin \beta_{i+1}}.$$

This relation can be obtained through solving the isoperimetric problem for the polygon $p_1, \cdots, p_n$ directly by means of elementary calculus and trigonometry. (11) contains the inequality (3), including the discussion of the equality sign, and yields thus a new elementary proof for the Brunn-Minkowski inequality for polygons. By a limit process the inequality for general convex curves is obtained, but the "only if" part in the condition for the equality sign is lost. For a regular n-gon (11) yields

[3] While this paper was in print a geometric construction for *all* curves C with symmetric $\perp_C$ was given by M. M. Day in: "Some characterizations of inner-product spaces," *Transactions of the American Mathematical Society*, vol. 62 (1947), pp. 320-337. The author seems to be unaware of the earlier work of Radon, Helly, and Blaschke on symmetric transversality.

$$L^2(D)/M(D) \geq 4\pi^{-1}n^2 \sin^2(\pi/n)$$

which becomes for $n \to \infty$ the ordinary isoperimetric inequality.

4. The discussion in Section **2** leads to the question: what is for arbitrary D the relation between $L(D)$ and $\bar{L}(D)$. This question will be answered completely in the present section for general n-dimensional Finsler spaces because it throws some light on the implications of the Lebesgue type of definition of length and area.

Let $F(x_1, \cdots, x_n; \alpha^1, \cdots, \alpha^n) = F(x, \alpha)$ be defined and continuous for all contravariant vectors (x, α) of an n-dimensional manifold of class 1, positive for $\alpha \neq 0$, and positive homogeneous of order 1 in the α^i. The Lebesgue F-length of a continuous curve $x(t)$ is defined as

$$\lambda_F^L(x(t)) = \inf [\liminf \int F(x_\nu(t), \dot{x}_\nu(t))\,dt]$$

for all sequences $\{x_\nu(t)\}$ of curves of class D' which tend to $x(t)$ in the sense of Fréchet.

With $F(x, \alpha)$ we associate a quasiregular integrand $\bar{F}(x, \alpha)$ as follows.[4] For fixed x (in a definite coordinate neighborhood) let $\bar{C}_x$ be the boundary of the convex closure of the indicatrix $F(x, \alpha) = 1$ in α-space. Then $\bar{F}(x, \alpha)$ is the integrand whose indicatrix (in the same x-system) is $\bar{C}_x$. The relation between $L(D)$ and $\bar{L}(D)$ is then determined by the following general theorem.

Theorem 2. *For any continuous curve* $x(t)$

$$\lambda_F^L(x(t)) = \lambda_{\bar{F}}^L(x(t)).$$

Since $\bar{F}$ is quasiregular the ordinary $\bar{F}$-length $\lambda_{\bar{F}}^L(x(t))$ of a curve[5] coincides with the Lebesgue length $\lambda_{\bar{F}}^L(x(t))$, so that Theorem 2 may also be formulated as

Theorem 2′. *For any continuous curve* $x(t)$

$$\lambda_F^L(x(t)) = \lambda_{\bar{F}}(x(t)).$$

---

[4] For further details see [3]. The following considerations follow closely Section 3 of that paper. The present Theorem 2 is a generalization of Theorem 1, [3, p. 184].

[5] This means again the limit of the Weierstrass sum $\sum \bar{F}(x(t_i), x(t_{i+1}) - x(t_i))$ which for absolutely continuous $x(t)$ coincides with the Lebesgue integral $\int \bar{F}(x, \dot{x})\,dt$. The F-length λ_F is defined correspondingly. Regarding these questions compare [5, p. 51].

Proof. Since $\bar{F}(x, \alpha) \leq F(x, \alpha)$ and $\lambda_{\bar{F}}$ is lower semicontinuous it follows that for a suitable sequence of curves $x_\nu(t)$ of class D' which tend to x

$$\lambda^{L}_{\bar{F}}(x) = \lim \lambda_{\bar{F}}(x_\nu) \geq \liminf \lambda_{\bar{F}}(x_\nu) \geq \lambda_{\bar{F}}(x).$$

Since $\lambda^{L}_{\bar{F}}$ is lower semicontinuous it suffices to prove $\lambda_{\bar{F}}(x) \geq \lambda^{L}_{\bar{F}}(x)$ for x which are of class D'. For then with a suitable sequence of curves of class D' which tend to x

$$\lambda_{\bar{F}}(x) = \lim \lambda_{\bar{F}}(x_\nu) \geq \liminf \lambda^{L}_{\bar{F}}(x_\nu) \geq \lambda^{L}_{\bar{F}}(x).$$

Since both lengths are additive it is also sufficient to consider a curve $x(t)$, $a \leq t \leq b$, of class C' which lies entirely in one coordinate neighborhood. If the norm of the partition $[t] = [t_0 = a < t_1 < t_2 < \cdots < t_{N+1} = b]$ tends to 0

$$\Sigma \bar{F}(x(t_i), x(t_{i+1}) - x(t_i)) \to \lambda_{\bar{F}}(x).$$

The direction of α is called quasiregular for $F(x, \alpha)$ if $F(x, \alpha)$ possesses at its point of the form $\delta\alpha$, $\delta > 0$, a supporting plane. *The relation $F(x, \alpha) = \bar{F}(x, \alpha)$ holds if and only if α is quasiregular.* If α is not quasiregular choose $\delta > 0$ such that $\delta\alpha$ lies on $\bar{F}(x, \alpha) = 1$. Then m, $2 \leq m \leq n$, regular directions $\beta_1, \cdots, \beta_m$ exist such that

$$\delta\alpha = \Sigma\delta_j\beta_j, \qquad \delta_j > 0, \qquad \Sigma\delta_i = 1 \tag{12}$$

For $F(x, \alpha) = 1$ is connected, hence n points $\beta_1, \cdots, \beta_n$ on $F(x, \alpha) = 1$ exist such that the $(n-1)$-simplex spanned by $\beta_1, \cdots, \beta_n$ contains $\delta\alpha$ (see [1, p. 9]). If $\delta\alpha$ is not in the interior of the simplex choose the notation so that the $(m-1)$-simplex spanned by $\beta_1, \cdots, \beta_m$ contains $\delta\alpha$ in its interior. There is a hyperplane through $\delta\alpha$ which is simultaneously a supporting plane P of $F(x, \alpha) = 1$ and $\bar{F}(x, \alpha) = 1$ (see [1, p. 6]). Then P cannot separate any two of the points $\beta_1, \cdots, \beta_n$ and must therefore contain all the points $\beta_1, \cdots, \beta_m$, so that all these points are regular. This proves (12).

Applying this result to $\alpha_i = x(t_{i+1}) - x(t_i)$ yields relations of the form

$$\alpha_i = \sum_j \delta_{ij}\beta_j, \qquad \delta_{ij} > 0.$$

Because F is homogeneous and the simplex spanned by $\beta_{i1} \cdots, \beta_{im_i}$ lies by construction on $\bar{F}(x, \alpha) = 1$,

$$\bar{F}(x(t_i), \alpha_i) = \sum_j \bar{F}(x(t_i), \delta_{ij}, \beta_{ij}) = \sum_k \bar{F}(x(t_i), y_{ik+1} - y_{ik})$$

where y_{ik} is the point $x(t_i) + \sum_{j=1}^{k} \delta_{ij}\beta_{ij}$. Since all the β_{ik} are quasiregular

$$F(x(t_i), y_{ik+1} - y_{ik}) = \bar{F}(x(t_i), y_{ik+1} - y_i).$$

If $p(t)$ is the euclidean polygon with vertices

$$y_{00} = x(a), y_{01}, \cdots, y_{0m_0} = x(t_1), y_1, \cdots, y_{1m_1} = x(t_2), \cdots, y_{Nm_N} = x(b)$$

then $\int F(p, \dot{p})dt$ will, for sufficiently small norm of $[t]$, differ arbitrarily little from

$$\sum_i \sum_k F(x(t_i), y_{ik+1} - y_{ik}) = \sum \bar{F}(x(t_i), x(t_{i+1}) - x(t_i)).$$

Hence, as the norm of $[t]$ tends to 0, the curve $p(t)$ tends to $x(t)$ in the sense of Fréchet and

$$\int F(p, \dot{p})\,dt \to \int \bar{F}(x, \dot{x})\,dt$$

which proves the theorem, because $\lambda \frac{L}{F}(x) \leq \lim \int F(p, \dot{p})\,dt.$

A consequence of Theorem 2, which is contained in the results of [3] and which can be generalized to planes as minimal surfaces in Minkowski spaces is this:

For any integrand $F(\alpha)$ in a Minkowski space the straight lines are shortest connections of their endpoints for Lebesgue F-length.

This is true since they are shortest connections for $\bar{F}$-length.

UNIVERSITY OF SOUTHERN CALIFORNIA.

REFERENCES

[1] T. Bonnesen and W. Frenchel, "Theorie der konvexen Körper," *Ergebnisse der Mathematik*, vol. 3, part 1, Berlin, 1934.
[2] H. Busemann, "Intrinsic area," *Annals of Mathematics*, vol. 48 (1947), pp. 234-267.
[3] H. Busemann and W. Mayer, "On the foundations of calculus of variations," *Transactions of the American Mathematical Society*, vol. 49 (1941), pp. 173-198.
[4] H. Minkowski, *Gesammelte Abhandlungen*, vol. II, Leipzig, 1911.
[5] C. Pauc, "La méthode métrique en calcul des variations," *Actualités Scientifiques et Industrielles*, No. 885, Paris, 1941.
[6] J. Radon, "Ueber eine besondere Art ebener konvexer Kurven," *Berichte über die Verhandlungen der Königlichen Sächsischen Gesellschaft der Wissenschaften zu Leipzig*, vol. 68 (1916), pp. 123-168.

A THEOREM ON CONVEX BODIES OF THE BRUNN-MINKOWSKI TYPE

By Herbert Busemann

University of Southern California, Los Angeles

Communicated by S. Lefschetz, October 7, 1948

The present note establishes a theorem on convex bodies in the n-dimensional Euclidean space E^n which is a counterpart to the Brunn-Minkowski Theorem. The following formulation of the latter [compare Bonnesen-Fenchel, *Theorie der konvexen Körper*, Berlin, 1934 (quoted as B. F.), pp. 71, 72] will exhibit the analogy:

In E^n let K be a convex body with interior points and P a two-dimensional half-plane bounded by the line L. Let the hyperplane normal to L intersect K in the non-empty set $H \cap K$ with $(n-1)$-dimensional volume $V(H \cap K)$. If $V^{1/(n-1)}(H \cap K)$ is laid off from the point $H \cap L$ on the ray $H \cap P$, then the resulting curve in P is convex (and turns its concavity towards L).

The present theorem is concerned with the intersection of K by pencils of non-parallel hyperplanes:

I. *In E^n let K be a convex body with interior points, L an $(n-2)$-dimensional linear space which intersects K, and P a 2-dimensional space normal to L at a point 0. Any half-hyperplane H bounded by L intersects K in a non-empty set $H \cap K$. If the $(n-1)$-dimensional volume $V(H \cap K)$ of $H \cap K$ is laid off from 0 on the ray $H \cap P$ then the resulting curve C is convex* (C contains 0 in its interior if L contains interior points of K, otherwise P lies on C).

Both theorems remain true, but are trivial if K has no interior points.

Notice the following Corollary of I:

II. *Let K be a convex body in E^n with 0 as interior point and center. If for any hyperplane H through 0 the volume $V(H \cap K)$ is laid off from 0 on the normal to H at 0 (in both directions), then the resulting surface S is convex.*

II follows from I because it suffices to know that the intersection of S with a two-dimensional plane P through 0 is a convex curve C'. But C' originates from the C of I by revolving C in P about 0 through $\pi/2$ and then dilating it in the ratio 2:1.

The result II is decisive for the theory of area in Finsler spaces and more specifically in Minkowski (or finite dimensional Banach) spaces. Its applications will be found in the author's series of articles on "Curvature and Angle in Non-Riemannian Spaces" which is appearing elsewhere.

Theorem I is trivial for $n = 2$, so that $n \geqslant 3$ will be assumed in the proof. It may also be assumed that L contains interior points of K, because the other case follows from this one by an obvious limit process. Then $V(H \cap K) > 0$ for every half-hyperplane H through L.

Let $\mathfrak{r}$ and $\mathfrak{s}$ be non-opposite rays in P with origin 0, and $\mathfrak{q}$ a ray with origin 0 in the angle with legs $\mathfrak{r}$ and $\mathfrak{s}$. Let the segment connecting the points $r \in \mathfrak{r}$, $s \in \mathfrak{s}$ $(r, s \neq 0)$ intersect $\mathfrak{q}$ at q. If $\angle(\mathfrak{r}, \mathfrak{q}) = \alpha'$, $\angle(\mathfrak{q}, \mathfrak{s}) = \beta'$ then the fact that the area of the triangle ors equals the sum of the areas of the triangles orq and oqs yields

$$or\, os \sin(\alpha' + \beta') = or\, oq \sin \alpha' + oq\, os \sin \beta',$$

hence

$$oq = or\, os \sin(\alpha' + \beta')/(or \sin \alpha' + os \sin \beta'). \tag{1}$$

Let H_r, H_s, H_q denote the half-hyperplanes bounded by L and through $\mathfrak{r}, \mathfrak{s}, \mathfrak{q}$, respectively. Then (1) shows that I is equivalent to the statement

$$V(H_q \cap K) \geqslant \frac{\sin(\alpha' + \beta')\, V(H_r \cap K)\, V(H_s \cap K)}{\sin \alpha' V(H_r \cap K) + \sin \beta' V(H_s \cap K)} \tag{2}$$

With the notations $\alpha = \sin \alpha'$, $\beta = \sin \beta'$, $\delta = \sin(\alpha' + \beta')$ and $V(H_q \cap K) = D$, $V(H_r \cap K) = A$, $V(H_s \cap K) = B$, the relation (2) becomes

$$D \geqslant \delta AB/(\alpha A + \beta B). \tag{3}$$

Let $\overline{A}_x(\overline{B}_y)$ denote the set at which the $(n - 2)$-dimensional space parallel to L through the point r with $or = x$ on $\mathfrak{r}$ (s with $os = y$ on $\mathfrak{f}$) intersects K. If $A_x(B_y)$ is the $(n - 2)$-dimensional measure of $\overline{A}_x(\overline{B}_y)$, then

$$A = \int A_x dx \qquad B = \int B_y dy.$$

Following the original idea of Brunn for the proof of the Brunn-Minkowski Theorem, we map the interval of $\mathfrak{r}$ for which $\overline{A}_x \neq 0$ on the interval of $\mathfrak{f}$ for which $B_y \neq 0$ through the relations

$$\rho A = \int_0^x A_t dt \qquad \rho B = \int_0^y B_t dt, \qquad 0 \leqslant \rho \leqslant 1 \tag{4}$$

so that

$$\frac{dx}{d\rho} = \frac{A}{A_x} \qquad \frac{dy}{d\rho} = \frac{B}{B_y}, \qquad 0 \leqslant \rho < 1. \tag{5}$$

Since K is convex it contains for every ρ the convex closure C_ρ of $\overline{A}_{x(\rho)}$ and $\overline{B}_{y(\rho)}$. Since $\overline{A}_x$ and $\overline{B}_y$ lie in $(n - 2)$-dimensional spaces parallel to L, the set C_ρ lies in the hyperplane H_ρ through these two spaces. H_ρ intersects H_q in an $(n - 2)$-dimensional space parallel to L and $\mathfrak{q}$ itself in a point whose distance $u(\rho)$ from 0 is by (1)

$$u(\rho) = \delta x(\rho)y(\rho)/[\alpha x(\rho) + \beta y(\rho)], \tag{6}$$

hence by (5)

$$\frac{du}{d\rho} = \delta\frac{\alpha x^2 y' + \beta y^2 x'}{(\alpha x + \beta y)^2} = \delta\frac{\alpha B x^2 B_y^{-1} + \beta A y^2 A x_x^{-1}}{(\alpha x + \beta y)^2}. \tag{7}$$

The intersection $\overline{C}_{u(\rho)}$ of C_ρ with H_q lies in $\overline{D} = H_q \cap K$ because $\overline{C}_{u(\rho)} = C_\rho \cap H_q \subset K \cap H_\rho \cap H_q = \overline{D} \cap H_\rho$. Therefore, if $C_{u(\rho)}$ is the $(n - 2)$-dimensional measure of $\overline{C}_{u(\rho)}$,

$$D \geqslant \int C_u du = \delta \int_0^1 C_{u(\rho)} \frac{\alpha B x^2 B_y^{-1} + \beta A y^2 A_x^{-1}}{(\alpha x + \beta y)^2} d\rho. \tag{8}$$

In the original triangle orq we find, if $V = \angle\, oqr$, that

$$or : rq = \sin V : \alpha, \qquad os : qs = \sin(\pi - V) : \beta,$$

hence

$$rq : qs = or \cdot \alpha : os \cdot \beta. \tag{9}$$

If (9) is applied to the points q, r, s with $oq = u(\rho)$, $or = x(\rho)$, $os = y(\rho)$ then $rq : qs$ equals the ratio of the distances d_r, d_s of $H_\rho \cap H_q$ from $H_\rho \cap H$ and $H_\rho \cap H_s$:

$$d_r : d_s = \alpha x : \beta y. \tag{10}$$

Therefore the Brunn-Minkowski Theorem (compare B. F., p. 88) yields applied to $A_{x(\rho)}$, $B_{y(\rho)}$ and $C_{u(\rho)}$ that

$$C_u^{1/(n-2)} \geqslant \frac{\beta y}{\alpha x + \beta y} A_x^{1/(n-2)} + \frac{\alpha x}{\alpha x + \beta y} B_y^{1/(n-2)} \tag{11}$$

hence by (8) and (11)

$$D \geqslant \delta \int_0^1 [\beta y A_x^{1/(n-2)} + \alpha x B_y^{1/(n-2)}]^{n-2} (\alpha B x^2 B_y^{-1} + \beta A y^2 A_x^{-1})(\alpha x + \beta y)^{-n} d\rho. \tag{12}$$

In order to prove (3) it suffices to show that *the integrand in (12) is for* $0 < \rho < 1$ *not less than* $AB(\alpha A + \beta B)^{-1}$. A similar procedure as in the proof of the Brunn-Minkowski Theorem in B. F., p. 89, will be used. If $\epsilon = (n-1)^{-1}$, then

$$\beta y A_x^{1/(n-2)} + \alpha x B_y^{1/(n-2)} = y^{1+\epsilon}\beta A^\epsilon \left(\frac{A_x}{A^{1-\epsilon} y^{1-\epsilon}}\right)^{1/(n-2)} + x^{1+\epsilon}\alpha B^\epsilon \left(\frac{B_y}{B^{1-\epsilon} x^{1-\epsilon}}\right)^{1/(n-2)}$$

$$\frac{\alpha B x^2}{B_y} + \frac{\beta A y^2}{A x} = y^{1+\epsilon}\beta A^\epsilon \frac{A^{1-\epsilon} y^{1-\epsilon}}{A_x} + x^{1+\epsilon}\alpha B^\epsilon \frac{B^{1-\epsilon} x^{1-\epsilon}}{B_y}$$

Putting $t = x^{1+\epsilon}\alpha B^\epsilon [x^{1+\epsilon}\alpha B^\epsilon + y^{1+\epsilon}\beta A^\epsilon]^{-1}$ the integrand in (12) may be written as

$$\left[(1-t)\left(\frac{A_x}{A^{1-\epsilon} y^{1-\epsilon}}\right)^{1/(n-2)} + t\left(\frac{B_y}{B^{1-\epsilon} x^{1-\epsilon}}\right)^{1/(n-2)}\right]^{n-2} \cdot \left[(1-t)\frac{A^{1-\epsilon} y^{1-\epsilon}}{A_x} + t\frac{B^{1-\epsilon} x^{1-\epsilon}}{B_y}\right] \cdot \frac{(x^{1+\epsilon}\alpha B^\epsilon + y^{1+\epsilon}\beta A^\epsilon)^{n-1}}{(\alpha x + \beta y)^n}. \tag{13}$$

By Jensen's Inequality (compare Hardy, Littlewood, Polya, *Inequalities*, Cambridge, 1934, Theorem 16, p. 26)

$$[(1-t)a_1^{1/\nu} + t a_2^{1/\nu}]^\nu \geqslant [(1-t)a_1^{-1} + t a_2^{-1}]^{-1} \quad \text{for } 0 < t < 1,\ a_i > 0,\ \nu > 0 \tag{14}$$

and the equality sign holds only when $a_1 = a_2$.

Application of (14) to (13) with $\nu = n - 2$ yields that the integrand in (12) is at least

$$\frac{(x^{1+\epsilon}\alpha B^\epsilon + y^{1+\epsilon}\beta A^\epsilon)^{n-1}}{(\alpha x + \beta y)^n} = \left(\frac{\alpha x \cdot x^\epsilon B^\epsilon + \beta y \cdot y^\epsilon A^\epsilon}{\alpha x + \beta y}\right)^{n-1} \cdot \frac{1}{\alpha x + \beta y}. \tag{15}$$

Applying (14) again with $\nu = n - 1 = \epsilon^{-1}$ and $t = \beta y(\alpha x + \beta y)^{-1}$ to the first factor on the right in (15) yields that (15), and therefore the integrand in (12), is at least

$$\left(\frac{\alpha x \cdot x^{-1}B^{-1} + \beta y . y^{-1}A^{-1}}{\alpha x + \beta y}\right)^{-1} \cdot \frac{1}{\alpha x + \beta y} = \frac{AB}{\alpha A + \beta B}$$

which proves Theorem I.

Conditions for the equality sign in (2) are important for several applications. Because of the continuity of the functions involved, the equality sign in (2) requires the equality sign in (11) and in the two cases of Jensen's Inequality. Moreover $\bar{C}_{u(\rho)}$ must coincide with $\bar{D} \cap H_\rho$.

The equality sign holds in (11) only if the sets $\bar{A}_{x(\rho)}$ and $\bar{B}_{y(\rho)}$ are homothetic (see B. F., pp. 72 and 88). The condition $a_1 = a_2$ for equality in (14) yields in the two cases

$$A_x/A^{1-\epsilon}y^{1-\epsilon} = B_y/B^{1-\epsilon}x^{1-\epsilon} \text{ and } xB = yA$$

so that $A_x = B_y$.

Therefore $\bar{A}_x$ and $x\bar{B}_y$ are not only homothetic but congruent. The relation $x:A = y:B$ shows then that $\bar{A} = H_r \cap K$ can be transformed into $\bar{B} = H_s \cap K$ by revolving $\bar{A}$ about L through $\alpha' + \beta'$ and dilating it at the ratio $B:A$ in the direction of $\mathfrak{f}$. If w' is the point of $\bar{B}$ into which the point w of $\bar{A}$ is mapped under this transformation, it is easily seen that the segments connecting w to w' form a convex set E when w tranverses $\bar{A}$, so that E is the convex closure of $\bar{A} \cup \bar{B}$. Clearly the equality sign holds in (2) for $K = E$. Thus we find

III. *The equality sign holds in (2) if and only if $H_r \cap K$ can be transformed into $H_s \cap K$ by a rotation about L and a subsequent dilation in the direction of $\mathfrak{f}$, and $H_q \cap K$ is the intersection of H_q with the convex closure of $H_r \cap K$ and $H_s \cap K$.*

From III and the fact that E is the union of the segments connecting w and w' we obtain the following addition to II:

IIa. *The surface S in II is strictly convex if K is strictly convex.*

But III shows that the strict convexity of K is not a necessary condition.

THE ISOPERIMETRIC PROBLEM FOR MINKOWSKI AREA.*

By HERBERT BUSEMANN.

The present paper centers about the following problem: A positive continuous function $\sigma(u)$ is defined on all unit vectors u of E^n. *To find among all simple closed surfaces D in a certain class Γ and with a given "area" $\int_D \sigma(u)dS$ those which bound the maximal (Euclidean) volume $V(D)$.* Here dS means the (Euclidean) surface element of D and u the exterior normal of D "at dS."

For the rather restricted class Γ which consists of all convex surfaces and the non-convex surfaces whose first partials piecewise satisfy Lipschitz conditions, the problem is solved by the *isoperimetric inequality*

$$(*) \qquad \int_D \sigma(u)\,dS \geq nV^{(n-1)/n}(D)\,V^{1/n}(K)$$

where K is the polar reciprocal (with respect to the unit sphere) to the boundary of the convex closure of the surface C: $\sigma^{-1}(u)u$; and the equality sign holds only for D homothetic to K.[1]

The author's interest in the problem is due to the fact that (*) *solves the isoperimetric problem for the intrinsic area in Minkowski (or finite dimensional Banach) spaces* (Section **6**). *The solution is in general not a Minkowski sphere, but a new convex surface which proves of great importance for the theory of Finsler spaces.*[2] The reader interested in these questions is referred to the author's lecture on "The Geometry of Finsler Spaces" which is going to appear in the Bulletin of the American Mathematical Society. The present paper shows only that Minkowski's definition of area in Euclidean spaces by means of parallel sets carries over to Minkowski spaces if the parallel sets are formed with solutions of the isoperimetric problem instead of spheres.

---

* Received February 11, 1948; revised January 31, 1949.

[1] If u means the interior normal, then K must be replaced by the surface symmetric to K with respect to the origin.

[2] For the plane the problem was solved in [3] and the methods of [3] are used here in Sections **1** and **5**. However, in the plane case the solution of the isometric problem reduces after a rotation through $\pi/2$ to the sphere of the dual space (in Banach space terminology), and is therefore not an essentially new geometric object.

Using (*) presupposes the validity of the representation $\int \sigma(u)\,dS$ for the intrinsic area of a surface of class Γ in Minkowski spaces. But [2], where this area is defined, establishes this representation only for surfaces of class D', and therefore not for all convex surfaces. The final Section **7** of the present paper fills this gap by proving the much more general theorem that $\int \sigma(u)\,dS$ *represents the intrinsic area of any rectifiable manifold in a Finsler space.* For rectifiable manifolds in Riemann spaces $\int \sigma(u)\,dS$ also represents the various other areas (of Lebesgue, Radó, Federer, . . .) none of which is intrinsic by definition. Thus the present theorem contributes to the theory of these areas the information that they are intrinsic in the rectifiable case.

The form of (*) indicates a close connection with the Brunn-Minkowski Theory, in fact, the area $\int \sigma(u)\,dS$ was first studied by Minkowski in [9, § 27], and (*) follows for convex C and D directly from Minkowski's inequality (Section **1**). For non-convex D the relation (*) follows from Lusternik's generalization [8] of the Brunn-Minkowski Theorem to non-convex sets, see Section **2**. However, Lusternik's method does not yield the condition for the equality sign in (*). Therefore Dinghas and Schmidt proved in [5] Lusternik's result again for the special case where C is a sphere, by adapting Kneser and Süss' proof of the Brunn-Minkowski Theorem (compare [1, pp. 88-90]), and showed how this method leads through an elegant additional device to the conditions for the equality sign in (*).

Sections **3** and **4** of the present paper use the method of [5], at times verbally, for arbitrary convex C instead of spheres, and go slightly farther, to yield *a necessary and sufficient condition for the equality sign in the isoperimetric problem for the Minkowski area corresponding to C.*

This result implies (*) for convex C. The passage to arbitrary C (Section **5**) is accomplished in a surprisingly simple manner by an idea which the author called regularity principle,[3] and which seems applicable to many similar problems.

1. Let $\sigma(u)$ be a positive continuous function defined on the unit vectors u of E_n, and such that the surface $C: \sigma^{-1}(u)u$ is convex. The definition of $\sigma(u)$ is extended to arbitrary vectors x by

$$\sigma(x) = |x|\,\sigma(x/|x|) \text{ for } x \neq 0, \qquad \sigma(x) = 0 \text{ for } x = 0.$$

Then C has the equation $\sigma(x) = 1$ and $\sigma(x)$ is a convex positive homogeneous

[3] Compare [4]. The author was incautious enough to state in [4], that the present problem for non-convex D was hopeless. He was then not aware of the paper [8] by Lusternik.

function of the first degree in x. Therefore it is supporting function of a convex body L with boundary K. It is well known [9, § 8] that K originates from C by a polar reciprocity in the unit sphere $|x| = 1$. If $\sigma(x)$ is analytic and D is any other analytic closed convex surface bounding the convex body E, then

$$\int_D \sigma(u)\, dS = nV(E, L, \cdots, L), \tag{1}$$

where (see [9, § 27] or [1, p. 64]) u is the exterior normal of D "at dS" and

$$V_1(E, L) = V(E, L, \cdots, L) \tag{2}$$

is the mixed volume of E and $(n-1)$-times L. If D and K are arbitrary closed convex surfaces, then sequences $\{D_\nu\}$ and $\{K_\nu\}$ of analytic convex surfaces exist, which tend to D and K respectively ([1, section 27]). Convergence of convex surfaces implies convergence of their supporting planes. Since $V_1(D, K)$ depends also continuously on D and K ([1, p. 40]) it follows from (1) and (2) that for any convex D and K

$$\int \sigma(u)\, dS = nV_1(E, L) = \lim \rho^{-1}(|E + \rho L| - |E|). \tag{3}$$

The integral on the left is extended over all points of D where the tangent plane exists, $|X|$ denotes generally the exterior Lebesgue measure of the set X, and $X + \rho Y$ means the set consisting of all points of the form $x + \rho y$ with $x \varepsilon X$ and $y \varepsilon Y$. For the last part of (3) compare [1, p. 47]. The number $nV_1(E, L)$ is called the area of D relative to K and was first studied by Minkowski ([9, § 27]). The inequality of Minkowski (see [1, p. 91] or the next section of this paper)

$$V_1^n(E, L) \geq |E|^{n-1} |L|$$

and (3) yield

$$\int_D \sigma(u)\, dS \geq n |E|^{(n-1)/n} |L|^{1/n} \tag{4}$$

and the equality sign holds only for E homothetic to L or D homothetic to K. Therefore

(5) *If $C: \sigma^{-1}(u)u$ is convex then the surfaces homothetic to the polar reciprocal K of C and only these minimize $\int_D \sigma(u)\, dS$ among all convex surfaces D with a given volume $|E|$.*

2. The discussion of the same problem for non-convex D rests on the expression which the right side of (3) gives for $\int_D \sigma(u)\,dS$. If H is the interior of K, then for any convex E

$$|E + \rho H| = |E + \rho L|.$$

Denoting the closure of the set X by $\bar{X}$, any set X satisfies the relation

$$X + \rho H = \bar{X} + \rho H. \tag{6}$$

Therefore we define generally as Minkowski area $B(M)$ relative to K of the boundary of a set X with $|\bar{X}| < \infty$, the number

$$B(X) = \liminf_{\rho \to 0+} \rho^{-1}(|X + \rho H| - |\bar{X}|). \tag{7}$$

Because of (6)

$$B(X) = B(\bar{X}). \tag{8}$$

For this area the isoperimetric problem can be solved *completely*. In order to formulate the solution adequately we introduce the following notation. For any bounded set M with $|\bar{M}| > 0$ and a given direction u, let a' be the least upper bound of those d for which the intersection of $\bar{M}$ with the set $xu \leq d$ has measure 0 and b' the greatest lower bound of the d for which $\bar{M}$ intersects $xu \geq d$ is a set of measure 0. We call $a' \leq xu \leq b'$ the *essential supporting strip* $W(u)$ of M normal to u. Obviously $W(u) = W(-u)$ and $|M - M \cap W(u)| = 0$. The intersection

$$M_E = M \cap \prod_u W(u)$$

is called the *essential part of* M. By definition $\overline{M_E} = (\bar{M})_E$. There is a countable set $\{u_\nu\}$ of directions u such that $\prod_u W(u) = \prod_\nu W(u_\nu)$. Since

$$M - M_E = M - M \cap \prod W(u_\nu) = \bigcup_\nu [M - M \cap W(u_\nu)]$$

it follows that

$$|M - M_E| = 0. \tag{9}$$

Corresponding to (4) we are going to prove the isoperimetric inequality:

Theorem 1. *For any set M with $0 < |\bar{M}| < \infty$,*

$$B(M) \geq n\,|\bar{M}|^{(n-1)/n}\,|L|^{1/n}, \tag{10}$$

and the equality sign holds if and only if $\bar{M}_E$ is homothetic to L and $B(M) = B(M_E)$.

Am. J. Math. 71 (1949), 743–762

Notice that the theorem is trivial for unbounded M since then $|M+\rho H|=\infty$ for any ρ, so that $B(M)=\infty$. In the proof we assume therefore that M is bounded. We call topological sphere a topological image D of the unit sphere $|x|=1$. If M is the interior of D then $M_E=M$. Therefore Theorem 1 yields

(11) *If C is convex then the surfaces homothetic to K and only these minimize $B(M)$ among all topological spheres D with a given (positive, finite) volume $|M|$ or $|\bar{M}|$, where M is the interior of D.*

(10) without the discussion of the equality sign follows from the Brunn-Minkowski Theorem $|M+\rho H|^{1/n}\geq|\bar{M}|^{1/n}+\rho|L|^{1/n}$ which yields

$$|M+\rho H|-|\bar{M}|\geq n\rho|\bar{M}|^{(n-1)/n}\cdot|L|^{1/n}.$$

The relation $|X+\rho R|^{1/n}\geq|X|^{1/n}+\rho|Y|^{1/n}$ was given for arbitrary measurable X and Y by Lusternik in [8]. For reasons explained in the introduction we follow very closely the method of [5] to establish Theorem 1 and the Brunn-Minkowski Theorem for arbitrary convex Y. With minor modifications the present proof works for any Y which contains an open set Z with $|Z|=|Y|$.

3. The purpose of the next two sections is then to prove simultaneously Theorem 1 and

THEOREM 2. *If M is a bounded set and K' is a bounded convex set with interior points, then for any $\rho>0$*

$$|M+\rho K'|^{1/n}\geq|\bar{M}|^{1/n}+\rho|K'|^{1/n} \tag{12}$$

and the equality sign holds if and only if $\bar{M}$ is homothetic to $\bar{K}'$.

Note. If either M or K' is not bounded or K' has no interior points (12) is trivially correct. The conditions for the equality sign are different, but equally trivial.

If H is the set of interior points of K' then $|H|=|K'|$, hence Theorem 2 follows, because of (6), from the special case where $K'=H$ and $M=\bar{M}$. Instead of (12) we therefore prove

$$|M+\rho H|^{1/n}\geq|M|^{1/n}+\rho|H|^{1/n} \text{ for closed } M. \tag{13}$$

Theorem 2 is trivial for $|M|=0$. For if M contains more than one point then $|M+\rho H|>\rho|H|$. And if M consists of one point, then $|M+\rho H|=\rho|H|$ and M is homothetic to H. Therefore we may assume that $|M| > 0$, and it means no restriction if we assume in addition that $|M| = |H|$, because the general case follows immediately from this case.

The proof proceeds by *induction.* The very simple case $n = 1$ is left to the reader, and Theorem 2 is assumed to hold for $n - 1$. In E_n introduce rectangular coordinates $x, y_1, \cdots, y_{n-1}$; let a and b (α and β) be the minimum and maximum of x in $M(\bar{H})$. Then $\alpha \leq x \leq \beta$ is both the ordinary and the essential supporting strip of H or $\bar{H}$ normal to the x-axis, but the essential supporting strip of M normal to the x-axis will in general have the form $a' \leq x \leq b'$ with $a \leq a' < b' \leq b$.

For $a \leq d \leq b$ ($\alpha \leq \delta \leq \beta$) denote by $M(d)$ ($H(\delta)$) the intersection of the halfspace $x \leq d$ ($x \leq \delta$) with $M(H)$, and by $m(d)$ ($h(\delta)$) the intersection of $M(H)$ with $x = d$ ($x = \delta$). Then $H(\alpha) = h(\alpha) = h(\beta) = 0$. If $|X|'$ is the $(n-1)$-dimensional measure of a set X in a hyperplane, then

$$(14) \qquad |M(d)| = \int_a^d |m(x)|' dx, \qquad H(\delta) = \int_a^\delta |h(z)|' dz.$$

Following the original idea of Brunn we introduce the function $z(x)$ by the relation

$$|H(z)| = |M(x)|, \qquad a \leq x \leq b.$$

$z(x)$ is a monotone increasing, absolutely continuous function of x with

$$(15) \qquad \begin{cases} z(x) = \alpha & \text{for } a \leq x \leq a' \\ z(x) = \beta & \text{for } b' \leq x \leq b \\ \alpha < z(x) < \beta & \text{for } a' < x < b'. \end{cases}$$

Because of $|h(z)|' > 0$ for $\alpha < z < \beta$ the relations (14) and (15) show that

$$(16) \qquad \begin{cases} \dfrac{dz}{dx} = 0 & \text{for } a \leq x < a' \text{ and } b' < x \leq b \\ \dfrac{dz}{dx} = \dfrac{|m(x)|'}{|h(z(x))|'} & \text{for almost all } x \text{ with } a' \leq x \leq b'. \end{cases}$$

Now construct the set M^* in the strip $a + \rho\alpha \leq d^* \leq b + \rho\beta$ by defining its intersection $m^*(d^*)$ with the plane $x = d^*$ as follows: Because $d + \rho z(d)$ is a strictly monotone and continuous function of d, a given d^* in $[a + \rho\alpha, b + \rho\beta]$ determines exactly one d in $[a, b]$ such that $d + \rho z(d) = d^*$. We then put

$$(17) \qquad \begin{aligned} m^*(d^*) &= m(d) + \rho h(z(d)) \text{ if } a' < d < b' \text{ and } m(d) \neq 0 \\ &= 0 \text{ for all other } d \text{ in } [a, b]. \end{aligned}$$

Since $h(z(d))$ lies in $x = z(d)$ and $m(d)$ in $x = d$, the set $m^*(d^*)$ lies really in $x = d + \rho z(d) = d^*$, and since $\rho h(z(d)) \subset \rho H$ and $m(d) \subset M$ the set M^* is contained in $M + \rho H$.

The inductive hypothesis yields

$$(18) \qquad | m^*(d^*) |' \geq \{ | m(d) |^{1/(n-1)} + \rho | h(z(d)) |^{1/(n-1)} \}^{n-1}$$

for the first set of d's in (17). The relations (16) and (18) show that for almost all x in $[a', b']$ with $| m(x) |' > 0$

$$(19) \quad | m^*(x^*) |' \cdot (1 + \rho dz/dx)$$
$$\geq | m(x) |' [1 + \rho \{ | h(z) |' / | m(x) |' \}^{1/(n-1)}]^{n-1} (1 + \rho (| m(x) |' / | h(z) |')).$$

But

$$(20) \qquad (1 + \rho \lambda^{1/(n-1)})^{n-1} (1 + \rho \lambda^{-1}) \geq (1 + \rho)^n \text{ if } \rho > 0,\ \lambda > 0$$

and the equality sign holds only for $\lambda = 1$ (compare [5] or [6, p. 61, Theorem 64]). (19) and (20) yield

$$(21) \qquad | m^*(x^*) |' (1 + \rho (dz/dx)) \geq | m\, x) |' \cdot (1 + \rho)^n$$

for almost all x in $[a', b']$ with $| m(x) |' > 0$. But (21) holds for all x with $| m(x) |' = 0$. The definition of $z(x)$ and (15) show that $| m(x) |' = 0$ a. e. in $[a, a']$ and $[b', b]$, so that (21) holds a. e. in $[a, b]$. Hence

$$(22) \quad | M + \rho H | \geq | M^* | = \int | m^*(x^*) |' \, dx^*$$
$$= \int_a^b | m^*(x^*) |' (1 + \rho (dz/dx)) \, dx$$
$$\geq (1 + \rho)^n \int_a^b | m(x) |' \, dx = (1 + \rho)^n | M | = \{ | M |^{1/n} + \rho | H |^{1/n} \}^n.$$

This proves the inequalities (10) and (12). We also observe that the equality sign can hold in (12) only if $| M + \rho H | = | M^* |$. This implies that $a' = a$ and $b' = b$, because M^* contains no points in $x < a' + \rho \alpha$ and $M + \rho H$ intersects any strip $a + \rho \alpha < x < c$ in a set of positive measure, because $m(a) \neq 0$. Varying the direction of the x-axis we find

(23) *If the equality sign holds in* (12), *then* $\bar{M}_E = \bar{M}$.

It is now quite simple to discuss the equality sign in (12), but (23) and Theorem 1 contain Theorem 2, because—as pointed out by Dinghas and Schmidt—the equality sign in (12) for some $\rho > 0$ implies equality for any $\rho' < \rho$. For if $\rho = \rho' + \delta$ then $M + \rho H = (M + \rho' H) \cdot + \delta H$ hence

$$|M|^{1/n}+\rho|H|^{1/n}=|M+\rho H|^{1/n}\geq|M+\rho' H|^{1/n}+\delta|H|^{1/n}$$
$$\geq|\bar{M}|^{1/n}+\rho'|H|^{1/n}+\delta|H|^{1/n}=|M|^{1/n}+\rho|H|^{1/n}.$$

Therefore the equality in (12) implies (for $\rho\to 0$) equality in (10).

4. That the equality sign holds in (10) when $\bar{M}_E$ is homothetic to L and $B(M)=B(M_E)$ is trivial. But it is worth noticing that the condition $B(M)=B(M_E)$ is not superfluous. For if N is a sequence of points which tends to ∞ and $M=L\cup N$, then $\bar{M}_E=L$ but $B(M)>B(L)$. Examples of W. H. and G. C. Young in [11, pp. 276-281] show that the same effect may be obtained with suitable convergent sequences.

(8) and the formulation of Theorem 1 show that we may restrict ourselves for the "only if" part in the discussion of the equality sign to the case of *closed* M. We show first:

(24) *If the equality sign holds in* (10) *then* $|m(d)|'>0$ *for* $a'<d<b'$.
First let $a<d<b$, put $M(d)=M'$, and call M'' the intersection of M with $x\geq d$. Let m'_ρ and m''_ρ be the intersections of $M'+\rho H$ and $M''+\rho H$ with $x=d$. If P denotes the orthogonal projection of H on $x=0$ then the intersection of $M'+\rho H$ $(M''+\rho H)$ with $x\geq d$ $(x\leq d)$ is contained in the cylinder with base $m'_\rho+\rho P$ $(m''_\rho+\rho P)$ and altitude $(\beta-\alpha)\rho$ (the width of ρH in the direction of the x-axis). Therefore

$$|(M'+\rho H)\cap(M''+\rho H)|\leq(\beta-\alpha)\rho(|m'_\rho+\rho P|'+|m''_\rho+\rho P|').$$

Since $M+\rho H=(M'+\rho H)+(M''+\rho H)$ it follows that

$$|M+\rho H|\geq|M'+\rho H|+|M''+\rho H|$$
$$-(\beta-\alpha)\rho(|m'_\rho+\rho P|'+|m''_\rho+\rho P|'),$$

hence because of $|M|=|M'|+|M''|$,

$$\rho^{-1}(|M+\rho H|-|M|)\geq\rho^{-1}(|M'+\rho H|-|M'|)$$
$$+\rho^{-1}(|M''+\rho H|-|M''|)$$
$$-(\beta-\alpha)(|m'_\rho+\rho P|'+|m''_\rho+\rho P|'). \tag{25}$$

Because $m'_\rho\supset m(d)$, $m''_\rho\supset m(d)$, and $m(d)$ is closed, both $m'_\rho+\rho P$ and $m''_\rho+\rho P$ tend for $\rho\to 0$ to $m(d)$, and it follows from (7) and (25) that

$$B(M)\geq B(M')+B(M'')-2(\beta-\alpha)|m(d)|'. \tag{26}$$

Assume now that $a'<d<b'$. Then $|M'|>0$, $|M''|>0$ and (10) and (26) yield

$$B(M) \geq n\,|\,L\,|^{1/n}(|\,M'\,|^{(n-1)/n} + |\,M''\,|^{(n-1)/n}) - 2(\beta-\alpha)\,|\,m(d)\,|$$
$$= n\,|\,M\,|^{(n-1)/n}\,|\,L\,|^{1/n}(|\,M'\,|^{(n-1)/n}\,|\,M\,|^{(1-n)/n}$$
$$+ |\,M''\,|^{(n-1)/n}\,|\,M\,|^{(1-n)/n}) - (\beta-\alpha)\,|\,m(d)\,|'.$$

Because of $|\,M'\,|\cdot|\,M\,|^{-1} + |\,M''\,|\cdot|\,M\,|^{-1} = 1$

$$|\,M'\,|^{(n-1)/n}\,|\,M\,|^{(1-n)/n} + |\,M''\,|^{(n-1)/n}\,|\,M\,|^{(1-n)/n} > 1,$$

so that

$$B(M) > n\,|\,M\,|^{(n-1)/n}\,|\,L\,|^{1/n} - (\beta-\alpha)\,|\,m(d)\,|'.$$

Hence if $m(d) = 0$ the equality sign cannot hold in (10), which proves (24).

Putting $\lambda = \lambda(x) = |\,h(z(x))\,|'/|\,m(x)\,|'$ it follows from (24) and (19) that equality in (10) implies that a. e. in $[a', b']$

$$|\,m^*(x^*)\,|'(1+\rho(dz/dx)) \geq |\,m(x)\,|'\{1+\rho[(n-1)\lambda^{1/(n-1)} + \lambda^{-1}]\};$$

hence (compare (22))

$$|\,M+\rho H\,| \geq |\,M^*\,| \geq \int_{a'}^{b'} |\,m(x)\,|\,dx$$
$$+ \rho\int_{a'}^{b'} |\,m(x)\,|'[(n-1)\lambda^{1/(n-1)} + \lambda^{-1}]\,dx$$

and

$$\rho^{-1}(|\,M+\rho H\,| - |\,M\,|) \geq \int_{a'}^{b'} |\,m(x)\,|'[(n-1)\lambda^{1/(n-1)} + \lambda^{-1}]\,dx.$$

Now

$$(n-1)\lambda^{1/(n-1)}(x) + \lambda^{-1}(x) \geq n$$

and the equality sign holds only for $\lambda(x) = 1$. This can be seen by differentiation or from the theorem of the arithmetic and geometric mean (see [6, p. 17]).

Therefore (since we assumed $|\,M\,| = |\,L\,|$)

$$B(M) > n\,|\,M\,| = n\,|\,M\,|^{(n-1)/n}\,|\,L\,|^{1/n}$$

unless $\lambda(x) = 1$ a. e. in $[a', b']$.

The absolute continuity of $z(x)$, the equality in (10), and (16) imply therefore that $z(x) = x + \alpha - a'$ for $a' \leq x \leq b'$. If we place L so that its center of gravity coincides with that M, then $\lambda(x) = |\,h(z)\,|'/|\,m(x)\,|' = 1$ a. e. shows that the essential supporting strip $a' \leq x \leq b'$ of M must be a supporting strip of L.

As a convex body, L is the intersection of its supporting strips. Varying the direction of the x-axis we see that $M_E \subset L$. If a point $p \,\varepsilon\, L - M_E$

existed, then a sphere S with center p and disjoint from M_E would exist because M_E is closed. Then $|S \cap L| > 0$ because L is convex, hence $|L| > |M_E|$, which contradicts (9) and $|L| = |M|$.

This proves that L and M_E are homothetic and

$$B(M_E) = n\,|\bar{M}|^{(n-1)/n}\,|L|^{1/n},$$

if the equality sign holds in (10). Therefore also $B(M) = B(M_E)$.

5. The statement (11) solves the main problem for convex C and those D for which $\int_D \sigma(u)\,dS \geq B(M)$. Unfortunately the class of surfaces for which this relation holds, even when $\sigma(u) \equiv 1$, is very restricted. Since the best possible results can doubtless not be obtained through using $B(M)$, we mention only one class Γ' for which the relation

$$\int_D \sigma(u)\,dS = B(M) \tag{27}$$

is very easily established: Γ' consists of topological spheres of class D' where in addition the set of points in which the first partial derivatives are continuous can be covered by a finite number of neighborhoods in which these partials satisfy a Lipschitz condition. The class Γ is defined to consist of Γ' and all closed convex surfaces.

(28) *If C is convex then the surfaces homothetic to K and only these minimize $\int_D \sigma(u)\,dS$ among all surfaces in Γ with a given volume.*

From this theorem we derive the corresponding theorem for *arbitrary* C. Let $\bar{C}$ denote the boundary of the convex closure of C and let $\bar{\sigma}^{-1}(u)\cdot u$ be the point of the form $\lambda \cdot u$, $\lambda > 0$, on $\bar{C}$. Then $\sigma(\bar{u}) \leq \sigma(u)$ hence

$$\int_D \bar{\sigma}(u)\,dS \leq \int_D \sigma(u)\,dS \tag{29}$$

for every surface D in Γ. We are going to show that the equality sign holds in (29) for surfaces D which are homothetic to the polar reciprocal K of $\bar{C}$ with respect to $|x| = 1$. It suffices, of course, to prove this for K itself.

Every point p of $\bar{C}$ lies on an $(n-1)$-simplex with vertices in C (see [1, p. 9]), and therefore on a simplex $T(p)$ of lowest dimension $\nu \leq n-1$ with vertices on C. Then $T(p) \subset \bar{C}$. This is obvious for $\nu = 0$. For $\nu > 0$ the point p is an interior point of T (considered as set in a ν-dimensional plane), and a hyperplane through p either contains $T(p)$ or separates at least two of the vertices of $T(p)$. A supporting plane π of $\bar{C}$ through p must therefore contain T, and since $\pi \cap \bar{C}$ is convex, $T(p) \subset \bar{C}$.

If $\nu \geq 1$ and $q_0, q_1, \cdots, q_\nu$ are the vertices of $T(p)$, then p has the form

$$p = \sum \epsilon_i q_i, \qquad \epsilon_i > 0, \qquad \sum \epsilon_i = 1, \tag{30}$$

hence, after defining $\bar{\sigma}(x)$ in the same way as $\sigma(x)$ for $|x| \neq 1$,

$$\begin{gathered} 1 = \bar{\sigma}(p) = \bar{\sigma}(\sum \epsilon_i q_i) \leq \sum \epsilon_i \bar{\sigma}(q_i) = 1 \text{ or} \\ \bar{\sigma}(p) = \sum \epsilon_i \bar{\sigma}(q_i) \end{gathered} \tag{31}$$

If $\bar{\sigma}(x)$ is now interpreted as supporting function of a body L with boundary K then (30) and (31) imply that at a point of K where the exterior normal is parallel to p, more than one, namely $\nu + 1$ linearly independent, supporting planes exist, so that K is not differentiable.

The crucial point is now that on the one hand for any point p of $\bar{C}$, but not on C, the dimension of $T(p)$ is obviously at least one, and on the other hand the points on $\bar{C}$ and not on C are exactly the points $p = \bar{\sigma}^{-1}(u)u$ for which $\bar{\sigma}(u) < \sigma(u)$. Therefore the preceding discussion shows that $\bar{\sigma}(u) = \sigma(u)$ at every point of K where a tangent plane exists, so that

$$\int_K \bar{\sigma}(u)\, dS = \int_K \sigma(u)\, dS. \tag{32}$$

If $V(D)$ denotes the volume bounded by the surface D in Γ, then Theorem 1, (27), (28), (29) and (32) contain the following main theorem:

THEOREM 3. *If $\sigma(u)$ is defined for $|u| = 1$, positive and continuous, then for any surface D in Γ*

$$\int_D \sigma(u)\, dS \geq n V^{(n-1)/n}(D)\, V^{1/n}(K)$$

where K is the polar reciprocal to the boundary $\bar{C}$ of the convex closure of C: $\sigma^{-1}(u)u$ with respect to the unit sphere. The equality holds only for D homothetic to K.

Although this theorem shows that the question whether C is convex or not is not of great importance as far as the isoperimetric problem for the Minkowski area $\int \sigma(u)\, dS$ is concerned, there are profound differences regarding other properties of this area. This is made clear by the following theorem which is due to Minkowski (see [9, § 27]):

(33) *The function* $\sigma(x)$ (*or the surface* C) *is convex if and only if the following holds: whenever the convex body* L_1 *contains the convex body* L_2 *then* $\int_{K_1} \sigma(u)\, dS \geq \int_{K_2} \sigma(u)\, dS$, *where* K_i *is boundary of* L_i.

An equivalent condition, which will be discussed elsewhere, is that the hyperplanes be minimal (that is area minimizing) surfaces.

6. Theorem 3 leads also to a solution of the *isoperimetric problem in Minkowski geometry.* Let $x_1, x_2, \cdots, x_n$ be a rectangular coordinate system in a Euclidean space associated with a given Minkowski metric μ (for this and the following see section 2 in [2]). Then a positive homogeneous function $f(x)$ of degree 1 exists, which is positive for $x \neq 0$ and such that $\mu(x, y) = f(x - y)$. Area and volume are defined only if μ is symmetric, that is $\mu(x, y) = \mu(y, x)$ or $f(x) = f(-x)$. The surface T: $f(x) = 1$ is the Minkowskian unit sphere. It bounds the closed convex set $S: f(x) \leq 1$.

If $|X|$ denotes the exterior Lebesgue measure of the set X with respect to the associated Euclidean space, then X has the *Minkowski measure*

(34) $$|X|_m = \sigma |X|$$

where

$$\sigma = \omega^{(n)} |S|^{-1} \text{ and } \omega^{(n)} = \pi^{n/2} \Gamma^{-1}(n + 2^{-1})\,;$$

that is, $\omega^{(n)}$ is the volume of the Euclidean unit sphere.

The intrinsic area of a surface in a Minkowski space is defined in terms of the one-dimensional metric μ, but admits for hypersurfaces D of class D' an integral representation of the form $\int \sigma(u)\, dS$, which is obtained as follows: Let the tangent hyperplane of D at a point where it exists, have normal u (because of the symmetry of μ or S we need not distinguish between interior and exterior normals). The hyperplane normal to u through the origin intersects S in an $(n-1)$-dimensional convex set $S(u)$. If $|X|'$ denotes exterior $(n-1)$-dimensional Lebesgue measure with respect to the associated Euclidean space, then

$$\sigma(u) = \omega^{(n-1)} / |S(u)|'.$$

The intrinsic area of a topological sphere D of class D' is then

(35) $$A_m(D) = \int_D \sigma(u)\, dS$$

where dS is, of course, the Euclidean surface element (see the next section).

Since convex surfaces are in general not of class D' this representation has not yet been established for the whole class Γ, but this gap will be filled in the next section.

The surface $C: \sigma^{-1}(u)u$ *is convex.* (A proof is found in the author's paper "A theorem on convex bodies of the Brunn-Minkowski type," *Proceedings of the National Academy of Sciences,* vol. 35 (1949), pp. 27-31). If K denotes the polar reciprocal of C with respect to $|x| = 1$ and $V_m(D)$ is the Minkowski volume (34) enclosed by the surface D, then Theorem 3, (34) and (35) yield

$$A_m(D) \geqq nV_m^{(n-1)/n}(D)\, V_m^{1/n}(K)\sigma^{-1}. \tag{36}$$

The factor σ^{-1} appears because K depends on the choice of the coordinate system x. But "homothetic" is independent of the coordinate system and has a Minkowski meaning. K can be characterized in terms of perpendicularity in a very similar manner as in the two dimensional case (see [3, p. 867]), but this will be discussed elsewhere.

For a satisfactory Minkowski isoperimetric inequality it is desirable to eliminate the factor σ^{-1} which depends on the associated Euclidean metric. This can be most easily accomplished by choosing among all the surfaces homothetic to K the surface $K^* = \sigma^{-1}K$. Then

$$V_m^{1/n}(K^*) = V_m^{1/n}(K)\sigma^{-1}. \tag{37}$$

A Minkowskian characterization of K^* is obtained by substituting K^* for D in (36):

$$A_m(K^*) = nV_m(K^*). \tag{38}$$

This relation may be considered as the Minkowskian analogue to the Euclidean fact that $n\omega^{(n)}$ is the surface of the unit sphere. (The relation $A_m(T) = nV_m(T)$ is in general not correct.) Since the associated Euclidean space to a given Minkowski metric is determined up to a non-degenerate affine transformation, *the class* Γ *is Minkowski invariant.* Therefore we may formulate the solution of the *Minkowski isoperimetric problem* as follows:

THEOREM 4. *For all surfaces* D *in* Γ

$$A_m(D) \geqq nV_m^{(n-1)/n}(D)\, V_m^{1/n}(K^*),$$

where K^* is homothetic to the polar reciprocal of C with respect to $|x| = 1$ and $A_m(K^*) = nV_m(K^*)$. The equality sign holds if and only if D is homothetic to K^*.

The choice of K^* has the great advantage that it permits us to generalize the fundamental relation (7) to Minkowskian geometry. It is not true that the intrinsic area of D is the limit of $\rho^{-1}(|M + \rho S|_m - |M|_m)$, where M is the interior of D. But

THEOREM 5. *If H^* is the interior of K^*, then for any surface D in Γ with interior M*

$$A_m(D) = \lim \rho^{-1}(|M + \rho H^*|_m - |\bar{M}|_m).$$

Proof. For the interior H of K the relations (27), (34) and (35) yield

$$\begin{aligned}\int_D \sigma(u)\,dS = A_m(D) &= \lim \rho^{-1}(|M + \rho H| - |\bar{M}|)\\ &= \lim (\sigma\rho)^{-1}(|M + \rho H|_m - |\bar{M}|_m)\\ &= \lim (\sigma\rho)^{-1}(|M + \rho\sigma(\sigma^{-1}H)|_m - |\bar{M}|_m)\end{aligned}$$

which proves the Theorem because $\sigma^{-1}H = H^*$.

In analogy to (7) we may now define as Minkowski area $B_m(X)$ of the boundary of a set X in a Minkowski space the number

$$B_m(X) = \liminf_{\rho \to 0+} \rho^{-1}(|X + \rho H^*|_m - |\bar{X}|_m), \tag{39}$$

and Theorem 1 yields

THEOREM 6. *For any set M with $0 < |\bar{M}|_m < \infty$*

$$B_m(M) \geq n\,|\bar{M}|_m^{(n-1)/n}\,|H^*|_m,$$

and the equality sign holds if and only if $\bar{M}_E$ is homothetic to $\bar{H}^*$ and $B_m(M) = B_m(M_E)$.

7. To make the preceding result valid, (35) must be established for convex surfaces. The results and notations of [2] will be used freely. Since convex surfaces in a Minkowski space are birectifiable, (35) is contained in the following general theorem:

THEOREM 7. *If (P, x) is a k-dimensional rectifiable manifold in a Finsler space F with local rectifiable representation $x(u)$, then*

(40) $$|P,x| = |P,x|_k^H = \int_p \sigma(u)^{(u,x)}\Delta^{(u,x)}(u)\,du_1\cdots du_k$$

where $\sigma^{(u,x)}$ is the term $\omega^{(k)}\Xi^{(u,x)}$ of [2, (7, 6)] (the k's in $\Xi^{(u,k)}$ and $\Delta^{(u,k)}$ are misprints) and means the following:

The derivatives $\partial x_i/\partial u_j$ exist for a definite u-system and a definite x-system almost everywhere (a. e.) in the u-coordinate neighborhood. If the matrix $\partial x/\partial u = \partial x_i/\partial u_j)_{\substack{i=1,\dots,n\\ j=1,\dots,k}}$ has rank k then the k-dimensional plane in the tangent space ξ at $x(u)$ spanned by the k vectors $\partial x/\partial u_j$ intersects the set $\Phi(x,\xi)=1$ in a k-dimensional convex set whose k-dimensional measure with respect to ξ considered as a rectangular coordinate system is $1/\Xi^{(u,x)}(u) = \omega^{(k)}/\sigma^{(u,x)}(u)$. Since $\Delta^{(u,k)}(u)\,du_1\cdots du_k$ is the Euclidean surface element of (P,x) at $x(u)$ (see [2, (7. 5)]) (40) has the form (35) (but u does not have the same meaning), moreover $\Delta^{(u,k)}(u)=0$ if $\partial x/\partial u$ has smaller rank than k. Therefore $\sigma^{(u,x)}(u)$ need not be defined in that case, we simply put the whole integrand $\sigma^{(u,x)}(u)\Delta^{(u,k)}(u)=0$.

Because of the additivity properties of $|P,x|$, $|P,x|_k^H$, and the integral, the general case can be reduced to the local case where P is a bounded closed convex set with positive measure $|P|'$ in E^k with rectangular coordinates $u=(u_1,\cdots,u_k)$ and $x(u)$ moves in one coordinate neighborhood (x) of P. We choose these coordinates once and for all, and can then replace $\sigma^{(u,x)}(u)$ and $\Delta^{(u,x)}(u)$ by $\sigma(u)$ and $\Delta(u)$. We show first that

(41) $$|P,x|_k^H = \int_P \sigma(u)\Delta(u)\,du.$$

The paper [2] contains a mistake regarding the measure $|P,x|_k^H$. The relation $|P_1,\pi_1^\epsilon|_n = \sum_K |x(P_1V_k),\delta|_n$ on p. 254 is in general not correct. Therefore the proof that $|P,x|_n$ is monotone does not hold, and, in fact, this measure is not monotone. This follows from another representation for $|P,x|_n^H$ which Prof. H. Federer communicated to the author. If P is the union of a countable number of compact sets in some metric space, $x(p)$ maps P continuously into a metric space with distance δ, and $\nu(x_0)$ denotes the number of (maximal connected) components of the set of points p in P with $x(p)=x_0$, then

(42) $$|P,x|_n^H = \int_{x(P)} \nu(x)\,d\,|M,\delta|_n,$$

where the right side means the integral of $\nu(x)$ over $x(P)$ with respect to

n-dimensional Hausdorff measure on $x(P)$. It follows from (42) that $|P, x|_n{}^H$ is invariant under the change of the parametrization, that is [2, (4.8)] (the proof given in [2] does not hold because it uses monotoneity.) It is easy to see that the integral in (42) is not monotone.

It is known that (41) holds if it has been established for an F_σ subset of P, on which the mapping $p \to x(p)$ is one-to-one, compare H. Federer and A. P. Morse, "Some properties of measurable functions," *Bulletin of the American Mathematical Society*, vol. 49 (1943), pp. 270-277, Theorem 3.1 and Section 5. This reference implies the mentioned reduction for the integral over the number $N(x_0)$ of points p with $x(p) = x_0$, but because of the rectifiability $N(x) \neq \nu(x)$ on a set M_0 with $|M_0, \delta|_n = 0$.

We therefore consider an F_σ subset $\bar{P}$ of P on which $p \to x(p)$ is one-to-one. It follows from (42) that $|\bar{P}, x|_k{}^H = |x(\bar{P}), \delta|_k$. Rectifiability means the existence of a constant β such that the distance δ in F satisfies the relation

(43) $$\delta(x(u), x(v)) \leq \beta\,|u - v| \quad \text{for } u, v \text{ in } P.$$

The partials $\partial x_i/\partial u_j$ exist therefore a. e. in P and are bounded. Consequently $\Delta(u)$ exists a. e. and is bounded. $\sigma(u)$ is bounded because P is closed.

For a given $\epsilon > 0$ there exists a compact subset P' of $\bar{P}$ with $|\bar{P} - P' < \epsilon$ such that $x(u)$ is uniformly totally differentiable on P'. By Kolmogoroff's Principle $|\bar{P} - P', x|_k{}^H \leq \beta^k |\bar{P} - P'|\,\beta^k\epsilon$ hence

(44) $$|x(P'), \delta|_k \leq |x(\bar{P}), \delta|_k = |\bar{P}, x|_k{}^H < |P', x|_k{}^H + |\bar{P} - P', x|_k{}^H \leq |x(P'), \delta|_k + \beta^k\epsilon.$$

The boundedness of $\Delta(u)\sigma(u)$ and (44) show that it suffices to establish (41) for P'.

The uniform total differentiability of $x(u)$ on P' implies: for every point $u_0 \,\varepsilon\, P'$ there is a $\rho_1 > 0$ such that for $u_i \,\varepsilon\, P' \cap S(u_0, \rho_1)$, $i = 1, 2$, ($S(u_0, \rho_1)$ is the set of all points u in E^k with $|u - u_0| < \rho_1$).

(45) $$x(u_1) - x(u_2) = y(u_1) - y(u_2) + z\,|u_1 - u_2| \text{ with } |z| < \epsilon.$$[4]

We put

(46) $$y(u) = x(u_0) + (u - u_0)\partial x(u)/\partial u.$$

If $\Delta(u_0) \neq 0$, then

---

[4] (45) is stated in [7, p. 363] and proved in [10, p. 751].

(47) $$| y(u_1) - y(u_2) | \geq 2k(u_0) | u_1 - u_2 | \ ; \ k(u_0) > 0.$$

There also is a $\rho_2 > 0$ such that for $u \in P' \cap S(u_0; \rho_2)$

(48) $$\sigma(u_0)\Delta(u_0)/(\sigma(u)\Delta(u)) = 1 + \theta \text{ with } |\theta| < \epsilon.$$

(45) and (47) imply the existence of a $\rho_3 > 0$, $\rho_3 \leq \rho_1$, such that

(49) $$| x(u_1) - x(u_2) | \geq k(u_0) | u_1 - u_2 | \text{ for } u_i \in P' \cap S(u_0, \rho_3).$$

With the notations

(50) $$w = x(u_1) - x(u_2), \ v = y(u_1) - y(u_2), \ \bar{w} = w/|w|, \ \bar{v} = v/|v|$$

we conclude from

$$|\bar{w} - \bar{v}| = \left| \frac{w(|v| - |w|) + |w|(w - v)}{|v| \cdot |w|} \right| \leq \frac{2|w - v|}{|v|},$$

(45), and (49) that

(51) $$|\bar{w} - \bar{v}| < 2|z|/k \text{ for } u_i \in P' \cap S(u_0, \rho_3).$$

We put $\Phi_0(\xi) = \Phi(x(u_0), \xi)$. Then for $|\xi| = 1$ and a suitable $\rho' > 0$

(52) $$\Phi_0(\xi)/\Phi(x, \xi) = 1 + \theta \text{ with } |\theta| \leq \epsilon \text{ when } |x - x(u_0)| < \rho'.$$

Because Φ is homogeneous in ξ the relation (52) holds for all $\xi \neq 0$. If δ_0 is the (Minkowski) distance which corresponds to Φ_0 as integrand, then it follows from (52) that

(53) $$\delta_0(x(u_1), x(u_2))/\delta(x(u_1), x(u_2)) = 1 + \theta \text{ with } |\theta| \leq \epsilon,$$

when $u_i \in S(u_0, \rho_4)$ with a suitable $\rho_4 > 0$. Moreover

$$\begin{aligned}\Phi_0(w)/\Phi_0(v) &= 1 + \frac{|w|\Phi_0(\bar{w}) - |v|\Phi_0(\bar{v})}{|v|\Phi_0(\bar{v})} \\ &= 1 + \frac{\Phi_0(\bar{w}) - \Phi_0(\bar{v})}{\Phi_0(\bar{v})} + \frac{|w| - |v|}{|v|\Phi_0(\bar{v})}\Phi_0(\bar{w}).\end{aligned}$$

Since $||w| - |v|| \leq |w - v|$ it follows from (45) and (51) that for a suitable positive $\rho_5 \leq \rho_3$

(54) $$\Phi_0(w)/\Phi_0(v) = 1 + \theta \text{ with } |\theta| \leq \epsilon \text{ for } u_i \in P' \cap S(u_0, \rho_5).$$

Let $\rho(u_0) = \min(\rho_2, \rho_4, \rho_5)$, then combination of the results (48), (53) and (54) yields for any measurable set $Q \subset P' \cap S(u_0, \rho(u_0))$

2

$$(55)\quad |x(Q),\delta|_k = (1+\theta_1)^k |x(Q),\delta_0|_k = (1+\theta_2)^{2k} |y(Q),\delta_0|_k$$
$$= (1+\theta_2)^{2k} |y(Q),\eta|_k \sigma(u_0)$$
$$= (1+\theta_2)^{2k} \int_Q \Delta(u_0)\sigma(u_0)\,du$$
$$= (1+\theta_3)^{2k+1} \int_Q \sigma(u)\Delta(u)\,du,$$

where $\eta(x,y)$ is the Euclidean distance $|x-y|$ and $|\theta_i| \leq \epsilon$.

Let P'' be the subset of P' where $\Delta(u) \neq 0$. There is a sequence of spheres $S(u^i, \rho(u^i))$ which cover P''. Putting $Q_1 = S_1 P''$ and generally $Q_n = P''S_n - S_n \cap (\bigcup_{i=1}^{n-1} Q_i)$ we have $P'' = \bigcup Q_i$ and $Q_i \cap Q_j = 0$. Because the mapping $x(u)$ is one to one on P'', $x(Q_1) \cap x(Q_2) = 0$ and the sets $x(Q_i)$ are measurable, therefore by (55)

$$(56)\quad |x(P'),\delta|_k = \sum |x(Q_i),\delta|_k = \sum (1+\theta^{(i)})^{2k+1} \int_{Q_i} \sigma(u)\Delta(u)\,du$$
$$= (1+\theta_4)^{2k+1} \int_{P''} \sigma(u)\Delta(u)\,du.$$

Since by [7, p. 362] or [10, p. 753]

$$|x(P'-P''),\delta|_k < \beta^k |x(P'-P''),\eta|_k = \beta^k \int_{P'-P''} \Delta(u)\,du = 0,$$

it follows that

$$|P',x|_k^H = \int_{P'} \sigma(u)\Delta(u)\,du.$$

By a method which illustrates convincingly the strength of Kolmogoroff's Principle, Theorem 7 can be reduced to (41) and Section 8 of [2]. As before or in [2, Section 8] the general case can be derived from the special case where P is a bounded closed convex set in E^k and x moves in one coordinate neighborhood in F, so that (P,x) is represented in the form $x(u) = (x_1(u_1,\cdots,u_k),\cdots,x_n(u_1,\cdots,u_k))$ with

$$(57)\qquad \delta(x(u_1),x(u_2)) \leq \beta\,|u_1-u_2| \quad \text{for } u_i \in P.$$

As auxiliary space we introduce the E^{n+k} with Cartesian coordinates $x_1,\cdots,x_n,u_1,\cdots,u_k$ and metrize (a portion of) it by the Finsler metric δ^* determined by the integrand

$$(58)\qquad \Phi((x,u),\xi_1,\cdots,\xi_n;\zeta_1,\cdots,\zeta_k) =)[\Phi(x,\xi) + |\zeta|]^{\frac{1}{2}}.$$

Then

$$(59)\qquad (x,u)_\epsilon = (x(u), \epsilon u)$$

is a birectifiable surface in the resulting Finsler space because by (57) and (58)

$$\epsilon | u_1 - u_2 | \leq \delta^*(x(u_1), x(u_2)) \leq (\beta^2 + \epsilon^2)^{\frac{1}{2}} | u_1 - u_2 | .$$

The matrix $\partial(x, u)/\partial u$ is

$$(60) \qquad \begin{pmatrix} \partial x_1/\partial u_1 \cdots \partial x_n/\partial u_1 & \epsilon \cdots 0 \\ \cdots \quad \cdots \quad \cdots & \cdots\cdots \\ \partial x_1/\partial u_k \cdots \partial x_n/\partial u_k & 0 \cdots \epsilon \end{pmatrix}.$$

Therefore the expressions $\Delta^\epsilon(u)$ and $\sigma^\epsilon(u)$ corresponding to $\Delta(u)$ and $\sigma(u)$ for the surface (59) have the following properties

$$(61) \qquad \Delta(u) \leq \Delta^\epsilon(u) \leq (\Delta^2(u) + \epsilon^2 B)^{\frac{1}{2}}$$

where B is a constant which depends only on n, k and β (because $| \partial x_i/\partial x_j | \leqq \beta$). For $\epsilon \to 0$ the expression $\Delta^\epsilon(u)$ tends therefore to $\Delta(u)$. The direction coefficients for the tangent space of (59) at a point u are the k-row determinants formed from (60) and divided by $\Delta^\epsilon(u)$. Hence, if $\Delta(u) \neq 0$, the tangent space tends to a k-dimensional space in $(\xi_1, \cdots, \xi_n, 0, \cdots, 0)$ and $\sigma^\epsilon(u) \to \sigma(u)$. Therefore

$$(62) \qquad \sigma^\epsilon(u)\Delta^\epsilon(u) \to \sigma(u)\Delta(u) \text{ when } \epsilon \to 0.$$

By (58) and (59)

$$(63) \qquad \delta^*[(x(u_1), u_1)_\epsilon, (x(u_2), u_2)_\epsilon] \geq \delta^*[x(u_1), u_1)_0, (x(u_2), u_2)_0] = \delta^*[(x(u_1), 0), (x(u_2),)].$$

If the set traversed by $(x; u)_\epsilon$, as u traverses P, is denoted by $(x(P), P)_\epsilon$ then by (41) and [2, (8.1)]

$$(64) \qquad | P, (x, u)_\epsilon | = | (x(P), P)_\epsilon, \delta | _k{}^H = \int_P \sigma^\epsilon(u)\Delta^\epsilon(u)\, du$$

and by (41)

$$(65) \qquad | P, (x, u)_0 | _k{}^H = \int_P \sigma^0(u)\Delta^0(u)\, du = \int_P \sigma(u)\Delta(u)\, du = | P, x | _k{}^H.$$

Kolmogoroff's Principle (see [2, p. 248]), (67) and [2, (4.10)] show that

$$(66) \qquad | P, (x, u)_\epsilon | \geq | P, (x, u)_0 | \geq | P, (x, u)_0 | _k{}^H.$$

Theorem 7 is now a consequence of (41), (64), (65) and (66).

UNIVERSITY OF SOUTHERN CALIFORNIA.

Am. J. Math. 71 (1949), 743–762

REFERENCES.

[1] T. Bonnesen and W. Fenchel, "Theorie der konvexen Körper," *Ergebnisse der Mathematik*, vol. 3, part 1, Berlin, 1934. .

[2] H. Busemann, "Intrinsic area," *Annals of Mathematics*, vol. 48 (1947), pp. 234-267.

[3] ———, "The isoperimetric problem in the Minkowski plane," *American Journal of Mathematics*, vol. 69 (1947), pp. 863-871.

[4] ———, "On the problem of Dido," *Studies and Essays Presented to R. Courant on his 60th Birthday*, New York, 1948, pp. 63-73.

[5] A. Dinghas and E. Schmidt, "Einfacher Beweis der isoperimetrischen Eigenschaft der Kugel im n-dimensionalen euklidischen Raum," *Abhandlungen der Preussischen Akademie der Wissenschaften*, 1943, Math. Nat. Klasse no. 7 (1944).

[6] G. H. Hardy, J. E. Littlewood, G. Polya, *Inequalities*, Cambridge, 1934.

[7] A. Kolmogoroff, "Beiträge zur Masstheorie," *Mathematische Annalen*, vol. 107 (1932), pp. 351-366.

[8] L. Lusternik, "Die Brunn-Minkowskische Ungleichnung für beliebige messbare Mengen," *Comptes Rendus (Doklady) de l'Académie des Sciences de l'URSS* (1935), vol. III (Nouvelle Série), pp. 55-58.

[9] H. Minkowski, "Theorie der konvexen Köper, insbesondre, Begründung ihres Oberflächenbegriffs," *Gesammelte Abhandlungen*, vol. II, Leipzig (1911), pp. 131-229.

[10] G. Nöbeling, "Über den Flächeninhalt dehnungsbeschränkter Flächen," *Mathematische Zeitschrift*, vol. 48 (1943), pp. 747-771.

[11] W. H. and G. C. Young, *The Theory of Point Sets*, Cambridge, 1906.

Am. J. Math. 71 (1949), 743–762

VOLUME IN TERMS OF CONCURRENT CROSS-SECTIONS

HERBERT BUSEMANN

1. Of the two expressions

$$|M| = \frac{1}{2}\int_0^{2\pi} r^2(\omega)\,d\omega = \frac{1}{2}\int_0^{2\pi}\left(\int_{-r(\omega-\pi/2)}^{r(\omega+\pi/2)} |\rho|\,d\rho\right)d\omega$$

for the area $|M|$ of a plane domain M, given in polar coordinates ρ, ω by the inequalities $0 \le \rho \le r(\omega)$, $0 \le \omega \le 2\pi$, the first has the well-known extension

$$(1) \qquad |M| = \frac{1}{n}\int_{\Omega_n} r^n(u)\,d\omega_u^n$$

to n dimensions. Here Ω_n is the surface of the unit sphere in the n-dimensional Euclidean space, $d\omega_u^n$ is its area element at the point u, and M is given by $0 \le \rho \le r(u)$, $u \in \Omega_n$.

In the second expression, $|\rho|$ may be interpreted as (1-dimensional) volume of the simplex with one vertex at the origin z and the other at a variable point $p = (\rho, \omega \pm \pi/2)$ in the cross-section of M with the line normal to ω. The purpose of the present note is *the proof and the application of the following extension of this second expression to* $n-1$ *sets* $M_1, \cdots, M_{n-1}$ *in* E_n:

$$(2) \qquad |M_1| \cdots |M_{n-1}|$$

$$= \frac{(n-1)!}{2}\int_{\Omega_n}\left(\int_{M_1(u)} \cdots \int_{M_{n-1}(u)} T(p_1, \cdots, p_{n-1}, z)\,dV_{p_1}^{n-1} \cdots dV_{p_{n-1}}^{n-1}\right)d\omega_u^n.$$

Here $M_j(u)$ is the cross-section of M_j with the hyperplane $H(u)$ through z normal to the unit vector u, the point p_j varies in $M_j(u)$, the differential $dV_{p_j}^{n-1}$ is the $((n-1)$-dimensional$)$ volume element of $M_j(u)$ at p_j, and $T(p_1, \cdots, p_{n-1}, z)$ is the volume of the simplex with vertices $p_1, \cdots, p_{n-1}, z$.

Received November 19, 1951.
Pacific J. Math. 3 (1953), 1-12

Replacing the sets $M_{n-r+1}, \cdots, M_{n-1}$ by the unit sphere U_n with center z yields expressions for $|M_1| \cdots |M_{n-r}|$ in terms of the volume $T(p_1, \cdots, p_{n-r}, z)$, in particular (1) for $r = n - 1$.

With the notation

$$\kappa_n = |U_n| = \pi^{n/2}/\Gamma\left(\frac{n}{2} + 1\right),$$

Steiner's symmetrization leads from (2) to the following result:

If $M_1, \cdots, M_{n-1}$ are convex bodies in E_n $(n \geq 3)$ with interior points, z is a given point in E_n, and $M_j(u)$ the cross-section of M_j with the plane normal to u and through z, then

$$(3)\quad |M_1| \cdots |M_{n-1}| \geq \frac{1}{n} \frac{\kappa_n^{n-2}}{\kappa_{n-1}^n} \int_{\Omega_n} |M_1(u)|^{n/(n-1)} \cdots |M_{n-1}(u)|^{n/(n-1)} d\omega_u^n,$$

and the equality sign holds only when the M_j are homothetic ellipsoids with center z.

It follows in particular for a convex body M that, for $n \geq 3$,

$$(4)\qquad |M|^{n-1} \geq \frac{1}{n} \frac{\kappa_n^{n-2}}{\kappa_{n-1}^n} \int_{\Omega_n} |M(u)|^n \, d\omega_u^n,$$

with the equality (if $|M| > 0$) only for ellipsoids with center z. The efforts to prove this inequality, which has applications in Finsler spaces, led to the present investigation. The—because of Jensen's inequality—weaker estimate

$$(5)\qquad |M| \geq \frac{1}{n} \kappa_n^{-n/(n-1)} \int_{\Omega_n} |M(u)|^{n/(n-1)} \, d\omega_u^n,$$

with equality sign (if $|M| > 0$) only for the spheres with center z, was found previously by L. A. Santaló who communicated it to the author. It is also the special case $M_{n-1} = M$, $M_j = U_n$ for $j < n - 1$, of (3).

2. Let $M_1, \cdots, M_{n-1}$ be bounded Jordan measurable sets in E_n, $n > 3$, such that the intersection of M_j with any ν-dimensional linear subspace (which in the future will be indicated by L_ν) through a fixed given point z possesses a

ν-dimensional Jordan measure. Since the subscripts run sometimes from 1 to n and other times from 1 to $n-1$, we agree to use α, β for the former type, and j, k for the latter, and may then omit mentioning the range.

Let x_α be rectangular coordinates in E_n with z as origin. Take $n-1$ copies E_n^j of E_n with coordinates x_α^j, and let M_j' be the image of M_j in E_n^j; that is, $x^j \in M_j'$ if and only if the point x with $x_\alpha = x_\alpha^j$ lies in M_j. Then x_α^j may be considered as rectangular coordinates in the product space

$$E = E_n' \times \cdots \times E_n^{n-1} = \prod E_n^j;$$

hence

$$\prod |M_j| = \prod |M_j'| = \int_{\prod M_j'} dx_1^1 \cdots dx_n^1 \cdots dx_1^{n-1} \cdots dx_n^{n-1}. \tag{6}$$

In E we introduce new coordinates

$$\bar{x}_1^1, \ldots, \bar{x}_{n-1}^1, v_1, \ldots, \bar{x}_1^{n-1}, \ldots, \bar{x}_{n-1}^{n-1}, v_{n-1}$$

through the relations

$$x_k^j = \bar{x}_k^j, \; x_n^j = v_1 \bar{x}_1^j + \cdots + v_{n-1} \bar{x}_{n-1}^j. \tag{7}$$

These equations fail to define v_j if $|x_k^j| = |\bar{x}_k^j| = 0$, i.e. if the points x^j in E_n are contained in an L_ν with $\nu < n-2$, or if the L_{n-2} spanned by the x^j is parallel to the x_n-axis. The geometric meaning of the right side of (2) shows that a special discussion of this case is superfluous.

To evaluate $\prod |(M_j')|$ in the new coordinates, observe that the first n rows in the $n(n-1)$-rowed Jacobian J of the transformation (7) are in blocks of $n \times n$ matrices:

$$\begin{matrix} 1 & 0 & \cdots & 0 & 0 & & 0 & 0 & \cdots & 0 & 0 & & 0 & 0 & \cdots & 0 & 0 \\ 0 & 1 & \cdots & 0 & 0 & & 0 & 0 & \cdots & 0 & 0 & & 0 & 0 & \cdots & 0 & 0 \\ \cdot & \cdot & \cdots & \cdot & \cdot & & \cdot & \cdot & \cdots & \cdot & \cdot & & \cdot & \cdot & \cdots & 0 & 0 \\ 0 & 0 & \cdots & 1 & 0 & & 0 & 0 & \cdots & 0 & 0 & & 0 & 0 & \cdots & 0 & 0 \\ v_1 & v_2 & \cdots & v_{n-1} & \bar{x}_1^1, & & v_1 & v_2 & \cdots & v_{n-1} & \bar{x}_2^1, & & v_1 & v_2 & \cdots & v_{n-1} & \bar{x}_{n-1}^1; \end{matrix}$$

hence

(8) $$J = |\bar{x}^j_k|.$$

The unit normal u in E_n to the plane

$$x_n = x_1 v_1 + \cdots + x_{n-1} v_{n-1}$$

is, with $w = (1 + v_1^2 + \cdots + v_{n-1}^2)^{1/2}$, either

$$u_j = v_j w^{-1/2},\ u_n = -w^{-1/2} \quad \text{or} \quad u_j = -v_j w^{-1/2},\ u_n = w^{-1/2}.$$

Then $w^{-1} = |\cos\theta|$, where θ is the angle between u and the x_n-axis, so that $d\omega_u^n = w\, du_1 \cdots du_{n-1}$ is the area element of Ω_n. Here we disregard again planes parallel to the x_n-axis. Now

$$\left|\frac{\partial u_j}{\partial v_k}\right| = w^{-3(n-1)} \begin{vmatrix} w^2 - v_1^2 & -v_1 v_2 & \cdots & -v_1 v_{n-1} \\ -v_2 v_1 & w^2 - v_2^2 & \cdots & -v_2 v_{n-1} \\ \cdot & \cdot & \cdots & \cdot \\ -v_{n-1} v_1 & -v_{n-1} v_2 & \cdots & w^2 - v_{n-1}^2 \end{vmatrix}$$

Since all principal minors of the determinant $|-v_j v_k|$ of order greater than 1 vanish, it follows (compare [4, pp. 125, 126]) that

$$\left|\frac{\partial u_j}{\partial v_k}\right| = w^{-3(n-1)} \left(w^{2(n-1)} - w^{2(n-2)}\sum v_j^2\right) = w^{-n-1}$$

and

(9) $$d\omega_u^n = w^{1-n-1}\, dv_1 \cdots dv_{n-1} = |\cos^n\theta|\, dv_1 \cdots dv_{n-1}.$$

The volume element dV_{xj}^{n-1} of the hyperplane $x_n^j = \sum x_k^j v_k$ is

(10) $$dV_{xj}^{n-1} = dx_1^j \cdots dx_{n-1}^j\, |\sec\theta|.$$

If we now interpret the points $x^1, \cdots, x^{n-1}$ as lying in the same E_n, then (8) shows that $|J|/(n-1)!$ is the volume of the projection of the simplex with vertices $x^1, \cdots, x^{n-1}$, z on the plane $x_n = 0$. Since these points determine a hyperplane $H(u)$ with normal u through z,

$$(11)\qquad (n-1)!\ T(x^1, \cdots, x^{n-1}, z) = |J \sec \theta|.$$

Replacing x^j by p_j, we briefly summarize the results (9), (10), (11) as

$$(12)\quad dV^n_{p_1} \cdots dV^n_{p_{n-1}} = (n-1)!\ T(p_1, \cdots, p_{n-1}, z)\, dV^{n-1}_{p_1} \cdots dV^{n-1}_{p_{n-1}}\, d\omega^n_u.$$

After observing that in (2) by integrating over Ω_n every $M^j(u)$ is counted twice (once for u and once for $-u$), we see that *the relation* (2) *follows from* (12).

For brevity we introduce, for sets $M_1, \cdots, M_r$ in E_s with $r \leq s$, the notation

$$\tau_s(M_1, \cdots, M_r, z) = \int_{M_1} \cdots \int_{M_r} T(p_1, \cdots, p_r, z)\, dV^s_{p_1} \cdots dV^s_{p_s},$$

and may then write (2) in the form

$$(13)\quad |M_1| \cdots |M_{n-1}| = \frac{(n-1)!}{2} \int_{\Omega_n} \tau_{n-1}(M_1(u), \cdots, M_{n-1}(u), z)\, d\omega^n_u.$$

3. In order to obtain *expressions for* $|M_1| \cdots |M_r|$ *with* $r < n-1$, we replace successively $M_{n-1}, \cdots, M_{n-r+1}$ by the unit sphere U_n. The contribution of the latter sets to the right side of (13) can then be integrated out by using the following fact:

Let an L_μ, $0 < \mu < n-1$, through the center z of the unit sphere U_{n-1} in E_{n-1}, intersect U_{n-1} in U_μ. For any point q in U_{n-1}, denote by r the distance qz, and by ϕ the angle between the ray qz and the L_μ. Then

$$(14)\qquad \int_{U_{n-1}} r|\sin\phi|\, dV^{n-1}_q = \frac{\omega_{\nu-1}}{\omega_\nu} \cdot \kappa_n, \quad \nu = n - \mu,$$

where $\omega_\nu = \nu \cdot \kappa_\nu = 2\pi^{\nu/2}\,\Gamma^{-1}(\nu/2)$ is the area of the surface Ω_ν of U_ν, in particular $\omega_1 = 2$.

To prove (14), let the $L_{\nu-1}$ normal to the L_μ through q intersect U_μ in p, and U_{n-1} in the sphere $S_{\nu-1}$. If $\rho = pz$ then $S_{\nu-1}$ has radius $\sigma = (1-\rho^2)^{1/2}$. Then $s = pq = r\ |\sin\phi|$; hence

$$\int_{U_{n-1}} r\,|\sin\phi|\ dV^{n-1}_q = \int_{U_\mu} \left(\int_{S_{\nu-1}} s\ dV^{\nu-1}_q \right) dV^\mu_p,$$

If $d\omega^{\nu-1}_q$ denotes the area element of the $\Omega_{\nu-1}$ with center p in the $L_{\nu-1}$ at the point $\bar{q}$ of the ray pq, then

$$\int_{S_{\nu-1}} s\, dV_q^{\nu-1} = \int_{\Omega_{\nu-1}} \int_0^{\sigma} s\, s^{\nu-2}\, ds\, d\omega_{\bar{q}}^{\nu-1} = \frac{\sigma^{\nu}}{\nu}\, \omega_{\nu-1} = (1-\rho^2)^{\nu/2}\, \frac{\omega_{\nu-1}}{\nu}.$$

Therefore, with a similar notation,

$$\int_{U_{n-1}} r\,|\sin\phi|\, dV_q^{n-1} = \frac{\omega_{\nu-1}}{\nu} \int_{\Omega_\mu} \int_0^1 (1-\rho^2)^{\nu/2}\, \rho^{\mu-1}\, d\rho\, d\omega_{\bar{p}}^{\mu}$$

$$= \frac{\omega_{\nu-1}\,\omega_\mu}{\nu}\, \frac{\Gamma(\mu/2)\,\Gamma(\nu/2+1)}{2\Gamma(\mu/2+\nu/2+1)} = \frac{\omega_{\nu-1}\,\omega_\mu}{\nu}\, \frac{\Gamma(\mu/2)\,\nu/2\,\Gamma(\nu/2)}{2\Gamma(n/2+1)} = \frac{\omega_{\nu-1}}{\omega_\nu}\, \kappa_n.$$

Returning to (13), we replace M_{n-1} by U_n. Then $M_{n-1}(u)$ becomes the $(n-1)$-dimensional unit sphere $U_n(u)$ in the hyperplane with normal u. If ϕ is the angle between the ray zp_{n-1} and the L_{n-2} spanned by $p_1, \cdots, p_{n-2}, z$, then, with $r = zp_{n-1}$,

$$T(p_1, \cdots, p_{n-1}, z) = (n-1)^{-1}\, r\,|\sin\phi|\; T(p_1, \cdots, p_{n-2}, z).$$

Hence, carrying out the integration over $U_n(u)$, by (14) we obtain

$$|M_1| \cdots |M_{n-2}| \cdot \kappa_n$$

$$= \frac{1}{2}(n-2)!\, \frac{\omega_1}{\omega_2}\, \kappa_n \int_{\Omega_n} \tau_{n-1}(M_1(u), \cdots, M_{n-2}(u), z)\, d\omega_u^n$$

or

(15) $|M_1| \cdots |M_{n-2}|$

$$= \frac{(n-2)!}{2\pi} \int_{\Omega_n} \tau_{n-1}(M_1(u), \cdots, M_{n-2}(u), z)\, d\omega_u^n.$$

If now M_{n-2} is replaced by U_n, then because of (14) the factor

$$\frac{1}{n-2}\, \frac{\omega_2}{\omega_3}\, \kappa_n$$

is introduced on the right. Continuing in this manner leads to *the general relation*

$$(16) \qquad |M_1| \cdots |M_{n-r}| = \frac{(n-r)!}{\omega_r} \int_{\Omega_n} \tau_{n-1}\,(M_1(u), \cdots, M_{n-r}(u), z)\, d\omega_u^n .$$

The integrand occurs in many $H(u)$, and it would be more natural to replace the integration over Ω_n by an integration over all L_{n-r}. The results of integral geometry [5] lead to such a reduction for general r; however, we restrict our attention here to the two simplest cases, where no new formulas of the type (12) are required.

It is clear that *the last formula in the sequence* (16),

$$|M_1| = \frac{1}{\omega_n} \int_{\Omega_n} \tau_{n-1}\,(M_1(u), z)\, d\omega_u^n ,$$

must be essentially identical with (1). Indeed, if M_1 can be represented in the form $0 \leq \rho \leq r(u)$, $u \in \Omega_n$, and we write the induced representation of $M_1(u)$ in the form

$$0 \leq \rho \leq r(v), \quad v \in \Omega_{n-2}(u) = H(u) \cap \Omega_n ,$$

then, with $pz = \rho$, we have

$$\tau(M_1(u), z) = \int_{M_1(u)} \rho\, dV_p^{n-1} = \int_{\Omega_{n-1}(u)} \int_0^{r(v)} \rho\, \rho^{n-2}\, d\omega_v^{n-1}$$

$$= \frac{1}{n} \int_{\Omega_{n-1}(u)} r^n(v)\, d\omega_v^{n-1}$$

and

$$|M_1| = \frac{1}{n \cdot \omega_{n-2}} \int_{\Omega_n} \left(\int_{\Omega_{n-1}(u)} r^n(v)\, d\omega_v^{n-2} \right) d\omega_u^n .$$

Now according to the results on cinematic measure on the sphere (see [5], for $n = 3$ already [3]), integrating over the v-normal to u first, and then over u, leads to the same result as integrating over the $H(w)$ that contain v, that is, those for which w is normal to v, and then over v. The first of the latter two integrations yields $\omega_{n-2} r^n(v)$, and (1) follows.

As second example, we indicate briefly the reduction of (15). Denoting by L^p_{n-2} the L_{n-2} spanned by $p_1, \cdots, p_{n-2}, z$, and by M^p_i the intersection of M_i with L^p_{n-2}, we obtain from (12) that if L^p_{n-2} lies in $H(u)$ and has there the normal v, then

$$\tau_{n-1}(M_1(u), \cdots, M_{n-2}(u), z)$$

$$= \int_{M_1(u)} \cdots \int_{M_{n-2}(u)} T(p_1, \cdots, p_{n-2}, z)\, dV^{n-1}_{p_1} \cdots dV^{n-1}_{p_{n-2}}$$

$$= \frac{(n-2)!}{2} \int_{\Omega_{n-1}(u)} \int_{M^p_1} \cdots \int_{M^p_{n-2}} T^2(p_1, \cdots, p_{n-2}, z)\, dV^{n-2}_{p_1} \cdots dV^{n-2}_{p_{n-2}}\, d\omega^{n-1}_v .$$

Substituting this in (15) leads besides the integrations over the M^p_i to integrations over $\Omega_{n-1}(u)$ and Ω_n. Similarly as in the preceding case, these latter two may be reduced to one integration by using the cinematic measure dL^p_{n-2} on Ω_n of the Ω_{n-2} in which L^p_{n-2} intersects Ω_n (compare [5]). The result (given without verification because it will not be used) is

$$(17)\quad |M_1| \cdots |M_{n-2}|$$

$$= \frac{[(n-2)!]^2}{2} \int_{\Omega_n} \int_{M^p_1} \cdots \int_{M^p_{n-2}} T^2(p_1, \cdots, p_{n-2}, z)\, dV^{v-2}_{p_1} \cdots dV^{n-2}_{p_{n-2}}\, dL^p_{n-2} .$$

4. To obtain the estimate (3) we use Steiner's symmetrization in the form suggested by Blaschke's treatment of Sylvester's Problem (compare [1, § 24]). In the following the subscripts i, h run from 1 to m.

Let $M_1, \cdots, M_m$ be convex bodies with interior points in E_m. In an arbitrary system of rectangular coordinates with origin z, symmetrize each M_i with respect to the $(x_1, \cdots, x_{m-1})$-plane P; that is, slide a segment in which a line L_1 parallel to the x_m-axis intersects M_i along L_1 such that its center falls on P. Call $\overline{M}_i$ the image of M_i under this transformation, and $\overline{p}_i$ the image in $\overline{M}_i$ of a given point p_i in M_i. The mapping preserves volume, $dV^m_{\overline{p}_i} = dV^m_{p_i}$. We are going to show that $\tau_m(M_1, \cdots, M_m, z)$ does not increase.

If $p_i \in M_i$, denote by p_i' the point symmetric to p_i with respect to the center of that chord of M_i parallel to the x_m-axis which goes through p_i. If $p_i^1, \cdots, p_i^m$ are the coordinates of p_i, then with $\eta = 1/m\,!$

$$\pm T(p_1, \cdots, p_m, z) = \eta\, |p_i^h|, \quad \pm T(p_1', \cdots, p_m', z) = \eta\, |p_i'^{\,h}| .$$

The images $\overline{p}_i$ of p_i and $\overline{p}_i'$ of p_i' satisfy the relation

$$|\overline{p}_i^h| = -|\overline{p}_i'^h|;$$

hence

$$2T(\overline{p}_1, \cdots, \overline{p}_m, z) = 2T(\overline{p}_1', \cdots, \overline{p}_m', z) = \eta \Big| |\overline{p}_i^h| - |\overline{p}_i'^h| \Big|.$$

But

$$p_i^h = p_i'^h = \overline{p}_i^h = \overline{p}_i'^h \text{ for } 1 \leq h \leq m-1, \text{ and } p_i^m - p_i'^m = \overline{p}_i^m - \overline{p}_i'^m.$$

so that

$$|\overline{p}_i^h| - |\overline{p}_i'^h| = |p_i^h| - |p_i'^h|;$$

hence

$$T(p_1, \cdots, p_m, z) + T(p_1', \cdots, p_m', z) \geq 2T(\overline{p}_1, \cdots, \overline{p}_m, z). \tag{18}$$

Since

$$\tau_m(M_1, \cdots, M_m, z) = \int_{M_1} \cdots \int_{M_m} T(p_1, \cdots, p_m, z)\, dV_{p_1}^m \cdots dV_{p_m}^m$$
$$= \int_{M_1} \cdots \int_{M_m} T(p_1', \cdots, p_m', z)\, dV_{p_1'}^m \cdots dV_{p_m'}^m,$$

we conclude from (18) and

$$dV_{\overline{p}_i}^m = dV_{p_i}^m,\ dV_{\overline{p}_i'}^m = dV_{p_i'}^m$$

that

$$\tau_m(M_1, \cdots, M_m, z) \geq \tau_m(\overline{M}_1, \cdots, \overline{M}_m, z). \tag{19}$$

To discuss the equality sign, consider points p_i in M_i which are centers of chords parallel to the x_n-axis. Then $p_i = p_i'$, and the points $\overline{p}_i = \overline{p}_i'$ lie in P, so that the right side of (18) vanishes. Therefore the equality sign can hold in (18) only when the points $p_1, \cdots, p_m, z$ are coplanar. Choosing $p_1, \cdots, p_{i-1}$,

$p_{i+1}, \cdots, p_m$ such that they and z do not lie in an L_{m-2} (the M_i have interior points!) we see that all centers of chords of M_i parallel to the x_m-axis must lie in the L_{m-1} spanned by $p_1, \cdots, p_{i-1}, p_{i+1}, \cdots, p_m, z$. Moreover, this same L_{m-1} must contain the centers of the chords parallel to the x_m-axis of all the other M_n. Thus we have proved:

(20) If $M_1, \cdots, M_m$ are convex bodies in E_m with interior points, then simultaneous symmetrization of the M_i with respect to any plane P through z decreases $\tau_m(M_1, \cdots, M_m, z)$ unless z and the centers of the chords perpendicular to P of all M_i are coplanar.

For given positive values $|M_1|, \cdots, |M_m|$, the expression $\tau_m(M_1, \cdots, M_m, z)$ can therefore be minimal only if the centers of every family of parallel chords of the different M_j lie on the same plane through z. This implies, first, that *each M_j is an ellipsoid with center z,*[1] *and then that all these ellipsoids are homothetic.*

That *the minimum is actually reached* in this case is proved by the following standard argument (see [1, § 24]). Using a suitable sequence P_ν of planes through z, and symmetrizing $M_1, \cdots, M_m$ successively in $P_1, P_2, \cdots$, yields a sequence $M_1^\nu, \cdots, M_m^\nu$ of convex bodies which tend to spheres $S_1, \cdots, S_m$ with center z and, of course, $|S_i| = |M_i^\nu| = |M_i|$ (compare [2, § 41]).

The functional $\tau_m(M_1, \cdots, M_m, z)$ is monotone [that is, $M_i' \subset M_i$ implies $\tau_m(M_1', \cdots, M_m', z) \leq \tau_m(M_1, \cdots, M_m, z)$] and positive homogeneous:

$$\tau_m(\lambda M_1, \cdots, \lambda M_m, z) = \lambda^{m(m+1)} \tau_m(M_1, \cdots, M_m, z) \quad \text{for } \lambda > 0.$$

For a given $\epsilon > 0$, choose $N(\epsilon) > 0$ such that $S_i \subset (1+\epsilon)M_i^\nu$ for $\nu > N(\epsilon)$ and all i. Then for $\nu > N(\epsilon)$, because of (20) and the two mentioned properties, we have

$$\begin{aligned}\tau_m(S_1, \cdots, S_m, z) &\leq (1+\epsilon)^{m(m+1)} \tau_m(M_1^\nu, \cdots, M_m^\nu, z)\\ &\leq (1+\epsilon)^{m(m+1)} \tau_m(M_1, \cdots, M_m, z),\end{aligned}$$

which proves $\tau_m(S_1, \cdots, S_m, z) \leq \tau_m(M_1, \cdots, M_m, z)$ and hence the mini-

[1] The proofs found in the literature all refer to the cases $m = 2, 3$; for references [2, § 70]. However, the extension to arbitrary m is immediate. A particularly simple proof, which works for all m and is not found in the literature, is obtained by using Loewner's result, that there is exactly one ellipsoid which has a given center, contains a given convex body, and has minimal volume.

mum property for homothetic ellipsoids with center z.

To evaluate $\tau_m(S_1, \cdots, S_m, z)$, denote the radius of S_i by r_i. Then the results of section 3 show that

$$\tau_m(S_1, \cdots, S_m, z) = \frac{1}{m} r_m^{m+1} \frac{\omega_1}{\omega_2} \kappa_{m+1} \tau_m(S_1, \cdots, S_{m-1}, z)$$

$$= \frac{1}{m(m-1)} \frac{\omega_1}{\omega_3} r_m^{m+1} r_{m+1}^{m+1} \kappa_{m+1}^2 \tau_m(S_1, \cdots, S_{m-2}, z) = \cdots$$

$$= \frac{1}{m!} \prod r_i^{m+1} \frac{\omega_1}{\omega_{m+1}} \kappa_{m+1}^m = \frac{1}{m!} \frac{2}{\omega_{m+1}} \frac{\kappa_{m+1}^m}{\kappa_m^{m+1}} \prod |S_i|^{(m+1)/m}.$$

Therefore we have:

(21) *If $M_1, \cdots, M_m$ are convex bodies in E_m with interior points, then*

$$\tau_m(M_1, \cdots, M_m, z) \geq \frac{2}{(m+1)!} \frac{\kappa_{m+1}^{m-1}}{\kappa_m^{m+1}} \prod |M_i|^{(m+1)/m},$$

and the equality sign holds only for homothetic ellipsoids with center z.

Applying this result to (13) yields the inequalities (3) and (4) with the conditions for the equality sign. The latter result may also be formulated as follows:

Among all convex bodies M with a given volume, the ellipsoids with center z (and only these) maximize $\int_{\Omega_n} |M(u)|^n d\omega_u^n$.

To ask for the minimum is senseless since for any convex body M the integral $\int_{\Omega_n} |M(u)|^n d\omega_u^n$ will tend to zero when M moves to infinity. However, it is a meaningful, but unsolved, problem to find the minimum of this integral for all convex bodies with a given volume and center z. This is equivalent to the problem of finding the smallest constant K such that for any convex M with center z the inequality

$$K \int_{\Omega_n} |M(u)|^n d\omega_u^n \geq |M|^{n-1}$$

holds. The existence of K follows readily from (2).

Finally (5) shows:

Among all convex bodies with center z the spheres (and only these) yield the maximum of

$$\min_u |M(u)|^n |M|^{1-n}.$$

The corresponding minimum maximum problem seems quite difficult.

References

1. W. Blaschke, *Vorlesungen über Differentialgeometrie*, vol. II, Berlin, 1923, and New York, 1949.

2. T. Bonnesen und W. Fenchel, *Theorie der konvexen Körper*, Berlin, 1934, and New York, 1948.

3. P. Funk, *Über Flächen mit lauter geschlossenen geodätischen Linien*, Math. Ann. 74 (1913), 278-300.

4. G. Kowalewski, *Einführung in die Determinantentheorie*, Leipzig, 1909.

5. B. Petkantschin, *Integralgeometrie 6. Zusammenhänge zwischen den Dichten der linearen Unterräume im n-dimensionalen Raum*, Abh. Math. Sem. Univ. Hamburg **11** (1936), 249-310.

University of Southern California

Pac. J. Math. 3 (1953), 1–12

PROBLEMS ON CONVEX BODIES

H. BUSEMANN and C. M. PETTY

The theory of Minkowski spaces leads to various problems concerning the plane sections through the center of a symmetric convex body. Many of these problems share with questions in number theory the great appeal of being understandable to a layman (at least in three dimensions) without being easy. Contributions to these problems would not only *advance the theory of Minkowski spaces, but lead the way to a new direction of research on convex bodies.*

This note lists some of the problems, stating what is known about them (if anything), and interpreting them in terms of Minkowskian geometry. These interpretations presuppose familiarity with the concepts created by one of the authors in [4] and may be omitted by readers interested only in convex bodies.

The following notations will be used throughout. We consider an n-dimensional euclidean space E^n, $n \geqq 3$, with rectangular coordinates $x_1, \ldots, x_n$ whose origin is denoted by z. The symbol u will always stand for a unit vector with origin z and components $u_1, \ldots, u_n$. The end point of u traverses the unit sphere Ω. If $f(u)$ is positive and defined on Ω the "surface $uf(u)$" is given by the equations

$$x_i = u_i f(u_1, \ldots, u_n), \qquad u \in \Omega .$$

The hyperplane $\Sigma x_i u_i = \Sigma x_i(-u_i) = 0$ is denoted by $H(u)$ or $H(-u)$. K is a convex body with z as interior point and *center*; V is the volume of K. The intersection of K and $H(u)$ is $K(u)$ and $A(u)$ is the $(n-1)$-dimensional measure or area of $K(u)$.

The first problem refers to two such convex bodies K and K' with center z. Let V' and $A'(u)$ be defined for K' as V and $A(u)$ for K.

PROBLEM 1: *Does $A'(u) \leqq A(u)$ for all $u \in \Omega$ imply $V' \leqq V$?*

The answer is affirmative if K' is an ellipsoid because, see [5],

$$V^{n-1} \geqq \varkappa_n^{\,n-2} \varkappa_{n-1}^{\,-n} n^{-1} \int_\Omega A(u)^n \, d\omega_u \tag{1}$$

Received March 31, 1956.

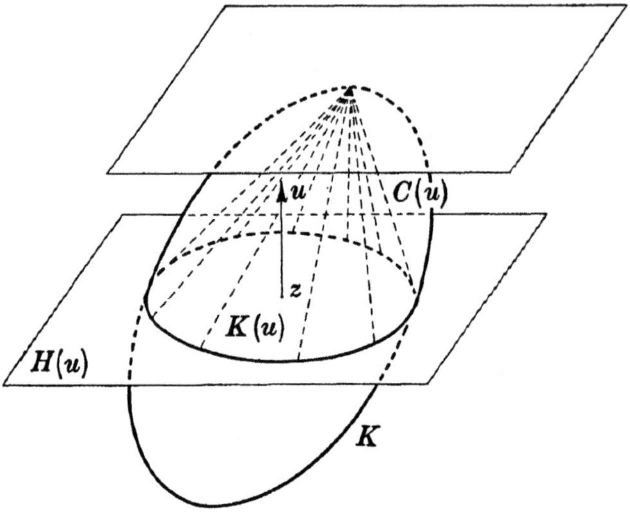

with equality only when K is an ellipsoid. Here $\varkappa_\nu = \pi^{\nu/2}/\Gamma(\nu/2+1)$ (equal to the volume of the solid unit sphere in E^ν) and $d\omega_u$ is the area element of Ω at u. The only other known special case is when $A'(u) = cA(u)$ for all $u \in \Omega$. It then follows from Funk's spherical integration theorem, see [2, pp. 136–138], that K' and K are homothetic.

This problem is equivalent to the following question on area in Minkowski spaces: Let $F(x-y)$, $F'(x-y)$ define two Minkowski metrics in E^n (i.e. $F(x)$ satisfies the conditions:

$$F(x) > 0 \text{ for } x \neq (0, \ldots, 0), \quad F(\mu x) = |\mu| F(x), \quad F(x) \text{ is convex;}$$

similarly for $F'(x)$, compare [6, p. 100]). If for some r, $2 \leqq r \leqq n-1$, the area of any (sufficiently smooth) r-dimensional surface with respect to the metric F does not exceed its area with respect to F', does then the same hold for s-dimensional surfaces with $s > r$? The answer is, in general, negative for $s < r$, even when both metrics are euclidean, i.e. K and K' are both ellipsoids, and is trivially affirmative if $r = 1$, because then $K' \subset K$.

The inequality (1) states that the ellipsoids maximize $V^{1-n} \int_\Omega A(u)^n \, d\omega_u$. This suggests

Problem 2: *To find a convex body K* (with center z) *which minimizes* $V^{1-n} \int_\Omega A(u)^n \, d\omega_u$.

The greatest lower bound of this expression is positive (see [5, p. 11]) and the existence of K can be deduced by an affine form of Blaschke's Selection theorem.

It is doubtful whether the solution of this problem would have interesting applications in Minkowskian geometry (unless the answer is an interesting body). However the following modification is important:

PROBLEM 3: *To find an estimate from above for V in terms of $A(u)$ such that the equality sign characterizes the ellipsoids.*

Experimenting with this problem led the authors to conjecture that such an estimate exists but involves the partial derivatives of

$$A(u) = A(u_1, \ldots, u_n) .$$

Any characterization of the ellipsoids is significant for Minkowskian geometry because it means a characterization of the euclidean geometry. A solution of Problem 3 would, most likely, in addition give important information about the solutions of the isoperimetric problem, which are the surfaces homothetic to the polar reciprocal of the surface $uA(u)$ with respect to Ω. These surfaces are, in general, not homothetic to the spheres (i.e. to the boundary of K), see the comments after Problem 5. However, they play in many Minkowskian theorems the role of spheres in the analogous euclidean theorems, see [4].

The inequality (1) implies further that

$$V^{n-1} \geqq \varkappa_n^{\,n-2} \varkappa_{n-1}^{\,-n} n^{-1} \min_u A(u)^n$$

with equality only for spheres, which therefore yield

$$\max_K \min_u A(u)^n V^{1-n} .$$

This leads to

PROBLEM 4: *Which K yields* $\min_K \max_u A(u)^n V^{1-n}$?

(The corresponding minmin and maxmax problems obviously have no solution.)

An affirmative answer to Problem 1 in the case where K is a sphere would solve Problem 4. For putting

$$r^{n-1} \varkappa_{n-1} = \max_u A'(u) ,$$

we would have

$$\varkappa_{n-1}^{\,n} \varkappa_n^{\,1-n} \leqq \max A'(u)^n (V')^{1-n}$$

for all K' so that Problem 4 would be solved by spheres. Its significance for Minkowskian geometry concerns again the solutions of the isoperimetric problem. The answer to Problem 4 would in particular decide the question whether the so-called isoperimetrix, i.e. the solution of the isoperimetric problem with center z and Minkowski area equal to n times its Minkowskian volume, can be properly contained in the unit sphere K. Problem 6, which is related to Problem 4, would also settle this question.

For a given u we construct the cone of maximal volume $C(u)$ with base $K(u)$ and apex in K. The apex of such a cone is any point of K on a supporting plane of K parallel to $H(u)$.

PROBLEM 5: *Are the ellipsoids characterized by the property that $C(u)$ is constant* (i.e. independent of u for fixed K)?

This problem is of particular interest owing to its connection with certain well known results of Radon [8] and Blaschke [1]. Radon discovered in [8] that symmetry of perpendicularity for lines in a Minkowski plane does not imply that the metric is euclidean. The plane geometries with symmetric perpendicularity are exactly those with unit circles for which the triangles, corresponding to $C(u)$, have constant area, see [4] and [6, p. 104]. Blaschke [1], compare also [6, p. 103], showed that symmetry of perpendicularity for lines in a Minkowski space of dimension $n \geqq 3$ implies that the metric is euclidean. This seemed to indicate that there is no analogue to Radon's curves in higher dimensions. However, in [4] it is shown that there is a natural way of defining perpendicularity of a line to a hyperplane and of a hyperplane to a line. The K with constant $C(u)$ are exactly those, which as unit spheres of Minkowskian geometries lead to the case where the two pairings between lines and hyperplanes are identical. Moreover, the surface $uC(u)^{-1}$ is a solution of the isoperimetric problem, so that Problem 5 also inquires whether in higher dimensional Minkowskian geometries, other than the euclidean, the solutions of the isoperimetric problem can be homothetic to the spheres.

PROBLEM 6: *To find a convex body for which $V^{-1} \max_u C(u)$ is minimal.*

The corresponding minmin and maxmax problems are simple and are solved by cylinders and double cones. The maxmin problem is also unsolved for higher dimensions, but in the plane it is solved by regular affine hexagons, see [7].

It is shown in [7] that the ellipse solves Problem 6 in two dimensions, so that the ellipsoid may be conjectured as answer to this problem. In terms of Minkowskian geometry the problem means: for each Minkowskian geometry we consider the maximum of the Minkowski sine between line and hyperplane, find the Minkowskian geometries for which this maximum is minimal.

PROBLEM 7: *Find a convex body K for which the integral $\int_{\partial K} A(u)^{-1}\, ds_u$*

is minimal, where ds_u is the area element of the boundary ∂K of K at a point of contact of a tangent plane parallel to $H(u)$.

For 2 dimensions, the anwer is a regular affine hexagon, see [7]. *The corresponding maximum problem is solved by the parallelepiped.* For a sketch of the proof, let $E(u)$ be the supporting function of K. Then, since $A(u)$ is maximal among the areas of parallel sections of K normal to u, we have

$$\int_{\partial K} A(u)^{-1}\, ds_u = \int_{\partial K} E(u)\,[E(u)A(u)]^{-1}\, ds_u \leqq 2V^{-1}\int_{\partial K} E(u)\, ds_u = 2n\,.$$

To find the condition on K for equality, we observe that there can be at most $2n$ distinct tangent planes to K for which $E(u)A(u) = V/2$. But since this condition must hold for almost all points of ∂K, K is a polyhedron with at most $2n$ faces. However, all polyhedra with center have at least $2n$ faces and the only polyhedron with center and $2n$ faces is a parallelepiped.

The Minkowskian interpretation of Problem 7 is to find a Minkowski metric for which the area of the unit sphere is minimal.

The difficulty in all these extremal problems derives from the fact that none of the symmetrization processes can be applied, because they do not transform plane sections into plane sections.

PROBLEM 8: *Are the ellipsoids characterized by the fact that the Gauss curvature at a point of contact with a tangent plane parallel to $H(u)$ is proportional to $A(u)^{-(n+1)}$?*

The answer is affirmative for two dimensions. The Minkowskian interpretation involves the Minkowski curvature of surfaces and asks: is the geometry euclidean if the Minkowski spheres have constant curvature?

For a given point p on the boundary ∂K of K, determine u such that the cone with apex p and base $K(u)$ has maximal volume (in general p is not an apex of $C(u)$). Call $P(u)$ the $(n-1)$-dimensional area of the projection of K on $H(u)$ parallel to the line pz. Clearly $P(u) \geqq A(u)$.

PROBLEM 9: *Can the ratio $P(u)/A(u)$ be constant for K that are not ellipsoids?*

If the value of this constant is one, the answer to Problem 9 is negative, because it is then merely a reformulation of Blaschke's theorem, see [1] and [6, p. 93], that the ellipsoids are the only convex bodies with plane shadow boundaries.

In the language of Minkowskian geometry with K as unit sphere, the plane $H(u)$ is normal to the line pz or pz is transversal to $H(u)$. The ratio $\varkappa_{n-1}P(u)/A(u)$ is the area of the tranversal projection of K on $H(u)$, hence the analogue to the "äussere Quermass" of K in the euclidean case, see [2, pp. 30–31]. Problem 9 inquires whether in a Minkowskian geometry, other than the euclidean, a sphere can have constant "äusseres Quermass". In Minkowski spaces, whose spheres solve Problem 9, there is an analogue to Cauchy's formula for area, see [2, p. 48], and hence to its implications, in particular in integral geometry.

The last problem is of somewhat different nature. It is proved in [3] that the surface $uA(u)$ is convex. Calling an r-dimensional linear subspace of E^n briefly an r-flat, our problem is this:

Problem 10: *If $n \geqq 4$ and $2 \leqq r < n-1$, does a result analogous to the convexity of the surface $uA(u)$ hold for the sections of K by r-flats through z?*

Here it is not even quite clear what the term "analogous" may mean, because the very formulation of convexity presupposes an underlying linear range, whereas the r-flats through z fail to form a linear variety, if $2 \leqq r < n-1$. However, the following suggests itself: The convexity of the surface $uA(u)$ may be reformulated by introducing for an arbitrary vector V with origin z the function

$$F(v) = |v|A(v/|v|)^{-1} \quad \text{if} \quad v \neq (0, \ldots, 0), \quad F(0) = 0 .$$

Then $F(v)$ is a convex function of v. Since $F(v)$ is positive homogeneous of degree 1, the inequality

$$F(v) \leqq \sum_{i=1}^{k} \lambda_i F(v^i) \quad \text{for} \quad v = \sum_{i=1}^{k} \lambda_i v^i, \quad \lambda_i \geqq 0 ,$$

is equivalent to the convexity of $F(v)$.

Introduce Plücker coordinates $p=(p_1, \ldots, p_N)$ for the r-flats in E^n (since we only consider r-flats through z, we may also use Plücker coordinates for the $(r-1)$-flats in $(n-1)$-dimensional projective space). If

$$|p| = \left(\sum_{j=1}^{N} p_j^2\right)^{\frac{1}{2}} = 1 ,$$

denote by $B(p)$ the r-dimensional area of the intersection of K with the r-flat p through z. If q represents any r-flat through z put

$$F(q) = |q| \, B(q/|q|)^{-1} \quad \text{and} \quad F(0) = 0 .$$

The following appears then as a natural analogue to the convexity of the surface $uA(u)$:

If q, and $q^1, \ldots, q^k$ represent r-flats through z and

$$q = \sum_{i=1}^{k} \lambda_i q^i, \qquad \lambda_i \geqq 0 ,$$

then

$$F(q) \leqq \sum_{i=1}^{k} \lambda_i F(q^i) .$$

The question on Minkowski area which leads to Problem 10 is this: A straight segment is a shortest connection of its endpoints in Minkowskian geometry. Let a sufficiently smooth surface S, homeomorphic to an $(r-1)$-sphere, in an r-flat L_r be given and bound in L_r the set W homeomorphic to the solid r-dimensional sphere. The convexity of $F(v)$ implies, see [4], that the r-dimensional area of W is not greater than that of any surface W' bounded by S and homeomorphic to W, provided W' lies in an $(r+1)$-flat. (The last condition is automatically satisfied for $r=n-1$.) Problem 10 inquires whether this statement remains true without the last restriction.

REFERENCES

1. W. Blaschke, *Räumliche Variationsprobleme mit symmetrischer Transversalitätsbedingung*, Ber. Sächs. Akad. Wiss. Leipzig 68 (1916), 50–55.
2. T. Bonnesen und W. Fenchel, *Theorie der konvexen Körper*, Berlin 1934, New York 1948.
3. H. Busemann, *A theorem on convex bodies of the Brunn-Minkowski type*, Proc. Nat. Acad. Sci. U.S.A. 35 (1949), 27–31.
4. H. Busemann, *The foundations of Minkowskian geometry*, Comment. Math. Helv. 24 (1950), 156–186.
5. H. Busemann, *Volume in terms of concurrent cross-sections*, Pacific J. Math. 3 (1953), 1–12.
6. H. Busemann, *The geometry of geodesics*, New York 1955.
7. C. M. Petty, *On the geometry of the Minkowski plane*, to appear in Rivista Mat. Univ. Parma.
8. J. Radon, *Über eine besondere Art ebener konvexer Kurven*, Ber. Sächs. Akad. Wiss. Leipzig 68 (1916), 131–134.

THE UNIVERSITY OF SOUTHERN CALIFORNIA, LOS ANGELES, CALIF., U.S.A.
LOCKHEED MISSILE SYSTEMS DIVISION, LOS ANGELES, CALIF., U.S.A.

MINKOWSKI'S AND RELATED PROBLEMS FOR CONVEX SURFACES WITH BOUNDARIES

Herbert Busemann

1. INTRODUCTION

The theory of convex bodies establishes uniqueness for general closed convex hypersurfaces in E^n, when an elementary symmetric function $R_1 R_2 \cdots R_m + \cdots$ ($1 \leq m \leq n - 1$) of the principal radii of curvature, or a suitable generalization of it in terms of set functions, is given as function of the normal. The method was initiated by Minkowski [8]. The general problem was solved independently by A. D. Alexandrov [1] and Fenchel and Jessen [5]. A convenient source in book form is provided by [3]. For surfaces of class C^2 in E^3 and given $R_1 R_2 < \infty$ Chern [4] gave a proof avoiding the Brunn-Minkowski Theory.

Also without using this theory, Hsiung [6, 7] proves the corresponding uniqueness theorems for smooth surfaces of positive curvature with a boundary in the cases: general n, m = 1, and n = 3, m = 2, following Chern's method in the latter case.

(Hsiung's historical remarks regarding the other methods cannot pass unchallenged. The decisive papers of Alexandrov, Fenchel and Jessen are not mentioned at all. Instead we find: uniqueness for closed surfaces ". . . was established by Minkowski and proved several decades later by Lewy for analytic surfaces . . . ." Lewy, as well as other cited authors, are concerned with the existence of smooth surfaces with smooth data and not with uniqueness.)

It is the purpose of this note to show that *the theory of convex bodies is applicable to surfaces with boundaries, just because it is not restricted to smooth surfaces.* In Theorem I we establish uniqueness under very wide conditions for surfaces in E^n with a given boundary and any given (generalized) $R_1 R_2 \cdots R_m + \cdots$. The completely general case would have required a modification of the theory, which will not be discussed here.

In addition we give *a uniqueness theorem and an existence theorem for convex caps* (Theorems II and III). A cap is a convex hypersurface with a closed hyperplane boundary, such that the normal projection of the surface on the plane of the boundary lies inside or on the latter. The interest in caps derives from their central role in the investigations of Pogorelov; see [3].

2. THE AREA FUNCTIONS

We use the term convex (hyper)surface as in [3] for a connected relative open subset K of a complete convex hypersurface K* in E^n. A normal to K is the unit normal to a supporting plane of K pointing into the exterior of the convex set bounded by K*. There is an ambiguity only in the trivial case, disregarded here, of a non-closed K lying in a hyperplane. The spherical image of K consists of the endpoints on the unit sphere Σ of the unit vectors beginning at the center of Σ and parallel to normals of K.

Received December 10, 1958.

Since the case where the boundary has several components requires merely more verbiage but no new ideas, we consider, besides certain closed surfaces, only surfaces K homeomorphic to E^{n-1} whose boundaries B are homeomorphic to S^{n-2}.

If K is of class C^2 and has positive curvature, then its principal radii of curvature $R_1, \cdots, R_{n-1}$ are defined and finite. $\{R_1 \cdots R_m\}$ $(1 \leq m \leq n-1)$ denotes the elementary symmetric function $R_1 R_2 \cdots R_m + \cdots$ of the R_i of which $R_1 \cdots R_m$ is one summand. Our first goal is a general form of the following statement:

I'. *Let* K, K' *be two convex hypersurfaces of class* C^2 *with positive Gauss curvature, with the same spherical image and the same boundary* B *of class* C^1. *If for some fixed* m *the function* $\{R_1 \cdots R_m\}$ *has the same value at points of* K *and* K' *with parallel normals, then* K = K'.

Consider a closed convex hypersurface K^*, and on K^* a set B homeomorphic to S^{n-2}. Let B decompose K^* into the sets K and C^0 homeomorphic to E^{n-1}, put $C = C^0 \cup B$, and assume that the spherical images X_K and X_C of K and C are disjoint.

We follow the method of [5], found also in [3, Sections 8, 9]. For each m $(1 \leq m \leq n-1)$, an area function $a_m(K^*, X)$ is defined for all Borel sets X on Σ. We agree to *consider only Borel sets*. For K^* of class C^2 with positive curvature,

$$a_m(K^*, X) = \int_X \{R_1 \cdots R_m\}\, d\Sigma_u .$$

For $X \subset \Sigma$, we define

$$a_m(K, X) = a_m(K^*, X \cap X_K), \qquad a_m(C, X) = a_m(K^*, X \cap X_C).$$

Because of $X_K \cap X_C = 0$ and $X_K \cup X_C = \Sigma$, we have

(1) $$a_m(K, X) + a_m(C, X) = a_m(K^*, X).$$

From the definition of $a_m(K^*, X)$ in [5] or [3], we deduce that the functions $a_m(K, X)$ and $a_m(C, X)$ are independent of K^* in this sense: If K lies on a second convex surface $K_1^* = K \cup C_1$ with $X_K \cap X_{C_1} = 0$, then $a_m(K, X)$ is the same function evaluated for K_1^* as for K^*. The same is true for $a_m(C, X)$ if $K_1^* = K_1 \cup C$, $X_{K_1} \cap X_C = 0$ and $X_K = X_{K_1}$, because then X_C is the same for K^* and K_1^*.

In particular, if no supporting plane of K contains a point of B, then the spherical images of K and C will be disjoint, no matter in which surface we imbed K. In that case, $a_m(K, X)$ depends therefore only on K, and $a_m(C, X)$ will, with the above notation, be the same function for K and any other K_1 which does not have supporting planes containing points of B, provided $X_K = X_{K_1}$.

We call $a_m(K, X)$ *the* m-*th area function of* K. For K of class C^2 and with positive curvature, we have

(2) $$a_m(K, X) = \int_{X \cap X_K} \{R_1 \cdots R_m\}\, d\Sigma_u ,$$

and for general K we see from [5] that $a_m(K, X)$ can be obtained by approximation of K^* with smooth convex surfaces as a limit of integrals (2).

3. UNIQUENESS FOR SURFACES WITH A GIVEN BOUNDARY

With the help of $a_m(K, X)$ we can now formulate our generalization of I':

THEOREM I. *Let* K, K' *be convex hypersurfaces in* E^n, *homeomorphic to* E^{n-1} *and with the same boundary* B *homeomorphic to* S^{n-2}, *such that the supporting planes of* K *and* K' *do not contain points of* B. *If* K *and* K' *have the same spherical image and the same* m-*th area function* (m *fixed*), *then* K = K'.

The assumption on the area function means that

$$a_m(K, X) = a_m(K', X) \qquad \text{for every } X \subset \Sigma.$$

If B lies in a hyperplane H, then it is a closed convex surface in H. If K coincides with the convex domain J bounded by B in H, then its spherical image consists of one point. Hence that of K' consists of the same point, so that K' lies in H and also coincides with J. For $B \subset H$ but $K \neq J$, the surfaces K, K' lie on the same side of H because they have the same spherical image. We denote by D and D' the closed convex surfaces obtained from K and K' by adding B and J.

We define D, D' similarly when B does not lie in a hyperplane, namely as the boundaries of the convex closures of $K \cup B$ and $K' \cup B$. Then $K \cup B \subset D$, because any subset of a convex surface lies on the boundary of its own convex closure. Put $C^0 = D - (K \cup B)$. A supporting plane H_p of D at a point p of C^0 is a supporting plane of B. For $H_p \cap (K \cup B)$ contains a point q (see [2, p. 6]), and the segment from p to q lies on H_p and contains a point of B; actually $q \in B$ because of our hypothesis on the supporting planes of K.

Thus C^0 is one of the two sets C^0, C_1^0 homeomorphic to E^{n-1} into which B decomposes the boundary of its own convex closure. By the same argument, D' is either $K' \cup B \cup C^0$ or $K' \cup B \cup C_1^0$. Since K and K' have the same spherical image, $D' = K' \cup B \cup C^0$.

By (1) and the ensuing discussion, the surfaces D and D' have the same m-th area function

$$a_m(D, X) = a_m(K, X) + a_m(C, X) = a_m(K', X) + a_m(C, X) = a_m(D', X).$$

Since D and D' do not lie in hyperplanes, we conclude from the results of [1] or [5] (see also [3, p. 70]) that D' originates from D by a translation. But D and D' have $C^0 \cup B$ in common; hence K = K'.

4. UNIQUENESS OF CAPS

The applicability of the theory of convex bodies to Theorem I rested essentially on the fact that this theory is not restricted to smooth surfaces. We give a second application of this fact.

A *convex cap* is a convex surface K homeomorphic to E^{n-1} whose boundary B is homeomorphic to S^{n-2} and lies in a hyperplane H, and whose normal projection on H falls on the domain J bounded by B, or on B. The cap is *spatial* when different from J, and *proper* when its projection is J. Certain, but not all, proper caps satisfy the hypothesis of Theorem I, so that their m-th area functions $a_m(K, X)$ are defined. Whether $a_m(K, X)$ determines a spatial K among all caps up to translations is not known and seems doubtful except for $m = n - 1$, but we shall define new functions $a^0_m(K, x)$, with $a^0_{n-1}(K, X) = a_{n-1}(K, X)$ for proper K, which determine K.

The restriction to spatial caps is essential, since caps in parallel hyperplanes and with the same area have the same (n - 1)-st area function. We consider improper caps because we do not only aim at a uniqueness theorem, but also at an existence theorem for given a^0_{n-1}, and there does not seem to exist a simple way of distinguishing general proper caps from certain improper ones in terms of a^0_{n-1}.

With the previous notations, let K be a spatial cap and K_1 its image under reflection in H. Then $K_* = K \cup B \cup K_1$ is a closed convex surface. Denote the unit normal to H towards the side of K by v, and by Σ^+_v, Σ^-_v, G_v respectively the subsets $x \cdot v > 0$, $x \cdot v < 0$, $x \cdot v = 0$ of Σ. We define

$$a^0_m(K, X) = a_m(K_*, \Sigma^+_v \cap X) + a_m(K_*, G_v \cap X)/2. \tag{3}$$

Then

$$a^0_m(K_1, X) = a_m(K_*, \Sigma^-_v \cap X) + a_m(K_*, G_v \cap X)/2$$

and

$$a^0_m(K, X) + a^0_m(K_1, X) = a_m(K_*, X). \tag{4}$$

In order to discuss the meaning of a^0_{n-1}, denote by Z the cylinder formed by the lines normal to H at the points of B. Then (see [5])

$$a_{n-1}(K_*, G_v) = \text{area } Z \cap K_*. \tag{5}$$

For proper caps, $Z \cap D = B$, hence $a_{n-1}(K_*, G_v) = 0$.

Also, $\Sigma^+_v \supset X_K$ for proper K, hence

$$a_m(K, X) = a_m(K_*, X_K \cap X) \leq a_m(K_*, \Sigma^+_v \cap X).$$

But

$$a_{n-1}(K_*, \Sigma^+_v - X_K) = 0,$$

because this number represents the Minkowski area (see [5]) of the points of K_* with normals in $\Sigma^+_v - X_K$, and all these points lie in B. Thus

$$a^0_{n-1}(K, X) = a_{n-1}(K, X) \qquad \text{for proper caps.}$$

The definition (3) of $a^0_{n-1}(K, X)$ as area function for improper caps is also entirely natural in view of (5), but (3) involves for $m < n - 1$ the behavior of K at the boundary in a non-obvious way. Our uniqueness theorem for caps is:

THEOREM II. *If two spatial caps* K, K' *have the same* m-*th area function* ($a^0_m(K, X) = a^0_m(K', X)$ *for all* $X \subset \Sigma$), *then* K' *originates from* K *by a translation.*

We define K'_1, K'_* for K' in the same way as K_* was defined for K. Then, for any set $X \subset \Sigma$ and its image X_1 under reflection in $x \cdot v = 0$,

$$a_m(K_*, X) = a_m(K_*, X_1), \qquad a_m(K_*', X) = a_m(K_*', X_1),$$

therefore also

$$a_m^0(K_1, X) = a_m(K_1', X),$$

and by (4),

$$a_m(K_*, X) = a_m(K_*', X) \qquad \text{for all } X \subset \Sigma;$$

hence it follows from the uniqueness theorem in [1] and [5] (see also [3, p. 70]), that K_*' originates from K_* by a translation. The planes H and H' containing the boundaries of K and K' and also the sides of H, H' containing K and K' correspond under the translation, because the images of Σ_v^+ on K_* and K_*' correspond. Moreover, $a_m^0(K, X) > 0$ for some $X \subset \Sigma_v^+$, $a_m^0(K, X) = 0$ for all $X \subset \Sigma_v^-$, and similarly for K', which defines the sides of H and H' on which K and K' lie.

As an application we observe that for $n = 3$ this theorem and the intrinsicness of the extrinsic curvature (see [3, p. 107]) imply that a spherical cap is, up to motions, determined by its intrinsic metric. This is a special case of a general theorem of Pogorelov (for references see [3, p. 166]). The same holds for spherical caps in E^n for odd $n > 3$ within the class of those smooth caps for which the intrinsicness of the Gauss-curvature $R_1^{-1} \cdots R_{n-1}^{-1}$ has been shown.

5. EXISTENCE OF CAPS WITH GIVEN $a_{n-1}^0(K, X)$

Any closed convex surface K_* satisfies

(6) $$\int_\Sigma u_i \, a_m(K_*, d\Sigma_u) = 0 \qquad (i = 1, \cdots, n).$$

For $m < n - 1$, it is known that the condition

$$\int_\Sigma u_i \, a(d\Sigma_u) = 0$$

is not sufficient for a non-negative completely additive set function a(X) on Σ to be the m-th area function of a convex surface; but necessary and sufficient conditions have not been found (see [3, Section 9]). For a_{n-1}^0 we have the following result:

THEOREM III. *Let* a(X) *be a non-negative completely additive set function defined on all Borel sets* X *on* Σ *satisfying the conditions*

(a) $$a(X) = 0, \textit{ when } X \textit{ lies in } x_n < 0,$$

(b) $$\int_\Sigma u_j \, a(d\Sigma_u) = 0 \textit{ for } j < n,$$

(c) $$a(G_u) < a(\Sigma) \textit{ for all } u$$

(*where* G_u *denotes, of course, the intersection of* Σ *with* $u \cdot x = 0$).

Then there exists, up to translations, exactly one spatial convex cap K *with* $a^0_{n-1}(K, X) = a(X)$ *for all Borel sets* $X \subset \Sigma$, *and its boundary lies in a plane* $x_n = \text{const.}$

The condition (c) is necessary to prevent degeneracy (compare [3, p. 63]). Denote by X_1 the image of the set $X \subset \Sigma$ under reflection in the hyperplane $H: x_n = 0$. We define a new set function $F(X)$ by

$$F(X) = a(X) + a(X_1),$$

then $F(X) = F(X_1)$, and $F(X)$ satisfies the conditions

$$\int_\Sigma u_i F(d\Sigma_u) = 0 \qquad (i = 1, \cdots, n).$$

If $u^* = (u_1, \cdots, u_{n-1}, -u_n)$, then for any u

$$F(G_u) = a(G_u) + a(G_{u^*}) < 2a(\Sigma) = F(\Sigma);$$

hence there exists a closed convex surface K_* with

$$a_{n-1}(K_*, X) = F(X) \qquad \text{for all } X \subset \Sigma$$

(see [1], [5] and [3, p. 64]).

We claim that K_* is symmetric with respect to a plane H' parallel to H. For if K_{*1} is the image of K_* under reflection in H, then

$$a_{n-1}(K_*, X_1) = a_{n-1}(K_*, X) = a_{n-1}(K_{*1}, X_1),$$

so that K_{*1} originates from K_* by a translation, whence the assertion follows.

Thus H' decomposes K_* into two spatial caps K, K_1 symmetric to H'. By construction, when $v = (0, \cdots, 0, 1)$ is a normal of K,

$$a^0_{n-1}(K, X) = a(X) \qquad \text{for } X \subset \Sigma_v^+,$$

$$a^0_{n-1}(K, X) = 0 = a(X) \qquad \text{for } X \subset \Sigma_v^-,$$

and also

$$a^0_{n-1}(K, X) = F(X)/2 = a(X) \qquad \text{for } X \subset G_v,$$

because $F(X)$ has for $X \subset G_v$ the following geometric meaning: consider the set Y of those points in $B = K_* \cap H'$ where the normal to B in H' falls in X, and let Z be the cylindrical set consisting of the lines normal to H at points of Y. Then $F(X)$ is the area of the set $Z \cap K_*$.

The last three equations and the additivity of $a^0_{n-1}(K, X)$ prove that

$$a^0_{n-1}(K, X) = a(X) \qquad \text{for } X \subset \Sigma.$$

The uniqueness, up to translations, is contained in Theorem II.

The question arises whether the cap K obtained in this way is smooth when the data are smooth. Nothing is known for $n > 3$. For $n = 3$ we can say the following: For any Borel subset X of the connected open subset Y of Σ_v^+, let

$$a_2^0(K, X) = \int_X f(u)\, d\Sigma_u,$$

where $f(u)$ is defined on Y and is positive and continuous. Then there is exactly one point $x \in K$ with normal u, so that x may be considered as function $x(u)$ on Y. Moreover, $f^{-1}(u)$ is the Gauss curvature of K at x if defined as the limit of the ratio of the area of a set on Σ shrinking to u and the area of the corresponding set on K. (This does *not* imply that the part of K corresponding to Y is of class C^2, see [3, p. 29]).

A theorem of Pogorelov [9], with an improvement due to Nirenberg (see [3, p. 36]) implies: *If* $f(u) = f(u_1, u_2, (1 - u_1^2 - u_2^2)^{1/2}$ *is on* Y *positive and of class* C^m $(m \geq 2)$, *then the components* $x_i(u_1, u_2, (1 - u_1^2 - u_2^2)^{1/2})$ *of* $x(u)$ *are* (all as functions of u_1, u_2) *at least of class* C^{m+1} (*analytic with* f). The following special case merits being formulated explicitly:

If $f(u_1, u_2, u_3)$ $(\Sigma u_i^2 = 1)$ is positive and continuous for $u_3 > 0$, vanishes for $u_3 \leq 0$, and satisfies

$$\int_\Sigma u_i f(u)\, d\Sigma_u = 0 \qquad (i = 1, 2).$$

then there is, up to translations, only one proper cap K with

$$a_2(K, X) = a_2^0(K, X) = \int_X f(u)\, d\Sigma_u.$$

The boundary of K lies in a plane parallel to $x_3 = 0$, and $f^{-1}(u)$ is the Gauss curvature of K defined as area limit.

The cap is of class C^{m+1} $(m \geq 2)$, or analytic, if $f(u_1, u_2, (1 - u_1^2 - u_2^2)^{1/2})$ is of class C^m or analytic.

All assertions have been proved before, except that K is proper when $f(u)$ is merely positive and continuous. This follows from a theorem of Alexandrov (see [3, p. 34, Theorem 5.4]).

REFERENCES

1. A. D. Alexandrov, *Zur Theorie der gemischten Volumina von konvexen Körpern, I*, Mat. Sbornik N.S. 2 (1937), 947-972; *II, ibid.*, 1205-1238; *III*, Mat. Sbornik N.S. 3 (1938), 27-46; *IV, ibid.*, 227-251 (Russian, German summaries).
2. T. Bonnesen, and W. Fenchel, *Theorie der konvexen Körper*, Berlin, 1934 or New York, 1948.
3. H. Busemann, *Convex surfaces*, New York, 1958.
4. S. S. Chern, *A proof of the uniqueness of Minkowski's problem for convex surfaces*, Amer. J. Math. 79 (1957), 949-950.

5. W. Fenchel and B. Jessen, *Mengenfunktionen und konvexe Körper*, Danske Vid. Selsk. Mat.-Fys. Medd. 16 No. 3 (1938), 1-31.

6. C. C. Hsiung, *On differential geometry of hypersurfaces in the large*, Trans. Amer. Math. Soc. 81 (1956), 243-252.

7. ———, *A uniqueness theorem for Minkowski's problem for convex surfaces with boundary*, Illinois J. Math. 2 (1958), 71-75.

8. H. Minkowski, *Volumen und Oberfläche*, Math. Ann. 57 (1903), 447-495.

9. A. V. Pogorelov, *Regularity of a convex surface with given Gaussian curvature*, Mat. Sbornik N.S. 31 (1952), 88-103 (Russian).

University of Southern California
and
Harvard University

MATHEMATICAL NOTES

EDITED BY ROY DUBISCH, Fresno State College

Material for this department should be sent to Roy Dubisch, Department of Mathematics, University of California, Berkeley 4, California.

VOLUMES AND AREAS OF CROSS-SECTIONS

HERBERT BUSEMANN, University of Southern California and Harvard University

Consider two bodies B_1, B_2 in ordinary space containing the point z and with positive volumes $V(B_i)$. For any unit vector w let $A(B_i, w)$ be the area of the intersection of B_i with the plane normal to w and through z.

Investigations on area lead to the problem, see [2]:

If B_1, B_2 are convex and have z as center, does $A(B_1, w) > A(B_2, w)$ for all w imply $V(B_1) > V(B_2)$?

This very simple sounding problem is still open. One would expect an affirmative answer, even if convexity is replaced by starshapedness with respect to z, or when the condition that z be the center of the B_i is dropped. Here we give examples which show that both assumptions are essential:

(*) *There are bodies of revolution B_1, B_2 starshaped with respect to their common center z with the property*

$$A(B_1, w) > A(B_2, w) \quad \textit{for all } w, \quad \textit{but } V(B_1) < V(B_2). \tag{1}$$

There are convex bodies of revolution B_i containing z and satisfying (1).

For the construction introduce geographic coordinates u, θ, $-\pi/2 \leqq u \leqq \pi/2$, $0 \leqq \theta < 2\pi$ on the unit sphere S^2. Let $f(u, \theta)$ be a rotation invariant, even, continuous, function on S^2, *i.e.*,

$$f(u, \theta) = f(u), \qquad f(u) = f(-u). \tag{2}$$

Putting $g(s, v) = f[\arcsin(\sin v \cdot \sin s)]$, the value of the integral of f with respect to the arc length over a great circle forming the angle v, $0 \leqq v \leqq \pi$, with the equator $u = 0$ is, because of (2),

$$I_{\pi-v} = I_v = \int_0^{2\pi} g(s, v)ds = 2\int_0^{\pi} g(s, v)ds$$

$$= 2\int_0^{\pi/2} g(s, v)ds + 2\int_{\pi/2}^{\pi} g(\pi - s, v)ds = 4\int_0^{\pi/2} g(s, v)ds, \qquad 0 \leqq v \leqq \pi/2.$$

In this v-interval $\sin v \cdot \sin s$ increases for each s, $0 < s \leqq \pi/2$, hence

$$I_v \geqq I_{\pi/2}, \text{ for } 0 \leqq v \leqq \pi, \text{ if } f(u) \text{ decreases in } [0, \pi/2]. \tag{3}$$

In spherical coordinates r, u, θ with z as the point $r = 0$ we define bodies B_h as bounded by $r = F(u, h) = e^{2|u|} + h(\cos 2u + 2^{-1})e^{-2|u|}$, $|u| \leqq \pi/2$. Then $F(u, h)$ satisfies (2) as function of u, hence B_h is star shaped with respect to z, has z as center and is a surface of revolution.

$$V(B_h) = 3^{-1}\int_0^{2\pi} d\theta \int_{-\pi/2}^{\pi/2} F^3(u, h)\cos u\,du = 4\pi 3^{-1}\int_0^{\pi/2} F^3(u, h)\cos u\,du,$$

$$V'(B_0) = \left.\frac{dV(B_h)}{dh}\right|_{h=0} = 4\pi\int_0^{\pi/2} F^2(u, 0)(\cos 2u + 2^{-1})e^{-2u}du$$

$$= 4\pi\int_0^{\pi/2} e^{2u}(\cos 2u + 2^{-1})du = \pi(-e^{\pi} - 1) + \pi(e^{\pi} - 1) = -2\pi.$$

Thus $V(B_h) < V(B_0)$ for small $h > 0$.

We want to show that $A(B_h, w) > A(B_0, w)$ for small h and all w. Instead of w we may use the previous angle v, because B_h is a body of revolution, so that $A(B_h, v)$ is the area of the intersection of B_h with a plane intersecting the plane $u=0$ at the angle v. Then as above

$$A(B_h, \pi - v) = A(B_h, v) = 2\int_0^{\pi/2} F^2[\arcsin(\sin v \cdot \sin s), h]ds, \qquad 0 \leqq v \leqq \pi/2,$$

$$A'(B_0, v) = \left.\frac{dA(B_h, v)}{dh}\right|_{h=0} = 4\int_0^{\pi/2} [\cos 2 \arcsin(\sin v \cdot \sin s) + 2^{-1}]ds.$$

Since $f(u) = \cos 2u + 2^{-1}$ decreases for $0 \leqq u \leqq \pi/2$ we conclude from (3) that $A'(B_0, v) \geqq A'(B_0, \pi/2) = \pi$ for $0 \leqq v \leqq \pi$. It follows that $A(B_h, v) > A(B_0, v)$ for small h and all v, which proves the first part of (*).

The second part is a consequence. For B_h is symmetric with respect to $u=0$. The body B_h^* bounded by $r = F(u, h)$ for $0 < u \leqq \pi/2$ and the disk $0 \leqq r \leqq F(0, h)$, $u=0$, is convex for small h. This is obvious for B_0 because $r = F(u, 0) = e^{2u}$, $0 \leqq u \leqq \pi/2$, is part of a spiral, and continuity shows the convexity of B_h for small $h > 0$.

On the other hand

$$v(B_h^*) = V(B_h)/2, \qquad A(B_h^*, v) = A(B_h, v)/2 \text{ for } v \neq 0, \pi,$$

$$A(B_h^*, v) = A(B_h, v) \qquad \text{for } v = 0, \pi.$$

This proves the second part of (*). In spite of the discontinuity of $A(B_h^*, v)$ at $v=0$ the relations (1) remain valid if z is replaced by a point on the normal to the disk at its center, inside B_h^* and close to that center.

For any two bodies which are star shaped with respect to z and have z as center, $A(B_1, w) = A(B_2, w)$ for all w, implies $B_1 = B_2$; hence all the more $V(B_1) = V(B_2)$. This fact, remarkable in view of our example, follows from a theorem of Funk [3]; see also [2].

If B_1 is an ellipsoid with center z and B_2 is any body containing z, then $A(B_1, w) < A(B_2, w)$ does imply $V(B_1) < V(B_2)$;

See [1]. The discussion there assumes that B_2 is convex, but the result of [1] on which this discussion is based, extends to very general sets B_2.

References

1. H. Busemann, Volume in terms of concurrent cross-sections. Pacific J. Math., vol. 3, 1953, 1–12.

2. H. Busemann and C. M. Petty, Problems on convex bodies, Math. Scand., vol. 4, 1956, 88–94.

3. P. Funk, Über Flächen mit lauter geschlossenen geodätischen Linien, Math. Ann., vol. 74, 1913, 278–300.

Am. Math. Mon. 67 (1960), 248–250

CORRECTIONS TO "VOLUMES AND AREAS OF CROSS-SECTIONS"*

HERBERT BUSEMANN, University of Southern California

In the second line from the bottom on page 249, the factor cos u is missing. Therefore $e^{2|u|}$ and $e^{-2|u|}$ in the definition of $F(u, h)$ must be replaced by $e^{t|u|}$ and $e^{-t|u|}$ with a suitable $t>2$ to attain

$$(4\pi)^{-1}V'(B_0) = \int_0^{1/2\pi} e^{tu}g(u)du < 0, \qquad g(u) = (\cos 2u + \tfrac{1}{2}) \cos u.$$

The inequalities $g(u) \leqq b = 3/2$, $g(u) \leqq 0$ in $[\frac{1}{3}\pi, \frac{1}{2}\pi]$, $g(u) \leqq -c < 0$ in $[\alpha, \beta]$ if $\frac{1}{3}\pi < \alpha < \beta < \frac{1}{2}\pi$ give

$$(4\pi)^{-1}V'(B_0) < b\int_0^{\alpha} e^{tu}du - c\int_{\alpha}^{\beta} e^{tu}du < t^{-1}e^{\beta t}[(b+c)e^{(\alpha-\beta)t} - c],$$

which tends to $-\infty$ for $t \to \infty$.

* This MONTHLY, vol. 67, 1960, pp. 248–250.

Am. Math. Mon. 67 (1960), 671

AREA AND NORMALITY

H. BUSEMANN AND E. G. STRAUS

1. Introduction. The simplest non-Riemannian a-dimensional area (concisely: a-area) is a translation invariant positive continuous measure (or area) defined on the a-dimensional linear subspaces, called a-flats, of an n-dimensional affine space A_n $(1 \leq a \leq n)$. Such areas have been studied by Wagner [15] and they are the subject of the present investigation which is in part related to Wagner's, but has no connection with the differential geometry of general area metrics persued principally in Japan by Kawaguchi, Iwamoto and others.

The simplest case, $a = 1$, is well known. In that case a segment with endpoints x, y has a translation invariant length $d(x, y)$. If the sphere $d(z, x) = 1$ (z fixed) has at x_0 a supporting $(n-1)$-flat (hyperplane) H_0 then H_0 is transversal to the 1-flat (line) L_0 through z and x_0, and L_0 is normal to H_0.

Therefore the existence of an $(n-1)$-flat transversal to a given line is equivalent to the convexity of the sphere $d(z, x) = 1$; which, in turn, is equivalent to the triangle inequality for $d(a, b)$, in other words, to the space being Minkowskian (normed linear).

If L_0 is normal to H_0 at x_0 then it is normal to every line L through x_0 in H_0 in the two-flat spanned by L_0 and L. A well-known theorem of Blaschke [2] states that for $n \geq 3$ normality between lines is symmetric only in euclidean space. However, as shown by Radon [13], this is not the case for $n = 2$.

Here we treat the analogous problems for arbitrary a, and then study the special case of Minkowski area.

We cannot give more than this vague hint without some definitions. Let $(x^1, \cdots, x^n)$ be affine coordinates of a point x in A^n with origin $z = (0, \cdots, 0)$. The *a-box* $[x_0, x_1, \cdots, x_a]$ consists of all points of the form $(1 - \theta_i)x_0 + \sum_{i=1}^{a} \theta_i x_i$ where $0 \leq \theta_i, \leq 1$; and hence is a (possibly degenerate) parallelepiped.

An *a-area* assigns to every Borel[1] set M in an a-flat a measure $\alpha(M)$ which is invariant under the translations of A^n, and continuous; that is, $\alpha([x_0, \cdots, x_a])$ depends continuously on $x_0, \cdots, x_a$. The invariance under translation applied to sets in the same a-flat A yields at once that the measure in A is determined up to a factor depending on A. If we introduce an auxiliary euclidean metric

Received, June 19, 1958. The first named author was supported by a grant from the National Science Foundation.

[1]All sets considered will be Borel sets.

$$e(x, y) = \left[\sum_{i,k=1}^{n} g_{ik}(x^i - y^i)(x^k - y^k)\right]^{1/2}$$

where the form $\sum g_{ik}x^i x^k$ is positive definite, then the a-dimensional Lebesgue measure, $|M|_a^e$, in A which results from this euclidean metric is invariant under translations so that

$$(1) \qquad \alpha(M) = f(A)|M|_a^e\,, \quad f(A) > 0\,.$$[2]

Translation invariance implies that $f(A) = f(A')$ if A and A' are parallel a-flats, and the continuity of α implies continuity of $f(A)$. Because of the invariance under translation we may also write.

$$\alpha([x_0, \cdots, x_a]) = F(x_1 - x_0, \cdots, x_a - x_0)$$

where the function $F(x_1, \cdots, x_a)$ satisfies some simple conditions $F_1, \cdots F_4$ listed at the end of § 2.

We call the area α convex if

$$(2) \qquad F(x_1' + x_1'', x_2, \cdots, x_a) \leq F(x_1', x_2, \cdots, x_a) + F(x_1'', x_2, \cdots, x_a)$$

and *strictly convex* if the strict inequality holds for independent x_1', x_1'', x_2, $\cdots$, x_a.

If an a-flat A and a b-flat B intersect in a d-flat D, where $0 \leq d < \min(a, b)$, then they span a q-flat Q with $q = a + b - d$. We call B *totally transversal to* A, or A *totally normal to* B (at D in Q, where ambiguities are possible) if $\alpha(M) \leq \alpha(M')$ for a projection[3] M parallel to B on A of any set M' which lies in an a-flat A' through D in Q. For $d = 0$, $b = n - a$ this is Caratheodory's concept of transversality[4]. If A is totally normal to B at D, $d > b + 1$, then A is totally normal to every b'-flat, $d < b' < b$ through D in B. We call A *normal to* B *at* D and B *transversal to* A, if A is totally normal to every $(d + 1)$-flat in B through D. For $d = 0$, $b = n - a$ this is Wagner's concept of transversality. Only for $d = \min(a, b) - 1$ does normality of A to B at D imply total normality. This is the only case with $d > 0$ which was studied previously in the literature, namely in [7] for Minkowski area.

We call α *totally convex* if an $(n - a)$-flat totally transversal to a given a-flat at a point exists. For totally convex α the a-flats minimize area in the sense that the α-area of the union of all but one face of a closed a-dimensional polyhedron is not less than the area of that face.

[2] Therefore the case $a = n$ is uninteresting as long as only areas for one definite A^n are considered. Hence we assume $1 \leq a \leq n - 1$ except in the last three sections.

[3] This concept needs clarification when $d > 0$. The precise form is found in § 2.

[4] Caratheodory treats more general a-dimensional variational problems. His ideas on transversality are easiest understood by consulting volume 1 of his Gesammelte Mathematische Schriften, München 1954; see in particular p. 364 and paper XX pp. 404–426.

However, the a-flats may minimize α-area for α which are not totally covex. On the other hand for $1 < a < n - 1$ the a-flats need not minimize area when α is merely convex. They will minimize a-area if α is *extendably convex* which means the following; α assigns an area $\phi(\mathfrak{a})$ to every simple a-vector, $\mathfrak{a}$, in the space V_a^n of all a-vectors, if $\phi(\mathfrak{a})$ can be extended to a convex function in all of V_a^n then α is extendably convex. The difference between extendable and total convexity has a very palpable interpretation in V_a^n.

If $F^2(x_1, \cdots, x_a)$ is a quadratic form in each set of variables $x_i^1, \cdots, x_i^n; i = 1, \cdots, a$; then we call $\alpha(M)$ *quadratic*. If $\alpha(M)$ is euclidean, that is if $\alpha(M) = |M|_a^e$ for a suitable choice of $e(x, y)$, then it is quadratic, but a quadratic area is not necessarily euclidean when $1 < a < n - 1$. The quadratic areas enter naturally as follows.

Let $0 \leq d < a \leq b < n$ and let a convex a-area α and a convex b-area β be defined in A^n. If normality (with respect to α) of an a-flat A to a b-flat B at a d-flat D is equivalent to normality (with respect to β) of B to A at D then both areas are quadratic unless $a + b = n, d = 0$. Whether the latter cases are really exceptional is not known except for $a = 1, b = n - 1$ (see below). If, in particular, $a = b$ and $\alpha \equiv \beta$, then equivalence of normality means that normality of two a-flats at a d-flat is a symmetric relation. Hence symmetry of normality implies—except for $a = n/2$, $d = 0$—that the area is quadratic. It will be euclidean only in special cases, for instance when $a < n/2$ and $d = 0$ or $a > n/2$ and $d = 2a - n$. For $a = b = 1, n > 2$ this becomes the above mentioned result of Blaschke [2].

All the results on symmetry and equivalence of normality also hold for total normality.

The a-dimensional Minkowski area (or measure), $2 \leq a \leq n$, in an n-dimensional Minkowski space with distance $F(x - y)$ is the area of the above type for which an a-dimensional unit ball in any a-flat A, that is the set $\{x | F(x - x_0) \leq 1; x, x_0 \in A\}$, has the euclidean volume $\pi^{a/2}/\Gamma(a/2 + 1)$. It is shown in [7] that these areas are convex and are strictly convex or differentiable if $F(x) = 1$ is strictly convex or differentiable.

We do not know whether Minkowski area is totally or extendably convex for $1 < a < n - 1$.

If the a-dimensional area $1 \leq a \leq n - 1$ of a Minkowski space is quadratic then the space is euclidean. Hence if normality of an a-flat A to a b-flat B at a d-flat D with respect to the a-area of one Minkowski space is equivalent to normality of B to A at D with respect to the b-area of another, then both Minkowski spaces are euclidean, unless $a + b = n, d = 0$. However only the case $a = 1, b = n - 1, d = 0$ is really known to be exceptional when the two spaces are different. When they are identical then already this case leads for $n > 2$ to an unsolved problem on convex bodies [10, Problem 5].

There are many interesting and difficult problems involving two areas in a Minkowski space of which we settle only a few. In the last section we obtain from the method and result of [8] a result of a different nature. If $b > a$ and $f_b(B), f_a(A)$ are the functions of (1) for a-and b-dimensional area of the same Minkowski space, we give an estimate from above for $f_b(B)$ in terms of $f_a(A)$ with $A \subset B$.

2. Normality. Our first objects are the relations between the various concepts of normality arising from different choices of d and b. In all that follows let $0 \leq d < \min(a, b)$; $q = a + b - d \leq n$. Moreover, A, B, D, Q with or without subscripts denote a-, b-, d-, q-flats respectively with $D \subset B \subset Q, A \subset Q, A \cap B = D$.

Choose in B a c-flat $C, c = b - d$, which intersects D in exactly one point and hence intersects A in this point only. The association of the points of A and A_0 which lie in the same c-flat parallel to C is a projection of A_0 on A, which depends on the choice of C. The restriction of this mapping to a subset M_0 of A_0 gives the projection of M_0 on a set M in A.

If C' is a second c-flat in B which intersects D in a point, and B^* is any b-flat in Q parallel to B, then the projection of $B^* \cap A_0$ on A with the use of C' is the product of the projection of $B^* \cap A_0$ on A with the use of C and of a translation parallel to D (which depends continuously on B^*).

This and (1) imply.

(2.1) Lemma. *If M_0 is a set in A_0 and M, M' are its projections on A with the use of C and C' respectively, then $\alpha(M) = \alpha(M')$.*

Thus the arbitrariness of C does not influence the measures of the projections. Moreover, if $0 < \alpha(M_0) < \infty$, then $\alpha(M)/\alpha(M_0)$ is according to (1) independent of the choice of M_0 in A_0.

We now define: A is totally *normal to B at D in Q*, or *B totally transversal to A at D in Q*, if for a fixed $M_0 \subset A_0$ with $0 < \alpha(M_0) < \infty$ and a fixed C the area $\alpha(M)$ of the projection of M_0 on A is minimal.

The preceding discussion shows that this definition is independent of the choice of A_0, M_0 and C; and hence depends only on D, B and Q.

The existence of an A normal to B at D in Q follows from two observations.

(i) The function $f(A)$ is continuous and has the same value for parallel A. Hence $f(A)$ attains its positive minimum f_1 and its finite maximum f_2 on the compact set of a-flats through z, so that

$$f_1 |M|_a^e \le \alpha(M) \le f_2 |M|_a^e .$$

(ii) $|M|_a^e \to \infty$ and hence $\alpha(M) \to \infty$ when A approaches a position for which $A \cap B$ is greater than D.

As previously observed, a B totally transversal to a given A at D in Q will in general fail to exist.

We now consider some properties of normality. In many of the following statements "totally" appears in parentheses, because they remain valid for the weaker concept of normality defined in the Introduction.

(2.2) *If A is* (totally) *normal toB at D in Q and the b'-flat B' lies in Q and contains B but does not contain A, then A is* (totally) *normal to B' at $D' = B' \cap A$ in Q.*

This is nearly obvious. A $(b-d)$-flat C in B which intersects D in exactly one point also intersects A and hence D' in this point only. Therefore the same C can be used for projection in both cases of normality.

(2.3) *If A is* (totally) *normal to B at D, $d < b' < b$, then A is* (totally) *normal to any b'-flat B' through D in B.*

Take a $(b'-d)$-flat C' in B' that intersects D in a point and choose a $(b-d)$-flat C in B which contains C' and intersects D in this point only. For any A' through D in the space spanned by A and B' the projection of A' on A parallel to B and B' respectively coincide if we use C and C'.

Proposition (2.3) implies in particular that A is totally normal to every $(d+1)$-flat through D in B. We shall see in § 5 that the converse is in general not true. It does hold in an important special case.

(2.4) Theorem. *If $A \cap B = D$, $a = d+1$, $b - d \ge 2$ and A is normal to every $(d+1)$-flat in B through D, then A is totally normal to B at D.*

For an indirect proof, assume that A is not totally normal to B and let $\overline{A} \ne A$ be totally normal to B at D in the space Q spanned by A and B. A suitable b-flat B' in Q parallel to B intersects A and $\overline{A}$ in two distinct d-flats D' and $\overline{D'}$ parallel to D. These lie therefore in a $(d+1)$-flat $D_+ \subset B'$. In D^+ take a line L which intersects D' in a point.

Consider a set M in A with $0 < \alpha(M) < \infty$. Since A is normal to D^+, the projection $\overline{M}$ of M on $\overline{A}$ parallel to L satisfies $\alpha(\overline{M}) \ge \alpha(M)$.

On the other hand, let C' be a $(b-d)$-flat in B' which contains L and intersects D' in $L \cap D'$ only. Projection of M on $\bar{A}$ parallel to B with the use of C' again yields the set $\bar{M}$. Since $\bar{A}$ is totally normal to B and A is not, we would have $\alpha(M) > \alpha(\bar{M})$, a contradiction.

Defining normality of A to B at D as in the Introduction we conclude from (2.4) that normality and total normality coincide for $d = \min(a, b) - 1$. Obviously (2.3) remains valid for normalily instead of total normality. To prove (2.2) in this case we observe that a $(d'+1)$-flat E through D' in B' intersects B in a $(d+1)$-flat $F \supset D$. For $b' - b = d' - d$ and $E \cup B$ spans B' so that

$$\dim E \cap B + b' = \dim E + \dim B = d' + 1 + b = b' + d + 1 .$$

By hypothesis A is totally normal to F at D, by (2.2) it is also totally normal to E at D' and hence normal to B'.

Moreover (2.2) and (2.3) also show that the case $b = n - a, q = n$ is decisive in the following sense.

(2.5) *If an $(n-a)$-flat (totally) transversal to A exists, then for given $D \subset A \subset Q$, $q = a + b - d$, a b-flat (totally) transversal to A at D in Q exists.*

By hypothesis there is an $(n-a)$-flat N transversal to A through a point $p \in D$. By (2.2) A is normal to the $(n-a+d)$-flat B' spanned by D and N. This settles the case $q = n$. If $q < n$ then according to (2.3) A is normal to the b-flat $B = Q \cap B'$.

For later purposes we note the following consequence of (2.4) and (2.5).

(2.6) LEMMA. *A b-flat B transversal to A at D in Q for any given $D \subset A \subset Q$ will exist if and only if*

(i) *For $p \in A$, every $(a+1)$-flat through A contains a line transversal to A at p.*

(ii) *The set formed by the totality of all transversals to A at p in the different $(a+1)$-flats though A contains an $(n-a)$-flat N.*

The flat N is then transversal to A.

Also for later application we notice as a consequence of the continuity of $f(A)$ the following.

(2.7) LEMMA. *If $A_\nu \to A$, $D_\nu \to D$, $B_\nu \to B$ and A_ν is (totally) normal to B_ν at D_ν then A is (totally) normal to B at D.*

We follow these considerations up analytically using Barthel [1]. The invariance of $\alpha(M)$ under translation implies that the area of the box $[x_0, x_1, \cdots, x_a]$ has the form $F(x_1 - x_0, \cdots, x_a - x_0)$ and

$$(2.8) \qquad F(x_1, \cdots, x_a) = F(x_1^1, \cdots, x_1^n, \cdots, x_a^1, \cdots, x_a^n) = f(A_x)|[z, x_1, \cdots, x_a]|_a^e ,$$

where A_x is the flat spanned by $x_1, \cdots, x_a$, if $x_1, \cdots, x_a$ are linearly independent and $F(x_1, \cdots, x_a) = 0$ otherwise. Thus $F(x_1, \cdots, x_a)$ has the following properties.

F_1: *$F(x_1, \cdots, x_a)$ is continuous in the $a \cdot n$ variables and symmetric in $x_1, \cdots, x_a$.*
F_2: *$F(x_1, \cdots, x_a) > 0$ if $x_1 \wedge \cdots \wedge x_a \neq 0$.*
F_3: *$F(\lambda x_1, x_2, \cdots, x_a) = |\lambda|\, F(x_1, \cdots, x_a)$.*
F_4: *$F(x_1 + \lambda x_j, x_2, \cdots, x_a) = F(x_1, \cdots, x_a)$ for $j > 1$.*

Conversely, if a function $F(x_1, \cdots, x_a)$ has the properties $F_1, \cdots, F_4$ then a well known argument (see e.g. [**14**, pp. 118, 124]) shows that $F(x_1, \cdots, x_a)$ has the form (2.8) with continuous $f(A_x)$ and vanishes for $x_1 \wedge \cdots \wedge x_a = 0$. Hence it defines an area function.

We now take definite independent vectors $u_1, \cdots, u_a$ and assume that $F(x_1, \cdots, x_d)$ possesses a differential as function of $x_1^1, \cdots, x_1^n$ at $x_i = u_i$. Then F_3 and F_4 yield for small $\lambda > 0$ and $j = 1, \cdots, a$

$$(2.9) \qquad \delta_j^1 \lambda F(u_1, \cdots, u_a) = F(u_1 + \lambda u_j, u_2, \cdots, u_a) - F(u_1, \cdots, u_a) = \sum_i \frac{\partial F(u_1, \cdots, u_a)}{\partial x_1^i} \lambda u_j^i + o(\lambda) .$$

For $\lambda \to 0$, using the symmetry of $F(x_1, \cdots, x_a)$ we obtain the following.

(2.10) *If $F(x_1, \cdots, x_a)$ possesses a differential as function of $x_k^1, \cdots, x_k^n$ at $u_1, \cdots, u_a$; $u_1 \wedge \cdots \wedge u_a \neq 0$, then*

$$(2.11) \qquad \sum_i \frac{\partial F(u_1, \cdots, u_a)}{\partial x_k^i} u_j^i = \delta_j^k F(u_1, \cdots, u_a) .$$

Let A be normal to B at D in Q, $z \in D$. Choose a non-degenerate q-box $[z, y_1, \cdots, y_b, u_{a+1}, \cdots, u_a]$ such that $y_1, \cdots, y_d$ lie in D; $y_{d+1}, \cdots, y_b$ in B and $u_{a+1}, \cdots, u_a$ in A. For any $\lambda_{d+1}, \cdots, \lambda_b$ the box $[z, y_1, \cdots, y_d, u_{d+1}, \cdots, u_a]$ originates from the box

$$\left[z, y_1, \cdots, y_d, u_{d+1} + \sum_{i=d+1}^{b} \lambda_i y_i, u_{d+2}, \cdots, u_a\right]$$

by projection parallel to B. If F possesses a differential at $y_1, \cdots, y_d$, $u_{d+1}, \cdots, u_a$ as function of $x_{d+1}^1, \cdots, x_{d+1}^n$, then normality of A to B implies that $F(y_1, \cdots, y_d, u_{d+1} + \sum \lambda_i y_i, u_{d+2}, \cdots, u_a)$ has a minimum for $\lambda_i = 0$. Hence

$$\sum \frac{\partial F(y_1, \cdots, y_d, u_{d+1}, \cdots, u_a)}{\partial x^i_{d+1}} y^i_j = 0 \qquad j = d+1, \cdots, b\,.$$

Thus we have found the following.

(2.12) *If the a-flat through z spanned by* $y_1, \cdots, y_d, u_{d+1}, \cdots, u_a$ *is normal to the b-flat B through z spanned by* $y_1, \cdots, y_b$ *and F is differentiable at* $y_1, \cdots, y_d, u_{d+1}, \cdots, u_a$ *as function of* $x^1_k, \cdots, x^n_k$ *for* $k = d+1, \cdots, a$ *then*

$$\text{(2.13)} \qquad \sum_i \frac{\partial F(y_1, \cdots, y_d, u_{d+1}, \cdots, u_a)}{\partial x^i_k} y^i_j = 0\,, \qquad \begin{array}{l} k = d+1, \cdots, a \\ j = d+1, \cdots b\,. \end{array}$$

We conclude from (2.11) that the matrix $\partial F(y, u)/\partial x^i_k$ has rank $a - d$. Therefore, if D, A, Q are given there can be at most one b-flat transversal to A at D in Q. For brevity we say that $F(x_1, \cdots, x_a)$ is *individually differentiable* at $u_1, \cdots, u_a$ if it possesses a differential at $u_1, \cdots, u_a$ with respect to each of the sets of variables $x^1_k, \cdots, x^n_k$; $k = 1, \cdots, a$.

With property (ii) of (2.6) in mind we state explicitly the following consequence of our discussion.

(2.14) Lemma. *If* $F(x_1, \cdots, x_a)$ *is individually differentiable at* $u_1, \cdots, u_a$ *with* $u_1 \wedge \cdots \wedge u_a \neq 0$, *and if in each* $(a+1)$*-flat containing the a-flat* A_u *spanned by* $u_1, \cdots, u_a$ *there exists a transversal to* A_u; *then this transversal is unique and the y corresponding to the different* $(a+1)$*-flats through* A_u *form the* $(n-a)$*-flat*

$$\sum_i \frac{\partial F(u_1, \cdots, u_a)}{\partial x^i_k} y^i = 0 \qquad k = 1, \cdots, a\,.$$

3. Convexity. Convexity, strict convexity and differentiability for the area α were determined in terms of the function $F(x_1, x_2, \cdots, x_a)$ in the introduction as follows.

(3.1) Definition. Writing $F(y, x) = F(y, x_2, \cdots, x_a)$ we say that α is *convex*, *strictly convex*, or *differentiable* according as the curve $F(\lambda_1 y_1 + \lambda_2 y_2, x) = 1$ has those properties in the plane spanned by y_1, y_2 for any linearly independent $y_1, y_2, x_2, \cdots, x_a$.

Thus for convex α we have

$$F(y_1 + y_2, x) \leq F(y_1, x) + F(y_2, x) \text{ for } y_1 \wedge y_2 \wedge \cdots \wedge x_a \neq 0$$

with strict inequality for strict convexity. If we do not exclude linear dependence of y_1, y_2, then setting $y_1 = \mu y_2$ we have

$$F(y_1 + y_2, x) = |1 + \mu| F(y_1, x) \begin{cases} = F(y_1, x) + F(y_2, x) \text{ if } \mu \geq 0 \\ < F(y_1, x) + F(y_2, x) \text{ if } \mu < 0 . \end{cases}$$

Thus we find the following.

(3.2) LEMMA. *The area function α is convex if and only if*

$$F(y_1 + y_2, x_2, \cdots, x_a) \leq F(y_1, x_2, \cdots, x_a) + F(y_2, x_2, \cdots, x_a)$$

for $y_i \wedge x_2 \wedge \cdots \wedge x_a \neq 0$; and is strictly convex if and only if equality implies $y_1 = \mu y_2$, $\mu > 0$.

Let α be convex and $u_1 \wedge \cdots \wedge u_a \neq 0$. The function F has a differential with respect to $x_1', \cdots, x_1^n$ at $u_1, \cdots, u_a$ if and only if the curve

$$F(\lambda u_1 + \mu v, u_2, \cdots, u_a) = 1$$

is differentiable at $\lambda = 1, \mu = 0$ for all v with $v \wedge u_1 \wedge \cdots \wedge u_a \neq 0$. We have thus proved the following.

(3.3) *A convex area function α is differentiable if and only if the corresponding function $F(x_1, \cdots, x_a)$ is individually differentiable for $x_1 \wedge \cdots \wedge x_a \neq 0$.*

The differentiability properties of convex functions imply that for every convex α the corresponding F has strong differentiability properties, of which we need only the following.

(3.4) LEMMA. *If α is convex and $u_1, \wedge \cdots \wedge u_a \neq 0$, then there exist sequences $\{u_{i\nu}\}$ such that $u_{i\nu} \to u_i$ $(i = 1, \cdots, a)$ and such that $F(x_1, \cdots, x_a)$ is individually differentiable at $u_{i\nu}, \cdots, u_{a\nu}$.*

Reformulation of these properties in terms of the function $f(A)$ will prove useful. Since $f(A)$ is defined relative to a definite euclidean metric $e(x, y)$ we may use euclidean concepts. In particular we will speak of "perpendicularity" when we mean normality with respect to $e(x, y)$.

Consider a plane P perpendicular at z to the $(a-1)$-flat L_{a-1} and choose in L_{a-1} an $(a-1)$-box $[z, x_2, \cdots, x_a]$ with euclidean $(a-1)$-volume 1. On each ray R in P with origin z choose y_R such that $F(y_R, x_2, \cdots, x_a) = 1$. The euclidean a-volume of this box is $e(z, y_R)$. Hence, if A_R is the a-flat containing R and L_{a-1} then

$$F(y_R, x_2, \cdots, x_a) = f(A_R) e(z, y_R) = 1 .$$

If the t-flat L_t, $2 \leq t \leq n - a + 1$ is perpendicular to L_{a-1} at z we denote by $S(L_{a-1}, L_t)$ the locus in L_t obtained by taking the point y_R

with $e(z, y_R) = f^{-1}(A_R)$ on a variable ray R in L_t with origin z. Then we can express our result as follows.

(3.5) LEMMA. *The area function α is convex, strictly convex, convex and differentiable if for any L_2, L_{a-1} and only if for all L_t, L_{a-1}, $2 \le t \le n - a + 1$ the surface $S(L_{a-1}, L_t)$ is convex, strictly convex, convex and differentiable.*

Following the arguments of [7] we now settle the case $d = \min(a, b) - 1$. The emphasis is not only on the result, but also on the *method* of constructing normal and transversal flats which the proof provides.

(3.6) THEOREM. *Let $d = \min(a, b) - 1, q = a + b - d \le n$. For given d-, a-, q-flats $D \subset A \subset Q$, there exists a b-flat B transversal to A at D in Q if and only if the area function α is convex. B is unique when α is differentiable. The normal to B at D in Q is unique for all given $D \subset B \subset Q$ if and only if α is strictly convex.*

Proof. There are two cases.

CASE I: $a = d + 1, b = q - 1$. If $z \in D \subset Q$ are given we take the $(q - d)$-flat L_{q-d} perpendicular to D in Q at z and construct the surface $S = S(D, L_{q-d})$ of (3.5). An a-flat A through D in Q intersects L_{q-d} in two rays, each containing a point of S. Let y_A be one of these points. We claim that *B is transversal to A at D in Q if and only if it is spanned by D and a $(q - d - 1)$-flat through z parallel to a supporting flat H of S in L_{q-d} at y_A.*

The additional remarks on strict convexity and differentiability are then obvious. For if $H \cap S$ contains more points than y_A then the normal A to B at D in Q is not unique, and if S has two different supporting flats at y_A then B is not unique.

To prove our assertion we take A_1 perpendicular to B through D in Q, and in A_1 we take a set M_1 with $0 < \alpha(M_1) < \infty$. If we use $C = L_{q-d} \cap B$ to define projection parallel to B, then we have for the projection M of M_1 on any A

$$(3.7) \qquad \alpha(M) = |M|^e_a f(A) = |M_1|^e_a |\sec(y_A z y_{A_1})| f(A) .$$

Therefore B is transversal to A if and only if $|\cos(y_A z y_{A_1})| f^{-1}(A)$ is maximal; or if and only if S has a supporting plane at y_A which is perpendicular to the ray from z through y_{A_1}, in other words is parallel to B.

The construction is easily freed from the intervening metric $e(x, y)$. Let $1 \le a = d + 1 < q \le n$ and let $z \in D \subset Q$ be given. Take a non-

Pac. J. Math. 10 (1960), 35–72

degenerate d-box $[z, x_1, \cdots, x_d]$ in D and a $(q-d)$-flat L_{q-d} in Q which intersects D at z only. In L_{q-d} construct the locus

$$S = \{x \mid F(x, x_1, \cdots, x_d) = 1\} .$$

Then the a-flat spanned by $x, x_1, \cdots, x_d$ with $x \in S$ is normal to the b-flat B in Q through D if and only if $B \cap L_{q-d}$ is parallel to a supporting $(q-d-1)$-flat of S at x.

CASE II: $b = d+1, a = q-1$. As in Case I take L_{q-d} perpendicular to D at z in Q. Instead of using S we now take the line perpendicular to a variable a-flat A through D in Q. The two points y_A with $e(z, y_A) = f^{-1}(A)$ generate a locus T. When α is convex, strictly convex, convex and differentiable then T has the corresponding property.

This time we claim that B is transversal to A at D in Q if and only if it is spanned by D and the perpendicular to a supporting $(q-d-1)$-flat of T in L_{q-d} at y_A. We define A_1 and M_1 as in Case I and use the line C perpendicular to A_1 at z for projection parallel to B. Then the projection M of M_1 on any A again satisfies (3.7) and $f^{-1}(A)|\cos(y_A z y_{A_1})|$ is maximal if and only if y_A lies on a supporting flat of T which is perpendicular to C. Since C is perpendicular to A_1 it lies in B. The additioned remarks follow as in Case I.

The definition of T cannot be entirely freed from extraneous concepts, but their role can be reduced.

If T is convex, let T' be the polar reciprocal in L_{q-d} of T with respect to the metric $e(x, y)$ (see [5, p. 28]). If T is strictly convex (differentiable) then T' is differentiable (strictly convex). In terms of T' we can interpret the normality relation in a manner similar to that of Case I; only the roles of normality and transversality are interchanged.

If $x \in T'$ then the $(d+1)$-flat spanned by x and D is transversal to the a-flat A through D in Q if and only if A is spanned by D and a $(q-d-1)$-flat parallel to a supporting flat of T' at x.

In the most interesting case, $d = 0$, the surface T' has a very interesting meaning. In $(Q = L_{q-d})$ take any $(q = a+1)$-measure invariant under translation. The only arbitrariness is then the unit of measure. Then T' is a *solution of the isoperimetric problem* to minimize the α-area among all closed convex hyper-surfaces in Q which bound a set of given $(a+1)$-measure. For details see [6]. Of course T' remains a solution even if we change the unit of $(a+1)$-measure.

Assume that α is convex and consider an a-flat A_u through z spanned by $u_1, \cdots, u_a$ and such that F is individually differentiable at $u_1, \cdots, u_a$. Then (3.6) (more particularly Case II) guarantees that in every $(a+1)$-flat containing A_u there exists a transversal to A_u at z. We conclude from (2.14) that the transversals at z to A_u in the different

$(a+1)$-flats form an $(n-a)$-flat N_A and from Theorem (2.6) that this N_A is transversal to A.

If F is not individually differentiable at $u_1, \cdots, u_a$ then we can find sequences $\{u_{i\nu}\}$ with $u_{i\nu} \to u_i$ $(i = 1, \cdots, a)$ such that F is individually differentiable at $u_{i\nu}, \cdots, u_{a\nu}$. Hence if A_ν contains $z, u_{i\nu}, \cdots, u_{a\nu}$ then there exists an $(n-a)$-flat N_ν transversal to A_ν at z. By the continuity of the area function every limit $(n-a)$-flat of a subsequence of N_ν is transversal to A. Thus if α is convex there exists an $(n-a)$-flat transversal to A. Using (2.14) and (2.5) we have proved

(3.8) THEOREM. *If the area function α is convex then, given an a-flat A, a d-flat $D \subset A$ and a q-flat $Q \supset A$ with $0 \leq d < a < q \leq n$; there exists a b-flat, $b = q - a + d$, transversal to A at D in Q, which is unique when α is differentiable.* (*Wagner* [15], *for* $d = 0$).

The conditions in (3.8) are also necessary, but we conclude from (2.5) and (3.6) that we need consider only fixed d and q.

(3.9) THEOREM. *With the notation of* (3.8); *if for fixed d, q and all A, D, Q a b-flat transversal to A at D in Q exists* (*and is unique*) *then α is convex* (*and differentiable*).

A normal to B at D in Q is in general not unique even for strictly and extendably convex α (as we shall see in (5.14)) when $d < \min(a, b) - 1$. For in that case normality is not equivalent to total normality. However, because total normals exist and are normal we have

(3.10) *If the a-flat A normal to B at D in Q is unique, then A is totally normal to B.*

Even the total normal is not necessarily unique for strictly and extendably convex α, see (5.14).

4. Area minimizing a-flats. Total and extendable convexity. The area $\alpha(\Delta)$ of an a-dimensional polyhedron Δ is defined as the sum of the a-areas of its a-faces. In the following we reserve Δ for the union of all a-faces but one, Δ_0, of an a-dimensional polyhedron in A^n which is abstractly a closed orientable a-dimensional manifold but may have self interersections in A^n. By A_Δ we denote the a-flat containing the face Δ_0 and hence the boundary of Δ.

We say that the a-flat A (strictly) minimizes α-area in the q-flat $Q \supset A$, $q > a$, if $\alpha(\Delta) \geq \alpha(\Delta_0)$ $(\alpha(\Delta) > \alpha(\Delta_0))$ for all choices of $\Delta \neq \Delta_0$ in Q for which $A_\Delta = A$. If this is true for all a-flats A in Q we say that the a-flats (strictly) minimize area in Q.

The case $a = 1$ is familiar; with the help of (3.6) we may formulate these results as follows.

The line L minimizes α-length in the q-flat Q if and only if a $(q-1)$-flat B transversal to L in Q at a point z exists. The line L strictly minimizes length in Q if and only if L is the only line normal to B at z.

The lines (strictly) minimize α-length in A^n if and only if α is (strictly) convex.

A few of these facts extend to the general case.

(4.1) *The a-flat A minimizes α-area in Q if a $(q-a)$-flat B totally transversal to A at a point z exists. Let B exist. Then A strictly minimizes α-area when A is the only a-flat totally normal to B at z or when a is strictly convex.*

Project Δ on A_Δ parallel to B. For topological reasons this projection covers Δ_0. Let σ be an a-dimensional face of Δ which lies in the a-flat A and let σ_0 be its projection on A_Δ. If $\dim(B \cap A) > 0$ then obviously $0 = \alpha(\sigma_0) < \alpha(\sigma)$. If $\dim(B \cap A) = 0$ then the transversality of B to A_Δ implies $\alpha(\sigma_0) \leq \alpha(\sigma)$. This proves $\alpha(\Delta) \geq \alpha(\Delta_0)$.

If α is strictly convex and $\Delta \neq \Delta_0$ then $\alpha(\Delta) > \alpha(\Delta_0)$ is obvious when $\dim(B \cap A) > 0$ for some A containing an a-face of Δ. Assume therefore $\dim(B \cap A) = 0$ for all such A. There is at least one pair of a-faces σ^1, σ^2 of Δ which have a common $(a-1)$-face and at least one of which is not parallel to A_Δ. If A^i is the a-flat containing σ^i then not both A^1, A^2 can be normal to B. For, if A_i is the a-flat parallel to A^i through $A_\Delta \cap B$ then $\dim(A_1 \cap A_2) = a-1$ and hence $A_1 \cup A_2$ spans an $(a+1)$-flat Q which intersects B in a line L through $A_\Delta \cap B$. Since α is strictly convex at least one of the two a-flats, say A_1, is by (2.3) and (3.6) not normal to B. Hence A' is not normal to B and $\alpha(\sigma') > \alpha(\sigma_0')$. Hence $\alpha(\Delta) > \alpha(\Delta_0)$.

If A is the only total normal to B at z then at least one a-face σ' of Δ is not totally normal to B and again $\alpha(\sigma') > \alpha(\sigma_0')$.

The case $q=a+1$ is completely known essentially through Minkowski (Theorie der konvexen Körper, § 27, Ges. Abh. 2, Leipzing 1911, 131–229). His terminology is so different that we give the argument here.

For each $(a+1)$-flat Q through z we construct the surface T_Q, analogous to T in the discussion of Case II in the proof of (3.6), as the locus T_Q of the points y_A with $e(z, y_A) = f^{-1}(A)$ on the perpendiculars to the a-flats A through z in Q.

(4.2) *The a-flat A minimizes α-area in the $(a+1)$-flat Q if and only if a line transversal to A in Q exists.*

A strictly minimizes area in Q if and only if a line transversal to in Q exists and y_A is not an interior point of an a-flat region on T_Q.

The sufficiency of the first part of (4.2) follows from (4.1) and the fact that a line transversal to A is totally transversal to A. We next prove the necessity statements in both parts of (4.2).

We choose rectangular coordinates such that Q is the flat $x^{a+2} = \cdots = x^n = 0$ and define, as usual,

$$H(0) = 0 \text{ for } x = 0, H(x) = |x| f(A_x) \text{ for } x \neq 0 ,$$

where A_x is the a-flat through z in Q with normal x and $|x| = (\sum x_i^2)^{1/2}$, so that T_Q has the equation $H(x) = 1$. The function $H(x)$ is convex with α.

If no transversal to A exists then, according to Case II in (3.6), T_Q does not possess a supporting a-flat at y_A; so that y_A is an interior point of the convex closure of T_Q. Hence independent points $x_1, \cdots, x_{a+1}$ on T_Q exist such that

$$(4.3) \qquad H(y_A) > \sum_{i=1}^{a+1} \lambda_i H(x_j) , \qquad y_B = \sum_{i=1}^{a+1} \lambda_i x_i , \qquad \lambda_i > 0 .$$

If y_A is an interior point of an a-flat set on T_Q then independent $x_1, \cdots, x_{a+1}$ on T_Q exist with

$$(4.4) \qquad H(y_A) = \sum_{i=1}^{a+1} \lambda_i H(x_i) , \qquad y_A = \sum_{i=1}^{a+1} \lambda_i x_i , \qquad \lambda_i > 0 .$$

Setting $\xi = y_A/|y_A|$, $\xi_i = x_i/|x_i|$ we have $-|y_A|\xi + \sum \lambda_i |x_i| \xi_i = 0$.

Therefore (see Bonnesen-Fenchel [5, p. 118]),[5] an $(a+1)$-simplex in Q exists whose faces have exterior normals, $-\xi, \xi_1, \cdots, \xi_a$ and area $|x|, \lambda_1|x_1|, \cdots, \lambda_a|x|$. The total area of the faces with normals $\xi_1, \cdots, \xi_a$ is

$$\sum \lambda_i |x_i| f(A_{x_i}) = \sum \lambda_i H(x_i)$$

and $|x| f(A_x) = H(x)$ is the area of the face with normal $-\xi$.

The relations (4.3), (4.4) prove the necessity statements in (4.2).

To establish sufficiency in the second part of (4.2) we resume the notation used in the last part of the proof of (4.1). We assume that Δ lies in Q and replace B by a line L transversal to $A = A_\Delta$.

For $\alpha(\Delta) = \alpha(\Delta_0)$ it is necessary that the mapping of Δ on Δ_0 by projection parallel to L be one-to-one and that all a-flats carrying a-faces of Δ be normal to L.

Now there are two supporting flats A', A'' of T_Q perpendicular to L. On the other hand the construction of the transversal in the discussion of Case II in (3.6) shows that at the points y_A which corresponds to an A normal to L the surface T_Q has supporting planes perpendicular to

[5] The proof there is involved but becomes very simple in the present case where the number of faces is $a + 1$.

L. Therefore A' and A'' each contain one of the two points y_A and one of the two points y_A for each A which carries an a-face of Δ.

Since projection of Δ on Δ_0 is one-to-one and $\Delta \neq \Delta_0$ it follows that among the points y_A, y_A in A' there are $a+1$ which do not lie in an $(a-1)$-flat. These points span an a-simplex which lies on T_Q.

(4.5) COROLLARY. *The a-flats minimize α-area in all $(a+1)$-flats if and only if α is convex. They strictly minimize area if and only if in addition the surface T_Q contains no a-flat piece for any $(a+1)$-flat Q.*

Our results are not as complete for $q>a+1$, $a\neq 1$. Consider the vector space V^n_a of all contravariant a-vectors $\mathfrak{A}$ in A^n. A simple a-vector $\mathfrak{A}\neq 0$ determines an oriented a-flat in A^n through the origin. For the α-area determined by $\mathfrak{A}$ we obtain a function $\Phi(\mathfrak{A})$ defined on all simple $\mathfrak{A}$-vectors whose relation to F is given by

$$a!\,\Phi(x_1\wedge\cdots\wedge x_a)=F(x_1,\cdots,x_a)\,.$$

Obviously Φ satisfies the conditions

$$\Phi_1 \qquad \Phi(\mathfrak{A})>0 \quad \text{for} \quad \mathfrak{A}\neq 0$$

$$\Phi_2 \qquad \Phi(\lambda\mathfrak{A})=|\lambda|\,\Phi(\mathfrak{A}) \quad \text{for all real } \lambda\,.$$

All a-vectors are simple only when $a=1$ and $a=n-1$. (If we exclude the trivial cases $a=0, n$). We shall prove at the end of this section that for $1<a<n-1$ and convex α it is in general impossible to extend $\Phi(\mathfrak{A})$ to a convex function defined for all a-vectors. An obviously necessary condition for extendability is

$$\Phi(\mathfrak{A})\leq\sum_{i=1}^{r}\Phi(\mathfrak{A}_i) \quad \text{for simple } \mathfrak{A}, \mathfrak{A}_i, \text{ with } \quad \mathfrak{A}=\sum_{i=1}^{r}\mathfrak{A}_i\,. \tag{4.6}$$

Condition (4.6) is also sufficient. The simple a-vectors form a basis of V^n_a. Hence if $\mathfrak{A}$ is any a-vector then simple a-vectors $\mathfrak{A}_i$ exist so that

$$\mathfrak{A}=\sum_{i=1}^{r}\mathfrak{A}_i\,, \tag{4.7}$$

since any scalar multiple of a simple vector is simple. We can now extend $\Phi(\mathfrak{A})$ to all of V^n_a by defining

$$\Phi(\mathfrak{A})=\inf\sum_{i=1}^{r}\Phi(\mathfrak{A}_i)$$

where the $\{\mathfrak{A}_i\}$ traverse all sets of simple vectors whose sum is $\mathfrak{A}$. Because of (4.6) $\Phi(\mathfrak{A})$ is not changed by this definition for simple $\mathfrak{A}$, and the extended function obviously is convex and satisfies Φ_1 and Φ_2.

We call α *extendably convex* if it satisfies (4.6). As before consider a polyhedron $\Delta \cup \Delta_0$. Orient it and let $\mathfrak{A}_1, \cdots, \mathfrak{A}_r$ be the simple a-vectors corresponding to the a-faces in Δ. Let $\mathfrak{A}_0$ correspond to Δ_0. Then

$$\sum_{i=0}^{r} \mathfrak{A}_i = 0 \quad \text{or} \quad \mathfrak{A} = -\mathfrak{A}_0 = \sum_{i=1}^{r} \mathfrak{A}_i$$

so that $\alpha(\Delta) \geq \alpha(\Delta_0)$ is equivalent to condition (4.6). In general the relation $\mathfrak{A} = \sum_{i=1}^{r} \mathfrak{A}_i$ for simple $\mathfrak{A}, \mathfrak{A}_i$ does not imply that $-\mathfrak{A}, \mathfrak{A}_1, \cdots, \mathfrak{A}_r$ correspond to the faces of a closed polyhedron. For example, the a-flats corresponding to $\mathfrak{A}, \mathfrak{A}_i$ through the origin z may intersect at z alone. However it is not unlikely that the validity of (4.6) for $\mathfrak{A}, \mathfrak{A}_i$ deriving from polyhedra implies its general validity. We have not been able to prove this. Thus we can only state:

(4.7) *If α is extendably convex then the a-flats minimize area.*

We call α *totally convex* if an $(n-a)$-flat totally transversal to a given a-flat at a point exists. If the condition in (4.7) is necessary then (4.1) shows that total convexity entails extendable convexity. We shall prove this directly, obtaining at the same time a very interesting geometric interpretation for the two types of convexity. The arguments are closely related to those of Wagner [15].

Denote by W_a the affine space associated with the vector space V_a^n, so that we may speak of hyperplanes etc. which do not pass through 0. The simple vectors in V_a^n form the *Grassmann cone* and the equation $\Phi(\mathfrak{A}) = 1$ defines on that cone the indicatrix I of the area α.

Extendable convexity of α means that I lies on the boundary of its convex closure in W_a; that is, that I possesses at every point a supporting hyperplane in W_a.

In order to interpret total convexity we provide A^n with the euclidean metric $g_{ik} = \delta_{ik}$. This metric induces a scalar product $\mathfrak{A} \cdot \mathfrak{B}$ for the simple a-vectors in A^n whose geomentric meaning, apart from sign, is the product of the (euclidean) area of one vector and the area of the orthogonal projection of the other on the a-flat of the first.

This scalar product for the vectors on the Grassmann cone can be extended to an inner product in V_a^n and hence induces a euclidean metric in W_a. To the projection of an a-flat A_1 on an a-flat A parallel to the $(n-a)$-flat B perpendicular to the a-flat B^* at a point there corresponds in V_a^n the projection of the line A_1 on the line A parallel to the hyperplane H_B perpendicular to the line B^*.

Assume now $\mathfrak{A} \in I$ and that I possesses at $\mathfrak{A}$ a simple supporting hyperplane H_B; that is a hyperplane H_B perpendicular to a line B^* on the Grassmann cone. If $\mathfrak{A}_1$ is a simple vector lying on H_B

(that is, interpreted in A^n, if $|\mathfrak{A}| = |\mathfrak{A}_1'|$ for the projection $\mathfrak{A}_1'$ of $\mathfrak{A}_1$ on the a-flat of $\mathfrak{A}$ parallel to the $(n-a)$-flat B which is perpendicular to B^*), then $\Phi(\mathfrak{A}_1) \geq \Phi(\mathfrak{A})$ since H_B is a supporting plane of I. Therefore B is totally transversal to A.

Conversely, if B is totally transversal to A at a point, then any simple $\mathfrak{A}_1$ whose projection parallel to H_B is $\mathfrak{A}$ satisfies $\Phi(\mathfrak{A}_1) \leq \Phi(\mathfrak{A}) = 1$, so that H_B is a supporting hyperplane of I. This could, of course, be formulated without the use of an auxiliary metric:

(4.10) *The area α is totally convex if and only if the indicatrix I posseses at every point $\mathfrak{A} = (a^\lambda)$ a simple supporting hyperplane $\sum a^\lambda b_\lambda = 1$, where $\mathfrak{B} = (b_\lambda)$ satisfies the conditions of a simple vector.*

If I is differentiable at $\mathfrak{A}$, so that the $a(n-a)$-flat, T, tangent to I at $\mathfrak{A}$ exists, then any supporting hyperplane of I at $\mathfrak{A}$ must pass through T. Through a given $a(n-a)$-flat there is exactly one simple hyperplane (see [15]). Since extendable convexity means only the existence of some supporting hyperplane of I at a given point we deduce from (4.10):

(4.11) *Total convexity implies extendable convexity but not conversely.*

That the converse is not valid does not follow from the preceding arguments, but in (5.13) we give an example of an extendably but not totally convex area.

We now show that convexity of α does not imply extendable convexity (Wagner [15] states this fact for $\min(a, n-a) > 2$; but, as it seems to us, he only proves that a certain definite extension of convex area is in general not convex). For this purpose we prove a lemma which seems to be of some independent interest.

(4.12) LEMMA. *Let S_a be a simple closed $(a-1)$-surface in an a-flat A so that at every point of S_a there is both an interior and an exterior supporting $(a-1)$-sphere of radius c in A. Let $z \in A$ be in the interior of S_a so that at the line zx from z to any $x \in S_a$ makes an angle no less than $\alpha > 0$ with the tangent $(a-1)$-flat of S_a at x.*

Then for every $\varepsilon > 0$ there exists a hypersurface $S \supset S_a$ such that every L_2 through z which contains a line that makes an angle greater than ε with A intersects S in convex curve.

Proof. For sufficiently small $\delta > 0$ the interior parallel surface S_a', which is the locus in the interior of S_a of points whose distance from S_a is δ, obviously satisfies the hypotheses of the lemma provided the constants c and α are replaced by suitable constants c' and α'. Let T_a' be the a-body bounded by S_a'.

Let S be the locus of points whose distance from T'_a is δ. Clearly $S_a \subset S$. Every $L_2 \ni z$ intersects S in a curve C. Assume that C is not convex; then there is an $x \in C$ at which C does not have a line of support in L_2 and therefore S does not have a plane of support at x. Thus the point x' nearest to x on T'_a must lie on S'_a and the line zx makes an angle less than $\tan^{-1}[d(x, x')/d(z, x')] \leq \tan^{-1}(\delta/d)$ with A, where d is the distance from z to S_a.

Now let L be the tangent line to C at x. Since L intersects the interior of C, the cylinder L_δ, which is the locus of points whose distance from L is δ, must intersect the interior of T'_a. Since the quadric $Q_\delta = L_\delta \cap A$ is tangent to S'_a at x' it follows that the minimal curvature of Q_δ at x' is less than $1/c'$. Let L' be the tangent line to Q_δ at x' in the direction of minimal curvature then the tangent of the angle between L and L' is less than $\sqrt{\delta/c'}$.

Thus for sufficiently small δ the two lines L and zx make arbitrarily small angles with the lines L' and zx' in A. Since the last named lines make an angle with each other which exceeds α' it follows that every line in L_2 makes an arbitrarily small angle with A.

Now, for example, in the space V_2^4 of 2-vectors in A^4 we can find a three-plane generated by simple vectors which contains no two-plane of simple vectors. Such a three-plans is L_3 generated by $e_1 \wedge e_2$, $e_3 \wedge e_4$ and $(e_1 + e_3) \wedge (e_2 + e_4)$. The simple vectors which it contains are all of the form $\lambda(e_1 + \mu e_3) \wedge (e_3 + \mu e_4)$. We can now define the area function F so that the indicatrix I does not lie on the boundary of its convex hull in L_3, for instance by $F(e_1, e_2) = F(e_3, e_4) = F(e_1 + e_3, e_2 + e_4) = 1$ and $F(e_1 + 2e_3, e_2 + e_2 + 2e_4) > 6$ in violation of (4.6); but so that $I \cap L_3$ satisfies all the conditions of Lemma (4.12) where z is the zero element of V_2^4. By Lemma (4.12) we can now extend I in such a way that its intersection with every two-plane of simple vectors is convex, in other words, so that F is convex. However, since I does not lie on the boundary of its convex hull, the area is not extendably convex.

5. Equivalence of normality. Example. Quadratic area. The normality relations determine the area up to a constant factor in the following sense.

(5.1) THEOREM. *Let α and α' be two a-dimensional convex area functions, $a + b - d \leq n$ and $d \leq \min(a, b) - 1$. For any d-flat D and any b-flat B through D let A be normal to B at D with respect to α' whenever this is the case with respect to α. Then $\alpha'(M)$ and $\alpha(M)$ differ only by a constant factor.*

The same holds for total normality if there exists a b-flat totally transversal with respect to α for any given a-flat at any given d-flat *in any given* $(a+b-d)$*-flat* (*in particular, when* α *is totally convex*).

Proof. Let $a=d+1$. With the notation of Case I in (3.6) we construct the surfaces S, S' belonging to α and α' respectively. The hypothesis of (5.1) means in terms of S, S': If H and H' are parallel supporting $(q-d-1)$-flats of S and S' then a line through z containing a point x of $S \cap H$ also contains a point of $S' \cap H'$. It folllows that S and S' are homothetic, and this conclusion remains valid when this condition on the line zx is assumed only for those $x \in S$ at which S is differentiable, that is H is unique.

This weakening of the hypothesis amounts to requiring that A be normal to B at D with respect to α' only when B is the unique transversal to A at D in $A \oplus B$ with respect to α.

The fact that S and S' are homothetic means that $\alpha'(M)/\alpha(M)$ is constant for all M lying in a-flats through a fixed $(a-1)$-flat in an $(a+b-d)$-flat. This yields the general answer, because two arbitrary a-flats A', A'' can be joined by a finite number of a-flats $A_1 = A', A_2, \cdots, A_r = A''$ such that $\dim A_i \cap A_{i+1} = a-1$ for $i=1, \cdots, r-1$.

Application of the result just obtained to the pencils determined by A_i and A_{i+1} proves the theorem.

The case $d < a-1$ is reduced to $d=a-1$ as follows. Let B^+, $\dim B^+ = b+a-d-1$ be the unique transversal to A at an $(a-1)$-flat D^+ in $A \oplus B^+$. In D^+ chose a d-flat D and an $(a-d-1)$-flat E such that $D^+ = D \oplus E$. Then $D^+ = D \oplus E$ where B is a b-flat and $A \oplus B = A \oplus B^+$ because $E \subset A$.

For normality we know, and for total normality we assume, that a b-flat B' totally transversal to A at D in $A \oplus B^+$ exists. By (2.2) $B' \oplus E$ is transversal to A at D^+ in $A \oplus B^+$ and $B' \oplus E = B^+$ because B^+ is unique. By hypothesis B' is transversal to A at D with respect to α', and again by (2.2) B^+ is transversal to A at D^+ with respect to α'.

This means that the hypothesis of the theorem is satisfied for $d=a-1$ and $b=a+b-d-1$, so that the assertion follows from the first part of the proof.

Let $0 \leq d < a < n$. For a given a-area α we say that normality at d-flats is *symmetric*, if normality of an a-flat A to an a-flat A' at a d-flat D implies that A' is normal to A at D.

If $0 \leq d < a < n$ and an a-area α and a b-area β are given, we say that α-normality and β-normality at d-flats are *equivalent*, if normality of an a-flat A to a b-flat B at a d-flat D with respect to α implies that B is normal to A at D with respect to β and conversely, normality of B to A at D implies that A is normal to B at D.

This formulation admits the possibility that $a=b$. If at the same time $\alpha = \beta$ then equivalence means symmetry. If $a = b$ but $\alpha \neq \beta$ then equivalence means that normality in one norm is equivalent to transversality in the other.

Symmetry and equivalence of total normality are defined in the same way by replacing everywhere normality and transversality by total normality or transversality.

In the next section we discuss the implications of symmetry or equivalence of normality. Here we give some examples where these phenomena occur and the area is not euclidean.

(5.2) *For $d = 0, a = 1, n = 2$ symmetry of normality does not imply that the length,* i.e. *the corresponding two-dimensional Minkowski metric, is euclidean.* All these metrics have been determined by Radon [13], (see also [9, p. 104]).

(5.3) *For any $(n-1)$-dimensional convex area function β there is a convex one-dimensional area,* i.e. *a Minkowski metric $F(x-y)$, such that normality of a hyperplane to a line for β is equivalent to normality of the line to the hyperplane for $F(x-y)$.*

To see this we construct the surface T' of Case II of (3.6) for β and $d = 0$. That is, on the perpendicular to a variable hyperplane $B \ni z$ at z we take the two points y_B with $e(y_B, z) = f^{-1}(B)$. These points y_B traverse a convex hypersurface T and T' is the polar reciprocal of T. As Minkowski metric $F(x-y)$ we take the metric with T' as unit sphere $F(x) = 1$. Then the discussion under (3.6) shows that the hyperplanes normal (for β) to a line zw at $w \in T'$ are the supporting planes of T' at w and these are exactly the planes transversal to zw at w for $F(x-y)$.

The a-area α, $1 \leq a \leq n-1$ is euclidean if $\alpha(M) = |M|_a^e$ for a suitable choice of $e(x, y)$. With the summation convention $\sum_k g_{ik}x^k = g_{ik}x^k$ this means for F that

$$(5.4) \qquad F^2(x_1, \cdots, x_a) = \sum_{i_1 < \cdots < i_a} \begin{vmatrix} x_1^{i_1} & \cdots & x_1^{i_a} \\ \cdot & & \cdot \\ \cdot & & \cdot \\ \cdot & & \cdot \\ x_a^{i_1} & \cdots & x_a^{i_a} \end{vmatrix} \begin{vmatrix} g_{i_1k}x_1^k & \cdots & g_{i_ak}x_1^k \\ \cdot & & \cdot \\ \cdot & & \cdot \\ \cdot & & \cdot \\ g_{i_1k}x_a^k & \cdots & g_{i_ak}x_a^k \end{vmatrix} .$$

We shall call α *quadratic* if F^2 is a quadratic form in each set of variables $x_i^1, \cdots, x_i^n$ $(i = 1, \cdots, a)$. A quadratic F^2 is a quadratic form in the Plücker coordinates.

$$P^{i_1 \cdots i_a} = \begin{vmatrix} x_1^{i_1} & \cdots & x_1^{i_a} \\ \cdot & & \cdot \\ \cdot & & \cdot \\ \cdot & & \cdot \\ x_a^{i_1} & \cdots & x_a^{i_a} \end{vmatrix}, \qquad 1 \leq i_1 < \cdots < i_a \leq n ,$$

of the a-flat through z spanned by $x_1, \cdots, x_a$, since $F(x_1, \cdots, x_a) = f(A)|[z, x_1, \cdots, x_a]|_a^e$ where the terms on the right depend only on the Plücker coordinates.

If F is quadratic then for any L_{a-1} and L_2 perpendicular to L_{a-1} at z the curve $S(L_{a-1}, L_2)$ of (3.5) is an ellipse and conversely. If Q is any $(a+1)$-flat through z we construct in Q the surface T of Case II of (3.6) for $D = z$. The section of T with any plane $L_2 \ni z$ is obtained from $S(_{a-1}, L_2)$, where L_{a-1} is perpendicular at z to L_2 in Q, by a rotation through $\pi/2$. Hence T is an ellipsoid. This implies that the area restricted to Q is euclidean Thus we have the following.

(5.5) THEOREM. *An a-area is quadratic if and only if it is euclidean in every $(a+1)$-flat; that is to say, if and only if normality of a-flats at $(a-1)$-flats in $(a+1)$-flats is symmetric.*

We now wish to determine under what conditions a quadratic area is euclidean.

(5.6) *A quadratic a-area is euclidean if $a = 1$ or $a = n - 1$, and in general is not euclidean if $1 < a < n - 1$.*

The first part of the statement is obvious since a quadratic length is euclidean by definition and a quadratic $(n-1)$-area is euclidean in n-space by (5.5).

A simple counting argument convinces us of the truth of the second part since a euclidean quadratic area is determined by the metric (g_{ij}) so that the manifold of euclidean quadratic areas is $n(n+1)/2$-dimensional, while the manifold of Plücker coordinates is of dimension $1 + a(n-a)$; or, in other words, there are $\binom{n}{a} - a(n-a) - 1$ independent (quadratic) identities satisfied by the $P^{i_1 \cdots i_a}$ (see e.g. [2]). The distinct quadratic form in the Plücker coordinates therefore have dimension

$$\frac{1}{2}\binom{n}{a}\left[\binom{n}{a} + 1\right] - \binom{n}{a} + a(n-a) + 1$$

which exceeds $\binom{n+1}{2}$ whenever $1 < a < n - 1$.

If, for example, we restrict our attention to a-areas for which

$$F^2(x_1, \cdots, x_a) = \sum_{i_1 < \cdots < i_a} A_{i_1} \cdots {}_{i_a}(P^{i_1 \cdots i_a})^2$$

then no two different forms can be identical. Thus the dimension of this set is $\binom{n}{a}$ while the dimension of each equivalence class is no greater than $\binom{n+1}{2}$.

In particular, the Cartesian form

$$F^2(x_1, \cdots, x_n) = \sum (P^{i_1 \cdots i_a})^2 \tag{5.7}$$

is preserved only under orthogonal transformations. For, if we assume the existence of a matrix $g_{ij} \neq \delta_{ij}$ which preserves the form (5.7) then, for a suitable choice of Cartesian coordinates, we have $g_{ij} = A_i \delta_{ij}$ and (5.7) becomes

$$F^2(x_1, \cdots, x_n) = \sum A_{i_1} \cdots A_{i_a} (P^{i_1 \cdots i_a})^2 \tag{5.8}$$

Now, if (5.8) is Cartesian in one Cartesian coordinate system then it is Cartesian in all. Thus $A_{i_1} \cdots A_{i_a} = 1$ for all $i_1 < \cdots < i_a$. Since $n > a$ this implies $A_i = 1$ and $g_{ij} = \delta_{ij}$.

Since every euclidean area can be brought to Cartesian form we have also proved the following (which also follows from Theorem 9.1).

(5.9) *If two metrics g_{ij} and g'_{ij} give rise to identical a-areas, $a < n$, then $g_{ij} = g'_{ij}$.*

We can now determine the relations which suffice to make a quadratic a-area euclidean:

(5.10) THEOREM. *A quadratic a-area is euclidean if it is euclidean in every $(a + 2)$-flat.*

Proof. We proceed by induction. Assuming the area is euclidean is every m-flat, $m \geq a + 2$, we wish to prove it euclidean in every $(m + 1)$-flat. Let the $(m + 1)$-flat L_{m+1} have the equations $x^{m+2} = \cdots = x^n = 0$. Since the area is euclidean in every sub-flat $x^i = 0$ $(i = 1, \cdots, m + 1)$, there exists a matrix $g^{(i)}_{pq}$ $(1 \leq p, q \leq m + 1; p, q \neq i)$ so that the area function has the form (5.4) in this sub-flat. By (5.9) we have $g^{(i)}_{pq} = g^{(j)}_{pq}$ if $p, q \neq i, j$ since that is the unique metric in the common sub-flat $x^i = x^j = 0$. Thus there exists a matrix $g_{pq} = g^{(i)}_{pq}$ $(i \neq p, q)$ that defines a euclidean a-area in L_{m+1} which coincides with the given a-area in every coordinate sub-flat.

Without loss of generality we may assume the coordinates in L_{m+1} chosen so that $g_{pq} = \delta_{pq}$. Then on L_{m+1} we have

$$F^2(x_1, \cdots, x_a) = \sum (P^{i_1 \cdots i_a})^2 + R \tag{5.11}$$

where R involves the products of distinct Plücker coordinates so that every index $1, \cdots, m + 1$ appears in every product (if $m + 1 > 2a$ then there are no such terms and the proof is complete).

Consider the sub-flat $x^{m+1} = \lambda x^m$ of L_{m+1} and introduce the coordinates $y^i = x^i$ $(i = 1, \cdots, m - 1)$, $y^m = (1 + \lambda^2)^{-1} x^m$. In terms of these coordinates (5.11) becomes

$$(5.12) \qquad F^2(x_1, \cdots, x_a) = \sum (P^{i_1 \cdots i_a})^2 + R'$$

where R' involves only products in which there appears every index $1, \cdots, m$. Now (5.12) is euclidean by hypothesis and the matrix g'_{ij} which represents it in the form (5.4) reduces to the identity matrix in every coordinate sub-flat. Hence $g'_{ij} = \delta_{ij}$ and $R' \equiv 0$. This means $R = 0$ in every sub-flat $x^{m+1} = \lambda x^m$, that is, $R \equiv 0$ so that (5.11) is euclidean.

The simplest case of an a-area with $1 < a < n-1$, namely quadratic 2-area in A^4, already povides examples to show that:

(5.13) *For $1 < a < n-1$ an extendably convex a-area need not be totally convex.*

For, denote the euclidean area in A^4 which belongs to $g_{ik} = \delta_{ik}$ by $E(x_1, x_2)$ and put $e_i = (\delta_{i1}, \cdots, \delta_{i4})$. For any $\varepsilon > 0$

$$F^2(x_1, x_2) = \varepsilon E^2(x_1, x_2) + \left(\begin{vmatrix} x_1^1 & x_1^2 \\ x_2^1 & x_2^2 \end{vmatrix} + \begin{vmatrix} x_1^3 & x_1^4 \\ x_2^3 & x_2^4 \end{vmatrix} \right)^2$$

defines a quadratic 2-area which obviously is extendably convex. The (x^1, x^2)-plane P^{12} is normal to every line in the (x^3, x^4)-plane P^{34}, because for arbitrary λ, μ, ρ we have

$$F^2(e_1 + \lambda e_3 + \mu e_4, e_2 + \rho\lambda e_3 + \rho\mu e_4) \geq \varepsilon E^2(e_1, e_2) + \left(\begin{vmatrix} 1 & 0 \\ 0 & 1 \end{vmatrix} + \begin{vmatrix} \lambda & \mu \\ \rho\lambda & \rho\mu \end{vmatrix} \right)^2$$
$$= \varepsilon + 1 = F^2(e_1, e_2) \,.$$

Thus P^{12} is normal to P^{34}. However, for small ε, P^{12} is not totally normal to P^{34}, since then

$$F^2(e_1 + e_3 + e_4/2, e_2 + e_3 - e_4/2) = \varepsilon E^2(e_1 + e_3 + e_4/2, e_2 + e_3 - e_4/2) < 1 + \varepsilon.$$

According to (3.10) the plane normal to P^{34} at z cannot be unique. Actually there is a one-parameter family of planes totally normal to P^{34} at z. To see this we observe that

$$F^2(e_1 + \lambda e_3 + \mu_4, e_2 + \rho e_3 + \sigma e_4)$$
$$= \varepsilon(1 + \lambda^2 + \mu^2 + \rho^2 + \sigma^2 + (\lambda\sigma - \mu\rho)^2) + (1 + \lambda\sigma - \mu\rho)^2.$$

For a given ε with $0 < \varepsilon < 1$ this expression attains the minimal value $4\varepsilon/(1+\varepsilon)$ for $\lambda = -\sigma = \delta \cos\theta$, $\mu = \rho = \delta \sin\theta$ where $\delta = (1-\varepsilon)^{1/2}(1+\varepsilon)^{-1/2}$ and θ is arbitrary. Hence

(5.14) *If $1<a<n-1$ then extendable strict convexity of an a-area does not imply that the a-flat totally normal to an $(n-a)$-flat at a point is unique. More generally, the a-flat totally normal to a b-flat at a d-flat is not necessarily unique when $d < \min(a, b) - 1$.*

6. Equivalence of normality. Implications. Equivalence of normality for two convex areas implies for most combinations of the dimensions a, b, d that both areas are quadratic.

(6.1) THEOREM. *Let $0 \leq d < a \leq b < n$ $a+b-d \leq n$, but not $a+b=n$ and $d=0$. If a convex a-area α and a convex b-area β have the property that (total) α-normality and (total) β-normality at d-flats are equivalent, then both α and β are quadratic.*

(6.2) COROLLARY. *If for a convex a-area (total) normality at d-flats is symmetric then the area is quadratic unless $n = 2a$ and $d = 0$.*

We know from (5.3) that $a = 1, b = n - 1$ is actually exceptional but no examples are known for $a > 1$. (See note at end of paper).

The following proof is arranged so that only the existence of normals and not of transversals is used. Since the total normals exist, the proofs remain valid when normality is replaced everywhere by total normality.

Since normals and total normals do exist for non-convex areas, it is possible that (6.1) also holds without the assumption that α and β be convex. However the present proof uses convexity.

The hypothesis on the dimensions means that either (1) $a + b < n + d$ or (2) $a + b = n + d$ and $d > 0$. We consider the two cases separately.

In case (1) we show first (denoting an i-flat by L_i):

(A) Given[6] $L_{a-1} \subset L_{a+1} \subset L_{a+2}$ there exists an $L_a \subset L_{a+2}$ with $L_a \cap L_{a+1} = L_{a-1}$ such that the a-flats through L_{a-1} in L_{a+1} are normal to L_a.

(A′) The same as (A) with b replacing a.[6]

The proofs are entirely analogous with a slight simplification for (A) which we shall point out.

To prove (A) take $L_{n-b+a} \supset L_{a+1}$ with $L_{n-b+a} \cap L_{a+2} = L_{a+1}$, then take B normal to L_{n-b+a} at $D \subset L_{a-1}$. If $d + 1 < a$ choose the $(a - 1 - d)$-flat C such that $D \oplus C = L_{a-1}$. Since $B \oplus L_{n-b-d} = A^n$ we can find L_a with $L_{a-1} \subset L_a \subset B \oplus C$ and $L_a \oplus L_{a+1} = L_{a+2}$. (Here we can take $L_a \subset B$, but in the proof of (A′) there would exist no $L_b \subset A$ for $b > a$, whereas $L_b \subset A \oplus C$ exists because C is a $(b-1-d)$-flat and hence dim $A \oplus C = b - 1 - d + a \geq b$.) B is normal, hence by hypothesis transversal at D to any a-flat A' in L_{n-b+a} through D. If $L_{n-b+a} \supset A' \supset L_{a-1}$ then A' is normal to $B \oplus C$ at L_{a-1} and hence is normal to L_a at L_{a-1}.

We now show that (A) implies that α is quadratic. Let $z \in L_a \subset L_{a+2}$ and take L_3 through a perpendicular to L_{a-1} in L_{a+2}. Construct the surface $S = S(L_3, L_{a-1})$ of (3.5). It follows from the discussion of (3.6) Case I that for two lines G, H through z in L_3 the a-flat $G \oplus L_{a-1}$ is normal to $H \oplus L_{a-1}$ if and only if H is parallel to a supporting line of S at one of the two points $G \cap S$.

[6] $a + 2 \leq b + 2 \leq n$ since $b < n + d - a \leq n - 1$.

Now it follows from (A): Given L_2 through z in L_3 there exists in L_3 a $G \ni z$ such that for $z \in H \subset L_2$ the a-flat $H \oplus L_{a-1}$ is normal to $G \oplus L_{a-1}$. In terms of S this means that every intersection of S with a plane through z lies in some circumscribed cylinder of S.

A well known theorem of Blaschke [3] (see also [4, p. 157]) states that a closed convex surface S' in A^3 is an ellipsoid if every cylinder touches S' in a plane curve. Blaschke assumes that S' is differentiable but not that S' has a center. The differentiability hypothesis is very easily removed (see e.g. [9, p. 93]).

Under the hypothesis that S' has a center z the hypothesis may be relaxed in two ways.

(B_1) *S' is an ellipsoid when every plane section of S' through z lies on a circumscribed cylinder.*

(B_2) *S' is an ellipsoid when every circumscribed cylinder contains a plane section of S' through z.*

(B_1) is proved by a trivial modification of the proof of Blaschke's theorem and is also well known from the theory of Banach spaces.

The proof of (B_2) requires a less obvious but far from difficult modification of Blaschke's proof. (B_1) and (A) show that S is an ellipsoid. It follows that $S(L_{a-1}, L_{n-a+1})$ is also an ellipsoid (compare for example [9, p. 91]).

In the same way we deduce from (A′) and (B_1) that the surface $S(L_{b-1}, L_3)$ constructed with the area β is an ellipsoid so that β is also quadratic.

We now turn to the case $a + b = n + d, d > 0$ and prove:

(C) *Given $z \in L_{a-2} \subset L_{a-1} \subset L_{a+1}$ there is an a-flat A in L_{a+1} with $A \cap L_{a-1} = L_{a-2}$ such that the a-flats A_G in L_{a+1} through L_{a-1} are normal to A. The same holds with b replacing a.*

Take B normal to L_{a-1} at an $L_{d-1} \subset L_{a-2}$. Such a B exists because $a - 1 + b = n + d - 1$, moreover $L_{a-1} \oplus B = A^n$.

For any line G through z in B the a-flat $A_G = L_{a-1} \oplus G$ is transversal to B at $D_G = L_{d-1} \oplus G$. Hence A_G is, by hypothesis, normal to B at D_G. If $a > d + 1$ choose an $(a - d - 1)$-flat C through z in L_{a-1} such that $L_{d-1} \oplus C = L_{a-2} \subset L_{a-1}$. Then A_G is normal to $B \oplus C$ at $L_{a-1}^G = D_G \oplus C$. Let $z \in L_2 \subset B, L_2 \cap L_{d-1} = z, L_2 \oplus L_{d-1} = L_{d-1}$. This L_2 exists because $L_{a-1} \cap B = L_{d-1}$ and $L_{a-1} \oplus B = A^n$. Then $A = L_2 \oplus L_{a-2} \subset B \oplus C$ and for $z \in G \subset L_2$ the a-flat A_G is normal to A at L_{a-1}^G.

We now construct a surface T as in Case II of (3.6). On the line perpendicular to a given a-flat A' through z in L_{a+1} we take $y_{A'}$ with $e(z, y_{A'}) = f^{-1}(A')$. The points $y_{A'}$ traverse T.

Also, for a given L_{a-2} with $z \in L_{a-2} \subset L_{a+1}$ we take the L_3

perpendicular to L_{a-2} through z ($L_3 = L_{a+1}$ if $a = 2$). If $A' \supset L_{a-2}$ then the perpendicular to A' at z lies in L_3. The perpendiculars to the $A' \supset L_{a-2}$ therefore intersect T in a surface T_0 and it suffices to prove that T_0, or its polar reciprocal T_0' in L_3, is an ellipsoid.

According to the discussion of Case II the a-flat spanned by $x \in T_0'$ and L_{a-2} is transversal to the a-flat $A' \supset L_{a-2}$ if and only if A' is spanned by L_{a-2} and a plane L_2 through z parallel to a supporting plane of T_0' at x. Then A' is normal also to every a-flat in L_{a+1} through L_{a-2} and x.

Statement (C) means in terms of T_0', that given a line H through z ($H \oplus L_{a-2}$ is the L_{a-1} in the hypothesis of (C)) the cylinder parallel to H circumscribed to T_0' touches T_0' in a set containing a section of T_0' by a plane L_2 through z ($H \oplus L_{a-2}$ is the L_a in the assertion of (C)). It now follows from (B_2) that T_0' is an ellipsoid.

The proof that β is quadratic for $a + b = n + d, d > 0$ is again entirely analogous.

The Corollary (6.2) can be improved in special cases as follows:

(6.3) THEOREM. *If $a < n/2$ and $d = 0$ or $a > n/2$ and $d = 2a - n$ and for a (totally) convex a-area α (total) normality at d-flats is symmetric, then α is euclidean.*

The area function is differentiable because, according to (6.2), it is quadratic (in other respects the present proof is independent of (6.2)).

Let $a < n/2, A \ni z$ and let B_A be the $(n - a)$-flat transversal to A at z. Then each a-flat $A' \ni z$ in B_A is transversal to A. Hence by hypothesis A' is normal to A so that $B_{A'} \supset A$. Thus $A' \supset B_A$ implies $B_{A'} \supset A$. The mapping $A \to B_A$ can therefore be extended to a correlation φ on itself of the bundle consisting of all i-flats $(1 \leq i \leq n - 1)$ through z (see [2, pp. 51–53]). Moreover φ is a polarity because $A\varphi^2 = A$, and if $L_1 \ni z$ then $L_1\varphi$ does not contain L_1. Thus φ coincides with the mapping which belongs to a suitable ellipsoid E with center z which associates $L_1 \ni z$ with its diametral hyperplane $L_1\varphi$. This nearly obvious fact may be seen as follows.

We extend A^n to a projective space P^n and the correlation φ to a correlation of P^n by first associating $z = (0, \cdots, 0,1)$ with the hyperplane at infinity $H = (0, \cdots, 0,1)$. With the intersection $L_1 \cap H = (x_1, \cdots, x_n, 0)$ we associate the hyperplane $L_1\varphi = (\xi_1, \cdots, \xi_n, 0)$. If T is the (symmetric) matrix of $(x_1, \cdots, x_n, 0) \to (\xi_1, \cdots, \xi_n, 0)$ then $\begin{pmatrix} T & 0 \\ 0 & -1 \end{pmatrix}$ is the matrix of a polarity in P^n which defines the ellipsoid E with the above property.

This ellipsoid taken as unit sphere defines a euclidean metric in A^n and also a euclidean a-area. By construction normality of a-flats at z for this area coincides with α-normality of a-flats at z. According to

(5.1) this shows that the two areas differ only by a constant factor so that α is also euclidean.

The case $a > n/2$, $d = 2a - n$ is very similar. If $A \ni z$ we take the $(n - a)$-flat B_A transversal to A at z. This time if $A' \supset B_A$ then A' is transversal to A at $A' \cap A$ where $\dim A' \cap A = 2a - n$. By hypothesis A' is normal to A so that $B_{A'} \subset A$. Since $A' \supset B_A$ implies $B_{A'} \subset A$, the mapping $A \to B_A$ can again be extended to a correlation of the bundle of all i-flats $(1 \leq i \leq n-1)$ through z on itself. From here on the proof proceeds exactly as in the first case.

7. Minkowski area. We now apply our results to the special cases from which the general theory originated.

Consider a symmetric Minkowski metric (or a 1-dimension convex area) $F(x)$ in A^n. We denote its unit ball $F(x) \leq 1$ by U and let $U(A)$ denote the intersection of U with the a-flat A through z. For any a-flat $\bar{A}$ parallel to A the intersection $(F(x - \bar{z}) \leq 1) \cap \bar{A}$, $\bar{z} \in \bar{A}$ originates from $U(A)$ by translation and is a unit ball in $\bar{A}$ for the metric induced by $F(x)$ in $\bar{A}$. Following [7] we define an a-dimensional area $1 \leq a \leq n$ in A^n by stipulating that the measure of $U(A)$ have the euclidean volume

$$\pi_a = \pi^{a/2}/\Gamma\left(\frac{a}{2} + 1\right),$$

(in particular $\pi_1 = 2$, $\pi_2 = \pi$) so that for a definite euclidean metric e we have

$$f_a(\bar{A}) = f_a(A) = \pi_a / |U(A)|_a^e .$$

The functions corresponding to our previous $\alpha(M)$ and $F(x_1, \cdots, x_a)$ will be denoted by $|M|_a$ and $F_a(x_1, \cdots, x_a)$ so that

$$|M|_a = f_a(A)|M|_a^e ,$$

$$F_a(x_1, \cdots, x_a) = f_a(A)|[z, x_1, \cdots, x_a]|_a^e$$

and $F_1(x) = F(x)$. Since we admitted $a = n$ we also have an n-dimensional measure

$$|M|_n = f_n |M|_n^e = \pi_n |M|_n^e / |U|_n^e .$$

For $a < n$ let $L_{a-1} \ni z$ be an $(a-1)$-flat and L_2 the plane perpendicular to L_{a-1} at z. On a variable ray R with origin z in L_2 take the point y_R with

$$e(z, y_R) = f_a^{-1}(A_R) = |U(A_R)|_a^e \pi_a^{-1} , \qquad A_R = L_{a-1} \oplus R .$$

That is the curve $S(L_{a-1}, L_2)$ for f_a as constructed in (3.5). It is a fundamental and non-trivial fact (see [7, p. 164]) that $S(L_{a-1}, L_2)$ for f_a is always convex and is strictly convex or differentiable when the unit sphere $F(x) = 1$ of the space is strictly convex or differentiable respectively. Thus we have the following.

(7.1) THEOREM. *The Minkowski areas* $|M|_a$, $(1 \leq a \leq n-1)$ *are all convex. They are strictly convex or differentiable if the unit sphere* $F(x) = 1$ *is strictly convex or differentiable.*

The question whether Minkowski areas are totally convex for $1 < a < n-1$ is equivalent to a difficult problem on convex bodies. Even extendable convexity is not known (see Problem 10 in [10]).

We mention the following further property of Minkowski area which is important for differential geometric investigations and was proved by Barthel [1].

(7.2) *If* $F(x)$ *is of class* C^r *for* $x \neq 0$ *then* $F_a(x_1, \cdots, x_a)$ *is of class* C^r *for* $x_1 \wedge \cdots \wedge x_a \neq 0$.

We also note:

(7.3) *If the a-area,* $1 \leq a \leq n-1$, *of a Minkowski space is quadratic then the space is euclidean.*

For, if $a > 1$ then we conclude from (6.3) that the area in any $(a+1)$-dimensional subspace is euclidean. It is easily seen and contained in Theorem (9.1) that therefore the metric in this subspace is euclidean. It is well known (see e.g. [9, (16.12) p. 91]) that then the metric of the whole space is euclidean. Therefore (6.1) and (6.2) yield the following.

(7.4) THEOREM. *Let* $0 \leq d < a \leq b < n$ *but not* $a + b = n$ *and* $d = 0$. *If* α, β *are Minkowski a-and b-areas respectively (not necessarily relative to the same Minkowski metric) and* α*-normality and* β*-normality at d-flats are equivalent then both Minkowski metrics are euclidean.*

If normality of a-flats at d-flats in a Minkowski space is symmetric then the space is euclidean unless $a = n/2, d = 0$.

We note in particular that for all $n > 2$ symmetry of normality of a-flats at $(a-1)$-flats suffices to make the Minkowski space euclidean. From (5.2) we know that the case $a = 1, b = n-1, d = 0$ is exceptional for two distinct Minkowski metrics. Whether this case is exceptional when α and β belong to the same Minkowski space amounts (unless $a = b = 1$) to an interesting open problem on convex bodies (see [10, Problem 5]).

Finally we see from Theorem (7.4) and the example at the end of § 6 that convex area functions—even quadratic area functions—are in general not Minkowskian. The problem of characterization of Minkowski areas among the convex area functions remains open, (see § 9).

The fact that area functions are now defined for all a leads to new concepts, in particular to a sine function. If $A \cap B = D \ni z$, where $0 \leq d \leq \min(a, b) - 1$ and $A \oplus B = Q$, take a non-degenerate q-box, $[z, y_1, \cdots, y_b, x_{d+1}, \cdots, x_a]$, such that $y_1, \cdots, y_d \in D; y_{d+1}, \cdots, y_b \in B - D$ and $x_{d+1}, \cdots, x_a \in A - D$. Now put

$$\text{(7.5)} \qquad \operatorname{sm}(A, B) = \frac{F_d(y_1, \cdots, y_d)F_q(y_1, \cdots, y_b, x_{d+1}, \cdots, x_a)}{F_a(y_1, \cdots, y_d, x_{d+1}, \cdots, x_a)F_b(y_1, \cdots, y_b)},$$

where $F_0 = 1$. The number $\operatorname{sm}(A, B)$ is called the *Minkowski sine* of the flats A, B because it depends only on the latter and not on the choice of the q-box. For example, if $d > 0$ then replacing $y_1, \cdots, y_d$ by other independent $\bar{y}_1, \cdots, \bar{y}_d \in D$ amounts to multiplying all four terms F_d, F_q, F_a, F_b in (7.5) by

$$|[z, \bar{y}_1, \cdots, \bar{y}_d]|_d^e / |[z, y_1, \cdots, y_d]|_d^e .$$

If D does not contain z, but $\bar{z} \in D$ then the vectors y_i, x_j in (7.5) must be replaced by $y_i - \bar{z}, x_j - \bar{z}$.

The sine function is *not* the function of a number, "the angle between A and B". Even in the euclidean case this angle is defined *only* for $d = \min(a, b)$-1. Hence the restriction to this case in [7] and [1]. The sine function for the euclidean metric will be denoted by se. Then obviously, with $f(L_0) = 1$, we have

$$\text{(7.6)} \qquad \operatorname{sm}(A, B) = \operatorname{se}(A, B)f_d(D)f_q(0)f_a^{-1}(A)f_b^{-1}(B) .$$

For any λ_j^k, $k = d + 1, \cdots, b$; $j = d + 1, \cdots, a$ put

$$y_k(\lambda) = \sum_{j=d+1}^{b} \lambda_j^k y_j .$$

Then the boxes of the form $[z, y_1, \cdots, y_d, x_{d+1} + y_{d+1}(\lambda), \cdots, x_a + y_a(\lambda)]$ have $[z, y_1, \cdots, y_d, x_{d+1}, \cdots, x_a]$ as projection in Q parallel to B on A. Since

$$F_a(y_1, \cdots, y_b, x_{d+1} + y_{d+1}(\lambda), \cdots, x_a + y_a(\lambda))$$

does not depend on the λ_j^k, the a-flat A is totally normal to B at D in Q if and only if

$$\operatorname{sm}(A, B) \geq \operatorname{sm}(A^*, B) \text{ for } A^* \cap B = D, A^* \subset Q .$$

We denote this maximal value of $\operatorname{sm}(A^*, B)$ for given B, D, Q by $\alpha(B, D, Q)$. If $q = n$ then $Q = A^n$ is unique and we write simply $\alpha(B, D)$.

(7.7) *If*

$$\operatorname{sm}(A_1, B_1) = \max_A \alpha(A, D, Q),$$

then

$$\operatorname{sm}(A_1, B_1) = \max_B \alpha(B, D, Q),$$

and conversely, hence A_2 *is totally normal to* B_2 *and* B_2 *is totally normal to* A_2.

Proof. If A is normal to B then

(7.8) $$\alpha(B, D, Q) = \operatorname{sm}(A, B) \leq \alpha(A, D, Q)$$

hence

(7.9) $$\max_B \alpha(B, D, Q) \leq \max_A \alpha(A, D, Q)$$

Similarly, if B' is totally normal to A' then

(7.10) $$\alpha(A', D, Q) = \operatorname{sm}(A', B') \leq \alpha(B', D, Q) .$$

Whence together with (7.9) we have

(7.11) $$\max_B \alpha(B, D, Q) = \max_A \alpha(A, D, Q) .$$

If

$$\operatorname{sm}(A_1, B_1) = \max_A \alpha(A, D, Q) = \alpha(A_1, D, Q)$$

then B_1 is totally normal to A_1. Hence (7.11) and (7.10) imply

$$\operatorname{sm}(A_1, B_1) = \alpha(B_1, D, Q) = \max(B, D, Q)$$

so that A_1 is totally normal to B_1.

(7.12) *If for given* A (B) *in* Q *through* D *there exists a b-flat (a-flat) totally transversal to* B (A) *at* D *in* Q *(which is always the case for* $\min(a, b) = d + 1$*) and*

$$\operatorname{sm}(A_2, B_2) = \min_A \alpha(B, D, Q)$$

then

$$\operatorname{sm}(A_2, B_2) = \min_B a(B, D, Q)$$

and A_2, B_2 are normal to each other.

For, if A is totally transversal to a given B, then

$$\alpha(A_2, D, Q) = \mathrm{sm}\,(A, B) \leq \alpha(B, D, Q)\,.$$

Hence

$$\min_A \alpha(B, D, Q) \leq \min_B \alpha(B, D, Q)\,.$$

The proof is analogous to that of (7.7).

As a consequence of (7.11) and (7.12) we have the following.

(7.13) COROLLARY. *If the function $\alpha(A, D, Q)$ is constant for fixed D, Q then $\alpha(B, D, Q)$ is constant and conversely. Moreover the constants have the same value. If $\alpha(A, D, Q)$ or $\alpha(B, D, Q)$ is constant then total normality of A to B and total normality of B to A are equivalent.*

The equivalence of total normality follows from the fact that for any A totally normal to B we have

$$\mathrm{sm}\,(A, B) = \max_A \alpha(A, D, Q)\,.$$

The equivalence of normality implies that B (A) totally transversal to A (B) at D in Q exist. Therefore both (7.11) and (7.12) apply.

Whether the converse of the second statement in (7.13) always holds is not known. However the proof of (3.6) yields the following special case.

(7.14) *If $d = \min(a, b) - 1$ and normality of A to B at D in Q is equivalent to normality of B to A, then $\alpha(A, D, Q)$ and $\alpha(B, D, Q)$ are constant.*

Proof. If $z \in D$ we take as in Case I of (3.6) the $(q-d)$-flat L_{q-d} perpendicular to D at z and construct, if $a \leq b$ say, the surface S by taking on each ray R in L_{q-d} with origin z the point y_R with $e(z, y_R) = f_a^{-1}(A_R)$ where $A_R = D \oplus R$.

For the b-area we construct T as in Case II by taking on the perpendicular in Q to a b-flat B through D in Q the two points y'_R with $e(z, y'_R) = f_b^{-1}(B)$, and denote by T' the polar reciprocal of T in L_{q-d} with respect to the metric $e(x, y)$.

If w_R is the point $R \cap T'$ then the supporting $(q-d-1)$-flat of T' at w_R spans together with the d-flat parallel to D through w_R a b-flat B normal to A_R. The reciprocity of T and T' implies that B has distance $f_b(B)$ from z. Hence by (7.6) we have[7]

[7] Because $d = \min(a, b) - 1$ the function se is the ordinary sine of the angle between A_R and B in the metric $e(x, y)$.

$$\alpha(A_R, D, Q) = \mathrm{sm}\,(A_R, B) = \mathrm{se}\,(A_R, B) f_d(D) f_q(Q) f_{\bar a}^{-1}(A_R) f_{\bar b}^{-1}(B) .$$

But

$$f_b(B) = \mathrm{se}\,(A_R, B) e(z, w_R), f_a(A_R) = e^{-1}(z, y_R) ,$$

so that we have the following nice *interpretation for* $\alpha(A_R, D, Q)$:

$$\alpha(A_R, D, Q) = f_{\bar d}^{-1}(D) f_{\bar q}^{-1}(Q) e(z, y_R)/e(z, w_R) .$$

If normality of A to B at D in Q is equivalent to that of B to A then S and T' are homothetic. Hence $e(z, y_R)/e(z, w_R)$ is constant, which proves (7.14).

8. The range of the sine functions. Problems regarding the ranges of $\alpha(B, D, Q)$ are important for Minkowskian geometry and are geometrically very attractive, but unfortunately often quite difficult—only in the simplest case $n = 2$ hence $a = b = 1, d = 0$ do we have complete answers owing to Petty [12] who found the following.

For any line L_1 *in* A^2 *through* z *we put* $\alpha(L_1, z, A^2) = \alpha(L)$ *and denote by* C_F *the unit circle* $F(x) = 1$. *Then*

$$\min_{L_1, F} \alpha(L_1) = \pi/4, \quad \max_{L_1, F} \alpha(L_1) = \pi/2 ,$$

and $\alpha(L_1) = \pi/4$ *or* $\alpha(L_1) = \pi/2$ *imply that* C_F *is a parallelogram and* L_1 *a suitable line (different in the two cases).*

Also

$$\max_F \min_{L_1} \alpha(L_1) = \pi/3 ,$$

where the maximum is attained only when C_F *is a hexagon which is regular for a suitable* $e(x, y)$.

Finally

$$\min_F \max_{L_1} \alpha(L_1) = 1 ,$$

where the minimum is attained only when C_F *is an ellipse, that is when the metric is euclidean.*

By (7.13) and (7.14) we have $\alpha(L_1) = k_F$, that is $\alpha(L_1)$ is independent of L_1, if and only if normality of lines in the plane is symmetric. This means that C_F is one of the curves discovered by Radon [13] which we encountered already several times implicitly and which we shall call Radon curves. Their construction is also found in Petty [12] and in [9, p. 104]. Since the regular hexagon is a Radon curve we find $1 \le k_F \le \pi/3$ with $k_F = 1$ only for the euclidean metric and $k_F = \pi/3$ only when C_F is a regular hexagon.

Under the hypothesis of (7.14), if $a = b$ and hence $d = a - 1$ then S and T' are Radon curves and we can derive the range of $\alpha(L_a, L_{a-1}, L_{a+1})$ (when constant) from Petty's results. Otherwise the ranges for $\alpha(A, D, Q)$ with D, Q fixed are not known. For variable D, Q we deduce from (7.13) and (7.14) the following.

(8.1) THEOREM. *If $0 \leq d < a \leq b < n$ but not $a + b = n$ and $d = 0$ then $\alpha(L_a, L_d, L_{a+b-d})$ is independent of L_a, L_d, L_{a+b-d} only in the euclidean geometry (where all α-functions are equal to 1).*

Beyond this result only very few facts on the ranges of the sine functions are known for $n > 2$, which we shall now discuss.

$$\min_{F,L_1} \alpha(L_1, L_0) = \min_{F,L_{n-1}} \alpha(L_{n-1}, L_0) = \pi_n/2\pi_{n-1} \tag{8.2}$$

$$\max_{F,L_1} \alpha(L_1, L_0) = \max_{F,L_{n-1}} \alpha(L_{n-1}, L_0) = n\pi_n/2\pi_{n-1}\,. \tag{8.3}$$

In the first of these relations equality is obtained only when the unit sphere S, that is $F(x) = 1$, is a *cylinder* and in the second only when S is a *double cone.*

The proof is very simple. The equality of the first two members in (8.2) or (8.3) follows from (7.12) and (7.7). Let H be a hyperplane through z and L_1 normal to H at z. If p, p' are the points $L_1 \cap S$ and $U_H = U \cap H$ then the hyperplanes parallel to H through p and p' are supporting planes of U. Moreover U_H has maximal $(n-1)$-dimensional volume among all sections U by hyperplanes parallel to H. Therefore

$$\pi_n = |U|_n \leq F(p - p')|U_H|_{n-1} \operatorname{sm}(L_1, H) = 2\pi_{n-1}\alpha(H, z)$$

with equality only for cylinders.

On the other hand U contains the double cone formed by the cones with apexes p, p' and bases U_H so that

$$\pi_n \geq n^{-1} 2\pi_{n-1}\alpha(H, z)$$

with equality only for double cones.

These relations successively provide bounds for all $\alpha(L_a, L_d)$, but these bounds are not sharp. We exemplify the procedure with $\alpha(L_{n-2}, L_0)$. If L_{n-2} is normal to L_2 at z then we consider in L_2 lines L_1' and L_1 through z such that L_1' is normal to L_1. Since L_{n-2} is normal to L_1 and L_1' we have, with $L_{n-1} = L_{n-2} \oplus L_1$,

$$\operatorname{sm}(L_{n-2}, L_2) \operatorname{sm}(L_1', L_1) = \operatorname{sm}(L_{n-2}, L_1) \operatorname{sm}(L_1', L_{n-1})$$

or

$$\begin{aligned} \operatorname{sm}(L_{n-2}, L_2) &= \alpha(L_1, z, L_{n-1}) \operatorname{sm}(L_1', L_{n-1}) \alpha^{-1}(L_1, z, L_2) \\ &\leq \frac{n-1}{2} \frac{\pi_{n-1}}{\pi_{n-2}} \frac{n}{2} \frac{\pi_n}{\pi_{n-1}} \frac{4}{\pi} = \frac{n(n-1)}{\pi} \frac{\pi_n}{\pi_{n-2}}\,. \end{aligned}$$

It is easily seen that with a proper choice of L_1, L_1' in L_2 the line L_1' is normal to L_{n-1}. Hence

$$\operatorname{sm}(L_{n-2}, L_2) \geq \frac{1}{2}\frac{\pi_{n-1}}{\pi_{n-2}}\frac{1}{2}\frac{\pi_n}{\pi_{n-1}}\frac{2}{\pi} = \frac{1}{2\pi}\frac{\pi_n}{\pi_{n-2}},$$

so that

$$\frac{1}{2\pi}\frac{\pi_n}{\pi_{n-2}} \leq \alpha(L_{n-2}, L_0) \leq \frac{n(n-1)}{\pi}\frac{\pi_n}{\pi_{n-2}}.$$

The only exact bound other than (8.2) and (8.3) which has been determined is the following.

(8.4) $$\max_{F, L_{n-1}} \alpha(L_{n-1}, L_{n-2}) = 2\pi_{n-2}\pi_n/\pi_{n-1}^2.$$

This equality holds only for a cylindrical unit sphere with $(n-2)$-dimensional generators and a parallelogram as 2-dimensional crosssection whose exact definition will emerge from the proof.

If an L_{n-2} is given we choose coordinates so that its equations are $x_{n-1} = x_n = 0$ and put $x_{n-1} = \rho\cos\varphi$, $x_n = \rho\sin\varphi$ so that $x_1, \cdots, x_{n-2}, \rho, \varphi$ are our coordinates. Set $U(L_{n-2}) = V$. For given x, φ with $x \in V$ let $(x, r(x, \varphi), \varphi)$ lie on the unit sphere S. Then, with $e^2(x, y) = \sum(x^i - y^i)^2$,

$$|U|_n^e = \frac{1}{2}\int_0^{2\pi}\int_V r^2(x, \varphi)dx\,d\varphi \geq \frac{1}{2|V|_{n-2}^e}\int_0^{2\pi}\left(\int_V r(x, \varphi)dx\right)^2 d\varphi$$

with equality only when $r(x, \varphi)$ is independent of x.

Now $\int_V r(x, \varphi_0)dx$ is the euclidean volume $A(\varphi_0)$ of the intersection of U with the half-hyperplane $\varphi = \varphi_0$. Hence if P_{φ_0} is the hyperplane containing $\varphi = \varphi_0$ we have

$$\begin{aligned}\operatorname{sm}(P_{\varphi_1}, P_{\varphi_2}) &= \frac{\sin|\varphi_1 - \varphi_2|2A(\varphi_1)2A(\varphi_2)}{|V|_{n-2}^e|U|_n^e}\cdot\frac{\pi_{n-2}\pi_n}{\pi_{n-1}^2} \\ &\leq \frac{4\sin|\varphi_1 - \varphi_2|A(\varphi_1)A(\varphi_2)}{(1/2)\int_0^{2\pi}A^2(\varphi)d\varphi}\cdot\frac{\pi_{n-2}\pi_n}{\pi_{n-1}^2}\end{aligned}$$

Considering the convex curve $\rho = A(\varphi)$ in $x_1 = \cdots = x_{n-2} = 0$ we see that the first factor on the right attains its maximum 2 when the curve is a parallelogram and φ_1, φ_2 fall in the diagonals. There will be equality in (8.4) if and only if in addition $r(x, \varphi)$ is independent of x. For $n = 3$ we have equality only for a parallelepiped.

The most important questions regarding the ranges of the sine functions concern

$$\min_F \max_{L_a} \alpha(L_a, L_a) = \min_F \max_{L_{n-a+a}} \alpha(L_{n-a+a}, L_a),$$

in particular whether, or for which a, d this number equals 1; and whether the value 1 characterizes euclidean geometry. The case $a = 1, d = 0$ is Problem 6 in [10].

9. Relations between the functions f_a. The Minkowski areas are derived from—and hence determined by—the Minkowski length. The question arises whether in a Minkowski geometry any of the areas $(1 < a < n)$ determine the remaining ones.

(9.1) THEOREM. *An a-dimensional area function $F_a(x_1, \cdots, x_a)$, $1 \leq a \leq n-1$, is an a-dimensional Minkowski area for at most one Minkowski geometry. In other words, if $F_a(x_1, \cdots, x_a)$ is known then $F(x)$ and hence the remaining $F_b(x_1, \cdots, x_b)$ are determined.*

This follows from a theorem of P. Funk [11]:

Let S_e be the sphere $e(z, x) = 1$ in B and let $S(A)$ be its intersection with $A \ni z$. Let $g_i(x)$, $i = 1, 2$ be an even continuous function on S_e and denote by $S(A, g_i)$ the integral of $g_i(x)$ over $S(A)$ with respect to $(a-1)$-dimensional area. If $S(A, g_1) = S(A, g_2)$ for each A with $z \in A \subset B$ then $g_1(x) \equiv g_2(x)$.

Induction reduces this statement to $a = b - 1$.

A proof for $b = 3$ is found in [5, p. 138]. A proof for general b is obtained by using expansion in terms of spherical harmonics. If $x \in S_e$ then $xF^{-1}(x)$ lies on $F(x) = 1$. Hence $|U(A)|_a^e = S(A, a^{-1}F^{-a}(x))$ so that by Funk's theorem this relation determines $F(x)$.

An explicit expression of $F(x)$ in terms of $f_a(A)$ can be found in [4, pp. 154, 155], and this yields, in principle, the value $f_b(B)$ for given B. Actually the expression thus obtained is much too involved to deduce pertinent information from it. There is however an inequality of a very simple form, although its proof is involved, which relates f_b and f_a and which we are now going to derive from the results of [8].

If $n \geq b > a > 1, B \ni z$ then

$$D(b, a) f_b^{-a}(B) \geq \int_{B \supset A \ni z} f_a^{-b}(A) dA \tag{9.2}$$

with equality only for the ellipsoid. In this formula dA is the kinematic density for a-flats in B, the quantity $D(b, a)$ is the measure of all a-flats through z in B and hence is a constant which depends only on a and b.

Since in (9.2) B acts as the whole space we may take $b = n$. The inequality is a special case of a relation between the functions

$$f_{i,a}(A) = \pi_a / |U_i(A)|_a^e, f_{i,n} = \pi_n / |U_i|_n^e \qquad i = 1, \cdots, a$$

belonging to different Minkowski metrics with unit spheres $U_1, \cdots, {}_aU$

with common center z:

$$(9.3) \qquad D(n, a)\prod_{i=1}^{a} f_{i,n}^{-1} \geq \int_{A \ni z} \prod_{i=1}^{a} f_{i,a}^{-n/a}(A)dA \ ,$$

with equality only when the U_i are homothetic ellipsoids, i.e. when the corresponding Minkowski metrics are proportional euclidean metrics.

The inequality (9.3) is in turn a consequence of a still more general inequality.

Let $M_1, \cdots, M_a$ be convex bodies in the n-dimensional euclidean space $E^n, n \geq 3, 2 \leq a \leq n-1$ then

$$(9.4) \qquad |M_1|_n \cdots |M_a|_n \geq \pi_n^a \pi_a^{-n} D^{-1}(n, a) \times \int_{A \ni z} |M_1 \cap A|_a^{n/a} \cdots |M_a \cap A|_a^{n/a} dA \ ,$$

with equality for $|M_i|_n > 0$ only when the M_i are homothetic ellipsoids with center z. The measure $|M|_i$ is of course, the i-dimensional Lebesgue measure in E^n.

We deduce (9.4) from the following relation for any closed bounded sets $M_1, \cdots, M_a$.

$$(9.5) \qquad |M_1|_n \cdots |M_a|_n = C^1(n, a) \times \int_{A \ni z} \int_{M_1 \cap A} \cdots \int_{M_a \cap A} T^{n-a}(P_1, \cdots, P_a, z) dV_{P_1}^a \cdots dV_{P_a}^a dA$$

where $T(P_1, \cdots, P_a, z)$ is the a-dimensional measure of the (possibly degenerate) simplex with vertices $P_1, \cdots, P_a, z$ and $dV_{P_i}^a$ is the area element of A at $P_i \in M_i \cap A$. The symbol $C^i(n, a)$ denotes a constant which depends only on n and a.

For $a = n-1$ and $a = n-2$ (9.5) is proved in [8, (2), (17)], hence we prove (9.5) by induction for decreasing a. Assume (9.5) to hold for some $a + 1 \leq n - 1$. As M_{a+1} we take the euclidean unit ball U with center z. Then if B denotes an $(a+1)$-flat we have

$$|M_1|_n \cdots |M_a|_n = \pi_n^{-1} C^1(n, a+1) \times \int_{B \ni z} \int_{M_1 \cap B} \cdots \int_{M_{a+1} \cap B} T^{n-a-1}(P_1, \cdots, P_{a+1}, z) dV_{P_1}^{a+1} \cdots dV_{P_{a+1}}^{a+1} dB \ .$$

Now $M_{a+1} \cap B$ is an $(a+1)$ — dimensional unit ball $\overline{U}$, and if φ is the angle between the a-flats spanned by $P_1, \cdots, P_a$ and the line through z and P_{a+1}, then

$$T(P_1, \cdots, P_{a+1}, z) = (a+1)^{-1} e(z, P_{a+1}) |\sin\varphi| T(P_1, \cdots, P_a, z) \ .$$

Since

$$\int_{\bar{U}} e^{n-a-1}(z, P_{a+1})|\sin^{n-a-1}\varphi| d V_{P+1}^{a+1}$$

depends only on n and a we obtain, after carrying out the integration ever U,

$$|M_1|_n \cdots |M_a|_n = C^2(n, a)$$
$$\times \int_{B \ni z} \int_{M_1 \cap B} \cdots \int_{M_a \cap B} T^{n-a-1}(P_1, \cdots, P_a, z) d V_{P_1}^{a+1} \cdots d V_{P_{a+1}}^{a+1} dB .$$

For a variable a-flat A through z in B we have (see [8, (12)])

$$d V_{P_1}^{a+1} \cdots d V_{P_a}^{a+1} = a!\, T(P_1, \cdots, P_a, z) d V_{P_1}^a \cdots d V_{P_a}^a dA .$$

Integration first over all A through z in B, and then over all B through z can, according to the properties of kinematic measure, be interpreted as an integration over all A through z (except for a factor which depends only on a), and this proves (9.5).

Steiner's symmetrization leads from (9.5) to (9.4). Consider a fixed a-flat A through z and let $M_1, \cdots, M_a$ be convex bodies. It is shown in [8, pp. 8-10] that under simultaneous symmetrization of the sets $M_i \cap A$ in any $(a-1)$-flat C through z in A the integral

$$\int_{M_1 \cap A} \cdots \int_{M_a \cap A} T^{n-a}(P_1, \cdots, P_a, z) d V_{P_1}^a \cdots d V_{P_a}^a$$

decreases unless the centers of all chords of all $M_i \cap A$ perpendicular to C are coplanar with z. Hence the $M_1 \cap A$ are homothetic ellipsoids with center z if the last integral is to be minimized. The minimum is actually attained for such ellipsoids [8, pp. 10, 11] and the integral has then the value

$$C^3(n, a)|M_1 \cap A|_a^{n/a} \cdots |M_a \cap A|_a^{a/a} .$$

This proves (9.4).

We note two consequences of these results.

(9.6) *The ellipsoids with center z maximize* $\int_{A \in z} |M \cap A|_a^n dA$ *among all convex bodies with a given volume.*

Application of (9.4) to the case $M_2 = \cdots = M_a = U$ yields

(9.7) $$|M|_n \geq \pi_n \pi_a^{-n/a} D^{-1}(n, a) \int_{A \in z} |M \cap A|_a^{n/a} dA ,$$

with equality only for the sphere. Hence the sphere gives the maximum of $\min_A |M \cap A|_a^n |M|_n^{a-n}$ for given volume $|M| > 0$.

References

1. W. Barthel, *Zum Inhaltsbegriff in der Minkowskischen Geometric*, Math. Z. **58** (1953), 358-375.

2. E. Bertini, *Einführung in die projektive Geometrice mehrdimensionaler Räume*, Wien 1924.

3. W. Blaschke, *Räumliche Variationsprobleme mit symmetrischer Transversalitätsbedingung*, Ber. Sächs. Akad. Wiss. Leipzig **68** (1916), 50-55.

4. ———, *Kreis and Kugel*, 2nd edition, Berlin 1956.

5. T. Bonnesen and W. Fenchel, *Theorie der konvexen Körper*, Ergebnisse Math. III 1, Berlin 1934 and New York 1958.

6. H. Busemann, *The isoperimetric problem for Minkowski area*, Amer. J. Math. **71** (1949), 743-762.

7. ———, *The foundations of Minkowskian geometry*, Comm. Math. Helv. **24** (1950), 156-186.

8. ———, *Volume in terms of concurrent cross-sections*, Pacific J. Math. **3** (1953), 1-12.

9. ———, *The geometry of geodesies*, New York 1955.

10. H. Busemann and C. M. Petty, *Problems on convex bodies*, Math. Scand. **4** (1956), 88-94.

11. P. Funk, *Über Flächen mit lauter geschlossenen geodätischen Linien*, Math. Ann. **74** (1913), 278-300.

12. C. M. Petty, *On the geometry of the Minkowski plane*, Riv. Mat. Univ. Parma **6** (1955), 269-292.

13. J. Radon, *Über eine besondere Art ebener konvexer Kurven*, Ber. Sächs. Akad. Wiss Leipzig **68** (1916), 131-134.

14. E. Sperner, *Einführung in die analytische Geometrie and Algebra I*, Göttingen 1948.

15. V. Wagner, *Geometry of a space with areal metric and its application to the calculus of variations*, (Russian), Mat. Sb. N. S. **19 (41)** (1946), 341-404.

University of Southern California and University of California, Los Angeles

Note: While this paper was in print it was shown in H. Busemann: *Areas in affine space* II (to appear in the Rend. Circ. Mat. Palermo) that the case $a+b=n, d=1$ in (6.1) is exceptional for all a and also that $a=n/2$, $d=0$ in (6.2) is always exceptional.

AREAS IN AFFINE SPACE I, II

by **Herbert Busemann** (*) (Los Angeles, California)

INTRODUCTION

Whereas areal metrics in general spaces have been studied since a long time, the elementary and more geometric aspects of the simple prototypes of general area have been neglected to the detriment of the general theory. Recently E. G. Straus and the author showed in [2] that there are indeed many concrete problems concerning the simplest area, many of which are still open, while others have quite unexpected answers. The present series of articles intends to complete the results of [2] and to widen their scope by considering more general, although still elementary areas, which we call a-areas in affine n-space X^n $(1 \leqslant a \leqslant n-1)$.

Such an area $\alpha(M)$ is *defined* as follows: *a non-empty open convex set D_α in X^n is given. For every a-flat A, whose intersection with D_α is not empty, $\alpha(M)$ is defined on all Borel sets $M \subset A \cap D_\alpha$, and is in each $A \cap D_\alpha$ a countably additive measure, which vanishes for M lying in $(a-1)$-flats and is continuous and positive*, i.e., $\alpha(S)$ depends continuously on the vertices of an a-simplex S (the vertices are not confined to one a-flat) and $\alpha(S) > 0$ when S is non degenerate. The restriction of α to a non-empty open convex subset of D_α also satisfies the conditions, because we did not make any completeness hypotheses. Completeness will be discussed in a subsequent article. We agree to consider *Borel sets only*.

(*) This work was supported by a grant from the National Science Foundation.

The α-area of an a-dimensional polyhedron in D_α is, of course, defined as the sum of the areas of its a-dimensional faces. The area of a face is invariant under subdivision of the face.

We say that the *a-flats (strictly) minimize area* if for any closed a-dimensional polyhedron P in D_α the sum of the areas of all but one face (exceeds) is not less than the area of that face. The special case, where P is the boundary of an $(a+1)$-simplex is important:

Simplex Inequality. The sum of the α-areas of a faces of an $(a+1)$-simplex in D_α is at least as large as the α-area of the remaining face.

In parts I, II we are interested in the case $D_\alpha = X^n$, where in addition $\alpha(M)$ is invariant under translations of M in the a-flat A in which M lies. (If M lies in more than one a-flat $\alpha(M)=0$, so that $\alpha(M)$ is invariant under all translations). We call such an area *weakly affine*, and reserve the term *affine* area for an a-area in X^n with $D_\alpha = X^n$ which is invariant under *all* translations of X^n.

If $x^1, \ldots, x^n$ are affine coordinates in X^n, we introduce auxiliary euclidean metrics

$$e(x, y) = [\Sigma g_{ik}(x^i - y^i)(x^k - y^k)]^{1/2},$$

where the form $\Sigma g_{ik} x^i x^k$ is positive definite, and speak of

$$\varepsilon(x, y) = [\Sigma (x^i - y^i)^2]^{1/2}$$

as the trivial euclidean metric.

The metric $e(x, y)$ induces in each a-flat an a-dimensional Lebesgue measure $\rho_a^e(M)$. Since there is, up to a constant factor, only one (countably additive) measure defined on all Borel sets in an a-flat A and invariant under translations of A, we see that for *any weakly affine a-area α*

$$(1) \qquad \alpha(M) = f(A)\rho_a^e(M), \qquad M \subset A,$$

where $f(A)$ is positive and continuous. If α is an affine area, then $f(A)$ has the same value on parallel a-flats. Conversely, if $f(A)$ is positive and continuous, then (1) defines a weakly affine area, and if $f(A)$ has the same value on parallel a-flats, then (1) is an affine area.

The affine areas are of special interest, because under certain differentiability hypotheses an *a-dimensional areal metric in a general n-dimensional space will locally,* or infinitesimally, *be an affine area.* From this point of view affine areas were first studied systematically in Wagner [5], and from a different point of view in [2]. We restate here those definitions and results of [2] which are necessary to understand the present series.

We consider oriented a-flats and put $f(A) = f(-A)$. If z is the origin of X^n and $z, x_1, \ldots, x_a$ are in this order the vertices of a non-degenerate a-simplex S in A with positive orientation relative to A, then

$$v^{(i,a)} = v^{i_1 \ldots i_a} = \begin{vmatrix} x_1^{i} & \ldots & x_1^{i_a} \\ \cdot & \cdots & \cdot \\ x_a^{i} & \ldots & x_a^{i_a} \end{vmatrix}, \qquad i_1 < i_2 < \cdots < i_a,$$

are the components of a simple a-vector v. After introducing $e(x, y)$ the length $|v|$ of v equals $a! \rho_a^e(S)$. It is convenient for us to *associate* v *with all sets* $M \subset A$ *for which* $\rho_a^e(M) = a! \rho_a^e(S)$ *or equivalently* $\alpha(M) = a! \alpha(S)$, i.e. v stands for a class of Borel sets in A with the same euclidean or α-area. The term «*perpendicular*» will be used for normality with respect to $e(x, y)$. If $v_{(i,a)}$ are the covariant components of v then the scaler product vv' of two simple vectors v, v' with $M \subset A$, $M' \subset A'$ as representatives has, except for sign, a simple geometric meaning (compare Cartan [3, Chapter I])

$$vv' = \sum_{(i,a)} v_{(i,a)} v'^{(i,a)} = \sum_{(i,a)} v^{(i,a)} v'_{(i,a)} = \pm \rho_a^e(M) \rho_a^e(M'_\pi) = \pm \rho_a^e(M_\pi) \rho_a^e(M')$$

where $M'_\pi(M_\pi)$ is the perpendicular projection of $M'(M)$ on $A(A')$. For the trivial metric ε, the covariant and contravariant components of v coincide.

We now pass to the vector space $Y_{n,a}$ of all a-vectors in X^n. The simple a-vectors fill, by definition, the Grassmann cone $G_{n,a}$ in $Y_{n,a}$. Put $Y_{n,a} = Y$, $G_{n,a} = G$. Each oriented a-flat A determines a ray R_A in G from the origin z_Y of Y, and a non-oriented a-flat determines a line $L_A = L_{-A}$ in G through z_Y. The scaler product vv' is defined on G but can be extended to a scaler product for all a-vectors in Y, providing Y with a euclidean metric. The sets M in the various oriented a-flats A in X^n for which $\alpha(M) = 1$ determine vectors $v(M)$ in G or, with Y interpreted as an affine space, a point *locus* I_α *on* G, *the indicatrix of the affine area* α, which has in Y the equation

$$I_\alpha : f(A)|v| = 1, \qquad v \in R_A.$$

Because if $f(A) = f(-A)$ the indicatrix is *symmetric* with respect to z_Y.

A Grassmann plane or a simple hyperplane is a hyperplane H in Y as affine space such that the coefficients in the linear equation representing H satisfy the conditions for a simple vector. In terms of the trivial metric $\varepsilon(x, y)$ a Grassmann plane is a hyperplane whose perpendicular through z_Y is a line on G.

In [2] three increasingly stronger types of convexity of an affine area α are discussed which coincide only for $a = 1$ and $a = n - 1$; they are called convexity, extendable convexity and total convexity and mean in terms of Y the following: 1) α is *convex* if the intersection of I_α with any two-flat or plane (through z_Y) contained in G is a convex curve. It follows that then the intersection of I_α with any flat through z_Y of dimension greater than 2 that lies in G is a convex hypersurface in this flat. 2) α is *extendably convex* if I_α possesses at every point a supporting hyperplane in Y. This is equivalent to saying that $f(A)|v|$, $v \in R_A$, can be extended to a convex function in all of Y. 3) α is *totally convex*, if I_α possesses at every point a simple supporting hyperplane.

In X^n extendable convexity guarantees that the a-flats minimize α-area, *convexity is equivalent to the Simplex Inequality.* For total convexity we need the concepts of normality and transversality.

The a-flat A is *totally normal* to the b-flat B $(1 \leqslant b \leqslant n-1)$ at the d-flat $D = A \cap B$ $(0 \leqslant d < \min(a, b))$ in $A + B$ (1), or B is *totally transversal* to A at D in $A + B$ (1), if for any a-flat $A' \subset A + B$ with $A' \cap B = D$ and any $M \subset A$ the projection M' of M parallel to B on A' satisfies $\alpha(M') \geqslant \alpha(M)$. For $d > 0$ projection is defined by projecting along a $(b-d)$-flat $C \subset B$ intersecting D in a point. It is true that M' depends on C, but $\rho_a^e(M')$ and hence $\alpha(M')$ do not, see [2].

α *is totally convex if an $(n-a)$-flat B totally transversal to a given a-flat A at a point exists.* Then for any b, $1 \leqslant b \leqslant n-1$ a b-flat totally transversal to a given a-flat at a d-flat $D \subset A$ in a given $(a + b - d)$-flat will also exist.

If A is totally normal to B at D, then it is also totally normal to any flat $B' \subset B$ through D $(\dim B' > d)$, in particular A is totally normal to every $(d+1)$-flat in B through D. If the latter is the case we call A *normal to B*, and B *transversal to A*, at D in $A + B$ (1). Normality and transversality are weaker than total normality and total transversality except for $d = \min(a, b) - 1$. In fact, the existence of a b-flat for all b, $1 \leqslant b \leqslant n-1$ transversal to a given a-flat at a given d-flat for all $d = \min(a, b) - 1$ is only equivalent to convexity of α.

I. A CHARACTERIZATION OF CONVEX AFFINE AREA

The content of this part is that translation invariance in each a-flat and the Simplex Inequality imply general translation invariance:

(1) «in $A + B$» will be omitted when no ambiguity can arise.

THEOREM. *A weakly affine a-area α which satisfies the Simplex Inequality is an affine area.*

For $a=1$, where the Simplex Condition becomes the triangle inequality, this theorem is proved in [1, Section 17]. For $1<a<n-1$ the interest of the extension lies partly in the fact that it was not a priori obvious that the analogue to the weakest convexity condition would produce general translation invariance. The theorem implies trivially that an area minimizing weakly affine a-area is affine, and also:

COROLLARY. *Let α be a weakly affine area and assume that for each a-flat A and $(n-a)$-flat B intersecting A in a point exists, such that the projection M parallel to B of any set M' in an a-flat A' satisfies $\alpha(M')\geqslant\alpha(M)$; then α is affine.*

For if P is any closed a-dimensional polyhedron with one face T in A, then the projection P^0 of the closure of $P-T$ on A parallel to B covers T, hence $\alpha(P-T)>\alpha(P^0)\geqslant\alpha(T)$. Therefore the a-flats minimize area.

We prove the theorem by induction with respect to a and assume therefore that the theorem holds for $a-1\geqslant 1$. After introducing an auxiliary euclidean metric $e(x, y)$ and writing ρ_a instead of ρ_a^e, the area $\alpha(M)$ takes the form

$$\alpha(M)=f(A)\rho_a(M),$$

and we have to prove that $f(A')=f(A'')$ if A' an A'' are parallel. A' and A'' lie in an $(a+1)$-flat and since $\alpha(M)$ restricted to this $(a+1)$-flat is still a weakly affine area, we may assume that X^n is this $(a+1)$-flat.

Let A_0 be an a-flat perpendicular to A' and A'', and T_0 a non-degenerate a-simplex in A_0. Take a ray R perpendicular to A_0 at one of the vertices, say q, of T_0, let $p\in R-q$. We denote by S_p the $(a+1)$-simplex spanned by p and T_0, by $T_1, \ldots, T_{a-1}$ the a-faces of S_p containing p and q, by T_a the remaining a-face, and by A_i the a-flat containing T_i. The flats $A_1, \ldots, A_{a-1}$ are the same for all $p\in R-q$, but A_a tends for $e(p, q)\to\infty$ to an a-flat A^+ perpendicular to A_0. By hypothesis

$$\sum_{i=0}^{a-1}\alpha(T_i)\geqslant\alpha(T_a)\quad\text{or}\quad\sum_{i=1}^{a-1}\rho_a^{-1}(T_a)\rho(T_i)f(A_i)+\rho^{-1}(T_a)\rho(T_0)f(A_0)\geqslant f(A_a).$$

$A_i\cap T_0$ is for $i=1, \ldots, a$ an $(a-1)$-face V_i of T_0, moreover $V_a=A_a\cap T_0=A^+\cap T_0$. If h_p denotes the euclidean distance of p from the $(a-1)$-face carrying V_a, then

$$\rho_a(T_i)=e(p, q)\rho_{a-1}(V_i),\qquad i=1, \ldots, a-1,\qquad \rho_a(T_a)=h_p\rho_{a-1}(V_a).$$

For $e(p, q) \to \infty$ we have

$$f(A_a) \to f(A^+), \qquad e(p, q)/h_p \to 1, \qquad \rho_a^{-1}(T_a)\rho_a(T_0) \to 0,$$

hence

$$\sum_{i=1}^{a-1} \rho_{a-1}(V_i) f(A_i) \geq \rho_{a-1}(V_a) f(A^+).$$

For any $(a-1)$-flat $A^- \subset A_0$ we define $f(A^-)$ as the value of $f(A)$ for the a-flat A perpendicular to A_0 at A^-. Then the last inequality shows that for $M^- \subset A^-$

$$\alpha^-(M^-) = f(A^-)\rho_{a-1}(M^-)$$

is a weakly affine $(a-1)$-area in A_0 which satisfies the Simplex Inequality. By the inductive hypothesis $f(A^-)$ has the same value for parallel $(a-1)$-flats in A_0. This implies that $f(A)$ has the same value $f(A') = f(A'')$ for the originally given a-flats.

II. NON-QUADRATIC AFFINE AREAS WITH EQUIVALENT NORMALITY

Let a (totally) convex a-area α and a (totally) convex affine b-area β be defined in X^n. If for some fixed $d < \min(a, b)$ (total) normality of an a-flat A to a b-flat B at a d-flat D with respect to α is equivalent to (total) normality of B to A at D with respect to β, we say that (total) α-normality and (total) β-normality at d-flats are equivalent. It is shown in [2] that in this case both α and β are quadratic ([1]) unless $a + b = n$ and $d = 0$.

Except for $a = 1$ the question whether the case $a + b = n$, $d = 0$ is really exceptional is left open in [2]. Our first aim here is to prove a theorem which implies that the answer is affirmative:

THEOREM 1. *For any affine a-area α there exists a totally convex affine $(n-a)$-area β such that whenever an a-flat A is totally normal to an $(n-a)$-flat B at a point (for α), then B is totally normal to A (for β). The area β is unique up to a constant factor when α is totally convex.*

The uniqueness is seen as follows: given an $(n-a)$-flat B z there exists an a-flat A totally normal to B at z. By hypothesis B is normal to A. If α is totally convex, then B transversal to a given A at z exists, so that the hypothesis becomes a statement about all a-flats. It is a special case of Theorem (5.1) in [2]

([1]) α is quadratic if for a simplex S with vertices $z, x^1, \ldots, x^a$ the function $\alpha^2(S)$ is quadratic in each set of variables $x_i^1, \ldots, x_i^n$, $i = 1, \ldots, a$.

that if two totally convex $(n-a)$-areas β, β_1 have the property that an $(n-a)$-flat B is normal to an a-flat A at z with respect to β whenever it is normal to A with respect to β_1, then $\beta_1(M) = k\beta(M)$.

We will actually obtain an explicit analytic expression for a β satisfying Theorem 1, when α is given.

For a simple a-vector $u_{(i,a)}$ let w be the supplementary $(n-a)$-vector, i.e.

(1) $$w^{(j,n-a)} = w^{j_1 \dots j_{n-a}} = \lambda_{ij} g^{-1/2} u_{i_1 \dots i_a} = \lambda_{ij} g^{-1/2} u_{(i,a)}$$

where $(i, j) = i_1, \dots, i_a, j_1, \dots, j_{n-a}$ is a permutation of $1, \dots, n$, $\lambda_{ij} = \text{sign}\,(i, j)(-1)^{a(n-a)}$, and $g = \det(g_{ik})$. The supplementary vector to w is u. Our definition differs in sign from the usual one (Cartan [3 Chapter I]) when both a and $n-a$ are odd.

SUPPLEMENT TO THEOREM 1: *If I_α is the indicatrix of α put*

(2) $$\psi(w) = H(u) = \max_{v \in I_\alpha} v u = \max_{v \in I_\alpha} \sum_{(i,a)} v^{(i,a)} u_{(i,a)},$$

where w is the supplementary vector (1) *to u. Then $\beta(M) = \psi(w)$, when M is a set in an $(n-a)$-flat representing w, satisfies Theorem* 1.

Since I_α is symmetric with respect to z_Y we have $H(u) = H(-u)$ hence $\psi(w) = \psi(-w)$. Also $H(\lambda u) = \lambda H(u)$ for $\lambda > 0$, hence $\psi(\lambda w) = \lambda \psi(w)$ so that $\beta(M)$ is an affine area. The fact that $\beta(M)$ satisfies our requirement rests on a geometric interpretation of projection in X^n in terms of Y.

Let a b-flat B $(1 \leqslant b \leqslant n-1)$ intersect the a-flat A at D $(d = \dim D < \min(a, b),\ z \in D)$ and consider an a-flat A' in $A+B$ with $A' \cap B = D$. Take the a-flat B_π in $A+B$ perpendicular to B at D and the $(n-a)$-flat N_B perpendicular to B_π at z, which intersects $A+B$ in a $(b-d)$-flat $C \subset D$. Projection of A on A' can be defined in terms of C, see above. Let $M' \subset A'$ be the projection parallel to C of the set $M \subset A$ $(0 < \alpha(M) < \infty)$, and v, v' the vectors in Y representing M and M'. Then by the meaning of the scalar product the points v, v' of Y as an affine space lie on the Grassmann plane H_B perpendicular to L_{B_π} (the line on G carrying the vectors corresponding to sets in B_π). Choose $M \subset A$ such that $\alpha(M) = 1$ or $v \in I_\alpha$. If B is totally transversal to A at D then $\alpha(M') \geqslant 1$, hence H_B must be a supporting hyperplane of the intersection of I_α with the subset $G(A+B, D)$ of G corresponding to the a-flats in $A+B$ through D. When D is a point we simply write $G(A+B)$. In the case $a+b=n$, $d=o$ we find that H_B is a supporting hyperplane of I_α at a.

If B is transversal (but not necessarily totally so) to A at a point, then for each $(a+1)$-flat A^+ containing A the hyperplane H_B must intersect $G(A^+)$ in a supporting a-flat of $G(A^+) \cap I_\alpha$ in $G(A^+)$, the latter being an $(a+1)$-flat in Y.

It is now clear that $\beta(M) = \psi(w)$ will satisfy Theorem 1: the hyperplane $vu = H(u)$ is a Grassmann supporting plane of I_α because $vu \leqslant H(u)$ for $v \varepsilon I_\alpha$. For a given u there is a $v_0 \varepsilon I_\alpha$ with $v_0 u = H(u)$ if and only if an $(n-a)$-flat transversal to the a-flat carrying the sets represented by v_0 exists. The set $H(u) \leqslant 1$ is the intersection of half spaces in Y bounded by Grassmann hyperplanes, therefore I_β as locus on $G_{n,n-a} \subset Y_{n,n-a}$ possesses at each point a supporting Grassmann plane in $Y_{n,n-a}$.

Constructing to β an a-area $\bar{\alpha}$ by using the analogue to (2) will not lead back to α unless α is totally convex. $I_{\bar{\alpha}}$ is what might called the totally convex closure of I_α. The area $\bar{\alpha}$ is characterized by the following facts: the intersections of I_α and $I_{\bar{\alpha}}$ with R_A coincide when an $(n-a)$-flat totally transversal to A at a point exists. If $\tilde{\alpha}(M)$ is any affine a-area $\tilde{\alpha}$ with this property then $\bar{\alpha}(M) \geqslant \tilde{\alpha}(M)$. The area (2) belonging to $\bar{\alpha}$ is again β. If we express the result on $\bar{\alpha}$ in terms of β we find the following characterization of β:

(3) *Select a non-degenerate $(n-a)$-simplex S. Among all $(n-a)$-areas $\tilde{\beta}$ satisfying Theorem* 1 *and with $\tilde{\beta}(S) = 1$ the area* (2) *is the smallest,* i.e., $\beta(M) \leqslant \tilde{\beta}(M)$ *for any M and any $\tilde{\beta}$.*

This discussion also leads to the solution of a second problem left open in [2]. Applied to $\alpha = \beta$ equivalence of normality becomes symmetry: If A is (totally) normal to another a-flat A' at D then A' is (totally) normal to A. The general theorem on equivalence of normality implies that unless $n = 2a$ and $d = 0$, α-normality at d-flats is symmetric only for quadratic α. That the case $n = 2a$, $d = 0$ is really exceptional had been shown long ago for $a = 1$ by Radon [4], but the case $a > 1$ is left open in [2]. Our previous discussion yields at once the first part of the following theorem:

THEOREM 2. *Let $n = 2a$ and let α be a totally convex affine a-area in X^n. Put $\Phi(v) = \alpha(M)$, where v is the simple a-vector representing M* (a Borel set in the oriented a-flat corresponding to the ray R_v in $Y_{n,a}$). *α-normality at points is symmetric if and only if*

$$\Phi(v) = \psi(v) \tag{4}$$

where $\psi(v)$ is the function defined by (2).

For every $a \geqslant 1$ there are non-quadratic totally convex α satisfying this condition.

Whereas Radon determines for $a = 1$ all $\Phi(v)$ satisfying (4), we construct for $a > 1$ only certain such α. We use the trivial metric $\varepsilon(x, y)$. The polar reciprocity R of Y with respect to the unit sphere $U: |v| = 1$ in Y associates with any $v \neq 0$ the hyperplane $vu = 0$ perpendicular to R_v at $v|v|^{-2}$, hence mates simple vectors and Grassmann planes.

Since α is totally convex it lies on a closed convex hypersurface C_α in Y symmetric to z_Y which possesses at each point of $I_\alpha = G \cap C_\alpha$ Grassmann supporting planes. If for any (not necessarily simple) u

$$H(u) = \max_{x \varepsilon C_\alpha} vu,$$

then $H(u) = 1$ is RC_α. If J denotes the mapping (1), then $JRC_\alpha = C_\beta$ is a convex hypersurface, and $I_\beta = C_\beta \cap G$ will define another a-area. Perpendicularity at points will be symmetric for α if $I_\beta = I_\alpha$ hence if $JRC_\alpha = C_\alpha$.

For brevity call U-surface a convex hypersurface in Y symmetric with respect to z_Y intersecting the coordinate planes $v_{i_1 \dots i_a} = 0$ in Y in the same points as the unit sphere U and having at these points unique supporting hyperplanes coinciding with those of U. The image of a U-surface under R or J is a U-surface.

A coordinate sector is one of the open sets into which the coordinate planes divide Y. With S the image $-S$ of S under $v' = -v$ is a coordinate sector, and so is JS, moreover, $J(-S) = -JS$, $J^2 S = S$, $J^2(-S) = -S$. We define C_α in each quadruple of sectors $S, -S, JS, -JS$. If one sector in a quadruple contains points of G, then all do ([1]). In a quadruple not containing points of G let C_α coincide with U.

Let S contain points of G. Then we take in S any piece C_S of a convex hypersurface satisfying the boundary conditions for a U-surface. Then RC_S also satisfies these boundary conditions. We define C_α in $-S$ as $-C_S$ and in JS, $-JS$ as JRC_S and $-JRC_S$ respectively. If we do this for each sector quadruple containing points of G, we obtain a convex hypersurface C_α with $JRC_\alpha = C_\alpha$, because R and J commute, and JR is an involution on C_α.

([1]) For $a = 2$ and $u = 4$ the sector $v_{ij} > 0$ for $ij \neq 13$, $v_{13} < 0$ contains no points of $G: v_{12} v_{34} - v_{13} v_{24} + v_{14} v_{23} = 0$.

We have to ascertain that C_S can be chosen so that it has Grassmann planes as supporting hyperplanes at points of G without being quadratic. But this is trivial. Let v a point of $U \cap G$ in S. A hyperplane H perpendicular to the vector at the point θv will for sufficiently large $\theta < 1$ intersect U in a small sphere U' of one dimension less than U which lies entirely in S. If the part of U on the other side of H from z_Y is replaced by the ball with boundary U', the hypersurface C_S obtained from $U \cap S$ by this modification obviously has the desired property. Clearly we can also find smooth C_S with this property.

This last construction also shows how to obtain for $n \neq 2a$ a non-quadratic totally convex a-area α and a non-quadratic totally convex $(n-a)$-area β such that α-normality and β-normality at points are equivalent.

Los Angeles, March 1960.

REFERENCES

[1] H. Busemann, *The Geometry of Geodesics*, New York 1955.

[2] H. Busemann and E. G. Straus, *Area and Normality*, Pacific Jour. of Math.

[3] E. Cartan, *Leçons sur la géométrie des espaces de Riemann*, 2nd ed., Paris 1946.

[4] J. Radon, *Über eine besondere Art ebener konvexer Kurven*, Ber. Sächs. Akad. Wiss. Leipzig 68 (1916), pp. 131-134.

[5] V. V. Wagner, *Geometry of spaces with areal metric and its applications to the calculus of variations*, Russian, (English Summary), Mat. Sbornik N. S. 19 (61) 1946, pp. 341-406.

Rend. Circ. Mat. Palermo, II. Ser. 9 (1960), 81–90

AREAS IN AFFINE SPACES III
THE INTEGRAL GEOMETRY OF AFFINE AREA (*)

by **Herbert Busemann** (Los Angeles, California)

1. INTRODUCTION

An a-dimensional affine area $\alpha(M)$ is a positive continuous translation invariant measure defined on all Borel sets M lying in a-flats ($1 \leqslant a \leqslant n$) of an n-dimensional affine space W^n. It induces a function $\alpha(A)$ on the simple a-vectors: if W^n is considered as a vector space with origin z then a vectors $x_1, \ldots, x_a$ span a (possibly degenerate) parallelepiped P whose area $\alpha(P)$ depends only on $A = x_1 \wedge \ldots \wedge x_a$ and may therefore be denoted by $\alpha(A)$. That α is positive means $\alpha(A) > 0$ for $A \neq 0$ and continuity means that $\alpha(A)$ depends continuously on A. The translation invariance implies $\alpha(kA) = |k|\alpha(A)$, and that $\alpha(A)$ determines $\alpha(M)$. For details see [11] and the introduction to this series in [10].

We use Z_a for a-flats passing through z. If a positive and continuous function $f(Z_a)$ is defined on all Z_a and a euclidean (translation invariant) metrization of W^n is given, and hence the absolute value $|A|$ (the euclidean analogue to $\alpha(A)$) we put

$$f(A) = |A| f(Z_a) \text{ for } A \text{ parallel to } Z_a, \; f(0) = 0,$$

and call $f(A)$ the *function induced by* $f(Z_a)$. Then $f(A)$ defines an affine area and any affine area $\alpha(A)$ defines an $f(Z_a)$, namely $\alpha(A^1) = \alpha(Z_a)$, when A^1 is a univector ($|A^1| = 1$) parallel to Z_a.

(*) This work was supported by a grant from the National Science Foundation.

The a-flats minimize α-area $(1 \leqslant a \leqslant n-1)$ if for any closed a-dimensional polyhedron the α-area of one face does not exceed the sum of the α-areas of the remaining faces. A sufficient condition for this property of α is that $\alpha(A)$ be *extendably convex* (see [10] or [11]):

$$(1) \qquad \alpha(A_1 + \cdots + A_k) \leqslant \alpha(A_1) + \cdots + \alpha(A_k).$$

Here and in the sequel it is always tacitly assumed that all vectors which occur are simple. In (1) it is therefore understood that not only the A_j but also $\sum A_j$ are simple. It is not known whether (1) is necessary for the a-flats to minimize area.

The classical theory of convex bodies deals exclusively with ordinary convexity, i.e. with the case $k = 2$, which is equivalent to (1) with general k only if $a = 1$, $n-1$ (or n) (see [11]). The first purpose of the present paper is to *establish extendable convexity for certain functions* $f(A)$ *derived from a convex body* in n-dimensional euclidean space E^n. The method consists in extending to general a and n results proved by Berwald and Varga in [3] for $a = 1$ or 2, $n = 3$.

Our second purpose is closely related to the first in method but not in content. We ask *whether for a given a-area* $\alpha(1 \leqslant a \leqslant n-1)$ *a density* dL_{n-a} *for the* $(n-a)$*-flats exists which makes the simplest Crofton formula valid,* namely

$$\int_{L_{n-a} \cap M \neq 0} dL_{n-a} = \alpha(M) \text{ for } M \text{ lying in } a\text{-flats } (^1).$$

Because the natural group for affine areas consists of the translations one would not expect such a density dL_{n-a} to exist. However Blaschke showed in [5] that for $a = 1$, $n = 2$ and for convex α a density for lines does exist satisfying Crofton's Formula. This also yields a kinematic density and many results which are true for E^2 (²).

Nevertheless our first impression is essentially correct, *the case* $a = 1$, $n = 2$ *is exceptional.* If $n > 2$ and L_{n-a} exists, then the area α must not only be extendably convex, which is a reasonable condition, but must in addition have a *special form.*

(¹) In the euclidean case and with the usual density for the $(n-a)$-flats $\alpha(M)$ appears with a factor depending on n and a. Here we use, even in the euclidean case, the normalization of dL_{n-a} for which this factor becomes 1.

(²) Blaschke defines a density for the extremals of a two-dimensional Finsler space with a regular integrand $F(x, \dot{x})$ which reduces to the present case if $F(x, \dot{x})$ does not depend on x.

We then ask whether for a given a-area of this special form, an $(n-a)$-area β and a density for a-flats exist such that a certain form of Cavalieri's Principle and Crofton's Formula $\int_{L_a \cap N \neq 0} dL_a = \beta(N)$ becomes true for sets N in $(n-a)$-flats. It turns out that this problem has so many solutions, that the arbitrariness detracts from the interest of the results derivable from these densities and areas. However, our analysis of the situation leads to a thorough understanding of the exceptional character of the case $a = b = 1$.

Moreover it provides us with the correct tools for the *integral geometry of affine area with respect to the translations* which we can now consider as the proper domain if $n > 2$. For E^3 such a study was carried out in the already mentioned paper [3] of Berwald and Varga. Here we indicate briefly a simple method for transcribing, as it were, euclidean results so that they become results for W^3 with preassigned affine length, area and volume.

Very few concrete integral geometric results are given here, although they are necessary to justify the present investigation, because the author learned that such results will be contained in work of Dr. G. H. Chakerian, who is, independently, studying the integral geometry of Minkowski spaces.

2. EXTENDABLY CONVEX FUNCTIONS BELONGING TO CONVEX BODIES

We introduce some notations which will be used throughout. The a-dimensional euclidean measure of a set M in an a-flat or the area of an a-dimensional surface in E^n is $\varepsilon_a(M)$. The area of the orthogonal projection of a set M on an r-flat parallel to Z_r, is $P_r(M, Z_r)$. As usual S^r is the r-dimensional unit sphere, $d_\varepsilon S^r$ its area element, $\pi_r' = (r+1)\pi_{r+1}$ its total area, so that π_{r+1} is the volume of the unit ball in E^{r+1}. For a given Z_a, $1 \leqslant a \leqslant n-1$ we consider the family $L(Z_a)$ of a-flats parallel to Z_a, but will for brevity, also use $\bar{L}$, if no ambiguity arises. $d\bar{L} = dL(Z_a)$ is the kinematic density for $L(Z_a)$. It equals, of course, the area element in an $(n-a)$-flat $Z_a^{\perp}$ orthogonal to Z_a. Using cartesian coordinates we denote by A^1, B^1 unit vectors and by A, B arbitrary a-and $(n-a)$-vectors respectively parallel to Z_a and $Z_a^{\perp}$.

Consider a convex body K in E^n with interior points and with boundary F. We assume at first that F is of class C^2 and has positive Gauss curvature. N_p is the unit normal to F at p. If $2 \leqslant a \leqslant n-1$ and $\bar{L} \cap F \neq 0$ for an a-flat parallel to Z_a which is not a supporting flat of K, then $\bar{L} \cap F$ is a closed

Rend. Circ. Mat. Palermo, II. Ser. 9 (1960), 226–242

convex hypersurface relative to $\bar{L}$ whose $(a-1)$-dimensional area element at p is $d_\varepsilon(\bar{L}\cap F)_p$. We may write

$$d_\varepsilon F = d_\varepsilon(\bar{L}\cap F)_p \wedge dM_p,$$

where dM_p corresponds to the $(n-a)$-flat M_p in the tangent hyperplane of F at p orthogonal to the tangent $(a-1)$-flat of $\bar{L}\cap F$ at p and the usual agreement of integral geometry is followed to consider all differentials as positive. If M_p^1 is a unit $(n-a)$-vector parallel to M_p, then

$$d\bar{L} = |A^1 \wedge M_p^1|\, dM_p = |B^1 \wedge N_p|\, dM_p,$$

therefore $d_\varepsilon(\bar{L}\cap F)_p \wedge d\bar{L} = |B^1 \wedge N_p|\, d_\varepsilon F_p$ and

$$\text{(2)}\qquad \bar{\varepsilon}_{a-1}(Z_a^{\perp}) = \int\limits_{\bar{L}\cap K\neq 0} \varepsilon_{a-1}(\bar{L}\cap F)\, dL(Z_a) = \int\limits_{F} |B^1 \wedge N_p|\, d_\varepsilon F_p.$$

This result yields others: using Cauchy's Formula for the area of a convex surface (see [6, p. 48]) we find

$$\pi_{a-2}\varepsilon_{a-1}(\bar{L}\cap F) = \int\limits_{S^{a-1}} P_{a-1}(\bar{L}\cap F,\ Z_{a-1}(\bar{L}))\, d_\varepsilon S^{a-1},$$

where $Z_{a-1}(\bar{L})$ traverses the different $(a-1)$-flats in $\bar{L}$ through a fixed point. If $Z_{a-1}(Z_a)$ is the $(a-1)$-flat parallel to $Z_{a-1}(\bar{L})$ through z we keep $Z_{a-1}(Z_a)$ fixed, vary $\bar{L}$ and denote by $H(\perp Z_a, \supset Z_{a-1}) = H$ the hyperplane orthogonal to Z_a and through $Z_{a-1}(Z_a)$, then

$$P_{n-1}(K, H) = \int\limits_{\bar{L}\cap K\neq 0} P_{a-1}(\bar{L}\cap F,\ Z_{a-1}(L_a))\, d\bar{L}$$

$$\int\limits_{\bar{L}\cap K\neq 0} d\bar{L} \int\limits_{S^{a-1}} P_{a-1}((\bar{L}\cap F),\ Z_{a-1}(L))\, d_\varepsilon S^{a-1} = \int\limits_{S^{a-1}} P_{n-1}(K, H)\, d_\varepsilon S^{a-1}$$

hence

$$\text{(3)}\quad \pi_{a-2}\bar{\varepsilon}_{a-1}(Z_a^{\perp}) = \int\limits_{S^{a-1}} P_{n-1}(K,\ H(\perp Z_a, \supset Z_{a-1}))\, d_\varepsilon S^{a-1} = \pi_{a-2}\int\limits_{F} |B^1 \wedge N_p|\, d_\varepsilon F_p.$$

Let $R_1, \ldots, R_{n-1}$ be the principal radii of curvature of F at p, ρ^k their k-th elementary symmetric function $R_1 \cdots R_k + \cdots$ and ν^k the corresponding function $R_1^{-1} \cdots R_k^{-1} + \cdots$ of the principal curvatures. If we pass to the exterior parallel surface F_h to F at distance h we have the well known relation

$$d_\varepsilon F_{hp} = (R_1 + h) \cdots (R_{n-1} + h) d_\varepsilon S^{n-1} =$$
$$= \rho^{n-1} d_\varepsilon S^{n-1} + h \rho^{n-2} d_\varepsilon S^{n-1} + \cdots + h^{n-2} \rho^1 d_\varepsilon S^{n-1} + h^{n-1} d_\varepsilon S^{n-1} =$$
$$= d_\varepsilon F_p + h \nu^1 d_\varepsilon F_p + \cdots + h^{n-2} \nu^{n-2} d_\varepsilon F_p + \cdots .$$

On the other hand, if $\partial P(K, H)$ is the area of the boundary of $P_{n-1}(K, H)$ and $\bar{B}P(F, H)$ the average width of $P_{n-1}(K, H)$ then, see [6, p. 63],

$$P_{n-1}(F_h, H) = P_{n-1}(K, H) + h \partial P(K, H) + \cdots + h^{n-2} \pi'_{n-1} 2^{-1} \bar{B}P(K, H) + \cdots .$$

Substituting these expressions into (3) applied to F_h and comparing the coefficients of h and h^{n-2} yields

(4) $$\int_{S^{a-1}} P(K, H(\perp Z_a, \supset Z_{a-1}(Z_a))) d_\varepsilon S^{a-1} = \pi_{a-2} \int_F \nu^1 |B^1 \wedge N_p| d_\varepsilon F_p$$

(5) $$\int_{S^{a-1}} \bar{B}P(K, H(\perp Z_a, \supset Z_{a-1}(Z_a))) d_\varepsilon S^{a-1} = \frac{2\pi_{a-2}}{\pi'_{n-1}} \int_F \nu^{n-2} |B^1 \wedge N_p| d_\varepsilon F_p .$$

The middle term in (3) and the left sides of (4) and (5) become after division by π'_{a-1} the averages of the areas, the areas of the boundaries, and the average widths of the projection of F or K on H:

$$\mathfrak{M}(P, K, Z_a^\perp) = \frac{1}{\pi'_{a-1}} \int_{S^{a-1}} P_{n-1}(K, H(\perp Z_a, \supset Z_{a-1}) d_\varepsilon S^{a-1}$$

$$\mathfrak{M}(\partial P, K, Z_a^\perp) = \frac{1}{\pi'_{a-1}} \int_{S^{a-1}} \partial P(K, H) d_\varepsilon S^{a-1}$$

$$\mathfrak{M}(\bar{B}P, K, Z_a^\perp) = \frac{1}{\pi'_{a-1}} \int_{S^{a-1}} \bar{B}P(K, H) d_\varepsilon S^{a-1} .$$

Following the agreement of the introduction we pass to the corresponding induced functions $\bar{\varepsilon}_{a-1}(B)$ etc. This amounts to replacing in each of the integrals $|B^1 \wedge N_p|$ by $|B \wedge N_p|$ with an arbitrary $(n-a)$-vector B parallel to $Z_a^\perp$. Now $|\Sigma B_j \wedge N_p| \leqslant \Sigma |B_j \wedge N_p|$, any convex body K can be approximated by bodies with boundaries of positive Gauss curvature and the limit of an extendably convex function is extendably convex. Therefore we have proved:

(6) *Theorem: For any convex body with interior points the functions* $\bar{\varepsilon}_{a-1}(B)$, $\mathfrak{M}(P, K, B)$, $\mathfrak{M}(\partial P, K, B)$, $\mathfrak{M}(\bar{B}P, K, B)$ *are extendably convex. Hence if any of these functions is used to define an affine* $(n-a)$*-area then the* $(n-a)$*-flats minimize* $(n-a)$*-area.*

Rend. Circ. Mat. Palermo, II. Ser. 9 (1960), 226–242

We proved (6) only for $a > 1$; for $a = 1$ the statements amount to ordinary convexity and are known: If u is the unit normal to the hyperplane H_u then

$$P_{n-1}(K, H_u) = P_{n-1}(F, H_u) = \frac{1}{2} \int_F |u \cdot N_p| \, d_e F_p \tag{7}$$

see [6, p. 49], which shows that the induced function $|v| P_{n-1}(F, H_{v/|v|})$ is convex. If we pass to the parallel surface F_h to F we find as above

$$P(F, H_u) = \frac{1}{2} \int_F \nu^1 |u \cdot N_p| \, d_e F_p, \qquad \dot{B} P(F, H_u) = \frac{2}{\pi'_{n-1}} \int_F \nu^{n-2} |u \cdot N_p| \, d_e F_p,$$

so that the induced functions are convex. If $h(v)$ is the supporting function of K, then:

$$2^{-1} \pi_{n-1} \bar{B} P(F, H_u) = \int_{S^{n-2}} h(v_u) \, d_e S^{n-2},$$

where v_u traverses the unit vectors orthogonal to u. Thus the integrals of $h(v)$ over the great S^{n-2} on S^{n-1} induce a convex function.

(Added during proof: While this paper was in print W. Fenchel and the *author discovered a general principle* which contains the preceding results, and in particular the theorem:

The integral of a convex function $h(v)$ over a variable great S^a on S^{n-1} induces an extendably convex function (3).

The present Theorem (6) contains this result only for $h(v) = |v| P_{n-1}(F, H_{v/|v|})$, i.e. for projection bodies, and these are not the most general convex bodies, see next section).

A very interesting open question is whether the projection

$$P_{n-a}(K, Z_a) = \int_{K \cap L \neq 0} dL(Z_a)$$

of K on the $(n-a)$-flat $Z_a^{\perp}$ orthogonal to Z_a induces an extendably convex function.

Our methods yield, however, an interesting expression for these projections (found in [3] for $n = 3$, $a = 2$). If $k_p(\bar{L})$ is the Gauss curvature of $\bar{L} \cap F$ at p and $k_p^n(\bar{L})$ the corresponding normal curvature (the Gauss curvature of the section of F by an a-flat containing N_p and the tangent $(a-1)$-flat of $\bar{L} \cap F$

(3) The result on the integrals over S^{n-2} implies, of course, for general a that the induced function is convex, i.e. satisfies (1) for $k = 2$.

at p) then

$$\int_{\bar{L}\cap K\neq 0} \bar{k}_p(L)\, d_\varepsilon (\bar{L}\cap F)_p = \pi'_{a-1}$$

therefore

(8) $$P_{n-a}(K, Z_a^{\perp}) = \frac{1}{\pi'_{a-1}} \int_{\bar{F}} k_p(\bar{L})\,|B^1 \wedge N_p|\, d_\varepsilon F_p = \frac{1}{\pi'_{a-1}} \int_{\bar{F}} k_p^n(\bar{L})\, d_\varepsilon F_p .$$

The bodies K for which $P_{n-1}(K, H)$ has the same value for all hyperplanes H play for $n = 3$ a considerable role as the bodies of constant « Helligkeit », [6, p. 140], Berwald [2] used the methods of [3] to characterize these bodies for $n = 3$. Extending his result to general n, we obtain a result which is striking in view of the original definition (2) of $\bar{\varepsilon}_{a-1}(Z_a^{\perp})$:

(9) *Theorem: $P_{n-1}(K, H)$ is constant if for any fixed a, $1 \leqslant a \leqslant n-1$, and only if for all these a, the function $\bar{\varepsilon}_{a-1}(Z_a^{\perp})$ is constant.*

For $a = 1$ the two statements are identical, so that we may assume $2 \leqslant a \leqslant n-1$. If $P_{n-1}(K, H)$ is constant, then (3) shows at once that $\bar{\varepsilon}_{a-1}(Z_a^{\perp})$ is constant for each a.

If $\bar{\varepsilon}_{a-1}(Z_a^{\perp})$ is constant for a fixed a then (3) implies that the integral of $P_{n-1}(K, H_u)$ over any great S^{a-1} on S^{n-1} has the same value and it follows from a theorem of Funk (compare [11, p. 35]) that $P_{n-1}(K, H)$ is constant.

3. INTEGRAL GEOMETRY FOR A SINGLE AREA

Let an affine a-area α be given $(1 \leqslant a \leqslant n-1)$ in W^n and put $b = n - a$. We ask whether, or under which conditions, a density dL_b for the b-flats exists such that the simplest integral geometric fact, namely

(10) *Crofton's Formula*

$$\int_{M\cap L_b\neq 0} dL_b = \alpha(M)$$

holds for a convex set M lying in an a-flat. We observed already that this normalization of dL_b does not correspond to the usual one in E^n, but we will use it even for euclidean densities.

The existence of dL_b depends, of course, on the requirements put on dL_b. Our first requirement is that dL_b be *positive*. Although obviously necessary, this turns out to be the stumbling block. One strong implication of this requirement can be seen at once. Consider a closed a-dimensional polyhedron with a-faces $F_0, F_1, \ldots, F_k$. An $(n-a)$-flat intersecting F_0 also intersects $F = \bigcup_{j=1}^{k} F_j$ and will, in general, intersect several F_j, $j = 1, \ldots, k$. Therefore

$$\sum_{j=1}^{k} \int_{L_b \cap F_j \neq 0} dL_b \geq \int_{L_b \cap F \neq 0} dL_b \geq \int_{L_b \cap F_0 \neq 0} dL_b,$$

and if (10) holds,

$$\sum_{j=1}^{k} \alpha(F_j) \geq \alpha(F_0) \ (^4).$$

This means that the a-flats minimize α-area, we will see that α is extendably convex.

We will not worry about differentiability, i.e., assume that dL_b is a true differential form, because the difficulties do not derive from a possible lack of smoothness of α. Also, we do not try to axiomatize our further requirements, since they are obviously reasonable.

The translation invariance of affine area suggests that the density of dL_b be *independent of the position* of L_b *in a family of parallel b-flats* $L(Z_b)$, we therefore only consider densities of the form

$$dL_b = dL(Z_b) \wedge dZ_b.$$

Moreover if M lies in an a-flat parallel to Z_a and Z_a, Z_b intersect in z only, we want $\int_{\delta} dL(Z_b)$ over some set δ of Z_b to be proportional to the area of the set M in a flat $L(Z_a)$ in which the $L(Z_b)$ in δ intersect $L(Z_a)$, and the factor of proportionality must be independent of the a-flat in $L(Z_a)$; or

$$\int_{M \cap L(Z_b) \neq 0} dL(Z_b) = h(Z_a, Z_b)\alpha(M).$$

If $dp(Z_a)$ is the element of α-area in a flat parallel to Z_a we find

$$dL(Z_b) = h(Z_a, Z_b)dp(Z_a), \qquad h(Z_a, Z_b) > 0 \text{ when } Z_a \cap Z_b = z.$$

If Z_a and Z_b have more points than z in common we want $h(Z_a, Z_b) = 0$.

We now introduce an auxiliary euclidean metric ε in W^n (invariant under the translations of W^n), and use our previous notations, in particular ε_a for

(4) The inequality holds when the polyhedron is non-degenerate, hence α is strictly convex, see [10] or [11].

euclidean a-measure. The euclidean densities are indicated by d_ε. Then $\alpha(A) = f_a(Z_a)\varepsilon_a(A)$ when A is parallel to Z_a, and

(11) $$dZ_b = g_1(Z_b)\,d_\varepsilon Z_b, \qquad g_1(Z_b) > 0 \text{ and continuous.}$$

When Z_a and Z_b intersect in a point and A^1, B^1 are euclidean unit a-and b-vectors parallel to Z_a and Z_b then

$$d_\varepsilon L_b = d_\varepsilon L(Z_b) \wedge d_\varepsilon Z_b = |A^1 \wedge B^1|\, d_\varepsilon p(Z_a) \wedge d_\varepsilon Z_b.$$

Thus

(12) $$\begin{cases} dL_b = h(Z_a, Z_b)\,dp(Z_a) \wedge dZ_b \\ \qquad = g_1(Z_b)\,h(Z_a, Z_b)\,|A^1 \wedge B^1|^{-1} f_a(Z_a)\,d_\varepsilon p(Z_a) \wedge d_\varepsilon(Z_b). \end{cases}$$

Now dL_b must be independent of the choice of Z_a, or

(12a) $$h(Z_a, Z_b) = |A^1 \wedge B^1|\, f_a(Z_a)\, g_2(Z_b), \qquad g_2(Z_b) > 0$$

and continuous.

Therefore

(13) $$dL_b = g_1(Z_b)\,g_2(Z_b)\,d_\varepsilon L_b = g(Z_b)\,d_\varepsilon L_b, \qquad g(Z_b) > 0 \text{ and continuous.}$$

Crofton's Formula (10) yields

$$f_a(Z_a)\varepsilon_a(M) = \alpha(M) = \int g(Z_b)\,d_\varepsilon L_b = \int_M d_\varepsilon p(Z_a) \int_{Y_b} |A^1 \wedge B^1|\, g(Z_b)\,d_\varepsilon Z_b,$$

where Y_b is the set of all b-flats Z_b.

Thus

(14) $$f_a(Z_a) = \int_{Y_b} g(Z_b)\,|A^1 \wedge B^1|\, d_\varepsilon Z_b$$

or

(14a) $$\alpha(A) = \int_{Y_b} g(Z_b)\,|A \wedge M^1|\, d_\varepsilon Z_b.$$

Since $|\Sigma A_j \wedge B^1| \leqslant \Sigma |A_j \wedge B^1|$ we find:

(15) *If a density dL_b exists (satisfying our various conditions) then $\alpha(A)$ is extendably and strictly convex.*

The condition (14) can, in general, not be satisfied, that is, *no positive function $g(Z_b)$ solving* (14) *will exist, even when α is extendably convex*. To see this we consider the case $a = n - 1$ and represent Z_a by its unit normal u, so that (14) takes the form

(16) $$\alpha(u) = \frac{1}{2}\int_{S^{n-1}} g(B^1)|u\cdot B^1|\,d_e S^{n-1}, \qquad g(B^1) = g(-B^1) > 0,$$

where B^1 is an ordinary unit vector.

Without the condition $g(B^1) > 0$ the equation (16) has a unique continuous solution, see Blaschke [4, pp. 154-157], but *the solution will for $n > 2$, in general take negative values and* hence not lead to a positive density, see (13). That $\alpha(u)$ must have special properties when $g(u) > 0$ can also be seen as follows: there is a closed convex hypersurface F whose Gauss curvature at the point with normal B^1 is $g^{-1}(B^1)$. Using (7) we find

$$P_{n-1}(F, H_u) = \alpha(u),$$

and not every convex function has this form.

The situation is the same for $a = 1$. We then represent Z_b by its unit normal u and A^1 is a unit vector, so that

$$\alpha(A^1) = \frac{1}{2}\int_{S^{n-1}} g(u)|A^1\cdot u|\,d_e S^{n-1}.$$

Uniqueness and existence have not been discussed for equations of the form (14) and $1 < a < n-1$, but it is more than probable that the situation is analogous.

4. INTEGRAL GEOMETRY FOR TWO AREAS

Assume that an a-area α is given and that (14) can be solved with a positive $g(Z_b)$. We ask whether *a b-dimensional area β ($b = n - a$) and a density dL_a for the a-flats can be introduced satisfying the Crofton Formula:*

(17) $$\int_{L_a \cap N \neq 0} dL_a = \beta(N)$$

for a set N lying in a b-flat.

The problem becomes most natural if we *also assume that an affine n-area or a volume $\nu(R)$ has been introduced in W^n*, which means that a constant f_n has been chosen such that $\nu(R) = f_n \varepsilon_n(R)$. Nothing would prevent us to choose the auxiliary metric such that $f_n = 1$, but we will need to see how the choice of f_n enters our results.

A minimal requirement is what might be called *Cavalieri's Principle:* If M, N are convex sets lying in Z_a and Z_b, $Z_a \cap Z_b = z$, and have z in common then $R = M + N$ (addition in the sense of the theory of convex bodies) is essentially the product of M and N. Using the previous notations we stipulate

$$\int_{L(Z_a)\cap R\neq 0} \alpha(L(Z_a)\cap R)\,dL(Z_a) = \int_{L(Z_b)\cap R\neq 0} \beta(L(Z_b)\cap R)\,dZ_b = h(Z_a, Z_b)\nu(R), \tag{18}$$

which is equivalent to

$$h(Z_a, Z_b)\alpha(M)\beta(N) = |A^1 \wedge B^1|\,\varepsilon_a(M)\varepsilon_b(N)f_n.$$

If $\beta(N) = f_b(Z_b)\varepsilon_b(N)$ then

$$h(Z_a, Z_b) = |A^1 \wedge B^1|\,f_a^{-1}(Z_a)f_b^{-1}(Z_b)f_n \text{ is intrinsic,} \tag{19}$$

which means that it is determined by α, β and ν and does not depend on the choice of the auxiliary euclidean metric. It corresponds to $|A^1 \wedge B^1|$ in the euclidean case and is called the Minkowski sine function in the special case of Minkowski area, see [11]. We define $h(A, B)$ for vectors A, B parallel to Z_a and Z_b by

$$h(A, B) = |A \wedge B|\,f_a^{-1}(Z_a)f_b^{-1}(Z_b)f_n. \tag{20}$$

If $A = x_1 \wedge \cdots \wedge x_a$, $B = x_{a+1} \wedge \cdots \wedge x_{a+b}$ then $h(A, B)$ is the ν-volume of the parallelepiped spanned by $x_1, \ldots, x_{a+b}$.

Because of (17) the following analogue to (14) must be satisfied

$$f_b(Z_b) = \int_{Y_a} g'(Z_a)|A^1 \wedge B^1|\,d_e Z_a \tag{21}$$

with a positive $g'(Z_a)$, and any positive continuous function $g'(Z_a)$ will provide an acceptable β. This means that *there is so much choice for the answer to our problem, that it loses much of its interest.* Nevertheless there are some further natural agreements possible, which relate these areas and densities mose closely.

We remember that $g(Z_b)$ was a product of two functions $g_1(Z_b)$ and $g_2(Z_b)$ and that only their product is determined. If we choose $g_2(Z_b) = f_b^{-1}(Z_b)f_n$ then

$$g_1(Z_b) = f_b(Z_b)f_n^{-1}g(Z_b), \qquad dZ_b = f_b(Z_b)f_n^{-1}g(Z_b)\,d_e Z_b. \tag{22}$$

Similarly

$$g_1'(Z_a) = f_a(Z_a)f_n^{-1}g'(Z_a), \qquad dZ_a = f_a(Z_a)f_n^{-1}g'(Z_a)\,d_e Z_a, \tag{23}$$

which yields because of (12a) and (14) the *desirable normalizations:*

$$1 = \int_{Y_b} h(Z_a, Z_b)\, dZ_b = \int_{Y_a} h(Z_a, Z_b)\, dZ_a.$$

We look again at the case $a = n - 1$. Proceeding similarly as after (16) we may consider $g'^{-1}(Z_{n-1})$ as the Gauss curvature of a hyper-surface T' at the points where the tangent hyper-planes are parallel to Z_a. Then $g'(Z_{n-1})d_e S^{n-1}$ will be the euclidean area element of T', hence $f_{n-1}(Z_{n-1})g'(Z_{n-1})d_e S^{n-1}$ will be its element of α-area, and (see (23)) dZ_a will be the element of α-area of the surface $T = f_n^{-1/(n-1)} T'$ at the points where the tangent plane is parallel to Z_{n-1}, which we write briefly as

(24) $$d\alpha(T, Z_{n-1}) = dZ_{n-1}.$$

The surface T is intrinsically determined by β and ν and solves the following isoperimetric problem (see [7]): among all hypersurfaces homeomorphic to S^{n-1} (and either convex or satisfying certain smoothness conditions) which bound the same volume as T the surface T minimizes α-area. Thus, in spite of all the arbitrariness, we stay close to the euclidean case. We can push the analogy slightly farther by a proper choice of ν. The surface T' is independent of f_n. If T_0, T_0' are the bodies bounded by T, T', we can solve the equation

$$n\varepsilon_n(f_n^{-1/(n-1)} T_0') = n f_n^{-n/(n-1)} \varepsilon_n(T_0') = f_n^{-1}\beta(T') = \beta(f_n^{-1/(n-1)} T')$$

for f_n and obtain

(25) $$n\nu(T_0) = \beta(T)$$

corresponding to $n\pi_n = \pi_{n-1}'$ in the euclidean case.

In a slightly less natural way we can also interpret dZ_1 as the area element of a hypersurface following an idea of Barthel [1]. Using a vector v instead of B take a supporting hyperplane $L(Z_{n-1})$ parallel to a given Z_{n-1} of the indicatrix $\beta(v) = 1$. Denote by N the segment from z to a point of contact of $L(Z_{n-1})$ with the indicatrix. For a set M in $L(Z_{n-1})$ we define

$$\alpha'(M) = \nu(M + N).$$

This $(n-1)$-area α' will in general be different from α, but $dv = dZ_1$ may be interpreted as the element of α'-area of a solution of the isoperimetric problem for α'.

If $h'(Z_a, Z_b)$ is defined for α' and β in the same way as $h(Z_a, Z_b)$ is for α and β, this amounts to choosing α' such that for each given Z_1 the relation $\max h'(Z_{n-1}, Z_1) = 1$ holds.

The case $b = a$ and $n = 2a$ is exceptional, because if (14) or (14a) has a positive solution $g(Z_b)$ we have a natural choice for β namely $\beta \equiv \alpha$.

In particular, for $a = 1$ we have a *Minkowski plane*. Since also $a = n - 1$ we can apply our previous results. Moreover (14) always has a positive solution: the dual curve to the indicatrix $\alpha(\nu) = 1$ is a curve with width $2f_1(Z_1)$ in the direction of Z_1 so that *$f_1(Z_1)$ is trivially a projection.* Because the vector u in (16) was orthogonal to Z_b, we obtain a solution to the isoperimetric problem for the length α by revolving this dual through $\pi/2$.

In order to come close to the usual notations we use an angle ψ instead of Z_1, and denote by $ds(T, \psi)$ the element of Minkowski arclength of T at the points where the tangent has direction ψ. Then we find as *density for the lines* L using (12) and (24):

$$dL = h(L_\varphi, L_\psi)\,dp(\varphi) \wedge ds(T, \psi). \tag{26}$$

The normalization (25) of ν is in this case equivalent to requiring that the unit circle have area π, see [8]. If dp in the element of ν-area, then $dp \wedge ds(T, \psi)$ is a natural candidate for kinematic density.

dL was in fact found by Blaschke [5] as a special case of a density for the extremals in a two-dimensional Finsler space with a line element $ds = F(x, dx) = F(x_1, x_2, dx_1, dx_2)$. Crofton's Formula holds. If $F(x, dx)$ is independent of x we have a Minkowski plane, and the uniqueness of the solution of (16) implies that Blaschke's density for lines must, except for a constant factor, coincide with ours. Blaschke expresses his density by means of the Weierstrass function $F_1(x, \dot{x})$ derived from $F(x, \dot{x})$ implicitly in his first expression [5, (6)] for dL, explicitly in the second [5, (30), (32)]. In our case F_1 is a function of $\dot{x}$ alone and may for $\dot{x}_1^2 + \dot{x}_2^2 = 1$ be written as $F_1(\cos\psi, \sin\psi)$. We know that

$$ds(T, \psi) = k_\varepsilon^{-1}(T, \psi) f_1(\psi)\, d_\varepsilon \psi,$$

where $k_\varepsilon(T, \psi)$ is the euclidean curvature of T at the points with tangent ψ, and it is shown in [9, formulae (10) and (14)] that

$$k_\varepsilon^{-1}(T, \psi) = F_1(\cos\psi, \sin\psi).$$

With the help of this relation, Blaschke's density is easily identified with ours.

5. INTEGRAL GEOMETRY FOR THE GROUP OF TRANSLATIONS

The preceding considerations show that except for $a = 1$, $b = 1$, $n = 2$, the natural domain for the integral geometry of affine area is the translation group.

Rend. Circ. Mat. Palermo, II. Ser. 9 (1960), 226–242

We saw in Section 1 how some of the results of Berwald and Varga [3] can be extended to n dimensions. We will restrict ourselves here to indicate how results of [3] *can, without deriving them again, be extended to a three dimensional affine space* W^3 *in which an affine* 1-*area or length* β, *an affine* 2-*area* (simply called area) α *and a volume* ν *are given. Convexity of* α *or* β *are not required.*

We denote lines by L, planes by P, ordinary vectors by small latin letters u, v, w, (simple) 2-vectors by A, B, C. It is clear from the previous discussion that the *sine functions* h will play a decisive role. We can *define these functions intrinsically* as follows:

$$h(v,\ w) = \alpha(v \wedge w)$$

$$h(A,\ v) = h(v,\ A) = \nu(v \wedge A)$$

$$h(A,\ B) = \nu(u \wedge v \wedge w)/\beta(w) \quad \text{if} \quad A = u \wedge w, \quad B = v \wedge w, \quad w \neq 0.$$

We introduce an auxiliary euclidean metric such that

$$f_3 = 1 \quad \text{or} \quad \varepsilon_3(u \wedge v \wedge w) = \nu(u \wedge v \wedge w).$$

If L_v, L_w are intersecting lines parallel to v and w, P_{vw} the plane containing these lines, then, see (19), with obvious notations

$$(27) \quad \begin{cases} h(L_v,\ L_w) = |\sin(L_v,\ L_w)|\, f_1^{-1}(L_v) f_1^{-1}(L_w) f_2(P_{vw}) \\ h(L_v,\ P_A) = |\sin(L_v,\ P_A)|\, f_1^{-1}(L_v) f_2^{-1}(P_A) \\ h(P_A,\ P_B) = |\sin(P_A,\ P_B)|\, f_2^{-1}(P_A) f_2^{-1}(P_B) f_1(L_w). \end{cases}$$

As *density for a family of parallel lines* $L(v)$ parallel to the line L_v we take for any plane P_A which intersects L_v in a point

$$d L(v) = h(L_v,\ P_A)\, d\alpha(P_A),$$

and as *density for the planes* $P(A)$ parallel to A

$$d P(A) = h(L_v,\ P_A)\, d\beta(L_v),$$

where $d\alpha(P_A)$ and $d\beta(L_v)$ are, of course, the elements of α-area in P_A and β-length on L_v respectively. We denote euclidean densities and area by d_ε.

If we use again $\bar{L}$ instead of $L(v)$ where ambiguities are impossible, and if $n(L)$ is the number of intersections of $\bar{L}$ with the surface F, then the simplest euclidean formula is (see (6.2) in [3]):

$$\int n(\bar{L})\, d_\varepsilon \bar{L} = \int |\sin(\bar{L},\ T_p)|\, d_\varepsilon F_p$$

where T_p is the tangent plane of F at p.

Now, using (27),

(28) $d_\varepsilon L = |\sin(L_v,\ P_A)|\, d\varepsilon_2(P_A) = h(L_v,\ P_A) f_1(L_v) f_2(P_A)\, d\varepsilon_2(P_A) = dL\, f_1(L_v)$,

hence

$$\int n(\bar{L})\, d\bar{L} = \int |\sin(\bar{L},\ T_p)|\, f_1^{-1}(L_v) f_2^{-1}(T_p) f_2(T_p)\, d_\varepsilon F_p$$

or

(29)
$$\int_{\bar{L}\cap F\neq 0} n(\bar{L})\, d\bar{L} = \int_F h(\bar{L},\ T_p)\, dF_p,$$

where dF_p is the element of α-area of F.

The analogue to (2) may be derived in a similar way. If F is a fixed surface (not necessarily convex), and $\bar{P}$ varies in the family $P(A)$, and the plane $\bar{P}$ intersects F in the curve $\bar{P}\cap F$, then we have as in Section 1 (also 3, (7.1)) $d_\varepsilon S_p \wedge d_\varepsilon \bar{P} = \sin(\bar{P},\ T_p)\, d_\varepsilon F_p$, where $d_\varepsilon s_p$ in the euclidean length element of $\bar{P}\cap F$. If the tangent of $\bar{P}\cap F$ at P is t_p, then the α-length element is $ds_p = f_1(t_p)\, d_\varepsilon s_p$. Exactly as in (28) we find

(30)
$$d_\varepsilon \bar{P} = dP f_2(P_A)$$

hence from (27)

$$\begin{aligned} ds_p \wedge d\bar{P} &= f_1(t_p) f_2^{-1}(P_A)\, d_\varepsilon s_p \wedge d_\varepsilon \bar{P} \\ &= f_1(t_p) f_2^{-1}(P_A) f_2^{-1}(T_p) |\sin(\bar{P},\ F_p)|\, f_2(T_p)\, d_\varepsilon F_p \\ &= h(\bar{P},\ T_p)\, dF_p. \end{aligned}$$

Therefore by integration

(31)
$$\int_{\bar{P}\cap F\neq 0} \beta(\bar{P}\cap F)\, d\bar{P} = \int_F h(\bar{P},\ T_p)\, dF_p,$$

which is the exact analogue of (2).

The density for the surface $\bar{F}$ originating from a fixed surface F by translations is measured by choosing a point on F and measuring its change, i.e. we simply take the volume element of the space, and since $\varepsilon_3(M) = \nu(M)$ we have $d_\varepsilon \bar{F} = d\bar{F}$.

If C is a fixed curve, t_p the tangent of C at $p \in C\cap \bar{F}$, and ds_p, $d_\varepsilon s_p$ the arclength elements of C, then it is proved in [3, (10.4)] that

$$d_\varepsilon \bar{F} = |\sin(t_p,\ \bar{T}_p)|\, d_\varepsilon s_p \wedge d_\varepsilon \bar{F}_p,$$

Geometries in which the planes minimize area ([1]).

Memoria di HERBERT BUSEMANN (a Los Angeles, U.S.A.)

To Enrico Bompiani on his scientific Jubilee.

Summary. - *It is shown that for $1 \leq a \leq n$ an open convex set in n-dimensional affine space can be provided with a complete a-dimensional area for which the a-flats minimize area, and that any such area can be extended to an area with the same properties in an open convex set in $(n+1)$ — space.*

1. Introduction

In his Problem IV HILBERT [3] suggested studying all geometries in which the straight lines are the shortest connections or minimize length. Although the problem in this form proved too general to be fruitful, it has given rise to many investigations dealing either with special geometries or with special questions. The higher dimensional analogue to Problem IV concerns *a-dimensional areal metrics in subsets of the n-dimensional projective space in which the a-flats minimize area.* Such geometries apparently have never been treated, at least in the large. The present paper represents a first step in this direction by answering two questions.

Given a non-empty open convex set in the n-dimensional affine space W^n, can D be provided with a complete a-dimensional area such that the intersections $A \cap D \neq 0$ of D with a-flats A in W^n minimize this area?

With completeness we mean — exact definitions will be given later — that an a-dimensional convex set K in D whose boundary $B(K)$ contains sufficiently many points of the absolute locus $B(D)$ has infinite area. The euclidean and hyperbolic geometries suggest distinguishing different degrees of completeness if $a > 1$. We speak of strong (or euclidean) completeness if the area of K is infinite whenever $B(K)$ contains a point of the absolute locus, and of hyperbolic completeness if the area is complete, but an a-simplex with one vertex on $B(D)$ has finite area.

It will be shown that *any D can be provided with a strongly complete a-dimensional area ($a \geq 1$), and that $D = W^n$ or any D with a strictly convex boundary in W^n can be provided with a hyperbolically complete area if $a > 1$.*

([1]) This work was supported by a grant from the National Science Foundation.

such that the a-flats minimize area. An idea suggested by integral geometry is essential for our construction.

Just as for $a = 1$ it will appear that for $a > 1$ the variety of such areas is so large that the problem of studying all such areas is not fruitful.

We then prove (answering our second question without formulating it first) the, in view of this variety non-obvious, fact that *a given complete a-dimensional area* α *in a set* $D \subset W^n$ *for which the a-flats minimize area can be extended to an a-area with the same properties defined in an open convex set* D^+ *in* $W^{n+1} \supset W^n$ *with* $D = W^n \cap D^+$. The proof is modeled after the case $a = 1$ which is treated in [1].

It gives a hyperbolic completion α^+ if α is hyperbolically complete but fails to yield a strongly complete α^+ if α is strongly complete unless $D = W^n$. (2)

Because of its usefulness we append a simple fact concerning *locally isometric maps* which is not connected with HILBERT's Problem IV, but was suggested by the considerations on order magnitude of volume in the first part of this paper.

2. Definition of the geometries and completeness.

An *a-dimensional area* ($1 \leq a \leq n$) *in the non-empty open convex set* D *of* W^n (without mentioning it every time it will always be understood that D has these properties) assigns to every BOREL set M lying in the intersection of D with an a-flat A ($A \cap D \neq 0$) an area $\alpha(M)$ such that $\alpha(M)$ is countably additive in each A, vanishes for M lying in $(a-1)$-flats, is finite and positive for a non-degenerate a-simplex S with vertices in D and depends continuously on the vertices of S.

The α-area of an a-dimensional polyhedron with a-faces $M_1, \ldots, M_r$ is $\sum_{j=1}^{r} \alpha(M_j)$. The properties of α guarantee that this area does not change under subdivision of the faces of the polyhedron. The analogue to HILBERT's Problem IV deals with those α for which the *a-flats strictly minimize area*. This means by *definition*: If $M^0, \ldots, M^r$ are the a-faces of a closed oriented (3) a-dimensional polyhedron then

$$\alpha(M^0) < \sum_{i>0} \alpha(M^i) \quad \text{for } M^0 \neq \bigcup_{i>0} M^i \tag{1}$$

(2) Whereas the author made no serious attempt in the first problem to produce hyperbolically complete a-areas for all D, because the problem does not seem sufficiently interesting, he did try to find a strongly complete α^+ for strongly complete α but without success.

(3) Orientability is inessential here but is necessary when areas depending on the orientation of the a-flats are considered.

If always $\alpha(M^0) \leq \sum_{i>0} \alpha(M^i)$ we say that the *a-flats minimize area.* Notice that (1) holds always for $a = n$.

The usual way of defining an a-dimensional areal metric in D is this: we introduce in W^n an auxiliary euclidean metric xy invariant under the translations of W^n. Terms like orthogonal, cartesian coordinates, etc. will refer to this metric. Consider an expression of the form $F(x, A)\, dV_a$, where $x \in D$, A is an a-flat through x and dV_a is the euclidean area element at x. The function $F(x, A)$ is at least continuous in x and A. The area of an a-dimensional surface which is piecewise of class C^1, in particular of a polyhedron, is then defined by integration of $F(x, A) dV_a$.

All the areas which we are going to construct in given D will in fact have this form with a smooth F. The extension theorem will apply to the general case and therefore the extended metric will in general not be smooth.

Let B_f be the boundary of D. It is empty only for $D = W^n$. If non-empty B_f is either homeomorphic to the $(n-1)$-sphere S^{n-1} or to W^{n-1} or to the topological product of S^r and $W^{n-r-1} (0 \leq r \leq n-2)$, see [2, p. 3].

We *complete* W^n not as usual to P^n, but to a *ball* by adding an $(n-1)$-sphere S_∞ whose points are in $(1-1)$-correspondence with the families of oriented parallel lines in W^n.

Then the boundary B of D is homeomorphic to S^{n-1} and we put $B_\infty = B - B_f$, with $B_\infty = S_\infty = B$ for $D = W^n$. If the flat A intersects D then the relative boundary $B(A \cap D)$ of $A \cap D$ is homeorphic to S^{a-1} and so is the relative boundary $B(K)$ of any a-dimensional convex set K in $A \cap D$.

We call the a-area α (not necessarily satisfying (1)) $(1 \leq a \leq n)$ *complete,* if $\alpha(K) = \infty$ whenever $B(K)$ contains a non-empty relative open subset of $B(A \cap D)$. A convex set K in A with the latter property has automatically dimension a.

We speak of *strong or euclidean completeness* if $\alpha(K) = \infty$ for any a-dimensional convex set K for which $B(K) \cap B(A \cap D) \neq 0$.

Denote as a-wedge the interior S of a non-degenerate a-simplex in D of which a vertices lie in D spanning an $(a-1)$-simplex S' and the remaining vertex v lies on B. If $\bar{S}$ is the closure of S relative to W^n, then $\alpha(\bar{S} - v) = \alpha(S)$ for any a-area in D. If $v \in B_\infty$ we mean, of course, the interior of a the prism formed by the rays beginning at points of S' and ending at v.

The a-area $\alpha (2 \leq a \leq n)$ is *hyperbolically complete* if it is complete, but all a-wedges have finite areas.

A complete a-area need not be either strongly or hyperbolically complete.

The case $a = 1$ requires some comments. A 1-wedge S is an open segment or an open ray in W^n. If L is the line containing S then $B(S) \cap B(L \cap D)$ is trivially open in $B(L \cap D)$, hence *a complete length cannot be hyperbolically complete and is,* in fact, *strongly complete.*

If the α-length of a segment is used as distance of its endpoints the resulting space is metric if the triangle inequality holds, or in our language, if the lines minimize length. Under this condition on α it is seen at once that the usual completeness in terms of distances coincides with the present.

Through yet finer distinctions between degrees of completeness the case $a=1$ would lose its exceptional character. Let F be a flat in A with $f=\dim F\leq a-2$. In F take a solid f-dimensional half cone C with apex in D and let $V=B(F\cap C)\cap B(D)$. With an appropriate definition of the convex hull S_V in D of V and an $(a-1)$-simplex S' in A we may require $\alpha(S_V)<\infty$. But the rigorous formulations are cumbersome and of little interest.

All problems are trivial for $a=n=1$ so that we will always assume $n>1$.

3. The case $n=a$.

With the euclidean distance xy and $xB=xB_f$ for $B_f\neq 0$ and $xB=\infty$ for $B_f=0$ put

$$f_\rho(x)=[1+(xB)^\rho](xB)^{-\rho},\ \rho>0,\ x\in D,$$

where $f_\rho(x)\equiv 1$ if $B_f=0$ or $D=W^n$. If dV is the euclidean volume element we define for any Borel set M in D

$$\alpha(M,\rho)=\int_M f_\rho(x)dV. \tag{1}$$

For $D=W^n$ this is the euclidean volume. Let $D\neq W^n$. Consider a wedge S in D and use the same notations as before. If $v\in B_\infty$ then $f_\rho(x)\geq 1$ yields

$$\alpha(S,\rho)=\infty\ \textit{for}\ v\in B_\infty,\ \rho>0. \tag{2}$$

If $v\in B_f$, let R be the segment from an interior point y of S' to v, and H a supporting hyperplane af B_f at v. Choose u on R so close to v, that a hyperplane H_z parallel to H through a point z of R between u and v intersects all edges of S ending at v. In all that follows $k_1, k_2, \ldots,$ *are suitably chosen positive constants.*

For $x\in H_z\cap S$ we have

$$f_\rho(x)>(xB)^{-\rho}\geq(xH)^{-\rho}=k_1 zv^{-\rho}$$

and the euclidean area of $H_z\cap S$ has the form $k_2 zv^{n-1}$.

Therefore

$$(3) \qquad \alpha(S,\ \rho) \geq k_3 \int_0^{uv} t^{n-1-\rho} dt,$$

which is infinite for $\rho \geq n$. Every a-dimensional convex set K with $B(K) \cap B \neq 0$ contains an a-wedge, hence

(4) *$\alpha(M,\ \rho)$ defines for $\rho \geq n$ a volume in D which is strongly complete.*

With the same notations and still $v \in B_f$ the intersection $H_u \cap S$ lies inside $H_u \cap B$, therefore the cone T_u formed by the rays from v and through pontis of $H_u \cap B$ (with obvious agreements when H contains other points of B than v) will contain S in its interior. If u is sufficiently close to v then

$$xB \geq xT_u \text{ for } x \in H_z \cap S,$$

hence for such x

$$f_\rho(x) \leq k_4 (xT_u)^{-\rho}, \qquad \beta(z) = \inf_{x \in H_z \cap S} xT_u = k_5 zv.$$

Therefore

$$f_\rho(x) \leq k_4 k_5 (zv)^{-\rho}$$

so that the part S^* of S cut out by the union of the planes H_z satisfies

$$(5) \qquad \alpha(S^*,\ \rho) \leq k_6 \int_0^{uv} t^{n-1-\rho} dt,$$

and $\alpha(S^*,\ \rho)$ and hence $\alpha(S,\ \rho)$ *are finite for* $\rho < n$.

If $B(K) \cap B_\infty \neq 0$ it follows from (2) that $\alpha(K,\ \rho) = \infty$ for any positive ρ. Consider a convex set K in D for which $B(K) \cap B_f$ contains a non-empty open subset G of B. The results of [2, pp. 23, 24] show that G contains a point w with the following properties: B_f has at w a tangent hyperplane T. If T is chosen as the plane $t = 0$ of a cartesian coordinate system $s_1, \dots, s_{n-1}, t$ with w as origin and the positive t-axis pointing into D, then a neighborhood of w on B is representable in the form

$$(6) \qquad t = h(s_1, \dots, s_{n-1}) = h(s) \geq 0,\ h(s) = 0(\mid s \mid^2),\ \mid s \mid < \varepsilon',$$

where $\mid s \mid = (\Sigma s_i^2)^{1/2}$. For a suitable $\varepsilon (0 < \varepsilon < \varepsilon')$ the intersection of the solid, cylinder $|s| \leq \varepsilon$ with B_f will lie in G and the part

$$Z: t \leq \varepsilon,\ t \geq h(s) \quad \text{of} \quad |s| \leq \varepsilon$$

lies in K. Because $xB \geq xT = t$ if $x = (s, t)$ we have

$$f_\rho(x) \geq k_7 t^{-\rho} \quad \text{for} \quad x \in Z.$$

(7) *The euclidean $(n-1)$-area of the intersection of Z with $t = t_0$ is by (6) for $0 < t_0 \leq \varepsilon$ at least $k_8 t_0^{(n-1)/2}$ hence*

$$\alpha(K, \rho) \geq k_7 k_8 \int_0^\varepsilon t^{(n-1)/2-\rho} dt$$

which is infinite for $\rho \geq (n+1)/2$.

This estimate is poor, but (7) will be useful later. The best estimate is obtained by observing that by going from a point (t, o) on a ray parallel to $t = 0$ we either hit the surface first, and then the integral over this ray of $f_\rho(x)$ will be infinite for $\rho \geq 1$ because $xB \leq t = h(x)$. Or we hit the cylinder first, and then the integral over $f_\rho(x)$ is at least $t^{-\rho}\varepsilon$, so that the contribution to $\alpha(K, \rho)$ for fixed t is at least $k_9 t^{-\rho}$.

Therefore

(8) *If $1 \leq \rho < n$, then $\alpha(M, \rho)$ defines for $B_f = B$ a volume which is hyperbolically complete.*
If $B_f \neq 0$ and $B_\infty \neq 0$ then $\alpha(M, \rho)$ is complete but neither strongly nor hyperbolically so.

In order to obtain a hyperbolically complete volume also for $B_\infty \neq 0$ we put

$$f_{\rho,\varepsilon}(x) = f_\rho(x)(1 + px)^{-1-\varepsilon},\ \rho > 0,\ \varepsilon > 0$$

and

(9) $$\alpha(M, \rho, \varepsilon) = \int_M f_{\rho,\varepsilon}(x) dV.$$

In the neighborhood of any point of B_f the volume $\alpha(M, \rho, \varepsilon)$ behaves like $\alpha(H, \rho)$ because $(1 + px)^{1+\varepsilon}$ is locally bounded with px. Consider a wedge S in D with $v \in \dot{B}_\infty$. We notice:

(10) *If $U(S', \delta) \subset D$, then $f_\rho(x) \leq (1 + \delta\rho)\delta^{-\rho}$ for $x \in S$.*

For let R be a ray beginning at a vertex q of S' and ending at v. Because D is convex any ray parallel to R and beginning at a point of $U(S', \delta)$ lies in D. Hence $U(S, D) \subset D$ and $xB \geq \delta$ for $x \in S$, moreover $(1 + t)t^{-1}$ is for $t > 0$ a decreasing function of t.

The intersection of S by hyperplanes parallel to S' and through points x of R all have the same euclidean $(n-1)$-area. If $qx = t$, then $px \geq t - pq$.

$$\alpha(S, \rho, \varepsilon) \geq k_{10} \int_0^\infty \max\,(1,\ t - pq)^{-1-\varepsilon} dt \tag{11}$$

which is finite for $\varepsilon > 0$.

Let K be convex set in D for which $B(K) \cap B$ is open in B and not empty. If $B(K)$ contains an open subset of B_f we conclude from (8) that $\alpha(K, \rho, \varepsilon) = \infty$ for $\rho \geq 1$. If $B(K)$ contains an open subset of B_∞, then K contains a solid circular cone with an apex a in D. If the ray R is the axis of this cone we find

$$\alpha(K, \rho, \varepsilon) > k_{11} \int_0^\infty t^{n-1} \max\,(1,\ t - pq)^{-1-\varepsilon} dt$$

which is infinite for $\varepsilon \leq n-1$. Thus

(12) *If* $1 \leq \rho < n$ *and* $0 < \varepsilon \leq n-1$ *then* $\alpha(M, \rho, \varepsilon)$ *defines in* D *(in particular for* $D = W^n$*) a volume which is hyperbolically complete.*

4. Minimizing areas for $a < n$. Behaviour at B_∞.

We use an idea suggested by integral geometry to derive a-areas in W^n from the preceding construction.

Select a point $p \in D$, denote by N a non-oriented $(n-a)$-flat $(1 \leq a \leq n-1)$ through p and by dN the euclidean kinematic density of these N. Further, A_N is the a-flat orthogonal to N through p and M_N is the projection of M parallel to N on A_N, in particular, D_N is the projection of D.

In each D_N we define the functions corresponding to $f_\rho(x)$ and $f_{\rho,\varepsilon}(x)$ and denote them by $f_\rho(x, N)$, $f_{\rho,\varepsilon}(x, N)$, and obtain in this may for each $M \subset D$ an a-dimensional area

$$\alpha(M_N, \rho) = \int_{M_N} f_\rho(x, N)\, dV_N, \qquad \alpha(M_N, \rho, \varepsilon) = \int_{M_N} f_{\rho,\varepsilon}(x, N)\, dV_N \tag{13}$$

in D_N, where dV_N is the euclidean area element in A_N.

Ann. Mat. Pura Appl., IV. Ser. 55 (1961), 171–189

We put

$$(14)\qquad \alpha(M,\rho)=\int\alpha(M_N,\rho)dN,\qquad \alpha(M,\rho,\varepsilon)=\int\alpha(M_N,\rho,\varepsilon)dN,$$

where the integration is extended over all N.

If M is a compact subset of D then a neighborhood $U(M,\delta)$ lies in D, hence the δ-neighborhood $U_N(M_N,\delta)$ of M_N relative to A_N lies in D_N. Thus the functions $f_\rho(x,N)$ and $f_{\rho,\varepsilon}(x,N)$ are bounded for $x\in M$ uniformly in N and $\alpha(M,\rho)$ and $\alpha(M,\rho,\varepsilon)$ are finite.

If $M\neq 0$ lies in an a-flat A and is open relative to A, then $\alpha(M_N,\rho)$ and $\alpha(M_N,\rho,\varepsilon)$ are positive for all N which intersect A in a point, so that $\alpha(M,\rho)>0$, $\alpha(M,\rho,\varepsilon)>0$, and these numbers vanish when M lies in an $(a-1)$-flat.

Therefore (14) *defines a-dimensional areas in D for all positive ρ, ε.* The advantage of the definition (14) is that it makes the minimizing property of the a-flats obvious which is usually very hard to establish for a given area.

Consider an orientable closed a-dimensional polyhedron with a-faces $M^0, M^1,\dots, M^r$ and $M^0\neq\bigcup_{i>0}M^i$. For topological reasons $M_N^0\subset\bigcup_{i>0}M_N^i$ for each N, and therefore

$$(15)\qquad \alpha(M_N^0,\rho)\leq\sum_{i>0}\alpha(M_N^i,\rho),\qquad \alpha(M_N^0,\rho,\varepsilon)\leq\sum_{i>0}\alpha(M_N^i,\rho,\varepsilon)$$

which shows after integration that the a-flats minimize area for $\alpha(M,\rho)$ and $\alpha(M,\rho,\varepsilon)$.

Because $M^0\neq\bigcup_{i>0}M^i$ there is an $(n-a)$-flat N_0' which intersects two different M^i, $i>0$ in an interior point each, and this remains true for all $(n-a)$-flats N' sufficiently close to N_0'. The flats parallel to these N' through p form an open subset Y of all N. An N' either intersects the a-flat carrying M^0 in a set of dimension ≥ 1 in which case $\alpha(M_N^0,\rho)=0$, $\alpha(M_N^0,\rho,\varepsilon)=0$ $(N\parallel N')$ or in exactly one point, which may or may not lie in M_N^0, whereas there are open sets in A_N covered by the projections of at least two different M^i $(i>0)$. Therefore

$$\alpha(M_N^0,\rho)<\sum_{i>0}\alpha(M_N^i,\rho),\qquad \alpha(M_N^0,\rho,\varepsilon)<\sum_{i>0}\alpha(M_N^i,\rho,\varepsilon)\quad\text{for}\quad N\in Y$$

which gives together with (15) by integration

$$\alpha(M^0,\rho)<\sum_{i>0}\alpha(M,\rho).\qquad \alpha(M^0,\rho,\varepsilon)<\sum_{i>0}\alpha(M,\rho,\varepsilon).$$

Thus we have proved

(16) *The functions $\alpha(M, \rho)$ and $\alpha(M, \rho, \varepsilon)$, $\rho > 0$, $\varepsilon > 0$ define a-areas in D for which the a-flats strictly minimize area,*

The preservation of completeness under the integration is obvious only at the points of B_∞. Consider an a-wedge S in an a-flat A intersecting D with $v \in B_\infty$. For any N which intersects A in a point S_N is an a-wedge in D_N whose vertex corresponding to v lies on S_∞. We conclude from (2) that $\alpha(S_N. \rho) = \infty$ for $\rho > 0$. Therefore

(17) $\alpha(S, \rho) = \infty$ *for* $\rho > 0$ *and* $v \in B_\infty$, S *an a-wedge.*

If the neighborhood $U(S', \delta)$ of S' (in the n-dimensional euclidean space) lies in D, then the neighborhood $U_N(S'_N, \delta)$ of S'_N in A_N lies in D_N. It follows from (10) that the functions $f_\rho(x, N)$ are bounded uniformly for $x \in S_N$ and all N, and conclude from the arguments proving (11) that $\alpha(S_N, \rho, \varepsilon)$ stays for a fixed $\varepsilon > 0$ below a bound independent of N.

(18) $\alpha(S, \rho, \varepsilon)$ *is finite for* $\rho > 0$, $\varepsilon > 0$, $v \in B_\infty$. S *an a-wedge.*

Consider a convex set K in $A \cap D$ for which $B(K)$ contains a non-empty open subset of $B(A)$. If N intersects A in a point then K_N contains a non-empty open subset of $B(A_N)$, therefore $\alpha(K_N, \rho, \varepsilon) = \infty$ for $\varepsilon \leq a -$ (see (12)).

(19) $a(K, \rho. \varepsilon) = \infty$ *if* $\varepsilon \leq a - 1$, $\rho > 0$, *and K is a convex set in A for which B(K) contains a non-empty open set of B(A).*

If $D = A_N$, then $f_\rho(x, N) \equiv 1$. If x_N is the projection of x on A_N parallel to N, our area has the form

(20) $$\alpha(M, \varepsilon) = \int dN \int_{M_N} (1 + px_N)^{-1-\varepsilon} d V_N$$

and we find

(21) *For* $0 < \varepsilon \leq a - 1$ *the function* $\alpha(M, \varepsilon)$ *defines in* W^n *a hyperbolically complete a-area,* $(a \geq 2)$, *which is invariant under all motions for the euclidean metric which leave p fixed.*

The surprising aspect of this fact is the arbitrariness of area, because we used only the order of magnitude of the function. If $h(x)$ is a measurable function satisfying

$$k'(1+px)^{-a} \leq h(x) \leq k''(1+px)^{-\lambda}, \qquad 1 < \lambda < a, \qquad k', k'' > 0,$$

then (20) with $(1+px)^{-1-\varepsilon}$ replaced by $h(x)$ also provides a hyperbolically complete a-area in W^n. When $h(x)$ has the form $h(px)$ then we have again rotation invariance. This shows as in the case $a=1$ that *the problem of determining all a-areas for which the a-flats strictly minimize area is too general to be interesting.*

In the case $a>1$ there are however some general questions of great interest which do not arise for $a=1$, for example: Let a length be defined in D for which the lines (strictly) minimize length. Is it possible (or under which condition on the length or D is it possible) to define an a-area α in D $(1 < a \leq n)$ for which the a-flats (strictly) minimize area and *α is intrinsic in each a-flat A*. This means that $\alpha(M)$ for $M \subset A$ is determined by the distances, i. e. lengths of segments in A.

For any reasonable length the a-dimensional HAUSDORFF measure in D (see [2, p. 86]) is intrinsic and it will solve a problem for $a=n$. But the question whether the a-flats (strictly) minimize area for HAUSDORFF measure leads to extremely diffilcult problems on convex bodies if $a<n$, which are unsolved even for MINKOWSKI length in $D=W^n$ when $1<a<n-1$.

5. Behaviour at B_f.

The points of B_f require more delicate estimates because *$v \in B_f$ is in general not projected on a point of $B(D_N)$*. We assume first that B_f does not contain an entire line.

If $v \in B_f$ and H is a supporting plane of B_f at v denote by H^+ the half-space bounded by H containing D. There is a half $(n-1)$-flat C^+ of H bounded by an $(n-2)$-flat H' in H through v and a number $\beta>0$ such that all points of $C^+ \cap B_f$ have distance less than β from v.

Let Y be the set of those $(n-a)$ flats N through p for which the $(n-a)$-flat N_v parallel to N through v has the property that $N_v \cap H^+$ falls in the halfspace bounded by the hyperplane through H' and p. For $N \in Y$ the projection w_N of a point w of B_f sufficiently close to v parallel to N on A_N satisfies

$$(22) \qquad w_N B(D_N) \leq w_N B(H \cap A_N) \leq \beta \mid \sin(N, H) \mid = \delta_N, \quad N \in Y.$$

Let S be an a-wedge in D with vertex v in the a-flat A, and $\mathfrak{A}_1$, $\mathfrak{N}_1$ unit a- and $(n-a)$-vectors respectively parallel to A and N. Then the estimate leading to (3) yields

$$\alpha(S_N, \rho) \geq k_{12} \mid \mathfrak{P}_1 \wedge \mathfrak{A}_1 \mid \int_0^{t_N} t^{a-1}(\delta_N + t)^{-\rho} dt, \tag{23}$$

where $t_N \geq t_0 > 0$ in any subset Y_1 of Y for which $\mid \mathfrak{P}_1 \wedge \mathfrak{A}_1 \mid \geq c > 0$. We choose c so small that Y_1 contains a non-empty open subset of all N. Then on the subset Y_2 of Y_1 where $\delta_N \leq t_0$,

$$\alpha(S_N, \rho) \geq k_{12} \int_0^{t_0} t^{a-1}(\delta_N + t)^{-\rho} dt \geq 2^{-\rho} k_{12} \int_0^{t_0} t^{a-1} \max{}^{-\rho}(\delta_N, t) dt \geq$$

$$\geq 2^{-\rho} k_{12} \int_0^{\delta_N} t^{a-1} \delta_N^{-\rho} dt = 2^{-\rho} k_{13} \delta_N^{a-\rho}$$

hence by (22)

$$\alpha(S, \rho) \geq \beta 2^{-\rho} k_{13} \int_{Y_2} \mid \sin (N, H) \mid^{a-\rho} dN$$

and this integral is infinite when ρ is sufficiently large, actually $\rho \geq a + 1$ suffices. Thus in conjunction with (17):

(24) *If B_f does not contain an entire line, then $\alpha(M, \rho)$ defines in D for $\rho \geq a + 1$ a strongly complete a-area.*

It will be seen presently that this result can be improved to $\rho \geq a$ if B_f is strictly convex.

The case where $B_x = 0$ is of special interest in view of hyperbolic geometry. In that case we may replace $f_\rho(x)$ by $(xB)^{-\rho}$. If B is the sphere with radius R about p then $xB = R - px$ and we find that

$$\alpha(M, \rho) = \int dN \int_{M_N} (R - xp)^{-\rho} d V_N$$

gives for $\rho \geq a$ a strongly complete a-area in $px < R$ which is invariant under all motions relative to the euclidean metric which leave p fixed. We observe again that only the order of magnitude of $(R - px)^{-\rho}$ enters and therefore many other functions can serve as integrands.

If in the preceding estimates the terms δ_N and $|\mathfrak{P}_1 \wedge \mathfrak{A}_1|$ are replaced by 0 and 1 respectively an upper bound for $\alpha(S_N, \rho)$ is obtained:

(25) $$\alpha(S_N, \rho) \leq k_{14} \int_0^{t_0} t^{a-1} t^{\rho} dt, \quad t_0 > 0,$$

so that $\alpha(S, \rho)$ is finite for $\rho < a$. For the further discussion of hyperbolically complete areas we restrict ourselves to *strictly convex (or empty)* B_f. For points w close to $v \in B_f$ we have

(26) $$w_N B(D_N) \leq k_{16} w H.$$

This estimate follows form the observation: if $y = h(x)$, $|x| < n$, is a strictly convex curve in the xy-plane with $h(0) = 0$ and $h(x) \geq 0$ and m is the slope of a supporting line of $y = h(x)$ at $(x_0, y_0 = h(x_0)$, $x_0 \neq 0$ then sign m = sign x_0 and the supporting line intersects the y-axis at $\bar{y} = y_0 - m x_0$, so that $|\bar{y}| < y_0$.

If a convex set K in A is given for which $B(K)$ contains an open subset of $B_f \cap B(A \cap D)$ choose v on $B(A \cap D)$ such that $B(A \cap D)$ satisfies at v the inequality (6). We then conclude from (7) and (26) because $wH \leq k_{16} t$ that with a suitable $t_0 > 0$

$$\alpha(K, \rho) \geq k_{17} \int_0^{t_0} t^{(a-1)/2}(t + w_N B(D_N))^{-\rho} dt \geq k_{18} \int_0^{t_0} t^{(a-1)/2} t^{-\rho} dt.$$

Therefore $\alpha(K, \rho) = \infty$ for $\rho \geq (a+1)/2$ and we find

THEOREM I. - *For* $(a+1)/2 \leq \rho < a$ *and* $\varepsilon \leq a - 1$ *the function* $\alpha(M, \rho, \varepsilon)$ *defines for a strictly convex or empty* B_f *a hyperbolically complete* a*-dimensional area for which the* a*-flats strictly minimize area.*

If B_f contains a line, then it contains a flat F^π of maximal dimension $d \geq 1$. Let F be the d-flat parallel to F^π through $p \in D$, and D' the intersection of D with an $(n-d)$-flat B' orthogonal to F through p. Then $B(D')$ does not contain a line.

We distinguish *two cases*: $d \leq n - a$ or dim $D' \geq a$ and $d > n - a$ or dim $D' < a$. In the first case we introduce in D' a strongly complete a-dimensional area α' using (4) or (24) and for any set M lying in the intersection of D with an a-flat A we denote by M' the projection of M parallel to F on B'. Then $M' \subset D'$ and we put $\alpha^1(M) = \alpha'(M')$. For any oriented closed

a-dimensional polyhedron with a-faces $M_0, \dots, M_r$ and $M_0 \neq \bigcup_{i>0} M_i$ we have

$$\alpha^1(M_0) \leq \sum_{i>0} \alpha^1(M_i).$$

although α^1 is not an area because $\alpha^1(M) = 0$ when $M \subset A$ and $A \cap F$ is not a point. Let $\alpha^2(M)$ be a strongly complete a-area in W^n, for which the a-flats strictly minimize area, then

$$\alpha^3(M) = \alpha^1(M) + \alpha^2(M)$$

is an a-area in D for which the a-flats strictly minimize area. It is complete only for $a = 1$. For if $a > 1$ an a-wedge S has $\alpha^3(S) = \infty$ if and only if either $v \in B_\infty$ or the a-flat $A_S \supset S$ intersects F in a point. Any convex set $K \subset D$ lying in an a-flat which intersects F in more than one point and with $B(K) \cap B_\infty = 0$ has finite area $\alpha^3(K)$. This phenomenon is interesting in itself.

To remedy the situation we denote by N_u an $(n - a)$-flat through p orthogonal to an a-flat containing an $(a - 1)$-flat in B' through p and a line in F through p, and by dN_u the euclidean density of these N_u. Defining $\alpha(M_{N_u}, \rho)$ by (13) we put

$$\alpha^4(M, \rho) = \int \alpha(M_{N_u}, \rho) dN_u .$$

Whereas $w_N B(D_N)$ in (22) is infinite for general N it will be finite for the N_u and we can apply the previous estimates yielding $\alpha^4(S, \rho) = \infty$ for $\rho \geq a + 1$ for all a-wedges S with $v \in B_f$ not parallel to B'. Therefore $\alpha^3(M) + \alpha^4(M, \rho)$ is a strongly complete area in D.

If $d > n - a$ we proceed differently. Take a $(d - n + a)$-flat C_j in F, denote by D_j' the intersection of D with the a-flat A_j formed by the $(d - n + a)$-flats parallel to C_j through points of D'. In D_j' define a strongly complete a-area α_j^1, and for any $M \subset D \cap A$ put as before $\alpha_j'(M) = \alpha_j^1(M_j')$ where M_j' is the projection of M on A_j parallel to the $(n - a)$-flat N_{A_j} orthogonal to A_j.

Then $\alpha^2(M) + \alpha_j'(M)$ is an area in D for which the a-flats strictly minimize area, but $\alpha^2(S) + \alpha_j'(S)$ is finite for a-wedges S with $v \in B_f$ whose a-flat A_S intersects N_{A_j} in more than one point. By choosing a finite number of suitable C_j, $j = 1, \dots, k$, we reach that A_S intersects at least one N_{A_j} in a point. Then

$$\alpha_1'(M) + \dots + \alpha_k^1(M) + \alpha^2(M)$$

defines in D a strongly complete a-area for which the a-flats strictly minimize area.

THEOREM II. – *Any non-empty open convex set D in W^n can for each a $(1 \leq a \leq n)$ be provided with a strongly complete a-area, for which the a-flats strictly minimize area.*

We note in passing that choosing $\alpha^2(M)$ as hyperbolically complete we obtain an a-area which is complete but neither hyperbolically nor strongly so.

6. Extension of given areas.

We now prove the extension theorem which was loosely formulated in the introduction.

THEOREM III. – *In the non-empty open convex set D in W^n let an a-area α $(1 \leq a \leq n)$ be given for which the a-flats (strictly) minimize area. Then an a-area α^+ exists which is defined in a set $D^+ \subset W^{n+1}$ with $W^n \subset W^{n+1}$ and $D^+ \cap W^n = D$ such that the a-flats (strictly) minimize area and $\alpha^+(M) = \alpha(M)$ for M lying in the intersection of an a-flat in W^n with D.*

Moreover, if α is (hyperbolically) complete, then so is α^+.

If α is only finitely additive then α^+ will be finitely additive, but if $a > 1$ then α^+ will *not* be strongly complete, even when α is, except for $D = W^n$. The interesting question, whether a strongly complete α^+ exists, remains open.

On the other hand, by using n-simplices instead of $(a+1)$-simplices which would have sufficed, the proof of III provides a D^+ which is independent of a and, consequently, the following important additional fact.

THEOREM III'. *If a-areas α_a are given in D for $a = 1,\ldots,n$ then $D^+ \subset W^{n+1}$ with $D^+ \cap W^n = D$ and areas $\alpha_1^+,\ldots,\alpha_n^+$ in D^+ exist such that the assertion of Theorem III' holds for each pair α_a, α_a^+, except for $a = n$ and $D \neq W^n$, where α_n^+ minimizes area but not strictly so.*

We first consider W^n as a vector space. If $\mathfrak{v} \in W^n$ and ρ a real number then the pair $(\mathfrak{v}, \rho)$ forms in a natural way an $(n+1)$-dimensional vector space. Let $\mathfrak{u}_1,\ldots,\mathfrak{u}_n$ be independent vectors in W^n. Then $(\mathfrak{v}, \rho)$ is isomorphic to the set $W = \{(\mathfrak{v}, \rho\mathfrak{u}_1,\ldots,\rho\mathfrak{u}_n)\}$ under

$$a(\mathfrak{v}, \rho) + a'(\mathfrak{v}', \rho') \to (a\mathfrak{v} + a'\mathfrak{v}', (a\rho + a'\rho')\mathfrak{u}_1,\ldots,(a\rho + a'\rho')\mathfrak{u}_n)$$

We think of W^{n+1} as the affine space obtained from W. The points of W^{n+1} may be interpreted as (possibly degenerate) simplices in W^n with vertices $\mathfrak{v}, \mathfrak{v} + \rho\mathfrak{u}_1, \ldots, \mathfrak{v} + \rho\mathfrak{u}_n$. The vertex $\mathfrak{v}$ is not actually distinguished; putting $\mathfrak{w} = \mathfrak{v} + \rho\mathfrak{u}_1$ the vertices become $\mathfrak{w} - \rho\mathfrak{u}_1, \mathfrak{w}, \mathfrak{w} + \rho(\mathfrak{u}_2 - \mathfrak{u}_1)\ldots, \mathfrak{w} + \rho(\mathfrak{u}_n - \mathfrak{u}_1)$. When we speak of simplices in this proof we always mean these simplices. Any two of them are homothetic with change of orientation if the corresponding ρ differ by signs. The degenerate simplices are those for which $\rho = 0$ and they may be identified with the points of W^n.

For any two distinct $(\mathfrak{v}, \rho)$, $(\mathfrak{v}', \rho')$ the $n+1$ lines through corresponding vertices $\mathfrak{v}$, $\mathfrak{v}'$; $\mathfrak{v} + \rho\mathfrak{u}_k$, $\mathfrak{v}' + \rho'\mathfrak{u}_k$, $(k \geq 1)$ are concurrent or parallel with an obvious agreement when two vertices coincide. Thus a line in W^{n+1} corresponds to a family of simplices in W^n which are perspective from a point, where we now use the completion of W^n to P^n.

We may formulate this fact differently: if two corresponding vertices of simplices on the line coincide then all vertices corresponding to these vertices coincide. Thus corresponding vertices determine a flat of lowest dimension (0 or 1 in this case) which contains all the vertices, but at least one is a line. The lines obtained in this way are not necessarily distinct. If there is only one, the center of perspectivity is undetermined.

Similarly, an a-flat A in W^{n+1} is determined by $a+1$ independent points $(\mathfrak{v}^0, \rho^0), \ldots, (\mathfrak{v}^a, \rho^a)$. Corresponding vertices $\mathfrak{v}^0, \ldots, \mathfrak{v}^a$ and $\mathfrak{v}^0 + \rho^a\mathfrak{u}_k, \ldots, \mathfrak{v}^a + \rho^a\mathfrak{u}_k$ $(k \geq 1)$ determine flats A_0, A_k respectively of minimal dimension, and each A_j contains all corresponding vertices of the simplices in A. At least one A_j is an a-flat. With a proper agreement any two simplices in A are perspective from a point in an $(a-1)$-flat in P^n, so that those A_j which have dimension a pass through this $(a-1)$-flat, which may be undetermined.

We now define D^+ as the set of those simplices all whose vertices lie in D. Then $D^+ \cap W^n = D$. By our discussion of lines in W^{n+1} the set D^+ is convex and is trivially non-empty and open in W^{n+1}. An a-flat A in W^{n+1} determines flats of lowest dimension A_0, A_1, .., A_n as above, and a Borel set $M \subset D^+ \cap A$ determines Borel sets $M_j \subset A_j$, $j = 0, \ldots, n$.
We put.

$$\alpha^+(M) = (n+1)^{-1} \sum_{j=0}^{n} \alpha(M_j).$$

Then obvioulsly $\alpha^+(M) = \alpha(M)$ for $M \subset W^n$. If M is open and non-empty, then at least one M_j is open and non-empty in A_j, hence $\alpha^+(M) > 0$. When M is compact then so are all M_j and $\alpha^+(M) < \infty$, if M lies in an $(a-1)$-flat then all M_j lie in $(a-1)$ – flats and $\alpha^+(M) = 0$. Finally $\alpha^+(M)$ is countably additive. *Thus α^+ is an a-area in D^+.*

To an a-wedge S with vertex v in A there correspond simplices $S_0, \ldots, S_{n+1}$ in D with vertices $v_0, \ldots, v_n$ corresponding to v. At least one of the v_j lies on $B(D)$, but unfortunately the corresponding S_j may be degenerate (this is easily verified with examples) and we cannot deduce from the strong completeness of α that $\alpha^+(S) = \infty$ except for $D = W^n$, where all v_j lie on $B(D)$. However we can deduce from $\alpha(S_j) < \infty$ that $\alpha^+(S) < \infty$.

A convex set K in A for which $B(K)$ contains a non-empty open subset of $B(A)$ produces convex sets $K_j \subset A_j$, and at least one K_j contains an open

Ann. Mat. Pura Appl., IV. Ser. 55 (1961), 171–189

subset of $B(D)$ and hence is a-dimensional. For if all A_i intersected $B(D)$ in an at most $(a-2)$-dimensional set, then A could not have dimension a. Thus at least one $\alpha(K_i) = \infty$ and $\alpha^+(K) = \infty$.

To finish the proof of Theorems III and III' we must show that α^+ (strictly) minimizes area if α does.

Let P be an a-dimensional oriented polyhedron in D^+ with a-faces $M^0, \dots, M^r$ and $M^0 \neq \bigcup_{i>0} M^i$. The sets M^i_j $(i = 0, \dots, r;\ j = 0, \dots, n)$ corresponding to the M^i satisfy

$$(27) \qquad \alpha(M^0_j) \leq \sum_{i<0} \alpha(M^i_j),\ j = 0, \dots, n.$$

Therefore

$$(28) \qquad \alpha^+(M^0) \leq \sum_{i>0} \alpha^+(M^i),$$

so that *the a-flats minimize area for α^+.*

If not all M^i lie in the same a-flat as M^0, then for $n > a$, a j exists such that $M^0_j, \dots, M^r_j$ do not lie in an a-flat of D and we have inequality in (27) for this j, if the a-flats strictly minimize area in D (although some M^i_j may be degenerate), so that α^+ satisfies (28) with the inequality for $n > a$.

To prove III for $a = n$ (and III' for $a = n$ and $D = W^n$) let F be a topological mapping of D on W^n and define for a Borel set $M \subset W^n$ a volume $\beta(M)$ by $\beta(M) = \alpha(F^{-1}M)$. For $D = W^n$ we let F be the identity. F will not map a-flats on a-flats for $a < n$ unless $D = W^n$ and can therefore not be used to establish III'. Using $\beta(M)$ we define $\beta^+(M)$ as before where D^+ is now all of W^{n+1} which we provide with a euclidean metric. An n-dimensional polyhedron for which all M^i lie in the same hyperplane A^0 as M^0 the inequality holds trivially in (28). We therefore assume that this is not the case. Also, the inequality holds in (28) when A^0 is parallel to W^n.

Take a line L in W^{n+1} not in W^n, through a point $p \in W^n$ let M_L be the projection of M on W^n parallel to L and put $\beta_L(M) = \beta(M_L)$. Again $\beta_L(M)$ is not an n-area but satisfies (28) and $\beta_L(M) = \beta(M)$ for $M \subset W^n$. For a given polyhedron we can find L such that $\beta_L(M^0) < \sum_{i>0} \beta_L(M^i)$. An integration over L would in general lead to infinite values, but the following procedure clearly leads to a usable area. Let R_n be the set of L whose angle with W^n is greater than n^{-1}, use the euclidean area element on the unitsphere about p to measure dL. If U_n is the open sphere with radius n about p, we put for M in a hyperplane W^{n+1}

$$\beta_n(M) = c_n \int_{R_n} \beta_L(M)\,dL$$

where c_n is chosen such that $\beta_L(M) \leq 2^{-n}$ for $M \subset U_n$. Then

$$(\beta^+(M) + \sum_{n=1}^{\infty} \beta_n(M))/(1 + \sum_{n=1}^{\infty} c_n \int_{R_n} dL)$$

will solve our problem.

7. Locally isometric maps.

A G-space is, except for differentiability hypotheses, a complete FINSLER space with symmetric distances, see [1] to which we will freely refer. The spherical neighborhood $px < \delta$ in a G-space R with distance xy will be denoted by $S(p, \delta)$. We assume that a *finitely additive volume* α is defined in R for which the $S(p, \delta)$ are measurable and that a point q with the following properties exists:

(a) $\alpha(S(q, \delta)) > 0$ *for* $\delta > 0$.

(b) *There are positive numbers* $t(q)$ *and* $\delta(q)$ such that

$$t(q)\alpha(S(q, \delta)) \leq \alpha(S(p, \delta))$$

if $\delta < \delta_0(q)$ *and* $S(p, \delta)$ *is isometric*[4]) *to* $S(q, \delta)$.

(c) $\alpha(S(q, v\beta)) = o(k+1)^v$ *for each* $\beta > 0$, k *a positive integer* (i. e. $\lim_{v\to\infty} \alpha(S(q, v\beta))(k+1)^{-v} = o$ *for fixed* β).

A locally isometric map J of R on itself is defined by the property that for a given $p \in R$ an $\varepsilon_p > 0$ exists such that J maps $S(p, \varepsilon_p)$ isometrically on $S(Jp, \varepsilon_p)$. It is shown in [1, p. 171] that a positive continuous function $\varepsilon(p)(= \rho(p)/2$ in the notation of [1]) exists in R such that J maps $S(p, \varepsilon(Jp))$ isometrically on $S(Jp, \varepsilon(Jp))$ and $pp_1 > 2\, \varepsilon(Jp)$ when $p \neq p_1$ and $Jp = Jp_1$.

If R is compact then a local isometry of R on itself is one-to-one, and hence is a global isometry, [1, p. 172]. The assertion becomes false when we omit the hypothesis that R is compact. The ordinary cylinder with a suitable hyperbolic metric possesses for each $k > 0$ locally isometric k to 1 maps on itself [1, p. 173]. We show here that under appropriate hypotheses on the order of magnitude of volumes of spheres a locally isometric map of a non-compact space on itself is $1 - 1$.

(4) This means that a distance preserving mapping of $S(p, \delta)$ on $S(q, \delta)$ exists.

THEOREM IV. *A locally isometric map J of a G-space R on itself is at most k to 1 (in particular, is $1 - 1$ for $k = 1$) if R possesses a volume satisfying (a), (b) and (c) for some point q.*

Assume for an indirect proof that $k+1$ distinct points are mapped on the same point. Then, see [1, p. 172], distinct point $q_1, \dots, q_{k+1}$ exist which by J are all mapped on q. We denote the set consisting of the q_j by Q_1 and have $JQ_1 = q$. We connect q to q by a segment T_j. Then

$$\max (\text{length } T_j) = \max qq_j = \beta > 0$$

because at most one q_j can coincide whith q. Put $\delta = \min (\delta(q), \varepsilon(q))$.

The sets $J^{-1}q_j$, $j = 1, \dots, k+1$ are disjoint and each contains at least $k+1$ distinct points, [1, p. 172].

Among the points in $J^{-1}q_j$ are those obtained from q_j by traversing the segment which lies over T_h $(h = 1, \dots, k+1)$ and begins at q_j, [1, p. 169]. None of these points has distance greater than β from q_j. The set of the $(k+1)^2$ points obtained in this way for $j = 1, \dots, k+1$ is denoted by Q_2. Then $J^2 Q_2 = q$, the spheres with radius δ about the points in Q_2 are disjoint and each is mapped isometrically on $S(q, \delta)$ by J^2.

Continuing in this manner we obtain a set Q_v consisting of $(k+1)^v$ points with $J^v Q_v = q$ where the spheres with radius δ about these points are disjoint and are isometric under J^v to $S(q, \delta)$. Moreover, for given point r_v in Q_v there is a point $r_{v-1} \in Q_{v-1}$ with $r_v r_{v-1} \leq \beta$, therefore

$$r_v q \leq r_v r_{v-1} + \dots + r_2 r_1 + r_1 q \leq v\beta$$

and

$$S(r_v, \delta) \subset S(q, v\beta + \delta).$$

Since the $S(r_v, \delta)$ are disjoint

$$\sum_{p \in Q_v} \alpha(S(p, \delta)) \leq \alpha(S(q, v\beta + \delta) \tag{29}$$

and by (a) and (b)

$$0 < t = t(q)\alpha(S(q, \delta)) \leq \alpha(S(p, \delta))$$

hence for $v \geq \delta/\beta$ by (29)

$$\alpha(S(q, v, 2\beta)) \geq \alpha(S(q, v\beta + \delta)) \geq (k+1)^v t$$

which contradicts $\alpha(S(q, \beta)) = o(k+1)^v$.

If R_e is an n-dimensional locally euclidean space, then it follows from the definition of $\varepsilon(p)$ that with the ordinary volume α_e in R_e we have for any point q

$$\alpha_e(S(q, \delta)) = \pi_n \delta^n \text{ for } \delta \leq \varepsilon(q), \text{ where } \pi_n = \pi^{n/2}/\Gamma(1 + n/2),$$

and that $\alpha_e(S(p, \delta)) = \alpha_e(S(q, \delta))$ when $S(p, \delta)$ isometric to $S(q, \delta)$. Therefore (b) is satisfied with $t(q) = 1$ and $\delta(q) = \varepsilon(q)$. Moreover, for any β

$$\alpha_e(S(q, v\beta)) \leq \pi_n(v\beta)^n = 0(2^v);$$

therefore

(30) *A local isometry of a locally euclidean space on itself is global.*

This result can be generalized. Assume the G-space R can be mapped on R_e with distance $e(x,y)$ by a map D satisfying

$$xy \leq se(Dx, Dy), \; s \text{ constant}.$$

(This implies that D is one-to-one.)

The n-dimensional HAUSDORFF measure (see [2, p. 86]) α of any BOREL set in R satisfies

$$\alpha(M) \leq s^n \alpha_e(DM),$$

in particular, because $DS(q, \delta) \subset S(Dq, s\delta)$,

$$\alpha(S(q, \delta)) \leq s^{2n}\pi_n\delta^n \text{ for any point } q,$$

so that α satisfies (c) with $k = 1$. The properties of HAUSDORFF measure give $\alpha(S(p, \delta)) = \alpha(S(q, \delta))$ for isometric $S(p, \delta)$ and $S(q, \delta)$ so that (b) holds. Thus we find.

COROLLARY. - *If the G-space R with distance xy can be mapped on the n-dimensional locally euclidean space R_e by a map D satisfying $xy \leq se(Dx, Dy)$ (s constant) and if for some point q the n-dimensional Hausdorff measures of the spheres $S(q, \delta)$ are positive, then a local isometry of R on itself is global.*

NOTE. - The condition $\alpha(S(q, \delta)) > 0$ is satisfied if for some positive δ_0, s_0 a relation $xy \geq s_0 e(Dx, Dy)$ holds for x, y in $S(q, \delta_0)$.

Because G-spaces are essentially FINSLER spaces it would seem more natural if a locally Minkowskian space R_m with distance $m(x, y)$ appeared in the Corollary instead of R_e. But R_m can be mapped on a suitable R_e by a mapping H satisfying

$$k^{-1}e(Hx, Hy) \leq m(x, y) \leq ke(Hx, Hy)$$

where $k > 1$ is constant, see [1, p. 192]. Therefore *the Corollary implies the analogous assertion with R_e replaced by R_m.*

REFERENCES

[1] H. BUSEMANN, *The Geometry of Geodesics*, New York 1955.
[2] H. BUSEMANN, *Convex Surfaces*, New York 1958.
[3] D. HILBERT, *Mathematische Probleme*, « Nachr. Akad. Wiss. Göttingen, Math. Phys. », 1900, pp. 253-297.

CONVEXITY ON GRASSMANN MANIFOLDS[1])

by Herbert Busemann

The definition of a convex function presupposes that the range of the independent variable is a convex subset of a linear space. However, the question of what convexity means on ranges of different types, in particular on Grassmann manifolds, arises naturally, for example in the search for an analogue to the Legendre Condition for a-dimensional variational problems in an n-dimensional space when $1 < a < n-1$. Thus the first new concept of convexity—total convexity in our language—occurs implicitly in the work of Carathéodory on such problems. A general and detailed discussion of convexity from the point of view of the calculus of variation was given by Wagner [11].

After the concept of convexity on Grassmann manifolds has been analyzed, it appears that *the theory of convex bodies is ready for a significant extension*, in that many theorems on convex bodies are probably special cases of more general theorems dealing with these new concepts of convexity.

Before entering a detailed discussion a *typical problem* concerning the meaning of convexity will be useful. Consider a convex body K with interior points in the n-dimensional euclidean space E^n. For a given simple a-vector $(1 \leqq a \leqq n-1)$ $\mathfrak{A} \neq 0$, let $P_1(\mathfrak{A})$ be the a-dimensional area of the projection of K on an a-flat parallel to $\mathfrak{A}$ and define

$$P(\mathfrak{A}) = |\mathfrak{A}| P_1(\mathfrak{A}), \quad P(O) = 0,$$

where $|\mathfrak{A}|$ is the absolute value (also called norm or area) of $\mathfrak{A}$. Clearly

$$P(k\mathfrak{A}) = |k| P(\mathfrak{A}) \quad \text{for all real } k.$$

For $a = 1, n-1$ the a-vectors form linear spaces; in both cases it was proved by Minkowski that $P(\mathfrak{A})$ is a convex function (compare [5, pp. 45, 51]).

---

1) Talk delivered at the Zurich Colloquium on Differential Geometry and Topology, June 1960.

For $1 < a < n - 1$ it has *a priori* no meaning to say that $P(\mathfrak{A})$ is convex, because the simple a-vectors form a Grassmann cone and not a linear manifold. We wish to find out whether we can *attribute a reasonable meaning to this statement.*

We will find that *there are three, possibly four, significant, increasingly stronger concepts of convexity*, the weakest is convexity in any linear subspace lying entirely on the Grassmann cone and follows for $P(\mathfrak{A})$ at once from Minkowski's result, the strongest is the total convexity found by Carathéodory.

First some *notations.* We use the same symbol for a vector space and the corresponding affine space. V^n is the n-dimensional affine space or vector space over the reals, V_a^n the space of a-vectors $1 \leqq a \leqq n-1$, Z_a the origin of V_a^n and G_a^n the Grassmann cone in V_a^n consisting of all simple a-vectors, so that $V^n = V_1^n = G_1^n$ and $V_{n-1}^n = G_{n-1}^n$. We will encounter explicitly only simple vectors so that symbols like $\mathfrak{A}_j$ and $\sum_{j=1}^{k} \mathfrak{A}_j$ *always mean simple vectors.*

Let V^n be realized as the set of n-tuples $x^1, \ldots, x^n$ of real numbers. A *euclidean metrization* of V^n has the form

$$[\Sigma g_{ik}(x^i - y^i)(x^k - y^k)]^{\frac{1}{2}},$$

where $\Sigma g_{ik} x^i x^k$ is, of course, positive definite. $[\Sigma (x^i - y^i)^2]^{\frac{1}{2}}$ is the *trivial* euclidean metric. Each euclidean metrization of V^n induces a euclidean metric in V_a^n, the trivial metric in V_a^n is induced by that in V^n.

A hyperplane in V_a^n considered as affine space is *simple* or a *Grassmann plane*, if the coefficients of its equation (except for the constant term) satisfy the Plücker identities or form a simple a-covector. In terms of the trivial metric this means that the *perpendicular through* Z_a *to the hyperplane falls on* G_a^n. A function defined in a subset of V_a^n is continuous if it is continuous with respect to one, and then all, euclidean metrics induced in V_a^n.

The sources for the following discussion are the already mentioned paper [11] of Wagner, [6], [7] and principally [8].

Consider a continuous function $f(\mathfrak{A})$ defined on G_a^n. For simplicity, and because this is the most important case, we assume that $f(\mathfrak{A})$ is positive and positive homogeneous of degree 1, i.e.,

$$f(\mathfrak{A})>0 \textit{ for } \mathfrak{A} \neq 0 \textit{ and } f(k\mathfrak{A}) = kf(\mathfrak{A}) \textit{ for } k \geqq 0 . \quad (1)$$

The weakest form of convexity was already mentioned: $f(\mathfrak{A})$ is *convex* if it is convex in each linear subspace lying entirely in G_a^n. This is equivalent to

$$f(\mathfrak{A}_1 + \mathfrak{A}_2) \leqq f(\mathfrak{A}_1) + f(\mathfrak{A}_2) , \quad (2)$$

because the simplicity of $\mathfrak{A}_1$, $\mathfrak{A}_2$, $\mathfrak{A}_1 + \mathfrak{A}_2$ implies that the a-flats parallel to $\mathfrak{A}_i$ through Z_a intersect in an $(a-1)$-flat. For $a = 1, n-1$ the relation (2) yields

$$f(\sum_{j=1}^{k} \mathfrak{A}_j) \leqq \sum_{j=1}^{k} f(\mathfrak{A}_j) \quad \text{for all} \quad k \geqq 2 , \quad (3)$$

but (2) does *not* imply (3) for $1 < a < n-1$. *If* f $(\mathfrak{A})$ *satisfies* (3) *we call it extendably convex*, because (3) is clearly necessary and sufficient for $f(\mathfrak{A})$ to be extendable to a convex function defined in all of V_a^n.

This condition may be rephrased. In V_a^n we consider the indicatrix I_f of $f(\mathfrak{A})$ which is the locus on G_a^n defined by $f(\mathfrak{A}) = 1$. The function $f(\mathfrak{A})$ is extendably convex if and only if I_f lies on the boundary of its own convex closure in V_a^n, hence if and only if I_f possesses at each point a supporting hyperplane in V_a^n.

Such a hyperplane will in general not be simple. *If* I_f *possesses at each point a simple supporting hyperplane, we call* f $(\mathfrak{A})$ *totally convex.*

These concepts would be of very little interest if they did not have *interesting geometric meanings in* V^n. It is wellknown that Minkowski interpreted $f(\mathfrak{A})$ for $a = 1$ as a length, less known that he also studied $f(\mathfrak{A})$ for $a = n-1$ as an area (see [5, p. 64]). We extend his ideas to general a.

Let $\rho_a(M)$ be the a-dimensional Lebesgue measure induced by a euclidean metrization of V^n of a set M in an a-flat. For a given a-vector $\mathfrak{A} \neq 0$ we consider the oriented a-flats A parallel to $\mathfrak{A}$ and in each A the open sets, or closures of open sets, M oriented as A and satisfying $\rho_a(M) = |\mathfrak{A}|$[1]. This condition

[1] For $0 < \rho_a(M) < \infty$ it is always assumed that M is open or the closure of an open set.

is independent of the choice of the euclidean metric. The totality of all sets M obtainable from $\mathfrak{A}$ in this way is denoted by $[M]$ and may be identified with $\mathfrak{A}$ because $\mathfrak{A}$ and $[M]$ determine each other uniquely. $[M] = \mathfrak{A}$ entails $[-M] = -\mathfrak{A}$. With the vanishing a-vector we identify the totality of all Borel sets M with $\rho_a(M) = 0$ and lying in a-flats.

A function $f(\mathfrak{A})$ (satisfying (1)) defines an affine a-area through

$$\alpha(M) = f(\mathfrak{A}) \quad \text{if} \quad [M] = \mathfrak{A} .$$

In general $\alpha(M) \neq \alpha(-M)$ because we do not assume $f(\mathfrak{A}) = f(-\mathfrak{A})$. If $f(\mathfrak{A}) \equiv f(-\mathfrak{A})$ we call f and the area α *symmetric.*

Consider an oriented closed a-dimensional polyhedron P in V^n with oriented a-faces $F_0, \ldots, F_k$. *If*

$$\alpha(-F_0) \leqq \sum_{1=j}^{k} \alpha(F_j) \tag{m}$$

for any P, we say that *the* a-*flats minimize* f-*area or* α-*area.* If $[F_j] = \mathfrak{A}_j$ then

$$\sum_{j=0}^{k} \mathfrak{A}_j = 0 \quad \text{or} \quad -\mathfrak{A}_0 = \sum_{j=1}^{k} \mathfrak{A}_j$$

so that (m) *is* equivalent to

$$f\left(\sum_{j=1}^{k} \mathfrak{A}_j\right) \leqq \sum_{j=1}^{k} f(\mathfrak{A}_j)$$

and is therefore *implied by extendable convexity. The converse is not known for* $1 < a < n-1$ and amounts to an involved semitopological question. The difficulties may be seen from the fact that for $a = 2$, $n = 4$ any 7 simple 2—vectors are dependent, but any two planes parallel to two of these 2-vectors will in general intersect in points only. This means that comparatively few sets of 2-vectors whose sum vanishes correspond to closed polyhedra, so that (m) provides rather little information on general sums $\Sigma\mathfrak{A}_j$.

In the cases $a = 1$, $n - 1$ extendable convexity becomes ordinary convexity and is easily seen to be equivalent to (m) [1]). The result goes back to Minkowski. Therefore, when $1 < a < n - 1$, convexity of $f(\mathfrak{A})$ is equivalent to (m) for P which lie in $(a + 1)$-flats.

For $1 < a < n - 1$ *the condition* (m) *is the natural analogue to the Legendre Condition.* It it should prove weaker than extendable convexity—the evidence available at present speaks against this possibility—it would therefore be the *fourth* significant form of convexity to which we alluded above. *Extendable convexity is probably the relevant concept for the theory of convex bodies.*

The interpretation of total convexity in V^n requires the concept of total normality or total transversality, which is at first difficult to grasp. In order to at least avoid the complications arising from orientation we assume that f or the area α is symmetric, i.e., that $f(\mathfrak{A}) = f(-\mathfrak{A})$ and indicate later the changes necessary for non-symmetric f.

In the simplest case $a = 1$ the line L is normal to the hyperplane H at the origin $z = L \cap H$ of V^n if for any set $M \subset L$ and any other line L' through $L \cap H$ the projection M' of M parallel to H on L' satisfies $\alpha(M') \geqq \alpha(M)$. (Figure 1.) If we choose as M a segment of α-length 1 beginning at z, its endpoint q lies on the indicatrix I_f of f, and we see that L is normal to H if and only if the hyperplane H_q parallel to H through q is a supporting hyperplane of I_f, because only then will the endpoint q of the projection of M lie on or outside of I_f for all choices of L'. Normality of L to H is the term used in Banach space theory, whereas in the calculus of variations one speaks of transversality of H to L. *Thus a length induces a concept of normality and transversality. We generalize this to arbitrary* a, *leaving as much free play for the dimensions as possible.*

If an a-dimensional affine area f or α is given, $1 \leqq a \leqq n - 1$, we wish to define when the a-*flat* A *is totally normal to the* b-*flat* B, $1 \leqq \mathrm{b} \leqq \mathrm{n} - 1$ *at the* d-*flat* $\mathrm{D} = \mathrm{A} \cap \mathrm{B}$, $\mathrm{d} < \min(\mathrm{a}, \mathrm{b})$, in

1) The conditions for the equality sign are different for $a = 1$ and $a = n - 1$, see [8, p. 49].

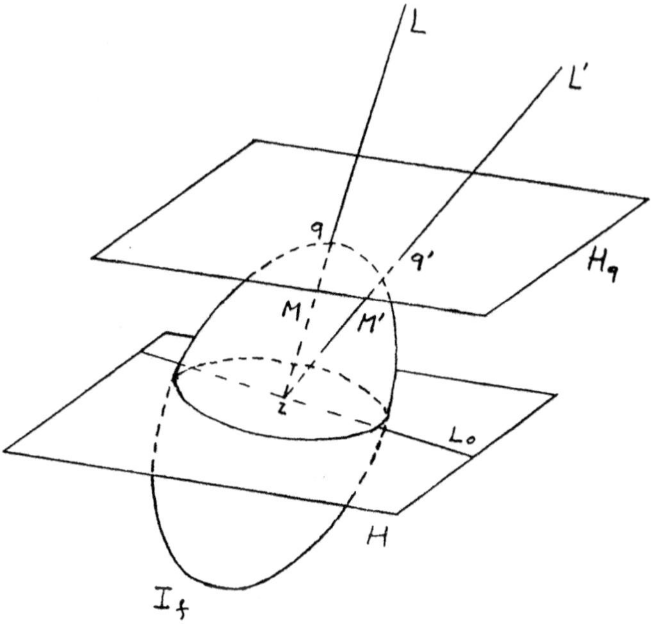

Fig. 1.

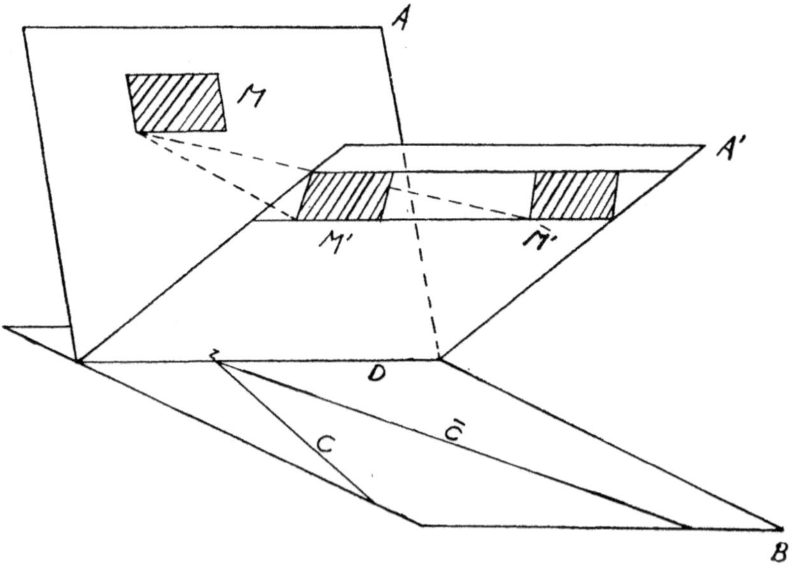

Fig. 2.

$A + B = Q$, $q = \dim Q = a + b - d \leqq n$. We write this relation as

$$A \overset{D,Q}{\underset{t}{\overline{\quad\quad}}}\Big| B \quad \text{or} \quad B \Big|\overset{D,Q}{\underset{t}{\overline{\quad\quad}}} A\,, \tag{4}$$

but read the second relation as: B *is totally transversal to* A *at* D *in* Q.

We distinguish two cases:

1) $d = 0$. Then $A \cap B$ is a point, it is convenient to always take the same point, say the origin z. Let M be a set in A with $0 < \alpha(M) < \infty$, and consider any a-flat A' in Q with $A' \cap B = z$ (then $A' + B = Q$) and the projection M' of M on A' parallel to B. *The relation* (4) *means that*

$$\alpha(M') \geqq \alpha(M) \tag{5}$$

for any choice of A$'$.

This inequality is independent of the choice of M.

2) $d > 0$. (The case $d = 1$, $a = b = 2$ is illustrated in Figure 2.) Assuming $D \ni z$ we choose in B a $(b - d)$-flat C with $C \cap D = z$. Then $C + A = Q$, $C \cap A = z$ and we define (4) by $A \overset{z,Q}{\underset{t}{\overline{\quad\quad}}}\Big| C$. This definition seems to depend on the choice of C, but does not. If C is replaced by another $(b - d)$-flat $\bar{C} \subset B$ with $\bar{C} \cap D = z$, then the projections M' and $\bar{M}'$ of $M \subset A$ on A' using C and $\bar{C}$ respectively, have the same measure: $\alpha(M') = \alpha(\bar{M}')$, because it is easily seen that $\rho_a(M') = \rho_a(\bar{M}')$. Therefore (5) for $\bar{M}'$ entails (5) for M' and conversely.

When A and B are known then D and Q are, and we omit them in (4). If $a + b = n$ then $Q = V^n$ is known and will be omitted etc.

The existence of A totally normal to B for given B, D, Q follows at once from the assumption that $f(\mathfrak{A})$ is positive and continuous. We saw already in the simplest case $d = 0$, $a = 1$, $b = n - 1$, $q = n$ that the existence of B totally transversal to A for given A, D, Q is a convexity property. In the general case we find:

f ($\mathfrak{A}$) *or* α *is totally convex if and only if a* B *totally transversal to* A *for given* $\mathrm{D} \subset \mathrm{A} \subset \mathrm{C}$ *exists for all dimensions* a, b, d, q *satisfying* $\mathrm{d} < \min(\mathrm{a}, \mathrm{b})$, $\mathrm{q} = \mathrm{a} + \mathrm{b} - \mathrm{d} \leqq \mathrm{n}$.

There are two simple, but very important facts, relating normality for different dimensions, the second of which was used already in a special case when reducing $d > 0$ to $d = 0$.

(6) *If* A *is (totally) normal to* B *at* D *and* $\mathrm{d} < \mathrm{b}' < \mathrm{b}$ *then* A *is (totally) normal to any* b'*-flat* B' *through* D *in* B. ($L \dashv L_0$ in Figure 1.)

(7) *If* A *is (totally) normal to* B *at* D *and the* b'*-flat* B' *lies in* $\mathrm{A} + \mathrm{B}$ *and contains* B *but not* A, *then* A *is (totally) normal to* B' *at* $\mathrm{D}' = \mathrm{B}' \cap \mathrm{A}$.

The parentheses about " totally " indicate that (6) and (7) are also valid for the ordinary normality to be defined presently.

Repeated application of (6) and (7) shows that $f(\mathfrak{A})$ *is totally convex if for a given* a*-flat* A *an* (n — a)*-flat transversal to* A *at a point exists.* This special case is Carathéodory's form of the Legendre Condition [9, paper XX], which is therefore *equivalent to total convexity.*

It is clear—and this is Carathéodory's starting point [9, p. 364] —that total convexity implies (m). With the previous notations, if A is the a-flat containing the face F_0 of P and B is an (n — a)-flat totally transversal to A at a point, then by the definition of total transversality the projection F_j' of F_j ($j \geqq 1$) parallel to B on A satisfies $\alpha(F_j') \leqq \alpha(F_j)$. For topological reasons $\bigcup_{j=1}^{k} F_j' \supset F_0$, hence ($m$).

If A is totally normal to B at D then A is by (6) totally normal to every ($d + 1$)-flat in B through D. The converse holds trivially if $b = d + 1$, less trivially, when $a = d + 1$; and $d = \min(a, b) - 1$ is the *only* case where the converse is always true. This leads us to the definition of normality.

A *is normal to* B *at* D *in* Q *or* B *is transversal to* A *at* D *in* Q, *in formulae*

$$A \left| \frac{D,Q}{} B \right. \quad or \quad B \left| \frac{D,Q}{} A \right. ,$$

if A *is totally normal to every* (d + 1)*-flat through* D *in* B.

Thus total normality and normality are in general equivalent only for $d = \min(a, b) - 1$. As stated before, (6) and (7) hold for normality, so that we can draw the analogous conclusion, namely that B transversal to A for given A, D, Q exists when an $(n - a)$-flat transversal to a given a-flat at a point exists. This special case is Wagner's concept of transversality.

B *transversal to* A *for given* A, D, Q *exists if and only if* f $(\mathfrak{A})$ *is convex. It suffices for convexity that* B *exists in the case* d = min) a, b) — 1 *or even only for* b = 1 (hence $d = 0$, $q = a + 1$).

The relations between the various concepts are summarized in the Table.

Table

Convexity	*Extendable Convexity*	*Total Convexity*
Def: $f(\mathfrak{A}_1 + \mathfrak{A}_2) \leqq f(\mathfrak{A}_1) + f(\mathfrak{A}_2)$	Def: $f\left(\sum_{j=1}^{k} \mathfrak{A}_j\right) \leqq \sum_{j=1}^{k} f(\mathfrak{A}_j)$	
$\downarrow\ \uparrow$	$\downarrow\ \uparrow$	
$f(\mathfrak{A})$ convex in each $(a+1)$-flat	$f(\mathfrak{A})$ extendable to a convex function in V_a^n	
$\downarrow\ \uparrow$	$\downarrow\ \uparrow$	
	I_f possesses supporting hyperplanes in V_a^n	Def: I_f possesses simple supporting hyperplanes in V_a^n
$\downarrow\ \uparrow$	$\downarrow\ \uparrow$?	$\downarrow\ \uparrow$
(m) for P in $(a+1)$-flats	(m) for general P	Existence of B with $B \frac{D,Q}{t}\Big\vert A$ for given A, D,
$\downarrow\ \uparrow$		
Existence of B with $B \frac{D,Q}{}\Big\vert A$ for given A, D, Q		
$\downarrow\ \uparrow$		
Existence of B with $B \frac{D,Q}{t}\Big\vert A$ if $d = \min(a, b) - 1$		

If f is not symmetric ($f(\mathfrak{A}) \neq f(-\mathfrak{A})$) then in the definition of (total) normality the flat B remains non-oriented, A is oriented and A' has the orientation induced by that of A through projection parallel to B. The values $\alpha(M)$ and $\alpha(M')$ are determined by these orientations.

Our exposition seems to justify the restriction to the case $b = n - a$, $d = 0$ considered by Carathéodory and Wagner, but in discussing symmetry and equivalence of normality we will find that the other cases prove most interesting. The answer to the equivalence problem requires the concept of quadratic area: *f* $(\mathfrak{A})$ *or* α *is quadratic if* $f^2(\mathfrak{A})$ *is a quadratic form in the components of* $\mathfrak{A}$. Clearly a quadratic area satisfies $f(\mathfrak{A}) = f(-\mathfrak{A})$ and is extendably convex, but need *not* be totally convex. A euclidean area, i.e. an area ρ_a induced by some euclidean metrization of V^n is quadratic, the converse holds only for $a = 1$ and $a = n - 1$. Therefore *a quadratic area is euclidean in every* $(a+1)$*-flat but not necessarily in every* $(a+2)$*-flat.* In fact, if the latter is the case, then the area is euclidean.

If a convex or totally convex area α is given then the relations

$$A \overset{D}{-}\!\Big|\, A' \quad \text{or} \quad A \underset{t}{\overset{D}{-}}\!\Big|\, A'$$

are defined for two a-flats A, A' with $D = A \cap A'$ provided $d < a$ and $2a - d \leq n$. It is natural to ask when these relations are symmetric. Since for $f(\mathfrak{A}) \neq f(-\mathfrak{A})$ the flat A has to be taken as oriented and A' is non-oriented the problem is natural only for symmetric areas. Our preceding remarks show that for $d = a - 1$ the normality relation is symmetric when the area is quadratic.

The problem is wellknown for $a = 1$ and was solved long ago (1916) by Radon and Blaschke. A convex 1-area is a Minkowski metric and our space is a finite-dimensional Banach space. The answer to the problem is as follows: if $n = 2$, then symmetry of normality between lines does not imply that the metric is euclidean (Radon [10]), but for $n > 2$ it does (Blaschke [3]). These results were rediscovered by the Banach space people in the 1940's and formulated in their own language.

Enseign. Math., II. Sér. 7 (1962), 139–152

In our case we have more free play and the problem presents itself in the following form:

Let a symmetric (totally) convex a-*area* α *and a symmetric (totally) convex* b-*area* β *be given, moreover fixed numbers* d, q *such that* $\mathrm{d} < \mathrm{a} \leqq \mathrm{b} < \mathrm{q} = \mathrm{a} + \mathrm{b} - \mathrm{d} \leqq \mathrm{n}$. *Then* α *defines the relation* $\mathrm{A} \frac{\mathrm{D}}{(\mathrm{t})}\Big| \mathrm{B}$. *When is this relation equivalent to* $\mathrm{B} \frac{\mathrm{D}}{(\mathrm{t})}\Big| \mathrm{A}$ *which is defined by* β ?

The answer is this:

If the two relations are equivalent, then both α *and* β *are quadratic, unless* $\mathrm{a} + \mathrm{b} = \mathrm{n}$ *and* $\mathrm{d} = 0$.

In the latter case, when a totally convex a-*area is given, an up to a constant factor unique totally convex* (n — a)-*area can be constructed such that the relations* $\mathrm{A} \frac{\mathrm{z}}{\mathrm{t}}\Big| \mathrm{B}$ *and* $\mathrm{B} \frac{\mathrm{z}}{\mathrm{t}}\Big| \mathrm{A}$ *are equivalent.*

The special case $\alpha \equiv \beta$ leads to symmetry of normality and we find:

If α *is a (totally) convex* a-*area and if* $\mathrm{A} \frac{\mathrm{D}}{(\mathrm{t})}\Big| \mathrm{A}'$ *implies* $\mathrm{A}' \frac{\mathrm{D}}{(\mathrm{t})}\Big| \mathrm{A}$ *then* α *is quadratic unless* $\mathrm{n} = 2\mathrm{a}$ *and* $\mathrm{d} = 0$. *In the latter case there are for each* $\mathrm{a} \geqq 1$ *non-quadratic totally convex areas for which* $\mathrm{A} \frac{\mathrm{z}}{\mathrm{t}}\Big| \mathrm{A}'$ *implies* $\mathrm{A}' \frac{\mathrm{z}}{\mathrm{t}}\Big| \mathrm{A}$.

For $a = 1$ *all* areas or Minkowski metrics with this property are known through Radon [10]; in such generality the problem has not been solved for $a > 1$.

In special cases we can assert that a (totally) convex area α is euclidean if normality of a-flats at d-flats is symmetric, namely when $2a < n$ and $d = 0$ or $2a > n$ and $d = 2a - n$.

Not all results for special dimensions have interesting extensions. An example is provided by the *integral geometry of (symmetric) affine areas* (for the following see [7]). Since affine areas are invariant only under translations, one would conjecture that a true integral geometry exists only for the translation group. Such a theory does exist, but is not essentially new, because there is a simple method of transcribing the results of the theory for the euclidean areas and this group, as it is developed in Berwald and Varga [2], directly without new proofs so that they become valid for affine areas. The only surprise is, that *results into which areas of different dimensions enter remain valid for completely unrelated, and not even necessarily convex affine areas.*

Contrary to our conjecture Blaschke showed in [3] that *the Minkowski plane possesses a much richer integral geometry*, in which most of the euclidean results for the full group of motions of E^2 hold. The central point is the *existence of a positive density* dL for all lines in the plane, such that for any segment S the relation

$$\alpha(S) = \int_{L \cap S \neq 0} dL \; ^{1)}$$

holds. Thus one might hope (and Blaschke predicted in [3]) that there is a rich integral geometry for a reasonably general a-dimensional area α. As criterion we take the existence of a positive density dB for the $(n - a)$-flats such that for any convex set S in an a-flat the relation

$$\alpha(S) = \int_{B \cap S \neq 0} dB$$

holds. *If* n $>$ 2 *such a density exists only if* α *is extendably convex* (which is a reasonable condition) *and has in addition a special form* which deprives the theory of much of its interest.

We come to the *new problems* arising *in the theory of convex bodies* through the new concepts. The classical theory proves for many functions derived from one or several convex bodies in E^n and depending on the position of a hyperplane, that they are convex. As our example of the projection function $P(\mathfrak{A})$ shows these functions are often naturally defined for a varying a-flat, and the classical results only contain the convexity of the new functions, and not extendable or total convexity.

---

1) The normalization of dL or dB differs from the usual one by a constant factor. We emphasize that dL and dB be *positive because this condition causes the trouble for* dB *when* $n > 2$.

No method for establishing total convexity is known.[1]) 7] shows *that in various cases extendable convexity can be proved with the help of integral geometry.* The following is a typical example: Let F be a closed convex hypersurface in E^n. For a given simple a-vector $\mathfrak{A} \neq 0$, $(a \geqq 2)$, take the family of oriented a-flats L_A parallel to $\mathfrak{A}$, let dL_A be their density and form, with ρ_{a-1} meaning ordinary $(a-1)$-dimensional area,

$$J(\mathfrak{A}) = |\mathfrak{A}| \cdot \int_{L_A \neq F \cap 0} \rho_{a-1}(L_A \cap F)\, dL_A\,, \quad J(O) = 0\,.$$

Then $J(k\mathfrak{A}) = |k|\, J(\mathfrak{A})$ and $J(\mathfrak{A})$ is extendably convex because

$$\int_{L_A \neq F \cap 0} \rho_{a-1}(L_A \cap F)\, dL_A = \int_F |\mathfrak{n}_p \wedge \mathfrak{A}^s|\, dF_p\,,$$

where $\mathfrak{n}_p$ is the exterior unit normal to F at p, dF_p is the area element of F and $\mathfrak{A}^s$ is the supplementary $(n-a)$-vector to $\mathfrak{A}$. Extendable convexity follows from $\left|\sum \mathfrak{n}_p \wedge \mathfrak{A}_j^s\right| \leqq \sum \left|\mathfrak{n}_p \wedge \mathfrak{A}_j^s\right|$.

W. Fenchel and the author noticed that the examples in [7] can be subordinated to a general principle, which is not discussed here because it is not yet clear how far it carries.

Whether the projection function $P(\mathfrak{A})$ is extendably convex is not known. Those familiar with the theory of convex bodies will recognize this as a *novel problem*, because

$$\mathrm{P}(\mathfrak{A}) = \binom{n}{a} \mathrm{V}_{n-a}(\mathrm{K}, \mathrm{M}), \ \textit{if}\ \mathrm{M},\ \textit{convex and}\ [\mathrm{M}] = \mathfrak{A}^s\,.$$

As usual (see [5, p. 40] $V_a(K, M)$ is the mixed volume of K taken $n-a$ times and M taken a times. The just mentioned general principle proves $V_{n-1}(K, M) = Q(\mathfrak{A})$ to be extendably convex.

We conclude with an interesting and probably quite dificult convexity problem. In its simplest form it is this: Let K be a convex body in E^n with z as interior point and center. For an a-vector $\mathfrak{A} \neq 0$ let A be the a-flat parallel to $\mathfrak{A}$ through z and put

$$\alpha(\mathfrak{A}) = |\mathfrak{A}|/\rho_a(A \cap K)\,, \quad \alpha(O) = 0\,.$$

---

1) This difficulty was already encountered by Carathéodory when he tried to establish his transversality condition for given variational problems.

This function is convex; the proof of convexity for the problem in a slightly more general form implies the Brunn-Minkowski Theorem (the relation between the two theorems is discussed in Barthel [1]). *This shows that the extendable convexity of* $\alpha(\mathfrak{A})$, *if true, is a deep theorem with many implications.*

REFERENCES

[1] Barthel, W., Zum Busemannschen und Brunn-Minkowskischen Satz. *Math. Z.*, 70 (1958-9), 407-429.

[2] Berwald, L. und O. Varga, Integralgeometrie 24, Ueber Schiebungen in Raum. *Math. Z.*, 42 (1937), 710-736.

[3] Blaschke, W., Räumliche Variationsprobleme mit symmetrischer Transversalitätsbedingung. *Ber. Sächs. Akad. Wiss. Leipzig*, 68 (1916), 50-55.

[4] —— Integralgeometrie 11, Zur Variationsrechnung. *Abh. Math. Sem. Univ. Hamburg*, 11 (1936), 359-366.

[5] Bonnesen, T. und W. Fenchel, Theorie der konvexen Körper. *Ergebnisse Math.*, III, Berlin, 1934 and New York, 1948.

[6] Busemann, H., Areas in affine space, II. *Rend. Circ. Mat. Palermo*, Serie II 9 (1960), 86-90.

[7] —— Areas in affine space, III. *Rend. Circ. Mat. Palermo*, *ibid.*, 226-242.

[8] —— and E. G. Straus, Area and normality. *Pacific J. Math.*, 10 (1960), 35-72.

[9] Carathéodory, C., *Gesammelte Mathematische Schriften*, vol. 1, München, 1954.

[10] Radon, J., Ueber eine besondere Art ebener konvexer Kurven. *Ber. Sächs. Akad. Wiss. Leipzig*, 68 (1916), 131-134.

[11] Wagner, V., Geometry of a space with areal metric and its applications to the calculus of variations (Russian, English Summary). *Mat. Sb. N.S.*, 19 (41) (1946), 341-404.

Department of Mathematics
University of Southern California
Los Angeles, California.

Enseign. Math., II. Sér. 7 (1962), 139–152

Convex Bodies and Convexity on Grassmann Cones

Dedicated to Reinhold Baer on his 60th birthday

By

H. Busemann and G. C. Shephard *)

Introduction. V. Totally convex functionals. In the preceding parts of this series [3] we were principally concerned with functions $f(R)$ which are defined on the Grassmann cone G_r^n of all simple r-vectors R over the n-dimensional affine space A^n, and are homogeneous (i.e. $f(\varrho R) = \varrho f(R)$ for $\varrho \geqq 0$). In particular we investigated whether certain functions of this type, which arose naturally from geometrical problems, were also convex, that is to say, could be extended to a convex function[1] $f(\tilde{R})$ defined in the space V_r^n of all r-vectors $\tilde{R}$ over A^n.

Though convexity defined in this way has interesting geometrical implications in A^n, it does not seem to be equivalent to any simple geometric property[2] in A^n. On the other hand, searching for an analogue to the usual Legendre condition for r-dimensional variational problems in n dimensions ($1 < r < n - 1$), Carathéodory created a stronger concept of convexity, namely *total convexity*. This proved too restrictive as a Legendre condition, but it has a simple and palpable geometric meaning: it is equivalent to the existence of an $(n - r)$-flat *totally transversal* to any given oriented r-flat. This term is defined as follows:

Let $\mathscr{R}^+$ be any given oriented r-flat and $\mathscr{S}$ any $(n - r)$-flat meeting $\mathscr{R}^+$ in a point. Let Q_0 be a variable r-vector, and Q be the projection of Q_0, parallel to $\mathscr{S}$, on $\mathscr{R}^+$. Then if $f(Q_0) \geqq f(Q)$ for all Q_0, the $(n - r)$-flat $\mathscr{S}$ is said to be *totally transversal* to $\mathscr{R}^+$.

It seems therefore worthwhile to continue the investigations of Parts II, III, IV by asking whether some of the functionals derived from convex bodies found there, as well as others, are totally convex. After a brief discussion of total convexity in a general context, the present Part V establishes total convexity in various cases by using a direct generalization of the ordinary supporting function.

---

*) This work was supported by a grant from the National Science Foundation, U.S.A.

1) We follow the conventions and notations of [3], considering A^n and V_r^n as both affine and vector spaces, and using R (with or without subscripts) for simple r-vectors and $\tilde{R}$ for general r-vectors. References such as IV (11) will mean statement (11) in Part IV of [3].

2) The conjecture in [2] that convexity of $f(R)$ with $f(R) > 0$ for $R \neq 0$ is equivalent to the area minimizing property of the r-flats with respect to $f(R)$ interpreted as area, is almost certainly false.

V.1. Total convexity. In Part I we derived the properties of convex functions on G_r^n from the more general theory of functions convex on a non-convex set in A^n. For reasons of consistency we give total convexity a similar general setting, although the theory is not really significant in this generality.

We assume throughout that M is a non-empty subset of A^n whose convex hull M^H is open in A^n. It was proved in I(9)II that a function $f(x)$ defined in M is convex if and only if, for a given point x_0, numbers $a_i(x_0)$ exist ($i = 1, \dots, n$) such that

$$(1) \qquad f(x) \geqq \sum_{i=1}^{n} a_i(x_0)(x^i - x_0^i) + f(x_0) \text{ for } x \in M.$$

Let N be a closed non-empty subset of an n-dimensional affine space. Sometimes it will be convenient to consider N as lying in a different space from M. We shall say that a function $f(x)$ defined in M is *N-convex* if, for each $x_0 \in M$, a point $p(x_0) = (p_1(x_0), \dots, p_n(x_0))$ in N exists such that

$$(2) \qquad f(x) \geqq \sum_{i=1}^{n} p_i(x_0)(x^i - x_0^i) + f(x_0) \text{ for } x \in M.$$

We begin with two trivial remarks.

(3) *If $\varrho N = N$ ($\varrho > 0$) and $f(x)$ is N-convex, so is $\varrho f(x)$.*

(4) *If N is convex and $f_1(x), \dots, f_m(x)$ are N-convex, then so is $\sum_{i=1}^{m} \lambda_i f_i(x)$.*

Here we adhere to the agreement of the previous parts that λ_i and μ_j denote non-negative numbers whose sum equals 1. The proof of (4) is clear:

$$f_j(x) \geqq \sum_{i=1}^{n} p_i^j(x_0)(x^i - x_0^i) + f_j(x_0), \quad j = 1, \dots, m,\ x \in M,$$

implies

$$\sum_{j=1}^{m} \lambda_j f_j(x) \geqq \sum_{i=1}^{n} \left(\sum_{j=1}^{m} \lambda_j p_i^j(x_0) \right) (x^i - x_0^i) + \sum_{j=1}^{m} \lambda_j f_j(x_0)$$

and the point $\sum \lambda_j p^j(x_0) \in N$ because N is convex.

In our application N will be taken to be a Grassmann cone, hence the condition $N = \varrho N$ is satisfied, but N is not convex. This leads to considerable differences between the theories of convexity and N-convexity.

From Part I we remember the following: If $\Phi(x)$ is convex in M then

$$g_\Phi(x) = \inf_{x = x_A} \sum_{i=1}^{n+1} \lambda_i \Phi(x_i), \quad x = \sum_{i=1}^{n+1} \lambda_i x_i, \quad x_i \in M$$

is a convex extension of $\Phi(x)$ to M^H.

If I is any index set and $f_i(x)$ ($i \in I$) is convex on M, and $s(x) = \sup_{i \in I} f_i(x)$ is finite, then we know that $s(x)$ is convex, see I (12). We need here that

$$(5) \qquad \sup_{i \in I} g_{f_i}(x) \leqq g_s(x).$$

For if $x \in M^n$ and $\varepsilon > 0$ are given, then points $x_j \in M$ and numbers λ_j ($j = 1, \dots, n+1$)

Arch. Math. 13 (1962), 512–526

exist such that $x = x_A$ and

$$\sum_{j=1}^{n+1} \lambda_j s(x_j) - g_s(x) < \varepsilon .$$

Because g_{f_i} is convex and is an extension of f_i,

$$g_{f_i}(x) \leqq \sum \lambda_j f_i(x_j),$$

hence

$$\sup g_{f_i}(x) \leqq \sum \lambda_j \sup f_i(x_j) = \sum \lambda_j s(x_j) < g_s(x) + \varepsilon .$$

Since ε is arbitrarily small, this proves (5). The same argument shows:

If $f_\nu(x)$ is convex on M and $f(x) = \limsup f_\nu(x)$ is finite, then,

(6) $$\limsup g_{f_\nu}(x) \leqq g_f(x) .$$

We now prove

(7) *If $f_i(x)$ $(i \in I)$ is N-convex and $s(x) = \sup_{i \in I} f_i(x)$ is finite, then $s(x)$ is N-convex.*

If the point $x_0 \in M$ is given, we can find a sequence of functions $f_\nu(x)$ among the $f_i(x)$ such that $f_\nu(x_0) \to s(x_0)$. By definition there is a point $p^\nu \in N$ such that

(7a) $$f_\nu(x) \geqq \sum_{i=1}^{n} p_i^\nu (x^i - x_0^i) + f_\nu(x_0) \text{ for } x \in M .$$

If the sequence of points $\{p^\nu\}$ is bounded, then some subsequence will converge to a limit point p^0, and $p^0 \in N$ since N is closed. In the limit (7a) becomes

$$s(x) = \lim f_\nu(x) \geqq \sum_{i=1}^{n} p_i^0 (x^i - x_0^i) + s(x_0),$$

and (7) is proved.

If, on the other hand the sequence of points $\{p^\nu\}$ were unbounded, then a subsequence $\{\mu\}$ of $\{\nu\}$ and a j would exist such that $p_j^\mu \to \infty$ or $p_j^\mu \to -\infty$. Since M^H is open the point u with coordinates $u^i = x_0^i$ for $i \neq j$, $u^j = x_0^j + (\operatorname{sign} p_j^\mu)\,\varepsilon$ will lie in M^H for a sufficiently small positive ε. But

$$g_{f_\mu}(u) \geqq \sum_{i=1}^{n} p_i^\mu (u^i - x^i) + f_\mu(x_0)$$

and $f_\mu(x_0) \to s(x_0)$. Hence $g_{f_\mu}(u) \to \infty$ which contradicts (5).

Using (6) instead of (5) the same reasoning yields:

(8) *If $f_\nu(x)$ is N-convex on M and $f(x) = \limsup f_\nu(x)$ is finite, then $f(x)$ is N-convex.*

Notice the following interesting application of (7):

(9) *A function $f(x)$ which is N-convex in M can be extended to an N-convex function in M^H.*

To each point $x_0 \in M$ there exists a function

$$L(x, x_0) = \sum_{i=1}^{n} p_i(x_0)\,(x^i - x_0^i) + f(x_0) \leqq f(x), \quad p \in N .$$

Each of these functions is N-convex in M^H and $L(x, x_0) \leqq g_f(x)$ for $x \in M^H$. Therefore $s(x) = \sup_{x_0 \in M} L(x, x_0)$ is finite in M^H and is an extension of $f(x)$. Now (7) applied to M^H shows that $s(x)$ is N-convex in M^H.

The original definition of total convexity (see [2]) of which N-convexity is a generalization, refers to homogeneous functions $f(x)$ with $f(x) > 0$ for $x \neq 0$, and uses the indicatrix I_f of f, that is the set of points in M for which $f(x) = 1$. Part c) of the following theorem states that in this case the two definitions are equivalent.

(10) *Let $f(x)$ be homogeneous in M where $M^H = A^n$ and $f(x) > 0$ for $x \neq 0$. Then*
a) *the convex hull I_f^H of I_f contains the origin in its interior,*
b) *$f(x)$ is convex if and only if I_f lies on the boundary of I_f^H or, equivalently, if I_f possesses at each point x_0 a supporting hyperplane $\sum a_i x^i = 1$,*
c) *$f(x)$ is N-convex if and only if I_f possesses at each point x_0 a supporting hyperplane $\sum p_i x^i = 1$, where $p = (p_1, \dots, p_n)$ is a point of N.*

If the origin were not an interior point of I_f^H then a hyperplane $\sum u_i x^i = 0$ would exist such that I_f^H lies in $\sum u_i x^i \geqq 0$. Because $M^H = A^n$ there is a point $y \in M$ with $\sum u_i y^i < 0$, hence $\beta^{-1} \underset{\text{def}}{=} f(y) > 1$. But then $f(\beta y) = \beta f(y) = 1$, hence $\beta y \in I_f$, which contradicts the fact that βy lies in $\sum u_i y^i < 0$.

Let $f(x)$ be convex and $x_0 \in I_f$ or $f(x_0) = 1$. There are numbers a_i such that

$$f(x) \geqq \sum a_i x^i, \quad \sum a_i x_i^0 = f(x_0) = 1,$$

(see IIa after I(9)). Thus not all a_i vanish and $f(x) \leqq 1$ implies $\sum a_i x^i \leqq 1$. This means that $\sum a_i x^i = 1$ is a supporting hyperplane of I_f, hence also of I_f^H, so that x_0 lies on the boundary of I_f^H.

Conversely let x_0 lie on the boundary of I_f^H. Since I_f^H contains the origin in its interior, it possesses at x_0 a supporting hyperplane $\sum a_i x^i = 1$ where $\sum a_i x_0^i = f(x_0) = 1$ and I_f^H lies in $\sum a_i x^i \leqq 1$. Therefore $x \in M$ and $f(x) = 1$ imply $f(x) \geqq \sum a_i x^i$. This relation extends to all $x \in M$ because $f(x) \geqq 0$ and $f(\varrho x) = \varrho f(x)$ for $\varrho \geqq 0$. Thus

$$f(x) \geqq \sum a_i x^i = \sum a_i (x^i - x_0^i) + f(x_0) \text{ for } x \in M.$$

This argument contains part c).

We come to the special case in which we are interested, namely $M = G_r^n \subset V_r^n$ and N is the Grassmann cone G_r^{n*} in the dual space. However, for the geometric reasons indicated in Part I, when A^n carries a euclidean metric, we prefer the old-fashioned interpretation of $R^* \in G_r^n$ as an r-vector in G_r^n written in covariant components. For these M, N we speak of *total convexity* rather than N-convexity, because for homogeneous $f(R)$ the concept is equivalent to the existence of total transversals (see the introduction and [2]). Precisely then, total convexity of a function $f(R)$ defined on G_r^n means: for a given $R_0 \in G_r^n$, there exists an $R^* \in G_r^n$ such that

$$f(R) \geqq R^* \cdot (R - R_0) + f(R_0) \text{ for all } R \in G_r^n, \tag{11}$$

34*

where

$$R^* \cdot (R - R_0) = \sum_{i_1 < \dots < i_r} R^*_{i_1 \dots i_r} (R^{i_1 \dots i_r} - R_0^{i_1 \dots i_r})$$

with lower subscripts referring to covariant components and upper subscripts to contravariant components.

Notice that total convexity is identical with ordinary convexity in the cases $r = 1$, $r = n - 1$.

If $f(R)$ is homogeneous, then $f(0) = 0$, $R \cdot R_0 = f(R_0)$ and (11) reduces to

$$(12) \qquad f(R) \geqq R^* \cdot R, \quad R^* \cdot R_0 = f(R_0).$$

A *simple* hyperplane in V_r^n is defined as one of the form $R^* \cdot \tilde{R} = c$ where R^* is a simple vector, i.e. lies in G_r^n.

If $f(R)$ is convex on G_r^n and differentiable in terms of local analytic parameters, and $f(R) > 0$ for $R \neq 0$, then I_f possesses, at each point, a unique tangent $(k-1)$-flat T in V_r^n, where $k = r(n-r) + 1 = \dim G_r^n$. Because of (10) the origin does not lie in T. If $R_0, \tilde{R}_1, \tilde{R}_2, \dots, \tilde{R}_{k-1}$ span T and the hyperplane $\sum a_i \tilde{R}^i = 1$ passes through T, then

$$\sum a_i (\tilde{R}_j^i - R_0^i) = 0 \;\; (j = 1, \dots, k-1) \quad \text{and} \quad \sum a_i R_0^i = 1.$$

This imposes k conditions on the a_i. It is therefore plausible (a formal proof can be found in [6]) that there is exactly one simple hyperplane through T. Thus total convexity requires that precisely one of the $\infty^{\binom{n}{r} - k}$ hyperplanes through T be a supporting hyperplane of I_f, so that it is, indeed, a much stronger condition than ordinary convexity.

Our results (4), (7) and (8) contain

(13) Theorem. *If $f(R)$ is totally convex then so is $\varrho f(R)$ for $\varrho > 0$.*

If $\{f_i(R)\}_{i \in I}$ is any family of totally convex functions in G_r^n and $s(R) = \sup_{i \in I} f_i(R)$ is finite, then $s(R)$ is totally convex.

If $f_1(R), f_2(R), \dots,$ is a sequence of totally convex functions and $f(R) = \limsup f_\nu(R)$ is finite, then $f(R)$ is totally convex.

On the other hand, as we have mentioned previously, the sum of a set of totally convex functions need not be totally convex. Consequently there is no process corresponding to the integration principle (Part II) in the theory of total convexity.

The following analogue to IV (7) holds:

(14) *If $0 < s < r < t < n$ and $f(\mathscr{R}^+)$ induces a totally convex function on G_r^n, then for variable*[3]) *s-flats $\mathscr{S}$, and t-flats $\mathscr{T}$*

$$F(\mathscr{S}) = \max_{\mathscr{R}^+ \supset \mathscr{S}} f(\mathscr{R}^+) \quad \textit{and} \quad F(\mathscr{T}) = \max_{\mathscr{R}^+ \subset \mathscr{T}} f(\mathscr{R}^+)$$

induce totally convex functions on G_r^n and G_t^n respectively.

---

[3]) Here all flats are assumed to pass through a fixed point say the origin z.

In the proof, as well as later, we need the following:

(15) Let B, C, D be simple b-, c- and d-vectors with $d = b + c \leqq n$. Then

$$(B \wedge C) \cdot D = B \cdot (C \cdot D), \quad D \cdot (B \wedge C) = (D \cdot B) \cdot C.$$

This can be deduced from [1, p. 105] and the identification of the vector space V with its dual V^* by $\langle x, y^* \rangle = x \cdot y$.

The assumption of (14) means that for each $R_0 \in G_r^n$, an inequality of type (11) exists with $R^* \in G_r^n$. Let C be a fixed $(r - s)$-vector, let S_0 be given, and put $R_0 = S_0 \wedge C$. Then R^* exists so that

$$\begin{aligned} f(S \wedge C) &\geqq (S \wedge C - S_0 \wedge C) \cdot R^* + f(S_0 \wedge C) = \\ &= [(S - S_0) \wedge C] \cdot R^* + f(S_0 \wedge C) = \\ &= (S - S_0) \cdot (C \cdot R^*) + f(S_0 \wedge C) \end{aligned}$$

and $C \cdot R^*$ is a simple s-vector. This shows that $f(S \wedge C)$ is a totally convex function of S on G_s^n, for fixed C.

If W is the set of all simple unit $(r - s)$-vectors C_u, then by (13),

$$F(S) = \max_{C_u \in W} f(S \wedge C_u)$$

is totally convex, and this is the function induced by $F(\mathscr{S})$, see the proof of IV(7).

For the second part let C be a fixed $(t - r)$-vector. Then $T \cdot C$ is a simple r-vector. Let T_0 be given, and put $R_0 = T_0 \cdot C$. Then R^* exists so that

$$\begin{aligned} f(T \cdot C) &\geqq (T \cdot C - T_0 \cdot C) \cdot R^* + f(T_0 \cdot C) = \\ &= (T - T_0) \cdot (C \wedge R^*) + f(T_0 \cdot C). \end{aligned}$$

Here $C \wedge R^*$ is a simple t-vector, hence $f(T \cdot C)$ is a totally convex function of T on G_t^n for fixed C. From here the proof proceeds as in IV(7).

As an application we mention that IV(8) can now be strengthened to:

(16) *If $f(x)$ is a homogeneous convex function in A^n and u denotes a unit vector, then $F(\mathscr{R}) = \max_{u \| \mathscr{R}} f(u)$ induces a totally convex function on G_r^n.*

V.2. Generalization of the supporting function. Let W be a non-empty bounded set in V_r^n with a euclidean metric. Since the functions $|R \cdot \tilde{R}|$ and $R \cdot \tilde{R}$ are convex in R for fixed $\tilde{R}$,

$$H(R, W) = \sup_{\tilde{R}' \in W} |R \cdot \tilde{R}'| \quad \text{and} \quad H_0(R, W) = \sup_{\tilde{R}' \in W} R \cdot \tilde{R}'$$

are convex on G_r^n. If $r = 1$, $H_0(R, W)$ is the ordinary supporting function of the set $W \subset V_1^n = A^n$. Trivially,

$$H(\varrho R, W) = |\varrho| \, H(R, W)$$

and

$$H_0(\varrho R, W) = \varrho \, H_0(R, W) \quad \text{for } \varrho \geqq 0.$$

Also, if $W' \subset W$,

$$H(R, W') \leqq H(R, W), \quad \text{and} \quad H_0(R, W') \leqq H_0(R, W).$$

If $\overline{W}$ is the closure of W, and W^H is the convex hull of W, then our general theory (see IV.1) yields the following facts:

$$(17)\qquad \begin{cases} H(R, W) = H(R, \overline{W}) = H(R, W^H) = H(R, \overline{W}^H), \\ H_0(R, W) = H_0(R, \overline{W}) = H_0(R, W^H) = H_0(R, \overline{W}^H). \end{cases}$$

If $W_1, \dots, W_k$ are non-empty bounded sets in V_r^n, and $\alpha_i \geqq 0$, then,

$$(18)\qquad \begin{cases} H\left(R, \sum_{i=1}^{k} \alpha_i W_i\right) \leqq \sum_{i=1}^{k} \alpha_i H(R, W_i), \\ H_0\left(R, \sum_{i=1}^{k} \alpha_i W_i\right) = \sum_{i=1}^{k} \alpha_i H_0(R, W_i) = \sum_{i=1}^{k} \alpha_i H_0(R, \overline{W}_i^H). \end{cases}$$

If R' is a simple vector, the function $R \cdot R'$ is totally convex because $R \cdot R' = (R - R_0) \cdot R' + R_0 \cdot R'$, and hence $|R \cdot R'| = \max[R \cdot R', R \cdot (-R')]$ is also totally convex. Therefore we conclude from (13):

(19) Theorem. *If $W \subset G_r^n$ then $H(R, W)$ and $H_0(R, W)$ are totally convex.*

We apply (19) to sets W of the following type. Let β_{ij} ($i = 1, \dots, r$; $j = 1, \dots, N$) be real numbers such that for each fixed j at least one β_{ij} is different from 0. If $M_1, \dots, M_N$ are non-empty bounded sets in A^n then we take $W = W(M_1, \dots, M_N) = W(M_i)$ to consist of the simple r-vectors

$$(20)\qquad R(x_1, \dots, x_N) = R(x_j) = \bigwedge_{i=1}^{r} \left(\sum_{j=1}^{N} \beta_{ij} x_j\right); \quad x_j \in M_j.$$

Because of (17),

$$H(R, W(M_i)) = H(R, W(\overline{M}_i)), \quad H_0(R, W(M_i)) = H_0(R, W(\overline{M}_i)).$$

Therefore, for a given r-vector R, we can find points $u_j \in \overline{M}_j$ so that

$$|R(u_j) \cdot R| = H(R, W(M_j)),$$

or alternatively, we can find such points so that

$$R(u_j) \cdot R = H_0(R, W(M_j)).$$

As in Part III we denote by $S(K, \mathscr{R})$ the shadow boundary of the convex body K, that is, the set of points which the $(n - r)$-dimensional supporting flats of K normal to $\mathscr{R}$ have in common with K or ∂K.

(21) Theorem. *If $u_i \in \overline{M}_i$ and*

$$|R(u_i) \cdot R| = H(R, W(M_i)) > 0,$$

or

$$R(u_i) \cdot R = H_0(R, W(M_i)) \neq 0,$$

then $u_i \in S(\overline{M}_1^H, \mathscr{R})$, where $\mathscr{R}$ is an r-flat parallel to the r-vector R.

We will prove $u_1 \in S(\overline{M}_1^H, \mathscr{R})$. Assume $\beta_{11} \neq 0$. The exterior product (20) defining $R(x_i)$ does not change if, for each $s > 1$, the first factor multiplied by β_{s1}/β_{11} is subtracted from the s-th factor. Therefore we may assume that x_1 occurs only in the first factor. If $T(x) = T\ (x_2, \ldots, x_N)$ is the product of the last $r-1$ factors, then, by (15),

$$R(u_i) \cdot R = \left\{\left(\beta_{11} u_1 + \sum_{j=2}^{N} \beta_{1j} u_j\right) \wedge T(u)\right\} \cdot R =$$
$$= \beta_{11} u_1 \cdot (T(u) \cdot R) + \left(\sum_{j=2}^{N} \beta_{1j} u_j\right) \cdot (T(u) \cdot R) =$$
$$= u_1 \cdot p + k,$$

where $p = \beta_{11}(T(u) \cdot R)$ is a 1-vector and k is a scalar, both depending only on $u_2, \ldots, u_N$. The hypothesis $R(u_i) \cdot R \neq 0$ shows that $T(u) \cdot R \neq 0$, hence $p \neq 0$ is a vector parallel to R.

As in the preceding parts let $P(\overline{M}_1, \mathscr{R})$ and $v_1 = P(u_1, \mathscr{R})$ be the projections of $\overline{M}_1$ and u_1 on $\mathscr{R}$. Then v_1 cannot be an interior point of $P(M_1, \mathscr{R})$. Otherwise $v_1 + \varepsilon p$ would lie in $P(\overline{M}_1, \mathscr{R})$ for small positive ε, hence a point $u_\varepsilon \in M_1$ with $v_1 + \varepsilon p = P(u_\varepsilon, \mathscr{R})$ would exist. Replacing u_1 by u_ε would increase $H(R, W(M_i))$ or $H_0(R, W(M_i))$ by $\varepsilon p^2 > 0$. This contradicts our choice of the u_j. Hence $v_1 \in \partial P\ (M_1, \mathscr{R})$.

We must show that $u_1 \in S(\overline{M}_1^H, \mathscr{R})$ or, equivalently, $v_1 \in \partial P(\overline{M}_1^H, \mathscr{R})$. If $y \in \overline{M}_1^H$ then

$$y = \sum_{i=1}^{n+1} \mu_j y_j, \quad \mu_j \geqq 0, \quad \sum \mu_j = 1, \quad y_j \in \overline{M}_1.$$

Since x_1 occurs only in the first factor of $R(x_i)$, we have,

$$R(y, x_2, \ldots, x_N) = \sum \mu_j R(y_j, x_2, \ldots, x_N) \in W^H$$

where $W = W(\overline{M}_1, M_2, \ldots, M_N)$. Therefore (17) yields

$$H(R, W) \leqq H(R, W(\overline{M}_1^H, M_2, \ldots, M_N)) \leqq H(R, W^H) = H(R, W).$$

So,

$$|R(u_i) \cdot R| = H(R, W) = H(R, W(\overline{M}_1^H, M_2, \ldots, M_N)).$$

The first half of the proof with $\overline{M}_1^H$ instead of $\overline{M}_1$ shows that $v_1 \in \partial P\ (\overline{M}_1^H, \mathscr{R})$. Hence $u_1 \in S(\overline{M}_1^H, \mathscr{R})$, which proves (21).

We now apply the general theory to different choices of the set $W(M_1, \ldots, M_N)$. First we take $N = r + 1$ and choose β_{ij} so that

$$R(x_1, \ldots, x_{r+1}) = (x_1 - x_{r+1}) \wedge \cdots \wedge (x_r - x_{r+1}).$$

Denote the possibly degenerate r-simplex with vertices $q_1, \ldots, q_{r+1}$ by $T(q_1, \ldots, q_{r+1})$ and its r-dimensional measure or volume by $\hat{T}(q_1, \ldots, q_{r+1})$. If $q_i = P(x_i, \mathscr{R})$, then for R parallel to $\mathscr{R}$,

$$|R(x_i) \cdot R| = r!\, \hat{T}(q_1, \ldots, q_{r+1}).$$

Therefore

$$H(R, W) = |R| \max_{q_i \in P(M_i, \mathscr{R})} \hat{T}(q_1, \ldots, q_{r+1})$$

is totally convex.

Applying the maximum principle (14) to this function we find that, for $r < t \leqq \leqq n - 1$, the maximum of $\hat{T}(q_1, \ldots, q_{r+1})$ with $q_i \in P(M_i, \mathscr{T})$ induces a totally convex function on G_t^n. Thus we have proved:

(22) Theorem. *If* $M_1, \ldots, M_{s+1}$ $(1 \leqq s \leqq r \leqq n - 1)$ *are non-empty bounded sets in* E^n, *then the supremum of the s-dimensional measures of all simplexes* $T(q_1, \ldots, q_{s+1})$ *where* $q_i \in P(M_i, \mathscr{R})$ *induces a totally convex function on* G_r^n.

The convexity of this function is contained in Theorem (9) of Part IV.

An alternative way of deducing this result for the case $s < r$ from the case $s = r$ (without using (14)) is to take W as the set of all r-vectors

$$(x_1 - x_{s+1}) \wedge \cdots \wedge (x_s - x_{s+1}) \wedge x_{s+2} \wedge \cdots \wedge x_{r+1}$$

where $x_i \in M_i$ $(i = 1, \ldots, s + 1)$, $x_i \in B^n$ (the unit ball with origin as center) for $i = s = 1, \ldots, r + 1$.

The special case $s = 1$, and $M = M_1 = M_2$ of (22) means that the diameter of $P(M, \mathscr{R})$ induces a totally convex function on G_r^n.

Before proceeding to the discussion of a second choice for W, we require a preliminary result. A box (= parallelepiped) is *properly circumscribed* to the n-dimensional convex body K if its $(n - 1)$-dimensional faces lie in supporting hyperplanes of K, and if, for any pair S_j, S_j' of parallel faces of the box, points $x_j \in S_j \cap K$, $x_j' \in S_j' \cap K$ can be found such that S_j and S_j' are parallel to $\bigwedge_{i \neq j} (x_i - x_i')$. This implies, clearly, that $x_j - x_j'$ is parallel to all the edges of the box which are not edges of S_j or S_j', and that the volume of the box is $\left|\bigwedge_{i=1}^{n} (x_i - x_i')\right|$.

There exists at least one box properly circumscribed to every convex body K. This is trivial if K is centrally symmetric, and is established in all cases by the following, probably well-known, fact:

(23) *The box B of minimal volume containing the n-dimensional convex body K is properly circumscribed to K.*

Let E be any edge of B and S, S' the two $(n - 1)$-faces of B not parallel to E. Suppose the chord of K of maximal length parallel to E meets ∂K in y, y' and S, S' in z, z' respectively, then $|z - z'| \geqq |y - y'|$. Let B^* be the box of which two $(n - 1)$-faces lie in parallel supporting hyperplanes of K at y, y' and the remaining faces lie in the hyperplanes containing the $(n - 1)$-faces of B parallel to E. Then clearly

$$1 \geqq \frac{|y - y'|}{|z - z'|} = (\text{volume of } B^*)/(\text{volume of } B).$$

Since B has minimal volume, equality must hold, $|y - y'| = |z - z'|$ and so y, y' lie in the faces S, S'. Thus B is properly circumscribed to K.

Let $M_1, \ldots, M_r, N_1, \ldots, N_r$ be non-empty bounded sets in E^n, and put

(23a) $$R(x_i, y_i) = \bigwedge_{i=1}^{r} (x_i - y_i), \quad x_i \in M_i,\ y_i \in N_i.$$

If W consists of the set of these $R(x_i, y_i)$ then $H(R, W) = H_0(R, W)$ if either $M_i = N_i$ for some i, or $M_i = M_k$, $N_i = N_k$ for some pair $i \neq k$. Because of (23a) with $K_i = \overline{M}_i^H$, $L_i = \overline{N}_i^H$,

$$H(R, W) = \begin{cases} |R| \sup \bigwedge_{i=1}^{r} (p_i - q_i), & p_i \in P(M_i, \mathscr{R}), \quad q_i \in P(N_i, \mathscr{R}), \\ |R| \max \bigwedge_{i=1}^{r} (p_i - q_i), & p_i \in P(K_i, \mathscr{R}), \quad q_i \in P(L_i, \mathscr{R}). \end{cases}$$

Moreover, if $u_i \in \overline{M}_i$ and $v_i \in \overline{N}_i$ satisfy

$$H(R, W) = \left| \left\{ \bigwedge_{i=1}^{r} (u_i - v_i) \right\} \cdot R \right| > 0$$

then $u_i \in S(K_i, \mathscr{R})$, $v_i \in S(L_i, \mathscr{R})$, R parallel to $\mathscr{R}$. We have, additionally:

If $p_j = P(u_j, \mathscr{R})$, $q_j = P(v_j, \mathscr{R})$, then p_j and q_j lie in supporting $(r-1)$-flats of $P(K_i, \mathscr{R})$ and $P(L_i, \mathscr{R})$ parallel to $\pm \bigwedge_{i \neq j} (p_i - q_i)$.

For suppose that the points $p_i \in P(K_i, \mathscr{R})$, $q_i \in P(L_i, \mathscr{R})$ are given for all $i \neq j$, and $\bigwedge_{i \neq j} (p_i - q_i) \neq 0$. Then $\left| \bigwedge_{i=1}^{r} (p_i - q_i) \right|$ is maximal when the absolute value of the component of $(p_j - q_j)$ normal to $\bigwedge_{i \neq j} (p_i - q_i)$ is maximal. This happens only when p_j and q_j behave as asserted.

We apply this to the case where the M_i and N_i are all equal to the same n-dimensional convex body K. If $R \neq 0$ and $p_i, q_i \in P(K, \mathscr{R})$ are chosen so that

$$H(R, W) = |R| \left| \bigwedge_{i=1}^{r} (p_i - q_i) \right| > 0$$

our discussion shows that the box circumscribed to $P(K, \mathscr{R})$ with faces parallel to $\pm \bigwedge_{i \neq j} (p_i - q_i)$ $(i = 1, \ldots, r)$ is a properly circumscribed one of maximal volume. We have proved:

(24) Theorem. *Let K be an n-dimensional convex body in E^n. The maximum of the volumes of the r-dimensional boxes properly circumscribed to $P(K, \mathscr{R})$ induces a totally convex function on G_r^n $(1 \leqq r \leqq n-1)$.*

If $r = 2$ then it is easily seen that p_1, p_2, q_1, q_2 are the vertices of a quadrilateral of maximal area contained in $P(K, \mathscr{R})$. That this area induces a convex function on G_2^n was demonstrated in IV(11). This theorem also asserted that the maximum area of a pentagon inscribed in $P(K, \mathscr{R})$ induces a convex function on G_2^n, but it is easy to see that, in general, this latter function is not totally convex.

In E^4 let T^4 be the simplex with vertices $p_0 = z$ (the origin) and $p_i = (\delta_i^1, \delta_i^2, \delta_i^3, \delta_i^4)$ $(i = 1, 2, 3, 4)$. With the definitions and notations of Part III, there is a direction R_1 for which the shadow boundary $S(T^4, \mathscr{R}_1)$ (R_1 parallel to $\mathscr{R}_1$) is the skew pentagon p_0, p_1, p_2, p_3, p_4 and R_1 lies inside one of the domains, say $G_1 \subset G_2^4$, where $\hat{P}(T^4, \mathscr{R})$ is linear. Moreover,

$$j_1(R) \leqq \hat{P}(T^4, R)$$

with equality for $R \in G_1$. Then $\zeta = j_1(\widetilde{R})$ is a supporting $\binom{n}{r}$-flat of the graph of $\zeta = \hat{P}(T^4, R)$ in an affine space of dimension $\binom{n}{r} + 1$. Because of III (20) and (21) this supporting hyperplane is unique. But $j_1(\widetilde{R}) = \widetilde{A} \cdot \widetilde{R}$ where

$$\widetilde{A} = (x_{14}, x_{24}, x_{34}, x_{23}, x_{31}, x_{12}) = (0, 0, 1, 1, 0, 1)$$

which is not a simple 2-vector because its components do not satisfy the Plücker relation $x_{12}x_{34} + x_{13}x_{42} + x_{14}x_{23} = 0$.

If T^4 is considered as a set in E^n $(n > 4)$ the same result follows for general n. Thus we find, in conjunction with (22) and IV (11):

(25) *Let K be an n-dimensional convex body in E^n with $n > 3$. If $r = 2$, then the function induced on G_2^n by the maximal area of an s-gon contained in $P(K, \mathscr{R})$ is totally convex for $s = 3, 4$, convex but not always totally convex for $s = 5$, and is, in general, non-convex for $s > 5$.*

In IV (11) we proved that for $3 \leqq r \leqq n - 2$, the maximal volume of a polyhedron with $r + 2$ vertices contained in $P(K, \mathscr{R})$ induces a convex function on G_r^n. (25) raises the question whether these functions are totally convex. This is so, but the proof is much less obvious and requires some preliminary considerations.

Let $\mathscr{S}$ and $\mathscr{T}$ be an s-flat and a t-flat intersecting in a single point v. If $T(p_1, \ldots, p_{s+1}) \subset \mathscr{S}$, and $T(q_1, \ldots, q_{t+1}) \subset \mathscr{T}$ are possibly degenerate simplexes containing v, then the convex hull of $T(p_1, \ldots, p_{s+1}) \cup T(q_1, \ldots, q_{t+1})$ is called the *(s, t)-bipyramid* $B(p_i, q_j)$. We first show

(26) *Every convex polyhedron in an r-flat $\mathscr{R}$ with $r + 2$ vertices is an (s, t)-bipyramid for suitable s, t with $s + t = r$.*

Let $p_1, \ldots, p_{r+2}$ be the vertices of the given polyhedron. Since $r + 2$ points in $\mathscr{R}$ are affinely dependent,

$$\sum_{i=1}^{r+2} \lambda_i p_i = 0 \text{ with } \sum \lambda_i = 0,$$

and not all $\lambda_i = 0$. We assume that $\lambda_1, \ldots, \lambda_{s+1}$ are non-negative and the remaining λ_i are negative. Put

$$\mu_i = -\lambda_{s+1+i}, \quad q_i = p_{s+1+i}, \quad i = 1, \ldots, t + 1 = r - s.$$

Then

$$\sum_{i=1}^{s+1} \lambda_i p_i = \sum_{i=1}^{t+1} \mu_i q_i, \quad \sum_{i=1}^{s+1} \lambda_i = \sum_{i=1}^{t+1} \mu_i \neq 0,$$

so that

$$v = \sum_{i=1}^{s+1} \lambda_i p_i / \sum \lambda_i = \sum_{i=1}^{t+1} \mu_i q_i / \sum \mu_i \in T(p_1, \ldots, p_{s+1}) \cap T(q_1, \ldots, q_{t+1})$$

and so the given polyhedron is $B(p_i, q_j)$.

(27) *Let* $C(p_i, q_j)$ *be the convex hull of the* $s + t + 2 = r + 2$ *points* $p_1, \ldots, p_{s+1}, q_1, \ldots, q_{t+1}$, *and*

$$R(p_i, q_j) = (p_1 - p_{s+1}) \wedge \cdots \wedge (p_s - p_{s+1}) \wedge (q_1 - q_{t+1}) \wedge \cdots \wedge (q_t - q_{t+1}).$$

Then $|R(p_i, q_j)| \leqq \frac{1}{r!} |C(p_i, q_j)|_r$ *with equality if* $C(p_i, q_j)$ *is the* (s, t)*-bipyramid* $B(p_i, q_j)$.

The last statement about equality in the case of a bipyramid is the result of a simple calculation (see [4]). To establish the inequality in general, we first notice that if either of the simplexes $T(p_1, \ldots, p_{s+1})$ or $T(q_1, \ldots, q_{t+1})$ is degenerate, then $|R(p_i, q_j)| = 0$ and the result is true. If neither is degenerate, choose any interior points e_1 and e_2 of $T(p_i)$ and $T(q_j)$ respectively. The system of bodies $C(p_i + t(e_2 - e_1), q_j)$ is, as t varies, a *linear parameter system* (see [5]) and so

$$F(t) = \frac{1}{r!} |C(p_i + t(e_2 - e_1), q_j)|_r$$

is a convex function of t. Further it is easy to see that, because e_1 and e_2 are interior points of the simplexes, $C(p_i + t(e_2 - e_1), q_j)$ is a bipyramid for t in some neighborhood of $t = 1$, and so $F(t)$ has the constant value $|R(p_i, q_j)|$ in this neighborhood. Hence in the (ζ, t)-plane, the line $\zeta = |R(p_i, q_j)|$ is a supporting line to the graph $\zeta = F(t)$. Consequently $F(t) \geqq |R(p_i, q_j)|$ for all t. Putting $t = 0$ proves (27).

We can now prove:

(28) *Let K be an n-dimensional convex body in* E^n $(n > 3)$. *Then the maximal volume of a polyhedron with* $r + 2$ *vertices contained in* $P(K, \mathscr{R})$ *induces a totally convex function on* G_r^n.

Let W consist of the set of r-vectors,

$$R(x_i, s, j, k) = (x_{i_1} - x_j) \wedge (x_{i_2} - x_j) \wedge \cdots \wedge (x_{i_s} - x_j) \wedge (x_{i_{s+1}} - x_k) \wedge \cdots \wedge (x_{i_r} - x_k)$$

where $x_i \in K$ $(i = 1, \ldots, r + 2)$, $1 \leqq j < k \leqq r + 2$, $0 < s < r$ and $(i_1, i_2, \ldots, i_r)$ is any permutation of $(1, 2, \ldots, j - 1, j + 1, \ldots, k - 1, k + 1, \ldots, r + 2)$. Since

$$|R(x_i, s, j, k) \cdot R| = |R(p_i, s, j, k) \cdot R|$$

with $p_i = P(x_i, \mathscr{R})$ we see that

$$H(R, W) = \max_{p_i \in P(K, \mathscr{R}), s, j, k} |R(p_i, s, j, k) \cdot R|.$$

From (27) this is exactly the volume of the maximal bipyramid with $r + 2$ vertices contained in $P(K, \mathscr{R})$ multiplied by $|R| r!$. By (26) this maximal bipyramid is the maximal polyhedron with $r + 2$ vertices, whose volume therefore induces the totally convex function $\frac{1}{r!} H(R, W)$ on G_r^n. This proves (28).

Following IV (13) we conjectured that the volume of the maximal polyhedron with $r + 3$ vertices in $P(K, \mathscr{R})$ $(2 \leqq r \leqq n - 2)$ induces a convex function on G_r^n. We now make the additional conjecture (by analogy with the case $r = 2$ discussed in (25)) that though convex, this function is not always totally convex.

Note that the proof of (28) contains the more general form analogous to IV (11), i.e. the vertices of the polyhedron may be chosen one from each of the sets $P(M_i, \mathscr{R})$ where the M_i are arbitrary non-empty compact sets.

V.3. Centrally symmetric bodies and area. Let K be a convex body in E^n with the origin z as interior point, $H(v)$ and $d(x)$ the supporting and distance functions of K. For a unit vector u, the supporting plane of K with exterior normal u has distance $H(u)$ from z and the ray ϱu, $\varrho \geqq 0$ intersects ∂K at the point $u/d(u)$. If z is the center of K and $z \in \mathscr{R}$, then $\max_{u \mid \mathscr{R}} H(u)$ is the radius of the circumsphere of $P(K, \mathscr{R})$ whereas $\max_{u \mid \mathscr{R}} d(u)$ is the reciprocal of the radius of the insphere of $K \cap \mathscr{R}$.

The reciprocal body K^* of K has $H(x)$ as distance function and $d(v)$ as supporting function. Therefore we conclude from (16):

(29) *If K is a convex body in E^n with z as interior point and center, then the radius of the circumsphere of $P(K, \mathscr{R})$ and the reciprocal of the radius of the insphere of $K \cap \mathscr{R}$ $(z \in \mathscr{R})$ induce totally convex functions on G_r^n.*

Under reciprocation, the supporting plane of ∂K^* normal to u (with distance $d(u)$ from z) corresponds to the point $u/d(u)$ of ∂K. A supporting $(n - r)$-flat of K^* normal to $\mathscr{R}$ corresponds to a supporting $(r - 1)$-flat of $K \cap \mathscr{R}$. These facts enable us to deduce properties of polyhedra related to $K \cap \mathscr{R}$ from the properties of polyhedra related to $P(K^*, \mathscr{R})$ which we proved in the last section. We discuss one result of this type which is of considerable significance for the area or volume problem in Minkowski spaces.

The problem is whether it is possible to define, simultaneously for all symmetric Minkowski spaces X^r $(r = \dim X^r \geqq 2)$, a volume $\alpha_r(M, X^r)$ with the following properties:

1. α_r is defined for all Borel sets M in X^r, is countably additive, and is positive and finite on non-degenerate r-simplexes.

2. If Φ is an isometry of X^r on another r-dimensional Minkowski space X'^r, then

$$\alpha_r(M, X^r) = \alpha_r(\Phi(M), X'^r).$$

Writing $|M|_r$ for the Lebesgue measure corresponding to a euclidean metrization of X^r invariant under the translations of X^r, then 1. and 2. imply that $\alpha_r(M, X^r) = k\,|M|_r$, $k > 0$.

3. If the metric of X^r is euclidean, then α_r coincides with the corresponding Lebesgue measure.

It is trivial to find α_r satisfying these conditions; the decisive requirement consists of inequalities relating these volumes. Let X^n be an n-dimensional Minkowski space $(n > r \geqq 2)$. The metric in X^n restricted to an r-flat $\mathscr{R}$ in X^n is a Minkowski metric for $X(\mathscr{R})$ in $\mathscr{R}$, such that $\alpha_r(M, X(\mathscr{R}))$ is defined for Borel sets M in $\mathscr{R}$.

4. If $F_0, \dots, F_k$ are the r-faces of an oriented closed r-dimensional (possibly skew) polyhedron in X^n and $F_i \subset \mathscr{R}_i$, then

(30)
$$\alpha_r(F_0, X(\mathscr{R}_0)) \leqq \sum_{i=1}^{k} \alpha_i(F_i, X(\mathscr{R}_i)).$$

It is known (see below) that α_r exist satisfying 1., 2., 3. and 4. for $n = r + 1$. Our present methods provide a solution for which **4.** holds without restrictions. The most convenient way of attacking the problem is this:

Consider a euclidean metrization of X^n invariant under translations of X^n and let $|M|_r$ be the induced r-dimensional Lebesgue measure for Borel sets M lying in r-flats. Then any α_r satisfying 1., 2., 3. has the form

$$\alpha_r(M) = \Phi(\mathscr{R}_z)\,|M|_r, \quad M \subset \mathscr{R}, \quad R \,\|\, \mathscr{R}_z \ni z \tag{31}$$

where $\Phi(\mathscr{R}_z)$ is positive.

With each member of a class of Borel sets M lying in oriented r-flats parallel to $\mathscr{R}_z^+$ and having equal measures $|M|_r$ we associate the r-vector $R \,\|\, \mathscr{R}_z^+$ with length $|R| = |M|_r$. The definition

$$f(R) = \alpha_r(M) \text{ if } R \text{ is associated with } M$$

is independent of the intervening euclidean metric. A sufficient condition for (30) is that $f(R)$ be convex. Thus the problem is reduced to finding $\Phi(\mathscr{R}_z)$ inducing convex functions on G_r^n.

In the solution we are going to give, the induced function is even totally convex. The intuitive meaning of this property was mentioned in the introduction. Our definition will be based on

(32) Theorem. *Let K be a convex body in E^n with z as interior point and center. For any r-flat $\mathscr{R} \ni z$ $(2 \leqq r \leqq n - 1)$ let $Q(\mathscr{R})$ be the minimum of the volumes of all boxes in $\mathscr{R}$ containing $K \cap \mathscr{R}$. Then $1/Q(\mathscr{R})$ induces a totally convex function on G_r^n.*

Proof. Let K^* be the reciprocal of K and Q^* any r-dimensional cross-polytope with center z contained in $P(K^*, \mathscr{R})$. The reciprocals of Q^* and $P(K^*, \mathscr{R})$ are cylinders with $(n - r)$-dimensional generators perpendicular to $\mathscr{R}$ whose sections by $\mathscr{R}$ are, respectively, an r-dimensional box Q and $K \cap \mathscr{R}$. Moreover Q contains $K \cap \mathscr{R}$ and a simple calculation shows:

$$r!\,|Q^*|_r = 1/|Q|_r. \tag{33}$$

We now apply (22) to $M_1 = \cdots = M_r = K^*$ and M_{r+1} as the set consisting of the point z only. Let $T(q_1, \ldots, q_r; z)$ be a simplex of maximal volume in $P(K^*, \mathscr{R})$ and take Q^* as the convex hull of the points $\pm q_1, \ldots, \pm q_r$. Then

$$|Q^*|_r = 2^r\,\hat{T}(q_1, \ldots, q_r; z)$$

shows that Q^* is a cross polytope of maximal volume contained in $P(K^*, \mathscr{R})$. By (33) the corresponding box has minimal volume $Q(\mathscr{R})$ among those containing $K \cap \mathscr{R}$ and having center z. Further, our proof of (23) shows that a box of minimal volume containing $K \cap \mathscr{R}$ has center z.

The solution of the volume problem is given in the following corollary of (32):

(34) *Let X^r be an r-dimensional Minkowski space and U its unit ball. Define the volume $\alpha_r(M, X^r)$ in X^r so that (it is invariant under translations and) the box of*

minimal volume containing U has volume 2^r. Then α_r satisfies the conditions 1. *to* 4. *Moreover the corresponding function $f(R)$ is totally convex.*

For, since the box of minimal volume containing U is affinely related to U, the conditions 1. and 2. are satisfied. 3. holds because of the normalization 2^r for the volume of the minimal box.

Let X^n be an n-dimensional Minkowski space ($n > r \geqq 2$) and consider any euclidean metrization of X^n invariant under the translations of X^n with corresponding r-dimensional Lebesgue measure $|M|_r$. Let K with center z be the unit ball of X^n. If $Q(\mathscr{R}_z)$ is the minimum of the euclidean volumes of the boxes in $K \cap \mathscr{R}_z$ ($\mathscr{R}_z \in z$) containing $K \cap \mathscr{R}_z$, then

$$\alpha_r(M, X(\mathscr{R})) = 2^r Q^{-1}(\mathscr{R}_z) \, |M|_r, \quad M \subset \mathscr{R}, \ R \parallel \mathscr{R}_z$$

so that

$$\Phi(\mathscr{R}_z) = 2^r Q^{-1}(\mathscr{R}_z).$$

But (32) contains that the function induced by $\Phi(\mathscr{R}_z)$ on G_r^n is totally convex.

This volume α_r has recently been studied by Dr. R. B. Benson in his thesis (University of Southern California), the principal results of which will be published later.

The volume to which we referred above is obtained by the normalization

$$\alpha_r(U, X^r) = |B^r|_r = \pi^{r/2}/\Gamma\left(1 + \frac{r}{2}\right).$$

The corresponding function is

$$\Phi(\mathscr{R}_z) = \frac{|B^r|_r}{|U \cap \mathscr{R}_z|_r}.$$

The question whether, or under what conditions, this function $\Phi(\mathscr{R}_z)$ induces a convex function on G_r^n when $n > r + 1$ is a deep and interesting problem on convex bodies.

References

[1] N. Bourbaki, Algèbre Multilinéaire. Éléments de Math. I, Livre II, Chap. 3. Paris 1948.
[2] H. Busemann, Convexity on Grassmann manifolds. Enseignment Math. **7**, 139–152 (1961).
[3] H. Busemann, G. Ewald and G. C. Shephard, Convex bodies and convexity on Grassmann cones, Parts I to IV. To appear in Math. Ann.
[4] C. A. Rogers and G. C. Shephard, Convex bodies associated with a given convex body. J. London Math. Soc. **35**, 270–281 (1958).
[5] C. A. Rogers and G. C. Shephard, Some extremal problems for convex bodies. Mathematika **5**, 93–102 (1958).
[6] V. Wagner, Geometry of a space with an areal metric and its applications to the calculus of variations. Mat. Sbornik, n. S. **19** (41), 341–404 (1946) (Russian).

Eingegangen am 23. 6. 1962

Anschrift der Autoren:

H. Busemann
Department of Mathematics
University of Southern California
Los Angeles (Cal.), USA

G. Ewald
Mathematisches Institut
der Universität Mainz

G. C. Shephard
Department of Mathematics
University of Birmingham
Birmingham, England

Convex Bodies and Convexity on Grassmann Cones*

Parts I to IV

By

H. BUSEMANN in Los Angeles, G. EWALD in Mainz
and G. C. SHEPHARD in Birmingham, England

Many statements in the theory of convex bodies assert the convexity of a functional depending on one or more convex bodies and a variable line or hyperplane. The present series of articles has the broad goal of initiating a systematic study of such functionals where the line or hyperplane is replaced by a variable r-flat $(1 < r < n - 1)$.

For the time being we limit ourselves to problems concerning r-flats passing through a fixed point. The more general case is most interesting but far too difficult, in fact the very concept of convexity needs clarification. For example, the natural generalization of the Brunn-Minkowski Theorem for non-parallel hyperplanes deals with pencils of half-hyperplanes rather than entire hyperplanes (see BARTHEL [2]).

The oriented r-flats $\mathfrak{R}^+$ in E^n passing through a fixed point correspond to the simple unit r-vectors on the Grassmann cone G_r^n formed by the simple r-vectors R in the linear space V_r^n of all r-vectors, the norm $|R|$ of R being derived from the euclidean norm in E^n. We say that a function $f(\mathfrak{R}^+)$ induces the function $f(R) = |R|\, f(\mathfrak{R}^+)$ in G_r^n, where R is parallel to $\mathfrak{R}^+$.

Since G_r^n is, for $1 < r < n - 1$, not a convex subset of V_r^n, the convexity of a function $f(R)$ on G_r^n remains to be defined. A systematic discussion of convexity on G_r^n was given in [8], and several concepts (all equivalent for $r = 1$ or $n - 1$) emerged. Here we will be principally concerned with one of these: $f(R)$ is *convex*[1]) if it can be extended to a convex function in V_r^n. The most incisive difference between the cases $r = 1, n - 1$ and $1 < r < n - 1$ lies in the fact that for $1 < r < n - 1$ convexity cannot be reduced to a local condition, hence cannot be expressed in terms of second differentials, even when the function in question is smooth.

We can now formulate our aim as follows: to find general principles yielding convex functions on G_r^n and to decide, for interesting special functions

* This work was supported by a grant from the National Science Foundation, U.S.A.

[1]) This is called extendable convexity in [8]. We change the name because convexity

$f(R)$ derived from convex bodies, whether, or under what conditions, they are convex. Preliminary results in this direction are found in [6] and two specific questions are raised which may serve as examples here. If $H(u)$ is the supporting function of a convex body, does the integral of $H(u)$, where u traverses all unit vectors ending on the intersection of the unit sphere in E^n with a variable r-flat $\mathfrak{R}$ through its center z, induce a convex function on G_r^n? Does the r-dimensional volume of the projection $P(K, \mathfrak{R})$ of a convex body K on $\mathfrak{R}$ induce a convex function? Part II contains an affirmative answer to the first question, whereas Part III answers the second question negatively, but provides conditions under which $P(K, \mathfrak{R})$ leads to a convex function.

A more detailed description of the content is found in the individual parts. Part V of this series, on "total convexity" is found in Arch. Math. **13** 512–526 (1962). We conclude this general introduction with some agreements on notation to which we shall adhere throughout.

λ_i and μ_i $(i = 1, \ldots, r)$ denote non-negative real numbers satisfying $\Sigma\,\lambda_i = 1$ and $\Sigma\,\mu_i = 1$. The point x_i in the n-dimensional (real) affine space A^n has affine coordinates $x_i^1, \ldots, x_i^n$. With the set $\Lambda = \{\lambda_i, \ldots, \lambda_r\}$ we associate the point $x_\Lambda = \sum_{i=1}^{r} \lambda_i x_i$ whose j-th coordinate is $\sum_i \lambda_i x_i^j$. If N is any set in A^n then N^H is its convex hull. It consists of all points x_Λ with $x_i \in N$. The set N is convex if $N^H = N$.

We consider real valued functions $f(x)$ defined on a not necessarily convex set M in A^n, but will for convenience always assume that $\dim M^H = n$. For convex M the function $f(x)$ is convex if, for $x_i \in M$ the inequality $f(x_\Lambda) \leqq \sum_{i=1}^{r} \lambda_i f(x_i)$ always holds, and it suffices that it holds for $r = 2$. We emphasise this point because the inequality for $r = 2$ will, for non-convex M, in general not imply the inequality for $r \geqq 3$.

A function defined on $M \subset A^n$ is *homogeneous* of degree $p \geqq 0$ if M contains the origin, $M = \varrho M$ for $\varrho > 0$ and $f(\varrho x) = \varrho^p f(x)$ for $\varrho \geqq 0$, in particular $f(0) = 0$. If $p = 1$ we simply call $f(x)$ homogeneous. The term *gauge function* used by some authors for homogeneous convex functions defined in all of A^n will not be employed here. $f(x)$ is *absolutely homogeneous* if M contains the origin, $M = \varrho M$ for $\varrho \neq 0$ and $f(\varrho x) = |\varrho| f(x)$ for all real ϱ.

The agreements on notations in the special case $M = G_r^n$ and $M^H = V_r^n$ are found in Section I.4.

As stated above, we will frequently be concerned with functions depending on a variable oriented r-flat $\mathfrak{R}^+$ or non-oriented r-flat $\mathfrak{R}$ through a fixed point. Without mentioning it every time $\mathfrak{R}^+$ and $\mathfrak{R}$ therefore always mean flats passing through the point z in A^n which we choose as origin. In one or two cases an r-flat will vary in a bundle of parallel r-flats, then we will express this by the

Part I. Convex Functions Defined on Non-convex Sets

The first two sections develop the theory of convexity on general sets M in A^n. With every function f defined on M, and bounded below by a linear function in M^H, we associate a convex function g_f. If f is convex, i.e. extendable to a convex function in M^H, then g_f is an extension of f. The properties of g_f readily yield the basic facts: conditions equivalent to convexity, the preservation of convexity under certain operations, etc.

The problem of finding the conditions under which f can be extended to a strictly convex function in M^H proves more delicate, and is the topic of Section 3.

Finally we apply the results of Sections 1 and 2 to the special case where M is the set of all simple r-vectors, and M^H the space of all r-vectors. This is the only important case for Parts II, III and IV.

I.1. The function g_f

A function $l(x)$ defined on the subset M of A^n is linear if it can be extended to a linear function in M^H. The extension is unique because we assume $\dim M^H = n$. If $p \in M$ and U is a neighborhood of p relative to M, then a function which is linear in U, i.e., can be extended linearly to the convex hull U^H of U in A^n, possesses a unique linear extension to A^n (or M^H) if and only if $\dim U^H = n$.

Let $f(x)$ be defined in M. If a convex function $c(x)$ in M^H exists satisfying $c(x) \leqq f(x)$ in M, we call $f(x)$ *bounded linearly below* (b. l. b.). This terminology is justified by the fact that, if this condition is satisfied, a linear function $l(x)$ in M (or M^H) also exists such that $l(x) \leqq f(x)$ in M. For if x_0 is an interior point of M^H, the graph $\zeta = c(x)$ of $c(x)$ in A^{n+1} (with coordinates $x^1, \ldots, x^n, \zeta$) possesses at $(x_0, c(x_0))$ a supporting hyperplane $\zeta = l(x) \leqq c(x)$; compare (4).

(1) *Let $f(x)$ be defined and bounded linearly below on a set M. Then*

(1a) $$g_f(x) = \inf_{x = x_A} \sum_{i=1}^{r} \lambda_i f(x_i) \qquad x_i \in M \quad 0 < r < \infty$$

is a convex function in M^H satisfying

(1b) $$g_f(x) \leqq f(x) \quad \text{for} \quad x \in M\,.$$

The infimum in (1a) is taken over all λ_i, and $x_i \in M$ for which $x = x_A$. Clearly $g_f(x)$ is defined in M^H, is finite since

$$\Sigma\, \lambda_i f(x_i) \geqq \Sigma\, \lambda_i c(x_i) \geqq c(x)\,,$$

and 1(b) is trivial. To prove that $g_f(x)$ is convex, let x, y be any given points in M^H and $\varepsilon > 0$. Then λ_i, $x_i \in M$ $(i = 1, \ldots, r)$ and μ_j, $y_j \in M$ $(j = 1, \ldots, s)$ exist such that $x = x_A$, $y = y_M$, and

$$g_f(x) + \varepsilon \geqq \Sigma\, \lambda_i f(x_i), \quad g_f(y) + \varepsilon \geqq \Sigma\, \mu_j f(y_j)\,.$$

Hence for $0 < \lambda < 1$

(2) *Let $f(x)$ be defined and bounded linearly below on a set M. Then,*

$$g_f(x) = \sup h(x) ,$$

where the supremum is taken over all convex functions $h(x)$ defined on M^H with $h(x) \leqq f(x)$ for $x \in M$.

For, if $x = \Sigma \lambda_i x_i$, $x_i \in M$, then for each $h(x)$

$$h(x) \leqq \Sigma \lambda_i h(x_i) \leqq \Sigma \lambda_i f(x_i)$$

and hence $h(x) \leqq g_f(x)$ and so $\sup h(x) \leqq g_f(x)$. On the other hand $g_f(x)$ is one of the convex functions $h(x)$, and so (2) follows.

(3) *Let $f(x)$ be defined and bounded linearly below on a set M. If it is also homogeneous of degree p, then $g_f(x)$ is homogeneous of degree p.*

Firstly, since $M = \varrho M$, $M^H = \varrho M^H$. Also

$$\begin{aligned} g_f(\varrho x) &= \inf_{x_A = \varrho x} \Sigma \lambda_i f(x_i) = \inf_{(x/\varrho)_A = x} \Sigma \lambda_i f(x_i) \\ &= \varrho^p \inf_{(x/\varrho)_A = x} \Sigma \lambda_i f(x_i/\varrho) = \varrho^p \inf_{x = x_A} \Sigma \lambda_i f(x_i) , \end{aligned}$$

with the last equality because $\varrho M = M$.

Any real valued function $\Phi(x)$ defined on a set S of A^n, where $x = (x^1, \ldots, x^n)$, can be represented by a graph consisting of the points $(x^1, \ldots, x^n, \Phi(x))$ in A^{n+1} defined by coordinates $(x^1, \ldots, x^n, \zeta) =_{\text{def}} (x, \zeta)$. If we denote by $S(\Phi)$ the set of all points (x, ζ) with $x \in S$ and $\zeta \geqq \Phi(x)$, then we clearly obtain:

(4) *Let S be a convex set of A^n and Φ a real valued function defined on S, then Φ is a convex function if and only if $S(\Phi)$ is a convex set.*

(5) *If $f(x)$ is defined and bounded linearly below on a set M, then*

$$\overline{[M(f)]^H} = \overline{M^H(g_f)} .$$

If f_1, f_2 are defined in M, then $f_1 \leqq f_2$ implies $M(f_2) \subset M(f_1)$; and for f defined in M_2 it follows from $M_1 \subset M_2$ that $M_1(f) \subset M_2(f)$ (using f also for its restriction to M_1). So we have $M(f) \subset M(g_f) \subset M^H(g_f)$, and since $M^H(g_f)$ is convex,

$$[M(f)]^H \subset M^H(g_f) .$$

On the other hand we can find, for an arbitrary closed convex set C of A^{n+1} with $M(f) \subset C$, a convex h such that $M^H(h)$ is the intersection of C with the cylinder over M^H (generators parallel to the ζ-axis). Using (2) therefore,

$$M^H(g_f) = \bigcap_h M^H(h) \subset \bigcap_{\text{all } C \supset M(f)} C = \overline{[M(f)]^H} .$$

(6) *Let $f(x)$ be defined and bounded linearly below on a set M. Then if M^H is open,*

$$g_f(x) = \sup l(x) = \inf \sum^{n+1} \lambda_i f(x_i)$$

The first part follows because $M^H(g_f)$ possesses at $(x_0, g_f(x_0))$ a supporting hyperplane of the form $\zeta = l_0(x)$, which means $l_0(x) \leqq g_f(x)$ for all $x \in M^H$ and $l_0(x_0) = g_f(x_0)$. Thus $g_f(x) \leqq \sup l(x)$; but by (2), $g_f(x) \geqq \sup l(x)$, and so $g_f(x) = \sup l(x)$.

We remark that if M^H is not open then $g_f(x) \neq \sup l(x)$ is possible. (Take, for example, $M^H = [-1, 1]$ in A^1, and $g_f(x) = -\sqrt{1-x^2}$ if $|x| < 1$, but $g_f(x) = 1$ for $|x| = 1$.)

The second part of (6) reduces the general r in (1a) to $n+1$ and is less trivial. Let $\{C_\nu\}$ be an increasing sequence of compact convex sets whose union is M^H and let f_ν be the restriction of $f(x)$ to $C_\nu \cap M$. Then $M(f_\nu) \subset M(f_{\nu+1})$ and $\bigcup_\nu M(f_\nu) = M(f)$, hence $[M(f_\nu)]^H \subset [M(f_{\nu+1})]^H \subset [M(f)]^H$, but in general not $\bigcup_\nu [M(f_\nu)]^H = [M(f)]^H$. This means that $g_{f_\nu}(x) \geqq g_f(x)$, and that $g_{f_\nu}(x)$ will, in general, not be a restriction of $g_f(x)$. However $g_f(x) = \lim_\nu g_{f_\nu}(x)$.

Let s be given and choose ν so large that C_ν contains x in its interior. If $g_{f_\nu}(x) = f_\nu(x) = f(x)$, then the assertion is trivial. If this is not the case we deduce from the compactness of C_ν that there is a simplex of minimal dimension, say r $(\geqq 1)$ with vertices $q_1, \ldots, q_{r+1}$ in $M(f_\nu)$ containing $q = (x, g_{f_\nu}(x))$ as an interior point, i.e. $q = \sum_{i=1}^{r+1} \lambda_i q_i$, $\lambda_i > 0$. If P is a supporting hyperplane of $[M(f_\nu)]^H$ at q, then P must contain the simplex, since otherwise it would separate at least two of the points q_j $(j = 1, \ldots, r+1)$. Therefore $r \leqq n$. If we put $q_i = (x_i, \zeta_i)$ then $\zeta_i = f(x_i)$, for the contrary would imply that P separates $(x_i, f(x_i))$ from $(x_i, \zeta_i + 1)$. It follows that for each ν, there exist sequences $\lambda_{i\varrho}^\nu$ and $x_{i\nu}^{(\varrho)} \in C_\nu$ $(i = 1, \ldots, n+1, \varrho = 1, 2, \ldots)$ with $x_{i\nu}^{(\varrho)} \to x_{i\nu}$, such that

$$g_{f_\nu}(x) = \lim_\varrho \sum_{i=1}^{n+1} \lambda_{i\varrho}^\nu f(x_{i\nu}^{(\varrho)}) .$$

Because $g_{f_\nu}(x) \to g_f(x)$ as $\nu \to \infty$,

$$g_f(x) = \lim_{\varrho,\nu} \sum_{i=1}^{n+1} \lambda_{i\varrho}^\nu f(x_{i\nu}^{(\varrho)}) \geqq \inf_{\varrho,\nu} \sum_{i=1}^{n+1} \lambda_{i\varrho}^\nu f(x_{i\nu})$$

$$\geqq \inf \sum_{i=1}^{r} \lambda_j f(x_j) = g_f(x) .$$

Note that the following is, in general not correct: If for some fixed $k < n+1$, each point x of M^H is representable in the form $x = \sum_{i=1}^{k} \lambda_i x_i$, $x_i \in M$, then $\inf_{x = x_\Lambda} \sum_{i=1}^{k} \lambda_i f(x_i)$ is convex. However, $n+1$ can be reduced to n in an important special case:

(7) *Let $f(x)$ be defined and bounded linearly below on an open set M. If f is*

From homogeneity of f we know the origin belongs to M, $\varrho M = M$ for $\varrho > 0$, $f(\varrho x) = \varrho f(x)$, and $f(0) = 0$. Applying (3) we have

$$g_f(\varrho x) = \varrho g_f(x) \quad \text{for} \quad \varrho > 0\,,$$

and since $g_f(x)$ is (as a convex function on an open set) continuous, $f(0) = g_f(0) = 0$.

The assertion is trivial for any x satisfying $g_f(x) = f(x)$. Take any point $q = (x, g_f(x))$ with $g_f(x) \neq f(x)$, and a supporting plane P of $M^H(g_f)$ at q. This hyperplane passes through $v = (0, 0)$ and contains the ray from v through q. As in the preceeding proof let $q_1, \ldots, q_{r+1}$ be a simplex of minimal dimension r $(1 \leq r \leq n)$ with vertices in $\overline{M(f)}$ containing q. Then the q_i lie in P, and for $x_i^{(\varrho)} \to x_i$, $q_i = (x_i, \lim_\varrho f(x_i^{(\varrho)}))$. If $\lim_\varrho f(x_i^{(\varrho)}) = 0$ for some i, say $i = r + 1$, then the assertion is trivial because

$$g_f(x) = \lim_\varrho \sum_{i=1}^{r} \lambda_i (1 - \lambda_{r+1})^{-1} f((1 - \lambda_{r+1}) x_i^{(\varrho)})\,.$$

If no $f(x_i)$ vanishes we take a supporting $(n-1)$-flat F through q, which may, or may not be unique, and project the points q_i parallel to the ζ-axis on F, obtaining points q_i'. These points lie in $\overline{M(f)}$ and their convex hull contains q. Since the q_i were vertices of a simplex of minimal dimension containing q we find $r \leq n - 1$.

I.2. Convex functions

A function defined in an arbitrary set M in A^n is called *convex* if it is the restriction to M of a convex function (in the usual sense) defined on M^H. Alternatively this can be expressed by saying that $f(x)$ may be extended to a convex function in M^H.

Such a function has the property that its restriction to any subset of M is also convex. We now list some basic properties of these functions.

(8) *A function $f(x)$ defined on M is convex if and only if it is bounded linearly below and the inequality*

$$f(x_\Lambda) \leq \sum_{i=1}^{r} \lambda_i f(x_i) \tag{8a}$$

holds for all $x_1, \ldots, x_r$ and $x_\Lambda = \sum \lambda_i x_i$ lying in M.

The conditions are clearly necessary and (8a) guarantees that for $x \in M$,

$$g_f(x) = \inf_{x = x_\Lambda} \sum \lambda_i f(x_i) = f(x)$$

so that $g_f(x)$ is a convex extension of $f(x)$ to M^H.

(9) *If M^H is open, then either of the following conditions is necessary and sufficient for $f(x)$ to be convex in M:*

(I) *$f(x)$ is bounded linearly below and the inequality*

$n+1$

(II) *for each point* $x_0 \in M$ *there exist numbers* $a_i(x_0)$ $(i = 1, \ldots, n)$ *such that*

$$f(x) \geqq \sum_{i=1}^{n} a_i(x_0)\,(x^i - x_0^i) + f(x_0) =_{\text{def}} l_{x_0}(x)\,.$$

The necessity and sufficiency of I follow immediately from (8) and (6).

If $f(x)$ is convex and $g(x)$ is a convex extension of $f(x)$ to M^H, then $M^H(g)$ is convex and so has a supporting hyperplane $\sum_{i=1}^{n} a_i(x_0)\,(x^i - x_0^i) + g(x_0) = 0$ at x_0. Hence for all x in M^H,

$$g(x) \geqq \sum_{i=1}^{n} a_i(x_0)\,(x_i - x_0^i) + g(x_0)\,.$$

In particular if $x \in M$, this reduces to the inequality of II, thereby proving necessity.

If II is satisfied, then for any x_0, $f(x) \geqq l_{x_0}(x)$ proves that $f(x)$ is b. l. b. By (2)

$$h(x) = \sup l_{x_0}(x) \leqq g_f(x) \quad \text{for} \quad x \in M\,.$$

Therefore $h(x_0) \geqq f(x_0)$ and $g_f(x_0) \leqq f(x_0)$ gives $h(x_0) = f(x_0)$.

If $f(x)$ is homogeneous and the set $M(f)$ contains a point $(x, f(x))$, $x \neq 0$, then it also contains the ray $(tx, tf(x)) = (tx, f(tx))$, $t \geqq 0$. A supporting hyperplane of $M(f)$ at $(x, f(x))$ must therefore contain the entire ray. Hence it suffices to establish II for one point different from 0 on each ray in M issuing from the origin. Moreover the constant term in II must vanish, so $f(x_0) = \Sigma\, a_i(x_0)\,x_0^i$, and II reads

(II a) $f(x) \geqq \sum_{i=1}^{n} a_i(x_0)\,x^i$ *with* $f(x_0) = \sum a_i(x_0)\,x_0^i$ *for* $x_0 \neq 0$.

From (3) and (7),

(10) *If $f(x)$ is a homogeneous function defined and bounded linearly below on M, and if M^H is open, then $f(x)$ is convex if and only if*

$$f(x_\Lambda) \leqq \sum_{i=1}^{n} \lambda_i f(x_i)$$

for all $x_1, \ldots, x_n$ *and* $x_\Lambda = \sum_{i=1}^{n} \lambda_i x_i$ *in* M.

The convex extension $g_f(x)$ of $f(x)$ will also, in this case, be homogeneous. We can rewrite (10) in the form

(11) *If $f(x)$ is a homogeneous function defined and bounded linearly below in M, and if M^H is open, then $f(x)$ is convex if and only if*

$$f(x) \leqq \sum_{i=1}^{n} f(x_i)$$

for all $x_1, \ldots, x_n$ *and* $x = \Sigma\, x_i$ *in* M.

Many of the usual theorems on the preservation of convexity under certain

If $f_1(x), \ldots, f_m(x)$ are non-negative and convex in M and $p \geqq 1$, then $\left[\sum_{i=1}^{m} f_i^p(x)\right]^{1/p}$ *is convex in M.*

If $\mathfrak{f} = \{f(x)\}$ is a set of convex functions defined on M, and $F(x) = \sup_{f \in \mathfrak{f}} f(x)$ is finite, then $F(x)$ is a convex function defined on M.

If $f_1(x), f_2(x), \ldots,$ is a sequence of convex functions defined on a set M and if $F(x) = \lim\sup f_i(x)$ is finite, then $F(x)$ is a convex function defined on M.

As a sample we produce a proof of the third statement. To each $f(x)$ in the family $\mathfrak{f}$ there exists a convex extension $g(x)$ to M^H. Let $G(x) = \sup g(x)$, the supremum extended over all $g(x)$. Clearly $F(x) = G(x)$ for $x \in M$. Given an $x = \Sigma\, \lambda_i x_i$ then $G(x) \leqq \Sigma\, \lambda_i \sup f(x_i) = \Sigma\, \lambda_i G(x_i) < \infty$. So $G(x)$ is a convex extension of $F(x)$ to M^H.

The integral analogues of the first and second assertions are:

(13) *Let $f(x, w)$ be a convex function of x in M for each w in a set W, in which a (non-negative) measure μ is defined. If $\alpha(x)$ is any non-negative function defined on W, then*

$$F_1(x) = \int_W \alpha(w)\, f(x, w)\, (d\mu)_W$$

is convex in M, provided it is bounded linearly below and the integral exists for each $x \in M$.

If, further, $f(x, w)$ is non-negative for all $x \in M$ and all $w \in W$, and $p > 1$, then

$$F_p(x) = \left[\int_W \alpha(w)\, f^p(x, w)\, (d\mu)_W\right]^{1/p}$$

is convex in M, provided the integral exists for each $x \in M$.

For a proof of the first part, we observe that $F(x)$ satisfies condition (8a). Considering the second part we know that $F_p(x)$ is b. l. b. by the constant function 0. Also, if $x_1, \ldots, x_r$ and $x_\Lambda = \Sigma\, \lambda_i x_i$ are in M, then

$$\begin{aligned} F_p(x) &= \left[\int_W \alpha(w)\, f^p(x, w)\, (d\mu)_W\right]^{1/p} \\ &\leqq \left[\int_W \alpha(w)\, [\Sigma\, \lambda_i f^p(x_i, w)]\, (d\mu)_W\right]^{1/p} \\ &\leqq \Sigma\, \lambda_i \left[\int_W \alpha(w)\, f^p(x_i, w)\, (d\mu)_W\right]^{1/p} \leqq \Sigma\, \lambda_i F_p(x_i) \end{aligned}$$

by Minkowski's inequality.

If $g_f(x, w)$ is the extension of $f(x, w)$ as a function of x, and $\alpha(w)\, g_f(x, w)$ or $\alpha(w)\, g_f^p(x, w)$ is integrable on W, then

$$\int_W \alpha(w)\, g_f(x, w)\, (d\mu)_W \quad \text{or} \quad \left[\int_W \alpha(w)\, g_f^p(x, w)\, (d\mu)_W\right]^{1/p}$$

We know from (3) that $g_f(\varrho x) = \varrho^p g_f(x)$. Since also $g_f(x) \geqq 0$, we can extend $\Phi(x)$ to M^H by putting $\Phi(x) = (g_f(x))^{1/p}$, and so we must show that $\Phi(x+y) \leqq \Phi(x) + \Phi(y)$ for each pair $x, y \in M^H$. The assertion is trivial if $\Phi(x) = \Phi(y) = 0$. If $\Phi(x) > 0$, $\Phi(y) > 0$, then

$$g_f(x/\Phi(x)) = g_f(x)/\Phi^p(x) = 1$$

and similarly $g_f(y/\Phi(y)) = 1$. Following a standard pattern,

$$g_f\left(\frac{x+y}{\Phi(x)+\Phi(y)}\right) = g_f\left(\frac{x}{\Phi(x)} \cdot \frac{\Phi(x)}{\Phi(x)+\Phi(y)} + \frac{y}{\Phi(y)} \cdot \frac{\Phi(y)}{\Phi(x)+\Phi(y)}\right)$$
$$\leqq \frac{\Phi(x)}{\Phi(x)+\Phi(y)} \cdot 1 + \frac{\Phi(y)}{\Phi(x)+\Phi(y)} \cdot 1 = 1\,.$$

Therefore also $\Phi\left(\frac{x+y}{\Phi(x)+\Phi(y)}\right) = \frac{\Phi(x+y)}{\Phi(x)+\Phi(y)} \leqq 1$. Finally if $\Phi(x) > 0$ and $\Phi(y) = 0$, we have, for $0 < t < 1$,

$$g_f\left(\frac{x+y}{\Phi(x)}\right) = g_f\left(t \cdot \frac{x}{t\Phi(x)} + (1-t) \cdot \frac{y}{(1-t)\,\Phi(x)}\right)$$
$$\leqq t\, g_f\left(\frac{x}{t\Phi(x)}\right) = t^{1-p}.$$

Hence $\Phi\left(\frac{x+y}{\Phi(x)}\right) \leqq t^{(1-p)/p}$. For $t \to 1$ we find $\Phi(x+y) \leqq \Phi(x)$ and so the assertion is true in this case also.

We have so far principally discussed the similarities between convex functions defined on a non-convex set, and convexity in the usual sense. The most profound difference, which makes many problems much more difficult, is that convexity on a non-convex set cannot be reduced to a local property.

For example, if M consists of two disjoint open sets S_1, S_2 in A^n, and $f(x) = i$ on S_i $(i = 1, 2)$, then $f(x)$ is locally convex, but not convex. Similar phenomena occur for functions on Grassmann cones, but we postpone the discussion of these until we can provide intrinsically interesting examples (III.4).

I.3. Strictly convex functions

We say that a function $f(x)$ defined on a convex set M of A^n is *strictly convex* if

(15) $\qquad f(x_A) < \sum_{i=1}^{r} \lambda_i f(x_i)$ for all $x_1, \ldots, x_r$ and $x_A = \sum \lambda_i x_i$

in M, $x_i \neq x_A$ $(i = 1, \ldots, r)$.

The restriction of $f(x)$ to an arbitrary subset of M is then also called *strictly convex*.

(16) *Let $f(x)$ be defined on a convex set M. Then the following statements are equivalent:*

(c) $f(x)$ *is convex on* M *and locally strictly convex, i.e. for each* $x \in M$ *there exists a spherical neighborhood* U *of* x *such that* $f(x)$ *is strictly convex in* $U \cap M$.

Clearly (a) implies (b) and (c). If $f(x)$ is convex but not strictly convex then $\zeta = f(x)$ contains a simplex and hence three distinct collinear points. Therefore (b) implies (a). If $f(x)$ is convex and $\zeta = f(x)$ has three distinct collinear points then $\zeta = f(x)$ contains a line segment and $f(x)$ is hence not locally strictly convex. So (a) follows from (c).

In the case where M is not convex, (15) is obviously necessary but not sufficient for strict convexity, as a simple example will show: Let M consist of the intervals $0 < x \leqq 1$ and $2 \leqq x < 3$ and let $y = f(x)$ be defined by

$$(x-1)^2 + (y-1)^2 = 1, y > 1 \quad \text{in} \quad 0 < x \leqq 1,$$

$$(x-2)^2 + (y-1)^2 = 1, y > 1 \quad \text{in} \quad 2 \leqq x < 3.$$

Then condition (15) is satisfied, but $f(x)$ cannot be extended to a strictly convex function in $M^H = (0, 3)$.

Another necessary condition is, for open M^H, that for each x_0 in M, numbers $a_i(x_0)$ exist such that

$$f(x) > a_i(x_0)(x^i - x_0^i) + f(x_0) \quad \text{for} \quad x \in M - \{x_0\}. \tag{17}$$

This condition is easily seen to be stronger than (15). The function $f(x)$ in the above example satisfies (15) but not (17). It will satisfy (17) as well if restricted to $0 < x < 1$, $2 < x < 3$, so that (17) is also not sufficient.

(18) **Theorem.** *Let* $f(x)$ *be defined on the set* M *in* A^n, *where* M^H *is open and* M *is closed in* M^H. *Then the following is a necessary and sufficient condition for* $f(x)$ *to be strictly convex on* M:

For each $x \in M$ *there exist* n *numbers* $a_i(x)$ *with the property that, for any given* $x_0 \in M$ *and* $1 > \delta > 0$, *we can find an* $\varepsilon = \varepsilon(x_0, \delta)$ *such that,*

$$f(x_0) > \Sigma a_i(x)(x_0^i - x^i) + f(x) + \varepsilon, \quad \textit{and} \tag{18a}$$

$$f(x) > \Sigma a_i(x_0)(x^i - x_0^i) + f(x_0) + \varepsilon|x_0 - x| \tag{18b}$$

whenever $|x - x_0| > \delta$, $x \in M$.

For any $x' \in M$, the equation

$$\zeta = a_i(x')(x^i - x'^i) + f(x') =_{\text{def}} l_{x'}(x)$$

defines in A^{n+1} a supporting hyperplane $l_{x'}$ of $M^H(g_f)$ at $(x', f(x'))$.

Let $g(x)$ be a strictly convex function on M^H with restriction f to M. The necessity of (18a) is obvious. To show the necessity of (18b) we join $(x_0, f(x_0)) =_{\text{def}} p_0$ by rays to all points $(y, g(y))$, $y \in M^H$ with $|y - x_0| = \delta$. These rays form a cone with vertex p_0. Clearly none of the rays is contained in l_{x_0}, and therefore the set of points $(y, g(y))$, being compact in $M^H(g)$, has a positive "vertical" distance d from l_{x_0}, i.e. a distance measured parallel to the ζ-axis. Putting $\varepsilon = d/\delta$ we obtain (18b).

Define $h(x) = \sup l_{x'}(x)$ where the supremum is taken over all $x' \in M$. Clearly $h(x) \leqq g_f(x)$ for $x \in M^H$, and $h(x) = g_f(x) = f(x)$ for $x \in M$. Suppose $h(x) = g_f(x)$ for some $x \in M^H - M$. Then $(x, h(x))$ is interior either to an r-dimensional simplex P or a semi-infinite "prism" Q with an s-dimensional simplex as basis, such that P or Q lies on the boundary of $M^H(g_f)$ $(1 \leqq r \leqq n)$ and all (finite) vertices of P or Q have x-coordinates in M.

In the case of a simplex P, let $q_i = (x_i, \zeta_i)$ be the vertices of P $(i = 1, \ldots, r+1)$. Then $x_i \in M$ since M is closed in M^H. There exists a convergent sequence $\{l_{x_\nu}\}$ of hyperplanes, where $\{x_\nu\} \in M$ is a Cauchy sequence in M or tends to infinity, such that $(x, g_f(x)) \in \lim_\nu l_{x_\nu}$ and hence $P \subset \lim_\nu l_{x_\nu}$. The limit of x_ν is different from at least one of the x_i and therefore we can find a $\delta > 0$ such that $(x_i, f(x_i))$ is not bounded away from the hyperplanes $l_{x'}$ with $|x' - x| > \delta$. This contradicts (18a).

In the case of an r-dimensional "half space" Q let the finite vertices of Q be $q_i = (x_i, \zeta_i)$, $i = 1, \ldots, s+1$, $0 \leqq s < r$. Again there exists a convergent sequence $\{l_{x_\nu}\}$ of hyperplanes with $Q \subset \lim_\nu l_{x_\nu}$, where $\{x_\nu\} \subset M$ is either a Cauchy sequence or tends to infinity. If $s \geqq 1$ or if $s = 0$ and $\{x_\nu\}$ does not tend to x_1 we conclude as before. Otherwise the limit hyperplane l_{x_1} of $\{l_{x_\nu}\}$ is not in the sense of (18b) bounded away from the points at infinity of Q. This completes the proof that $h(x) \neq g_f(x)$ for $x \in M^H - M$ and we conclude that $M^H(h) - M^H(g_f)$ is open in $M^H(h)$.

Let $p_1 = (x_1, \zeta_1)$ be a point with $x_1 \in M^H - M$ and $h(x_1) < \zeta_1 < g_f(x_1)$. We join p_1 with all points of $M^H(g_f)$ by rays starting at p_1, obtaining a cone C with apex p_1. If the boundary of C contained an entire line, then $M^H(g_f)$ would also contain an entire line contrary to what we have shown above. Hence we can find a hyperplane E through p_1 intersecting C only in p_1, and a closed segment (cap) S of the surface of an n-dimensional sphere tangent to E at p_1, such that

(a) p_1 is interior to S (on the surface of the respective sphere),
(b) S is contained in $M^H(h) - M^H(g_f)$,
(c) no tangent plane of S meets $M^H(g_f)$, and
(d) S represents a convex function $\zeta = s(x)$.

We restrict $\zeta = s(x)$ to a closed spherical neighborhood N_1 of x_1 contained in M^H for which $s(x)$ is defined, obtaining a function $\zeta = s_1(x)$. Putting

$$f_1(x) = \begin{cases} f(x) & \text{if } x \in M \\ s_1(x) & \text{if } x \in N_1 \end{cases}$$

we arrive at a function f_1 that satisfies all conditions of (18) for $M \cup N_1$ instead of M.

instead of M. It remains to be shown that we can select the N_ν such that

$$M^H = M \cup N_1 \cup \cdots \cup N_\nu \cup \cdots$$

and f_ν tends to strictly convex limit function.

Let $\varepsilon > 0$ be given. We can proceed in such a way that after finitely many, say ν_1 steps, $M^H - (M \cup N_1 \cup \cdots \cup N_{\nu_1})$ contains no segment of length $\geqq \varepsilon$ except possibly inside an ε-neighborhood of $M \cup \partial M^H$ (here ∂ denotes the boundary) or outside a ball of radius 1 about a fixed point $x^* \in M$. After $\nu_2 - \nu_1$ further steps $M^H - (M \cup N_1 \cup \cdots \cup N_{\nu_2})$ can be assumed to have no segment of length $\geqq \varepsilon/2$ except possibly inside an $(\varepsilon/2)$-neighborhood of $M \cup \partial M^H$ or outside a ball of radius 2 about x^*, and so on. $\lim_l f_{\nu_l}$ clearly exists and is strictly convex on M^H.

It is possible to find conditions for strict convexity of $f(x)$ on an arbitrary set of A^n. But all cases which will occur in the following sections are covered by (18). The condition that M be closed in M^H, may be omitted if (18a, b) are strengthened:

(19) *Let M^H be open. Then a necessary and sufficient condition for a function $f(x)$ defined on M to be strictly convex is:*

For each $x \in M$ there exist n numbers $a_i(x)$ with the property that, for any given $x_0 \in M$ and $1 > \delta > 0$, we can find an $\varepsilon = \varepsilon(x_0, \delta)$ such that

(19a) $$f(x_1) > a_i(x)\,(x_1^i - x^i) + f(x) + \varepsilon,$$

and

(19b) $$f(x) > a_i(x_1)\,(x^i - x_1^i) + f(x_1) + \varepsilon\,|x_1 - x|$$

whenever $x, x_1 \in M$, $|x - x_0| > \delta$ and $|x_1 - x_0| < \delta/2$.

The necessity of the conditions is seen as in the proof of (18). To prove the sufficiency we define $f(x)$ for x in the relative closure of M in M^H by $f(x) = \lim_{x_\varrho \to x} f(x_\varrho)$ where $x_\varrho \in M$ (this limit exists since $g_f(x)$ is continuous on M^H). Let a δ_0 be given simultaneously for all x_ϱ and let $\varepsilon_\varrho = \varepsilon_\varrho(x_\varrho, \delta_0)$ be the numbers assumed in our theorem. For a sufficiently large ϱ, say ϱ_0, we find a $\delta' < \delta_0/2$ such that $|x - x_{\varrho_0}| < \delta'/2$ and $|x_\varrho - x_{\varrho_0}| < \delta'/2$ for all $\varrho \geqq \varrho_0$. This implies that $\inf_\varrho \varepsilon_\varrho > 0$. If we choose $\delta < \delta'/2$ and put $\varepsilon(x, \delta) = \frac{1}{2} \inf_\varrho \varepsilon_\varrho$ then conditions (19a) and (19b) are satisfied, and so, in particular, are (18a) and (18b). Now (19) follows from (18).

Note. It can be shown by counter-examples that (18a) and (18b) are in general not sufficient if M is not closed in M^H.

I.4. Convexity on Grassmann Cones

Considering A^n also as an n-dimensional vector space over the reals, we denote by V_r^n the set of all r-vectors over A^n, and by G_r^n the Grassmann cone of all simple (or decomposable) r-vectors. It will be convenient to also interpret

Clearly $\varrho G_r^n = G_r^n$ for $|\varrho| > 0$. Almost all functions $f(R)$ which we consider will be homogeneous. A necessary condition for $f(R)$ to be convex is that $f\left(\sum_{i=1}^{k} R_i\right) \leq \sum_{i=1}^{k} f(R_i)$, whenever ΣR_i is simple, and it suffices by (10) that this inequality holds for $k \leq \binom{n}{r}$.

The note after the proof of (6) was made against the following fact as background: any 2-vector R in V_2^n $(n \geq 3)$ can be represented as the sum of $n-2$ or fewer simple 2-vectors. This is true for $n = 3$ because every 2-vector is simple in that case. The general case is proved by induction: If the statement is true for $n-1 \geq 3$ and

$$e_i = (\delta_i^1, \delta_i^2, \ldots, \delta_i^n) \qquad (i = 1, \ldots, n)$$

are the coordinate vectors in A^n, then

$$\tilde{R} = \sum_{i<j\leq n} \alpha_{ij} e_i \wedge e_j = \left(\sum_{i=1}^{n-1} \alpha_{in} e_i\right) \wedge e_n + \sum_{i<j\leq n-1} \alpha_{ij} e_i \wedge e_j .$$

By the induction hypothesis the last term can be represented as the sum of at most $n-3$ simple vectors, and the first term on the right is also a simple vector.

As the note after the proof of (6) pointed out, in this case

$$g(\tilde{R}) =_{\text{def}} \inf \sum_{i=1}^{n-2} f(R_i), \quad \tilde{R} = \sum_{i=1}^{n-2} R_i$$

will not, unfortunately, in general provide a convex function $g(\tilde{R})$ in V_r^n.

A euclidean norm in A^n induces a norm or metric in V_r^n. We call the norm in A^n cartesian if $|x| = [\Sigma (x^i)^2]^{1/2}$. The induced norm in V_r^n is also cartesian. The theory of convex bodies always uses cartesian norms. For the more general norm

$$|x| = (\Sigma g_{ik} x^i x^k)^{1/2}$$

the supporting function $H(u)$ of a convex body K must be defined as

$$H(u) = \sup_{x\in K} \Sigma g_{ik} x^i u^k ,$$

or, if we want to preserve the simple original form, as

$$H(u) = \sup_{x\in K} \Sigma x^i u_i = \sup_{x\in K} \Sigma x_i u^i ,$$

where the x_i and u_i are the covariant components of x and u. We use this old-fashioned parlance[2]) because interpreting u as a vector in the dual space

[2]) Its relation to BOURBAKI [4] terminology is this:

A scalar product $x \cdot y$ in an n-dimensional real vector space V defines an isomorphism Φ_1 of the dual space V^* onto V by $x \cdot y = \langle x, y^* \rangle$, which canonically induces an isomorphism Φ_r of $\Lambda^r(V^*)$ onto $\Lambda^r(V)$. With dual bases in V and V^* the components of $\tilde{R} \in \Lambda^r(V)$ $(1 \leq r \leq n-1)$ are the contravariant components and those of $\Phi_r^{-1}(\tilde{R})$ are the covariant components of $\tilde{R}$. Concerning the scalar product $\tilde{R} \cdot \tilde{S}$ of an r-vector and an s-vector, if obscures the geometrically relevant picture of u as a vector in the original space normal to a supporting hyperplane of K.

The distinction between covariant and contravariant components can be avoided by using cartesian norms only, but it is often useful to know the formulae for general norms.

Let $1 \leqq s < r \leqq n$, $b = r - s$ and R be a non-vanishing r-vector parallel to the oriented r-flat $\mathfrak{R}^+$. If S is a (simple) s-vector parallel to $\mathfrak{S}^+ \subset \mathfrak{R}$, then the *supplementary* (simple) *b-vector B to S relative to R* (or $\mathfrak{R}^+$) is defined as follows: $B = 0$ for $S = 0$. If $S \neq 0$, then B is parallel to the b-flat in $\mathfrak{R}^+$ normal to $\mathfrak{S}^+$, $|B| = |S|$, and $R \wedge S = \lambda B$ with $\lambda > 0$. For $r = n$ this yields the standard definition of the supplementary vector depending on the orientation of A^n.

The scalar product $\tilde{R} \cdot \tilde{S}$ of an r-vector $\tilde{R}$ $(1 \leqq r \leqq n-1)$ and an s-vector $\tilde{S}$ is a scalar for $r = s$ and an $(|r-s|)$-vector for $r \neq s$. Let $r \geqq s$ and $b = r - s$. If $R^{i_1 \ldots i_r}$ are the contravariant components of $\tilde{R}$ and $S_{j_1 \ldots j_s}$ are the covariant components of $\tilde{S}$, then the contravariant components of $\tilde{R} \cdot \tilde{S}$ are

$$(20) \qquad (\tilde{R} \cdot \tilde{S})^{i_1 \ldots i_b} = \sum_{j_1 < \cdots < j_s} R^{i_1 \ldots i_b j_1 \ldots j_s} S_{j_1 \ldots j_s}$$

where $i_1 < \cdots < i_b, j_1, \ldots, j_s$ are r different numbers among $1, 2, \ldots, n$, and, when $b = 0$, $(\tilde{R} \cdot \tilde{S})^{i_1 \ldots i_b} = \tilde{R} \cdot \tilde{S}$.

For simple vectors R, S the vector $R \cdot S$ $(r > s)$ is simple. If $R \neq 0$ then $R \cdot S$ is the supplementary vector relative to R of the projection S' of S on $\mathfrak{R}$ multiplied by $|R|$. To verify this, since $R \cdot S$ is known to be independent of the coordinates, we can use a cartesian norm and choose the coordinates so that

$$R = R^{1 \ldots r} e_1 \wedge \cdots \wedge e_r, \qquad R^{1 \ldots r} = |R| > 0\,.$$

Then $(R \cdot S)^{i_1 \ldots i_b} = 0$ if one $i_k > r$ and

$$(21) \qquad (R \cdot S)^{i_1 \ldots i_b} = R^{1 \ldots r} \operatorname{sign}(i_1 \ldots i_{r-s} j_1 \ldots j_s) S_{j_1 \ldots j_s}\,,$$

if all $i_k \leqq r$ and $(i_1 \ldots i_b j_1 \ldots j_s)$ is a permutation of $(1, \ldots, r)$. The projection S' of S on $\mathfrak{R}$, $x^{r+1} = \cdots = x^n = 0$, has the covariant components

$$S'_{j_1 \ldots j_s} = \begin{cases} S_{j_1 \ldots j_s} & \text{if all } j_k \leqq r\,, \\ 0 & \text{if one } j_k > r\,. \end{cases}$$

Then the right side of (21) is in fact the supplementary vector in the usual sense to S' in $(x^1, \ldots, x^r)$-space (compare [9, pp. 16, 17]).

Besides its geometric meaning the following properties of the scalar product are important for us. $\tilde{R} \cdot \tilde{S}$ is bilinear on $V^n_r \times V^n_s$, hence $|\tilde{R} \cdot \tilde{S}|$ is convex as a function of $\tilde{R}$ (or $\tilde{S}$) in V^n_r (or V^n_s). Therefore,

(22) *$|R \cdot S|$ is convex on G^n_r as a function of R, and on G^n_s as a function of S. For $r = s$, the function $R \cdot R'$ is bilinear on $G^n_r \times G^n_r$ and hence $R \cdot R'$ is convex on G^n_r in either variable.*

For a function $F(\mathfrak{R}^+)$ defined on all oriented r-flats $\mathfrak{R}^+$ through the origin we define the function $F(R)$ *induced by* $F(\mathfrak{R}^+)$ in G_r^n as follows: $F(0) = 0$; if R is any non-vanishing r-vector parallel to $\mathfrak{R}^+$ (including orientation) then

$$F(R) = |R| F(\mathfrak{R}^+) .$$

Trivially $F(\varrho R) = \varrho F(R)$ for $\varrho \geqq 0$.

If $F(\mathfrak{R})$ depends on the non-oriented r-flat $\mathfrak{R}$, i.e. has the same value for both orientations of $\mathfrak{R}$, then

$$F(\varrho R) = |\varrho| F(R) .$$

If $F(R)$ is convex, we say that $F(\mathfrak{R})$ induces a convex function on G_r^n. Here is a simple, but typical example:

In E^3 with a cartesian norm let $\mathfrak{L}$, $\mathfrak{E}$ be a line and a plane through the origin, and let u, E_u be a unit vector and a unit 2-vector parallel to $\mathfrak{L}$ and $\mathfrak{E}$ respectively. For a compact set M with a mass-distribution of positive continuous density α:

$$m(M, \alpha, \mathfrak{L}) = \int_M |u \wedge x|^2 \alpha(x)\, d\mu_x ,$$

and

$$m(M, \alpha, \mathfrak{E}) = \int_M |E_u \wedge x|^2 \alpha(x)\, d\mu_x ,$$

are the moments of inertia of M with respect to $\mathfrak{L}$ and $\mathfrak{E}$ if μ is the ordinary Lebesgue measure.

In E^n with a cartesian norm we may define analogously

$$m(M, \alpha, \mathfrak{R}) = \int_M |R_u \wedge x|^2 \alpha(x)\, d\mu_x$$

as the moment of inertia of M with respect to the r-flat $\mathfrak{R}$ through 0 parallel to the unit r-vector R_u. Then $m^{\frac{1}{2}}(M, \alpha, \mathfrak{R})$ induces, by (13), a convex function on G_r^n, namely

$$|R|\, m^{\frac{1}{2}}(M, \alpha, \mathfrak{R}) = \left[\int_M |R \wedge x|^2 \alpha(x)\, d\mu_x\right]^{\frac{1}{2}} .$$

Part II. An Integration Principle

in collaboration with W. Fenchel

The purpose of this brief part is the proof of the following theorem: Let $f(R)$ be homogeneous and convex on G_r^n. If $1 \leqq r < s \leqq n-1$ then $\int\limits_{R_u \| \mathfrak{S}} f(R_u)\, dR_u$ induces a convex function on G_s^n. Here R_u means a simple unit r-vector, $R_u \| \mathfrak{S}$ indicates that R_u is parallel to the s-flat $\mathfrak{S}$ and dR_u is the kinematic density of the oriented r-flats in $\mathfrak{S}$. The proof is obtained by applying a result of integral geometry to a simple general fact.

The latter may be stated as follows: Let U, V be vector spaces with be defined. If $g(u)$ is a convex function on U then $g \circ \Phi_x(v) = g(\Phi_x(v))$ is defined on V and is convex because

$$g \circ \Phi_x[(1-\lambda)v_1 + \lambda v_2] = g \circ [(1-\lambda)\Phi_x(v_1) + \lambda \Phi_x(v_2)]$$
$$\leqq (1-\lambda) g \circ \Phi_x(v_1) + \lambda g \circ \Phi_x(v_2) .$$

If, for each fixed v the function $g \circ \Phi_x(v)$ is integrable over Q then

(1) $$\int_Q g \circ \Phi_x(v)\, (d\mu)_x$$

is a convex function of v. (Here it need not be assumed, as in I(13) that the integral is bounded linearly below because V is convex.)

Let b and c be positive integers with $d =_{\text{def}} b + c \leqq n$. We apply (1) to $U = V_c^n$, $V = V_d^n$. $Q = \{B_u\}$ is the set of all simple unit b-vectors, the measure is given by the kinematic density dB_u of the oriented b-flats, and $\Phi_x = \Phi_{B_u}$ is defined by

$$\Phi_{B_u}(\tilde{C}) = B_u \wedge \tilde{C} .$$

Since $B_u \wedge C$ is simple for simple C, this map induces a map of G_c^n into G_d^n. For any convex (not necessarily homogeneous) function $g(\tilde{C})$ in V_c^n

(2) $$G(\tilde{C}) = \int_{\{B_u\}} g(B_u \wedge \tilde{C})\, dB_u$$

trivially exists (is finite) and is, by (1), a convex function on V_d^n.

Assume now that $g(D)$ is defined and convex, i.e. is extendable to a convex function $g(\tilde{D})$ in V_d^n. Then (2) provides a convex extension of $\int_{\{B_u\}} g(B_u \wedge C)\, dB_u$. Thus

(3) **Theorem.** *If b and c are positive integers with $d = b + c \leqq n$ and $g(D)$ is convex on G_d^n then*

$$F(C) = \int_{\{B_u\}} g(B_u \wedge C)\, dB_u \, .$$

is a convex function on G_c^n.

If $g(D)$ is homogeneous and $C = |C|C_u$, where C_u is a unit vector then,

$$F(C) = |C| \int_{\{B_u\}} g(B_u \wedge C_u)\, dB_u \, .$$

Assume that $g(D)$ is absolutely homogeneous ($g(\varrho D) = |\varrho|\, g(D)$). If $\mathfrak{B}, \mathfrak{C}, \mathfrak{D}$, are the non-oriented b-, c- and d-flats parallel to B_u, C_u, D_u then $g(B_u \wedge C_u)$ depends only on $\mathfrak{B}$ and $\mathfrak{C}$. Moreover, if $B_u \wedge C_u = |B_u \wedge C_u|\, D_u$ ($|D_u| = 1$), then

$$g(B_u \wedge C_u) = |B_u \wedge C_u|\, g(D_u)$$

where $g(D_u)$ depends only on $\mathfrak{D}$. Therefore we may write without ambiguity

$$F(C) = 2|C| \int g(B_u \wedge C_u)\, d\mathfrak{B}$$

The integration over $\{\mathfrak{B}\}$ can be carried out by integrating first over all b-flats $\mathfrak{B}$ lying in a d-flat $\mathfrak{D}$ containing a given c-flat $\mathfrak{C}$ and then over all $\mathfrak{D}$ that contain $\mathfrak{C}$. This means there is a relation of the form

$$d\mathfrak{B} = \Phi(\mathfrak{B} \subset \mathfrak{D}, \mathfrak{C})\, d(\mathfrak{D} \supset \mathfrak{C}) \wedge d(\mathfrak{B} \subset \mathfrak{D})$$

where $d(\mathfrak{D} \supset \mathfrak{C})$ or $d(\mathfrak{B} \subset \mathfrak{D})$ is the kinematic density for the d-flats $\mathfrak{D}$ containing $\mathfrak{C}$ or the b-flats $\mathfrak{B}$ contained in $\mathfrak{D}$ (see [13, § 3, section 2]). Therefore

$$\int_{\{\mathfrak{B}\}} g(B_u \wedge C_u)\, d\mathfrak{B} = \int_{\{\mathfrak{D} \supset \mathfrak{C}\}} d(\mathfrak{D} \supset \mathfrak{C})\, g(D_u) \int_{\{\mathfrak{B} \subset \mathfrak{D}\}} |B_u \wedge C_u|\, \Phi(\mathfrak{B} \subset \mathfrak{D}, \mathfrak{C})\, d(\mathfrak{B} \subset \mathfrak{D}) .$$

Because the notation is very clumsy we will here, and at similar occasions in the future (III.1) replace $d(\mathfrak{D} \supset \mathfrak{C})$ and $d(\mathfrak{B} \subset \mathfrak{D})$ by $d\mathfrak{D}$ and $d\mathfrak{B}$ when the sets written below the integral sign clearly suggest the correct interpretation.

Because the kinematic densities are invariant under motions of E^n, the second integral on the right is a constant depending only on b, c, n, and we find

$$F(C) = k|C| \int_{\{\mathfrak{D} \supset \mathfrak{C}\}} g(D_u)\, d\mathfrak{D} \text{ is convex on } G_c^n.$$

This is essentially the theorem which we wanted to prove. More useful forms are obtained by passing from D to its supplementary r-vector R $(r = n - d)$. If $f(R)$ is defined and homogeneous on G_r^n then

$$g(D) = f(R) + f(-R) \qquad (|D| = |R|)$$

is absolutely homogeneous and

$$|C| \int_{\{\mathfrak{D} \supset \mathfrak{C}\}} g(D_u)\, d\mathfrak{D} = |C| \int_{R_u \perp C} f(R_u)\, dR_u = |S| \int_{R_u \| S} f(R_u)\, dR_u,$$

where S is the supplementary $(n - c)$-vector to C, the last two integrals are defined as 0 for $C = S = 0$, and $R_u \perp C$ means $R_u \cdot C = 0$. Thus we have proved:

(4) **Theorem.** *If $f(R)$ is convex and homogeneous on G_r^n and $0 < c < n$, $0 < r < s \leq n - 1$, then*

$$F(C) = |C| \int_{R_u \perp C} f(R_u)\, dR_u,\ F(0) = 0, \textit{ is convex on } G_c^n, \textit{ and}$$

$$G(S) = |S| \int_{R_u \| S} f(R_u)\, dR_u,\ G(0) = 0, \textit{ is convex on } G_s^n.$$

Sometimes (4) occurs in an equivalent form:

(4a) *If $f(\mathfrak{R})$ induces a convex function on G_r^n and if $1 \leq t < r$, $1 \leq r < s \leq n - 1$, then*

$$\int_{\{\mathfrak{R} \supset \mathfrak{T}\}} f(\mathfrak{R})\, d\mathfrak{R} \quad \textit{and} \quad \int_{\{\mathfrak{R} \subset \mathfrak{S}\}} f(\mathfrak{R})\, d(\mathfrak{R})$$

induce convex functions on G_t^n and G_s^n respectively.

The special case $r = 1$ of (4) answers the question raised in [6]:

If $h(u)$ is the supporting function of a convex body K and $\pi_s = \pi^{s/2}/\Gamma(1+\frac{1}{2}s)$ is the volume of the unit ball in E^s, then

$$[s\,\pi_s]^{-1} \int_{u \parallel \mathfrak{S}} (h(u) + h(-u))\,du$$

is the average width of the orthogonal projection of K on $\mathfrak{S}$. Therefore:

(6) *The average width of the projection of a convex body in E^n on a variable s-flat induces a convex function on G^n_s.*

Part III. The Projections of Convex Bodies

We turn to a detailed study of the convexity properties of the r-dimensional volume $\hat{P}(K, \mathfrak{R})$ of, and other functions related to, the orthogonal projection $P(K, \mathfrak{R})$ of a convex body K in E^n on a variable r-flat $\mathfrak{R}$. It is well known (and we shall here give further proofs) that the projection function $\hat{P}(K, R) = |R|\,\hat{P}(K, \mathfrak{R})$ is convex for $r = 1$ and $r = n-1$. We will see that this is, in general, not the case for $1 < r < n-1$.

First we derive the convexity properties obtainable from the theory of mixed volumes. We will show that the projection functions are convex for bodies which can be approximated by sums of line segments, and that, even if $\hat{P}(K, R)$ is not convex, the integral of $\hat{P}(K, \mathfrak{R})$ over the r-flats lying in a hyperplane induces a convex function on G^n_{n-1}.

In the second section we give a sufficient condition for the convexity of $\hat{P}(K, R)$ in terms of the shadow boundaries $S(K, \mathfrak{R})$. The latter is the locus in which the cylinder with $(n-r)$-dimensional generators normal to $\mathfrak{R}$ and circumscribed to K touches K. We shall also show that this condition is necessary for polyhedra.

The third section provides various, sometimes surprising, examples. In particular, for $n > 3$ there are convex polyhedra with $n+2$ vertices in E^n whose projection function fails to be convex for all r other than 1 and $n-1$.

Finally we shall show that all bodies whose boundaries are of class C^2 and have non-vanishing Gauss curvature have locally convex (but in general not convex) projection functions, a fact to which we referred at the end of Section I.2.

III.1. Mixed volumes and projections

For the definition of mixed volumes and all statements for which no reference or proof is given here, we refer to sections 29–32 of [3]. α_i always denotes a non-negative number. The mixed volume of bodies $K_1, \ldots, K_n$ in E^n is denoted by $V(K_1, \ldots, K_n)$. We put

$$V_{r,n-r}(K, L) = V(\underbrace{K, K, \ldots, K}_{r}, \underbrace{L, L, \ldots, L}_{n-r})$$

and

$$V_{r,s,n-r-s}(H, K, L) = V(\underbrace{H, \ldots, H}_{r}, \underbrace{K, \ldots, K}_{s}, \underbrace{L, \ldots, L}_{n-r-s})\,.$$

(1) $V(K_1, \ldots, K_n) > 0$ if and only if a line segment $L_i \subset K_i$ for each i exists so that $L_1 + \cdots + L_n$ does not lie in a hyperplane. In particular,

(2) If F is a convex body contained in an s-flat, then

$$V(K_1, \ldots, K_{n-j}, \underbrace{F, F, \ldots, F}_{j}) = 0 \text{ for } j > s .$$

The simple s-vector S_F is said to be *associated* with the convex body F in the oriented s-flat $\mathfrak{S}^+$ if S_F is parallel to $\mathfrak{S}^+$ and the norm $|S_F|$ of S_F equals the s-dimensional area $|F|_s$ of F. If $\mathfrak{S}$ is not oriented — as will always be the case here — then S_F and $-S_F$ are associated with F.

(3) *If F_1 and F_2 are convex bodies lying in an r-flat and an $(n-r)$-flat respectively, and R_{F_1}, S_{F_2} are associated with F_1 and F_2, then*

$$\tbinom{n}{r} V_{r,n-r}(F_1, F_2) = |R_{F_1} \wedge S_{F_2}| .$$

For, even if the flats have more than one common point, $V(F_1 + F_2) = |R_{F_1} \wedge S_{F_2}|$ and by (2),

$$V(F_1 + F_2) = \sum_{i=1}^{n} \tbinom{n}{i} V_{i,n-i}(F_1, F_2) = \tbinom{n}{r} V_{r,n-r}(F_1, F_2) .$$

Let $F_1 = L_1 + \cdots + L_r$, where the L_i are line segments and write F for F_2. If v_i is a vector associated with L_i, then (2) and (3) yield

$$\begin{aligned} |v_1 \wedge \cdots \wedge v_r \wedge S_F| &= \tbinom{n}{r} V_{r,\,n-r}(L_1 + \cdots + L_r, F) \\ &= \tbinom{n}{r} r!\, V(L_1, \ldots, L_r, \underbrace{F, \ldots, F}_{n-r}) . \end{aligned}$$

Therefore if $L_1, \ldots, L_m$ are line segments and $v_1, \ldots, v_m$ the associated vectors,

$$\begin{aligned} \tbinom{n}{r} V_{r,n-r}\left(\sum_{i=1}^{m} L_i, F\right) &= \tbinom{n}{r} \Sigma V(L_{i_1}, \ldots, L_{i_r}, \underbrace{F, \ldots, F}_{n-r}) \\ &= (r!)^{-1} \Sigma |v_{i_1} \wedge \cdots \wedge v_{i_r} \wedge S_F| , \end{aligned} \tag{4}$$

where the summation is extended over all sets of r different subscripts.

With the previous notation we prove a fact that is well known for $r = n - 1$.

(5) *Let F be a convex body lying in an $(n-r)$-flat $\mathfrak{R}^\perp$ orthogonal to $\mathfrak{R}$. Then for any convex body K,*

$$\tbinom{n}{r} V_{r,\,n-r}(K, F) = |F|_{n-r} \hat{P}(K, \mathfrak{R}) .$$

For a proof we place $\mathfrak{R}$ and $\mathfrak{R}^\perp$ so that they pass through the origin and put $P = P(K, \mathfrak{R})$, $P^\perp = P(K, \mathfrak{R}^\perp)$. Then for a suitable convex body $Q \subset \mathfrak{R}^\perp$,

$$P + P^\perp \supset K, \quad K + Q \supset P = P(K + Q, \mathfrak{R}) . \tag{6}$$

Then (3) and (6) yield, for any $\alpha > 0$,

$$\begin{aligned} V(\alpha K + F) \leq V(\alpha P + (\alpha P^\perp + F)) &= \tbinom{n}{r} V_{r,\,n-r}(\alpha P, \alpha P^\perp + F) \\ &= \alpha^r \tbinom{n}{r} V_{r,\,n-r}(P, \alpha P^\perp + F) , \end{aligned}$$

so that, by (3) and (4),

On the other hand, we deduce from (3) and (6),

$$\binom{n}{r} V_{r,n-r}(K,F) = \binom{n}{r} \sum_{i=0}^{r} \binom{r}{i} V_{r-i,i,n-r}(K,Q,F)$$

$$= \binom{n}{r} V_{r,n-r}(K+Q,F) \geq \binom{n}{r} V_{r,n-r}(P,F).$$

We generalize (5) slightly:

(7) *If $K_1, \ldots, K_r$ are convex bodies in E^n and F is contained in an $(n-r)$-flat $\mathfrak{R}^{\perp}$ normal to $\mathfrak{R}$, then*

$$\binom{n}{r} V(K_1, \ldots, K_r, \underbrace{F, \ldots, F}_{n-r})$$

$$= |F|_{n-r} V(P(K_1, \mathfrak{R}),\ P(K_2, \mathfrak{R}), \ldots, P(K_r, \mathfrak{R})).$$

For using

$$P(\Sigma \alpha_i K_i, \mathfrak{R}) = \Sigma \alpha_i P(K_i, \mathfrak{R}), \tag{8}$$

$$\binom{n}{r} \Sigma \alpha_{i_1} \ldots \alpha_{i_r} V(K_{i_1}, \ldots, K_{i_r}, \underbrace{F, \ldots, F}_{n-r})$$

$$= \binom{n}{r} V_{r,n-r}(\Sigma \alpha_i K_i, F) = |F|_{n-r} \hat{P}(\Sigma \alpha_i K_i, \mathfrak{R}) = |F|_{n-r} |\Sigma \alpha_i P(K_i, \mathfrak{R})|_r$$

$$= |F|_{n-r} \Sigma \alpha_{i_1} \ldots \alpha_{i_r} V(P(K_{i_1}, \mathfrak{R}), P(K_{i_2}, \mathfrak{R}), \ldots, P(K_{i_r}, \mathfrak{R})).$$

Comparing the coefficients of $\alpha_1 \ldots \alpha_r$ yields (7).

Since, with the above notation, $|F|_{n-r} = |S_F|$, $V(K_1, \ldots, K_r, F, \ldots, F)$ depends only on S_F and not on the shape of F. It will be convenient later to use this fact by putting

$$V(K_1, \ldots, K_r, \underbrace{S, \ldots, S}_{n-r}) = V(K_1, \ldots, K_r, \underbrace{F, \ldots, F}_{n-r})$$

for any $(n-r)$-vector S and any set F with which S is associated. Secondly, if R is any r-vector parallel to $\mathfrak{R}$ and S_R is its supplementary $(n-r)$-vector, then

$$V(K_1, \ldots, K_r, S_R, \ldots, S_R) = |R|\, V(P(K_1, \mathfrak{R}), \ldots, P(K_r, \mathfrak{R}))$$

is the function induced by $V(P(K_1, \mathfrak{R}), \ldots, P(K_r, \mathfrak{R}))$ on G_r^n.

Since for any vectors $v_1, \ldots, v_r$,

$$|v_1 \wedge \cdots \wedge v_r \wedge S_R| = |(v_1 \wedge \cdots \wedge v_r) \cdot R|,$$

we conclude from (4) and (5) that

$$|R|\, \hat{P}\left(\sum_{i=1}^{m} L_i, \mathfrak{R}\right) = (r!)^{-1} \Sigma\, |(v_{i_1} \wedge \cdots \wedge v_{i_r}) \cdot R|.$$

Each summand on the right being convex, the left side is also convex, so that any finite sum of segments has a convex projection function and more generally:

(9) **Theorem.** *If $K = \lim K_i$ where K_i is a finite sum of segments, then*

convex body for which the projection function $\hat{P}(K, R)$ is convex. In particular, if K is a convex body and dA_w the area element of its boundary at the point with unit normal w (this can be defined rigorously even for arbitrary K, compare [5, section 8]) then for a hyperplane $\mathfrak{H}$ with unit normal u, $v = |v| u$, and S_v a supplementary $(n-1)$-vector[3]),

$$|S_v| \hat{P}(K, \mathfrak{H}) = |v| \hat{P}(K, \mathfrak{H}) = \tfrac{1}{2} \int_{\Omega} |w \cdot v| \, dA_w$$

(Ω is the unit sphere $|w| = 1$) has this form. It is the function $\hat{P}(K, S_v)$. It is the supporting function of the so called projection body of K, and (9) implies that projection bodies have convex projection functions for all r.

We remember Kubota's Formula,

$$(10) \qquad \int_{\{R\}} \hat{P}(K, \mathfrak{R}) \, d\mathfrak{R} = k(r, n) \, V_{r, n-r}(K, B^n) \qquad 1 \leq r \leq n-1 ,$$

where B^n is the unit ball in E^n. A proof of (10) as well as the explicit value of the constant $k(r, n)$ can be found in [12, pp. 232—234]. Because

$$P(P(K, \mathfrak{H}), \mathfrak{R}) = P(K, \mathfrak{R}) \qquad \text{for } \mathfrak{R} \subset \mathfrak{H} ,$$

(10) applied to $\mathfrak{H}$ as space yields, for $1 \leq r \leq n-2$

$$\int_{\mathfrak{R} \subset \mathfrak{H}} \hat{P}(K, \mathfrak{R}) \, d\mathfrak{R} = k(r, n-1) \, V_{r, n-r-1}(P(K, \mathfrak{H}), B(\mathfrak{H})) .$$

where $B(\mathfrak{H})$ denotes the unit ball in $\mathfrak{H}$.

On the other hand (7) for $r = n-1$ and a vector v normal to $\mathfrak{H}$ gives,

$$n \, V_{r, n-r-1, 1}(K, B^n, v) = |v| \, V_{r, n-r-1}(P(K, \mathfrak{H}), B(\mathfrak{H}))$$

so that

$$|S_v| \int_{\mathfrak{R} \subset \mathfrak{H}} \hat{P}(K, \mathfrak{R}) \, d\mathfrak{R} = n^{-1} k(r, n-1) \, V_{r, n-r-1, 1}(K, B^n, v) .$$

Replacing K by $\Sigma \alpha_i K_i$ and proceeding as in the proof of (7) shows

$$|S_v| \int_{\mathfrak{R} \subset \mathfrak{H}} V(P(K_1, \mathfrak{R}), \ldots, P(K_r, \mathfrak{R})) \, d\mathfrak{R}$$

$$= n^{-1} k(r, n-1) \, V(K_1, \ldots, K_r, B^n, \ldots, B^n, v) .$$

---

[3]) This relation establishes again that $|S_v| P(K, \mathfrak{H})$ is convex.

It yields for $1 < r < n-1$ that $\hat{P}(K, R)$ is convex on the set of r-vectors parallel to a fixed $(r+1)$-flat. R_1, R_2 and $R_1 + R_2$ are simple if and only if they are parallel to an $(r+1)$-flat ([4], p. 99). Therefore

$\hat{P}(K, R_1 + R_2) \leq \hat{P}(K, R_1) + \hat{P}(K, R_2)$ if $R_1 + R_2$ is simple, which does not imply (as it would for $r = 1$ or $n-1$) that $\hat{P}\left(K, \sum_{i=1}^{k} R_i\right) \leq \sum_{i=1}^{k} \hat{P}(K, R_i)$ for simple $\sum_{i=1}^{k} R_i$ and $k > 2$, because $\hat{P}(K, R)$ need not be convex.

This fact elucidates our statement in I.4: If $r = 2$, $n = 4$, then any 2-vector $\widetilde{R} = R_1 +$ $+ R_2$ hence

It is well known that the right side is a convex function of v, therefore we have proved:

(11) **Theorem:** *For any convex bodies $K_1, \ldots, K_r$ the integral*

$$\int_{\mathfrak{R}\subset\mathfrak{H}} V(P(K_1, \mathfrak{R}), \ldots, P(K_r, \mathfrak{R}))\, d\mathfrak{R}$$

induces a convex function on G^n_{n-1}.

Notice that the integrand represents $\hat{P}(K, \mathfrak{R})$ when $K_i = K$, and the $(r-1)$-dimensional volume of the boundary of $P(K, \mathfrak{R})$ when $K_i = K$ for $i = 1, \ldots, r-1$ and $K_r = B^n$. The interest of (11) lies, of course, in the fact that the integrand does not, in general induce a convex function on G^n_r and our integration principle of Part II is not applicable.

But we use the principle to derive some other convexity properties. A lemma is needed. Let $\mathfrak{R}$ be a fixed r-flat through the origin and $\mathfrak{R}_p$ a variable r-flat parallel to $\mathfrak{R}$. Then for any set $K \neq \emptyset$ and any set $M \subset \mathfrak{R}$, $M \neq \emptyset$

$$(12)\qquad \begin{aligned} &(K \cap \mathfrak{R}_p) + M = (K + M) \cap \mathfrak{R}_p \quad \text{if } K \cap \mathfrak{R}_p \neq \emptyset\,,\\ &(K + M) \cap \mathfrak{R}_p = \emptyset \qquad\qquad\qquad\;\; \text{if } K \cap \mathfrak{R}_p = \emptyset\,,\end{aligned}$$

where $\emptyset$ is the empty set. The second statement is trivial.

Choose the coordinates so that $\mathfrak{R}$ is given by $x^i = 0$ for $i > r$. Then $M + \mathfrak{R}_p = \mathfrak{R}_p$ and the general relation

$$(A \cap B) + C \subset (A + C) \cap (B + C)$$

with the agreement $D + \emptyset = \emptyset$ yields

$$(K \cap \mathfrak{R}_p) + M \subset (K + M) \cap \mathfrak{R}_p$$

including $K \cap \mathfrak{R}_p = \emptyset$ for $(K + M) \cap \mathfrak{R}_p = \emptyset$.

Let $x = y + z \in (K + M) \cap \mathfrak{R}_p$, $y \in K$, $z \in M$. If $x^i = c^i$ $(i > r)$ are the equations of $\mathfrak{R}_p$, then $z^i = 0$ for $i > r$ gives $y^i = c^i$, hence $y \in \mathfrak{R}_p$ and

$$x \in (K \cap \mathfrak{R}_p) + M\,.$$

(13) *If $\mathfrak{R}$ passes through the origin and the convex body F lies in $\mathfrak{R}$ then for any convex body K and $i \leq r < n$,*

$$\binom{n}{i} V_{n-i,i}(K, F) = \binom{r}{i} \int_{\mathfrak{R}_p \cap K \neq \emptyset} V_{r-i,i}(K \cap \mathfrak{R}_p, F)\, d\mathfrak{R}_p$$

where $d\mathfrak{R}_p$ is the kinematic density in the bundle of r-flats parallel to $\mathfrak{R}$.

For, by (2) and (12),

$$\binom{n}{i} \alpha^i V_{n-i,i}(K, F) = V(K + \alpha F) = \int_{\mathfrak{R}_p \cap (K + \alpha M) \neq \emptyset} |(K + \alpha F) \cap \mathfrak{R}_p|_r d\mathfrak{R}_p$$

$$= \int_{\mathfrak{R}_p \cap K \neq \emptyset} |(K \cap \mathfrak{R}_p) + \alpha F|_r d\mathfrak{R}_p$$

r

(14) **Theorem:** *If $K_1, \ldots, K_{n-s}$ are convex bodies in E^n and S_u is a variable unit s-vector parallel to the r-flat $\mathfrak{R}$ $(s < r)$ then*

$$\int_{S_u \| \mathfrak{R}} V(K_1, \ldots, K_{n-s}, \underbrace{S_u, S_u, \ldots, S_u}_{s})\, dS_u$$

$$= k(r, s, n)\, V(K_1, \ldots, K_{n-s}, \underbrace{B(\mathfrak{R}), \ldots, B(\mathfrak{R})}_{s}) .$$

This reduces (compare the proof of (7)) to

$$\int_{S_u \| \mathfrak{R}} V_{n-s,s}(K, S_u)\, dS_u = k(r, s, n)\, V_{n-s,s}(K, B(\mathfrak{R})) .$$

By (13),

$$\int_{S_u \| \mathfrak{R}} V_{n-s,s}(K, S_u)\, dS_u = \binom{r}{s} \binom{n}{s}^{-1} \int_{\mathfrak{R}_p \cap K \neq 0} d\mathfrak{R}_p \int_{S_u \| \mathfrak{R}} V_{r-s,s}(K \cap \mathfrak{R}_p, S_u)\, dS_u .$$

If $\mathfrak{D}$ denotes the $(r-s)$-flat normal to S_u through a fixed point of $\mathfrak{R}_p$ then since S_u is oriented and $\mathfrak{D}$ is not, and by (5) and (10),

$$\tfrac{1}{2}\binom{r}{s} \int_{S_u \| \mathfrak{R}} V_{r-s,s}(K \cap \mathfrak{R}_p, S_u)\, dS_u = \int_{\mathfrak{D} \| \mathfrak{R}} P(K \cap \mathfrak{R}_p, \mathfrak{D})\, d\mathfrak{D}$$

$$= k(s, r)\, V_{r-s,s}(K \cap \mathfrak{R}_p, B(\mathfrak{R})) .$$

Using (13) once more,

$$\int_{\mathfrak{R}_p \cap K \neq 0} V_{r-s,s}(K \cap \mathfrak{R}_p, B(\mathfrak{R}))\, d\mathfrak{R}_p = \binom{n}{s} \binom{r}{s}^{-1} V_{n-s,s}(K, B(\mathfrak{R})).$$

The last three equations contain the assertion.

From (14) we obtain in particular for $s = 1$,

(15) $$\int_{u \| \mathfrak{R}} V(K_1, \ldots, K_{n-1}, u)\, du = \Phi(r, n)\, V(K_1, \ldots, K_{n-1}, B(\mathfrak{R})) .$$

Since $V(K_1, K_2, \ldots, K_{n-1}, v)$ is a convex function of v, our integration principle shows

(16) **Corollary:** $V(K_1, \ldots, K_{n-1}, B(\mathfrak{R}))$ *induces a convex function on G_r^n.*

In the special case of (13),

$$V_{n-1,1}(K, B(\mathfrak{R})) = n^{-1} r \int_{\mathfrak{R}_p \cap K \neq 0} V_{r-1,1}(K \cap \mathfrak{R}_p, B(\mathfrak{R}))\, d\mathfrak{R}_p$$

the integrand is the $(r-1)$-dimensional volume of the boundary of $K \cap \mathfrak{R}_p$. This fact and the convexity of $\int_{u \| \mathfrak{R}} V_{n-r,r-1,1}(K, B^n, u)\, du$ yields the results of Section 2 in [6].

Kubota's Formula (10) applied to all s-flats $\mathfrak{S}$ lying in an r-flat $\mathfrak{R}$ yields

$$\int_{\mathfrak{S} \subset \mathfrak{R}} \hat{P}(K, \mathfrak{S})\, d\mathfrak{S} = k(s, r)\, V_{s,r-s}(P(K, \mathfrak{R}), B(\mathfrak{R})) .$$

Therefore we conclude from II (4):

(16a) *If $\hat{P}(K, S)$ is convex and $s < r \leq n-1$ then $V_{s,r-s}(P(K, \mathfrak{R}), B(\mathfrak{R}))$ induces a convex function on G_r^n.*

III.2. A test for the convexity of $\hat{P}(K, R)$

The projection function $\hat{P}(K, R)$ of a convex body, is, in general, not convex. This will be illustrated in the next section by examples whose discussion is based on the criterion (19) derived here. We require some preliminary results and definitions.

Let K be an n-dimensional convex body and $\mathfrak{R}$ an r-flat. Denote by $Z(K, \mathfrak{R})$ the cylinder with $(n-r)$-dimensional generators normal to $\mathfrak{R}$ and circumscribed to K. For obvious geometrical reasons we call the set

$$S(K, \mathfrak{R}) = Z(K, \mathfrak{R}) \cap K$$

a *shadow boundary* of K, and call the shadow boundary *sharp* if every generator of $Z(K, \mathfrak{R})$ touches it in exactly one point. Since

$$P(S(K, \mathfrak{R}), \mathfrak{R}) = \partial P(K, \mathfrak{R}) = Z(K, \mathfrak{R}) \cap \mathfrak{R},$$

the shadow boundary is sharp if and only if it is homeomorphic to the $(r-1)$-dimensional sphere S^{r-1}, or the projection of $S(K, \mathfrak{R})$ on $\mathfrak{R}$ is a topological map. All shadow boundaries are sharp if and only if K is strictly convex. The final example in the next section will show that $S(K, \mathfrak{R})$ need not contain any subset homeomorphic to S^{r-1}.

Assume $S(K, \mathfrak{R})$ is sharp. Then an orientation $\mathfrak{R}^+$ of $\mathfrak{R}$ induces an orientation $P(K, \mathfrak{R}^+)$ of $P(K, \mathfrak{R})$, $\partial P(K, \mathfrak{R}^+)$ of $\partial P(K, \mathfrak{R})$ and $S(K, \mathfrak{R}^+)$ of $S(K, \mathfrak{R})$.

If K is a polyhedron and $H_1, H_2, \ldots, H_p$ are simple vectors associated with the faces of all dimensions d $(1 \leq d \leq n-1)$ of K, then $S(K, \mathfrak{R})$ is sharp if $R \cdot H_j \neq 0$ for $j = 1, \ldots, p$, where $R \neq 0$ corresponds to $\mathfrak{R}$. These equations represent a finite number of linear relations for R and hence determine, on G_r^n, a set E of measure 0 (with respect to the standard $(r(n-r)+1)$-dimensional measure on G_r^n). The complement $G_r^n - E$ of E consists of a finite number of open regions $G_1, \ldots, G_t$ and $S(K, \mathfrak{R}^+)$ is the same, and sharp, for all $\mathfrak{R}^+$ corresponding to r-vectors R in a fixed G_k. (Note that $R \in G_k$ implies $R \neq 0$ and $-R \notin G_k$.) Let $\mathfrak{R}_0^+$ be any one of these $\mathfrak{R}^+$, then we put $S(K, \mathfrak{R}_0^+) = S_k$.

Projecting S_k on any oriented r-flat $\mathfrak{R}^+$ gives an oriented cycle C_k in $\mathfrak{R}^+$. Then every point of $\mathfrak{R}^+ - C_k$ has, with respect to C_k, a definite order or index $j(p)$. Since the number of components of $\mathfrak{R} - C_k$ is finite and $j(p)$ vanishes in the unbounded component, the index integral

$$\int_{\mathfrak{R}-C_k} j(p)\,dA$$

is defined and finite, where dA is the volume element in $\mathfrak{R}^+$. We denote the index integral by $j(\mathfrak{R}_0^+, \mathfrak{R}^+)$ or briefly by $j_k(\mathfrak{R}^+)$ (where $\mathfrak{R}_0^+$ corresponds to $R_0 \in G_k$) and observe that if $\mathfrak{R}_0^-$ and $\mathfrak{R}^-$ are the opposite orientations of $\mathfrak{R}_0^+$ and $\mathfrak{R}^+$, then

$$j(\mathfrak{R}_0^-, \mathfrak{R}^+) = j(\mathfrak{R}_0^+, \mathfrak{R}^-) = -j(\mathfrak{R}_0^+, \mathfrak{R}^+). \tag{17}$$

(18) *The index function* $j(\mathfrak{R}_0^+, R) = |R|\, j(\mathfrak{R}_0^+, \mathfrak{R}^+)$, *where* R *is a vector corresponding to* $\mathfrak{R}^+$, *induced by the index integral is (defined when* $S(K, \mathfrak{R}_0)$ *is sharp and) linear; moreover*

$$j(\mathfrak{R}_0^+, R_0) = \hat{P}(K, R_0)$$

when R_0 *corresponds to* $\mathfrak{R}_0^+$.

Proof. We have to show that if $R_0 \in G_k$ then the function $j_k(R) = j(\mathfrak{R}_0^+, R)$ is linear. We join a vertex of S_k to the remaining points of S_k by segments obtaining an r-dimensional polyhedron Q homeomorphic to an r-ball which we orient so that $S_k = S(K, \mathfrak{R}_0^+)$ is its boundary. Let $F_1, \ldots, F_m$ be the simple r-vectors associated with the r-dimensional faces of Q. If R_u is the unit vector corresponding to $\mathfrak{R}^+$ then

$$\sum_{i=1}^{m} F_i \cdot R_u$$

may be called the *area of the oriented projection* of Q on $\mathfrak{R}^+$ because it has the following geometric interpretation: With the exception of a set of measure 0, a point p of $\mathfrak{R}^+$ either does, or does not, lie in the interior of the projection of an r-dimensional face of Q. In the latter case, we say that its multiplicity $m(p) = 0$, and in the former we put $m(p) = s - t$ when p is covered s times positively and t times negatively by the interiors of the projections of the r-dimensional faces of Q ($\min(r, s) > 0$). Then

$$\sum_{i=1}^{m} F_i \cdot R_u = \int_{\mathfrak{R}^+} m(p)\, dA\,.$$

On the other hand (see [1, p. 420]) $m(p) = j(p)$ and so

$$j_k(R) = \sum_{i=1}^{m} F_i \cdot R$$

which proves linearity. If R_u is parallel to R_0, then every point of $P(K, \mathfrak{R}_0^+)$, with the exception of a set of measure 0 is covered with multiplicity $+1$, and $m(p) = 0$ for $p \in \mathfrak{R}_0^+ - P(K, \mathfrak{R}_0^+)$. This proves (17).

We can now state our main result:

(19) **Theorem:** *Let* K *be an* n*-dimensional convex polyhedron in* E^n. *Then* $\hat{P}(K, R)$ *is convex if and only if, for each* $\mathfrak{R}_0$ *for which the shadow boundary is sharp, the index integral satisfies the inequality*

$$|j(\mathfrak{R}_0^+, \mathfrak{R}^+)| \leq \hat{P}(K, \mathfrak{R})\,,$$

where the orientations $\mathfrak{R}_0^+, \mathfrak{R}^+$ *of* $\mathfrak{R}_0$ *and* $\mathfrak{R}$ *are arbitrary.*

Note. Since the absolute value of the index integral enters, (17) shows that the arbitrariness of the orientation imposes no extra condition. Also, because of (17) and $\hat{P}(K, R) = \hat{P}(K, -R)$ the condition is equivalent to

The sufficiency is very simple. If $R_0 \in G_r^n - E$ then $R_0 \in G_k$ for a suitable k and by (18),

$$\hat{P}(K, R_0) = j_k(R_0)\,.$$

This relation also holds for $R_0 \in E$. It is trivial for $R_0 = 0$. For $R_0 \neq 0$ there is a G_k and a sequence of r-vectors $R_1, R_2, \ldots,$ in G_k which tend to R_0. Then

$$j_k(R_i) = \hat{P}(K, R_i)$$

and the assertion follows because both sides are continuous.

Applying the hypotheses of (19) we find that, for all R,

$$\hat{P}(K, R) = \sup_k |j_k(R)|$$

and since the j_k are linear, $\hat{P}(K, R)$ is convex.

Two lemmas are needed to prove the necessity. For the proof of the first we are indebted to Professors B. L. van der Waerden and K. Leichtweiss.

(20) *Let* $R_0 \in G_r^n$ $(1 < r < n-1)$ *and let* U *be a neighborhood of* R_0 *on* G_r^n. *Then* R_0 *is an interior point of* U^H, *the convex hull of* U *in* V_r^n.

It suffices to prove that any hyperplane in V_r^n through R_0 separates at least two points of U. For then U^H cannot possess a supporting hyperplane at R_0. This is obvious for $R_0 = 0$, so assume $R_0 \neq 0$.

The homogeneity properties of G_r^n permit us to take R_0 as the special point $e_1 \wedge \cdots \wedge e_r = (1, 0, \ldots, 0)$. If y_i $(1 \leqq i \leqq r)$ denotes the point with coordinates δ_i^α for $\alpha \leqq r$ and x_i^h for $h > r$, then $\varepsilon > 0$ exists such that

$$R_y = y_1 \wedge \cdots \wedge y_r \in U \quad \text{for} \quad |x_i^h| < \varepsilon, \quad i \leqq r, h > r\,.$$

Our proof is indirect. Suppose the lemma is false and there exists a hyperplane

$$L(\tilde{R}) = \Sigma\, a_{i_1 \ldots i_r} R^{i_1 \cdots i_r} - a = 0$$

through R_0 such that $L(R) \geqq 0$ for $R \in U$. Because G_r^n contains the line joining R_0 to the origin, the hyperplane must pass through the origin, so $a = 0$. Since it passes through the point $R_0 = e_1 \wedge \cdots \wedge e_r$, also $a_{12 \ldots r} = 0$. We show that $L(R_y) \geqq 0$ for $|x_i^h| < \varepsilon$ implies that all the remaining coefficients $a_{i_1 \ldots i_r}$ are zero so that $L(\tilde{R})$ does not represent a hyperplane.

Write a given sequence $i_1 < \cdots < i_r$ in the form $j_1 < \cdots < j_h < k_1 < \cdots < k_{r-h}$ where $j_\alpha \leqq r$, $k_\beta > r$ ($h = 0$ not excluded) and denote by $m_1, \ldots, m_{r-h}$ those among the numbers $1, 2, \ldots, r$ which are different from $j_1, \ldots, j_h$. Then

$$R_y^{i_1 \ldots i_r} = \varepsilon(j_\alpha, k_\beta) \begin{vmatrix} x_{m_1}^{k_1} & \ldots & x_{m_1}^{k_{r-h}} \\ \cdot & \ldots & \cdot \\ x_{m_{r-h}}^{k_1} & \ldots & x_{m_{r-h}}^{k_{r-h}} \end{vmatrix}$$

where $\varepsilon(j_\alpha, k_\beta) = \pm 1$. Therefore $L(R_y)$ has the following form

$$L(R_y) = \sum_{1 \leqq m \leqq r} L_m(x_m) + \sum_{1 \leqq m_1 \leqq m_2 \leqq r} L_{m_1 m_2}(x_{m_1}, x_{m_2}) + \cdots$$

where $L_{m_1 m_2 \cdots m_h}(x_{m_1}, x_{m_2}, \ldots, x_{m_h})$ is a multilinear form, namely for each s $(1 \leqq s \leqq h)$ linear in $x_{m_s}^{r+1}, x_{m_s}^{r+2}, \ldots, x_{m_s}^{n}$.

For arbitrary x_α^h with $|x_\alpha^h| < \varepsilon$ and $x_i^h = 0$ for $i \neq \alpha$ we obtain $L_\alpha(x_\alpha) \geqq 0$. Replacing x_α by $-x_\alpha$ we get also $-L_\alpha(x_\alpha) \geqq 0$, hence $L_\alpha = 0$ or $a_{1\,2 \cdots \alpha-1\,\alpha+1 \cdots r\,i} = 0$ for $\alpha \leqq r$ and $i > r$.

Putting $x_i^h = 0$ for $i \neq \alpha, \beta$ $(\alpha < \beta)$ we obtain also $L(x_\alpha, x_\beta) \geqq 0$ and replacing x_α by $-x_\alpha$ also $-L(x_\alpha, x_\beta) \geqq 0$. Hence

$$L_{\alpha\beta} = 0 \text{ or } a_{1\,2 \cdots \alpha-1\,\alpha+1 \cdots \beta-1\,\beta+1 \cdots r\,i\,j} = 0 \quad \begin{matrix} \text{for } \alpha < \beta \leqq r \\ \text{and } j > i > r\,. \end{matrix}$$

Proceeding in this way we find that all the $L_{j_1 \ldots j_h}$ vanish identically and hence that all the coefficients $a_{i_1 \ldots i_r}$ of $L(\tilde{R})$ vanish. Hence (20) is true.

As a corollary of (20) and our remark at the beginning of Part I, we mention:

(21) *If $f(R)$ is linear in a non-empty relative open subset of G_r^n, then it possessses a unique linear extension to V_r^n.*

(22) *If $f(R)$ is a linear function of R in a neighborhood U of R_0 on G_r^n, then f can be extended to a convex function $f(\tilde{R})$ in U^H in a unique way, namely as a linear function of $\tilde{R}$.*

$f(R)$ can clearly be extended as a linear function to U^H, so we need only prove that this is the only convex extension.

Let $f(\tilde{R})$ be any convex extension of f to U^H. Write $\tilde{R}_1$ for any point of U^H distinct from R_0. Since, by (20), R_0 is an interior point of U^H, we can find $\tilde{R}_2$ in U^H, also distinct from R_0, such that $\tilde{R}_1, R_0, \tilde{R}_2$ are collinear in that order, hence

$$R_0 = \theta \tilde{R}_1 + (1 - \theta)\tilde{R}_2 \qquad (0 < \theta < 1)\,.$$

We have,

$$\tilde{R}_i = \sum_{j=1}^{m} \lambda_{ij} R_{ij} \qquad \left(i = 1, 2;\, m = \binom{n}{r} + 1\right)$$

where the R_{ij} are simple r-vectors in U, and the λ_{ij} are positive with sum 1 for each i. Since $f(\tilde{R})$ is convex,

$$\text{(23)} \qquad \begin{aligned} f(R_0) &\leqq \theta f(\tilde{R}_1) + (1 - \theta) f(\tilde{R}_2) \leqq \\ &\leqq \theta \sum_{j=1}^{m} \lambda_{1j} f(R_{1j}) + (1 - \theta) \sum_{j=1}^{m} \lambda_{2j} f(R_{2j})\,. \end{aligned}$$

But

$$R_0 = \theta \sum_{j=1}^{m} \lambda_{1j} R_{1j} + (1 - \theta) \sum_{j=1}^{m} \lambda_{2j} R_{2j}\,,$$

and so, by linearity of $f(R)$ on $U \cap G_r^n$, the extreme left and extreme right of (23) are equal. Thus all three terms are equal, and

$$f(\tilde{R}_i) = \Sigma \lambda_{ij} f(R_{ij}) \qquad (i = 1, 2)\,.$$

Thus $f(\tilde{R})$ is linear in U^H, it is a linear extension of $f(R)$ and (22) is proved.

$j_k(\tilde{R})$ to V_r^n, and from (22),

$$\hat{P}(K, \tilde{R}) = j_k(\tilde{R}) \qquad \text{for } R \in U^H.$$

In the notation of I (4) the hyperplane $(\tilde{R}, j_k(\tilde{R}))$ is a supporting hyperplane of the convex set $U^H(\hat{P}(K, \tilde{R}))$ and hence must be a supporting hyperplane of $V_r^n(\hat{P}(K, \tilde{R}))$ because this set is convex. This means that $j_k(\tilde{R}) \leqq P(K, \tilde{R})$ for all $\tilde{R}$, and so the necessity of condition (19) is established.

We now consider the possibility of formulating a condition similar to (19) for an arbitrary convex body K (and not just for a polyhedron). As we have already stated, $S(K, \mathfrak{R}_0)$ need not contain a subset $\Gamma(K, \mathfrak{R}_0)$ homeomorphic to S^{r-1}. If it does we write $\Gamma(K, \mathfrak{R}_0^+)$ for this with the orientation induced by $\mathfrak{R}_0^+$ and $C(\mathfrak{R}_0^+, \mathfrak{R}^+)$ for the oriented continuous cycle which is its projection on $\mathfrak{R}^+$. Each point P of $\mathfrak{R}^+ - C$ has a definite order $j(p)$ with respect to C which is constant in each component of $\mathfrak{R}^+ - C$, and in particular is zero in the unbounded component (see [1, l.c.]). But since $\mathfrak{R}^+ - C$ may have infinitely many components and $j(p)$, although finite, may be unbounded, the integral $\int_{\mathfrak{R}^+ - C} j(p)\, dA$ may not exist.

So that we can formulate conditions corresponding to those of (19) for all bodies K, we approximate K by a sequence of polyhedra $K_1, K_2, \ldots,$ for which $S(K_i, \mathfrak{R}_0)$ is a sharp shadow boundary. The corresponding index function (calculated as $j_k(R)$ with appropriate k, for each of the polyhedra K_i) will be denoted by $j_i(\mathfrak{R}_0^+, R)$. Because $j_i(\mathfrak{R}_0^+, R_0) = \hat{P}(K_i, R_0)$ and $\hat{P}(K_i, R_0) \to \hat{P}(K, R_0)$, we have

$$j_i(\mathfrak{R}_0^+, R_0) \to \hat{P}(K, R_0) \qquad \text{as } i \to \infty . \tag{24}$$

In view of the preceding remarks it is clear that $j_i(\mathfrak{R}_0, R)$ will not in general converge for all R. If however

$$j^*(\mathfrak{R}_0^+, R) = \lim_{i \to \infty} j_i(\mathfrak{R}_0^+, R)$$

exists for a suitable sequence of polyhedra $K_i \to K$ for which $S(K_i, \mathfrak{R}_0)$ is sharp, we shall call $j^*(\mathfrak{R}_0^+, R)$ an index function of K with respect to $\mathfrak{R}_0^+$. This function is, in general, not unique but as a limit of linear functions it is linear. Moreover

$$j^*(\mathfrak{R}_0^-, R) = -j^*(\mathfrak{R}_0^+, R) = j^*(\mathfrak{R}_0^+, -R) .$$

If ∂K is of class C^2 and has non-vanishing Gauss curvature, it is easily seen that $j^*(\mathfrak{R}_0^+, R)$ always exists and is unique for all $\mathfrak{R}_0^+$.

(25) **Theorem:** *Let K be an n-dimensional convex body in E^n. If K possesses index functions $j^*(\mathfrak{R}_0^+, R)$ for a dense set of r-vectors R_0, and for each such R_0 and for each R,*

$$j^*(\mathfrak{R}_0^+, R) \leqq \hat{P}(K, R)$$

then $\hat{P}(K, R)$ is convex.

hyperplane, and the R_0 are dense among the R, the set $G_r^n(\hat{P}(K, R))$ possesses a supporting hyperplane everywhere. Hence $\hat{P}(K, R)$ is convex.

It is probable that any convex set K possesses index functions for a dense set of vectors R_0 and that the condition in (25) is also necessary.

Our results and the convexity of $\hat{P}(K, R)$ for $r = n - 1$ indicate that in this case the index function can never exceed the projection function. In fact the multiplicities $j(p) = m(p)$ in the bounded components of $\mathfrak{R}^+ - C$ are always ± 1. Since $C \subset P(K, \mathfrak{R})$ it follows at once that $|j^*(\mathfrak{R}_0^+, R_u)| \leqq \hat{P}(K, \mathfrak{R})$, ($R_u$ a unit vector corresponding to $\mathfrak{R}$).

To prove the statement regarding $m(p)$ assume that $S(K, \mathfrak{R}_0)$ is sharp. $\mathfrak{R}_0$ and $\mathfrak{R}$ intersect in an $(n-2)$-flat $\mathfrak{S}$, and any 2-flat or plane normal to $\mathfrak{S}$ intersects $P(K, \mathfrak{R}_0)$, if at all, in a line segment $T(x_1, x_2)$. Let x_i' be the point of $S(K, \mathfrak{R}_0)$ projected into x_i, and x_i'' the projection of x_i' on $\mathfrak{R}$.

The union of segments $T(x_1', x_2')$ is a surface Q homeomorphic to the $(n-1)$-dimensional ball, which spans $S(K, \mathfrak{R}_0)$. The segment $T(x_1', x_2')$ is projected on $T(x_1'', x_2'')$ and different $T(x_1', x_2')$ project into different $T(x_1'', x_2'')$. Therefore the map of Q on R by projection is $1-1$ except when $x_1' = x_2'$ and then $x_i'' = T(x_1'', x_2'') \subset C$. Thus $m(p) = \pm 1$ and the statement is proved.

Theorem (25) has a pointwise version which is frequently useful:

(26) *For the convex body K and the r-flat $\mathfrak{R}_0^+$, let an index function $j^*(\mathfrak{R}_0^+, R)$ exist such that $j^*(\mathfrak{R}_0^+, R) \leqq \hat{P}(K, R)$ for all R. Then,*

$$\hat{P}(K, R_0) \leqq \sum_{j=1}^{r} \hat{P}(K, R_j) \quad \text{if} \quad \sum_{j=1}^{r} R_j = R_0.$$

This is contained in the more general statement:

If $f(x)$ is defined in M and the numbers a_i exist such that

$$f(x) \geqq \Sigma a_i (x^i - x_0^i) + f(x_0),$$

then $f(x_0) \leqq \sum_{j=1}^{r} \lambda_j f(x_j)$ if $x_0 = x_\Lambda$.

For,

$$\Sigma \lambda_j f(x_j) \geqq \sum_j \lambda_j \sum_i a_i (x_j^i - x_0^i) + f(x_0) = f(x_0).$$

Notice the corollary,

(27) *If $S(K, \mathfrak{R}_0)$ lies in an r-flat, then*

$$\hat{P}(K, R_0) \leqq \sum_{j=1}^{r} \hat{P}(K, R_j) \quad \text{if} \quad \sum_{j=1}^{r} R_j = R_0.$$

For in this case, the index $j(p)$ in any $\mathfrak{R}^+$ must have the values $0, 1, -1$ everywhere; $j(p) = 0$ for $p \in \mathfrak{R} - P(S(K, \mathfrak{R}_0), \mathfrak{R})$ and $P(S(K, \mathfrak{R}_0), \mathfrak{R}) \subset P(K, \mathfrak{R})$.

A consequence of (27) is that an n-dimensional solid ellipsoid C^n in E^n has a convex projection function $\hat{P}(C^n, R)$. Further, since any s-dimensional ellipsoid C^s $(s < n)$ is the limit of a sequence of n-dimensional ellipsoids, and the pro-

Another consequence of (27) is that $\hat{P}(K, R)$ is convex if $\dim K = r$, which also follows from $\hat{P}(K, R) = |R_0 \cdot R|$ where R_0 is an r-vector associated with K. This implies:

(28) *$\hat{P}(K, R)$ is convex if* $\dim K = r + 1$.

It suffices to prove this for $(r + 1)$-dimensional convex polyhedra K. If $H_1, \ldots, H_p$ are the r-dimensional faces of K, then any $(n - r)$-flat through an interior point of K normal to $\mathfrak{R}$ intersects ∂K in two points. Therefore

$$2\hat{P}(K, \mathfrak{R}) = 2\hat{P}(\partial K, \mathfrak{R}) = \sum_{i=1}^{p} \hat{P}(H_i, \mathfrak{R})$$

and each summand on the right induces a convex function on G_r^n.

III.3. Examples of projection functions

This section provides various examples of projection functions illustrating some of the phenomena that do not occur for $r = 1, n - 1$. In particular several examples will be given of non-convex projection functions.

(a) *A polyhedron K^4 in E^4 with six vertices whose projection function $\hat{P}(K^4, R)$ $(r = 2)$ is not convex.*

In (c) we shall show that the simplex T^4 in E^4 has a convex projection function $\hat{P}(T^4, R)$ $(r = 2)$ and so this is the simplest example of a non-convex projection function that can arise.

It is convenient to represent a point (x_1, x_2, x_3, x_4) in E^4 by two complex coordinates (z_1, z_2) with $z_1 = x_1 + ix_2$, $z_2 = x_3 + ix_4$. We shall use $\mathfrak{R}_1$, $\mathfrak{R}_2$ to represent the 2-dimensional planes $z_2 = 0$, $z_1 = 0$ respectively.

K^4 is defined as the convex hull of the six points

$$p_1 = (1, 1) \qquad p_2 = (\omega, -\omega^2) \qquad p_3 = (\omega^2, \omega)$$
$$p_4 = (1, -1) \qquad p_5 = (\omega, \omega^2) \qquad p_6 = (\omega^2, -\omega)$$

where ω is a primitive cube root of unity. With appropriate orientation, the projection of K^4 on $\mathfrak{R}_2^+$ is a regular hexagon with vertices $1, -\omega^2, \omega, -1, \omega^2, -\omega$, so that $S(K^4, \mathfrak{R}_2^+) = S_2$ is a sharp shadow boundary, namely the hexagon $p_1p_2p_3p_4p_5p_6$. The union of the four triangles $p_1p_2p_3$, $p_1p_3p_4$, $p_1p_4p_5$, $p_1p_5p_6$ forms a surface Q spanning S_2 and hence we can calculate $j(\mathfrak{R}_2^+, \mathfrak{R}_1^+)$. Two of the triangles have zero projection and the other two each project into the triangle $1, \omega, \omega^2$ with area Δ. Hence

$$|j(\mathfrak{R}_2^+, \mathfrak{R}_1^+)| = 2\Delta > \hat{P}(K^4, \mathfrak{R}_1) = \Delta\,.$$

Hence, by (19) $\hat{P}(K^4, R)$ is not convex.

(b) *A polyhedron K^n in E^n with $(n + 2)$ vertices whose projection functions $\hat{P}(K, R)$ are not convex for any r, $1 < r < n - 1$.*

This example is a generalization of (a) and the latter will be used in its construction. Assume that $n > 4$, and let $\mathfrak{S}$ be any 4-flat in E^n passing through points are the vertices of a regular simplex in $\mathfrak{S}^{\perp}$. K^n is defined as the convex hull of the points $p_1, \ldots, p_{n+2}$.

For any integer r satisfying $1 < r < n-1$, consider only those r-flats $\mathfrak{R}$ which contain the simplex with vertices $z, p_7, \ldots, p_{r+4}$ and which therefore intersect $\mathfrak{S}$ in a 2-flat $\mathfrak{R}^*$. For such $\mathfrak{R}$,

$$\hat{P}(K^n, \mathfrak{R}) = k\hat{P}(K^4, \mathfrak{R}^*)$$

where the constant k depends only on r and n. But $\hat{P}(K^4, R^*)$ is not convex on G_2^4, so $\hat{P}(K^n, R)$ is not convex on G_r^n.

(c) *The projection functions of the four-dimensional simplex T^4 are all convex.*

We need only consider $P(T^4, \mathfrak{R})$ where $r = 2$. T^4 is the convex hull of five points in general position in E^4, and twelve of the sharp shadow boundaries S_i are skew pentagons, whose edges and vertices are edges and vertices of T^4. The projection of such a pentagon S_1 on to a plane $\mathfrak{R}_2$ in general position is a general pentagon C dividing $\mathfrak{R}_2$ into regions as explained in III.2. In calculating $|j(\mathfrak{R}_1^+, \mathfrak{R}_2^+)|$ these regions are counted with various multiplicities which can be determined by examining the index numbers $j(p)$ of C. In the diagrams below, two such pentagons are illustrated and the appropriate indexes inserted.

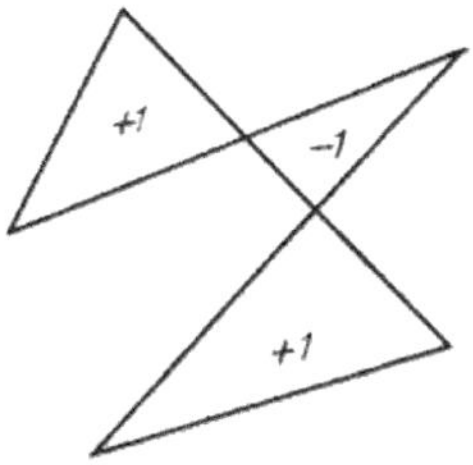

Fig. 1

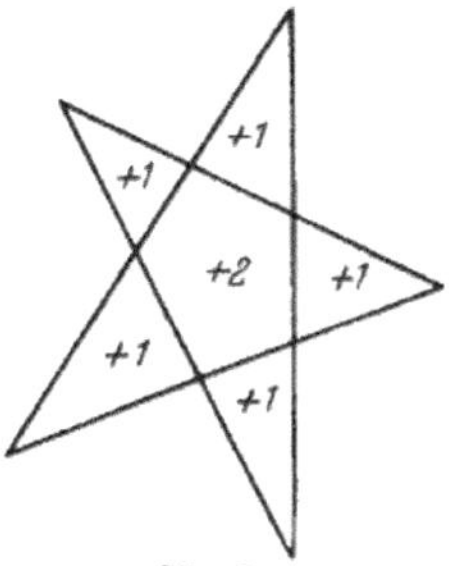

Fig. 2

By examining all possible pentagons it is easy to verify that, unless the projection is a pentagram (fig. (2)) the only indexes that can occur are 0, ± 1, and hence

$$|j(\mathfrak{R}_1^+, \mathfrak{R}_2^+)| \leq |C^H|_2 = \hat{P}(T^4, \mathfrak{R}_2)\,.$$

For the pentagram an index ± 2 occurs as in fig (2) but a simple calculation, which we omit, shows that the above inequality still holds.

By (19), the function $\hat{P}(T^4, R)$ is convex.

(d) *For $1 < r < n-2$, the projection functions $\hat{P}(T^n, R)$ of the n-dimensional simplex T^n are not convex.*

Consider first the case $r = 2$, $n = 5$. The coordinates of the vertices of a (regular) T^5 may be written

$$\left(\cos\frac{2k\pi}{3}, \sin\frac{2k\pi}{3}, \cos\frac{k\pi}{3}, \sin\frac{k\pi}{3}, \frac{(-1)^k}{2}\right) \qquad k = 0, 1, \ldots, 5\,,$$

and so $\hat{P}(T^5, R) = \hat{P}(P(T^5, \mathfrak{H}), R)$. However, from the above coordinates it is easy to see that $P(T^5, \mathfrak{H})$ is the body K^4 of example (a). Thus $\hat{P}(P(T^5, \mathfrak{H}), R)$ is not convex for $\mathfrak{R} \subset \mathfrak{H}$ and so $\hat{P}(T^5, R)$ is not a convex function of R.

In a similar way, for $r > 2$ it can be readily verified that the body K^n of (b) with $(n + 2)$ vertices is the projection of a simplex T^{n+1} on to a suitable hyperplane $\mathfrak{H}$. Hence for $\mathfrak{R} \subset \mathfrak{H}$,

$$\hat{P}(T^{n+1}, R) = \hat{P}(P(T^{n+1}, \mathfrak{H}), R) = \hat{P}(K^n, R)$$

is not a convex function of R.

Since $\hat{P}(T^n, R)$ is convex for $r = n - 1$, or 1 and is not convex for $1 < r \leqq n - 3$, only one case remains, namely $1 < r = n - 2$. We have not been able to determine whether $\hat{P}(T^n, R)$ $(r = n - 2)$ is convex or not, except when $n = 4$, $r = 2$ (see (c)) but we conjecture that it is always convex[4]).

Our next example illustrates the fact, at first sight surprising, that even if two bodies K_1 and K_2 have convex projection functions for a certain value of r, their vector sum can have a non-convex projection function for the same value of r.

(e) *Two simplexes T_1^4 and T_2^4 in E^4 with the property that $\hat{P}(T_1^4 + T_2^4, R)$ $(r = 2)$ is not convex.*

We represent a point of E^4 as (z_1, z_2) as in (a). Let θ be a primitive fifth root of unity, let T_1^4 be the convex hull of $p_1 = (1, 1)$, $p_2 = (\theta^2, \theta)$, $p_3 = (\theta^4, \theta^2)$, $p_4 = (\theta, \theta^3)$, $p_5 = (\theta^3, \theta^4)$ and let T_2^4 be the convex hull of $q_1 = (\theta^4, -\theta^2)$, $q_2 = (\theta, -\theta^3)$, $q_3 = (\theta^3, -\theta^4)$, $q_5 = (1, -1)$, $q_5 = (\theta^2, -\theta)$. Both T_1^4 and T_2^4 are regular simplexes (see [10, p. 245]) and $T_1^4 + T_2^4$ is the convex hull of the 25 points $p_i + q_j$ (see [14]). Now $P(T_1^4, \mathfrak{R}_1)$ is the pentagon with vertices 1, $\theta, \theta^2, \theta^3, \theta^4$ and $P(T_2^4, \mathfrak{R}_1)$ is the pentagon with vertices $-1, -\theta, -\theta^2, -\theta^3, -\theta^4$, and so $P(T_1^4 + T_2^4, \mathfrak{R}_1) = P(T_1^4, \mathfrak{R}_1) + P(T_2^4, \mathfrak{R}_1)$ is the decagon with vertices $1 - \theta^2$, $1 - \theta^3$, $\theta - \theta^3$, $\theta - \theta^4$, $\theta^2 - \theta^4$, $\theta^2 - 1$, $\theta^3 - 1$, $\theta^3 - \theta$, $\theta^4 - \theta$, $\theta^4 - \theta^2$. Thus $S_1(T_1^4 + T_2^4)$ is a sharp shadow boundary consisting of the skew decagon with vertices $p_1 + q_1$, $p_1 + q_2$, $p_2 + q_2$, $p_2 + q_3$, $p_3 + q_3$, $p_3 + q_4$, $p_4 + q_4$, $p_4 + q_5$, $p_5 + q_5$, $p_5 + q_1$ in this order (with appropriate orientation of $\mathfrak{R}_1$). From this we can calculate the value of $|j(\mathfrak{R}_1^+, \mathfrak{R}_2^+)|$ as twice the area of the pentagon with vertices $1 + \theta^4$, $\theta + 1$, $\theta^2 + \theta$, $\theta^3 + \theta^2$, $\theta^4 + \theta^3$.

Now $\hat{P}(T_1^4 + T_2^4, \mathfrak{R}_2)$ is the pentagon with vertices 2, 2θ, $2\theta^2$, $2\theta^3$, $2\theta^4$, and an elementary calculation shows that

$$|j(\mathfrak{R}_1^+, \mathfrak{R}_2^+)| > \hat{P}(T_1^4 + T_2^4, \mathfrak{R}_2)$$

and so, by (19), $\hat{P}(T_1^4 + T_2^4, R)$ is not convex.

It is not difficult to find examples of convex bodies K such that, for a suitable 2-dimensional plane $\mathfrak{R}_1$, $S(K, \mathfrak{R}_1)$ does not contain a subset homeomorphic to a circle. The following is an example with an additional interesting property:

[4]) *Added in proof:* In the meantime G. C. Shephard has given a proof of this conjecture

(f) *A convex body K in E^4 with the property that the projection of a subset of $S(K, \mathfrak{R}_1)$ onto a second plane $\mathfrak{R}_2$ winds infinitely often around a fixed triangle.*

Write $\omega_n = \exp(2\pi i/n)$ and consider the convex hull of the infinite sequence of points in E^4 whose coordinates are, in the notation of (a),

$$(\omega_n, \omega_3^n) \qquad n = 1, 2, \ldots, .$$

$P(K, \mathfrak{R}_1)$ is an infinite polygon Π inscribed in a semicircle, whose vertices are ω_n. Hence $S(K, \mathfrak{R}_1)$ is an infinite polygon with vertices (ω_n, ω_3^n) $(n = 1, 2, \ldots)$ in this order. It is of infinite length. The only closed subset of $S(K, \mathfrak{R}_1)$ which projects into the polygon Π is $S(K, \mathfrak{R}_1)$ itself. Further, $P(S(K, \mathfrak{R}_1), \mathfrak{R}_2)$ is a curve winding infinitely often around the triangle $1, \omega_3, \omega_3^2$.

III.4. Local convexity

At the end of Section I.2 it was briefly mentioned that convexity on non-convex sets is not a local property. We discuss this important point now, showing in particular that a non-convex projection function may be locally convex.

If $f(R)$ is defined on G_r^n and is homogeneous, then convexity in a neighborhood of the origin on G_r^n evidently implies convexity in all of G_r^n. We therefore call $f(R)$ *locally convex* if every simple r-vector $R \neq 0$ has a neighborhood U_R on G_r^n such that $f(R)$ is convex in U_R, that is to say, can be extended to a convex function $f(R)$ in U_R^H, which, by (20), is an open subset of V_r^n. It suffices, of course, to require this for every simple unit vector. We observe first,

(29) *If $f(x)$ is homogeneous in A^n $(n > 1)$ and is convex in a neighborhood of each $x \neq 0$, then $f(x)$ is convex.*

If y_1, y_2 are distinct points of A^n, $\lambda_1 + \lambda_2 = 1$, $\lambda_i > 0$, then

$$f(y_\lambda) \leqq \lambda_1 f(y_1) + \lambda_2 f(y_2)$$

is trivial when one of the points y_i, say y_1, is the origin z, because $f(x)$ is linear on the ray αy_2 $(\alpha \geqq 0)$. Assume that neither point is z and that the segment T from y_1 to y_2 does not pass through z. Then $f(x)$ is locally convex, hence convex on T.

For $z \in T$ it suffices to consider the case $y_\lambda = z$. Let w be any point not on the line $y_1 y_2$ (here we use $n > 1$), $w_t = (1 - t)\, y_1 + t w$ and $y'_\lambda = \lambda_1 w_t + \lambda_2 y_2$. Then $f(y'_\lambda) \leqq \lambda_1 f(w_t) + \lambda_2 f(y_2)$ and $f(y'_\lambda) \to f(y_\lambda) = f(z)$, $f(w_t) \to f(y_1)$ as $t \to 0$.

Thus local convexity of $f(R)$ on G_1^n or G_{n-1}^n $(n > 1)$ implies convexity, and hence is on G_r^n for $1 < r < n - 1$ another true generalization of ordinary convexity. It fits into our scale as follows:

$$\text{total convexity}^{5)} \Rightarrow \text{convexity} \Rightarrow \text{local convexity} \Rightarrow$$
$$\Rightarrow \text{convexity on convex subsets}.$$

We will see presently in (30) that local convexity is, for $1 < r < n - 1$, weaker than convexity. For this reason it is not very interesting. Properties

(30) Theorem: *Let K be a convex body in E^n whose boundary ∂K has class C^2 and non-vanishing Gauss curvature. Then $\hat{P}(K, R)$ is, for $1 < r < n-1$, locally convex, but, in general, not convex.*

Since $K_i \to L$ implies $\hat{P}(K_i, R) \to \hat{P}(L, R)$ and since any convex body L can be approximated by bodies K_i satisfying the hypotheses of (30), the $\hat{P}(K_i, R)$ cannot all be convex if $\hat{P}(L, R)$ is not convex.

If ∂K has the properties required in (30), then $S(K, \mathfrak{R}) = S(\mathfrak{R})$ is homeomorphic to S^{r-1} and possesses at every point an $(r-1)$-dimensional tangent flat $\mathfrak{T}_p$ which depends continuously on p. The tangent hyperplane $\mathfrak{H}_p$ of ∂K at p contains $\mathfrak{T}_p$ and the $(n-r)$-dimensional generators $\mathfrak{S}_p$ of $Z(K, \mathfrak{R})$ through p. Moreover $\mathfrak{T}_p$ and $\mathfrak{S}_p$ are orthogonal in the euclidean metric in $\mathfrak{H}_p$ for which the Dupin Indicatrix I_p of ∂K at p is the unit sphere.

Let S_u and T_{up} be the unit $(n-r)$- and $(r-1)$-vectors corresponding to $\mathfrak{S}_p$ and $\mathfrak{T}_p$. Because I_p depends continuously on p and ∂K is compact there exists a positive $\alpha(K)$ such that $|T_{up} \wedge S_u| \geqq \alpha(K)$ for all $S(\mathfrak{R})$ and all $p \subset S(\mathfrak{R})$.

Let S'_u be a simple unit $(n-r)$-vector and $\mathfrak{S}'_p$ the corresponding $(n-r)$-flat through $p \in S(\mathfrak{R})$. We claim that $\varepsilon > 0$ exists such that for $|S_u - S'_u| < \varepsilon$ no $\mathfrak{S}'_p$ intersects $S(\mathfrak{R})$ at a point different from p.

If this were not correct, then a sequence of unit $(n-r)$-vectors $S_u^{(\nu)} \to S_u$, points $p_\nu \in S(\mathfrak{R})$ and $(n-r)$-flats $\mathfrak{S}^{(\nu)}$ corresponding to the $S_u^{(\nu)}$ through p_ν would exist such that $\mathfrak{S}^{(\nu)}$ intersects $S(\mathfrak{R})$ in a point $q_\nu \neq p_\nu$. We may assume that p_ν and q_ν converge to points p, q (on $S(\mathfrak{R})$). Then $\mathfrak{S}^{(\nu)} \to \mathfrak{S}_p$ and $q \in \mathfrak{S}_p$, hence $q = p$. But the line through p_ν and q_ν tends to a line in $\mathfrak{T}_p$ (because $\mathfrak{T}_p$ depends continuously on p) which would imply $|T_{up} \wedge S_u| = 0$.

It is clear that in this argument we could also have let S_u vary by replacing it with a sequence of simple $(n-r)$-vectors tending to S''_u, and thus establishing uniformity:

There exists an $\varepsilon > 0$ such that for $|S'_u - S''_u| < \varepsilon$ an $(n-r)$-flat corresponding to S'_u through a point of $S(\mathfrak{R}'')$, where $\mathfrak{R}''$ is normal to S''_u, does not intersect $S(\mathfrak{R}'')$ a second time.

Let Q be the r-dimensional surface spanning $S(\mathfrak{R}'')$ formed by joining one point of $S(\mathfrak{R}'')$ to every other point, then it is clear from the above argument that each $(n-r)$-flat corresponding to S'_u meets Q in at most one point. Hence the projection of Q onto $\mathfrak{R}'$ (where $\mathfrak{R}'$ is the r-flat normal to S'_u) is $1-1$. With the same agreements on orientation and notation as in III.2, we have, therefore,

$$j^*(\mathfrak{R}'', \mathfrak{R}') \leqq \hat{P}(K, \mathfrak{R}') \qquad \text{for } |R''_u - R'_u| < \varepsilon\,,$$

which, if ε is small, is equivalent to

$$j^*(\mathfrak{R}'', R') \leqq \hat{P}(K, R') \qquad \text{for } |R''_u - R'_u| < \varepsilon/2\,. \tag{31}$$

We can now show that $\hat{P}(K, R)$ is convex in an $(\varepsilon/4)$-neighborhood of any given unit vector R_u. If both R' and R'' lie in this neighborhood, then $|R' - R''| <$

In the course of our investigations we encountered an expression for $\hat{P}(K, R)$ as an integral over the tangent bundle of ∂K, which seems worth nothing. Since it has not been used, we give it here without proof and only for the simplest interesting case $r = n - 2$.

Let ∂K have non-vanishing curvature, M_p be the mean curvature of ∂K at p. As usual put

$$M = \int_{\partial K} M_p d A_p = n V_{n-2,2}(K, B^n) .$$

For a fixed $(n-2)$-plane $\mathfrak{F}$, let $\mathfrak{E}$ be the 2-plane normal to $\mathfrak{F}$ through $p \in \partial K - S(K, \mathfrak{F})$. Then $\mathfrak{E}$ intersects the tangent hyperplane $\mathfrak{H}_p$ of ∂K at p in a line. Denote by v_p a unit vector with origin p along this line, and by u_p a variable unit tangent vector of ∂K at p, and by $k(u_p)$ the corresponding normal curvature. Then

$$\hat{P}(K, \mathfrak{F}) = \frac{1}{2\pi_n} \int_{(\partial K, u_p)} k(u_p) \, |u_p \cdot v_p| \, d u_p \wedge d A_p - \frac{n-1}{2\pi} M , \tag{32}$$

where dA_p is the area element of ∂K, and the integral is extended over the tangent bundle of K. Using (5) we obtain an interesting expression of this integral in terms of mixed volumes ($\mathfrak{F}^{\perp}$ denoting the plane normal to $\mathfrak{F}$)

$$\int_{(\partial K, u_p)} k(u_p) \, |u_p \cdot v_p| \, d u_p \wedge d A_p = 2(n-1)\pi_{n-2} \, [V_{n-2,2}(K, B(\mathfrak{F}^{\perp})) + \\ + V_{n-2,2}(K, B^n)] .$$

(32) shows that the integral induces a convex function on G^n_{n-2} whenever $P(K, \mathfrak{F})$ does.

Part IV. Suprema of convex functions

Forming the supremum of a set of convex functions, each defined on the same set M, yields a convex function. This idea underlies the usual definition of the supporting function, but it has many other applications which form the topic of Part IV.

We first sketch the general theory, and then use it to show, as a first application: if $f(\mathfrak{R}^+)$ induces a convex function on G^n_r and $0 < s < r$, $0 < r < t \leq n-1$, then $\sup_{\mathfrak{R}^+ \supset \mathfrak{S}} f(\mathfrak{R}^+)$ and $\sup_{\mathfrak{R}^+ \subset \mathfrak{T}} f(\mathfrak{R}^+)$ induce convex functions on G^n_s and G^n_t respectively.

In the second section we use the properties of the projection functions of simplexes to establish results of the following type. For $2 \leq r \leq n-1$ let $v(r, n)$ be the supremum of integers s with the property that, for every convex body K in E^n, the maximum of the r-dimensional volumes of all polyhedra with s vertices inscribed in $P(K, \mathfrak{R})$ induces a convex function on G^n_r. Then $v(n-1, n) = \infty$, but it is clear that $v(r, n)$ must be finite for $r < n-1$ because otherwise we could deduce that $\hat{P}(K, R)$ is convex for all K. We prove $v(2, n) = 5$, and $r + 2 \leq v(r. n) \leq r + 3$ for $2 < r \leq n - 2$. but conjecture that

IV.1. Basic theory

Our first statement lists some obvious but useful facts. 1 (d) is the basic result which we shall use frequently in Part IV. It has already been proved in I (12).

(1) *Let a function $f(x, w)$ be defined for each w in an arbitrary set W and each $x \in M \subset A^n$. For each x let*

$$F(x, W) = \sup_{w \in W} f(x, w) < \infty .$$

Then,

(a) $F(x, W_0) \leq F(x, W)$ *for* $W_0 \subset W$,

(b) *if* $W = \bigcup_j W_j$ $(j \in J)$ *then* $F(x, W) = \sup_{j \in J} F(x, W_j)$,

(c) *if each $f(x, w)$ is (absolutely) homogeneous of degree $p \geq 0$ in x, then so is $F(x, W)$, and*

(d) *if $f(x, w)$ is convex in M for each $w \in W$, then $F(x, W)$ is convex in M.*

Statement (1) is true whatever the nature of the set W. Further properties emerge if, as in IV.3, we take W to be a subset of A^n.

(2) *Let $F(x, W)$ be defined as in (1), $W \subset A^n$, and $f(x, w)$ be convex in x for each $w \in W$, and convex in w for each $x \in M$. Then,*

(a) *if* $W = W_0^H$ *then* $F(x, W_0) = F(x, W)$,

(b) *if* $W_0 = \bigcup_{j \in J} W_j$ *and* $W = W_0^H$, *then* $F(x, W) = \sup_{j \in J} F(x, W_j)$,

(c) *if $W = A^m$ and $f(x, w)$ is, for fixed x, bounded in any bounded subset of A^m, then for $\alpha_i \geq 0$ and any bounded sets W_i $(i = 1, \ldots, r)$,*

$$F\left(x, \sum_{i=1}^{r} \alpha_i W_i\right) \leq \sum_{i=1}^{r} \alpha_i F(x, W_i) .$$

Proof. If $W = W_0^H$ then a given $w \in W$ is representable as $w = \Sigma \mu_i w_i$, $w_i \in W_0$ and

$$f(x, w) \leq \Sigma \mu_i f(x, w_i) \leq \Sigma \mu_i \sup_{w \in W_0} f(x, w) = F(x, W_0) .$$

Therefore,

$$F(x, W) = \sup_{w \in W} f(x, w) \leq F(x, W_0) .$$

Assertions (a) and (b) now follow from 1 (a) and 1 (b). The inequalities,

$$\sup_{w_i \in W_i} f(x, \Sigma \alpha_i w_i) \leq \sup_{w_i \in W_i} \Sigma \alpha_i f(x, w_i) \leq \Sigma \alpha_i \sup_{w_i \in W_i} f(x, w_i) \tag{3}$$

contain (c).

(4) *If $F(x, W)$ is defined as in (1), $W = A^m$ and $f\left(x, \sum_{i=1}^{r} \alpha_i w_i\right) = \sum_{i=1}^{r} \alpha_i f(x, w_i)$ for $\alpha_i \geq 0$, then for any bounded subsets W_i,*

$$F\left(x, \sum^{r} \alpha_i W_i\right) = \sum^{r} \alpha_i F(x, W_i) = \sum^{r} \alpha_i F(x, W_i^H) .$$

Then,

$$F(x, \Sigma\alpha_i W_i) \geqq f(x, \Sigma\alpha_i w_i) = \Sigma\alpha_i f(x, w_i) \geqq$$
$$\geqq \Sigma\alpha_i F(x, W_i) - \varepsilon\Sigma\alpha_i .$$

This together with (3) proves the first equality of (4). The second is obtained from the first, 2 (b) and the identity[6]), $(\Sigma\alpha_i W_i)^H = \Sigma\alpha_i W_i^H$.

If $f(R)$ is convex on G_r^n, $t + s = r$, $t > 0$, $s > 0$ and T, S are variable simple t-vectors and s-vectors then $f(T \wedge S)$ is defined and is convex in T on G_t^n and in S on G_s^n. If W is any bounded set of simple t-vectors (with respect to any euclidean norm) then by 1 (d),

(5) $$F(S) = \sup_{T\in W} f(T \wedge S) \text{ is convex on } G_s^n .$$

Similarly if $t - s = r$, $0 < s < t < n$ and S, T are variable simple s-vectors and t-vectors, then for any bounded set W of simple s-vectors,

(6) $$F(T) = \sup_{S\in W} f(T \cdot S) \text{ is convex on } G_t^n .$$

From these facts we deduce:

(7) **Theorem:** *Let $f(\mathfrak{R}^+)$, where $\mathfrak{R}^+$ is a variable oriented r-flat, induce a convex function on G_r^n. If $0 < s < r$, $0 < r < t < n$, then for variable s-flats and t-flats $\mathfrak{S}$ and $\mathfrak{T}$,*

(a) $F(\mathfrak{S}) = \max_{\mathfrak{R}^+\supset\mathfrak{S}} f(\mathfrak{R}^+)$ *induces a convex function $F(S)$ on G_s^n, and,*

(b) $F(\mathfrak{T}) = \max_{\mathfrak{R}^+\subset\mathfrak{T}} f(\mathfrak{R}^+)$ *induces a convex function $F(T)$ on G_t^n.*

Proof. Since $f(0) = 0$, and the function $f(R)$ induced by $f(\mathfrak{R}^+)$ is convex, $f(R) + f(-R) \geqq 0$. Because $\mathfrak{R}^+ \supset \mathfrak{S}$ and $\mathfrak{R}^+ \subset \mathfrak{T}$ implies $\mathfrak{R}^- \supset \mathfrak{S}$ and $\mathfrak{R}^- \subset \mathfrak{T}$, where $\mathfrak{R}^-$ is the same r-flat as $\mathfrak{R}^+$ but with the opposite orientation,

$$F(\mathfrak{S}) = \max_{\mathfrak{R}^+\supset\mathfrak{S}} f(\mathfrak{R}^+) \geqq 0,\ \ F(\mathfrak{T}) = \max_{\mathfrak{R}^+\subset\mathfrak{T}} f(\mathfrak{R}^+) \geqq 0 .$$

To prove 7 (a) let W be the set of all simple unit $(r - s)$-vectors C_u. For a unit s-vector S_u, choose C_u' so that

$$f(S_u \wedge C_n') = \max_{C_u\in W} f(S_u \wedge C_u) ,$$

[6]) Since the literature seems to contain only the case of compact W_i, we give a proof of this last relation for any bounded sets W_i. If $w_1 \in W_1^H$, then $w_1 = \Sigma\mu_j y_j$ with $y_j \in W_1$. If $w_i \in W_i$ for $i > 1$, then

$$\alpha_1 y_j + \sum_{i=2}^{r} \alpha_i w_i \in \sum_{i=1}^{r} \alpha_i W_i .$$

Hence,

$$\sum_j \mu_j \left(\alpha_1 y_j + \sum_{i=2}^{r} \alpha_i w_i\right) = \sum_{i=1}^{r} \alpha_i w_i \in (\Sigma\alpha_i W_i)^H ,$$

so that,

$$\alpha_1 W_1^H + \sum_{i=2}^{r} \alpha_i W_i \subset (\Sigma\alpha_i W_i)^H .$$

and so $f(S_u \wedge C'_u) \geqq 0$. If $f(S_u \wedge C'_u) > 0$ then C'_u must be normal to S_u. For otherwise we can choose C''_u normal to S_u with

$$S_u \wedge C''_u = \lambda(S_u \wedge C'_u) \qquad (\lambda > 0)\,.$$

Then $\lambda > 1$ and,

$$f(S_u \wedge C''_u) = \lambda f(S_u \wedge C'_u) > f(S_u \wedge C'_u)$$

which is impossible, by choice of C'_u. Therefore even if $f(S_u \wedge C'_u) = 0$

$$\max_{C_u \in W} f(S_u \wedge C_u) = \max_{C^*_u \perp S_u} f(S_u \wedge C^*_u)\,.$$

But $S_u \wedge C^*_u$ is a unit vector R_u, and the corresponding r-flats traverse all the r-flats containing $\mathfrak{S}$. Thus $F(S) = \max\limits_{C_u \in W} f(S_u \wedge C_u)$, which, by (5), is convex. This proves 7 (a).

For a proof of 7 (b) let W be the set of all simple unit $(t-r)$-vectors C_u. If T_u is a unit t-vector then $T_u \cdot C_u$ is a simple r-vector parallel to T_u. As in the proof of 7 (a) we see that

$$\max_{C_u \in W} f(T_u \cdot C_u) = \max_{C^*_u \parallel \mathfrak{T}} f(T_u \cdot C^*_u)$$

where $\mathfrak{T}$ is the non-oriented flat corresponding to T_u. Moreover $T_u \cdot C^*_u$ is a unit r-vector and traverses all r-vectors parallel to $\mathfrak{T}$. Applying (6), the function $F(T)$ is convex.

(8) **Corollary:** *If $f(x)$ is a homogeneous convex function in A^n and u denotes a unit vector, then*

$$F(\mathfrak{R}) = \max_{u \parallel \mathfrak{R}} f(u)$$

induces a convex function on G^n_r.

If, for example, we put $f(u) = h(u) + h(-u)$ where $h(u)$ is the supporting function of a convex body K, then $F(\mathfrak{R})$ is the diameter of $P(K, \mathfrak{R})$, and this therefore induces a convex function on G^n_r. Actually this function is totally convex (see Part V).

As a typical application of 7 (a) we mention: Let K be a convex body in E^n with z as interior point and center, $\mathfrak{H}$ a variable hyperplane through z. Then $1/|K \cap \mathfrak{H}|_{n-1}$ induces a convex function on G^n_{n-1} (see [2] or [8, section 7]). We conclude from 7 (a) that for an s-flat $\mathfrak{S}$ $(0 < s < n-1)$, $1/\min\limits_{\mathfrak{H} \supset \mathfrak{S}} |K \cap \mathfrak{H}|_{n-1}$ induces a convex function on G^n_r.

IV.2. Extremal polyhedra in projections

Let $T(x_1, \ldots, x_{r+1})$ be a (possibly degenerate) r-simplex with vertices $x_1, \ldots, x_{r+1}$. Denote by $T^s_k(x_1, \ldots, x_{r+1})$ $\left(k = 1, \ldots, \binom{r+1}{s+1}\right)$ the s-faces of $T(x_1, \ldots, x_{r+1})$ and put

(9) **Theorem:** *Let $M_1, \ldots, M_{r+1}$ be non-empty bounded sets in E^n and $0 < s \leqq r \leqq t < n$. If $P(M_i, \mathfrak{T})$ denotes the projection of M_i on the t-flat $\mathfrak{T}$, then for $\alpha \geqq 1$*

$$\sup_{q_i \in P(M_i, \mathfrak{T})} \left(\sum_{k}^{\binom{r+1}{s+1}} [V_k^s(q_i, \ldots, q_{r+1})]^\alpha \right)^{1/\alpha}$$

induces a convex function on G_t^n.

Proof. Let W be the set of simplexes $w = T(x_1, \ldots, x_{r+1})$ where $x_i \in M_i$. From III (27) and (7) we know that $|P(T(x_1, \ldots, x_{s+1}), \mathfrak{T})|_s$ induces a convex function on G_t^n. Therefore, with $q_i = P(x_i, \mathfrak{T})$,

$$f(\mathfrak{T}, w) = \left(\sum_{k=1}^{\binom{r+1}{s+1}} [|P(T_k^s(x_1, \ldots, x_{r+1}), \mathfrak{T})|_s]^\alpha \right)^{1/\alpha}$$

$$= \left(\sum_{k=1}^{\binom{r+1}{s+1}} [V_k^s(q_1, \ldots, q_{r+1})]^\alpha \right)^{1/\alpha}$$

induces a convex function on G_t^n for each w, (see I.12). Further, by 1 (d), $F(\mathfrak{T}) = \sup_{w \in W} f(\mathfrak{T}, w)$ also induces a convex function on G_t^n. However this supremum is the supremum occurring in the statement of the theorem, since for every $q_i \in P(M_i, \mathfrak{T})$ there exists an $x_i \in M_i$ such that $q_i = P(x_i, \mathfrak{T})$.

This theorem also follows by the methods of Part V, where special cases of it are strengthened to give total convexity.

The following corollary is the special case $\alpha = 1$, $s = r$, $r - 1$, and $M_i = K$ (all i) of the theorem:

(10) **Corollary:** *If K is a convex body in E^n and $n > t \geqq r$, then the maximum of the volume and the maximum of the area (i.e. the sum of the volumes of the $(r-1$-faces) of an r-simplex contained in $P(K, \mathfrak{T})$ induce convex functions on G_t^n.*

For $r = 1$, this maximal volume is the diameter of $P(K, \mathfrak{T})$ whose convexity has already been mentioned in the preceding section.

Because the projection function of an $(r+1)$-simplex on an r-flat is also convex, the argument leading to (9) also proves the following:

(11) **Theorem:** *Let $M_1, \ldots, M_{r+2}$ be non-empty bounded sets in E^n. Then the maximum volume of the polyhedra with $(r+2)$ vertices, one chosen from each of the sets $P(M_i, \mathfrak{R})$ induces a convex function on G_r^n.*

If we put all the sets M_i equal to a convex body K we get:

(12) **Corollary:** *If K is a convex body in E^n, then the maximal volume of a polyhedron with $(r+2)$ vertices inscribed in $P(K, \mathfrak{R})$ induces a convex function on G_r^n.*

When $r = 2$, this implies that the maximum area of a quadrilateral inscribed in $P(K, \mathfrak{R})$ induces a convex function on G_2^n. However if we use the fact that the projection function of a 4-dimensional simplex on to a 2-flat is convex (see (c)

We can now formulate our principal result in this direction:

(14) **Theorem:** *Let $v(r, n)$ $(2 \leqq r \leqq n-1)$ denote the supremum of integers s with the property that, for every convex body K in E^n, the maximum of the r-dimensional volumes of all polyhedra with s vertices contained in $P(K, \mathfrak{R})$ induces a convex function on G_r^n. Then $v(n-1, n) = \infty$, $v(2, n) = 5$ for $n > 3$, and $r + 2 \leqq v(r, n) \leqq r + 3$ for $3 \leqq r \leqq n-2$.*

The assertion $v(n-1, n) = \infty$ is trivial since for any convex polyhedron K' in K, the projection $P(K', \mathfrak{H})$ ($\mathfrak{H}$ a hyperplane) induces a convex function on G_{n-1}^n.

(12) and (13) imply $v(r, n) \geqq r + 2$ and $v(2, n) \geqq 5$. It remains to show that $v(r, n) \leqq r + 3$ for $2 \leqq r \leqq n - 2$, that is to say, that the maximal volume of a polyhedron with $s \geqq r + 4$ vertices contained in $P(K, \mathfrak{R})$ does not, in general, induce a convex function. In a fixed $(r + 2)$-flat take the polyhedron K^{r+2} of example (b) of III.3, which has $r + 4$ vertices. $P(K^{r+2}, \mathfrak{R})$ is a polyhedron with at most $r + 4$ vertices, so the maximum volume of a polyhedron with $s \geqq r + 4$ vertices inscribed in $P(K^{r+2}, \mathfrak{R})$ is exactly $\hat{P}(K^{r+2}, \mathfrak{R})$. However this does not induce a convex function on G_r^n, which proves the assertion.

We have already stated the conjecture that $v(r, n) = r + 3$ for $2 \leqq r \leqq n-2$. This is, of course, equivalent to the conjecture of III.3 example (d) that $\hat{P}(T^n, R)$ is convex when $r = n - 2$[7]). It should be mentioned here that if K has a center and if, instead of taking arbitrary polyhedra with s vertices in $P(K, \mathfrak{R})$ we restrict our attention to those having a center, further statements are possible, see (17).

Returning now to our discussion of the general method, we notice that in the proof of (9) we only needed the fact that for any s-dimensional simplex T^s and any t-flat $\mathfrak{T}$, $(s \leqq t)$, the function $|P(T^s, \mathfrak{T})|_s$ induces a convex function on G_t^n, and not that the summands were faces of a simplex. Therefore we can show, in the same manner:

(15) *Let $M_1, \ldots, M_m$ be non-empty bounded sets in E^n, $0 < s \leqq r < n$, and $s \leqq m - 1$. If $\alpha \geqq 1$ and C is the class of all sets $(i_1, \ldots, i_{s+1})$ of $(s + 1)$ different numbers selected from $1, 2, \ldots, m$, with $i_{k+1} > i_k$ $(k = 1, 2, \ldots, s)$ then*

$$\sup_{q_i \in P(M_i, \mathfrak{R})} \Big(\sum_C |T(q_{i_1}, \ldots, q_{i_{s+1}})|_s^\alpha \Big)^{1/\alpha}$$

induces a convex function on G_r^n.

The particular case $r = 2$, $s = 1$, $\alpha = 1$, $M_i = K$ (all i) gives:

(16) **Corollary:** *Let K be a convex body in E^n and let $m \geqq 2$ be any given integer. Then the maximum of the sum of the lengths of the sides and diagonals of an m-gon contained in $P(K, \mathfrak{R})$ $(r = 2)$ induces a convex function on G_2^n.*

By an argument similar to that used for proving (9), taking K to be a

(17) *Let K be a convex body with z as interior point and center. Then the maximum volume of a polyhedron with $2(r+1)$ vertices, with center z, inscribed in $P(K, \mathfrak{R})$ induces a convex function on G_r^n.*

If $r = 2$ we may relax the condition that the polygon is centrally symmetric with center z. The maximal hexagon in $P(K, \mathfrak{R})$ necessarily has center z (see [11]).

In conclusion, we mention that the integration principle (Part II) may be applied to any of the functions $F(R)$ whose convexity we have established. This enables us to assert that the average value of $F(\mathfrak{R})$ taken over all $\mathfrak{R}$ through a given t-flat $\mathfrak{T}$ also induces a convex function on G_t^n.

References

[1] Alexandroff, P., and H. Hopf: Topologie I, Berlin 1935, (reprinted Ann Arbor, Mich. 1945).
[2] Barthel, W.: Zum Busemannschen und Brunn-Minkowskischen Satz. Math. Z. **70**, 407—420 (1959).
[3] Bonnesen, T., and W. Fenchel: Theorie der konvexen Körper. Berlin 1934 and New York 1948.
[4] Bourbaki, N.: Algèbre Multilinéaire, Éléments de Math. I, Livre II Chap. 3. Paris 1958.
[5] Busemann, H.: Convex surfaces. New York 1958.
[6] — Area in affine spaces III. Rend. Circ. Mat. Palermo, Ser. II, **9**, 226—242 (1960).
[7] — Convexity on Grassmann manifolds. Enseign. Math. **7**, 139—152 (1961).
[8] —, and E. G. Strauss: Area and normality. Pacific J. Math. **10**, 35—72 (1960).
[9] Cartan, É.: Leçons sur la géometrie des espaces de Riemann. 2nd Ed. Paris 1946.
[10] Coxeter, H. S. M.: Regular polytopes. New York 1948.
[11] Dowker, C. H.: On minimum circumscribed polygons. Bull. Amer. Math. Soc. **50**, 120—122 (1944).
[12] Hadwiger, H.: Vorlesungen über Inhalt, Oberfläche und Isoperimetrie. Berlin-Göttingen-Heidelberg: Springer 1957.
[13] Petkantschin, B.: Integralgeometrie 6, Zusammenhänge zwischen den Dichten der linearen Unterräume im n-dimensionalen Raum. Abhandl. Math. Seminar Hamburg. Univ. **11**, 249—310 (1936).
[14] Shephard, G. C.: Inequalities between mixed volumes of convex sets. Mathematika **7**, 125—138 (1960).

(Received June 8, 1962)

Convexity problems arising in the theory of area

Herbert Busemann (USA)

Proc. Fourth All-Union Math. Congr. (Leningrad, July 3–12 1961)
Vol. II, Leningrad: Nauka, 1964, pp. 177–182.

translated from the Russian by Dmitriy Slutskiy,
Institut de Recherche Mathématique Avancée,
Université de Strasbourg,
7 rue René-Descartes, 67084 Strasbourg, France
email: slutskiy@math.unistra.fr

In my report, I will deal with issues related to convexity and arising in the study of those geometries where the planes are minimal surfaces.

Our space is a convex domain, i.e. a non-empty open set G in an n-dimensional affine space A^n. Let $1 \leq a \leq n-1$; an a-plane is a non-empty intersection: $A_G = A \cap G$, where A is an a-dimensional plane in A^n. We consider measures α of the following type:

1) $0 \leq \alpha(M) \leq \infty$ and $\alpha(M)$ is defined for any Borel set M lying in some A_G;

2) $\alpha(M) = 0$ if M is contained in an $(a-1)$-dimensional plane;

3) $0 < \alpha(S) < \infty$ if S is an a-dimensional simplex in G and $\alpha(S)$ depends continuously on the vertices of S;

4) $\alpha(M)$ is a completely additive function in each A_G.

Such a measure will be called an a-dimensional measure; we will always assume that M lies inside some A_G. We say that a-planes are minimal surfaces for the measure α, or that α satisfies the triangle inequality, if for every closed a-dimensional polyhedron with a-dimensional faces $F_0, \ldots, F_k$ the following polyhedron inequality holds:

$$\alpha(F_0) \leq \alpha(F_1) + \cdots + \alpha(F_k);$$

a-planes are strictly minimal if α satisfies the strict polyhedron inequality

$$\alpha(F_0) < \alpha(F_1) + \cdots + \alpha(F_k) \quad \text{for} \quad F_0 \neq \textstyle\bigcup_{i>0} F_i.$$

We will also need a special case — the simplex inequality

$$\alpha(F_0) \leq \alpha(F_1) + \cdots + \alpha(F_{a+1}) \quad \text{if} \quad \textstyle\bigcup_{i=0}^{a+1} F_i = \partial(S^{a+1}),$$

i.e. $\bigcup F_i$ is the boundary of an $(a+1)$-dimensional simplex.

The question of whether the polyhedron inequality holds or not for any given measure α is very difficult even in very simple cases. Therefore, we must first make sure that interesting, complete in some sense, measures exist for given a and G.

Let K be the cone formed by the rays with origin at a vertex P of an a-dimensional simplex and passing through its opposite face. We say that a measure α is complete if for any such cone K with vertex in G, the intersection of K with G has an infinite measure: $\alpha(K \cap G) = \infty$. A measure α is complete in the sense of Euclid, if every a-dimensional convex set $B \subset G$ whose boundary contains a point of the boundary Γ of G, has an infinite measure $\alpha(B) = \infty$. A measure α is complete in the sense of Lobachevsky, if it is complete and the interior of each a-dimensional simplex S with one vertex lying on Γ and all other vertices belonging to G has a finite measure.

One can prove [Bus61] that there is a measure α, which is complete in the sense of Euclid for given G and a, that satisfies the strict polyhedron inequality. If the boundary Γ is either empty or strictly convex, then there is also a measure which is complete in the sense of Lobachevsky. Perhaps the condition of strict convexity in unnecessary here, but I did not consider this specific case in detail.

These measures depend on some arbitrary functions; therefore, there are many such measures. This fact gives rise to a second question that I immediately formulate in the form of a theorem answering this question (see [Bus61]).

Let G be a convex domain in A^n and $\alpha_1, \ldots, \alpha_{n-1}$ measures on G. Here α_a is an a-dimensional measure on G, satisfying the strict polyhedron inequality. These measures can be simultaneously extended to the $(n+1)$-dimensional space A^{n+1}. Hence, there is a convex domain $G^* \subset A^{n+1}$, where $G = A^n \cap G^*$, and also there are measures $\alpha_1^*, \ldots, \alpha_{n-1}^*$ defined on G^* such that $\alpha_a^*(M) = \alpha_a(M)$ if $M \subset A_G \subset G$. From that, α_a^* satisfies the strict polyhedron inequality. Also α_a^* is complete if α_a is complete, and α_a^* is complete in the sense of Lobachevsky if it is so for α_a.

I could not prove a similar result for the measures that are complete in the sense of Euclid, except for the case $G = A^n$.

In case $a = 1$ we can determine the distance between the points a and b as the length of the segment with endpoints a and b. Here the simplex inequality transforms into the triangle inequality and also implies the general polyhedron inequality. Then we can define intrinsic measures with the help of the distance. For example, an a-dimensional Hausdorff measure is such a measure if we choose a length reasonably.

The question of when the measures satisfy the polyhedron inequality leads to difficult problems that I mentioned at the beginning.

Further we will consider only the case $G = A^n$ and the measures that are invariant with respect to all translations in A^n. It would seem to be interesting

Proc. Fourth All-Union Math. Congr. (Leningrad, July 3–12 1961) Vol. II, Leningrad: Nauka (1964), 177–182

to study a-dimensional measures $\alpha(M)$, $M \subset A$, invariant only with respect to translations that send a plane A to itself, but it turns out [Bus60a] that such a weak invariance and the simplex inequality imply the invariance with respect to all translations.

Let us introduce an auxiliary Euclidean metric

$$|x - y| = \sqrt{\sum g_{ik}(x^i - y^i)(x^k - y^k)},$$

where $x^1, \dots, x^n$ are affine coordinates. It induces an ordinary a-dimensional measure $|M|_a$. In order to distinguish these measures from the measures α, we will call them volumes.

According to the theory of Haar measure, a measure $\alpha(M)$, $M \subset A$ is of the form $\alpha(M) = f(A)|M|_a$, where $f(A)$ is a continuous function in A which has the same value for parallel a-planes.

For $a = 1$ the length of the segment $\overline{xy}$ can be written as

$$\alpha(\overline{xy}) = f(x - y) = f(x^1 - y^1, \dots, x^n - y^n) = f(u)|x - y| = f(A)|x - y|,$$

where u is a unit vector which is parallel to the segment $\overline{xy}$, $f(u) = f(-u)$, and A is a straight line passing through x and y.

The polyhedron inequality is equivalent to the convexity of the function $f(x)$

$$f(x_1 + x_2) \le f(x_1) + f(x_2).$$

Therefore, we have a linear space, or a Minkowski space.

Let $a = n - 1$ and $G(u)$ be a hyperplane with a unit normal vector u. If $M \subset G(u)$, then we can write

$$F(G(u)) = F(u) \quad \text{and} \quad \alpha(M) = F(u)|M|_{n-1} = F(v),$$

where $v = u|M|_{n-1}$, $F(0) = 0$. According to Minkowski's Theorem ([Min11, §27] and [BS60]), the simplex inequality is equivalent to the polyhedron inequality, and is also equivalent to the convexity of the function $F(v)$, or of the surface $F(v) = 1$

$$v = \tfrac{u}{F(G(u))}, \quad |u| = 1.$$

The case where $1 < a < n - 1$ is much more difficult; we will consider it later.

Let us now consider, for example, a Hausdorff measure in Minkowski space. Let S be the unit ball, where $f(x) \le 1$. Then the section of S by an a-dimensional plane A passing through the origin $z = (0, \dots, 0)$ is a unit ball in A. The Hausdorff measure leads to the normalization

$$\alpha(A \bigcap S) = \pi_a = \tfrac{\pi^{\frac{a}{2}}}{\Gamma(\frac{a}{2}+1)},$$

which is equal to the volume of a unit sphere in E^a.

It should be mentioned that there are different definitions of the Hausdorff measure, but they are all equivalent because of a theorem which says that a ball has the largest volume among all sets of a fixed diameter in Minkowski space.

Therefore, a Hausdorff measure of $M \subset A'$ is equal to

Proc. Fourth All-Union Math. Congr. (Leningrad, July 3–12 1961) Vol. II, Leningrad: Nauka (1964), 177–182

$$\alpha(M) = \frac{\pi_a}{|A \bigcap S|_a} |M|_a, \quad A' \| A \ni z.$$

Hence, the polyhedron inequality for $a = n - 1$ is equivalent to the convexity of the surface

$$T \; : \; r = u|G(u) \bigcap S|_{n-1}, \quad |u| = 1,$$

that we obtain by drawing segments of length $|G(u) \bigcap S|_{n-1}$ starting at z in the directions of the normals to $G(u)$.

The convexity of the surface T follows from a more general theorem [Bus49]. Let K be an arbitrary convex body, L a $(n-2)$-dimensional plane, and P a two-dimensional plane perpendicular to L at the point z. If a half-hyperplane G with boundary L intersects K then we draw a segment of length equal to the volume $|G \bigcap K|_{n-1}$ starting at z along the ray where G intersects P. By varying G, we get a convex curve in P.

This theorem contains Brunn's Theorem as a limit case. When L moves to infinity, the half-planes turn into a pencil of parallel hyperplanes. Thus, the polyhedron inequality for the $(n-1)$-dimensional Hausdorff measure in Minkowski space leads to a theorem that is stronger than Brunn's Theorem. It is clear that every informative measure in Minkowski space leads to new interesting properties concerning convexity.

Absolutely new problems arise when comparing Hausdorff measures. First we note that each a-dimensional Hausdorff measure uniquely determines other measures, in particular, length in the initial Minkowski geometry [BS60].

Let us now consider two different Minkowski metrics in A^n, and let α_a, β_a be their a-dimensional Hausdorff measures. Suppose that for each M

$$\alpha_a(M) \leq \beta_a(M).$$

Does this imply that

$$\alpha_b(N) \leq \beta_b(N), \quad b > a,$$

for any N lying in a b-dimensional plane? If $b < a$, it is, generally speaking, not true even for two different Euclidean geometries. The question is equivalent to the following one.

Let S and S' be convex bodies with center z and $|G \bigcap S|_{n-1} \leq |G \bigcap S'|_{n-1}$ for every hyperplane $G \ni z$. Does this imply that

$$|S|_n \leq |S'|_n?$$

I spent a long time trying to prove this. At last, five years ago I published it as a problem, and I know that good mathematicians were studying it without success. It is very possible that the answer is negative, because it is so for star-shaped bodies of rotation in a three-dimensional space.

If S is an ellipsoid then the answer is affirmative. This follows from one general inequality.

Let $M_1, \ldots, M_{n-1}$ be closed bounded sets of positive volume $|M_i|_n > 0$ in E^n. Let $G(u)$ be a hyperplane passing through the origin and with a unit normal vector u. Suppose that

Proc. Fourth All-Union Math. Congr. (Leningrad, July 3–12 1961) Vol. II, Leningrad: Nauka (1964), 177–182

$$M_i(u) = |G(u) \bigcap M_i|_{n-1}, \quad \widetilde{M_i}(u) = G(u) \bigcap M_i.$$

Then [Bus53] the product of volumes

$$|M_1|_n \dots |M_{n-1}|_n \geq k \int\limits_{\Omega} M_1^{\frac{n}{n-1}}(u) \dots M_{n-1}^{\frac{n}{n-1}}(u) dm(u),$$

where $k = \frac{1}{n} \frac{\pi_n^{n-2}}{\pi_{n-1}^n}$; Ω is a unit sphere $|u| = 1$; $m(u)$ is a spherical measure on Ω. Equality holds if and only if the sets M_i are homothetic ellipsoids with center z.

Recently Petty [Pet61] found an equality that contains the above-mentioned inequality. Let $|N|_n > 0$ and $s(u)$ be the center of gravity of a part $C(u)$ of the set N that lies on the positive side of the plane $G(u)$.[1] A point $t(u)$ is defined as the center of gravity of the points $s(u)$ and $s(-u)$ with masses $|C(u)|_n |N|_n^{-1}$ and $|C(-u)|_n |N|_n^{-1}$. The points $t(u)$ form a surface F, and if we denote by $T(N)$ a convex body bounded by the surface F (expanded with the help of a similarity),

$$T(N) : \frac{(n+1)\pi_n}{\pi_{n-1}}, \quad F \text{ is the boundary of } T(N),$$

then the following statements are true.

1) $|T(N)_n| \geq |N|_n$, and the equality holds only if N is an ellipsoid with the center z;

2) $|M_1|_n \dots |M_{n-1}|_n = k \int\limits_{\Omega} M_1(u) \dots M_{n-1}(u) V(T(\widetilde{M_1}(u)), \dots, T(\widetilde{M}_{n-1}(u))) dm(u)$.

Here $T(\widetilde{M_i}(u))$ is defined as $T(N)$; $G(u)$ appears as a space; V is a mixed volume.

From the Aleksandrov-Fenchel Theorem [Ale37] it follows that

3) $V^{n-1}(T(\widetilde{M_1}(u)), \dots, T(\widetilde{M}_{n-1}(u))) \geq |T(\widetilde{M_1}(u))|_{n-1} \dots |T(\widetilde{M}_{n-1}(u))|_{n-1}$.

Thus, the equality 2) with the inequalities 1) and 3) imply my previous inequality.

We now come back to the polyhedron inequality for $1 < a < n-1$. A difficulty arises from the fact that a-dimensional planes passing through z do not form a linear manifold. Consider the open oriented sets and their closures in the a-dimensional planes. We say that two sets of this type are equivalent if they lie in parallel a-dimensional planes and they have the same volume and the same orientation. A class of equivalent sets is a simple multivector (a-vector) V_a. All the sets of zero volume in the a-dimensional planes are represented by a zero a-vector. The simple a-vectors form a so-called Grassmann cone G_a^n. It is an algebraic cone in a linear space of all a-vectors V_a^n of dimension C_n^a.

If $\alpha(M)$ is an a-dimensional measure in A^n invariant under all translations and if M belongs to the class of the a-vector V_a, then we can write:

$$\alpha(M) = \alpha(V_a) = \alpha(-V_a), \quad \text{where } |V_a| = |M|_n.$$

Since the measure $\alpha(V_a)$ is defined only in G_a^n but not in the whole space V_a^n, it does not a priori make sense to talk about the convexity of the function

[1] $s(u) = z$ if $|C(u)|_n = 0$.

Proc. Fourth All-Union Math. Congr. (Leningrad, July 3–12 1961) Vol. II, Leningrad: Nauka (1964), 177–182

$\alpha(V_a)$. However it is natural to say that the function is convex if for simple a-vectors $V_a^1, \dots, V_a^k$ and for a simple sum $\sum V_a^i$ the following inequality holds

$$\alpha(V_a^1 + \cdots + V_a^k) \leq \alpha(V_a^1) + \cdots + \alpha(V_a^k). \tag{0.1}$$

In fact, this convexity provides the polyhedron inequality. The inequality

$$\alpha(V_a^1 + V_a^2) \leq \alpha(V_a^1) + \alpha(V_a^2)$$

is equivalent to the simplex inequality; it is also equivalent to the polyhedron inequality only in the case where the polyhedron lies in an $(a+1)$-dimensional plane; i.e. when $1 < a < n-1$, the case $k = 2$ is not equivalent to the case $k > 2$ (see [Vag46] and [BS60]).

We saw that the function $\alpha(M)$ for the Hausdorff measure is equal to

$$\alpha(M) = \tfrac{\pi_a}{|A \bigcap S|_a} |M|_a, \quad M \subset A' \| A \ni z,$$

and we know that this function satisfies the inequality (0.1) for $k = 2$, but the question for the general k is rather difficult, and I have not yet found any idea to attack it. An affirmative answer would include most of the current theory of convex bodies as a very special case.

Obviously, this is only one of many such questions that lead to a systematic and important extension of the theory of convex bodies. In many statements it is said that a function depending on one or more bodies and on a variable hyperplane, is convex or concave. Such statements often preserve their meaning if you replace a hyperplane with a variable a-dimensional plane, and a classical result contains the inequality for $k - 2$. Is the corresponding inequality true for a general k?

Fenchel and I found a common principle (not yet published, see special cases in [Bus60b]) by which we can affirm that in simple cases. The essence of the principle is that convex functions of linear functions are convex and that linear combinations of convex functions are convex. Instead of formulating the principle, I will explain it by an example.

Let $f(u)$ be a support function of a convex body K, and A an a-dimensional plane passing through the origin. Let also

$$B(A) = \tfrac{2}{2\pi_a} \int\limits_{A \bigcap \Omega} f(u) dm(u),$$

where $m(u)$ is an $(a-1)$-dimensional spherical measure on $A \bigcap \Omega$. Then $B(A)$ is the average width of the projection of the body K onto A. For an a-vector V_a parallel to A we set

$$B(V_a) = |V_a| B(A), \quad B(0) = 0.$$

Then

$$B(V_a^1 + \cdots + V_a^k) \leq B(V_a^1) + \cdots + B(V_a^k).$$

Our principle is applicable because integration is a linear operation.

Instead of the average width of the projection let us consider the volume $P(A)$ of the projection of a body K onto A and let us set

Proc. Fourth All-Union Math. Congr. (Leningrad, July 3–12 1961) Vol. II, Leningrad: Nauka (1964), 177–182

$$P(V_a) = |V_a|P(A)$$

if V_a is parallel to an a-dimensional plane A.

It is not yet proved that

$$P(V_a^1 + \dots + V_a^k) \leq P(V_a^1) + \dots + P(V_a^k)$$

for $k > 2$.[2] For $k = 2$ the inequality follows from a well-known result of Minkowski ([BF34, § 29] and [Min11, § 26]). Here our principle is not applicable because we deal with a non-linear operation. In fact, let M be an $(n-a)$-dimensional convex body lying in an $(n-a)$-dimensional plane which is normal to an a-vector V_a, and let $|M|_{n-a} = |V_a|$; then

$$P(V_a) = C_n^a V(\underbrace{K, \dots, K}_{a \text{ times}}, \underbrace{M, \dots, M}_{(n-a) \text{ times}}),$$

and this is linear only for $n - a = 1$. Despite the overall impression, the question is not related to the Aleksandrov-Fenchel theorem that generalizes the Minkowski inequality.

I wanted to show by this report that questions arising from the concept of area lead to a number of deep problems that are natural within the theory of convex bodies, and interesting regardless of their origin.

References

[Ale37] A. D. Aleksandrov. Zur Theorie der gemischten Volumina von konvexen Körpern. II. Neue Ungleichungen zwischen den gemischten Volumina und ihre Anwendungen. *Rec. Math. Moscou, n. Ser.*, 2:1205–1238, 1937.

[BES62] H. Busemann, G. Ewald, and G. C. Shephard. Convex bodies and convexity on Grassmann cones. V. Totally convex functionals. *Arch. Math. (Basel)*, 13:512–526, 1962.

[BES63] H. Busemann, G. Ewald, and G. C. Shephard. Convex bodies and convexity on Grassmann cones. I–IV. *Math. Ann.*, 151:1–41, 1963.

[BF34] T. Bonnesen and W. Fenchel. *Theorie der konvexen Körper.* (Erg. Math. u. ihrer Grenzgebiete 3, No.1) Berlin: Julius Springer. VII, 164 S., 8 Fig., 1934.

[BS60] H. Busemann and E. G. Straus. Area and normality. *Pac. J. Math.*, 10:35–72, 1960.

[Bus49] H. Busemann. A theorem on convex bodies of the Brunn-Minkowski type. *Proc. Natl. Acad. Sci. USA*, 35:27–31, 1949.

[2]This inequality is not true for a general convex body K, where $1 < a < n-1$. A detailed study of the function $P(V_a)$ and of the conditions on the body K guaranteeing that $P(V_a)$ satisfies the above-mentioned inequality, will be given in a common paper by G. Ewald, G. C. Shepard, and of the speaker. (*Addendum of the lecturer, December 1961*; here the author speaks about the articles [BES62] and [BES63] – translator's note.)

Proc. Fourth All-Union Math. Congr. (Leningrad, July 3–12 1961) Vol. II, Leningrad: Nauka (1964), 177–182

[Bus53] H. Busemann. Volume in terms of concurrent cross-sections. *Pac. J. Math.*, 3:1–12, 1953.

[Bus60a] H. Busemann. Areas in affine space. I, II. *Rend. Circ. Mat. Palermo, II. Ser.*, 9:81–90, 1960.

[Bus60b] H. Busemann. Areas in affine spaces. III: The integral geometry of affine area. *Rend. Circ. Mat. Palermo, II. Ser.*, 9:226–242, 1960.

[Bus61] H. Busemann. Geometries in which the planes minimize area. *Ann. Mat. Pura Appl., IV. Ser.*, 55:171–189, 1961.

[Min11] H. Minkowski. Theorie der konvexen Körper. *Gesammelte Abhandlungen*, 2:131–229, 1911.

[Pet61] C. M. Petty. Centroid surfaces. *Pac. J. Math.*, 11:1535–1547, 1961.

[Vag46] V. V. Wagner. Geometry of a space with an areal metric and its applications to the calculus of variations. 1946.

Proc. Fourth All-Union Math. Congr. (Leningrad, July 3–12 1961) Vol. II, Leningrad: Nauka (1964), 177–182

Convex Bodies and Convexity on Grassmann Cones (X): Projection Functions of Parallel Convex Bodies.

H. BUSEMANN (*) (Los Angeles) and G. C. SHEPHARD (Birmingham)

In memory of Guido Castelnuovo, in the recurrence of the first centenary of his birth.

Summary. - *It is shown that if K is a convex polyhedron or a smooth convex body, then for sufficiently large positive ρ, the body parallel to K at distance ρ has convex projection functions. An example is given of a convex body which does not have this property.*

Introduction.

The results of the first part of this paper form a contribution to the growing collection of geometric phenomena which are at variance with our three-dimensional experience. As such, they may be of interest to a wider circle of geometers, so we have tried to make the introduction understandable to the non-specialist.

If K is a convex body in n-dimensional Euclidean space E^n, with a cartesian coordinate system whose origin is z, then $P(K, \mathfrak{R})$ denotes the orthogonal projection of K on the r-flat $\mathfrak{R}$ through z, and $\widehat{P}(K, \mathfrak{R})$ denotes its r-area. For a simple r-vector R parallel to $\mathfrak{R}$, we put $\widehat{P}(K, R) = |R| \widehat{P}(K, \mathfrak{R})$, $\widehat{P}(K, 0) = 0$, and call $\widehat{P}(K, R)$ the *r-dimensional projection function* of K. MINKOWSKI showed that if $r = 1$ or $n - 1$, the function $\widehat{P}(K, R)$ is always convex. Hence we shall assume in the following that $1 < r < n-1$, and so $n \geq 4$.

A function $f(R)$, where R is a simple r-vector, is called *homogeneus* if $f(\lambda R) = \lambda f(R)$ for all real $\lambda \geq 0$. Such a function is defined on the GRASSMANN cone G_r^n of all simple r-vectors and is said to be *convex* if either

(i) it can be extended to a convex function on the linear space of all r-vectors, i.e. the space spanned by G_r^n,

or, in terms of simple vectors only:

(ii) whenever $R = R_1 + \dots + R_k$ $(k \leq \binom{n}{r})$, then

$$f(R) \leq f(R_1) + \dots + f(R_k),$$

(see I) [1].

(*) Supported by a grant from the National Science Foundation, U. S. A.

[1] References in Roman numerals, such as this, are to earlier parts of this series of papers, listed at the end of this paper.

A projection function $\widehat{P}(K, R)$ of a convex body K may or may not be convex in this sense, depending on the particular body K and the values of n and r (see III). If we write B for the unit ball in E^n, center z, then $K+\rho B$ (in the sense of vector addition) is the parallel body to K at distance $\rho > 0$. As ρ increases, the shape of $K+\rho B$ becomes more and more spherical in as much as $\rho^{-1}K+B$ tends to B. Therefore it seems highly plausible that for any body K, $\widehat{P}(K+\rho B, R)$ will be convex for sufficiently large ρ. Very surprisingly this is false. There is a pseudo-polyhedron (2) Π_∞ in E^4, i.e. the convex hull of a denumerable number of points, such that $\widehat{P}(\Pi_\infty + \rho B, R)$ $(r=2)$ is not convex for $\rho > 0$.

This phenomenon cannot be attributed alone either to the fact that the boundary of Π_∞ is predominantly flat, or to the fact that it has infinitely many vertices. For we shall show in §§ 3 and 5 that $\widehat{P}(K+\rho B, R)$ is convex for large ρ and all r, when K is a (finite) polyhedron, or when it can «roll» in some sphere; in the latter case it may have infinitely many vertices. In fact, in § 5, we shall prove the following strong uniformity properties:

(a) There exists a constant $\alpha(N, n, r)$ such that $\widehat{P}(K+\rho B, R)$ is convex for all $\rho \geq M\alpha(N, n, r)$, whenever K is a convex polyhedron with at most N vertices and diameter M.

(b) There exists a constant $\beta(n, r)$ such that $\widehat{P}(K+\rho B, R)$ is convex for all $\rho \geq M\beta(n, r)$, whenever K is a convex body that can roll in a sphere of radius M.

It is known (see III(30)) that for $1 < r < n-1$ (in contrast to the cases $r=1, n-1$) local convexity of a function $f(R)$ does not imply convexity. Nevertheless, local convexity has some bearing on our problem. Using the notation that R stands for a simple r-vector with unspecified r, and R^i for a simple i-vector, we will prove in § 1 that the local convexity of $\widehat{P}(K, R^i)$ for $2 \leq i \leq s$, implies the convexity of $\widehat{P}(K+\rho B, R^s)$ for large ρ. This does not contain the result about the convexity of $\widehat{P}(\Pi+\rho B, R)$ for a polyhedron Π and large ρ, because, as we shall show in § 2, in general $\widehat{P}(\Pi, R)$ is not locally convex. In particular, if T^5 is a 5-dimensional simplex in E^5, then $\widehat{P}(T^5, R^2)$ is not locally convex. In this conection we recall the fact, proved elsewhere, that although $\widehat{P}(T^5, R^3)$ is convex for every 5-simplex T^5 (see VI), given any $\rho_0 > 0$, we can find a 5-simplex T_0^5 with the property that $\widehat{P}(T_0^5 + \rho_0 B, R^3)$ is not convex (see [4]).

The last part of the paper deals with closely related but more technical

(2) The existence of this was conjectured by G. Ewald.

questions. For r convex bodies $K_1, \dots, K_r$ in E^n, and an r-vector R parallel to the r-flat $\mathfrak{R}$ we define te *mixed projection function* $\widehat{P}(K_1, \dots, K_r; R)$ to be the mixed volume $|R| V(P(K_1, \mathfrak{R}), \dots, P(K_r, \mathfrak{R}))$. While this function need not be convex, its convexity is affine-invariant in the sense that for any affinity φ, the function $\widehat{P}(\varphi K_1, \dots, \varphi K_r; R)$ is convex when $\widehat{P}(K_1, \dots, K_r; R)$ is. Because of their geometric meaning, the cases

$$P_{i,r-i}(K, B; R) = \widehat{P}(\underbrace{K, \dots, K}_{i}, \underbrace{B, \dots, B}_{r-i}; R) \qquad (i = 1, \dots, r)$$

are of special interest. $\widehat{P}_{r,0}(K, B; R) = \widehat{P}(K, R)$ is in general not convex, but $\widehat{P}_{1,r-1}(K, B; R)$ always is (see III pp. 23 and 31). This leads quite naturally to the problem posed in [1], whether there are numbers $1 < i < r < n-1$ such that $\widehat{P}_{i,r-i}(K, B; R)$ is always convex. Using the methods of VIII and [4] we will show that none of these functions is convex for all K. In particular, the $(r-1)$-area of the boundary of $P(K, \mathfrak{R})$ yields $r\widehat{P}_{r-1,1}(K, B; R)$ and hence a non-convex function, in general, if $2 < r < n-1$.

1. Implications of Local Convexity.

We begin by proving

(1) *If $f(R)$ is a homogeneous locally convex function on G_r^n, then a ρ_0 exists such that $f(R) + \rho|R|$ is convex for $\rho \geq \rho_0$.*

Throughout we shall use a notation previously introduced: the letter R, with or without suffixes and superscripts, stands for a simple r-vector, whilst $\tilde{R}$ means a general r-vector which may or may not be simple. We recall that a function $f(R)$ is locally convex if for each $R_0 \neq 0$ there exists a neighbourhood $U(R_0)$ on G_r^n such that $f(R)$ can be extended to a convex function $f_{R_0}(\tilde{R})$ on the convex hull $U^H(R_0)$ of $U(R_0)$. If $f(R)$ is homogenous, as here, $f_{R_0}(\tilde{R})$ can also be chosen homogeneous. Further there exists $\varepsilon(R_0)$ such that $U^H(R_0)$ contains the ball $|\tilde{R} - R_0| < \varepsilon(R_0)$.

Because $f(R)$ is locally convex it is continuous on G_r^n, so if R_u denotes a simple unit r-vector, then $\min f(R_u) = m > -\infty$. Further, also by compactness, there exists a finite set of simple unit vectors $R_u^1, \dots, R_u^k$ and a $\delta > 0$ such that for each R_u, a ν $(1 \leq \nu \leq k)$ exists for which the ball $|\tilde{R} - R_u| \leq \delta$ lies in $|\tilde{R} - R_u^\nu| \leq \varepsilon(R_u^\nu)$.

Let $f_\nu(\tilde{R})$ be a convex extension of $f(R)$ to $|\tilde{R} - R_u^\nu| < 3\varepsilon(R_u^\nu)$. Then for a suitable b_ν,

$$|f_\nu(\tilde{R}_1) - f_\nu(\tilde{R}_2)| \leq b_\nu |\tilde{R}_1 - \tilde{R}_2|$$
$$\text{if } |\tilde{R}_i - R_u^\nu| \leq 2\varepsilon(R_u^\nu),\ i = 1,\ 2.$$

Therefore, with $b = \max_\nu b_\nu$, we have, for any unit vector R_u^0:

$$|f(R_u) - f(R_u^0)| \leq b|R_u - R_u^0| \quad \text{if } |R_u - R_u^0| \leq \delta.$$

For given numbers $A_{i_1 \dots i_r} (1 \leq i_1 < i_2 < \dots < i_r \leq n)$ we put, for brevity,

$$|A| = (\sum_{i_1, \dots, i_r} A^2_{i_1 \dots i_r})^{\frac{1}{2}}$$

and $A \cdot R = \Sigma A_{i_1 \dots i_r} R^{i_1 \dots i_r}$ where the $R^{i_1 \dots i_r}$ are the components of the r-vector R. Since $f_\nu(\tilde{R})$ is convex, there are, for a given R_u^0 with $|R_u^0 - R_u^\nu| \leq \varepsilon(R_u^\nu)$, numbers $A_{i_1 \dots i_r}(R_u^0)$ such that for $|\tilde{R} - R_u^0| \leq \delta$,

$$f_\nu(\tilde{R}) \geq A(R_u^0) \cdot \tilde{R}, \quad f_\nu(R_u^0) = A(R_u^0) \cdot R_u^0.$$

Hence we have, for $|\tilde{R} - R_u^0| \leq \delta$, that

$$b|\tilde{R} - R_u^0| \geq f_\nu(\tilde{R}) - f_\nu(R_u^0) \geq A(R_u^0) \cdot (\tilde{R} - R_u^0).$$

We choose $\tilde{R}$ such that $|\tilde{R} - R_u^0| = \delta$ and

$$\tilde{R} - R_u^0 = \lambda A(R_u^0)$$

with $\lambda \geq 0$. Then

$$b\delta \geq |A(R_u^0)|\delta \quad \text{or} \quad |A(R_u^0)| \leq b.$$

The condition $|R_u - R_u^0| \leq \delta$ is equivalent to $R_u \cdot R_u^0 \geq 1 - \eta$ with a suitable $\eta(\delta) > 0$. Then for $R_u \cdot R_u^0 \geq 1 - \eta$,

$$f(R_u) + \rho|R_u| - A(R_u^0) \cdot R_u - \rho R_u \cdot R_u^0 \geq 0. \tag{2}$$

If, on the other hand, $R_u \cdot R_u^0 \leq 1 - \eta = |R_u| - \eta$ then

$$f(R_u) + \rho|R_u| - A(R_u^0) \cdot R_u - \rho R_u \cdot R_u^0 \geq m - b + \rho\eta,$$

and so (2) holds for all R_u if $\rho \geq (b - m)/\eta$.

Since (2) is homogeneous in R_u, the assumption that R_u is a unit vector is superfluous, and and we may replace it by R throughout. If R_0 is any non-zero vector, replace $\overset{0}{R}_u$ by $R_0/|R_0|$ and we obtain

$$f(R) + \rho|R| \geq A(R_0/|R_0|) \cdot R + \rho R \cdot R_0/|R_0|,$$

with $f(R_0) + \rho|R_0| = A(R_0/|R_0|) \cdot R_0 + \rho R_0 \cdot R_0/|R_0|$, which proves, see I p. 7, that $f(R) + \rho|R|$ is convex for $\rho \geq (b-m)/\eta$.

The following lemma is also needed:

(3) Lemma. - *If a convex body K has the property that a number τ exists such that*

$$\widehat{P}(K,\ R^i) + \tau|R^i|$$

is convex for $1 \leq i \leq s$, then a number ρ_0 exists such that

$$\widehat{P}(K + \rho B,\ R^s)$$

is convex for all $\rho \geq \rho_0$.

Proof. - Since all the functions considered are homogeneous, it will suffice to cousider their values taken on unit vectors only. Thus we shall say that

$$\widehat{P}(K,\ R^i_u) + \tau \qquad 1 \leq i \leq s$$

is convex (meaning precisely that the extension of this function to the whole of G^n_r by homogeneity is convex). By the integration principle (II) and Kubota's formula,

$$\int\limits_{\mathfrak{R}^i \subset \mathfrak{R}^s} (\widehat{P}(K,\ R^i_u) + \tau) d\mathfrak{R}^i = k_1(i,\ s)\widehat{P}_{i,s-i}(K,\ B;\ R^s_u) + k_2(i,\ s)\tau$$

is convex. (Kubota's formula, as well as the values of the positive constants k_1, k_2 can be found in [2].) Hence if we put

$$\rho_i = k_2(i,\ s)\tau/k_1(i,\ s), \qquad 1 \leq i < s,$$

$$\rho_s = \tau$$

we deduce that

$$\widehat{P}_{i,s-i}(K,\ B;\ R^s_u) + \rho_i \qquad 1 \leq i \leq s$$

is convex. Now consider

$$\widehat{P}(K+\rho B, R_u^s)$$

$$= \widehat{P}(K, R_u^s) + \sum_{i=1}^{s-1}\binom{s}{i}(\widehat{P}_{i,s-i}(K, B; R_u^s)\rho^{s-i} + \widehat{P}(B, R_u^s)\rho^s$$

by VIII(2). We rewrite this in the form

$$(\widehat{P}(K, R_u^s) + \rho^s) + \sum_{i=1}^{s-1}\binom{s}{i}(\widehat{P}_{i,s-i}(K, B, R_u^s) + \rho_i)\rho^{s-i}$$

$$+\left(\rho^s - \sum_{i=1}^{s}\binom{s}{i}\rho^{s-i}\rho_i\right)\widehat{P}(B, R_u^s).$$

All the terms except the last we have shown to be convex, and the last is convex if

$$\rho^s - \sum_{i=1}^{s}\binom{s}{i}\rho^{s-i}\rho_i \geq 0.$$

Clearly we can choose ρ_0 so that this holds for $\rho \geq \rho_0$, and so (3) is proved.

Combining (1) and (3) yields

(4) THEOREM. - *If $\widehat{P}(K, R^i)$ is locally convex for $2 \leq i \leq s$, then a number ρ_0 exists such that $\widehat{P}(K+\rho B, R^i)$ $(1 \leq i \leq s)$ is convex for $\rho \geq \rho_0$.*

In III(30) we have shown that if the boundary of K is smooth and nowhere flat, then the hypotheses of (4) are satisfied for all s. In precise form we obtain:

(5) COROLLARY. - *If the boundary of K is of class C^2 and has non-vanishing Gauss curvature, then a number ρ_0 exists such that $\widehat{P}(K+\rho B, R)$ is convex for all $\rho \geq \rho_0$, and all r.*

In fact, it follows from IX, that $\widehat{P}(K+\rho B, R)$ is locally strongly convex.

2. Polyhedra with Locally Non-convex Projection Functions.

The assertion of (5) is correct if K is a polyhedron (instead of being smooth), but the proof cannot be based on (4) because, as we shall now show, $\widehat{P}(T^3, R^2)$ is not locally convex (see the introduction). We begin with some general remarks about the projections of shadow boundaries, which will also be used later.

So far, all projections considered have been orthogonal. It is convenient to introduce parallel (oblique) projections. Let X be any set, and $\mathfrak{R}$, $\mathfrak{R}_0$ any two r-flats. Then we write $P_0(X, \mathfrak{R})$ for the parallel projection of X on to $\mathfrak{R}_0$ in a direction orthogonal to $\mathfrak{R}$, and $\widehat{P}_0(X, \mathfrak{R})$ for the r-area of this projection. In particular, if we take $\mathfrak{R}_0$ to be one of the coordinate r-flats, and notice that

$$\widehat{P}(X, R) = \widehat{P}_0(X, \mathfrak{R}) \tag{6}$$

when $|R_0| = |R \cdot R_0| = 1$ (see [3], p. 13), the calculation of the projection functions is much simplified. Take, for example, $\mathfrak{R}_0$ to be $x_{r+1} = \dots = x_n = 0$ in some rectangular cartesian coordinate system, and R_1 to be

$$\begin{array}{l}(1, 0, \dots 0, x_{r+1,1}, x_{r+2,1}, \dots, x_{n,1}) \wedge \\ (0, 1, \dots 0, x_{r+1,2}, x_{r+2,2}, \dots, x_{n,2}) \wedge \\ \dots\dots\dots\dots\dots\dots \wedge \\ (0, 0, \dots 1, x_{r+1,r}, x_{r+2,r}, \dots, x_{n,r})\end{array} \tag{7}$$

so that R_0 is given by the same expression with all the $x_{i,j} = 0$. If $t = (t_1, \dots, t_n)$ then clearly

$$P(t, \mathfrak{R}_0) = P_0(t, \mathfrak{R}_0) = (t_1, \dots, t_r).$$

To find $P_0(t, \mathfrak{R}_1)$ we notice that the $(n-r)$-vector

$$\begin{array}{l}(-x_{r+1,1}, -x_{r+1,2}, \dots, -x_{r+1,r}, 1, 0, \dots, 0) \wedge \\ (-x_{r+2,1}, -x_{r+2,2}, \dots, -x_{r+2,r}, 0, 1, \dots, 0) \wedge \\ \dots\dots\dots\dots\dots\dots\dots\dots \wedge \\ (-x_{n,1}, -x_{n,2}, \dots, -x_{n,r}, 0, 0, \dots, 1)\end{array}$$

is perpendicular to the vector (7) (and so is parallel to the direction of projection). From this we obtain

$$P_0(t, \mathfrak{R}_1) = (t_1 + \sum_{i=r+1}^{n} t_i x_{i,1}, \dots, t_r + \sum_{i=r+1}^{n} t_i x_{i,r}).$$

In other words, the effect of changing the direction of projection on to $\mathfrak{R}_0$

from that perpendicular to $\mathfrak{R}_0$ to that perpendicular to $\mathfrak{R}_1$, is to displace the projection of t by an amount

(9) $$P_0(t, \mathfrak{R}_1) - P_0(t, \mathfrak{R}_0) = \Big(\sum_{i=r+1}^{n} t_i x_{i,1}, \ldots, \sum_{i=r+1}^{n} t_i x_{i,r}\Big).$$

Working out actual coordinates of R_0 and R_1 from (7) we see that

$$|R_0 - R_1| \geq \Big(\sum_{i=r+1}^{n} \sum_{j=1}^{n} x_{i,j}^2\Big)^{\frac{1}{2}}$$

and so if $T = \max_{r+1 \leq i \leq n} t_i$, then $|R_0 - R_1| < \varepsilon$ implies that

(10) $$|P_0(t, \mathfrak{R}_1) - P_0(t, \mathfrak{R}_0)| < T\varepsilon.$$

We shall apply this to the case when $X = S$ is some (polygonal) shadow boundary of a polyhedron K that, as we may suppose without loss of generality lies inside the ball MB, for some number M. Then the polygons $P_0(S, \mathfrak{R}_1)$ and $P_0(S, \mathfrak{R}_0)$ have the property that when $|R_0 - R_1| < \varepsilon$, one can be obtained from the other by displacing its vertices by a distance not exceeding $M\varepsilon$.

We can now prove

(11) $\widehat{P}(T^5, R^2)$ *is not locally convex.*

Let T^5 be the convex hull of the six points

$$p_0(0, 0, 0, 0, 0),\ p_1(1, 0, 0, 0, 0),\ \ldots,\ p_5(0, 0, 0, 0, 1).$$

Consider $P_0(T^5, \mathfrak{R}_i)$ $(i = 0, 1, 2)$ where

$$R_0 = (1, 0, 0, 0, 0) \wedge (0, 1, 0, 0, 0)$$

so that $\mathfrak{R}_0$ is $x_3 = x_4 = x_5 = 0$,

$$R_1 = (1, 0, \xi_3, \xi_4, \xi_5) \wedge (0, 1, \eta_3, \eta_4, \eta_5)$$

where the points (ξ_3, η_3), (ξ_4, η_4), (ξ_5, η_5) are chosen near to $(0, 0)$ and so that the hexagon of figure 2 is convex, and

$$R_2 = (1, 0, 0, \zeta, 0) \wedge (0, 1, 0, 0, \zeta)$$

where $0 < \zeta < 1$.

Writing $P_0(p_i, \mathcal{R}_0) = q_i$, $P_0(p_i, \mathcal{R}_1) = q_i'$, $P_0(p_i, \mathcal{R}_2) = q_i''$ $(i = 0, 1, \dots, 5)$, using (9) we see that the positions of these 18 points in $\mathcal{R}_0$ are as shown in the figures 1, 2 and 3. Since, by construction, the hexagon in figure 2 is convex, it follows that the shadow boundary $S_1 = S(T^5, \mathcal{R}_1)$ is the skew hexa-

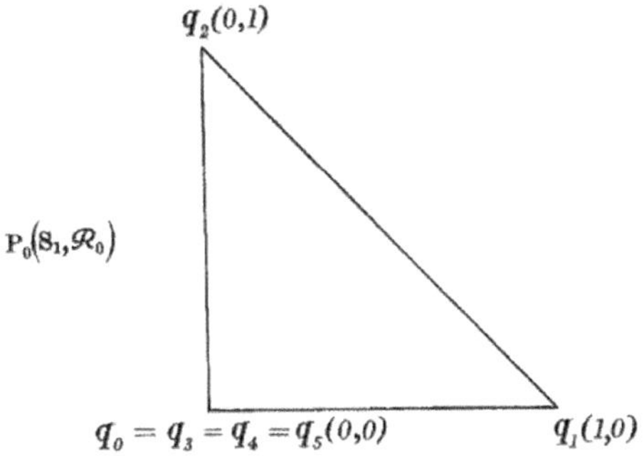

Fig. 1

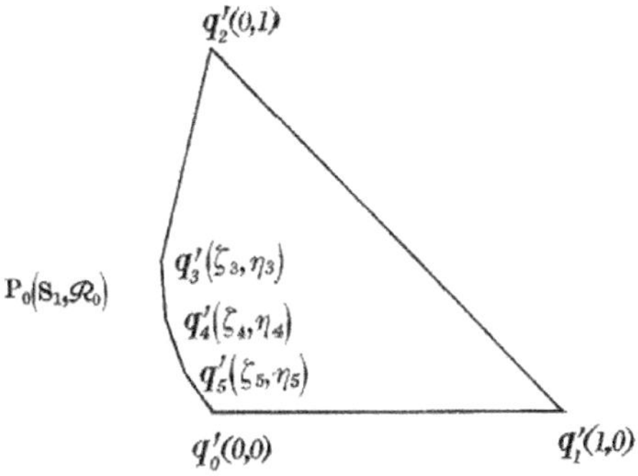

Fig. 2

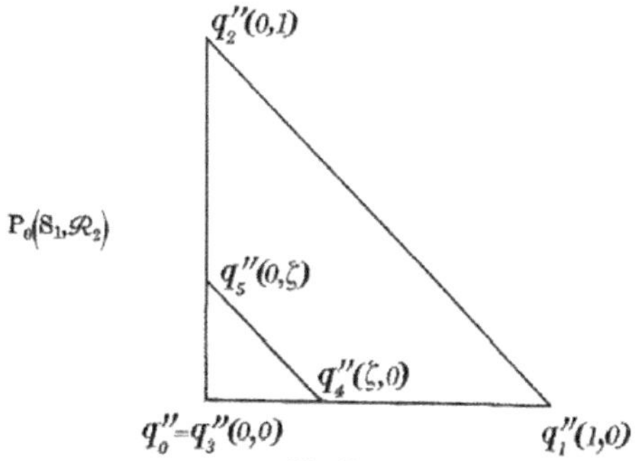

Fig. 3

gon $p_0p_1p_2p_3p_4p_5$ and hence, by constructing a surface Q spanning this hexagon as in III p. 25, we can calculate the corresponding index function, namely, with appropriate orientations

$$j(\mathcal{R}_1^+, R) = R \cdot \left(\frac{1}{2}(e_1 \wedge e_2 + e_2 \wedge e_3 + e_3 \wedge e_4 + e_4 \wedge e_5)\right)$$

where e_i is the unit vector along the i–th coordinate axis. Now

$$R_2 = e_1 \wedge e_2 + \zeta(e_1 \wedge e_5 + e_4 \wedge e_2) + \zeta^2(e_4 \wedge e_5)$$

from which

$$j(\mathcal{R}_1^+, R_2) = \frac{1}{2}(1 + \zeta^2).$$

(This could have been seen immediately from figure 3 for the area of the cycle $q''_0 q''_1 \dots q''_5$, regions being counted with the multiplicities shown, is clearly $\frac{1}{2}(1+\zeta^2)$. The statement about the value of the index function follows from the analogue of (6) since $R_2 \cdot R_0 = 1$). However $\widehat{P}(T^5, R_2) = \frac{1}{2}$ (since $P_0(T^5, \mathfrak{R}_2)$ is the triangle $q''_0 q''_1 q''_2$ in figure 3, of area $\frac{1}{2}$, and then the value of $\widehat{P}(T^5, R_2)$ follows from (6)), and so

$$j(\mathfrak{R}_1^+, R_2) = \widehat{P}(T^5, R_2) + \frac{1}{2}\zeta^2.$$

This implies

$$j(\mathfrak{R}_1^+, R_2) > \widehat{P}(T^5, R_2)$$

however small ξ_i, η_i, ζ are (i.e. however close R_1 and R_2 are to R_0). Consequently, by an obvious extension of III (25), $\widehat{P}(T^5, R)$ fails to be locally convex at the point R_0.

3. Parallel Bodies to Polyhedra.

The following lemma concerning the convexity of $f(R) + \rho|R|$ will be needed in the further discussion of polyhedra. We shall use U_r^n for the set of all simple unit r-vectors in E^n. As previously, a tilde signifies that a vector is not necessarily simple.

(12) Lemma. - *Let $f(R)$ be a homogeneous, continuous, function defined on G_r^n. On a dense subset D of U_r^n, let an r-vector $\tilde{R}(R_D) = \tilde{R}_D$ ($R_D \in D$) be defined with $f(R_D) = \tilde{R}_D \cdot R_D$. For any $R_u \in U_r^n - R_D$ put*

(13) $$C(R_u, R_D) = \{\tilde{R}_D \cdot R_u - f(R_u)\}/\{1 - R_D \cdot R_u\}$$

and

$$C_f = \max\Big(\sup_{R_D \in D,\, R_u \in U_r^n - D} C(R_u, R_D),\ 0\Big).$$

Then:

(a) If C_f is finite, $f(R) + \rho|R|$ is convex for $\rho \geq C_f$.

(b) If $C_f = \infty$ and

(14) $$\begin{cases} \text{for each } R_D, \text{ an } \varepsilon(R_D) > 0 \text{ exists such that} \\ f(R) \geq \tilde{R}_D \cdot R \text{ when } |R - R_D| < \varepsilon(R_D), \end{cases}$$

then $f(R) + \rho|R|$ is not convex for any ρ.

PROOF. – Let C_r be finite. Then for $\rho \geq C_r$,

$$\tilde{R}_D \cdot R_u - f(R_u) = C(R_u, R_D)(1 - R_D \cdot R_u) \leq \rho(1 - R_D \cdot R_u),$$

and so

$$R_u \cdot (\tilde{R}_D + \rho R_D) \leq f(R_u) + \rho$$

with equality for $R_u = R_D$. This means, in the notation of I, p. 4, that the hyperplane

$$\zeta = \tilde{R} \cdot (\tilde{R}_D + \rho R_D)$$

is a supporting hyperplane to the graph $\zeta = f(R) + \rho|R|$ at R_D. Since D is dense in U_r^n, and $f(R)$ is continuous, a supporting plane of this graph exists for every unit vector and hence for every R. So the sufficient conditions for convexity in IIa (I, p. 7) hold, and statement (a) of the lemma follows.

If $C_r = \infty$, then for any ρ, a pair R_u, R_D exists such that

$$(15) \qquad R_u \cdot (\tilde{R}_D + \rho R_D) > f(R_u) + \rho .$$

But the hypothesis (14) means that $\zeta = \tilde{R}_D \cdot \tilde{R}$ is a local supporting plane of $\zeta = f(R)$ at R_D, and therefore $\zeta = (\tilde{R}_D + \rho R_D) \cdot \tilde{R}$ is a local supporting plane of $\zeta = f(R) + \rho|R|$ at R_D. But (15) shows that this hyperplane is not globally a supporting plane, and hence $f(R) + \rho|R|$ cannot be convex.

(16) COROLLARY. – *If for each* $R_D \in D$, $\tilde{R}_D$ *is chosen so that* (14) *holds, then* $C_r = 0$ *if and only if* $f(R)$ *is convex.*

For this is precisely the condition that a local supporting plane to the graph $\zeta = f(R)$ should be global.

(17) THEOREM. – *Given a convex polyhedron K, there is a number* ρ_0 *such that* $\widehat{P}(K + \rho B, R)$ *is convex for* $\rho \geq \rho_0$ *and all* r.

To prove this it suffices to show:

(18) *If K is a polyhedron, then for each* $r (1 < r < n - 1)$ *the number* C_r *of* (12) *is finite for* $f(R) = \widehat{P}(K, R)$.

For if (18) is true, let τ be the maximum of the C_r for $1 < r < n - 1$. By (12) the functions $\widehat{P}(K, R) + \rho|R|$ are convex for $\rho \geq \tau$, so that (17) follows from (3).

To establish (18) we consider, for simplicity, only the case $r = 2$; other values of r can be treated similarly. Figures 1, 2 and 3 illustrate the discussion here also.

A polyhedron K has a finite number of sharp shadow boundaries $S_1, \ldots$ $\ldots, S_k$ (for this and the following, see III pp. 24, 25). To each S_i belongs an open region $G_i \subset G_2^n$ (which does not contain 0, but if $R \in G_i$, then $\lambda R \in G_i$ for $\lambda > 0$) such that the shadow boundary $S(K, \mathfrak{R})$ equals S_i when $\mathfrak{R}$ is parallel to a vector in G_i. Moreover $\bigcup_i \bar{G}_i = G_2^n$. For each i there is a 2-vector $\tilde{F}_i$ such that

$$(19) \qquad \widehat{P}(K, R) = R \cdot \tilde{F}_i \quad \text{for} \quad R \in \bar{G}_i .$$

(see III pp. 25-26).

If we put $D = (\bigcup_i G_i) \cap U_r^n$, then (19) shows that the hypothesis (14) is satisfied with $\tilde{R}_D = \tilde{F}_i$, when $R_D \in G_i$. Suppose $C_f = \infty$, then there must be a sequence of pairs of unit vectors $(R_u^{(j)}, R_D^{(j)})$ $j = 1, 2, \ldots$ such that $C_f(R_u^{(j)}, R_D^{(j)}) \to \infty$ as $j \to \infty$. Since the numerator of (13) is bounded, this implies that the denominator $1 - R_D^{(j)} \cdot R_u^{(j)} \to 0$, and further that $R_u^{(j)}, R_D^{(j)}$ belong to different regions, say G_0, G_1, for infinitely many values of j. (If $R_u^{(j)}$ and $R_D^{(j)}$ belong to the same region G_i, then $C_f(R_u^{(j)}, R_D^{(j)}) = 0$). We may, by picking out appropriate subsequences, assume that $R_u^{(j)} \in G_0$ and $R_D^{(j)} \in G_1$ for all j, and so the sequences $\{R_u^{(j)}\}$ and $\{R_D^{(j)}\}$ both converge to a vector $R_0 \in \bar{G}_0 \cap \bar{G}_1$, with $|R_0| = 1$.

Let us write $\bar{R} = R/|R \cdot R_0|$ for any non-zero simple vector R, with suffixes and superscripts added as necessary. Thus $\bar{R}_u^{(j)} \cdot R_0 = \bar{R}_D^{(j)} \cdot R_0 = 1$, and the discussion leading to (6) and (10) applies. So $P_0(S_0, \mathfrak{R}_u^{(j)})$ is a convex polygon of area $\bar{R}_u^{(j)} \cdot \tilde{F}_0$ and $P_0(S_0, \mathfrak{R}_D^{(j)})$ is a general polygon obtained by displacing the vertices of $P_0(S_0, \mathfrak{R}_u^{(j)})$ by a distance not exceeding $\varepsilon =$ $= M|\bar{R}_u^{(j)} - \bar{R}_D^{(j)}|$ (see (10)). We are only interested in the case where the numerator of (13) is positive, that is,

$$R_D^{(j)} \cdot \tilde{F}_0 > \widehat{P}(K, R_D^{(j)})$$

or, equivalently,

$$\bar{R}_D^{(j)} \cdot \tilde{F}_0 > \widehat{P}(K, \bar{R}_D^{(j)}) .$$

This means that the cycle $P_0(S_0, \mathfrak{R}_D^{(j)})$ must enclose regions of multiplicity at least two (in absolute value).

The total area of the parts convered more than once, each counted with its non-signed, and hence also signed multeplicity, has the order of ε^2.

For, multiplicities greater than one can occur only in the ε-neighbourhoods of vertices, or $(r-2)$-faces in the general case. The area of the union of the these neighbourhoods is $0(\varepsilon^2)$. Because the multiplicity at a point equals its winding number or index with respect to the boundary, no multiplicity can exceed $N/2$ if N is the number of $(r-1)$-faces.

In fact we can estimate

$$\bar{R}_D^{(j)} \cdot \tilde{F}_0 - \widehat{P}(K, \bar{R}_D^{(j)}) = \bar{R}_D^{(j)} \cdot (\tilde{F}_0 - \tilde{F}_1)$$

more exactly. Let

$$\bar{R}_D^{(j)} = (1, 0, \xi_3, \ldots, \xi_n) \wedge (0, 1, \eta_3, \ldots, \eta_n)$$

and F_{ij} be the components of $\tilde{F}_0 - \tilde{F}_1$. Then

$$\bar{R}_D^{(j)} \cdot (\tilde{F}_0 - \tilde{F}_1) = \sum_{i=3}^{n} F_{1i}\eta_i - \sum_{j=3}^{n} F_{2j}\xi_j + \sum_{\substack{i,j=3,\ldots,n \\ i \neq j}} F_{ij}(\xi_i\eta_j - \xi_j\eta_i).$$

The above remark shows that this is at most a constant multiple of ε^2, so that the two linear terms in η_i, ξ_j (the first two) must be negative or zero for $R_D^{(j)} \in G_1$. It follows that

$$\begin{aligned} R_D^{(j)} \cdot \tilde{F}_0 - \widehat{P}(K, R_D^{(j)}) &= R_D^{(j)} \cdot (\tilde{F}_0 - \tilde{F}_1) \\ &\leq \bar{R}_D^{(j)} \cdot (\tilde{F}_0 - \tilde{F}_1) \\ &< 2FM^2\varepsilon^2, \end{aligned} \tag{20}$$

where F is the sum of the components $|F_{ij}|$. But if $|\bar{R}_u^{(j)} - \bar{R}_D^{(j)}| < \varepsilon$, then $|R_u^{(j)} - R_D^{(j)}| < \varepsilon$, and so for sufficiently small ε, $1 - R_u^{(j)} \cdot R_D^{(j)} > \frac{1}{4}\varepsilon^2$. Substituting these inequalities in (13) we see

$$\begin{aligned} C_r(R_u^{(j)}, R_D^{(j)}) &< 2FM^2\varepsilon^2/\tfrac{1}{4}\varepsilon^2 \\ &< 8FM^2, \end{aligned}$$

and so is uniformly bounded. This contradicts our assumption that $C_r(R_u^{(j)}, R_D^{(j)}) \to \infty$, and (18) is proved.

4. Projection of Pseudo-polyhedra.

We now establish that:

(21) *There exist convex bodies K with the property that $\widehat{P}(K + \rho B, R)$ is not convex for any value of of ρ.*

The example we give of such a body is a pseudo-polyhedron in E^4, i. e. the convex hull of a denumerable number of points. It is similar to example (*f*) of III.3, but a number of modifications are introduced in order to simplify the calculations. We take for K the convex hull of

$$p_n = \left(\cos\frac{\pi}{n},\ \sin\frac{\pi}{n},\ \alpha_n,\ \beta_n\right),\ n = 1, 2, \ldots,$$

(22)

$$q_n = \left(\cos\frac{\pi}{m},\ -\sin\frac{\pi}{m},\ \beta_m,\ \alpha_m\right),\ m = 2, 3, \ldots,$$

where $\alpha_n = 1$ if $n \equiv 2 \pmod 3$ and $= 0$ otherwise,

$\beta_n = 1$ if $n \equiv 0 \pmod 3$ and $= 0$ otherwise.

The projection on the plane $\mathfrak{R}_0 : x_3 = x_4 = 0$ is the «infinite polygon» illustrated in figure 4, and that on the plane $\mathfrak{R}_1 : x_1 = x_2 = 0$ is the triangle $\triangle$:

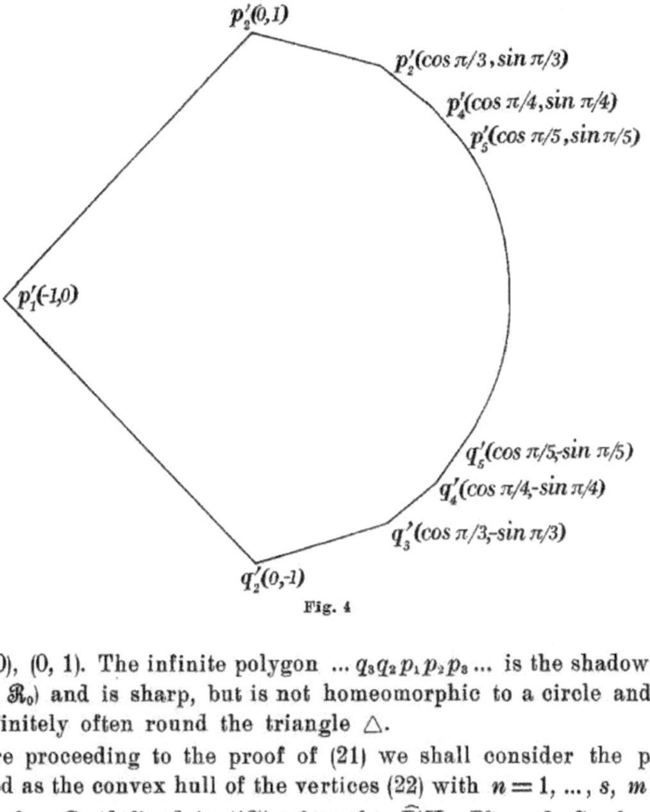

Fig. 4

(0, 0), (1, 0), (0, 1). The infinite polygon ... $q_3 q_2 p_1 p_2 p_3$ ... is the shadow boundary $S_0 = S(K, \mathfrak{R}_0)$ and is sharp, but is not homeomorphic to a circle and $P(S_0, \mathfrak{R}_1)$ winds infinitely often round the triangle $\triangle$.

Before proceeding to the proof of (21) we shall consider the polyhedron K_s defined as the convex hull of the vertices (22) with $n = 1, \ldots, s$, $m = 2, \ldots, s$. Write C_s for C_f (defined in (12)) when $f = \widehat{P}(K_s, R^2)$ and C_∞ for C_f when

$f = \widehat{P}(K, R^2)$. Then it is easily verified (using III(19)) that the projection function $\widehat{P}(K_s, R^2)$ is not convex and that $C_s \to \infty$ as $s \to \infty$. It is therefore not suprising that K has the property (21), but this cannot be established by taking limits as $s \to \infty$ since the function $C_f(R_u, R_D)$ $(f = \widehat{P}(K, R))$ is not continuous in K. Thus $K_s \to K$ does not imply that $C_s \to C_\infty$, and therefore we cannot deduce that $C_\infty = \infty$ as required.

Our approach is therefore to consider K itself and choose a sequence of unit vectors $\{R_0^{(j)}\}$ and a fixed unit vector R_1 such that $C_\infty(R_1, R_0^{(j)}) \to \infty$ as $j \to \infty$. Take $\mathfrak{R}_1$ to be the plane $x_1 = x_2 = 0$ as above and

$$R_0^{(j)} = (1, 0, \xi_j, \xi_j) \wedge (0, 1, -\eta_j, \eta_j)$$

where ξ_j and η_j are chosen small and positive (the appropriate choices will be apparent from the following discussion). By (9) the polygon $P_0(S_0, \mathfrak{R}_0^{(j)})$ is formed from $P_0(S_0, \mathfrak{R}_0)$ by displacing the vertices as follows:

$$\begin{aligned}
p'_n \to\ & p'_n + (\xi_j, -\eta_j) && \text{if } n \equiv 2 \pmod 3\\
& p'_n + (\xi_j, \eta_j) && \text{if } n \equiv 0 \pmod 3\\
& p'_n && \text{if } n \equiv 1 \pmod 3\\
q'_m \to\ & q'_m + (\xi_j, \eta_j) && \text{if } m \equiv 2 \pmod 3\\
& q'_m + (\xi_j, -\eta_j) && \text{if } m \equiv 0 \pmod 3\\
& q'_m && \text{if } m \equiv 1 \pmod 3
\end{aligned}$$

where p'_n, q'_m are the orthogonal projections of p_n, q_m on to $\mathfrak{R}_0$. Taking the convex hull of these displaced vertices gives us $P_0(K, \mathfrak{R}_0^{(j)})$. (See figure 5 where the displaced vertices are denoted by double primes). It is clear that ξ_j can be chosen small, and η_j compared with ξ_j, so that at least $6j+1$ consecutive vertices

$$q''_{3j+1} q''_{3j} \cdots q''_2 p''_1 p''_2 \cdots p''_{3j+1}$$

will lie on the boundary of $P_0(K, \mathfrak{R}_0^{(j)})$, i.e. on the convex hull of the displaced vertices. Consider now the shadow boundary $S_0^{(j)} = S(K, \mathfrak{R}_0^{(j)})$. This is sharp, is an infinite polygon, but, as is easily seen, is homeomorphic to a circle. Further it is unchanged for small variations of ξ_j, η_j and so there exists a vector $\widetilde{F}_0^{(j)}$ (constructed from a surface spanning $S_0^{(j)}$ (see III, p. 25)) such that $\zeta = \widetilde{F}_0^{(j)} \cdot R$ is a local supporting hyperplane to the graph $\zeta = \widehat{P}(K, R)$ at $R = R_0^{(j)}$. Further, the projection of $S_0^{(j)}$ on $\mathfrak{R}_1$ winds at least $2j$ times round the triangle $\triangle$, so that $\tilde{F}_0^{(j)} \cdot R_1 \geq 2j \cdot \frac{1}{2} = j$. Hence

$$C_\infty(R_1, R_0^{(j)}) = \frac{R_1 \cdot \tilde{F}_0^{(j)} - \widehat{P}(K, R_1)}{1 - R_0^{(j)} \cdot R_1} > \frac{j - \frac{1}{2}}{2}.$$

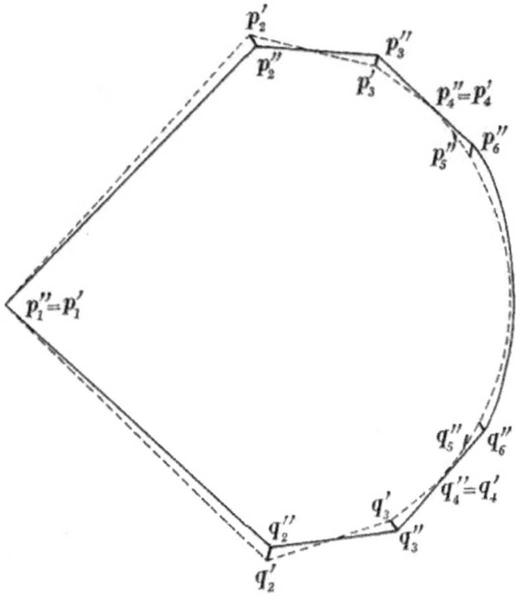

Fig. 5

Thus as $j \to \infty$, $C_\infty(R_1, R_0^{(j)}) \to \infty$, and so $C_\infty = \infty$. This proves, incidentally, that $\widehat{P}(K, R^2) + \rho|R^2|$ is not convex for any value of ρ.

To complete the proof we write

$$\widehat{P}(K + \rho B, R) = \widehat{P}(K, R) + 2\rho\widehat{P}(K, B: R) + \rho^2\widehat{P}(B, R).$$

Since $\widehat{P}(K, B; R)$ and $\widehat{P}(B, R)$ are both convex, they possess supporting hyperplanes, say

$$\zeta = \tilde{D}_0^{(j)} \cdot R \qquad \zeta = \tilde{E}_0^{(j)} \cdot R$$

to their graphs at $R_0^{(j)}$, respectively. Hence if we put $f(R) = \widehat{P}(K + \rho B, R)$,

$\tilde{R}_D = \tilde{F}_0^{(j)} + 2\rho\tilde{D}_0^{(j)} + \rho^2\tilde{E}_0^{(j)}$, then

$$C_f(R_1,\ R_0^{(j)}) = \frac{R_1 \cdot (\tilde{F}_0^{(j)} + 2\rho\tilde{D}_0^{(j)} + \rho^2\tilde{E}_0^{(j)})}{1 - R_0^{(j)} \cdot R_1}$$
$$> \frac{j - \left(\frac{1}{2} + \rho(2 + \sqrt{2}) + \pi\rho^2\right)}{2}$$

and so, whatever the value of ρ, j may be chosen large enough so that $C_f(R_1,\ R_0^{(j)}) > 0$. Thus $C_f > 0$ and since $\tilde{R}_D$ defined above satisfies condition (14), we deduce from (16) that $\widehat{P}(K + \rho B,\ R)$ is not convex. This completes the proof of (21).

5. Uniformity Theorems.

We now prove the two uniformity theorems mentioned in the introduction.

Denote by $\Lambda_M (M > 0)$ the class of all convex bodies K in E^n which can «roll» in a sphere of radius M. Precisely this means that every point of the boundary ∂K of K lies on the boundary of some ball of radius M that contains K. Then:

(23) THEOREM. – *There exists a number $\beta(n,\ r)$ depending on n and r only, such that*

$$\widehat{P}(K + \rho B,\ R)$$

is convex for $K \in \Lambda_M$ and $\rho \geq M\beta(n,\ r)$.

Let Λ_M^2 be the set of those K in Λ_M for which ∂K is of class C^2. Notice that $K \in \Lambda_M$ or $K \in \Lambda_M^2$ implies that $M^{-1}K \in \Lambda_1$ or $M^{-1}K \in \Lambda_1^2$ respectively. Λ_M^2 consists of those K for which ∂K has class C^2 and the maximal normal radius of curvature of ∂K does not exceed M.

For $K \in \Lambda_1^2$, the normal radii of curvature of $\partial(K + \rho B)$ lie in the interval $[\rho,\ 1 + \rho]$, and hence those of $K_\rho = \rho^{-1}(K + \rho B) = \rho^{-1}K + B$ lie in $[1,\ 1 + \rho^{-1}]$.

In the tangent hyperplane $\mathcal{H}_p$ of K_ρ at $p \in \partial K_\rho$, consider an $(n - r)$-flat $\mathfrak{N}_p$ through p, and let $\mathfrak{J}_p$, $\mathfrak{J}'_p$ be the $(r - 1)$-flats in $\mathcal{H}_p$ such that $\mathfrak{J}_p$ is normal to $\mathfrak{N}_p$ in the sense of E^n, and $\mathfrak{J}'_p$ is normal to $\mathfrak{N}_p$ with respect to the Euclidean metric in $\mathcal{H}_p$ with the DUPIN indicatrix of ∂K_ρ at p as unit sphere. Since the normal radii of curvature of K_ρ lie in $[1,\ 1 + \rho^{-1}]$, there is a $\beta(n,\ r,\ \varepsilon)$ such that unit $(r - 1)$-vectors T_u, T'_u parallel to $\mathfrak{J}_0$, $\mathfrak{J}'_p$ satisfy

(24) $$|T_u \cdot T'_u| \geq 1 - \varepsilon \quad \text{for } K \in \Lambda_1^2, \text{ and } \rho \geq \beta(n,\ r,\ \varepsilon).$$

If we place K_ρ so that $B \subset K_\rho$ then

$$B \subset K_\rho \subset (1 + 2\rho^{-1})B \tag{25}$$

For, the diameter of a plane convex curve whose radius of curvature does not exceed δ is at most 2δ, so that the diameter of any plane section of ∂K_ρ is at most $2(1+\rho^{-1})$. If ∂K_ρ contained a point a not in $(1+2\rho^{-1})B$, then the second point in which the line through a and z intersects ∂K_ρ, would have a distance greater than $1+(1+2\rho^{-1}) = 2(1+\rho^{-1})$ from a, and this contradicts the above statement about the diameter.

For a given $(n-r)$-flat $\mathfrak{N}$ through z we construct the cylinder $Z(K_\rho, \mathfrak{N})$ whose generators are the $(n-r)$-flats parallel to $\mathfrak{N}$ and tangent to K_ρ. It touches K_ρ along the shadow boundary $S(K_\rho, \mathfrak{R})$ where $\mathfrak{R}$ is the r-flat through z normal to $\mathfrak{N}$, and intersects $\mathfrak{R}$ in the boundary of $P(K_\rho, \mathfrak{R})$ (see III, p. 18). If $\mathfrak{N}_p$ is the generator of $Z(K_\rho, \mathfrak{N})$ through $p \in S(K_\rho, \mathfrak{R})$, then the above defined $(r-1)$-flat $\mathfrak{T}'_p$ is the tangent flat of $S(K_\rho, \mathfrak{R})$ at p.

Connecting z to all $p \in S(K_\rho, \mathfrak{R})$ by line segments produces a surface $L(K_\rho, \mathfrak{R})$ homeomorphic to an r-ball. Because of (25) the lengths of these segments tend to 1 for $\rho \to \infty$ uniformly for all $K \in \Lambda_1^2$, and the existence of $\beta(n, r, \varepsilon)$ shows that $\beta_1(n, r)$ can be chosen so that

$$\text{Area } L(K_\rho, \mathfrak{R}) - \pi_r < 1 \text{ for } \rho \geq \beta_1(n, r). \tag{26}$$

The function $\beta(n, r, \varepsilon)$ also yields a $\beta_2(n, r)$ with the following property. If $\mathfrak{R}$ is any r-flat and $\mathfrak{N}$ is any $(n-r)$-flat passing through a point p of $S(K_\rho, \mathfrak{R})$ such that unit vectors R_u, N_u parallel to $\mathfrak{R}$ and $\mathfrak{N}$ satisfy $|R_u \wedge N_u| \geq \frac{1}{2}$, then $\mathfrak{N}$ will not intersect $S(K_\rho, \mathfrak{R})$ at a point other than p, when $K \in \Lambda_1^2$ and $\rho \geq \beta_2(n, r)$.

Consider a pair of r-flats $\mathfrak{R}$, $\mathfrak{R}'$ through z and project $S(K_\rho, \mathfrak{R})$ on $\mathfrak{R}'$. Denote by $h(p)$ the (non-signed) multiplicity with which the projection of $L(K_\rho, \mathfrak{R})$ on $\mathfrak{R}'$ covers p. If N_u is a unit $(n-r)$-vector normal to $\mathfrak{R}'$ and $|R_u \wedge N_u| < \frac{1}{2}$, then (26) shows that

$$\int_{\mathfrak{R}'} h(p)dp \leq \frac{1}{2}(\pi_r + 1) < \pi_r \text{ for } \rho \geq \beta_1(n, r).$$

Since (25) implies that $\widehat{P}(K_\rho, \mathfrak{R}') \geq \pi_r$ we have, with the notation of III p. 28, that

$$|j^*(\mathfrak{R}, \mathfrak{R}')| \leq \int_{\mathfrak{R}'} h(p)dp \leq \widehat{P}(K, \mathfrak{R}')$$

for $K \in \Lambda_1^2$, $\rho \geq \beta_1(n, r)$ and $|R_u \wedge N_u| < \frac{1}{2}$.

If $|R_u \wedge N_u| \geq \frac{1}{2}$ and $\rho \geq \beta_2(n, r)$ then projecting $S(K_\rho, \mathfrak{R})$ on $\mathfrak{R}'$ is an injection, moreover the projection lies in $P(K_\rho, \mathfrak{R}')$. Therefore

$$|j^*(\mathfrak{R}, \mathfrak{R}')| \leq \widehat{P}(K_\rho, \mathfrak{R}') \text{ for } K \in \Lambda_1^2, \ \rho \geq \beta_2(n, r)$$

and $|R_u \wedge N_u| \geq \frac{1}{2}$.

The sufficient condition for the convexity of $\widehat{P}(K_\rho, R)$ and hence of $\widehat{P}(K + \rho B, R)$ given in III (25) is therefore satisfied for all $K \in \Lambda_1^2$ and $\rho \geq \beta(n, r) = \max(\beta_1(n, r), \beta_2(n, r))$.

Because $K \in \Lambda_M^2$ implies $M^{-1}K \in \Lambda_1^2$, we see that $\widehat{P}(K + \rho MB, R)$ is convex for all $\rho \geq \beta(n, r)$.

Finally we eliminate the hypothesis that ∂K is of class C^2. If $K \in \Lambda_1$, then $K + d^{-1}B$ lies in $\Lambda_{1+1/d}$, and $\partial(K + d^{-1}B)$ is of class C^1. By any of several well-known methods, for example that of TONELLI [5], $\partial(K + d^{-1}B)$ can be approximated by analytic surfaces for which the maximal normal radius of curvature is at most $1 + 2/d$. This implies that the surfaces are convex. Therefore $K_d \in \Lambda_{1+2/d}^2$ with $K_d \to K$ and $d \to \infty$ exists. The convexity of $\widehat{P}(K_d + \rho B, R)$ for $\rho \geq (1 + 2/d)\beta(n, r)$ yields the convexity of $\widehat{P}(K + \rho B, R)$ for $\rho \geq \beta(n, r)$. Thus we have proved (23).

Our second uniformity theorem relates to polyhedra:

(27) THEOREM. - *If K is a polyhedron with N vertices and diameter M, there exists a constant $\alpha(N, n, r)$ such that $\widehat{P}(K + \rho B, R)$ is convex for all $\rho \geq M\alpha(N, n, r)$.*

The proof of this statement arises by refining the argument of § 3 leading to (17). The term $R_D^{(j)} \cdot (\tilde{F}_0 - \tilde{F}_1)$ in (20) is precisely the amount by which the area of the cycle $P_0(S_0, \mathfrak{R}_D^{(j)})$ exceeds that of the convex polygon $P_0(S_1, \mathfrak{R}_D^{(j)})$. If K has N vertices, its projection is a convex polygon with at most N vertices and then a displacement ε of each vertex can lead to, at most, N regions of multiplicity $N/2$. (In fact this estimate is much too large, but it serves our purpose here.) Hence we deduce

$$R_D^{(j)} \cdot (\tilde{F}_0 - \tilde{F}_1) \leq \frac{1}{2} N^2 \varepsilon^2 M^2$$

since the diameter of K being M, it can be put in a ball of radius M. Consequently, proceeding as in the proof of (18)

$$C_f(R_u^{(j)}, R_D^{(j)}) \leq 2N^2 M^2.$$

We deduce that $\widehat{P}(K, R^2)+\rho|R^2|$ is convex for $\rho \geq NM$, and so $\widehat{P}(K+\rho B, R^2)$ is also convex for $\rho \geq NM$. Hence in this case (27) holds with $\alpha(N, n, 2)=N$. A similar, though more complicated argument can be applied for $r>2$, and (27) established generally.

6. Mixed Projection Functions.

The preceding parts of this series used, without proof, the fact that for any affinity φ, the function $\widehat{P}(\varphi K, Q)$ is convex when $\widehat{P}(K, R)$ is. This is indeed intuitively clear. For r convex bodies $K_1, \ldots, K_r$ in E^n, and an r-flat $\mathfrak{R}$ through z, we put, for R parallel to $\mathfrak{R}$,

$$\widehat{P}(K_1, \ldots, K_r; R)=|R| V(P(K_1, \mathfrak{R}), \ldots, P(K_r, \mathfrak{R})),$$

$$\widehat{P}(K_1, \ldots, R_r; 0)=0.$$

This is called the *mixed projection function* of the r bodies (see VIII). Clearly $\widehat{P}(K_1, \ldots, K_r; R)$ is invariant under translations of the K_i, but for affinities $\varphi_1, \ldots, \varphi_r$ the function $\widehat{P}(\varphi_1 K_1, \ldots, \varphi_r K_r; R)$ may not be convex even if $\widehat{P}(K_1, \ldots, K_r; R)$ is. (A simple example for the case $n=4$, $r=2$ can be deduced from (*e*) of III.3. Using T_1^4, T_2^4 for the two 4-simplexes introduced there, we see that $\widehat{P}(T_1^4, T_1^4; R)$, $\widehat{P}(T_2^4, T_2^4; R)$ are convex and $\widehat{P}(T_1^4+T_2^4, R)$ is not, whence it is almost immediate that $\widehat{P}(T_1^4, T_2^4: R)$ is not convex). However we can prove

(28) *If $\widehat{P}(K_1, \ldots, K_r; R)$ is convex and φ is an affinity, then $\widehat{P}(\varphi K_1, \ldots, \varphi K_r; R)$ is also convex.*

The proof is simple, but not entirely obvious. Because of the invariance under translation we may assume that φ is a central affinity, and hence identify it with a linear transformation of the vector space belonging to E^n. Then φ induces an automorphism of the Grassmann algebra, we denote the image of an r-vector R by φR. It will be convenient in the proof to use the notations

$$\widehat{Q}(K, T)=|T| \widehat{P}(K, \mathfrak{R})$$

$$\widehat{Q}(K_1, \ldots, K_r; T)=|T| V(P(K_1, \mathfrak{R}), \ldots, P(K_r, \mathfrak{R})),$$

where T is any vector parallel to an $(n-r)$-flat perpendicular to $\mathfrak{R}$.

We first show that

(29) $\widehat{Q}(\varphi K, \varphi T) = \Delta \widehat{Q}(K, T)$ *where* φ *is any affinity with* $|\det \varphi| = \Delta > 0$. (³)

For let $Z(K, T)$ $(T \neq 0)$ be the solid cylinder formed by all the $(n-r)$-flats parallel to T intersecting K. On one of these flats take a parallelepiped C whose volume $V_{n-r}(C)$ is equal to $|T|$, and in each of the others take a translate of C such that the union M of all these translates in measurable. (In fact we can choose these translates so that M is convex.) Then

$$V_n(M) = \widehat{P}(K, \mathfrak{R}) V_{n-r}(C) = |T| \widehat{P}(K, \mathfrak{R}) = \widehat{Q}(K, T).$$

Since $\varphi Z(K, T) = Z(\varphi K, \varphi T)$ and $V_{n-r}(\varphi C) = |\varphi T|$ we find in the same way that

$$V_n(\varphi M) = \widehat{Q}(\varphi K, \varphi T).$$

Since $V_n(\varphi M) = \Delta V_n(M)$ we deduce that

$$\widehat{Q}(\varphi K, \varphi T) = \Delta \widehat{Q}(K, T)$$

proving (29). (An alternative proof equivalent to the above can also be deduced from [3] (20).)

To prove (28) we proceed as follows. Since φ is a central affinity, for $\lambda_i \geq 0$ $(i = 1, \ldots, r)$

$$\varphi(\Sigma \lambda_i K_i) = \Sigma \lambda_i \varphi K_i.$$

Therefore, using (29),

$$\widehat{Q}(\Sigma \lambda_i \varphi K_i, \varphi T) = \widehat{Q}(\varphi(\Sigma \lambda_i K_i), \varphi T) = \Delta \widehat{Q}(\Sigma \lambda_i K_i, T).$$

Expanding in terms of $\widehat{Q}(K_{i_1}, \ldots, K_{i_r}; T)$ and equating coefficients of $\lambda_1 \ldots \lambda_r$ yields

$$\widehat{Q}(\varphi K_1, \ldots, \varphi K_r; \varphi T) = \Delta \widehat{Q}(K_1, \ldots, K_r; T).$$

Using supplementary vectors it is clear from the definition that the convexity of $\widehat{Q}(K_1, \ldots; K_r; T)$ and that of $\widehat{P}(K_1, \ldots, K_r; R)$ are equivalent. Therefore, if $\widehat{P}(K_1, \ldots, K_r; R)$ is convex, and $T = \sum_{i=1}^{k} T_i$,

$$\begin{aligned} \widehat{Q}(\varphi K_1, \ldots, \varphi K_r; T) &= \Delta \widehat{Q}(K_1, \ldots, K_r; \varphi^{-1} T) \\ &\leq \Delta \sum_{i=1}^{k} \widehat{Q}(K_1, \ldots, K_r; \varphi^{-1} T_i) = \sum_{i=1}^{k} \widehat{Q}(\varphi K_1, \ldots, \varphi K_r; T_i) \end{aligned}$$

and (28) follows.

(³) $|\det \varphi|$ means the absolute value of the determinant of φ.

This argument also yields

(30) *If* $\widehat{P}(K_1, \ldots, K_r; R)$ *is locally convex and* φ *is an affinity, then* $P(\varphi K_1, \ldots, \varphi K_r; R)$ *is locally convex.*

Let $\widehat{P}_{i,r-i}(K, B; R)$ be defined as in the introduction. We can now show:

(31) *If* $1 < i < r < n-1$, *there exists convex bodies* K *in* E^n *such that* $\widehat{P}_{i,r-i}(K; B; R)$ *is not convex.*

The method of proof is similar to that used in [4].

Since $1 < i < n-2$, we can find a convex body K such that $\widehat{P}(K, R^i)$ is not convex. (In fact, by III, K may be chosen to be a simplex). It follows from VIII (7) [4] that if B^{r-i} represents any $(r-i)$-dimensional unit ball in E^n, then $\widehat{P}_{i,r-i}(K, B^{r-i}; R)$ is also not convex.

Let $B(\varepsilon)$ be an n-dimensional ellipsoid with $r-i$ semi-axes of length 1 coinciding with radii of B^{r-i}, and the remaining $n-r+i$ semi-axes of length ε. Thus if d represents the Hausdorff metric, $d(B(\varepsilon), B^{r-i}) = \varepsilon$. Now $\widehat{P}(K_1, \ldots, K_r; R)$ is a continuous function of each argument K_i, and so if ε is sufficiently small, the non-convexity of $\widehat{P}_{i,r-i}(K, B^{r-i}; R)$ implies the non-convexity of $\widehat{P}_{i,r-i}(K, B(\varepsilon); R)$.

Let φ be an affinity such that $\varphi(B(\varepsilon)) = B$, and let $\varphi K = K^*$. Then by (28) the non-convexity of $\widehat{P}_{i,r-i}(K, B(\varepsilon); R)$ implies the non-convexity of $\widehat{P}_{i,r-i}(\varphi K, \varphi B(\varepsilon); R)$ and so of $\widehat{P}_{i,r-i}(K^*, B; R)$. Thus the body K^* has the required property and (31) is established.

7. Problems.

We conclude with the statement of some unsolved problems closely connected with the subject of this paper.

(1) *If* $\widehat{P}(K, R)$ *is convex, does a number* ρ_0 *exist such that* $\widehat{P}(K + \rho B; R)$ *is convex for* $\rho \geq \rho_0$?

The answer to this question is in the affirmative if $r = 2$, as can be deduced from (4), or also from VIII. For $r > 2$, the question is still open.

(2) *If* K *can roll in a sphere, is* $\widehat{P}(K, R)$ *locally convex?*

More generally one could ask whether the conditions for locally convexity given in III (30) can be significantly weakened.

---

[4] Corrigendum: In this statement the words «and only if» sould be omitted.

(3) *If $K_1, \ldots, K_r$ are convex bodies whose boundaries are of class C^2 and with non-vanishing Gauss curvature, is $\widehat{P}(K_1, \ldots, K_r; R)$ locally convex? In particular, if K is such a body, is $\widehat{P}_{i, r-i}(K, B; R)$ locally convex for $1 < i < r-1$?*

REFERENCES

The previous papers in the series « Convex Bodies and Convexity on Grassmann Cones » by H. Busemann, G. Ewald, and G. C. Shephard appear in the following journals:

Parts I - IV	Math. Annalen 151 (1963), 1-41.
Part V	Arch. Math., 13 (1962), 512-526.
Part VI	J. London Math. Soc., 39 (1964), 307-319.
Part VII	Abh. Math. Sem. Univ. Hamburg, 27 (1964), 167-170.
Part VIII	J. London Math. Soc., 39 (1964), 417-423.
Part IX	Math. Annalen 157 (1964), 219-230.

[1] H. Busemann, *Un nuovo tipo di problemi nella teoria dei corpi convessi*, Seminari dell'Istituto Nazionale di Alta Matematica 1962-63, 178-205.

[2] H. Hadwiger, *Vorlesungen über Inhalt, Oberfläche und Isoperimetrie*, Berlin-Göttingen-Heidelberg: Springer 1957,

[3] G. C. Shephard, *Shadow systems of convex sets*, Israel J. of Math. 2 (1964), 229-236.

[4] — —, *A property of projection functions of parallel convex bodies*, J. London Math. Soc., 40 (1965).

[5] L. Tonelli, *Sulla rappresentazione analitica delle funzioni di più variabili reali*, Rend. Circ. Mat. Palermo, 29 (1910), 1-36.

Ann. Mat. Pura Appl., IV. Ser. 70 (1965), 271–293

UN NUOVO TIPO DI PROBLEMI NELLA TEORIA DEI CORPI CONVESSI. (*)

HERBERT BUSEMANN

1. Introduzione.

Sono molto grato per l'opportunità di parlare nel Seminario di Algebra, Geometria e Topologia dell'Istituto Nazionale di Alta Matematica, perchè voglio riferire su un nuovo campo relativo alla geometria pluridimensionale, della quale esiste una lunga e fortissima tradizione italiana. Potrebbe darsi che qualcuno dei miei ascoltatori possa risolvere taluna delle numerose questioni aperte o procedere in una delle molte nuove direzioni strettamente connesse con queste conferenze.

I concetti e i problemi che tratteremo provengono da due campi: dal calcolo delle variazioni e dalla teoria dei corpi convessi. Prima di entrare in dettagli, spiegherò come nacquero tali questioni e descriverò concisamente il contenuto delle conferenze.

Consideriamo un problema variazionale in forma parametrica ad una dimensione:

$$\int f(x, \dot{x})\, dt = \min,$$

dove è $f(x, v) = f(x^1, \dots, x^n; v^1, \dots, v^n) > 0$ se $v \neq 0$, e $f(x, \lambda v) = \lambda f(x, v)$ se $\lambda \geq 0$. Il problema è quasiregolare nel punto x, o soddisfa la condizione di Legendre, se la funzione $f(x, v)$ è convessa come funzione di v, oppure, il che è lo stesso, se la cosidetta indicatrice di $f(x, v)$ nel punto x, cioè la ipersuperficie $f(x, v) = 1$

(*) Conferenze tenute al Seminario di Algebra, Geometria e Topologia nei giorni 5, 6, 13, 14 dicembre 1962.

nello spazio di v, è convessa. Spesso occorre considerare un concetto più debole, la quasiregolarità in una sola direzione (x, v_0), $v_0 \neq 0$. Questo significa che l'indicatrice possiede nel punto $v_0 / f(x, v_0)$ un iperpiano di appoggio. Tale iperpiano è trasversale a v_0 nel linguaggio del calcolo delle variazioni.

Per formulare i problemi variazionali a r dimensioni ($1 \leq r \leq \leq n-1$) ci occorre la nozione di r-vettore semplice, che sostituisce quella del vettore ordinario nel caso precedente. Per metterci d'accordo sulle notazioni riprendiamo brevemente questo concetto. Gli r vettori $v_1, \ldots, v_r$ nello spazio vettoriale reale n-dimensionale, V^n, determinano l'r-vettore semplice

$$v_1 \wedge \ldots \wedge v_r = \mathop{\Lambda}_{i=1}^{r} v_i,$$

loro prodotto esterno. Tale prodotto si può definire implicitamente dando le condizioni di uguaglianza per due r-vettori. Precisamente diamo che è $\mathop{\Lambda}_{i=1}^{r} v_i = \mathop{\Lambda}_{i=1}^{r} w_i$ se

1) i vettori v_i e w_i giacciono in uno stesso S_r

2) i volumi dei parallelepipedi, determinati dai vettori v_i e dai w_i, sono uguali (¹)

3) se quei volumi non svaniscono, le due r-ple ordinate di vettori $v_1, \ldots, v_r$ e $w_1, \ldots, w_r$ determinano la stessa orientazione di S_r.

Una metrica euclidea, definita da una forma quadratica $\Sigma g_{ik} v^i v^k$ (> 0 se $v \neq 0$), induce una norma $|R|$ per l'r-vettore $R = \mathop{\Lambda}_{i=1}^{k} v_i$, data dal volume — relativo a questa metrica — del parallelepipedo determinato da $v_1, \ldots, v_r$.

Gli r-vettori semplici individuano lo spazio lineare, V_r^n, di *tutti* gli r-vettori dove, come è noto, è $\dim V_r^n = \binom{n}{r}$. Gli r-vettori semplici formano in V_r^n il cono grassmanniano G_r^n ed è $\dim G_r^n = r(n-r) + 1$. Posto $V_1^n = V^n$ si ha, come è ben noto, che è $V_r^n = G_r^n$ se, e soltanto se, è $r = 1$ oppure $r = n-1$.

(¹) Si osservi che, mentre la nozione di volume di un insieme di un S_r non è di natura affine (tranne nel caso in cui il volume sia zero), il concetto d'uguaglianza tra volumi di insiemi situati in S_r paralleli è di natura affine (in quanto il rapporto tra tali volumi è una nozione affine).

Se invece è $1 < r < n-1$, il numero $K_r^n = \dim V_r^n - \dim G_r^n$ è positivo e G_r^n è l'intersezione di K_r^n coni quadratici in V_r^n. Il cono è una ipersuperficie, cioè $K_r^n = 1$, soltanto se $n = 4$, $r = 2$. Se è $1 < r < n-1$ il cono G_r^n non è mai un insieme convesso in V_r^n.

Ora ritorniamo ai problemi variazionali di dimensione r in spazi di dimensione n. L'integrando ha la forma

$$f(x, R) = f(x^1, \ldots, x^n;\ R^1, \ldots, R^{\binom{n}{r}}) \qquad R \in G_r^n,$$

$$f(x, R) > 0 \text{ se } R \neq 0 \text{ e } f(x, \lambda R) = \lambda f(x, R) \text{ se } \lambda \geq 0.$$

L'analogo alla quasiregolarità (od alla condizione di Legendre) sarebbe la convessità di $f(x, R)$ come funzioni di R.

Una funzione $g(y)$ definita in un insieme M nello spazio affine $A^n (= V_1^n = V^n)$ è convessa se è

$$g((1-t)\, y_1 + t y_2) \leq (1-t)\, g(y_1) + t g(y_2), \qquad 0 < t < 1.$$

Questo presuppone che $(1-t)\, y_1 + t y_2 \in M$, cioè che M sia convessa. Siccome G_r^n $(1 < r < n-1)$ non è un insieme convesso in V_r^n, la convessità di una funzione su G_r^n deve esser definita. Nel caso $r = 1$ la convessità di $f(x, v)$, o della indicatrice $f(x, v) = 1$ equivale all'esistenza di iperpiani di appoggio, oppure all'esistenza di un iperpiano trasversale a una direzione, v_0, data. Carathéodory — uno dei primi matematici che si sono occupati seriamente del caso $1 < r < n-1$ — ebbe l'idea che la condizione di Legendre dovrebbe ancora equivalere all'esistenza di un S_{n-r} trasversale ad un dato r-vettore non nullo. Tali trasversali sono trasversali totali nella nostra terminologia. Wagner [11] ha studiato la convessità e la trasversalità dal punto di vista del calcolo delle variazioni, usando il termine di trasversale in un senso più debole, che sarà quello da noi qui adoperato.

L'esistenza di una trasversale totale a un dato vettore R (nel punto x) equivale alla convessità totale di $f(x, R)$ su G_r^n, che definiremo nel capitolo seguente. Siccome ci sono molti problemi variazionali che hanno soluzioni senza che $f(x, R)$ sia totalmente convessa, la condizione di Carathéodory è troppo restrittiva.

Infatti esistono parecchie nozioni di convessità su G_r^n che coincidono se $r = 1$ o $r = n-1$, ma tali che l'una implica la successiva se $1 < r < n-1$.

Ora vediamo come lo stesso problema, cioè il significato di convessità su G_r^n, nasce nella teoria dei corpi convessi. Molti teoremi ([2]) asseriscono che una funzione che dipende da uno o parecchi corpi convessi e da una retta oppure da un iperpiano variabile è convessa. Tali funzioni rimangono quasi sempre ben definite se un S_r variabile sostituisce la retta o l'iperpiano. Prendiamo un esempio: Sia K un corpo convesso nello spazio E^n euclideo, $\mathcal{R}$ un S_r variabile per un punto fisso, e $P_r(K, \mathcal{R})$ il volume (r-dimensionale) della proiezione $P(K, \mathcal{R})$ di K su $\mathcal{R}$. Da $P_r(K, \mathcal{R})$ deriviamo una funzione omogenea, la cosidetta funzione indotta da $P_r(K, \mathcal{R})$ su G_r^n ponendo, $R \in G_r^n$,

$$P(K, R) = |R| \, P_r(K, \mathcal{R}) \text{ se } R \parallel \mathcal{R}, \quad P(K, 0) = 0.$$

Minkowski ha provato che $P(K, R)$ è convessa se $r = 1$ e $r = n - 1$ (²). Così incontriamo di nuovo la questione se $P(K, R)$, ed altre funzioni analoghe, sono convesse su G_r^n, se $1 < r < n - 1$, in un senso o in un altro, o di trovare condizioni sotto le quali sono convessi in uno fissato dei vari sensi.

Discuteremo prima i differenti concetti di convessità, poi le loro relazioni con l'esistenza di trasversali, ed alcuni teoremi con-

(²) La convessità di $P(K, R)$ per $r = 1$ segue senz'altro dal fatto che la funzione d'appoggio di K è convessa (cfr. cap. 5).

Se $r = n - 1$ sia ∂K la frontiera di K, e dS_u l'elemento d'area di ∂K, u denotando il vettore unitario normale esterno all'elemento d'area dS_u. Se l'iperpiano $\mathfrak{R}$ é parallelo al $(n-1)$-vettore unitario R_u oppure normale al vettore unitario u_R (normale a R_u), è chiaro che

$$\underset{n=1}{P}(K, \mathfrak{R}) = P(K, R_u) = \frac{1}{2} \int_{|u|=1} |u \cdot u_R| \, dS_u = \frac{1}{2} \int_{|u|=1} |R_u \wedge u| \, dS_u .$$

Se $R \parallel \mathfrak{R}$, cioè $R = |R| \cdot R_u$

$$P(K, R) = \frac{1}{2} \int |R \wedge u| \, dS_u .$$

Quindi abbiamo

$$P(K, \sum_{i=1}^{k} R_i) \leq \sum_{i=1}^{k} P(K, R_i),$$

cioè la convessità di $P(K, R)$ se $r = n - 1$. Un'altra dimostrazione si trova in [2] § 29.

cernenti trasversalità. Finalmente ci occuperemo della convessità di funzioni derivate da corpi convessi.

Il tempo non basta per dar le relative dimostrazioni, però rimanderemo volta per volta alla bibliografia posta alla fine.

2. Convessità su insiemi non-convessi.

Sia M un insieme non vuoto nello spazio affine A^n, denoteremo con M^H lo «inviluppo convesso» di M, cioè l'intersezione di tutti i convessi contenenti M. Supporremo inoltre che M sia tale che M^H sia aperto. Vogliamo esaminare diverse possibilità di definire la convessità di una funzione $f(x) = f(x^1, \ldots, x^n)$ su M. Ci interessa sopratutto il caso $M = G_r^n$ cioè $M^H = V_r^n$, e $f(\lambda x) = \lambda f(x)$ se $(\lambda \geq 0$ cfr. [4] parte I, e [6]).

La condizione minima per la convessità di $f(x)$ è che $f(x)$ sia convessa su ogni sottoinsieme convesso di M. Tali funzioni le chiameremo *debolmente convesse*. È chiaro che la

$$f((1-t)x_1 + tx_2) \leq (1-t)f(x_1) + tf(x_2)$$

con

$$x_i \in M, \quad (1-t)x_1 + tx_2 \in M$$

equivale a la convessità debole.

Diciamo che $f(x)$ è omogenea se è $M = \lambda M$ per $\lambda > 0$; $0 \in M$, e $f(\lambda x) = \lambda f(x)$ per $\lambda \geq 0$. In questo caso la

$$f(x_1 + x_2) \leq f(x_1) + f(x_2)$$

esprime la convessità debole.

Osserviamo subito che questa convessità *non è interessante nella teoria dei corpi convessi* (8). Se R_1, R_2 e $R_1 + R_2$ sono r-vettori semplici, gli S^r per z paralleli a R_1 e R_2 si intersecano in un S_{r-1}, quindi giacciono in un S_{r+1} e sono iperpiani in questo spazio. Dunque, per esempio, la disuguaglianza

$$P(K, R_1 + R_2) \leq P(K, R_1) + P(K, R_2), \quad R_i \in G_r^n, \quad R_1 + R_2 \in G_r^n$$

(8) Questa è la ragione per cui usiamo il termine di «debolmente convesso» invece di «convesso» che è usato in [6]. Noi seguiremo [4], riservando la parola «convessità» per un concetto più restrittivo ma più importante, che in [6] è chiamato «convessità estensibile».

segue dal teorema di Minkovski e non ci dà niente di nuovo. Però la convessità debole è interessante nella teoria della trasversalità.

La funzione $f(x)$ è *convessa* se si può estendere ad una funzione in M^H convessa nel senso ordinario. È evidente che l'esistenza di una funzione lineare $L(x) \leq f(x)$ e la *disuguaglianza*:

$$(1) \qquad f\left(\sum_{i=1}^{r} \lambda_i x_i\right) \leq \sum_{i=1}^{r} \lambda_i f(x_i),$$

con

$$\lambda_i \geq 0, \quad \Sigma \lambda_i = 1, \quad x_i \in M, \quad \Sigma \lambda_i x_i \in M,$$

è una condizione necessaria e sufficiente perchè $f(x)$ sia convessa.

Inoltre non è difficile dimostrare che perchè la $f(x)$ sia convessa basta che valga la (1) per $r = n + 1$.

Se $f(x)$ è omogenea, l'esistenza di $L(x) \leq f(x)$ e la disuguaglianza

$$(1') \qquad f\left(\sum_{i=1}^{n} x_i\right) \leq \sum_{i=1}^{n} f(x_i), \text{ con } x_i \in M, \ \sum_{i=1}^{n} x_i \in M.$$

equivalgono alla convessità di $f(x)$. Osserviamo che non si può diminuire il numero di termini nelle sommatorie che compaiono in (1'), anche se avviene che, per un k fissato minore di n, ogni punto x di M possa rappresentarsi nella forma $x = \sum_{i=1}^{k} x_i, x_i \in M$. Per esempio, se $M = G_2^4$, e quindi $M^H = V_2^4$, ogni bivettore $\tilde{R} \in V_2^4$ è la somma di due bivettori semplici, $\tilde{R} = R_1 + R_2$. Ma la disuguaglianza

$$f(R_1 + R_2) \leq f(R_1) + f(R_2) \qquad (R_i, R_2, R_1 + R_2 \text{ semplici})$$

esprime soltanto la convessità debole, che non implica la convessità, come vedremo nel caso di $P(K, R)$ (cfr. cap. 5). Dunque l'esistenza di $L(x) \leq f(x)$ e *la disuguaglianza*

$$(1'') \qquad f\left(\sum_{i=1}^{\binom{n}{r}} R_i\right) \leq \sum_{i=1}^{\binom{n}{r}} f(R_i), \text{ con } R_i \in G_r^n, \ \Sigma R_i \in G_r^n$$

equivalgono alla convessità di una funzione omogenea su G_r^n, e non si può sostituire $\binom{n}{r}$ con un numero minore.

Ritorniamo ad una funzione convessa $f(x)$ in M. Sia $g(x)$ è un'estensione convessa di $f(x)$ da M a M^H, l'insieme dei punti

$x \in M^H$, $z \geq g(x)$ in A^{n+2} è convessa; quindi possiede in ogni punto $(x_0, g(x_0))$ un iperpiano di appoggio

$$z = \sum_{i=1}^{n} a_i(x_0)(x^i - x_0^i) = a(x_0) \cdot (x - x_0)$$

cioè

$$g(x) \geq a(x_0) \cdot (x - x_0) + g(x_0), \text{ con } x, x_0 \in M^H, \text{ e}$$

(2) $$f(x) \geq a(x_0) \cdot (x - x_0) = f(x_0), \text{ con } x, x_0 \in M.$$

Viceversa, se per ogni punto $x_0 \in M$ esistono numeri $a_i(x_0)$ soddisfacenti la (2), allora $f(x)$ è convesso: Se $x_i \in M$ e $x_0 = \sum_{i=1}^{r} \lambda_i x_i \in M$ segue dalla (2) che

$$f(x_i) \geq a(x_0) \cdot (x_i - x_0) + f(x_0)$$

$$\Sigma \lambda_i f(x) \geq a(x_0)(\Sigma \lambda_i x_i - x_0) + f(x_0) = f(x_0).$$

Se $f(x)$ è omogenea, concludiamo dalla (2) per $\lambda \geq 0$ che è

$$\lambda f(x_0) = f(\lambda x_0) \geq a(x_0)(\lambda - 1) x_0 + f(x_0),$$

$$(\lambda - 1) f(x_0) \geq (\lambda - 1) a(x_0) \cdot x_0 .$$

Ponendo una volta $\lambda > 1$ e una seconda volta $\lambda < 1$, troviamo $f(x_0) = a(x_0) \cdot x_0$. Perciò la (2) prende la forma

(2′) $$f(x) \geq a(x_0) \cdot x, \quad f(x_0) = a(x_0) \cdot x_0 \quad \textit{se} \quad f(x) \textit{ è omogenea.}$$

Abbiamo discusso questo fatto banale perchè il terzo concetto di convessità restringe i numeri $a_i(x_0)$.

Sia N un insieme non-vuoto chiuso in uno spazio affine n-dimensionale, diremo che una funzione è N-convessa se per ogni punto $x_0 \in M$ esiste un punto $p(x_0) \in N$ tale che sia

$$f(x) = p(x_0) \cdot (x - x_0) + f(x_0), \qquad x, x_0 \in M.$$

Il caso interessante è quello in cui sia $M = G_r^n$ ed N il cono grassmanniano, G_r^{n*}, nello spazio V_r^{n*} duale di V_r^n, e $f(R)$ omogenea. Una tale funzione si chiama *totalmente convessa*. In forza della (2′) la $f(R)$ è totalmente convessa se, per ogni R_0, esiste un $R^*(R_0) \in G_r^{n*}$ tale che si abbia

(2″) $$f(R) \geq \Sigma R_i^*(R_0) R_0^i, \quad f(R_0) = \Sigma R_i(R_0) R_0^i .$$

Nelle applicazioni, $A^n (= V^n)$ è provvisto di una metrica euclidea ed è preferibile di considerare R^* non come un r-covettore ma come un r-vettore in G_r^n e $\Sigma R_i^* R_0^i$ come il prodotto scalare $R^* \cdot R_0$. Allora gli R_i^* sono i componenti covarianti di R^* nella vecchia terminologia. [4] La (2'') diventa

$$(2''') \qquad f(R) \geq R^*(R_0) \cdot R, \quad f(R_0) = R^*(R_0) \cdot R_0 .$$

Se $r = 1$ o $r = n - 1$ i tre concetti di convessità coincidono. Se $1 < r < n - 1$ la convessità totale di $f(x, R)$ come funzione di R é la generalizzazione di Carathèodory della condizione di Legendre.

Sia $f(x)$ una funzione definita in un insieme M. Se ogni punto $x_0 \in M$ ha un intorno $U(x_0) \subset M^H$ tale che la restrizione di $f(x)$ a $M \cap U(x_0)$ si può estendere a una funzione convessa in $U(x_0)$ diciamo che $f(x)$ è *localmente convessa.*

Se $M = M^H$, ossia nel caso della convessità ordinaria, la convessità locale implica la convessità. Perciò *una funzione localmente convessa in un M arbitrario è debolmente convessa, però in generale non è convessa.* Per esempio, se M è l'insieme di due aperti M_1, M_2 disgiunti ed è $f(x) = i$ in M_i, allora $f(x)$ è localmente convessa senza di esser convessa.

Per funzioni omogenee l'origine è un punto speciale. Una funzione convessa in un intorno di o è convessa. Perciò una fun zione omogenea è chiamata localmente convessa, se soddisfa la condizione di convessità locale in ogni punto diverso da o. Vedremo che anche nel caso di una funzione omogenea su G_r^n la convessità locale non garantisce la convessità.

Per conseguenza, anche se $f(R)$ è di classe C^2 in coordinate locali analitiche su G_r^n, *la convessità non è esprimibile mediante derivate seconde.*

Questa è la differenza più profonda e più incisiva tra la convessità su insiemi convessi e quella su insiemi non-convessi.

Le funzioni convesse ordinarie $f(x)$, $f_i(x)$ rimangono tali effettuando su esse certe operazioni; ricordiamo le seguenti:

I $\quad \lambda f(x)$ è convessa, se $\lambda > 0$

II $\quad \sum_{i=1}^{k} f_i(x)$ è convessa

[4] Naturalmente la differenza fra componenti covarianti e controvarianti sparisce se la metrica è cartesiana, vuol dire, è data dalla forma $\Sigma (x^i)^2$.

III Se $\sup_{i \in I} f_i(x)$ è finito, allora esso è convesso

IV Se $\lim_m \sup f_m(x)$ è finito, allora esso è convesso.

Importa sapere quali di queste proprietà valgono per i nostri nuovi concetti. È chiaro che la risposta dipende da M ed N. Perciò consideriamo soltanto il caso più interessante: $M = G_r^n$, $N = G_r^n$. Si ha allora il seguente quadro:

$M = G_r^n$, $N = G_r^n$	I	II	III	IV
convessità debole	si	si	si	si
convessità locale	si	si	no	no
convessità	si	si	si	si
convessità totale	si	no	si	si

Giacché la convessità locale non implica una proprietà globale e per essa non sono valide le proprietà III e IV, non è interessante.

Anche gli altri tre concetti varrebbero poco, se non avessero applicazioni ed interpretazioni interessanti nello spazio A^n, delle quali ci occuperemo ora.

3. Trasversalità.

In questo capitolo supporremo sempre che $f(R)$ sia una funzione omogenea e continua su G_r^n e per evitare discussioni di casi particolari e questioni di orientazione che

$$f(R) > 0 \text{ se } R \neq 0, \quad f(\lambda R) = |\lambda| f(R) \text{ per ogni } \lambda \text{ reale.}$$

L'r-vettore semplice R_F dicesi associato all'unione, F, di un numero finito di insiemi compatti convessi situati nell'S_r orientato $\mathcal{R}^+$, se $R_F \parallel \mathcal{R}^+$ ed in una, e quindi in tutte le metriche euclidee $\Sigma g_{ik} x^i x^k$, la norma $|R_F|$ di R_F è uguale al volume r-dimensionale di F.

Interpretiamo $f(R_F)$ come un'area $\alpha(F)$ di F, ponendo:

$$\alpha(F) = f(R_F).$$

Con «poliedro» intenderemo sempre una superficie poliedrale di dimensione r, e con «facce» del poliedro, quelle di dimensione r.

Siano $F_0, \dots, F_{r+1}$ le facce del bordo di un $(r+1)$-simplesso. Allora la

$$(3) \quad \alpha(F_0) \leq \sum_{i=1}^{r+1} \alpha(F_i) \text{ chiamasi } \textit{la disuguaglianza simpliciale.}$$

Se $F_0, \dots, F_k$ sono le facce di un poliedro chiuso orientabile, la

$$(3') \quad \alpha(F_0) \leq \sum_{i=1}^{k} \alpha(F_i) \text{ chiamasi } \textit{la disuguaglianza poliedrale.}$$

Ovviamente la (3') segue dalla (3), se $r = 1$. Minkowski ha provato ciò anche per $r = n - 1$, inoltre ha dimostrato che nei precedenti due casi la (3) o la (3') equivalgono a la convessità della funzione $f(R)$ ([8] § 27). Ne segue:

Per $1 < r < n - 1$ la convessltà debole di $f(R)$ equivale a la disuguaglianza simpliciale ed alla disuguaglianza poliedrale per poliedri giacenti in spazi di dimensione $r + 1$. (Ma non è noto, si implica la (3') per poliedri sghembi).

I riferimenti bibliografici per quanto segue si trovano in [6] e [11].

Se $F_0, \dots, F_k$ sono le facce di un poliedro chiuso orientato gli r-vettori associati R_i agli F_i soddisfano la $\sum_{i=1}^{k} R_i = 0$ o la $-R_0 = \sum_{i>0} R_i$. Perciò la (3') prende la forma

$$f\left(\sum_{i>0} R_i\right) = f(-R_0) = \alpha(F_0) \leq \sum_{i>0} \alpha(F_i) = \sum_{i>0} f(R_i).$$

Quindi:

La convessità di $f(R)$ garantisce la disuguaglianza poliedrale per l'area corrispondente, α.

È molto probabile che l'asserzione inversa non sia vera. Allora la disuguaglianza poliedrale sarebbe un quarto concetto interessante di convessità, che potrebbe essere importantissimo. Però sembra che tale concetto non sia per niente maneggevole, specialmente dal punto di vista algebrico. Non esiste un criterio per riconoscere quando gli r-vettori $F_0, \dots F_k$ sono associati alle facce di un poliedro.

Denoteremo in seguito con $\mathcal{R}, \mathcal{T}, \mathcal{D}, \dots$ rispettivamente un $S_r, S_t, S_d, \dots$.

Per arrivare ad una interpretazione geometrica della convessità totale diamo la seguente definizione (per $1 \leq r \leq n-1$):

$\mathcal{T}(t=n-r)$ *è totalmente trasversale* ad $\mathcal{R}$ (rispetto all'area α o alla funzione $f(R)$), nel punto p, se $\mathcal{R} \cap \mathcal{T} = p$ e se l'area $\alpha(K)$ di qualsiasi corpo convesso K r-dimensionale non è minore dell'area $\alpha(K_\pi)$ della proiezione K_π di K su $\mathcal{R}$ parallela a $\mathcal{T}$.

Scriveremo allora

$$\mathcal{T} \overset{p}{\underset{t}{|\!-}} \mathcal{R} \quad \text{o} \quad \mathcal{R} \overset{p}{\underset{t}{-\!|}} \mathcal{T}$$

e leggeremo la seconda formula così: R è totalmente normale a $\mathcal{T}$ nel punto p (cfr. fig. 1).

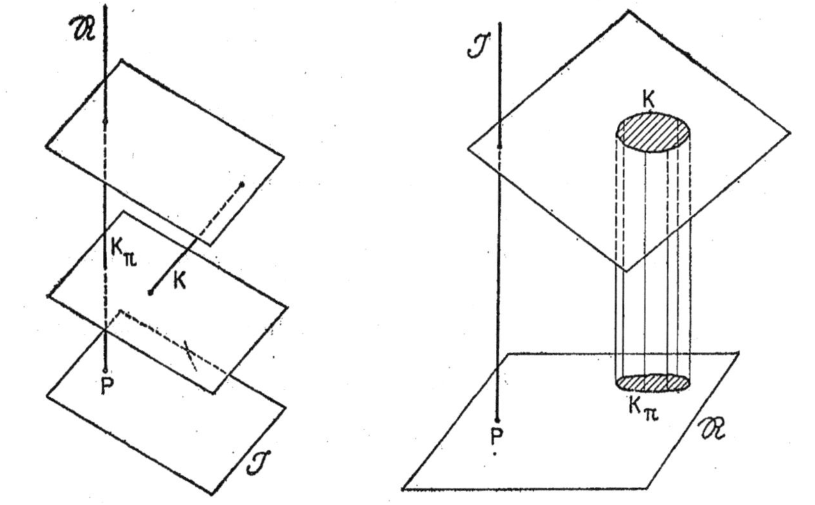

caso $n=3$, $r=1$ caso $n=3$, $r=2$

Fig. 1

Dati $\mathcal{T}$ e un punto $p \in \mathcal{T}$, l'esistenza di $\mathcal{R} \overset{p}{\underset{t}{-\!|}} \mathcal{T}$ segue senz'altro dalla continuità di $f(R)$ e dalla $f(R) > 0$, se $R \neq 0$.

L'esistenza di un $\mathcal{T}$ *totale trasversale ad un dato* $\mathcal{R}$ *in un punto* p *equivale a la convessità totale di* $f(R)$.

Così la convessità totale, che sembra poco interessante come proprietà in V_r^n, ha un interessante significato molto intuitivo in A^n. Questo ha indotto, o piuttosto sedotto, il Carathéodory a considerare la convessità totale di $f(x, R)$, x fissato, come una forma appropriata della condizione di Legendre in x.

Dalla convessità totale segue facilmente la disuguaglianza poliedrale senza esser costretti ad usare V_r^n. Se F_0 giace in $\mathcal{R}$ e $\mathcal{T}$ è trasversale ad $\mathcal{R}$ in un punto, e F_i' è la proiezione di F_i su $\mathcal{R}$ parallela a $\mathcal{T}$, risulta dalla definizione stessa di $\mathcal{T}$ che è $\alpha(F_i') \leq \leq \alpha(F_i)$ e da argomenti topologici semplici che $\bigcup F_i' \supset F_0$. Perciò si ha:

$$\sum_{i>0} \alpha(F_i) \geq \sum_{i>0} \alpha(F_i') \geq \alpha(\bigcup_{i>0} F_i') \geq \alpha(F_0).$$

Se Q è un S_q $(q > r)$ la restrizione di f oppure di α a Q è un'area in Q, per conseguenza definisce una normalità ed una trasversalità in Q. Se è dim $\mathcal{T} = q - r$, con $\mathcal{T} \subset Q$, e $\mathcal{T}$ totalmente trasversale ad $\mathcal{R}(\subset Q)$ nel punto p scriveremo

$$\mathcal{T} \mid\!\frac{p,\,Q}{t}\, \mathcal{R} \quad \text{o} \quad \frac{p,\,Q}{t}\!\mid \mathcal{T}$$

Generalizziamo questa nozione: sia $0 \leq d \leq \min(r, t)$, $q = r + + t - d \leq n$ e siano $\mathcal{D} \subset \mathcal{T} \subset Q$ spazi delle dimensioni d, t, q e sia $\mathcal{C}$ $(c = t - d)$ un S_c in $\mathcal{T}$ che interseca in un punto p.

Diremo che $\mathcal{T}$ è totalmente trasversale a $\mathcal{R}$ lungo $\mathcal{D}$ in Q e scriveremo

$$\mathcal{C} \mid\!\frac{\mathcal{D},\,Q}{t}\, \mathcal{R} \quad \text{o} \quad \mathcal{R}\,\frac{\mathcal{D},\,Q}{t}\!\mid \mathcal{T}$$

se per ogni corpo convesso K giacente in un S_r contenente $\mathcal{D}$ e contenuto in Q la proiezione K_π di K su $\mathcal{R}$ parallela a $\mathcal{C}$ soddisfa la $\alpha(K_\pi) \leq \alpha(K)$.

Pare che questa definizione dipenda dalla scelta di $\mathcal{C}$ in $\mathcal{T}$ perchè la forma della proiezione K_π di un corpo convesso K r-dimensionale su $\mathcal{R}$ varia con $\mathcal{C}$. Però è facile vedere che l'area $\alpha(K_\pi)$ non dipende da $\mathcal{C}$.

Siano dati i tre spazi $\mathcal{D}, \mathcal{T}, Q$, con $\mathcal{D} \subset \mathcal{T} \subset Q$. Siccome $f(R)$ è continua e positiva, se $R \neq 0$, esiste un $\mathcal{R}$ con $\mathcal{R}\,\frac{\mathcal{D},\,Q}{t}\!\mid \mathcal{T}$.

Se sono dati $\mathcal{D}, \mathcal{R}, \mathcal{D}$, con $\mathcal{D} \subset \mathcal{R} \subset Q$, esiste un $\mathcal{T} \mid\!\frac{\mathcal{D},\,Q}{t}\, \mathcal{R}$, se $f(R)$ è totalmente convessa.

Sappiamo che la debole e la totale convessità sono equivalenti se $r = 1$ o $r = n - 1$. Nel primo caso è $d = r - 1$ e nel secondo $d = t - 1$. Un poco più generalmente si ha che:

La debole convessità di $f(R)$ equivale all'esistenza di $\mathcal{G} \,|\!\overset{\mathcal{D},\,Q}{\underset{t}{\longrightarrow}}\, \mathcal{R}$ *se* $d = \min(r, t) - 1$ (5).

Se $\mathcal{R} \overset{\mathcal{D},\,Q}{\underset{t}{\longrightarrow}}\!|\, \mathcal{G}$ e $\mathcal{D} \subset \mathcal{S} \subset \mathcal{G}$ $(d < s < t)$, allora $\mathcal{R}$ è anche totalmente normale a $\mathcal{S}$ nell'S_{r+s-d} individuato da $\mathcal{R}$ e $\mathcal{S}$ (6).

In particolare $\mathcal{R}$ è totalmente normale ad ogni S_{d+1} in $\mathcal{G}$ contenente $\mathcal{D}$. L'inverso non è vero, cioè $\mathcal{R}$ può esser totalmente normale a tutti gli S_{d+1} in $\mathcal{G}$ contenenti $\mathcal{D}$ senza esser totalmente normale a $\mathcal{G}$.

Diremo che $\mathcal{R}$ *è normale a* $\mathcal{G}$ *lungo* $\mathcal{D}$ *in* Q e scriveremo

$$\mathcal{R} \overset{\mathcal{D},\,Q}{\longrightarrow}\!|\, \mathcal{G} \quad \text{o} \quad \mathcal{G} \,|\!\overset{\mathcal{D},\,Q}{\longrightarrow} \mathcal{R}$$

se $\mathcal{R} \overset{\mathcal{D},\,Q}{\underset{t}{\longrightarrow}}\!|\, S_{d+1}$ per ogni $S_{d+1} \subset \mathcal{G}$, contenente $\mathcal{D}$.

Nel caso speciale $d = 0$, $t = n - r$ il concetto di trasversalità si deve a Wagner, [11].

L'esistenza di un $\mathcal{G}$ *trasversale a un dato* $\mathcal{R}$ *lungo un dato* $\mathcal{D} \subset \mathcal{R}$ *in un dato* Q $(d, q = r + t - d \leq n$ arbitrari) *equivale alla convessità debole di* $f(R)$,

Basta per la debole convessità che esista un S_{n-r} trasversale a un dato $\mathcal{R}$ in un punto.

Da ciò si prova adesso facilmente l'affermazione

$\mathcal{R} \overset{\mathcal{D},\,Q}{\longrightarrow}\!|\, \mathcal{G}$ e $\mathcal{R} \overset{\mathcal{D},\,Q}{\underset{t}{\longrightarrow}}\!|\, \mathcal{G}$ *sono equivalenti se* $d = \min(r, t) - 1$.

In generale non sono equivalenti in altri casi.

Oltre f e α consideriamo una seconda funzione $g(S)$ $(1 \leq s \leq \leq n - 1)$ e l'area corrispondente β. Se $0 \leq d < \min(r, s)$ $q = r + + s - d \leq n$, f, od α, definisce la relazione $\mathcal{R} \overset{\mathcal{D},\,Q}{\underset{(t)}{\longrightarrow}}\!|\, \mathcal{S}$ e g, o β, definisce $\mathcal{S} \overset{\mathcal{D},\,Q}{\underset{(t)}{\longrightarrow}}\!|\, \mathcal{R}$.

(5) Per esempio se $r = 2$, $n = 4$ e $f(R)$ è debolmente convessa, esistono una retta totalmente trasversale ad un dato piano in un dato S_3, un S_3 totalmente trasversale (in A^4) a un dato piano lungo una data retta, un piano totalmente trasversale ad un altro lungo una data retta in un dato S_3, ma in generale non esiste un piano totalmente trasversale a un dato piano in un punto.

(6) Per esempio una retta totalmente normale a un iperpiano $\mathcal{H}$ in un punto p è totalmente normale a ogni S_r $(1 \leq r \leq n - 2)$ in $\mathcal{H}$ che passa per p.

Ci domandiamo quanto si possa dedurre dall'ipotesi che queste due relazioni siano *equivalenti*. Nel caso speciale $f \equiv g$ le relazioni diventano

$$\mathcal{R} \frac{\mathcal{D}, Q}{(t)}| \mathcal{R}' \quad \text{e} \quad \mathcal{R}' \frac{\mathcal{D}, Q}{(t)}| \mathcal{R} \text{ (7)}$$

ci domandiamo cioè, quando la normalità di un S_r ad un altro rispetto a f) lungo un S_d ($d < r$, $q = 2r - d$) è *simmetrica*.

L'unico caso esaminato prima è $f \equiv g$ e $r = 1$; ossia *la normalità simmetrica di rette*. Se $f(R)$ è convessa, cioè se lo spazio è Minkowskiano, Blaschke [1] ha provato nel 1916 che: *Se* $n > 2$ *e la normalità fra rette è simmetrica la metrica è euclidea.*

Invece Radon [9] ha mostrato (anche nel 1916) che *la normalità di rette in un piano* ($n = 2$) *Minkowskiano può esser simmetrica, senza che la metrica sia euclidea.*

Chiamando $f(R)$ *quadratica* se $f^2(R) = f^2(R^1, \dots, R^{\binom{n}{r}})$ è una forma quadratica, si dimostra, per quanto concerne il nostro problema in tutta la sua generalità, che:

Se $f(R)$ *e* $g(S)$ *sono debolmente convesse e se per fissati interi* d, r, s, ($d < \min(r, s)$, $q = r + s - d \leq n$) *le relazioni*

$$\mathcal{R} \frac{\mathcal{D}. Q}{(t)}| \mathcal{S} \quad \text{e} \quad \mathcal{S} \frac{\mathcal{D}, Q}{(t)}| \mathcal{R} \text{ (7)}$$

sono equivalenti, ambedue le funzioni $f(R)$ *e* $g(S)$ *sono quadratiche a meno che* $r + s = n$ *e* $d = 0$ (cfr. [6]).

Se $r + s = n$ *e* $d = 0$, *data una* $f(R)$ *debolmente convessa esiste una* $g(S)$ *totalmente convessa tale che le due relazioni siano equivalenti* (cfr. [3]).

Applicando questo teorema al caso $f \equiv g$ troviamo che f *è quadrica se la normalità* (*totale*) *di un* S_r *ad un altro lungo un* S_d (*d fissato*) ($q = 2r - d \leq n$) *è simmetrica a meno che* $n = 2r$ e $d = 0$.

In quest'ultimo caso esistono per ogni $r \geq 1$ *funzioni non-quadratiche totalmente convesse per cui la normalità* (*in un punto*) *è simmetrica* (cfr. [3]).

(7) Le parentisi indicano che tutto ciò che segue vale tanto per normalità quanto per normalità totale.

Se $r = 1$ o $r = n - 1$ una $f(R)$ quadratica è l'area corrispondente ad una opportuna metrica euclidea $\Sigma g_{ik} x^i x^k$. Se $1 < r < n - 1$ ci sono più $f(R)$ quadratiche che aree derivabili da metriche euclidee.

4. Funzioni derivate da corpi convessi.

Consideriamo lo spazio euclideo E^n ad n dimensioni provvisto di un sistema di coordinate cartesiane con origine z. Una funzione $f(\mathcal{R}^+)$ dipendente da un S_r orientato variabile per z induce una funzione omogenea $f(R)$ (su G_r^n) data da:

$$f(R) = |R| f(\mathcal{R}^+) \text{ se } R \parallel \mathcal{R}^+, \; f(0) = 0 \; (\text{è } f(\lambda R) = \lambda f(R) \text{ se } \lambda \geq 0).$$

Se $f(\mathcal{R}^+)$ non dipende dall'orientazione di $\mathcal{R}$, ossia ha lo stesso valore per le due orientazioni di $\mathcal{R}$, scriveremo $f(\mathcal{R})$ invece di $f(\mathcal{R}^+)$. Allora la funzione indotta soddisfa la relazione

$$f(\lambda R) = |\lambda| \, f(R) \text{ per ogni } \lambda \text{ reale.}$$

Si tratta di decidere se certe funzioni $f(\mathcal{R}^+)$ inducono, o sotto quali condizioni inducono, funzioni convesse o totalmente convesse su G_r^n. I risultati si trovano nei lavori [4] parti II, III, IV, e [5], che ho scritto insieme con G. Ewald e C. G. Shephard.

Cominciamo con un esempio molto semplice

(4) Sia $1 \leq a < r \leq n - 1$ e $g(A)$

una funzione omogenea (totalmente) convessa su G_a^n. Designando con A_u un a-vettore semplice unitario e per $\mathcal{R}$ un S_r per z l'estremo

$$f(\mathcal{R}) = \sup_{A_u \parallel \mathcal{R}} g(A_u)$$

induce una funzione (totalmente) convessa su G_r^n.

Osserviamo dapprima che è $g(A_u) + g(-A_u) \geq g(0) = 0$ a causa della convessità di $g(A)$; quindi si ha $\max(g(A_u), g(-A_u)) \geq 0$. Se R è un r-vettore semplice, e C_u è un $(r - a)$-vettore unitario semplice, $R . C_u$ è un a-vettore semplice parallelo a R. [8]

([8]) Se $r > a$ ricordiamo che il prodotto scalare $R \cdot A$ è nullo se $R = 0$, $A = 0$ oppure A è normale a R. In caso contrario, sia A' la proiezione di A

Per ogni C_u fisso, $g(R \cdot C_u)$ è una funzione convessa su G_r^n, perchè è:

$$g((\Sigma R_i) \cdot C_u) = g(\Sigma (R_i \cdot C_u)) \leq \Sigma g(R_i \cdot C_n).$$

Perciò $\sup_{C_u} g(R \cdot C_u)$ è convessa su G_r^n. Risulta

$$\sup_{C_u} g(R_u \cdot C_u) = \sup_{C_u \parallel \mathfrak{R}} g(R_u \cdot C_u), \quad |R_u| = 1$$

in quanto, se C_u non è parallelo a $\mathcal{R}$ esiste un C_u' con $R_u \cdot C_u' = \lambda (R_u \cdot C)$, $\lambda > 1$, ed allora

$$g(R_u \cdot C_u') = \lambda g(R_u \cdot C_u);$$

e vale la $\sup_{C_u} g(R_u \cdot C_u) \geq 0$. Ma se è $C_u \parallel \mathcal{R}$, il vettore $R_u \cdot C_u$ è unitario e percorre tutti glį a-vettori semplici unitari paralleli a $\mathcal{R}$.

Per questo la

$$\sup_{C_u} g(R_u \cdot C_u) = f(\mathcal{R})$$

induce la funzione convessa $\sup_{C_u} g(R \cdot C_u)$. Occorre una ulteriore semplice argomentazione — che omettiamo per brevità — per provare che la funzione indotta è totalmente convessa, se $g(A)$ lo è.

Sussiste un teorema analogo concernente l'integrazione di $g(A)$:

(5) *Se $g(A)$ è convessa e omogenea su G_a^n, allora*

$$f(\mathcal{R}) = \int_{A_u \parallel \mathfrak{R}} g(A_u)\, dA_u$$

induce una funzione convessa su G_r^n.

Qui dA_u è la densità cinematica nel senso della geometria integrale degli S_a orientati in $\mathcal{R}$.

sul S_r orientato $\mathfrak{R}^+$ parallelo ad $\mathfrak{R}$, allora $R \cdot A$ è un $(r-a)$-vettore semplice parallelo a $\mathfrak{R}$ e normale ad A'. La norma e la direzione di $R.A$ sono determinate da

$$|R \cdot A| = |R| \cdot |A'|, \qquad A' \wedge (R \cdot A) = \lambda R, \qquad \lambda > 0.$$

Nazionale di Alta Matematica (1962–1963), 178–205

La dimostrazione è più difficile della precedente ed usa un teorema della geometria integrale. Inoltre, se $g(A)$ è totalmente convessa non segue che $f(\mathcal{R})$ induca una funzione totalmente convessa, in quanto la somma di due funzioni totalmente convesse non è, in generale, totalmente convessa.

Ci interessano anzitutto funzioni derivate da corpi convessi. Sia K un corpo convesso n-dimensionale in E^n. La sua funzione di appoggio

$$H(v) = \sup_{x \in K} v \cdot x$$

è convessa, essendo l'estremo superiore di funzioni convesse (lineari), ne segue che è $H(v) > 0$ per $v \neq 0$ se z è un punto interno di K.

Per un vettore unitario la distanza da z del piano d'appoggio, con normale esterna u, è $H(u)$. Perciò

$$H(u) + H(-u) = P_1(K, \mathcal{L})$$

è la lunghezza della proiezione di K su la retta $\mathcal{L}$ parallela ad u. (Come precedentemente designamo con $P_r(K, \mathcal{R})$ il volume r-dimensionale della proiezione $P(K, \mathcal{R})$ di K su $\mathcal{R}$). Conviene supporre che $\mathcal{R}$ passi per z. Allora

$$P(K, v) = |v| P_1(K, \mathcal{L}) = H(v) + H(-v), \qquad v \parallel \mathcal{L},$$

è la funzione indotta da $P_1(K, \mathcal{L})$ su $G_1^n = V^n$ ed è totalmente convessa. L'estremo

$$\sup_{u \parallel \mathcal{R}} P(K, u)$$

è il diametro di $P(K, \mathcal{R})$. Perciò deduciamo dalla (4):

Il diametro di $P(K, \mathcal{R})$ induce una funzione totalmente convessa.

$B(u) = H(u) + H(-u)$ e la distanza degli iperpiani di appoggio di K normali a $\pm u$ e si chiama la larghezza di K nelle direzioni $\pm u$. Se $u \parallel \mathcal{R}$ il numero $B(u)$ è anche la larghezza di $P(K, \mathcal{R})$. Poichè

$$\pi_n = \pi^{\frac{n}{2}} / \Gamma\left(1 + \frac{r}{2}\right), \qquad n\pi_n$$

sono il volume della sfera piena unitaria, B_n, e l'area $(n-1)$-dimensionale della sua frontiera Σ^{n-1}, allora

$$\frac{2}{r\pi_r}\int_{u\,\|\,\mathfrak{R}} H(u)\,du = \frac{1}{r\pi_r}\int_{u\,\|\,\mathfrak{R}} B(u)\,du$$

è la cosiddetta larghezza media di $P(K, \mathfrak{R})$. Ne deduciamo dalla (5) che:

La larghezza media di $P(K, \mathfrak{R})$ induce una funzione convessa.

È chiaro che per un fissato r-vettore semplice R_0, la funzione $R \cdot R_0$ è totalmente convessa. Perciò la funzione

$$|R \cdot R_0| = \max\,[R \cdot R_0\,, R_0 \cdot (-R_0)]$$

è anche essa totalmente convessa. Se W è un insieme di r-vettori semplici limitato (rispetto a la norma $|R|$), allora

$$H_0(R, W) = \sup_{R_0 \in W} R \cdot R_0 \quad \text{e} \quad H(R, W) = \sup_{R_0 \in W} |R \cdot R_0|$$

sono totalmente convesse (9). Scegliendo differenti insiemi W si ottengono diverse funzioni totalmente convesse. Qui consideriamo il solo esempio seguente.

Siamo $C_0, \ldots, C_r$ $(1 \leq r \leq n-1)$ insiemi non-vuoti compatti in E^n e sia W l'insieme degli r-vettori

$$(x_1 - x_0) \wedge \ldots \wedge (x_r - x_0)\,, \quad x_i \in C_i\,.$$

Se $T(q_0, \ldots, q_r)$ è il volume r-dimensionale dell'r-simplesso coi vertici $q_0, \ldots, q_r$ e R_u è un r-vettore unitario parallelo a $\mathfrak{R}$, si ha:

$$|R_u \cdot [(x_1 - x_0) \wedge \ldots \wedge (x_r - x_0)]| = r!\; T(P(x_0, \mathfrak{R}), \ldots, P(x_r, \mathfrak{R}))\,.$$

Perciò è:

$$\frac{1}{r!} H(R_u, W) = \max_{q_i \in P(C_i, \mathfrak{R})} T(q_0, \ldots, q_r)$$

e

$$\frac{1}{r!} H(R, W) = |R| \max_{q_i \in P(C_i, \mathfrak{R})} T(q_0, \ldots, q_r)\,, \quad R = |R|\,R_u$$

(9) Se $r = 1$ e W è l'insieme dei vettori x i cui punti terminali giacciono nel corpo convesso K (scriviamo $x \in K$), la funzione $H_0(R, W)$ è la funzione d'appoggio di K.

Dunque:

Il volume massimo di un r-simplesso con vertici $q_i \in P(C_i, \mathcal{R})$ $(i = 0, \ldots, r)$ induce una funzione totalmente convessa su G_r^n.

Applicando la (4) ne deduciamo:

Se $1 \leq r \leq t \leq n-1$ *e* $C_0, \ldots, C_r$ *sono insiemi compatti* $\sup_{q_i \in P(C_i, \mathcal{S})} T(q_0, \ldots q_r)$ *induce una funzione totalmente convessa su* G_t^n.

Se tutti i C_i coincidono con lo stesso corpo convesso K otteniamo (usando s, r invece di r, t):

Il volume (s-dimensionale, $1 \leq s \leq r$*) massimo degli s-simplessi contenuti in* $P(K, \mathcal{R})$ *induce una funzione totalmente convessa su* G_r^n.

Per $s = 1$ otteniamo di nuovo la convessità totale della funzione indotta dal diametro di $P(K, \mathcal{R})$.

Vedremo che, in generale, la funzione $P(K, \mathcal{R}) = |R| P_r(K, \mathcal{R}) (R \parallel \mathcal{R})$ non è convessa se $1 < r < n-1$. Siano $(V_t(r, n))$ $V(r, n)$ gli estremi superiori di tutti gli interi s con la proprietà che il volume massimo r-dimensionale dei poliedri con s vertici contenuti in $P(K, \mathcal{R})$ induce una funzione (totalmente) convessa su G_r^n per ogni corpo convesso K.

Allora è chiaro che $V^t(r, n) \leq V(r, n)$ e che $V(r, n)$ deve esser finito, perchè $P(K, \mathcal{R})$ può esser approssimato con poliedri contenuti in $P(K, \mathcal{R})$. Abbiamo il seguente teorema: *Per* $1 < r < n-1$ *valgono le uguaglianze:*

$$V_t(r, n) = r + 2, \qquad V(r, n) = r + 3.$$

(Il caso $r = 2$ e la disuguaglianza

$$r + 2 \leq V_t(r, n) \leq V(r, n) \leq r + 3$$

per $r > 2$ si trovano in [4] e [5], il resto in [10]).

La dimostrazione per poliedri con $s = r + 2$ o $s = r + 3$ vertici è basata sulle proprietà delle proiezioni di un simplesso n-dimensionale T^n in E^n. La funzione $P(T^n, R)$ su G_r^n non è convessa per $1 < r < n-2$ e per $r = n-2$ è convessa ma non totalmente convessa (vedere il capitolo seguente). Poichè $r \leq n-2$ abbiamo $s \leq n+1$. La proiezione di un $T^{s-1} \subset K$ su un S_r è un poliedro con s vertici in $P(K, S_r)$ e ogni tale poliedro in $P(K, S_r)$ è la proiezione di un $T^{s-1} \subset K$. Per questo si tratta della funzione

$$f_s(K, R) = \sup_{T^{s-1} \subset K} P(T^{s-1}, R).$$

Si prova ([10]) che la $P(T^{s-1}, R)$ è (totalmente) convessa su G_r^{s-1}, se, e soltanto se $P(T^{s-1}, R)$ è (totalmente) convesso su G_r^n. Dunque segue della III nel capitolo precedente e della convessità totale della $P(T^{r+1}, R)$ su G_r^{r+1} che la $f_{r+2}(K, R)$ è totalmente convessa su G_r^n e del fatto che la $P(T^{r+2}, R)$ è convessa ma non totalmente convessa su G_r^{r+2} che lo stesso vale per la $f_{r+3}(K, R)$ su G_r^n.

Che il volume massimo di un poliedro con $r+4$ vertici contenuti in $P(K, \mathcal{R})$ non conduce ad una funzione convessa su G_r^n segue in modo analogo della esistenza di un poliedro Π con $n+2$ vertici in E^n che ha $P(\Pi, R)$ non convesse per $1 < r < n-1$ (vedesi il terzo capitolo).

Menzoniamo un teorema un poco differente che si ottiene coi nostri metodi.

Sia K un corpo convesso n-dimensionale avente il punto z come centro. Per un S_r variabile $\mathcal{R}$ per z $(2 \leq r \leq n-1)$ sia $Q(K, \mathcal{R})$ il volume minimo dei parallelepipedi r-dimensionali in $\mathcal{R}$ contenenti $K \cap \mathcal{R}$. Allora $Q^{-1}(K, \mathcal{R})$ induce una funzione totalmente convessa.

5. L'area delle proiezioni dei poliedri.

Sia K un corpo convesso (n-dimensionale) in E^n e $\mathcal{R}$ un S_r per un fissato punto z.

La frontiera $\partial P(K, \mathcal{R})$ di $P(K, \mathcal{R})$ è l'insieme dei punti in cui gli S_{n-r} di appoggio di K normali a $\mathcal{R}$ intersecano $\mathcal{R}$. L'insieme dei punti in cui questo S_{n-r} toccano K o ∂K si chiama la «frontiera d'ombra» $S(\mathcal{R})$ di K corrispondente a $\mathcal{R}$, perchè nel caso $n=3$, $r=2$ la $S(\mathcal{R})$ è proprio la frontiera della parte illuminata di K da una sorgente di luce che si trova all'infinito nella direzione di un raggio normale a $\mathcal{R}$. La frontiera d'ombra $S(\mathcal{R})$ dicesi «*netta*» se ogni S_{n-r} di appoggio normale a $\mathcal{R}$ tocca K in un sol punto. Allora la proiezione di $S(\mathcal{R})$ su $\partial P(K, \mathcal{R})$ è una rappresentazione topologica; cosicchè $S(\mathcal{R})$ è omeomorfa a Σ^{r-1}. Tutte le $S(\mathcal{R})$ sono nette se, e soltanto se, ∂K non contiene segmenti retti o è strettamente convessa.

Se $S(\mathcal{R})$ è netta, un'orientazione $\mathcal{R}^+$ di $\mathcal{R}$ induce un'orientazione $P(K, \mathcal{R}^+)$ di $P(K, \mathcal{R})$, $\partial P(K, \mathcal{R}^+)$ di $\partial(P, \mathcal{R})$ ed $S(\mathcal{R}^+)$ di $S(\mathcal{R})$,

Supponiamo adesso che K sia un poliedro. Allora una $S(\mathcal{R})$ netta è un poliedro di dimensione $r-1$ omeomorfo a Σ^{r-1}.

Esiste un insieme E di misura zero su G_r^n (che contiene o) tale che $G_r^n - E$ consista di un numero finito di aperti connessi e $S(\mathcal{R})$ sia netto se $\mathcal{R}$ è parallelo a un r-vettore in $G_r^n - E$.

Inoltre $S(\mathcal{R})$ è la stessa per tutti gli $\mathcal{R}$ paralleli a R in uno degli aperti.

Prendiamo adesso un $\mathcal{R}_0$ per cui $S(\mathcal{R}_0)$ è netta. Sia $\mathcal{R}_0^+$ una orientazione di $\mathcal{R}_0$ e $\partial P(K, \mathcal{R}_0^+)$, $S(\mathcal{R}_0^+)$ le corrispondenti orientazioni di $\partial P(K, \mathcal{R}_0)$ e $S(\mathcal{R}_0)$. Proiettando $S(\mathcal{R}_0^+)$ su un altro r-piano orientato $\mathcal{R}^+$, otteniamo un ciclo orientato $C = C(\mathcal{R}_0^+, \mathcal{R})$ in $\mathcal{R}$ ed ogni punto p di $\mathcal{R}^+ - C$ ha un indice finito $j(p)$ rispetto a C, che svanisce nella componente illimitata di $\mathcal{R}^+ - C$. Se dA è l'elemento d'area in $\mathcal{R}$, l'integrale

$$j(\mathcal{R}_0^+, \mathcal{R}^+) = \int_{\mathfrak{R} - C_k} j(p)\, dA$$

esiste ed è finito. Se $\mathcal{R}_0^-$, $\mathcal{R}^-$ sono le altre orientazioni di $\mathcal{R}_0$ e $\mathcal{R}$ è chiaro che è

$$(6)\quad j(\mathcal{R}_0^-, \mathcal{R}^+) = j(\mathcal{R}_0^+, \mathcal{R}^-) = -j(\mathcal{R}_0^+, \mathcal{R}^+) = -j(\mathcal{R}_0^-, \mathcal{R}^-).$$

Dimostriamo che:

(7) *La funzione* $j(\mathcal{R}_0^+, R) = |R|\, j(\mathcal{R}_0^+, \mathcal{R}^+)$ $(R \parallel \mathcal{R}^+)$ *indotta da* $j(\mathcal{R}_0^+, \mathcal{R}^+)$ *è lineare e*

$$j(\mathcal{R}_0^+, R_0) = P\,(K, R_0) \ \textit{se}\ R_0 \parallel \mathcal{R}_0^+ .$$

L'equazione equivale alla seguente altra:

$$j(\mathcal{R}_0^+, \mathcal{R}_0^+) = P_r(K, \mathcal{R}_0),$$

ed è banale poichè

$$j(p) = 1 \text{ se } p \in P(K, \mathcal{R}_0) - \partial P(K, \mathcal{R}_0) \text{ e } j(p) = 0 \text{ se } p \in \mathcal{R}_0 - P(K, \mathcal{R}_0).$$

Per provare la linearità uniamo (con segmenti) un vertice di $S(\mathcal{R}_0)$ a tutti gli altri punti di $S(\mathcal{R}_0)$. Otteniamo un poliedro omeomorfo a B_r (sfera piena unitaria r-dimensionale). Orientiamo poi il poliedro in modo che $S(\mathcal{R}_0^+)$ sia la sua frontiera. Siano $F_1, \ldots, F_k$ gli r-vettori associati alle facce (r-dimensionali) del poliedro. Se R_u è il vettore unitario parallelo a $\mathcal{R}^+$, l'area dotata di segno della proiezione della faccia corrispondente a F_i è $R_u \cdot F_i$.

Un punto p di $\mathcal{R}^+$ che non giace nella proiezione delle frontiere delle facce di un poliedro ha una multiplicità di ricoprimento $m(p)$, data dalla differenza tra il numero dei ricoprimenti positivi e quello dei ricoprimenti negativi, porremo poi $m(p) = 0$ se p non è la proiezione di un punto interno di una faccia. Allora è

$$\sum_{i=1}^{k} F_i \cdot R_u = \int_{\mathfrak{R}^+} m(p)\, dA.$$

Ma è un fatto topologico ben noto che $m(p) = j(p)$.

Per conseguenza si ha:

$$j(\mathcal{R}_0^+, R) = \sum_{i=1}^{k} F_i \cdot R,$$

e $j(\mathcal{R}_0^+, R)$ è lineare.

Dimostriamo ora il seguente fondamentale risultato:

(8) *Sia K un poliedro convesso n-dimensionale in E^n. Perchè l'area $P_r(K, \mathcal{R})$ della proiezione di K induce una funzione convessa, è necessario e sufficiente che ogni coppia $\mathcal{R}_0^+$, $\mathcal{R}$ di r-piani orientati per cui $S(\mathcal{R}_0)$ è netta soddisfa la diseguaglianza*

$$|j(\mathcal{R}_0^+, \mathcal{R}^+)| \leq P_r(K, \mathcal{R}).$$

In forza della (6) e della uguaglianza $P(K, R) = P(K, -R)$, la condizione precedente equivale a

$$j(\mathcal{R}_0^+, R) \leq P(K, R)$$

per tutti gli R (semplici). La sufficienza è facile da dimostrare. Infatti:

Sia C una componente di $G_r^n - E$ e $R_0 \in C$, $R \subset C$. Poichè $S(\mathcal{R})$ è lo stesso insieme, per tutti gli $\mathcal{R}$ corrispondenti a punti di C, abbiamo

$$P(K, R) = j(\mathcal{R}_0^+, R)\, R_0, \; R \subset C.$$

Questo vale anche se R giace sulla frontiera di C perchè ambedue le funzioni sono continue.

L'ipotesi enunciata in (8) implica che è

$$P(K, R) = \sup_{\mathfrak{R}_0^+} |j(\mathcal{R}_0^+, R)|,$$

quindi $P(K, R)$ è convessa, come estremo superiore di funzioni convesse.

La dimostrazione della *necessità* si fonda sul seguente Lemma che non dimostreremo per brevità.

LEMMA. Sia $f(R)$ una funzione lineare definita in un aperto U (relativo) su G_r^n, cioè $f(R)$ può estendersi ad una funzione lineare nell'inviluppo convesso U^H di U in V_r^n. Allora $f(R)$ possiede una sola estensione a U^H che sia convessa, la quale, perciò, coincide con quella funzione lineare.

Per ipotesi attualmente $P(K, R)$ può estendersi ad una funzione convessa $P(K, \tilde{R})$ in V_r^n. Se R_0 è un punto di una delle componenti, C, di $G_r^n - E$ già sappiamo che, se $\mathcal{R}_0^+ \parallel R_0$, è

$$j(\mathcal{R}_0^+, R) = P(K, R) \text{ per tutti gli } R \in C.$$

$j(\mathcal{R}_0^+, R)$ possiede una sola estensione lineare $j(\tilde{R})$ a V_r^n ne segue dal Lemma che è $j(\tilde{R}) = P(K, \tilde{R})$ per tutti gli $R \in C^H$. L'iperpiano $(\tilde{R}, j(\tilde{R}))$ nello spazio affine A^+ consistente dei punti $(\tilde{R}, \zeta) = (\tilde{R}^1, \ldots, \tilde{R}^{\binom{n}{r}}, \zeta)$ è un iperpiano di appoggio dell'insieme convesso $\tilde{R} \in C^H, \zeta \geq P(K, \tilde{R})$ in A^+. Poichè l'insieme più vasto $(\tilde{R}, \zeta)$, $\tilde{R} \in V_r^n, \zeta \geq P(K, \tilde{R})$ è anche convesso (per ipotesi), l'iperpiano $(\tilde{R}, j(\tilde{R}))$ deve essere un iperpiano di appoggio del secondo insieme. Questo significa che è

$$j(\tilde{R}) \leq P(K, \tilde{R}) \text{ per ogni } \tilde{R} \in V_r^n,$$

ne segue in particolare che è

$$j(R) = j(R_0^+, R) \leq P(K, R) \text{ su } G_r^n \text{ o } |j(\mathcal{R}_0^+, \mathcal{R})| \leq P_r(K, \mathcal{R}).$$

Vediamo come si applica il criterio nel caso più semplice che sia $r = 2$, $n \geq 4$ e K un simplesso, T^n in E^n. Siccome la convessità di $P(K, R)$ è invariante rispetto alle affinità di K, supponiamo che T^n sia un simplesso regolare.

Ogni curva di Jordan chiusa che consiste di spigoli di T^n è una frontiera d'ombra (se $r = 2$). Ci sono piani $\mathcal{R}$ per i quali $P(T^n, \mathcal{R})$ è un poligono regolare.

Consideriamo prima il caso $n = 6$. Per un certo piano, $\mathcal{R}_1$, la proiezione $P(T^6, \mathcal{R}_1)$ sarà un ettagono regolare $q_1 q_2 q_3 q_4 q_5 q_7$. Per un altro piano $\mathcal{R}_0$ la proiezione di $S(\mathcal{R}_0)$ su $\mathcal{R}_1$ sarà l'ettagono

$q_1 q_3 q_5 q_7 q_2 q_4 q_6$. Con opportune orientazioni $\mathcal{R}_0^+$, $\mathcal{R}_1^+$ l'indice $j(p)$ di un punto p di $\mathcal{R}_1$, rispetto al secondo ettagono ha i valori o,

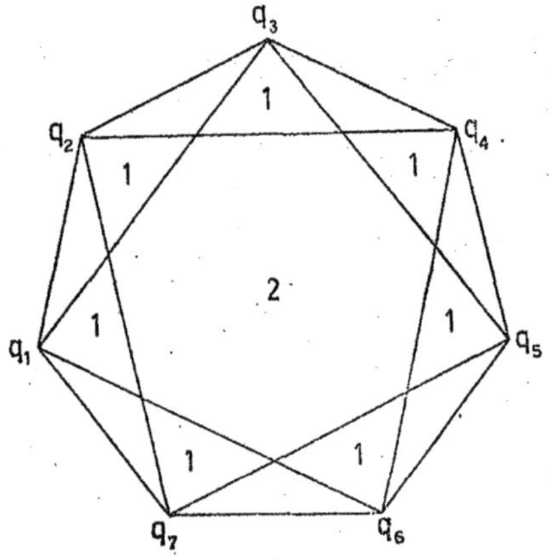

Fig. 2

1, 2. I valori sono indicati nella figura 2, dove sono positivi. È facile vedere che è

$$j(\mathcal{R}_0^+, \mathcal{R}_1^+) = \int_{\mathcal{R}_1} j(p)\, dA > \text{ area } (\bigcup q_i)^H = P_2(T^6, \mathcal{R}_1).$$

($(\bigcup q_i)^H$ è l'inviluppo convesso dei punti q_i). Quindi la condizione per la convessità di $P(T^6, R)$ ($r = 2$) non è soddisfatta e $P(T^6, R)$ *non è convessa.*

È chiaro che la situazione è peggiore per $n > 6$.

Nel caso di un T^4, ci sono cinque punti $q_1, \ldots, q_5$ ed abbiamo da considerare pentagoni (non necessariamente regolari) di due tipi

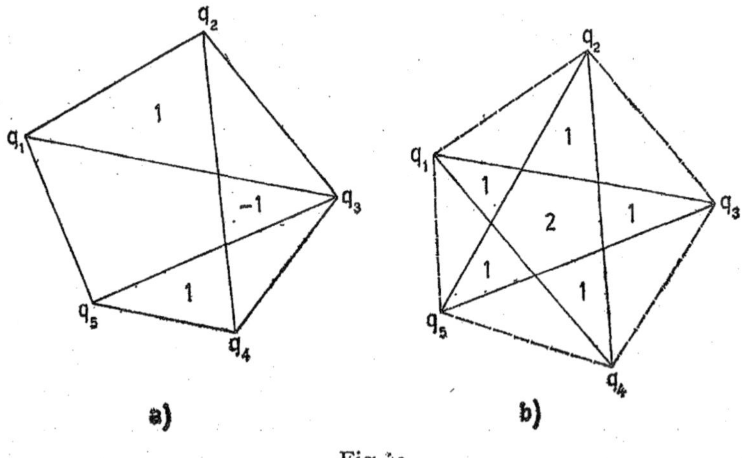

Fig. 3

Gli indici $j(p)$ rispetto ad orientazioni opportune sono dati nelle Figure 3 (a), (b). Nel caso (a) è banale che è

$$|j(\mathcal{R}_0^+, \mathcal{R}_1^+)| < \text{area}\,(\dot{\bigcup}\, q_i)^H = P(T^4, \mathcal{R}_1)$$

Un facile calcolo dimostra che questa disuguaglianza vale anche nel caso (b), cioè che $P(T^4, R)$ è convessa. Invece $P(T^5, R)$ $(r = 2)$ non è convessa.

Il risultato completo è il seguente ([4] e [10]).

$P(T^n, R)$ non è convessa se $2 \leq r \leq n - 3$ $(n \geq 5)$ ed è convessa ma non totalmente convessa se $r = n - 2$ $(n \geq 4)$.

Si può provare che ([4]):

Per ogni $n > 3$ ci sono poliedri Π convessi con $n + 2$ vertici tali che $P(\Pi, R)$ non sia convessa per nessun r con $l < r < n - 1$.

Ci sono altri fenomeni sorprendenti quando si pensa alla teoria classica. Per esempio, benchè $P(T^4, R)$ $(r = 2)$ sia convessa, ci sono simplessi regolari T_1^4, T_2^4 in E^4 tali che $P(T_1^4 + T_2^4, R)$ non sia convessa, ($T_1^4 + T_2^4$ significa la somma nel senso vettoriale, cioè consiste di tutti i punti $x_1 + x_2$ con $x_i \in T_i^4$). Invece, un corpo K convesso che è la somma di segmenti induce funzioni $P(K, R)$ convesse per ogni r.

6. **Corpi rotondi e convessità locale. Problemi.**

Finora abbiamo parlato soltanto di poliedri. Infatti, non siamo riusciti a provare un teorema analogo per corpi convessi arbitrari. Le difficoltà hanno cause diverse.

In primo luogo, le frontiere d'ombra di un corpo convesso qualunque possono essere patologiche. Per esempio, se $S(R)$ non è netta può avvenire che non contiene un sottoinsieme omeomorfo alla sfera Σ^{r-1}. Se poi $S(R)$ è netta l'indice $j(p)$ di un punto $p \in R^+ - C$ rispetto alla proiezione $C = C(\mathcal{R}_0^+, \mathcal{R}^+)$ di $S(\mathcal{R}_0^+)$ su $\mathcal{R}^+$ è finito malgrado ciò può accadere che l'integrale $j(\mathcal{R}^+, \mathcal{R}^+) = \int j(p)\, dA$ non esista.

Per brevità diremo che il corpo K è *rotondo* se ∂K è di classe C^2 e la curvatura gaussiana di ∂K è dappertutto diversa da zero. Le frontiere d'ombra di un corpo rotondo sono tutte nette e l'integrale $j(\mathcal{R}_0^+, \mathcal{R}^+)$ esiste. Dalla stessa argomentazione fatta per i poliedri, risulta che:

Se K è rotondo basta per la convessità di $P(K, R)$ che sia

$$|j(\mathcal{R}_0^+, \mathcal{R}^+)| \leq P_r(K, \mathcal{R})$$

per ogni coppia $\mathcal{R}_0^+, \mathcal{R}^+$ di S_r orientati.

È molto probabile — ma non l'abbiamo dimostrato — che questa condizione sia anche necessaria. Ci mancano delle informazioni sulla struttura dei coni G_r^n dal punto di vista della geometria differenziale: Il Lemma, su cui abbiamo fondato la dimostrazione nel caso dei poliedri, dovrebbe esser sostituito con un lemma molto più fino.

Abbiamo sottinteso che la funzione $P(K, R)$ di un corpo rotondo, in generale, non è convessa, ciò che è facile vedere: Sia K un corpo convesso per cui le funzioni $P(K, R)$ $(1 < r < n - 1)$ non sono convesse. Si sa che K si può approssimare con una successione $K_1, K_2, \ldots$ di corpi rotondi (persino analitici). Poichè $P(K_i, R)$ tende a $P(K, R)$, ed il limite di funzioni convesse è convesso, segue che a partire da un certo indice i_0 le $P(K_i, R)$ non sono convesse.

Però le $P(K, R)$ di un corpo K rotondo sono localmente convesse. Infatti, sia $S(\mathcal{R}_0)$ una frontiera d'ombra di K. Si prova facilmente che:

Se $\mathcal{R}$ è abbastanza vicino a $\mathcal{R}_0$, un S_{n-r} normale a $\mathcal{R}$ per un punto di $S(\mathcal{R}_0)$ non interseca $S(\mathcal{R}_0)$ in un secondo punto. Per conseguenza, proiettando $S(\mathcal{R}_0)$ su $\mathcal{R}$, otteniamo una rappresentazione topologica di $S(\mathcal{R}_0)$ su la sua proiezione $C(\mathcal{R}_0, \mathcal{R})$ su $\mathcal{R}$. Ne risulta che $C(\mathcal{R}_0, \mathcal{R})$ è omeomorfa alla sfera Σ^{r-1} e delimita un insieme aperto D omeomorfo alla sfera aperta r-dimensionale.

Un'orientazione $\mathcal{R}_0^+$ di $\mathcal{R}_0$ induce, come dianzi, orientazioni $\partial P(K, \mathcal{R}_0^+)$ di $\partial P(K, \mathcal{R}_0)$, $S(\mathcal{R}_0^+)$ di $S(\mathcal{R}_0)$ e $C(\mathcal{R}_0^+, \mathcal{R}^+)$ di $C(\mathcal{R}_0, \mathcal{R})$ nella orientazione $\mathcal{R}^+$ di $\mathcal{R}$. Poichè la proiezione di $S(\mathcal{R}_0)$ su $C(\mathcal{R}_0, \mathcal{R})$ è topologica, l'indice $j(p)$ di un punto p di $\mathcal{R}^+ - C(\mathcal{R}_0, \mathcal{R})$ rispetto a $C(\mathcal{R}_0^+, \mathcal{R})$ è zero al di fuori di $C(\mathcal{R}_0^+, \mathcal{R})$ ed è 1 o -1 in D. Perciò ne risulta dalla $D \subset P(K_0 \mathcal{R})$ che è

$$|j(\mathcal{R}_0^+, \mathcal{R}^+)| = \left| \int_{\mathfrak{R}} j(p)\, dA \right| = \text{volume } D \leq P_r(K, \mathcal{R}),$$

Dunque la nostra condizione sufficiente per la convessità di $P(K, R)$ vale per $\mathcal{R}$ abbastanza vicini a $\mathcal{R}_0$. È chiaro che questo fatto deve implicare la convessità locale di $P(K, R)$. Abbiamo quindi provato che:

Esistono corpi convessi rotondi tali che $P(K, R)$ $(1 < r < n - 1)$ *sia localmente convessa senza essere convessa.*

Perciò il fenomeno di una funzione localmente ma non globalmente convessa non è affatto patologico.

Concludiamo queste conferenze con alcuni *problemi* uno dei quali abbiamo già precedentemente incontrato.

1. Sia $V_{i,r-i}(L_1, L_2)$ $(0 \leq i \leq r)$ il volume misto (cf [2], § 29) dei corpi L_i, L_2 in un S_r in cui L_1 compare i volte e L_2 compare $r - i$ volte. Allora $P_r(K, \mathcal{R}) = V_{r,0}(P(K, \mathcal{R}), B(\mathcal{R}))$ e la larghezza media di $P(K, \mathcal{R})$ è $cV_{1,r-1}(PK, \mathcal{R}), B(\mathcal{R}))$, dove c è una costante e $B(\mathcal{R})$ è la sfera piena unitaria in $\mathcal{R}$. (Cfr. [2], p. 63). Trovare condizioni sotto cui $V_{i,r-i}(P(K, R), B(\mathcal{R}))$ è convessa, se $1 < i < r$.

2. Decidere se per una $f(R)$ omogenea con $f(R) > 0$, per $R \neq 0$, la disuguaglianza poliedrale è meno forte della convessità o più forte della convessità debole.

3. In particolare, soddisfa $P(K, R)$ la disuguaglianza poliedrale?

4. Sia K un corpo convesso n-dimensionale col centro z, e $A(\mathcal{R})$ il volume r-dimensionale di $K \cap \mathcal{R}$, $\mathcal{R} \ni z$. $A^{-1}(\mathcal{R})$ induce la funzione $A'(R) = |R| / A(\mathcal{R})(R \| \mathcal{R})$. È noto che $A'(R)$ è debolmente convessa (cfr. [6], p. 62).

Domandiamo se $A'(R)$ è convessa o soddisfa la disuguaglianza poliedrale.

Questo problema conduce ad una serie di questioni importantissime, su cui non sappiamo ancora niente, nemmeno le formulazioni appropriate. Si tratta di descrivere mediante opportuni concetti di convessità il comportamento del volume r-dimensionale di $K \cap \mathcal{R}$ per un corpo convesso qualunque e per un $\mathcal{R}$ variabile che non passa per un punto fisso.

BIBLIOGRAFIA

[1] W. Blaschke, *Räumliche Probleme mit symmetrischer Transversalitäts-bedingung,* Ber Sächs. Akad. Wiss. Leipzig 68 (1916), 50-55.

[2] T. Bonnesen e W. Fenchel, *Theorie der konvexen Körper,* Ergebnisse Math. III I Berlin 1934 (New York 1958).

[3] H. Busemann, *Areas in affine space I, II,* Rend. Circ. Mat. Palermo Serie II 9 (1960), 81-90.

[4] H. Busemann, G. Ewald e G. C. Shephard, *Convex Bodies and Convexity on Grassmann Cones,* Parti I-IV Math. Ann. 151 (1963), 1-41.

[5] H. Busemann, G. Ewald e G. C. Shephard, 151 (1963), 1-41. Parte V Archiv der Mathematik, 13 (1962), 512-526.

[6] H. Busemann e E. G. Straus, *Area and Normality,* Pacific J. Math. 10 (1960) 35-72.

[7] C. Carathéodory, *Math. Schriften vol. I,* München 1954.

[8] H. Minkowski, *Theorie der konvexen Körper, insbesondre Begründung ihres Oberflächenbegriffs,* Ges. Abb. vol. 2 (Leipzig und Berlin 1911), 131-229.

[9] J. Radon, *Über eine besondere Art ebener konvexer Kurven,* Ber. Sächs. Akad. Wiss. Leipzig 68 (1916), 131-134.

[10] G. C. Shephard, *Convex Bodies and Convexity on Grassmann Connes.* Parte VI in Jour. London Math. Soc. 39 (1964).

[11] V. Wagner, *Geometry of a space with areal metric and its applications to the calculus of variations* (Russo, riassunto inglese) Mat. Sbornik N. S. 19 (41) (1946), 341-404.

CONVEX BODIES AND CONVEXITY ON GRASSMANN CONES XI
SUBLINEAR FUNCTIONS

HERBERT BUSEMANN

1.

A real valued function $f(x)$ defined on a non-empty set M of the n-dimensional affine space A^n is *sublinear* if

$$(1)\qquad f(x) \leqq \sum_{j=1}^{k} \lambda_j f(x_j) \qquad \text{for } x_j \in M,\ x = \sum_{j=1}^{k} \lambda_j x_j \in M,\ \lambda_j \geqq 0, \sum_{j=1}^{k} \lambda_j = 1\,,$$

where k is arbitrary. The function $f(x)$ is *sublinear of order* m if (1) holds for $k=m$ and hence $k \leqq m$.

If M is a convex set then sublinearity of order 2 means that $f(x)$ is a convex function and therefore has the two properties:

1) $f(x)$ is sublinear and

2) $f(x)$ is linearly bounded below, i.e., an affine (a not necessarily homogeneous linear) function $L(x)$ in A^n exists with $L(x) \leqq f(x)$ on M.

For arbitrary M the function $f(x)$ is called *convex* if it has the properties 1) and 2), which is equivalent to requiring that $f(x)$ can be extended to a convex function on the convex hull M^H of M, see [2].

Here we treat the basic questions with a new method and study in particular the implications of 1) alone. For open M^H it is shown in [2] that for an $f(x)$ which is linearly bounded below sublinearity of order $n+1$ implies sublinearity. Here we prove

THEOREM 1. *If $f(x)$ is sublinear of order $n+1$ then it is sublinear.*

THEOREM 2. *For some fixed t, $1 \leqq t \leqq n-1$, and each point $x \in M$ let a t-flat T_x through x exist such that $M \cap T_x = M^H \cap T_x$. If $f(x)$ is defined on M and sublinear of order $n-t+1$ then it is sublinear.*

Note that $M \cap T_x = M^H \cap T_x$ holds when $T_x \subset M$.

A simple observation which was overlooked in [2] shows that 1) implies 2) in all interesting cases.

Received January 27, 1968.

This work was supported by a grant from the National Science Foundation, U.S.A.

Theorem 3. *If M contains a relative interior point of M^H and $f(x)$ is sublinear on M then $f(x)$ is convex.*

Thus it would be preferable to define convexity only for functions defined on M with this property. We refrain from doing so in order to conform with the other parts of this series.

If M is a union of lines through a fixed point then it satisfies this hypothesis as well as that of Theorem 2 with $t=1$. Therefore any $f(x)$ which is absolutely homogeneous of degree $p \geqq 1$ and sublinear of order n is convex.

The analogous question for positive homogeneous functions requires some case distinctions, see Corollaries 2, 3.

2.

For brevity we write $\varrho \in \Delta_m$ resp. $\varrho \in \Delta_m{}^+$ to indicate that ϱ is an m-tuple of reals $\varrho_1,\ldots,\varrho_m$ with $\sum_{i=1}^m \varrho_i = 1$ and $\varrho_i \geqq 0$ resp. $\varrho_i > 0$. The proofs will be based on the following

Lemma. *In A^n let*

$$y = \sum_{j=1}^{k} \lambda_j y_j, \quad \lambda \in \Delta_k .$$

Then for a suitable r there are numbers $\theta \in \Delta_r{}^+$ and $\mu^i \in \Delta_k$, $i=1,\ldots,r$, such that

$$\sum_{i=1}^{r} \theta_i \mu_j^i = \lambda_j, \quad j = 1,\ldots,k , \tag{2}$$

and for each i

$$y = \sum_{j=1}^{k} \mu_j^i y_j \quad \textit{and the } y_j \textit{ with } \mu_j^i > 0 \textit{ are independent} . \tag{3}$$

The set C in R^k of all $\sigma \in \Delta_k$ for which $y = \sum_{j=1}^k \sigma_j y_j$ is compact and convex, hence is the convex hull of its extreme points. In particular

$$\lambda = \sum_{i=1}^{r} \theta_i \mu^i, \quad \theta \in \Delta_r{}^+, \quad \mu^i \text{ an extreme point of } C .$$

We claim that the μ^i satisfy the assertion. Since $\mu^i \in C$ implies the equation in (3) we must show that the y_j with $\mu_j^i > 0$ are independent. For simplicity let $\mu_j^i > 0$ for $j \leqq s$ and $\mu_j^i = 0$ for $j > s$. If, for example,

$$y_s = \sum_{j=1}^{s-1} \varrho_j y_j, \quad \sum_{j=1}^{s-1} \varrho_j = 1 ,$$

then

$$\mu_j^i + \varepsilon\varrho_j \geqq 0, \quad j=1,\ldots,s-1, \quad \mu_s^i - \varepsilon \geqq 0 \text{ for } |\varepsilon| \leqq \varepsilon_0$$

with a suitable $\varepsilon_0 > 0$ and

$$y = \sum_{j=1}^{s-1} (\mu_j^i + \varepsilon\varrho_j)y_j + (\mu_s^i - \varepsilon)y_s .$$

But this means that μ^i is not an extreme point of C.—For this proof, which is shorter than the original one, I am indebted to W. Fenchel.

3.

Theorem 1 follows at once from the Lemma. For, if $y = \sum_{j=1}^k \lambda_j y_j$, $y \in M$, $y_j \in M$, $\lambda \in \Delta_k$, we introduce θ, μ^i according to the Lemma. For a given i there are at most $n+1$ positive μ_j^i. If $f(x)$ is sublinear of order $n+1$ we conclude using (2) that

$$f(y) = \sum \theta_i f(y) \leqq \sum_{i,j} \theta_i \mu_j^i f(y_j) = \sum \lambda_j f(y_j) .$$

This proof shows that Theorem 2 will follow if

$$f(y) \leqq \sum_{j=1}^{s+1} \lambda_j f(y_j) \tag{4}$$

for $y = \sum_{j=1}^{s+1} \lambda_j y_j \in M$, $y_j \in M$, $\lambda \in \Delta_{s+1}^+$ and independent y_j.

The hypothesis of Theorem 2 contains (4) for $s \leqq n-t$. Let $n-t < s$ $(\leqq n)$. The y_j span a non-degenerate s-simplex S with y as interior point and S lies in an s-flat L. The t-flat T_y intersects L in a flat D with dimension

$$d \geqq s+t-n \geqq 1, \quad \text{whence} \quad s-d \leqq n-t . \tag{5}$$

For $d=t$ the flats T_y and L coincide and the assertion (4) follows from the convexity of $M \cap T_y$ and $n-t+1 \geqq 2$.

Let $d<t$. Then $T_y \cap S = D \cap S$ is a convex polyhedron, whose vertices $z_1,\ldots,z_q$ lie in $(s-d)$-faces of S and also in M, because $S \subset M^H$ and $M \cap T_y = M^H \cap T_y$. Then

$$y = \sum_{h=1}^{q} \mu_h z_h, \quad \mu \in \Delta_q , \tag{6}$$

and

$$f(y) \leqq \sum \mu_h f(z_h) \tag{7}$$

because $M \cap T_y$ is convex. Now, z_h lies on an $(s-d)$-face of S, so that

$$z_h = \sum_{j=1}^{s+1} \varrho_j^h y_j, \quad \varrho^h \in \Delta_{s+1}, \text{ with at most } s-d+1 \text{ positive } \varrho_j^h .$$

The hypothesis, (5), (6) and (7) yield

$$f(y) \leqq \sum_{h=1}^{q} \mu_h \sum_{j=1}^{s+1} \varrho_j^h f(y_j), \quad y = \sum_{j=1}^{s+1} \left(\sum_{h=1}^{q} \mu_h \varrho_j^h \right) y_j .$$

The λ_j in (4) are uniquely determined by y and the y_j, whence $\sum_h \mu_h \varrho_j^h = \lambda_j$, and (4) follows.

4.

Let $f(x)$ be defined on the arbitrary nonempty set M in A^n. For $x \in M^H$ put as in [2]

$$(8) \qquad g_f(x) = \inf_{x=\Sigma\lambda_j x_j} \sum_{j=1}^{k} \lambda_j f(x_j), \quad \lambda \in \Delta_k, \; x_j \in M ,$$

where k is arbitrary, possibly $g_f(x) = -\infty$. Applying the Lemma to x and x_j we find

$$\sum_{j=1}^{k} \lambda_j f(x_j) = \sum_{j=1}^{k} \sum_{i=1}^{r} \theta_i \mu_j^i f(x_j) \geqq \min_i \sum_j \mu_j^i f(x_j)$$

and therefore

$$(9) \quad g_f(x) = \inf_{x=\Sigma\lambda_j x_j} \sum_{j=1}^{k} \lambda_j f(x_j), \quad \lambda \in \Delta_k^+, \; x_j \in M, \; \textit{with independent } x_j .$$

In particular we see, as proved in [2], that it suffices in (8) to take $k = n+1$.

It is easy to see (comp. [2]) that $g_f(x)$ is convex if finite. On the other hand $g_f(x)$ is an extension of $f(x)$ if $f(x)$ is sublinear. To prove Theorem 3 it therefore suffices to show that $g_f(x)$ is finite if $f(x)$ is sublinear and M contains a relative interior point y of M^H.

Because of sublinearity $g_f(y) = f(y) > -\infty$. Let x be any other point of M^H. Since y is a relative interior point of M^H a $z \in M^H$ and $0 < \theta < 1$ with $y = (1-\theta)x + \theta z$ exist. If

$$x = \sum_{i=1}^{h} \lambda_i x_i, \quad x_i \in M_j, \; \lambda \in \Delta_h ,$$

$$z = \sum_{j=1}^{k} \mu_j z_j, \quad z_j \in M, \; \mu \in \Delta_k ,$$

then by sublinearity

$$f(y) \leqq (1-\theta) \textstyle\sum_i \lambda_i f(x_i) + \theta \sum_j \mu_j f(z_j)$$

or

$$\Sigma_i \lambda_i f(x_i) \geqq (1-\theta)^{-1}[f(y)-\theta \Sigma_j f(z_j)] ,$$

so that $g_f(x)$ is finite.

5.

The hypothesis that M contains relative interior points of M^H is trivially satisfied in the frequently occurring case where M^H is open. It also holds when M is a union of t-flats through a fixed point w. For, assuming, without loss of generality, that $\dim M^H = n$, no supporting hyperplane of M at w exists, so that w is an interior point of M^H. Moreover M satisfies the hypothesis of Theorem 2. Therefore:

COROLLARY 1. *If M is a union of t-flats through a fixed point and $f(x)$ is defined on M and sublinear of order $n-t+1$, then $f(x)$ is convex.*

THEOREM 4. *Let M be a union of rays with origin w and let $f(x)$ be defined on M and sublinear of order n.*

If M possesses a supporting hyperplane then $f(x)$ is sublinear but in general not convex. If M possesses no supporting hyperplane and

$$(10) \qquad f(w) \leqq \sum_{j=1}^{n+1} \mu_j f(y_j) \quad \text{for} \quad w = \sum_{j=1}^{n+1} \mu_j y_j,\ y_j \in M,\ \mu \in \Delta_{n+1}^{+}$$

and independent y_j, then $f(x)$ is convex. The condition (10) *cannot be omitted.*

Note that (10) holds when $f(w) = \min f(x)$.

A supporting plane of M passes through w so that in the first case w is not an interior point of a nondegenerate n-simplex with vertices in M. Because of the Lemma it suffices to prove (4) which holds for $s<n$ since we assume sublinearity of order n. Let $s=n$. Then y is an interior point of the n-simplex S_n with vertices $y_1,\ldots,y_{n+1}$. The ray from w through y intersects the boundary of S_n in two points z_1, z_2 which lie on $(n-1)$-faces of S_n and the proof proceeds as that of Theorem 2.

That $f(x)$ need not be convex (when M does not contain an interior point of M^H) is shown in Section 6.

In the second case w is an interior point of M. It suffices again to prove (4) for $s=n$. For $y=w$ this is (10). Let $y \neq w$ and use the previous notation. If w is not an interior point of S_n, then (4) follows as before.

If w lies in the interior of S_n the ray from w through y intersects an $(n-1)$-face of S_n in a point z, so that

Math. Scand. 24 (1969), 93–101

$$z = \sum_{j=1}^{n+1} \varrho_j y_j, \quad \varrho_j \in \Delta_{n+1}, \text{ at least one } \varrho_j = 0\,.$$

We express w in the form (10) and obtain with a suitable $0 < \theta < 1$

$$\begin{aligned} y &= (1-\theta)w + \theta z = (1-\theta)\,\textstyle\sum_j \mu_j y_j + \theta\,\textstyle\sum_j \varrho_j y_j\,, \\ f(y) &\leqq (1-\theta)f(w) + \theta f(z) \\ &\leqq (1-\theta)\,\textstyle\sum_j \mu_j f(y_j) + \theta\,\textstyle\sum_j \varrho_j f(y_j) = \textstyle\sum_j \lambda_j f(y_j)\,, \end{aligned}$$

because y and y_j determine the λ_j uniquely.

That $f(x)$ need not be convex without (10) is clear: Let M consist of the nonnegative x^i-axes R_i, $i=1,\ldots,n$, and the ray R: $x^1 = \ldots = x^n$, $x^i \leqq 0$. Put $f(x)=0$ on $\bigcup R_i$ and $f(x)=x^1$ on R.

The function $f(x)$ is *absolutely homogeneous of degree* $p>0$ if $0 \in M \neq \{0\}$, $\alpha M = M$ for $\alpha \neq 0$ and $f(\alpha x) = |\alpha|^p f(x)$ for all α. For $p<1$ the function is not convex on any line through 0 unless it vanishes; therefore $f(x)$ is not sublinear of order 2 unless $f(x) \equiv 0$.

The function $f(x)$ is *positive homogeneous of degree* $p>0$ if $0 \in M \neq \{0\}$, $\alpha M = M$ for $\alpha > 0$ and $f(\alpha x) = \alpha^p f(x)$ for $\alpha \geqq 0$. For $p<1$ the function is not sublinear of order 2 if it takes positive values and for $p>1$ if it takes negative values. In most applications also $\alpha M = M$ for $\alpha < 0$, but not always $f(x) = f(-x)$. If $\alpha M = M$ for $\alpha \neq 0$ then $p<1$ and sublinearity imply $f(x) \equiv 0$.

The most important special case of Corollary 1 is

Corollary 2. *$f(x)$ is convex if it is sublinear of order n and absolutely homogeneous of degree $p \geqq 1$ or is positive homogeneous of degree $p \geqq 1$ and $\alpha M = M$ also for $\alpha < 0$.*

Theorems 3, 4 imply

Corollary 3. *$f(x)$ is convex if it is sublinear of order n, positive homogeneous of degree $p>0$ and M possesses a supporting hyperplane and contains an interior point of M^H.*

$f(x)$ is convex if it is sublinear of order n and positive homogeneous of degree $p>1$ and M possesses no supporting plane.

$f(x)$ is convex if sublinear of order n, positive homogeneous of degree 1 and linearly bounded below.

In the second part we have $f(x) \geqq 0$ so that (13) holds. The third part contains Theorem (11) in [2, p. 7] where it is assumed that $M^H \setminus \{0\}$ (owing to a misprint [2] states M^H) is open. To prove it we observe that by hypothesis numbers $u_1,\ldots,u_n,c$ exist such that

$$u \cdot x - c = \textstyle\sum_i u_i x_i - c \leqq f(x) \quad \text{on } M\,.$$

Then $u \cdot x \leqq f(x)$, because $u \cdot x_0 = f(x_0) - \delta$, $\delta > 0$, would imply

$$\alpha u \cdot x_0 - \alpha\delta = \alpha f(x_0) = f(\alpha x_0) \geqq u \cdot (\alpha x_0) - c \quad \text{for all } \alpha > 0\,.$$

Therefore, if $0 = \sum_j \mu_j y_j$,

$$0 = u \cdot \textstyle\sum_j \mu_j y_j = \sum_j \mu_j (u \cdot y_j) \leqq \sum_j \mu_j f(y_j)$$

and (10) holds. If M possesses a supporting hyperplane we apply the second part of Theorem 4 and that $f(x)$ is linearly bounded below.

6.

The order of sublinearity in the hypotheses of Theorems 1, 2 *and Corollaries* 1, 2, 3 *cannot be lowered*, although one might have expected the contrary in Corollaries 2, 3 because their proofs do not fully use the homogeneity of $f(x)$. We show this with some simple *examples*.

Let M_0^n consist of the vertices $x_1, \ldots, x_{n+1}$ and an interior point x of a nondegenerate simplex. Put $f(x_j) = 0$, $f(x) = 1$; then $f(x)$ is sublinear of order n on M_0^n but not sublinear. For Theorem 2 take any $(n-t)$-flat F and a t-flat T' intersecting F in a point. In F take a set

$$M_0^{n-t} = \{y_1, \ldots y_{n-t}, y\}$$

and let M_t^n consist of the t-flats T_i, T parallel to T' through y_i, y. Put $f(x) = 0$ on $\bigcup T_i$ and $f(x) = 1$ on T. Then $f(x)$ is sublinear of order $n-t$ but not $n-t+1$.

Further examples are based on the following observation. Let $f(y)$ be defined on a set M' in the hyperplane $x^n = 1$ in A^n and extend $f(y)$ to the cone M consisting of the rays from 0 through points of M' by

$$f(\tau y) = \tau f(y) \quad \text{for } \tau \geqq 0,\ y \in M'\,.$$

For $0 = \sum_{j=1}^k \lambda_j x_j$, $\lambda \in \Delta_k^+$ and $x_j \in M$ we have $x_j = 0$. If $0 \neq x = \sum_{j=1}^k \lambda_j x_j$, $\lambda \in \Delta_k$, $x \in M$, $x_j \in M$ and $x = \tau y$, $x_j = \tau_j y_j$, $y \in M'$, $y_j \in M'$ then

$$y = \sum_{j=1}^k \tau^{-1} \lambda_j \tau_j y_j, \qquad \sum_{j=1}^k \tau^{-1} \lambda_j \tau_j = 1\,,$$

because $y^n = y_j^n = 1$. From the definition of $f(x)$

$$f(x) \lesseqqgtr \sum_{j=1}^k \lambda_j f(x_j)$$

is equivalent resp. to

$$f(y) \gtreqqless \sum_{j=1}^k \tau^{-1} \lambda_j \tau_j f(y_j)\,.$$

Thus, if $f(x)$ is linear resp. sublinear of order k on M' then so is $f(x)$ on M.

We use this first to show that $f(x)$ need not be convex in the first part of Theorem 4. Take M' as an $(n-2)$-dimensional ellipsoid and define $f(x)$ on M' in any way such that $\inf f(y) = -\infty$. Then $f(y)$ is sublinear on M', but not linearly bounded below. The same holds for $f(x)$.

Next take as M' the set M_{t-1}^{n-1}, $t \geqq 1$; then $f(x)$ is sublinear of order $n-t$ but not $n-t+1$. Thus we obtain an example for Corollary 1, and for $t=1$, since $f(x)$ is positive homogeneous of degree 1, also for the last part of Corollary 3.

Let L_i be the x^i-axes and L the line $x^1 = \ldots = x^n$, $n > 2$. Put

$$M = L_1 \cup \ldots \cup L_n \cup L ,$$

$f(x)=0$ on $\bigcup L_i$, $f(x) = |x^1|^p$, $p \geqq 1$, on L. Then $f(x)$ is sublinear of order $n-1$ but not n. This takes care of Corollary 2 and the second part of Corollary 3. Let M^* be the subset of M in $\{x \mid x^i \geqq 0\}$. Then the restriction of $f(x)$ to M^* settles the first part of Corollary 3 with $p \geqq 1$, and for $0 < p < 1$ we take $f(x) = -(x^i)^p$ on $L_i \cap M^*$, $f(x)=0$ on $L \cap M^*$.

7.

The theory of functions which are convex on a nonconvex set was developed specifically for the case, where the space is the linear space V_r^n (also considered as affine space) of all r-vectors over A^n (considered as vector space) and M is the Grassmann cone G_r^n of all simple r-vectors R (compare [2]). The maximal dimension of a flat contained in G_r^n is

$$t_r^n = \max(r+1, n-r+1) ,$$

see [1, p. 300], and each point R of G_r^n lies on a t_r^n-flat through the origin. Since $\dim V_r^n = \binom{n}{r}$ we deduce from Corollary 1:

Corollary 4. *A function $f(R)$ which is defined on G_r^n and sublinear of order $g_r^n = \binom{n}{r} - \max(r, n-r)$ is convex.*

In the applications $f(R)$ is positive (frequently absolutely) homogeneous of degree 1. According to a previous observation this alone would not lead us to expect that g_r^n can be replaced by a smaller number. However, there are infinitely many t_r^n-flats in G_r^n through each R and G_r^n has in other respects a very special structure. If we denote by o_r^n the smallest integer such that sublinearity of order o_r^n guarantees convexity for a function on G_r^n which is positive homogeneous of degree 1, then most likely $o_r^n < g_r^n$. Possibly o_r^n has a smaller order of magnitude than

g_r^n, that is, $o_r^n/g_r^n \to 0$ when $n \to \infty$ and $\max(r, n-r)/n < \theta < 1$. For various questions in the theory of convex bodies it would be important to know the precise value of o_r^n.

In the simplest case $n=4$, $r=2$ it is known that $o_2^4 > 2$ (see [2, p. 21[3]]) and Corollary 4 gives $o_2^4 \leqq g_2^4 = 4$. It seems probable that $o_2^4 = 3$.

REFERENCES

1. W. Burau, *Mehrdimensionale projektive und höhere Geometrie* (Math. Monographien 5), VEB Deutch Verlag d. Wissenschaften, Berlin, 1961.
2. H. Busemann, G. Ewald and G. S. Shephard, *Convex bodies and convexity on Grassmann cones*, Parts I–IV, Math. Ann. 151 (1963), 1–41.

UNIVERSITY OF SOUTHERN CALIFORNIA, LOS ANGELES, CALIF., U.S.A.

PART III
BUSEMANN'S PAPERS ON HILBERT'S PROBLEM IV

UDC 513.73+513.82

On Hilbert's fourth problem

Herbert Busemann

Uspehi Mat. Nauk **21** *(1966), no. 1, pp. 155-164*
Received on December 12, 1964.

translated from the Russian by Dmitriy Slutskiy,
Institut de Recherche Mathématique Avancée,
Université de Strasbourg,
7 rue René-Descartes, 67084 Strasbourg, France
email: slutskiy@math.unistra.fr

1 Accurate statement of the problem

By the time of Hilbert's report [Hil00] the following results related to Problem IV were known.

1) Beltrami's Theorem which states that a Riemannian metric on an open connected set in n-dimensional projective space P_n has constant curvature if its geodesics belong to projective (straigth) lines. Therefore, among all spaces of this type, the only spaces that are complete in their internal metrics are elliptic spaces defined on the whole P_n, Euclidean space defined on an affine space $A_n \subset P_n$, and hyperbolic spaces defined in the interiors of ellipsoids in A_n (see Cartan [Car46, Part IV] or [Bus55, §15]).

2) Darboux's local solution to the problem of determination of all subintegral functions (integrands) $f(x, y, y')$ of two-dimensional variational problems for which straight lines are extreme, but not necessarily the shortest [Dar94, p. 58].

3) Two special complete geometries, one of which was invented by Minkowski, the other — by Hilbert. They will be considered in §§ 2 and 3.

Hilbert considers the analysis of the concept of distance as part of the problem. If we accept the axiom of continuity (we will see that even in this case the problem remains extremely general), the distance will be a real number and it will be possible to use Fréchet's concept of a metric space R, modified so as to include non-symmetric metrics:

a) there is a nonnegative function xy defined on $R \times R$ and such that $xy = 0$ only if $x = y$;

b) $xy + yz \geq xz$;

c) $xx_\nu \to 0$ if and only if $x_\nu x \to 0$.

The space R or its metric are called symmetric if $xy = yx$ for all pairs x, y.

It is convenient to introduce the following notation:

$K(p, \varrho)$ is a sphere $\{x \mid px = \varrho\}$, $\varrho > 0$;
$B(p, \varrho)$ is a ball $\{x \mid px \leq \varrho\}$, $\varrho > 0$.

In the non-symmetric case, we also need to define

$$K'(p, \varrho) = \{x \mid xp = \varrho\},\ B'(p, \varrho) = \{x \mid xp \leq \varrho\}.$$

Since we are interested in the metrizations of subsets of P_n, the completeness is expressed in the fact that

d) *the balls* $B(p, \varrho)$ *are compact for all* p *and* ϱ. This *does not imply* the compactness of all $B'(p, \varrho)$.

An oriented *segment* $S(a, b)$ going from a to b is a curve with starting point a and endpoint b, and of the length ab. Therefore it can be represented in the form $x(s)$, $\alpha \leq s \leq \alpha + ab$, $x(s_1)x(s_2) = s_2 - s_1$ for $s_2 > s_1$ (see [Bus55] and [Zau59]).

The supports of our spaces R are nonempty open connected subsets D_n of the space P_n with the property that the intersection L of a projective line with D_n (if it is not empty) is either the whole projective line, or the projective line with a point or a closed interval removed. Such a point and the ends of such an interval are called the ends of L; they do not belong to our space. Let us call L *a (straight) line* in D_n.

The spaces serving as the subject of research in Problem IV are called *Desarguesian*. These spaces can be now defined as follows:

A Desarguesian space is a metric space R satisfying the conditions a) — d) *and*

e) *any two points can be connected by a segment* $S(a, b)$;

f) R *is a subset* D_n *of the space* P_n;

g) $xy + yz > xz$ *if* x, y, z *do not belong to the line* L *in* D_n.

This implies that the segment $S(a, b)$, $a \neq b$, either lies on a (unique) straight line $L(a, b)$ in D_n passing through a and b or $S(a, b)$ is the only segment leading from a to b, if L is not the entire projective line. Then it follows from d) that the line $L(a, b)$ oriented so that the point a is followed by b can be written in the form

$$x(s),\ \alpha < s < \infty,\ x(s_1)x(s_2) = s_2 - s_1 \text{ for } s_2 > s_1,$$

where α can take the value $-\infty$, and moreover, α must be equal to $-\infty$ if the ball $B'(p, \varrho)$ is also compact.

The problem of *determination of all possible D_n is fully resolved in a symmetric case*: according to Hamel [Ham03][1], either D_n is the whole P_n and all

[1]This and several other facts mentioned below (restricted sometimes to the symmetric situation or to the dimension condition $n > 2$) are special cases of more general theorems

straight lines are isometric to one (ordinary) circle, or D_n is a convex subset of the space A_n and all straight lines are isometric to an ordinary straight line. Any open convex subset of A_n can serve as D_n (see [Bus55], (18.14)). It follows from Hamel's Theorem that the subset D_2 of the space P_2, which consists of the points of intersection of two different tangent lines of a fixed conic section, can not be metrized by a symmetric distance satisfying conditions a) — g).

It would be nice to *characterize our spaces with the help of some internal property* instead of immersing them into P_n. In the two-dimensional situation, the geodesics satisfy Desargues' Theorem as well as the inverse theorem in such cases when the intersections required for the formulation of Desargues' Theorem exist. For a space where geodesics exist in some general sense, the above-mentioned condition is also sufficient for immersing. In the case $n > 2$, it suffices that planes containing three given points exist. In the symmetric case the proofs are given in [Bus55] (§§13, 14). This justifies the name "Desarguesian space".

There is *a wide variety* even of symmetric metrizations of a given (admissible) D_n. For instance, if a given domain is A_2 then there are infinitely many essentially different ways of metrization of A_2 viewed as a Desarguesian space with a symmetric metric when the additional restrictions are satisfied: all rotations exist around a fixed point, and, for any pair of parallel straight lines L, L', the distance xL' goes to 0 (or always tends to ∞), if x runs over L in an arbitrary direction [Bus55, §11a].

There can be no doubt that Hilbert did not know about a wide variety of Desarguesian spaces when he stated Problem IV and that the "determination and study of" all Desarguesian spaces (or of all such spaces with symmetric metrics) proposed by him is not a fruitful task. That is why the major researchers in the field, except Hamel alone, considered special Desarguesian spaces.

2 The Minkowski geometry

A Desarguesian metrization of a space A_n which is invariant under parallel translations of the space A_n is called a *Minkowski* geometry, metric or space. It suffices to require that the distance between two different points x, y to be invariant with respect to a translation along the straight line $L(x, y)$ (see [Bus55], §17; the proof is also valid for a non-symmetric case). The balls $B'(p, \varrho)$ are then compact.

about the spaces that are either not Desarguesian or their Desarguesity follows from other hypotheses. These more general results are given in [Bus55] and also in Zaustinsky's paper [Zau59]. This concerns here Theorem (31.2) of the book [Bus55].

The Minkowski spaces are the most important of Desarguesian spaces. They were introduced by Minkowski [Min68] because of their applications in number theory; in the case of symmetric metrics and under some weakening of the condition g) they are finite-dimensional Banach spaces. In the differential geometry of Finsler spaces with a linear element $ds = F(x^1, ..., x^n, dx^1, ..., dx^n) = F(x, dx)$ satisfying certain standard conditions (see, for instance, [Bus55], §14), the Minkowski spaces play role of *local spaces* which is similar to the role of Euclidean spaces in Riemannian geometry: the linear element $ds_m = F(x_0, dx)$ defines the Minkowski metric that approximates the Finsler metric in the neighborhood of x_0. Thus, all smooth Desarguesian spaces locally behave as Minkowski spaces.

We can only mention a few of the many characteristics of Minkowski spaces. They are the only Desarguesian metrizations of A_n with compact $B'(p, \varrho)$ such that the spheres $K(p, \varrho)$ and $K'(p, \varrho)$ are convex hypersurfaces (the proof is very easy in the symmetric case[2]; see [Bus55], (24.1); and slightly more difficult in the general case; see [Hil00], (10.3)).

The existence of *similarities*, i.e. of applications $x \to x'$ of a space onto itself such that $x'y' = \varepsilon xy$, $\varepsilon > 0$, $\varepsilon \neq 1$, characterizes Minkowski geometry only under certain assumptions of differentiability[2] (see [Bus55], §34a).

In an arbitrary Desarguesian space defined in $D_n \subset A_n$, let us consider an oriented straight line L^+, a fixed point q on L^+, and a variable point moving along L^+ in a positive direction to the endpoint u of L^+. Then there exist the *"limit spheres"* [2]

$$\Lambda'(q, L^+) = \lim_{x \to u} K'(x, qx), \ \Lambda(q, L^+) = \lim_{x \to u} K(x, xq).$$

Note that $qx \to \infty$, although $xq \to \infty$, in general, only for a compact $B'(p, \varrho)$.

In Minkowski geometry, the limit spheres $\Lambda'(q, L^+)$ and $\Lambda(q, L^+)$ are both hyperplanes if the spheres $K(p, \varrho)$ and, consequently, $K'(p, \varrho)$ are differentiable, and this property characterizes the Minkowski geometries with differentiable spheres[2]. It is interesting that just the fact that $\Lambda'(q, L^+)$ is a plane does not imply that the space we work with is a Minkowski geometry; see §4.

It is important to keep in mind that *the concept of perpendicularity* in non-Riemannian geometries is, in general, *not symmetric.* For simplicity, let us consider only the symmetric case; the general case is described in [Zau59].

A straight line L_1 is perpendicular to a straight line L_2 at a point f if $L_1 \cap L_2 = f$ and $xf < xy$ for $x \in L_1$ and $y \in L_2 - \{f\}$. *The existence of a straight line L_1 perpendicular to a given straight line L_2 and passing through a given point $x \in L_2$ is equivalent to the (straight) convexity of the spheres* $K(p, \varrho)$footrefnote1 and, under the appropriate conditions of smoothness, it is equivalent to the absence of focal points in terms of the calculus of variations (see [Bus55], §§20, 25).

According to well-known Blaschke theorems ([Bla16], see also [Bus55], (18,25)), a Minkowski space where the perpendicularity of straight lines is sym-

[2] see Footnote[1]

metric, is Euclidean for $n > 2$, whereas there exist two-dimensional Minkowski geometries with symmetric perpendicularity that are not Euclidean. They were found by Radon ([Rad16], see also [Bus55], §17).

3 The Hilbert geometry

Before his report, Hilbert discovered a non-Riemannian analogue of hyperbolic geometry (see [Hil30], Annex I). Let C be a closed convex hypersurface in A_n with interior D_n. Given two different points x and y in D_n, let us denote the ends of the straight line $L(x,y)$ by the symbols u and v so that the order of points on $L(x,y)$ is $vxyu$ and let us suppose

$$h(x,y) = \alpha \ln R(x,y,u,v),\ \alpha > 0,$$

where R means the cross-ratio, i.e., if we introduce an auxiliary Euclidean metric $e(x,y)$ in A_n,

$$h(x,y) = \alpha\Big[\ln \tfrac{e(x,u)}{e(y,u)} + \ln \tfrac{e(y,x)}{e(x,v)}\Big] = h(y,x).$$

The function $h(x,y)$ satisfy Conditions a) — f), and Condition g) is verified if and only if a two-dimensional plane passing through a point in D_n never intersects C along a curve that contains two proper noncollinear segments.

The spheres are convex ([Bus55], §18), *therefore perpendiculars exist, but even for $n = 2$ perpendicularity of straight lines is symmetric only in the hyperbolic case* (Kelly and Page, [KP52]), i. e. when C is an ellipsoid.

The sets $\{x | xS \leq \delta\}$, where S is a segment in D_n, are convex. If this property is postulated locally to geodesic segments, then in the Riemannian case it is equivalent to the requirement of nonpositive curvature (Pedersen [Ped52], see also [Bus55], (41,6)). However, the condition that the segment joining the midpoints of the two sides of a triangle is not longer than half of the third side is valid only for the hyperbolic geometry (Kelly and Straus, [KS58]), although if we require this condition to be verified locally for the geodesic triangles in Riemannian spaces, it is equivalent to the non-positivity of curvature (see [Bus55], (41,6)).

The Minkowski geometry and the Hilbert geometry have a common elementary characteristic that was found by Berwald [Ber29] and Funk [Fun35]. Let us denote by P the following property of a Desarguesian space:

P. An isometric mapping from one oriented straight line L_1 to another line L_2 is induced by a projective mapping from L_1 to L_2.

The geometries of Minkowski and Hilbert are the only non-compact Desarguesian spaces with compact $B'(p,\varrho)$ which satisfy Property P.

The proof of Berwald and Funk uses the notion of curvature and hence it uses differentiability, whereas a geometric proof can be obtained immediately. The Minkowski and Hilbert geometries, with evidence, satisfy the property P.

An isometric map from L_1 onto L_2 must send an endpoint of the straight line L_2 to one of the endpoints of L_1, therefore P guarantees that either all straight lines have one endpoint or all lines have two endpoints, therefore, if we choose an appropriate plane at infinity then D_n either coincides with the whole A_n or D_n is the interiour of some convex hypersurface in A_n. Applying Property P to two intersecting lines L_1 and L_2 and also to an isometric map which sends the lines L_1 and L_2 to themselves, we obtain a perspectivity with center lying at infinity if $D_n = A_n$. This immediately leads to the desired result.

4 The Funk geometries

Using the same notation as in the definition of the function $h(x, y)$, Funk [Fun29] specifies an asymmetric distance in D_n by the formulas

$$f(x, y) = \alpha \ln \tfrac{e(x,u)}{e(y,u)},$$

$$f(y, x) = \alpha \ln \tfrac{e(y,v)}{e(x,v)}.$$

He calls this geometry "Geometrie der spezifischen Massbestimmung" (a geometry with a specialized definition of measure). The function $f(x, y)$ satisfies Conditions a) — f), but Condition g) is verified if and only if the hypersurface C is strictly convex; for more details see [Zau59] (Annex I). The spheres $B'(p, \varrho)$ are not compact.

If the hypersurface C is differentiable, then the limit spheres $\Lambda'(q, L^+)$ are flat, i. e. they lie on hyperplanes, more precisely, on the hyperphanes that are parallel to the hyperplane which is tangent to C at the positive endpoint u of the straight line L^+. The limit spheres $\Lambda(q, L^+)$ are homothetic to C.

From the definition of the sets $\Lambda'(q, L^+)$ as limits of spheres it follows that two distinct limit spheres

$$\Lambda_1'(q_1, L^+) = H_1 \text{ and } \Lambda_2'(q_2, L^+) = H_2$$

(q_1 lies between u and q_2) are equidistant in the sense that for a variable point $x_2 \in H_2$ the distance x_2H_1 is constant ($f(x_2, x_1) = x_2H_1$ if x_1 lies on the straight line with the endpoint u and which passes through x_2). In [Fun30] Funk proved (under some regularity conditions) for $n = 2$ (although the theorem is valid for any n) that only *the Minkowski metrics and also Funk metrics $f(x, y)$ lead to Desarguesian spaces where the locus of points equidistant to a hyperplane is a hyperplane.*

Let us now consider strictly convex hypersurfaces C, C' in A_n with interiors D, D' such that D and D' have nonempty intersections with D_n. Let us set Funk metrics $f(x, y)$ in D and $f'(x, y)$ in D' with the same α. Then the function

$$f_0(x, y) = f(x, y) + f'(x, y)$$

metrizes D_n to be a Desarguesian space which also becomes a Hilbert space if $C = C'$. We can obtain the distance $f(x, y)$ from $f_0(x, y)$ if we force C' to tend to the hyperplane at infinity.

5 Spaces of constant curvature

In this and the following sections, we assume that some appropriate regularity conditions are verified and some convex hypersurface has positive Gaussian curvature.

For each oriented linear element, or direction, (x, δ) of a Finsler space, there exists a contact Riemannian metric that approximates the Finsler metric in a neighborhood of the element (x, δ) (see Finsler [Fin18] and Berwald [Ber29]). A flat element π passing through x but not necessarily through δ determines the sectional curvature of the Riemannian metric, which is called "a measure of curvature" (Krümmungsmaß) of a Finsler space in (x, δ, π). Using a convolution we get *a scalar curvature* $K'(p, \delta)$ as in the Riemannian case. Berwald [Ber29] proved that *if the straight lines are extremals (not necessarily the shortest) then*

$$K(x, \delta, \pi) = K'(p, \delta)$$

for all $\pi \supset \delta$. *If the curvature* $K'(p, \delta)$ *does not depend on* δ *then it is constant, i.e.* $K'(p, \delta)$ *does not depend on* x *as well.*

For the spaces of constant curvature, Property P is different (see Berwald [Ber29] and Funk [Fun35]):

If the extremals are straight lines, then $K'(p, \delta)$ is constant if and only if the length-preserving mapping of one extremal to another (i.e. $\int F(x, dx)$) is induced by a projective transformation.

Therefore, *the Minkowski and Hilbert spaces are the only Desarguesian spaces in* $D_n \subset A_n$ *of constant curvature and with compact* $B'(p, \varrho)$.

The Minkowski space has zero curvature; the curvature of the Hilbert space is equal to $-\frac{1}{4}\alpha^2$. If we do not require the sets $B'(p, \varrho)$ to be compact, then we get (see Funk [Fun29] and Berwald [Ber29]):

Desarguesian geometries of constant negative curvature $-\frac{1}{4}\alpha^2$ *are the Funk geometries* $f(x, y)$ *and* $f_0(x, y)$. Thus, the Hilbert geometries are the only Desarguesian spaces with symmetric distance and of constant negative curvature.

There are many Desarguesian spaces of zero curvature with non-compact $B'(p, \varrho)$. They were found by Funk [Fun29] and Berwald [Ber29] and do not have any remarkable geometric properties; however, the relationship between the zero curvature and the arbitrariness in existence of spaces is interesting. Let D_n be the interior of a closed convex hypersurface C in A_n. Let us intruduce in A_n an auxiliary Euclidean metric $e(x, y)$ with a linear element ds_e. If (x, δ) is a given direction in D_n and the ray starting in x with the direction δ intersects C in u and has a (Euclidean) angle β with the tangent hyperplane C, then

$$ds(x,\delta) = e^{-2}(x,u)\sin^{-2}\beta g(u)ds_e,$$

where $g(u) > 0$ is an arbitrary function, if we only require the straight lines in D to be extremals. In case the space is Desarguesian, $g(u)$ satisfies some additional inequality. In the two-dimensional case, if we introduce a Euclidean arc length s_e in C as a parameter instead of u and if we denote by $R(s_e)$ a radius of curvature of C, this inequality has the form (Funk [Fun29]):

$$(g'' + gR^{-2})12g - (R'R^{-1}g - 3g')^2 > 0$$

(here the primes denote differentiation with respect to s_e).

Desarguesian spaces of constant positive curvature were found only for $n = 2$ by Funk [Fun29]. $D_2 = P_2$, but there are many non-symmetric as well as symmetric metrics besides the elliptic one, although $B'(p,\varrho)$ are compact with evidence. They do not have interesting geometric properties, and we do not list them.

6 The Hamel construction

Hamel's PhD thesis [Ham01], written under the supervision of Hilbert, appeared shortly after Hilbert's report and it was the first contribution to the solution of Problem IV. We consider the Hamel construction the latest, because it is based on an application to straight lines as extremals, and also on the Weierstrass theory, which was particularly interesting for Hilbert at that time, as some observations of his report show. In [Ham01] Hamel admits even more general distances than non-symmetric. For instance, in the plane corresponding to the linear element $ds(x,\beta)$ (here β is an angle) it is not required for $ds(x,\beta)$ to satisfy the "weak axiom of monodromy"

$$ds(x,\beta) = ds(x,\beta + 2\pi).$$

It corresponds to the optionality of the symmetry of the distance in our sense. But in [Ham03] Hamel is interested almost exclusively in the "strong axiom of monodromy"

$$ds(x,\beta) = ds(x,\beta + \pi)$$

or

$$F(x,dx) = F(x,-dx)$$

in the general case. This leads to a symmetric distance.

For $n = 2,3$, Hamel constructed *all linear elements for which straight lines are the shortest lines*, but he did not give conditions of the compactness of $B(p,\varrho)$ (except for the case $D_2 = P_2$ in [Ham03]). Apparently, this problem is extremely difficult. *The precise degree of arbitrariness* is the most interesting result of this research.

Suppose that in the plane (x, y) with the Euclidean line element

$$ds_e = (dx^2 + dy^2)^{\frac{1}{2}}$$

there is a closed convex curve, an indicatrix,

$$r = g^{-1}(\alpha, b)$$

for all points $(0, b)$ of an open subset of the y-axis. Here r, α are polar coordinates with the pole $(0, b)$ and

$$g(\alpha + \pi, b) = g(\alpha, b)$$

in the symmetric case. Let

$$w(\alpha, b) = \frac{d^2 g(\alpha, b)}{d\alpha^2} + g(\alpha, b).$$

Then in the symmetric case the linear element

$$ds = \frac{ds_e}{2} \int_{\beta-\pi}^{\beta} \sin(\beta - \alpha) w(\alpha, y - x \tan \alpha) d\alpha \tag{6.1}$$

defines a metric where the straight lines are the shortest lines, and any such metric can be written in the form (6.1). In the non-symmetric case this expression contains a function of two variables with a large arbitrariness. The case $D_2 = P_2$ can be obtained by passing to the limit.

Hamel solved the problem also for $n = 3$, and again, the precise degree of arbitrariness is the most interesting part here. In the space (x, y, z), for each point $(0, 0, c)$ of an open set on the z-axis we can specify an indicatrix, i.e. a closed convex surface, and for any point $(0, b, c)$ on the plane (y, z) there is an intersection of the indicatrix with the plane $z = c$ satisfying some obvious requirements of consistency for $b = 0$. In the symmetric case, the points $(0, 0, c)$ and $(0, b, c)$ must be the centers of symmetry of the surfaces and the lines, respectively.

Then, in the symmetric case, the linear element is uniquely determined by an expression that is structurally similar to the expression (6.1), but more complicated. In the non-symmetric case, this expression also includes a function of three variables with a wide arbitrariness.

Translated from English by A. I. Sirota.[3]

References

[Ber29] L. Berwald. Über die n-dimensionalen Geometrien konstanter Krümmung, in denen die Geraden die kürzesten sind. *M. Z.*, 30:449–469, 1929.

[3] The initial English manuscript is lost.

[Bla16] W. Blaschke. Räumliche Variationsprobleme mit symmetrischer Transversalitätsbedingung. *Leipz. Ber.*, 68:50–55, 1916.

[Bus55] H. Busemann. *The geometry of geodesics.* Pure and applied Mathematics, 6. New York: Academic Press, Inc. X, 422 p. , 1955.

[Car46] E. Cartan. *Lecons sur la géométrie des espaces de Riemann. 2 me ed.* Paris: Gauthier-Villars. VIII, 378 p. , 1946.

[Dar94] G. Darboux. *Leçons sur la théorie générale des surfaces et les applications géométriques du calcul infinitésimal. IIIe Partie. Lignes géodésiques et courbure géodésique. Paramètres différentiels. Déformation des surfaces. IVe Partie. Déformation infiniment petite et représentation sphérique.* Paris. Gauthier-Villars et Fils (1894 et 96) , 1894.

[Fin18] P. Finsler. *Über Kurven und Flächen in allgemeinen Räumen.* PhD thesis, Göttingen, Zürich: O. Füssli, 120 S. 8° , 1918.

[Fun29] P. Funk. Über Geometrien, bei denen die Geraden die Kürzesten sind. *Math. Ann.*, 101:226–237, 1929.

[Fun30] P. Funk. Über Geometrien, bei denen die Geraden die kürzesten Linien sind und die Äquidistanten zu einer Geraden wieder Gerade sind. *Monatshefte f. Math.*, 37:153–158, 1930.

[Fun35] P. Funk. Über zweidimensionale Finslersche Räume, insbesondere über solche mit geradlinigen Extremalen und positiver konstanter Krümmung. *Math. Z.*, 40:86–93, 1935.

[Ham01] G. Hamel. *Über die Geometrien, in denen die Geraden die Kürzesten sind.* Diss. Göttingen: Universitäts-Buchdruckerei (W. Fr. Kaestner). 92 S , 1901.

[Ham03] G. Hamel. Über die Geometrien, in denen die Geraden die Kürzesten sind. *Math. Ann.*, 57:231–264, 1903.

[Hil00] D. Hilbert. Mathematische Probleme. Vortrag, gehalten auf dem internationalen Mathematiker-Congress zu Paris 1900. 1900.

[Hil30] D. Hilbert. *Grundlagen der Geometrie.* 7. Aufl. VII + 326 S. 100 Fig. Leipzig, B. G. Teubner (Wissenschaft und Hypothese Bd. 7) , 1930.

[KP52] P.J. Kelly and L.J. Paige. Symmetric perpendicularity in Hilbert geometries. *Pac. J. Math.*, 2:319–322, 1952.

[KS58] Paul Kelly and Ernst Straus. Curvature in Hilbert geometries. *Pac. J. Math.*, 8:119–125, 1958.

[Min68] Hermann Minkowski. *Geometry of numbers. (Geometrie der Zahlen.).* Bibliotheca Mathematica Teubneriana. 40. New York, NY: Johnson Reprint Corp. vii, 256 p. , 1968.

[Ped52] Flemming P. Pedersen. On spaces with negative curvature. *Mat. Tidsskr. B*, 1952:66–69, 1952.

[Rad16] J. Radon. Über eine besondere Art ebener konvexer Kurven. *Leipz. Ber.*, 68:123–128, 1916.

[Zau59] E.M. Zaustinsky. Spaces with non-symmetric distance. *Mem. Am. Math. Soc.*, 34:91 p., 1959.

Uspehi Mat. Nauk 21 (1966), 155–164

Journal of Geometry. Vol. 4/2 1974. Birkhäuser Verlag Basel

Spaces with homothetic spheres

by

Herbert Busemann

1. The Minkowskian geometry and the less familiar Funk geometry (see below) have homothetic balls $\{x; px \leq \rho\}$. It is the purpose of the present note to show that they are the only spaces with this property.

The precise definitions and hypotheses are as follows: The space is denoted by R.

a) A function xy is defined on $R \times R$ such that $xx = o$, $xy > o$ for $x \neq y$, $xy + yz \geq xz$, $x_\nu x \to o$ and $xx_\nu \to o$ are equivalent, the balls $B^+(p,\rho) = \{x; px \leq \rho\}$ (but not necessarily the $B^-(p,\rho) = \{x; xp \leq \rho\}$) are compact. Any two distinct points x,y can be connected by a (metric) segment T(x,y), i.e. a curve z(t) $(o \leq t \leq xy)$ satisfying $z(o) = x$, $z(xy) = y$ and

(*) $z(t_1)z(t_2) = t_2 - t_1$ for $t_1 < t_2$.

(b) R is a nonempty open convex set D in the n-dimensional affine space A^n (with affine coordinates x^i), and, of course, $x_\nu x \to o$ meaning $x^i_\nu \to x^i$, $i = 1,\ldots,n$) moreover $xy + yz > xz$ unless x,y,z are collinear. It follows that T(x,y) is unique and falls on the affine segment S(x,y) (i.e. $(1 - \beta)x + \beta y$, $o \leq \beta \leq 1$).

Calling line resp. hyperplane in D the intersection of an affine line resp. hyperplane with D if not empty, each oriented line L^+ can be written in the form z(t), $\alpha(z) < t < \infty$, where

$\alpha(z)$ may or may not equal $-\infty$, such that $z(t_2)$ follows $z(t_1)$ for $t_2 > t_1$ and (*) holds. $\alpha(z)$ depends on the specific representation $z(t)$ of L^+; the only matter of importance is that for a given L^+ either all $\alpha(z)$ are finite or all equal $-\infty$.

The conditions a) and b) define the noncompact desarguesian spaces. The compact desarguesian metrics are defined in entire projective spaces, see [4,p 37].

c) A map $x \to x'$ of A^n on itself is called homothetic resp. antihomothetic, if

$$x' = \lambda x + a \text{ with } \lambda > o \text{ resp. } \lambda < o$$

For $\lambda = 1$ it is a translation, for $\lambda > o$ but $\lambda \neq 1$ it is a dilation with "focus"[1] $f = (1-\lambda)^{-1}a$ and factor λ. If $\lambda < o$ it is a dilation with factor $-\lambda$ and focus f followed by the reflection in f.

We say that the spheres of a noncompact desarguesian space R defined in D are homothetic, if a compact set B in D exists such that the set of all balls $B^+(p,\rho)$ coincides with the set of all homothetic images of B in D.

d) In A^n introduce an auxiliary euclidean metric $\varepsilon(x,y)$ by choosing a positive definite quadratic form $\sum g_{ik}x^ix^k$ and putting $\varepsilon(x,y) = [\sum g_{ik}(x^i - y^i)(x^k - y^k)]^{1/2}$. For pairs of points x,y and x',y' which lie on parallel lines in A^n the ratio $\varepsilon(x,y):\varepsilon(x',y')$ is an affine invariant, in particular independent of the choice of the g_{ik}.

---

1) This unusual term is employed here instead of "center" to avoid confusion with centers of balls or spheres.

To obtain a Minkowski metric $\mu(x,y)$ in A^n select an n-dimensional, strictly convex, compact set B and an interior point p of B. Put $\mu(x,x) = o$ and for $x \neq y$ let the ray from p parallel to the ray $R(x,y)$ from x through y intersect the boundary of B in z. Then

$$\mu(x,y) = \varepsilon(x,y) : \varepsilon(p,z).$$

Obviously $B = B^+(p,1)$ and the balls are homothetic. The $B^-(p,\rho)$ are also compact and $\alpha(z) = -\infty$ for each line. $\mu(x,y) = \mu(y,x)$ for all x,y if and only if B has an affine center and this coincides with p.

e) Let D be the interior of a strictly convex closed convex hypersurface $C \subset A^n$. A Funk metric $\varphi(x,y)$, see [5], is defined in D by $\varphi(x,x) = o$ and for $x \neq y$ by

$$\varphi(x,y) = k \log \frac{\varepsilon(x,z)}{\varepsilon(y,z)}, \quad z = R(x,y) \cap C, \quad k > o .$$

A proof that $xy + yz > xz$ for noncollinear x,y,z can be found in [7, Appendix I], It is again obvious that the $B^+(p,\rho)$ are homothetic. The $B^-(p,\rho)$ are compact for fixed p and small ρ, but not for large ρ. Therefore $\alpha(z)$ is finite for all lines and $xy = yx$ for all x,y is impossible.

Whereas in a Minkowskian geometry the homothety mapping $B^+(p,\rho)$ on $B^+(q,\sigma)$ sends p into q, this is in general not the case in a Funk geometry.

In both geometries an isometry of one oriented line on another is (the restriction of) a projectivity, which is according to a theorem of Berwald [1], a simpler proof is found in [6], equivalent to constant curvature (inneres Krümmungsmass) when the customary regularity and smoothness conditions hold.

Our aim is:

<u>Theorem.</u> <u>A desarguesian space with homothetic spheres is a Minkowski or a Funk space.</u>

"Noncompact" is omitted because "homothetic" is meaningless in a compact desarguesian space.

2. We first show that the $B^+(p,\rho)$ are strictly convex (which would be trivial under the usual assumptions in Finsler spaces, because then small spheres, and hence by homothety all spheres are strictly convex, compare [3, p 162]).

The equivalence of $xx_\nu \to o$ and $x_\nu x \to o$, and the compactness of the $B^+(p,\rho)$ imply:

(1) <u>For each p there is a $\beta(p) > o$ such that $B^-(p,\rho)$ is compact for $\rho < \beta(p)$.</u>

A terminal foot or briefly t-foot (initial foot or briefly i-foot) f of a point p on a set M is defined by

$$f \in M \text{ and } pf = \min_{x \in M} px \quad (fp = \min_{x \in M} xp).$$

Because the $B^+(p,\rho)$ are compact each point has a t-foot on a set M which is closed in R, but not necessarily an i-foot. If $x \neq z$ and y is an interior point of S(x,z) we write (xyz). Then trivially, see [7, (8.5)] or [2, (7.9)]:

(2) <u>If f is a t-foot (i-foot) of p on M and (pqf) then f is the unique t-foot (i-foot) of q on M.</u>

Also trivial, but important are the following two statements:

(3) <u>If q,r lie on opposite sides of the hyperplane H and f is a t-foot of q on H and an i-foot of r on H then (qfr),</u>

(4) <u>If (pqb) then $B^+(q,qb) - \{b\} \subset \{x;px < \rho\}$ and $B^-(q,bq) - \{b\} \subset \{x;xp < \rho\}$</u> (compare [3,(22.12)]).

Putting, for $\rho > 0$,

$K^+(p,\rho) = \{x;px = \rho\}$, $K^-(p,\rho) = \{x;xp = \rho\}$

we have the following converse of the first part of (4) :

(5) <u>If $B^+(q,\sigma) - \{b\} \subset \{x;px < \rho\}$ where $b \in K^+(p,\rho) \cap K^+(q,\sigma)$, $\rho \geq \sigma$ and $p \neq q$ then (pqb).</u>

For assume the ray R(p,q) intersects $K^+(p,q)$ in $r \neq b$. Then (pqr) and R(p,q) intersects $K^+(q,\sigma)$ in a point s with (qsr) (there may be other intersections). But $pq + sr = pr - qs = \rho - \sigma$, hence $pq < \rho - \sigma$ and $pq + qb = pq + \sigma > \rho$.

(6) <u>Every point of a hyperplane H is t-foot of some point on a given side Q^+ of H</u> (i.e. one of the two open sets bounded by H in D) <u>and the points in H which are i-feet of points in Q^+ are dense in H.</u>

The second assertion follows from (1). For the first let $p \in Q^+$ and f be a t-foot of p on H. Then H is a supporting (hyper)plane of $B^+(p,pf)$. For a given $x \in H$ there is a set homotheric to $B^+(p,pf)$ in $Q^+ \cup H$ touching H at x. This set is a ball $B^+(q,qx)$, $q \in Q^+$, so that x is a t-foot of q.

(7) <u>If $f \in H$ is i-foot (t-foot) on H of p and q in Q^+ then $p = q$, (pqf) or (qpf).</u>

Denote the other side of H by Q^-. If f is i-foot of p on H then by (6) there is an $r \in Q^-$ whose t-foot on H is f. By (3) the points p,q,r,f are collinear

Assume (7) were false for t-feet or that p and q in Q^+ with the same t-foot f and not collinear with f exist. By (6) a homothety arbitrarily close to the identity will take $B^+(p,pf) \cup B^+(q,qf)$ into $B^+(p',p'x) \cup B^+(q',q'x)$, where $x \in H$ is close to f and i-foot of a point s in Q^-. Then p',q',x,s would be collinear, which is impossible since they are close to p,q,f respectively. A proof without this continuity argument can be obtained from (5).

(8) The t-foot of a point p on a hyperplane H is unique.

Assume p had two t-feet f and f' on H. The dilation with focus f taking f' into $f'' = (1-\delta)f + \delta f'$ $(0 < \delta < 1)$ takes $B^+(p,pf)$ into a ball $B^+(q,\sigma)$ which touches H at f and f". Choose δ so small that $\sigma < pf$. Then f and f" are t-feet of q on H, and (7) would yield (pqf), but f is by (2) the only t-foot of q.

(9) The balls $B^+(p,\rho)$ are strictly convex.

For any nonconvex or not strictly convex ball $B^+(p,\rho)$ would have a supporting plane touching it in two distinct points f,f' and these would be t-feet of p on H.

(10) Concentric balls $B^+(p,\rho)$ and $B^+(p,\sigma)$ are homothetic with focus p.

Let $\rho > \sigma$, take any line L through p and denote its intersection with $K = K^+(p,\rho)$ and $K' = K^+(p,\sigma)$ by r,r' and s,s' where (ps's) and pr'r). On S(p,r) there is by continuity a point q with qs' = s'r. Then $K^+(q,qr)$ passes through s' and by (4) contains $K^+(p,\sigma)$ in its interior except for s'. The homothety Ω sending K into $K^+(q,qr)$ has r as focus and takes L into itself.

The homothety Ω' taking $K^+(q,qr)$ into K' has focus s' and also leaves L fixed. Thus $\Omega'\Omega$ maps K on K' and L on itself.

Therefore K and K' have at s and s' parallel supporting planes, and this holds generally for the endpoints of any common radius of K and K', whence the assertion.[2)]

(11) Either $D = A^n$ or D is the interior of a strictly convex closed hypersurface C.

We show that there cannot be a ray $R \subset D$ and a segment $S(p,b)$ connecting the origin p of R to a boundary point b of D.

For, if (pxb) and $\varepsilon(x,b) \to 0$ then $px \to \infty$, so that $B^+(p,px)$ tends to the entire space and $\varepsilon(p,x') \to \infty$ when $x' = K^+(p,px) \cap R$. On the other hand $\varepsilon(p,x):\varepsilon(p,x')$ is independent of x because of (10) and $\varepsilon(p,x) < \varepsilon(p,b)$.

This proves that either $C = \emptyset$ hence $D = A^n$ or D is bounded. For any $p \in D$ the $K^+(p,\rho)$ tend for $\rho \to \infty$ to C, hence are homothetic to C which is therefore strictly convex.

Thus for bounded D the fact that $B^+(p,\rho)$ is homothetic to $D \cup C$ and hence convex is obvious to start with, and the convexity of the $B^+(p,\rho)$ trivially implies their strict convexity, see [2] or [7,p 37].

3. Let L^+ be an oriented line and let x,y follow p on L^+. By (10) the spheres $K^+(p,px)$ and $K^+(p,py)$ have at x and y parallel supporting planes H_x and H_y, moreover x,y are the t-feet of p

---

2) The underlying theorem on convex bodies reduces to the plane case; a method of proof is given in [2,p 414,footnote 3 to Section 16].

on H_x and H_y. The arguments for (7) show that for (qpx) the spheres $K^+(p,px)$ and $K^+(q,qx)$ have the same supporting planes at x. Thus:

(12) For a given oriented line L^+ there is at least one family H_x of parallel hyperplanes ($H_x \cap L^+ = x$) such that x is the t-foot on H_x of any point $p \in L^+$ preceding x.

In the case of bounded D the H_x are parallel to a supporting plane of C at $R(p,x) \cap C$.

With the same notations, if $y \in H_x$ there is a $K^+(q,qy)$ with q on the same side of H_x as p which has H_x as supporting plane at y. Let L_1^+ be the line through q and y so oriented that y follows q.

(13) L^+ and L_1^+ do not intersect.

This is clear for bounded D. For a proof in either case assume that L^+ and L_1^+ intersect at $r \in D$ and take the hyperplane H_r (parallel to H_x) through r and $a \in L^+$, $a_1 \in L_1^+$ preceding r on L^+ resp. L_1^+. Then r would by (12) be the foot of both a and a_1 on H_r contradicting (8). This implies

(14) No point has two i-feet on a hyperplane.

For if f and f' were i-feet of r on H, let (afr) and (a'f'r). By (3) the only points which have f(f') as t-foot on H lie on the ray R(f,a) (R(f',a')) which violates (13). Therefore:

(15) The compact $B^-(p,\rho)$ or $K^-(p,\rho)$ are strictly convex.

It also follows that, if $K^-(p,\sigma)$ is compact, H is a supporting plane of $K^+(p,\rho)$ at x and if (ypx) with $yp = \sigma$ then $K^-(p,\sigma)$ has at y a supporting plane parallel to H. Consequently:

(16) A ball $B^+(p,\rho)$ and a compact $B^-(p,\sigma)$ are antihomothetic with focus p.

Consider a hyperplane H, one of its sides, Q^+, and the lines which connect points in Q^+ to their t-feet on H. In the case of bounded D we know that the affine lines carrying the considered lines are concurrant, namely at the point of C where the supporting plane parallel to H on the opposite side of Q^+ touches C. We want to establish the analogous fact for $D = A^n$.

(17) For $D = A^n$, if H is a hyperplane, Q^+ a side of H then the lines connecting points of Q^+ to their t-feet on H are parallel.

Let L^+ be an oriented line and $p \in L^+$. For x preceding p on L^+ the spheres $K^+(x,xp)$ all have the same tangent cone T at p. This cone is enveloped by the hyperplanes on which p is a t-foot of x and hence is independent of x. If $\varepsilon(p,x) \to \infty$ then $K^+(x,xp)$ converge (because of (4)) monotonically to T, which is therefore the so-called limit sphere $K_\infty(L^+,p)$. For $y \in K^+(x,xp)$ and $y \to q \in T$ the segments $S(y,x)$ converge to the ray from q parallel to L^+ and the general theory of limit spheres[3] states that any point z on the same side of T as x has the intersection of the parallel to L^+ through z with T as t-foot on T.

If T is a hyperplane, i.e. if $K^+(x,xp)$ is differentiable at p, then this proves (17) for H = T and proves (17) in general if the $K^+(x,xp)$ are everywhere dofferentiable.

---

3) The theory is found in [2] and [7] under the hypothesis that the $B^-(p,\rho)$ are also compact, but this assumption does not enter the proof of the mentioned facts at all.

If not, we use that they are almost everywhere so (in the sense of the ordinary measure induced by $\varepsilon(x,y)$ on $K^+(p,\rho)$). Consider $K_p = K^+(p,\rho)$ and $K_q = K^+(q,\sigma)$. Map K_p on K_q by associating points in which parallel rays issuing from p and q (radii) intersect K_p resp K_q. Let S_p and S_q be the sets of $x_p \in K_p$ and $x_q \in K_q$ for which K_p, K_q are simultaneously differentiable at x_p resp. y_q. Then $K_p - S_p$ and $K_q - S_q$ have measure 0, hence S_p, S_q are everywhere dense on K_p, K_q.

By (12) and the first part of this proof K_p and K_q have parallel supporting planes at corresponding points of S_p and S_q. The continuity properties of supporting planes now yield (17). This implies:

(18) If $D = A^n$ then the balls $B^+(p,\rho)$ and $B^+(q,\sigma)$ as well as the compact $B^-(p,\rho)$ and $B^-(q,\sigma)$ are homothetic with p corresponding to q. Each $B^+(p,\rho)$ is antihomothetic to each compact $B^-(q,\sigma)$ with p corresponding to q.

Therefore the balls $B^+(p,\rho)$ and compact $B^-(p,\rho)$ are balls $B^+(p,\rho')$, $B^-(p,\rho'')$ for a suitable Minkowski metric if $D = A^n$ and balls $B^+(p,\rho')$, $B^-(p,\rho'')$ for a Funk metric in D if bounded. It remains to show that $\rho' = k\rho$ with k independent of ρ. To comprise both cases we use $\delta(x,y)$ for $\mu(x,y)$ resp. $\varphi(x,y)$ and first prove the lemma:

(19) If a,b,c,d,e lie in this order on a line, ab = de, cb = ab and $K^-(b,cb)$ is compact then $\delta(a,b) = \delta(d,e)$.

Choose p with (cpd) and pd = pc. Then $e \in K^+(p,pb)$, $c \in K^+(p,pd)$ and $a \in K^-(b,cb)$.

Since $K^+(p,pb)$ is also a $K^+_\delta(p,\rho)$ etc. we conclude $\delta(p,e) = \delta(p,b)$, $\delta(p,c) = \delta(p,d)$, hence $\delta(d,e) = \delta(c,b)$. Because $K^-(b,cb)$ is also $K^-_\delta(b,\sigma)$ we find $\delta(c,b) = \delta(a,b)$.

We can now show:

(20) If (abc) and $ab = bc$ then $\delta(a,b) = \delta(b,c)$; or, a midpoint b of a and c for xy is also one for $\delta(x,y)$.

On $T(a,b)$ choose points $a = x_1, x_2, \ldots, x_n = b$ $(n \geq 3)$ in this order such that $K^-(x_{i+1}, x_i x_{i+1})$ is compact and if $y_1 = b, y_2, \ldots, y_n = c$ are the points on $T(b,c)$ with $y_i y_{i+1} = x_i x_{i+1}$, the sphere $K^-(x_{i+1}, x_i x_{i+1})$ intersects $T(x_{i+1}, c)$ in a point z_i with $(x_{i+1} z_i y_i)$.

Then the points $x_i, x_{i+1}, z_i, y_i, y_{i+1}$ satisfy the hypotheses of (19), whence $\delta(x_i, x_{i+1}) = \delta(y_i, y_{i+1})$ and $\delta(a,b) = \delta(b,c)$.

Because of (20) there is for each oriented line L^+ a number $\gamma(L^+)$ such that

$$xy = \gamma(L^+)\,\delta(x,y), \text{ if } y \text{ follows } x \text{ on } L^+.$$

Since a $K^+(p,\rho)$ is also a $K^+_\delta(p,\rho')$ the value of $\gamma(L^+)$ is the same for L^+ through p, and since any two nonintersecting lines can be intersected by a third line, $\gamma(L^+)$ is constant. This proves the Theorem.

References

[1] L. Berwald, Über eine characteristische Eigenschaft der allgemeinen Räume konstanter Krümmung mit geradlinigen Extremalen, Monath.Math. Phys. 36 (1929) pp 315-330

[2] H. Busemann, Local metric geometry, Trans. Am. Math. Soc.56 (1944) pp 200-274

[3] H. Busemann, The Geometry of Geodesics, New York 1955

[4] H. Busemann, Recent Synthetic Differential Geometry, Ergebnisse Math. vol 54 (1970)

[5] P. Funk , Über Geometrien, bei denen die Geraden die kürzesten sind, Math. Ann 101 (1929) pp 226-237

[6] P. Funk, Über zweidimensionale Finslersche Räume, insbesondre über solche mit geradlinigen Extremalen und positiver konstanter Krümmung, Math. Zeitschr. 40 (1935) pp 86-93

[7] E. M. Zaustinsky, Spaces with nonsymmetric distance, Mem. Am. Math. Soc. No 34 (1959)

Herbert Busemann
Department of Mathematics
University of Southern California
Los Angeles, Cal. 90007
U. S. A.

(Eingegangen am 5. November 1973)

PLANES WITH ANALOGUES TO EUCLIDEAN ANGULAR BISECTORS

HERBERT BUSEMANN

To Werner Fenchel on his 70th birthday.

1.

In the euclidean plane the bisector of an angle in a triangle divides the opposite side in the ratio of the adjacent sides. Avoiding angular measure this may be expressed as the following property:

(P) *Inside a (nonstraight) convex angle with legs N_1, N_2 and vertex v there is a ray M with origin v such that any segment $T(a_1, a_2)$ with $a_i \in N_i$, $a_i \neq v$, intersects M in a point $b = b(a_1, a_2)$ for which the distances satisfy*

$$va_1 : va_2 = ba_1 : ba_2 .$$

P is so strong that one might expect it to distinguish the euclidean plane at least among all Desarguesian planes, i.e. the planes whose geodesics fall on the ordinary affine lines. Actually one verifies easily that *every Minkowski plane has property* P, see Note 1.

It is the purpose of this paper to show that P *characterizes the Minkowski planes among all Desarguesian planes and, if the* (by P convex) *circles are differentiable, even among all two-dimensional straight G-spaces,* which we will briefly call straight planes[1].

Whether the differentiability hypothesis is necessary is an open question closely related to a problem which has remained unsolved for more than forty years, namely, whether the Minkowski planes are the only straight planes with convex circles satisfying the parallel axiom.

2.

A straight plane R is given by a metrization xy of the ordinary plane such that any two distinct points a, b lie on exactly one curve, denoted as the *line* $L(a, b)$, which is isometric to the real axis, i.e. representable, given an arbitrary t_0, as $z(t)$ with

$$(1) \qquad z(t_0) = a, \quad z(t_0 + ab) = b, \quad z(t_1)z(t_2) = |t_1 - t_2|$$

[1] All concepts occurring here can be found in [1], but for the convenience of the reader we will briefly recapitulate in Section 2 the properties of straight planes used later.

Received August 26, 1974.

and $-\infty < t < \infty$. The arc from a to b on $L(a,b)$ is the *segment* $T(a,b)$ and is represented by (1) restricted to $[t_0, t_0 + ab]$.

Orient the line L and let $p \notin L$. If x traverses L in the positive direction then the line $L(p,x)$ tends to the *asymptote* A^+ to the positive orientation of L and is *independent of* p in the sense that any $q \in A^+$ also leads to A^+. The asymptote A^- to the negative orientation of L is defined analogously. If A^+ and A^- lie on the same line M, then M is called *parallel* to L. But notice that this will, in general, not imply that L is parallel to M, see [1, p. 141].

Convexity of curves or sets in R is defined in the usual way, but, of course, with respect to the $T(x,y)$ as segments. The *circle* $K(p,\varrho)$, $\varrho > 0$, is the locus $px = \varrho$. *The convexity of all circles is equivalent to the existence of exactly one foot f of a given point p on a given line L* (i.e. $pf = \min_{x \in L} px$, see [1, p. 121]).

If L^+ is an oriented line, $p \in L^+$ and x traverses L^+ in the positive sense then $K(x, xp)$ converges for $px \to \infty$ to the so-called *limit circle* $K_\infty(L^+, p)$ through p with central line L^+. Limit-circles $K_\infty(L^+, p_1)$ and $K_\infty(L^+, p_2)$, $p_i \in L^+$, are *equidistant*; they intersect any asymptote A to L^+ in points q_1, q_2 with $p_1p_2 = q_1q_2$ and q_i is the foot of q_j $(i \neq j)$ on $K_\infty(L^+, p_i)$, see [1, p. 138].

An *angular domain* D in R is defined as the convex set bounded by two distinct rays N_1, N_2 (a ray is a halfline) with the same origin v which do not form a line. A ray M in D with origin v which has the property P is called the *bisector of* D or also of N_1 *and* N_2. We say that R has the property P if every angular domain in R possesses a bisector.

3.

We begin by showing that P implies another property P′ from which our assertions will follow.

(2) *If the straight plane R has property* P *then it also satisfies*:

(P′) *The parallel axiom holds. For two distinct parallel lines L_1, L_2 there is a line L (parallel to the L_i) which contains the centers of all segments $T(p_1, p_2)$ with $p_i \in L_i$.*

Let L_1 be given and $p_2 \notin L_1$, $p_1 \in L_1$ and orient L_1. If x traverses L_1 in the positive direction, $L(p_2, x)$ tends to the asymptote A_2^+ to the positive orientation L_1^+ of L_1. The bisector of the domain bounded by the rays from x and through p_1, p_2 intersects $T(p_1, p_2)$ in a point y with

$$xp_1 : xp_2 = yp_1 : yp_2 .$$

The inequality $|xp_1 - xp_2| \leqq p_1p_2$ implies

$$xp_1 : xp_2 \to 1 \quad \text{as } p_1x \to \infty ,$$

therefore y tends to the center p of $T(p_1,p_2)$ and $L(y,x)$ to the asymptote A^+ to L_1^+ through p, see [1, p. 138].

If $q_1 \in L_1$, $q_2 \in A_2^+$ then by the same argument A^+ passes through the center q of $T(q_1,q_2)$. For, A^+ does not depend on p_1 and the asymptote to L_1^+ through q_2 is also A_2^+.

Also by the same argument, if x traverses L_1 in the negative direction then $L(p_2,x)$ tends to the asymptote A_2^- to the opposite orientation L_1^- of L_1 and $L(y,x)$ to the asymptote A^- to L_1^- through p.

Choose q_1 on L_1 so close to (but different from) p_1 that $L(q_1,p)$ intersects A_2^+ and A_2^- in points q_2^+, q_2^-. Then $T(q_1,q_2^+)$ and $T(q_1,q_2^-)$ have p as common center, so that $q_2^+ = q_2^-$ and A_2^+ and A_2^- fall on the same line L_2 parallel to L_1. Obviously L_2 is the only line through p_2 not intersecting L_1, and since L_1 and p_2 were arbitrary this implies the parallel axiom, in particular L_1 is also parallel to L_2.

The asymptotes A^+, A^- to L_1 through p then also fall on the same line L parallel to the L_i and L contains by the preceding discussion the centers of all $T(q_1,q_2)$ with $q_i \in L_i$.

(3) *If a straight plane has one of the properties* P, P′ *then its circles are convex.*

Because of (2) it suffices to prove that P′ implies the convexity of the circles, or according to Section 2, that a given point $p \notin L$ has exactly one foot f on L.

Let f be a foot of p on L and L_1 the parallel to L through p. The parallel L_0 to L through the center c of $T(p,f)$ bisects by (2) every segment leading from L to L_1, therefore p is a foot of c on L_1. Choose f_2 such that p is the center of $T(f,f_2)$. Since f is a foot of p on L, the point f_2 is a foot on the parallel L_2 to L through f_2. If c_2 is the center of $T(p,f_2)$ then, as before, p is a foot of c_2 on L_1.

If now p had another foot g on L then $T(p,g)$ would intersect L_0 in the center c_0 of $T(p,g)$ and p would be a foot of c_0 on L_1. This leads to a contradiction because c_2, p, c_0 are not collinear so that $T(c_2,c_0)$ intersects L_1 in $q \neq p$ and

$$c_2c_0 = c_2q + qc_0 < c_2p + pc_0 ,$$

but $c_2p \leqq c_2q$ and $pc_0 \leqq qc_0$.

THEOREM 1. *A Desarguesian straight plane is Minkowskian if, and only if, it has one of the properties* P, P′.

This follows from the fact, see [1, p. 144], that Desarguesian spaces with convex spheres satisfying the parallel axiom are Minkowskian.

Without the assumption that the space be Desarguesian we derive from the preceding arguments the following additional fact:

(4) *If a straight plane has one of the properties* P,P′, *then a family of parallel lines has a family of parallels lines as common perpendiculars.*

A *perpendicular* to L is a line H intersecting L at a point f such that every point of H has f as foot on L. The existence of a unique perpendicular to a given line through a given point (in two dimensions) follows from the convexity of the circles, see [1, pp. 121, 122].

With the same notation as in the proof of (3) the center of $T(c,f)$ has c as unique foot on L_0. By iterated bisecting and doubling of distances on $L(p,f)$ we obtain a set S dense on $L(p,f)$ such that $L(p,f)$ is perpendicular to every parallel to L through a point of S, so that by continuity $L(p,f)$ is perpendicular to every parallel to L.

Since perpendiculars to the same line do not intersect and the parallel axiom holds, the common perpendiculars must be parallel.

4.

This is as far as we have been able to proceed without differentiability assumptions or the Desargues property. The circles being convex they are differentiable (i.e. have a unique supporting line) except at an at most countable number of points.

(5) *In a straight plane with the parallel axiom and differentiable convex circles parallel lines are equidistant.*

If L is perpendicular to H we call H transversal to L. If $H \cap L = p$ and $x \in L$, $x \neq p$ then H is the supporting line of $K(x,xp)$ at p and hence the only transversal to L at P. This implies that transversals to a given line L are parallel. For, if two distinct transversals H_1, H_2 to L intersected at a point q then the parallel L_0 to L through q would be perpendicular to both H_1 and H_2 at q, and L_0 would have two transversals at q.

Let H be given and L perpendicular to H at p. Orient L obtaining L^+ and let x follow p on L^+. For $px \to \infty$ the circle $K(x,xp)$ tends to the limitcircle $K_\infty(L^+,p)$. On the other hand it follows from the general theory (see [1, p. 147]) that $K_\infty(L^+,p) = H$.

The limitspheres with L^+ as central line are therefore transversals of L. Consequently, see Section 2, they are equidistant and have the perpendiculars to H as common perpendiculars.

Math. Scand. 36 (1975), 5–11

THEOREM 2. *If the straight plane R has one of the properties* P, P′ *and differentiable circles then it is Minkowskian.*

The proof which follows can be simplified if it is known beforehand that a pair of mutually perpendicular lines exists, as it does in every Minkowskian geometry.

Let $u(x)$, $-\infty<x<\infty$, represent an arbitrary line M in the form (1). Denote the two sides of M by π^+ and π^-. To any point p we assign coordinates x,y as follows: x is determined by the intersection $u(x)$ of the transversal to M through p. For $p \in M$ put $y=0$. For $p \notin M$ let f be the foot of p on M and put $y=pf$ if $p \in \pi^+$, $y=-pf$ if $p \in \pi^-$. There is exactly one point with given coordinates x,y, because for a fixed x_0 the ordinate y increases monotonically from $-\infty$ to ∞ on the transversal $x=x_0$ to M, since the parallels to M are by (5) the lines $y=\text{const.}$

Let L be any line which is neither transversal nor parallel to M and $a_i \in L$ $(i=1,2,3)$ with $a_1a_2=a_2a_3$. If $a_i=(x_i,y_i)$ it is no restriction to assume that $x_1<x_2<x_3$ and $y_1<y_2<y_3$.

For $i=1,3$ let the parallels to M through a_i intersect the perpendicular to M through a_2 at b_i and let the transversals to M through a_i intersect the parallel to M through a_2 at c_i. Then $b_i=(w_i,y_i)$, $c_i=(x_i,z_i)$ with suitable w_i,z_i.

Because of (4) and P′ we have

$$b_1a_2 = a_2b_3 \quad \text{and} \quad c_1a_2 = a_2c_3 .$$

Moreover by (5) $c_1a_2=u(x_1)u(x_2)=x_2-x_1$ and similarly $a_2c_3=x_3-x_2$.

By definition $b_1a_2=y_2-y_1$, $a_2b_3=y_3-y_2$, so that

$$(y_1-y_2):(x_1-x_2) = (y_3-y_2):(x_3-x_2) . \tag{6}$$

Bisecting and doubling segments on L and using (6) we obtain a dense set S on L such that for $(x,y) \in S$ the relation

$$\frac{y-y_2}{x-x_2} = \frac{y_1-y_2}{x_1-x_2} \tag{7}$$

holds. Therefore (7) is the equation of L. The fact that all lines have linear equations is obviously equivalent to the Desarguesian character of R and the assertion follows from Theorem 1.

NOTE 1. To show that *a Minkowski plane has property* P denote (using the notations of P) by u_i the point $u_i \in N_i$ with $vu_i=1$ and select an ellipse with center v passing through u_1 and u_2. The ellipse is the unit-circle of a euclidean metric $\varepsilon(x,y)$ (invariant under translations). Let M be the angular bisector of N_1 and N_2 with respect to $\varepsilon(x,y)$. Then

$$\varepsilon(v,a_1):\varepsilon(v,a_2) = \varepsilon(b,a_1):\varepsilon(b,a_2) .$$

Now $\varepsilon(v,u_i)=vu_i=1$ implies $\varepsilon(v,a_i)=va_i$ and $\varepsilon(b,a_1):\varepsilon(b,a_2)=ba_1:ba_2$ holds because a_1,b,a_2 are collinear.

Note 2. Since the question probably occurred to the reader we observe:

(8) *If in a straight plane the angles of a triangle possess bisectors then these are concurrent.*

The proof is obvious: If abc is the triangle and the bisectors of the angles at a and b intersect at p, put $L(c,p)\cap T(a,b)=q$. Then

$$cp:qp = ac:aq=bc:bq ,$$

hence

$$ac:bc = aq:bq ,$$

so that q and hence p lies on the bisector of the angle at c.

Note 3. The condition P′ can be replaced by a seemingly much weaker one.

(9) *In a straight plane* P′ *is equivalent to*:
(P″) *For any line L and any point $p \notin L$ the centers of the segments $T(p,x)$, $x \in L$, lie on a line L_1.*

That P′ implies P″ is obvious. Let P″ hold and orient L obtaining L^+. When x traverses L^+ in the positive direction L_1 intersects $T(p,x)$ in its center x_1 and the line $L(p,x)$ tends to the asymptote M to L^+ through p. Either M intersects L_1 or M is asymptote to the induced orientation $L_1{}^+$ of L_1. The first case is impossible, because px_1 would tend to a finite limit whereas $px \to \infty$.

Repeating the argument yields a sequence of lines L_i, $i=1,2,\ldots$, with induced orientations $L_i{}^+$ such that L_{i+1} contains the centers x_{i+1} of all $T(p,x_i)$, $x_i \in L_i$, and M is asymptote to all $L_i{}^+$. Consequently L_i converges to M. Since L_i is independent of the orientation L^+ of L, the line M is also an asymptote to the opposite orientation of L, hence parallel to L. Because L and p were arbitrary, the parallel axiom holds, in particular L_1 is parallel to L and M.

Let $x \in L$, then the centers of all $T(x,y)$, $y \in M$, lie on a line $L_1{}'$, so $L_1{}'$

passes through the center x_1 of $T(p,x)$, whence $L_1'=L_1$. Therefore L_1 contains the centers of all $T(x,y)$, $x \in L$, $y \in M$, and P′ follows.

Thus we have as a corollary of Theorem 2 and (9)

THEOREM 3. *If, in a straight plane, for a given line L and a given point* $p \notin L$ *the centers of the segments* $T(p,x)$, $x \in L$, *lie on a line and the* (by (3) and (9) convex) *circles are differentiable, then the metric is Minkowskian.*

REFERENCE

1. H. Busemann, *The Geometry of Geodesics*, Academic Press, New York, 1955.

UNIVERSITY OF SOUTHERN CALIFORNIA,
LOS ANGELES, CALIFORNIA, U.S.A.

REMARK ON "PLANES WITH ANALOGUES TO EUCLIDEAN ANGULAR BISECTORS"

HERBERT BUSEMANN

The paper [1] mentioned in the title ends with a theorem which is suggested by its context but unsatisfactory by itself, because it distinguishes the value $\frac{1}{2}$, whereas any positive $\varrho \neq 1$ seems equally natural. We will confirm this here by briefly indicating how [1], which we assume to be at the reader's elbow, must be modified to yield the

THEOREM. *Let $\varrho > 0$, $\varrho \neq 1$, and a straight plane R be given. For any point p and line L in R, $p \notin L$, denote for $x \in L$ by x' the point on the ray from p through x with $px' : px = \varrho$. If the x' lie on a line and the* (automatically convex) *circles are differentiable then the metric is Minkowskian.*

The assumption that the circles be differentiable can be replaced by requiring that R be *Desarguesian*, see [1].

Since the cases ϱ and $1/\varrho$ are easily seen to be equivalent we assume $0 < \varrho < 1$, so that x' lies on the segment $T(p,x)$. The assertion will follow through obvious modifications of [1] from a generalization P_ϱ' of P' in [1].

P_ϱ': *R satisfies the parallel axiom. If L_0 and L are* (distinct) *parallel lines then for $x_0 \in L_0$ and $x \in L$ the x' on the segments $T(x_0,x)$ with $x_0x' : x_0x = \varrho$ lie on a line L'.*

(P' is the case $P_{\frac{1}{2}}'$). The parallel axiom is proved exactly as in [1, (9)].

With $p \notin L$ and L' the line belonging to p and L by the hypothesis, let L_0 be parallel to L through p. We must show that $x_0x' : x_0x = \varrho$ when $x_0 \in L_0$, $x \in L$, and $T(x_0,x)$ intersects L' at x'. Here is the *only instance* where [1] uses $\varrho = \frac{1}{2}$ in an essential way. This can be avoided by applying the parallel axiom. When q traverses the ray from x' through x_0 then $qx' : qx$ increases from 0 to 1, so that L' is the line of the hypothesis corresponding to q and L for exactly one position of q on the ray. We show that this must be x_0.

Let $q \notin L_0$. Then $L(p,q)$ intersects L and L' (by the parallel axiom) in points u and u' with $pu' : pu = \varrho$. As y traverses the ray from u' through

Received October 15, 1975.

p the value $yu':yu$ increases from 0 to 1 and hence equals ϱ only at p and not at q.

NOTE ADDED IN PROOF. A paper by B. B. Phadke which will appear in this journal entitled *The theorem of Desargues in planes with analogues to Euclidean angular bisectors* implies that the assumption that the circles be differentiable can be omitted in the above theorem.

REFERENCE

1. H. Busemann, *Planes with analogues to Euclidean angular bisectors*, Math. Scand. 36 (1975), 5–11.

UNIVERSITY OF SOUTHERN CALIFORNIA,
LOS ANGELES, CALIFORNIA, U.S.A.

PROBLEM IV: DESARGUESIAN SPACES

Herbert Busemann

The Fourth Problem concerns the geometries in which the ordinary lines, i.e. lines of an n-dimensional (real) projective space P^n or pieces of them are the shortest curves or geodesics. Specifically, Hilbert asks for the construction of all these metrics and the study of the individual geometries. It is clear from Hilbert's comments that he was not aware of the immense number of these metrics, so that the second part of the problem is not a well posed question and has inevitably been replaced by the investigation of special, or special classes of, interesting geometries. Already the historically first (non-Riemannian) of these metrics, the Minkowskian, exhibited the phenomenon of non-symmetric distances.

In giving a precise modern version of Problem IV we therefore begin with the definition of a not necessarily symmetric metric space R:

A distance xy is defined on $R \times R$ with $xx = 0$, $xy > 0$ if $x \neq y$, $xy + yz \geqq xz$, and $xx_\nu \to 0$ if and only if $x_\nu x \to 0$.

There are two types of balls:

$$B^+(p,\rho) = \{x : px \leqq \rho\}, \quad B^-(p,\rho) = \{x : xp \leqq \rho\} \quad (\rho > 0).$$

For a symmetric distance we use $B(p,\rho)$. A segment $T(x,y)$ from x to y is a curve from x to y isometric to an interval of length xy on the real axis, i.e. representable as

$$z(t), \ \alpha \leqq t \leqq \beta = \alpha + xy, \ z(\alpha) = x, \ z(\beta) = y \text{ and}$$
$$z(t_1)\, z(t_2) = t_2 - t_1 \text{ if } t_1 < t_2.$$

R is an n-dimensional Desarguesian space if R is a nonempty open subset of P^n, metrized by a not necessarily symmetric distance xy (with the correct topology, of course) such that

(a) The $B^+(p,\rho)$, but not necessarily the $B^-(p,\rho)$, are compact.

(b) $T(x,y)$ exists for any x, y.

(c) $xy + yz > xz$ unless x, y, z are collinear in P^n.

Since $T(x,y)$ exists it must by (c) fall on the intersection of a projective line L with R and $L \cap R$ must be connected. Because R is an open subset of P^n either $L \cap R = L$ or $L \cap R$ is an open connected subset of L. In the sequel a line, plane, hyperplane etc. in R will always mean a nonempty intersection of a projective line, plane, hyperplane, etc. with R.

The subject is clearly related to three fields: the foundations of geometry on which

Hilbert put the principal emphasis, the calculus of variations and differential geometry.

Hilbert does not mention the latter for an obvious historical reason. The only metrics considered in differential geometry before 1900 were Riemannian, and for these the problem had been solved long before (1866) by Beltrami who showed that a Riemannian metric whose geodesics fall on ordinary lines has constant curvature. This in conjunction with the compactness of the $B(p, \rho)$ yields the three well known solutions: the entire P^n with an elliptic metric, the P^n with a hyperplane omitted, or the n-dimensional affine space A^n, with a euclidean metric and (with a proper choice of the ideal locus) the interior of an ellipsoid in A^n with a hyperbolic metric.

As to the calculus of variations, we have a so-called inverse problem. Usually an integrand F depending on point and direction is given and the curves giving the integral over F extremal values are sought. In an inverse problem the extremals are given (in our case the lines) and the corresponding F are to be found. The first contribution to Problem IV, Hamel's Thesis [1], was written under Hilbert's guidance and concentrates on the calculus of variations aspect with nonsymmetric distances. The later version [2] emphasizes principally symmetric distances.

From the point of view of the foundations of geometry it is mandatory to have an intrinsic geometric property of the geodesics which guarantees that they may be regarded as parts of projective lines, in fact, Hilbert's formulation implies this approach.

We keep all the axioms except the strong congruence axiom and replace this by the requirement that ordinary lines be the shortest connections. For $n > 2$ the situation is very simple. One axiom is that three points lie in a plane and this suffices. The space can then be mapped into P^n such that the given lines fall on projective lines. It is not necessary to postulate that any $r + 1$ points lie in a r-flat when $n - 1 \geqq r > 2$ because this follows.

For $n = 2$ everything lies in a plane and an additional requirement becomes necessary. The validity of the Theorem of Desargues and its converse in those cases where the necessary intersections exist (which we denote as the Desargues Property) is clearly necessary and proves also sufficient. Detailed proofs of these facts can be found, for instance, in [3, Sections 13 and 14]. It is a curious fact that Hilbert forgot to mention Desargues in his formulation of Problem IV, although he himself discovered and emphasized this role of the Desargues Theorem. The omission is repeated in [1], but corrected in [2].

The two-dimensional result is strongly used in the proof for $n > 2$, because a plane in a space of dimension exceeding 2 has automatically the Desargues Property. This justifies the brief term "Desarguesian Spaces" used here as title.

There are very few general theorems on Desarguesian spaces. The most important goes back to Hamel [2] and is formulated there only for symmetric distances, but it also holds in the general case, see [4, p. 37]:

If R is Desarguesian ($n \geqq 2$) then either $R = P^n$ and all lines have the same length (traversed either way) or R leaves out an entire hyperplane and may therefore be regarded as an open convex set in A^n. The theorem implies, for example, that the exterior of a conic in P^2 (i.e. the set of points lying on two tangents) cannot be metrized as a Desarguesian space.

This theorem exhibits one of two phenomena which apply to many statements in the theory, namely that the Desarguesian character is either not necessary or deducible from other hypothses. We give an example for the first.

Define a geodesic in a symmetric metric space as a locally isometric map of the entire real axis. If the $B(p,\rho)$ are compact and the geodesic through two distinct points is unique, then all geodesics are either isometric to the real axis or to one circle C (with its intrinsic distance). Moreover, in the latter case, if the dimension exceeds 1, the space has a two-sheeted universal covering space, whose geodesics are isometric to a circle with twice the length of C and share with the ordinary sphere the property that all geodesics through one point meet again at a second, see [3, p. 201]. The omission of a hyperplane requires, of course, the Desarguesian character. We will not return to this point later.

A second general theorem on Desarguesian spaces which is obvious in the Riemannian, but not the general case and has so far only been proved for symmetric distances is this (see [4, p. 32]):

An n-dimensional Desarguesian space can be embedded (with preservation of the metric) as a hyperplane in an (n+1)-dimensional Desarguesian space.

Our next task is to construct some interesting special Desarguesian metrics and to exhibit general methods of generating an abundance of Desarguesian metrics of which very few have interesting geometric properties.

For this purpose it is useful, although unsatisfactory from the purist's point of view, to introduce an auxiliary euclidean metric $\epsilon(x,y)$ in an affine space A^n with affine coordinates $x^1,\dots,x^n$ by

$$\epsilon(x,y) = [\Sigma(x^i - y^i)^2]^{1/2}.$$

Besides the topology given by $\epsilon(x,y)$ we use only the following simple, but fundamental fact: If the pairs of points a, b and c, d lie on parallel lines then

$$\epsilon(a,b) : \epsilon(c,d) \text{ is an affine invariant,}$$

i.e. does not change if we replace the x^i by other affine coordinates.

The first non-Riemannian Desarguesian space and by far the most important one was constructed around 1890 by Minkowski for applications in number theory. Let C be a bounded nonempty open convex set in A^n with the (compact) boundary B and $p \in C$. For any two points x, y define the Minkowski distance $m(x,y)$ by $m(x,x) = 0$ and for $x \neq y$ as follows: let the ray with origin p parallel to the ray from x through y intersect B at q. Then

$$m(x,y) = \epsilon(x,y) : \epsilon(p,q).$$

If C has an affine center and this is p, then always $m(x,y) = m(y,x)$ and we obtain what is now known as a finite-dimensional Banach space. The condition c) is satisfied if and only if B is strictly convex. Both the $B^+(p,\rho)$ and $B^-(p,\rho)$ are compact.

The importance of Minkowski spaces is for non-Riemannian smooth spaces (Finsler spaces) the same as that of euclidean spaces for Riemannian geometry: every Finsler space behaves locally like a Minkowski space.

Consequently there are very many characterizations of the Minkowskian spaces. Besides the obvious one by translation invariance we mention here only two. A third and fourth will appear later in connection with the Funk geometry and the concept of curvature. The Minkowskian geometries are the only Desarguesian metrics defined in all of A^n and compact convex $B^{\pm}(p,\rho)$, see Zaustinsky [5, p. 46]. They are the only Desarguesian spaces in which the loci equidistant to a hyperplane H (i.e. $xH = \rho$ and $Hx = \rho$) are pairs of hyperplanes, see Phadke [6]. The principal point in the far from simple proof is showing that the metric must be defined in all of A^n. It is rather easily seen that the boundary cannot contain a strictly convex piece and is therefore polyhedral. In the differential geometric approach to Problem IV it is always assumed that the boundary has nonvanishing curvature, so that the problem becomes rather trivial. This applies to many situations and explains why the contributions of the differential geometers, which will be discussed later, have little relation to the foundations of geometry.

A non-Riemannian analogue $h(x,y)$ to hyperbolic geometry was discovered by Hilbert in 1894, see [7, Anhang 1]. With the same notation as above and R( ) denoting cross-ratio let x, y be points of C. Put $h(x,x) = 0$ and for $x \neq y$ let the line through x and y intersect B at u and v in the order v, x, y, u. Then with a positive constant k

$$h(x,y) = k \log R(x,y,u,v) = k \log\left[\frac{\epsilon(x,u)}{\epsilon(y,u)} + \log\frac{\epsilon(y,v)}{\epsilon(x,v)}\right] = h(y,x)$$

The triangle inequality follows from the convexity of C, but $xy + yz > xz$ for noncollinear x, y, z requires only that a plane which intersects C, intersects B in a curve which does not have two noncollinear segments. The set $\{x : xT \leq \rho\}$ is convex for every segment T (see Pedersen [8] and [3, Section 18]). In Riemannian geometry this is equivalent to nonpositive curvature, see [8] and [3, Section 41].

The requirement that in a triangle with vertices a, b, c and midpoints b' of a, b and c' of a, c always $2h(b', c') \leq h(b, c)$ is in Riemannian geometry also equivalent to nonpositive curvature [3, l. c.], but implies for a Hilbert geometry that it is hyperbolic (see Kelly and Straus [9]).

Finally we come to a most interesting metric $f(x,y)$ discovered by Funk [10] in 1929. It is so-to-say half a Hilbert geometry. Its main features are that not all $B^-(p,\rho)$ are compact and that it resembles euclidean geometry in some and hyperbolic geometry in other respects. Keeping the previous notations put $f(x,x) = 0$ and for $x \neq y$

$$f(x,y) = k \log \frac{\epsilon(x,u)}{\epsilon(y,u)}.$$

For $\epsilon(y,u) \to 0$ we have $f(x,y) \to \infty$ so that the $B^+(p,\rho)$ are compact, but for $\epsilon(x,v) \to 0$

$$f(x,y) \to k \log \frac{\epsilon(v,u)}{\epsilon(y,u)} < \infty,$$

so that $B^-(p,\rho)$ is not compact for fixed p and large ρ.

The condition c) is equivalent to the strict convexity of B. The proof uses the

Theorem of Menelaus (compare [5, Appendix I, where the triangle fa'z must be replaced by xyz]).

Obviously the $B^+(p,\rho)$ are all homothetic to $B \cup C$ and hence to each other. Requiring that the $B^+(p,\rho)$ be homothetic distinguishes the geometries of Minkowski and Funk among all Desarguesian spaces; see [11].

With the above notation the spheres through x about y (i.e. $f(y,x) = \rho$) tend for $\epsilon(y,u) \to 0$ to a curve homothetic to B and touching B at u. Thus we encounter the hyperbolic phenomenon of limitspheres. There are several other facts showing this ambivalent behaviour of the Funk geometry.

There does not seem to be any special interesting non-Riemannian geometry defined in all of P^n, although we will see that there are very many.

The abundance of Desarguesian metrics is perhaps best understood by beginning with a very simple example. We modify $\epsilon(x,y)$:

$$xy = \epsilon(x,y) + |e^{x^n} - e^{y^n}|.$$

This defines a Desarguesian metric in all of A^n because the absolute value satisfies the triangle equality and e^t is monotone. Any other monotone function would have done, many such functions can be added, and watching convergence even infinitely many. Also, the addition can be replaced by an integration over a continuous family.

Of course, it is very easy to generate nonsymmetric distances by this method, for example by modifying a nonsymmetric Minkowski distance.

It seems clear that Hilbert was not aware of these simple possibilities. For, speaking of Desarguesian geometries defined in all of A^n he says: "One finds that such a geometry really exists and is no other than that which Minkowski constructed in his book, Geometrie der Zahlen...".

The method of addition of quasimetrics was not invented merely for destructive purposes, but to answer a variety of questions. For example, it allows us to show in a few lines, that infinitely many essentially different Desarguesian metrics exist for any given open convex set in A^n, see [3, p. 111]. Here one example will suffice to exhibit the method : the strip $-\pi/2 < x^n < \pi/2$ is metrized by

$$xy = \epsilon(x,y) + |\tan x^n - \tan y^n| .$$

The only property used is that tan t increases monotonically from $-\infty$ to ∞ in $(-\pi/2, \pi/2)$.

It is very easily seen that a metric in a set $C \subset A^2$ which possesses all rotations about two distinct points is euclidean or hyperbolic, but there is much liberty in constructing Desarguesian metrics with all rotations about one point. The above method applied with integration yields among others Desarguesian metrics of this type in all of A^2 such that for any two parallel lines L_1, L_2 always $xL_2 \to 0$ (or always $xL_2 \to \infty$) when x traverses L_1 in either direction, see [4, p. 29].

Integral geometry leads to a completely different way of generating Desarguesian spaces. In integral geometry one considers measures on sets of geometric objects other than points, for example sets of hyperplanes. For significant results, in E^n say, congruent sets of hyperplanes must have the same measure, but this aspect is irrevelant in the present context. Important is that the number or measure of hyperplanes intersecting a curve, each counted as often as it meets the curve, equals the length of the curve, usually with a constant factor depending on the unit of the measure. The author observed in [12] that this method can be reversed as it were and then automatically yields Desarguesian metrics.

Since we concentrated on A^n so far, let us consider P^n. Take a countably additive nonnegative set function $w(S)$ defined on all Borel sets S of hyperplanes H in P^n. For any pointset M put

$$H_M = \{H : H \cap M \neq \emptyset\}.$$

and require

(1) $0 < w(G) < \infty$ for a nonempty open set G of H.

(2) $w(H_p) = 0$ for any point p.

Then, for any line L,

$$0 < w(H_L) = w(H_{P^n}) = 2k < \infty,$$

because $w(\{H \supset L\}) = 0$ by (2). If A is an arc of a projective line divided by the point q into the subarcs A_1, A_2 then, by (2),

$$w(H_{A_1}) + w(H_{A_2}) = w(H_A).$$

Consider a triangle with sides A, B, C such that $A \cup B \cup C$ bounds a simply connected domain in the plane of the triangle. Then

$$w(H_A) + w(H_B) > w(H_C).$$

For, any hyperplane intersecting C intersects A or B, and there is an open set of hyperplanes cutting $A \cup B$ without intersecting C.

This means that the projective lines are locally shortest connections and hence geodesics. We now define a distance xy as follows: $xx = 0$ and if $x \neq y$ we take that arc A (or one of the two) on the line L through x and y for which $w(H_A) \leqq k$ and put

$$xy = w(H_A). \qquad (*)$$

Then xy is a Desarguesian metrization of P^n. Thus each w yields a Desarguesian geometry, but notice that the distance is always symmetric.

The same can be done in A^n provided we admit ∞ as possible value for $w(S)$. A domain for which the distances are finite is automatically convex. For, choose p arbitrarily and consider the set of all points x for which $px < \infty$. This set is open if it is n-dimensional and is then convex. For, if $px < \infty$, $py < \infty$ then $xy \leqq xp + py < \infty$. Hence, if z lies on the segment from x to y then $zx < \infty$ and $zp \leqq px + xp < \infty$.

This leads naturally to the question of how much more general a symmetric Desarguesian space can be. In [13], entitled "A complete solution of Hilbert's Fourth Problem", Pogorelov announced the very beautiful result, that each symmetric

Desarguesian space can be obtained in this way. The proof is very sketchy and a detailed version does not seem to be available as yet.

The argument consists of two parts: in the first a smooth function $g(H)$ is determined such that for a smooth Desarguesian distance $w(S) = \int_S g(H)$ and (*) hold. Then for a given not necessarily smooth distance xy a sequence w_ν, derived from smooth distances, converging to a set function w is constructed by an averaging process such that (*) holds.

The second part, although many details are missing, seems clear enough. In the first part the case $n = 2$ is treated in some detail, but that there must be a solution for $n = 2$ is evident from an old paper of Blaschke [14] who solved the analogous problem for general two-dimensional variational problems, i.e. also those in which the extremals are not necessarily straight lines, giving a measure for the extremals such that a given curve has a length (in terms of the variational problem) which equals the number of extremals intersecting it.

Unfortunately Pogorelov passes the cases $n > 2$ over by a remark to the effect that they are treated in the same way as the plane case. However, it seems to the author that $n > 2$ offers essentially new difficulties and he was not able to reconstruct a complete proof from the brief observation of Pogorelov.

The title of [13] is somewhat misleading in as much as the integral geometric method yields only symmetric distances, whereas in Problem IV nonsymmetric distances have been considered from the very beginning (already by Minkowski). For these the first paper [1] by Hamel (which is restricted to $n = 2, 3$ and smooth metrics) provides until now the only solution. We do not reproduce it here, since it is not based on a general idea like integral geometry and the expression given by Hamel is not illuminating. Also, inevitably in view of the time it was written, the discussion of completeness does not satisfy our modern requirements, the necessary concepts did not yet exist. However, the immense number of possibilities becomes quite clear from [1].

Finally we discuss the contributions of differential geometry to Problem IV. These could not begin before a general theory of non-Riemannian spaces was created by Finsler [15] in 1918. Taking all necessary smoothness conditions for granted one considers on a differentiable manifold a function $F(x, \xi) = F(x^1, \ldots, x^n; \xi^1, \ldots, \xi^n)$ where ξ is a contravariant or tangent vector. The length of a curve $x(t)$ is given by $\int F(x, \dot{x})dt$. To yield meaningful results F must satisfy certain conditions. $F(x, \xi) > 0$ for $\xi \neq 0$ to make length positive, $F(x, k\xi) = kF(x, \xi)$ for $k \geqq 0$ to assure independence of the parametrization. This homogeneity implies with the summation convention

$$F^2(x, \xi) = \tfrac{1}{2} F^2_{\xi^i \xi^k}(x, \xi)\, \xi^i \xi^k = g_{ik}(x, \xi)\, \xi^i \xi^k \quad \text{for } \xi \neq 0.$$

In the Riemannian case the $g_{ik}(x, \xi)$ are independent of ξ. Our previous considerations show the enormous increase in generality caused by this at first sight innocuous looking difference.

The Legendre Condition that the matrix $(g_{ik}(x, \xi))$ be positive definite for $\xi \neq 0$

expresses that the locus $F(x,\xi) = 1$ in ξ-space has for fixed x nowhere vanishing curvature. It is always assumed in the theory of Finsler spaces. The Jacobi Condition, which plays a considerable role in the calculus of variations and hence in some aspects of Finsler geometry is irrelevant in our case, because it deals with conjugate points, which do not occur when the straight lines are the extremals.

Unfortunately there are many different theories of Finsler spaces; for an excellent detailed discussion see Rund [16]. It turns out that many Riemannian concepts can be generalized to Finsler spaces in different ways so that they coincide with the Riemannian when the g_{ik} do not depend on ξ. Fortunately for our subject, only one of these theories has been seriously applied to Problem IV, namely the historically first with Berwald and Funk as principal representatives, which flourished roughly in the decade from 1928 to 1938 in the then existing German University at Prague.

Following Berwald [17] (which generalizes results obtained by Funk [10] for $n = 2$ and gives a brief introduction to Berwald's general theory), we choose the parameter such that $F(x,\dot{x}) = 1$. Then the extremals are the same for F and F^2. The Euler equations for F^2 have the form

$$\ddot{x}^k + 2\Gamma^k(x,\dot{x}) = 0, \quad k = 1,\ldots,n,$$

where the Γ^k are positive homogenous of order 2 in $\dot{x}$. With the notations

$$\frac{\partial\Gamma^k}{\partial\dot{x}^l} = \Gamma^k_l, \quad \frac{\partial^2\Gamma^k}{\partial\dot{x}^l\partial\dot{x}^m} = \Gamma^k_{lm}$$

the curvature tensor is given by

$$K^{\ i}_{k\cdot lm} = \left(\frac{\partial\Gamma^i_{kl}}{\partial x^m} - \frac{\partial\Gamma^i_{kl}}{\partial\dot{x}^r}\Gamma^r_m\right) - \left(\frac{\partial\Gamma^i_{km}}{\partial x^l} - \frac{\partial\Gamma^i_{km}}{\partial\dot{x}^r}\Gamma^r_l\right) + \Gamma^r_{kl}\Gamma^i_{rm} - \Gamma^r_{km}\Gamma^i_{rl}.$$

As in Riemannian geometry we put

$$K_{iklm} = g_{kp}K^{\ p}_{i.lm}, \quad K_{il} = K^{\ p}_{i.lp}$$

and define

$$s(x,\dot{x},\pi) = \frac{K_{iklm}\dot{x}^i\dot{x}^l\pi^k\pi^m}{(g_{il}g_{km} - g_{im}g_{kl})\dot{x}^i\dot{x}^l\pi^k\pi^m}$$

as the sectional curvature (Krümmungsmass) of the space at the line element $x,\dot{x}$ and the plane element π. The curvature scalar (or Ricci curvature) at $x,\dot{x}$ is

$$q(x,\dot{x}) = (n-1)^{-1}K_{ik}\dot{x}^i\dot{x}^k.$$

Among the many results of Berwald we mention in particular: If the extremals are straight lines then $s(x,\dot{x},\pi)$ has the same value for each plane π containing $\dot{x}$ and hence coincides with $q(x,\dot{x})$:

$s(x,\dot{x},\pi) = q(x,\dot{x})$ if $\dot{x}$ is parallel to π.

If the extremals are straight lines and $q(x,\dot{x})$ is independent of $\dot{x}$ (or $s(x,\dot{x},\pi)$ is independent of $\dot{x}$) then $q(x)$ is constant.

A particularly attractive result of Berwald is, that the Desarguesian spaces of constant curvature are characterized by the property

P: An isometry of one line on another or itself is (the restriction of) a projectivity.

The three special geometries of Minkowski, Hilbert and Funk discussed above have this property and constant curvatures 0, $-k^{-2}/4$, $-k^{-2}/4$.

The Minkowskian and Hilbert geometries are jointly characterized by the properties: they are defined in subsets of A^n, have compact $B^{\pm}(p, \rho)$ and constant curvature.

If we replace the last property by P we obtain a theorem whose assertion does not involve differentiability and can, in fact, be proved very easily without smoothness assumptions, compare [4, p. 38].

It should be emphasized that both the hypotheses that the space be noncompact and the $B^{\pm}(p, \rho)$ compact are important. If either of them is omitted there are many solutions, most of them uninteresting. Berwald determined in [17] all n-dimensional Desarguesian spaces with curvature 0 and Funk [18] all two-dimensional Desarguesian metrics with constant positive curvature. Even the latter depend on an essentially arbitrary function, where "essential" means that the function is restricted by an inequality only.

We conclude with some remarks on the future of Problem IV. The problem of generating all not necessarily symmetric Desarguesian metrics, preferably in some elegant way like Pogorelov's, remains still open. Also, it is quite possible that some new interesting special metrics will be discovered. A desideratum is, in particular, a nonelliptic metric in P^2 or P^n with intuitively appealing properties.

The present development seems, however, to point in a different direction. In spite of the generality of Desarguesian spaces some very interesting spaces, which ought to be considered under this heading, do not fall under our difinition, namely those where the lines are geodesics, but not necessarily the only ones. A general theory comprising this case is outlined in [19].

The simplest problem, where this case appears naturally, arises when we ask, in which symmetric Minkowski planes the unit-circle has maximal or minimal length. The maximal length is 8 and obtained when the circles are parallelograms; the minimal length is 6 and obtained when the circles are affinely equivalent to regular euclidean hexagons, see Petty [20].

There are many Minkowski planes in which the perpendicularity of lines is symmetric, compare [2, p. 104]. The only Hilbert planes with symmetric perpendicularity are those in which the boundary is either an ellipse (i.e. the metric is hyperbolic) or a triangle; see Kelly and Page [21].

A most interesting new two-dimensional (generalized Desarguesian) metric where the domain is a triangle and the circles are hexagons was recently discovered by Phadke [22]. It is besides the Minkowskian the only metric in which the loci equidistant to lines are again pairs of lines.

Added in proof.

At the time of my lecture (referred to below as R) only the Doklady note [13] by Pogorelov, claiming to solve Problem IV for symmetric distances, existed, with no proofs for dimension $n > 2$. In the meantime a detailed treatment by Pogorelov has appeared as the booklet

(1) The Fourth Problem of Hilbert, Moscow 1974

which, unfortunately, I only succeeded to see too late to amend R.

This is to inform the reader about a serious discrepancy between [13] and (1).

The most important tool is a "density" function $g(H)$ defined on the hyperplanes. In [13] it is asserted to be positive for all n. However, I had proved in

(2) Areas in affine spaces III — The integral geometry of affine area, Rend. Cont. Circ. Mat. Palermo Ser. II, 9(1960), 1–17

that a positive $g(H)$ solving Problem IV, which is translation invariant does not exist for $n > 2$ in every Minkowskian geometry, but that a unique $g(h)$ which may also take negative values does (it is positive if and only if the spheres are projection bodies). For the purposes of (2) which aims at general integral geometry, densities which take negative values had to be rejected.

Now Pogorelov's $g(H)$ is translation invariant in the Minkowski case, so that his claim in [13] could not be valid, but (2) suggested that $g(H)$ with some negative values might work for the general Problem IV. Very misleadingly I stated in R only that I could not complete Pogorelov's "proof" for $n > 2$, trying to give Pogorelov the privilege of revealing his error.

However, (1) does not mention [13] at all and replaces the correct reference [12] by an incorrect one to [3]. In (1) Pogorelov allows $g(H)$ indeed to take negative values giving necessary and sufficient conditions for the extent to which $g(H)$ may be negative. Thus he constructs, in fact, all symmetric Desarguesian spaces.

Nonsymmetric distances which entered the problem from its inception are completely disregarded in (1) so that the earliest contribution to Problem IV, Hamel's thesis [1] (written under Hilbert) remains to this day the only one in the general case.

Perhaps the two omissions of Pogorelov are due to the fact that the series in which (1) appeared is aiming at an elementary level.

Herbert Busemann

Proc. Symp. Pure Math. 28, De Kalb 1974 (1976), 131–141

REFERENCES

[1] G. Hamel, Ueber die Geometrien, in denen die Graden die Kürzesten sind, Dissertation, Göttingen, 1901.

[2] G. Hamel, same title, Math. Ann. 57 (1903), 231-264.

[3] H. Busemann, The Geometry of Geodesics, New York, 1955.

[4] H. Busemann, Recent Synthetic Differential Geometry, Ergebnisse der Mathematik, Vol. 54, 1970.

[5] E.M. Zaustinsky, Spaces with nonsymmetric distance, Mem. Amer. Math. Soc., No 34, 1959.

[6] B.B. Phadke, Equidistant loci and the Minkowskian geometries, Canad. J. Math. 24 (1972), 312-327.

[7] D. Hilbert, Grundlagen der Geometrie, 8th ed. Stuttgart, 1956.

[8] F.P. Pedersen, On spaces with negative curvature, Mat. Tidsskrift B (1952), 66-89.

[9] P. Kelly and E.G. Straus, Curvature in Hilbert geometry, Pacific J. Math. 8 (1958), 119-125.

[10] P. Funk, Über Geometrien, bei denen die Geraden die Kürzesten sind, Math. Ann. 101 (1929), 226-237.

[11] H. Busemann, Spaces with homothetic spheres, Journal of Geometry 4-2(1974), 1-12.

[12] H. Busemann, Geometries in which the planes minimize area, Ann. Mat. Pure Appl. (IV) 55 (1961), 171-190.

[13] A.V. Pogorelov, A complete solution of Hilbert's Fourth Problem, Soviet Math. Dokl. 14 (1973), 46-49.

[14] W. Blaschke, Integralgeometrie 11, Zur Variationsrechnung, Abh. Math. Sem. Univ. Hamburg 11 (1936), 359-366.

[15] P. Finsler, Über Kurven und Flächen in allgemeinen Räumen, Dissert. Göttingen 1918 and Basel 1951.

[16] H. Rund, The Differential Geometry of Finsler Spaces, Berlin-Göttingen-Heidelberg, 1959.

[17] L. Berwald, Über die n-dimensionden Geometrien konstanter Krümmung in denen die Geraden die Kürzesten sind, Math. Zeitschr. 30 (1929), 449-469.

[18] P. Funk, Über zweidimensionale Finslersche Räume, insbesondre über solche mit geradliningen Extremalen und positiver konstanter Krümmung, Math. Zeitschr. 40 (1935), 86-93.

[19] H. Busemann, Spaces with distinguished shortest joins, A spectrum of Mathematics, Auckland, 1971, 108-120.

[20] C.M. Petty, On the geometry of the Minkowski plane, Rivista di Mat. Univ. Parma 6 (1955), 269-292.

[21] P. Kelly and L.J. Page, Symmetric perpendicularity in Hilbert geometries, Pacific J. Math. 2 (1952), 319-322.

[22] B.B. Phadke, A triangular world with hexagonal circles, Geometriae Dedicata 3(1975), 511-520.

MINKOWSKIAN GEOMETRY, CONVEXITY CONDITIONS AND THE PARALLEL AXIOM

H. Busemann

The problem of characterising Minkowskian spaces is an important problem of that branch of differential geometry in which spaces more general than the complete Riemann and Finsler spaces are studied axiomatically using synthetic geometric methods. The fundamental theorem in this field is the result that a Desarguesian straight *G*-space in which the parallel axiom holds and the spheres are convex is Minkowskian. However the question as to whether the hypothesis of the space being Desarguesian is necessary or not has remained unsolved for over forty years. It is therefore natural to investigate conditions stronger than the mere convexity of spheres. In this paper such geometric conditions derived from functions which measure the distance between lines and points on lines are studied. Besides characterising the Minkowskian spaces these investigations also bring out the interplay between the parallel axiom and the convexity and linearity conditions.

1. INTRODUCTION. This paper deals with the relations between Minkowskian Geometry (M), various convexity and linearity conditions (A, B, C and D) and the Parallel Axiom (P) in straight spaces. A straight space is a metric space satisfying the Bolzano-Weierstrass Theorem in which any two distinct points a,b lie on exactly one curve, called the line L(a,b), isometric to the real axis.[1)]

In a Desarguesian straight space, i.e., a space defined in a convex subset of the real n-dimensional affine space A^n whose lines fall on the affine lines (but need not coincide with them) the convexity of the spheres (C) and P guarantee M, see [1, (24.1), p. 144]. (When the meaning is clear from context, we omit the parentheses around

[1)] We repeat some standard definitions of *G*-space theory in the introduction to make it widely understandable; in the text and for results we refer to the book [1].

the names of the conditions.)

Whether in general straight spaces $C \cup P \sim M$ is an open question whose answer has eluded us since over forty years and still does. However, various conditions, some only slightly stronger, characterising M are known. We mention first the linearity requirement:

A triangle abc in a straight space and the midpoints b' of a,b and c' of a,c satisfy $2b'c' = bc$.

That this makes the metric Minkowskian is proved in [1, (36.2), (39.12)]. Here we show first that a similar hypothesis (A) which is often easier to verify does the same (i.e. $A \sim M$).

(A): If L is a line, $p \in L$, $a \notin L$ and a' is the midpoint of p and a then $2a'L = aL$.

We then turn to convexity conditions. For $n = 2$, it is known, see [1, (25.6)], that $B_\mu \cup P \sim M$ where

(B_μ): A fixed $\mu \geq 1$ exists such that whenever $a \neq b$ and m is the midpoint of a,b then for $p \notin L(a,b)$

$$2pm^\mu - pa^\mu - pb^\mu \leq 0.$$

(B_μ) becomes with increasing μ (in straight spaces) only seemingly weaker, but leads for $\mu \to \infty$ to the actually weaker C.

If instead of μ we let pm tend to ∞ when p traverses a ray (not on L(a,b)) with origin m we obtain

($\overline{B}$): $$\lim_{pm\to\infty} (2pm - pa - pb) \leq 0.$$

It will be shown that in any straight space (B_1), (B_μ) and $(\overline{B})$ are equivalent and we use (B) for all three.

On the other hand, if $(\overline{B})$ is strengthened to

$$(D): \qquad \lim_{pm\to\infty} (2pm - pa - pb) = 0,$$

then the space becomes Minkowskian with differential spheres (S_d). It is known that $C \cup P \cup S_d \sim M \cup S_d$ for $n > 2$, see [1, (24.6)]. The condition D has the advantage that $D \sim M \cup S_d$ works also for $n = 2$.

This leaves the question whether in straight spaces without P the condition B can be stronger than C $(B > C)$. We will see that suitable Hilbert geometries (see [1, §18]) have this property, so that in general straight, even Desarguesian, spaces

$$D > A > B > C$$

observing that $A > B$ follows from hperbolic geometry.

For the convenience of the reader we repeat the frequently occurring Divergence Property [1, p. 230]: for any two distinct noncollinear rays R_1 and R_2 having the same origin the distance $xR_1 \to \infty$ as $px \to \infty$, $x \in R_2$.

2. LINES WITH BOUNDED DISTANCE. For any lines A, B we say that A has bounded distance from B and write $A\beta B$ if a constant β exists such that $xB \leqslant \beta$ for all $x \in A$. This concept is evidently reflexive and transitive. It is also symmetric but this requires proof. If B contains a segment $T(f,f')$ whose end points f, f' are feet on B of points on A and whose interior J contains no such feet then $uA \leqslant 2\beta$ for $u \in J$. To see this let W and W' be the subrays of B which are the

components of $B - J$ and let $S(S')$ be the (closed) set of those points on A which have feet on $W(W')$. If $v \in S \cap S'$ and w, w' are feet of v on $A = S \cup S'$ in W and W' respectively, then $T(f,f') \subset T(w,w')$ and $ww' \leqslant wv + vw' \leqslant 2\beta$. We may assume without loss of generality that $uw \leqslant uw'$. Then $uw \leqslant \beta$ and $uA \leqslant uw + wv \leqslant 2\beta$. Using an argument essentially based on the above method we can show that every subray of B contains feet of points of A. Thus every point u of B is either a foot of a point of A or is contained in the interior of a segment T as above. This proves that $A\beta B$ implies $B\gamma A$, with $\gamma = 2\beta$.

If the space has the Divergence Property [1, p. 230] then through any point $p \notin B$ there is at most one line A through p with bounded distance from B. For let C be any line through p distinct from A. For $z \in C$ let f be a foot of z on B. As above there is a point $h \in A$ for which we have $hf \leqslant 2\beta$ if $A\beta B$. Then $zA \leqslant zh \leqslant zf + fh \leqslant zf + 2\beta$. Hence $C\gamma B$ implies $zA \leqslant \gamma + 2\beta$ violating the Divergence Property.

The above considerations yield the following:

THEOREM 1: If for any line A and point p a line B through p with bounded distance from A exists and the Divergence Property holds, then each line lies in a unique family of lines with bounded distance from each other and this family covers the space simply.

The hypothesis of the Divergence Property is essential in the above theorem. This is seen from the metrization $xy = e(x,y) + |\tan x_1 - \tan y_1|$ of the strip $|x_1| < \pi/2$ in the cartesian (x_1,x_2)-plane with the cartesian distance $e(x,y) = [(x_1-y_1)^2 + (x_2-y_2)^2]^{\frac{1}{2}}$. In this space there are exactly two families of lines with mutually bounded distances:

x_1 = constant and the rest of the lines.

We do not know if the conclusion of the above theorem also entails the Parallel Axiom. Our condition (A) does entail the Divergence Property, existence of lines with bounded distance (in fact every asymptote of A is equidistant from (A)) and the Parallel Axiom. This is proved in the next section.

3. THE LINEARITY CONDITION. In this section we consider the implications of the condition (A) for a straight space. We prove that the condition (A) implies that the space satisfies the Parallel Axiom, any three points are contained in a 2-dimensional straight subspace, the theorem of Desargues holds and that the metric indeed is Minkowskian.

A first consequence of (A) is the Divergence Property. If R_1, R_2 are distinct (non-opposite) rays with the same origin p, let H be the line carrying R_2 and $q \in R_1$, $q \neq p$. Then $xR_2 \geqslant xH = (px/pq).qH$ by (A), so that $xR_2 \to \infty$ when $px \to \infty$.

Secondly, the spheres are convex or, equivalently, see [1, (20.9)], any point p has exactly one foot on a given line H. Let f be a foot of p on H, $p \notin H$. Then any point x on the ray R(p,f) through f containing p satisfies xf : pf = xH : pH and pH = pf gives xH = xf. Now [1, (20.6)] applied to (fpx) shows that f is the only foot of p on H.

Thirdly the Parallel Axiom holds. For if H^+ is an orientation of the line H and $p \notin H$, $\eta > 0$ and x is so far out on H^+ that $px > \eta$, let $pr_i = \eta$, (i = 1,2) with (pr_1x), (r_1pr_2). The oriented line $L^+(p,x)$ tends for $px \to \infty$ to the oriented asymptote Q^+ to H^+ through p, see [1, (23.2)] and r_i to a point q_i on Q^+ with (q_1pq_2) and $q_ip = \eta$. By condition (A) $r_iH : pH = r_ix : px \to 1$, so that $q_iH = pH$. Hence Q^+ is equidistant from H.

Therefore the hypotheses of Theorem 1 are satisfied with the further property that the line A of that theorem is an asymptote to B. Consequently the Parallel Axiom holds as formulated in [1, p. 141] with the additional properties that parallelism is an equivalence relation and, if Q and H are parallels, then $qH = hQ$ for all $q \in Q$ and $h \in H$.

We next show that given three non-collinear points a, b, c there is a 2-dimensional subspace which is a *G*-space containing a, b, c. Let H be the line joining b, c and L be the perpendicular to H from a. Consider the metric subspace P consisting of all points on all parallels to H through different points of L.

To show that P is a *G*-space we have to show that with any two points a', b' the entire line joining a', b' also lies in P. Let a' lie on the parallel $H_{a'}$ to H and b' lie on the parallel $H_{b'}$ to H. We may assume that $H_{a'} \neq H_{b'}$. Let c' be a point with $(a'c'b')$. We first show that c' lies on a line which is perpendicular to both $H_{a'}$ and $H_{b'}$. Let γ be the distance between $H_{a'}$ and $H_{b'}$ and let $f_{a'}$ and $f_{b'}$ be respectively the feet of c' on $H_{a'}$ and $H_{b'}$. Then by condition (A) we have $c'f_{a'} : \gamma = a'c' : a'b'$ and $c'f_{b'} : \gamma = c'b' : a'b'$. Hence $c'f_{a'} + c'f_{b'} = \gamma$. Thus $(f_{a'}c'f_{b'})$ i.e. c' lies on the common perpendicular $L(f_{a'},f_{b'})$ to $H_{a'}$, $H_{b'}$.

If c' is a point with $(c'a'b')$ we have $a'b' : c'b' = \gamma : c'f_{b'}$ and $c'a' : c'b' = c'f_{a'} : c'f_{b'}$; on considering the parallel C through

c' to $H_{a'}$, observing that C and $H_{b'}$ are also parallels and using condition (A) and the fact that parallels are equidistants.

The above ratios yield, in this case, $\gamma + c'f_{a'} = c'f_{b'}$ so that $(c'f_{a'}f_{b'})$ which shows that the perpendicular from c' to $H_{a'}$ meets $H_{b'}$.

To show that $c' \in P$ we must show that the parallel $H_{c'}$ to $H_{a'}$ meets L. Let $H_{a'}$, $H_{b'}$ meet L in p and q respectively. Let r be the foot of q on $H_{c'}$ and s the foot of r on $H_{a'}$. Consider first the case when $(a'c'b')$. Then remembering that parallels are equidistants, $qs \leqslant qr + rs = f_{b'}c' + c'f_{a'} = f_{b'}f_{a'} = qp$. Thus $s = p$ as q has a unique foot p on $H_{a'}$. Hence (qrs) and the line $L(q,a) = L$ meets $H_{c'}$ in r. The case $(c'a'b')$ can be treated similarly. This completes the proof that P is a 2-dimensional straight G-space.

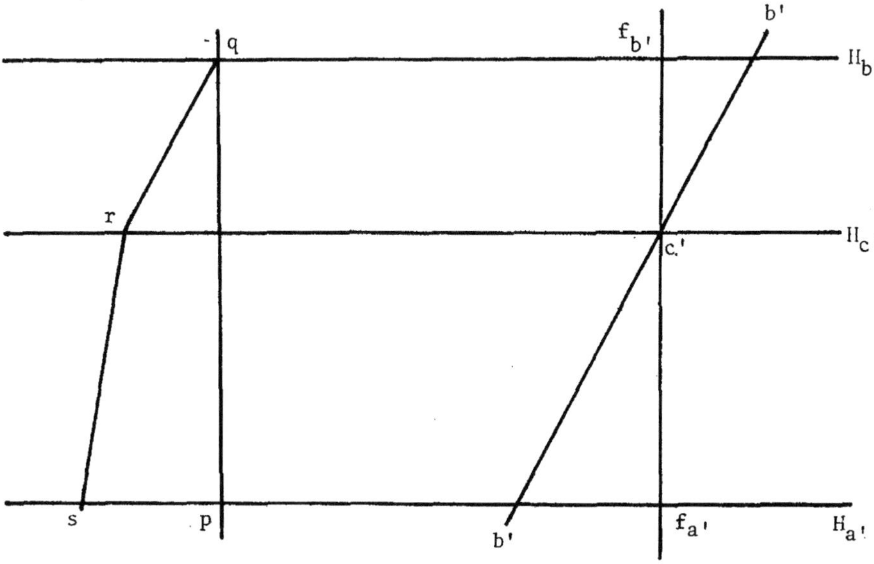

Figure 1

We now prove that the metric in this 2-dimensional subspace P is Minkowskian. To see this consider a triangle with vertices a, b, c in P and let d be a point on the side T(a,b). Draw through d, a line parallel to L(b,c) and let it meet T(a,c) in e. Applying condition (A) and remembering that the parallel lines L = L(d,e) and L' = L(b,c) are equidistants, we have

$$ad : db = aL : bL = aL : cL = ae : ce.$$

Thus we have proved the following.

LEMMA: In a 2-dimensional straight G-space with condition (A) a line parallel to the base of a triangle divides the other two sides in equal ratios.

The fact that the metric in such a space P is Minkowskian can now be deduced from the results contained in the two papers [2, 4]. Thus in the given straight G-space R any plane P is a Minkowskian plane. From this it follows that any plane P of R and a point outside P lie in a Minkowskian 3-dimensional subspace of R and that any k-dimensional subspace and a point outside it span a (k+1)-dimensional Minkowskian subspace in R. As R is finitely compact, R cannot contain Minkowskian subspaces of dimension k for every k. This can be proved by using an argument similar to [1, p. 261] and by applying the Riesz Lemma of Functional Analysis, see, for example, [5, p. 96]. Therefore R itself is finite dimensional and Minkowskian. That any Minkowskian space satisfies condition (A) is easy to prove, see, for example, [1, §17, pp. 94-96]. Hence we have the following:

THEOREM 2: A straight G-space is Minkowskian if and only if it satisfies the condition (A).

4. THE CONVEXITY CONDITIONS. We first prove the equivalence of the conditions B_1, B_μ and $\bar{B}$. If a continuous and positive function $t \to f(t)$ is convex then so also is the function $t \to (f(t))^\alpha$ for $\alpha > 1$. Hence B_1 implies B_μ for $\mu > 1$. Similarly B_μ for $\mu > 1$ implies B_σ for $\sigma > \mu$ so to show that B_μ implies $\bar{B}$ it suffices to consider the case when μ is an integer greater than one.

Let now a, b, c be three collinear points with b as midpoint of a and c. Let M be a line through b represented by $t \to x(t)$, $-\infty < t < \infty$, with t as arc length parameter on $M \neq L(a,b)$. Denoting by q_t the distance $qx(t)$ for any point q we have, by condition (B_μ),

$$b_t^\mu \leqslant \tfrac{1}{2}a_t^\mu + \tfrac{1}{2}c_t^\mu \text{ so that } \tfrac{1}{2}(b_t^\mu - a_t^\mu) + \tfrac{1}{2}(b_t^\mu - c_t^\mu) \leqslant 0.$$

Remembering that μ is an integer > 1 we have, on factorizing, dividing and simplifying,

$$\tfrac{1}{2}(b_t - a_t) + \left[\tfrac{1}{2}(b_t - c_t).\left(\textstyle\sum^\sigma (c_t/b_t)^\sigma\right)/\left(\textstyle\sum^\sigma (a_t/b_t)^\sigma\right)\right] \leqslant 0$$

where the summations are taken over σ, $0 \leqslant \sigma \leqslant \mu-1$. As $t \to \infty$ or $-\infty$, $b_t - q_t$ tends to a finite number. Hence $(q_t/b_t)^\sigma \to 1$ for all integers σ because $b_t \to \infty$ as $|t| \to \infty$. Hence taking limits as $t \to \infty$ or $-\infty$ we have

$$\lim\left(2bx(t) - ax(t) - cx(t)\right) \leqslant 0.$$

This proves that condition (B_μ) implies condition $(\bar{B})$.

To show that $(\bar{B})$ implies (B_1) we consider the function

$$\Delta(t) = 2bx(t) - ax(t) - cx(t).$$

Since $\Delta(t) = \big(bx(t) - ax(t)\big) + \big(bx(t) - cx(t)\big) \leqslant ba + bc = ac$ we see that $\Delta(t)$ is bounded above. Now if p_1, p_2 are points on M with $p_i = x(t_i)$ and (bp_1p_2) then

$$\Delta(t_2) = 2bp_2 - ap_2 - cp_2$$

$$> 2bp_1 + 2p_1p_2 - (ap_1 + p_1p_2) - (cp_1 + p_1p_2)$$

$$= 2bp_1 - ap_1 - cp_2 = \Delta(t_1).$$

Thus the limits $\Delta(\infty)$ and $\Delta(-\infty)$ of $\Delta(t)$ as $t \to \infty$, $t \to -\infty$ respectively exist, and if $x(t_o) = b$ then $\Delta(\infty) > \Delta(t)$ for any $t \geqslant t_o$ and $\Delta(-\infty) > \Delta(t)$ for any $t \leqslant t_o$. If $\overline{B}$ holds then both $\Delta(\infty)$ and $\Delta(-\infty)$ are $\leqslant 0$; hence $\big(2bx(t) - ax(t) - cx(t)\big) < 0$ for any t. Thus $(\overline{B})$ implies (B_1).

In fact we have deduced from $(\overline{B})$ the condition (B_1) with strict inequality. Thus we have, as an additional property of straight spaces the following:

REMARK. Let a line L be represented by $t \to x(t)$, $-\infty < t < \infty$ with t as arc length parameter on L in a straight space R and let p be a point of R not on L. Then if the function $t \to px(t)$ is a convex function of t for all positions of p not on L then it is actually a strictly convex function of t.

The result of this lemma is an analogue of the result [1, (20.9)] where it is proved that in straight spaces convexity of spheres implies their strict convexity.

Since the conditions (B_1), (B_μ) and $(\overline{B})$ are equivalent we refer to them collectively as conditions (B). Defining for an ellipse with foci p, q and semimajor axis α (i.e. the locus with equation $px + xq = 2\alpha$)

the corresponding sphere as the sphere with centre at the midpoint of p, q and radius α we have the following geometric interpretation of the condition (B): The condition (B) holds if and only if no part of an ellipse lies outside any corresponding sphere.

The important open problem can now be phrased as: Is $C \cup P \sim B \cup P$? This raises the question as to whether in straight spaces (B) is strictly stronger than (C). As Hilbert geometries [1, §18] have convex spheres we answer this question by showing that in suitable Hilbert geometries the condition (B) is not satisfied.

As the condition (B) implies that the ellipses are convex [1, p. 161] it suffices to show that in suitable Hilbert geometries the ellipses are not convex.

Now the Hilbert metric inside a closed convex curve C in the real affine plane defines a straight Desarguesian plane if and only if C does not contain two non-collinear segments, see [1, §18]. But if a sequence C_n of such curves tends to a curve C_0 and the Hilbert metric inside C_0 has a nonconvex ellipse then there must exist a C_n such that not all the ellipses in the corresponding geometry are convex.

We take as C_0 a triangle but with one side at infinity and the other two along the positive axes of the Cartesian (x_1,x_2)-plane. (The curves C_n then are the branches $x_1 > 0$ of the hyperbolas $x_1x_2 = 1/n$.) Simple observations on similar triangles show that the Hilbert metric inside C_0 then is the metric

$$xy = \max\{|\log \frac{x_1}{y_1}|, \; |\log \frac{x_2}{y_2}|, \; |\log \frac{x_1y_2}{x_2y_1}|\}$$

defined inside the quadrant $\mathcal{Q} = \{x = (x_1,x_2), x_i > 0\}$. This metric is discussed in [3], but at the time of the writing of [3] it was not realised that the metric can be identified with the Hilbert metric inside a triangle.

From the properties of this metric proved in [3] we can directly plot the ellipse with foci $f_1 = (1,1)$ and $f_2 = (1,2)$ and eccentricity 1/2; i.e. the locus with $f_1x + xf_2 = \log 4$ and verify that this ellipse is not convex. The ellipse is plotted in figure 2 below. In this figure $a = (1/\sqrt{2},1)$, $b = (1/\sqrt{2},1/\sqrt{2})$, $c = (1,1/2)$, $d = (2,1/\sqrt{2})$, $e = (2/\sqrt{2},1)$, $f = (2/\sqrt{2},\sqrt{2})$, $g = (2,2)$ and $h = (1,\sqrt{2})$. The arcs from c to d and f to g are concave. This example is the first instance confirming explicitly the statement in [1, p. 161] that the convexity of circles does not imply the convexity of ellipses.

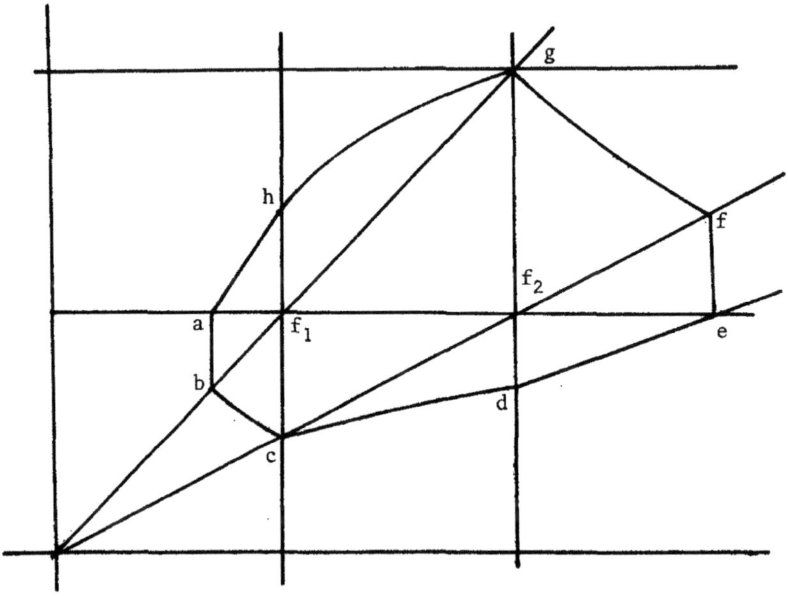

Figure 2

Another example is obtained from Hilbert geometries whose limiting case is the Hilbert metric defined in a square in the Cartesian (x,y)-plane. With the notation of figure 3, with $0 < c < d < b$, and taking the line L^+ as oriented downwards and represented in terms of the arc length parameter $t \to u(t)$, $u(0) = q = (s,s)$, we have, when u follows q, the following relations:

$$\frac{dt}{dy} = -\frac{(d-c)}{(d-y)(y-c)}$$

$$\frac{d^2(pu(t))}{dt^2} = \frac{b}{(d-c)} \cdot \frac{\Delta}{y^2(b-y)^2} \cdot \frac{dy}{dt}$$

with $\Delta = y^2(c + d - b) - 2cdy + bcd.$

Hence taking $b = 1$, $c = (1/2)$, $d = (3/4)$ we have $s = .6$ and $4\Delta = y^2 - 3y + (3/2)$. Thus $\Delta = 0$ when $y = y_1 = (3+\sqrt{3})/2 > 1$ and when $y = y_2 = (3-\sqrt{3})/2 > .63$. Choosing p sufficiently away from q we can ensure that $w_y < (1/2) + \varepsilon$. Then for y satisfying $w_y < y < s$ we have $\Delta > 0$ and $\frac{dy}{dt} < 0$. Therefore

$$\frac{d^2pu(t)}{dt^2} < 0 \text{ for } w_y < y < s.$$

Thus the condition (B) is not satisfied for the Hilbert metric inside the unit square. By projective equivalence the same conclusion holds for Hilbert metrics defined inside any convex quadrangle. By approximating the quadrangle by suitable convex curves (approximation is needed in only very small parts, see [1, §18]) we obtain straight G-planes which then also furnish examples of (B) > (C).

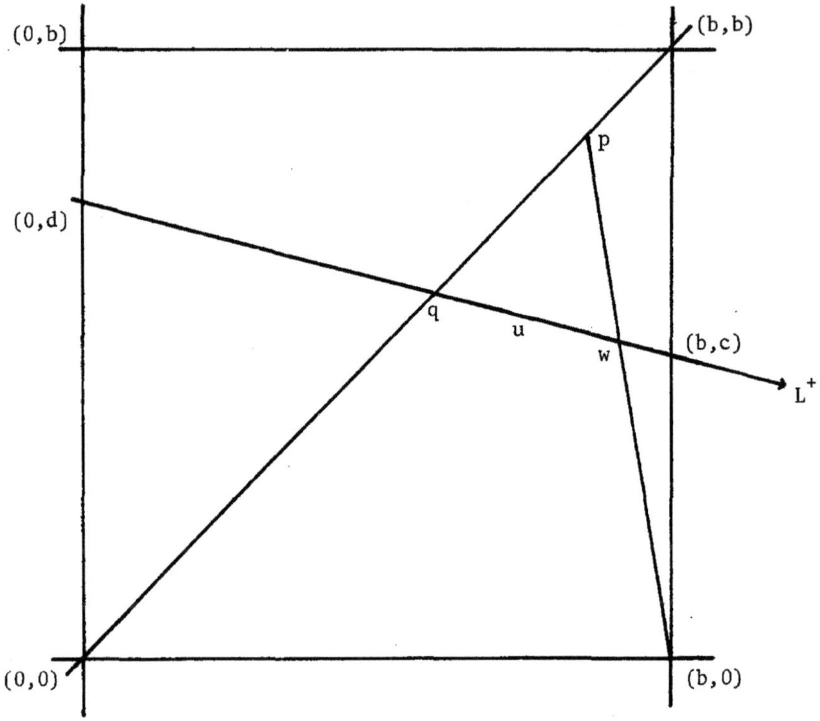

$q = (s,s)$, $p = (a,a)$, $u(t) = \big(x(t), y(t)\big)$, $w = (w_x, w_y)$

Figure 3

5. THE MEANING OF CONDITION (D). In view of the fact that the condition (B) is satisfied on any simply connected surface of negative curvature and hence does not imply the Parallel Axiom it is interesting to see that the condition (D) does entail the Parallel Axiom and that it actually implies that the space is Minkowskian. We prove this result in this section.

The main idea of the proof is to show that the condition (D) implies that the limit spheres are flat. (For definition and properties of limit spheres see [1, §22].) To prove this we first obtain a geometric interpretation of condition (D). For a limit sphere K and a point s of the space R we denote by sgn sK the distance of s from K if s lies in the exterior of K and the negative of this distance if s lies in the interior of K or in K. Then if K is a limit sphere with axis L^+ emanating from y we have sgn sK = lim(sx - yx) as $yx \to \infty$ with x following y on L^+. See [1, (23.2)]. Now lim(2py - px - pz) = lim(py - px) + lim(py - pz) as both the latter limits exist. Thus we have the following geometric meaning of the conditions (B) and (D).

LEMMA. Let K be a limit sphere and let T be a segment whose mid-point lies in K. Then the condition (B) implies that the sum of the signed distances of the ends of T from K is non-negative. The condition (D) implies that this sum is always equal to zero.

An immediate consequence of the above lemma and of (D) is that if one end point of T is exterior to K then the other end point of T is interior to K and also that if one end point of T lies in K then so also does the other end point. As no point of a supporting line of K is interior to it this shows that all supporting lines of K lie in K.

We are now ready to prove that the condition (D) implies that the limit spheres are flat, i.e. a limit sphere contains with any two points in it the entire line joining these two points. We first observe that as the condition (D) implies the condition (B) the spheres and hence the limit spheres are convex and also that any limit sphere decomposes the space. Now let R_1 and R_2 be the two rays beginning at p in which a line through p is divided and let K_j be the limit sphere through p with axis R_j, j = 1,2. Let I_j denote the interior of K_j, $j = 1,2$. We are going to show that $K_1 = K_2$.

In fact if K_2 contains a point u which is not in K_1 then the point v such that (vpu) and vp = pu also lies in K_2. But as K_1 is closed, K_1 and K_2 do not intersect the interiors of each other and hence $u \notin K_1$ implies $uK_1 > 0$. Hence sgn vK_1 = -sgn uK_1 = $-uK_1 < 0$. Thus $v \in I_1 \cap K_2$. This is a contradiction to $K_2 \cap I_1 = \phi$. Hence $K_2 \subset K_1$. Similarly $K_1 \subset K_2$. This proves that $K_1 = K_2$.

As K_1 and K_2 both bound disjoint convex sets we have proved that $K_1 = K_2$ is flat. These considerations enable us to prove the following:

THEOREM 3: A straight *G*-space is a Minkowskian space with differentiable spheres if and only if it satisfies the condition (D).

A Minkowskian space with differentiable spheres has flat limit spheres, see [1, (17.21), p. 102]. Hence with the above notation; since $K_1 = K_2 = K$ and the interior of K_1 is exterior of K_2 we have sgn xK_i = -sgn xK_j; $i,j = 1,2$; $i \neq j$. Now any Minkowskian space satisfies condition (B), see [1, (17.23), p. 102]. Therefore applying condition (B) to both orientations of a line L and using the above result that sgn xK_i = -sgn xK_j we have sgn xK + sgn yK = 0 whenever the midpoint p of x and y lies in K. Thus condition (D) holds. This proves the necessity of condition (D).

For the sufficiency proof we recall that the condition (D) implies that the limit spheres are flat and that this property, in turn, implies the Parallel Axiom and the differentiability of spheres. The theorem is now contained in (24.7) [1, p. 146] for dim $R > 2$ and in (25.6) [1, p. 157] for dim R = 2 because (D) implies (B).

REFERENCES.

[1] Busemann, H.: The Geometry of Geodesics, Academic Press Inc., New York, 1955.

[2] Busemann, H.: Remark on "Planes with analogues to euclidean angular bisectors", Math. Scand. 38, (1976), 81-82.

[3] Phadke, B.B.: A triangular world with hexagonal circles, Geometriae Dedicata, 3, (1975), 511-520.

[4] Phadke, B.B.: The theorem of Desargues in planes with analogues to euclidean angular bisectors, to appear in Math. Scand. vol. 39.

[5] Taylor, A.E.: Introduction to Functional Analysis, John Wiley & Sons, Inc., New York, 1964.

The University of Southern California, Los Angeles
The Flinders University of South Australia, Bedford Park

(Eingegangen am 10. Mai 1977)

LOCALLY DESARGUESIAN SPACES

H. Busemann and B.B. Phadke

In Riemannian spaces, locally Desarguesian spaces have constant curvature and are therefore locally symmetric. This does not hold for Finsler spaces, so that locally Desarguesian spaces represent a generalization other than the obvious one we studied previously of (certain) Riemannian symmetric spaces. In this paper we discuss them in detail; as an example of the results obtained we mention that a simply connected locally Desarguesian space without conjugate points is globally Desarguesian. Applications are then given to spaces which are locally symmetric in a wider sense. We also study (and in Minkowski spaces determine exactly) the properties of functions which measure the distance of a point from those on a line.

1. INTRODUCTION. According to Beltrami's theorem [1,p.86] the locally Desarguesian (wherein the geodesics behave locally like projective or affine lines) Riemannian spaces are the spaces of constant curvature, which are obviously locally symmetric and (when complete) have the Euclidean, hyperbolic and spherical spaces as their universal covering spaces. On the other hand there is an abundance of nonsymmetric straight Desarguesian *G*-spaces (and Finsler spaces) so that Beltrami's theorem does not have a direct analogue in more general spaces. However, from the results proved in [1,Section 52] and the theorem

that 'a straight locally symmetric G-space is globally symmetric' (which was proved in [3]) it follows that the only straight locally symmetric Desarguesian spaces are the Minkowskian and hyperbolic spaces.

In this paper we first investigate the question as to when a locally Desarguesian space is globally Desarguesian. We show that this holds for all simply connected spaces without conjugate points[1], for all spaces of the elliptic type and also for all noncompact simply connected G-surfaces. We also show that all sphere-like locally Desarguesian spaces and all compact locally Desarguesian G-surfaces are metrizations of the Euclidean sphere (of the same dimension as the space) with geodesics along the great circles. (For a definition of the sphere-like spaces and the spaces of the elliptic type, see respectively [1, p.128] and [1,p.124].) These results are considerable generalizations of the mentioned Riemannian result.

We apply these results to generalized locally symmetric (g.l.s.) G-spaces which are defined by postulating that for every p a symmetric $S(p,\rho)$ exists. We recall, see [3,p.3], that a locally symmetric space is a G-space R together with a continuous positive function $\sigma(p)$ defined on R such that each $S(p,\sigma(p))$ is symmetric. In the Riemannian case the definition of a locally symmetric (l.s.) space is always such that $\sigma(p)$ exists; whether the axioms for a G-space imply that a g.l.s. space is l.s.; we do not know, but doubt it, because there are very

[1]By absence of conjugate points we mean that the universal covering space is straight [1,p.38] as the concepts coincide in Finsler spaces. Noteworthy among these spaces are in particular the Riemann spaces of non-positive curvature or, more generally, the G-spaces with convex capsules, see [1,p.254].

trivial examples of spaces which are "almost" G-spaces where the concepts differ.

We prove that g.l.s. locally Desarguesian straight spaces with the parallel axiom are Minkowskian. We also show that the distance g(t) = px(t) from a point p to a variable point x(t) of a line which plays a considerable role in [1] and in [2] and in [3] and which is always convex in straight l.s. spaces, remains so in straight g.l.s. spaces, when the absence of conjugate points is strengthened to that of focal points (i.e. convexity of the spheres, see [1,p.163]).

We conclude the paper by giving exact conditions under which a convex function g(t) occurs as px(t) in straight, in particular, Minkowskian spaces. Except for this last theorem which includes a more general situation, all spaces are assumed to be G-spaces.

2. LOCALLY DESARGUESIAN SPACES. A G-space R is said to be locally Desarguesian[2] if each point p has a neighborhood $\mathcal{U}(p)$ which can be mapped topologically onto an open set V in affine space such that the segments are mapped on affine segments.

Since V has open convex subsets containing the image of p we may and will assume that $\mathcal{U}(p)$ is convex in the sense of G-spaces.

Two observations are in order. The condition implies that R has finite dimension (a G-space always has the same dimension at different points, in the present case this follows from the definition). The second remark is that in contrast to l.s. spaces we do not require that $\mathcal{U}(p)$

[2]The reason for this terminology can be found in [1, Chapter II].

is uniform in the sense that it contains a sphere $S(p,\lambda(p))$ with a continuous positive $\lambda(p)$. For this reason the results of this section are applicable to the generalized l.s. spaces which will be discussed in the last section. The space is (globally) Desarguesian if it can be mapped geodesically into a projective space.

The principal result is analogous to our main theorem (2.7) in [3].

(2.1) THEOREM. A locally Desarguesian straight space is globally Desarguesian.

This is clearly a consequence of

(2.2) An open convex subset T of a locally Desarguesian space R is Desarguesian.

We consider first the case when R is 2-dimensional.

(a) The first step in the proof of this result is the observation that a convex subset contained in the union of two Desarguesian open sets is itself Desarguesian. In fact if $\delta_{\mathcal{U}}$ and δ_V are two Desarguesian charts on $\mathcal{U}$ and V respectively and if $W = \mathcal{U} \cap V \neq \phi$ then $\delta_{\mathcal{U}} \circ \delta_V^{-1} : \delta_V(W) \to \delta_{\mathcal{U}}(W)$ has a unique extension $g_{\mathcal{U}V}$ to a collineation of the real projective space $\mathbb{P}^n$. Hence $\delta_V^* = g_{\mathcal{U}V} \circ \delta_V$ is also a Desarguesian chart for V. Further δ_V^* has the property that $\delta_{\mathcal{U}} = \delta_V^*$ on W.

Hence if X is a convex subset of R such that $X \subset \mathcal{U} \cup V$ then we define $\delta_X : X \to \mathbb{P}^n$ by $\delta_X(p) = \delta_{\mathcal{U}}(p)$ if $p \in \mathcal{U}$ and $\delta_X(p) = \delta_V^*(p)$ if $p \in V$. Then the map δ_X is well defined and by applying the arguments in [1,p.72,73] we see that δ_X is injective and preserves geodesics. Therefore X is Desarguesian.

(b) Now if points a,b,c,d belong to a convex Desarguesian neighborhood $\mathcal{U}$ and points c,d,e,f belong to a convex Desarguesian neighborhood V and if (ade) and (bcf), see figure 1, then a convex subset X can be chosen such that $X \subset \mathcal{U} \cup V$ and a,b,e,f lie in X.

J. Geom. 14 (1980), 123–138

Consequently, applying the observation in (a), we conclude that whenever quadruples a,b,c,d and c,d,e,f lie in Desarguesian convex neighborhoods, the quadruple a,b,e,f also lies in a Desarguesian convex neighborhood provided (ade) and (bcf). We note that this result holds even when any of the points a,b,c,d are coincident.

(c) Continuing with the proof, let us say that a triple a,b,c of points in T forms a narrow triangle if points $a = a_0, a_1, a_2, \ldots, a_k = c$, $(a_i a_{i+1} a_{i+2})$, $i = 0$ to $k-2$ on T(a,c) and points $b = b_0, b_1, \ldots, b_k = c$, $(b_i b_{i+1}, b_{i+2})$, $i = 0$ to $k-2$ on T(b,c) can be found such that each quadruple $a_i, a_{i+1}, b_{i+1}, b_i$, $i=0$ to $k-2$, and the triple a_{k-1}, c, b_{k-1} is contained in a convex open Desarguesian neighborhood, See figure 2.

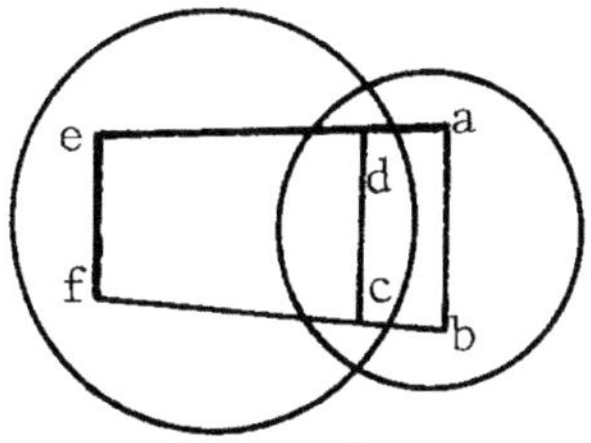

Figure 1

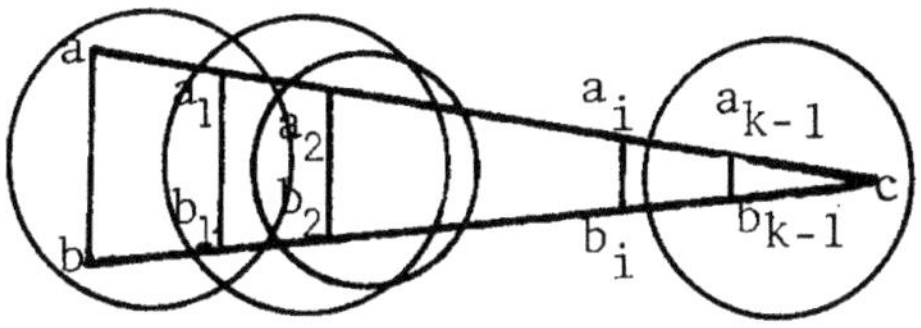

Figure 2

Then the steps (a) and (b) show that all "narrow" triangles lie in convex open Desarguesian neighborhoods.

(d) We now make the transition to any arbitrary triple of points a,b,c in T. Given any such triple we choose points $b = b_0, b_1, \ldots, b_k = c$ on T(b,c) such that each triple a, b_i, b_{i+1}, $i = 0$ to $k-1$ forms a narrow triangle. From the steps (c) and (d) it follows that the triple a,b,c lies in a Desarguesian neighborhood.

The result can be extended to any convex polygon Π

because, by joining a vertex of Π to all the other vertices we can divide Π into a finite number of triangles $\Delta_1, \Delta_2, \ldots, \Delta_k$ such that $\bigcup_{1 \leqslant i \leqslant j} \Delta_i$, $1 \leqslant j \leqslant k$ is convex for each j. The preceding argument can now be applied repeatedly to deduce that Π is contained in a Desarguesian open set.

Since every Desargues configuration in T is contained in a convex polygon, this proves that the Desargues Property [1,p.67] holds in T. A slight modification of [1, Theorem (13.1), p.68] hence shows that T is itself a Desarguesian set.

This completes the proof of (2.2) in the case when the dimension of R is 2.

To complete the proof we consider the case when dimension of R exceeds 2. We show that an open convex subset of a straight locally Desarguesian G-space satisfies a property similar to the existence of 2-dimensional planes.

More precisely, let three points a,b,c in T be given. Let $\mathcal{U}$ be a Desarguesian neighborhood of a in T and let K be a 2-dimensional plane element of $\mathcal{U}$ such that K contains in its relative interior a and points d, e with (adb) and (aec).

Let H be the union of all segments in T which intersect K in segments having a as an interior point. Clearly H is 2-dimensional and a,b,c belong to H. We show that H is convex.

To show this it suffices to prove that H contains with any two sufficiently close points p and x the whole segment T(p,x). In fact given $p \in H$, cover T(a,p) with Desarguesian neighborhoods Q_1, Q_2, ..., Q_k such that $a \in Q_1$, $p \in Q_k$ and $Q_i \cap Q_{i+1}$ is nonempty, $1 \leqslant i \leqslant k-1$. Then

$\bigcup_{1\leq i\leq k} Q_i$ covers a neighborhood of T(a,p) and in particular a narrow triangle formed by segments lying in H. Arguments similar to (b) and (c) above, applied repeatedly, show that the narrow triangle contains a plane piece in Q_k through p and hence H contains T(p,x) for x close to p.

Slight modifications of the theorem (14.1) of [1,p.76] now show that T is a Desarguesian set. This completes the proof of (2.2).

(2.2) also implies

(2.3) A simply connected n-dimensional locally Desarguesian G-space defined on a subset of the n-dimensional euclidean sphere S^n admits a geodesic map either on the entire S^n or a convex open subset of an open hemisphere of S^n (or of the n-dimensional real affine space A^n).

This result implies, in particular, that an n-dimensional locally Desarguesian spherelike space (see [1, p.128]) is a metrization of the euclidean sphere S^n as a G-space R such that the great circles of S^n are geodesics of R. A G-space of the elliptic type has a spherelike space as two-sheeted universal covering space [1, (31.2), p.201] and whenever the affine theorem of Desargues holds in an affine space, the projective theorem of Desargues holds in the projective completion of that affine space. Combining these two facts with (2.2) and (2.3) we have:

(2.4) A locally Desarguesian space of the elliptic type is globally Desarguesian.

We now apply these results to obtain more explicit analogues of the Riemannian theorem that the universal covering space of a locally Desarguesian space is geodesically spherical or globally Desarguesian.

To begin with we observe that if a space is locally

Desarguesian then so are its covering spaces. Also the absence of conjugate points is equivalent to the straightness of the universal covering space. Hence we have:

(2.5) The universal covering space of a locally Desarguesian space without conjugate points is globally Desarguesian.

This is the Finsler space analogue of the above mentioned Riemannian theorem.

Using again the fact that the universal covering space of a space of the elliptic type, of dimension greater than 1, is a spherelike space, we have, from (2.3) and (2.4) the following:

(2.6) The universal covering space of a locally Desarguesian space R (of dimension $n > 1$) of the elliptic type is a metrization of the euclidean sphere S^n as a G-space such that the geodesics of R coincide with the euclidean great circles of S^n.

For 2-dimensional spaces we can generalize both the results (2.5) and (2.6). We prove:

(2.7) A 2-dimensional simply connected locally Desarguesian space is either a straight Desarguesian space defined inside a convex subset of the real affine plane A^2 or is a metrization of the entire euclidean sphere S^2 as a G-space R such that the geodesics of R coincide with the great circles of S^2.

To prove this we need the following result which is also of some interest on its own:

(2.8) If an open non-empty subset R of S^n is metrized as a G-space with the geodesics of R falling on the great circles of S^n then either R lies in an open hemisphere of S^n (and is a straight Desarguesian space) or else R is defined on all of S^n.

To prove this result first note that whenever points a,b are not S^n-antipodes the S^n-great circle, and hence the R-geodesic, through a,b is unique. Hence for such points a,b the R-geodesic through them contains a segment T(a,b). The same holds even for S^n-antipodal points a,b by continuity of distance. Thus every R-geodesic contains with any two points a segment containing those two points. It follows from [1,(9.6),p.46] that each R-geodesic is an R-straight line or an R-great circle.

This also shows that R-antipodes and S^n-antipodes are identical.

Thus to prove (2.8) it suffices to prove that if R contains two antipodes a and b then $R = S^n$. For such points a segment T(a,b) must be a semigreatcircle T and hence a geodesic containing T(a,b) is a great circle. R being open, all semigreatcircles close to T lie in R and so do the great circles containing them. Let Q be the point set formed by all great circles in R through a and b. If a point p of S^n not in Q existed, choose c,d in Q such that they are not antipodes and the S^n-greatcircle C through c and d passes through p. By the construction of Q it follows that C contains antipodes lying in Q, hence in R and so p lies in R as above.

This proves (2.8). Since 2-dimensional G-spaces are manifolds and since the plane and the sphere are the only 2-dimensional simply connected manifolds, we obtain (2.7) by combining (2.3), (2.5) and (2.8)

In higher dimensions there are many simply connected spaces whose geodesics are great circles but probably no Desarguesian ones.

As a consequence of [3,(2.7),(4.5)] and (2.7) we have the following analogue of Beltrami's theorem for G-surfaces:

(2.9) A locally symmetric and locally Desarguesian G-surface is either locally Minkowskian or locally hyperbolic or locally spherical.

Because of (4.2) in [3] and (2.1) here we find:

(2.10) The locally Desarguesian and locally symmetric straight spaces are Minkowskian or hyperbolic.

Together with (4.5) in [3] this is a generalization to G-spaces of Beltrami's theorem in the large as formulated in [1,(15.4)]. A strictly local form like [1,(15.3)] does not fit into our setup, because we consider only G-spaces which are complete.

Also (2.7), (4.5) in [3] and (2.3) here yield:

(2.11) A locally symmetric, locally Desarguesian metrization of a simply connected subset of the n-dimensional Euclidean sphere is Minkowskian, hyperbolic or spherical.

We expect that for algebraic reasons a simply connected locally Desarguesian manifold is topologically the Euclidean sphere or an open subset of it. If this is true then (2.11) suggests that the twin properties of locally symmetric and locally Desarguesian provide the analogue in Finsler or G-spaces to the property of constancy of curvature in Riemann spaces.

3. GENERALIZED LOCALLY SYMMETRIC SPACES.

According to the introduction a g.l.s. space is defined as a G-space R where every point p has a symmetric neighborhood $S(p,\kappa(p))$ and no condition except positivity is imposed on $\kappa(p)$.

We stated already that we do not know, but doubt, whether the axioms for a G-space make a g.l.s. space automatically l.s. The example of a circular cone, where every point has a symmetric neighborhood and the axioms for a G-space are satisfied except at the apex, at which the space is symmetric, shows that the concepts of a g.l.s. and l.s. space can differ in more general spaces. We show:

(3.1) THEOREM. A g.l.s. locally Desarguesian straight space with the parallel axiom is Minkowskian.

In the proof we will need the following easily proved lemma:

LEMMA. Let the t-axis be metrized as a straight line by $|t_1-t_2|$ as well as $\rho(t_1, t_2)$ such that each t_0 has a neighborhood $(t_0-\varepsilon, t_0+\varepsilon)$ where $t_0-\varepsilon < t_1' < t_0 < t_2' < t_0+\varepsilon$ and $t_0-t_1' = t_2'-t_0$ implies $\rho(t_1', t_0) = \rho(t_0, t_2')$ $[=\frac{1}{2}\rho(t_1', t_2')]$ then there exists a constant k such that $\rho(t_1, t_2) = k|t_1-t_2|$ for any t_1, t_2.

To prove (3.1) we first observe that by (2.1) the space is globally Desarguesian and since the parallel axiom holds, it is a metrization of the entire affine space such that the affine lines are metric lines. In view of the above lemma, therefore, a reflection of a sphere $S(p,\rho)$ in p is the restriction of an affine reflection. This implies that in any 2-dimensional subspace small circles have parallel supporting lines at the ends of any diameter. Consequently the space is locally Minkowskian [1,pp.89 and 94]. It is globally Minkowskian because it is its own universal covering space, see [1, (30.1),p.189].

We saw in [3,(3.7] that in straight symmetric spaces the functions px(t) are convex. We now show

(3.2) THEOREM. In a g.l.s. straight space R with convex spheres for any point p not on a line represented by $x(t)$ the function $px(t)$ is strictly convex. The ellipses in R are, therefore, also strictly convex.

The proof uses the easily verified lemma (which at first sight seems to require differentiability):

(3.3) LEMMA. If a real valued function $g(t)$, $-\infty < t < \infty$, is continuous and satisfies, with a suitable $\varepsilon(t_0) > 0$ for each t_0, $g(t_0+h) + g(t_0-h) - 2g(t_0) \geqslant 0$, $0 \leqslant h \leqslant \varepsilon(t_0)$ then $g(t)$ is convex.

In view of (3.3) to prove (3.2) it suffices to exhibit for each t_0 an $\varepsilon(t_0) > 0$ such that for $g(t) = px(t)$, $g(t_0+h) + g(t_0-h) - 2g(t_0) > 0$ for $0 < h \leqslant \varepsilon(t_0)$.

Put $x(t_0) = m$ and let $y(t)$ be a representation of the line H joining m and p such that $y(0) = m$ and the positive ray H^+ of H starting at m contains p. Let $K = K_\infty(H^+,m)$ be the limit sphere with H^+ as an axis. Since the spheres are convex, K is convex and the transversals G to H at m form a set V which decomposes the space into two arcwise connected sets, one of which contains K; see [1,(20.12),p. 123].

Since R is locally symmetric at m we can find $\rho > 0$ such that $S = S(m,\rho)$ is symmetric and since spheres are convex we can find $\varepsilon(t_0) = \varepsilon > 0$ such that $\varepsilon < \rho$ and whenever $0 \leqslant h \leqslant \varepsilon$ the feet of $x(t_0 \pm h)$ on V lie in S.

Denoting by sgn xK the signed distance of a point x from K as defined in [2,p.15] and using the inequalities proved in [2,p.10] we have: $g(t_0+h) + g(t_0-h) - 2g(t_0) >$ limit of $[\{px(t_0+h) - px(t_0)\} + \{px(t_0-h) - px(t_0)\}]$, as $pm \to \infty$ with $p \in H^+$, and this limit equals $\operatorname{sgn} x(t_0+h)K + \operatorname{sgn} x(t_0-h)K$.

Thus it suffices to show that the sum of signed distances of $x(t_0 \pm h)$ from K is nonnegative.

Since K is convex at most one of the points $x(t_0 \pm h)$ lies inside K. If neither of these points lies inside K we have $\operatorname{sgn} x(t_0 \pm h)K = x(t_0 \pm h)K \geqslant 0$ and we are done.

If only one, say $x(t_0 - h)$, lies in K let q be the foot of $x(t_0 + h)$ on V and let G be the transversal to H which joins m to q. Then as q lies in S and S is symmetric we have $x(t_0+h)G = x(t_0-h)G$. As $G \subset V$ and K is contained in that side of V which contains $x(t_0-h)$ we have:
$\operatorname{sgn} x(t_0+h)K = x(t_0+h)K \geqslant x(t_0+h)V = x(t_0+h)G = x(t_0-h)G \geqslant x(t_0-h)V \geqslant x(t_0-h)K = -\operatorname{sgn} x(t_0-h)K$. Thus $\operatorname{sgn} x(t_0+h)K + \operatorname{sgn} x(t_0-h)K \geqslant 0$ in this case also. Hence g(t) is strictly convex. This completes the proof of (3.2)

Considering the importance of the convexity of the functions px(t) it is desirable to know precisely which convex px(t) occur in straight spaces; it would, of course, be even more satisfactory to know the exact properties of all px(t), not only the convex ones, but this problem seems very difficult.

The case when p lies on the line x(t), say $p = x(t_0)$, is trivial and therefore uninteresting: $px(t) = |t-t_0|$. Disregarding this case in the future the triangle inequality applied to p, $x(t_1)$, $x(t_2)$ gives with $g(t) = px(t)$,

(3.4) $g(t_1) + g(t_2) \geqslant |t_1 - t_2| \geqslant |g(t_1) - g(t_2)|, \quad g(t) > 0.$

The second half shows that even in general straight spaces g(t) has bounded difference quotients and hence is rectifiable and almost everywhere differentiable. We close this paper by:

(3.5) THEOREM. Given a convex function g(t) satisfying (3.4) there is a unique Minkowskian plane with a point p and a line x(t) for which $px(t) = g(t)$.

The plane is straight (i.e. the circles are strictly convex) if and only if

(3.4') $g(t_1) + g(t_2) > |t_1 - t_2| > |g(t_1) - g(t_2)|$ for

$t_1 \neq t_2$, $g(t) > 0$.

The result is somewhat surprising because one would expect such a special Desarguesian geometry to impose special conditions on $g(t)$ beyond convexity.

In the proof we only consider (3.4), the reader can easily check the ensuing arguments to verify that (3.4') implies the strict triangle inequality for noncollinear points.

First we establish uniqueness. In a Minkowskian plane take affine coordinates x, y such that $x(t) = (t,0)$ and $p = (0,-1)$. If the metric is given by the function $f(x,y)$ with $f(kx,ky) = |k| f(x,y)$, $f(x_1+x_2, y_1+y_2) \leqslant f(x_1,y_1) + f(x_2,y_2)$ then $g(t) = f(t,1)$; hence

(3.6) $f(x,y) = |y| g(\frac{x}{y})$ if $y \neq 0$, $f(x,0) = |x|$.

To prove the existence define $f(x,y)$ by (3.6). Then $f(x,y)$ is continuous; this is clear for $y \neq 0$, but also for $y = 0$ because for $\varepsilon \neq 0$, $f(x,\varepsilon) = |\varepsilon| g(\frac{x}{\varepsilon}) = |x| g(\frac{x}{\varepsilon}) |\frac{x}{\varepsilon}|^{-1} \to |x| = f(x,0)$ for $\varepsilon \to 0$ since (3.4) implies $g(t)|t|^{-1} \to 1$ for $|t| \to \infty$.

The homogeneity $f(kx,ky) = |k| f(x,y)$ is obvious. To prove the convexity of $f(x,y)$ we distinguish several cases which are all simple except for the last.

a) $y_1 + y_2 = 0$. Then either $y_1 = y_2 = 0$ or $y_1 = y > 0$, $y_2 = -y$. The first case is trivial, in the second we have using the first part of (3.4), $f(x_1,y) + f(x_2,-y) = y[g(\frac{x_1}{y}) + g(\frac{-x_2}{y})] \geqslant y|\frac{x_1}{y} - (-\frac{x_2}{y})| = |x_1 + x_2| = f(x_1+x_2, 0)$.

b) $y_1 = 0$ or $y_2 = 0$, say $y_1 = 0$, $y_2 \neq 0$. The second half of (3.4) yields $g(\frac{x_1+x_2}{y_2}) - g(\frac{x_2}{y_2}) \leqslant |\frac{x_1+x_2}{y_2} - \frac{x_2}{y_2}| = |\frac{x_1}{y_2}|$, hence

$f(x_1+x_2,y_1+y_2) = f(x_1+x_2,y_2) = |y_2|g\left(\frac{x_1+x_2}{y_2}\right) \leqslant$
$|y_2|\left[\left|\frac{x_1}{y_2}\right| + g\left(\frac{x_2}{y_2}\right)\right] = f(x_1,0) + f(x_2,y_2).$

c) $y_1 > 0$, $y_2 > 0$ (which because of $f(x,y) = f(-x,-y)$ also covers the case $y_1 < 0$, $y_2 < 0$). Since $g(t)$ is convex, $f(x_1+x_2,y_1+y_2) = (y_1+y_2)g\left(\frac{x_1}{y_1}\frac{y_1}{y_1+y_2} + \frac{x_2}{y_2}\frac{y_2}{y_1+y_2}\right)$
$\leqslant (y_1+y_2)\left[\frac{y_1}{y_1+y_2} g\left(\frac{x_1}{y_1}\right) + \frac{y_2}{y_1+y_2} g\left(\frac{x_2}{y_2}\right)\right] = f(x_1,y_1) + f(x_2,y_2).$

d) $y_1 > 0$, $y_2 < 0$, $y_1+y_2 < 0$. The segment from (x_1,y_1) to (x_1+x_2,y_1+y_2) intersects the x-axis in a point $(x_3,0)$. Hence there is θ, $0 < \theta < 1$, with $x_3 - x_1 = \theta x_2$, $-y_1 = \theta y_2$, $x_1 + x_2 - x_3 = (1-\theta)x_2$ and $y_1 + y_2 = (1-\theta)y_2$. Therefore by the homogeneity of $f(x,y)$ we have $f(x_3-x_1,-y_1) + f(x_1+x_2-x_3, y_1+y_2) = f(\theta x_2,\theta y_2)+f\big((1-\theta)x_2,(1-\theta)y_2\big)=f(x_2,y_2)$. Using a) and b) we find $f(x_1,y_1) + f(x_2,y_2) = f(x_1,y_1) + f(x_3-x_1,-y_1) + f(x_1+x_2-x_3,y_1+y_2) \geqslant f(x_3,0) + f(x_1+x_2-x_3,y_1+y_2) \geqslant f(x_1+x_2,y_1+y_2)$. The case $y_1+y_2 > 0$ is reduced to this by $f(-x_1,-y_1) + f(-x_2,-y_2) \geqslant f(-x_1-x_2,-y_1-y_2)$. The proof is now complete.

If $g(t) = g(-t)$ then $(0,0)$ is the foot of p on $x(t)$ and the plane possesses the reflection in the y-axis and hence all lines $x = \text{const}$.

In the hyperbolic plane with the distance $h(u,v)$ if the point $p\big(\notin x(t)\big)$ has $x(0)$ as foot on $x(t)$ (with $h\big(x(t_1), x(t_2)\big) = |t_1-t_2|$) then $g(t) = h\big(p,x(t)\big) = \text{Area}\cosh[\cosh h(p,x(0)) \cosh t]$ is convex and hence, unexpectedly, also occurs as distance $g(t)$ in a Minkowskian plane which possesses the reflection in each of a family of parallel lines.

REFERENCES

[1] Busemann, H. : The Geometry of Geodesics, Academic Press, New York, 1955.

[2] Busemann, H. and Phadke, B.B. : Minkowskian Geometry, Convexity Conditions and the Parallel Axiom, Journal of Geometry 12/1(1979)17 - 33.

[3] Busemann, H. and Phadke, B.B.: Symmetric Spaces and Ellipses, Proceedings International Christoffel Symposium, Birkhäuser, Basel 1980.

The University of Southern California, Los Angeles, California, 90007, U.S.A.
The Flinders University of South Australia, Bedford Park, South Australia, 5042, Australia.

(Eingegangen am 20.Februar 1979)

BOOK REVIEWS

Hilbert's fourth problem, by A. V. Pogorelov, Wiley, New York; Holt, Rinehart and Winston, New York, 1979, vi + 97 pp., \$16.00.

Hilbert's Problem 4 stripped of its comments[1] is this: Omit from his axioms for the foundations of geometry besides the parallel axioms all those which contain the concept of angle, and replace them by the triangle inequality, which follows from the congruence axiom for triangles (*CT*).

(1) Determine all geometries satisfying these conditions.

(2) Study the individual ones.

This is not quite Hilbert's formulation because with respect to angles he only omits the *CT*. The remaining angle axioms have no significant applications without *CT*. Pogorelov uses the preferable form given above.

Only two such geometries, besides the elementary ones, were known at the time, the Minkowskian satisfying the euclidean parallel axiom and Hilbert's geometry generalizing in a similar way the hyperbolic situation. It seems that Hilbert did not think of a mixed situation like a half plane.

Nor is it clear, whether he wanted to include nonsymmetric (n.s.) distances (a Minkowski metric may be symmetric or not). Hilbert was admittedly probing and considered, in fact, an analysis of the distance concept as one of the tasks connected with the problem. Since absence of symmetry violates one of the axioms Pogorelev does not admit n.s. distances.

That Hilbert was interested in them is evident from Hamel's thesis [**2**], which he directed immediately after his lecture. It dwells largely on n.s. distances. They were deemphasized in the later version [**3**] of [**2**], most probably because from the great variety of Desarguesian metrics (both symmetric and not) which Hamel exhibited, no appealing new one emerged. The situation changed in 1929, when Funk discovered a very interesting, always n.s. geometry which resembles euclidean geometry in some respects and hyperbolic in others. In addition it led to the proper definition of completeness: The balls $\{x|px \leqslant \rho\}$ are compact, but not necessarily the $\{x|xp \leqslant \rho\}$, where xy is the distance.

Hamel approached the problem through the Weierstrass Theory of the calculus of variations, which requires smoothness properties alien to the foundations of geometry, the framework visualized by Hilbert for the problem. This is less surprizing than it appears because Hilbert was very interested

---

[1]These are translated in *Hilbert's mathematical problems* **8** (1902) of this Bulletin. The translation is reprinted in [**1**], a report on the symposium on Hilbert's Problems held in 1974. Problem 4 appears there on pp. 131–141 with the title *Desarguesian spaces*. Hilbert's heading, slightly modernized, is: *The geometries in which the ordinary lines provide the shortest connections*. We refer the reader to the article in [**1**] for all facts concerning Problem 4 omitted here, because they are not connected to Pogorelev's book.

in the Weierstrass Theory at the time; in fact, he contributed his important invariant integral in his comments to Problem 23.

The discovery of the great variety of solutions showed that part (2) of Problem 4 is not feasible. *It is therefore no longer considered as part of the problem.* But many interesting special cases have been studied since 1929.

[2] remains to this day the only attack on the n.s. case. Therefore I only give the modern version of Problem 4 for symmetric distances. The length of a curve in a metric space is defined in strict analogy to the elementary procedure. The *Desarguesian* spaces R are given by the following conditions.

(1) *R is a nonempty open subset of the real projective space P^n metrized by a distance xy* (providing the correct topology) such that

(2) *The balls $\{x|px \leqslant \rho\}$ are compact* (or the Bolzano-Weierstrass Theorem holds).

(3) *Any two distinct points x, y can be connected by a segment, i.e. a curve of length xy.*

(4) *$xy + yz > xz$ when x, y, z are not collinear in P^n.*

In the spirit of the time Hilbert restricted himself to $n = 2, 3$ and so does Pogorelev. However, this has doubtless pedagogical reasons, because he addresses a wide class of readers. The real difference is between $n = 2$ and $n > 2$. Pogorelev's method works for $n > 3$, but requires greater technicalities. For the same reason he *assumes* that R is either all of $P^{2,3}$ or all of the affine $A^{2,3}$ or a convex subset of $A^{2,3}$, although Hamel *proved* in **[3]** that R must have one of these forms (this holds actually for all n and even n.s. distances). The proof becomes simple by using covering spaces, which Pogorelev avoids. Thus he accomplishes more than appears at first sight.

I will explain in some detail his principal contribution, because for my report in **[1]** the Russian book (price: 24 kopeks) was not yet at my disposal.

Led by integral geometry I had observed (references are found in the translation) that many Desarguesian spaces can be obtained as follows, taking $R = P^n$ as an example. Denote hyperplanes by H, and for a point set M put $H_M = \{H|H \cap M \neq \varnothing\}$. Define on the set of all hyperplanes a nonnegative measure ε with (a) $\varepsilon(H_{\{p\}}) = 0$ for each point p, (b) $\varepsilon(H_M) > 0$ when $M \neq \varnothing$ is open, (c) $\varepsilon(H_{P^n}) = 2k < \infty$. Notice that $\varepsilon(H_L) = 2k$ for any line L because $H_L = H_{P^n}$. Two distinct points x, y divide the line L through them into two arcs A_1, A_2 and $\varepsilon(H_{A_1}) + \varepsilon(H_{A_2}) = \varepsilon(H_L) = 2k$ because of (a), so that at least one A_i, say A_1, satisfies $\varepsilon(H_{A_1}) \leqslant k$. Putting $xy = \varepsilon(H_{A_1})$ defines a Desarguesian metrization of P^n with A_1 as segment.

Without essential changes this construction can be used when $R \subset A^n$ by admitting the value ∞ for $\varepsilon(H_M)$.

Pogorelev proves that in detail. Now the question arises whether this method yields all Desarguesian metrics. *The answer is affirmative for $n = 2$, but $n > 2$ requires measures δ, which may take negative values on many sets.* Unexpectedly, this proves compatible with $\delta(H_T) > 0$ for every segment T. I had noticed this situation previously in the very special case of the Minkowskian geometries.

We denote a metric given by a δ with $\delta(H_T) > 0$ as a δ-metric. *The construction of* δ (*or* ε *for* $n = 2$) *for a given Desarguesian metric is Pogorelev's prime contribution.*

He shows first that every δ-metric can be approximated, uniformly on every compact set, with Desarguesian metrics derived from integrands $F(x, dx)$ satisfying the standard conditions of the calculus of variations ($n = 2, 3$).

In cartesian coordinates the euclidean densities d_eL, d_eH (i.e. $H_e(M) = \int_{H \cap M \neq \varnothing} d_eH$) for a line L in P^2 or a plane H in normal form $x \cdot u - p = 0$, is $d_eL = dudp$, $d_eH = dudp$ (u is a unit vector normal to L, resp. H, and p the distance from the origin). It is clear that in the general case the densities must have the form $\gamma(L)d_eL$ or $\gamma(H)d_eH$. *Thus the problem becomes finding* γ *in terms of* $F(x, dx)$.

By an ingenious use of a special form which the Euler equations take when the straight lines are the extremals, this task is reduced to Funk's problem of representing the supporting function $G(y)$ of a convex body K with the origin as center in the form $G(y) = \int_U \gamma(u)|y \cdot u|du$ with even $\gamma(u)$. The problem is easily solved with $\gamma(u) \geqslant 0$ for $n = 2$, but is far from trivial for $n = 3$ and $\gamma(u)$ may take negative values.[2]

The solution is applied for each x to $F(x, \dot{x})$ with $y = \dot{x}$ in $\dot{x}$-space. $F(x, y)$ is, in fact, because of $F(x, kx) = |k|F(x, x)$ and the Legendre condition, the supporting function of a convex body with the origin as center in y-space. This defines $\gamma(u, x \cdot u)$ and $dH = \gamma(u, xu)dpdu$.

The final answer for $\delta(n = 3)$ is expressed in terms of the measure $\delta'(Q')$ of the point set Q' obtained by a correlation from a set Q of planes where $\delta'(Q') = \delta(Q)$. *The conditions are* $\delta'(P) = 0$ *for every plane* P (corresponding to (a)), $\delta'(C) \geqslant 0$ *for every cone* C *and* $\delta'(C) > 0$ *when* C *has interior points.*

The very well written book begins with an introduction to the most elementary facts concerning $P^{2,3}$ whose points are defined as equivalence classes of triples, resp. quadruples of reals. It then gives topologies for the lines and planes. The detailed exposition of the results sketched above follows. Finally the tract turns to the foundations of geometry (with Pogorelev's own variation of Hilbert's axioms) where it is clearly undesirable to take the analytic definition of $P^{2,3}$ for granted. The special geometries of Minkowski and Hilbert are discussed. As all this is rather known territory, I will not dwell on it.

The blurb on the cover states that the book is accessible to advanced undergraduates. This seems unlikely if for no other reason than the breadth of knowledge which is taken for granted and is certainly so in the Russian original, which provides practically no references. The excellent translation by R. A. Silverman contains a good list of these, for all knowledge assumed, and a very judicious four pages of notes, both contributed by E. Zaustinsky, which will greatly help the reader (not only students) to locate the material,

---

[2]For readers familiar with the theory of convex bodies I mention that $\gamma(u) > 0$ means that K is a projection body, so that the δ-metric is an ε-metric, when $F(x, y)$ supports for each fixed x a projection body in y-space (see below). It is trivial that a two-dimensional K is always a projection body, so that γ is easy to find, but for $n > 2$ projection bodies are very special cases.

and beyond this explains at crucial points the reasons for certain steps which could baffle a newcomer to the field.

References

1. Felix E. Browder (ed.), *Mathematical developments arising from Hilbert problems*, Proc. Sympos. Pure Math., vol. 28, Amer. Math. Soc., Providence, R. I., 1976.
2. G. Hamel, *Ueber die Geometrien, in denen die Geraden die Kürzesten sind*, Dissertation, Göttingen, 1901, 90 pp.
3. ______, same title, Math. Ann. **57** (1903), 231–264.

Herbert Busemann

Bull. Amer. Math. Soc. 4 (1981), 87–90

A GENERAL VERSION OF BELTRAMI'S THEOREM IN THE LARGE

H. BUSEMANN AND B. B. PHADKE

Beltrami's Theorem (1865) determines the Riemann spaces whose geodesics behave locally like affine lines. Its global form (proved much later) states that a complete simply connected Riemann space with this property is a euclidean, hyperbolic or spherical space. Here we establish what we believe to be the most general meaningful version of this theorem. We define a general class of complete metric spaces, called chord spaces, which possess distinguished extremals. In our case these must behave locally like affine lines, but they need not be the only extremals. This situation occurs in many important spaces.

Our principal result is that such a space (of n dimensions, $n > 1$) can be mapped topologically and geodesically either on the entire n-sphere S^n or on an arbitrary open convex subset of an open hemisphere of S^n, considered as the affine space A^n. Examples of such spaces with some unexpected phenomena and the significance of chord spaces in general are discussed.

1. Introduction. The classical theorem of Beltrami was published in 1865 in [**1**]. It states that an n-dimensional Riemann space (Beltrami assumes $n = 2$, but this case implies the result for $n > 2$. We take $n > 1$ throughout) in which the geodesics behave locally like projective lines has constant curvature and is therefore, according to Riemann, locally euclidean, hyperbolic or spherical.

The corresponding theorem in the large which was proved only in this century is: *If the space R is (in addition) complete and simply connected then it is a spherical space S^n or is defined in an open hemisphere of S^n, i.e. the affine space A^n and either $R = A^n$ is the euclidean space or, with a proper choice of the hemisphere R is a hyperbolic space defined in the interior of an ellipsoid in A^n.*

The proofs use strongly the properties of these three metrics.

The purpose of the present paper is to give a Beltrami's Theorem (B.T.) in the large the most general form which seems reasonable. Because the Theorem of Desargues plays a fundamental role we call for brevity a space locally desarguesian (l.d.)[1] when each point has a neighborhood

[1]Our paper [**9**] treats special cases of interest and some independent topics. At the time we were not aware of the monodromy argument. We used it in a very similar form in [**11**].

which can be mapped topologically and geodesically on a convex set V in the projective space P^n, where convex means that there is a hyperplane $H \subset P^n$ with $V \cap H = \varnothing$ and V is convex as set in the affine space $P^n - H$. Although it proves convenient to use a metric it turns out to be immaterial which model of P^n or S^n we use and we agree to always take P^n or S^n with curvature 1.

The usual setting for generalizations of the present type has been G-spaces, see [**3** quoted as G]. But some of the most interesting and important desarguesian spaces are not G-spaces. [**8**], which is not readily accessible, gives axioms for a wider class of spaces where certain shortest joins, called chords, are distinguished. They comprise the l.d. spaces where locally the projective lines furnish shortest connections but not necessarily the only ones. We give here a different, more concise, set of axioms for these "*chord spaces*" and discuss in §2 those of their properties which are needed here.

We split the B.T. in the large into two parts of which the first is of a topological nature and our main contribution. The types of simply connected spaces become very numerous for large n, so that it is not at all obvious that the very general metrics admitted in chord spaces do not lead to other image spaces than S^n. We prove, keeping in mind that chord spaces are by definition complete:

MAIN THEOREM. *A locally desarguesian simply connected chord space R is either defined in all of S^n or is an arbitrary open convex set of an open hemisphere of S^n (considered as A^n).*

That each convex subset R of A^n can be metrized as a straight G-space in infinitely many essentially different ways is contained in the proof of G(18.14). This is one reason why an analogue to the proof in the Riemannian case is not feasible.

Pogorelov [**17**] determined all globally desarguesian G-spaces. We will see that his results can be used to also determine all l.d. chord metrics so that we have the following

METRIC COMPLEMENT TO THE MAIN THEOREM. *All simply connected locally desarguesian chord spaces are also known metrically.*

The general l.d. chord spaces are those covered by the simply connected ones. But they are not all known even in the Riemannian case, in particular our knowledge on the hyperbolic case is very lacunary.

This material forms the content of §3. Finally we owe the reader cogent reasons which force us to go beyond G-spaces. In §4 we provide examples showing the importance of admitting chord spaces. In addition it will become clear that chord spaces are interesting quite independently of their significance for desarguesian spaces.

2. Chord spaces. Our space R is always a metric space with distance between x and y denoted simply by xy satisfying the metric space axioms:

$$xy \text{ is real valued and } > 0 \text{ if } x \neq y,$$
$$xx = 0, \quad xy = yx \quad \text{and} \quad xy + yz \geq xz$$

for any x, y, z in R. Convergence of sets in R is understood in terms of Hausdorff's closed limit, see G §3. When points a, b, c satisfy $ab + bc = ac$ we say that b lies between a and c and denote this by the symbol (abc); here a, b, c are assumed to be distinct. Length of curves in R is defined in the usual manner and a segment $T(a, b)$, $a \neq b$, is an isometric map $x(t)$ of an interval of length ab of the real t-axis $\mathbf{R}$ into R i.e. $x(t)$ is defined in an interval $\alpha \leq t \leq \beta = \alpha + ab$, $x(\alpha) = a$, $x(\beta) = b$,

$$x(t_1)x(t_2) = |t_1 - t_2|.$$

We note that if $a_\nu \to a$, $b_\nu \to b \neq a$ and if segments $T(a_\nu, b_\nu)$ exist and converge then the limit is a segment $T(a, b)$.

For a given subset $\overline{\mathscr{C}}$ of the set of all segments in R a partial $\overline{\mathscr{C}}$-extremal is a class of maps $\{x(t), M\}$, where M is a closed connected subset with more than one point of the reals $\mathbf{R}$, $t \in M$, $x(t) \in R$ and for each t in M numbers $t' < t < t''$ exist with

$$(*) \qquad x(t_1)x(t_2) = |t_1 - t_2| \quad \text{for } t_1, t_2 \text{ in } [t', t''] \cap M$$

and

$$(**) \qquad x(t)|[t', t''] \cap M \in \overline{\mathscr{C}}.$$

In the cases $M = \mathbf{R}$ or $M = [\alpha, \beta]$ we use the terms $\overline{\mathscr{C}}$-extremal or $\overline{\mathscr{C}}$-extremal curve respectively. If $M = [\alpha, \beta]$ and $(*)$ and $(**)$ hold for all t_1, t_2 in $[\alpha, \beta]$ we speak of a $\overline{\mathscr{C}}$-segment from $a = x(\alpha)$ to $b = x(\beta)$.

If the class is restricted to maps $t \to t + \eta$ we obtain an oriented partial $\overline{\mathscr{C}}$-extremal,...,oriented $\overline{\mathscr{C}}$-segment.

A $\overline{\mathscr{C}}$-extremal $\{x(t), \mathbf{R}\}$ is a $\overline{\mathscr{C}}$-straight line or simply a $\overline{\mathscr{C}}$-line if $(*)$ holds for any t_1, t_2. It is a $\overline{\mathscr{C}}$-great circle of length $\beta > 0$ if

$$x(t_1)x(t_2) = \min_{\nu = 0, \pm 1, \pm 2, \dots} |t_1 - t_2 + \nu\beta|.$$

We now state the axioms for a space R to be a chord space.

I. *R satisfies the conditions* I, II, III *of G-spaces which are* (*see* G p. 37):

(i) *The space is metric with distance xy.*

(ii) *The space is finitely compact, i.e. a bounded infinite set has at least one accumulation point.*

(iii) *Given two distinct points x, z then a point y with (xyz) exists.*

Before stating the second set II of axioms we note that it follows from I that any two points can be joined by a segment, this is proved in G(6.8).

The second set of axioms is:

II. *There is a set $\mathscr{C}$ of segments termed chords $Q(a, b)$. Calling $\mathscr{C}$-extremals and $\mathscr{C}$-extremal curves geodesics* resp. *geodesic curves we require*:

(a) *Each class $\mathscr{H}$ of homotopic curves form a to b (for given a, b with $\mathscr{H} \neq \emptyset$ for $a = b$) contains a shortest curve which is a geodesic curve.*

(b) *The limit of a converging sequence $Q(a_\nu, b_\nu)$ with $a_\nu \to a$, $b_\nu \to b \neq a$ is a $Q(a, b)$.*

(c) *If two chords $Q(a, b)$ and $Q(a, b')$ with $ab \geq ab'$ have a common point other than a then $Q(a, b') \subset Q(a, b)$ or $Q(a, b) \cap Q(a, b') = \{a, b\}$. (Trivially $b = b'$ in the latter case.)*

(d) *Each point has a neighborhood $S(p, \sigma_p)$, $0 < \sigma_p$, such that for a, b in $S(p, \sigma_p)$ a $Q(a, c) \Supset Q(a, b)$ exists.*

The notations $A \Subset B$ or $B \Supset A$ mean that A is a proper subset of B.

We note some consequences of the axioms. From (a) alone (I is always assumed) we have

(1) *$Q(a, b)$ exists for $a \neq b$. Every subarc of $Q(a, b)$ is a chord.*

From (a) and (c)

(2) *When b' is an interior point of $Q(a, b)$ then $Q(a, b')$ is unique and a subarc of $Q(a, b)$.*

(3) *If $Q(a, c_i) \Supset Q(a, b)$ and $bc_1 = bc_2$ (or equivalently $ac_1 = ac_2$) then $Q(a, c_1) = Q(a, c_2)$.*

If not then (c) would imply $Q(a, c_1) \cap Q(a, c_2) = \{a, c_1\}$ but $Q(a, b) \subset Q(a, c_i)$, $i = 1, 2$.

Let $\sigma(p)$, $\sigma'(p)$ be resp. the supremum of the ρ for which (d) holds resp. $Q(a, b)$ is unique for $a \neq b$ in $S(p, \rho)$.

Since $Q(a, c) \Supset Q(a, b)$ implies by (2) that $Q(a, b)$ is unique we have first that $Q(a, b)$ is unique in $S(p, \sigma_p)$. Then we have using G(7.5) that $\sigma'(p) \geq \sigma(p) > 0$. (d) holds in $S(p, \sigma(p))$ and

(4) *$\sigma(p) \equiv \infty$ or $0 < \sigma(p) < \infty$ and $|\sigma(p) - \sigma(q)| \leq pq$.*

The proof is similar to G(7.5) and the same proof shows that σ' satisfies the relations (4) too.

We note, without using it, that in our case $\sigma(p) \equiv \sigma'(p)$ because a l.d. space is a topological manifold, see [7, p. 19].

(5) *If $pa < \sigma(p)$, a chord $Q(a, b) \ni p$ with $pa = pb$ exists.*

The proof of (5) is given explicitly in G(7.8).

(6) *A chord can be extended uniquely to a geodesic. $\sigma(p) \equiv \infty$ characterizes the straight chord spaces, i.e. those where all geodesics are straight lines.*

The proof is a simple combination of the arguments for G(7.9) and (8.4, 5).

(7) *If a geodesic contains with any $a \neq b$ a $Q(a, b)$ then it is either a straight line or a great circle.*

The proof is identical with that of G(9.6). Although we always assume dim $R > 1$ we mention for justification the following corollary of (7) (whose proof is as in G(9.7)).

(8) *A 1-dimensional chord space is a straight line or a great circle.*

REMARK. *Dimension of a chord space is understood as its dimension as a metric space as defined, for example, in Hurewicz and Wallman* [**14**].

The following Theorem (10) is known for G-spaces but not found in G or [7] but in [**9**, (2.8)]. Because it is clearly of prime importance for B.T. we provide a proof for chord spaces, in particular because it is partly superior to, and more complete than the argument in [**9**, (2.8)].

But first we note:

(9) *Let $x_\nu(t)$, $\nu = 0, 1, 2, \ldots$, represent geodesics and $\lim_{\nu \to \infty} x_\nu(t) = x_0(t)$ for $t \in [\alpha, \beta]$, $\alpha \neq \beta$. Then $x_\nu(t)$ converges to $x_0(t)$ uniformly on any bounded set of* **R**.

The proof is like that of G(8.11, 12).

(10) THEOREM. *If an open set $R \neq \emptyset$ of S^n is metrized as a chord space whose geodesics fall on the great circles in S^n then R is either a convex set in an open hemisphere (i.e. A^n) and straight or $R = S^n$ and S^n-antipodes are R-antipodes.*

When $a \neq b$ are points of R and not antipodes in S^n, then the great circle on S^n and hence the R-geodesic through a and b is unique. It contains all $Q(a, b)$ because each lies on a geodesic. When a, b are antipodes, choose a $Q(a, b)$ and a sequence $b_\nu \to b$, $b_\nu \neq b$, $b_\nu \in Q(a, b)$. Then a and b_ν are not antipodes hence $Q(a, b_\nu) \subset Q(a, b) \subset R$ and (9) shows that there is a geodesic containing $Q(a, b)$; moreover $Q(a, b)$ is an S^n-semigreat circle. It follows from (7) that each geodesic in R is a great circle or a straight line.

A chord in R which is a semigreat circle $Q(a, b)$ cannot be prolonged to a chord $Q(a, c) \supsetneqq Q(a, b)$. For if d is an interior point of the subchord $Q(b, c)$ then both arcs from a to d of the S^n-great circle containing $Q(a, b)$ would be a $Q(a, d)$ and unique, one because a and d are not antipodes, the other because a proper subchord of a chord is unique.

To finish the proof we show: if R contains two antipodes a, b then $R = S^n$. We proved that with any other point q of R the entire S^n-great circle through a, b, q lies in R. For an arbitrary p ($\neq a, b$) in S^n the S^n-great circle C' through a, b, p contains other points of R because a neighborhood of a lies in (the open) R. Therefore $p \in C' \subset R$.

If in our S^n as given chord space we identify antipodes we obtain a chord metric in P^n, see G p. 129.

We will see in §4 that often fundamentally different chord spaces may be constructed with the same metric. But if for some two points $a \neq b$ the segment $T(a, b)$ is unique, then it must be a $Q(a, b)$ in any compatible chord space. This implies that if the given metric defines a G-space then the only chord space compatible with it is the space with all $T(a, b)$ as chords. For (abc) in a G-space the segment $T(a, b)$ is unique, so that the $Q(a, b)$ are determined locally.

Therefore

(11) *A chord space is a G-space* iff *all segments are chords. In this case the metric determines the chord space uniquely.*

The same argument shows

(12) *If a space is metrized such that $T(a, b)$ always exists and is unique, then the space is either a chord space with the $T(a, b)$ as $Q(a, b)$ or there is no compatible chord space.*

The latter may happen even when for given $a \neq b$ and $\rho > 0$ a c with $ba = \rho$ and (abc) exists. For example, consider the following metrization of the (x^1, x^2)-plane. For $x = (x^1, x^2)$ and $y = (y^1, y^2)$ we define

$$xy = \sqrt{(x^1 - y^1)^2 + (x^2 - y^2)^2} + |x^1 - y^1| \quad \textit{if } x^2, y^2 \geq 0$$

$$xy = 2\sqrt{(x^1 - y^1)^2 + (x^2 - y^2)^2} \quad \textit{if } x^2, y^2 \leq 0$$

$$xy = \min_t \left(x(t,0) + (t,0)y \right) \quad \text{if } x^2 \le 0 \text{ and } y^2 \ge 0.$$

Then the prolongation of, for example, the segment along the x^2-axis from $a = (0,1)$ to $z = (0,0)$ is not unique. None of these many continuations of $T(a, z)$ can be omitted because otherwise no $T(a, c)$ for some $c = (c^1, c^2)$ with $c^2 < 0$ would exist.

(For suitable conditions which prevent this nonuniqueness see [**12**] where we construct a class of examples of G-spaces).

3. Locally desarguesian spaces. In this paper we call a subset Q of a chord space R "convex" if the following condition holds: when x and y lie on Q there is one and only one chord $Q(x, y)$ in R joining x and y and this chord is contained in Q.

When R is the euclidean space, this concept of convexity coincides with the usual affine space convexity. We note also that our definition of convexity above yields the following notion of convexity in the projective space P^n: a set Q in the real projective space P^n is called convex if Q is convex in the G-space sense above when P^n is metrized as the elliptic space. (The elliptic space is a G-space). This notion of convexity, of course, coincides with the definition in the Introduction. Finally we mention the case when R is the spherical space S^n. Here our definition amounts to the following: $Q \subset S^n$ is convex if Q lies in an open hemisphere and if for x, y in Q *the* shorter of the two arcs of *the* great circle containing x and y lies in Q.

We now define and explain the notion of a locally desarguesian chord space. A chord space R is said to be locally desarguesian if each point p of R is contained in a nonempty open convex set U which can be mapped homeomorphically onto an open convex set $\tilde{U}$ in the real affine space A^n in such a way that the geodesic curves contained in U are mapped onto affine segments in $\tilde{U}$. (Such a map is called a geodesic map.) This is equivalent to saying that each point p of R is contained in an open convex set which can be mapped homeomorphically and geodesically onto an open convex set of the sphere S^n, or the projective space P^n.

In the sequel we use throughout a fixed sphere S^n (and neither P^n nor A^n) as the "model space" i.e. the space into which we map.

If the open convex set U of R is mapped homeomorphically and geodesically onto an open convex subset of S^n by means of a map ϕ we say that U is a "desarguesian domain" and ϕ is a "desarguesian chart". The pair (U, ϕ) is called a "desarguesian element".

The connection between our use of the word "desarguesian" and the theorem of Desargues derives from the "*Desargues Property*" formulated for G-spaces by H. Busemann and the implications of this property for G-spaces; see [G, Chapter II, §§12, 13 and 14].

The arguments in the proof of our Main Theorem are to a great extent simplified because the following properties hold for our desarguesian domains and charts:

(A) *Standardization of the size of a domain: There exists a positive continuous function $\gamma(p)$ such that each closed ball $\bar{S}(p, \gamma(p))$ lies in a desarguesian domain $W(p) \ni p$.*

(B) *Uniqueness of charts: If (U, ϕ) and (V, ψ) are two desarguesian elements and $\phi = \psi$ on a nonempty open set $Z \subset U \cap V$ then $\phi = \psi$ on the entire intersection $U \cap V$.*

(C) *Freedom of transition from one chart to another: If (W, ψ) and (Z, ϕ) are desarguesian elements and $W \supset Z$ then there exists a desarguesian element (W, Φ) such that $\Phi|Z = \phi$ i.e. the restriction of Φ to Z coincides with ϕ.*

We now establish these properties. First consider (A). The proof of this uniformity of standardized sizes is based on the frequently used idea of [G (7.5) p. 33]. In fact, let $\mu(p)$ be the supremum of positive numbers α such that $S(p, \alpha)$ is contained in some desarguesian domain V. Since R is locally desarguesian, $\mu(p)$ is positive. When $S(p, \alpha)$ is contained in V, $S(q, \alpha - pq)$ is also contained in V if $\alpha > pq$. Therefore, $\mu(p)$ is either identically equal to ∞ or else $|\mu(p) - \mu(q)| \leq pq$ for all p, q in R. Hence if we put, for example, $\gamma(p) = \frac{1}{2}\min(\mu(p), \sigma(p))$, then each closed $\bar{S}(p, \gamma(p))$ is contained in some desarguesian domain $W(p)$. We note that since $\gamma(p)$ is continuous and positive, it is bounded away from zero on any compact set.

We now consider properties (B) and (C). The proof of these properties is, in turn, based on the following facts about geodesic maps of the sphere S^n onto itself. (Such bijective homeomorphisms of S^n onto S^n sending great circles to great circles are also called spherical collineations.)

I. *A spherical collineation sends a convex set onto a convex set.*

II. *If f and g are spherical collineations and if $f = g$ on a nonempty open set U then $f = g$ on the whole of S^n.*

III. *Let $\tilde{f}$ be a bijection of a nonempty open convex set $Q \subset S^n$ onto a nonempty open convex set $Q' \subset S^n$ such that $\tilde{f}$ sends arcs of great circles in Q onto arcs of great circles in Q'. Then $\tilde{f}$ can be extended to a collineation f of S^n. The extension f of $\tilde{f}$ is unique.*

The proofs of these three properties are fairly standard and can be obtained by using information about projective collineations and the standard covering of P^n by S^n.

We now derive the properties (B) and (C).

For (B) we simply have to observe that if $\tilde{U} = \phi(U)$ and $\tilde{V} = \psi(V)$ and $\tilde{Z} = \phi(Z) = \psi(Z)$ then $\phi \circ \psi^{-1}$ sends $\tilde{Z}$ to $\tilde{Z}$ and is the identity on $\tilde{Z}$ and hence its unique extension to S^n is the identity. Hence $\phi \circ \psi^{-1}$ is the identity on $\psi(U \cap V)$ also. Hence $\psi(U \cap V) = \phi(U \cap V)$ and $\phi = \psi$ on $U \cap V$.

Finally we consider (C). Here the map $\tilde{g} = \psi \circ \phi^{-1}$ from $\phi(Z)$ to $\psi(Z)$ satisfies the hypothesis in III above and hence has an extension g to S^n. We now define $\Phi = g^{-1} \circ \psi$ so that $\Phi | Z = \phi$.

From now on we assume that all our desarguesian domains have the property that if U is a desarguesian domain then there exists at least one $p \in U$ such that $\bar{S}(p, \gamma(p))$ is contained in U.

When desarguesian elements (U, ψ) and (V, ϕ) satisfy the condition that $\psi = \phi$ on $U \cap V$ and $U \cap V$ is nonempty we say that (U, ψ) and (V, ϕ) are immediate continuations of each other.

Let (U, ϕ) be a desarguesian element and let C be a continuous curve parametrized by $\alpha(t)$, $a \le t \le b$, with $\alpha(a) \in U$. We say that (U, ϕ) is continued along C to (V, ψ) if there exists a partition Δ: $a = t_0 < t_1 < t_2 < t_3 < \cdots < t_k = b$ and a family (U_i, ϕ_i), $i = 0, 1, 2, \dots, k$, of desarguesian elements such that:

(i) $\alpha([t_i, t_{i+1}]) \subset S(\alpha_i, \gamma_\alpha/2) \subset \bar{S}(\alpha_i, \gamma_\alpha/2) \subset U_i$ where $\alpha_i = \alpha(t_i)$ and $\gamma_\alpha = \inf\{\gamma(\alpha(t)), a \le t \le b\} > 0$. ($\gamma_\alpha > 0$ since γ is continuous and the trace of, i.e. the actual pointset occupied by, $\alpha(t)$ is a compact set).

(ii) (U_i, ϕ_i) and (U_{i+1}, ϕ_{i+1}) are immediate continuations of each other, $i = 0, 1, \dots, k - 1$.

(iii) $U \subset U_0$, $\phi_0 = \phi$ on U, $V \subset U_k$ and $\psi = \phi_k$ on V.

If a curve C joins p to q and lies completely in a single desarguesian domain W then from properties (B) and (C) we conclude that continuation of any element (U, ϕ), $p \in U$ along C to (V, ψ) exists; in fact we can take $V = W$ and $\phi = \psi$ on $W \cap U$.

Since any general continuous curve $\alpha(t)$, $a \le t \le b$, has a compact trace which can be covered by the union of a finite number of desarguesian domains of appropriate sizes satisfying condition (i) above we conclude that continuations of desarguesian elements along all curves exist.

We now show that continuation is "unique". First consider the case when C_1 and C_2 are two curves both joining p to q ($C_1 = C_2$ is not excluded) and both lying in a single desarguesian domain W. Then if (U_i, ϕ_i) are continuations of (U, ϕ) along C_i, $i = 1, 2$, then the properties (B) and (C) above show that $\phi_1 = \phi_2$ on $U_1 \cap U_2$. Thus, in single desarguesian domains continuation is independent of the path. In other words a single desarguesian domain W is always a "neutral" neighborhood for continuation. (Compare arguments on p. 43 of [**11**].)

We remember now that our chord space R is simply connected and consider the case when C_1 and C_2 are any two curves both joining p to q. ($C_1 = C_2$ is not excluded.) Then since C_1 and C_2 are homotopic we can subdivide the deformation square of the deformation homotopy into a finite number of small subsquares in such a way that the image of each subsquare lies in a "neutral" neighborhood. A familiar argument (similar to those occurring in some proofs of the monodromy theorem, for example, in complex analysis) then shows that if (U, ϕ) is continued to (U_i, ϕ_i) along C_i, $i = 1, 2$, then $\phi_1 = \phi_2$ on $U_1 \cap U_2$. (Again compare [**11**, pp. 43, 44].)

Thus we have:

LEMMA. *Let R be a simply connected locally desarguesian chord space and let (U, ϕ) be a desarguesian element. Then (U, ϕ) can be continued along all curves beginning in U. If C_1 and C_2 are two curves starting at p and ending at q and if (V_i, ϕ_i), $i = 1, 2$, are continuations of (U, ϕ) along C_i, $i = 1, 2$, respectively, then $\phi_1 = \phi_2$ in the nonempty neighborhood $V_1 \cap V_2$ of q.*

Choose now a fixed point z of R and a fixed desarguesian element (U, ϕ) with $z \in U$. For an arbitrary point $x \in R$ let C be a curve from z to x and (U_C, ϕ_C) be a desarguesian element obtained by continuing (U, ϕ) along C. Then from the lemma above we see that $x' = \phi_C x$ is independent of the choice of the curve C and the partition used during continuation. Hence the association of x' with x yields a desarguesian chart in the neighborhood $W(x)$ of x. We denote by Δ the map which sends $x \in R$ to $x' \in S^n$ as above.

Since Δ is a desarguesian chart on each $W(x)$, the image under Δ of any geodesic in R lies on (i.e. is contained in) a great circle in S^n. Using this and the local injectivity of Δ we show that Δ is actually globally injective i.e. if $a \neq b$ then $\Delta a \neq \Delta b$.

For an indirect proof assume that $\Delta a = \Delta b$ and $a \neq b$. Let Q be any chord joining a to b. Then Δ maps Q onto a great circle Γ, say, in $R' = \Delta R \subset S^n$. Hence on Q there exist points p, q such that $p' = \Delta p$, $q' = \Delta q$; p, q are distinct from a, b and p', q' are antipodal on Γ. Since $R' = \Delta R$ is an open subset of S^n we can find another great circle $\Sigma \neq \Gamma$, $\Sigma \subset R'$, $p', q' \in \Sigma$. Choose now sequences p'_n, q'_n on Σ such that p'_n is not antipodal to q'_n, $p'_n \to p'$, $q'_n \to q'$ as $n \to \infty$. As Δ is injective in neighborhoods of p, q we can find sequences p_n, q_n in R such that $p_n \to p$, $q_n \to q$

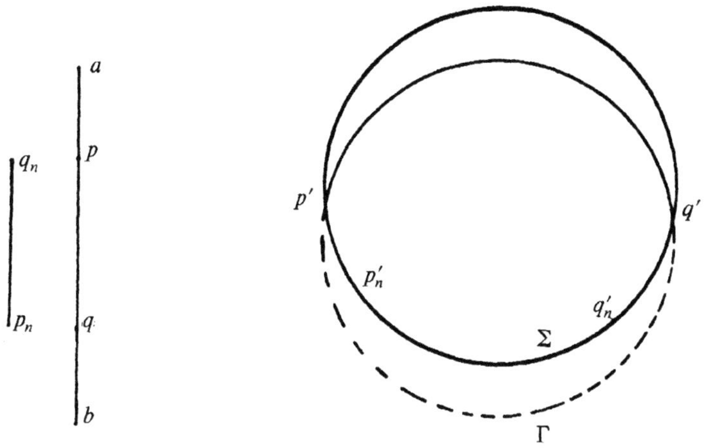

FIGURE 1

and $\Delta p_n = p_n'$, $\Delta q_n = q_n'$. Let Q_n be any chord joining p_n and q_n. See Figure 1.

$Q(p, q)$ is unique since $p, q \in Q - \{a, b\}$, hence there is a subsequence, which we may again denote by Q_n such that $Q_n \to Q_0 = Q(p, q) \subset Q$ as $n \to \infty$. Now p_n', q_n' are not antipodal, hence the great circle joining p_n', q_n' is unique, namely Σ. Hence $\Delta Q_n \subset \Sigma$ for all n. Therefore $\Delta Q_0 = \Delta \lim Q_n = \lim \Delta Q_n \subset \Sigma$. Thus $\Delta Q_0 \subset \Gamma \cap \Sigma$. This is a contradiction to $\Gamma \neq \Sigma$ and hence if $a \neq b$ then $\Delta a \neq \Delta b$ i.e. Δ is globally injective.

We have thus proved that Δ maps R homeomorphically onto an open subset R' of S^n in such a way that Δ sends the geodesics of R onto portions of great circles of S^n. We may thus identify R with R' and regard the latter as a chord space. Hence we can apply Theorem (10) of §2 to conclude that either R' lies in an open hemisphere of S^n (i.e. A^n) and is a straight desarguesian chord space or else R' is the entire S^n and antipodes in R' and S^n coincide. This completes the proof of the Main Theorem as stated in the Introduction.

Pogorelov **[17]** determined all globally desarguesian (g.d.) G-spaces (in P^n). This allows us to construct all chord metrics in the sets occurring in the Main Theorem. In the case $R \subset A^n$ our spaces are g.d. If xy metrizes R as a g.d. chord space and $\varepsilon(x, y)$ is a Euclidean metric in A^n (invariant under translations), then R is with $xy + (1/k)\varepsilon(x, y)$ a G-space. Because of Theorem (10) chords in R and segments on S^n coincide (as point sets) so that if $\sigma(x, y)$ is the spherical metric $xy + (1/k)\sigma(x, y)$ is a G-metric on S^n with the great circles as geodesics.

Pogorelov's procedure is the same in both cases, so let $R \subset A^n$. He proves that each g.d. metric in R can be approximated by explicitly given g.d. metrics $\rho_\nu(x, y)$ of class C^∞, uniformly on any compact set. Applying this to $xy + (1/k)\varepsilon(x, y)$ we find that any g.d. chord metric can be approximated uniformly on compact sets by definite $\rho_\nu(x, y)$ of class C^∞. Evidently the limit of such ρ_ν is the given xy, thus all chord metrics can be constructed.

The result can be formulated in a more satisfactory way, because the $\rho_\nu(x, y)$ can evidently be varied. If K_M is the set of all hyperplanes intersecting the point set M, then $\rho_\nu(x, y)$ can be written in the form

$$\rho_\nu(x, y) = \mu_\nu\left(K_{T(x,y)}\right) = \int_{K_{T(x,y)}} f_\nu(H)\, dH$$

where f_ν is smooth and dH is the euclidean density of hyperplanes. For $\nu \to \infty$ we obtain a set function μ with the property that

$$xy = \mu\left(K_{T(x,y)}\right).$$

Pogorelov determines for G-metrics precisely the μ with this property (he considers only $n = 2, 3$, but his method for $n = 3$, where μ taking negative values must be admitted, works for $n > 3$). We call these set functions briefly P-functions. If W_R is the totality of all P-functions for R and W_R^n the subset of W_R corresponding to the G-metrics of the form $xy + \delta\varepsilon(x, y)$, $0 < \delta \leq 1/n$, then $W_R^n \supset W_R^{n+1}$ and $\bigcap W_n = W$ contains all g.d. G-metrics in R and in addition precisely the g.d. chord metrics.

That a metric analogue to the B.T. singling out three basic geometrics cannot exist without additional conditions is evident from our preceding considerations. All that can be expected is a condition weaker than constant curvature and meaningful in G-spaces. The following version may be of interest.

THEOREM. *A simply connected locally desarguesian and locally symmetric G-space is Minkowskian, hyperbolic or spherical.*

We indicate the proof briefly. Locally symmetric G-spaces which generalize locally symmetric Riemann and Finsler spaces are defined, see [**11**], as G-spaces in which a positive continuous $\beta(p)$ exists such that each $S(p, \beta(p))$ is symmetric in p. In [**10**, (4.2), (4.4)] we proved that a locally symmetric globally desarguesian G-space is Minkowskian or hyperbolic and that a locally symmetric spherelike G-space is spherical. These results when combined with the Main Theorem give the theorem stated above.

4. The significance of chord spaces. Taking axioms I to IV of G-spaces we obtain extremals, but they may behave so pathologically that no theory of truly geometric interest can be built on these axioms alone. (We note here that the axioms I, II and III of G-spaces were stated in §2 above. The axiom IV, see G p. 37, states: To every point p of the space there corresponds a positive ρ_p, such that for any two distinct points x, y in $S(p, \rho_p)$ (i.e. $xp < \rho_p$, $yp < \rho_p$) a point z with (xyz) exists. The axiom IId of chord spaces is an analogue to this axiom).

It is shown in [7, p. 7] that there are metrics in the plane where each segment can be continued to a straight line and which permit nevertheless closed extremals and also $x(t)$ defined in $(-\infty, \alpha)$, α finite, satisfying (1) locally, but which cannot be continued to extremals. In [**6**] one finds for any plane Minkowski metric whose circles are polygons extremals which are transitive in a very strong sense.

All this holds for one of the most important Minkowski planes, namely those whose circles are parallelograms (which are all isometric by G(17.10)). We denote this metric by m_p. In addition to m_p there is another very important plane Minkowski metric m_h whose circles are affine images of euclidean regular hexagons.

These metrics occur as solutions of various extremum problems of which we will discuss a few. They are all found in Petty [**16**]. The very simplest problem is: which are the plane Minkowski geometries yielding the longest or the shortest unit circle? The answer is that m_p (and only m_p) yields the maximum 8, similarly only m_h yields the minimum 6.

Let A, B be affine lines in A^2 intersecting at p, A', B' subrays of A, B with origin p and A'', B'' the opposite rays. A Minkowski metric is invariant under translations and the theory of Haar measure tells us that up to a factor only one measure invariant under translation exists which we normalize for each Minkowski metric such that the unit disk has measure π. If $a' \in A' - p$, $b' \in B' - p$ we define the Minkowski sine of the angle formed by A' and B' by

$$\mathrm{sm}(A', B') = \frac{2\Delta}{\sqrt{pa' \cdot pb'}}$$

where Δ is the area of the triangle $pa'b'$. This is independent of the choice of a', b' and trivially $\mathrm{sm}(A'', B'') = \mathrm{sm}(A', B')$. But corresponding to $\sin\alpha = \sin(\pi - \alpha)$ we also have

$$\mathrm{sm}(A', B'') = \mathrm{sm}(A'', B') = \mathrm{sm}(A', B').$$

Thus sm depends only on A, B and we write $\mathrm{sm}(A, B)$. This can, in general, not be interpreted as a function of a number, but is evidently invariant under translation of A or B, which we can therefore assume to pass through a fixed p.

If B is kept fixed and parallel to the direction ω at p then A is perpendicular to B or B transversal to A iff

$$\mathrm{sm}(A', B) \leq \mathrm{sm}(A, B)$$

for any $A' \ni p$. We set $\alpha(B) = \mathrm{sm}(A, B) = \alpha(\omega) = \alpha(\omega + \pi)$. The function α plays an important role, for example in the solution of the isoperimetric problem, but that does not interest us here.

All metrics satisfy

$$\pi/4 \leq \alpha(\omega) \leq \pi/2;$$

in this case both bounds are attained only by m_p. Also always

$$\max \alpha(\omega) \geq 1,$$

with equality only for the ellipse, i.e. the euclidean geometry. But $\min \alpha \leq \pi/3$ with the equality only for m_h.

The geometries for which perpendicularity is symmetric are those whose circles are Radon curves and they are characterized by $\alpha(\omega) =$ *constant*. The minimum and maximum for constants α are 1 and $\pi/3$ because both in the euclidean metric and m_h perpendicularity is symmetric.

As an example of a different type we consider the Hilbert geometries (see G §18 and Hilbert [**13**, Appendix 1]). Let S be a closed convex hypersurface in A^n and $\mathscr{I}$ its interior. In $\mathscr{I}$ we define distance just as in hyperbolic geometry. If the affine line through $a \neq b$ in $\mathscr{I}$ intersects S in u and v, the distance

$$h(a, b) = |\log \rho(a, b, u, v)|$$

where ρ is cross ratio. Hilbert shows that the triangle inequality holds and that the intersections of $\mathscr{I}$ with affine lines are straight extermals and the only ones (so that $\mathscr{I}$ becomes a G-space) iff no 2-plane intersecting $\mathscr{I}$ intersects S in a curve containing two non-collinear affine segments. Taking affine segments in $\mathscr{I}$ as chords yields therefore a desarguesian chord space.

The most degenerate case is where $S \cup \mathscr{I}$ is a simplex. For $n = 2$ this geometry and the hyperbolic (where S is an ellipse) are the only ones with symmetric perpendicularity [**15**].

The case of general n is most interesting. The (unique) Hilbert geometry possessing a transitive abelian group of motions where the affine segments are the chords (motion means that both distance and chords are preserved) is given by a simplex S, see [**5**, p. 35]. If we realize $\mathscr{S}$ as the first quadrant $x_i > 0$ of an affine coordinate system, the group is given by $x'_i = \beta_i x_i$, $\beta_i > 0$.

Denoting by ϕ the map $y_i = \log x_i$ of $\mathscr{S}$ on A^n the group becomes the group of translations of A^n and putting $m(y', y'') = m(\phi x', \phi x'') = h(x', x'')$ we get a chord-metric in A^n invariant under the translations, whose chords are the images of the affine segments in $\mathscr{S}$. The geodesics are straight and satisfy the hyperbolic parallel axiom, moreover the Theorem of Desargues. m is a Minkowski metric because it is invariant under the translations and we can take the affine segments as chords. Transforming this into $\mathscr{S}$ we obtain in $\mathscr{S}$ another chord metric (based on h) which satisfies both the parallel axiom and Theorem of Desargues.

From a general point of view we arrive at the very interesting question: If R is a metric space satisfying axioms I, II, III and $\mathscr{C}(R)$ is the set of all chord spaces compatible with the given metric, how different can the elements of $\mathscr{C}(R)$ be? For example $\mathscr{C}(m_p)$ contains straight chord spaces whose geodesics (but not the metric) are invariant under the translations where the theory of parallels (into which we do not want to enter here) has none of the usual properties of euclidean or hyperbolic geometry. The examples given in G p. 140 confirm this.

This leads to the further question whether special metrics like the Minkowski metrics in A^n can be characterized by intrinsic properties instead of defining the chords. In general this is clearly impossible owing to the arbitrariness in defining chords, but easy in the case mentioned.

A theorem of Pontrjagin [**18**, p. 170] states that a simply connected topological manifold which is the group space of a transitive abelian group of motions is equivalent to A^n with the translations as group. The affine lines are the orbits of the one-parameter subgroups. This contains as special case that among the chord metrics in A^n with the translations as motions the Minkowskian are those whose geodesics are the orbits of A^n considered as group space. Applied to our example it determines in the Hilbert simplex at once the geodesics of the desarguesian chord metric with the parallel axiom as the orbits of $x'_i = \beta_i x_i$, $1 \leq i \leq n$.

We hope to have convinced the reader that a generalization of G-spaces is necessary and possible to obtain a complete analogy to the topological aspect of B.T. i.e. our Main Theorem, and that chord spaces are one possibility which leads in addition to many interesting new problems and often to unexpected phenomena.

Surprisingly, one can clearly deduce from a paragraph (which is too long to be produced here in toto) in Beltrami's paper [2] that he would have welcomed our generalization of his theorem. He says that a theorem which holds under weaker hypotheses than stated has not been fully understood and that the proper generalization may involve the disappearance of some of the original concepts (in our case the Riemannian metric).

References

[1] E. Beltrami, *Risoluzione del problema: riportare i punti di una superficie sopra un piano in modo che le linee geodetiche vengano rappresentate da linee rette*, Opere, **I** (1865), 262–280.

[2] _____, *Saggio di interpretazione della Geometria non-Euclidea*, Giornale di Matematiche, **VI** (1868), 284–312.

[3] H. Busemann, *The Geometry of Geodesics*, (1955), Academic Press, New York.

[4] _____, *Geometries in which the planes minimize area*, Ann. Mat. Pura. Appl. (IV), **55** (1961), 171–190.

[5] _____, *Timelike spaces*, Dissert. Math. No. 53, (1967), Warsaw.

[6] _____, *Transitive geodesics in Minkowski planes*, Math. Chronicle, **1** (1969), 27–29.

[7] _____, *Recent Synthetic Differential Geometry*, Ergeb. Math. Band 54, (1970), Springer-Verlag, New York.

[8] _____, *Spaces with Distinguished Shortest Joins*, A Spectrum of Mathematics (essays presented to H. G. Forder, edited by J. C. Butcher; University Press, Auckland, New Zealand; Oxford University Press), (1971), 108–120.

[9] H. Busemann and B. B. Phadke, *Locally desarguesian spaces*, J. Geometry, **14** (1980), 123–138.

[10] _____, *Symmetric Spaces and Ellipses*, Proc. Int. Christoffel Symposium, A collection of articles dedicated to E. B. Christoffel on the occasion of his 150th birthday, (1981), Birkhäuser, Basel, 626–635.

[11] _____, *Two theorems on general symmetric spaces*, Pacific J. Math., **92** (1981), 39–48.

[12] _____, *Nonconvex spheres in G-spaces*, J. Indian Math. Soc., **44** (1980), 39–50.

[13] D. Hilbert, *Grundlagen der Geometrie*, 10. Auflage, (1968), B. G. Teubner, Stuttgart.

[14] W. Hurewicz and H. Wallman, *Dimension Theory*, (1941), Princeton University Press, Princeton.

[15] P. J. Kelly and L. J. Paige, *Symmetric perpendicularity in Hilbert geometries*, Pacific J. Math., **2** (1952), 319–322.

[16] C. M. Petty, *On the geometry of the Minkowski plane*, Rivista di Matematica della Universita di Parma, **6** (1955), 269–292.

[17] A. V. Pogorelov, *Hilbert's Fourth Problem*, (Russian), (1974), Izdat. "Nauka", Moscow. The English translation (same title, (1979), V. H. Winston & Sons, Washington D.C.) contains, in contrast to the original, a complete bibliography and 4 pages of useful notes.

Pac. J. Math. 115 (1984), 299–315

[18] L. S. Pontrjagin, *Topological Groups*, (1939), Princeton University Press, Princeton.

Received March 2, 1983.

THE UNIVERSITY OF SOUTHERN CALIFORNIA
LOS ANGELES, CA 90007

AND

THE FLINDERS UNIVERSITY OF SOUTH AUSTRALIA
BEDFORD PARK, AUSTRALIA

PART IV
MISCELLANEOUS PAPERS

Zur Differentiation der Lebesgueschen Integrale.

Von

H. Busemann und W. Feller (Kopenhagen).

Die folgende Untersuchung greift die Frage auf, nach welchen Mengen die totalstetigen additiven Mengenfunktionen oder, was dasselbe ist, die unbestimmten Lebesgueschen Integrale differenziert werden können, d. h. wenn R ein System messbarer Mengen positiven Masses ist, von denen sich auf jeden Punkt Q des Raumes eine Teilfolge $\varrho_1, \varrho_2, \ldots = \{\varrho_\nu\}$ zusammenzieht, so wird gefragt: Wie muss das System R beschaffen sein, damit für jede summierbare Funktion $f(P)$ und alle Punkte Q mit Ausnahme einer Nullmenge

$$\lim_{k \to \infty} \frac{1}{m(\varrho_k)} \int_{\varrho_k} f(P)\, dP = f(Q)$$

ist [1]).

Man beweist gewöhnlich [2]) mit Hilfe des Vitalischen Satzes, dass das System der achsenparallelen Würfel (oder ein gleichwertiges) von dieser Art ist. Ein einfacher Schluss zeigt weiter, dass man dann auch nach den Mengen jeder in bezug auf das zuerst behandelte System regulären Mengenfamilie [3]) differenzieren kann. Ob es auch bezüglich der achsenparallelen Würfel *nicht* reguläre Mengensysteme gibt, nach welchen man differenzieren kann, ist nicht bekannt. Insbesondere weiss man in dieser Hinsicht nichts über belie-

[1]) $m(\ )$ bezeichnet immer das Lebesguesche Mass.

[2]) Vgl. z. B. C. Carathéodory: Reelle Funktionen, 2. Auflage, Kap. IX.

[3]) Die Definition derselben, sowie die erwähnte Schlussweise findet man in der Bemerkung zu § 1. dieser Arbeit.

bige Intervalle [4]) [5]). Für die Funktion

$$F(x, y) = \int_0^x \int_0^y f(\xi, \eta)\, d\xi\, d\eta, \quad f(x, y) \text{ summierbar,}$$

ist also nur bewiesen, dass der Differenzenquotient

$$\frac{1}{h \cdot k} \int_x^{x+h} \int_y^{y+k} f(\xi, \eta)\, d\xi\, d\eta =$$

$$= \frac{1}{h \cdot k} [F(x+h, y+k) - F(x+h, y) - F(x, y+k) + F(x, y)]$$

gegen einen Grenzwert strebt, wenn h und k so gegen 0 gehen, dass das Verhältnis $\left|\frac{h}{k}\right|$ zwischen zwei positiven Konstanten bleibt. Die Existenz des entsprechenden Doppellimes ist hierin nicht enthalten.

Bei der Behandlung der allgemeinen Frage nach der Beschaffenheit der Mengensysteme, nach welchen man differenzieren kann, zeigt sich ein in den bisherigen Ergebnissen nicht auftretender Unterschied zwischen dem Verhalten der Integrale beschränkter und der nicht beschränkter Funktionen. Insbesondere werden wir sehen, dass erstere nach den Intervallen differenziert werden können, letztere dagegen nicht immer.

Der Fall des beschränkten Integranden lässt sich zurückführen auf die Behandlung der Integrale über solche Funktionen, welche nur die Werte 0 und 1 annehmen. Daher ist hier die Gültigkeit des Dichtesatzes bezüglich R für die Differenzierbarkeit notwendig und hinreichend. Wir sagen dabei, *für R gilt der Dichtesatz, wenn für jede messbare Menge $\varkappa$ in fast allen Punkten Q des Raumes für jede sich auf Q zusammenziehende Folge $\{\varrho_k\}$ aus R der Grenzwert*

$$\lim_{k \to \infty} \frac{m(\varrho_k \varkappa)}{m(\varrho_k)}$$

existiert, und zwar für $Q \subset \varkappa$ gleich 1, *und sonst gleich* 0 *ist.*

[4]) Unter einem *Intervall* verstehen wir hier stets ein achsenparalleles Parallelepiped des Euklidischen Raumes E^r.

[5]) Für dieses System gilt der Vitalische Satz nicht, wie Banach und H. Bohr gezeigt haben. Vgl. Carathéodory, a. a. O. Anhang. Übrigens folgt das auch beiläufig aus den Ergebnissen des § 4 dieser Arbeit.

15*

Für den Dichtesatz lässt sich ein für das Folgende grundlegendes Kriterium angeben, welches sich einfach ausspricht, wenn man die Vereinigungsmenge $\sigma_\alpha(\varkappa)$ derjenigen Mengen ϱ aus R einführt, in welchen die „mittlere Dichte" von $\varkappa$ grösser als eine Zahl α zwischen 0 und 1 ist; d. h. für die

$$\frac{m(\varrho\varkappa)}{m(\varrho)} > \alpha$$

ist. Für die Gültigkeit des Dichtesatzes (§ 1) erweist sich dann als hinreichend, wenn für jede beschränkte messbare Menge K und jedes α zwischen 0 und 1 eine Beziehung der Form

$$(*) \qquad m(\sigma_\alpha(\varkappa)) \leqslant C(\alpha)\, m(\varkappa)$$

besteht, wo $C(\alpha)$ eine nur von α und R, aber nicht von $\varkappa$ abhängige Konstante bedeutet. (*) ist im allgemeinen nicht nötwendig, wohl aber, wenn R mit irgend einer Menge auch alle dazu ähnlich gelegenen [6]) enthält. In diesem Falle ist (*) gleichbedeutend mit der Forderung, dass für jede Menge $\varkappa$ endlichen Masses auch $\sigma_\alpha(\varkappa)$ endliches Mass habe. Es zeigt sich nun (§ 2) merkwürdigerweise, dass dass System der Intervalle der Bedingung (*) genügt, dagegen das System a l l e r rechtwinkligen Parallelepipede nicht, so dass für ersteres der Dichtesatz gilt, für letzteres aber nicht.

Man kann demnach (§ 3) die unbestimmten Integrale beschränkter messbarer Funktionen $f(x_1,\dots,x_r)$ nach Intervallen differenzieren. Hieraus folgt, dass die Funktion

$$F(x_1,\dots,x_r) = \int_0^{x_1}\int_0^{x_2}\dots\int_0^{x_r} f(\xi_1,\dots,\xi_r)\, d\xi_1 \dots d\xi_r$$

für $s \leqslant r$ bis auf eine Nulmenge die gemischten Ableitungen

$$F_{(x_{\nu_1}\dots x_{\nu_s})}, \quad \nu_i \neq \nu_k \quad \text{für} \quad i \neq k,$$

im Sinne des s-fachen Limes (des entsprechenden Differenzenquotienten) besitzt, und dass dieser daher für jede Permutation $\mu_1,\dots,\mu_s$ der Zahlen $\nu_1, \nu_2,\dots,\nu_s$ gleich

$$\frac{\partial^s F}{\partial x_{\mu_1}\dots\partial x_{\mu_s}}$$

[6]) Unter ä h n l i c h g e l e g e n verstehen wir hier und im folgenden „ähnlich und ähnlich gelegen".

ist. Ein besonderer Beweis der Vertauschbarkeit der Differentiationsreihenfolge ist in diesem Falle also nicht erforderlich.

Schliesslich (§ 4) stellen wir eine zu (*) analoge notwendige und hinreichende Bedingung dafür auf, dass bezüglich eines Systems, das mit irgendeiner Menge alle dazu ähnlich gelegenen enthält, die unbestimmten Integrale über beliebige summierbare Funktionen differenziert werden können. Man sieht sofort, dass z. B. die Würfel und die Kugeln dieser Bedingung genügen, die Intervalle dagegen nicht. Es gibt also totalstetige additive Mengenfunktionen, die sich nach Intervallen *nicht* differenzieren lassen, und wo die gemischten Ableitungen der zugehörigen Punktfunktionen im Sinne des mehrfachen Limes im allgemeinen *nicht* existieren. (Vgl. den *Zusatz bei der Korrektur* am Ende der Arbeit).

§ 1.

Kriterien für die Gültigkeit des Dichtesatzes.

Es sei $R = \{\varrho\}$ ein System von beschränkten offenen Mengen des r-dimensionalen Euklidischen Raumes E^r von der Art, dass sich auf jeden Punkt des Raumes eine Folge $\{\varrho_k\}$ von Mengen aus R zusammenzieht. Die Voraussetzung der Offenheit der Mengen ϱ bedeutet keine wesentliche Einschränkung, da man die Resultate mühelos auf Systeme beliebiger meßbarer Mengen positiven Maßes überträgt. Wenn $\varkappa$ eine messbare Menge ist, so verstehen wir unter der *mittleren Dichte von* $\varkappa$ *in einer Menge* ϱ den Quotienten

$$\frac{m(\varkappa \cdot \varrho)}{m(\varrho)}.$$

Wir bilden zunächst den Limes inferior der mittleren Dichte für eine beliebige sich auf einen Punkt P zusammenziehende Folge $\{\varrho_k\}$ von Mengen aus R. Die untere Grenze dieser Limites inferiores für alle sich auf P zusammenziehenden Folgen $\{\varrho_k\}$ aus R nennen wir mit Lebesgue die *untere Dichte von* $\varkappa$ *bezüglich* R *in* P. Entsprechend nennen wir die obere Grenze der Limites superiores derselben Folgen die *obere Dichte von* $\varkappa$ *bezüglich* R *in* P. Sind in einem Punkte diese beiden Dichten gleich, so sprechen wir von *Dichte* schlechthin. Die Dichte existiert insbesondre in allen Punkten, in denen entweder die obere Dichte gleich Null oder die untere Dichte gleich 1 ist.

Wir sagen, *für das System R gilt der Dichtesatz, wenn für jede messbare Menge die (untere) Dichte in fast allen Punkten der Menge gleich 1 ist, oder, was dasselbe ist, wenn die (obere) Dichte für jede messbare Menge in fast allen Punkten ihrer Komplementärmenge verschwindet.*

Wir führen nun einige Bezeichnungen ein. $d(\lambda)$ bedeute stets den Durchmesser der Punktmenge λ; der Buchstabe N mit oder ohne Indizes ist für Nullmengen reserviert. Es sei $0 < \alpha < 1$, $\delta < 0$ und $\varkappa$ eine messbare Menge. *Unter $\sigma_{\alpha,\delta}(\varkappa)$ verstehen wir die Vereinigungsmenge aller Mengen ϱ aus R, in denen die mittlere Dichte von $\varkappa$ grösser als α ist und deren Durchmesser kleiner als δ sind, für die also*:

$$1)\quad m(\varrho\varkappa) > \alpha \cdot m(\varrho),$$
$$2)\quad d(\varrho) < \delta$$

ist. $\sigma_\alpha(\varkappa)$ *ist*, wie in der Einleitung angegeben, *die Vereinigungsmenge aller Mengen aus R, welche nur der ersten Einschränkung unterworfen sind.*

Aus der Definition sieht man, dass für die Vereinigungsmenge beliebiger messbarer Mengen λ stets

$$\sigma_{\alpha,\delta}\left(\sum \lambda\right) \supset \sum \sigma_{\alpha,\delta}(\lambda), \tag{1}$$

und ferner, dass für $\alpha_1 \leqslant \alpha$, $\delta_1 \geqslant \delta$ und $\varkappa_1 \supset \varkappa$ stets

$$\sigma_{\alpha_1,\delta_1}(\varkappa_1) \supset \sigma_{\alpha,\delta}(\varkappa) \tag{2}$$

ist. Hiernach enthält der Durchschnitt $\prod\limits_{\delta>0} \sigma_{\alpha,\delta}(\varkappa)$ bei festem α alle Punkte des Raumes, in denen die obere Dichte von $\varkappa$ grösser als α und darüber hinaus höchstens solche Punkte, in denen sie gleich α ist.

Wenn der Dichtesatz gilt, folgt hieraus

$$\prod_{\delta>0} \sigma_{\alpha,\delta}(\varkappa) + N_1 = \varkappa + N_2; \tag{3}$$

wenn umgekehrt (3) für jedes α zwischen 0 und 1 gilt, so ergibt sich daraus wegen (2)

$$\prod_{0<\alpha<1} \prod_{\delta>0} \sigma_{\alpha,\delta}(\varkappa) + N_3 = \varkappa + N_4,$$

d. h. die obere Dichte von $\varkappa$ ist in fast allen Punkten von $E^r - \varkappa$ gleich 0.

Wir betrachten jetzt eine monoton abnehmende Folge beschränkter messbarer Mengen $\varkappa_\nu$ mit leerem Durchschnitt, und eine monoton gegen 0 abnehmende Folge positiver Zahlen δ_ν. Wenn der Dichtesatz gilt, so strebt für jedes α zwischen 0 und 1

$$m(\sigma_{\alpha,\delta_\nu}(\varkappa_\nu)) \to 0; \tag{4}$$

denn nach (2) ist für $\nu > n$

$$m(\sigma_{\alpha,\delta_\nu}(\varkappa_\nu)) \leqslant m(\sigma_{\alpha,\delta_\nu}(\varkappa_n))$$

und nach (3)

$$\lim_{\nu\to\infty} m(\sigma_{\alpha,\delta_\nu}(\varkappa_n)) = m(\varkappa_n),$$

also

$$\lim_{\nu\to\infty} m(\sigma_{\alpha,\delta_\nu}(\varkappa_\nu)) \leqslant m(\varkappa_n)$$

für alle n, woraus (4) wegen $m(\varkappa_n) \to 0$ folgt.

Die somit für die Gültigkeit des Dichtesatzes notwendige Bedingung (4) ist schwächer als (3), aber trotzdem *hinreichend*. Dazu haben wir zu zeigen, dass für jede beschränkte messbare Menge Λ die Teilmenge der Punkte, in denen sie bezüglich R nicht die Dichte 1 hat, eine Nullmenge ist. Es sei λ die Teilmenge von Λ, in welcher die untere Dichte von Λ bezüglich R kleiner als $1-\alpha$ ist. Es ist zu beweisen, dass jede abgeschlossene Teilmenge φ von λ das Mass 0 hat. Auf jeden Punkt von φ zieht sich nach Voraussetzung eine Folge $\{\varrho_i\}$ von Mengen aus R zusammen mit

$$m(\varphi\cdot\varrho_i) \leqslant m(\lambda\cdot\varrho_i) < (1-\alpha)\, m(\varrho_i). \tag{5}$$

Wir wählen unter diesen Mengen endlich viele, etwa $\varrho_1^1, \dots, \varrho_{s_1}^1$, mit kleinerem Durchmesser als 1 so, dass sie φ überdecken und setzen $\sum_{i=1}^{s_1} \varrho_i^1 = \gamma_1$. Sind $\gamma_1, \dots, \gamma_{n-1}$ schon definiert, so wählen wir endlich viele der Ungleichung (5) genügende Mengen, etwa $\varrho_1^n, \dots, \varrho_{s_n}^n$, die in γ_{n-1} enthalten und deren Durchmesser kleiner als $1/n$ sind, so, dass sie φ überdecken, und setzen $\sum_{i=1}^{s_n} \varrho_i^n = \gamma_n$. Es gilt dann $\gamma_n \subset \gamma_{n-1}$ und ferner

$$m(\varphi) \leqslant m(\gamma_n) = m(\varphi) + \epsilon_n \quad \text{mit} \quad \epsilon_n \to 0.$$

Die Mengen $\varkappa_n = \gamma_n - \varphi$ bilden also eine monoton abnehmende Folge mit leerem Durchschnitt. Es ist wegen $\varrho_i^n \subset \gamma_n$

$$m(\varrho_i^n) = m(\varkappa_n\cdot\varrho_i^n) + m(\varphi\cdot\varrho_i^n);$$

ferner gilt nach (5)

$$m(\varphi \cdot \varrho_i^n) < (1-\alpha)\, m(\varrho_i^n),$$

also

$$m(\varkappa_n \cdot \varrho_i^n) > \alpha \cdot m(\varrho_i^n);$$

daher ist

$$\varphi \subset \sigma_{\alpha, \frac{1}{n}}(\varkappa_n).$$

Da die Mengen $\varkappa_n$ die Voraussetzungen für (4) erfüllen, ergibt sich aus (4): $m(\sigma_{\alpha, \frac{1}{n}}(\varkappa_n)) \to 0$, also $m(\varphi) = 0$. Wir haben damit gefunden:

I. *Für das System R gilt der Dichtesatz dann und nur dann, wenn für jedes α zwischen* 0 *und* 1, *für jede monoton abnehmende Folge beschränkter messbarer Mengen $\varkappa_n$ und jede Folge $\delta_1 > \delta_2 > \ldots, \delta_\nu \to 0$ die Bedingung*

$$m(\sigma_{\alpha, \delta_\nu}(\varkappa_\nu)) \to 0 \tag{4}$$

erfüllt ist.

Insbesondre gilt also der Dichtesatz, wenn für jede beschränkte messbare Menge $\varkappa$ die Abschätzung

$$m(\sigma_\alpha(\varkappa)) < C(\alpha) \cdot m(\varkappa) \tag{6}$$

besteht, wobei $C(\alpha)$ nur vom System R und von α abhängt, nicht aber von $\varkappa$. Dass die Bedingung (6) im allgemeinen nicht notwendig sein kann, ist trivial. Hingegen ist sie es, wenn das System R mit irgend einer Menge ϱ auch alle zu ihr ähnlich gelegenen Mengen enthält. Dem Beweise dieser Behauptung schicken wir folgenden Hilfsatz voraus:

Ist $\varkappa$ eine messbare beschränkte Menge positiven Masses, so gibt es zu jedem $\varepsilon > 0$ in jeder offenen Menge Γ abzählbar viele zu $\varkappa$ ähnlich gelegene Mengen $\varkappa_1, \varkappa_2, \ldots$ derart, dass

$$1) \quad m\left(\sum \varkappa_\nu\right) = m(\Gamma)$$

$$2) \quad \sum m(\varkappa_\nu) < m(\Gamma) + \varepsilon$$

ist. Dabei kann man die $\varkappa_\nu$ der Einschränkung unterwerfen, dass ihre Durchmesser kleiner sind als eine vorgegebene Zahl δ [7]).

[7]) Wenn $\varkappa$ nach aussen quadrierbar ist, kann $\varepsilon = 0$ gemacht werden, was dann ein Spezialfall des Vitalischen Satzes ist.

I sei ein offenes Intervall dass $\varkappa$ enthält, und

$$\frac{m(\varkappa)}{m(I)} = a;$$

dabei kann $a < 1$ vorausgesetzt werden, da sonst nichts zu beweisen ist. Wir bemerken zunächst folgendes: λ sei eine beliebige Teilmenge von Γ, welche Vereinigungsmenge von zu $\varkappa$ ähnlich gelegenen Mengen ist, und es werde

$$\frac{m(\lambda)}{m(\Gamma)} = q$$

gesetzt. Dann gibt es in Γ zu jedem $\eta > 0$ eine Folge von paarweise fremden zu $\varkappa$ ähnlich gelegenen Mengen $\varkappa_\nu$ derart, dass für ihre Vereinigungsmenge $\Lambda = \Sigma\, \varkappa_\nu$ die beiden Ungleichungen

$$\text{a)} \quad \frac{m(\lambda + \Lambda)}{m(\Gamma)} > q + a\frac{1-q}{2}$$

$$\text{b)} \quad m(\lambda\Lambda) < \eta$$

gelten. Zum Beweise überdecken wir $\Delta = \Gamma - \lambda$ durch eine offene Menge Δ' so dass

$$m(\Delta' - \Delta) > \operatorname{Min}\left[\eta,\, a\frac{1-q}{2} m(\Gamma)\right]$$

wird. Wir stellen Δ' bis auf eine Nullmenge als Vereinigungsmenge von abzählbar vielen zu I ähnlich gelegenen, paarweise fremden Intervallen I_ν dar. Bei der Dilatation, die I in I_ν überführt, gehe $\varkappa$ in $\varkappa_\nu$ über: $\Lambda = \Sigma\, \varkappa_\nu$ sei ihre Vereinigungsmenge. Dann ist

$$m(\Lambda) = m\left(\sum \varkappa_\nu\right) = \sum m(\varkappa_\nu) = a \sum m(I_\nu) = a\, m(\Delta')$$

und

$$m(\lambda \cdot \Lambda) \leqslant m(\lambda \cdot \Delta') < \operatorname{Min}\left[\eta,\, a\frac{1-q}{2} m(\Gamma)\right].$$

Um auch (a) zu beweisen, bemerke man, dass

$$m(\lambda + \Lambda) = m(\lambda) + m(\Lambda) - m(\lambda\Lambda) \geqslant$$
$$\geqslant m(\lambda) + a \cdot m(\Delta') - a\frac{1-q}{2} m(\Gamma).$$

Beachtet man $m(\lambda) = q \cdot m(\Gamma)$ und

$$m(\Delta') \geqslant m(\Delta) = m(\Gamma) - m(\lambda),$$

so ergibt sich

$$m(\lambda + \Lambda) > \left(q + a\,\frac{1-q}{2}\right) m(\Gamma).$$

Aus dieser Bemerkung leitet man den Hilfssatz so ab:

$\varkappa^1$ sei eine zu $\varkappa$ ähnlich gelegene Teilmenge von Γ. Wir nehmen zur Induktion an, dass die Menge $\varkappa^n$ bereits definiert sei, und zwar als Vereinigungsmenge von gewissen zu $\varkappa$ ähnlich gelegenen Teilmengen $\varkappa_\nu^n$ von Γ, derart dass

$$\sum_\nu m(\varkappa_\nu^n) < m(\varkappa^n) + \left(1 - \frac{1}{2^n}\right)\varepsilon. \tag{7}$$

Es sei

$$q_n = \frac{m(\varkappa^n)}{m(\Gamma)}.$$

Wir wenden das soeben beschriebene Verfahren an auf

$$\lambda = \varkappa^n, \quad q = q_n, \quad \eta = \frac{\varepsilon}{2^{n+1}}.$$

Dadurch wird eine Menge $\Lambda = \Sigma \varkappa_\nu$ von zu $\varkappa$ ähnlich gelegenen Mengen $\varkappa_\nu$ bestimmt, und zwar ist wegen $\varkappa_\nu \cdot \varkappa_\mu = 0$ für $\nu \neq \mu$ und wegen (7)

$$\sum_\nu m(\varkappa_\nu^n) + \sum m(\varkappa_\nu) \leqslant m(\varkappa^n) + \left(1 - \frac{1}{2^n}\right)\varepsilon + m(\Lambda) + \frac{\varepsilon}{2^{n+1}},$$

sodass $\varkappa^{n+1} = \varkappa^n + \Lambda$ der Ungleichung (7) mit $n+1$ und außerdem der Beziehung

$$m(\varkappa^{n+1}) > \left(q_n + a\,\frac{1-q_n}{2}\right) m(\Gamma)$$

genügt. Setzt man wieder

$$q_{n+1} = \frac{m(\varkappa^{n+1})}{m(\Gamma)},$$

so ist nach Konstruktion

$$q_{n+1} > q_n + a\,\frac{1-q_n}{2}.$$

Die Zahlen q_n wachsen monoton und streben offenbar gegen 1. Die Grenzmenge der $\varkappa^n$ ist daher mit Γ massgleich; sie ist Vereinigungsmenge von zu $\varkappa$ ähnlich gelegenen Teilmengen von Γ und genügt wegen (7) auch der zweiten Bedingung des Hilfssatzes. Schliesslich ist es klar, dass man sich bei der ganzen Konstruktion auf solche zu $\varkappa$ ähnlich gelegene Mengen beschränken kann, deren Durchmesser kleiner als δ sind. Damit ist der Hilfsatz bewiesen.

Unsere Behauptung über die Ungleichung (6) bestand im folgenden: Wenn das System R mit jeder Menge ϱ auch alle zu ϱ ähnlich gelegenen Mengen enthält, und der Abschätzung (6) *nicht* genügt, so gilt bezüglich R der Dichtesatz *nicht.* Wir zeigen, dass es unter den angegebenen Voraussetzungen in jedem offenen Intervall I eine messbare Menge $\varkappa$ mit

$$m(\varkappa) < \frac{2}{3} m(I)$$

gibt, die in allen Punkten von I bezüglich R eine positive obere Dichte hat.

Da die Ungleichung (6) für R nicht erfüllt ist, gibt es für ein festes $\alpha > 0$ und für jede positive Zahl C eine beschränkte messbare Menge $\varkappa$ mit

$$m(\sigma_\alpha(\varkappa)) > C \cdot m(\varkappa).$$

Es sei $\{\varkappa_n\}$ eine Folge von solchen Mengen mit

$$m(\sigma_\alpha(\varkappa_n)) > 4^n \cdot m(\varkappa_n).$$

$\sigma_\alpha(\varkappa_n)$ ist Vereinigungsmenge von Mengen ϱ des Systems R, in denen die mittlere Dichte von $\varkappa_n$ grösser als α ist. Wir können unter diesen endlich viele so herausgreifen, dass ihre Vereinigungsmenge π_n beschränkt ist, und dass für ihr Mass die Ungleichung

$$m(\pi_n) > 4^n \cdot m(\varkappa_n)$$

erfüllt ist. Nun konstruieren wir nach unserem Hilfssatz in I für jedes n eine Folge von zu π_n ähnlich gelegenen Mengen π_n^ν, deren Durchmesser $d(\pi_n^\nu) < 1/n$ sind, und für die

$$m(I) = m\left(\sum_\nu \pi_n^\nu\right) \leqslant \sum_\nu m(\pi_n^\nu) < 2 \cdot m(I)$$

ist. $\varkappa_n^\nu$ sei das Bild von $\varkappa_n$ in π_n^ν; ferner sei

$$\pi^n = \sum_\nu \pi_n^\nu, \quad \varkappa^n = \sum_\nu \varkappa_n^\nu.$$

Es ist dann

$$m(\pi^n) = m(I),$$

$$m(\varkappa^n) \leqslant \sum_\nu m(\varkappa_n^\nu) < \frac{1}{4^n} \sum_\nu m(\pi_n^\nu) < \frac{2}{4^n} m(I).$$

Für die Menge $\varkappa = \sum_n \varkappa^n$ ist also

$$m(\varkappa) < \frac{2}{3} m(I);$$

diese Menge hat aber in fast allen Punkten des Intervalls I bezüglich R mindestens die obere Dichte α. Jede Menge π_n^ν ist Vereinigungsmenge von Mengen ϱ des Systems R, in denen $\varkappa_n^\nu$ und wegen $\varkappa \supset \varkappa^n \supset \varkappa_n^\nu$ erst recht die Menge $\varkappa$ eine grössere mittlere Dichte als α hat. Nun liegen fast alle Punkte von I für jedes n in einem $\pi_n^{\nu_n}$. Wegen $d(\pi_n^\nu) < 1/n$ zieht sich wirklich auf fast jeden Punkt von I eine Folge von Mengen ϱ unsres Systems zusammen, in denen die mittlere Dichte von $\varkappa$ grösser als α ist. Wir haben damit folgenden Satz bewiesen:

II. *Dem System R mögen mit jeder Menge ϱ auch alle zu ϱ ähnlich gelegenen Mengen angehören; dann ist für die Gültigkeit des Dichtesatzes bezüglich R notwendig und hinreichend, dass für jedes α zwischen 0 und 1 und alle beschränkten messbaren Mengen $\varkappa$ die Ungleichung*

$$m(\sigma_\alpha(\varkappa)) < C(\alpha) \cdot m(\varkappa)$$

besteht, wobei $C(\alpha)$ nur von α und vom System R, nicht aber von $\varkappa$ abhängt.

Man kann diesem Kriterium eine etwas andere Fassung geben:

II′. *Der Dichtesatz gilt bezüglich des Systems R, das wieder mit jeder Menge alle dazu ähnlich gelegenen enthalte, dann und nur dann, wenn für jedes α zwischen 0 und 1 und jede messbare Menge $\varkappa$ endlichen Masses* [8]) *auch die Menge $\sigma_\alpha(k)$ endliches Mass hat.*

Es gebe eine Menge $\varkappa$ endlichen Maßes mit $\sigma_\alpha(\varkappa) = \infty$. Wir wählen bei vorgegebenem C aus den Mengen ϱ, welche $\sigma_\alpha(\varkappa)$ bilden —

[8]) Das Wesentliche ist, daß $\varkappa$ nicht beschränkt vorausgesetzt wird.

für die also $m(\varrho\cdot\varkappa) > \alpha\cdot m(\varrho)$ ist — endlich viele, etwa $\varrho_1,\ldots,\varrho_n$, so aus, dass

$$m(\varrho_1+\ldots+\varrho_n) > C\cdot m(\varkappa)$$

wird, und bezeichnen den Durchschnitt $\varkappa\cdot\sum_{i=1}^{n}\varrho_i$ mit $\varkappa'$. Dann ist $\varkappa'$ beschränkt, und andrerseits $\sigma_\alpha(\varkappa')\supset\varrho_1+\ldots+\varrho_n$, also

$$m(\sigma_\alpha(\varkappa'))\geqslant m(\varrho_1+\ldots+\varrho_n)\geqslant C\cdot m(\varkappa)\geqslant C\cdot m(\varkappa').$$

Jetzt zeigen wir, dass die Bedingung hinreichend ist. Wenn der Dichtesatz bezüglich R nicht gilt, so gibt es bei passendem α eine Folge beschränkter messbarer Mengen $\varkappa_n$, für die

$$m(\sigma_\alpha(\varkappa_n)) > n^2\cdot m(\varkappa_n).$$

Wir verstehen unter $\varkappa_n'$ eine zu $\varkappa_n$ ähnlich gelegene Menge mit $m(\varkappa_n')=\frac{1}{n^2}$ und unter s_n eine solche beschränkte Teilmenge von $\sigma_\alpha(\varkappa_n')$, für die ebenfalls

$$m(s_n) > n^2\cdot m(\varkappa_n') = 1$$

gilt. Wir verschieben die Mengen $\varkappa_n'$ so, dass die zugehörigen Mengen s_n paarweise fremd werden. Dann ist

$$m\left(\sum\varkappa_n'\right)\leqslant\sum m(\varkappa_n')=\sum\frac{1}{n^2},$$

$$m\left(\sum s_n\right)=\infty$$

und [9])

$$\sigma_\alpha\left(\sum\varkappa_n'\right)\supset\sum\sigma_\alpha(\varkappa_n')\supset\sum s_n.$$

Bemerkung.

Wenn der Dichtesatz für ein System R gilt, so gilt er nach

[9]) Es gibt Systeme R, für welche das Kriterium I erfüllt ist, II aber nicht. Aus § 2 wird hervorgehen, dass dies zutrifft für das System aller Rechtecke ϱ der Ebene, für die die Verhältnisse der Seitenlängen grösser oder gleich $\frac{1}{1+t}$ sind, wobei t die Entfernung des Mittelpunktes von ϱ vom Ursprung ist. Dieses System hat allerdings die Eigenschaft, in jedem beschränkten Bereich der Bedingung (6) zu genügen; es gibt jedoch auch einfache Beispiele, von Systemen, die (4) erfüllen, und wo es trotzdem eine Menge positiven Masses derart gibt, dass die Bedingung (6) in keiner Umgebung eines Punktes dieser Menge erfüllt ist.

Lebesgue auch für alle bezüglich des Systems R regulären Mengenfamilien. Dabei heisst die Mengenfamilie π bezüglich R regulär, wenn sich auf jeden Punkt P eine Mengenfolge $\{\pi_k\}$ aus π zusammenzieht mit der Eigenschaft, dass jedes π_k in einem solchen ϱ_k aus R enthalten ist, für das

$$\frac{m(\pi_k)}{m(\varrho_k)} > \vartheta(P) > 0$$

gilt, wobei $\vartheta(P)$ eine beliebige von P, aber nicht von der speziellen Folge $\{\pi_k\}$ abhängige positive Funktion ist. Etwas allgemeiner nennt man die Mengenfamille π *„in P regulär bezüglich R"*, wenn es zu jeder gegen P konvergierenden Folge von Menge π_k aus π eine sich auf P zusammenziehende Folge $\{\varrho_k\}$ aus R gibt, mit

$$\varrho_k \supset \pi_k, \quad \frac{m(\pi_k)}{m(\varrho_k)} > \vartheta > 0.$$

(Zum Unterschied gegen eben wird $\pi_k \supset P$ nicht gefordert).

Wenn dann für eine beliebige messbare Menge $\varkappa$ und eine Folge $\{\varrho_k\}$, die sich auf P zusammenzieht,

$$\lim_{k\to\infty} \frac{m(\varkappa\varrho_k)}{m(\varrho_k)} = 1$$

ist, so gilt auch

$$\lim_{k\to\infty} \frac{m(\varkappa\pi_k)}{m(\pi_k)} = 1.$$

Denn ist C die Komplementärmenge von $\varkappa$, so ist

$$1-\frac{m(\pi_k\varkappa)}{m(\pi_k)} = \frac{m(C\pi_k)}{m(\pi_k)} \leqslant \frac{m(C\varrho_k)}{m(\varrho_k)}\cdot\frac{m(\varrho_k)}{m(\pi_k)} \leqslant \frac{1}{\vartheta}\left(1-\frac{m(\varrho_k\varkappa)}{m(\varrho_k)}\right) \to 0.$$

Wenn also eine messbare Menge in irgend einem Punkte des Raumes bezüglich R die Dichte 1 oder 0 hat, so hat sie an dieser Stelle dieselbe Dichte für jede dort bezüglich R reguläre Mengenfamilie.

§ 2.

Spezielle Mengensysteme, für die der Dichtesatz gilt oder nicht gilt.

A. Die Intervalle.

$R = \{\varrho\}$ sei nun das System aller offenen Intervalle [4]) des E^r.

Nach (6) gilt bezüglich der Intervalle der Dichtesatz, wenn wir zeigen können:

Für jede beschränkte messbare Menge $\varkappa$ des E^r ist

(8) $$m(\sigma_\alpha(\varkappa)) \leqslant 2^{(r^2)} \cdot \alpha^{-r} \cdot m(\varkappa).$$

Wir führen den Beweis durch Induktion nach r. Vorher bemerken wir, dass $\varkappa$ als *offene* Menge angenommen werden darf. Denn aus (8) für offene Mengen ergibt sich dieselbe Beziehung für beliebige messbare Mengen durch Grenzübergang.

1). Es sei $r=1$ und γ eine beschränkte offene Menge auf der Geraden E^1. $\sigma_\alpha(\gamma)$ ist die Vereinigungsmenge der offenen Intervalle ϱ mit

$$\alpha \cdot m(\varrho) < m(\varrho \cdot \gamma).$$

Aus diesen ϱ wählen wir endlich viele, etwa $\varrho_1, \ldots, \varrho_h$ so aus, dass $m(\sigma_\alpha(\gamma)) - m(\sum_1^h \varrho_i) < \varepsilon$ wird. Haben drei dieser ϱ_i einen nichtleeren Durchschnitt, so ist wenigstens eins in der Summe der beiden anderen enthalten. Wir können demnach $\sum_{i=1}^h \varrho_i$ so als Vereinigungsmenge gewisser ϱ_i, etwa $\varrho_{i_1}, \ldots, \varrho_{i_s}$, darstellen, dass jeder Punkt von $\sigma_\alpha(\gamma)$ in höchstens zwei der ϱ_{i_k} vorkommt. Dann liegt auch jeder Punkt von $\gamma \cdot \Sigma \varrho_i$ in höchstens zwei der offenen Mengen $\gamma \cdot \varrho_{i_k}$; daher gilt

$$m(\sigma_\alpha(\gamma)) - \varepsilon < m\left(\sum_{i=1}^h \varrho_i\right) \leqslant \sum_{k=1}^s m(\varrho_{i_k}) \leqslant \frac{1}{\alpha} \sum_{k=1}^s m(\gamma \cdot \varrho_{i_k}) \leqslant$$
$$\leqslant \frac{2}{\alpha} m\left(\sum_{k=1}^s \gamma \cdot \varrho_{i_k}\right) \leqslant \frac{2}{\alpha} m(\gamma).$$

Eine etwas eingehendere Überlegung liefert [10]), die genaue Abschätzung

$$m(\sigma_\alpha(\gamma)) \leqslant \left(\frac{2}{\alpha} - 1\right) m(\gamma),$$

welche wir aber nicht brauchen.

[10]) vgl. z. B. H. Lebesgue: Ann. de l'École Norm. sup. (3) **27**, 1910.

2) Für den weiteren Beweis führen wir einige Bezeichnungen ein: $x_1,\dots,x_r$, z seien die Koordinaten des E^{r+1}, E^r sei die $(x_1,\dots,x_r)$-Ebene. $m^1(\ )$ bedeutet das lineare, $m^r(\ )$ das r-dimensionale, $m^{r+1}(\ )$ das $(r+1)$-dimensionale Mass. Ist λ irgendeine Menge des E^{r+1}, so bezeichne λ_t den Durchschnitt von λ mit der Ebene $z=t$, und λ^P den Durchschnitt von λ mit der Parallelen zur z-Achse durch den Punkt P. Ist λ_t r-dimensional und λ^P eindimensional messbar, so sei $\sigma^r_\beta(\lambda_t)$ die Vereinigungsmenge aller r-dimensionalen Intervalle e der Ebene $z=t$ mit

$$m^r(\lambda_t\cdot e) > \beta\cdot m^r(e),$$

und $\sigma^1_\beta(\lambda^P)$ die Vereinigungsmenge aller linearen Intervalle i der Parallelen zur z-Achse durch P mit

$$m^1(\lambda^P\cdot i) > \beta\cdot m^1(i).$$

3) Es sei (8) für den Index r schon bewiesen. γ sei irgend eine beschränkte offene Menge des E^{r+1}. Wir behaupten, dass $\sigma_\alpha(\gamma)$ in der folgendermassen konstruierten Menge Γ enthalten ist: wir ersetzen jedes γ_t durch die Menge $\sigma^r_{\frac{\alpha}{2}}(\gamma)$ und bilden

$$\Gamma_* = \sum_{t=-\infty}^{\infty} \sigma^r_{\frac{\alpha}{2}}(\gamma_t).$$

Γ_* ist offen [11]). Dann ersetzen wir jeden linearen Schnitt Γ^P_* von Γ_* durch die Menge $\sigma^1_{\frac{\alpha}{2}}(\Gamma^P_*)$ und definieren

$$\Gamma = \sum_{P\subset E^r} \sigma^1_{\frac{\alpha}{2}}(\Gamma^P_*).$$

[11]) Ist nämlich $Q\subset E_t$ ein Punkt von Γ_*, so gibt es in E_t ein r-dimensionales Intervall ϱ_t mit

$$m^r(\gamma_t\,\rho_t) > \frac{\alpha}{2}\, m^r(\rho_t).$$

Nun besteht $\rho_t\cdot\gamma_t$ aus inneren Punkten von γ; zu jedem Punkt von $\rho_t\cdot\gamma_t$ gibt es also ein ihn enthaltendes $(r+1)$-dimensionales Intervall, das in γ liegt. Aus diesen suchen wir endlich viele, $I^1,\dots,I^h$, aus, deren Vereinigungsmenge auch noch der Ungleichung

$$m^r(\rho_t(I^1+\dots+I^h)_t) > \frac{\alpha}{2}\, m^r(\rho_t)$$

genügt. Die Schnitte $(I^1+\dots+I^h)_\tau$ sind für hinreichend kleine $|t-\tau|$ kongruent, so dass auch die entsprechenden $\sigma^r_{\frac{\alpha}{2}}(I^1+\dots+I^h)_\tau$ kongruent sind. Andrerseits liegen sie alle in Γ_*, also gehört mit $Q\subset\rho_t$ eine ganze Umgebung zu Γ_*.

Γ ist wieder offen [11]). Um $\sigma_\alpha(\gamma) \subset \Gamma$ zu zeigen, haben wir nachzuweisen, dass jedes $(r+1)$-dimensionale Intervall ϱ mit

$$m^{r+1}(\gamma \cdot \varrho) > \alpha \cdot m^{r+1}(\varrho)$$

in Γ liegt. Das Intervall j sei die Projektion von ϱ auf die z-Achse, ξ sei die Projektion von ϱ auf E^r, ferner e diejenige Teilmenge des Intervalls j, für welche

(9) e: $$m^r(\gamma_t \cdot \varrho_t) > \frac{\alpha}{2} \cdot m^r(\varrho_t)$$

gilt. e ist offen, und zwar gilt

(10) $$m^1(e) > \frac{\alpha}{2} m^1(j);$$

denn wegen $m^r(\varrho_t) = m_r(\xi)$ für $t \subset j$ und $m^r(\gamma_t \cdot \varrho_t) \leqslant \frac{\alpha}{2} m^r(\varrho_t)$ für $t \subset j - e$ ist

$$\alpha \cdot m^1(j) \cdot m^r(\xi) = \alpha \cdot m^{r+1}(\varrho) < m^{r+1}(\varrho\gamma) = \int_{j-e} m^r(\gamma_t \cdot \varrho_t)\, dt +$$
$$+ \int_e m^r(\gamma_t \cdot \varrho_t)\, dt \leqslant \frac{\alpha}{2} \int_{j-e} m^r(\varrho_t)\, dt + \int_e m^r(\varrho_t)\, dt \leqslant$$
$$\leqslant m^r(\xi) \left(\frac{\alpha}{2} m^1(j) + m^1(e)\right).$$

Nach (9) gilt für $t \subset e$

$$\sigma^r_{\frac{\alpha}{2}}(\gamma_t) \supset \sigma^r_{\frac{\alpha}{2}}(\gamma_t \cdot \varrho_t) \supset \varrho_t,$$

daher

(11) $$\sum_{t \subset e} \varrho_t \subset \Gamma_*.$$

Jede Parallele zur z-Achse durch einen Punkt P aus ξ schneidet $\sum_{t \subset e} \varrho_t$ in einer zu e kongruenten Menge. Daher besteht für die Schnitte ϱ^P und $\Gamma^P_* \varrho^P$ nach (10) und (11) die Beziehung

$$\frac{\alpha}{2} \cdot m^1(\varrho^P) = \frac{\alpha}{2} \cdot m^1(j) < m^1(e) \leqslant m^1(\Gamma^P_* \cdot \varrho^P) \quad (P \subset \xi).$$

Fundam. Math. 22 (1934), 226–256

Demnach ist

$$\varrho^P \subset \sigma^1_{\frac{\alpha}{2}}(\Gamma^P_*) \qquad \text{für} \quad P \subset \xi$$

und damit

$$\varrho = \sum_{P \subset \xi} \varrho^P \subset \sum_{P \subset \xi} \sigma^1_{\frac{\alpha}{2}}(\Gamma^P_*) \subset \Gamma,$$

also auch $\sigma_\alpha(\gamma) \subset \Gamma$.

4) Wir zeigen nun noch

$$m^{r+1}(\Gamma) \leqslant 2^{(r+1)^2} \cdot \alpha^{-(r+1)} \cdot m^{r+1}(\gamma).$$

Nach der Induktionsvoraussetzung ist

$$m^r(\sigma^r_{\frac{\alpha}{2}}(\gamma_t)) \leqslant \frac{2^{r^2} \cdot 2^r}{\alpha^r} \cdot m^r(\gamma_t),$$

also

$$m^{r+1}(\Gamma_*) = \int_{-\infty}^{\infty} m^r(\sigma^r_{\frac{\alpha}{2}}(\gamma_t))\, dt \leqslant 2^{r^2+r} \cdot \alpha^{-r} \cdot m^{r+1}(\gamma).$$

Ferner ist, wie unter 1) gezeigt wurde,

$$m^1(\sigma^1_{\frac{\alpha}{2}}(\Gamma^P_*)) \leqslant \frac{4}{\alpha}\, m(\Gamma^P_*),$$

daher

$$m^{r+1}(\sigma_\alpha(\gamma)) \leqslant m^{r+1}(\Gamma) = \int_{E^r} m^1(\sigma^1_{\frac{\alpha}{2}}(\Gamma^P_*))\, dP \leqslant \frac{4}{\alpha}\, m^{r+1}(\Gamma_*) \leqslant$$
$$\leqslant 2^{(r+1)^2} \cdot \alpha^{-(r+1)} \cdot m^{r+1}(\gamma).$$

Damit ist (8) bewiesen, und wir haben gefunden:

Bezüglich des Systems aller achsenparalleler Parallelepipede des E^r gilt der Dichtesatz.

Hierbei ist, wie aus dem Beweis ersichtlich, unwesentlich, dass die Achsen aufeinander senkrecht stehen, sodass der Dichtesatz überhaupt für Systeme von Parallelepipeden gilt, die zu einem festen ähnlich gelegen sind.

Allgemeiner zeigt man, etwa in der Ebene, leicht, dass *der Dichtesatz für das System aller konvexen Polygone gilt, deren Seiten zu einer von endlich vielen fest vorgegebenen Richtungen parallel sind.*

B. Erstes Beispiel eines Systems, für das der Dichtesatz nicht gilt.

Sehr leicht zu behandeln ist das System aller der Γ-förmigen Mengen der Ebene, die aus zwei achsenparallelen Rechtecken zusammengesetzt sind:

$$\varrho = (x_0 < x < X_0,\ y_0 < y < Y_0;\ X_0 < x < X_1,\ y_0 < y < Y_1 < Y_0).$$

Betrachtet man etwa das Einheitsquadrat W, so sieht man ohne weiteres, dass für $0 < \alpha < 1$ die Menge $\sigma_\alpha(W)$ den ganzen Halbstreifen $0 < y < 1$, $x > 0$ enthält. Da unser System mit jeder Menge zugleich alle ähnlich gelegenen enthält, folgt aus dem Kriterium II' unmittelbar, dass dafür der Dichtesatz nicht gilt.

C. Das System aller Parallelepipede.

Wir beweisen nun, dass der Dichtesatz bereits für das System aller Rechtecke der Ebene nicht gilt, woraus man sofort die entsprechende Tatsache für alle höheren Dimensionen ableitet.

Da das System mit jeder Menge auch die dazu ähnlich gelegenen enthält, genügt es nach II', eine Menge $\varkappa$ endlichen Masses anzugeben, für die

$$m(\sigma_{\frac{1}{2}}(\varkappa)) = \infty$$

ist [12]).

1) Es sei ABC eine offene Dreiecksfläche. Wir verdoppeln die Seiten AB und AC über B und C hinaus und bekommen so die Punkte B' und C'. Dann liegt für jeden Punkt Q des offenen Dreiecks $AB'C'$ die Strecke QA mehr als halb in ABC; also gibt es zu Q ein offenes Rechteck, dessen eines Seitenpaar parallel zu QA ist, und das halb von ABC überdeckt wird. Daher ist

$$AB'C' \subset \sigma_{\frac{1}{2}}(ABC).$$

Die folgende Konstruktion gibt eine Folge von Dreiecken $A_\nu B_\nu C_\nu$ so an, dass ihre Vereinigungsmenge endliches, die Vereinigungsmenge der entsprechenden Dreiecke $A_\nu B'_\nu C'_\nu$ aber unendliches Mass hat.

[12]) Das folgende Beispiel ist einer Konstruktion nachgebildet, welche O. Perron zur Vereinfachung der von Besicovitch gegebenen Lösung des Kakeyaschen Problems gemacht hat (siehe Math. Zeitschr. **28**, 1928, S. 383/6). Übrigens kann man diese selbst auch für unsere Zwecke verwenden.

16*

2) Es bezeichne D das offene Dreieck ABC (siehe Fig.), dessen Grundlinie BC auf und dessen Spitze A oberhalb der x-Achse liegt. Die Länge der Höhe sei h, die der Grundlinie BC sei a, die Punkte B' und C' mögen dieselbe Bedeutung haben wie unter 1).

Es sei $0 < h' < h$. Wir bringen die Geraden $y = h + h'$, $y = h - h'$ und $y = -(h + h')$ zum Schnitt mit den Geraden AB und AC. Die Schnittpunkte nennen wir in der aus der Figur ersichtlichen Reihenfolge P, Q, R, S, B'', C''. Die beiden parallelen Geraden PQ und RS schneiden die Strecken BC und $B''C''$ in U, V bzw. U'', V''. Die (offenen) Dreiecksflächen PUC und SVB bezeichnen wir mit D_1 und D_2.

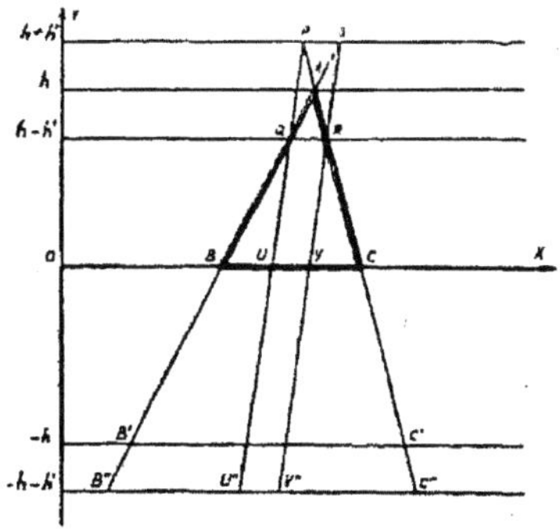

Von dieser Konstruktion werden wir folgendes gebrauchen:

Die Höhe der Dreiecke D_1 und D_2 beträgt $h + h'$. Die Länge der Strecke QR ist $\frac{h'}{h} \cdot a$. Die beiden Dreiecke SAR und PAQ bilden das halbe Parallelogramm $PQRS$, und daher ist

$$m(D_1 + D_2) - m(D) = h' \cdot \frac{h'}{h} a.$$

Die Summe der Längen der Grundlinien BV und UC von D_1 und D_2 beträgt

$$a + \frac{h'}{h} a = \frac{a}{h} \cdot (h + h').$$

Die Vereinigungsmenge der zu D_1 und D_2 gehörenden Dreiecke $SB''V''$ und $PU''C''$ enthält $AB''C''$. Der Flächeninhalt des Trapezes $B'C'C''B''$ ist grösser als $2a \cdot h'$. Für das Folgende ist wesentlich, dass dieser Inhalt linear mit h' wächst, während $m(D_1 + D_2) - m(D)$ zu h'^2 proportional ist.

3) Wir gehen von der soeben beschriebenen Konfiguration aus und setzen zur Induktion (ν bedeute in a^ν u. a. stets Indizes):

$$D = \Delta^1 = k^1, \quad a = a^1;$$

$$D_1 = \Delta_1^2, \quad D_2 = \Delta_2^2, \quad \Delta_1^2 + \Delta_2^2 = \varkappa^2;$$

$$h = h^1, \quad B' = B^1, \quad C' = C^1, \quad B'' = B^2, \quad C'' = C^2.$$

Ferner sei speziell $h' = \frac{h}{2}$, also $h + h' = \frac{3}{2} h = h^2$. Die Grundlinie von Δ_k^i bezeichnen wir mit a_k^i. Wir nehmen an, die Dreiecke

$$\Delta_k^n, \quad k = 1, 2, \ldots, 2^{n-1},$$

seien schon definiert. Es werde $\varkappa^n = \sum_k \Delta_k^n$ gesetzt. Die Grundlinie von Δ_k^n liege auf der x-Achse und ihre Länge sei a_k^n. Die Δ_k^n haben alle dieselbe Höhe h^n, und folgende Gleichungen (die für $n = 1$, $n = 2$ richtig sind) mögen bestehen:

a) $h^n = h^1 \left(1 + \frac{1}{2} + \ldots + \frac{1}{n}\right)$,

b) $a_1^n + a_2^n + \ldots + a_{2^{n-1}}^n = a^1 \frac{h^n}{h^1}$,

c) $m(\varkappa^n) = m(\Delta_1^n + \ldots + \Delta_{2^{n-1}}^n) \leqslant a^1 \cdot h^1 \left(1 + \frac{1}{2^2} + \ldots + \frac{1}{n^2}\right) - \frac{a^1 \cdot h^1}{2}$,

d) $\sigma_{\frac{1}{2}}(\varkappa^n)$ enthält das Dreieck $AB^n C^n$, wobei B^n und C^n die Schnittpunkte der Geraden AB und AC mit $y = -h^n$ sind.

Der $(n+1)$-te Schritt besteht dann im folgenden: Wir wenden die Konstruktion aus 2) an auf jedes Δ_k^n, wobei wir $h' = \frac{h^1}{n+1}$ setzen. Den Dreiecken D_1 und D_2 entsprechen dabei Δ_{2k-1}^{n+1} und Δ_{2k}^{n+1} mit den Grundlinien a_{2k-1}^{n+1} und a_{2k}^{n+1}. Die gemeinsame Höhe $h + h'$ bezeichnen wir mit h^{n+1}. Dann ist

$$h^{n+1} = h^n + h' = h^1 \left(1 + \frac{1}{2} + \ldots + \frac{1}{n+1}\right).$$

Wir zeigen, dass auch b) bis d) für den Index $n+1$ erfüllt sind. Wir setzen wieder $\varkappa^{n+1} = \sum_{k=1}^{2^n} \Delta_k^{n+1}$; es ist, wie unter 2) gezeigt wurde,

$$a_{2k-1}^{n+1} + a_{2k}^{n+1} = \frac{a_k^n}{h^n} h^{n+1},$$

also

$$\sum_{k=1}^{2^n} a_k^{n+1} = \frac{h^{n+1}}{h^n} \sum_{k=1}^{2^{n-1}} a_k^n = a^1 \cdot \frac{h^{n+1}}{h^1}.$$

Ferner ist

$$m(\Delta_{2k-1}^{n+1} + \Delta_{2k}^{n+1}) - m(\Delta_k^n) = \left(\frac{h^1}{n+1}\right)^2 \cdot \frac{a_k^n}{h^n},$$

also

$$m(\varkappa^{n+1}) = m\left(\sum_{k=1}^{2^n} \Delta_k^{n+1}\right) = m\left(\sum_{k=1}^{2^{n-1}} \Delta_k^n\right) +$$

$$+ m\left(\sum_{k=1}^{2^{n-1}} [(\Delta_{2k-1}^{n+1} + \Delta_{2k}^{n+1}) - \Delta_k^n]\right) \leqslant a^1 h^1 \left(1 + \frac{1}{2^2} + \ldots + \frac{1}{n^2}\right) - \frac{a^1 h^1}{2} +$$

$$+ \left(\frac{h^1}{n+1}\right)^2 \cdot \frac{1}{h^n} \cdot \sum_{k=1}^{2^{n-1}} a_k^n = a^1 \cdot h^1 \left(1 + \frac{1}{2^2} + \ldots + \frac{1}{(n+1)^2}\right) - \frac{a^1 h^1}{2}.$$

Schliesslich seien B^{n+1} und C^{n+1} die Schnittpunkte der Geraden $y = -h^{n+1}$ mit den Geraden AB und AC. Offenbar enthält $\sum_{k=1}^{2^n} \sigma_{\frac{1}{2}}(\Delta^{n+1})$ das ganze Dreieck $AB^{n+1}C^{n+1}$, und daher ist auch

$$AB^{n+1}C^{n+1} \subset \sigma_{\frac{1}{2}}\left(\sum_{k=1}^{2^n} \Delta_k^{n+1}\right) = \sigma_{\frac{1}{2}}(\varkappa^{n+1}).$$

Damit sind die Mengen $\varkappa^n$ für alle n definiert, und zwar ist wegen $\varkappa^n \subset \varkappa^{n+1}$ für $\varkappa = \lim \varkappa^n$

$$m(\varkappa) = \lim m(\varkappa^n) \leqslant a^1 h^1 \sum_{n=1}^{\infty} \frac{1}{n^2} - \frac{a^1 h^1}{2}.$$

Andrerseits ist

$$\sigma_{\frac{1}{2}}(\varkappa^n) \subset \sigma_{\frac{1}{2}}(\varkappa),$$

und da $\sigma_{\frac{1}{2}}(\varkappa^n)$ das Dreieck AB^nC^n enthält, dessen Höhe

$$h^1 + h^n = h^1\left(1 + \sum_{\nu=1}^{n} \frac{1}{\nu}\right)$$

gegen ∞ strebt, ist der ganze innere Winkelraum der Halbstrahlen AB und AC in $\sigma_{\frac{1}{2}}(\varkappa)$ enthalten. Wir haben also gefunden:

Für das System aller rechtwinkligen Parallelepipede des E^r $(r \geqslant 2)$ gilt der Dichtesatz nicht.

Bemerkungen.

a) Die von uns zum Beweise der Notwendigkeit der Bedingung (6) verwandte Schlussweise liefert ein Verfahren, aus den Mengen $\varkappa^n$ eine beschränkte messbare Menge zu konstruieren, die in einer zu ihr fremden Menge positiven Masses in bezug auf die Rechtecke eine positive Dichte hat.

b) Indem wir nochmals auf 1) und 2) zurückgreifen, stellen wir fest, dass jeder Punkt der Dreiecke $AB^\nu C^\nu$ in einem solchen Rechteck liegt, dessen eines Seitenpaar parallel zu einer Richtung des Winkelraumes BAC ist. Wenn (W) eine in BAC dichte Menge von Geraden durch A ist, so lässt sich dieses Rechteck natürlich auch so wählen, dass das erwähnte Seitenpaar parallel zu einer passenden Richtung aus (W) ist. Da der Winkel BAC keiner Beschränkung unterlag, sehen wir: Der Dichtesatz gilt bereits *nicht* bezüglich des Systems der Rechtecke, deren eines Seitenpaar parallelel zu einer beliebigen unter den Geraden einer in irgend einem Winkelraum dichten Geradenmenge ist.

§ 3.

Die Differentiation der unbestimmten Integrale mit beschränkten Integranden.

Die Mengenfunktion $\varphi(\varkappa)$ heisse im Punkte P bezüglich R differenzierbar, wenn für jede sich auf P zusammenziehende Folge $\{\varrho_k\}$ aus R der Grenzwert

$$\lim_{k\to\infty} \frac{\varphi(\varrho_k \varkappa)}{m(\varrho_k)}$$

existiert.

Wir setzen voraus, dass für das System R der Dichtesatz gilt, und wollen feststellen, was sich daraus über die Differenzierbarkeit der totalstetigen absolut additiven Mengenfunktionen, oder was dasselbe ist [12]), der unbestimmten Lebesgueschen Integrale entnehmen lässt. Der Dichtesatz selbst erledigt diese Frage für die Integrale solcher messbarer Funktionen, welche nur die Werte 0 und 1 annehmen.

Es sei $f(P)$ eine summierbare Funktion und $M_{\alpha,\beta}$, $\alpha < \beta$, die Menge der Punkte, für die

$$M_{\alpha,\beta}: \qquad \alpha \leqslant f(P) < \beta$$

ist. $N_{\alpha,\beta}$ sei die Menge der Punkte von $M_{\alpha,\beta}$, in welchen die untere Dichte von $M_{\alpha,\beta}$ kleiner als 1 ist. $N_{\alpha,\beta}$ hat das Mass 0. Wir setzen

$$N_f = \sum_{\alpha<\beta} N_{\alpha,\beta},$$

wo (α, β) alle rationalen Zahlenpaare mit $\alpha < \beta$ durchläuft,

Als einfache *Folge des Dichtesatzes* ergibt sich: *Wenn $f(P)$ beschränkt ist, so ist das Integral $\int\limits_{x} f(P)\cdot dP$ in allen Punkten Q, die nicht der Nullmenge N_f angehören, bezüglich R differenzierbar, und seine Ableitung hat dort den Wert $f(Q)$.*

Es sei nämlich $|f(P)| < C$, α, β rational, Q ein Punkt aus $M_{\alpha,\beta}$, der nicht in N_f enthalten ist; $M'_{\alpha,\beta}$ sei die Komplementärmenge von $M_{\alpha,\beta}$ und ϱ_k irgend eine sich auf Q zusammenziehende Folge aus R. Dann ist

$$-C\cdot m(M'_{\alpha,\beta}\varrho_k) + \alpha\cdot m(M_{\alpha,\beta}\varrho_k) \leqslant \int\limits_{\varrho_k} f(P)\cdot dP \leqslant$$
$$\leqslant \beta\cdot m(M_{\alpha,\beta}\varrho_k) + C\cdot m(M'_{\alpha,\beta}\varrho_k),$$

woraus

$$\alpha \leqslant \overline{\underline{\lim}}_{k\to\infty} \frac{\int_{\varrho_k} f(\varrho)\, dP}{m(\varrho_k)} \leqslant \beta$$

folgt.

Hiernach sind die Integrale über beschränkte Funktionen speziell nach den Intervallen differenzierbar. Dieses Ergebnis formulieren wir in üblicher Weise als einen Satz über die Punktfunktion

$$F(x_1, \ldots, x_r) = \int_0^{x_1}\int_0^{x_2}\ldots\int_0^{x_r} f(X)\, dX, \quad |f(x)| < C.$$

[12]) Bekanntlich benötigt man, um das zu beweisen, nur die äusserst einfache Differentiation nach den Mengen eines sich verfeinernden quadratischen Gitters. Vgl. etwa de la Vallée Poussin, Intégrales de Lebesgue..., Paris 1916, S. 61 ff.

Wir betrachten den r-ten gemischten Differenzenquotienten der Funktion $F(x_1, \dots, x_r)$ im Punkte $X = (x_1, \dots, x_r)$; er lässt sich in der Integralform

$$\frac{\Delta^r F(x_1, \dots, x_r)}{h_1 \dots h_r} = \frac{1}{h_1 \dots h_r} \int_{x_1}^{x_1+h_1} \dots \int_{x_r}^{x_r+h_r} f(X)\, dX$$

schreiben; für $r = 2$ ist z. B.

$$\frac{\Delta^2 F(x_1, x_2)}{h_1 \cdot h_2} =$$

$$= \frac{F(x_1 + h_1, x_2 + h_2) - F(x_1 + h_1, x_2) - F(x_1, x_2 + h_2) + F(x_1, x_2)}{h_1 \cdot h_2}.$$

Zur Bildung dieses Differenzenquotienten wird, geometrisch gesprochen, ein Intervall benutzt, von dem X ein Eckpunkt ist. Diese Intervalle bilden im Punkte X ein bezüglich des Systems der Intervalle reguläre Mengenfamilie (vgl. S. 238), so dass man nach ihnen differenzieren kann. Das heisst aber, dass *die r-te gemischte Ableitung $F_{(x_1 x_2 \dots x_r)}$ als Grenzwert des betrachteten Differenzenquotienten fast überall existiert und zwar im Sinne des r-fachen Limes.*

Bei dem üblichen Vorgehen kommt die Existenz dieses Limes nicht heraus, da man voraussetzt, dass die Intervalle, nach denen man differenziert, bezüglich der Würfel regulär sind, so dass die Verhältnisse $h_i : h_k$ sämtlich beschränkt bleiben müssen.

Die Existenz des Doppellimes

$$\lim_{\substack{h_1 \to 0 \\ h_2 \to 0}} \frac{1}{h_1 h_2} [F(x_1 + h_1, x_2 + h_2) - F(x_1 + h_1, x_2) - F(x_1, x_2 + h_2) + F(x_1, x_2)]$$

macht, wenn die Existenz der ersten partiellen Ableitungen $\frac{\partial F}{\partial x}$ und $\frac{\partial F}{\partial y}$ bekannt ist, einen besondren Beweis für die Existenz und Gleichheit der beiden gemischten Ableitungen $\frac{\partial^2 F}{\partial x \partial y}$ und $\frac{\partial^2 F}{\partial y \partial x}$ unnötig.

Entsprechend existieren in r Dimensionen die gemischten Ableitungen s-ter Ordnung ($s \leqslant r$) fast überall, und es ist

$$\text{(12)} \qquad \frac{\partial^s F}{\partial x_{\nu_1} \dots \partial x_{\nu_s}} = F_{(x_{\nu_1} \dots x_{\nu_s})}, \quad \nu_i \neq \nu_k \text{ für } i \neq k,$$

wobei die Ableitung rechts als s-facher Limes des entsprechenden Differenzenquotienten definiert ist. (Darin ist die Vertauschbarkeit der Ableitungsreihenfolge links natürlich enthalten).

Wenn nämlich bekannt ist, dass die Ableitungen $F_{(x_{\nu_1} \dots x_{\nu_s})}$ ($\nu_i \neq \nu_k$ für $i \neq k$) für alle $s \leqslant r$, abgesehen von einer r-dimensionalen Nullmenge existieren, so folgt aus bekannten Grenzwertsätzen auch die Existenz der iterierten Ableitungen und die Gleichung (12). Um aber auch für $s < r$ zu zeigen, dass $F_{(x_{\nu_1} \dots x_{\nu_s})}$ fast überall existiert, genügt es zu wissen, dass die Menge, in der $F_{(x_{\nu_1} \dots x_{\nu_s})}$ nicht existiert, r-dimensional messbar ist. Denn ihr Schnitt mit allen s-dimensionalen Hyperebenen parallel zu $(x_{\nu_1}, \dots, x_{\nu_s})$, in denen F totalstetig ist, hat das s-dimensionale Mass 0. Wir betrachten den s-ten Differenzenquotienten, durch dessen Grenzwert $F_{(x_{\nu_1} \dots x_{\nu_s})}$ definiert ist, und konstruieren die Menge der Punkte, in denen sein Limes superior vom Limes inferior verschieden ist. Dabei ist es offenbar gestattet, sich auf rationale Zuwächse $h_{\nu_1}, \dots, h_{\nu_s}$ zu beschränken. Wir ordnen alle rationalen Kombinationen $(h_{\nu_1}, \dots, h_{\nu_s})$ mit $\sum_{i=1}^{s} h_{\nu_i}^2 < \frac{1}{n}$ in eine Folge, und bilden die entsprechende Folge der Differenzenquotienten. Ihr Limes superior sei $L_n(P)$, der Limes inferior $l_n(P)$. Da die beiden Folgen $L_n(P)$ und $l_n(P)$ monoton ab-, bzw. zunehmen, existiert $L(P) = \lim L_n(P)$ und $l(P) = \lim l_n(P)$. Die Funktionen $L(P)$ und $l(P)$ sind messbar, und daher auch die Menge, in der $L(P) > l(P)$ ist; sie besteht offenbar genau aus den Punkten, in denen der Differenzenquotient keinem Grenzwert zustrebt.

§ 4.

Differentiation beliebiger unbestimmter Integrale.

In diesem Paragraphen soll gezeigt werden, dass aus dem Dichtesatz die Differenzierbarkeit der Integrale beliebiger summierbarer Funktionen, z. B. schon für das System der Intervalle, *nicht* folgt.

Zu diesem Zweck beweisen wir den folgenden, dem Kriterium II analogen Satz:

Das System R enthalte mit jeder Menge ϱ alle dazu ähnlich gelegenen Mengen. Bezüglich R ist dann und nur dann jede totalstetige additive Mengenfunktion differenzierbar, wenn für beliebige positive Zahlen $a_1, \ldots, a_n$ und beliebige fremde, beschränkte messbare Mengen $\varkappa_1, \ldots, \varkappa_n$ die Vereinigungsmenge s aller Mengen ϱ aus R mit

$$\sum_{\nu=1}^{n} a_\nu \cdot m(\varkappa_\nu \varrho) > m(\varrho) \tag{13}$$

der Bedingung

$$m(s) < C \cdot \sum_{\nu=1}^{n} a_\nu \cdot m(\varkappa_\nu) \tag{14}$$

genügt, wo C lediglich von System R abhängt.

1) Hinreichend. Wir bemerken, dass (14) wesentlich schärfer ist als (6), so dass der Dichtesatz bezüglich R gilt. Es genügt offenbar Integrale über positive summierbare Funktionen zu betrachten;

$$\varphi(\varkappa) = \int_{\varkappa} f(P) \cdot dP, \quad f(P) > 0,$$

sei die zu untersuchende Funktion; $\varkappa_\nu$ sei die Menge der Punkte mit

$$\varkappa_\nu : \qquad \nu - 1 \leqslant f(P) < \nu.$$

Dann konvergiert $\sum_{\nu=1}^{\infty} \nu \cdot m(\varkappa_\nu)$. Es sei $s_{n_0, \varepsilon}$ die Vereinigunsmenge der Menge ϱ aus R, für die

$$\sum_{n_0}^{\infty} \nu \cdot m(\varkappa_\nu \varrho) > \varepsilon \cdot m(\varrho), \quad \varepsilon > 0,$$

ist. $s_{n_0, \varepsilon}$ nimmt bei festem ε monoton ab; der Durchschnitt ist wegen (14)[14]) eine Nullmenge. Wir bilden die Vereinigungsmenge N dieser Nullmengen für $\varepsilon = \frac{1}{2}, \frac{1}{3}, \ldots$ N_f sei die auf S. 248 definierte, zu f gehörige Nullmenge. Wir zeigen nun: Wenn $P \not\subset N_f + N$,

[14]) Aus der Gültigkeit von (14) für eine beliebige endliche Zahl von Mengen $\varkappa_\nu$ folgt offenbar die Gültigkeit für abzählbar viele $\varkappa_\nu$ (eventuell mit „$\leqslant$" statt „$<$").

so ist für jede sich auf P zusammenziehende Folge $\{\varrho_k\}$ aus R

$$\lim_{k=\infty}\frac{\varphi(\varrho_k)}{m(\varrho_k)}=f(P).$$

Wir wählen bei festem ε die Zahl n_0 so gross, dass P nicht mehr in $s_{n_0,\varepsilon}$ enthalten ist. Dann gilt

$$\frac{\varphi(\varrho_k)}{m(\varrho_k)}=\frac{\sum\limits_{\nu=1}^{n_0-1}\varphi(\varrho_k\cdot\varkappa_\nu)}{m(\varrho_k)}+\frac{\sum\limits_{\nu=n_0}^{\infty}\varphi(\varrho_k\cdot\varkappa_\nu)}{m(\varrho_k)}.$$

Die Wahl von n_0 hat zur Folge, dass der zweite Summand rechts höchstens die Grösse ε hat. Andrerseits folgt aus dem Dichtesatz, dass der erste Summand mit wachsendem k gegen $f(P)$ strebt.

2) Notwendig. Der Beweis verläuft genau wie der Nachweis der Notwendigkeit von (II). Wenn (14) nicht erfüllt ist, so gibt es zu jedem n gewisse Zahlen $a_1^n,\ldots,a_{\nu_n}^n$, die offenbar grösser als 1 vorausgesetzt werden dürfen, und fremde beschränkte, messbare Mengen $\varkappa_1^n,\ldots,\varkappa_{\nu_n}^n$ derart, dass eine beschränkte messbare Menge s^n existiert, welche Vereinigungsmenge von Mengen ϱ aus R mit

$$\sum_i a_i^n\cdot m(\varrho\,\varkappa_i^n)>m(\varrho) \tag{15}$$

ist und der Ungleichung

$$m(s^n)>4^n\cdot\sum_i a_i^n\cdot m(\varkappa_i^n) \tag{16}$$

genügt. Es sei

$$h(P;s^n)=\begin{cases}a_i^n & \text{für}\quad P\subset s^n\cdot\varkappa_i^n\\ 0 & \text{sonst.}\end{cases} \tag{17}$$

Dann ist

$$\int\limits_{s^n}h(P;s^n)\cdot dP=\sum_i a_i^n\cdot m(s^n\varkappa_i^n)<\frac{1}{4^n}\cdot m(s^n), \tag{18}$$

und jeder Punkt von s^n liegt in einer Menge ϱ aus R mit

$$\int\limits_{\varrho}h(P;s^n)\cdot dP>m(\varrho). \tag{19}$$

Durch Ähnlichkeitsübertragung werde $h(P;\tau)$ für alle zu s^n ähnlich gelegenen Mengen τ definiert. Die Gleichungen (18), (19) übertragen sich auf $h(P;\tau)$, da ja R mit ϱ alle zu ϱ ähnlich gelegenen Mengen enthält. Wir überdecken nach dem Hilfssatz von S. 232 den Einheitswürfel W mit zu s^n ähnlich gelegenen Mengen s_j^n so, dass

$$d(s_j^n) < \frac{1}{n}, \quad \sum_j m(s_j^n) \leqslant 2, \quad m\Big(\sum_j s_j^n\Big) = 1$$

wird. Wir setzen weiter

$$f^n(P) = \sum_j h(P; s_j^n).$$

Wegen (16) und (18) ist f^n summierbar und

$$\int_W f^n \, dP < \frac{1}{4^n} \sum_j m(s_j^n) = \frac{2}{4^n}.$$

Die Funktion $f = \sum\limits_{n=1}^{\infty} f^n$ ist dann ebenfalls summierbar mit

$$\int_W f \, dP < 2 \sum_{n=1}^{\infty} \frac{1}{4^n} = \frac{2}{3}.$$

Da wir alle $a_i^n > 1$ vorausgesetzt haben, hat f, soweit es nicht verschwindet, wenigstens den Wert 1. Deshalb hat die Menge der Punkte mit $f > 0$ höchstens das Mass 2/3. Andrerseits liegt, von einer Nullmenge abgesehen, jeder Punkt Q von W für jedes n in einem s_j^n, also auch in einer Menge ϱ aus R mit (vergl. (19))

$$\int_\varrho f(P) \cdot dP \geqslant \int_\varrho h(P; s_j^n) \cdot dP > m(\varrho).$$

Da der Durchmesser

$$d(s_j^n) < \frac{1}{n}$$

ist, ziehen sich diese Mengen ϱ auf Q zusammen; die obere Ableitung der Funktion f bezüglich R ist also fast überall in W

gleich 1, während f mindestens in einer Teilmenge des Masses 2/3 von W verschwindet, womit die Behauptung bewiesen ist.

Bekanntlich kann man bezüglich R jede totalstetige Funktion differenzieren, wenn für dass System R der **Vitalische Überdeckungssatz** gilt. Aus diesem muss also die Bedingung (14) folgen. Unser Notwendigkeitsbeweis zeigt nun gerade, dass das System R dem Vitalischen Satze nicht genügen kann, wenn (14) nicht erfüllt ist. Von den beim Beweise verwandten Mengen ϱ zieht sich nämlich auf fast jeden Punkt von W eine Teilfolge zusammen. Trotzdem: Wie man auch abzählbar viele paarweise fremde unter den ϱ herausgreift, stets hat ihre Vereinigungsmenge höchstens das Mass 2/3. Es seien ϱ^ν irgend welche fremde unter den Mengen ϱ, als deren Vereinigungsmenge s^n definiert ist. Für sie gilt (15), woraus nach (16)

$$m\left(\sum_\nu \varrho^\nu\right)=\sum_\nu m(\varrho^\nu)<\sum_i\sum_\nu a_i^n\cdot m(\varrho^\nu x_i^n)<\sum_i a_i^n\cdot m(x_i^n)<$$
$$<\frac{1}{4^n}\cdot m(s^n)$$

folgt. Dieselbe Abschätzung gilt für jedes System der Mengen ϱ in irgend einem s_j^n. Für ein beliebiges System von fremden Mengen ϱ gilt also

$$m\left(\sum \varrho\right)<\sum_n\frac{1}{4^n}\sum_j m(s_j^n)=\sum\frac{2}{4^n}=\frac{2}{3}.$$

Der Umweg über den Vitalischen Satz zum Nachweis von (14) ist jedoch nicht erforderlich. Denn man sieht sehr leicht:

Wenn ϱ eine Menge des Systems R ist, $d(\varrho)$ ihr Durchmesser, ferner $U(\varrho)$ die Menge aller Punkte, deren Abstand von ϱ kleiner als $2\cdot d(\varrho)$ ist, und wenn

$$\frac{m(U(\varrho))}{m(\varrho)}<\beta, \tag{20}$$

gilt wo β eine von ϱ unabhängige Zahl bedeutet, so genügt das System R der Bedingung (14). Solche Systeme bilden z. B. die **Kugeln**, die **Würfel** und ferner alle **rechtwinklingen Parallelepipede**, **bei denen die Verhältnisse der Kantenlängen beschränkt sind.**

Es sei s die wie oben definierte Vereinigungsmenge der ϱ mit (13). Wir beweisen, dass (14) mit $C=\beta$ erfüllt ist, wobei wir eine auch beim üblichen Beweis des Vitalischen Satzes vorkommende Idee benutzen. Aus den ϱ, welche s bilden, wählen wir eine Folge $\varrho_1', \varrho_2', \ldots$, welche s überdeckt. Wir setzen $\varrho_1=\varrho_1'$, und bezeichnen mit ϱ_2 das erste ϱ_ν' der Folge, das nicht in $U(\varrho_1)$ enthalten ist. Allgemein sei ϱ_{n+1} das erste ϱ_j' der Folge ϱ_ν', das nicht in $\sum_{\nu=1}^{n} U(\varrho_\nu)$ enthalten ist. So erhalten wir eine Folge $U(\varrho_n)$, deren Vereinigungsmenge s überdeckt. Aus dieser Folge denken wir uns jedes $U(\varrho_i)$ bereits gestrichen, das ganz in einem anderen enthalten ist. Dann sind die ϱ^ν paarweise fremd. Denn für $n>m$ enthält ϱ_n einen Punkt ausserhalb von $U(\varrho_m)$. Hätte es einen Punkt P mit ϱ_m gemeinsam, so wäre $d(\varrho_n)\geqslant 2\cdot d(\varrho_m)$. Daher enthielte $U(\varrho_n)$ alle Punkte, die von P um weniger als $4\cdot d(\varrho_m)$ entfernt sind; ein Punkt von $U(\varrho_m)$ hat aber von P höchstens den Abstand $3\cdot d(\varrho_m)$, und es wäre $U(\varrho_m)\subset U(\varrho_n)$. Daher ist

$$m\Big(\sum_i \varrho_i\Big)=\sum_i m(\varrho_i)>\frac{1}{\beta}\sum_i m(U(\varrho_i))\geqslant\frac{1}{\beta}m\Big(\sum_i U(\varrho_i)\Big)\geqslant\frac{1}{\beta}m(s).$$

Andrerseits ist wegen (13) und der Fremdheit der ϱ_i

$$\sum_i m(\varrho_i)<\sum_\nu\sum_i a_\nu\cdot m(\varkappa_\nu\,\varrho_i)\leqslant\sum_\nu a_\nu\cdot m(\varkappa_\nu),$$

womit

$$m(s)<\beta\sum_\nu a_\nu\cdot m(\varkappa_\nu)$$

bewiesen ist.

Wenn die Beziehung (14) für ein System R nicht gilt, so lässt sich das im allgemeinen noch leichter feststellen, z. B. unmittelbar für das System der achsenparallelen Rechtecke der Ebene. Bedeutet W ein achsenparalleles Quadrat, so ist für dieses System

$$m(\sigma_{\frac{1}{n}}(W))=(4n\cdot\log n+1)\cdot m(W).$$

(14) ist also schon für $\varkappa_1=W$, $a_1=n$, $\varkappa_\nu=0$ für $\nu>1$, nicht erfüllt [15]).

[15]) Damit ist nochmals gezeigt, dass für die achsenparallelen Rechtecke der Vitaliche Satz nicht gilt.

Man kann also die unbestimmten Integrale nicht beschränkter Funktionen im allgemeinen nach achsenparallelen Rechtecken nicht differenzieren, oder, anders ausgedrückt: Für die Funktion

$$F(x, y) = \int_0^x \int_0^y f(\xi, \eta)\, d\xi\, d\eta$$

braucht, wenn $f(x, y)$ nicht beschränkt ist, die zweite gemischte Ableitung $f_{(x,y)}$ im Sinne des Doppellimes nicht zu existieren.

Zusatz bei der Korrektur.

Nach Vollendung der Arbeit bemerken wir, dass einige Behauptungen der Einleitung durch das neue Buch von S. Saks: Théorie de l'Intégrale, Warszawa 1933, überholt sind. Der Dichtesatz für das System der Intervalle wurde bereits von Herrn Saks bewiesen (vgl. S. 231 f.). Ferner hat Herr Saks dort auch schon bemerkt, dass die Integrale über unbeschränkte Integranden nach Intervallen nicht differenzierbar zu sein brauchen; wir verdanken einen Beweis und eine, auf dem Begriff der Baireschen Kategorien beruhende, Verschärfung dieser Behauptung einer freundlichen brieflichen Mitteilung von Herrn Saks. Schliesslich sei erwähnt, dass man nach einer Bemerkung von O. Nikodym bereits aus einem, zu anderen Zwecken konstruierten Beispiel von A. Zygmund schliessen kann, dass der Dichtesatz für das System aller Rechtecke der Ebene nicht gilt. (vgl. S. 232 des zitierten Buches).

Kopenhagen, den 30. XI. 1933.

Fundam. Math. 22 (1934), 226–256

Lokale Eigenschaften der zu Variationsproblemen gehörigen metrischen Räume.

Von

Herbert Busemann (Princeton, U.S.A.).

1. Einem positiv definiten, quasiregulären Variationsproblem mit hinreichend glattem Integranden kann in bekannter Weise ein Finslerscher Raum zugeordnet werden. Wenn der Integrand $F(P,\varphi)$ nicht genügend regulär ist, sondern etwa wie bei den Mengerschen [1] [2] Untersuchungen nur stetig von Punkt und Richtung abhängt, so gelangt man, wenn man den Wert des Integrals über $F(P,\varphi)$ längs einer gerichteten Minimanten von A nach B als Abstand von A und B einführt (die Existenz solcher Minimanten ist durch die erwähnten Mengerschen Resultate gesichert) zu einem nicht notwendig symmetrischen metrischen Raum R. $F(P,\varphi)$ definiert an jeder Stelle die Indikatrix des Variationsproblems, und Quasiregularität ist als (schwache) Konvexität der Indikatrix erklärt. Gołąb [3]) hat kürzlich unter den Mengerschen Differenzierbarkeitsannahmen gezeigt, dass die Konvexität der Indikatrix notwendig ist, damit der Raum R metrisch (bezw. fastmetrisch) wird.

In dieser Note möchte ich einige Zusätze zu den Ergebnissen von Menger und Gołąb machen, die, wie ich glaube, zur Klärung der Zusammenhänge geeignet sind. Ich gehe aus von einem metrischen Raum der obigen Art, d. h. ich betrachte einen auf einen Euklidischen Raum topologisch abgebildeten, nicht notwendig

[1]) K. Menger, *Metrische Geometrie und Varationsrechnung*, Fund. Math. **25** (1935), pp. 441–458.

[2]) K. Menger, *Die metrische Methode in der Variationsrechnung*, Erg. Math. Kolloq. **8** (1937), pp. 1–32.

[3]) St. Gołąb, *Ein Beitrag zum Mengerschen Begriff des fastmetrischen Raumes*, Fund. Math. **31** (1938), pp. 67–73.

symmetrischen metrischen Raum R (vgl. die Bedingungen (2, 3) abc), und zwar beschränke ich mich der Einfachheit halber auf den zweidimensionalen Fall. Indem man den Euklidischen Abstand zweier Punkte mit dem Abstand der entsprechenden Punkte in R in Beziehung bringt, sieht man in einfacher Weise, dass schon unter diesen allgemeinen Voraussetzungen in einem gewissen Sinne von einer konvexen Indikatrix gesprochen werden kann. Da nämlich keine Differenzierbarkeitsannahmen gemacht werden, muss die eine Indikatrix durch zwei, eine *obere* und eine *untere*, ersetzt werden. *Die untere Indikatrix ist dann stets konvex.* Wenn eine Indikatrix existiert, d. h. wenn obere und untere Indikatrix zusammenfallen, ist sie daher konvex. Bei gegebener unterer Indikatrix muss die obere einer einfachen Beziehung (b) genügen. Dies ist merkwürdigerweise *genau dieselbe Beziehung, in der die obere Krümmungsindikatrix einer konvexen Fläche in einem Punkt zur unteren stehen muss.* Auf Grund der bereits vorliegenden ausführlichen Untersuchung dieser Krümmungsindikatrizen [4]) zeigt es sich, dass jedesmal, wenn eine beliebige einen gegebenen Punkt P umschliessende oder enthaltende konvexe Kurve i und eine der Beziehung (b) genügende Kurve I gegeben ist, i und I auftreten können als untere und obere Indikatrix von P (dem Aufpunkt von i und I) in einem konvexen metrischen Raume, dessen topologische Beziehung auf die euklidische Ebene ausserhalb P beliebig glatt gewählt werden kann.

In unserer Definition der oberen und unteren Indikatrix ist enthalten, dass, wenn beide für alle Punkte eines Gebietes G zusammenfallen, die Indikatrix in G stetig von ihrem Aufpunkt abhängt. Setzt man ausserdem voraus, dass die Indikatrix im scharfen Sinne konvex und überall beschränkt ist und ihren Aufpunkt im Innern enthält, so ist dies bereits mehr, als nach Menger für die Existenz von rektifizierbaren Minimanten nötig ist. Meines Wissens ist bisher nicht untersucht worden, *ob die Minimanten unter den angegebenen Voraussetzungen überall* (und nicht nur fast überall) *differenzierbar sind.* Wir zeigen an einem Beispiel, dass das im allgemeinen *nicht der Fall* ist. Wohl aber kommt den Minimanten *überall eine eigenartige, der Differenzierbarkeit ähnliche Eigenschaft* zu, die schwächer ist als stetige Differenzierbarkeit,

---

[4]) H. Busemann und W. Feller, *Bemerkungen zur Differentialgeometrie der konvexen Flächen II, Über die Krümmungsindikatrizen*, Mat. Tidsskrift **B** (1935), pp. 87–115. Im folgenden zitiert mit „B. F.“.

gewöhnliche Differenzierbarkeit aber (wie jenes Beispiel zeigt) weder enthält, noch (wie an anderen Beispielen zu sehen ist) in ihr enthalten ist. Man bekommt also *auch für Punkte, wo eine Tangente existiert, eine zusätzliche Aussage*. Unter gewissen Umständen, z. B. bei Existenz einer rechten oder einer linken Halbtangente, folgt die Differenzierbarkeit aus unserer Eigenschaft.

2. π sei die euklidische Ebene mit einem kartesischen Koordinatensystem (x,y) und einem Polarkoordinatensystem (r,φ), wobei $x=r\cos\varphi$, $y=r\sin\varphi$ sein soll. Den euklidischen Abstand zweier Punkte $A=(x_1,y_1)=(r_1,\varphi_1)$ und $B=(x_2,y_2)=(r_2,\varphi_2)$ bezeichnen wir mit

$$(2.1)\qquad e(A,B)=[(x_1-x_2)^2+(y_1-y_2)^2]^{\frac{1}{2}}=[r_1^2+r_2^2-2r_1r_2\cos(\varphi_1-\varphi_2)]^{\frac{1}{2}}.$$

Wir notieren zunächst eine einfache, später oft gebrauchte Beziehung: Seien $\overrightarrow{PB_1}$, $\overrightarrow{PB_2}$ zwei von P ausgehende, nicht entgegengesetzt gerichtete Strahlen und $\bar{B}$ ein Punkt auf der Strecke $\overline{B_1B_2}$. Setzt man dann $\sphericalangle B_1P\bar{B}=\gamma_1$, $\sphericalangle \bar{B}PB_2=\gamma_2$ so ist

$$(2.2)\qquad \frac{\sin(\gamma_1+\gamma_2)}{e(P,\bar{B})}=\frac{\sin\gamma_2}{e(P,B_1)}+\frac{\sin\gamma_1}{e(P,B_2)}.$$

Ausser $e(X\cdot Y)$ liege in π eine zu $e(\;)$ topologisch äquivalente und nicht notwendig symmetrische Metrik vor. D. h., es sei zu je zwei Punkten A,B ein Abstand $r(A,B)$ erklärt mit den Eigenschaften:

(2.3)
a) $r(A,B)\geqslant 0$;
b) $r(A,B)=0$ dann und nur dann, wenn $A=B$;
c) $r(A,B)+r(B,C)\geqslant r(A,C)$;

Die topologische Äquivalenz von $r(\;)$ und $e(\;)$ bedeutet, dass aus $r(A,A_\nu)\to 0$ oder $r(A_\nu,A)\to 0$ stets $e(A,A_\nu)\to 0$, und aus $e(A,A_\nu)\to 0$ stets $r(A,A_\nu)+r(A_\nu,A)\to 0$ folgen soll.

Bezeichnet man bei einem hinreichend glatten, positiv definiten Variationsproblem mit $r(X,Y)$ die kürzeste Entfernung im Sinne des Variationsproblems von X nach Y (d. h. den Wert des Integrals längs einer gerichteten Minimanten von X nach Y), so bekommt man die Indikatrix in einem Punkt O folgendermassen: es sei $\vec{s}$ ein von O ausgehender Strahl mit der Richtung φ, und A_ν,B_ν seien Punktepaare derart, dass die X Strahlen $\overrightarrow{A_\nu B_\nu}\,\|\,\vec{s}$ sind und

$A_\nu \to O$, $B_\nu \to O$ gilt. Man bilde dann

$$\Psi(A_\nu, B_\nu) = \frac{e(A_\nu, B_\nu)}{r(A_\nu, B_\nu)}$$

und trage von O aus auf $\underset{\to}{s}$ den Punkt $(g(\varphi), \varphi)$ mit

$$g(\varphi) = \lim_{\nu \to \infty} \Psi(A_\nu, B_\nu)$$

ab. $r = g(\varphi)$ stellt dann die Indikatrix des Variationsproblems für O als Aufpunkt dar.

Um in unserem allgemeinen Falle zu einer Indikatrix zu gelangen, führen wir daher auch die Funktion

$$\Psi(x, y) = \frac{e(x, y)}{r(x, y)} \tag{2.4}$$

ein und versuchen, die Dreiecksungleichung (2.3 c) als Ungleichung für $\Psi(X, Y)$ zu schreiben. Dazu seien A, A_1, A_2 irgend drei nicht auf einer Geraden gelegene Punkte und $\sphericalangle A_1 A A_2 = \gamma$, $A A_1 A_2 = \gamma_1$, $A_1 A_2 A = \gamma_2$. Dann ist

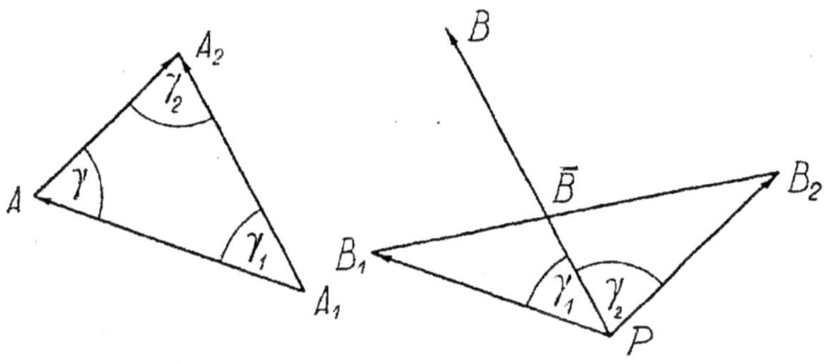

$$\frac{1}{e(A_1, A_2)} = \frac{1}{e(A_1, A)} \frac{\sin\gamma_2}{\sin\gamma} = \frac{1}{e(A, A_2)} \frac{\sin\gamma_1}{\sin\gamma}$$

und daher

$$\frac{r(A_1, A_2)}{e(A_1, A_2)} \leqslant \frac{r(A_1, A)}{e(A_1, A_2)} + \frac{r(A, A_2)}{e(A_1, A_2)} = \frac{r(A_1, A)}{e(A_1, A)} \frac{\sin\gamma_2}{\sin\gamma} + \frac{r(A, A_2)}{e(A, A_2)} \frac{\sin\gamma_1}{\sin\gamma}$$

oder wegen $\sin\gamma = \sin(\gamma_1 + \gamma_2)$

$$\frac{\sin(\gamma_1 + \gamma_2)}{\Psi(A_1, A_2)} \leqslant \frac{\sin\gamma_2}{\Psi(A_1, A)} + \frac{\sin\gamma_1}{\Psi(A, A_2)}. \tag{2.5}$$

Zieht man von P aus Strahlen $\overrightarrow{PB_1} \| \overrightarrow{A_1 A}$, $\overrightarrow{PB_2} \| \overrightarrow{A A_2}$, und $\overrightarrow{PB} \| \overrightarrow{A_1 A_2}$, wo B_1, B_2, B so gewählt sein mögen, dass

$$e(P,B_1)=\Psi(A_1,A),\quad e(P,B_2)=\Psi(A,A_2),\quad e(P,B)=\Psi(A_1,A_2)$$

wird, so fällt $\overrightarrow{PB}$ ins Innere des von $\overrightarrow{PB_1}$ und $\overrightarrow{PB_2}$ gebildeten Winkels, trifft also die Strecke $\overline{B_1 B_2}$ in einem Punkt $\bar{B}$. (2.2) und (2.5) besagen dann, dass B nicht ins Innere des Dreiecks $B_1 P B_2$ fallen kann.

Wir definieren nun die Indikatrizen. Wir fixieren einen Punkt P und eine Richtung φ, betrachten dann alle Folgen $\{A_\nu, B_\nu\}$ von Punktepaaren $A_\nu, B_\nu (A_\nu \neq B_\nu)$ derart, dass 1) $\overrightarrow{A_\nu B_\nu}$ die Richtung φ hat, 2) $A_\nu \to P$, $B_\nu \to P$ gilt. $W(P,\varphi)$ sei die Gesamtheit dieser Folgen. Wir bilden dann

$$(2.6)\quad \begin{aligned} &\text{a)}\quad g(P,\varphi)=\underset{\{A_\nu,B_\nu\}\subset W(P,\varphi)}{\text{Untere Grenze}}\ \underset{\nu\to\infty}{\underline{\lim}}\,\Psi(A_\nu,B_\nu),\\ &\text{b)}\quad G(P,\varphi)=\underset{\{A_\nu,B_\nu\}\subset W(P,\varphi)}{\text{Obere Grenze}}\ \underset{\nu\to\infty}{\overline{\lim}}\,\Psi(A_\nu,B_\nu). \end{aligned}$$

Aus dieser Definition folgt sofort, dass für festes φ und $P_n \to P$

$$(2.7)\quad \begin{aligned} g(P,\varphi) &\leqslant \underset{n\to\infty}{\underline{\lim}}\ g(P_n,\varphi),\\ G(P,\varphi) &\geqslant \underset{n\to\infty}{\overline{\lim}}\ G(P,\varphi) \end{aligned}$$

gilt, sodass $g(P,\varphi)$ bei festem φ als Funktion von P nach unten und $G(P,\varphi)$ nach oben halbstetig ist. Tragen wir nun auf dem von P ausgehenden Strahl mit der Richtung φ die Punkte $\underline{R}_\varphi$ und $\bar{R}_\varphi$ ab, für die

$$e(P,\underline{R}_\varphi)=g(P,\varphi)\qquad\text{bezw.}\qquad e(P,\bar{R}_\varphi)=G(P,\varphi)$$

ist, und variiert man φ, so entstehen Punktmengen i und I, die wir als untere und obere Indikatrix von $r(X,Y)$ für P als Aufpunkt bezeichnen. Da ∞ als Wert für $g(P,\varphi)$ und $G(P,\varphi)$ zugelassen werden muss und dann die Punkte $\underline{R}_\varphi$ und $\bar{R}_\varphi$ ohne eine neue Konvention keinen Sinn haben, ist es häufig bequemer, statt der Mengen i und I die bezüglich P sternförmigen Mengen γ und Γ zu betrachten, die aus den P mit den Punkten $\underline{R}_\varphi$ bezw. $\bar{R}_\varphi$ verbindenden Strecken bestehen, worunter für $\underline{R}_\varphi=\infty$ oder $\bar{R}_\varphi=\infty$ die ganzen von P ausgehenden Strahlen zu verstehen sind. Wir wollen zunächst zeigen:

I: *γ ist konvex.*

Beweis: Es seien B_1' und B_2' irgend zwei zu γ gehörige Punkte, wir müssen zeigen, dass jeder Punkt B' von $\overline{B_1'B_2'}$ zu γ gehört. Hierbei darf angenommen werden, dass die Gerade $\overline{B_1'B_2'}$ nicht durch P geht, da dieser Fall trivial ist. $\underrightarrow{PB_1'}$, $\underrightarrow{PB_2'}$, $\underrightarrow{PB'}$ mögen die Richtungen $\varphi_1, \varphi_2, \varphi$ haben. Wir wählen nun eine Folge von Punktepaaren $\{A_1^\nu, A_2^\nu\}$ in $W(P, \varphi)$ derart, dass

$$\lim \Psi(A_1^\nu, A_2^\nu) = g(P, \varphi)$$

wird. Dann konstruieren wir den Punkt A^ν, für den

$$\underrightarrow{A_1^\nu A^\nu} \parallel \underrightarrow{PB_1'}, \qquad \underrightarrow{A^\nu A_2^\nu} \parallel \underrightarrow{PB_2'}$$

wird, und wählen eine solche Teilfolge $\{\nu_n\}$ aus $\{\nu\}$ aus, dass

$$s_{\varphi_1} = \lim \Psi(A_1^{\nu_n}, A^{\nu_n}) \geqslant g(P, \varphi) \quad \text{und} \quad s_{\varphi_2} = \lim \Psi(A^{\nu_n}, A_2^{\nu_n}) \geqslant g(P, \varphi)$$

existieren. Tragen wir dann auf $\underrightarrow{PB_1'}$, $\underrightarrow{PB_2'}$, $\underrightarrow{PB'}$ die Punkte S_{φ_1}, und S_{φ_2} mit $e(P, S_{\varphi_1}) = s_{\varphi_1}$, $e(P, S_{\varphi_2}) = s_{\varphi_2}$ ab, so folgt aus (2.5), dass der Punkt $\underline{R}_\varphi$ nicht im Inneren des Dreiecks $S_{\varphi_1}PS_{\varphi_2}$ liegen kann. Da aber jeder Punkt von $\overline{B_1'B_2'}$ dem abgeschlossenen Dreieck $S_{\varphi_1}PS_{\varphi_2}$ angehört, so muss $B' \subset \overline{P\underline{R}_\varphi}$ und damit $B' \subset \gamma$ gelten.

II: *Wenn $B_1' \subset \gamma$ und $B_2' \subset \Gamma$, so gilt $\overline{B_1'B_2'} \subset \Gamma$.*

Beweis: Wir zeigen wieder, dass jeder Punkt B' von $\overline{B_1'B_2'}$ in Γ liegt, und dürfen dazu wieder annehmen, dass $\overline{B_1'B_2'}$ nicht durch P geht. Sind $\varphi_1, \varphi_2, \varphi$ definiert wie oben, so wählen wir diesmal eine Folge $\{A^\nu, A_2^\nu\}$ in $W(P, \varphi)$ derart dass

$$\lim \Psi(A^\nu, A_2^\nu) = G(P, \varphi_2)$$

wird, und bestimmen dann A_1^ν so, dass

$$\underrightarrow{A_1^\nu A^\nu} \parallel \underrightarrow{PB_1'}, \qquad \underrightarrow{A_1^\nu A_2^\nu} \parallel \underrightarrow{PB'}$$

wird. Dann werde $\{\nu_n\}$ in $\{\nu\}$ so bestimmt, dass

$$s_{\varphi_1} = \lim \Psi(A_1^{\nu_n}, A^{\nu_n}) \geqslant g(P, \varphi_1)$$

existiert. S_{φ_1} sei der Punkt auf $\underrightarrow{PB_1'}$ mit $e(P, S_{\varphi_1}) = s_{\varphi_1}$ und S_φ der Punkt auf $\underrightarrow{PB'}$ mit $e(P, S_\varphi) = \varliminf_{n\to\infty} \Psi(A_1^{\nu_n}, A_2^{\nu_n})$.

Dann folgt aus (2.5), dass der Punkt S_φ nicht im Inneren des Dreiecks $S_{\varphi_1} P \bar{R}_{\varphi_2}$ liegen kann, sodass B' der Strecke $\overline{PS_\varphi}$, also erst recht der Strecke $\overline{P\bar{R}_\varphi}$ und damit Γ angehört.

Aus I und II ergibt sich [5]):

III. *Wenn* $g(P,\varphi_0)>0$, *so existieren (bei festem* P*) für alle Richtungen, ausser möglicherweise für* $\varphi=\varphi_0$ *und* $\varphi=\varphi_0+\pi$, *die Grenzwerte* $G(P,\varphi-)$ *und* $G(P,\varphi+)$, *und es ist:*

$$G(P,\varphi+)\leqslant G(P,\varphi)\leqslant G(P,\varphi-) \qquad \text{für} \quad \varphi_0<\varphi<\varphi_0+\pi,$$

$$G(P,\varphi-)\leqslant G(P,\varphi)\leqslant G(P,\varphi+) \qquad \text{für} \quad \varphi_0+\pi<\varphi<\varphi_0+2\pi.$$

Als Spezialfall von I und III haben wir:

III a. *Wenn* $g(P,\varphi)>0$ *für alle* φ, *so sind die Funktionen* $g(P,\varphi)$ *und* $G(P,\varphi)$ *stetig für alle* φ.

Bislang wurde noch nicht benutzt, dass die Metrik $r(\,)$ topologisch äquivalent zu $e(\,)$ ist. Dies ist nicht so befremdend, wie es zunächst scheint, da II für $g(P,\varphi)\equiv 0$ leer ist und jede Voraussetzung über das Nichtverschwinden von $g(P,\varphi)$ in Richtung der topologischen Äquivalenz der beiden Metriken geht. Z. B. besagt $g(P,\varphi)>0$, dass $r(x,y)/e(x,y)$ in einer Umgebung von P beschränkt ist. Dies wird noch deutlicher, wenn man Aussagen über $G(P,\varphi)$ bei beliebigem $g(P,\varphi)$ machen will. Wir wollen zeigen:

IV. $G(P,\varphi)$ *ist Limes einer monoton fallenden Folge nach unten halbstetiger Funktionen* $\Lambda_m(\varphi)$. [6])

Zum Beweise betrachten wir alle Punktepaare A,B derart, dass $\overrightarrow{AB}$ die Richtung φ hat und

$$\frac{1}{m}\geqslant e(A,B)\geqslant\frac{1}{m\cdot 2^\nu}, \quad e(A,B)\leqslant M; \qquad m,\nu=1,2,3,\ldots$$

ist. Wir bilden

$$\Lambda_{m,\nu}(\varphi)=\text{obere Grenze } \Psi(A,B)$$

[5]) Siehe B. F., p. 103.

[6]) Funktionen dieses Typus wurden zuerst von W. H. Young untersucht. Literatur und einige ihrer Eigenschaften sind in B. F., pp. 90/1, zusammengestellt. Dort ist auch gezeigt, dass IV für $g(P,\varphi)\not\equiv 0$ aus III folgt.

über alle diese Punktepaare. Wenn nun $r(A,B)$ topologisch äquivalent zu $e(A,B)$ ist, so ist für $A \neq B$ die Funktion $\Psi(A,B)$ stetig in A und B, sodass stets ein Punktepaar (A_ν, B_ν) mit

$$\Psi(A_\nu, B_\nu) = \Lambda_{m,\nu}(\varphi)$$

existiert. Daher ist ferner $\Lambda_{m,\nu}(\varphi)$ eine stetige Funktion von φ. Da $\Lambda_{m,\nu}(\varphi) \leqslant \Lambda_{m,\nu+1}(\varphi)$, so existiert

$$\lim_{\nu \to \infty} \Lambda_{m,\nu}(\varphi) = \Lambda_m(\varphi)$$

und ist nach unten halbstetig. Ferner ist $\Lambda_{m+1}(\varphi) \leqslant \Lambda_m(\varphi)$ und

$$G(P,\varphi) = \lim_{m \to \infty} \Lambda_m(\varphi),$$

womit IV bewiesen ist. Man konstruiert leicht Beispiele, die zeigen, dass IV nicht zu stimmen braucht, wenn $e(\ )$ und $r(\ )$ nicht äquivalent sind.

3. *Wir geben uns nun einen Punkt O und zwei bezüglich O sternförmige Mengen γ und Γ vor, die den Bedingungen I, II und IV genügen. Wir fragen dann, ob eine zu $e(\ )$ äquivalente Metrik $r(\ )$ existiert, mit den durch γ und Γ festgelegten Indikatrizen.* Es ist natürlich wünschenswert, $r(\ )$ so zu bestimmen, dass es ausserhalb O möglichst glatt ist. Um einen kurzen Ausdruck zu haben, sagen wir, $r(\ )$ sei im Punkte Q normal, wenn in Q die untere Indikatrix mit der oberen zusammenfällt, beschränkt ist und Q im Inneren enthält. Wenn $r(\ )$ in allen Punkten eines Gebietes normal ist, so ist die Indikatrixfunktion $g(P,\varphi)$ wegen (2.6) stetig in P. Wegen der Konvexität ist sie sodann stetig in P und φ zusammen.

Wir wollen zeigen, dass $r(\ )$ so gewählt werden kann, dass die Ebene π oder, was genügt, ein O im Inneren enthaltender Teilbereich von π ein konvexer, metrischer („metrisch" soll heissen: ein den Bedingungen (2.3 a, b, c) genügender) Raum wird, der überall, ausser vielleicht in O, normal ist und in O die vorgegebenen Indikatrizen hat. Die Konstruktion eines solchen $r(\ )$ ist an sich recht mühsam, jedoch ist sie im Wesentlichen in B. F. bereits durchgeführt. Sind nämlich $r = g(O,\varphi) = g(\varphi)$ und $r = G(O,\varphi) = G(\varphi)$ die Gleichungen der Indikatrizen, so ist dort gezeigt, dass nach Einführung einer räumlichen Koordinate z eine konvexe Fläche

$$z = \Phi(r,\varphi)$$

gefunden werden kann, wo für kleine r

$$r^3 \leqslant \Phi(r,\varphi) \leqslant r^{\frac{3}{4}} \tag{3.1}$$

ist, und für die $r=g(\varphi)$ die untere und $r=G(\varphi)$ die obere Krümmungsindikatrix im Punkte O ist. Dies bedeutet, dass für jedes feste φ

$$\varliminf_{r\to 0} \frac{r}{\sqrt{2\Phi(r,\varphi)}} = g(\varphi), \qquad \varlimsup_{r\to 0} \frac{r}{\sqrt{2\Phi(r,\varphi)}} = G(\varphi) \tag{3.2}$$

ist. Man kann nun auf Grund hiervon wie folgt eine für unser Problem geeignete Metrik konstruieren: Die Kurve $\Phi(r,\varphi)=k>0$ ist wegen (3.1) konvex und beschränkt und enthält O im Inneren. Ist $r=\Psi_k(\varphi)$ ihre Projektion auf die Ebene $z=0$, so ordnen wir einen Punkt P dieser Projektion mit den kartesischen Koordinaten a,b diejenige Kurve k_P als Indikatrix zu, die aus $r=\Psi_k(\varphi)$ durch die Ähnlichkeitstransformation

$$x' = \frac{x}{\sqrt{2k}} + a, \qquad y' = \frac{y}{\sqrt{2k}} + b$$

hervorgeht. Der Integrand des zugehörigen Variationsproblems bestimmt sich demnach durch

$$F(P,\varphi) = \frac{\sqrt{2k}}{\Psi_k(\varphi)}, \qquad k=\Phi(P). \tag{3.3}$$

Ausserhalb O ist $F(P,\varphi)$ stetig in den beiden Variablen P und φ [7]) zusammen.

Der Funktion $F(P,\varphi)$ ordnen wir nun eine Metrik zu, wie folgt: Wir bezeichnen den Wert des Integrals über $F(P,\varphi)$ längs einer gerichteten, stückweise stetig differenzierbaren Kurve c (wir nennen solche c zulässig) mit $I(c)$, insbesondere bezeichnen wir den Wert dieses Integrals längs der Strecke mit dem Aufangspunkt A und dem Endpunkt B mit $I(A,B)$. Wegen (3.1) ist $I(O,P)$ endlich für jedes P. Daher ist $I(A,B)$ stets endlich. Um $r(A,B)$ zu definieren, verbinden wir A mit B durch zulässige Kurven mit A als Anfangspunkt und setzen

$$r(A,B) = \text{untere Grenze}_{c}\ I(c). \tag{3.4}$$

---

[7]) Wie $F(O,\varphi)$ definiert wird, ist gleichgültig. Bequem für manche Zwecke ist es $F(P,\varphi) = \frac{1}{g(\varphi)}$ zu setzen.

Fundam. Math., Warszawa 32 (1939), 265–287

Dann ist $r(A,B)$ stets endlich und genügt den Axiomen (2.3a, b, c), ferner ist klar, dass $r(\ )$ topologisch äquivalent zu $e(\ )$ ist.

Es ist nun leicht zu sehen, dass $r(A,B)$ in jedem Punkte $P_0 \neq O$ normal ist und k_{P_0} zur Indikatrix hat. Es sei nämlich $\{A_\nu, B_\nu\} \subset W(P_0, \varphi)$. Dann folgt zunächst aus dem ersten Mittelwersatz der Integralrechnung, dass

$$\frac{I(A_\nu, B_\nu)}{e(A_\nu, B_\nu)} \to F(P_0, \varphi) \tag{3.5}$$

also wegen $r(A_\nu, B_\nu) \leqslant I(A_\nu, B_\nu)$

$$\overline{\lim} \frac{r(A_\nu, B_\nu)}{e(A_\nu, B_\nu)} \leqslant F(P_0, \varphi). \tag{3.6}$$

Ferner sei die gerichtete, von A_ν nach B_ν führende, zulässige Kurve c_ν so gewählt, dass

$$\left| \frac{r(A_\nu, B_\nu)}{e(A_\nu, B_\nu)} - \frac{I(c_\nu)}{e(A_\nu, B_\nu)} \right| \leqslant \frac{1}{\nu} \tag{3.7}$$

wird. Da $F(P,\varphi)$ in P und φ stetig ist, gilt

$$F(P,\varphi) = F(P_0,\varphi) + \vartheta(P,\varphi) \cdot \varepsilon(e(P,P_0)) \tag{3.8}$$

mit $|\vartheta(P,\varphi)| < 1$ und $\varepsilon(e(P,P_0)) \to 0$ für $P_0 \to P$. Ist $e(c_\nu)$ die euklidische Länge von c_ν, so muss

$$\frac{e(c_\nu)}{e(A_\nu, B_\nu)} < R < \infty \tag{3.9}$$

sein, da sonst offenbar für eine passende Teilfolge $\{\nu_n\}$ aus $\{\nu\}$

$$\frac{I(c_{\nu_n})}{I(A_{\nu_n}, B_{\nu_n})} \to \infty$$

und damit

$$\frac{r(A_{\nu_n}, B_{\nu_n})}{I(A_{\nu_n}, B_{\nu_n})} \to \infty$$

gälte. Aus (3.789) und $c_\nu \to P_0$ folgt

$$\frac{r(A_\nu, B_\nu)}{e(A_\nu, B_\nu)} - \frac{\int_{c_\nu} F(P_0, \varphi)\, ds}{e(A_\nu, B_\nu)} \to 0. \tag{3.10}$$

Für den Integrand $\bar{F}(P,\varphi)=F(P_0,\varphi)$ sind die Geraden Minimanten ($\bar{F}(P,\varphi)$ definiert eine Minkowskische Metrik). Deshalb ist

(3.11) $$\int_{C_\nu} F(P_0,\varphi)\,ds \geqslant \int_{\overline{A_\nu B_\nu}} F(P_0,\varphi)\,ds = e(A_\nu,B_\nu)\,F(P_0,\varphi).$$

Aus (3.10) und (3.11) folgt

(3.12) $$\underline{\lim}\,\frac{r(A_\nu,B_\nu)}{e(A_\nu,B_\nu)}=F(P_0,\varphi).$$

Die Formeln (3.6) und (3.7) zeigen, dass $r(\ )$ in P_0 die Indikatrix k_{P_0} hat.

Nun sei $\{A_\nu,B_\nu\}\subset W(O,\varphi)$. Aus (3.2) folgt

(3.13) $$\underline{\lim}\,\frac{e(A_\nu,B_\nu)}{r(A_\nu,B_\nu)}\geqslant \underline{\lim}\,\frac{e(A_\nu,B_\nu)}{I(A_\nu,B_\nu)}\geqslant g(\varphi).$$

Es seien dann gemäss (3.2) die Folgen $r_\nu\downarrow 0$ und $r'_\nu\downarrow 0$ so gewählt, dass

(3.14) $$\frac{r_\nu}{\sqrt{2\Phi(r_\nu,\varphi)}}\to g(\varphi),\qquad \frac{r'_\nu}{\sqrt{2\Phi(r'_\nu,\varphi)}}\to G(\varphi).$$

Da $r(\ )$ in $P_\nu=(r_\nu,\varphi)$ die Indikatrix k_{P_ν} hat, kann $h_\nu>O$ so klein bestimmt werden, dass für $Q_\nu=(r_\nu+h_\nu,\varphi)$ die Beziehung

(3.15) $$\left|\frac{e(P_\nu,Q_\nu)}{r(P_\nu,Q_\nu)}-\frac{r_\nu}{\sqrt{2\Phi(r_\nu,\varphi)}}\right|<\frac{1}{\nu}$$

gilt. Ebenso kann $h'_\nu\downarrow 0$ so bestimmt werden, dass für $P'_\nu=(r'_\nu,\varphi)$ und $(Q'_\nu=(r'_\nu+h'_\nu,\varphi)$ die Beziehung

(3.16) $$\left|\frac{e(P'_\nu,Q'_\nu)}{r(P'_\nu,Q'_\nu)}-\frac{r_\nu}{\sqrt{2\Phi(r'_\nu,\varphi)}}\right|<\frac{1}{\nu}$$

gilt. Aus (3.13), (3.14) und (3.15) folgt, dass $r=g(\varphi)$ die untere Indikatrix von $r(\ )$ in O ist. Aus (3.14) und (3.16) folgt für die obere Indikatrix $r=G^*(\varphi)$ von $r(\ )$ in O, dass

(3.17) $$G^*(\varphi)\geqslant G(\varphi)$$

ist. Der vollständige Beweis von $G^*(\varphi)=G(\varphi)$ erfordert ein näheres Eingehen auf die Struktur der Fläche $z=\Phi(r,\varphi)$. Um diese hier nicht noch einmal auseinandersetzen zu müssen, werde ich mich mit Andeutungen des Beweises begnügen. Man betrachte zunächst

Fundam. Math., Warszawa 32 (1939), 265–287

den Fall $G(\varphi)=g(\varphi)$. Wenn für alle φ die Ungleichung $0<g(\varphi)<\infty$ gilt, so ist die Minkowskigeometrie mit $r=g(\varphi)$ als Einheitskreis um O eine wohlbekannte Lösung. Übrigens liefert auch die in B. F. (p. 94) für diesen Fall angegebene Fläche $z=r^2/2g^2(\varphi)$ vermöge (3.3) gerade diese Lösung. Wenn nun $g(\varphi)$ nicht obiger Ungleichung genügt, so zerlege man den Bereich $0\leqslant\varphi<2\pi$ in die drei Teile: 1) $g(\varphi)=\infty$, 2) $0<g(\varphi)<\infty$, 3) $g(\varphi)=0$. Aus (3.17) folgt dass $(G^*(\varphi)=\infty$ für $g(\varphi)=\infty$. Die Konstruktion von Φ zeigt, dass es im Falle, 2) stets ein $\varrho(\varphi)>0$ gibt derart, dass

$$\Phi(r,\varphi)=r^2/2g^2(\varphi) \qquad \text{für } 0<r<\varrho(\varphi)$$

und hieraus schliesst man leicht, dass $G(\varphi)=g(\varphi)$ im Falle 2). Schliesslich ist der Teil 3) der Fläche aus Torsenstücken und dem Stück einer Rotationsfläche zusammengesetzt, deren Form nach B. F. explizite angegeben kann, und daraus kann dann $G(\varphi)=g(\varphi)$ auch in diesem Falle geschlossen werden.

Wenn $G(\varphi)$ nicht mit $g(\varphi)$ identisch ist, so ist die Konstruktion derart, dass die Fläche $\Phi(r,\varphi)$ aus der zu $r=g(\varphi)$ als Krümmungsindikatrix konstruierten hervorgeht, indem letztere in Zonen um O herum abgeändert wird und diese Zonen beliebig selten gewählt werden können. Dann zeigt der schon erledigte Teil, dass die (3.6) zugrunde liegende Wahl sehr kleiner Strecken $\overline{P'_\nu Q'_\nu}$, die gerade in diese Zonen fallen, in gewissem Sinne die günstigste ist.

Damit ist dann alles gezeigt, ausser dass der Raum mit der Metrik $r(\ )$ konvex ist. Dies folgt nun aus Mengers Untersuchungen [8]). Es wird dort ein Integralbegriff benutzt, der für zulässige Kurven mit dem üblichen (Riemannschen oder Lebesgueschen) Kurvenintegral zusammenfällt. Ferner wird gezeigt, dass zwei Punkte A,B unter den hier gültigen Voraussetzungen stets durch eine rektifizierbare gerichtete Kurve c verbunden werden können, für die

$$\int_c F(P,\varphi)\,ds=r(A,B)$$

wird, und für keine solche Kurve dieses Integral einen kleineren Wert bekommt. Daher gilt für jeden Punkt C auf c

$$r(A,C)+r(C,B)=r(A,B)$$

dies enthält aber die Konvexität der Metrik.

[8]) l. c. [2]), insbesondere § 56, Erstes Existenztheorem für Minimanten.

Bemerkungen. Da die Fläche $z=\Phi(r,\varphi)$ im allgemeinen nicht im scharfen Sinne konvex ist, werden es die ebenen Schnitte $z=$konst. auch nicht immer sein, d. h. unser Integrand $F(P,\varphi)$ wird ausserhalb O nicht überall regulär, sondern nur quasiregulär sein. Die Regularität kann jedoch durch eine triviale Änderung von Φ erreicht werden, indem man $\Phi(r,\varphi)+r^4$ anstatt $\Phi(r,\varphi)$ betrachtet. Die Indikatrizen in O ändern sich dabei nicht. Ferner ist leicht zu sehen, dass $\Phi(r,\varphi)$ ausserhalb O bei gegebenem d mindestens d-mal stetig differenzierbar gewählt werden kann, ohne dass die Indikatrizen in O sich ändern. *Man bekommt so schliesslich einen ausserhalb O regulären, beliebig glatten Integranden, und jeder Punkt kann mit O durch eine rektifizierbare, ausserhalb O beliebig glatte Extremale verbunden werden.*

4. Wir kommen nun zu den *Differenzierbarkeitseigenschaften der Extremalen.* In einer Umgebung von O (inklusive O) sei zu jedem Punkt P eine beschränkte konvexe Kurve k_P definiert, die den Punkt im Inneren enthält und sich stetig mit P ändert. Wie schon in der vorangehenden Nummer gezeigt wurde, gehört zu dieser Kurvenfamilie ein Integrand $F(P,\varphi)$ und eine Metrik $r(\ )$, die im Punkte P die Kurve k_P zur Indikatrix haben. Nach dem unter [8]) zitierten Satz von Menger können je zwei hinreichend nahe bei O gelegene Punkte P,Q durch eine Minimante c verbunden werden derart, dass $\int_c F(P,\varphi)\,ds=r(P,Q)$ wird und für keine von P nach Q führende Kurve dieses Integral einen kleineren Wert bekommt. Wir nehmen an, dass c durch O geht ($P=Q$ und $Q=O$ zugelassen), und wollen c in O auf Differenzierbarkeit hin untersuchen. Wenn man über *jede* Extremale etwas aussagen will, muss vorausgesetzt werden, dass k im scharfen Sinne konvex ist. Sonst gilt nämlich folgender, wie wir sehen werden, wesentlicher Hilfsatz nicht.

Hilfssatz. *Für Folgen*

$$A_\nu\to O,\quad B_\nu\to O,\quad C_\nu\to O\qquad (A_\nu\neq B_\nu\neq C_\nu\neq A_\nu)$$

gilt

$$R(A_\nu,B_\nu,C_\nu)=\frac{r(A_\nu,B_\nu)+r(B_\nu,C_\nu)}{r(A_\nu,C_\nu)}\to 1 \tag{4.1}$$

dann und nur dann, wenn

$$E(A_\nu,B_\nu,C_\nu)=\frac{e(A_\nu,B_\nu)+e(B_\nu,C_\nu)}{e(A_\nu,C_\nu)}\to 1 \tag{4.2}$$

gilt.

Setzt man:

$$\alpha_\nu = \sphericalangle C_\nu A_\nu B_\nu, \qquad \beta_\nu = \sphericalangle A_\nu B_\nu C_\nu, \qquad \gamma_\nu = \sphericalangle B_\nu C_\nu A_\nu,$$

so zeigt eine leichte Rechnung, dass (4.2) dann und nur dann gilt, wenn (mindestens) eins der beiden Paare von Gleichungen gilt:

$$\text{(4.3)} \qquad \begin{aligned} &\alpha_\nu \to 0 \quad \text{und} \quad \frac{\alpha_\nu}{\beta_\nu} \to 0, \\ &\gamma_\nu \to 0 \quad \text{und} \quad \frac{\gamma_\nu}{\beta_\nu} \to 0. \end{aligned}$$

Es genügt den Hilfsatz für solche Folgen $\{A_\nu\}, \{B_\nu\}, \{C_\nu\}$ zu beweisen, für die die Folgen $\left\{\frac{\sin\alpha_\nu}{\sin\gamma_\nu}\right\}$, $\{\overrightarrow{A_\nu B_\nu}\}$, $\{\overrightarrow{B_\nu C_\nu}\}$, $\{\overrightarrow{A_\nu C_\nu}\}$ konvergieren und zwar gegen η (∞ zugelassen) bezw. gegen Strahlen mit den Richtungen $\varphi_1, \varphi_2, \varphi_3$. Es existieren dann auch

$$\alpha = \lim \alpha_\nu, \qquad \beta = \lim \beta_\nu, \qquad \gamma = \lim \gamma_\nu.$$

Stellt dann $r = g(\varphi)$ die Kurve k_0 dar, so folgt aus den Überlegungen der vorigen Nummer, falls wieder

$$\Psi(x,y) = \frac{e(x,y)}{r(x,y)}$$

gesetzt wird, dass

$$\text{(4.4)} \qquad \Psi(A_\nu, B_\nu) \to g(\varphi_1), \qquad \Psi(B_\nu, C_\nu) \to g(\varphi_2), \qquad \Psi(A_\nu, C_\nu) \to g(\varphi_3)$$

gilt. Durch eine triviale Umformung bekommt man

$$\text{(4.6)} \qquad \begin{aligned} &\frac{R(A_\nu, B_\nu, C_\nu)}{E(A_\nu, B_\nu, C_\nu)} \cdot \frac{1}{\Psi(A_\nu, C_\nu)} = \\ &= \left[\Psi(A_\nu, B_\nu)\left(1 + \frac{e(B_\nu, C_\nu)}{e(A_\nu, B_\nu)}\right)\right]^{-1} + \left[\Psi(B_\nu, C_\nu)\left(1 + \frac{e(A_\nu, B_\nu)}{e(B_\nu, C_\nu)}\right)\right]^{-1}. \end{aligned}$$

Nun gelte zunächst $E(A_\nu, B_\nu, C_\nu) \to 1$ und es sei etwa

$$\eta = \lim \frac{e(B, C)}{e(A, B)} = \frac{\sin\alpha_\nu}{\sin\gamma_\nu} \leqslant 1.$$

Aus (4.3) folgt $\alpha_\nu \to 0$, also $\varphi_2 = \varphi_3$. Ist nun $\eta = 0$, so liefert (4.6) sofort die Behauptung (4.1). Ist $\eta > 0$, so muss auch $\gamma_\nu \to 0$, daher $\varphi_1 = \varphi_2 = \varphi_3$ gelten, woraus unter Benutzung von (4.6) wieder (4.1) folgt. In diesem Teil des Hilfsatzes wurde übrigens die starke Konvexität noch nicht benutzt.

Es gelte jetzt $R(A_\nu, B_\nu, C_\nu) \to 1$. Dann ist $E(A_\nu, B_\nu, C_\nu)$ wegen

$$E(A_\nu, B_\nu, C_\nu) = \frac{1}{\Psi(A_\nu, C_\nu)} \frac{r(A_\nu, B_\nu)\,\Psi(A_\nu, B_\nu) + r(B_\nu, C_\nu)\,\Psi(B_\nu, C_\nu)}{r(A_\nu, C_\nu)}$$

beschränkt. Wir wollen zuerst zeigen, dass nicht $\alpha > 0$ *und* $\gamma > 0$ sein kann. Wenn dies gälte, so wäre auch $\beta > 0$, weil sonst

$$E(A_\nu, B_\nu, C_\nu) = \frac{\sin\alpha_\nu + \sin\gamma_\nu}{\sin\beta_\nu} \tag{4.7}$$

unbeschränkt wäre. Für $\alpha > 0$, $\beta > 0$, $\gamma > 0$ würde aber aus (4.6 71) folgen, dass

$$\frac{\sin(\alpha+\gamma)}{g(\varphi_3)} = \frac{\sin\gamma}{g(\varphi_1)} + \frac{\sin\alpha}{g(\varphi_2)}$$

ist, was nach (2.2) der Tatsache widerspricht, dass k_0 im *scharfen* Sinne konvex ist.

Demnach muss $\alpha = 0$ oder $\gamma = 0$ sein. Sei etwa $\gamma = 0$. Wenn dann

$$\overline{\lim}\, \frac{\gamma_\nu}{\beta_\nu} > 0$$

wäre und wir annehmen, dass die Folge

$$\frac{e(A_\nu, B_\nu)}{e(A_\nu, C_\nu)} = \frac{\sin\gamma_\nu}{\sin\beta_\nu}$$

bereits gegen einen Grenzwert ζ (>0) strebt, so gälte $\alpha_\nu \to \pi$ daher nach (4.3)

$$E(B_\nu, A_\nu, C_\nu) \to 1$$

und damit

$$\frac{e(B_\nu, C_\nu)}{e(A_\nu, C_\nu)} \to 1 + \zeta.$$

Dann gälte aber wegen $\gamma = 0$

$$\lim \Psi(A_\nu, C_\nu) = \lim \Psi(B_\nu, C_\nu) = g(\varphi_3)$$

und daher

$$\underline{\lim}\, R(A_\nu, B_\nu, C_\nu) = \underline{\lim}\, \frac{\dfrac{r(A_\nu, B_\nu)}{e(A_\nu, C_\nu)} + \Psi^{-1}(B_\nu, C_\nu)\, \dfrac{e(B_\nu, C_\nu)}{e(A_\nu, C_\nu)}}{\Psi^{-1}(A_\nu, C_\nu)} \geqslant 1 + \zeta.$$

Damit ist der Hilfssatz bewiesen.

Bevor wir weitergehen, wollen wir uns an einem Beispiel davon überzeugen, dass die scharfe Konvexität von k zu seiner Gültigkeit nötig ist. Dazu sei k_0 die Kurve

$$y = \begin{cases} \pm 1 & \text{für } |x| \leqslant 2, \\ \pm \sqrt{1-(x+2)^2} & \text{für } -3 \leqslant x < -2, \\ \pm \sqrt{1-(x-2)^2} & \text{für } 2 < x \leqslant 3 \end{cases}$$

und k_P gehe aus k_0 durch die Translation hervor, die O in P überführt. Wir bekommen so eine symmetrische Minkowskische Metrik $m(X,Y)$, in der die Geraden Minimanten sind. Ausser den Geraden gibt es aber auch andre Kurven, die Minimanten sind, z. B. ist die Summe der beiden Strecken die für $a>0$ den Punkt (a,a) mit $(0,0)$ und $(0,0)$ mit $(a,-a)$ verbinden eine Extremale, da

$$m\big((a,a),0\big)+m\big(0,(a,-a)\big)=m\big((a,a),(a,-a)\big)$$

ist. Setzt man für a die Werte $1/n$ ein und setzt:

$$A_n=(1/n,1/n), \qquad B_n=(0,0), \qquad C_n=(1/n,-1/n),$$

so ist demnach:

$$\frac{m(A_n,B_n)+m(B_n,C_n)}{m(A_n,C_n)}=1, \quad \text{jedoch} \quad \frac{e(A_n,B_n)+e(B_n,C_n)}{e(A_n,C_n)}=\sqrt{2}.$$

Man kann auch Extremalen mit O als Anfangspunkt konstruieren, die den Hilfsatz widerlegen, und zwar so: Wir nehmen die Strahlen w_1: $\varphi=\pi/8$ und w_2: $\varphi=5\pi/8$. Ausgehend von einem Punkt P_0 auf w_1 verfolgen wir den Strahl parallel zu $\varphi=7\pi/4$ bis zum Schnittpunkt P_1 mit w_2 gehen von dort parallel zu $\varphi=5\pi/4$ bis zum Schnittpunkt P_2 mit w_1 von da parallel zu $\varphi=7\pi/4$ bis zum Schnittpunkt P_3 mit w_2 u. s. w. Dann ist der unendliche Streckenzug $\sum_{i=0}^{\infty}\overline{P_iP_{i+1}}+O$ eine Minimante von P_0 nach O (und daher wegen der Symmetrie von $m(\ )$ auch eine von O nach P), da für alle i

$$m(P_{i-1},P_i)+m(P_i,P_{i+1})=m(P_{i-1},P_{i+1})$$

ist. Demnach gilt:

$$\frac{m(P_{2i-1},P_{2i})+m(P_{2i},P_{2i+1})}{m(P_{2i-1},P_{2i+1})}=1, \quad \text{jedoch} \quad \lim E(P_{2i-1},P_{2i},P_{2i+1})>1.$$

Wir kehren zum allgemeinen Fall zurück, und wenden den Hilfsatz wie folgt an: Unter denselben Voraussetzungen wie oben sei c eine von P nach Q führende Minimante, die O enthält ($O=P$ oder $O=Q$ zugelassen). Es sei dann (A_ν, B_ν, C_ν) eine Folge von Punktetripeln auf c, und zwar treffe man von P kommend erst A_ν dann B_ν, dann C_ν, sodass

$$R(A_\nu, B_\nu, C_\nu)=\frac{r(A_\nu, B_\nu)+r(B_\nu, C_\nu)}{r(A_\nu, C_\nu)}=1$$

ist. Gilt nun $A_\nu \to 0$, $B_\nu \to 0$, $C_\nu \to 0$, so folgt aus dem Hilfsatz, dass

$$E(A_\nu, B_\nu, C_\nu)\to 1 \tag{4.8}$$

strebt. In wie weit dies eine eine Differenzierbarkeitsaussage ist, folgt aus (4.3). (4.8) gilt sicher für jede Kurve c, die in O stetig differenzierbar ist; denn dann gilt $\beta_\nu = \sphericalangle A_\nu B_\nu C_\nu \to \pi$, sodass sogar beide Paare von Gleichungen in (4.3) erfüllt sind. Andrerseits gilt (4.8) nicht für jede in O differenzierbare Kurve, wie man an der Kurve

$$y=x^2\cdot \sin e^{1/x} \qquad (y=0 \quad \text{für } x=0)$$

nachrechnen kann. Ferner ist leicht zu sehen, dass eine Kurve c, die O im Inneren enthält und dort eine rechte Halbtangente besitzt, wenn sie die Eigenschaft (4.8) hat, eine Tangente überhaupt besitzt. Es fragt sich daher, ob (4.8) *stets die Differenzierbarkeit von c in O nach sich zieht.* Wir wollen an einem Beispiel zeigen, dass das *nicht der Fall* ist.

5. Ein solches Beispiel liefert die Kurve bestehend aus dem Punkt O und den Kurven

$$\begin{aligned}\varphi&=\sin\log|\log r|,\\ \varphi&=\sin\log|\log r|+\pi.\end{aligned} \tag{5.1}$$

Wir setzen

$$f(r)=\sin\log|\log r|$$

und stellen vor dem Beweis von (4.8) einige Eigenschaften der Kurve zusammen. Wenn $r_\nu^1\to 0$ und $r_\nu^2\to 0$ und $1/a<r_\nu^1/r_\nu^2<a$, $1<a<\infty$, dann gilt

$$|f(r_\nu^1)-f(r_\nu^2)|=\left|\sin\log|\log r_\nu^1|-\sin\log|\log r_\nu^2|\right|\leqslant 2\cdot\left|\sin\frac{\log\left|\dfrac{\log r_\nu^1}{\log r_\nu^2}\right|}{2}\right|$$

und da $\left|\dfrac{\log r_\nu^1}{\log r_\nu^2}\right|\to 1$, so hat man also

$$f(r_\nu^1)-f(r_\nu^2)\to 0 \quad \textit{für} \quad \frac{1}{a}<\frac{r_\nu^1}{r_\nu^2}<a \quad 1<a<\infty \quad \textit{und} \quad r_\nu^1\downarrow 0,\ r_\nu^2\downarrow 0. \tag{5.2}$$

Es ist

$$(5.3)\qquad f'(r)=\frac{\cos\log|\log r|}{r\log r},\qquad 0<r<1.$$

Bezeichnet k den Ast von (5.1) bestehend aus O und $\varphi=f(r)$, ist ferner s die von O aus auf k gemessene Bogenlänge von k, $P(s)$ der zum Wert s gehörige Punkt von k, so gilt

$$ds=\sqrt{1+r^2f'^2(r)}\,dr$$

und daher wegen (5.3) für $s_\nu^1\to 0$, $s_\nu^2\to 0$ $(s_\nu^1\neq s_\nu^2)$

$$(5.4)\qquad \frac{|s_\nu^1-s_\nu^2|}{e\left(P(s_\nu^1),\,P(s_\nu^2)\right)}\to 1.$$

Ist $\mu(s)$ der mit passendem Vorzeichen gerechnete Winkel zwischen dem Strahl $\overrightarrow{OP(s)}$ und der Tangente an k in $P(s)$, so gilt wegen (5.3)

$$(5.5)\qquad \operatorname{tg}\mu(r)=r\frac{d\varphi}{dr}\to 0\qquad \text{für}\quad r\to 0.$$

Aus (5) und (5.5) sieht man, dass für $0<r_1<r_3<1$ eine Abschätzung

$$(5.6)\qquad \int_{r_1}^{r_3}\left|\frac{d\mu(r)}{dr}\right|dr\leqslant\int_{r_1}^{r_3}\left|\frac{d\operatorname{tg}\mu(r)}{dr}\right|dr\leqslant\text{konst.}\frac{|\log r_3-\log r_1|}{|\log r_3|}$$

gilt.

Wir wollen nun (4.8) für die Kurve (5.1) beweisen. Wir orientieren die Kurve so, dass O auf den Punkt $(1,f(1)+\pi)$ folgt. Es seien dann $A_\nu^i=(r_\nu^i,f(r_\nu^i))$, $i=i,2,3$, drei gegen O konvergierende Punktfolgen auf (5.1) und zwar folge A_ν^2 auf A_ν^1 und A_ν^3 auf A_ν^2. Zum Beweis von

$$(5.7)\qquad E(A_\nu^1,A_\nu^2,A_\nu^3)\to 1$$

dürfen wir annehmen, dass die Grenzwerte

$$e_{ij}=\lim_{\nu\to\infty}\frac{r_\nu^i}{r_\nu^j},\qquad i,j=1,2,3,$$

alle existieren (∞ zugelassen) und dass jede Folge $\{A_\nu^i\}$ entweder ganz vor O liegt oder ganz dem abgeschlossenen Bogen $\langle O,(1,f(1))\rangle$

von (5.1) angehört. Aus Symmetriegründen genügt es ferner, die folgenden beiden Fälle zu unterscheiden: I. $\{A_\nu^1\}$ liegt vor O, und $\{A_\nu^2\}$ und $\{A_\nu^3\}$ nicht. II. Weder $\{A_\nu^1\}$ noch $\{A_\nu^2\}$ noch $\{A_\nu^3\}$ liegt vor O.

Im Falle I ist $r_\nu^2 < r_\nu^3$, daher $e_{21} \leqslant e_{31}, e_{23} \leqslant 1$. Es ist

$$E(A_\nu^1, A_\nu^2, A_\nu^3) =$$

$$= \frac{\left[(r_\nu^1)^2 + (r_\nu^2)^2 + 2r_\nu^1 r_\nu^2 \cos\left(f(r_\nu^1) - f(r_\nu^2)\right)\right]^{\frac{1}{2}} + \left[(r_\nu^2)^2 + (r_\nu^3)^2 - 2r_\nu^1 r_\nu^2 \cos\left(f(r_\nu^2) - f(r_\nu^3)\right)\right]^{\frac{1}{2}}}{\left[(r_\nu^1)^2 + (r_\nu^3)^2 + 2r_\nu^1 r_\nu^3 \cos\left(f(r_\nu^1) - f(r_\nu^2)\right)\right]^{\frac{1}{2}}}.$$

Ist nun $e_{31} = 0$, so ist auch $e_{21} = 0$ und (5.7) folgt sofort, indem Zähler und Nenner durch r_ν^1 dividiert werden. Daher werde $0 \leqslant e_{13} < \infty$ angenommen. Dann bleiben die Fälle 1) $e_{23} = 0$ und $e_{13} = 0$; 2) $e_{23} = 0$ und $0 < e_{13} < \infty$; 3) $0 < e_{23} \leqslant 1$ und $e_{13} = 0$; 4) $0 < e_{23} \leqslant 1$, $0 < e_{13} < \infty$. Man kürze nun in allen 4 Fällen durch r_ν^3. Wenn $0 < e_{ij} < \infty$, so folgt aus (5.2), dass

$$\cos\left(f(r_\nu^i) - f(r_\nu^j)\right) \to 1$$

geht, sodass in allen Fällen

$$\lim E(A_\nu^1, A_\nu^2, A_\nu^3) = \frac{(e_{13} + e_{23}) + (1 - e_{23})}{e_{13} + 1} = 1$$

wird.

Im Falle II ist $r_\nu^1 < r_\nu^2 < r_\nu^3$, daher $0 \leqslant e_{13} \leqslant e_{23} \leqslant 1$ und

$$E(A_\nu^1, A_\nu^2, A_\nu^3) =$$

$$= \frac{\left[(r_\nu^1)^2 + (r_\nu^2)^2 - 2r_\nu^1 r_\nu^2 \cos\left(f(r_\nu^1) - f(r_\nu^2)\right)\right]^{\frac{1}{2}} + \left[(r_\nu^2) + (r_\nu^3) - 2r_\nu^2 r_\nu^3 \cos\left(f(r_\nu^2) - f(r_\nu^3)\right)\right]^{\frac{1}{2}}}{\left[(r_\nu^1)^2 + (r_\nu^3)^2 - 2r_\nu^1 r_\nu^3 \cos\left(f(r_\nu^1) - f(r_\nu^3)\right)\right]^{\frac{1}{2}}}.$$

Man kürze nun durch r_ν^3 und verwende dieselbe Schlussweise wie im Falle I. Man sieht dann, dass für $e_{13} \neq 1$ stets

$$\lim E(A_\nu^1, A_\nu^2, A_\nu^3) = \frac{(e_{23} - e_{13}) + (1 - e_{23})}{1 - e_{13}} = 1$$

wird.

Der ausgelassene Fall $e_{13} = 1$ (es sind dann alle e_{ij} gleich 1) spielt geometrisch wirklich eine Sonderrolle. Wenn die Kurve bestehend aus O und den Ästen $\varphi = h(r)$ und $\varphi = h(r) + \pi$ nämlich in O differenzierbar ist, d. h. wenn $h(r)$ für $r \to 0$ einen Grenzwert zustrebt, so gilt für $r_\nu^i \to 0$ und $r_\nu^j \to 0$ stets

$$\cos\left(f(r_\nu^i) - f(r_\nu^j)\right) \to 0.$$

Unsere vorangehenden Überlegungen gelten, während bereits festgestellt war, dass (5.7) nicht für jede differenzierbare Kurve gilt. Der ausstehende Fall erledigt sich so: Es ist $\sphericalangle A_\nu^1 A_\nu^2 A_\nu^3$ mindestens gleich π minus der Gesamtvariation von $\mu(r)$ von A_ν^1 bis A_ν^3. Da nun für $e_{13}=1$

$$\frac{|\log r_\nu^3 - \log r_\nu^1|}{|\log r_\nu^3|} \to 0$$

strebt, folgt aus (5.6) dass $\sphericalangle A_\nu^1 A_\nu^2 A_\nu^3 \to \pi$ strebt, und daraus unter Benutzung von (4.3) die Behauptung.

6. Es ist nun leicht, mit Hilfe der Kurve k eine Metrik $r(\ )$ zu konstruieren, die überall in der Ebene normal ist, deren Indikatrix k_0 in O im scharfen Sinne konvex ist, und deren Extremalen in O nicht differenzierbar sind. Wir werden nämlich ein $r(\ )$ mit folgenden Eigenschaften angeben:

1) $r(\ )$ *ist euklidisch.*

2) $r(\ )$ *ist ausserhalb O analytisch, d. h. der Integrand des zu $r(\ )$ gehörigen Variationsproblems hängt ausserhalb O analytisch von Punkt und Richtung ab.*

3) *Die Indikatrix von $r(\ )$ in O ist der Einheitskreis mit O als Mittelpunkt.*

4) *Die Extremalen, die hier wegen der Euklidizität durch je zwei Punkte eindeutig bestimmt sind, sind in O nicht differenzierbar.*

Durch jeden von O verschiedenen Punkt $P=(r,\varphi)$ der Ebene geht genau eine Kurve der Schar

$$\varphi = f(r) + \lambda, \qquad 0 \leqslant \lambda < 2\pi.$$

Bezeichnet s die auf dieser Kurve von O aus gemessene Bogenlänge, so bestimmt daher P die Werte s und λ eindeutig und wird umgekehrt eindeutig durch diese Werte bestimmt. Deshalb setzen wir auch $P=(s,\lambda)$. Ist dann $P_1=(s_1,\lambda_1)$, $P_2=(s_2,\lambda_2)$, so setzen wir

$$r(P_1,P_2) = [s_1^2 + s_2^2 - 2 s_1 s_2 \cos(\lambda_1 - \lambda_2)]^{\frac{1}{2}}. \tag{6.1}$$

Die Metrik $r(\ )$ ist dann euklidisch, sie genügt daher gewiss den Axiomen (2.3abc). Da die Transformation, die O sich selbst und dem Punkt (r,φ) den Punkt $(s=r,\ \lambda=\varphi)$ zuordnet, ausserhalb O analytisch ist, so ist der Integrand des zu $r(\ )$ gehörigen Variationsproblems ausserhalb O analytisch in Punkt und Richtung, die Indikatrix existiert und ist daher konvex, sie ist ferner beschränkt und enthält ihren Aufpunkt im Inneren. Es wäre auf Grund der Euklidizität von $r(\ )$ leicht zu sehen, dass die Indikatrix ausserhalb O überall im scharfen Sinne konvex ist.

Wir wollen nun zeigen, dass für $\overset{i}{P}_\nu=(\overset{i}{r}_\nu,\overset{i}{\varphi}_\nu)=(\overset{i}{s}_\nu,\overset{i}{\lambda}_\nu)\to 0$, $i=1,2$, stets

$$(6.2)\qquad \frac{e(\overset{1}{P}_\nu,\overset{2}{P}_\nu)}{r(\overset{1}{P}_\nu,\overset{2}{P}_\nu)}=\Psi(\overset{1}{P}_\nu,\overset{2}{P}_\nu)\to 1$$

gilt. Da beide Metriken gegenüber Drehungen $r'=r,\ \varphi'=\varphi+t$ um O invariant sind, darf angenommen werden, dass $\overset{2}{\lambda}_\nu=0$ für alle ν gilt. Ist dann auch $\overset{1}{\lambda}_\nu=0$, so folgt (6.2) aus (6.1) und (5.4), ebenso, wenn $\overset{1}{s}_\nu=\overset{2}{s}_\nu$ für alle ν gilt. (Für $\overset{1}{s}_\nu=\overset{2}{s}_\nu$ ist $\overset{1}{\lambda}_\nu=\overset{1}{\varphi}_\nu-\overset{2}{\varphi}_\nu$). Im allgemeinen Falle darf angenommen werden, dass $\overset{2}{s}_\nu\geqslant\overset{1}{s}_\nu$, $\overset{1}{\lambda}_\nu\geqslant 0$ ist. Es sei dann P_ν der Punkt $(\overset{1}{s}_\nu,0)=(\overset{1}{r}_\nu,\overset{1}{\varphi}_\nu-\overset{1}{\lambda}_\nu)$. Es gilt dann, wie soeben gezeigt wurde,

$$(6.3)\qquad \Psi(\overset{1}{P}_\nu,P_\nu)\to 1 \qquad \text{und} \qquad \Psi(P_\nu,\overset{2}{P}_\nu)\to 1.$$

Es genügt (6.2) in dem Fall zu beweisen, wo die Grenzwerte

$$(6.4)\qquad \begin{gathered}\lambda=\lim\overset{1}{\lambda}_\nu\\ a=\lim\frac{e(P_\nu,\overset{2}{P}_\nu)}{e(\overset{1}{P}_\nu,P_\nu)}=\lim\frac{r(P_\nu,\overset{2}{P}_\nu)}{e(\overset{1}{P}_\nu,P_\nu)}=\lim\frac{r(P_\nu,\overset{2}{P}_\nu)}{r(\overset{1}{P}_\nu,P_\nu)}\end{gathered}$$

existieren. Da $r(\ )$ euklidisch ist, hat man

$$(6.5)\qquad r(\overset{1}{P}_\nu,\overset{2}{P}_\nu)=[r^2(\overset{1}{P}_\nu,P_\nu)+r^2(P_\nu,\overset{2}{P}_\nu)+2r(\overset{1}{P}_\nu,P_\nu)r(P_\nu,\overset{2}{P}_\nu)\sin\overset{1}{\lambda}_\nu/2]^{\frac{1}{2}}.$$

Ist $a=0$ bezw. $a=\infty$, so gilt

$$\frac{e(\overset{1}{P}_\nu,P_\nu)}{e(\overset{1}{P}_\nu,\overset{2}{P}_\nu)}\to 1 \qquad \text{bezw.} \qquad \frac{e(P_\nu,\overset{2}{P}_\nu)}{e(\overset{1}{P}_\nu,\overset{2}{P}_\nu)}\to 1$$

und (6.2) folgt aus (6.3) und (6.5).

Es sei daher $0<\alpha<\infty$. Bezeichnet ψ_ν den Winkel, den der Strahl $\overrightarrow{OP}$ mit dem Strahl $\overrightarrow{P_\nu P_\nu^2}$ bildet (mit geeignetem Vorzeichen) so ist

$$\sphericalangle P_\nu^2 P_\nu P_\nu^1 = \frac{\pi}{2} + \frac{\lambda_\nu^1}{2} + \psi_\nu$$

daher

$$(6.6)\quad e(P_\nu^1, P_\nu^2) = \left[e^2(P_\nu^1, P_\nu) + e^2(P_\nu, P_\nu^2) + 2e(P_\nu^1, P_\nu)\, e(P_\nu, P_\nu^2) \sin\left(\frac{\lambda^1}{2} + \psi_\nu\right)\right]^{\frac{1}{2}}.$$

Wir wollen zunächst zeigen, dass $\psi_\nu \to 0$. Dazu unterscheiden wir zwei Fälle:

1) $\lambda \neq 0$. Dann ist

$$(6.7)\qquad \lim \frac{e(P_\nu^1, P_\nu^2)}{e(P_\nu, O)} = \sin\frac{\lambda}{2} > 0$$

und daher wegen $0<\alpha<\infty$

$$(6.8)\qquad 0 < \lim \frac{e(P_\nu, P_\nu^2)}{e(P_\nu, O)} < \infty.$$

Da andrerseits O, P_ν, P_ν^2 in dieser Reihenfolge auf k: $\varphi = f(r)$ liegen (es ist $s_\nu^2 > s_\nu^1$), gilt wie in der vorigen Nummer gezeigt wurde,

$$E(O, P_\nu, P_\nu^2) \to 1$$

und daher infolge (6.8)

$$\sphericalangle P_\nu^2 P_\nu O \to \pi,$$

woraus $\psi_\nu \to 0$ folgt.

2) $\lambda = 0$. Hier ist das Argument ganz anders. Es gilt dann

$$\frac{e(P_\nu^1, P_\nu)}{e(P_\nu, O)} \to 0$$

und daher wegen $0<\alpha<\infty$ auch

$$(6.9)\qquad \lim \frac{e(P_\nu^2, O)}{e(P_\nu, O)} = \lim \frac{r(P_\nu^2, O)}{r(P_\nu, O)} = \lim \frac{s_\nu^2 - s_\nu^1}{s_\nu^1} = 0.$$

Auf dem Bogen $\overline{P_\nu P_\nu^2}$ von k gibt es einen Punkt $P(s_\nu^0)$ derart, dass die Tangente an k in $P(s_\nu^0)$ parallel zu der Sehne $\underline{P_\nu P_\nu^2}$ ist. Aus $s_\nu^1 \leqslant s_\nu^0 \leqslant s_\nu^2$ und (6.9) folgt $\frac{s_\nu^0}{s_\nu^1} \to 1$ und daraus nach (5.2), dass

(6.10) $$\varphi\left(r(s_\nu^0)\right)-\varphi(r_\nu^1)\to 0$$

strebt. Wenn $\mu(s)$ wie in (5.5) erklärt ist, so ist

$$\psi_\nu=\mu(s_\nu^0)-\varphi\left(r(s_\nu^0)+\varphi(r_\nu^1)\right),$$

und aus (5.5) und (6.10) folgt $\psi_\nu\to 0$.

Aus $\psi_\nu\to 0$ und (6.3456) folgt nun:

$$\lim\frac{r(P_\nu^1,P_\nu^2)}{e(P_\nu^1,P_\nu)}=\left[1+\alpha^2+2\alpha\sin\frac{\lambda}{2}\right]^{\frac{1}{2}},$$

$$\lim\frac{e(P_\nu^1,P_\nu^2)}{e(P_\nu^1,P_\nu)}=\left[1+\alpha^2+2\alpha\sin\frac{\lambda}{2}\right]^{\frac{1}{2}},$$

womit (6.2) in allen Fällen bewiesen ist.

Fundam. Math., Warszawa 32 (1939), 265–287

NON-EUCLIDEAN GEOMETRY

Herbert Busemann

1. *Introduction.* The root of non-euclidean geometry is the problem whether our ordinary geometry[1] is the only possible one.

For centuries mathematicians were convinced that the answer must be positive and tried to prove this. Since all postulates of Euclid seemed self evident, except for the *parallel axiom*,[2] the efforts were directed towards establishing this axiom as a consequence of the others. *The failure of all these efforts led Gauss, Bolay and Lobachewsky during the first third of the 19th century to the conclusion that the parallel axiom is not a logical necessity but that consistent so-called non-euclidean geometries exist, in which this axiom does not hold.*

Since then mathematical research has enlarged our concepts regarding geometry and space to such an extent that the non-euclidean geometries appear now as very special examples of general geometric structures which are not euclidean (so that the terms non-euclidean and not euclidean are not synonymous). From the modern point of view the euclidean and the non-euclidean geometries resemble each other very closely. They can be derived jointly from a principle which was discovered by the physicist Helmholtz.[3]

In an implicit form the principle appears in Euclid's "proof" of the first congruence theorem: the triangles ABC and $A'B'C'$ are congruent if $AB = A'B'$, $AC = A'C'$ and $\angle BAC = \angle B'A'C'$. Euclid *moves* $\angle B'A'C'$ until it falls on $\angle BAC$. Then B' must fall on B and C' on C, etc. Clearly this argument is not the same type of proof as Euclid's other deductions. In fact the presupposed mobility of figures is equivalent to the congruence theorem and should therefore have been stated as an axiom. Helmholtz emphasized that *the mobility of rigid objects, the fact that we recognize an object independently of its position in space, is one of our basic physical experiences.* He therefore put, and partially solved, the mathematical problem of finding those spaces or geometries in which figures can be moved freely.

The answer to the problem is that the euclidean and the non-euclidean geometries have that property and that no other geometry does. A complete proof of this fundamental fact is beyond the scope of the present article, but one typical argument will indicate the procedure by which the result is obtained.

2. *Spherical geometry.* The reader is familiar with one non-euclidean geometry without realizing it, namely with the geometry on a sphere. Let Σ be (the surface of) a sphere with radius ρ and center O in the ordinary or euclidean space. Distances on Σ will be measured not in the space but along Σ. Airplane travel has made everyone familiar with the fact that shortest connections on Σ are arcs of great circles. (A great circle is a circle in which a plane through O intersects Σ.)

From the point of view of spherical geometry going along a great circle means choosing the most direct or straight route from one point to another. The great circles play therefore on Σ the role of straight lines and we shall frequently call them so.

If A,B are points on Σ and α is the radian measure of $\angle AOB$ then the length of the great circle arc (or the segment) from A to B is $\rho\alpha$, the spherical distance of A and B.

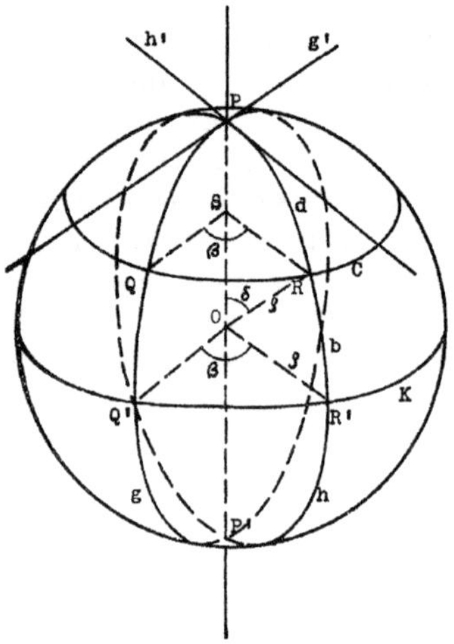

Figure 1

Let the two great circles g and h intersect at P and P'. These points are the intersections with Σ of the line which is common to the planes in which g and h lie. P' and P are diametrically opposite points of Σ or antipodes.

The circle C with radius d and center P on Σ is by definition the locus of those points which have spherical distance d from P. Let $R(Q)$ be one of the two intersections of $h(g)$ with C. If $d = \rho.\delta$ then $\delta = \angle QOP = \angle ROP$. Now C is also the euclidean circle whose radius is the perpendicular distance RS of R from the line PP'. The figure shows that $RS/RO = RS/\rho = \sin\delta$ or $RS = \rho\sin\delta$ so that C has length

$$L = 2\pi\,\rho\,\sin\delta = 2\pi\,\rho\,\sin\frac{d}{\rho} \tag{1}$$

This is the first of a series of non-euclidean formulas to be derived here. It solves the problem: given the radius d of a circle, find its length. The euclidean answer is $2\pi d$, the spherical answer is $2\pi\rho\sin d/\rho$. The constant ρ is typical for the specific spherical geometry under consideration. *It is a length and can be determined by purely geometric means* (for instance, the distance of any two antipodes is $\pi\rho$), *whereas in euclidean geometry no definite length* *can be determined by geometry alone.* We must use physical aids like the standard meter at Paris. The existence of absolute lengths is one of the main differences between the euclidean and the two non-euclidean geometries.

The relation $\sin x/x \to 1$ for $x \to 0$, which is familiar from calculus, and (1) show that

$$2\pi\ \rho \sin \frac{d}{\rho}/2\pi d = \sin \frac{d}{\rho}/\frac{d}{\rho} \to 1 \text{ for } d \to 0.$$

This means that for d which are small compared to ρ the spherical length of a circle is approximately equal to the euclidean length $2\pi d$. We recognize here a special case of a procedure which we use all the time, namely, *the application of euclidean geometry to small domains on the earth.*

Among the circles about P there is a spherical straight line k, the equator for P and P' as poles, which corresponds to the value $\delta = \pi/2$. Viewed from k the previously considered circle C appears as the locus of those points which have the same distance $b = QQ' = RR' = \rho(\pi/2 - \delta)$ from k (see Figure 1). Hence, *in spherical geometry the locus of the points which are equidistant to a straight line is not a straight line.* Also the lengths of the corresponding arcs $Q'R'$ and QR of k and C are not equal. for if β is the radian measure of the angle $<Q'OR'$ and $<QSR$ then $Q'R'$ has length $s = \rho\beta$ and the arc QR of C has length $t = \beta\rho \sin \delta$ so that

$$t = s \sin \delta = s \cos(\pi/2 - \delta) = s \cos b/\rho \tag{2}$$

That t is proportional to s was obvious beforehand, (2) determines the factor of proportionality.

Hitherto only distances or lengths of curves were considered. In euclidean geometry angle and area are two other fundamental concepts. The same goes for spherical geometry. There is no question regarding angles, because locally, or close to a fixed point P, the spherical geometry is almost euclidean, so that angles with vertices at P can be measured in terms of the local euclidean geometry. This amounts to the following: instead of operating on Σ we operate on the tangent plane π to Σ at P. The two great circles g and h have tangents g' and h' at P which lie in π. As measure of the angle between g and h we take the measure of the euclidean angle of g' and h' in π (any unit can be used. Here we always use radian measure). Then the previously used β is also the measure of one of the angles formed by g and h, and we find

$$C = \beta\rho \sin \frac{d}{\rho} \tag{3}$$

as spherical formula for the length of a circular arc with radius d which subtends the spherical angle β at its center P.

Everyone knows the importance of *Pythagoras' Theorem* in euclidean geometry. Unfortunately it is frequently formulated as a relation between areas. The reader, who remembers the applications of the theorem (for instance in trigonometry) will realize that its importance is due to the fact that *it expresses one side of a right triangle in terms of the other two sides.* The corresponding problem is just as important for spherical geometry. To solve it consider on Σ the triangle ABC with a right angle at C. Denote the (spherical) lengths of the sides AB, BC, CA by c, a, b respectively (all sides are assumed to be shortest connections of their endpoints). Let the tangent plane π of Σ at A (see Figure 2) intersect the line OB in B' and OC in C'. The plane is perpendicular to the radius OA and therefore to the plane OAC. The plane OBC is also perpendicular to OAC because the angle at

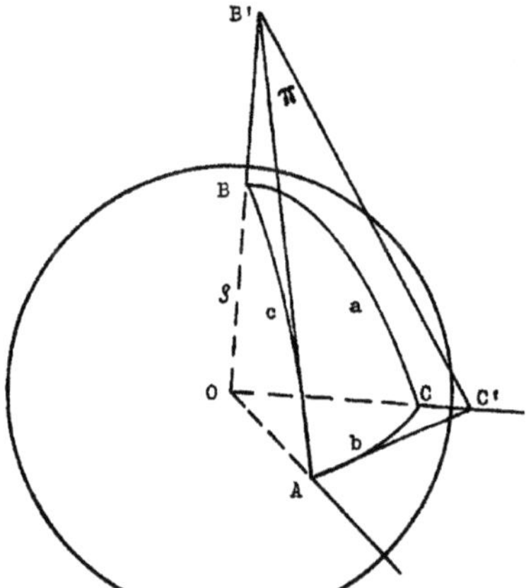

Figure 2

C is a right angle. Therefore the line $B'C'$ is perpendicular to the plane OAC, hence

$$\angle B'C'A' = \angle B'C'O = \angle B'AO = \angle C'AO = \pi/2$$

Since $\angle AOB' = c/\rho$, $\angle B'OC' = a/\rho$, $\angle C'OA = b/\rho$ we find

$$OA/OB' = \cos c/\rho, \quad OC'/OB' = \cos a/\rho, \quad OA/OC' = \cos b/\rho$$

and because of $OA/OB' = (OC'/OB')\cdot(OA/OC')$

(4) $\cos c/\rho = \cos a/\rho \cdot \cos b/\rho$ (*Spherical Theorem of Pythagoras*)

Those who are familiar with the expansion of $\cos x$ see that

$$1-\frac{c^2}{2\rho^2}+\ldots=(1-\frac{a^2}{2\rho^2}+\ldots)(1-\frac{b^2}{2\rho^2}+\ldots)=1-\frac{a^2}{2\rho^2}-\frac{b^2}{2\rho^2}-\frac{a^2b^2}{4\rho^4}+\ldots$$

Math. Mag. 24 (1950), 19–34

$$c^2 = a^2 + b^2 + \dots,$$

where the omitted terms contain at least ρ^2 in the denominator. This confirms again that the *euclidean geometry holds for triangles whose sides are small compared to* ρ.

The angle at A of the spherical triangle ABC equals (by our agreement) the euclidean angle $\angle B'AC'$. Therefore

$$\text{(5)} \qquad \cos A = \frac{AC'}{AB'} = \frac{AC'/OA}{AB'/OA} = \frac{\tan b/\rho}{\tan c/\rho} = \frac{\sin b/\rho}{\sin c/\rho}\cdot\frac{\cos c/\rho}{\cos c/\rho}$$

By (4) the last factor on the right equals $\cos a/\rho$. To evaluate the preceding factor observe that

$$\text{(6)} \qquad \sin A = \frac{B'C'}{AB'} = \frac{B'C'/B'O}{AB'/B'O} = \frac{\sin a/\rho}{\sin c/\rho}$$

Since A and B play the same role in the triangle ABC, the relation

$$\text{(7)} \qquad \sin B = \frac{\sin b/\rho}{\sin c/\rho}$$

must also hold. (5) and (7) yield

$$\text{(8)} \qquad \cos A = \sin B\cdot\cos a/\rho$$

Contrary to all our euclidean experiences (8) reveals that *in spherical geometry the angles of a triangle determine its sides*. A theory of similar figures does therefore not exist. *There is no such concept as shape independent of size*. This is, of course, closely related to the previously discussed existence of an absolute length.

Finally we turn our attention to area. A two-gon on Σ is a figure bounded by two semi great circles h, k with diametrically opposite points P,P' as vertices. Obviously the area A of a two-gon is proportional to the angle α at its vertices:

$$A = \omega\cdot\alpha$$

where ω is the factor of proportionality. Consider a triangle PQR, where Q and R lie on h and k. Let β and γ be the angles at Q and R of the spherical triangle PQR and Q',R' the antipodes to Q and R. Clearly any triangle is congruent to the triangle formed by its antipodes and has therefore the same area so that, for instance, $PQR = P'Q'R'$ and $PQR = P'Q'R$.

Now $PQR + PQR'$ is a two-gon with angle γ, hence

$$\text{(9)} \qquad PQR + PQR' = \omega\gamma$$

Similarly

$$\text{(10)} \qquad PQR + PRQ' = \omega.\beta$$

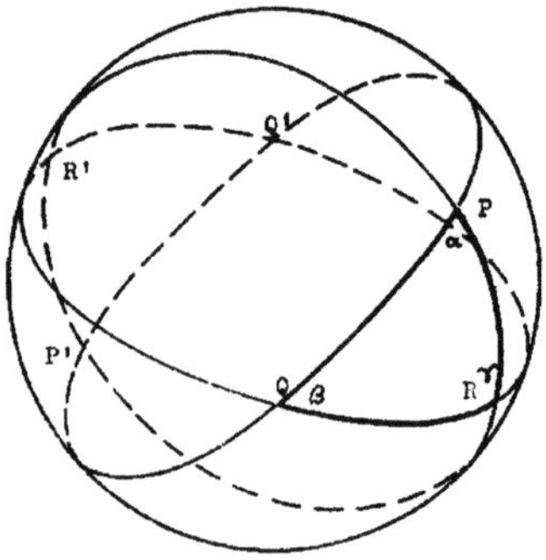

Figure 3

But $PQR' = P'Q'R$ and $PRQ' + P'Q'R$ is a two-gon with angle $\pi - \alpha$ so that

$$PQR' + PRQ' = P'Q'R + PRQ' = \omega(\pi - \alpha) \tag{11}$$

Adding (9) and (10) and subtracting (11) yields

$$2\,PQR = \omega(\alpha + \beta + \gamma - \pi) \tag{12}$$

The area of a triangle is therefore proportional to the excess of the angle sum over π. This confirms that the size of a triangle is determined by its angles. The constant ω depends on the unit of area used. The usual agreement is to make the area of a spherical triangle as object of spherical geometry equal to the area of the same triangle as curved surface in euclidean space. This amounts to putting $\omega = 2\rho^2$.

A noteworthy consequence of (12) is that *the sum of the angles of a spherical triangle is always greater than* π.

Spherical geometry was, of course, well known prior to 1800, in fact, it was known to Ptolemy and the Arabs. Why then, did the discovery of non-euclidean geometry come so late? Here is the reason. Spherical geometry does not satisfy some of the euclidean axioms. Two straight lines (or great circles) intersect twice. There is however a geometry, closely related to the spherical, the so-called elliptic geometry, in which straight lines intersect only once. But in elliptic geometry only figures which are not too large can be moved freely. Moreover, neither in spherical nor in elliptic geometry can we continue walking straight ahead without eventually retracing our steps. Euclid assumes that this does not happen. The pioneers looked only for geometries in which the straight lines have the euclid-

Math. Mag. 24 (1950), 19–34

ean property of infinite extension in both directions. Thus they discovered the other non-euclidean geometry which is called hyperbolic. Since this geometry cannot be interpreted as geometry on a surface in ordinary space it is much more recondite.

After hyperbolic geometry had been fully developed its formulas proved amazingly similar to spherical geometry and unification became natural.

3. *The hyperbolic functions.* The analogy between spherical and hyperbolic geometries becomes apparent when the so-called hyperbolic trigonometric functions are introduced.

The hyperbolic sine and cosine of x are defined by the formulas:

$$\sinh x = \frac{e^x - e^{-x}}{2} \qquad \cosh x = \frac{e^x + e^{-x}}{2} \tag{13}$$

Clearly

$$\frac{d \sinh x}{dx} = \cosh x \text{ and } \frac{d \cosh x}{dx} = \sinh x \tag{14}$$

Since $e^x > 0$ for all x, the function $\cosh x$ is positive, therefore $\sinh x$ increases monotonically over the whole x-range. Because $\sinh 0 = 0$ it follows that

$$\sinh x < 0 \text{ for } x < 0 \text{ and } \sinh x > 0 \text{ for } x > 0$$

Therefore (14) and the elementary theory of maxima and minima yield that $\cosh x$ reaches its minimum 1 for $x = 0$. From $e^{x} \to 0$ for $x \to \infty$ and (13) we see that $\sinh x \to \infty$ for $x \to \infty$.

Most formulas for the ordinary trigonometric functions hold without change for the hyperbolic functions. For instance

$$\begin{aligned} (15)\quad \cosh(x+y) + \cosh(x-y) &= (½)(e^{x+y} + e^{-x-y}) + (½)(e^{x-y} + e^{-x+y}) \\ &= (½)(e^x e^y + e^{-x}e^{-y} + e^x e^{-y} + e^{-x}e^y) = (½)(e^x + e^{-x})(e^y + e^{-y}) \\ &= 2 \cosh x \cosh y. \end{aligned}$$

But the reader is warned that

$$\cosh^2 x - \sinh^2 x = 1.$$

The formulas of hyperbolic geometry can be obtained from those of spherical geometry by a simple formal procedure: The angles are in both cases determined by the local euclidean geometry and are therefore treated alike. This means that *all functions of angles are left as ordinary trigonometric functions. But the ordinary trigonometric functions of the sides are replaced by the corresponding hyperbolic functions.* For instance the hyperbolic Theorem of Pythagoras is

$$\cosh c/\rho = \cosh a/\rho \cosh b/\rho,$$

whereas (8) becomes

$$\cos A = \sin B \cosh a/\rho.$$

Thus the angles of a triangle determine also in hyperbolic geometry its sides. **ρ is again a positive constant typical for the specific hyperbolic geometry under consideration, but it does not have such a simple euclidean interpretation as in spherical geometry.**

That *straight lines in hyperbolic geometry intersect at most once* can be seen as follows: Let the lines g and h form the angle α at P. On g lay off from P the distance x to the point X and let d be the perpendicular distance of X from h. The hyperbolic formula which is analogous to (7) shows that $\sin\alpha = \dfrac{\sinh d/\rho}{\sinh x/\rho}$ or $\sinh d/\rho = \sin\alpha\cdot\sinh x/\rho$. It was proved above that $\sinh x/\rho$ increases monotonically from 0 to ∞ as x goes from 0 to ∞. Therefore $\sinh d/\rho$ tends monotonically to ∞ and so does d because of the monotonicity of $\sinh x$. This implies of course that X will never lie on h when $x > 0$. (In the spherical case X lies on h for $x = \rho\pi$, because $\sin \rho\pi/\rho = 0$.)

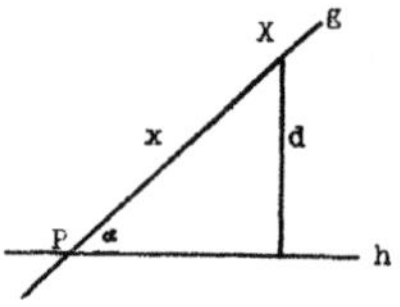

Figure 4

The preceding discussion is nothing but a purely formal procedure for deriving hyperbolic facts from spherical ones. No geometric argument was advanced for justifying this procedure. The next section will indicate how the mobility principle leads to such arguments.

4. *Mobility of figures. A motion of the euclidean plane Π or of the sphere Σ is a mapping f of Π (or Σ) on itself which preserves distances*, which means that the images $f(P)$ and $f(Q)$ of any two points P,Q of Π (or Σ) have the same distance as P and Q:

$$f(P)f(Q) = PQ \tag{16}$$

As example we consider *rotations*. On Π or Σ choose an arbitrary point R and an oriented angle α. A rotation f about R through α is

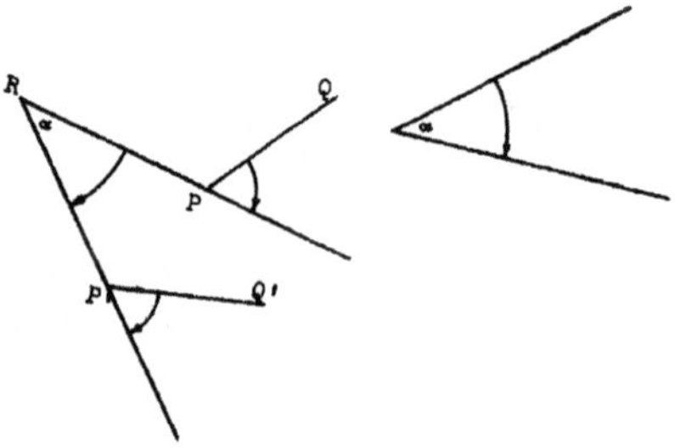

Figure 5

Math. Mag. 24 (1950), 19–34

defined as follows: 1) $f(R) = R$ 2) For any point $P \neq R$ we lay off the angle $\angle PRP' = \alpha$ with RP as initial and RP' as terminal side and choose $P' = f(P)$ such that $RP' = RP$. The reader will easily prove that (16) holds for rotations of Π or Σ.

Reflections in a line give a second example of motions. Let g be a straight line on Π or Σ. For any R on g we put $f(R) = R$. If P is not on g let F be a point of g closest to P (the point F is uniquely determined unless (on Σ) P happens to be a pole to g as equator). On the prolongation of PF beyond F we choose $P' = f(P)$ such that $P'F = PF$. The reader will again easily confirm (16). The present definition of rotations and reflections carry over to hyperbolic geometry.

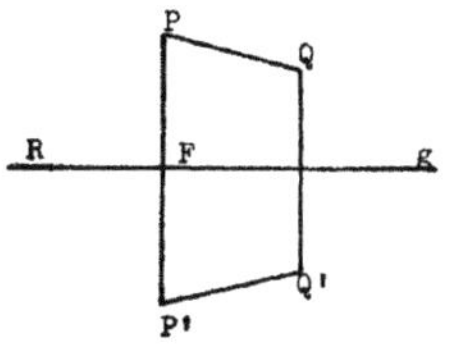

Figure 6

The most general motion can be composed from rotations and reflections. Let ABC and $A'B'C'$ be any two congruent triangles in Π or Σ. If R is any point which has the same distance from A and A', then a rotation f_1 about R through a suitable angle will carry A' into A or $f_1(A') = A$.

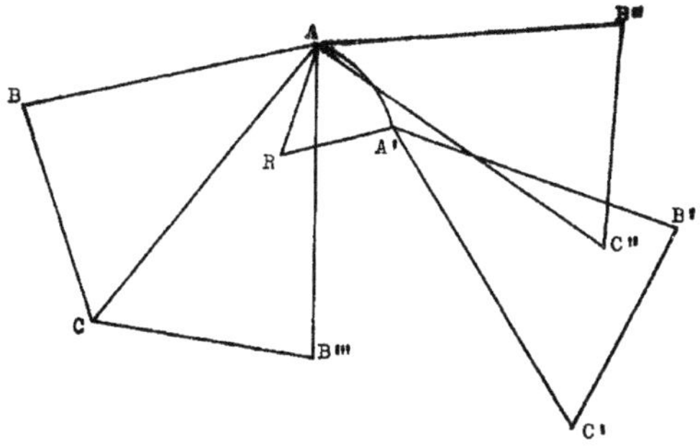

Figure 7

Let $f_1(B') = B''$, $f_1(C') = C''$. Then the triangle $AB''C''$ is congruent to $A'B'C'$ and therefore to ABC. Since $AC = AC''$ a rotation f_2 about A (through $C''AC$) will carry C'' into C or $f_2(C'') = C$. Let $f_2(B'') = B'''$. Since ACB''' is congruent to ACB, the point B''' either coincides with B or (as in the figure) B can be obtained from B''' by a reflection in the line AC. In any case we can move the triangle $A'B'C'$ into the triangle ABC by combining at most two rotations and a reflection.

The spherical case exhibits a phenomenon which is foreign to the other two geometries: a rotation f about R also moves a straight line h, the equator to R as pole, into itself. Or in the standard terminology: f is a *translation* of Σ along h. In the other two geometries translations can be combined from rotations (or reflections) but they are not themselves rotations. They are defined as follows: Let an oriented **straight** line h and a positive number x be given.

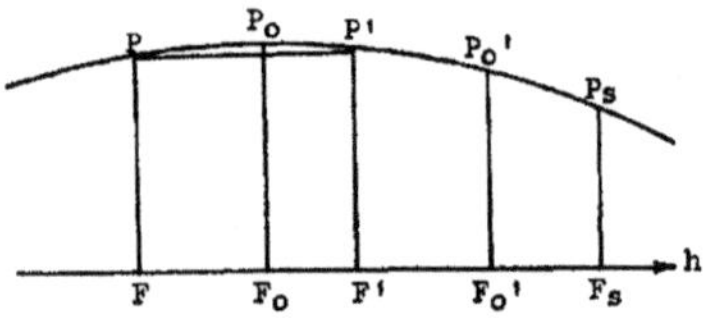

Figure 8

To find the image $P' = f(P)$ of a given point P drop the perpendicular PF from P on h ($P = F$, if P lies on h). From F lay off x in the positive direction on h to the point F'. On h erect at F' a perpendicular $F'P'$ on the same side as P and such that $F'P' = FP$. Then $P' = f(P)$. If s changes $f(P)$ varies on an equidistant curve k to h because every point of k has distance PF from h.

Only in the euclidean case is k a straight line, f is then also a translation along k. In the hyperbolic and spherical cases f cannot be interpreted as translation along any other line but h.

The mobility principle is the requirement that congruent figures can be moved into each other. Here we restrict ourselves to the two-dimensional case. It then suffices to require mobility for triangles. A rigorous form of the principle is this:

(17) *If* $AB = A'B'$, $BC = B'C'$, $CA = C'A'$

then a motion f exists such that $f(A') = A$, $f(B') = B$, $f(C') = C$. Since a combination of motions is again a motion the preceding considerations show that euclidean and spherical geometries satisfy the mobility principle. We want to understand why hyperbolic geometry is the only other geometry in which this principle holds.

The problem presupposes a set of general geometries from which the mobility principle selects the three geometries. How can ''general geometries'' be defined? (17) shows that the concept of distance is essential. We therefore assume that our space is such that for any two points x,y a distance xy is defined and has the following three properties whose content will be clear without further comment:

1) $xy = 0$ means $x = y$

2) If $x = y$, then $xy = yx > 0$

3) $xy + yz > xz$

These conditions are, of course, still much too weak as a foundation.

For instance, take any set of points and define $xy = yx = 1$ for $x \neq y$ and $xx = 0$. Then properties 1), 2), 3), and (17) are satisfied (to see the latter we define f at the points A', B', C' as in (17) and $f(P) = P$ for $P \neq A'$, B', C', A, B, C, finally $f(A) = A'$, $f(B) = B'$, $f(C) = C'$). It is clear what is wrong with the example: it is impossible to go continuously from one point to another. For our problem *the distance of x and y must be the length of the shortest route from x to y*. In exact language we require

4) If $x = y$ then a curve $z(t)$, $0 \leq t \leq xy$ from x to y (that is $z(0) = x$, $z(xy) = y$) exists such that $z(t_1)z(t_2) = |t_1 - t_2|$

This implies that for $t_1 < t_2 < t_3$

$$z(t_1)z(t_2) + z(t_2)z(t_3) = |t_1 - t_2| + |t_2 - t_3|$$

$$t_2 - t_1 + t_3 - t_2 = t_3 - t_1 = |t_1 - t_3| = z(t_1)z(t_3)$$

so that distances along a shortest route add up (instead of only 3)). Requirement 4) excludes the above example, but it also excludes Σ, if the distances of ordinary space are used instead of the spherical distances along great circles (going through the earth instead of along its surface).

4) guarantees only that y can be reached from s by a curve of length xy. It is also necessary to know that we can go beyond y, at least if y is not too far from x. (On Σ the antipode P' to a given P is a point beyond which it is not possible to go if we start at P). Moreover we need to know that there is only one shortest route from x to y, again only if y is not too far from x. (There are infinitely many different shortest routes from P to P' on Σ). All this can be easily expressed by conditions similar to 4). But 1) to 4) will suffice to give the reader an idea how the problem of defining ''general geometries'' is approached in modern mathematics.

Without going into further details we assume that *every point P is contained in a plane piece in which shortest connections are unique and can be prolonged*. If P is on Σ, a hemisphere with P as pole would be such a piece. It is true that a hemisphere is not a piece of a plane in the usual sense, but it can be bent and stretched into one if we imagine it to be of rubber. After this bending process we define the distance of two atoms x and y as equal to the distance of the same atoms in their original position on Σ. Thus we obtain a plane piece with a spherical geometry. Intrinsically the hemisphere does not differ from the plane piece. A difference appears only if we try to imbed the piece in ordinary space, that is derive its geometry from the geometry of the surrounding space as we did in section 1. In the plane form it cannot be imbedded. The reader will wee that the following argument is entirely intrinsic.

Consider a straight line h and an equidistant curve k to h obtained by translation of a point P along h (see Figure 8). If the feet F,F' of P and P' are such that they have the same distance (and orientation) as the feet F_0, F'_0 of P_0 and P'_0, then the figures $PFF'P'$ and $P_0F_0F'_0P'_0$ consisting of three straight segments and an arc of k each are congruent, because one figure can be obtained from the other by translations along h. Therefore the arcs PP' and $P_0P'_0$ of k have equal lengths. We see that equal arcs of k correspond to equal distances of their feet on h. Therefore, if F_s is the point on h obtained by laying off the distance s from F_0 in the positive direction and P_s is the corresponding point of k, then the length t of the arc P_0P_s must be proportional to s; the factor of proportionality depends on the distance $b = PF = P_0F_0 = P_sF_s$, or in a formula,

$$t = \phi(b)\cdot s. \tag{18}$$

We have to determine the function $\phi(b)$. For that purpose construct the equidistant curves to h (all on the same side of h) at distances x, $x + y$, and $x - y$ from h where $x > y > 0$. Divide the segment from F_0 to F_s into n equal parts, where n is any even positive integer.

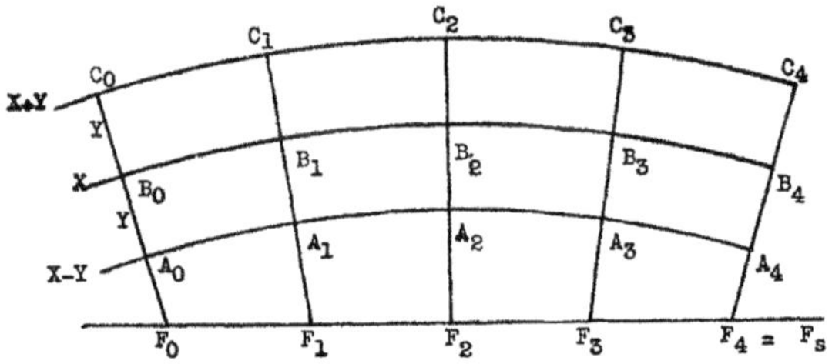

Figure 9

Denote the points of division by $F_0, F_1, F_2, \ldots, F_n = F_s$, and the points corresponding to F_i on the equidistant curves by A_i, B_i, C_i as indicated in the figure which corresponds to $n = 4$. The relation (18) yields for the length of the arc of the equidistant curve

$$\text{length } A_0A_n = \phi(x - y)\cdot s, \quad \text{length } B_0B_n = \phi(x)\cdot s, \tag{19}$$
$$\text{length } C_0C_n = \phi(x + y)\cdot s$$

The quadrangles $A_iC_iC_{i+1}A_{i+1}$ bounded by the arcs A_iA_{i+1} and C_iC_{i+1} of equidistant curves and the straight segments A_iC_i and $A_{i+1}C_{i+1}$ are all congruent. By using the mobility principle we put these quadrangles together in a new way, namely we put every second one upside down, so that A_1 coincides with C_1, etc. as shown in Figure 10.

In the new figure the points $B_0, \ldots, B_n$ lie on a straight line. The length of the middle arc is the same as before, that is $z = \phi(x)\cdot s$ because it consists of the same arcs. The upper and lower arcs consist each of half the previous upper and lower arcs and have therefore

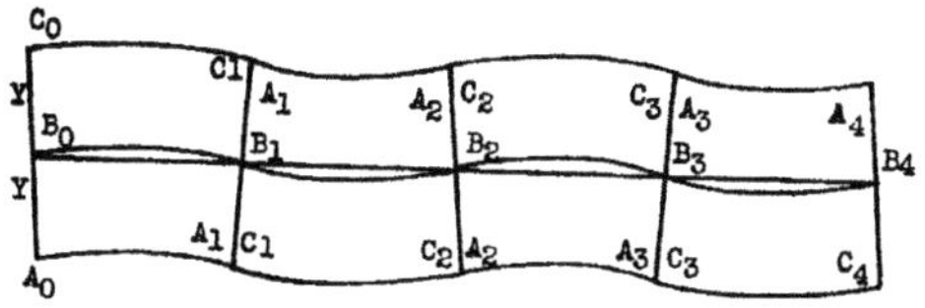

Figure 10

length (compare (19))

$$w = \tfrac{1}{2}(\textit{length } A_0A_n + C_0C_n) = \tfrac{1}{2}[\phi(x - y) + \phi(x + y)]\cdot s$$

The segments C_iA_i are still perpendicular to the new middle curve. Therefore as n increases the middle curve tends to a straight segment of length z and the upper and lower curves tend to the equidistant curves to the middle curve at distance y. By using (18) we find

$$w = \phi(y)\cdot z,$$

and substituting the values of w and z and multiplying by 2

$$\phi(x + y) + \phi(x - y) = 2\phi(x)\cdot\phi(y). \tag{20}$$

We have reduced the problem to solving this so-called functional equation for ϕ. Everyone notices at once the trivial solution $\phi(x) \equiv 0$, but this is excluded because $\phi(0) = 1$, which expresses that for $b = 0$ or $x = 0$ the points P_s and F_s coincide so that $t = s$. Another obvious solution is $\phi(x) \equiv 1$. This leads to the euclidean geometry in which P_s moves on a straight line and $t = s$.

From trigonometry the reader is familiar with the identity

$$\cos(x + y) + \cos(x - y) = 2\cos x \cos y$$

Therefore cos x and, more generally, cos ax, where a is any constant, is a solution of (20). Formula (15) shows that cosh ax is also a solution of (20).

It is quite within the reach of a sophomore to see that 0, 1, cos ax, and cosh ax are the *only continuous solutions* of (20), but the proof will be omitted due to lack of space. If we replace a by $1/\rho$ we can say (compare (2)):

If the mobility principle holds, then the length t of the arc of an equidistant curve to a straight line at distance b has one of the following three forms:

$$t = s, \quad t = \cos b/\rho\cdot s, \quad t = \cosh b/\rho\cdot s$$

The third expression originates from the second by the formal procedure explained in Section 2. But now it was obtained by geometric arguments. In the same way, not only all the hyperbolic formulas can be derived, but the reader sees that the euclidean and spherical formulas appear at the same time and that *no other formulas are compatible with the mobility principle.*

5. *Hyperbolic geometry.* After the procedure of Section 2 has been justified hyperbolic geometry can be legitimately based on it. We describe here, without proofs, some of its properties.

An equidistant curve k to a straight line h is a straight line in euclidean geometry. In spherical geometry it is a circle whose center is on the other side of k from g. Therefore k turns its convex side towards g. In hyperbolic geometry k is neither a circle nor a straight line, but a new kind of curve which turns its concave side towards g as in Figure 8. The segment $P_s F_s$ forms with k (and h) a right angle. Therefore the straight quadrangle $PP'F'F$ has at P and P' angles which are (equal and) smaller than $\pi/2$, so that the angle sum in $PP'F'F$ is less than 2π. This leads to the fact that

In hyperbolic geometry the sum of the angles of a triangle is less than π.

The area of a hyperbolic triangle with angles α, β, γ is proportional to $\pi - \alpha - \beta - \gamma$, and remains therefore below a fixed number although triangles with arbitrary large sides exist (for instance, equilateral ones).

Consider a straight line g in the hyperbolic plane and a point P not on g. Drop the perpendicular PF from P on g and let X traverse one of the rays of g determined by F. The line PX turns monotonically

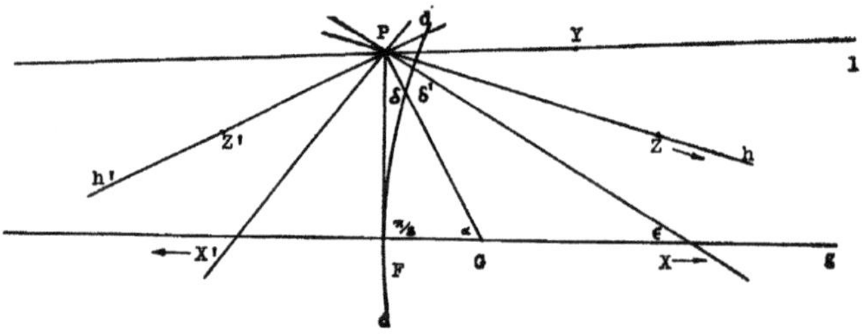

Figure 11

and tends to a limit position h. If G is a fixed point on the same ray as X we find, with the notation of figure 11,

$$\pi/2-(\delta+\delta')>\pi/2-(\delta+\delta'+\epsilon)=\pi-(\alpha+\delta+\pi/2)+\pi-(\delta'+\epsilon+\pi-\alpha).$$

The parentheses on the right are the angle sums in the triangles PFG and PGX respectively. Since these are smaller than π

$$\pi/2 - (\delta + \delta') > \pi - (\alpha + \delta + \pi/2) > 0.$$

Because P, F, G do not depend on X we have

$$\lim \angle FPX = \lim(\delta + \delta') = \angle FPZ < \pi/2$$

By reflecting this figure in the line PF we see: if X' traverses the other ray of g then

$$\angle FPZ' = \lim \angle PFX' = \lim \angle PFX < \pi/2$$

Therefore $\angle ZPZ' < \pi$, so that the limit position h' of PX' is different from h. The lines through P inside $\angle ZPZ'$ intersect g, all the other lines through P do not intersect g. *The parallel axiom does not hold.*

A line l inside the supplementary angle to $\angle ZPZ'$ is usually called a hyper parallel to g. The distance of a variable point y on 1 from g reaches a minimum at some point and tends to ∞ as y tends in either direction to ∞.

If Z traverses h in the indicated direction its distance from g tends to 0. The lines h and h' are called parallels to g. The angle $\beta(d) = \angle FPZ = \angle FPZ'$ which depends only on the distance $b = PF$, is called the *parallel angle* belonging to b. Its value is given by the relation

$$\tan \tfrac{1}{2}\beta(d) = e^{-d/\rho}$$

In spherical geometry a straight line is at the same time a circle (about the pole to the line as equator). In euclidean geometry a straight line is not a circle but it is the limit of circles whose radii tend to ∞. *In hyperbolic geometry a circle with center X through F (Figure 11) does not tend to a straight line as X tends on g to* ∞, but to an important new kind of curve d which is orthogonal to all parallels to the orientation of g in which X recedes.

These facts explain why the discovery of hyperbolic geometry had such a deep influence on mathematics. Because of its many new phenomena it freed us from the shackles of euclidean tradition. *It is somewhat richer than the other two geometries.* The following fact elucidates this contention: In three dimensional hyperbolic space the geometry on a sphere is spherical (but its hyperbolic radius is not equal to the ρ belonging to the same sphere if imbedded in euclidean space). As the hyperbolic sphere becomes large it does not tend to a plane, but to a curved surface, whose geometry is euclidean. Thus both the other geometries can be derived from hyperbolic geometry as the spherical from euclidean. On the other hand, three dimensional spherical geometry does not contain models of either of the other two geometries, a result which is made plausible by the finite extent of spherical geometry.

FOOTNOTES

1) The ordinary geometry, with which the reader was first acquainted in high school, is the main subject of Euclid's Elements. This book remained the standard textbook on geometry until the end of the last century. Therefore ordinary geometry is also called euclidean geometry.

2) The parallel axiom states: if a line g and a point P not on g are given, then the plane determined by P and g contains exactly one line through P which does not intersect g. The formulation of the axiom in Euclid's Elements is slightly different, but equivalent.

3) There are other unifying principles; for instance, projective geometry furnishes such a principle.

University of Southern California

Math. Mag. 24 (1950), 19–34

positive integers given above.

THEOREM (*Peano's postulate of finite induction*). *Let A be the set of positive integers, and let $B \subset A$ such that*

(1) *B includes the first element of A, and*

(2) *$a \in B$ implies that its immediate successor $a' \in B$.*

Then we can conclude that $A = B$.

Proof. Hypothesis (1) of transfinite induction is the same as hypothesis (1) of finite induction. Next suppose that $a' \in A$ implies that all elements preceding a' are in B. Then, since the immediate predecessor of a' is in B by hypothesis (2), $a' \in B$; thus hypothesis (2) of transfinite induction is satisfied. Thus by the transfinite induction theorem, $A = B$.

The rest of Peano's axioms follow almost trivially, and the development of the real numbers can now proceed in the classical form.

Reference

1. Henry P. Thielman, Theory of Functions of Real Variables, Englewood Cliffs, N. J., 1953, Chs. I and II, pp. 1–51.

MATHEMATICAL EDUCATION NOTES

EDITED BY JOHN A. BROWN, University of Delaware, AND
JOHN R. MAYOR, AAAS and University of Maryland

All material for this department should be sent to John R. Mayor, 1515 Massachusetts Avenue, N.W., Washington 5, D. C.

THE ROLE OF GEOMETRY FOR THE MATHEMATICS STUDENT*

HERBERT BUSEMANN, University of Southern California

The importance of geometry in college mathematics has declined for some time, in fact there are now quite a few colleges which offer no course at all in geometry proper. While this development does not worry all mathematicians, it caused concern to sufficiently many to start the present discussion. Here we first deal with the present status of geometry in college, *i.e.*, with the courses which are actually offered, then try to evaluate these courses in view of the aims of geometry teaching, and will arrive at some definite proposals regarding courses to be eliminated or added. The aims, and consequently the evaluation and the proposals, necessarily depend partly on the speaker's views and taste.

The presentation of the actual situation is principally based on the information furnished by the catalogues of 182 colleges, including all the better known ones. The catalogues usually fail to provide all the relevant facts. Course de-

* Presented at the Fortieth Summer Meeting of the Association, Salt Lake City, September 1, 1959.

scriptions are often inadequate or absent. For example, it would have been important for our present purposes to know whether a course in projective geometry uses an algebraic, a geometric or a mixed approach. There is often no indication, which may have the very good reason that the choice of method is left to the teacher. Also, the frequency with which a course is given is not always listed.

Nevertheless the following two tables extracted from the catalogues are believed to be useful; they give the total number of courses in geometry per college and for each individual course the number of colleges teaching it. However it must first be stated which courses are taken into account. Since this lecture is concerned with undergraduate teaching only, graduate courses in geometry do not figure in the tables unless they are explicitly listed as open to undergraduates. This point principally concerns differential geometry. Vector analysis is not counted although it almost always comprises some solid and some differential geometry. Above all, the standard course: "Analytic Geometry and Calculus" is disregarded. This course originated from the waning interest in geometry, almost always proceeds at the expense of geometry,* and is largely responsible for perpetuating the condition.

TABLE 1

Total number of geometry courses per college

Number n of courses in college	Number of colleges offering n courses
0	13
1	23
2	75
3	39
4	24
5	4
6	4

TABLE 2

Types of Geometry Courses

Course	Number of colleges offering this course	
Analytic Geometry		58
Projective Geometry		
undefined	69	99
algebraic	20	
geometric-axiomatic	10	
Foundations of Geometry		9
Noneuclidean Geometry		29
Higher Geometry		33
Modern Euclidean Geometry (triangle and circle)		52
Differential Geometry		61
Other		21

The sum of the numbers in the right column does not equal $1\times 23+2\times 75+\cdots+6\times 4$ because two semester courses in the same subject, like Higher Geometry I, II, appear only once on the right and twice on the left.

Examining the individual courses, we notice that Differential Geometry is in a different position from the rest. Its importance is generally admitted, the subject flourishes. But this leads naturally to the tendency of introducing the student to the now active parts of the field. The course then becomes an advanced graduate course. On the other hand the theory of surfaces in E^3 is a very rich field still quite active, and many of its results are not contained in the

* Since this talk was given Angus E. Taylor's Calculus with Analytic Geometry appeared. It contains most of the material taught in analytic geometry.

type of advanced course just mentioned. Perhaps every undergraduate ought to have a chance (but should not be required) to learn the classical theory, which is also useful in many applications.

There are influential mathematicians doubting the value of any of the remaining courses. We will discuss them from three points of view which may be briefly listed as: factual information, axiomatic thinking, training of the geometric intuition.

Nobody denies that some analytic geometry is necessary, but geometers doubt that "Analytic Geometry and Calculus" suffices. Nevertheless, since saving time for newer developments is essential, this situation may be accepted, provided the student becomes familiar with projective coordinates. Our table shows that projective geometry is by far the most popular course in geometry. The usual objection to the course (which also turned up in the discussion after this lecture) is that the topic is no longer an active subject of research; and that it can easily be introduced (but not adequately treated) as an appendix to the now popular vector spaces. The principles underlying this attitude are highly dangerous. The subject is beautiful, it required major efforts of some of the best mathematicians during and after the last century to understand its structure thoroughly, it still gives a deep insight into geometry. The projective duality principle, besides being constantly used by geometers, is an exciting experience for any interested student. The duality in vector spaces is a very meager substitute. But most of all, do we have the right to completely disrupt historical continuity whenever a subject moves out of the focus of contemporary interests? Do we really expect or agree that our present mathematical efforts will be altogether junked (at least from courses) as soon as the interests change? This disregard for historical continuity obviously leads to dangerous absurdities if taken seriously and must be fought. As far as geometry is concerned the speaker believes that projective geometry is a subject on which to make a stand. A mathematics major should not obtain his degree without knowing projective geometry. There remains the question how the subject should be taught. It will be seen later why the axiomatic approach seems less important now, hence a course should preferably contain both algebraic and geometric arguments. This freedom in method also has the advantage of yielding results most quickly.

Noneuclidean geometry played, historically, the important role of an eye opener. It will still have this effect on the student, moreover it lends itself readily for an exciting undergraduate course and should therefore be offered once in a while either by itself or as a part of another course.

The value of higher geometry as a course depends on its content. It often consists mainly of projective and noneuclidean geometry; then the preceding arguments justify its existence. All too often the content is left to the instructor. Instances are not rare, where the course is assigned to a nongeometer, and then degenerates, for example, into the pure algebra connected with constructions by ruler and compass.

Modern euclidean geometry appears under various synonyms, for example

"Modern (or Advanced) Geometry of Circles and Triangles." It deals with those facts of elementary euclidean geometry which were discovered after Apollonius, so that the definition "modern" used here means "not more than 2100 years old." The course consists of individual niceties entirely unrelated to anything else. It has a slight merit as a training of the intuition and for high school teachers, but we will see later that this can be accomplished far better. The course should therefore disappear.

We now come to our second point, namely axiomatics and in particular to the foundations of geometry as a course, which was passed over above. Not so very long ago this course and axiomatic projective geometry constituted (for historical reasons) the only opportunity for introducing the student to axiomatic thinking, which greatly appeals to many minds with a mathematically critical or a logical bent. Now large parts of mathematics are treated axiomatically, and, what is worse, the axioms for a group, for a field, or for a Hausdorff space, are simpler by far than those for euclidean geometry. Therefore the foundations of geometry no longer satisfy a real need; the same applies to the axiomatic approach to noneuclidean or projective geometry, although the axioms for the latter are simple. Many geometers will find it hard to admit this, because axiomatics is very close to their hearts and is still a rather active field of research. Students interested in the subject turn up now and then, but they can very well be helped by a reading course. In this connection it should be considered whether the modern development of the foundations of geometry on the basis of group theory* might not be preferable to Hilbert's approach.

The shift in mathematics towards the axiomatic approach and the concomitant abstractness make it imperative that some course in geometry strengthen the geometric intuition which is in great danger of being lost, but has always been one of the main inspirations in mathematical creation. The abstract approach not only fails to develop the intuition, but the unexpected phenomena arising in very general situations often completely destroy the confidence in intuitive arguments just in the best students. When "moving a hyperplane in E^n continuously" (in the course of a proof), the author was repeatedly interrupted with the question: "Which topology for hyperplanes do you use?" Needless to say that a student who has qualms regarding this operation is forever lost to geometry and geometric reasoning.

The course in "Modern Euclidean Geometry" does train the intuition to a certain extent, but is objectionable for the reasons already given and, in the present context, because it seems to confirm the rather prevalent feeling that modern mathematics is too advanced or intricate for the intuition.

A glance at *Mathematical Reviews* shows that the theory of convex bodies is quite active. It is applied in many other fields, for example in game theory; its proofs are often purely synthetic. Didactically, convexity has the advantage

---

* A book on this topic is: F. Bachmann, Aufbau der Geometrie aus dem Spiegelungsbegriff, Berlin, 1959.

Amer. Math. Mon. 67 (1960), 281–285

over the classical subjects of euclidean geometry that it is defined essentially by an inequality and hence requires much deeper arguments. The proof that convexity can be treated in a form accessible to sophomores is provided by the book of I. M. Yaglom and V. G. Boltyanski on convex figures, which was first published in 1951 in Russia, has since appeared in a second Russian edition and in a German translation (*Konvexe Figuren*, Verlag der deutschen Wissenschaften, 1956), and will soon appear in an English translation (P. J. Kelly and L. F. Walton). It deals with plane convex figures, but leads there to very subtle topics, providing training in continuity arguments, applications of Helly's Theorem, addition of convex figures, maximum and minimum problems with and without solutions, curves of constant width and their generalizations.

The material is divided into theorems or lemmas posed as problems occupying one third, and the solutions occupying the remaining two thirds, of the book, but it can be used as a text.

This book was discussed at some length because there is at present no other book satisfying our requirements, namely, to make strong demands on the intuition, to furnish a useful training for high school teachers as well as to those contemplating research, and to be accessible to sophomores, say. Thus the book constitutes a constructive existence proof, but we do not contend that the problem has only one solution. It is up to the geometers here and elsewhere to discover other possibilities. The more we try the better the chance will be of arriving at a synthesis of ideas which will reverse the present adverse trend in geometry.

Amer. Math. Mon. 67 (1960), 281–285

TWO APPLICATIONS OF GEOMETRY

Lectures by Herbert Busemann

(Lecture notes by Melvin Hausner)

Lecture I: The Simultaneous Approximation of n Real Numbers by Rationals.

Of all the problems in mathematics in which geometry can be used, I prefer those in which geometric insight makes the situation absolutely clear. The most famous example is Riemann's application of topology to the study of algebraic functions. Another example is Poincaré's application of hyperbolic geometry to the theory of automorphic functions. In this lecture and the next I shall consider two topics which need very little background but in which the application of geometry is most forceful. The first topic is a problem in approximation by rationals. The other is a problem in the calculus of variations, which will be discussed in the next lecture.

If $q > 0$ is an integer, and if $\rho_1,\ldots,\rho_{n-1}$ are $n-1$ real numbers, then there exist integers $p_1,\ldots,p_{n-1}$ such that

$$\left|\frac{p_i}{q} - \rho_i\right| \leqq \frac{1}{2q}, \qquad i = 1,\ldots,n-1.$$

To see this, we mark off on the real axis the points $0, \pm\frac{1}{q}, \pm\frac{2}{q},\ldots$. Each point ρ_i falls in one of the intervals determined by the points $\frac{n}{q}$ and it is sufficient to take the nearest endpoint to obtain $\frac{p_i}{q}$ Indeed the distance to the nearest endpoint of an interval is never more than half the length of the interval.

Hermite proved the following stronger result. Suppose $\rho_1,\ldots,\rho_{n-1}$ are given real numbers. Then there exist integers $p_1,p_2,\ldots,p_n$ $(p_n > 0)$ such that

$$\left|\frac{p_i}{p_n} - \rho_i\right| \leqq \frac{1}{p_n^{n/(n-1)}}, \qquad i = 1,\dots,n-1.$$

In this case, the error for large p_n is of course smaller than the $\frac{1}{2p_n}$ guaranteed by the previous argument. Actually, Hermite's result had another factor, larger than 1, on the right-hand side. We shall give Minkowski's geometric approach in what follows. It is worthy of note that the result certainly does not appear to have anything to do with geometry.

We consider a second problem also due to Hermite. Let

$$E^2(x) = \Sigma_{i,k=1}^n g_{ik}x_ix_k$$

be a positive quadratic form. Thus $E(x) > 0$ provided $x \neq 0$. Hermite considered the question of how small can $E(p)$ be, if p is a non-zero vector with integral coordinates. Hermite found a lower bound for such $E(p)$ which depends only on $D = \det(g_{ik})$ and the dimension n. This result also appears to have no geometric significance. However, we can see geometry entering by recalling that the volume of the ellipsoid given by $E(x) \leqq 1$ can be expressed in terms of D. If we let V_E be the volume of the ellipsoid $E(x) \leqq 1$, we have $V_E = \pi_n/\sqrt{D}$, where π_n is the volume of the unit ball ($\pi_n = \pi^{\frac{1}{2}n}/\Gamma(1+\frac{1}{2}n)$).

We can verify this in the simple case of the ellipse $\frac{x^2}{a^2} + \frac{y^2}{b^2} = 1$. Here

$$D = \begin{vmatrix} \frac{1}{a^2} & 0 \\ 0 & \frac{1}{b^2} \end{vmatrix} = \frac{1}{a^2b^2} \quad \text{and} \quad \pi_2 = \pi.$$

Thus the formula above gives $V_E = \pi_2/\sqrt{D} = \pi ab$.

In the general theory it is the volume V_E which enters, rather than the specific formula for it. Therefore we shall not concern ourselves with the proof of this formula. Using V_E, rather than the equivalent expression in terms of D, Hermite showed that it was possible to find a lattice point p such that

$$0 < E(p) \leqq 2/\sqrt[n]{V_E}.$$

(A lattice point is a point with integral coordinates.)

We can rephrase the theorem in the following way. If $V_E \geqq 2^n$, then the ellipse $E(x) \leqq 1$ must contain a lattice point other than the origin 0. In the diagram, any ellipse with center 0 and area $4 = 2^2$ will contain

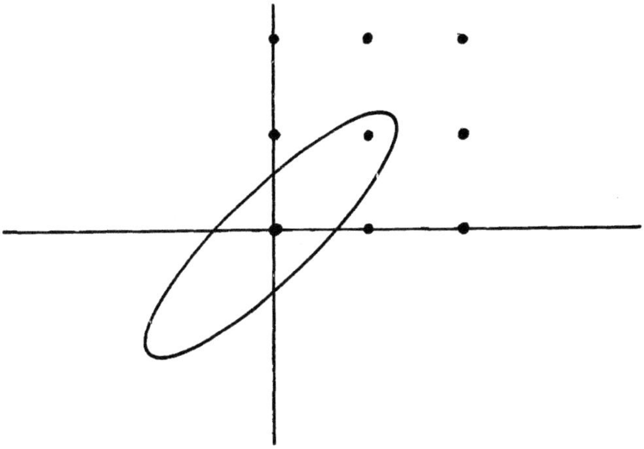

a nontrivial lattice point, even if the ellipse is long and thin.

Minkowski observed that the same theorem is true if the ellipsoid is replaced by any centrally symmetric convex body with the origin as its center. In this case we also have the result that if $V_K \geqq 2^n$, then there exists a lattice point p in K with $p \neq 0$. Here V_K is the volume of K. In this general formulation the number 2^n is best possible. To see this we need only consider a cube centered at 0 with sides slightly less than 2. Minkowski also noted that this general theorem on convex bodies implies Hermite's result on rational approximations. We shall prove this, but we first consider a result of Minkowski on convex bodies.

We shall work in n-space. Suppose that K is a given (closed and bounded) convex body with center at the origin. We introduce a function $r(u)$, defined for all unit vectors in R^n: $r(u)$ is the largest number λ such that λu is in K. It is not hard to verify that $r(u)$ is continuous and that $r(-u) = r(u)$. Since $r(u)$ is continuous on the unit sphere it has a minimum value $r_1 > 0$ and a maximum value $r_2 < \infty$.

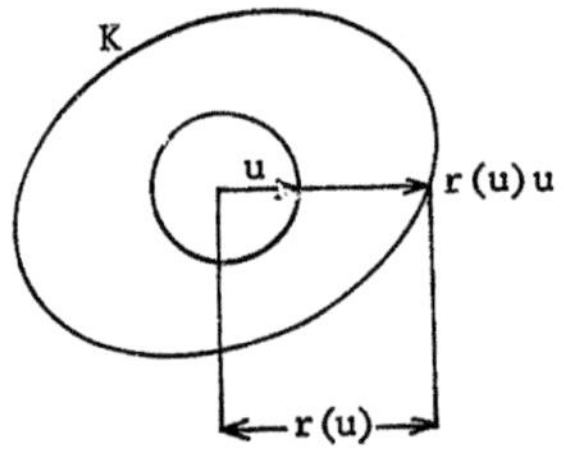

Corresponding to the function $E(x)$ for the ellipsoid, we now define the real-valued function $K(x)$ for all points of space:

$$K(0) = 0,$$

$$K(x) = \frac{|x|}{r(x/|x|)}, \quad \text{if} \quad x \neq 0.$$

Theorem 1. The function $K(x)$ has the following properties:

1. $K(x) > 0$ for $x \neq 0$.
2. $K(tx) = |t|K(x)$.
3. $K(x)$ is a convex function.

The proof of statement 1 is immediate. To prove statement 2, we note that it is trivial if $t = 0$ or if $x = 0$. Otherwise,

$$K(tx) = \frac{|tx|}{r(tx/|tx|)} = \frac{|t|\,|x|}{r(\pm x/|x|)} = |t|K(x),$$

since r satisfies the equation $r(u) = r(-u)$. We shall find it more convenient to prove the following alternate formulation of statement 3:

3'. $K(x+y) \leqq K(x) + K(y)$.

We first show the equivalence of statements 3 and 3' (assuming statement 2). First, assuming 3', we have

$$K((1-\theta)x + \theta y) \leqq K((1-\theta)x) + K(\theta y) = (1-\theta)K(x) + \theta K(y)$$

if $0 \leqq \theta \leqq 1$. This proves that K is convex, and we have statement 3. To go the other way, assuming that K is convex, we have

$$\tfrac{1}{2}K(x+y) = K(\tfrac{1}{2}x + \tfrac{1}{2}y) \leqq \tfrac{1}{2}K(x) + \tfrac{1}{2}K(y).$$

Thus statement 3' follows from statement 3.

Before proceeding to the proof of 3', we mention two important facts about $K(x)$. These are:

4. $K(x) \leqq 1$ if and only if x is in the convex set K.
5. $\frac{|x|}{r_2} \leqq K(x) \leqq \frac{|x|}{r_1}$.

Both of these statements follow immediately from the definitions of $r(u)$ and of $K(x)$.

We now prove 3'. The statement is clearly true if $x = 0$ or if $y = 0$, so we may assume $x \neq 0$ and $y \neq 0$ with no loss in generality. The points $x/K(x)$ and $y/K(y)$ are in K, since

$$K(x/K(x)) = 1 = K(y/K(y)),$$

by property 2. Thus, using property 4, we see that they are points in K. We now express the point $\frac{x+y}{K(x)+K(y)}$ as a convex combination of these

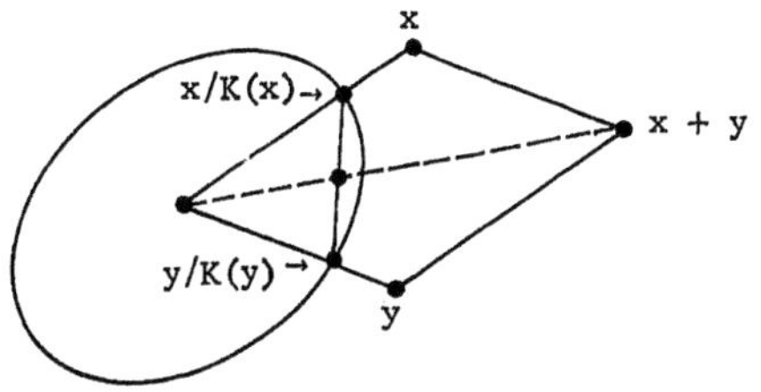

two points. We have

$$\frac{x+y}{K(x)+K(y)} = \frac{K(x)}{K(x)+K(y)}\,\frac{x}{K(x)} + \frac{K(y)}{K(x)+K(y)}\,\frac{y}{K(y)}\ .$$

Thus $\frac{x+y}{K(x)+K(y)}$ is an element of K. Again by 4, this implies

$$K\left(\frac{x+y}{K(x)+K(y)}\right) \leqq 1.$$

Finally, using 2, we obtain the required result after multiplying the last inequality by $K(x) + K(y)$. This completes the proof of the theorem.

We remark that, conversely, if $K(x)$ is a real-valued function defined in n-space satisfying conditions 1, 2, and 3 (or 3'), it is an easy matter to show that the inequality $K(x) \leqq 1$ defines a centrally symmetric convex body. Thus there is a one-to-one correspondence between such functions and convex bodies with the origin as center. To see that $K(x) \leqq 1$ defines a convex body, we note that if $K(x) \leqq 1$ and $K(y) \leqq 1$, then

$$K((1-\theta)x + \theta y) \leqq (1-\theta)K(x) + \theta K(y) \leqq (1-\theta) + \theta = 1,$$

for $0 \leqq \theta \leqq 1$. Thus the inequality is satisfied by every convex combination of points which satisfy it.

For an ellipsoid the expression

$$E(x-y) = \sqrt{\Sigma_{ik} g_{ik}(x_i - y_i)(x_k - y_k)}$$

gives the ordinary Euclidean distance in oblique coordinates. One of Minkowski's two ideas in this development was the idea of generalizing this formula to introduce what we now call the Minkowski distance $m(x,y) = K(x-y)$. (The other main idea was the recognition that only the volume was involved in the formulation of the problem.) We note that while this distance function was introduced before Fréchet's general formulation, Minkowski proved the three conditions for a distance. Thus, defining $m(x,y) = K(x-y)$, we have

a) $m(x,x) = 0$,

b) $m(x,y) = m(y,x) > 0$ if $x \neq y$,

c) $m(x,y) + m(y,z) \geqq m(x,z)$ (the triangle inequality).

The verification is straight-forward:

a) $m(x,x) = K(x-x) = K(0) = 0$.

b) $m(x,y) = K(x-y) = K(-1(y-x)) = |-1|K(y-x) = m(y,x) > 0$ if $y \neq x$.

c) $m(x,z) = K(x-z) = K((x-y)+(y-z)) \leqq K(x-y) + K(y-z) = m(x,y) + m(y,z)$.

The distance $m(x,y)$ satisfies some further properties which we shall need. One of these properties is that it is translation invariant: only the difference $x-y$ enters into the definition. Thus

d) $m(x+a,y+a) = m(x,y)$.

More generally, if $x' = \beta x + a$ is a similitude, the Minkowski distance changes by the factor $|\beta|$.

e) $m(x',y') = |\beta| m(x,y)$ for dilations with a factor β.

A special case is a reflection in a point. This occurs when $\beta = -1$. Thus the metric is invariant under reflections in points.

We also remark that the closed ball of radius ρ about p given by the inequality $m(p,x) \leqq \rho$ can be transformed into the ball with center p and radius σ by a dilation with the factor σ/ρ. Thus the Minkowski metric has some of the features of the Euclidean metric. Of course some features are lost. For example, it is not invariant under rotations.

With this preparation, we are ready to solve the problem of Minkowski.

Theorem 2. If $K(x)$ is a function satisfying the hypothesis of Theorem 1, and if V_K is the Euclidean volume of the convex set K consisting of all points x with $K(x) \leqq 1$, then there exists a lattice point $p = (p_1, \ldots, p_n)$ such that

$$0 < K(p) \leqq \frac{2}{\sqrt[n]{V_K}} .$$

Another way of stating this result asserts that if the volume of the convex body K is at least 2^n, then there is a non-zero lattice point p in K. By symmetry, of course, there are at least two. In what follows p will be used to denote a lattice point. We now prove Theorem 2.

We first note that the inequality 5, $|x|/r_2 \leqq K(x)$, implies that $K(x) \to \infty$ as $|x| \to \infty$. Therefore the minimum of K(p) over all non-zero lattice points exists and is a positive number. We shall let M be this minimum:

$$0 < M = \min_{p \neq 0} K(p).$$

We note further that M is the minimum Minkowski distance between two different lattice points: $m(p,p') = K(p-p')$. It follows that the closed balls of radius $\frac{1}{2}M$ about distinct lattice points p and p' do not have a common interior point. For if y were such a point, we would have $m(p,y) < \frac{1}{2}M$

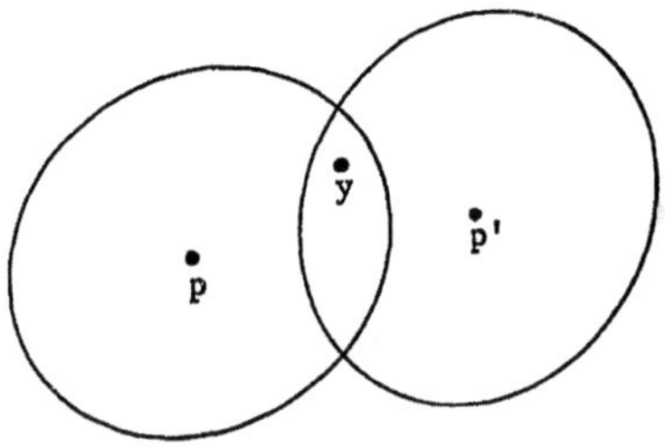

and $m(p',y) < \frac{1}{2}M$. It would follow that

$$m(p,p') \leqq m(p,y) + m(y,p') = m(p,y) + m(p',y) < \tfrac{1}{2}M + \tfrac{1}{2}M = M.$$

Thus we would have $m(p,p') < M$, which contradicts the definition of M as the least Minkowski distance between distinct lattice points.

Now let ω be any positive even integer. (In the end we shall let $\omega \to \infty$, but for the present, keep it fixed.) We now consider all of the lattice points whose coordinates are chosen from among the numbers $\{0,\pm 1,\ldots,\pm\frac{1}{2}\omega\}$, the lattice points inside the closed cube of length ω centered at 0. Since there are $\omega + 1 = 1 + 2(\frac{1}{2}\omega)$ choices for each coordinate, there are

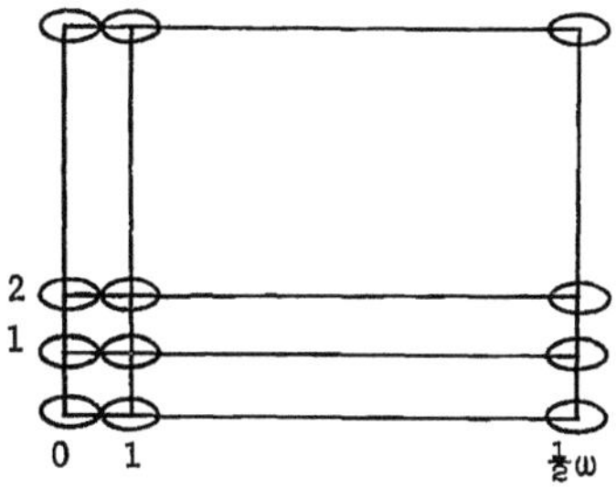

exactly $(\omega + 1)^n$ such lattice points. We call the set of these lattice points S_ω. We now cover each point p of S_ω with S_p, the closed ball (using the Minkowski metric) centered at p with radius $\frac{1}{2}M$, and we consider the volume of the union $\bigcup_{p\in S_\omega} S_p$. First, each ball has the same volume since the balls are translates of one another. Next, the ball centered at 0 with radius $\frac{1}{2}M$ is obtained from the unit ball by a dilation of factor $\frac{1}{2}M$. Thus the volume of S_p is equal to $(\frac{1}{2}M)^n V_k$, since dilations in n-space magnify volumes by the $n^{\underline{th}}$ power of their factors. Finally, since there are $(\omega + 1)^n$ lattice points in S_ω and since the spheres have no common interior points, it follows that the total volume of this set of balls is

$$V = (\omega + 1)^n (\tfrac{1}{2}M)^n V_K,$$

where V_K is the volume of the unit ball.

We shall now obtain an upper bound for this volume by enclosing the set in a cube. Since the spheres at the boundary of the cube $|x| \leqq \omega$ stick out for a Minkowski distance $\frac{1}{2}M$, it is necessary to estimate this as a Euclidean distance and to enlarge this cube by an appropriate amount to contain the union of our spheres. If $K(x) \leqq \frac{1}{2}M$ then, using inequality 5, we have

$$|x| \leqq r_2 K(x) \leqq r_2(\tfrac{1}{2}M).$$

Therefore, if we extend the cube for a Euclidean distance of $r_2(\frac{1}{2}M)$ in each direction, the resulting cube will enclose all of our spheres. The side of this

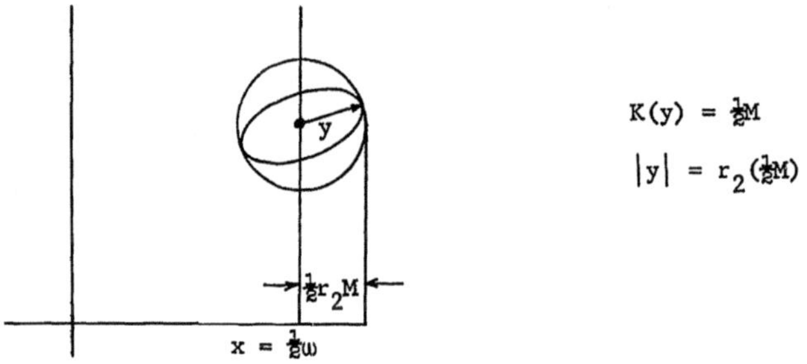

larger cube has Euclidean length $2(\frac{1}{2}\omega + r_2(\frac{1}{2}M)) = \omega + r_2M$, so the volume is $(\omega + r_2M)^n$. We therefore obtain the volume inequality

$$(\omega + 1)^n(\tfrac{1}{2}M)^n V_K \leqq (\omega + r_2M)^n.$$

Dividing by $(\omega + 1)^n$, we have

$$(\tfrac{1}{2}M)^n V_K \leqq \left(\frac{\omega + r_2M}{\omega + 1}\right)^n.$$

Finally we let $\omega \to \infty$ and obtain $(\frac{1}{2}M)^n V_K \leqq 1$ or $M \leqq 2/\sqrt[n]{V_K}$. Since M is the minimum value of $K(p)$, for $p \neq 0$, we have the result.

Again we rephrase the result: A convex body in n-space, symmetric about the origin, with volume at least 2^n, contains a non-zero lattice point.

We started with the theorem of Hermite concerning the approximation of real numbers by rationals. As we shall see this is an easy consequence of Minkowski's result.

Suppose $L_i(x)$ is a non-trivial linear function of x:

$$L_i(x) = a_{i1}x_1 + \cdots + a_{in}x_n,$$
$$(a_{i1},\ldots,a_{in}) \neq (0,\ldots,0), \quad i = 1,\ldots,\nu,$$

where $\nu \geqq n$. We shall suppose that the matrix (a_{ik}) has rank n.

We now consider the system of ν linear inequalities given by $|L_i(x)| \leqq 1$. Each of these inequalities has as its solution an infinite strip, symmetric about the origin, and bounded by the hyperplanes $L_i(x) = \pm 1$, $i = 1,\ldots,\nu$. The intersection of these strips is a convex set with the origin as center. The condition that the matrix of coefficients have rank n assures us that this set is a bounded convex set. (Geometrically, the rank condition states that the normals to the bounding hyperplanes span n-space.) If we let K denote this intersection, the point x is in K if and only if x satisfies each of the ν inequalities $|L_i(x)| \leqq 1$. Therefore, applying Minkowski's theorem, we can make a statement about non-trivial integral solutions of this system of inequalities.

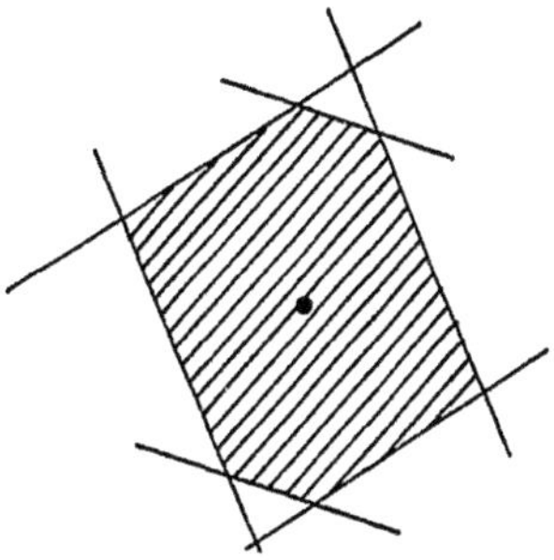

Theorem 3. Let $L_i(x) = \Sigma_k a_{ik}x_k$, $i = 1,\ldots,\nu$, and suppose that the matrix (a_{ik}) has rank n. Let V_K be the volume of K, the set of all points satisfying the inequalities $|L_i(x)| \leqq 1$. Then there is a lattice point

p satisfying

$$|L_i(p)| \leqq 2/\sqrt[n]{V_K}, \quad p \neq 0, \quad \text{for} \quad i = 1,\ldots,\nu.$$

With these preliminaries we turn to the theorem on approximation by rationals.

Theorem 4. Let $\rho_1,\ldots,\rho_{n-1}$ be given real numbers. Then there exist integers $p_1,\ldots,p_n$ with $p_n > 0$ such that

$$\left|\frac{p_i}{p_n} - \rho_i\right| \leqq \frac{1}{p_n^{n/(n-1)}} .$$

To prove this theorem, we specialize the discussion above to the case $\nu = n$. In this case we have n strips, meeting in a parallelepiped whose volume can be found easily. We suppose then that $L_i(x) = \Sigma_k a_{ik}x_k$, $i = 1,\ldots,n$, $\Delta = \det(a_{ik})$, and K is the convex set consisting of all points satisfying the inequalities $|L_i(x)| \leqq 1$. To find the volume of K we change coordinates, taking $x_i' = \Sigma_{k=1}^n a_{ik}x_k$. The convex set K, given by the inequalities $|\Sigma_{k=1}^n a_{ik}x_k| \leqq 1$, goes into the convex set K', given by $|x_i'| \leqq 1$. But K' is simply a cube of side 2, whose volume is 2^n. The Jacobian of the transformation is $\det(a_{ik}) = \Delta$. Thus, the formula for changing variables gives

$$2^n = \iint_{K'} dx_1'\cdots dx_n' = \iint_{K} |\Delta| dx_1\cdots dx_n = |\Delta| V_K.$$

Hence $V_K = 2^n/|\Delta|$. Consequently the Minkowski theorem, applied to this special case, shows that there is a non-zero lattice point p satisfying

$$|L_i(p)| \leqq 2/\sqrt[n]{V_K} = |\Delta|^{\frac{1}{n}}, \quad \text{for} \quad i = 1,\ldots,n.$$

We now choose particular linear functions as follows:

$$L_i(x) = x_i - \rho_i x_n, \quad i = 1,\ldots,n-1,$$
$$L_n(x) = x_n t^{-n}$$

where t may be any number greater than 1. We can compute Δ for this system easily:

$$\Delta = \begin{vmatrix} 1 & & & & -\rho_1 \\ & 1 & & & \cdot \\ & & \cdot & & \cdot \\ & 0 & & \cdot & \\ & & & & t^{-n} \end{vmatrix} = \frac{1}{t^n}.$$

Thus $\Delta = 1/t^n$ and $|\Delta|^{1/n} = 1/t$. Applying our general results, we obtain a system of integers p_i, $i = 1,\ldots,n$, not all zero, for which

$$|p_i - \rho_i p_n| \leq \frac{1}{t}, \quad i = 1,\ldots,n-1,$$

$$\left|\frac{p_n}{t^n}\right| \leq \frac{1}{t}.$$

Hence p_n cannot vanish, since the first $n-1$ relations would then imply that each p_i vanished. By considering $-p_i$ instead of p_i, we may assume $p_n > 0$. We therefore conclude that for each $t > 1$, there are integers $p_1,\ldots,p_n$, $p_n > 0$, satisfying

$$\left|\frac{p_i}{p_n} - \rho_i\right| \leq \frac{1}{tp_n}, \quad 0 < p_n \leq t^{n-1}.$$

This result is somewhat stronger than the result we set out to prove. We obtain Hermite's result by noting that $p_n^{1/(n-1)} \leq t$:

$$\left|\frac{p_i}{p_n} - \rho_i\right| \leq \frac{1}{p_n^{1/(n-1)} \cdot p_n} = \frac{1}{p_n^{n/(n-1)}}.$$

The result can be improved somewhat. For example, the factor $(n-1)/n$ can be introduced into the upper bound. This was done by Minkowski, also geometrically.

Before leaving the field of Minkowskian geometry, I should like to give another striking application. We defined a function $K(x)$ associated with any compact centrally symmetric convex body K; then we defined the Minkowski distance $m(x,y) = K(x-y)$. Suppose H is a hyperplane and p is a point not on H. Then it is an easy matter to show that there is a shortest Minkowski

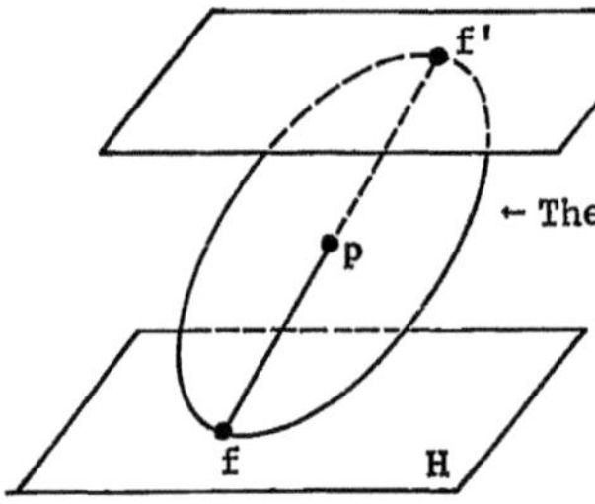

distance from p to H. (Here we use p to denote any point, not just a lattice point.) To see this, we note that the Euclidean distance from p to a point h on H approaches infinity as h approaches infinity on H. But the Minkowski distance is greater than some positive multiple of the Euclidean distance. Hence, in order to find the shortest distance, we may assume that h is in some large ball with center p. Then we can apply the usual compactness argument to show that the shortest distance is achieved. (If K has a (k-1)-dimensional face, the nearest point may not be unique.) Let us call a nearest point f on H the foot, as in Euclidean geometry, and let us call the line through p and f a perpendicular from p to H. Then, because of the properties of the Minkowski distance under dilations, it is easy to prove that this line is a perpendicular to H from any point on it. If a Minkowski ball centered at p with radius m(p,f) is constructed, then H will be a hyperplane supporting this ball at f. Because of central symmetry, the hyperplane H' obtained by reflecting H through p will support the ball at f', the reflection of f through p. The line through p and f is perpendicular to H' too. Any line parallel to pf is also perpendicular to H.

We now discuss a problem which was considered by many people and solved only in a few special cases until A. E. Taylor* observed that its solution is a trivial consequence of the ideas we have been discussing.

Problem. Let K be a compact convex centrally symmetric body. Is it possible to circumscribe a box (i.e., a parallelepiped) about K so that the center of each $(n-1)$-face is in K?

To solve this problem, we consider first the following problem, which will turn out to be useful for solving the original problem. Given $\rho > 0$, construct the box of maximum volume, each of whose sides has Minkowski length ρ.

This problem clearly has a solution, since, if one corner is kept fixed at the origin, the other vertices range over a compact set. (Note that we use the term box in an extended sense, including possibly degenerate parallelepipeds.) We claim that an edge a_1a_2 is perpendicular to the hyperplanes H_1 and H_2 which carry the $(n-1)$-faces that intersect the edge in its endpoints a_1, a_2. For if a_2 is not a foot of a_1 in H_2, let f be such a foot. Then $m(a,f) < \rho$. Prolong a_1f beyond f to the point a_2' for which $m(a_1, a_2') = \rho$. Then translate the $(n-1)$-face in H_2 so that a_2 falls on a_2'. The box spanned by the translated face and a_1 then has greater volume than the original box, but all its edges have Minkowski length ρ.

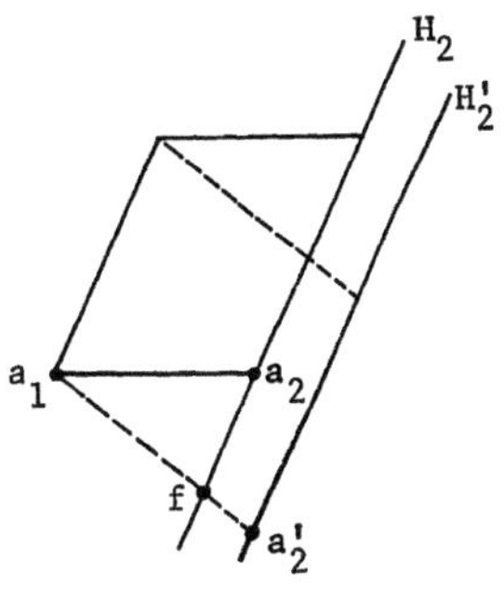

*Bulletin of the American Mathematical Society, 53(1947), 614-616.

Finally, to solve the original problem, we take $\rho = 2$. The box of maximum volume will be the required box. We merely translate it so that its center is at the origin and take K as the unit ball in Minkowski geometry. Since

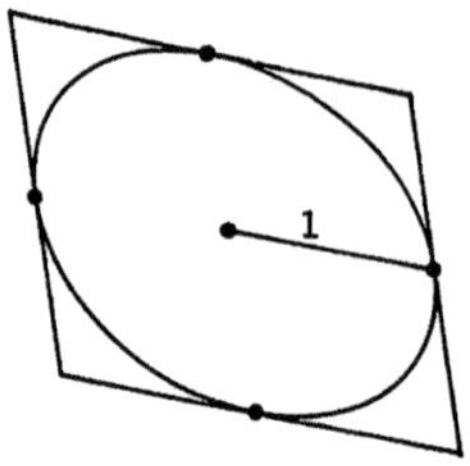

each edge of the box has Minkowski length 2, it follows that each face is a supporting hyperplane of K. For the line from the origin, parallel to the edges and meeting a face in one point, is perpendicular to that face, since it is a translate of an edge. But this line meets the hyperplane in its center, and it meets at a Minkowski distance 1 from the origin. Hence the center of any face is in K, and the proof is finished.

Lecture II: An Application of Integral Geometry to the Calculus of Variations.

My second example of an application of geometry to another field of mathematics uses an idea from integral geometry but does not actually use integral geometry itself. This idea provides the basis for a simple proof of a theorem which had previously been considered difficult; but nobody would have thought of this proof unless he had seen integral geometry. We therefore give a brief description of integral geometry in a very simple setting.

In the plane it is reasonable to regard the area of a domain D as "the number of points in the domain." If we write $p = (x_1,x_2)$ and $dp = dx_1dx_2$, we simply call

$$\int_D dx_1dx_2 = \int_D dp$$

the number of points in D. In integral geometry we count lines and other objects in similar ways. That is, we introduce a suitable measure or density dL in the space of lines and evaluate $\int_X dL$ over some set X of lines. The answer is called "the number of lines in X." However, to give geometrically significant results this measure must be invariant under rotations and translations. Abstractly we have a set of objects and a group acting on these objects. We are required to introduce a density which is invariant under this group. It might be thought that one can always be found if the group is nice enough. But it is not so. For example, the linear subspaces of a fixed dimension in a projective space do not have a density invariant under the projective group. (A theorem of Chern gives general conditions for such a density to exist.)

However, we can easily find a suitable density for the lines in the plane. We first find suitable coordinates to express these lines and then introduce a density in the space of these coordinates. Natural coordinates are introduced if we use the normal form for the line:

$$x \cos \varphi + y \sin \varphi - p = 0, \qquad 0 \leqq \varphi < 2\pi, \qquad p \geqq 0.$$

We shall use φ and p as coordinates for the line and we shall show how to find a density which is invariant under the group of Euclidean motions.

We suppose that $f(p,\varphi)\, d\varphi\, dp$ is such a density. That is, if X is a set of lines in (φ,p) space, the number of lines in X will be given by $\int_X f(p,\varphi)\, d\varphi\, dp$. If X is a set of lines which is transformed by a motion into a set X^*, invariance requires that

$$\int_X f(p,\varphi)\, d\varphi\, dp = \int_{X^*} f(p^*,\varphi^*)\, d\varphi^*\, dp^*.$$

(It will turn out here that we can choose $f(p,\varphi) \equiv 1$ identically. However, this is an accident which depends on our choice of coordinates to describe the lines.)

We now compute what a motion does to lines and see how the motion transforms a set X. Any motion (on points) is given by the equations

$$x = x^* \cos \alpha - y^* \sin \alpha + a,$$

$$y = x^* \sin \alpha + y^* \cos \alpha + b.$$

The equation $x \cos \varphi + y \sin \varphi - p = 0$ is transformed into

$$(x^* \cos \alpha - y^* \sin \alpha + a) \cos \varphi + (x^* \sin \alpha + y^* \cos \alpha + b) \sin \varphi - p = 0,$$

which after simplification becomes

$$x^* \cos (\varphi-\alpha) + y^* \sin (\varphi-\alpha) - (p - a \cos \varphi - b \sin \varphi) = 0.$$

The coordinates φ^*, p^* of the transformed line are therefore given by the equations

$$\varphi^* = \varphi - \alpha,$$

$$p^* = p - a \cos \varphi - b \sin \varphi,$$

if $p - a \cos \varphi - b \sin \varphi \geqq 0$. If not, we do not have normal form, so we would have

$$\varphi^* = \varphi - \alpha + \pi,$$

$$p^* = -(p - a\cos\varphi - b\sin\varphi).$$

Thus

$$d\varphi^* = d\varphi$$

$$dp^* = \pm(dp + (a\sin\varphi - b\cos\varphi)\,d\varphi).$$

The Jacobian of this transformation is ± 1. But since it is the absolute value of the Jacobian which appears in the formula for a change of variables in a double integral, we have

$$d\varphi^*\,dp^* = d\varphi\,dp.$$

Therefore group invariance, together with the formula for changing variables in a double integral, gives

$$\int_X f(p,\varphi)\,dp\,d\varphi = \int_{X^*} f(p^*,\varphi^*)\,dp^*\,d\varphi^* = \int_X f(p^*,\varphi^*)\,dp\,d\varphi.$$

Then since X is arbitrary and any line may be moved into any other line, it follows that $f(p,\varphi) = f(p^*,\varphi^*)$ or $f(p,\varphi) = \text{constant}$. Since we can always multiply by a constant, we shall normalize so that $f(p,\varphi) = 1$. Thus $\int_X dp\,d\varphi$ is the number of lines in a set X of lines, coordinatized by p and φ. This number is left invariant if a motion is applied to X.

We now consider a typical problem in integral geometry. Given a curve C (assumed sufficiently differentiable), how many lines intersect the curve C? In the counting process we count each line as many times as it intersects the curve C. More formally, let $n(L \cap C)$ be the number of points in $L \cap C$. We wish to compute $\int n(L \cap C)\,dL$.

In order to make this computation, we choose as coordinates for a line which intersects the curve, the arc-length s along the curve and θ, the angle which the line makes with the tangent vector, measured positively. Here $0 \leqq s \leqq \lambda$, where λ is the length of C, and $0 \leqq \theta \leqq \pi$. It is necessary

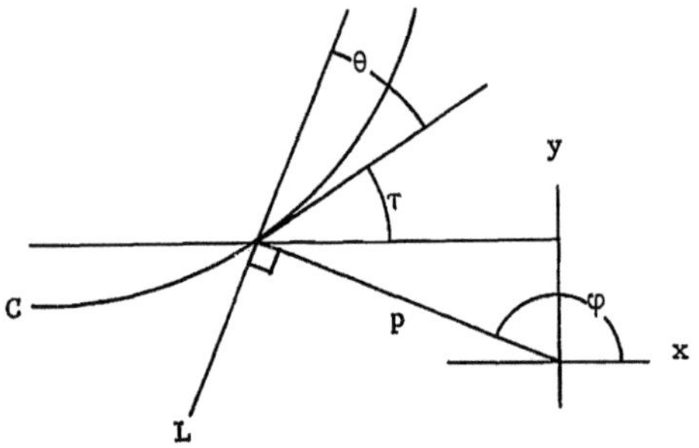

to relate these coordinates with p and φ before we can make our computation. We introduce the angle τ which the curve makes with the x-axis. Then the coordinate φ of the line is clearly $\varphi = \theta + \tau \pm \frac{1}{2}\pi$. Therefore,

$$d\varphi = d\theta + \frac{d\tau}{ds}\, ds.$$

If the point $(x(s), y(s))$ lies on the line $x \cos\varphi + y \sin\varphi - p = 0$,

$$x(s) \cos\varphi + y(s) \sin\varphi - p = 0.$$

Taking differentials and recalling that $\frac{dx}{ds} = \cos\tau$, $\frac{dy}{ds} = \sin\tau$, we obtain after some simplification

$$dp = (-x \sin\varphi + y \cos\varphi)\, d\varphi + \cos(\varphi - \tau)\, ds.$$

We now compute $\frac{\partial(s,\theta)}{\partial(p,\varphi)}$ in two stages:

$$\frac{\partial(s,\theta)}{\partial(s,\varphi)} = \begin{vmatrix} 1 & 0 \\ \frac{d\tau}{ds} & 1 \end{vmatrix} = 1,$$

while

$$\frac{\partial(s,\varphi)}{\partial(p,\varphi)} = \begin{vmatrix} \cos(\varphi-\tau) & - \\ 0 & 1 \end{vmatrix} = \cos(\varphi-\tau).$$

But $\varphi = \theta + \tau \pm \frac{1}{2}\pi$. So $\cos(\varphi-\tau) = \cos(\theta\pm\frac{1}{2}\pi) = \pm \sin\theta$. Since absolute values of Jacobians are desired the sign is of no importance and

$$\left|\frac{\partial(s,\theta)}{\partial(p,\varphi)}\right| = \left|\frac{\partial(s,\theta)}{\partial(s,\varphi)}\right| \left|\frac{\partial(s,\varphi)}{\partial(p,\varphi)}\right| = |\cos(\varphi-\tau)| = |\sin\theta|.$$

Now we use the change of variables formula to transform $\int n(L \cap C)\, dL$ and obtain

$$\int n(L \cap C)\, dL = \int_0^\lambda \int_0^\pi |\sin\theta|\, ds\, d\theta = 2\lambda.$$

This result is very surprising. Except for a constant factor, the number of lines which intersect the curve is the length of the curve. For example, if a curve of a given length is crushed, to make it wiggle, there will be fewer lines meeting the curve, but these lines will meet it more often to compensate.

We make a few observations concerning this result. For a segment of length λ, each line which intersects the segment intersects it only once. The one line which is determined by the segment meets it infinitely often, of course, but one line by itself is of measure zero and can be ignored. For the boundary of a convex curve of length λ, each intersecting line will meet it

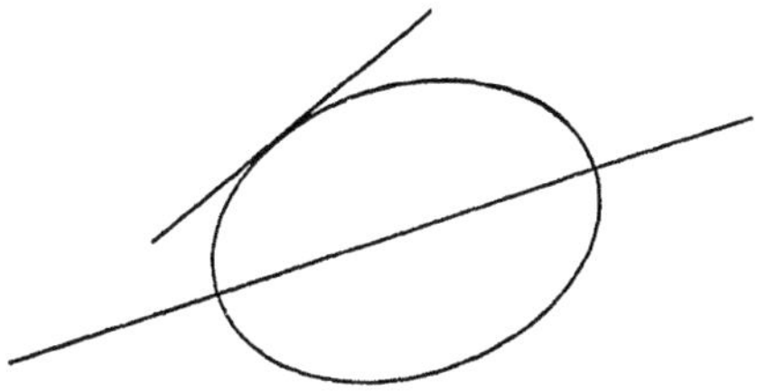

twice, with the exception of the supporting lines. But the supporting lines form a set of measure zero, so they may be ignored. Thus for a convex set in the plane, the number of lines which meet its boundary K is λ, the length of K: $\int_{L \cap K \neq \emptyset} dL = \lambda$.

Before leaving the subject of integral geometry, I would like to mention that this topic would make an excellent course. The problems are always well liked by all the students who can be interested at all. There are also generalizations to space. For example, the density of planes in space. The number of planes which intersect a space curve, counted as many times as they intersect the curve, is again an absolute constant times the length of the curve. Similarly the number of lines which intersect a surface, each counted as many times as it intersects the surface, is an absolute constant times the area of the surface. These results generalize to r-dimensional surfaces in n-space, where we count the number of (n-r)-dimensional planes which intersect the surface, with the correct multiplicity.

Another example is the following one. Suppose C_1 and C_2 are two smooth curves in the plane. Suppose C_2 is kept fixed, but C_1 is rigidly moved to all possible positions. We obtain a density for the various images of C_1, which we may denote dC_1. Then we find

$$\int n(C_1 \cap C_2)\, dC_1 = 4\lambda(C_1)\lambda(C_2).$$

As far as the possibilities for teaching this subject are concerned, the only difficulty is the use of the Jacobian in the change of variables formula for a multiple integral. But if we stay in the plane, even this amounts at most to a 3×3 determinant. If the elements of multilinear algebra are known, then the derivation of the formula for a change of variables is very easy and all of these results can be proved very simply.

We shall now see how these ideas can be used in the calculus of variations. Recall that if $f(x,y,y')$ is a function of three variables, one of the problems in the calculus of variations is to find a function $y = y(x)$ which is an extremal for $\int f(x,y,y')\,dx$ where the endpoints of the curve $y(x)$ are given. The classical Euler equation gives a necessary condition that y be an extremal:

$$\frac{\partial^2 f}{(\partial y')^2}\, y'' + \frac{\partial^2 f}{\partial y \partial y'}\, y' + \frac{\partial^2 f}{\partial y \partial x} - \frac{\partial f}{\partial y} = 0.$$

where x, y, and y' are taken to be independent variables for the computation of the partial derivatives. The solutions $y(x)$ of this equation may maximize the given functional, minimize it, or do neither. They are called extremals in all cases.

Darboux posed the following problem. Suppose that in the plane we are given a two parameter system of curves. Then can this system be regarded as the extremals of a variational problem? His method was roughly as follows. We can consider this system as the solutions of a differential equation $y'' = \varphi(x,y,y')$. If $\varphi(x,y,y')$ is substituted for y'' in Euler's equation, we obtain a partial differential equation for f, as a function of the three independent variables x,y,y'. Finally, using the appropriate existence theorem, we can solve the equation, at least locally for f. Some of the details are omitted, of course. Geometrically we would want to consider vertical lines. Also, the exact meaning of a two parameter system of curves has been left open. But it is clear that the solution by this method is local. Furthermore, all that is known is that the given curves, even locally, are extremals of the problem. Nothing is known about the question of whether they are minimal curves, etc. (Darboux was actually interested in minima.)

Before going into the precise statement of the result, I would like to sketch the idea. Our method will give <u>global</u> results, and furthermore, the given curves will turn out to be unique <u>minimal</u> curves for a variational problem.

Assume that the curves are parametrized by points in the (u,v)-plane. Thus each (u,v) is mapped into a curve $L(u,v)$ of the given system, in a one-to-one way. Since the curves are not necessarily graphs of functions $y = y(x)$, we should consider curves in parametric form, and similarly express the variational problem as a minimization of $\int F(x,y,\dot{x},\dot{y})\, dt$. What we shall find is not F itself, but the value of F integrated along a curve of the given system. Then, by differentiating, we shall be able to recover the function F. Therefore, we shall find

$$\lambda(A) = \int_A F(x,y,\dot{x},\dot{y})\, dt$$

where A is any curve, or portion of a curve, of the given system. We shall arrange to have $\lambda(A)$ equal to the <u>length</u> of A in a suitable geometry, and the curves of the system the "lines". In this case, the triangle inequality "the line is the shortest path between two points" will show that the curve A minimizes the length between two points. Thus we shall need to have the triangle inequality. Now suppose that a_1 and a_2 are joined by a curve of the given system. Take an arbitrary point b and join b to a_i by the curve B_i of the system. What is required is the inequality:

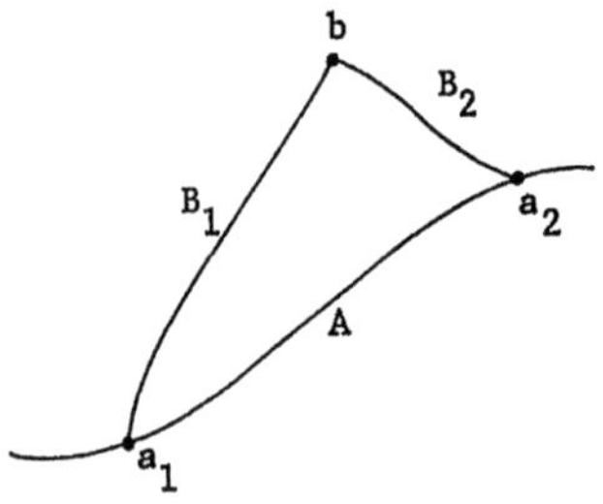

$$\text{Length } B_1 + \text{Length } B_2 \geqq \text{Length } A$$

with equality if and only if b is a point of A.

It is here that integral geometry gives us the leading idea. Although integral geometry is not used, no one could get the idea of this approach unless he has seen some of the ideas of the subject. We remember that the length of a segment is, up to a constant factor, the number of lines which intersect that segment. Therefore, all we need is a density for all the curves of the system, and we could compute this number, and hence the length of any of the curves of the system. Since we are not concerned with rotations and translations, we may choose *an arbitrary positive continuous function* $\varphi(u,v)$ as a density, and use $dL(u,v) = \varphi(u,v)\, du\, dv$. We now *define* the length $\lambda(A)$ by the formula

$$\lambda(A) = \int_{L \cap A \neq \emptyset} dL(u,v)$$

and we maintain that this gives a length function in the sense described above.

To see this, note that if b is on the curve A, then any curve which meets A meets either arc a_1b or ba_2. However, the curves which meet both

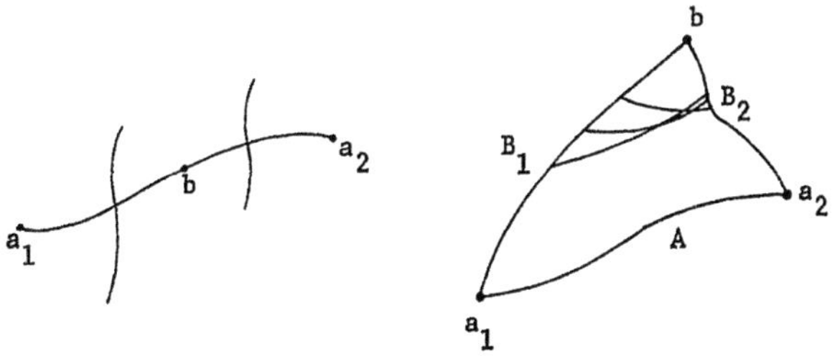

of these arcs form a set of measure zero in this density, and we have $\lambda(A) = \lambda(B_1) + \lambda(B_2)$. On the other hand, if b is not in A, every line which intersects A will intersect B_1 or B_2. This proves the triangle inequality. To obtain the strict inequality, we observe that there is an open set of lines which intersect both B_1 and B_2 and not A, and this set has positive measure. This proves the strict inequality in this case.

Thus the idea is quite a simple one, but it probably would not have been conceived unless one had already seen integral geometry.

To make the argument more rigorous, we must describe the curve system Σ more accurately. We first suppose that the curves are given by coordinates in the plane. We suppose that each curve of a system is an open Jordan curve, and may be parametrized by a function

$$P = P(t), \quad -\infty < t < \infty,$$

where

$$P(t_1) \neq P(t_2) \quad \text{if} \quad t_1 \neq t_2.$$

Furthermore, we require that $P(t_\nu)$ diverge as $|t_\nu| \to \infty$. We also require that any two distinct points belong to exactly one curve in Σ.

A flaw in the reasoning is the assumption that the curves are parametrized by the points in the plane. If $a \neq b$ and if $a_\nu \to a$ and $b_\nu \to b$ it is easy to prove that, for the curves, $L_\nu \to L$. Indeed the curves form a topological space which can be shown to be a two-dimensional manifold. But what sort of manifold is Σ? The curves of the system through a point cannot be contracted to a point. Thus Σ cannot be a plane. However, it can be proved that this space is topologically the projective plane with one point removed, just as it is in the simple case when all the curves are lines. We can then use the measure dL_p in the projective plane, and define the length

$$\lambda(A) = \int_{L \cap A \neq \emptyset} g(L)\, dL_p,$$

where $g(L)$ is an arbitrary continuous positive function, and the argument goes on as before.

Having gone this far, we can answer some other questions. First, we can make every line of finite length if g(L) is bounded. Can we make every line have infinite length in both directions? This can be done with some effort by adjusting g(L).

We can consider a similar problem for curves in the projective plane. Here, the projective lines are homeomorphic to a circle, and any two distinct points determine a line. Suppose we consider in the projective plane P^2 a system Σ of curves in P^2 in which each curve is homeomorphic to a circle and for which any two distinct points lie on exactly one curve in the system. Is it then possible to find a metric in P^2 in which these curves are geodesics? (Since the projective line is not the shortest distance in the large, we can only expect a local result.) The method used to prove the result in the plane broke down in this case, and was saved with great effort by Skornyakov. However, this result is entirely trivial using the method suggested by integral geometry. Here we prove that the curve system is homeomorphic to the projective plane. We can thus use the projective density and weight it with any positive continuous function. What is the length of an entire curve? Because every curve must intersect it, the integral determining the length will be taken over the entire space of curves, and therefore the length has a constant value 2k. This illustrates the general theorem that for any metrization of the projective plane the length of a line is a constant. An arc of a curve in Σ of length less than k strictly minimizes length.

Now assume that the system Σ has some mobility. That is, suppose there are some collineations. Of course, by a collineation in this case we mean a topological map of the underlying space of points (the plane or projective space) which sends the system Σ into itself. Is it then possible to find a metric which is invariant under all of these conditions?

In general the answer is no. For example, in the projective plane, it cannot be true of the projectivities, since a projectivity can send a segment into a proper subset of itself. However, it is possible to give a simple answer if the group is a compact group. In this case, choose one metric and then average it over the group. These considerations solve the problem for the projective plane. For if a compact space has a metric, then the group of all its metric preserving collineations must be compact. Therefore, for a system of curves Σ in the projective plane, it is necessary and sufficient that a group of collineations be contained in a compact collineation group in order for there to exist a metric invariant under these collineations.

We now give an example of a system of curves in the plane, which shows that there is no analogue for this result in the plane. The system will be invariant under all translations, but there will be no metric (for which the curves are geodesics) which is also invariant under translations. The system includes

1) All lines with slope $\leqq 0$.

2) All lines whose equations are of the form $x = \text{constant}$.

3) All translates of the branch of the hyperbola $xy = -1$, $-\infty < x < 0$.

It may be verified that this system has the property that any two distinct points of the plane lie on exactly one of these curves. But it is known that the only geometry invariant under all translations is a Minkowskian geometry, which has the straight lines as geodesics. Thus there is no translation invariant metric on the plane with this system of curves as its geodesics.

The example also illustrates a point in the theory of parallel lines which was considered here by Coxeter and which we shall need later. (We can also use the term asymptotes instead of parallels for the same idea.) It is a usual result, if $A \parallel B$, then $B \parallel A$. I should like to point out that this result depends on the mobility of the plane, that there is a sufficiently large group of motions on it. To find a parallel to a given line A in a given direction from a point p, we consider the lines pq, where q traverses A in one direction. Then the limiting line exists and is defined to be the

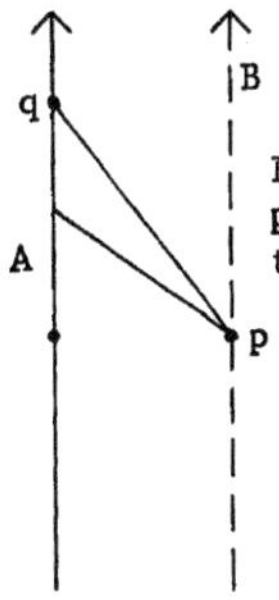

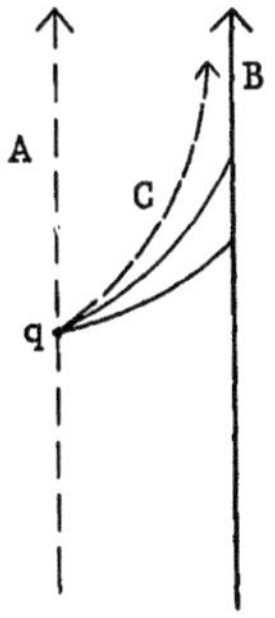

parallel B to A through p. The diagram shows that this is not a symmetric relation in this geometry. The usual proofs from the axioms of hyperbolic geometry use mobility in some way.

If we consider the analogue of this problem on more general surfaces, we encounter a discrete group of motions. For example, consider a system of curves on a cylinder which, in the language of the calculus of variations, "has no conjugate points." This means that if we take the plane to be the universal covering space of the cylinder, then the system of curves in the plane over the given system on the cylinder is a curve system in the plane of the type we have been considering. After the cylinder is "unrolled" we have a curve system Σ in the plane. But we must obtain a metric in the plane which is invariant under the translations which correspond to the identity transformation of the cylinder. This is a discrete group, generated by one translation T. What is the general situation? Can one always find such a metric?

The answer is negative. The condition for the existence of a metric invariant under a discrete group of translations can be nicely stated, although there is not enough time to give the proof. We first define an axis of a translation T as a line mapped on itself by T.

Then, if T does not have an axis, the problem has a solution. For example, if we revolve a branch of a hyperbola about one of its asymptotes, we obtain a surface of negative curvature, which topologically is a cylinder. The rotation of this surface through 2π can be represented in the plane by a translation in the direction of the x-axis of length 2π. But an axis corresponding to this translation

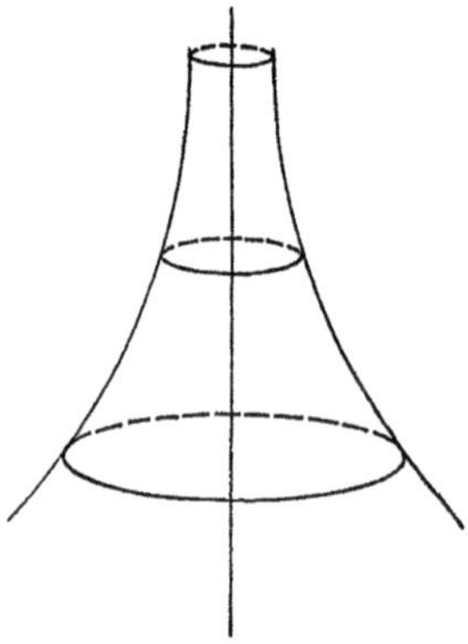

is any of the orbits of a point on the hyperbola. These circles are not lines on the surface of negative curvature. Thus, for example, T has no axis.

But if T has an axis, a necessary and sufficient condition that there is a metric invariant under T is that for any point P the lines L_n through P and T^nP should converge towards a line which is parallel (or asymptotic) to the axis of T.

This property may be illustrated nicely by referring to the example given above of lines and hyperbolas in the plane. We may easily verify that if a translation has a positive x-component and a positive y-component, there is no axis. For this case, therefore, there is a metric invariant under T. If the x-component is positive, but the y-component is negative, then there is an axis, but the above condition is satisfied. However, for a translation with no y-component or with no x-component, the above condition fails, and there will not exist a metric for this system of curves invariant under this translation.

Suppose, instead of a discrete group of translations generated by one translation, we have a continuous one parameter group of translations. In this case we can show that there is a metric invariant under this group which has the given family of curves in the plane as its geodesics if, for one of these translations not equal to the identity, the above criterion is satisfied. (This is clearly also a necessary condition.) The reason is that after we factor out this translation we are left with a compact group.

For my concluding example, I would like to consider a system Σ of curves on a surface of genus $p > 1$, and ask when such a system constitutes the geodesics on this surface without conjugate points. In this case, we can represent the universal covering space by the interior of the unit circle. We shall use the Poincaré model for hyperbolic geometry in order to obtain the appropriate theorem for the system Σ.

Suppose that the system Σ goes over into a system Σ' in the unit circle. Suppose that for Σ', <u>any two points determine a unique curve</u>. Just as certain translations of the plane correspond to the identity transformation of the cylinder and certain translations of the plane correspond to the identity transformation of the torus which it covers, so there are certain <u>covering transformations</u> of the hyperbolic geometry which correspond to the identity transformation of the given surface of genus p on which the curves of the system Σ lie. The system Σ' is invariant under the covering transformations. We postulate that <u>for each covering transformation which has an axis, the curves L_n determined by p and $T^n p$ must approach an asymptote to this axis</u>. We can prove then that each curve of the system meets the circle $|z| = 1$ in two distinct points. (This is to be interpreted in a limiting sense.) We then further assume that <u>a curve of Σ' is uniquely determined by its end-points</u>. With these assumptions, we can state that the given system is a system of geodesics for a suitable metric.

To see this, we note that there is a density for the lines of the hyperbolic plane which is invariant under all motions of this geometry. Since we are postulating that a curve of the system Σ' is determined by its end-points, we can transfer this density onto the curves of the system Σ'. This density can then be used to define length for curves in the system Σ'. It has the proper invariance and hence this length may be transferred onto the given surface of genus p. This solves the problem.

We conclude with the remark that the problem in the large for surfaces of genus p would be quite hopeless without this idea borrowed from integral geometry.

Discussion.

The question of generalizing this type of problem to space was considered. It was observed that in this case the situation is more complicated. It is known, for example, that it is not sufficient that two points uniquely determine a curve. For in the plane, curves separate the plane, and many arguments about crossings go through. But this is not true in space. In 3-space, Douglas gave necessary and sufficient conditions for a solution, locally, in the sense that the curves are extremals for a problem in the calculus of variations.

Math. Assoc. Amer. (1967), 73–105

IRREDUCIBLE SUMS OF SIMPLE MULTIVECTORS

HERBERT BUSEMANN AND D. EDWARD GLASSCO II

Denoting by $V^n(F)$ the n-dimensional vector space over the field F of characteristic 0, let $V_r^n(F)$ be the linear space of all r-vectors $\tilde{R}$ over $V^n(F)$ and $G_r^n(F)$ the Grassmann cone of the simple r-vectors R in $V_r^n(F)$. The sum $\tilde{R} = \sum_{i=1}^k R_i (R_i \in G_r^n(F))$ is irreducible if $\tilde{R}$ is not the sum of fewer than k elements of $G_r^n(F)$. (Duality reduces the interesting cases to $2 \leqq r \leqq n/2$.) Such sums are trivial only for $r = 2$, because $\bigwedge_{i=1}^k R_i \neq 0$ while always sufficient for irreducibility is then also necessary. Extension of F does not influence irreducibility if $r = 2$ but it can for $r > 2$.

The sets $W_r^n(F, k)$ of those $\tilde{R}$ in $V_r^n(F)$ which are irreducible sums of k terms behave as expected when $r = 2$, but have the most surprising properties for larger r. Although $V_3^6(F) = \bigcup_{k=1}^3 W_3^6(F, k)$ and $W_3^6(F, 3) \neq \phi$, the sets $W_3^6(\boldsymbol{R}$ or $\boldsymbol{C}, 2)$ have interior points as sets in $V_3^6(\boldsymbol{R}$ resp. $\boldsymbol{C})$ and so does $W_3^6(\boldsymbol{R}, 3)$ but $W_3^6(\boldsymbol{C}, 3)$ does not.

The paper is based on the thesis [1] with the same title by the second author.

The smallest number k for which $V_r^n(F, k) = \bigcup_{i=1}^k W_r^n(F, i)$ coincides with $V_r^n(F)$ is denoted by $N(F, n, r)$ which by duality equals $N(F, n, n - r)$. Obviously $N(F, n, r) \leqq \binom{n}{r}$. But in spite of various inequalities relating these numbers which show that $\binom{n}{r}$ is much too large, the precise value of $N(F, n, r)$ is known only in the two cases implied by the above statements: namely $N(F, n, 2) = [n/2]$ and $N(F, 6, 3) = 3$. The values $N(\boldsymbol{C}, 7, 3) = 5$, $N(\boldsymbol{C}, 8, 3) = 7$, and $N(\boldsymbol{C}, 9, 3) = 10$ have been claimed but questioned, see Schouten [3, p. 27] and [1].

The purpose of our investigation is to elucidate why the case $r = 2$ is so much simpler than $2 < r < n - 2$. In addition to the already mentioned facts we show that $V_2^n(F, k)$ is an algebraic variety, because, if $\tilde{R}^{(i)}$ is the ith exterior power of $\tilde{R}$, then $\tilde{R}^{(k+1)} = 0$ is necessary and sufficient for $\tilde{R} \in V_r^n(F, k)$ when $r = 2$, but merely necessary when $r > 2$. This implies $\dim V_2^n(\boldsymbol{R}$ resp. $\boldsymbol{C}, k) < \dim V_2^n(\boldsymbol{R}$ resp. $\boldsymbol{C}, k + 1)$ for $1 \leqq k < [n/2]$ in contrast to the case $n = 6, r = 3$. In fact we show that $V_r^n(\boldsymbol{R}$ or $\boldsymbol{C}, k)$ is for $r > 2, k > 1$, and $n \geqq (k - 1)r + 3$ not even a closed set.

An irreducible representation $\tilde{R} = \sum_{i=1}^k R_i$, $k > 1$, is for $r = 2$ never unique, but for $r > 2$ it is (up to a permutation) if $\bigwedge_{i=1}^k R_i \neq 0$ and $k \leqq r$. The condition $k \leqq r$ is probably superfluous but enters —like $n \geqq (k - 1)r + 3$ (instead of $n \geqq r + 3$) above—because we use the Plücker relations for simple vectors which get out of hand for large k. A coordinate-free approach would therefore be preferable, but in many cases we were not able to devise one.

We will continue using capitals (R, S, T) with a tilde and with or without subscripts for general multivectors and omit the tilde only when the vectors are known or assumed to be simple.

2. Results for general F, n, r, k. The following agreement will prove convenient. $e_1, e_2, \cdots$ are used for elements of a base. If two spaces $V^m \subset V^n$ occur, then the base $e_1, \cdots, e_n$ of V^n is chosen so that $e_1, \cdots, e_m$ is a base of V^m. We begin with some simple remarks.

(2.1) *If* $R \in G_r^n$ *then* $R = R' + S \wedge e_n$ *with* $R' \in G_r^{n-1}$ *and* $S \in G_{r-1}^{n-1}$.

For, with suitable $v_i \in V^{n-1}$ and β_i

$$R = \bigwedge_{i=1}^{r} (v_i + \beta_i e_n)$$
$$= \bigwedge_{i=1}^{r} v_i + [\sum_{i=1}^{r} (-1)^{n-i} \beta_i v_1 \wedge \cdots \wedge v_{i-1} \wedge v_{i+1} \wedge \cdots \wedge v_r] \wedge e_n .$$

If the v_i are dependent, the bracket reduces to one term; if not, the bracket is an $(r-1)$-vector in the r-space spanned by $v_1, \cdots, v_r$ and hence is simple.

We apply (2.1) to prove

(2.2) $\tilde{R} \in W_r^n(F, k)$ *if and only if* $\tilde{R} \wedge e_{n+1} \wedge \cdots \wedge e_{n+m} \in W_{r+m}^{n+m}(F, k)$.

It suffices to prove this for $m = 1$. We show if $\tilde{R} \in W_r^n(k)$ and $\tilde{R} \wedge e_{n+1} \in W_{r+1}^{n+1}(l)$, then $l = k$. Trivially $\tilde{R} \wedge e_{n+1} \in V_{r+1}^{n+1}(k)$, whence $l \leq k$. By (2.1) and the hypothesis $\tilde{R} \wedge e_{n+1} = \sum_{i=1}^{l} R_i = \sum_{i=1}^{l} (R_i' + S_i \wedge e_{n+1})$ with $R_i \in G_{r+1}^{n+1}$, $R_i' \in G_{r+1}^n$, and $S_i \in G_r^n$. Therefore, $\Sigma R_i' = 0$ and $\tilde{R} \wedge e_{n+1} = (\Sigma S_i) \wedge e_{n+1}$, which implies $\tilde{R} = \sum_{i=1}^{l} S_i$ and $k \leq l$.

Corollary 2.3. $N(F, n+1, r+1) \geqq N(F, n, r)$.

Anticipating $N(F, n, 2) = [n/2]$ we see that both equality and inequality occur. $N(2m, 2m-2) = N(2m, 2) > N(2m-1, 2) = N(2m-1, 2m-3)$. Similarly $N(2m+1, 2m-1) = N(2m, 2m-2)$. Also $N(n, r) \geqq [(n-r+2)/2]$, but this lower bound is for $r > 2$ too small to be useful.

A consequence of (2.1) is the generalization

(2.4) *If* $\tilde{R} \in V_r^n(k)$, *then* $\tilde{R} = \tilde{R}' + \tilde{S} \wedge e_n$ *with* $\tilde{R}' \in V_r^{n-1}(k)$ *and*

$\tilde{S} \in V_{r-1}^{n-1}(k)$.

By hypothesis $\tilde{R} = \sum_{i=1}^{l} R_i (l \leq k,\ R_i \in G_r^n)$. Applying (2.1) to each R_i yields $\tilde{R} = \sum_{i=1}^{l} (R_i' + S_i \wedge e_n) = \sum_{i=1}^{l} R_i' + (\sum_{i=1}^{l} S_i) \wedge e_n$ with $R_i' \in G_r^{n-1}$ and $S_i \in G_{r-1}^{n-1}$, which is the assertion.

With $k = N(F, n, r)$ we deduce from (2.4):

(2.5) $N(F, n, r) \leqq N(F, n-1, r) + N(F, n-1, r-1)$.

For $r = 2$ equality holds when n is even and inequality holds when n is odd.

A linear map $f: U^m \to V^n$ induces a homomorphism $f^*: U_r^m \to V_r^n$ given by $f^*(u_1 \wedge \cdots \wedge u_r) = f(u_1) \wedge \cdots \wedge f(u_r)$. The map f^* is surjective when f is. We note

(2.6) *If* $f^*(R_1) + \cdots + f^*(R_k)$ *is irreducible in* V_r^n, *then so is* $R_1 + \cdots + R_k$ *in* U_r^m.

We apply this first to the projection $f: V^{n+i} \to V^n$ defined by

$$f: \sum_{i=1}^{n+k} \alpha^i e_i \longrightarrow \sum_{i=1}^{n} \alpha^i e_i$$

and find:

(2.7) *If* $R_i \in G_r^n(F)$ *and* $\sum_{i=1}^{k} R_i$ *is irreducible in* $V_r^n(F)$, *then it is irreducible in* $V_r^{n+k}(F)$.

Hence

(2.8) $N(F, n+1, r) \geqq N(F, n, r)$.

The case $r = 2$ shows again that both inequality and equality can occur in (2.8). Next we apply (2.6) to the map $f: V^{n+k} \to V^{n+1}$ given by

$$f: \sum_{i=1}^{n+k} \alpha^i e_i \longrightarrow \sum_{i=1}^{n} \alpha^i e_i + (\sum_{i=n+1}^{n+k} \alpha^i) e_{n+1}$$

and find using (2.2):

(2.9) *If* $\sum_{i=1}^{k} R_i$ *is irreducible in* $V_r^n(F)$, *then* $\sum_{i=1}^{k} R_i \wedge e_{n+i}$ *is irreducible in* $V_{r+1}^{n+k}(F)$.

Two important facts will now be proved together:

THEOREM 2.10. *If* $\bigwedge_{i=1}^{k} R_i \neq 0$, *then* $\sum_{i=1}^{k} R_i$ *is irreducible. The converse holds only for* $r = 2$.

THEOREM 2.11. *If* $\tilde{R} \in V_r^n(F, k)$ *then* $\tilde{R}^{(k+1)} = 0$. *The converse holds only for* $r = 2$.

If r is odd then $\tilde{R}^{(i)} = 0$ for any $i > 1$ so that $\tilde{R}^{(k+1)} = 0$ imposes no condition. If r is even the relation $(\sum_{i=1}^k R_i)^{(k+1)} = 0$ is obvious, so that the first part of (2.11) holds. Since

(2.12) $(\sum_{i=1}^k R_i)^{(k)} = k! \bigwedge_{i=1}^k R_i$ *for even* r

it follows that $\sum_{i=1}^k R_i \in W_r^n(k)$ when $\bigwedge_{i=1}^k R_i \neq 0$. Applying (2.9) we see that this also holds for odd r.

If $\bigwedge_{i=1}^k R_i = 0$, $r = 2$, and $R_i = v_i \wedge w_i$ then one of the v_i or w_i depends on the rest, say $v_k = \sum_{i=1}^{k-1} \lambda_i v_i + \sum_{i=1}^k \mu_i w_i$ so that

$$\sum_{i=1}^{k} R_i = \sum_{i=1}^{k-1} [v_i \wedge w_i + (\lambda_i v_i + \mu_i w_i) \wedge w_k] .$$

Each bracket represents a simple vector because it is a 2-vector in the space spanned by v_i, w_i, and w_k.

That $\bigwedge_{i=1}^k R_i \neq 0$ is necessary for irreducibility only when $r = 2$ follows from (2.2). This establishes (2.10).

It remains only to prove the second part of (2.11). Let $r = 2$ and $\tilde{R}^{(k+1)} = 0$. Then $\tilde{R} \in W_r^n(k + i)$ with $i \geqq 1$ is impossible because (2.10) and (2.12) would imply $\tilde{R}^{(k+i)} \neq 0$. That $\tilde{R}^{(k+1)} = 0$ is not sufficient for $\tilde{R} \in V_r^n(F, k)$ is obvious for odd r and follows from (2.2) for even $r > 2$.

Corollaries of (2.10) resp. (2.11) are:

(2.13) $N(F, n, 2) = [n/2]$.

(2.14) *If* $\tilde{R} \in W_2^n(F, k)$ *then also* $\tilde{R} \in W_2^n(F_0, k)$ *for any extension field* F_0 *of* F. *This is not true for* $r > 2$.

The latter means that for each $n - 2 > r > 2$ there are $\tilde{R}$, $k' < k$, $F \subset F_0$ with $\tilde{R} \in W_r^n(F, k)$ and $\tilde{R} \in W_r^n(F_0, k')$, and follows from (5.9) and (2.2). Note: The first part of (2.14) does not mean, for example, that $\tilde{R} \in V_2^n(F)$, $\tilde{R} \in W_2^n(F_0, 2)$, hence $\tilde{R} = R_1 + R_2$ with $R_i \in G_2^n(F_0)$, imply $R_i \in G_2^n(F)$, but only that $R_i' \in G_2^n(F)$ with $\tilde{R} = R_1' + R_2'$ exist, compare (4.3).

Whereas in (2.2) and (2.9) the number of summands is the same in hypothesis and assertion, it is different in the next theorem which is therefore harder to prove.

THEOREM 2.15. *Let* $\tilde{R} \in W_r^n(F, k)$, $E_i = \bigwedge_{l=1}^r e_{n+(i-1)r+l} (i = 1, \cdots, j)$, *then* $\tilde{R} + \sum_{i=1}^j E_i \in W_r^{n+rj}(F, k + j)$.

Evidently it suffices to prove this for $j = 1$, or with $E = E_1$ that $\tilde{R} + E \in W_r^{n+r}(k + 1)$. Let $\tilde{R} + E = \sum_{i=1}^m S_i$, $S_i \in G_r^{n+r}$, and denote by S_i' the projection of S_i on V_r^n. Then S_i' is simple and $\tilde{R} = \sum_{i=1}^m S_i'$. Therefore, $\tilde{R} \in W_r^n(k)$ implies $m \geqq k$ and that for $m = k$ all $S_i' \neq 0$. We show that $m = k$ is impossible.

There are at least two S_i which do not lie in G_r^n. For, $S_i \in G_r^n$, if $i > 1$, would entail $S_1 = S_1' + E$ with $S_1' \wedge E \neq 0$, but $S_1' + E$ is not simple by (2.10). Assume that S_1 and S_2 do not lie in G_r^n. For $w = \sum_{i=1}^{n+r} \alpha^i e_i$, put $w' = \sum_{i=1}^n \alpha^i e_i$ and $w'' = \sum_{i=n+1}^{n+r} \alpha^i e_i$. Then

$$S_i = \bigwedge_{j=1}^r w_{ij} = \bigwedge_{j=1}^r (w_i' + w_j'')$$

and we may assume further that $w_{11}'' \neq 0$ and $w_{21}'' \neq 0$.

There are subscripts i, j, k, l with $i \neq k$ such that $w_{ij}'' \wedge w_{kl}'' \neq 0$. Otherwise $w_{11}'' \wedge w_{kl}'' = 0$ for $k \neq 0$ so that $w_{kl}'' = \lambda_{kl} w_{11}''$ for $k \neq 1$. Similarly $w_{kl}'' = \mu_{kl} w_{21}''$ for $k \neq 2$, so that $w_{1l}'' = \mu_{1l}\lambda_{21} w_{11}''$.

This, with $\lambda_{11} = 1$ and $\lambda_{1l} = \mu_{1l}\lambda_{21}$, gives

$$S_i = \bigwedge_{j=1}^r (w_j' + \lambda_{ij} w_{11}'') .$$

But then ΣS_i cannot produce E. Thus we may assume (with a possible change of notation) that $w_{11}'' \wedge w_{21}'' \neq 0$. Then $e_1 \wedge \cdots \wedge e_n \wedge w_{11} \wedge w_{21} = e_1 \wedge \cdots \wedge e_n \wedge w_{11}'' \wedge w_{21}'' \neq 0$ and there is a base $\{e_i'\}$ of V^{n+r} with $e_i' = e_i$ for $i \leqq n$, $e_{n+1}' = w_{11}$, and $e_{n+2}' = w_{21}$. Then with the original $\tilde{R}$, $E, S_1, \cdots, S_k$,

$$(\tilde{R} + E) \wedge e_{n+1}' \wedge e_{n+2}' = (S_3 + \cdots + S_k) \wedge e_{n+1}' \wedge e_{n+2}' ,$$

i.e.,

$$\tilde{R} \wedge e_{n+1}' \wedge e_{n+2}' \in W_{r+2}^{n+r}(k - 1)$$

contradicting (2.2) and (2.7),

3. The sets $V_r^n(F_t, k)$. Let F_t be a topological field. Obviously $G_r^n(F_t) = V_r^n(F_t, 1) = W_r^n(F_t, 1)$ is a closed set in $V_r^n(F_t)$. It is clear that for $k < N(F_t, n, r)$ the set $V_r^n(F_t, k)$ cannot be open, but one might expect it to be closed. This is true for $r = 2$, see below, but in general not for $r > 2$. To show the latter it is not necessary to study general n and $r > 2$ because of the following:

THEOREM 3.1. *If for a topological field* F_t *the set* $V_r^n(F_t, k)$ *is not closed in* $V_r^n(F_t)$ *then for* $m \geqq n$, $s \geqq r$, $m - s \geqq n - r$ *and* $j \geqq 0$ *the set* $V_s^{m+js}(F_t, k+j)$ *is not closed in* $V_s^{m+js}(F_t)$.

First let $j = 0$, $m \geqq n$ and $\tilde{R} \in V_r^n(F_t, k)$. By (2.7) $\tilde{R} \in V_r^m(F_t, k)$ so that the latter is not closed. For any m we conclude from $\tilde{R} \in V_r^m(F_t, k)$ and (2.2) that

$$\tilde{R} \wedge e_{m+1} \wedge \cdots \wedge e_{m+h} \in V_{r+k}^{m+k}(F_t, k) .$$

Since $V_r^m(F_t, k)$ is not closed there are $\tilde{R}_\nu$ in $V_r^m(F_t, k) (\nu = 1, 2, \cdots)$ such that $\tilde{R}_\nu \to \tilde{R} \in W_r^m(F_t, k')$ with $k' > k$.
Then by (2.2)

$$\tilde{R}_\nu \wedge e_{m+1} \wedge \cdots \wedge e_{m+h} \longrightarrow \tilde{R} \wedge e_{m+1} \wedge \cdots \wedge e_{m+h} \in W_{r+h}^{m+h}(F_t, k')$$

so that $V_{r+h}^{m+h}(F_t, k)$ is not closed. This settles the case $j = 0$ or that $V_s^m(F_t, k)$ is not closed.

With the notation of (2.15) we see with the same argument

$$V_s^{m+js}(F_t, k+j) \ni \tilde{R}_\nu + \sum_{i=1}^{j} E_i \longrightarrow \tilde{R} + \sum_{i=1}^{j} E_i \in W_s^{m+js}(F_t, k'+j)$$

which proves (3.1).

In §5 it will be shown that $N(F, 6, 3) = 3$ and $V_3^6(\boldsymbol{R}$ resp. $\boldsymbol{C}, 2)$ is not closed in $V_3^6(\boldsymbol{R}$ resp. $\boldsymbol{C})$. Probably no $V_r^n(\boldsymbol{R}$ resp. $\boldsymbol{C}, k)$ with $3 \leqq r \leqq n-3$ and $1 < k < N(\boldsymbol{R}$ resp. $\boldsymbol{C}, n, r)$ is closed, but from (3.1) we obtain (with $2 + j = k$) this best possible result only for $k = 2$.

THEOREM 3.2. *The sets* $V_r^n(\boldsymbol{R}, k)$ *and* $V_r^n(\boldsymbol{C}, k)$ *are not closed in* $V_r^n(\boldsymbol{R})$ *resp.* $V_r^n(\boldsymbol{C})$ *when* $k \geqq 2$, $r \geqq 3$, *and* $n \geqq (k-1)r + 3$.

The mentioned best result would require a direct treatment of the case $k > 2$ instead of reduction to $k = 2$. The fact that we use Plücker relations in §5, which become very involved for large n, r, k, is responsible for our incomplete result in the case $k > 2$.

We now discuss the case $r = 2$. The by (2.11) necessary and sufficient condition $\tilde{R}^{(k+1)} = 0$ for $\tilde{R} \in V_2^n(F, k)$ amounts to polynomial conditions on the components α^{ik} of $\tilde{R} = \sum_{1 \leqq i < k \leqq n} \alpha^{ik} e_i \wedge e_k$. *The set* $V_2^n(F, k)$ *is therefore an algebraic cone in* $V_2^n(F)$ and hence closed when F carries a topology.

It is also clear that for $1 \leqq k < k' \leqq [n/2]$ the set $V_2^n(F, k)$ is a proper subset of $V_2^n(F, k')$ and plausible but, since we do not know whether $V_2^n(F, k)$ is an irreducible manifold, not *a priori* certain, that the *dimension in the sense of algebraic geometry* (denoted by *a-dim*) and consequently in the case of $\boldsymbol{R}$ resp. $\boldsymbol{C}$ also the *topological dimension* (=dim), of $V_2^n(F, k)$ is less than that of $V_2^n(F, k')$. That a proof is necessary may be seen from the case $r = 3$ (see §§5 and 6). In spite of $N(F, 6, 3) = 3$ the sets $W_3^6(\boldsymbol{R}$ resp. $\boldsymbol{C}$, 2) and $W_3^6(\boldsymbol{R}, 3)$ have nonempty interiors in $V_3^6(\boldsymbol{R}$ resp. $\boldsymbol{C})$ so that

$$\dim V_3^6(\boldsymbol{R} \text{ resp. } \boldsymbol{C},\ 2) = \dim V_3^6(\boldsymbol{R} \text{ resp. } \boldsymbol{C})$$
$$\dim W_3^6(\boldsymbol{R}, 2) = \dim W_3^6(\boldsymbol{R}, 3) = \dim V_3^6(\boldsymbol{R}) = 20\ .$$

But $W_3^6(\boldsymbol{C}, 3)$ has no interior points and hence by a theorem in dimension theory (see [2, p. 46])

$$\dim W_3^6(\boldsymbol{C}, 3) < \dim W_3^6(\boldsymbol{C}, 2) = \dim V_3^6(\boldsymbol{C}) = 40\ .$$

Although we need only the expression for $\tilde{R}^{(k)}$ in the case $r = 2$, we give, owing to its potential usefulness, the *expression of* $\bigwedge_{i=1}^k \tilde{R}_i$ *of k different r-vectors in terms of the components of the* $\tilde{R}_i$. The rather long proof can be found in [1, p. 51].

Put $J = \{j_1, \cdots, j_r\}$ where $1 \leqq j_1 < \cdots < j_r \leqq n,\ n \geqq kr$.

Let $\alpha_i^J = \alpha_i^{j_1 \cdots j_r} (i = 1, \cdots, k)$ be indeterminates and define for a permutation π of $\{1, \cdots, r\}$

$$\alpha_i^{\pi(J)} = \alpha_i^{j_{\pi(1)} \cdots j_{\pi(r)}} = \operatorname{sgn} \pi \alpha_i^J\ .$$

If $\{H = h_1, \cdots, h_{kr}\}$ with $1 \leqq h_1 < \cdots < h_{kr} \leqq n$ and $J_1 \cup \cdots \cup J_k = H$ (disregarding order) then $J_\nu \cap J_\mu = \phi$ for $\nu \neq \mu$ and $J_1, \cdots, J_k$ in this order is a permutation of H whose sign is denoted by

$$\begin{bmatrix} J_1 \cdots J_k \\ H \end{bmatrix}.$$

We then define

$$F^H(\alpha_1, \cdots, \alpha_k) = \sum_{J_1 \cup \cdots \cup J_k = H} \begin{bmatrix} J_1 \cdots J_k \\ H \end{bmatrix} \alpha_1^{J_1} \cdots \alpha_k^{J_k}$$

where α_ν stands for $\{\alpha_\nu^J : J \subset H\}$. If π is a permutation of $\{1, \cdots, k\}$, then

$$F^H(\alpha_{\pi(1)}, \ldots, \alpha_{\pi(k)}) = (\operatorname{sgn} \pi)^r F^H(\alpha_1, \cdots, \alpha_k)\ .$$

THEOREM 3.3. *If* $\tilde{R}_i = \sum_{J \subset N} \alpha_i^J e_J$ *with* $N = \{1, \cdots, n\}$, *then* $\bigwedge_{i=1}^k \tilde{R}_i = \sum_{H \subset N} F^H(\alpha_1, \cdots, \alpha_k) e_H$.

Consequently, if $Q^H(\alpha)$ originates from $F^H(\alpha_1, \cdots, \alpha_k)$ through replacing each α_i by the same $\alpha = \{\alpha^J\}$, then we obtain

COROLLARY 3.4. *If* $\tilde{R} = \sum_{J \subset N} \alpha^J e_J$, *then* $\tilde{R}^{(k)} = k! \sum_{H \subset N} Q^H(\alpha) e_H$.

From (3.4) one deduces with the conventions $\binom{0}{2} = \binom{1}{2} = 0$,

$$a\text{-dim } V_2^n(F, k) = \binom{n}{2} - \binom{n-2k}{2}, \ 1 \leqq k \leqq \left[\frac{n}{2}\right]$$

(see [1, p. 65]). Hence

$$\dim V_2^n(\boldsymbol{R}, k) = \binom{n}{2} - \binom{n-2k}{2}, \ \dim V_2^n(\boldsymbol{C}, k) = 2\left[\binom{n}{2} - \binom{n-2k}{2}\right],$$

and so

$$\dim V_2^n(\boldsymbol{R} \text{ resp. } \boldsymbol{C}, k) < \dim V_2^n(\boldsymbol{R} \text{ resp. } \boldsymbol{C}, k+1) \text{ for } 1 \leqq k < \left[\frac{n}{2}\right]$$

in contrast to the case $r = 3$.

4. Uniqueness. Let $R_i \in G_r^n(F)$, $(i = 1, \cdots, k)$. The sum $\sum_{i=1}^k R_i$ is called *unique* in $V_r^n(F)$ if $S_i \in G_r^n(F)(i = 1, \cdots, k)$ and $\Sigma R_i = \Sigma S_i$ imply that $S_{\pi(i)} = R_i (i = 1, \cdots, k)$ for a suitable permutation π of $\{1, \cdots, k\}$.

Obviously:

(4.1) *If* $\sum_{i=1}^k R_i$ *is unique then it is irreducible.*

(4.2) *If* $\sum_{i=1}^k R_i$ *is irreducible resp. unique then so is* $\sum_{i=1}^j R_i$ *for* $j < k$.

The converse of (4.1) does not hold; in particular:

(4.3) *If* $r = 2$, $k > 1$, $\bigwedge_{i=1}^k R_i \neq 0$ *then* $\sum_{i=1}^k R_i$ *is not unique, i.e., no irreducible sum of 2-vectors is unique.*

Because of (4.2) it suffices to observe that

$$e_1 \wedge e_2 + e_3 \wedge e_4 = (e_1 + e_3) \wedge e_2 + e_3 \wedge (-e_2 + e_4) .$$

However, if $r > 2$ and $\bigwedge_{i=1}^k R_i \neq 0$, then $\sum_{i=1}^k R_i$ probably is unique. Because the Plücker relations are hard to handle for large k we were able to prove only:

THEOREM 4.4. *If* $r > 2$, $k \leqq r$, *and* $\bigwedge_{i=1}^k R_i \neq 0$, *then* $\sum_{i=1}^k R_i$ *is unique.*

Here both the field F and the dimension n of the space (except that $n \geqq rk$ is, of course, implied) are deliberately omitted because they are immaterial.

First we convince ourselves that n is unimportant and at the end of the proof we indicate why F is.

Let $R_i = v_{(i-1)r+1} \wedge \cdots \wedge v_{ir} (i = 1, \cdots, k)$, $\bigwedge v_i \neq 0$. It suffices to prove (for a given F) that ΣR_i is unique in the space V spanned by $v_1, \cdots, v_{kr}$. For, let also

$$\tilde{R} = \sum_{i=1}^{k} R_i = \Sigma R_i^*,\ R_i^* \in V_r^n,\ V^n \supset V .$$

Under projection of V^n on V let $R_i^* \to R_i'$. Then $\Sigma R_i^* \to \Sigma R_i'$, $\tilde{R} \to \tilde{R}$, $\bigwedge R_i^* \to \bigwedge R_i'$ so that $\tilde{R} = \Sigma R_i'$ and if (4.4) holds in V then $\{R_i'\}$ is a permutation of $\{R_i\}$. Therefore, $\bigwedge R_i' \neq 0$ and hence $\bigwedge R_i^* \neq 0$. If $R_i^* = v_{(i-1)r+1}^* \wedge \cdots \wedge v_{ir}^*$ then because (4.4) holds in the space spanned by $v_1^*, \cdots, v_{rk}^*$ we have $R_{\pi(i)}^* = R_i$ for a suitable permutation π of $\{i, \cdots, k\}$.

In the proof of (4.4) we therefore assume that $n = rk$ and $\tilde{R} = \sum_{i=1}^k R_i$ with

$$R_i = e_{(i-1)r+1} \wedge \cdots \wedge e_{ir} = e_{L(i)} ,$$

where

$$L(\nu) = \{(\nu - 1)r + 1, \cdots, \nu r\} \quad (\nu = 1, \cdots, k) .$$

We further put

$$I = \{i_1, \cdots, i_k\} \quad \text{with} \quad 1 \leqq i_1 < \cdots < i_k \leqq rk ,$$

also

$$I(\nu) = I/\{i_\nu\},\ I(\nu, \mu) = I/\{i_\nu, i_\mu\},\ \text{etc} .$$

It will prove convenient and causes no ambiguities to use $I(\nu)$ for $\{i_1, \cdots, i_{\nu-1}, i_{\nu+1}, \cdots, i_k\}$ even if i_ν is not defined.

Ω is the set of all I with $i_\nu \in L(\nu) (\nu = 1, \cdots, k)$, and $I(\nu) \in \Omega$ means $i_\mu \in L(\mu)$ for $\mu \neq \nu$.

We also use

$$e_{L(\nu)} = \bigwedge\nolimits_{i \in L(\nu)} e_i, \quad e_I = \bigwedge\nolimits_{i \in I} e_i, \ \text{etc.}$$

The sign depends on the order but will prove irrelevant. Finally $E(\nu)$ and $F(\nu)$ are the spaces spanned by the e_i with $i \in L(\nu)$ or $i \notin L(\nu)$ respectively.

From now on we will often use the *Plücker relations* (see [3, p. 23] and [4, p. 27]) which in our type of notation may be stated as follows:

Let $P = \{p_1, \cdots, p_r\}$, $1 \leqq p_i \leqq n$, $P(i) = P/\{p_i\}$,

$$\alpha^P = \alpha^{p_1 \cdots p_r} = \operatorname{sgn} \pi \alpha^{p_{\pi(1)} \cdots p_{\pi(r)}}$$

for a permutation π of $\{1, \cdots, r\}$ and similarly for Q. The vector

$$\frac{1}{r!}\sum_P \alpha^P e_P = \sum_{1\leqq p_1<\cdots<p_r\leqq n} \alpha^{p_1\cdots p_r} e_{p_1}\wedge\cdots\wedge e_{p_r}\in V_r^n$$

is simple if and only if for any P, Q

Plücker: $\alpha^P\alpha^Q + \sum_{i=1}^r (-1)^i \alpha^{P(r)q_i}\alpha^{p_r Q(i)} = 0.$

We prove several lemmas beginning with

(4.5) *Let $\tilde{T} = \sum \gamma^I e_I$ and suppose $\tilde{T}\wedge\tilde{R} = 0$. If $k < r$, or $k = r$ but $\tilde{T}$ is simple, then $\gamma^I \neq 0$ only if $I \in \Omega$. Thus simple $\tilde{T} \neq 0$ implies $\gamma^I \neq 0$ for at least one $I \in \Omega$.*

If $k < r$ the assertion follows from

$$\tilde{R}\wedge\tilde{T} = \sum_{\nu=1}^k \left[e_{L(\nu)}\wedge\left(\sum_{I\cap L(\nu)=\phi}\gamma^I e_I\right)\right],$$

$$e_{L(\nu)}\wedge e_I \neq e_{L(\mu)}\wedge e_I \quad\text{for}\quad \nu\neq\mu \quad\text{and}\quad k<r,$$

and the observation that $I\cap L(\nu) = \phi$ for some ν is equivalent to $I\notin\Omega$.

If $k = r$ then the terms in $\tilde{R}\wedge\tilde{T}$ with $e_{L(1)}$ as a factor are

$$e_{L(1)}\wedge\left[\sum_{i_1>r}\gamma^I e_I + (-1)^r\gamma^{L(1)}(e_{L(2)}+\cdots+e_{L(k)})\right].$$

Therefore, $\gamma^I = 0$ if $i_1 > r$ (hence $I\notin\Omega$) and $\gamma^{L(\nu)} + (-1)^r\gamma^{L(1)} = 0$ for $\nu > 1$. Generally $\gamma^I = 0$ if, $I\notin\Omega$ and I is no $L(\nu)$; moreover,

$$\gamma^{L(\mu)} + (-1)^r\gamma^{L(\nu)} = 0 \quad\text{if}\quad \mu\neq\nu.$$

If r is even then $\gamma^{L(\nu)} = 0$ for all ν so that $\gamma^I\neq 0$ only for $I\in\Omega$.

If r is odd then $\gamma^{L(\nu)} = \lambda$ for all ν and thus

$$\tilde{T} = \lambda\tilde{R} + \sum_{I\in\Omega}\gamma^I e_I.$$

We show $\tilde{T}$ is simple only if $\lambda = 0$ which completes the assertion.

Let $I\in\Omega$ and assume $i_1\neq r$. Then with

$$L(1, r)i_s = \{1, \cdots, r-1, i_s\}$$

one of the Plücker relations for the simplicity of $\tilde{T}$ is

$$0 = \gamma^{L(1)}\gamma^I + \sum_{s=1}^r(-1)^s\gamma^{L(1,r)i_s}\gamma^{rI(s)} = \lambda\gamma^I,$$

for $\gamma^{L(1,r)i_s} = 0$ because $L(1, r)\in\bigcup_{\nu=1}^r L(\nu)\cup\Omega$ for $s > 1$, and $L(1, r)i_1$ contains a repeated index (since $i_1\neq r$). If $i_1 = r$ just permute $L(1)$ so that r is not the last element. Thus $\gamma^I = 0$ for all $I\in\Omega$, or

$\lambda = 0$. Since $\tilde{T} = \lambda\tilde{R}$ would not be simple we must have $\lambda = 0$.

Let $H = \{h_1, \cdots, h_{k-1}\}$ with $1 \leqq h_1 < \cdots < h_{k-1} \leqq rk$.

(4.6) *If* $\tilde{S} = \sum \beta^H e_H$ *and* $\tilde{R} \wedge \tilde{S}$ *is simple then, for*

$$I \in \Omega,\ \beta^{I(s)}\beta^{I(t)} = 0$$

if $s \neq t$.

The terms in the expansion

$$\tilde{R} \wedge \tilde{S} = \Sigma \alpha^{i_1 \cdots i_{r+k-1}} e_{i \cdots i_{r+k-1}}$$

which contain e_I as a factor are given by

(4.7) $e_I \wedge [\pm \beta^{I(1)} e_{L(1,i_1)} \pm \cdots \pm \beta^{I(k)} e_{L(k,i_k)}]$

where $L(\nu, i_\nu) = L(\nu)/\{i_\nu\}$. Consider the Plücker relation for $\tilde{R} \wedge \tilde{S}$ beginning with

$$\alpha^{I,L(s,i_s)}\alpha^{I,L(t,i_t)} = \beta^{I(s)}\beta^{I(t)}\ .$$

The terms not written down all vanish. The first $k-1$ that follow vanish because the first factor has a repeated superscript. From the $(k+1)$st term on, the last element of $L(i, i_s)$ is the first superscript of the second factor α which then vanishes because it does not appear in (4.7). (This requires $r \geqq 3$. The first α also vanishes and for a similar reason.)

The following is the decisive step in our long argument:

(4.8) *If both* $S = \Sigma\beta^H e_H$ *and* $\tilde{R} \wedge S$ *are simple and some* $\beta^{I(1)} \neq 0$ $(I(1) \in \Omega)$ *then* $\beta^{i_1\nu_1\cdots\nu_{k-2}} = 0$ *for* $i_1 \in L(1)$ *and any* $\nu_s (s = 1, \cdots, k-2)$. *Briefly* $S \in F(1)_{k-1}$.

Take any $i_1 \in L(1)$ and join it to $I(1)$. This produces an $I \in \Omega$. We prove inductively.

$$\beta^{\nu_1\cdots\nu_\lambda I(k-\lambda,\cdots,k)} = 0 \quad \text{for all} \quad \nu_s \quad \text{and} \quad \lambda \leqq k-2\ .$$

If $k = 2$ we have $\beta^{i_1} = \beta^{I(2)} = 0$ by (4.6) and $\beta^{I(1)} \neq 0$.

If $k \geqq 3$ we make

Step 1. Consider the Plücker relation

$$0 = \beta^{I(1)}\beta^{\nu I(k-1,k)} - \beta^{I(1,k)\nu}\beta^{i_k I(k-1,k)} + \beta^{I(1,k)i_1}\beta^{i_k\nu I(1,k-1,k)}$$
$$- \beta^{I(1,k)i_2}\beta^{i_k\nu I(2,k-1,k)} \pm \cdots \pm \beta^{I(1,k)i_{k-2}}\beta^{i_k\nu I(k-2,k-1,k)}\ .$$

Except for order

$$i_k I(k-1,k) = I(k-1), \quad \text{and} \quad I(1,k)i_1 = I(k)$$

so that the second and third terms vanish by (4.6). The remaining terms vanish because the sets $I(1,k)i_2, \cdots, I(1,k)i_{k-2}$ contain repeated elements. If $k = 3$ we are finished. If $k > 3$ we make

Step 2. Take the Plücker relation

$$\begin{aligned} 0 = {} & \beta^{I(1)}\beta^{\nu\mu I(k-2,k-1,k)} - \beta^{I(1,k)\nu}\beta^{i_k\mu I(k-2,k-1,k)} \\ & + \beta^{I(1,k)\mu}\beta^{i_k\nu I(k-2,k-1,k)} - \beta^{I(1,k)i_1}\beta^{i_k\nu\mu I(1,k-2,k-1,k)} \\ & + \beta^{I(1,k)i_2}\beta^{i_k\nu\mu I(2,k-2,k-1,k)} - \cdots \pm \beta^{I(1,k)i_{k-3}}\beta^{i_k\nu\mu I(k-3,k-2,k-1,k)} . \end{aligned}$$

Except for order

$$i_k\mu I(k-2,k-1,k) = \mu I(k-2,k-1)$$

and

$$i_k\nu I(k-2,k-1,k) = \nu I(k-2,k-1)$$

so that the second and third terms vanish by Step 1 (that $k-1, k$ are replaced by $k-2, k-1$ is immaterial since the argument of Step 1 is the same for any permutation of $\{2, \cdots, k\}$). The fourth term vanishes because of (4.6) and $I(1,k)i_1 = I(k)$. In all following terms the sets of superscripts in the first factor β contain repeated elements and these terms vanish also. This completes the argument in case $k = 4$. If $k > 4$, the process clearly continues.

(4.9) *If both* $S = \Sigma\beta^H e_H$ *and* $\tilde{R} \wedge S$ *are simple and* $\beta^{I(t)} \neq 0$ *for some* $I(t) \in \Omega$ *then* $S \in F(t)_{k-1}$. *If* $S = w_1 \wedge \cdots \wedge w_{k-1}$ *then each* $w_i \in F(t)$.

The first part is a consequence of (4.8). The second statement follows from the general lemma.

(4.10) *If* $m < n$, $V^m \subset V^n$, $v_j \in V^n$, $\bigwedge_{j=1}^s v_j \neq 0$ *and* $\bigwedge_{j=1}^s v_j \in V_s^m$ *then* $v_j \in V^m (j = 1, \cdots s)$.

Setting $v_i = v_i' + \beta_i e_n$ $(i = 1, \cdots, s)$ in the proof of (2.1) yields the case $m = n-1$ from which the general case follows.

THEOREM 4.11. *If* $k \leq r$, $S = \bigwedge_{i=1}^{k-1} w_i \neq 0$ *and* $\tilde{R} \wedge S$ *is simple then* $w_i \in F(t)$ *for a suitable* t.

If $S = \Sigma\beta^H e_H$ then it suffices by (4.9) to show that $\beta^{I(t)} \neq 0$ for a suitable $I(t) \in \Omega$. Because $\tilde{R} \wedge S$ is a simple $(r+k-1)$-vector there is a vector $v = \sum_{i=1}^{rk} \delta^i e_i$ such that $\tilde{R} \wedge S \wedge v = 0$ and $S \wedge v \neq 0$. If

$T = S \wedge v = \Sigma\gamma^J e^J$ then

$$\gamma^J = \Sigma \pm \beta^{J(t)}\delta^{j_t} .$$

By (4.5) there is at least one $I \in \Omega$ with $\gamma^I \neq 0$, hence $\beta^{I(t)} \neq 0$ for some t.

After these preparations we are ready to prove (4.4). First observe

(4.12) *If* $\tilde{R} = \sum_{\nu=1}^{k} v_{L(\nu)} (v_i \in V)$ *then* $\bigwedge_{i=1}^{rk} v_i \neq 0$.

Each $v_i \neq 0$ because $\sum_{i=1}^{k} e_{L(\nu)} = \tilde{R}$ is irreducible. Assme $\bigwedge v_i = 0$ and let $\{w_j\}, 1 \leqq j \leqq \lambda < kr$, be a maximal set of independent v_i. Since the w_i span a proper subspace V' of V, an e_μ with $w_1 \wedge \cdots \wedge w_\lambda \wedge e_\mu \neq 0$ exists, and the w_i together with e_μ span a space V'' with $V' \subset V'' \subset V$. Now $\Sigma v_{L(\nu)}$ is irreducible in V and V', and therefore (see (2.2) and (2.7)) $\Sigma_{L(\nu)} \wedge e_\mu$ is irreducible in V'' and in V. But if $\mu \in L(s)$ then $\tilde{R} \wedge e_\mu = \sum_{\nu \neq s} e_{L(\nu)} \wedge e_\mu$.

(4.13) *If* $\tilde{R} = \sum_{\nu=1}^{k} v_{L(\nu)} (v_i \in V)$, $k \leqq r$, $I = \{i_1, \cdots, i_k\} \in \Omega$ *then* $v_{i_t} \in E(\pi(t))$ *for a suitable permutation* π *of* $\{1, \cdots, k\}$.

First $v_{I(s)} \neq 0$ by (4.12). Next

$$\tilde{R} \wedge v_{I(s)} = v_{L(s)} \wedge v_{I(s)} \neq 0$$

is simple. Therefore (4.11) yields $v_{i_\nu} \in F(\pi(s))$, $(\nu \neq s)$ for a suitable number $\pi(s)$, $(1 \leqq \pi(s) \leqq k)$. We must show that $\pi(s)$ defines a permutation of $\{1, \cdots, k\}$ or that $\pi(s) \neq \pi(t)$ for $s \neq t$. Assume $\pi(s) = \pi(t)$ for some $s \neq t$. Then $v_{i_\nu} \in F(\pi(s))$ for $\nu = 1, \cdots, k$ because $I(s) \cup I(t) = I$ whence $v_I \in F(\pi(s))_k$ and $\tilde{R} \wedge v_I = e_{L(\pi(s))} \wedge v_I \neq 0$ contradicting $\tilde{R} \wedge v_I = 0$. Thus $v_{i_t} \in \bigcap_{i_t \in I(s)} F(\pi(s)) = \bigcap_{s \neq t} F(\pi(s)) = E(\pi(t))$.

This establishes the uniqueness of $\Sigma e_{L(\nu)}$. For, consider $I \in \Omega$ and put $I' = \{j_1, i_2, \cdots, i_k\}$ with $j_1 \in L(1)$. Then $I' \in \Omega$. Since $v_{i_\nu} \in E(\pi(\nu))$ for $\nu > 1$ it follows from (4.13) that $v_{j_1} \in E(\pi(1))$. Thus $v_{j_1} \in E(\pi(1))$ for all $j_1 \in L(1)$ and $v_{L(1)} = \alpha_1 e_{L(\pi(1))}$.

Generally, $v_{L(\nu)} = \alpha_\nu e_{L(\pi(\nu))}$ whence $\alpha_\nu = 1 (\nu = 1, \cdots, k)$ and uniqueness follows.

The condition $v_i \in V$ which entered the proof of (4.12) because we applied (2.2) can now be eliminated; $\Sigma e_{L'\nu)}$ retains its form after extension of the underlying field and therefore remains unique after the extension. This justifies the formulation of (4.4) which does not mention a field.

5. The case $n = 6, r = 3$. The remainder of the paper deals with the case $n = 6, r = 3$ whose importance was noted in connection with (3.2). We first show $N(F, 6, 3) = 3$ which may be new for $F \neq C$. Our inequalities (2.3) and (2.5) give only

$$2 = N(F, 5, 2) \leqq N(F, 6, 3) \leqq N(F, 5, 3) + N(F, 5, 2) = 4 .$$

With $e_{ijk} = e_i \wedge e_j \wedge e_k$ we prove:

(5.1) $\tilde{S} = e_{145} + e_{246} + e_{356}$ *is irreducible; whence* $N(F, 6, 3) \geqq 3$.

This proof rests on the observation:

(5.2) *If* $R = \tilde{S} \wedge \sum_{i=1}^{6} \beta^i e_i$ *is simple then* $\beta^1 = \beta^2 = \beta^3 = 0$.

(The converse is trivial but not needed.) If

$$R = \sum_{1 \geqq i < j < k < l < \geqq 6} \alpha^{ijkl} e_{ijkl}$$

then

$$\sum_{1 < j < k < 6} \alpha^{1jk6} e_{1jk6} = e_{16} \wedge [\beta^6 e_{45} - \beta^1(e_{24} + e_{35})] .$$

Therefore one of the Plücker relations for R is

$$0 = \alpha^{1624}\alpha^{1635} - \alpha^{1623}\alpha^{4165} + \alpha^{1625}\alpha^{4613} = (\beta^1)^2 .$$

Similarly, $\beta^2 = \beta^3 = 0$.

Assume $\tilde{S}$ were reducible, $\tilde{S} = v_{123} + v_{456}$ (where again $v_{ijk} = v_i \wedge v_j \wedge v_k$) with $v_i = \sum_{k=1}^{6} \beta_i^k e_k$. Then $\tilde{S} \wedge v_i$ is simple, so that by (5.2) $\beta_i^k = 0$ for $k \leqq 3$, whence

$$\tilde{S} = [\det(\beta_k^{3+i}) + \det(\beta_{3+k}^{3+i})] e_{456} ,$$

which is false because $\tilde{S} \wedge e_6 = e_{1456} \neq 0$.

To show $N(F, 6, 3) \leqq 3$ we need the lemma:

(5.3) *Given* $\tilde{R}_i \in V_2^4(F)(i = 1, \cdots, m)$ *there are* $\lambda_i \in F$ *and* $R_i \in G_2^4(F)(i = 0, \cdots m)$ *such that* $\tilde{R}_i = R_i + \lambda_i R_0(i = 1, \cdots, m)$.

If $\tilde{R}_i$ is simple then $R_i = \tilde{R}_i$, $\lambda_i = 0$ will do, so we assume that no $\tilde{R}_i$ is simple. G_2^4 is a quadratic cone and a hypersurface in $V_2^4(F)$. Therefore, $R_0 \in G_2^4$ exists such that the tangent hyperplane of G_2^4 at R_0 does not contain any $\tilde{R}_i$, and no line through $\tilde{R}_i$ and R_0 intersects G_2^4 (as a locus in V_2^4 completed to a projective space) at infinity. Then the line through $\tilde{R}_i$ and R_0 intersects G_2^4 in a second point R_i' so that

$$\tilde{R}_i = (1 - \lambda_i)R_i' + \lambda_i R_0 = R_i + \lambda_i R_0 .$$

This argument does not require extending F because it amounts to solving a quadratic equation of which one root is F.

Now let $\tilde{R} = \sum_{1 \leq i<j<k \leq 6} \alpha^{ijk} e_{ijk} \in V_3^6(F)$ be given. A simple calculation shows that either $\tilde{R} \in V_3^6(F, 2)$ or a base $\{\bar{e}_i\}$ exists in terms of which

$$\tilde{R} = \sum_{1 \leq i<j \leq 4} \beta^{ij5} \bar{e}_{ij5} + \sum_{1 \leq i<j \leq 4} \beta^{ij6} \bar{e}_{ij6} = \tilde{S}_1 \wedge \bar{e}_5 + \tilde{S}_2 \wedge \bar{e}_6$$

with $\tilde{S}_i \in V_2^4$. By (5.3) there are $S_i \in G_2^4$ and $\lambda_i \in F$ such that

$$\begin{aligned}\tilde{R} &= (S_1 + \lambda_1 S_0) \wedge \bar{e}_5 + (S_2 + \lambda_2 S_0) \wedge \bar{e}_6 \\ &= S_1 \wedge \bar{e}_5 + S_2 \wedge \bar{e}_6 + S_0 \wedge (\lambda_1 \bar{e}_5 + \lambda_2 \bar{e}_6) \in V_3^6(F, 3) .\end{aligned}$$

Thus:

$$N(F, 6, 3) = 3 . \tag{5.4}$$

By a similar argument we prove

$$3 \leqq N(F, 7, 3) \leqq 5 . \tag{5.5}$$

The left inequality follows from $3 = N(6, 3) \leqq N(7, 3)$, see (2.8). For the right inequality one shows (see [1, p. 90]) that either $\tilde{R} \in V_3^7(2)$ or with a suitable base $\{\bar{e}_i\}$

$$\tilde{R} = \sum_{1 \leq i<j \leq 4} \beta^{ij5} \bar{e}_{ij5} + \sum_{1 \leq i<j \leq 4} \beta^{ij6} \bar{e}_{ij6} + \sum_{1 \leq i<j \leq 4} \beta^{ij7} \bar{e}_{ij7} + \sum_{i=1}^{5} \beta^{i67} \bar{e}_{i67} .$$

The last sum is simple and applying (5.3) to the first three terms on the right yields $N(F, 7, 3) \leqq 5$. This method does not extend to $N(n, 3)$ with $n > 7$.

We now study a *special type* of $\tilde{R} \in V_3^6(C)$ which will confirm some of the important assertions made previously.

Let Y be the set of triples

$$Y = \{123, 126, 135, 156, 234, 246, 345, 456\}$$

and suppose that the α^I, $I \in Y$, satisfy the inequalities

$$\begin{aligned} &\alpha^{123}\alpha^{156} + \alpha^{126}\alpha^{135} \neq 0 , \qquad \alpha^{123}\alpha^{246} + \alpha^{126}\alpha^{234} \neq 0 , \\ &\alpha^{123}\alpha^{345} + \alpha^{135}\alpha^{234} \neq 0 , \qquad \alpha^{234}\alpha^{456} + \alpha^{246}\alpha^{345} \neq 0 , \\ &\alpha^{135}\alpha^{456} + \alpha^{156}\alpha^{345} \neq 0 , \qquad \alpha^{126}\alpha^{456} + \alpha^{156}\alpha^{246} \neq 0 , \end{aligned} \tag{5.6}$$

and that the roots λ, μ of

$$(\alpha^{123} x - \alpha^{234})(\alpha^{156} x - \alpha^{456}) + (\alpha^{126} x + \alpha^{246})(\alpha^{135} x + \alpha^{345}) = 0 \tag{5.7}$$

are distinct. They are different from zero.

Theorem 5.8. *If* $\tilde{R} = \sum_{I \in Y} \alpha^I e_I \in V_3^6(C)$ *and the* α^I *satisfy* (5.6), *then* $\tilde{R} = R_\beta + R_\gamma$, $R_\beta = \sum_{I \in Y} \beta^I e_I$, $R_\gamma = \sum_{I \in Y} \gamma^I e_I$, *where* R_β *and* R_γ *are simple with* $R_\beta \wedge R_\gamma \neq 0$. *Hence the representation* $R_\beta + R_\gamma$ *is* *unique (by* (4.4)).

If λ, μ *are the solutions of* (5.7) *then the* β^I *and* $\gamma^I(I \in Y)$ *are given by*

$$\beta^{1ij} = \frac{\mu\alpha^{1ij} - \alpha^{4ij}}{\mu - \lambda}, \quad \gamma^{1ij} = \frac{\alpha^{4ij} - \lambda\alpha^{1ij}}{\mu - \lambda},$$
$$\beta^{4ij} = \lambda\beta^{1ij}, \qquad \gamma^{4ij} = \mu\gamma^{1ij}.$$

No β^I *or* $\gamma^I(I \in Y)$ *vanishes.*

This representation was found by using Plücker relations (see [1, pp. 98–106]), but after it is explicitly given one readily verifies that R_β and R_γ are simple and that $R_\beta \wedge R_\gamma \neq 0$. In fact, it is easy to factor R_β and R_γ, see [4, p. 21]: Since $\beta^I \neq 0$ if $I \in Y$, letting $\nu = (\beta^{123})^{-2/3}$ we find

$$R_\beta = u \wedge v \wedge w = \Sigma u^i e_i \wedge \Sigma v^i e_i \wedge \Sigma w^i e_i$$

with

$$u^i = \nu\beta^{23i},\ v^i = -\nu\beta^{13i},\ w^i = \nu\beta^{12i}$$

(see also [1, p. 102]), and similarly for R_γ.

First we confirm the statement in the introduction that *irreducibility may depend on the field.*

(5.9) *If the* α^I *in* (5.8) *are real and* λ, μ *are not, then* $\tilde{R} \in W_3^6(\boldsymbol{C}, 2)$ *but* $\tilde{R} \in W_3^6(\boldsymbol{R}, 3)$.

For because $R_\beta + R_\gamma$ is unique, $\tilde{R} \in V_3^6(\boldsymbol{R}, 2)$ is impossible, and this with $N(\boldsymbol{R}, 6, 3) = 3$ entails the assertion.

Next we observe that the vector

$$\tilde{R}(\eta) = e_1 \wedge (e_2 + e_5) \wedge e_6 + e_1 \wedge e_3 \wedge e_5 + \eta e_2 \wedge e_4 \wedge e_6$$
$$+ (e_2 + e_5) \wedge e_3 \wedge e_4 \qquad (\eta \neq 0)$$

is a special case of (5.8) and that λ, μ are real when $\eta < 0$. Letting $\eta \to 0^-$ we find

$$\tilde{R}(0^-) = e_1 \wedge (e_2 + e_5) \wedge e_6 + e_1 \wedge e_3 \wedge e_5 + (e_2 + e_5) \wedge e_3 \wedge e_4$$

which by (5.1) lies in $W_3^6(\boldsymbol{R}$ or $\boldsymbol{C}, 3)$. Therefore:

(5.10) *The sets* $V_3^6(\boldsymbol{R}, 2)$ *resp.* $V_3^6(\boldsymbol{C}, 2)$ *are not closed in* $V_3^6(\boldsymbol{R})$ *resp.* $V_3^6(\boldsymbol{C})$.

Theorem (3.2) whose proof used (5.10) is therefore completely established.

Pac. J. Math. 49 (1973), 13–32

We now prove a surprising fact for $\boldsymbol{C}$ which has no analogue for $\boldsymbol{R}$ (see (6.3)):

(5.11) *The interior of* $W_3^6(\boldsymbol{C}, 3)$ *as a set in* $V_3^6(\boldsymbol{C})$ *is empty.*

We show that if $\tilde{R} = R_1 + R_2 + R_3$, $R_i \in G_3^6(\boldsymbol{C})$, is irreducible then it is the limit of elements in $V_3^6(\boldsymbol{C}, 2)$.

R_i and $R_j (i \neq j)$ have no nonvanishing 2-vector as a common factor, because $R_i + R_j$ would then be simple. Thus two cases are to be considered:

(1) $R_i \wedge R_j \neq 0$ for some i, j, say $R_1 \wedge R_2 \neq 0$,

(2) R_i and R_j have for $i \neq j$ a vector $v_k \neq 0$ (but no 2-vector $\neq 0$) as a common factor where (i, j, k) is a permutation of $(1, 2, 3)$.

In the latter case the v_i are either parallel or no two v_i are parallel. If they were parallel we could choose e_6 parallel to the v_i so that $\tilde{R} = \tilde{S} \wedge e_6$ with $\tilde{S} \in V_2^5(\boldsymbol{C})$, and ΣR_i would be reducible since $N(F, 5, 2) = 2$. If no two v_i are parallel then with suitable u_i

$$R_1 = u_1 \wedge v_2 \wedge v_3,\ R_2 = u_2 \wedge v_1 \wedge v_3,\ R_3 = u_3 \wedge v_1 \wedge v_2\ .$$

The vectors u_i, v_j are independent, for otherwise ΣR_i would be a 3-vector in a space of dimension less than 6 and by $N(F, 5, 3) = 2$ reducible. The proof of (5.10) shows $\tilde{R}$ can be approximated by elements of $V_3^6(\boldsymbol{C}, 2)$.

In case (1) there are vectors $w_1, \cdots, w_6, v_1, v_2, v_3$ such that w_1, w_2, w_3 are parallel to R_1, w_4, w_5, w_6 are parallel to R_2, $R_3 = v_1 \wedge v_2 \wedge v_3$ and

$$v_1 = a_1 w_1 + a_4 w_4,\ v_2 = a_2 w_2 + a_5 w_5,\ v_3 = a_3 w_3 + a_6 w_6\ .$$

If $\bigwedge_{i=1}^6 w_i \neq 0$ then $\tilde{R} = \sum_{I \in Y} \alpha^I w_I$ and $\prod_{i=1}^6 a_i \neq 0$ is equivalent to (5.6), so we have a special case of (5.8) (see [1, p. 83]) and hence $\tilde{R} \in V_3^6(\boldsymbol{C}, 2)$ contrary to the hypothesis. If $\bigwedge_{i=1}^6 w_i = 0$ and/or $\prod_{i=1}^6 a_i = 0$ we can choose w_i' and a_i' arbitrarily close to w_i resp. a_i such that $\bigwedge_{i=1}^6 w_i' \neq 0$ $\prod_{i=1}^6 a_i' \neq 0$ and $\lambda \neq \mu$, so that $\tilde{R}$ is the limit of elements in $V_3^6(\boldsymbol{C}, 2)$.

For $kr \leqq n$ let $Z_r^n(F, k)$ *be the set of* $\tilde{R} = \sum_{i=1}^k R_i$ *with* $\bigwedge_{i=1}^k R_i \neq 0$. Then $Z_r^n(F, k) \subset W_r^n(F, k)$ by (2.10).

(5.12) $Z_r^n(\boldsymbol{R}$ *resp.* $\boldsymbol{C}, k)$ *is dense in* $V_r^n(\boldsymbol{R}$ *resp.* $\boldsymbol{C}, k)$.

This is nearly obvious: If $\tilde{R} = \sum_{i=1}^j R_i \in Z_r^n(j)$, $j < k$, then $R_{j+1}, \cdots, R_k$ exist with $\bigwedge_{i=1}^k R_i \neq 0$ and

$$\tilde{R} = \lim\,(\tilde{R} + \delta \sum_{i=j+1}^{k} R_i) \text{ as } \delta \to 0\ .$$

If $\tilde{R} = \sum_{i=1}^j R_i \in W_r^n(j)$, $j \leqq k$, $\bigwedge_{i=1}^j R_i = 0$ and $R_i = \bigwedge_{h=1}^r v_{(i-1)r+h}$ then

$w_i \to v_i$ with $\bigwedge_{i=1}^{jr} w_i \neq 0$ exist and $\sum_{i=1}^{j} \bigwedge_{h=1}^{r} w_{(i-1)r+h} \to \tilde{R}$.

Because

$$V_3^6(C) = Z_3^6(C, 2) \cup V_3^6(C, 2)/Z_3^6(C, 2) \cup W_3^6(C, 3) ,$$

(5.11, 12) show that $Z_3^6(C, 2)$ is dense in $V_3^6(C)$, so that $V_3^6(C)/Z_3^6(C, 2)$ has no interior points, and hence has dimension less than 40 (= dim $V_3^6(C)$), see [2, p. 46]. In the next section we will see that $Z_3^6(C, 2)$ is open. Thus

(5.13) *The set* $Z_3^6(C, 2)$ *is open and dense in* $V_3^6(C)$, *hence*

$$V_3^6(C)/Z_3^6(C, 2)$$

is closed and $\dim Z_3^6(C, 2) = 40$, $\dim V_3^6(C)/Z_3^6(C, 2) < 40$.

Note that $W_3^6(C, 3) \subset V_3^6(C)/Z_3^6(C, 2)$ and that therefore *the closure of* $W_3^6(C, 3)$ *has dimension less than* 40.

6. The sets $Z_3^6(\boldsymbol{R}$ resp. $\boldsymbol{C}, 2)$ and $W_3^6(\boldsymbol{R}, 3)$. We now prove that $Z_3^6(\boldsymbol{R}$ resp. $\boldsymbol{C}, 2)$ is open. Actually our next theorem provides much more information which will allow us to show that $W_3^6(\boldsymbol{R}, 3)$ has a nonempty interior.

THEOREM 6.1. *Let* $F = \boldsymbol{R}$ or $\boldsymbol{C}$. *If* $R_1, R_2 \in G_3^6(F)$ *and* $R_1 \wedge R_2 \neq 0$ *then there is a neighborhood* $\tilde{U}(\tilde{R}_0)$ *of* $\tilde{R}_0 = R_1 + R_2$ *in* $V_3^6(F)$ *such that for* $\tilde{R} \in \tilde{U}(\tilde{R}_0)$ *there are simple* R_1', R_2' *with* $\tilde{R} = R_1' + R_2'$. *Furthermore, given neighborhoods* $U_i(R_i)$ *of* R_i *in* $G_3^6(F)$ *there is a neighborhood* $\tilde{U}'(\tilde{R}_0) \subset \tilde{U}(\tilde{R}_0)$ *such that* $\tilde{R} \in \tilde{U}'(\tilde{R}_0)$ *implies* $R_i' \in U_i(R_i)$ *and* $R_1' \wedge R_2' \neq 0$. *Consequently* $\tilde{R} \in Z_3^6(F, 2)$ *and by* (4.4) $R_1' + R_2'$ *is unique.*

Necessary for (6.1) to work is that $G_3^6 \times G_3^6$ and V_3^6 have the same dimension, which is the case because a-dim $G_r^n = r(n-r) + 1$, a-dim $V_r^n = \binom{n}{r}$ and $2[3(6-3)+1] = 20 = \binom{6}{3}$. But this argument is far from sufficient as similar situations for other dimensions show; the structure of G_3^6 enters.

Since $R_1 \wedge R_2 \neq 0$ we can choose a base so that $R_1 = e_{123}$, $R_2 = e_{456}$. A neighborhood of R_1 on G_3^6 consists of the simple $R_1' = \Sigma\beta^I e_I = \sum_{1\le i<j<k\le 6} \beta^{ijk} e_{ijk}$ with β^{123} close to 1 and the remaining β^I close to 0, so $\beta^{123} \neq 0$ may be assumed. Similarly for $R_2' = \Sigma\gamma^I e_I$ and $\gamma^{456} \neq 0$. The components of $\tilde{R}_0$ are $1, 0, \cdots, 0, 1$.

The special properties of G_3^6 arise from the Plücker relations which (see [1, p. 69]) with $\lambda = (\beta^{123})^{-1}$ are equivalent to

$$\beta^{ijk} = \pm\lambda \begin{vmatrix} \beta^{i\rho j} & \beta^{i\rho k} \\ \beta^{i\sigma j} & \beta^{i\sigma k} \end{vmatrix} 1 \leqq i \leqq 3,\ 4 \leqq j < k \leqq 6,\ (i, \rho, \sigma) = \pi(1, 2, 3)\,,$$

$$\beta^{456} = \lambda^2 \begin{vmatrix} \beta^{124} & \beta^{125} & \beta^{126} \\ \beta^{134} & \beta^{135} & \beta^{136} \\ \beta^{234} & \beta^{235} & \beta^{236} \end{vmatrix} = B\,.$$

Similarly with $\mu = (\gamma^{456})^{-1}$

$$\gamma^{ijk} = \pm\mu \begin{vmatrix} \gamma^{i\rho k} & \gamma^{j\rho k} \\ \gamma^{i\sigma k} & \gamma^{j\sigma k} \end{vmatrix} 1 \leqq i < j \leqq 3,\ 4 \leqq k \leqq 6,\ (\rho, \sigma, k) = \pi(4, 5, 6)\,,$$

$$\gamma^{123} = \mu^2 \begin{vmatrix} \gamma^{145} & \gamma^{245} & \gamma^{345} \\ \gamma^{146} & \gamma^{246} & \gamma^{346} \\ \gamma^{156} & \gamma^{256} & \gamma^{356} \end{vmatrix} = C\,.$$

If now $\Sigma\alpha^I e_I = \tilde{R} = R_1' + R_2' = \Sigma\beta^I e_I + \Sigma\gamma^I e_I$ then $\alpha^I = \beta^I + \gamma^I$, and substitution gives

$$\alpha^{123} = \beta^{123} + C\,,$$

$$\alpha^{ijk} = \beta^{ijk} \pm \mu \begin{vmatrix} \gamma^{i\rho k} & \gamma^{j\rho k} \\ \gamma^{i\sigma k} & \gamma^{j\sigma k} \end{vmatrix} 1 \leqq i < j \leqq 3,\ 4 \leqq k \leqq 6,\ (\rho, \sigma, k) = \pi(4, 5, 6)\,,$$

$$\alpha^{ijk} = \pm\lambda \begin{vmatrix} \beta^{i\rho j} & \beta^{j\rho k} \\ \beta^{i\sigma j} & \beta^{i\sigma k} \end{vmatrix} + \gamma^{ijk},\ 1 \leqq i \leqq 3,\ 4 \leqq j < k \leqq 6,\ (i, \rho, \sigma) = \pi(1, 2, 3),$$

$$\alpha^{456} = B + \gamma^{456}\,.$$

Thus the 20 components of $\tilde{R}$ are expressed in terms of β^{123}, the nine β^I with $1 \leqq i \leqq 3$, $4 \leqq j < k \leqq 6$, the nine γ^I with $1 \leqq i < j \leqq 3$, $4 \leqq k \leqq 6$, and γ^{456}. Evaluation of the functional determinant at $(1, 0, \cdots, 0, 1)$ gives the value 1. Therefore, the implicit function theorem is applicable and yields the assertion. The details of the calculation may be found in the thesis, [1, pp. 93–97].

As a corollary we have

(6.2) *The set* $Z_3^6(\boldsymbol{R}$ *resp.* $\boldsymbol{C}, 2)$ *is open in* $V_3^6(\boldsymbol{R}$ *resp.* $\boldsymbol{C})$.

But in contrast to (5.11):

(6.3) *The interior of* $W_3^6(\boldsymbol{R}, 3)$ *is not empty.*

For take any $\tilde{R}_0 = \sum_{I\in Y} \alpha^I e_I$ of (5.8) for which the α^I are real but λ, μ are not. Then

$$\tilde{R}_0 = R_\beta + R_\gamma \quad \text{with} \quad R_\beta \wedge R_\gamma \neq 0\,.$$

By (6.1) for $\tilde{R} \in \tilde{U}'(\tilde{R}_0) \subset V_3^6(\boldsymbol{C})$

$\tilde{R} = R'_1 + R'_2$ with $R'_1 \wedge R'_2 \neq 0$, R'_1 close to R_β and R'_2 close to R_γ so that R'_1 and R'_2 cannot be real either. Also $R'_1 + R'_2$ is unique by (4.4) and this implies as in the proof of (5.9) that

$$\tilde{R} \in W_3^6(\boldsymbol{R}, 3) \quad \text{for} \quad \tilde{R} \in \tilde{U}'(\tilde{R}_0) \cap V_3^6(\boldsymbol{R}) ,$$

where we consider $V_3^6(\boldsymbol{R})$ as a subset of $V_6^3(\boldsymbol{C})$.

References

1. D. E. Glassco, *Irreducible Sums of Simple Multivectors*, Dissertation, University of Southern California, 1971.

2. W. Hurewicz and H. Wallmann, *Dimension Theory*, 1st edition, Princeton, 1941.

3. J. Schouten, *Tensor Analysis for Physicists*, 2nd edition, Oxford, 1954.

4. B. L. Van der Waerden, *Einführung in die Algebraische Geometrie*, Berlin, 1938, New York, 1945.

Received July 17, 1972.

University of Southern California
and
California Polytechnic State University at San Luis Obispo

GPSR Compliance

The European Union's (EU) General Product Safety Regulation (GPSR) is a set of rules that requires consumer products to be safe and our obligations to ensure this.

If you have any concerns about our products, you can contact us on ProductSafety@springernature.com

In case Publisher is established outside the EU, the EU authorized representative is:

Springer Nature Customer Service Center GmbH
Europaplatz 3
69115 Heidelberg, Germany

Batch number: 10755308

Printed by Printforce, the Netherlands